Structure in
Protein Chemistry

Dedicated to my teachers

Richard Ramette
Charles Carlin
John Edsall
Gustav Lienhard
Klaus Weber
Guido Guidotti
Jon Singer
Russell Doolittle
Charles Perrin

Structure in Protein Chemistry

Jack Kyte

Professor of Chemistry
University of California at San Diego

Garland Publishing, Inc.
New York & London
1995

*"to place before mankind the common sense of the subject,
in terms so plain and firm as to command their assent"*

Thomas Jefferson
Letter to Henry Lee, 1825

Library of Congress Cataloging-in-Publication Data

Kyte, Jack.
    Structure in protein chemistry / Jack Kyte.
    p. cm.
    Includes bibliographical references and index.
    ISBN 0-8153-1701-8
    1. Proteins—Structure. 2. Proteins—Analysis. I. Title.

QP551.K98 1994
547.7′50442—dc20   94-23417
                                      CIP

Artwork for the cover was supplied by Dr. Anita Sielecki,
University of Alberta

15 14 13 12 11 10 9 8 7 6 5 4 3 2 1

Manufactured in the United States of America

Contents

Preface

All of the refined crystallographic molecular models of proteins that are now available were obtained over the last two decades. These molecular models have revolutionized our understanding of the structures of proteins, and any book devoted to this subject must rely heavily on them. At the same time, the other chemical and biophysical methods that are used to study the structures of proteins have retained their importance, both in the laboratory and in the classroom. This book is a detailed description of the crystallographic molecular models themselves and a discussion of those other chemical and biophysical approaches in the context of those crystallographic molecular models. The result is a comprehensive view of how the structures of proteins are studied and what has been learned.

This book is designed for a senior undergraduate or graduate course covering the structures of proteins. I have used it as the text in such a course over the last two academic years. Because it is designed to be a formal text, there are problems after each section to reinforce the concepts that were just presented.

The course created by this text is intended to bridge the gap between the introductory chemistry and biochemistry classes and the research literature. There are suggested readings in which the concepts discussed in each section are applied in an experimental setting, and there are also citations within the text that should direct the student to the research literature. At the completion of this course, the student should be equipped to take charge of her own education by reading critically the experimental literature. To do this she must be able to understand the experiments performed and be able to reach the same conclusions as do the authors or to realize that the authors are mistaken. It is my intention to develop in the student the ability to draw her own conclusions from only the experimental results. To this end, the problems at the end of each section are usually based on actual experimental results which are to be evaluated by the student.

To present a comprehensive view of proteins, this text combines concepts of bonding and chemical reactivity, descriptions of macromolecular structure, principles of thermodynamics, and explanations of biophysical methods. The concepts of bonding and chemical reactivity are presented in structural drawings of individual molecules or of chemical reactions in which electronic and mechanistic aspects are emphasized. The descriptions of macromolecular structure are highlighted by the stereo views of the crystallographic molecular models. The principles of thermodynamics are illustrated with relationships among the equilibrium constants and fundamental state functions. The explanations of biophysical methods rely on the mathematical equations defining the physical properties. The results of the experiments themselves are found in graphs and tables. It is this combination of structural drawings, stereo images, mathematical equations, and graphs and tables that makes this book both unique and comprehensive. It also places severe demands on the student. He must have a firm background in physics, mathematics, analytical chemistry, organic chemistry, and physical chemistry to understand the material. In the broadest sense, the intention of the course is to educate protein chemists. A protein chemist should be able to evaluate critically the results of any of the methods applied to the study of proteins.

To further assist the student in her understanding of the material, there is available the Kinemage program, which displays crystallographic molecular models on a computer monitor. The drawings can be rotated on the screen to afford the impression of three dimensions and can be expanded to focus on particular regions. Kinemage files have been prepared as a supplement to *Introduction to Protein Structure* (C. Branden and J. Tooze, Garland Publishing: New York, 1991). This supplement is available through Garland Publishing, 1000A Sherman Avenue, Hamden, Connecticut 06514 (800-627-6273). Most of the crystallographic molecular models found in this supplement are also discussed in the present volume. Cross-references to the Kinemage Supplement to *Introduction to Protein Structure* are presented in parentheses in the text. The cross-references are presented in the format (cn Kin.m), where the integer n refers to the number of the chapter and the integer m to the number of the image in that chapter of the Kinemage Supplement.

Over the last few months I have had the onerous task of reading the proofs of this book. I have been impressed by the fact that each time I have read through a proof, I find errors I missed before. I am certain that many still exist and would be grateful to any reader who draws them to my attention.

It is a pleasure to thank everyone who has helped me in the preparation of this book. My wife, Francey, out of the goodness of her heart, typed the first draft of the manuscript. Ian Glynn provided me with enthusiastic support and an office next to the library in the Physiological Laboratories at Cambridge University during the sabbatical year in which I wrote the first draft. My secretary, Brenda Leake, transferred the manuscript onto the computer, a major task because of the mathematical and chemical equations. Scott Ralph drew all of the structural formulas on the computer. Russell Doolittle and Harvey Itano read large portions of the manuscript and provided excellent suggestions. Individual sections of the manuscript were reviewed critically by Frank Huennekens, Bruno Zimm, Charles Perrin, Steven Clarke, Ajit Varki, David Matthews, John Edsall, Cyrus Chothia, Arthur Lesk, David DeRosier, Nigel Unwin, Steven Harrison, Fred Hartman, John Simon, George Fortes, Rachel Klevit, Ken Dill, Robert Baldwin, Howard Schachman, Dennis Haydon, and Guido Guidotti. Each of them provided many helpful comments.

Jack Kyte
August 1994
La Jolla, California

List of Stereo Drawings

The following is a list of the stereo drawings included in the text. It is now standard practice to publish drawings of crystallographic molecular models in this format, sometimes in color. To appreciate the results of crystallographic studies, one must be able to view these images. Although a few individuals can view them effortlessly by crossing their eyes, the rest of us require a stereo viewer. The best I have found is the stereo viewer distributed by Luminos Photo Corporation, P.O. Box 158, Yonkers, New York 10705 (800-586-4667). It has been my experience that a novice will usually complain that although everyone else can learn to use one of these viewers, he cannot. It is also my experience that everyone learns to use one. It is essential that anyone interested in the structures of proteins learn to view crystallographic molecular models in stereo. The drawings have been placed vertically rather than in their usual horizontal orientation and each has been placed on the outside edge of a page. This has been done to allow each image to be spread as flat as possible for the best viewing.

Figures

CHAPTER 1

Purification

The living world that teems around us, the world of species, individual organisms, organs, tissues, and cells, can be viewed as the manifestation of a vast fluid array of protein molecules, each appearing and disappearing in the proper place at the proper time. One fruitful approach in our attempt to understand this phenomenon has been to study these proteins individually or in small groups to gain insight into the role of each one in the overall scheme. An argument could be made that a cell does seem to be no more than the sum of its parts and that a significant understanding of life itself can be gained by studying those parts individually. Because the proteins are the parts of a cell that perform all of the chemical and structural transformations that occur within it, they have attracted the most attention. The most dynamic region in a living organism is the cytoplasm of the cells or cell from which it is made. About 20–30% of the total mass of cytoplasm is protein; the majority of the remainder is water.

The strategy that has been applied most frequently to the study of proteins is to identify a particular biological feature of a living organism and then purify the protein or proteins responsible for it. Typically, when a complex, intricately organized biological specimen, such as a tissue or a suspension of cells, is submitted to the first step in any purification procedure, it is immediately sundered beyond recognition and becomes a nondescript jumble of broken fragments suspended in an aqueous solution of proteins, nucleic acids, metabolites, and salts. This event is referred to as **homogenization**. The homogenization is usually accompanied by the dilution of the proteins in the initial specimen by the addition of a buffered aqueous solution. Following the homogenization, insoluble fragments are removed by centrifugation to produce a clear solution, the protein concentration of which is 1–10%. This solution contains most of the proteins that were once the living cytoplasm of the specimen. It is from this solution that particular proteins can be isolated. The separation of one protein from all of the others in a homogenate is referred to as **purification**. A particular protein must be purified before its molecular structure can be studied.

Usually, the only interest that one has in a particular protein arises from its participation in some process of biological importance. It might be an enzyme responsible for catalyzing a particular reaction; it might be a structural protein creating the macroscopic shape of the cell; it might be a protein that binds a hormone or neurotransmitter; or it might be a protein that binds to DNA and controls its transcription. To distinguish one protein from the others in a complex mixture, an assay for the protein of interest, based on its particular function, is required.

The most widely used procedure for purifying proteins is chromatography. This technique separates molecules of protein by differences in the rate at which they move along a cylinder of a porous solid phase as a liquid phase percolates through it. If the solid phase is properly chosen, each protein travels through the cylinder at a different rate and each emerges at a different time. In this way, one can be separated from the others. In order to distinguish the protein of interest from the others as they emerge from the chromatographic column, the assay for that protein is used. As the protein becomes purified, the preparation displays greater and greater activity in the specific assay for a given amount of total protein.

Once the protein has been purified, analytical methods must be used to demonstrate that only one protein is the major component in the final preparation and that this protein is responsible for the biological function of interest. The analytical procedure most suited to this demonstration is electrophoresis. Electrophoresis separates proteins by both their charge and their shape; and, if used with discontinuous stable boundaries, electrophoresis can have very high resolution.

Once a protein of known function has been purified to homogeneity, it can be crystallized. As in organic chemistry, crystallization is a way of harvesting a particular substance in a highly purified form. Ideally, every protein that was purified would be crystallized and stored in this form, as are organic molecules. In this form, each suspension of crystals would represent a pure chemical compound. In practice, because crystals are often difficult to make and yields in crystallizations are poor, purified proteins are usually left in solution or precipitated for storage. It is these solutions, precipitates, or suspensions of crystals that are the raw material for studies of the structures and functions of the proteins they contain. The purpose of this chapter is to describe how a particular protein is purified from a complex mixture of proteins such as the homogenate of a tissue or suspension of cells.

Adsorption to stationary phases and chromatography are the bases for both the purification of proteins and many of the assays used to identify particular proteins, so these processes will be considered first.

Adsorption to Stationary Phases and Chromatography

The goal of any procedure used to purify a particular substance from a complex mixture is to separate that substance from all of the other components in the mixture. When adsorption or chromatography is used for this purpose, differences in the preferences of solutes in a solution for another phase separate from the solution are exploited. The simplest example of such a strategy is an affinity adsorbent. Suppose a small molecule that could be tightly bound by only one particular protein in a solution was covalently attached to a solid surface. This affinity adsorbent would collect molecules of only that one protein on its surface. The rest of the molecules of protein in the solution could be washed away, and the molecules of the desired protein could then be released. Unfortunately, such highly specific adsorbents are not usually available, so small differences in affinity among proteins or among other molecules for separate phases are amplified by the process of chromatography.

When a chemical substance A, which will be referred to as the solute, is added to a vessel containing two immiscible phases and the system is allowed to come to equilibrium, the solute A will distribute between the two phases in a characteristic manner. The solute can be an inorganic ion, a small organic molecule, a protein, a nucleic acid, a polysaccharide, or any other similar substance. The two phases can be two immiscible liquids, a liquid and a solid, or a gas and a liquid; the only requirement is that they be brought into sufficient contact to permit the distribution of solute A between them to reach equilibrium and that they then be separated in some way that does not redistribute the solute. The simplest examples are a two-phase, solvent–solvent extraction or the suspension of some finely divided solid in a liquid followed by its removal from the liquid by filtration.

After the equilibration and separation, the moles of solute A in each phase can be determined. In the cases that are generally encountered, at least one of the phases is a liquid that can be freed entirely of the other phase. This liquid will be arbitrarily called the mobile phase. In the special case when a protein is solute A, the mobile phase is invariably an aqueous solution of moderate ionic strength buffered at a specific pH. In any situation, however, the molar concentration of solute A in the mobile phase can be readily measured. The second phase, arbitrarily referred to as the stationary phase, can be an immiscible liquid, a solid, or a solid in which a liquid is entrapped. Because of the peculiarities of this phase, the best way to express the concentration of solute A that has become physically associated with the stationary phase, $[A]'_S$, is in moles (liter of bed)$^{-1}$, where the volume of the bed is the volume filled by the stationary phase when it has settled.

Three general types of behavior[1] have been observed in such a partition (Figure 1–1). The simplest behavior,

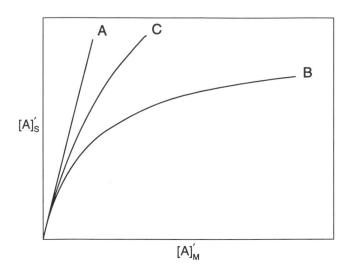

Figure 1–1: Partition of a solute between two phases referred to as the stationary phase, S, and the mobile phase, M. The concentration of solute A at equilibrium in the stationary phase, $[A]'_S$, is presented as a function of its concentration in the mobile phase, $[A]'_M$, for three types of behavior designated A, B, and C.

type A, occurs when the concentration of the solute A in the stationary phase increases in direct proportion to its concentration in the mobile phase. This type of behavior is encountered in solvent–solvent extractions or in molecular exclusion chromatography. In the latter example, it results from the fact that the stationary phase is nothing more than trapped, and thereby immobilized, mobile phase. Behavior of type B (Figure 1–1) is encountered when the stationary phase can be saturated with solute A. It results from the presence of only a finite number of sites on the stationary phase with which the molecules of solute A can associate, if these sites are all equivalent in their individual affinities for solute A and distributed over the stationary phase so that they do not interact with each other at saturation. Behavior of this type is encouraged by choosing microscopically uniform stationary phases. It is advantageous because at low concentrations of solute A the partition of solute closely approximates the direct proportionality of behavior of type A. Stationary phases showing this type of behavior are highly uniform ion-exchange resins or uniform, inert matrices to which molecules of a small organic compound displaying an affinity for solute A have been randomly and sparsely attached. In more heterogeneous stationary phases, specific sites with which molecules of solute A associate are distributed in such a way that they have an array of different affinities. This means that the small number of sites with very high affinities for solute A are occupied first, followed by those with lower and lower affinities sequentially. This produces behavior of type C (Figure 1–1), which is unpredictable and not uniform.

Examples of a stationary phase of this type are hydroxylapatite or a matrix to which a polyclonal antibody had been attached.

The earliest use of such distributions of solutes between two phases was selective adsorptions. **Selective adsorption** is a technique in which conditions are sought that promote the almost complete confinement of the molecule of interest to one phase while other, unwanted molecules distribute into the other phase and can thus be removed. When a protein is being isolated in this way, the ionic strength, pH, temperature, and choice of stationary phase is varied until conditions are found that permit the protein of interest to distribute almost completely into one phase while the maximum amount of the other, undesired proteins distribute into the other. An example is one of the steps in the purification of the protein fumarate hydratase.[2] Calcium phosphate gel was added to a crude mixture of proteins containing fumarate hydratase and dissolved in 0.1 M sodium acetate, pH 5.2. All of the fumarase (> 95%) associated tightly with the calcium phosphate gel. After the gel was washed, the adsorbed fumarase was then eluted with 5% $(NH_4)_2SO_4$ and 0.1 M sodium phosphate, pH 7.3, in 97% yield, even though only 20% of the original protein remained in the final solution.

Selective adsorption is a rather unsophisticated use of the distribution of a solute between a mobile and a stationary phase. It can be remarkably improved upon by operating at concentrations of solutes low enough that $[A]'_S$ is directly proportional to the molar concentration of A in the mobile phase and by causing the mobile phase to move through or across the stationary phase. This process is known as chromatography.

Chromatography is the process by which solutes are separated from one another on the basis of differences in the rate at which they pass over a bed of stationary phase through which a liquid mobile phase is continuously flowing. A chromatographic system is designed so that the mobile phase passes by the stationary phase in such a way that the contact between the two is maximized and equilibration of the solute between them is encouraged. Examples are paper chromatography, in which the liquid mobile phase moves down the paper while flowing among the cellulose fibers that form the stationary phase; column chromatography, in which the liquid mobile phase percolates through a finely divided, solid stationary phase compacted in a cylinder; or thin-layer chromatography, in which the liquid mobile phase creeps up a thin layer of the solid, dry stationary phase drawn by the capillary force arising from its movement between finely divided particles.

Zonal chromatography is chromatography in which the mixture to be separated is introduced in a thin zone at one end of the bed of stationary phase and the mobile phase is then set in motion. The molecules of solute in the mixture meander through the system, drawn forward by the movement of the mobile phase but retarded by the

stationary phase in which each spends a certain fraction of its time. The fraction of the time each solute spends in the stationary phase is determined by its affinity for the stationary phase, and this is determined by its bulk distribution behavior (Figure 1–1). Since the molecules of each solute spend a different fraction of their time in the immobility of the stationary phase, each solute moves through the system at a different rate and the components of the mixture are isolated one from the other into separate zones. The separated solutes are collected either by dividing the stationary phase itself and extracting them, as in paper chromatography, thin-layer chromatography, or countercurrent distribution chromatography, or by continuously collecting the mobile phase as it emerges at the opposite end of the bed of the chromatographic system, as in column chromatography or gas–liquid chromatography.

The important properties of the chromatogram are the relative mobilities, the widths of the peaks of the concentrations of the solutes at their half heights, and the resolution. The **relative mobility**, ^AR_f, of a particular solute A is either (1) the distance that the peak of its distribution has traveled through the system divided by the distance traveled by the mobile phase or (2) the total volume of the mobile phase in the bed of the system, referred to as the **void volume**, V_0, divided by the total volume that has passed through the system before the peak of the distribution of solute A emerges, referred to as the **elution volume**, ^AV_e. These are two different ways to measure the same parameter. The **width** of the distribution of solute A **at half height**, $^Aw_{1/2}$, is the width, in units of distance for definition (1) or volume for definition (2) above, between the two points at which the concentration of solute A is half of its maximum concentration at the peak. The **resolution** between two solutes is the completeness with which they are separated, a property that increases as the difference in their relative mobility increases and decreases as their widths at half height increase. The larger the differences in the various ^iR_f and the smaller the various $^iw_{1/2}$, the more successful will be the separation of the different solutes; expressions for ^AR_f and $^Aw_{1/2}$, as functions of parameters that can be manipulated, are of value in the understanding and design of chromatographic separations.

There are two approaches to describing the phenomenon of chromatography in theoretical terms.[3] It can be treated as the continuous process that it is, and differential equations can be formulated to describe the differential changes in solute positions and concentrations with time. These differential equations, however, do not have simple solutions, nor do they lead to an intuitive understanding of the process. The alternative approach is based on the concept of the theoretical plate, which was developed originally to describe the separation performed by a fractional distillation column.[4,5] Although this is a discontinuous model for a continuous process, the treatment is formulated in terms of an easily understood process and does provide, in at least one case, that of countercurrent distribution chromatography, an exact solution to the problem. Martin and Synge[6] were the first to apply this model to the process of chromatography.

Suppose that a chromatographic separation always operates at concentrations of solute A such that the amount associated with the stationary phase and the mobile phase in the chromatographic system is a linear function of its concentration in the mobile phase (behavior of type A, Figure 1–1). If so, at equilibrium

$$[A]'_S = \alpha_A [A]'_M \tag{1-1}$$

where the concentration $[A]'_M$ is the concentration of the solute A in the mobile phase in units of moles (liter of bed)$^{-1}$ where now the volume of the bed, V_T, is the volume filled by the stationary and mobile phases together as they are packed into the chromatographic system, and α_A is a partition coefficient. The units for $[A]'_S$ are the same, moles (liter of bed)$^{-1}$.

The stationary phase is formally divided into a series of $p + 1$ equivalent theoretical plates. A set of **theoretical plates** is a set of contiguous compartments formed by a set of evenly spaced planes normal to the direction in which the mobile phase flows. The **height equivalent to a theoretical plate**, h, is the distance the mobile phase must flow past the stationary phase until the concentration of solute in the fluid emerging from the theoretical plate is equal to the concentration the solute would have had if the fluid entering the theoretical plate had come into equilibrium with the stationary phase that fills the theoretical plate. For example, if the fluid entering the theoretical plate had a concentration of solute A equal to $[A]'_{M,ent}$ and the stationary phase had solute A immobilized within it at a concentration of $[A]'_{S,im}$, the formal boundary of the theoretical plate would occur downstream at the point where the concentration of the solute in the mobile phase, $[A]'_{M,lv}$, had reached a value

$$[A]'_{M,lv} = \frac{[A]'_{M,ent} + [A]'_{S,im}}{1 + \alpha_A} \tag{1-2}$$

where all concentrations are expressed in moles (liter of bed)$^{-1}$. If V_M is the volume of the mobile phase within a theoretical plate, then an apparent volume seen by the solute, V_A, can be defined such that

$$V_A \equiv (1 + \alpha_A) V_M \tag{1-3}$$

This apparent volume is the volume in which the total amount of solute A in the theoretical plate would have to be dissolved to produce a concentration equal to its concentration in the mobile phase within the theoretical plate. The apparent volume seen by the solute is always equal to or greater than V_M.

With these definitions, the continuous process of zonal chromatography is equivalent to the following discontinuous sequence of events. A number of moles of solute A equal to $^A m_{TOT}$ is added to the first theoretical plate and allowed to come to equilibrium between the stationary and mobile phases. The entire mobile phase of each plate is then moved to its neighbor downstream. Mobile phase containing no solute is added to the first plate, and the mobile phase in plate $p + 1$ is collected if necessary. After the new situation is allowed to come to equilibrium, the same transfers of mobile phases are made. The cycle of equilibrium and transfer is repeated n times. A machine[7] that performs countercurrent distribution chromatography mechanically completes this exact sequence of transfers. The theoretical plates in the countercurrent machine are individual glass vials, the mobile phases are equal volumes of an aqueous solution, the stationary phases are equal volumes of an immiscible organic solvent, and the steps of equilibration and transfer are discrete. Normally, however, this sequence of events is only a theoretical treatment equivalent to what actually happens.

At the conclusion of n steps, the total amount of solute A, $^A m_{TOT}$, will be distributed over the set of theoretical plates according to the formula[6]

$$^A m_{i+1} = {}^A m_{TOT} \left[\frac{n! \left(1 - \dfrac{V_M}{V_A}\right)^{n-i} \left(\dfrac{V_M}{V_A}\right)^i}{i!(n-i)!} \right] \tag{1-4}$$

where $^A m_{i+1}$ is the number of moles of solute A in the theoretical plate $i + 1$, when $i < n$. This equation is exact and applicable to situations where samples are collected from the bed, as in thin-layer chromatography, paper chromatography, or countercurrent distribution chromatography. It describes a binomial distribution of the solute A over the first n plates, or a binomial distribution truncated at the last plate, $p + 1$, if $n > p + 1$.

There are several characteristic features of this binomial distribution.[3] The maximum concentration of solute A coincides with the mean of the distribution, $\overline{i+1}$. The position of the mean, and hence the position of the peak of concentration, can be approximated by the relationship

$$\overline{i+1} \cong \frac{nV_M}{V_A} = \frac{n}{1 + \alpha_A} \tag{1-5}$$

which leads to

$$^AR_f = \frac{\overline{i+1}}{n} = \frac{1}{1+\alpha_A} \qquad (1\text{--}6)$$

because the mobile phase has moved a distance equal to n theoretical plates, nh, while the maximum of the distribution of solute A has only moved a distance equal to $\overline{i+1}$ theoretical plates, $(\overline{i+1})h$. The width of the distribution of solute A as it moves through the chromatographic system is governed by the standard deviation, $^A\sigma_n$, of this binomial distribution after n steps

$$^A\sigma_n = \frac{h\sqrt{n\alpha_A}}{1+\alpha_A} \qquad (1\text{--}7)$$

and, since $^Aw_{1/2} = \sqrt{8(\ln 2)}\,(^A\sigma_n)$ for the normal curve of error that coincides with this binomial distribution, the width of the distribution of solute A after n steps is

$$^Aw_{1/2} = \frac{h\sqrt{8(\ln 2)n\alpha_A}}{1+\alpha_A} = h\sqrt{n}\left[\sqrt{8(\ln 2)^AR_f\left(1-{}^AR_f\right)}\right] \qquad (1\text{--}8)$$

When mobile phase is collected from the end of the chromatographic system, the effluent collected is, by definition, the mobile phase emerging from plate $p + 1$. The concentration of solute A in this effluent is monitored as a function of the total volume that has emerged since the chromatogram was begun. As fluid emerges from the end of the system, the concentration of solute A that it contains increases, reaches a maximum, and then declines. This results from the approach of the mean of the binomial distribution to plate $p + 1$, its arrival at plate $p + 1$, and its passage through plate $p + 1$. The volume at which the maximum passes through plate $p + 1$ is the elution volume of solute A, ^AV_e. It corresponds to the volume of mobile phase that must pass through the system to bring the maximum of the distribution defined by Equation 1–4 into plate $p + 1$. When $\overline{i+1} = p + 1$, $n = n_e$, the number of steps necessary to bring the peak of solute A into plate $p + 1$. These definitions and Equation 1–5 lead to the relationship

$$n_e = (p+1)(1+\alpha_A) \qquad (1\text{--}9)$$

Since ^AV_e equals the number of steps, n_e, times the volume of each step, V_M, and the total volume of the mobile phase in the chromatographic system, or the void volume, V_0, equals $(p + 1)V_M$, it follows that

$$^AV_e = V_0(1+\alpha_A) \qquad (1\text{--}10)$$

and

$$^AR_f \equiv \frac{V_0}{{}^AV_e} = \frac{1}{1+\alpha_A} \qquad (1\text{--}11)$$

This is the fundamental relationship governing chromatography. It connects the volume at which the solute A emerges from the end of the chromatographic system with its bulk partition coefficient for the material composing the stationary phase and was verified experimentally by Martin and Synge.[6] In terms of the idea of apparent volume, solute A emerges at an elution volume that is greater than the void volume by the same factor that the apparent volume it saw in each plate was greater than the volume of the mobile phase in the theoretical plate.

The other variable of importance in chromatography, the width of the peak of concentration at half height, can be shown to be a function of the number of theoretical plates[3,6,8,9]

$$^Aw_{1/2} = \frac{{}^AV_e\sqrt{8(\ln 2)}}{\sqrt{p+1}} \qquad (1\text{--}12)$$

The theory as described leads to a number of insights into the important variables in chromatography.[6] The distance that solute A has moved through a given chromatographic system, $(\overline{i+1})h$, is directly proportional to the number of theoretical plates through which the mobile phase has moved (Equation 1–5), but the width of its distribution, $^Aw_{1/2}$, is proportional to the square root of the number of theoretical plates through which the mobile phase has moved (Equation 1–8). As a result, as the chromatography progresses, solutes can separate from each other more rapidly than they spread, and it is this property that permits chromatography to perform separations. The farther the mobile phase is allowed to move, the greater will be the separation observed. If the mobile phase has moved a certain distance, d, through the system, where $d = nh$, then the width of the distribution of solute A within the bed, $^Aw_{1/2}$, will be defined by

$$^Aw_{1/2} = \sqrt{dh}\left[\sqrt{8(\ln 2)^AR_f\left(1-{}^AR_f\right)}\right] \qquad (1\text{--}13)$$

Since ^AR_f is not a function of the height equivalent to a theoretical plate, h, if d remains constant, the width of the distribution of solute A will decrease as h is decreased, and consequently the number of theoretical plates that have been passed through is increased. As a result, resolution is enhanced by decreasing the height equivalent to a theoretical plate.

Similar logic can be applied to the situation in which the solute is collected as it emerges from the chromatographic system. At constant void volume, the elution position of the solute A, $^A V_e$, is only a function of the partition coefficient, α_A, of the solute A between the mobile phase and the stationary phase (Equation 1–10). If the solvent, ionic strength, temperature, and pH of the mobile phase and the volume and chemical structure of the stationary phase remain the same, the resolution of the separation can be improved by increasing the number of theoretical plates, $p + 1$, that it contains (Equation 1–12).

Because the height of a theoretical plate, h, is defined as the distance of passage required for equilibrium to be reached, h decreases and p increases as the flow rate is decreased, until diffusion between the plates becomes a significant factor. In most cases, however, diffusion is severely hindered by the structure of the stationary phase itself and almost never becomes important, and the slower the flow, the better the resolution. This is particularly important in the chromatography of proteins, especially when they are denatured, because their slow rates of diffusion significantly decrease rates of equilibration with the stationary phase.

The height of the theoretical plate decreases as the diameter of the particles in a solid stationary phase decreases.[6] A realization of this fact has led to the recent development of the high-pressure liquid chromatography foreseen by Martin and Synge.[6] **High-pressure liquid chromatography** improves upon earlier methods in several respects. In these systems, the high pressure is required only to force the liquid mobile phase through the finely divided solid particles of the stationary phase at a realistic rate. The particles themselves are spherical in shape and of uniform diameter to promote uniform flow of as rapid a rate as possible over the bed. Because the smaller particles of the solid phase decrease the height of a theoretical plate, more theoretical plates can exist in a given length of bed. This advantage can be exploited either to increase the resolution or to decrease the length of the chromatographic system or both.

Because high pressures are used to increase the rate of flow through a shorter column, the major advantage of high-pressure liquid chromatography is the speed with which the chromatograms can be run. For example, when peptides are separated by cation-exchange chromatography on sulfonated polystyrene,[10] at low pressure (< 500 psi), the chromatography takes about 25 h; when peptides are separated by reverse-phase adsorption chromatography,[11] at high pressure (> 1000 psi), the chromatography takes only 1 h, even though the resolution in each case is about the same. With reverse-phase adsorption chromatography, the solvents used are also more transparent to ultraviolet light, so peptides can be followed simply by their absorbance with a continuous-flow spectrophotometer.

Improvements in the size, uniformity, and rigidity of the particles of the stationary phase have permitted similar increases in the rate at which chromatography of

proteins can be performed. These developments are referred to commercially as **fast protein liquid chromatography**. In both high-pressure liquid chromatography and fast protein liquid chromatography, the principles remain the same as before, often the solid phases remain the same as before, and the technological improvements of the original techniques are based on previously noted predictions of the original theory.

The theory presented here for chromatography has been developed for regions of the partition curves (Figure 1–1) where solute A distributes with a constant partition coefficient, α_A. It turns out that the most usual deviation from such ideal behavior is for the stationary phase to display saturation (curves B and C, Figure 1–1). The more prominent this behavior becomes, the poorer the resolution of the chromatogram becomes.[3] As a rule, uniform stationary phases of high capacity, by promoting the linearity of the partition function, provide the highest resolution.

The fact that peak height decreases in almost inverse proportion to α_A precludes the use of conditions where the solute has a high affinity for the stationary phase. Usually, conditions such as solvent, temperature, ionic strength, and pH of the mobile phase and the chemical structure of the stationary phase are manipulated to bring the values of α_A for the solutes to be separated into a useful range. A variation in one of these properties of the mobile phase, however, can also be incorporated into the chromatography itself.

To this point, only isocratic zonal chromatography has been described. **Isocratic zonal chromatography** is chromatography in which the mobile phase introduced continuously into the first plate remains of constant composition. It is also possible to vary continuously and monotonically the composition of the mobile phase entering the column. This systematic variation produces a **gradient** of one or more properties of the mobile phase. For example, the ionic strength of the entering mobile phase can be increased continuously over time so that it is a linear function of the volume introduced into the system. Mechanical devices are available to produce linear gradients or gradients that are exponential or logarithmic or some other function of the volume by mixing two or more solutions that differ in the property to be varied. When a gradient of pH is required, the situation becomes somewhat more complicated because the pH of a solution is usually controlled with a buffer. Not only is the pH a logarithmic function of the concentrations of the conjugate acid and base of the buffer, but changing the concentrations of conjugate acid and base often affects the ionic strength. There is no requirement, however, that the gradient be some particular function of a particular property; the only requirement is that the property be varied continuously and monotonically. Therefore, this is not a serious problem.

The method of gradient chromatography is an important tool because it permits the partition coefficient,

α_A, to be decreased over the chromatographic run. This often is essential because if the partition coefficient for a particular solute is too large, it emerges from the system with such a large elution volume, ^AV_e, that the width of its band is unacceptably large (Equation 1–12). To produce satisfactory chromatography, the partition coefficient must be less than 10 in most situations, but frequently the values of the partition coefficient of solutes in a complicated mixture can spread over a large range for one particular mobile phase of constant composition. By using a gradient formulated so that all of the partition coefficients decrease continuously, even the solutes with the highest affinity for the stationary phase eventually have low enough partition coefficients to emerge from the system within a reasonable time. Usually, a gradient of ionic strength, cosolvent, or pH is employed. It is constructed in such a way that the chosen property continuously changes in a direction that will cause the solutes to have smaller and smaller affinities for the stationary phase and elute earlier than they would under isocratic conditions. For example, if a solute is being adsorbed to a nonpolar stationary phase, a gradient that increases in the concentration of a miscible nonpolar solvent in water is used to decrease gradually the affinity of the solute for the stationary phase.

The stationary phase in a chromatographic system is referred to as the **chromatographic medium**. It determines the molecular property used by the chromatographic system to separate the solutes.

Adsorption chromatography is chromatography that uses as a chromatographic medium a solid phase with which the solutes physically associate by noncovalent forces. Certain amorphous or heterogeneous solids such as silica gel and hydroxylapatite have been used as chromatographic media for adsorption chromatography. Silica gel has the unfortunate property of adsorbing hydrogen-bonding solutes very tightly and is generally confined to nonpolar solutes. Hydroxylapatite has been used extensively in protein purification. Unfortunately, it is heterogeneous, which causes it to have nonlinear distribution behavior, and it saturates readily. Both of these properties limit its resolution.

Currently, more carefully designed media are being produced for adsorption chromatography of proteins by synthetically coupling defined organic molecules to beaded hydrophilic matrices such as agarose. Although the intention in these syntheses has often been to produce a chromatographic medium with a specific affinity for one particular protein or class of proteins, many of these products have turned out to be simple adsorption media with useful and unexpected affinities for proteins in general.[12] Ironically, this makes them more valuable than they were originally intended to be. Successful purification of a minor component from a complex mixture requires that the set of distribution coefficients, α_i, for the components present assume a new and randomly permuted sequence of increasing magnitude as each new

chromatographic medium is employed. A series of chromatographic steps should be designed so that all of the components that had similar values of α_i in the preceding step, and that were not separated, have very different values of α_i in the next step, and are separated. The availability of a collection of microscopically uniform adsorption media with peculiar and unexpected affinities for proteins in general assists in this strategy.

Reverse-phase chromatographic media are used for adsorption chromatography at high pressure. These solid phases are spherical beads of silica gel that have been alkylated with hydrocarbons of uniform length, for example, octadecyl or octyl groups. This creates a very apolar surface on the beads that adsorbs apolar functional groups on the solutes. These media, however, show no affinity for completely polar solutes.

In media for adsorption chromatography, the affinity of molecules of the solute for the stationary phase arises from their direct physical attachment to the molecular surface of the stationary phase. These transient associations are noncovalent in nature and can be considered as ionic, hydrophobic, or hydrogen bonding—designations that imply direct molecular contact between solid phase and solute. It is this feature that distinguishes adsorption chromatography and ion-exchange chromatography.

Media for **molecular exclusion chromatography** separate molecules on the basis of differences in their size and shape. The beaded solids used as stationary phases, dextran, agarose, and polyacrylamide, are tangled webs of hydrophilic, linear polymers cross-linked among themselves randomly along their length. These matrices can be produced in two ways. Polysaccharides such as agarose and dextran spontaneously imbibe water and swell when the dry solid is exposed to an aqueous solution. The degree to which the linear polymers are cross-linked among themselves determines how much water they will imbibe at saturation. This is designated as their water regain, W_r, in milliliters (gram of polysaccharide)$^{-1}$. This in turn determines the fraction of the volume of the stationary phase occupied by solid polymer, f_{poly}

$$f_{poly} = \frac{V_{poly}}{V_{H_2O} + V_{poly}} = \frac{\bar{v}_{poly}}{W_r + \bar{v}_{poly}} \qquad (1–14)$$

where V_{poly} is the volume occupied by polysaccharide, V_{H_2O} is the volume occupied by water, and $\bar{v}_{poly}$ is the partial specific volume of the polysaccharide in milliliters gram^{-1}. Polyacrylamide does not swell readily but can be polymerized from acrylamide monomers and a small amount of the cross-linker N,N'-methylenebis-(acrylamide), both dissolved at a certain concentration in an aqueous solution. This produces a rigid gel that can be fragmented.

The majority of the internal volume of any of these stationary phases is occupied by water. When fluid is within the tangled web of the bead, however, it is no

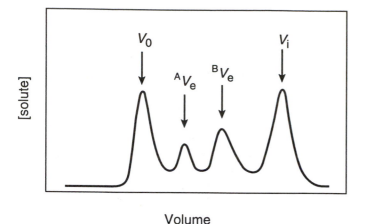

Figure 1–2: Molecular exclusion chromatography. All molecules larger than a certain size move at the void volume, V_0; all molecules below a certain size travel at the included volume, V_i; and solutes A and B travel at ^AV_e and ^BV_e, respectively, and are separated from each other because a molecule of solute A is larger than a molecule of solute B.

longer mobile but stationary. The mobile phase percolates around the beads and flow occurs only in the interstices among the beads. Only fluid between the beads is mobile phase.

The larger the molecule of solute, the less of the open space in the stationary phase within the bead is available to it. In terms of Equation 1–3, the larger is solute A, the smaller will be V_A, the apparent volume seen by solute A, and the smaller will be α_A. If solute A is very large, it cannot enter the beads at all; it has an $\alpha_A = 0$, and its peak emerges from the system at V_0 because $V_A = V_M$. Therefore, the elution position of the completely excluded molecules marks the position of V_0. A small molecule (in theory, water itself or something equivalent to it) can enter the entire open space in each bead and its elution position marks V_i, the included volume. Unlike with most other chromatographic separations, there is an end to a molecular exclusion chromatogram because no solute can see a larger volume than V_i. The only useful separation that occurs in such a system is of those solutes that emerge between V_0 and V_i, because all solutes larger than a certain size travel together at V_0 and all solutes smaller than a certain size travel together at V_i (Figure 1–2). Between V_0 and V_i, the larger solutes are the first to emerge.

The fraction of the total volume of the stationary phase, $V_T - V_0$, where V_T is the total volume of the bed, that is available to solute A is designated $^AK_{av}$

$$^AK_{av} \equiv \frac{^AV_e - V_0}{V_T - V_0} = \frac{\alpha_A}{\dfrac{V_T}{V_0} - 1} \qquad (1\text{--}15)$$

Another parameter is often used to describe the elution during molecular exclusion chromatography. This is the fraction of the volume of the stationary phase available to a small reference solute, solute R, that is also available to solute A, and it is designated ^AK_D

$$^AK_D \equiv \frac{^AV_e - V_0}{^RV_e - V_0} \qquad (1\text{--}16)$$

where ^RV_e is the volume at which solute R elutes. If the reference solute were able to enter the entire aqueous phase within the stationary phase, V_i, then ^RV_e would be equal to V_i and ^AK_D would equal $K_{av}(1 - f_{poly})^{-1}$.

Usually, media for **ion-exchange chromatography** are solids formed from neutral polymers to which charged organic functional groups have been covalently attached (Figure 1–3).[8] The most commonly used polymers are polystyrenedivinylbenzene copolymers, cellulose, and agarose. The fixed charges are added synthetically to the solid phase. **Basic media** are solid phases to which functional groups of positive charge at neutral pH have been covalently attached, and **acidic media** are solid phases to which functional groups of negative charge at neutral pH have been attached. A distinction can be made between weakly basic or acidic and strongly basic or acidic media based on whether the fixed charges, be they positive or negative, can or cannot be neutralized by variation of the pH within the ranges normally employed for chromatography. This is an important distinction because the density of charge, and hence the capacity of the medium, can be changed by changing the pH when weakly basic or acidic media are used but not when strongly basic or acidic media are used. Examples of weakly basic functional groups are tertiary amines such as those on (diethylaminoethyl)cellulose; examples of strongly basic functional groups are quarternary ammonium cations such as those on (trimethylammonium) methylpolystyrene; examples of weakly acidic functional groups are carboxylates, such as those on (carboxymethyl)cellulose, or phosphates, such as those on phosphocellulose; and examples of strongly acidic functional groups are sulfonates, such as those on sulfonated polystyrene (Figure 1–3).

The fixed charges on the stationary phase are responsible for the tendency of solutes of an opposite charge to associate with it. The simple univalent ionic functional groups on these solutes do not form physical contact with the isolated fixed charges of opposite sign that are attached to the stationary phase when ion-exchange chromatography is performed in aqueous solution. Rather, charged solutes (for example, nucleotides, amino acids, or proteins) can be considered to be trapped as mobile counterions to the covalently attached charges in an ionic double layer.[13] The two layers in an **ionic double layer** are an inner layer of covalently attached charges on the surface of the polymer forming the stationary phase and an outer layer of solution, adjacent to that surface, that is enriched in mobile, dissolved counterions of charge opposite to those fixed charges.

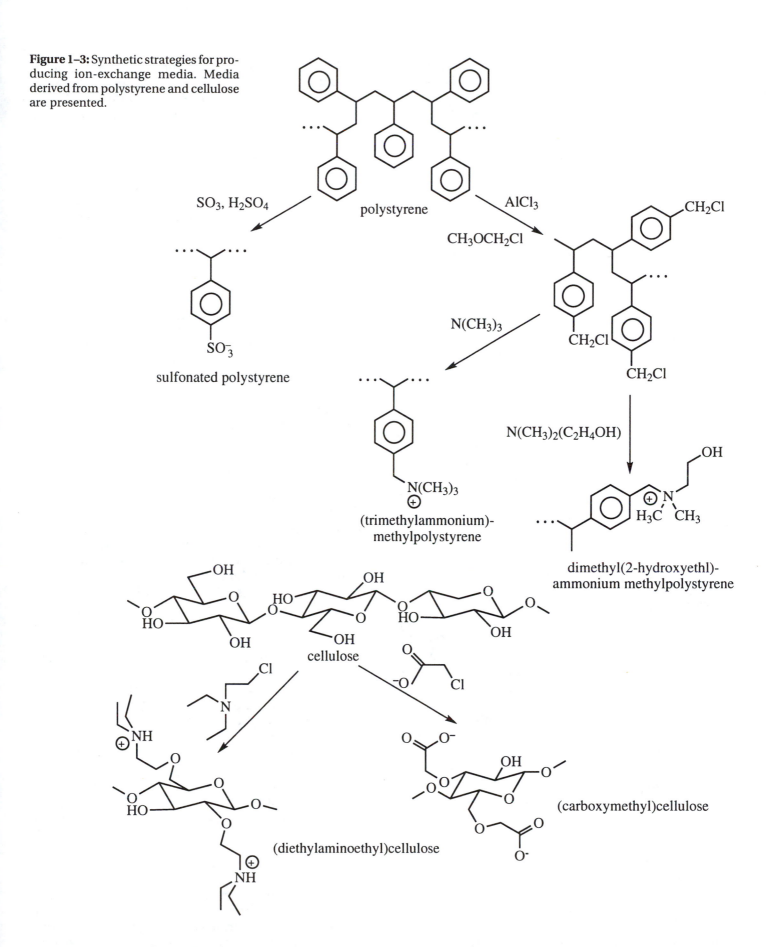

Figure 1–3: Synthetic strategies for producing ion-exchange media. Media derived from polystyrene and cellulose are presented.

polystyrene

SO₃, H₂SO₄

AlCl₃

CH₃OCH₂Cl

SO₃⁻

sulfonated polystyrene

N(CH₃)₃

CH₂Cl

CH₂Cl

CH₂Cl

CH₂Cl

N(CH₃)₃
⊕

(trimethylammonium)-
methylpolystyrene

N(CH₃)₂(C₂H₄OH)

OH

⊕N

H₃C   CH₃

dimethyl(2-hydroxyethl)-
ammonium methylpolystyrene

cellulose

Cl

N

⊕NH

O

HO

O

O

NH

(diethylaminoethyl)cellulose

O

Cl

O

O⁻

OH

O

O

O

O⁻

(carboxymethyl)cellulose

This enrichment results from the requirement for maintaining electroneutrality. The outer layer of solution contains solutes of both net positive and net negative charge but has an excess of solutes of net charge opposite to that of the functional groups in the layer of covalently attached charges. The inner layer of covalently attached charges is usually considered to be localized in a geometric surface representing the physical surface of the polymer, and the outer layer of solution enriched in the respective counterions is considered to have the properties of a space charge extending into the surrounding solvent.[13] The geometric surface of fixed charge is the inner boundary of the double layer. The reason that the diffuse space charge extends a significant distance into the solvent surrounding this boundary is that the positive and negative charges in the solution are on mobile, dissolved cations and anions, and the enthalpic tendency of the counterions to gather at the charged surface of the boundary is counterbalanced by the entropic tendency for them to diffuse randomly throughout the surrounding solution. Because the imbalance in charge that defines the ionic double layer falls off exponentially, it has no outer boundary. It is, however, arbitrarily assigned a thickness which is approximately that distance, from the surface of fixed charges, at which the space charge has decreased by a factor of e^{-1}. Under the normal conditions of chromatography, the thickness of the double layer would be less than 10 nm.[13] It can be assumed that the boundary that separates the stationary phase from the mobile phase during the chromatography lies at a greater distance than this from the molecular surface of the charged strands of polymer because flow occurs around particles of dimensions at least a thousand times larger. Therefore, the entire ionic double layer must be within the chromatographic stationary phase.

If this assumption is made, the distribution of counterions between the stationary phase and the mobile phase becomes formally equivalent to the distribution of permeant counterions across a permeable membrane when a charged, impermeant macromolecule is present on only one side of the membrane. If this is the case, the sum of the fixed charges and dissolved charges of the same sign in the stationary phase must equal the sum of the dissolved charges of the opposite sign in the stationary phase. It follows that the concentration within the stationary phase of any solute of charge opposite to the fixed charges must always be greater than its concentration within the mobile phase, and it is this bias that can produce significant values of α_i. This bias can be treated by the Donnan formalism.[14]

Consider the situation of an anion-exchange medium of univalent fixed positive charges [an example would be (trimethylammonium) methylpolystyrene, Figure 1–3], N^+, and a univalent anionic solute, A^-, in the presence of a dissolved univalent salt, K^+Cl^-, referred to as the electrolyte. The original stationary phase was the chloride

salt of N^+, and the solute before it was added to the stationary phase was the potassium salt of A^-. All concentrations are expressed in terms of moles (liter of phase)$^{-1}$, hence the unprimed values. From the requirement for electroneutrality

$$[K^+]_S + [N^+]_S = [Cl^-]_S + [A^-]_S \qquad (1\text{–}17)$$

$$[K^+]_M = [Cl^-]_M + [A^-]_M \qquad (1\text{–}18)$$

where the subscripts refer to the stationary and mobile phases. Since the electrolytes are at equilibrium within the theoretical plate

$$[K^+]_M[Cl^-]_M = [K^+]_S[Cl^-]_S \qquad (1\text{–}19)$$

$$[K^+]_M[A^-]_M = [K^+]_S[A^-]_S \qquad (1\text{–}20)$$

In the particular circumstance where the concentration of solute A^- is significantly less than the concentration of Cl^- so that $[A^-]$ becomes negligible in both Equations 1–17 and 1–18 and the concentration of fixed charges in the stationary phase, $[N^+]_S$, is so large that $[K^+]_S$ in Equation 1–17 becomes negligible

$$\alpha''_{A^-} \equiv \frac{[A^-]_S}{[A^-]_M} \cong \frac{[N^+]_S}{[K^+]_M} \qquad (1\text{–}21)$$

The partition coefficient, α''_{A^-}, for solute A should be inversely proportional to the concentration of K^+ in the mobile phase. Because the internal volumes of the stationary phases in ion-exchange chromatography are fairly small and the capacities of most media are large even in terms of equivalents (liter of bed)$^{-1}$, the situation in which $[K^+]_S \geq [N^+]_S$ is probably rarely approached, and Equation 1–21 should govern most concentrations of salt employed. The effect of adding a univalent salt to the mobile phase is to decrease the value of the partition coefficient for the anion A^- between the cationic stationary phase and the aqueous mobile phase. In this way, the value of α''_{A^-} can be adjusted by varying the concentration of electrolyte to optimize an isocratic separation, or a gradient of the electrolyte can be used to vary α''_{A^-} continuously. If the concentration of electrolyte is very low, α''_{A^-} will be large and the mobility of A^- will be negligible. Therefore, the solute can be gathered at the origin of the chromatographic system from a large volume of a dilute solution, and chromatography can then be initiated by increasing the concentration of the electrolyte.

In a weakly basic or acidic ion-exchange medium, the titration of the charges that occurs upon adding acid or base, respectively, occurs over a broad range of pH because of electrostatic repulsion among the fixed cations or anions. This permits $[N^+]_S$ (or $[O^-]_S$) to be continuously decreased by incorporating a gradient of pH into

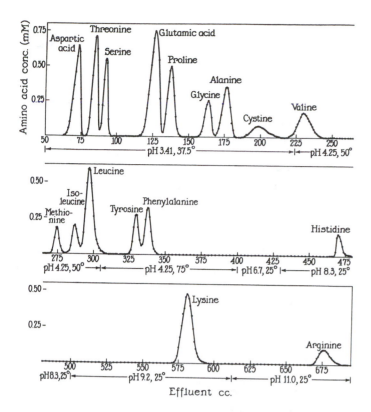

Figure 1–4: Separation of amino acids on sulfonated polystyrene (see Figure 1–3).[15] A mixture of amino acids in the ratios typical of those found in a protein was submitted to chromatography on a column (0.90 cm × 100 cm) of sulfonated polystyrene in the sodium form. The values of the pH and temperatures of the buffered mobile phases are noted below the axis of the abscissa, which is the volume of the mobile phase that has passed through the column (in centimeters[3]) since initiation of the chromatography. Changes from one mobile phase to the next were made discontinuously at the times noted. Individual fractions of the effluent emerging from the bottom of the column were collected and assayed for their concentration of amino acids (millimolar). Reprinted with permission from ref 15. Copyright 1951 *Journal of Biological Chemistry.*

the matrix.[15] An ion-exchange medium can also participate in adsorbing simple cations or anions by chelation, such as occurs in the binding of alkali metal cations to polygalacturonic acid.[17,18]

Proteins are macromolecular polyelectrolytes whose effective charge is a function of pH and varies over a wide range. If the pH is changed, their partition coefficients upon ion exchange can be varied, and gradients of pH as well as gradients of ionic strength are used in their chromatography. In the case of polyelectrolytes of this type, interactions between the solute and the stationary phase may also lead to direct adsorption to the stationary phase. Although simple univalent ions almost certainly do not physically associate with each other in aqueous solution, polyelectrolytes of opposite charge, such as proteins and ion-exchange media, often do. This results from a cooperative association of the opposite charges on the polymers that arises from the fact that those on the ion-exchange medium are covalently fixed and those of the opposite sign on the protein are also covalently fixed. It is always possible that there is a population of sites on the solid phase where the distribution of charge complements the distribution of charge on the protein. This is probably a rarer phenomenon, however, than the situations in which the molecules of proteins are simply trapped as mobile counterions in the ionic double layer.

Suggested Reading

Moore, S., & Stein, W.H. (1951) Chromatography of Amino Acids on Sulfonated Polystyrene Resins, *J. Biol. Chem.* 192, 663–681.

Problem 1–1: Assume that the total volume of mobile phase in the column described in Figure 1–4 is 45 cm^3. Calculate α_i for aspartic acid, threonine, glutamic acid, proline, glycine, alanine, and valine. Calculate the number of theoretical plates in the column used for the separation shown in Figure 1–4 from the peaks for threonine, serine, proline, glycine, alanine, and valine.

Problem 1–2: List the following amino acids in order of their elution from a column of sulfonated polystyrene.

the entering mobile phase. For example, if the stationary phase has fixed, protonated tertiary ammonium cations, a gradient of increasing pH would decrease $[R_3NH^+]_S$ as it progresses. The decrease in $[N^+]_S$ (or $[O^-]_S$) produces a decrease in α''_{A-} (or α''_{A+}), causing the solutes to emerge sooner than they would under isocratic conditions. When the solutes themselves are weak acids or bases, however, their ionization may also vary as the gradient of pH progresses, but in the opposite sense to the stationary phase; their effective charge will be increasing as the gradient progresses.

There is no question that Equation 21, although intuitively informative, does not describe real ion-exchange processes. At face value it predicts that α''_{A-} should only be a function of the charge density on the stationary phase and the concentration of electrolyte, and this is simply not the case. Even simple solutes upon ion exchange display affinities for the supporting polymeric medium that sometimes differ greatly from this expectation. The reason for these deviations is almost certainly due to the fact that solutes, brought to high concentration within the double layer by ion exchange, adsorb physically to the matrix, and this adsorption is superimposed upon the basic process of ion exchange. The clearest example of this is found in the separation of amino acids on sulfonated polystyrene (Figure 1–4).[15,16] Even though the solutes in the series alanine, valine, leucine, and phenylalanine have very similar acid dissociation constants, and hence ionic charge, they are cleanly separated. There is little doubt that the separation observed in this series is due to adsorption chromatography performed by the styrene–divinylbenzene copolymer of

$(R = {}^-OOCCHNH_3{}^+)$

$R-CH_2CH_2OH$ homoserine

$R-CH_2CH_2-\overset{+}{S}-CH_3$ [(aminocarbonyl)
$\quad\quad\quad\quad\quad CH_2$ methyl]-methionine
$\quad\quad\quad\quad\quad C$
$\quad\quad\quad H_2N \quad O$

$R-CH_2CH_2CH_2CH_3$ norleucine

$R-CH_2CH_2SH$ homocysteine

$R-CH_2$ (indole ring) tryptophan

$R-CH_2$ (imidazole ring) NCH_2C N-(carboxymethyl)-
histidine

Assay

Before any one of the proteins in the complex mixture produced by the homogenization of a biological specimen can be purified, there must be a way to identify it. An **assay** serves this purpose. An assay is the only connection between a specific biological phenomenon, which is one of the myriad features that, along with others, produces life, and the clear solution that contains the protein responsible for this phenomenon and that is otherwise indistinguishable from another clear solution containing some other protein. During the purification, separations of high resolution are performed that produce large numbers of separate samples, and the need to locate the protein of interest within these separated fractions requires that they be individually assayed (see points in Figure 1–4).

In its simplest form, an assay could be the monitoring of a chemical reaction catalyzed by an enzyme. For example, one of the phenomena that occurs in living organisms is the conversion of fumarate to malate

$${}^-OOC\quad H \quad + \quad H_2O \rightleftharpoons {}^-OOC\quad H\quad\quad H\quad OH \quad\quad COO^- \quad (1\text{–}22)$$

When the homogenate from a porcine heart is added to a solution of fumarate, which is otherwise quite stable, the fumarate begins to disappear and malate appears.[2] Because fumarate has significant absorbance at 300 nm (A_{300}), the decrease in A_{300} with time can be followed.

Conveniently, this wavelength is in a region where proteins and nucleic acids do not absorb strongly. This is a particularly important point. Because the enzymatic activity in the initial homogenate is usually quite low, large amounts of heterogeneous mixtures must be added to the assay at early steps in the purification. The other proteins, or the nucleic acids, in these mixtures can interfere with many otherwise useful enzymatic assays. This can happen because these components interfere either passively, for example, by absorbing light, or actively, for example, by catalytically converting essential reactants in the assay into products other than the product of the enzymatic reaction being followed.

The purpose of an assay is to produce a signal that is directly proportional to the molar concentration of the particular protein in the sample analyzed, and a necessary and sufficient condition for this to be the case is that the quantity measured, for example, the rate of decrease in A_{300} in the first few minutes, be itself directly proportional to the amount of sample added. The range over which this direct proportionality between the measurement and the added sample occurs must be fairly large to avoid the problem of having to assay every sample at a series of different dilutions. It is also helpful if the quantity measured, such as A_{300} in the case of fumarase, changes as a linear function of time within the range of times chosen to monitor the reaction.

The most unambiguous and unimaginative assay of a chemical transformation is one in which the reactants and products are **chromatographically separated** after the reaction and the quantities of each are determined. The recent introduction of rapid, high-pressure liquid chromatography with associated monitoring systems of high sensitivity makes this possible, but the expense of the equipment and the appreciable turnaround times of the chromatographic columns make this approach tedious. If, however, **radioactive reactants** are available that can be turned into radioactive products, reactants and products from a large number of assays can be separated in large arrays of simple, inexpensive chromatographic systems and their respective quantities can be determined by scintillation counting.

Examples of such assays have been used for the purifications of the proteins farnesyldiphosphate farnesyltransferase, methylamine-glutamate methyltransferase, nicotinate phosphoribosyltransferase, and S-adenosylmethionine: N^ε-L-lysine methyltransferase. Farnesyldiphosphate farnesyltransferase catalyzes the reaction

$$2\text{-farnesyldiphosphate} + NADPH \rightleftharpoons$$
$$squalene + NADP^+ + pyrophosphate \quad (1\text{–}23)$$

A sample to be assayed for this enzymatic activity can be mixed with [^{14}C]farnesylpyrophosphate and incubated for a set time. The reaction can then be terminated by adding ethanol. After hydrolysis and extraction, the resulting [^{14}C]farnesol and [^{14}C]squalene in each sample can be separated on small plates by thin-layer chromatography and separately quantified.[19] Methylamine-glutamate methyltransferase catalyzes the reaction

$$L\text{-glutamate} + [^{14}C]\text{methylamine} \rightleftharpoons$$
$$\text{ammonia} + N\text{-}[^{14}C]\text{methyl-L-glutamate} \quad (1\text{-}24)$$

The [^{14}C]methylammonium cation and the [^{14}C]methyl-L-glutamate can be separated by isocratic, cation-exchange chromatography.[20] Nicotinate phosphoribosyltransferase catalyzes the reaction

$$5\text{-phospho-}\alpha\text{-D-ribose 1-diphosphate} + [^{14}C]\text{nicotinate} \rightleftharpoons$$
$$[^{14}C]\text{nicotinate D-ribonucleotide} + \text{pyrophosphate}$$
$$(1\text{-}25)$$

The [^{14}C]nicotinate and [^{14}C]nicotinate D-ribonucleotide can be separated by descending paper chromatography.[21] L-Lysine, N^ε-methyl-L-lysine, and $N^\varepsilon,N^\varepsilon$-dimethyl-L-lysine are converted in the presence of S-[^{3}H]methyl-S-adenosylhomocysteine into mixtures of N^ε-[^{3}H]methyl-L-lysine, $N^\varepsilon,N^\varepsilon$-[^{3}H]dimethyl-L-lysine, and $N^\varepsilon,N^\varepsilon,N^\varepsilon$-[^{3}H]trimethyl-L-lysine by S-adenosylmethionine:N^ε-L-lysine methyltransferase. After removal of unreacted S-[^{3}H]methyl-S-adenosylhomocysteine with activated charcoal, the three radioactive products can be separated by thin-layer chromatography and quantified individually.[22]

As in the previous example, where unreacted S-[^{3}H]methyl-S-adenosylhomocysteine was removed by adsorption to activated charcoal, the chemical transformation performed by an enzyme often produces a product that can be exclusively transferred to a **separable phase**. For example, tryptophan-tRNA ligase catalyzes the reaction

$$\text{MgATP} + L\text{-}[^{14}C]\text{tryptophan} + \text{tRNA}^{\text{Trp}} \rightleftharpoons$$
$$\text{AMP} + \text{Mg-pyrophosphate} + L\text{-}[^{14}C]\text{tryptophan-tRNA}^{\text{Trp}}$$
$$(1\text{-}26)$$

The L-[^{14}C]tryptophan-tRNA^{Trp} can be isolated as a precipitate, free of L-[^{14}C]tryptophan, by treatment with acid and filtration through glass fiber filters.[19] The [^{14}C]CO$_2$ released from L-[1-^{14}C]glutamate by glutamate decarboxylase[24] or from 4-hydroxyphenyl[1-^{14}C]pyruvate by 4-hydroxyphenylpyruvate dioxygenase[25] can be released as a gas from the assay solutions by treatment with acid and collected in a separate well containing a strong base.

As in the assay for fumarate hydratase, the **change in absorbance** of a reactant or product is often followed. In such cases, it is usually desirable to be able to make observations at 290 nm or above to avoid the absorbance of the crude solutions of protein and nucleic acid themselves. For example, tryptophan 2,3-dioxygenase catalyzes the reaction

$$L\text{-tryptophan} + O_2 \rightleftharpoons L\text{-formylkynurenine} \quad (1\text{-}27)$$

and the appearance of the product can be followed by its absorbance at 321 nm.[26] The reaction catalyzed by protocatechuate 3,4-dioxygenase

$$3,4\text{-dihydroxybenzoate} + O_2 \rightleftharpoons$$
$$3\text{-carboxy-}cis,cis\text{-muconate} \quad (1\text{-}28)$$

can be followed by the increase in absorbance at 290 nm as the reaction proceeds.[23]

This latter example also illustrates one of the difficulties of using an assay to follow the purification of a protein. In crude homogenates, the 3-carboxy-*cis,cis*-muconate was converted by a contaminating enzyme into 3-carboxy-*cis,cis*-muconolactone, and absorbances had to be corrected for this further transformation until the contaminating enzyme had been lost at an intermediate step in the purification.[23] All of the assays described so far have been used during total purifications of the respective proteins, but as has already been noted, many assays that work quite well with purified proteins are unsuitable for following that protein in crude mixtures.

Of all of the changes of absorbance that are employed in enzymatic assays, none is more heavily used than the decrease in absorbance at 340 nm of dihydronicotinamide adenine dinucleotide (NADH; $\varepsilon_{340} = 6220 \text{cm}^{-1}\text{M}^{-1}$)[28]—or its phosphate (NADPH; $\varepsilon_{340} = 6100 \text{cm}^{-1}\text{M}^{-1}$)—when it is oxidized to nicotinamide adenine dinucleotide (NAD$^+$)—or to its phosphate (NADP$^+$)—or the increase in absorbance that occurs in the reverse reactions. As each of these redox pairs is always oxidized or reduced at the expense of the reduction or oxidation of another redox pair of reactant and product, the assay can be made very specific by following the change of absorbance in the absence and then in the presence of the specific reactant.

There is a large class of enzymes, known as dehydrogenases, that use the redox pair of NADH and NAD$^+$ or NADPH and NADP$^+$, and they can be assayed directly. For example, 3-hydroxyacyl-CoA dehydrogenase catalyzes the reaction

$$S\text{-acetoacetylpantetheine} + \text{NADH} \rightleftharpoons$$
$$(S)\text{-}S\text{-(3-hydroxybutyryl)pantetheine} + \text{NAD}^+ \quad (1\text{-}29)$$

The loss of NADH can be followed at 340 nm.[29] In the opposite sense, the increase in A_{340} can be used to follow the reaction catalyzed by glyceraldehyde-3-phosphate dehydrogenase[30]

$$D\text{-glyceraldehyde 3-phosphate} + \text{HOPO}_3^{2-} + \text{NAD}^+ \rightleftharpoons$$
$$3\text{-phospho-D-glycerol phosphate} + \text{NADH} \quad (1\text{-}30)$$

The absorbance change produced by a dehydrogenase can also be used in a **coupled assay** to monitor enzymatically catalyzed reactions that do not involve NADH. Examples of such coupled assays have been used for the purifications of 5-aminopentanamidase, pyruvate carboxylase, and phosphomevalonate kinase. The ammonium ion released during the reaction catalyzed by 5-aminopentanamidase

$$5\text{-aminopentanamide} + H_2O \rightleftharpoons \\ 5\text{-aminopentanoate} + NH_4^+ \quad (1\text{-}31)$$

can be monitored continuously and quantitatively as it is produced by adding NADH, 2-oxoglutarate, and excess glutamate dehydrogenase. As the ammonium is converted to glutamate by the glutamate dehydrogenase,[27] the NADH is stoichiometrically converted to NAD^+

$$2\text{-oxoglutarate} + NH_4^+ + NADH \rightleftharpoons \\ NAD^+ + H_2O + L\text{-glutamate} \quad (1\text{-}32)$$

The oxaloacetate produced by pyruvate carboxylase

$$MgATP + pyruvate + HCO_3^- \rightleftharpoons \\ MgADP + HOPO_3^{2-} + oxaloacetate \quad (1\text{-}33)$$

can be monitored by adding NADH and excess malate dehydrogenase[32]

$$NADH + oxaloacetate \rightleftharpoons NAD^+ + (S)\text{-malate} \quad (1\text{-}34)$$

Phosphomevalonate kinase catalyzes the reaction

$$MgATP + (R)\text{-5-phosphomevalonate} \rightleftharpoons \\ MgADP + (R)\text{-5-diphosphomevalonate} \quad (1\text{-}35)$$

The MgADP can be monitored continuously as it is produced by adding phosphoenolpyruvate, NADH, and an excess of both pyruvate kinase and L-lactate dehydrogenase[33]

$$MgADP + phosphoenolpyruvate \rightleftharpoons MgATP + pyruvate \quad (1\text{-}36)$$

$$pyruvate + NADH \rightleftharpoons (S)\text{-lactate} + NAD^+ \quad (1\text{-}37)$$

One of the more subtle uses of a coupled assay based on the absorbance of NADH is the one devised[34] for hydroxymethylglutaryl-CoA lyase

$$(S)\text{-3-hydroxy-3-methylglutaryl-SCoA} \rightleftharpoons \\ acetyl\text{-SCoA} + acetoacetate \quad (1\text{-}38)$$

This coupled assay takes advantage of the fact that the equilibrium of the malate dehydrogenase reaction (Equation 1–34) lies in the direction of NAD^+ and (S)-malate so that if NAD^+, malate, and malate dehydrogenase are mixed together, very little oxaloacetate and NADH are formed. With this in mind, it can be seen that if (S)-malate, NAD^+, and excesses of citrate (*si*) synthase and malate dehydrogenase are added during the progress of the reaction catalyzed by hydroxymethylglutaryl-CoA lyase, the conversion of the acetyl-SCoA

$$acetyl\text{-SCoA} + H_2O + oxaloacetate \rightleftharpoons citrate + HSCoA \quad (1\text{-}39)$$

into citrate consumes oxaloacetate and pulls the unfavorable equilibrium of the malate dehydrogenase reaction in the direction of NADH production, and hence an increase in the absorbance of the solution is observed.

The two or more enzymatic steps in a coupled assay are sometimes disconnected rather than allowed to proceed simultaneously. An example would be an assay[35] for ribose–phosphate pyrophosphokinase

$$MgATP + D\text{-ribose 5-phosphate} \rightleftharpoons \\ MgAMP + 5\text{-phospho-}\alpha\text{-D-ribose 1-diphosphate} \quad (1\text{-}40)$$

The reaction is quenched by boiling, and the 5-phospho-α-D-ribose 1-diphosphate that has accumulated is determined by adding orotate, orotate phosphoribosyltransferase, and orotidine-5'-phosphate decarboxylase

$$5\text{-phospho-}\alpha\text{-D-ribose 1-diphosphate} + orotate \rightleftharpoons \\ orotidine 5'\text{-phosphate} + pyrophosphate \quad (1\text{-}41)$$

$$orotidine 5'\text{-phosphate} \rightleftharpoons UMP + CO_2 \quad (1\text{-}42)$$

The total decrease in A_{295} due to the loss of orotate is proportional to the phosphoribosylpyrophosphate originally present in the quenched samples. The decarboxylation draws the reactions to completion.

Many assays depend on a chromogenic chemical reaction in which one of the products will participate. Examples of such **colorimetric assays** were used for the purification of Na^+/K^+-transporting ATPase,[36] glutamine–pyruvate aminotransferase, galactonate dehydratase, and selenocysteine lyase. Na^+/K^+-transporting ATPase catalyzes the reaction

$$3Na^+ + 2K^+ + MgATP \rightleftharpoons \\ 3Na^+ + 2K^+ + MgADP + HOPO_3^{2-} \quad (1\text{-}43)$$

After a certain interval the enzymatic reaction is quenched and the phosphate produced is determined colorimetrically by the addition of ammonium molydate in dilute sulfuric acid and a strong reductant, which together produce a blue color proportional in magnitude to the phosphate present.[37,38] Glutamine–pyruvate aminotransferase will also catalyze the reaction

$$\begin{array}{c}\text{L-glutamine} + \text{glyoxylate} \rightleftharpoons \\ \text{2-oxoglutaramate} + \text{glycine}\end{array} \quad (1\text{-}44)$$

The glycine produced and the L-glutamine remaining will react with o-phthalaldehyde and a thiol, after the enzymatic conversion has been terminated, to produce complexes that absorb in the near ultraviolet.[39] The glycine complex, however, absorbs at a higher wavelength ($\lambda_{max} = 330$ nm). Galactonate dehydratase catalyzes the reaction

$$\text{D-galactonate} \rightleftharpoons \text{2-oxo-3-deoxy-D-galactonate} + H_2O \quad (1\text{-}45)$$

After the reaction is quenched, the ketonic product is reacted with semicarbazide[40] to produce a semicarbazone that absorbs at 250 nm.[40] Selenocysteine lyase catalyzes the reaction

$$\text{selenocysteine} + 2RSH \rightleftharpoons \text{L-alanine} + H_2Se + RSSR \quad (1\text{-}46)$$

After the enzymatic reaction is stopped, the H_2Se can be assayed colorimetrically by its reaction with lead acetate, a reaction that yields a yellow color.[41]

In such colorimetric assays, the enzymatic reaction is usually quenched after a certain time, and the amount of product formed is determined colorimetrically. In at least one case, however, the colorimetric reaction was compatible with the enzymatic reaction. It is possible to monitor the production of coenzyme A by citrate (si) synthase (Equation 1–39) continuously[42] by the addition of 5,5'-dithiobis(2-nitrobenzoate). This reagent reacts with the thiol of the coenzyme A as it is formed to release the bright yellow 2-nitro-5-thiolatobenzoate dianion.

Certain proteins, known loosely as **receptors**, often do not catalyze a chemical reaction but respond to specific small molecules, referred to as **agonists**, by binding them and then undergoing a change in structure. Receptors are assayed by their ability to bind either these agonists or similar molecules that also bind but do not elicit the response, referred to as **antagonists**. In such **binding assays**, the receptor and a suitable radioactive agonist or antagonist are mixed together, the binding is allowed to come to equilibrium, and the receptor-agonist or receptor-antagonist complex is separated from unbound agonist or antagonist, respectively. The amount of bound radioactivity is then determined by scintillation counting.

Agonists or antagonists of very high affinity for the receptor are required to ensure that the binding is at saturation, so that all receptors are counted, and to prevent dissociation of receptor and agonist or receptor and antagonist during the separation of bound and free radioactivity. These reagents are often produced by the synthesis of analogues of the natural compounds. For example, [³H]dihydroalprenolol is a radioactive synthetic compound that binds very tightly (dissociation constant = 2 nM)[43] to the β-adrenergic receptor, which physiologically responds to epinephrine, and its binding has been used as an assay during the purification of this receptor.[44] Often a synthetic compound whose binding to a receptor is very strong has been obtained during a search for pharmaceutically useful agents. An example of this kind of product is prazosin, which was developed as a drug specific for α_1-adrenergic receptors and whose binding (dissociation constant = 1 nM) could be used as an assay during the purification of the α_1-adrenergic receptor.[45] At times, the naturally occurring agonist can be rendered radioactive and has an affinity great enough to be used in an assay during the purification of the receptor. Examples of this practice would be the use of the binding of ¹²⁵I-epidermal growth factor[46] (dissociation constant = 20 nM) and the binding of [1,2-³H₂]progesterone[47] (dissociation constant = 1 nM) as assays for their receptors.

In all binding assays for receptors, the difficulty is to determine how much of the added agonist has bound to the receptor. With weakly bound agonists (and antagonists), the complex between receptor and agonist cannot be separated from the free agonist in solution without dissociation of the agonist, and some technique such as equilibrium dialysis must be used as an assay. With agonists and antagonists that bind tightly, the complex can be separated rapidly with little loss of bound radioactivity by rapid molecular exclusion chromatography on small, disposable columns.[48]

Binding assays have also been developed for proteins that associate with specific nucleotide sequences in DNA,[49] such as promoters or other regulatory elements. A short fragment of DNA labeled with radioactive phosphate at one end and containing the sequence of interest is used as a reagent. When such a fragment is digested with deoxyribonuclease I, a characteristic pattern of shorter segments of DNA of various lengths is obtained as a result of the random cleavage by the nuclease of the phosphodiesters along the double-stranded DNA. The presence of a protein that binds specifically to a particular nucleotide sequence in a short fragment of end-labeled DNA results in prevention of cleavage of the DNA by the nuclease at that site. The fragments resulting from cleavages in this region disappear from the display, and this **footprint** demonstrates that the DNA-binding protein is present. Such an assay can be used to determine the relative concentration of the DNA-binding protein by examining the patterns produced as a series of dilutions is performed in the solution of the protein added to the end-labeled DNA.

Biological assays are assays in which the ability to evoke a complex biological response by samples added to cells or whole organisms is determined. For example, the assay for a protein referred to as the Hurler corrective factor measures the ability of this protein to prevent the accumulation of sulfated mucopolysaccharide in lysosomes of intact cells. It is this accumulation that causes the Hurler syndrome. Samples are added to a series of petri dishes on which fibroblasts from a patient with the Hurler syndrome have been grown and $[^{35}S]SO_4$ is added. After several days, the accumulation of ^{35}S-sulfated mucopolysaccharide is assessed by washing the cells and submitting them to scintillation counting.[50] In this particular assay, the decrease in accumulation of radioactivity was not directly proportional to the amount of sample added, and this problem was overcome by constructing a dose–response curve.

A biological assay was also used for the maturation-promoting factor, which is a protein involved in controlling the cell cycle.[51] Samples containing this protein could be assayed for its activity by injecting sequentially diluted aliquots into individual oocytes from the frog *Xenopus laevis* and scoring the cells for the disappearance of geminal vesicles.[52] With the use of this assay, the protein could be followed during a purification procedure[52] and the remarkable fluctuation of its concentration during the cell cycle could be documented.[51]

Suggested Reading

McClure, W.R. (1969) A Kinetic Analysis of Coupled Enzyme Assays, *Biochemistry 8*, 2782–2786.

Problem 1–3: Design a coupled assay, based on the release of $[^{14}C]CO_2$, for the enzyme *cis*-aconitase, which catalyzes the reaction

$$citrate \rightleftharpoons isocitrate$$

Problem 1–4: Design a coupled assay based on the reduction of NAD^+ for the enzyme fumarate hydratase.

Problem 1–5: Design a coupled assay for phosphofructokinase, the enzyme that catalyzes the reaction

$$fructose\ 6\text{-}phosphate + MgATP \rightleftharpoons$$
$$MgADP + HOPO_3^{2-} + fructose\ 1,6\text{-}diphosphate$$

Purification of a Protein

The goal of an attempt to purify a protein is to isolate that protein, the presence and relative molar concentration of which can be followed by a specific assay, from all of the other proteins in the homogenate. To do this, advantage is taken of the properties that distinguish a molecule of one protein from a molecule of another. Proteins are macromolecules of molar mass 10,000–10,000,000 g mol^{-1}. Unlike synthetic polymers, each molecule of a given protein has the same covalent structure,

the same distribution of polar and nonpolar functional groups over its surface, and the same shape as every other molecule of the same protein. Different proteins are distinguished from each other by differences in these properties.

A molecule of protein is a **globular macromolecule**, the shape of which resembles a hollow metal sphere that has been dented at random by blows from a hammer. The diameters of globular proteins vary from about 2 to 10 nm. Positively and negatively charged functional groups are distributed in a characteristic array over the surface of each molecule of a particular protein. In addition to guanidinium cations, these charged functional groups are carboxylate anions and ammonium and imidazolium cations that can be neutralized by lowering or raising the pH, respectively. As a result, the net charge on a molecule of protein varies with the pH and can be negative or positive within normal ranges. Patches of nonpolar functional groups are distributed in a characteristic array over the surface of each molecule of protein. The affinity of these patches for nonpolar solid phases can be exploited to separate molecules of one protein from those of another. Chromatography is used to separate molecules of protein by differences in their size, their shape, their charge as a function of pH, and the unique character of their surfaces.

The strategy for the purification of a protein is tailored to the particular problems faced in each instance. Usually it includes a series of steps, each involving chromatography or adsorption. Each step produces a series of fractions, from two to several hundred, each contained in a small volume of aqueous solution. The fractions containing the protein of interest are identified by the assay, they are pooled together, and the protein they contain is submitted to the next step. At each step in the procedure, the material in the pooled fractions is examined by electrophoresis to determine how many proteins remain. Electrophoresis is used analytically to separate proteins and provide a catalogue of the number of different proteins in a given solution. As the purification progresses, the number of other proteins should decrease as the protein of interest becomes the dominant component.

The three requirements for the successful purification of a protein are an assay for the protein, a protein whose function is stable, and a source of the protein of sufficient abundance. To satisfy the last requirement, different tissues and different species are scanned with the assay to find a source in which the particular protein is present at the highest relative concentration. Often it is possible to increase the amount of a given protein in a given cell by selective pressure. For example, when human cells of the HeLa line are grown on media with increasing concentrations of ouabain, a drug that inhibits Na^+/K^+-transporting adenosinetriphosphatase, mutant cells that can survive under these conditions are selected. These mutants counter the inhibition by producing 10 times as much of the protein as normal cells.[53]

Table 1–1: Purification of Mevalonate Kinase from Porcine Liver[55]

purification step	total protein (mg)	total activity[a] (units)	specific activity[a] (units mg^{-1})	volume (mL)	purification	yield of units (%)
extract	136,500	3,969	0.029	4,010		100.0
ammonium sulfate	60,050	3,120	0.052	1,274	1.78	78.8
calcium phosphate gel	9,398	2,270	0.242	3,530	8.34	57.4
ammonium sulfate	7,855	2,035	0.259	335	8.93	51.4
DEAE-cellulose	833	724	0.87	760	30.0	18.3
ammonium sulfate	562	622	1.105	33	38.1	13.3
DEAE-cellulose	29	258	8.99	153	310.0	6.5
Sephadex G-150	5	87	17.5	30	605.0	2.2

[a]A unit of enzymatic activity is defined as the amount of enzyme required to incorporate 1 μmol of phosphate into (R)-5-phosphomevalonate min^{-1}.

Once a particular protein has been purified and the complementary DNA for that protein has been cloned, it is often possible to express that complementary DNA under the control of a strong promoter to create cells that produce high levels of the protein. For example, the catalytic subunit of cyclic AMP-dependent protein kinase was originally purified from bovine heart, and its amino acid sequence was used to obtain complementary DNA encoding the protein from the mouse. This complementary DNA was incorporated into a plasmid next to the T7 promoter. When this plasmid was transfected into *Escherichia coli*, a strain of the bacteria resulted that produced the murine catalytic subunit as 30% of its total protein.[54] This method provided an abundant source of this protein for further studies.

The progress of the purification of a protein is evaluated by examination of both the **total activity** recovered, which is a measure of the yield of the particular protein at each step, and the **specific activity**, the units of activity (milligram of protein)$^{-1}$, which is a measure of the enrichment of the protein of interest relative to the other proteins present (Table 1–1). There is a conventional order in which the various steps of the purification are carried out. This order is usually determined by the volume of material a certain procedure can accommodate, because the volumes that must be processed, if the protein has been concentrated after each step, always decrease as the purification proceeds because of the decrease in the total amount of protein.

Precipitations can be carried out on very large volumes and are often the first step in a purification. **Ammonium sulfate** at high concentrations causes most proteins to precipitate from solution. Each protein precipitates in a given range of ammonium sulfate concentration. Extraneous proteins that precipitate at lower concentrations can be removed first, and then the protein being purified can be precipitated by raising the concentration of ammonium sulfate and thus be separated from proteins that remain soluble at the higher concentration. In the example of mevalonate kinase

$$MgATP + (R)\text{-}mevalonate \rightleftharpoons$$
$$MgADP + (R)\text{-}5\text{-}phosphomevalonate \qquad (1\text{–}47)$$

from porcine liver (Table 1–1),[55] the enzyme was precipitated between 15% and 45% ammonium sulfate for a purification of 1.8-fold. Purification by ammonium sulfate precipitation is usually not very great, but the procedure is a mild one, usually of high yield, which can be used to concentrate rapidly and gently a solution of protein between later steps in a purification (Table 1–1). Poly(ethylene glycol) has also been used to precipitate proteins selectively and reversibly. Tryptophan 5-monooxygenase can be purified 5-fold after precipitation with poly(ethylene glycol) and redissolution in aqueous buffer.[56] Purifications in excess of 3–4-fold are rarely accomplished by differential precipitation. Another precipitation method for purifying proteins is **isoelectric precipitation**. At the pH at which a given protein bears no net charge, its isoelectric pH, it is least soluble in water. If the pH is adjusted to this value and the salts in the solution are removed by dialysis, the protein will often precipitate, while other proteins, which have different isoelectric points, do not. An isoelectric precipitation has been used in the purification of aspartate transcarbamylase[57] and in the purification of fibrinogen.[58]

Ammonium sulfate precipitation, poly(ethylene glycol) precipitation, and isoelectric precipitation are reversible, and the protein is readily redissolved by decreasing the concentration of precipitant or changing the pH, but precipitation by **acid** or **heat**, two other widely used procedures, is usually not reversible. In these situations advantage is taken of the ability of the protein of interest to remain in solution while other proteins precipitate irreversibly. An example of the use of precipitation with heat occurs in the purification of 6-phosphofructokinase, and during this step a 2.5-fold increase in specific activity was recorded.[59] These techniques are quite harsh and can lead to proteolytic degradation of the protein being purified or to deamidation of glutamine and asparagine residues even though very little loss of enzymatic activity is recorded.

During their purification, proteins may be submitted to **selective adsorption**, **ion-exchange chromatography**, **adsorption chromatography**, and **molecular exclusion chromatography**. The procedures are usually performed in this order because the media used for the first procedure are inexpensive, and the third and fourth procedures require smaller samples than the second procedure. The matrices from which the ion-exchange media and media for adsorption chromatography are made are usually cellulose or dextran, which do not by themselves adsorb proteins. The matrices used as molecular exclusion media are agarose, dextran, or polyacrylamide, for the same reason.

The purification of mevalonate kinase (Table 1–1) illustrates this systematic strategy. In each step of the purification the specific enzymatic activity [units (milligram of protein)$^{-1}$] increases as extraneous proteins are separated from the desired protein, and the yield of enzymatic activity after each step is high. Nevertheless, because there are so many steps, the overall yield is low. In this example, molecular exclusion chromatography (on Sephadex G-200 or Sephadex G-150) is used in the latter steps when total amounts of protein (560 and 30 mg) are small enough to permit the samples to be concentrated to the small volumes required by this procedure. Anion-exchange chromatography (DEAE-cellulose), however, can be used earlier in the purification because large volumes (335 mL) at low concentration of electrolyte can be passed through the solid phase, which adsorbs the desired protein as the solution passes and in the process concentrates it. The chromatography itself is then initiated by increasing the concentration of electrolyte. Selective precipitation (ammonium sulfate) and selective adsorption (calcium phosphate gel) are used as initial steps.

Proteins are separated by these techniques because of differences among them in particular properties. Different proteins have different sizes and shapes and can be separated by molecular exclusion chromatography. Different proteins also have different charges at a given pH and can be separated by ion-exchange chromatography. In the case of ion exchange, a pH is usually chosen at which the protein to be purified has a net charge opposite to the fixed charge on the stationary phase so that it will participate in ion exchange with the stationary phase as the chromatography progresses. The elution of the protein is usually performed with a gradient of increasing concentration of a simple monovalent salt such as KCl. If a gradient of pH is used, the change in pH is usually in the direction that would decrease the magnitude of the net charge on the protein.

The separation of three different proteins in the ammonium sulfate precipitate from a crude homogenate by molecular exclusion chromatography[60] is illustrated in Figure 1–5. The three proteins that were assayed were glyceraldehyde-3-phosphate dehydrogenase (GDH), phosphoglycerate mutase (PGM), and phosphoglycerate kinase (PGK). Each of the three enzymes migrates with a characteristic elution volume, V_e, and the glyceraldehyde-3-phosphate dehydrogenase is cleanly separated from the other two enzymes by the column of Sephadex G-150. The noted fractions (IV-K-M) were combined and submitted directly to chromatography on (diethylaminoethyl)cellulose developed with a gradient of sodium chloride (Figure 1–5, bottom). In this step the phosphoglycerate mutase was cleanly separated from the phosphoglycerate kinase. These examples illustrate the use of column chromatography, monitored by enzymatic assay, to separate proteins.

An example of the use of a sequence of steps of column chromatography to purify a particular protein is found in the purification[61] of α-ketoisocaproate oxygenase from rat liver (Figure 1–6). Aside from an initial ammonium sulfate precipitation, only the three consecutive steps, ion-exchange chromatography (Figure 1–6A), adsorption chromatography (Figure 1–6B), and molecular exclusion chromatography (Figure 1–6C), were necessary to purify the enzyme to homogeneity.

When proteins are purified by column chromatography, the increase in specific activity seen in each of the chromatographic steps is usually around 5-fold. Extreme examples of purification, such as the 100-fold purification of phospho-2-dehydro-3-deoxyheptonate aldolase on phosphocellulose[62] or the 100-fold purification of methylcrotonyl-CoA carboxylase on (diethylaminoethyl)cellulose,[63] are rare. There are two methods, however, that do not rely on chromatography and that often produce even greater degrees of purification. They are based on the selective adsorption to or selective elution from a stationary phase and can be referred to as affinity adsorption or affinity elution, respectively.

When a protein is purified by **affinity elution**, it is first adsorbed to a stationary phase, such as a chromatographic medium; after all unabsorbed proteins have been washed away, a compound that binds with high specificity to the protein of interest is added. The presence of this compound can sometimes cause only that protein to which it binds to elute from the stationary phase. For example, when (carboxymethyl)cellulose is added to a crude, clarified homogenate from liver at pH 6, all of the fructose-1,6-bisphosphatase is adsorbed along with many other proteins. The (carboxymethyl)cellulose is then collected and washed well with 5 mM sodium malonate, pH 6. When it is then rinsed with 0.06 mM fructose-1,6-bisphosphate in 5 mM sodium malonate, pH 6, only the fructose-1,6-bisphosphatase elutes. In one step the enzyme can be purified 400-fold, to homogeneity.[64] In

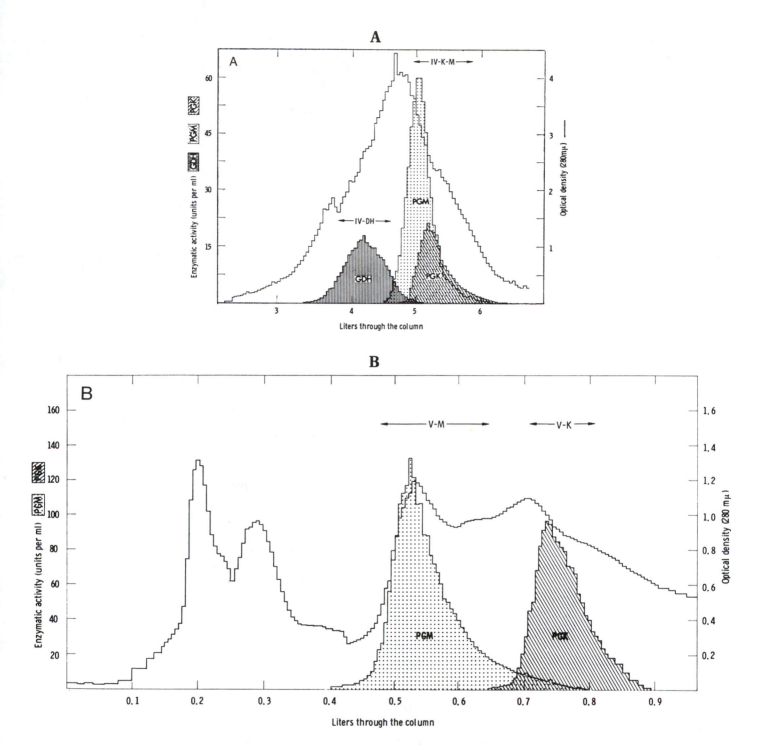

Figure 1–5: Molecular exclusion chromatography[60] (A) and anion-exchange chromatography (B) of proteins in a homogenate from the bacterium *Escherichia coli*. The clarified homogenate was submitted to precipitation with ammonium sulfate (30–45%). The precipitate (7.2 g of protein) was redissolved in a minimum volume (120 mL) of aqueous buffer and submitted to zonal chromatography on a column (10 cm × 120 cm) of Sephadex G-150 (A). Fractions were assayed for protein (optical density at 280 mμ) and enzymatic activity (units milliliter⁻¹) for glyceraldehyde 3-phosphate dehydrogenase (GDH), phosphoglycerate mutase (PGM), and phosphoglycerate kinase (PGK), respectively. The units of enzymatic activity are micromoles minute⁻¹. The proteins contained in the fractions in pool IV-K-M from the chromatogram in panel A were combined and submitted to anion-exchange chromatography (B). The ionic strength of the buffer used for the molecular exclusion chromatography was low enough that the sample (900 mL) could be passed directly through the column (2.2 cm × 25 cm) of (diethylaminoethyl) (DEAE-)cellulose while the proteins gathered at the top of the ion-exchange medium. Chromatography was then initiated with a gradient of NaCl (0–0.15 M in the same buffer at pH 8). Fractions were again assayed for protein and enzymatic activity. Reprinted with permission from ref 60. Copyright 1971 *Journal of Biological Chemistry*.

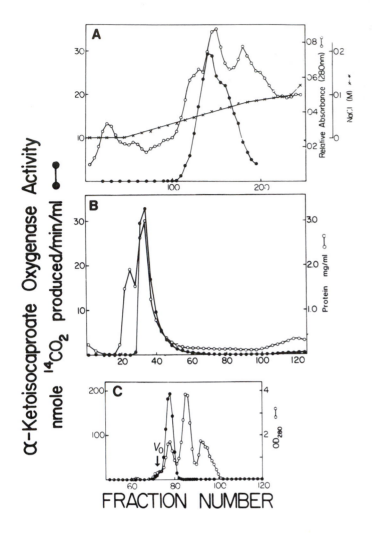

Figure 1–6: Column chromatography of rat liver α-ketoiso-caproate oxygenase.[61] An ammonium sulfate (45–75%) precipitate (35 g of protein) of the clarified homogenate was redissolved, dialyzed to remove salt, and applied to a column (5 cm × 80 cm) of DEAE-cellulose (A). Fractions containing enzymatic activity (4 g of protein) were pooled, concentrated, brought to 2.5 M NaCl, and applied to a column (4 cm × 40 cm) of agarose to which phenyl groups had been attached covalently. The proteins were eluted with a gradient between 2.5 M NaCl and buffer without added NaCl (B). The fractions containing enzymatic activity (500 mg of protein) were pooled, concentrated, and applied to a column (5 cm × 80 cm) of the molecular exclusion medium Sephacryl S-200 (C). In each panel, α-ketoisocaproate oxygenase activity (nanomoles minute^{-1} milliliter^{-1}) is presented as a function of fraction number. The total protein in each fraction was also monitored by absorbance (optical density) at 280 nm. The final yield was 70 mg of protein in the peak of enzymatic activity. Reprinted with permission from ref 61. Copyright 1982 *Journal of Biological Chemistry.*

cleotides or dinucleotides. Deoxythymidine 3,5-bisphosphate is a specific inhibitor of the nuclease that binds to it very tightly. A *p*-aminophenyl derivative of this inhibitor was synthesized and attached covalently to agarose through its aniline nitrogen to produce a stationary phase displaying deoxythimidine 3,5-bisphosphate (Figure 1–7). When a crude supernate containing micrococcal nuclease was passed over this affinity medium, none of the nuclease emerged but almost all of the protein did. The nuclease could then be eluted nonspecifically with dilute acetic acid in greater than 90% yield. It was completely purified in this one step.

Since this early report, the technical aspects of affinity adsorption have been exhaustively explored. The main difficulty to which many of these investigations have been directed is positioning the ligand far enough from the polymeric matrix of the agarose to interact effectively with the protein.[69,70] This problem may explain many of the failed attempts to use the technique of affinity adsorption. Several long, hydrophilic connecting links, usually referred to as arms, that serve the purpose of the *p*-aminophenyl in the above example (Figure 1–7), have been developed to solve this problem. Often a long hydrophilic arm is created during the set of reactions used to attach the ligand to the solid phase (Figure 1–8). Many different strategies for attaching ligands of various structures to the stationary phase have been developed.

The cases in which affinity adsorption has been successful in the purification of proteins provide a provocative collection of examples (Table 1–2). Because purifications of 100-fold in one step are not unusual, this approach has obvious advantages over the usual strategy that combines selective precipitation, selective adsorption, and chromatography (Table 1–1), where several steps are required to achieve the same degree of purification. It remains the case, however, that affinity adsorption often requires a greater investment than assembling a se-

another example, hexokinase in homogenates of the brain remained bound to a particulate fraction which could be washed extensively and from which the hexokinase could be eluted selectively with its substrate glucose 6-phosphate.[65]

Affinity adsorption is more widely used and has been very successful in a number of instances. The basic idea in affinity adsorption is to synthesize a stationary phase to which has been covalently attached a chemical compound that adsorbs specifically the protein to be purified. The compound synthetically attached to the stationary phase is usually an analogue or a derivative of a reactant or product in the reaction catalyzed by an enzyme, an inhibitor of the enzyme, an allosteric activator of the enzyme, or an agonist or antagonist of a receptor. This compound, when attached to the stationary phase, is referred to as an immobilized ligand for the protein. Agarose[66,67] is the stationary phase to which the immobilized ligand is usually attached.

One of the original examples of this technique[68] can serve to illustrate the idea. Micrococcal nuclease is an enzyme from *Staphylococcus aureus* that can hydrolyze the phosphodiesters of either single-stranded RNA or double-stranded DNA to produce 3'-phosphomononu-

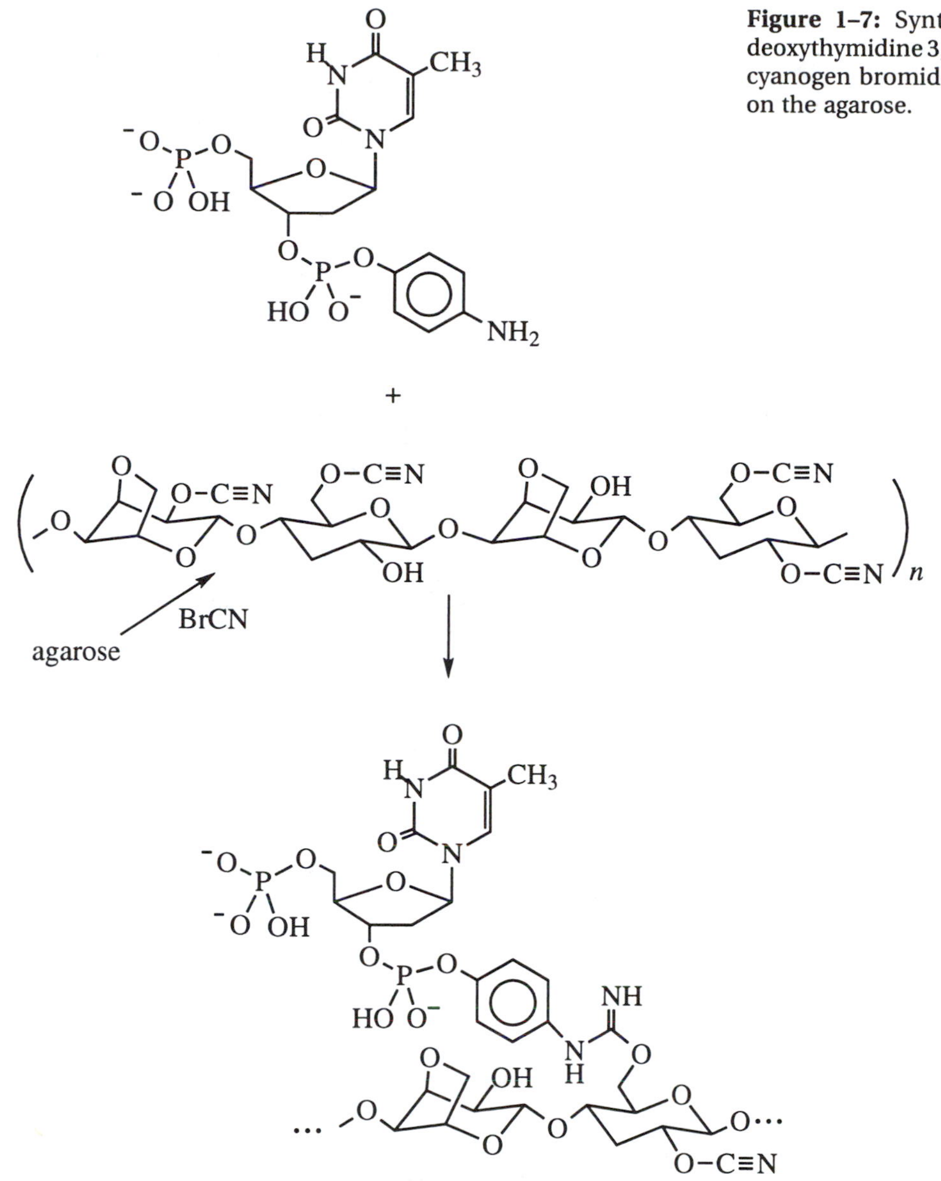

Figure 1–7: Synthetic strategy used to couple covalently deoxythymidine 3,5-bisphosphate to agarose by activation with cyanogen bromide (BrCN). The cyanylation occurs randomly on the agarose.

quence of simple chromatographic steps and has a higher risk of failure. Often the affinity adsorbent produces only a modest purification of 10-fold or less under conditions that suggest that the process occurring is either nonspecific ion exchange[86,87] or simple adsorption[88] or affinity elution from a nonspecific stationary phase.[89,90] Often the protein adsorbs so tightly to the affinity medium that it can be eluted only in very low yield.[91]

The central, defining feature of affinity adsorption is the design of the stationary phase, but the conditions used to elute the bound protein are also characteristic. Often they are merely the application of a mobile phase of extreme pH or ionic strength such as in the original example of micrococcal nuclease. The ideal approach, however, is to combine affinity adsorption with affinity elution to gain an advantage in each of the two steps, and the protein is often eluted with a solution of the soluble ligand from which the immobilized ligand was derived (Figure 1–9, Table 1–2).

Affinity adsorption has been used to purify proteins that bind to particular nucleotide sequences in DNA.[92] In this instance, the spacer arm holding the DNA recognized by the protein away from the surface of the agarose can be produced by simply polymerizing short fragments of DNA containing the target sequence to produce a long repeating double strand of DNA and then attaching this long repeating polymer to the agarose through one of its ends. The DNA closest to the surface of the agarose acts as a spacer arm for the more peripheral segments.

Figure 1–8: Use of a hydrophilic arm to connect a specific ligand to a polymeric support.[45] *N,N*-Di-(3-aminopropyl)amine was attached to agarose by activating the polysaccharide with cyanogen bromide (Figure 1–7). 1-(4-Amino-6,7-dimethoxy-2-quinazolinyl)piperazine, which is a portion of prazosin, a specific antagonist for α_1-adrenergic receptors, was succinylated and then attached to the aliphatic amine by activation of the resulting carboxylic acid with 1-[(*N,N*-dimethyl-amino)propyl]-3-ethylcarbodiimide. This produced an arm of 14 atoms connecting an oxygen of the polysaccharide with the nitrogen of the ligand. The arm is hydrophilic by virtue of the *O*-alkyl-*N*-alkyl urea, the amine, and the two *N*-alkyl amides. This affinity medium was used to purify the α_1-adrenergic receptor.

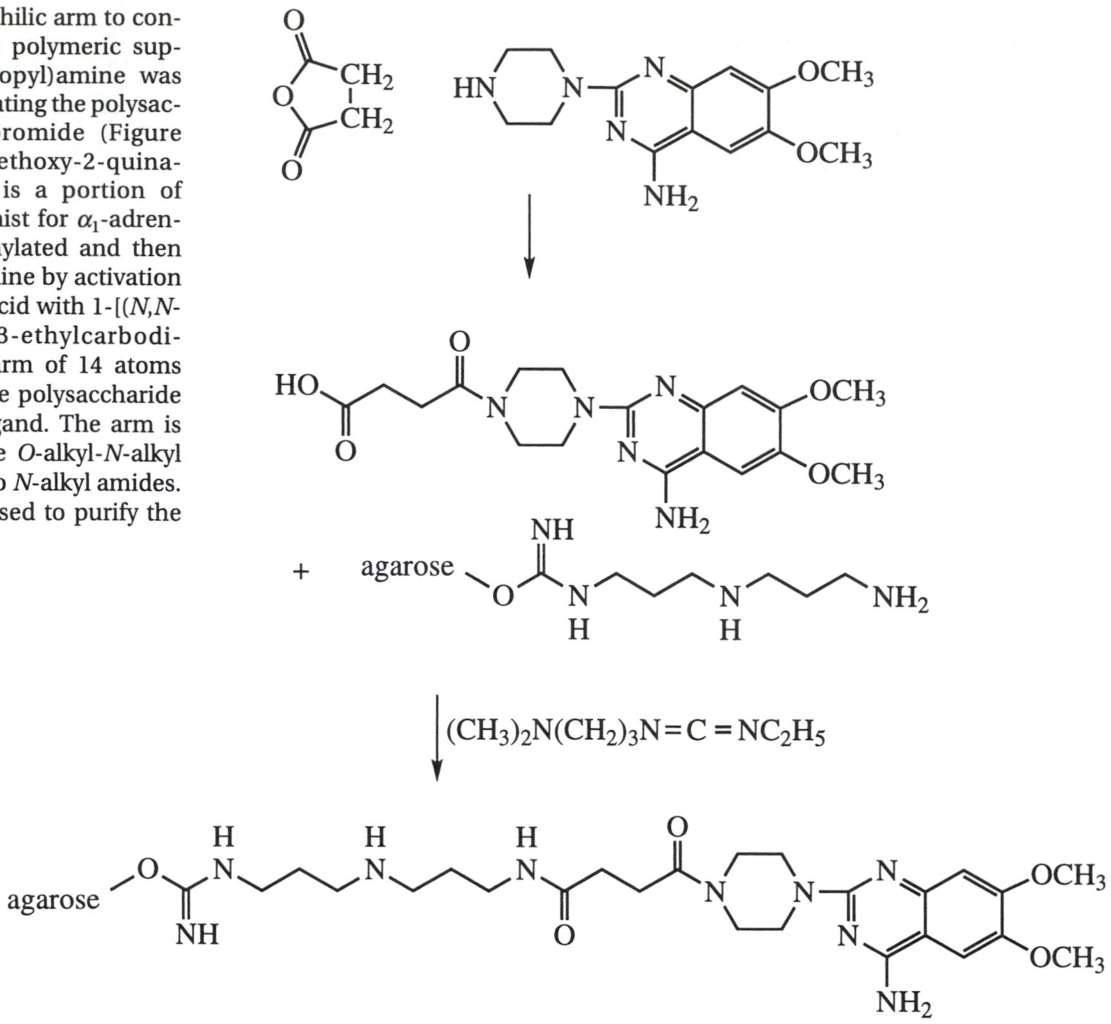

Such an affinity adsorbent was used to purify the promoter-specific transcription factor,[49] which is a protein referred to as Sp1. This protein binds to the nucleotide sequence GGGGCGGGGC in double-stranded DNA, and its concentration in a particular solution can be assayed by observing its footprint on DNA containing this specific sequence. An extract of nuclei from HeLa cells was purified by molecular exclusion chromatography, adsorption chromatography on heparin bound to agarose, cation-exchange chromatography on sulfated dextran, and affinity adsorption on agarose to which the specific DNA was attached. In the last step, the protein was eluted with a high concentration (0.5 M) of KCl. The first three steps produced 100-fold purification with a 20% yield, and the last step alone produced a further 100-fold purification with a 50% yield.

The goal of purification is to obtain the protein of interest isolated from all of the other proteins that were originally in the homogenate derived from the biological specimen. That this has been achieved is often suggested by the coelution of the protein present and the biological or enzymatic activity in the last chromatographic step of

the purification (Figure 1–10).[93] This is only an indication of purity, and the absolute purity of the final preparation must always be demonstrated independently by electrophoresis.

Suggested Reading

Grimshaw, C.E., Henderson, G.B., Soppe, G.G., Hansen, G., Mathur, E.J., & Huennekens, F.M. (1984) Purification and Properties of 5,10-Methenyltetrahydrofolate Synthetase from *Lactobacillus casei*, *J. Biol. Chem.* 259, 2728–2733.

Problem 1–6: Calculate the number of theoretical plates in the column used for the separation displayed in Figure 1–5 from the width of the peak of phosphoglycerate mutase. Use the number of theoretical plates to calculate the width the peak of glyceraldehyde-3-phosphate dehydrogenase should have. Why might its peak be wider than the width calculated?

Problem 1–7: Calculate the number of theoretical plates in the column used in Figure 1–6C.

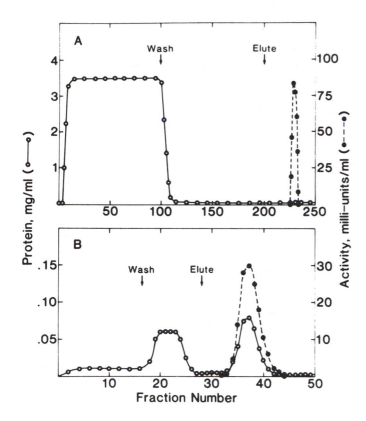

Figure 1–9: Affinity adsorption and affinity elution used in combination to purify 5-formyltetrahydrofolate cyclo-ligase.[83] (A) A crude extract (7.3 g of protein in 2 L) from the bacterium *Lactobacillus casei* was passed over a column (4 cm × 18 cm) of agarose to which 5-formyltetrahydropteroylglutamate had been attached. After the affinity medium was washed with 2 L of buffer until no more protein emerged, the bound enzyme was eluted with a solution of 5-formyltetrahydrofolate, a reactant in the enzymatic reaction. (B) A purified fraction (0.7 mg of protein in 40 mL from a later step in the procedure) was passed over a column (2 cm × 13 cm) of agarose to which ATP had been attached. After the affinity medium was washed with 100 mL of buffer, the bound enzyme was eluted with a solution of ATP, another reactant in the enzymatic reaction. Protein concentration (milligrams milliliter^{-1}) and enzymatic activity (milliunits milliliter^{-1}) were measured for each fraction collected from each column. A unit of activity is defined as a micromole minute^{-1}. Reprinted with permission from ref 83. Copyright 1984 *Journal of Biological Chemistry*.

Problem 1–8: Write a chemical mechanism for the second step in the reaction shown in Figure 1–8.

Problem 1–9: The table on the page 25 describes the purification of α-ketoisocaproate oxygenase (Figure 1–6). Calculate the specific activity, fold purification, and yield at each step.

Problem 1–10: Design an affinity adsorbent for the enzyme nicotinate phosphoribosyltransferase.

Problem 1–11: Alprenolol (A) binds tightly and specifically to β-adrenergic receptor (βAR), which is a protein in the plasma membranes of certain animal cells. The dissociation constant for this binding is the equilibrium constant defined by the equation

$$K_d = \frac{[A][\beta AR]}{[A \cdot \beta AR]}$$

where all concentrations are in moles (liter)$^{-1}$. They are the concentration of free alprenolol, [A], the concentration of uncomplexed β-adrenergic receptor, [βAR], and the concentration of the complex between the alprenolol and β-adrenergic receptor [A · βAR]. The value for K_d is 8 nM.

Alprenolol was covalently attached to agarose to produce an affinity adsorbent for the purification of β-adrenergic receptor. The final concentration of the alprenolol covalently bound to the solid phase, $[A_B]'_{TOT}$, was 2 mM in units of millimoles (liter of bed)$^{-1}$. All molar concentrations designated with primes are in moles (liter of bed)$^{-1}$. Assume that the dissociation constant between covalently bound alprenolol and β-adrenergic receptor is the same as that for unbound alprenolol (8 nM).

Consider what happens when a solution containing β-adrenergic receptor is added to a chromatographic column containing the affinity adsorbent. If, as is reasonable, $[\beta AR]' << [A_B]'_{TOT}$, where $[A_B]'_{TOT}$ is the molar concentration of covalently bound alprenolol (2 mM), then $[A_B]'_{TOT} = [A_B]'$, the molar concentration of covalently attached alprenolol to which β-adrenergic receptor is not bound; and, from the equation for K_d

$$\frac{[A_B]'_{TOT}}{K_d} = \frac{[A_B \cdot \beta AR]'}{[\beta AR]'} \cong \alpha_{\beta AR}$$

where α is the partition coefficient for β-adrenergic receptor between the mobile phase, βAR, and its complex with alprenolol covalently bound to the stationary phase, $A_B \cdot \beta$AR.

(A) If the chromatographic column has a volume of mobile phase, V_0, of 2.0 mL, calculate the elution volume, $^{\beta AR}V_e$, of β-adrenergic receptor.

One way to decrease the elution volume of β-adrenergic receptor would be to add free alprenolol to the mobile phase at a particular molar concentration $[A_M]'$ in moles (liter of bed)$^{-1}$. Again, if $[\beta AR]' << [A_B]'_{TOT}$, then

$$\alpha_{\beta AR} \cong \frac{[A_B \cdot \beta AR]'}{[\beta AR]' + [A_M \cdot \beta AR]'}$$

(B) Derive an equation for $\alpha_{\beta AR}$ in terms of $[A_M]'_{TOT}$, $[A_B]'_{TOT}$, and K_d, if

Table 1–2: Examples of the Use of Affinity Adsorption in the Purification of Proteins

protein	ligand	point of connection to ligand	elution conditions	purification (fold)	ref.
phospho-2-dehydro-3-deoxyheptonate aldolase	tyrosine (allosteric inhibitor)	N^α		100	71
procollagen-proline, 2-oxoglutarate 4-dioxygenase	(Pro-Gly-Pro)$_n$ ($n \cong 10$)	amino terminus	solution of (Pro-Gly-Pro)$_n$	1500	72
UDPglucose 4-epimerase	UDP	β-phosphate	UMP	100	73
L-lactate dehydrogenase	NAD$^+$	adenosine N^6	phosphate	4	74
	NAD$^+$	adenosine C^8	pyruvyl NAD$^+$	40	
	AMP	adenosine C^8	pyruvyl NAD$^+$	40	
isocitrate dehydrogenase	AMP	ribose	NAD$^+$	40	75
choline acetyltransferase	coenzyme A		gradient of NaCl	100	76
cathepsin D	pepstatin	carboxyl group	pH 8.5	100	77
N-acetylglucosamine kinase	glucosamine	N^2	glucose	20	78
hexokinase	glucosamine	N^2	glucose	40	79
dihydrofolate reductase	methotrexate	carboxyl group	dihydrofolate	< 200	80
N-acetylgalactosaminidemucin-β-1,3-galactosyltransferase	asialomucin	bound to DEAE-cellulose	EDTA	20	81
ornithine decarboxylase	pyridoxamine phosphate	–NH$_2$	pyridoxal phosphate	1000	82
5-formyltetrahydrofolate cyclo-ligase	5-formyltetrahydro-pteroylglutamate	carboxyl group	5-formyltetra-hydrofolate	4000	83
β-adrenergic receptor	alprenolol	olefin addition	isoproterenol	100	84
α_1-adrenergic receptor	analogue of prazosin	carboxyl group	prazosin	200	45
plasminogen	L-lysine	amino group	ε-aminocaproic acid	200	85

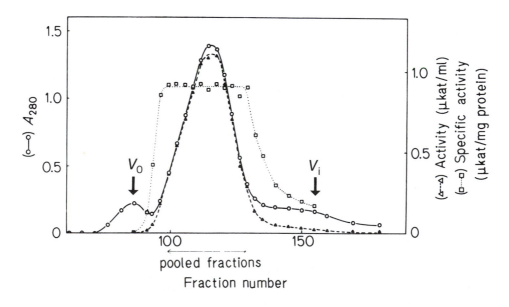

Figure 1–10: Elution of malate synthase upon molecular exclusion chromatography.[93] A solution (280 mg of protein in 7 mL) of malate synthase, from the penultimate step in the purification procedure from bakers' yeast, was loaded onto a column (1.8 L) of Sephadex G-100. The fractions (7 mL) collected from the bottom of the column were assayed for protein (A_{280}) and enzymatic activity, and specific activity was calculated by dividing enzymatic activity (milliliter)$^{-1}$ by A_{280}. Reprinted with permission from ref 93. Copyright 1981 Springer-Verlag.

Table for Problem 1–9

step	fraction	volume (mL)	total protein (mg)	total activity (nmol min^{-1})
1	10,000 g supernatant	7,200	104,200	58,600
2	45–75% $(NH_4)_2SO_4$ fraction	1,200	34,700	43,100
3	pre-DEAE-dialysate	1,400	20,900	34,800
4	DEAE-cellulose pool	1,770	4,000	30,500
5	concentrated DEAE pool	250	4,200	29,200
	116 mL of the concentrated DEAE pool was used for the remainder of the purification			
6	pre-phenyl-Sepharose	120	1,800	12,400
7	phenyl-Sepharose pool	675	513	9,790
8	phenyl pool concentrated	20	474	10,700
9	Sephacryl S-200 pool	60	71	7,450

$$K_d = \frac{[A_M]'\,[\beta AR]'}{[A_M \cdot \beta AR]'}$$

(C) Calculate the elution volume of β-adrenergic receptor from the same chromatographic column ($V_0 = 2$ mL) if the concentration of alprenolol in the mobile phase, $[A_M]'_{TOT}$, is 0.10 mM.

Net Molecular Charge

Before electrophoresis can be understood, the property of a macromolecule that permits electrophoresis to occur, namely, its net molecular charge, must be described. A molecule of protein is a polyelectrolyte bearing a net electrostatic charge that is the difference between the sum of its many positive charges and the sum of its many negative charges. These individual charges are the charges of loosely adsorbed electrolytes from the solution, the tightly bound coenzymes and metallic cations, and the covalent posttranslational modifications of the protein as well as the positive charges of the guanidinium, ammonium, and imidazolium cations and the negative charges of the carboxylates, thiolates, and phenolates that are incorporated into the covalent molecular structure of the protein itself. Each of these latter functional groups is the conjugate acid or base of a weak neutral base or acid, respectively, and the degree to which they are each ionized is a function of the pH of the solution.

A molecule of a protein responsible for one biological function differs in its structure from a molecule of a protein responsible for another biological function; for example, a molecule of fumarate hydratase differs in its structure from a molecule of citrate (*si*) synthase. When two proteins differ from each other in structure, they will differ from each other in the number of each type of these weak bases and weak acids that they contain. These differences in composition cause each protein to have a characteristic net proton charge at a given pH. The **net proton charge**, $^H\bar{Z}_i$ [moles of charge (mole of protein)$^{-1}$], of protein i is the summation, at a particular pH, of all the positive and negative charges on a molecule of that protein that arise from ions or functional groups that remain affixed to the protein, covalently or noncovalently, in pure water in the absence of other dissolved electrolytes in the solution. Because the rates of protonic equilibria are extremely rapid, all molecules of the protein i, even though each has a different integral charge at any instant, will have the same average net proton charge, $^H\bar{Z}_i$, if this is averaged over a time as long as a second.

The change in the average net proton charge on a protein as a function of pH is measured by performing a simple acid–base titration.[94] The number of moles of protons or hydroxide ions necessary to adjust an unbuffered solution containing a known molar concentration of protein i to a given final pH from a given initial pH is measured. The number of moles of protons or hydroxide ions necessary to adjust an identical unbuffered solution, lacking the protein, to the same final pH from the same initial pH is then determined. The solution containing the protein will always consume more equivalents of protons or hydroxide ions than the control, and this additional amount is the moles of positive charge gained by the protein upon association of the protons or positive charge lost upon dissociation of the protons and their combination with hydroxide anion, respectively, as the pH of the solution is changed from the initial value to the final value.

To determine the actual average net proton charge on a protein, rather than only its relative net proton charge, as a function of pH, the isoionic point of the protein must be determined.[94] The **isoionic point**, $pH_i^{isoionic}$, is the pH of a solution containing only protein i and water. Since the only cations and anions in such a solution, other than the charges bound to the protein, are protons and hydroxide ions

$$(^H\bar{Z}_i^{\ isoionic})[\text{protein } i] + [H^+]^{isoionic} = [OH^-]^{isoionic} \quad (1\text{–}48)$$

where [protein i] is the molar concentration of the protein, $^H\bar{Z}_i^{isoionic}$ is the average net proton charge [expressed in units of moles of net atomic charges (mole of protein)$^{-1}$] on the protein at its isoionic point, and $[H^+]^{isoionic}$ and $[OH^-]^{isoionic}$ are the molar concentrations of protons and hydroxide ions in this isoionic solution. This equation can be combined with the expression for the ionization of water ($K_w = [H^+][OH^-]$) and the definition of pH ($[H^+] = 10^{-pH}$) to give

$$^H\overline{Z}_i^{\,isoionic} = \frac{K_w - 10^{-2pH_i^{isoionic}}}{[protein\ i]10^{-pH_i^{isoionic}}} \qquad (1\text{--}49)$$

where K_w is the ionization constant of water. Equation 1–49 can be used to calculate the average net proton charge on the protein i at its isoionic point, and this provides a measurement of the absolute average net proton charge on the protein i at one pH in the absence of electrolytes. A solution containing only protein i and water, an isoionic solution, is usually obtained by passing a solution of protein i over a mixed-bed ion-exchange medium to remove all salts, and the pH of the resulting solution is then measured.[95]

The isoionic point, $pH_i^{isoionic}$, should be formally distinguished from the pH at which the protein has zero average net proton charge because at the isoionic point the protein does bear an average net proton charge. It is clear from Equation 1–49, however, that if [protein i] is significant and $pH_i^{isoionic}$ is between pH 5 and 9, there is very little difference between the isoionic point and the point of zero net proton charge.

The acid–base titration, which is a measurement of the relative average net proton charge on protein i at a series of pH values, and the isoionic point, which is a measurement of the absolute average net proton charge on protein i at one specific pH, can be combined to give the average net proton charge, $^H\overline{Z}_i$, as a function of pH (Figure 1–11).[95] From such a relationship, $^H\overline{Z}_i$ at any given pH can be determined by interpolation.

The net proton charge on protein i, $^H\overline{Z}_i$, differs from the net molecular charge on protein i, Q_i (moles of charge mol^{-1}), because proteins have a tendency to bind electrolytes, even ones as simple as halides[96] and alkali metal ions.[17] This binding can occur even at the point of zero net proton charge and is reflected as a decrease or increase in the $pH_i^{isoionic}$ as a neutral salt is added to an isoionic solution.[94] For example, if protein i in an isoionic solution binds more of the anions than the cations of a neutral salt that is added, the increase in its negative charge will indirectly cause it to take up more protons, increasing $pH_i^{isoionic}$. The reverse effect on the isoionic point is observed when the cations are preferentially bound. These shifts in $pH_i^{isoionic}$ with ionic strength must be noted to correct the absolute values of net proton charge for changes in ionic strength.

This binding of small simple ions, such as halides and alkali metal cations, to proteins probably results from chelation. Two or more fixed charges on the protein, of

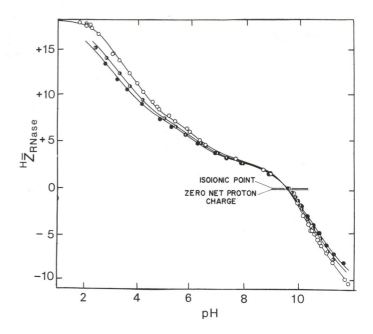

Figure 1–11: Average net proton charge on ribonuclease as a function of pH. Solutions of ribonuclease at ionic strengths 0.01 M (●), 0.03 M (◑), and 0.15 M (○), produced with KCl, were titrated with either KOH or HCl.[95] The changes in pH as a function of the equivalents of acid or base added (mole of protein)$^{-1}$ were recorded. The isoionic point was determined by passing a solution of the protein over a mixed-bed ion-exchange medium to remove all electrolytes except H$^+$ and OH$^-$. The net proton charge, $^H\overline{Z}_{RNase}$, is presented as a function of the pH. Reprinted with permission from ref 95. Copyright 1956 American Chemical Society.

opposite sign to the bound ion, would have to be properly oriented to perform such chelation. If this is the case, the amount of binding at the isoionic point and its ionic preference would be a unique and unpredictable property of each protein, as has been observed. In deoxyhemoglobin, a site at which chloride binds to the protein has been identified, and it sits between two functional groups, an ammonium cation and a guanidinium cation, that both bear a positive charge and presumably chelate the chloride.[97] Another binding site for chloride on deoxyhemoglobin has been tentatively assigned to a location among an imidazolium cation and two ammonium cations on the surface of the protein.[97] Proteins with high densities of negative charge seem to bind cations more readily than those with low densities of negative charge,[17] and this tendency would presumably result from the increase in the probability of proper juxtaposition for chelation with the increase in the density of negative charge. As the pH is lowered from the point of zero net proton charge, the density of positive charge on a protein rarely increases; rather, the density of negative charge decreases as carboxylates are neutralized. It has been observed that the number of bound anions increases as the pH is lowered,[98] which results from the decrease in electrostatic repulsion, due to these carboxylates, that at neutral pH was inhibiting the chelation of dissolved anions by the fixed positive charges on the protein. For reasons that are not well understood, but that may include the differences in ionic radii, proteins seem to bind halides more readily than they do alkali metal cations.

The **net molecular charge**, Q_i [in moles of atomic charge (mole of protein)$^{-1}$], on protein i in a solution containing simple neutral salts such as NaCl or KCl is the sum of the net proton charge and the net charge contributed by these bound ions:

$$Q_i = {}^H\overline{Z}_i + \sum_{j=1}^{m} \overline{v}_j z_j \qquad (1\text{–}50)$$

where $\overline{v}_j$ is the mean number of ions of species j and valence z_j bound by the protein. It is this net charge on protein i that determines its behavior on ion-exchange chromatography or electrophoresis.

Suggested Reading

Tanford, C., & Wagner, M.L. (1954) Hydrogen Ion Equilibria of Lysozyme, *J. Am. Chem. Soc.* 76, 3331–3336.

Problem 1–12: At a protein concentration of 3×10^{-4} M, the isoionic pH of ribonuclease is 9.60. Calculate ${}^H Z_{\text{RNase}}^{\text{isoionic}}$.

Electrophoresis

When a molecule of protein i at a given pH in an aqueous solution of electrolytes is placed in an electric field, it will experience a force, F_{el} (in dynes), in the direction x such that

$$F_{\text{el}} = \frac{\mathcal{F} Q_i E_x}{N_{\text{Av}}} \qquad (1\text{–}51)$$

where Q_i is the net molecular charge on protein i under these circumstances in number of atomic charges (molecule of protein)$^{-1}$; $\mathcal{F}$ is Faraday's constant (9.65×10^{11} erg V^{-1} mol^{-1}), N_{Av} is Avogadro's number (6.02×10^{23} molecules mol^{-1}), and E_x is the electrical field (volts centimeter^{-1}) or gradient of the electric potential ($\partial V/\partial x$) in the direction x. Electrophoresis is usually run in an apparatus designed so that ($\partial V/\partial y$) and ($\partial V/\partial z$) are zero, and the force F_{el} will cause the molecule of protein i to move only in the x direction. For the moment, it will be assumed that only the molecule of protein i and its physically bound ions move. As the molecule of protein i moves, a frictional force, F_{fric}, exerted by the surrounding stationary liquid is experienced by the molecule. The frictional force is proportional to the velocity of movement

$$F_{\text{fric}} = -f_i \left(\frac{\partial x}{\partial t} \right)_E \qquad (1\text{–}52)$$

where the constant of proportionality, f_i, is the frictional coefficient (grams second^{-1}) of the molecule.

At this point a digression is necessary to explain the frictional coefficient before continuing with a discussion of electrophoresis. The most direct way to determine the frictional coefficient of a molecule of protein is from its diffusion coefficient, D. This parameter measures the net tendency of any particular substance to move from a region of high concentration to a region of low concentration; the driving force behind this movement is not a function of any intrinsic feature of the molecule such as its charge or its mass. The diffusion coefficient (centimeters^2 second^{-1}) of any substance i in solution is defined by Fick's law

$$\left(\frac{\partial m_i}{\partial t} \right)_x = -D_i \left(\frac{\partial C_i}{\partial x} \right)_t \qquad (1\text{–}53)$$

where $(\partial m_i/\partial t)_x$ is the flux [moles (centimeter^2 second)$^{-1}$] of substance i through a planar surface of unit area, C_i is the concentration [moles centimeter^{-3}] of the substance i at any point, and x is the distance (centimeters) along an axis normal to the planar surface. The greater $(\partial C_i/\partial x)_t$, the greater the diffusive force and the greater the net flux. The diffusion coefficient, D_i, is simply the constant of this proportionality. It can be shown that, for any substance i

$$f_i = \frac{k_B T}{D_i} \qquad (1\text{–}54)$$

where k_B is Boltzmann's constant (1.38×10^{-16} erg K^{-1}) and T is the temperature (Kelvins).

The diffusion coefficient of a protein is usually measured by creating a sharp boundary between two solutions, one of which contains the protein at a given initial concentration and the other of which is otherwise identical to the first but does not contain the protein (Figure 1–12). At any time after initiating the experiment, $(\partial C_i/\partial x)_t$, where x is normal to the original boundary, will be a Gaussian function. Its width will increase with time as diffusion spreads the boundary, and

$$D = \frac{1}{4\pi t} \left(\frac{A}{H} \right)^2 \qquad (1\text{–}55)$$

where A is the area (concentration) of the curve of $(\partial C_i/\partial x)_t$ against x and H is its maximum height (concentration centimeter^{-1}).

The frictional coefficients of spheres or ellipsoids of revolution can be calculated. For a sphere

$$f = 6\pi\eta r \qquad (1\text{–}56)$$

where η is the viscosity [grams (second · centimeter)$^{-1}$] of the solution and r is the radius (centimeters) of the sphere. This formalism has led to the concept of the **effective sphere** or Stokes' sphere representative of protein i, the radius of which, a_i, is defined as

$$a_i = \frac{k_B T}{6\pi\eta D_i} \qquad (1\text{-}57)$$

This radius, a_i, can be considered to be the radius of a sphere whose diffusion coefficient is the same as the diffusion coefficient of protein i. It is usually referred to as the Stokes' radius of the protein.

It is now possible to return from the digression defining the frictional coefficient to the molecule of protein in the electric field. When the electric field is turned on, a steady state[3] is rapidly reached in which $F_{el} = F_{fric}$ and which is characterized by a constant net velocity $(\partial x_i/\partial t)_E$ of the molecules of protein i in the direction of the electric field. At steady state, because $F_{el} = -F_{fric}$

$$\left(\frac{\partial x_i}{\partial t}\right)_E = \frac{\mathcal{F}Q_i E_x}{N_{Av} f_i} \qquad (1\text{-}58)$$

Although electrophoresis is usually carried out in an apparatus in which the current passes through a complicated path, the region of the apparatus over which the proteins are actually separated is uniform in its dimensions and in its specific conductance, σ_{sp} [in (ohm · centimeter)$^{-1}$]. If the current, I (in amperes) through the apparatus, remains constant, then from Ohm's law

$$E_x = \left(\frac{\partial V}{\partial x}\right) = \frac{I}{A\sigma_{sp}} \qquad (1\text{-}59)$$

where A is the cross-sectional area (square centimeters) of the region of the apparatus where separation occurs normal to the direction of the flow of current. The **free electrophoretic mobility**, $u_i°$ [in centimeters^2 (volt · second)$^{-1}$] is defined as

$$u_i° \equiv \frac{(\partial x_i/\partial t)_E}{(\partial V/\partial x)} = \frac{d_i \sigma_{sp} A}{tI} \qquad (1\text{-}60)$$

where d_i is the distance (centimeters) moved by protein i in time t (seconds) in a particular electric field. This definition causes the electrophoretic mobility to be only a function of the molecule of protein and the medium through which it is moving.

It follows that, if the assumptions that have been made were correct, the relationship governing electrophoresis would be

$$u_i° = \frac{\mathcal{F}Q_i}{N_{Av} f_i} \qquad (1\text{-}61)$$

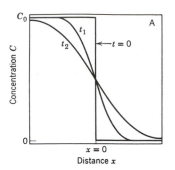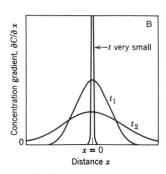

Figure 1–12: (A) Spreading of a boundary of concentration at the interface formed between two solutions, one containing the solute and the other not containing the solute.[99] A solution containing the solute is brought in contact with a solution otherwise identical but lacking the solute. At the initial time the concentration function (C) is discontinuous at the boundary, but as time progresses the solute diffuses into the vacant solution and a gradient of concentration develops. (B) The first derivative of the concentration with respect to the distance in the direction x [$(\partial C/\partial x)_t$] at any instant is always a Gaussian function, the width of which increases and the height of which decreases with time, t. Reprinted with permission from ref 99. Copyright 1961 John Wiley.

This relationship, however, is an incomplete description of electrophoresis and fails to explain actual behavior.[100] Equation 1–61 states that electrophoretic mobility will be affected by ionic strength only insofar as Q_i is affected by ionic strength. In general, Q_i increases gradually but not impressively with ionic strength, as ionic shielding permits the molecule of protein i to bear a greater net charge (Figure 1–11), yet it is observed that electrophoretic mobility declines precipitously as ionic strength is increased (Figure 1–13).[100]

The inadequacy of Equation 1–61 is due to the erroneous assumption that only the molecule of protein and its bound ions move through the solution. What actually occurs is that the molecule of protein draws along some of the counterions in the solution surrounding it. These counterions, moving under the influence of the moving macromolecule, have the effect of decreasing the effective charge of the macromolecule. This decrease arises from the fact that the layer of solution surrounding a charged molecule of protein dissolved in an aqueous solution of electrolytes bears, as does the solution surrounding a charged solid phase for ion-exchange chromatography, a net charge of opposite sign due to the existence of an ionic double layer.[13] The layer of solution surrounding the molecule of protein is enriched, as is the layer of solution surrounding a charged solid phase, in counterions to the net charge on the protein and depleted in ions of like sign. A region of solution large enough to contain the molecule of protein and the entire ionic double layer would be electrically neutral, and if the molecule of protein moving through the solution drew along the entire double layer, it would have no net charge.

If the molecule of protein were the sphere of Stokes radius a (centimeters) and that sphere were of uniform surface charge density and of net molecular charge, Q, the radial distribution of electrostatic potential, $\psi(r)$, in volts, through the outer layer of the ionic double layer would be approximated by

$$[\psi(r)]_{r>a} = -\frac{e_a}{\varepsilon r}\left[\frac{Qe^{\kappa(a-r)}}{1+\kappa a}\right] \qquad (1\text{--}62)$$

where r is the distance (centimeters) from the center of the sphere, ε is the dielectric constant of the solvent, and e_a is the elementary charge (4.80×10^{-10} esu). Because an electrostatic unit is defined by the relationship

$$\text{dyne} \equiv \frac{\text{esu}^2}{\text{cm}^2} \qquad (1\text{--}63)$$

the units on an electrostatic unit are gram^{½} centimeter$^{3⁄2}$ second^{-1}. Because 299.8 V is equal to an erg (electrostatic unit)$^{-1}$

$$1\,\frac{\text{esu}}{\text{cm}} = 299.8\ \text{V} \qquad (1\text{--}64)$$

The parameter κ(centimeters^{-1}) in Equation 1–62 is defined by the relationship

$$\kappa^2 \equiv \frac{4\pi e_a^2}{\varepsilon k_B T}\sum_{j=1}^{m} n_j z_j^2 \qquad (1\text{--}65)$$

where z_j is the valence of the ion of species j composing the electrolyte and n_j is the number of ions j for each cubic centimeter of the solution. The units on e_a^2 can be directly converted (Equation 1–63) from (electrostatic units)^2 to dynes $\cdot$ centimeter^2, which also explains how

$$e_a = \frac{\mathcal{F}}{N_{Av}} \qquad (1\text{--}66)$$

The term $\sum n_j z_j^2$ is related to the ionic strength, $\Gamma/2$ (moles liter^{-1}), defined as

$$\Gamma/2 \equiv \frac{1}{2}\sum_{j=1}^{m}[J]z_j^2 \qquad (1\text{--}67)$$

where [J] is the molar concentration (moles liter^{-1}) of the ion of species j, by the relationship

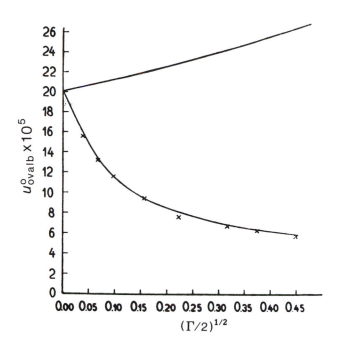

Figure 1–13: Free electrophoretic mobility (u°_{ovalb}) of the protein ovalbumin at pH 7.1 as a function of the square root of the ionic strength [$(\Gamma/2)^{½}$].[100] The upper curve is the behavior of the ideal electrophoretic mobility calculated with Equation 1–61; the points are the observed mobilities. The line through the points is the behavior of the electrophoretic mobility calculated with Equation 1–71. Reprinted with permission from ref 100. Copyright 1940 Royal Society of Chemistry.

$$\sum_{j=1}^{m}n_j z_j^2 = 2(\Gamma/2)N_{Av} \qquad (1\text{--}68)$$

and

$$\kappa^2 = \frac{8\pi e_a^2 N_{Av}}{\varepsilon k_B T}(\Gamma/2) \qquad (1\text{--}69)$$

The term within the brackets on the right side of Equation 1–62 can be considered to be the effective charge of the sphere expressed at a distance r from its center. If there were no electrolytes in the solution so that $\kappa = 0$, the potential would decrease radially only as the inverse of the distance, r, as expected for a sphere of charge in a medium of uniform dielectric ε, and the full charge, Q, would contribute to the potential at all values of r. If κ does not equal zero, however, the effective charge also decreases as r increases due to the presence of the ionic double layer. Because the term κ has the dimensions of centimeters^{-1}, its inverse, κ^{-1}, is used as a measure of the "thickness of the double layer."[13]

Equations 1–62 and 1–69 define the dimensions of the double layer and state that the thickness of the double

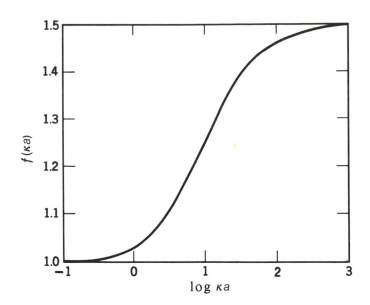

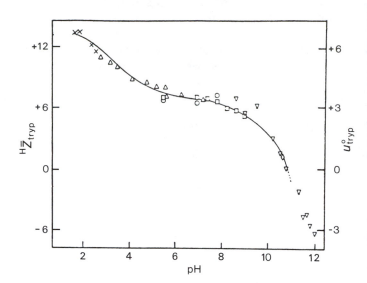

Figure 1–14: Graphic presentation of Henry's function.[99] Reprinted with permission from ref 99. Copyright 1961 John Wiley.

Figure 1–15: Comparison of the electrophoretic mobilities of trypsin at 0 °C ($u°_{tryp}$, points) with the acid–base titration curve of trypsin determined at 20 °C ($^H\bar{Z}_i$, continuous curve).[103] The respective scales on the two variables compared, electrophoretic mobility and net proton charge, both with respect to pH, were adjusted to produce maximum coincidence. The value for $^H\bar{Z}_i = 0$ was arbitrarily set to coincide with the isoelectric point. The coincidence displayed is in shape rather than absolute value or excursion. Reprinted with permission from ref 103. Copyright 1952 Academic Press.

layer will decrease as ionic strength is increased. As the thickness of the ionic double layer decreases, the layer of counterions tightens around the molecule of protein and the molecule of protein draws more of these counterions with it as it moves through the solution. This decreases its effective charge. On the basis of these assumptions, an equation has been derived[100–102] to describe the electrophoretic mobility of protein i if its shape is approximately that of a sphere

$$u_i° = \frac{\mathcal{F}Q_i}{f_i N_{Av}}\left[\frac{1+\kappa a_j}{1+\kappa a_j + \kappa a_i}\right]f(\kappa a_i) \qquad (1\text{–}70)$$

where a_j is the radius of the ions in the supporting electrolyte, a_i is the effective radius of the protein, and $f(\kappa a_i)$ is a function in κa_i for which there is no exact expression[101] but which can be expressed graphically (Figure 1–14).[99] The value of this function varies between 1.0 and 1.5. It can be seen that, when $\kappa a_j < 1$, as is usually the case

$$u_i° \cong \frac{\mathcal{F}Q_i}{f_i N_{Av}}\left[\frac{1}{1+\kappa a_i}\right]f(\kappa a_i) \qquad (1\text{–}71)$$

This equation predicts that the electrophoretic mobility will decrease as the ionic strength increases (Figure 1–13) because κ increases as the ionic strength increases (Equation 1–69).

The points in Figure 1–13 are the observed electrophoretic mobilities of the protein ovalbumin at various ionic strengths as measured by Tiselius and

Svensson.[100] The top line is their calculation of the mobilities from Equation 1–61 using independent measurements of $Q_{ovalbumin}$ and $f_{ovalbumin}$. The lower line is their calculation of the mobilities from Equation 1–71. The agreement between calculated values and observed values is quite satisfactory. As the authors point out, the calculated value from Equation 1–61, in the absence of electrolyte, comes very close to the extrapolated value of the actual mobilities.

According to Equation 1–70, at a constant ionic strength, the electrophoretic mobility of protein i should be directly proportional to Q_i, and this proportionality is reflected in the direct proportionality that obtains between $^H\bar{Z}_i$ and $u_i°$ (Figure 1–15) when $^H\bar{Z}_i$ is varied by varying the pH at a constant ionic strength.[103] The absolute values of the electrophoretic mobilities of several proteins have been calculated from experimental values of their net proton charges, $^H\bar{Z}_i$, using Equation 1–70 with the assumption that $^H\bar{Z}_i = Q_i$ or a more complicated equation derived from a cylindrical model rather than a spherical one. The agreement between calculated values of u_i and experimental values of u_i was within a factor of 2 or less.[102] The lack of exact agreement between calculated and experimental values was assumed to be due to the difference between $^H\bar{Z}_i$ and Q_i caused by the binding of inorganic ions to the proteins. In this case, the proportionality between $^H\bar{Z}_i$ and u_i observed in Figure 1–15 could still be explained, if the binding of counterions increases proportionately as $^H\bar{Z}_i$ increases in magnitude.[96]

The **isoelectric point** of protein i, pI_i, is the pH at which its net molecular charge, Q_i, is zero.[104] At pI_i, the electrophoretic mobility of protein i becomes zero (Equation 1–70), and this fact permits this quantity to be measured by electrophoresis. To determine this pH, electrophoretic mobilities are measured at values of pH greater than and less than pI_i and the pH of zero mobility is determined by interpolation (Figure 1–15). The effect of ionic strength on the isoelectric point of a protein in the absence of actual binding of the ions in the electrolyte to the protein has been calculated to be smaller than the experimental error in measurement.[94] Nevertheless, significant variations in isoelectric point with ionic strength are generally observed (Figure 1–16), and these depend on the particular neutral salt chosen to adjust the ionic strength.[105] The explanation for this behavior can only be the preferential binding of particular ions—in Figure 1–16, always that of the anions—in the chosen electrolyte. The advantage of following the binding of ions by observing changes in the isoelectric point is that the binding is directly measured because, from Equation 1–50, when $Q_i = 0$

$$^H\bar{Z}_i = -\sum_{j=1}^{m}\bar{v}_j z_j \qquad (1-72)$$

Since $^H\bar{Z}_i$ is available from titration data (Figure 1–11), the net binding of ions at the isoelectric pH can be calculated directly.

To this point, only the free electrophoretic mobility, u_i°, has been discussed. The **free electrophoretic mobility** is the electrophoretic mobility displayed by a protein in free solution. This property of the protein is measured by moving boundary electrophoresis[106] in an apparatus developed by Tiselius.[107] This technique has been supplanted by electrophoresis in continuous gels of cross-linked polyacrylamide. A **gel of cross-linked polyacrylamide** is a hydrated plastic cast in a mold from solutions of acrylamide and the cross-linker N,N'-methylenebis(acrylamide) dissolved in aqueous solutions of buffers and other salts. The concentration of acrylamide in the final gel can be varied from 3% to 20%.

It has been demonstrated experimentally by Morris[108] that the relative electrophoretic mobilities of proteins in polyacrylamide gels vary regularly with the concentration of acrylamide used to cast the gel (Figure 1–17) and

$$u_i = u_i^\circ e^{-{}^iK_r T_a} \qquad (1-73)$$

where u_i is the electrophoretic mobility of protein i observed on a gel cast from a solution whose total concentration of acrylamide, in percent, was T_a and ^iK_r is a constant unique to protein i. Such behavior was first

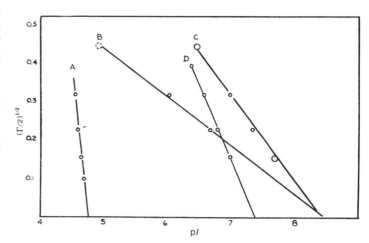

Figure 1–16: Variations in the electrophoretic isoelectric points as a function of the square root of the ionic strength.[105] Line A, ovalbumin in acetate; line B, fructose-bisphosphate aldolase in phosphate; line C, fructose-bisphosphate aldolase in acetate; line D, carboxy-hemoglobin in phosphate. Reprinted with permission from ref 105. Copyright 1949 American Chemical Society.

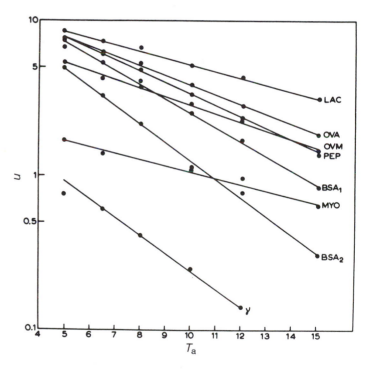

Figure 1–17: Electrophoretic mobility (u), presented on a logarithmic scale, of various proteins on gels of various concentrations of polyacrylamide (T_a).[108] The gels were cast from solutions of pH 8.88 with total concentrations of acrylamide plus N,N'-methylenebis(acrylamide) (T_a) varying between 5% and 15%. The concentration of N,N'-methylenebis(acrylamide) was always 20-fold less than the concentration of acrylamide. The proteins were β-lactoglobulin (LAC), ovalbumin (OVA), ovomucoid (OVM), pepsin (PEP), bovine serum albumin monomer (BSA$_1$) and dimer (BSA$_2$), myoglobin (MYO), and immunoglobulin G (γ). Reprinted with permission from ref 108. Copyright 1966 Elsevier.

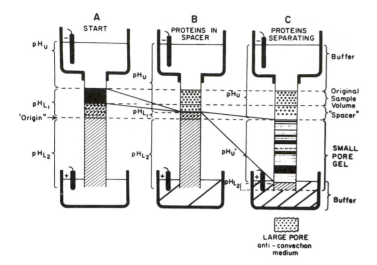

Figure 1–18: Disc electrophoresis.[111] At the start (A) the proteins in the original sample are in a large volume and at a low pH (pH$_{L1}$). They are compressed to a small volume, or disc, as they move through the spacer by being trapped in the boundary between the upper solution (glycine buffer) and the solution (pH$_{L1}$) of the original sample and the spacer. (B) Upon fusion of this boundary and the solution (pH$_{L2}$) originally in the small-pore gel, the pH at the boundary increases and the new more rapidly moving boundary outstrips the proteins and deposits a newly created solution of higher pH (pH$'_u$) behind it as it moves ahead of the separating proteins (C). The proteins also escape the first boundary because, at about the same time as the jump in pH at the fusion of the two boundaries, they also encounter the small pore gel, which decreases their mobility. Reprinted with permission from ref 111. Copyright 1964 New York Academy of Sciences.

noted by Ferguson[98] on gels cast from starch. The same equation applies (Equation 1–73), but the concentration in this case is T_s, the concentration of the starch.[98]

According to Equation 1–73, $u_i°$ should be the free electrophoretic mobility of protein i, and this has been shown to be the case.[108] It follows that

$$u_i \cong \frac{\mathcal{F}Q_i}{f_i N_{Av}}\left[\frac{f(\kappa a_i)}{1+\kappa a_i}\right]e^{-iK_rT_a} \qquad (1\text{–}74)$$

Examination of this relationship reveals that the electrophoretic mobilities of the proteins in a complex mixture upon a gel of polyacrylamide are directly proportional to their respective charges, which are determined by complex functions of pH (Figure 1–11); are functions of their respective frictional coefficients, which are determined by their sizes and shapes; and are exponentially proportional to the product of a constant, which is unique for each, and the concentration of acrylamide. At a given pH, ionic strength, and concentration of polyacrylamide, each of the proteins in this mixture will have a characteristic electrophoretic mobility (Figure 1–17) and they can

be separated one from the other. In this way, electrophoresis can provide a catalogue of the number of proteins present in the mixture and the relative amounts of each.

Electrophoresis is also the most reliable method available for assessing the homogeneity of a sample of purified protein. A sample of pure protein should display only one component upon electrophoresis. Because the electrophoretic mobilities of two proteins change disproportionately as either the pH (Figure 1–15) or the concentration of acrylamide (Figure 1–17) is changed, the possibility that the single component observed under one set of conditions results from the accidental coelectrophoresis of two or more proteins can be dismissed by running electrophoresis at several values of pH[109,110] or several concentrations of acrylamide.[108]

If they are to be used in the roles of cataloguing mixtures and establishing purity, electrophoretic separations on polyacrylamide gels must have as high a resolution as possible. This resolution is achieved by using a discontinuous buffer system and performing what has been referred to as **disc electrophoresis**,[111,112] the pun apparently intended. This technique relies upon the creation of three stable moving boundaries (Figure 1–18),[111] each between two solutions of different ionic composition. The first of these boundaries is used to trap the proteins and sweep them into an extremely narrow band prior to electrophoretic separation. This process has been called **stacking**. It significantly improves the resolution of the subsequent separation by shrinking the original sample to a hairline so that all of the molecules of the proteins begin the electrophoretic separation at very nearly the same point.

The stacking occurs because the proteins are initially placed as a sample that is sandwiched between an upper solution and a lower solution adjusted so their respective ionic compositions will form a stable boundary of a particular type. To describe this boundary, it will be assumed that the direction of electrophoretic movement of both the proteins and the stable moving boundary is downward and a pH has been chosen such that the proteins are all anionic. In this case, both the upper solution and the lower solution are prepared from salts of the same cationic weak acid, but an anion whose mobility is less than the mobilities of all the proteins is used to make the upper solution and an anion whose mobility is greater than those of all the proteins is used to make the lower solution. If a molecule of one of the proteins is in the lower solution, it is surrounded by anions that are moving faster than it is, and it is overtaken by the boundary. If a molecule of one of the proteins is in the upper solution, it is surrounded by anions that are slower than it is, and it outstrips them and returns to the boundary. The result of these events is that the proteins all gather at the descending boundary itself, which remains extremely sharp if the two solutions have the proper ionic compositions.

The stacking process is able to compress the proteins to a disc, but in order for electrophoretic separation to occur, they must be released from the boundary after they have been stacked. This can be done if the upper solution of the descending boundary has been made with an anion, α, that is slower than the protein only because it is the conjugate anionic base of a weak neutral acid and the pH of the upper solution has been chosen to be significantly lower than the pK_a of that weak acid. Under these conditions, the anion α is slow because only a small fraction of the weak acid is anionic at any instant. The acid–base equilibrium has the effect of decreasing the mobility of the anion α from its value in the absence of its conjugate acid to a lower value, and

$$u_\alpha = u_\alpha^\circ \, f_\alpha \qquad (1\text{--}75)$$

where u_α is the mobility of the upper anion at the ratio of conjugate base to acid chosen, u_α° is its mobility in the absence of conjugate acid, and f_α is the fraction of the total weak acid that is ionized at the ratio chosen. The proteins are released from the descending boundary by abruptly increasing the pH, and hence the value of f_α, so that u_α becomes greater than the mobilities of the proteins, and the new stable, rapidly descending boundary that results leaves the proteins behind at the origin of the electrophoretic separation.

The abrupt increase of the pH of the upper solution is accomplished by the arrival of a stable ascending boundary behind which is a solution of a cationic weak acid of a much higher pK_a than the cationic weak acid used as the counterion in the first descending boundary. This ascending boundary has been constructed so that the anion in both its upper and lower solutions is the same. Because its upper solution is the lower solution of the descending boundary, this anion is the fast anion of the upper boundary. The cationic weak acid of the upper solution of the ascending boundary must be the cationic weak acid of the two solutions used to make the first descending boundary. The cation of the lower solution of the ascending boundary is chosen to be the cationic conjugate acid of a neutral base strong enough to adjust the final pH behind the new descending boundary to a value higher than the pK_a of the neutral conjugate acid of the anion α and release the proteins from the first descending boundary. If the release is unsuccessful, or only partially successful, the proteins, or some of the proteins, remain trapped in the new descending boundary and are never separated. These trapped unseparated proteins form an extremely sharp but uninformative band at the bottom of the final electrophoretogram.[113]

To ensure that as many proteins are released as possible, shortly after the fusion of the ascending boundary and the first descending boundary, the descending band of proteins encounters a much higher concentration of polyacrylamide, the running gel, which decreases all of the mobilities of the proteins by virtue of the relationship in Equation 1–73. This frictional deceleration of the proteins increases the probability that all of their mobilities will be less than that of the now accelerated anion of the upper solution. This is the requirement for their escape from the new descending boundary.

The polyacrylamide gel is poured in two stages: the running gel, the polyacrylamide concentration of which is high and upon which the separation will occur, and the stacking gel, the polyacrylamide concentration of which is as low as possible to keep the mobilities of the proteins as high as possible and upon which the stacking will occur. A stacking gel is used to prevent convective disintegration of the stable moving boundaries.

Three stable moving boundaries must be constructed. The first descending boundary between the slow anion and the fast anion that compresses the proteins is initially the boundary between the upper electrode solution and the solution in the stacking gel at the start of the electrophoresis. The ascending boundary between the upper cation and the cation of the weak base that will deliver the pH jump is the boundary between the running gel and the stacking gel at the start of the electrophoresis. The second descending boundary that deposits behind it the solution in which the proteins are actually separated forms upon fusion of the other two.

As the first descending boundary moves, it must maintain a constant pH behind it to maintain the low and constant mobility of the slow anion in the upper solution. As the second descending boundary moves, it must deposit behind itself a solution of constant pH and ionic composition to form a uniform electrophoretic field upon which the proteins can be separated. The constant pH deposited behind each of these descending boundaries is established by the weak cationic acid found on both sides of the boundary and its conjugate base, which together buffer the deposited solution, and it is their pK_a that determines the value of the deposited pH.

The equations that govern the creation of a stable moving boundary and the ability of that boundary to deposit a solution of uniform pH and ionic composition were derived by Ornstein[111] from the regulating functions described by Kohlrausch.[114] On the basis of these equations, Jovin[115] has developed a more elaborate theoretical description of discontinuous electrophoresis, and he and his colleagues have provided the necessary recipes for a large number of discontinuous systems.[116]

Suggested Reading

Tiselius, A., & Svensson, H. (1940) The Influence of Electrolyte Concentration on the Electrophoretic Mobility of Egg Albumin, *Trans. Faraday Soc. 36*, 16–22.

Problem 1–13: The uptake of protons by 1 mol of horse carboxyhemoglobin in the range of pH 6–8 is about 9 mol of protons for each drop of 1 unit in pH.[117] Use this value to estimate the moles of phosphate bound by a mole of horse carboxyhemoglobin at its isoelectric point at the phosphate concentration of the last point in curve D of Figure 1–16 ([phosphate] = 0.12 M). Assume that no cations other than protons are binding to the protein under these conditions.

Problem 1–14: The following facts describe a set of proteins:

protein	isoionic point (pH^{isoionic})	molar mass
chymotrypsinogen	9.4	23,200
hemoglobin	7.0	64,500
lysozyme	11.1	14,100
ovalbumin	4.9	45,000

(A)   Draw the elution profile for a mixture of these four proteins passed through a molecular exclusion chromatographic medium.

(B)   If chymotrypsinogen, hemoglobin, and ovalbumin are mixed and spotted on a block of starch at a centrally located origin and an electric field is applied across the block, draw the final distribution pattern of the proteins if the pH of the buffer soaking the block of starch is (1) pH 7.0; (2) pH 9.4.

(C)   What would be the order of elution of the same three proteins under similar conditions from a column of (carboxymethyl)cellulose at (1) pH 4.0; (2) pH 6.0; (3) pH 8.0.

Problem 1–15: The isoelectric point of normal hemoglobin, hemoglobin A, is 6.87, and that of sickle hemoglobin, hemoglobin S, is 7.09 when electrophoresis is carried out under the same conditions. In the vicinity of the isoelectric point, the charge on either of these hemoglobins changes by about 13 equiv for every mole of protein for every change of 1 unit in pH. At the same pH, anywhere between their respective isoelectric points, what is the difference in charge between hemoglobin A and hemoglobin S?

Problem 1–16: The frictional coefficient of trypsin at 10 °C is 5.5×10^{-8} g s^{-1}. Assume the molecule to be a sphere and calculate its free electrophoretic mobility at 10 °C and at pH 6 and $\Gamma/2$ = 0.13 M using the results of

the acid–base titration in Figure 1–15, which are for 20 °C, and Equation 1–71. Assume that $Q_{tryp} = {}^H\bar{Z}_{tryp}$ and that ${}^H\bar{Z}_{tryp}$ at pH 6 is the same at 10 °C as at 20 °C.

Problem 1–17: The frictional coefficient of ribonuclease at 25 °C is 2.6×10^{-8} g s^{-1}. Assume the molecule to be a sphere and calculate its free electrophoretic mobility at pH 6 and [KCl] = 0.15 M using the results presented in Figure 1–11 and Equation 1–71 with the assumption that $Q_{RNase} = {}^H\bar{Z}_{RNase}$. In a field of 20 V cm^{-1}, how far would ribonuclease travel in 3 h if it had this mobility?

Problem 1–18: The table below contains information about five imaginary proteins where a is the Stokes' radius, Q is the charge on the protein at pH 7, $(\partial^H\bar{Z}_i/\partial pH)_{\Gamma/2}$ is the change in net proton charge with pH, K_r is the retardation coefficient for polyacrylamide, and u° is the free electrophoretic mobility for a temperature of 25 °C, an ionic strength of 0.1 M, and a pH of 7.0.

(A)   Assume that, at constant ionic strength, $(\partial^H\bar{Z}_i/\partial pH)_{\Gamma/2}$ is equivalent to $(\partial Q_i/\partial pH)_{\Gamma/2}$ for each of the five proteins, and calculate the electrophor-etic mobilities of these five proteins, at 25 °C and an ionic strength of 0.1 M, under each of the following conditions: (1) pH 7.0 on 5% polyacrylamide; (2) pH 7.0 on 10% polyacrylamide; (3) pH 5.0 on 5% polyacrylamide; (4) pH 5.0 on 10% polyacrylamide.

(B)   What is the order of the migration of these five proteins under each of these four conditions?

(C)   What will happen to protein E at pH 7.0 that would not happen at pH 5.0 if a mixture of the proteins is run on vertical polyacrylamide gels with the cathode at the bottom and the anode at the top?

(D)   Assume that Q_i does not change as ionic strength changes and calculate the mobilities of the five proteins at an ionic strength of 0.2 M at pH 5 and at 25 °C on 5% polyacrylamide. How does the increase in ionic strength affect the mobilities?

(E)   How should the charge on these proteins change with increasing ionic strength?

protein	a (nm)	Q (at pH 7)	$\left(\dfrac{\partial^H\bar{Z}_i}{\partial pH}\right)_{\Gamma/2}$	K_r ([% acrylamide]$^{-1}$)	u° (cm^{2}V^{-1}s^{-1})
A	2.4	+1.4	−0.2	0.045	1.7×10^{-5}
B	5.3	+9.8	−1.8	0.152	3.2×10^{-5}
C	4.9	+6.2	−2.7	0.146	2.3×10^{-5}
D	2.6	+0.9	−0.5	0.048	1.0×10^{-5}
E	3.4	−3.4	−2.3	0.073	-2.4×10^{-5}

Criteria of Purity

When the purification of a particular protein is monitored analytically by disc electrophoresis (Figure 1–19), the array of other proteins present at the early states of the purification is seen gradually to become less complex in the later stages as one component emerges from the background and becomes more prominent until it alone remains.[118,119] The random loss or copurification of the various components during the successive steps creates a new array of proteins at each step. Such patterns differ characteristically from patterns obtained by mixing a crude homogenate and a pure protein in various ratios.[116] To be certain that the single component observed at the last step of the purification is the only one present in the purified preparation, electrophoresis should be run at a variety of protein concentrations in addition to a few different acrylamide concentrations and values of pH.[120] At high concentrations of protein, minor impurities are most easily recognized, while at low concentrations, two closely running components can be resolved. Also, by running polyacrylamide gels loaded with a series of protein concentrations, the number and relative amounts of any minor impurities can be quantitatively assessed.[121]

To demonstrate that the protein which has been purified is the one responsible for the biological function, the single component observed upon electrophoresis of a sample from the final step of the purification is submitted to the assay. This can be done either by slicing the polyacrylamide gel and performing the assay on each slice (Figure 1–20)[33,61,122] or by staining the intact polyacrylamide gel for enzymatic activity. The latter is accomplished by placing the gel in a solution that gives a colored product from the enzymatic reaction. For example, by adding lead acetate, the SeH_2 produced in a polyacrylamide gel from the action of selenocysteine lyse can be made to form a yellow band where the enzyme is located.[35] The most widely used stain for enzymatic activity is based on the ability of NADH to reduce p-nitrotetrazolium blue to give a blue color.[123,124] It is obvious that through coupled assays this reaction can be used to visualize a large array of different enzymatic activities. At times, the protein being purified is itself colored, by virtue of a bound chromophore, such as the coenzyme B_{12} associated with D-lysine 5,6-aminomutase,[125] and the coelectrophoresis of the purified protein and that color can be observed directly.

Several artifacts can produce misleading results on electrophoresis. For example, aggregation of individual

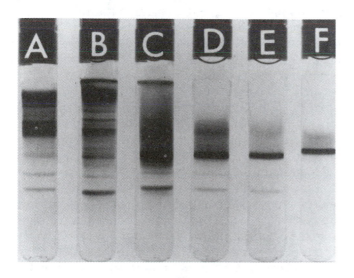

Figure 1–19: Disc electrophoresis on gels of polyacrylamide of (acyl carrier protein) malonyltransferase from *Escherichia coli* at various stages of purification.[119] Electrophoresis was performed on polyacrylamide gels cast from 15% solutions of acrylamide in a tris(hydroxymethyl)aminomethane–glycylglycine discontinuous buffer system. The different gels represent samples from successive steps in a complete purification of the enzyme, seen in its final purified state on gel F. Reprinted with permission from ref 119. Copyright 1973 *Journal of Biological Chemistry.*

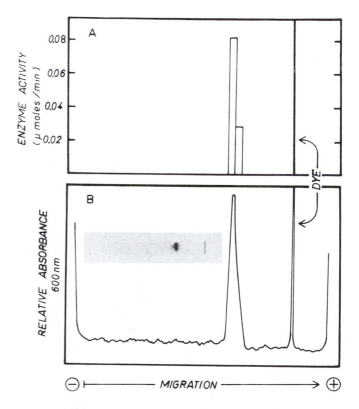

Figure 1–20: Electrophoresis of purified porcine phosphomevalonate kinase (20 μg) on a gel cast from a 10% solution of acrylamide.[33] Following the electrophoresis, the cylindrical gel was divided in half longitudinally. One half was cut into slices laterally, and the slices were assayed individually for enzymatic activity (A). The other half was stained for protein and then scanned for the resulting absorbance (B). The inset in panel B is a photograph of the stained gel. Reprinted with permission from ref 33. Copyright 1980 American Chemical Society.

Figure 1-21: Separation of proteins from the cytoplasm of the bacterium *Escherichia coli* by electrophoresis in two dimensions.[136] A sample (10 μg of protein) from a homogenate of bacteria grown in the presence of [^{14}C]amino acids was submitted to isoelectric focusing (pH 3-10), under conditions where the proteins were unfolded (9 M urea), on a cylindrical (0.25 cm × 13 cm) gel of polyacrylamide. After the unfolded proteins had reached their respective isoelectric points, the gel was removed from its tube, soaked in a solution of sodium dodecyl sulfate (SDS) to coat the unfolded polypeptides with this anionic detergent, and the cylinder was laid across the top of a flat slab (14 cm × 16 cm × 0.3 mm). The proteins separated by isoelectric focusing (IF) in the first dimension were then separated by electrophoresis (SDS) in the second dimension. [14C]Proteins were located by placing the slab on photographic film and exposing the film for a long enough time that the radioactive disintegrations in each spot of protein produced the dark spots seen in the figure. Reprinted with permission from ref 136. Copyright 1975 *Journal of Biological Chemistry*.

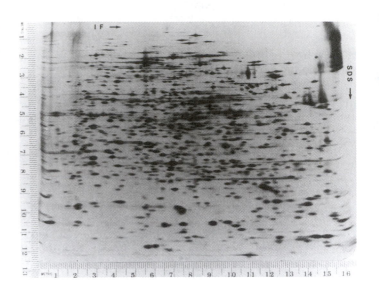

molecules of the same protein can occur[126] during either the purification or the stacking process, and this produces an array of complexes, each with a different f_i and K_i. The amides of glutamines and asparagines on the protein can hydrolyze randomly and in low yield to produce carboxylates, and this modification leads to variations in Q_i that produce multiple components from the same protein. Because these or other similar alterations are integral processes, the components that result from them are usually evenly spaced upon the electrophoretogram, and the nature of the artifact can be recognized by this pattern.[126-128] Each component, however, should be biologically active if the protein is pure.[127,128]

Although the coelectrophoresis of the purified protein and the biological activity is the most convincing criterion of purity, occasionally the electrophoresis itself destroys the activity.[129] For this reason, or simply for personal satisfaction, other criteria of purity can be used. Antibodies raised against the purified enzyme should behave on immunodiffusion and immunoelectrophoresis as expected of antibodies directed against a single antigen and also should precipitate all of the protein and the biological activity.[130] Activity and protein should comigrate on chromatography (Figures 1-6 and 1-10)[131] or co-sediment upon gradients of sucrose.[132] The grams of protein for every mole of binding site is between 15,000 and 100,000 g mol^{-1} for most proteins. The concentration of binding sites for a ligand, such as an agonist or antagonist, known to be specific for a desired protein, such as the respective receptor, can be determined. If the value obtained lies within the expected range and if only one protein can be discerned on electrophoresis, these observations are taken to be convincing criteria of purity, especially if the value of grams mole^{-1} agrees with the measured molar mass of the protomer of the protein that has been purified. For example, purified histidinol-phosphate aminotransferase binds 1 mol of pyridoxal phosphate for every 37,000 g of protein,[120] purified

methylmalonyl-CoA mutase contains 1 mol of adenosylcobalamine for every 73,000 g of protein,[133] and purified α_1-adrenergic receptor binds 1 mol of [^{3}H]prazosin for every 69,000 g of protein.[45]

Isoelectric focusing is a method for assessing purity that is based on electrophoresis. A gel of polyacrylamide is cast from a solution containing a mixture of polyelectrolytes known as ampholytes. The isoelectric points of the ampholytes in the mixture vary over a continuous range of pH values. Upon application of an electric field, this mixture forms a stable gradient of pH in the gel. Each protein migrates through this gradient until it reaches a pH equal to its isoelectric point where it can no longer move, and the proteins in a mixture are spread upon the field in order of their respective isoelectric points. It is a technique that is less flexible than disc electrophoresis because it separates molecules on the basis of only one property rather than three. It also seems to be more sensitive to minor heterogeneities of charge than is electrophoresis.[134] The coisoelectrofocusing of protein and biological activity,[29,130,135] however, is a criterion of purity independent from the observation of coelectrophoresis. Isoelectric focusing has been combined with electrophoresis to resolve complex mixtures of proteins in two dimensions.[136] When the clarified homogenate produced from the cytoplasm of the bacterium *Escherichia coli* was submitted to such a procedure, more than 1000 different proteins were represented upon the field (Figure 1-21). This display should indicate the complexity of the mixture of proteins in a cell. From such a mixture, a single protein with a single biological activity is purified.

Heterogeneity

Often heterogeneity of a purified protein is detected by one of these criteria, even though all of the various components are biologically active; often heterogeneity is discovered in later experiments. This heterogeneity may have a biological origin, and the various forms of the protein producing this heterogeneity may coexist in the tissue prior to homogenization, but usually the heterogeneity arises from harsh treatments during the purification itself. This often becomes apparent when a new, more rapid, less debilitating method of purification is devised for a certain protein, and the heterogeneity noted previously, the subject of many publications, simply disappears. When fructose bisphosphatase was purified by a shorter method,[137] the previously studied requirement of the enzyme for alkaline conditions no longer held. When aconitate hydratase was purified by a more rapid procedure,[138] it was isolated with its iron still attached. When glyceraldehyde-3-phosphate dehydrogenase from yeast was purified rapidly by affinity chromatography,[139] the heterogeneous behavior in its binding of ligands[140] was no longer observed.

One of the most publicized causes of heterogeneity or artifactual alteration of a protein during its purification is **proteolytic degradation**. Proteins whose biological role is to degrade other proteins are known as proteolytic enzymes. Under normal circumstances, they are usually segregated from the cytoplasm of the cell in which they are located or in which they were produced. This is accomplished by enclosing them in tight, membrane-sealed packages, the lysosomes, or excreting them into the extracellular surroundings. Upon homogenization, the natural boundaries between the cytoplasm and the proteolytic enzymes are destroyed and artifactual degradation of the proteins being purified can commence.

Proteolytic degradation is not always a problem. Most native proteins are remarkably resistant to digestion, and in most instances, proteins can be purified without being degraded. Harsh treatments, however, such as heat, the use of detergents, and extremes of pH encourage proteolytic degradation, and proteins purified by procedures employing these conditions often display evidence of deterioration. Because a protein can be nicked by a proteolytic enzyme and remain almost unaltered, the cumulative effects of proteolytic degradation become most obvious when they are assessed by electrophoresis in solutions of sodium dodecyl sulfate,[140] a technique that is used to catalogue the number and lengths of the polypeptides present in a given preparation. In the absence of such assessment, proteolytic degradation often goes unnoticed.

There are four major classes of proteolytic enzymes,[141–143] and their properties determine the precautions that can be taken to inhibit them during a purification procedure. Acid proteolytic enzymes are active only at acidic ranges of pH, and if the purification is carried out at neutral or slightly alkaline pH, their action can be avoided. Sulfhydryl proteolytic enzymes contain a thiol necessary for activity and can be permanently inactivated by treatment with iodoacetamide or iodacetate. Metalloproteolytic enzymes require transition metal cations or alkaline earth cations and can be inactivated by adding chelating agents such as ethylenediaminetetraacetate or *o*-phenanthroline. Serine proteolytic enzymes are invariably inactivated by diisopropyl fluorophosphate, but this compound is extremely toxic. They are often inactivated by phenylmethanesufonyl fluoride or by chloromethyl ketones of various specificities.

There is a vast array of natural and synthetic inhibitors of proteolytic enzymes[141–143] that are more or less specific for one or several members of a particular class. Some have been used successfully as additives during the purification of proteins. For example, acetyl-CoA carboxylase has been purified from chicken liver in the presence of parotid trypsin inhibitor,[144] and glyceraldehyde-3-phosphate dehydrogenase, from the same source in the presence of leupeptin.[145] Often, however, inhibitors of proteolytic activity are used prophylactically in the absence of any evidence that they are effective.

Crystallization

As in the isolation of a natural product in organic chemistry, the production of a crystalline preparation (Figure 1–22)[146] was once considered to be the final step in any isolation of a protein. Although the time-consuming search for the proper conditions necessary to crystallize a given protein has gone out of fashion, the exhilarating gallery of photographs of crystalline enzymes compiled by Dixon and Webb[147] testifies to the pleasure that such a conclusion to a long purification must inspire.

Crystallization as a method of purification is usually less effective than chromatography. Some examples of crystallization as the last step in a purification are the

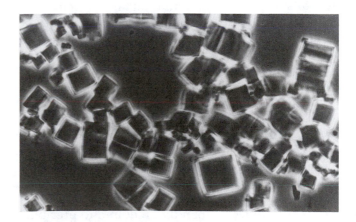

Figure 1–22: Crystals of α-galactosidase isolated from *Mortierella vinacea*.[146] Homogenates of cells of *M. vinacea* were submitted sequentially to ammonium sulfate fractionation, anion-exchange chromatography on (diethylaminoethyl)dextran, and molecular exclusion chromatography. The final protein was crystallized from a solution of ammonium sulfate. The largest crystal in the field is 10 μm across. Reprinted with permission from ref 146. Copyright 1970 *Journal of Biological Chemistry*.

purification of 1.5-fold seen upon recrystallization of phosphoenolpyruvate carboxykinase (ATP),[148] the purification of 1.4-fold with a 40% yield seen upon recrystallization of acylphosphatase,[149] and the 2-fold purification with a 90% yield seen upon recrystallization of nicotinate–nucleotide pyrophosphorylase.[150] Recrystallization has been observed to eliminate some of the heterogeneous behavior displayed by a purified protein,[151] presumably due to an increase in its homogeneity.

Crystals of a protein, aside from their intrinsic beauty, are the specimens required to determine the molecular structure of the purified protein by x-ray crystallography, and there is considerable interest in crystallizing proteins.[152] For crystallographic studies, single, untwinned crystals of 1 or 2 mm in size are required, and to produce suitable crystals is a tedious process involving a good deal of trial and error. Because of the number of attempts that are necessary and because suitable crystals only form in concentrated solutions of the protein, hundreds of milligrams to several grams of protein are required. Furthermore, crystals usually will grow only from homogeneous solutions of monodisperse protein, so the protein must be as pure as possible and it must have suffered as little damage as possible during the procedures used to purify it. For all of these reasons, proteins purified rapidly in one or two steps from overproducing microorganisms, in which more than 10% of the cellular protein can be the protein of interest, are usually the ideal starting material for crystallizations. The construction of such a microorganism is usually the first step in an attempt to crystallize a protein for crystallographic studies.

Crystals of proteins are produced by slowly and continuously increasing the concentration of both the protein and a solute that promotes crystallization. Originally, salts such as ammonium sulfate, which induce proteins to precipitate from solution, were used to promote crystallization in a process referred to as salting out. Recently, it has been discovered that the aliphatic polyether poly(ethylene glycol) is a more effective agent. Both ammonium sulfate[153] and poly(ethylene glycol)[154] are thought to decrease the solubility of proteins by the same mechanism. Because either solute is excluded from the water surrounding a protein, the system containing the dissolved protein becomes unstable relative to the two separated phases of solvent and crystal.

Initially, the solute promoting crystallization is added to a large drop on a cover slip of a solution of the protein at a concentration less than that required to precipitate the protein. The cover slip is inverted over a well containing an aqueous solution in which the activity of water is less than that of the solution of protein and solutes. The system is sealed and left in the cold for several weeks. Slowly, water evaporates from the drop and condenses in the well, and if one is lucky, crystals of protein form in the drop. As this is a rare event, hundreds of hanging drops are made, each with a different pH, ionic strength, or concentration of protein over wells with different solutions in them. Small molecules, for example, substrates or ligands, that are known to bind to the protein are also added to some of the drops in the hope that they might encourage crystallization. The entire process is a gamble, but an inescapable one.

Suggested Reading

Noyes, B.E., & Bradshaw, R.A. (1973) Purification and Characterization of Beef Liver Dihydrofolate Reductase, *J. Biol. Chem. 248*, 3052–3059.

Kaufman, B.T., & Kemerer, V.F. (1976) L-3-Hydroxyacyl Coenzyme A Dehydrogenase from Pig Heart Muscle, *Arch. Biochem. Biophys. 172*, 289–300.

References

1. Cassidy, H.G. (1957) in *Fundamentals of Chromatography; Techniques of Organic Chemistry* (Weissberger, A., Ed.) Vol. 10, pp 31–33, Interscience, New York.
2. Massey, V. (1952) *Biochem. J. 51*, 490–494.
3. Morris, C.J.O.R., & Morris, P. (1963) *Separation Methods in Biochemistry*, pp 47–82, Pitman and Sons, London.
4. Lewis, W.K. (1922) *Ind. Eng. Chem. 14*, 492–497.
5. Peters, W.A. (1922) *Ind. Eng. Chem. 14*, 476–479.
6. Martin, A.J.P., & Synge, R.L.M. (1941) *Biochem. J. 35*, 1358–1368.
7. Craig, L.C., Hausmann, W., Ahrens, E.H., Jr., & Harfenist, E.J. (1951) *Anal. Chem. 23*, 1236–1244.
8. Said, A.S. (1956) *Am. Inst. Chem. Eng. J. 2*, 477–481.
9. Keulemans, A.I.M., & McNair, H.M. (1961) in *Chromatography* (Heftmann, E., Ed.) pp 169–171, Reinhold, New York.
10. Bradshaw, R.A., Garner, W.H., & Gurd, F.R.N. (1969) *J. Biol. Chem. 244*, 2149–2158.
11. Mahoney, W.C., & Hermodson, M.A. (1980) *J. Biol. Chem. 255*, 11199–11203.
12. Hutchens, T.W., & Porath, J. (1986) *Anal. Biochem. 159*, 217–226.
13. Overbeck, J.T.G., & Lijklema, J. (1959) in *Electrophoresis* (Bier, M., Ed.) pp 1–34, Academic Press, New York.
14. Stein, W.D. (1967) *The Movement of Molecules Across Cell Membranes*, pp 60–62, Academic Press, New York.
15. Moore, S., & Stein, W.H. (1951) *J. Biol. Chem. 192*, 663–681.
16. Spackman, D.H., Stein, W.H., & Moore, S. (1958) *Anal. Chem. 30*, 1190–1202.
17. Carr, C.W. (1956) *Arch. Biochem. Biophys. 62*, 476–484.
18. Walkinshaw, M.D., & Arnott, S. (1981) *J. Mol. Biol. 153*, 1055–1073.
19. Qureshi, A.A., Beytia, E., & Porter, J.W. (1973) *J. Biol. Chem. 248*, 1848–1855.
20. Pollock, R.J., & Hersh, L.B. (1971) *J. Biol. Chem. 246*, 4737–4743.
21. Kosaka, A., Spivey, H.O., & Gholson, R.K. (1971) *J. Biol. Chem. 246*, 3277–3283.
22. Borum, P.R., & Broquist, H.P. (1977) *J. Biol. Chem. 252*, 5651–5655.
23. Joseph, D.R., & Meunch, K.H. (1971) *J. Biol. Chem. 246*, 7602–7609.
24. Wu, J., Matsuda, T., & Roberts, E. (1973) *J. Biol. Chem. 248*, 3029–3034.

25. Roche, P.A., Moorhead, T.J., & Hamilton, G.A. (1982) *Arch. Biochem. Biophys. 216*, 62–73.
26. Poillon, W.N., Maeno, H., Koike, K., & Feigelson, P. (1969) *J. Biol. Chem. 244*, 3447–3456.
27. Bull, C., & Ballou, D.P. (1981) *J. Biol. Chem. 256*, 12673–12680.
28. Horecker, B.L., & Kornberg, A. (1948) *J. Biol. Chem. 175*, 385–390.
29. Noyes, B.E., & Bradshaw, R.A. (1973) *J. Biol. Chem. 248*, 3052–3059.
30. Duggleby, R.G., & Dennis, D.T. (1974) *J. Biol. Chem. 249*, 162–166.
31. Reitz, M.S., & Rodwell, V.R. (1970) *J. Biol. Chem. 245*, 3091–3096.
32. McClure, W.R., Lardy, H.A., & Kneifel, H.P. (1971) *J. Biol. Chem. 246*, 3569–3578.
33. Bazaes, S., Beytia, E., Jabalquinto, A.M., Solis de Ovando, R., Gomez, I., & Eyzaguirre, J. (1980) *Biochemistry 19*, 2300–2304.
34. Kramer, P.R., & Miziorko, H.M. (1980) *J. Biol. Chem. 255*, 11023–11028.
35. Switzer, R.L. (1969) *J. Biol. Chem. 244*, 2854–2863.
36. Kyte, J. (1971) *J. Biol. Chem. 246*, 4157–4165.
37. Nakao, T., Nagano, K., Adachi, K., & Nakao, M. (1963) *Biochem. Biophys. Res. Commun. 13*, 444–448.
38. Leloir, L.F., & Cardini, C.E. (1957) *Methods Enzymol. 3*, 843–844.
39. Cooper, A.J.L., & Meister, A. (1972) *Biochemistry 11*, 661–671.
40. Donald, A., Sibley, D., Lyons, D.E., & Dahms, A.S. (1979) *J. Biol. Chem. 254*, 2132–2137.
41. Esaki, N., Nakamura, T., Tanaka, H., & Soda, K. (1982) *J. Biol. Chem. 257*, 4386–4391.
42. Moriyama, T., & Srere, P.A. (1971) *J. Biol. Chem. 246*, 3217–3223.
43. Caron, M.G., & Lefkowitz, R.J. (1976) *J. Biol. Chem. 251*, 2374–2384.
44. Shorr, R.G.L., Strohstacker, M.W., Lavin, T.N., Lefkowitz, R.J., & Caron, M.G. (1982) *J. Biol. Chem. 257*, 12341–12350.
45. Graham, R.M., Hess, H., & Homcy, C.J. (1982) *J. Biol. Chem. 257*, 15174–15181.
46. Cohen, S., Ushiro, H., Stoscheck, C., & Chinkers, M. (1982) *J. Biol. Chem. 257*, 1523–1531.
47. Kuhn, R.W., Schrader, W.T., Smith, R.G., & O'Malley, B.W. (1975) *J. Biol. Chem. 250*, 4220–4228.
48. Penefsky, H.S. (1977) *J. Biol. Chem. 252*, 2891–2899.
49. Briggs, M.R., Kadonaga, J.T., Bell, S.P., & Tjian, R. (1986) *Science 234*, 47–52.
50. Barton, R.W., & Neufeld, E.F. (1971) *J. Biol. Chem. 246*, 7773–7779.
51. Gerhart, J., Wu, M., & Kirschner, M. (1984) *J. Cell Biol. 98*, 1247–1255.
52. Wu, M., & Gerhart, J.C. (1980) *Dev. Biol. 79*, 465–477.
53. Ash, J.F., Fineman, R.M., Kalka, T., Morgan, M., & Wire, B. (1984) *J. Cell Biol. 99*, 971–983.
54. Slice, L.W., & Taylor, S.S. (1989) *J. Biol. Chem. 264*, 20940–20946.
55. Beytia, E., Dorsey, J.K., Marr, J., Cleland, W.W., & Porter, J.W. (1970) *J. Biol. Chem. 245*, 5450–5458.
56. Tong, J.H., & Kaufman, S. (1975) *J. Biol. Chem. 250*, 4152–4158.
57. Gerhart, J.C., & Holoubek, H. (1967) *J. Biol. Chem. 242*, 2886–2892.
58. Fuller, G.M., & Doolittle, R.F. (1971) *Biochemistry 10*, 1305–1311.
59. Uyeda, K., & Kurooka, S. (1970) *J. Biol. Chem. 245*, 3315–3324.
60. D'Alessio, G., & Josse, J. (1971) *J. Biol. Chem. 246*, 4319–4325.
61. Sabourin, P.J., & Bieber, L.L. (1982) *J. Biol. Chem. 257*, 7460–7467.
62. Nimmo, G.A., & Coggins, J.R. (1981) *Biochem. J. 197*, 427–436.
63. Lau, E.P., Cochran, B.C., & Fall, R.R. (1980) *Arch. Biochem. Biophys. 205*, 352–359.
64. Sarngadharan, M.G., Watanabe, A., & Pogell, B.M. (1970) *J. Biol. Chem. 245*, 1926–1929.
65. Chou, A.C., & Wilson, J.E. (1972) *Arch. Biochem. Biophys. 151*, 48–55.
66. March, S.C., Parikh, I., & Cuatrecasas, P. (1974) *Anal. Biochem. 60*, 149–152.
67. Araki, C., & Arai, K. (1957) *Bull. Chem. Soc. Jpn. 30*, 287–293.
68. Cuatrecasas, P., Wilchek, M., & Anfinsen, C.B. (1968) *Proc. Natl. Acad. Sci. U.S.A. 61*, 636–643.
69. Cuatrecasas, P. (1970) *J. Biol. Chem. 245*, 3059–3065.
70. Steers, E., Cuatrecasas, P., & Pollard, H.B. (1971) *J. Biol. Chem. 246*, 196–200.
71. Chan, W.W.C., & Takahashi, M. (1969) *Biochem. Biophys. Res. Commun. 37*, 272–277.
72. Berg, R.A., & Prokop, D.J. (1973) *J. Biol. Chem. 248*, 1175–1182.
73. Geren, C.R., & Ebner, K.E. (1977) *J. Biol. Chem. 252*, 2082–2088.
74. Lee, C., Lappi, D.A., Wermuth, B., Everse, J., & Kaplan, N.O. (1974) *Arch. Biochem. Biophys. 163*, 561–569.
75. Nealon, D.A., & Cook, R.A. (1979) *Biochemistry 18*, 3616–3622.
76. Ryan, R.L., & McClure, W.O. (1979) *Biochemistry 18*, 5357–5365.
77. Huang, J.S., Huang, S.S., & Tang, J. (1979) *J. Biol. Chem. 254*, 11405–11417.
78. Allen, M.B., & Walker, D.G. (1980) *Biochem. J. 185*, 565–575.
79. Magnani, M., Serafini, G., Stocchi, V., Bossù, M., & Dachà, M. (1982) *Arch. Biochem. Biophys. 216*, 449–454.
80. Kautman, B.T., & Pierce, J.V. (1971) *Biochem. Biophys. Res. Commun. 44*, 608–613.
81. Mendicino, J., Sivakami, S., Davila, M., & Chandrasekaran, E.V. (1982) *J. Biol. Chem. 257*, 3987–3994.
82. Kitani, T., & Fujisawa, H. (1983) *J. Biol. Chem. 258*, 235–239.
83. Grimshaw, C.E., Henderson, G.B., Soppe, G.G., Hansen, G., Mathur, E.J., & Huennekens, F.M. (1984) *J. Biol. Chem. 259*, 2728–2733.
84. Caron, M.G., Srinivasan, Y., Pitha, J., Kociolek, K., & Lefkowitz, R.J. (1979) *J. Biol. Chem. 254*, 2923–2927.
85. Nold, J.G., Kang, A.H., & Gross, J. (1970) *Science 170*, 1095–1099.
86. Sugden, B., & Keller, W. (1973) *J. Biol. Chem. 248*, 3777–3788.
87. Niedel, J., & Dietrich, L.S. (1973) *J. Biol. Chem. 248*, 3500–3505.
88. Hsu, Y., & Kohlhaw, G.B. (1980) *J. Biol. Chem. 255*, 7255–7260.

89. Cvetanovic, M., Moreno de la Garza, M., Dommes, V., & Kunau, W. (1985) *Biochem. J. 227*, 49–56.

90. Holden, J.A., Meredith, G.S., & Kelley, W.N. (1979) *J. Biol. Chem. 254*, 6951–6955.

91. Payne, M.E., Schworer, C.M., & Soderling, T.R. (1983) *J. Biol. Chem. 258*, 2376–2382.

92. Kadonaga, J.T., & Tjian, R. (1986) *Proc. Natl. Acad. Sci. U.S.A. 83*, 5889–5893.

93. Durchschlag, H., Biedermann, G., & Eggerer, H. (1981) *Eur. J. Biochem. 114*, 255–262.

94. Tanford, C. (1962) *Adv. Protein Chem. 17*, 69–165.

95. Tanford, C., & Hauenstein, J.D. (1956) *J. Am. Chem. Soc. 78*, 5288–5291.

96. Carr, C.W. (1953) *Arch. Biochem. Biophys. 46*, 417–423.

97. Matthew, J.B., Hanania, G.I.H., & Gurd, F.R.N. (1979) *Biochemistry 18*, 1928–1936.

98. Ferguson, K.A. (1964) *Metabolism 13*, 985–1002.

99. Tanford, C. (1961) *Physical Chemistry of Macromolecules*, John Wiley, New York.

100. Tiselius, A., & Svensson, H. (1940) *Trans. Faraday Soc. 36*, 16–22.

101. Henry, D.C. (1931) *Proc. R. Soc. London, A 133*, 106–129.

102. Brown, R.A., & Timasheff, S.N. (1959) in *Electrophoresis* (Bier, M., Ed.) pp 317–367, Academic Press, New York.

103. Duke, J.A., Bier, M., & Nord, F.F. (1952) *Arch. Biochem. Biophys. 40*, 424–436.

104. Edsall, J.T., & Wyman, J. (1958) *Biophysical Chemistry*, pp 504–510, Academic Press, New York.

105. Velick, S.F. (1949) *J. Phys. Colloid Chem. 53*, 135–149.

106. Longsworth, L.G. (1959) in *Electrophoresis* (Bier, M., Ed.) pp 137–177, Academic Press, New York.

107. Tiselius, A. (1937) *Trans. Faraday Soc. 33*, 524–531.

108. Morris, C.J.O.R. (1966) in *Protides of the Biological Fluids* (Peeters, H., Ed.) Vol. 14, pp 543–561, Elsevier, Amsterdam.

109. Stahl, P.D., & Touster, O. (1971) *J. Biol. Chem. 246*, 5398–5406.

110. Philippov, P.P., Shestakova, I.K., Tikhomirova, N.K., & Kochetov, G.A. (1980) *Biochim. Biophys. Acta 613*, 359–369.

111. Ornstein, L. (1964) *Ann. N.Y. Acad. Sci. 121*, 321–349.

112. Davis, B.J. (1964) *Ann. N.Y. Acad. Sci. 121*, 404–427.

113. Laemmli, U.K. (1970) *Nature 227*, 680–685.

114. Kohlrausch, F. (1897) *Ann. Phys. Chem. 62*, 209–239.

115. Jovin, T.M. (1973) *Biochemistry 12*, 871–898.

116. Spector, M., O'Neal, S., & Racker, E. (1980) *J. Biol. Chem. 255*, 5504–5507.

117. Cohn, E.J., Green, A.A., & Blanchard, M.H. (1937) *J. Am. Chem. Soc. 59*, 509–517.

118. Katze, J.R., & Konigsberg, W.J. (1970) *J. Biol. Chem. 245*, 923–930.

119. Ruch, F.E., & Vagelos, P.R. (1973) *J. Biol. Chem. 248*, 8086–8094.

120. Henderson, G.B., & Snell, E.E. (1973) *J. Biol. Chem. 248*, 1906–1911.

121. Zampighi, G., Kyte, J., & Freytag, W. (1984) *J. Cell Biol. 98*, 1851–1864.

122. Kolhouse, J.F., Utley, C., & Allen, R.H. (1980) *J. Biol. Chem. 255*, 2708–2712.

123. Schachter, H., Sarney, J., McGuire, E.J., & Roeman, S. (1969) *J. Biol. Chem. 244*, 4785–4792.

124. Li, J.J., Ross, C.R., Tepperman, H.M., & Tepperman, J. (1975) *J. Biol. Chem. 250*, 141–148.

125. Morley, C.G.D., & Stadtman, T.C. (1970) *Biochemistry 9*, 4890–4900.

126. Yu, C., Gunsalus, I.C., Katagiri, M., Suhara, K., & Takemori, S. (1974) *J. Biol. Chem. 249*, 94–101.

127. Olsen, A.S., & Milman, G. (1974) *J. Biol. Chem. 249*, 4030–4037.

128. Scott, W.A., & Tatum, E.L. (1972) *J. Biol. Chem. 246*, 6347–6352.

129. Warnick, G.R., & Burnham, B.F. (1971) *J. Biol. Chem. 246*, 6880–6885.

130. Fernandez-Sorenson, A., & Carlson, D.M. (1971) *J. Biol. Chem. 246*, 3485–3493.

131. Reed, B.C., & Rilling, H.C. (1975) *Biochemistry 14*, 50–54.

132. Beytia, E., Dorsey, J.K., Marr, J., Cleland, W.W., & Porter, J.W. (1970) *J. Biol. Chem. 245*, 5450–5458.

133. Fenton, W.A., Hack, A.M., Willard, H.F., Gertler, A., & Rosenberg, L.E. (1982) *Arch. Biochem. Biophys. 214*, 815–823.

134. Arnold, W.J., & Kelley, W.N. (1971) *J. Biol. Chem. 246*, 7398–7404.

135. Norton, I.L., Pfuderer, P., Stringer, C.D., & Hartman, F.C. (1970) *Biochemistry 9*, 4952–4958.

136. O'Farrell, P.H. (1975) *J. Biol. Chem. 250*, 4007–4021.

137. Traniello, S., Melloni, E., Pontremoli, S., Sia, C.L., & Horecker, B.L. (1972) *Arch. Biochem. Biophys. 149*, 222–231.

138. Kennedy, C., Rauner, R., & Gawron, O. (1972) *Biochem. Biophys. Res. Commun. 47*, 740–745.

139. Gennis, L.S. (1976) *Proc. Natl. Acad. Sci. U.S.A. 73*, 3928–3932.

140. Pringle, J.R. (1970) *Biochem. Biophys. Res. Commun. 39*, 46–52.

141. Lorand, L. (1970) *Methods Enzymol. 19*.

142. Lorand, L. (1970) *Methods Enzymol. 45*.

143. Lorand, L. (1970) *Methods Enzymol. 80*.

144. Mackall, J.C., Lane, M.D., Leonard, K.R., Pendergast, M., & Kleinschmidt, A.K. (1978) *J. Mol. Biol. 123*, 595–606.

145. Grant, G.A., Keefer, L.M., & Bradshaw, R.A. (1978) *J. Biol. Chem. 253*, 2724–2726.

146. Suzuki, H., Li, S., & Li, Y. (1970) *J. Biol. Chem. 245*, 781–786.

147. Dixon, M., & Webb, E.C. (1964) *Enzymes*, pp 794–808, Longmans, Green & Co., London.

148. Cannata, J.J.B. (1970) *J. Biol. Chem. 245*, 792–798.

149. Shiokwa, H., & Noda, L. (1970) *J. Biol. Chem. 245*, 669–673.

150. Iwai, K., & Taguchi, H. (1974) *Biochem. Biophys. Res. Commun. 56*, 884–891.

151. Montielhet, C., & Blow, D.M. (1978) *J. Mol. Biol. 122*, 407–417.

152. McPherson, A. (1990) *Eur. J. Biochem. 189*, 1–23.

153. Arakawa, T., & Timasheff, S.N. (1982) *Biochemistry 21*, 6545–6552.

154. Arakawa, T., & Timasheff, S.N. (1982) *Biochemistry 21*, 6536–6544.

CHAPTER 2

Electronic Structure

When crystals of protein are submitted to chemical analysis, it can be shown that, in addition to the solvent included within them, they are composed of 20 amino acids: aspartic acid, asparagine, threonine, serine, glutamine, glutamic acid, proline, glycine, alanine, cysteine, valine, methionine, isoleucine, leucine, tyrosine, phenylalanine, lysine, histidine, tryptophan, and arginine. Each protein has different relative amounts of each of these amino acids. The amino acids a protein contains are coupled together in a particular order to create polymers 50–5000 amino acids in length, referred to as polypeptides. To understand the structure of molecules of protein, one must understand the structures of the amino acids, the order in which they are connected, and the way that these long polymers are folded up to produce the globular conformation of the molecule. The first level of understanding is grounded in a firm knowledge of the bonding and molecular structure of small molecules. The second level of understanding requires a description of the complete covalent structure of the polymers composing proteins. The third level of understanding proceeds from crystallographic molecular models of proteins that are the products of X-ray crystallography.

It is a remarkable fact that each molecule of a particular protein has the same precisely defined covalent structure and that the polypeptides from which it is composed always assume the same few precisely dictated conformations even though the complete molecule of the protein is very large. These two properties are foreign to a synthetic chemist. Molecules produced synthetically are either precise but small or large but heterogeneous. Large heterogeneous polymers produced synthetically never have defined structures. Yet a molecule of protein is a molecule made from atoms held together by the same covalent chemical bonds holding together the smaller molecules to which one is already accustomed. All of the rules of bonding exerted with such inescapability in small molecules are as inescapable in a molecule of protein.

The covalent bonds holding the atoms together in any molecule are pairs of electrons confined to molecular orbitals. The molecular orbitals are either localized σ molecular orbitals or delocalized π molecular orbitals. A distinction between these two types of molecular orbitals is crucial to an understanding of bond lengths, bond angles, and rotational motions about bonds.

In addition to the covalent bonds, molecules of protein are filled with lone pairs of electrons. Because σ lone pairs of electrons are the only valence electrons that do not participate in covalent bonds and because there are also lone pairs of electrons participating in π molecular orbitals, to understand the details of molecular structure one must be able to distinguish localized σ lone pairs of electrons from delocalized π lone pairs of electrons. The distinction between these two types of electrons is reflected in their basicity, or their ability to house a proton.

The lone pairs of electrons in a molecule of protein are potential bases, and the hydrogens in a molecule of protein are potential acids. Which lone pair will act as a base and which hydrogen will act as an acid is determined by the acid dissociation constant for the conjugate acid or the acid, respectively. Every lone pair is basic and every hydrogen is acidic, but most lone pairs are such weak bases and most hydrogens are such weak acids that their basicity or acidity can be ignored. To understand the atomic structure of a molecule of protein, the acids and bases within it must be identified and categorized. It is also necessary to distinguish an acid dissociation, in which protons leave the molecule of protein, from a tautomerization, in which protons redistribute among lone pairs of electrons within the molecule of protein.

The chemical capacities available to a protein are a reflection of the amino acids from which it is constructed. Each of the 20 amino acids has a peculiar set of chemical properties. These properties are mixed in a unique way by the amino acid sequence and the resulting native structure to produce the distinct protein; but, to understand the combination, the properties of the ingredients must be understood. These properties include the bonding and acid–base behavior of each of the 20 side chains of the amino acids. With the exception of the regular polyamide of the polymer; the covalent bonds, acidic hydrogens, and basic lone pairs of electrons that fill a molecule of protein are contributed by these side chains.

π and σ

Molecules, including proteins, are arrays of atomic nuclei required to maintain particular distances and angular dispositions relative to each other by electrons confined to particular regions of space known as **orbitals**. Every electron in a molecule is confined to a specific orbital, and almost every orbital is occupied by two electrons. Each orbital is either confined exclusively to one nucleus or distributed between or among particular nuclei. The electrons, in their occupation of these orbitals, create the distribution of electron density observed by a crystallographer. The electrons present in a molecule can be divided into three categories, core electrons, π electrons, and σ electrons, that reflect the degree to which they are confined and define their chemical reactivity.

A

B

Figure 2–1: Two ways of representing the electronic structure of *N*-acetylglutamate α-amide. (A) In the Lewis dot formula, each main atom is surrounded by an octet of electrons and the total number of electrons represented equals the sum of the number of valence electrons in each neutral atom and the molecular charge. The negative sign surrounded by a circle locates the formal charge. (B) In a stereochemical representa-

tion distinguishing electrons, a σ bond is designated by a line, a localized σ lone pair of electrons is designated by two dots surrounded by a circle, a π bond is indicated by a second or third line between two atoms, and a π lone pair of electrons is shown by two uncircled dots. The atoms are arranged in space to represent the tetrahedral or trigonal geometry dictated by their respective hybridizations.

Core electrons are the electrons that are immediately adjacent to a nucleus. Aside from hydrogen, almost all of the atoms present in molecules of protein are either carbon, oxygen, or nitrogen. Each of these atoms has two core electrons spherically confined about the nucleus. Occasionally, sulfur or phosphorus occurs in a protein, and these atoms each have 10 core electrons. Because they are confined close to the nucleus, the core electrons provide the greatest electron density and are prominent features in a map of electron density. They are, however, chemically inert.

Valence electrons are the outermost electrons surrounding each atom. All of the chemistry of a molecule, its chemical bonds and its reactive locations, results from the valence electrons. Unless one electron is missing, as in the case of a radical, or two electrons are momentarily missing, as in a carbocation, every carbon, nitrogen, oxygen, sulfur, or phosphorus in a molecule of protein can be logically associated with eight valence electrons, but often the association is a weak one. By convention, these octets are assigned by Lewis structures. This formalism divides valence electrons into **bonding electrons** and **lone pairs of electrons** and assigns **formal charge** to certain atoms. An example would be the Lewis structure of the model compound for glutamic acid in a polypeptide, *N*-acetylglutamate α-amide (Figure 2–1A). The intent of a Lewis structure is to count valence electrons.

A pair of bonding electrons occupies a **bonding molecular orbital** that is formed from the overlap of two or more atomic orbitals, each contributed by a different atom in the molecule. These bonding electrons must be clearly distinguished as occupants of either π molecular

orbitals, forming π bonds, or σ molecular orbitals, forming σ bonds.

The overlap of two or more adjacent and parallel *p* atomic orbitals creates a π molecular orbital system. Two adjacent *p* orbitals can overlap only above and below the line of centers between the two atoms from which they are contributed (Figure 2–2). This has two consequences: it prevents rotation about axes connecting the nuclei of adjacent atoms, and it permits a series of overlaps to occur simultaneously. Because rotation is prevented, structures containing a π molecular orbital system are rigid. The fact that a series of overlaps can occur permits the electrons occupying a π molecular orbital system to be delocalized.

Delocalization of a pair of electrons occupying one π molecular orbital in such a system results from the fact that each π molecular orbital is a linear combination of the *p* orbitals that overlap. Each π molecular orbital is spread over and shared by every atom that contributed a *p* orbital to the system unless a node is located at that atom. When a pair of electrons occupies a π molecular orbital, it cannot be assigned to a particular atom, notwithstanding the formal requirement of the Lewis dot structure that it be so localized for the purposes of bookkeeping. Confusion between actuality and accounting sometimes leaves the impression that π electrons are localized.

An example of a combination of *p* atomic orbitals is the π molecular orbital system that forms when four parallel *p* orbitals mix (Figure 2–2). The number of π molecular orbitals that result from any combination of this type is always equal to the number of *p* orbitals

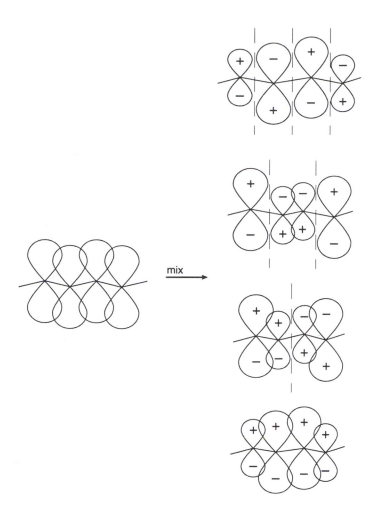

Figure 2–2: A π molecular orbital system formed by the combination of four parallel *p* orbitals on four adjacent atoms held together by σ bonds. The four *p* orbitals overlap, and four linear combinations of these four atomic orbitals are permitted. In the combination of lowest electronic energy, all four *p* orbitals overlap in phase (indicated arbitrarily with + and –). In each of the higher π molecular orbitals, nodes are present on either side, of which the constituent *p* orbitals have opposite phase so overlap is not possible. The four individual π molecular orbitals (to the right) are arranged in order of increasing electronic energy from bottom to top. In all linear π molecular orbital systems, such as the one represented, the number of nodes increases by one upon moving to the next higher energy level. The nodes are evenly distributed from one end of the structure to the other. In this π molecular orbital system formed from four *p* orbitals there are usually four electrons occupying the two π molecular orbitals of lowest energy. The sizes of the atomic orbitals approximately represent the magnitude of their contribution to each π molecular orbital.

combined; in this case there are four π molecular orbitals in the system. Each *p* orbital can be mixed in one of two phases, and adjacent *p* orbitals can be either in phase, in which case they overlap—a favorable interaction—or out of phase, in which case a node—an unfavorable interaction—occurs. A node is a position at which the phase inverts. In a linear system such as the one shown in Figure 2–2 the number of nodes increases by one for each molecular orbital in the series.

Each of these four π molecular orbitals has an energy level associated with it that is equal to the electrostatic potential energy one electron would experience were it confined within that orbital. These energy levels are distributed symmetrically above and below the potential energy an electron would have in one of the isolated *p* orbitals from which the system was created. The more nodes the molecular orbital has, the higher its energy and the less likely that an electron will occupy it. A π molecular orbital is designated as bonding, nonbonding, or antibonding depending on whether its energy level is less than, equal to, or greater than the energy level of an isolated *p* orbital, respectively. If the π molecular orbital is bonding, the equivalent of a covalent bond is formed because a pair of electrons has a lower energy in the

molecular orbital than it would have were it split between two isolated atoms. That covalent bond is spread over the atoms contributing the *p* orbitals, notwithstanding the impression left by the Lewis structure that it is a second localized bond.

The general structure of each π molecular orbital is determined only by the number of *p* orbitals that have mixed together and their topology; however, the nature of the atom—carbon, nitrogen, or oxygen—that has contributed each of the *p* orbitals does affect the shape and energy of the molecular orbital through the coulomb effect. Roughly, this causes the region of a bonding π molecular orbital over a more electronegative atom, such as oxygen or nitrogen, to swell at the expense of the region over the less electronegative atoms, usually carbon. A π molecular orbital that has a significant volume over an oxygen or a nitrogen before the coulomb effect is considered significantly lowered in energy by the coulomb effect relative to the same π molecular orbital in a molecule in which that same position is occupied by carbon. A π molecular orbital that has a node at an oxygen or nitrogen has about the same energy after the coulomb effect is considered as the same π molecular orbital in a molecule in which that same position is occupied by carbon. The mean energy of an electron in an isolated *p* orbital is also lowered for a π molecular orbital system containing electronegative atoms because one or more of the *p* orbitals came from the electronegative atoms.

The number of π molecular orbitals in a given system is determined solely by the number of *p* orbitals that have been mixed together, but the number of those molecular orbitals that are occupied by pairs of electrons is determined by other properties of the molecule as well. The two decisions, how many *p* orbitals have combined and how many π electrons have occupied the π molecular orbital system, are made by examining all valid resonance structures for the molecule. Drawing resonance structures is nothing more than making this decision. Any electrons that are active participants in resonance have been explicitly designated as π electrons by the person who drew those resonance structures, and any atom whose bonding changes among the resonance structures has been explicitly designated as an atom that has contributed a p orbital to the π molecular orbital system. All double bonds and triple bonds in a molecule are necessarily participants in π molecular orbital systems.

The amide is a simple, relevant example of this process of designation (Figure 2–3). The two resonance structures for the amide state that the oxygen, the carbon, and the nitrogen each contribute a p orbital to the π molecular orbital system because their bonding changes between the two structures. The resonance structures state that the π molecular orbital system contains four π electrons because two pairs of electrons shift between the two resonance structures. When three adjacent p orbitals are mixed, three π molecular orbitals are created (Figure 2–3). That four electrons occupy these three molecular orbitals places one pair in each of the two molecular orbitals of lowest energy. The two electrons in the lowest bonding level, if the coulomb effect is disregarded, would have half of their density each distributed over carbon and one-quarter of their density distributed over oxygen and nitrogen. The two electrons in the middle nonbonding level have half of their density distributed over nitrogen and half over oxygen, as is implied by the resonance structures. Usually the resonance structures provide information about the distribution of electrons in the highest occupied molecular orbital or the distribution of electron deficiency in the lowest unoccupied molecular orbital. In the case of the amide, the resonance structures indicate that the pair of electrons in the highest occupied molecular orbital can only occupy locations over the nitrogen and the oxygen. This illustrates the fact that resonance structures and molecular orbitals should agree in their assessment of electron distribution.

Figure 2–3: Electronic structure of an amide. In the lower left corner of the figure the two resonance structures for the amide are presented. The molecular orbital system is formed from the linear combination of three atomic p orbitals: one from nitrogen, one from carbon, and one from oxygen. They are combined to produce three π molecular orbitals presented in order of increasing energy. In the combination of lowest energy, all three of the constituent p orbitals overlap in phase. In linear π molecular orbital systems with an odd number, n, of atoms, the number of nodes in the central nonbonding molecular orbital is equal to $\frac{1}{2}(n - 1)$. This means that there are lobes on the two end atoms and nodes on every other atom in between, alternating lobe, node, lobe, node, and so forth. In the three-atom system displayed here, there is a lobe at nitrogen, a node at carbon, and a lobe at oxygen in the central nonbonding π molecular orbital.

Resonance theory has always incorporated the fact that the several structures drawn do not have independent existence, but occasionally, by mistake, it is implied that they do. In the extreme, the double-headed arrow of resonance becomes replaced with the two arrows of a chemical equilibrium, a mistake that engenders serious confusion. That only one, undivided π molecular orbital system represents the resonance hybrid is a reaffirmation of the absence of independent existence. Unfortunately, while π molecular orbitals present a more accurate picture of the molecular structure and avoid the confusion with equilibrium, they do not have the accounting capability of formal resonance structures, and each view, whether molecular orbitals or resonance structures, has its appropriate use.

The first decision that must be made about the electronic structure of any molecule is the location of all π molecular orbital systems. Any carbon, nitrogen, or oxygen that has contributed a $2p$ atomic orbital to a π molecular orbital system has only two $2p$ atomic orbitals remaining to hybridize with its lone $2s$ atomic orbital, but any carbon, nitrogen, or oxygen that is not involved in a π molecular orbital system has three $2p$ atomic orbitals to hybridize with its $2s$ atomic orbital. It is these hybrids between s orbitals and p orbitals that overlap to form σ **bonds**. These bonds lie along the line of centers between the two respective atoms that they connect, and they are localized. They are stronger covalent bonds than π bonds, and as a result every pair of atoms joined by more than one covalent bond must be joined

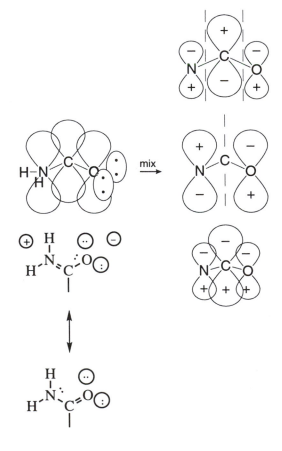

by one and only one σ bond. These σ bonds form the molecular skeleton defining the structure of the molecule, in particular its bond angles. This skeleton is the σ structure of the molecule. Each σ bond is also an occupied molecular orbital, but this realization is not informative in issues of molecular structure. In the particular instance of molecules in biological situations, when an atom has contributed one p orbital to a system of π molecular orbitals, it will almost always be hybridized [p, sp^2, sp^2, sp^2]. At that point in the σ structure, the molecule will be planar, and the σ covalent bonds and σ lone pairs radiate within that plane in three directions from the atom at approximately 120° angles. When an atom has not contributed a p orbital to a system of π molecular orbitals, it will almost always be hybridized [sp^3, sp^3, sp^3, sp^3]. At that point in the σ structure, σ covalent bonds or σ lone pairs should radiate in four directions tetrahedrally.

Because the σ structure incorporates these bond orders and bond angles, it dictates the details of molecular structure. These details cannot be appreciated until decisions on hybridization can be made correctly. To pursue an earlier example, the oxygen, carbon, and nitrogen of an amide are each contributing a p orbital to the π molecular orbital system, and each is hybridized [p, sp^2, sp^2, sp^2]. In the σ structure each of these three atoms and all of the σ bonds and σ lone pairs of electrons radiating from them are in a plane, and each bond angle is approximately 120° (Figure 2–3).

Lone pairs of electrons are identified by writing a Lewis structure of the molecule. Thereafter, it is conventional to ignore them, on the assumption that everyone knows that they are there. Because lone pairs of electrons are of paramount importance in biochemistry and because an understanding of a biologically important molecule is incomplete if ever they are forgotten, it is safer to include them explicitly in any structure drawn.

As a result of electron repulsion, a lone pair of electrons on any oxygen or nitrogen unconjugated to a π molecular orbital system will occupy one of the sp^3 orbitals of that atom. This is a σ lone pair of electrons, and it resides at one of the tetrahedral vertices of the atom. A σ **lone pair** of electrons is a lone pair confined to a single atom because it resides in a hybridized or unhybridized atomic orbital that does not overlap with any other atomic orbital from another atom. A σ lone pair of electrons is designated in a drawing by enclosing it within a circle. If an oxygen or nitrogen containing a lone pair of electrons is sterically able both to rotate until that lone pair is parallel to an immediately adjacent π molecular orbital system and to rehybridize to sp^2 at its three remaining bonded positions, the lone pair of electrons can enter the π molecular orbital system. To do this, the atom carrying the lone pair must rehybridize, which requires sufficient energy to overcome the electron repulsion that originally placed the lone pair in an sp^3 orbital. The favorable energy resulting from the de-

localization of the lone pair into the π molecular orbital system must exceed this deficit. If it does, the lone pair of electrons becomes a delocalized π **lone pair** of electrons, occupying a π molecular orbital spread over two or more atoms. It is so designated in a drawing by not enclosing it within a circle (Figures 2–1B and 2–3).

When either oxygen or nitrogen has contributed only one of its p orbitals to a π molecular orbital system and is left with three valence orbitals, one $2s$ orbital and two $2p$ orbitals, it is usually assumed that they mix to form three sp^2 orbitals that lie together within a plane normal to the π molecular orbital system and are arrayed at 120° angles. If there are two or three covalent σ bonds to the heteroatom, the hybridization is usually [p, sp^2, sp^2, sp^2] because sp^2 orbitals provide maximum overlap in a σ bond. Thus a single lone pair left on a nitrogen that has contributed only one p orbital and one or two valence electrons to a π molecular orbital system and also participates in two σ bonds is always a σ lone pair in an sp^2 orbital, and it is designated as such by surrounding it with a circle. An example of such a lone pair is the lone pair on a nitrogen in an imine

2–1

The situation becomes ambiguous, however, in the case of an oxygen that has contributed a p orbital and one valence electron to a π molecular orbital system, participates in one σ bond, and remains with two lone pairs of electrons. An example of such an oxygen would be the oxygen of a carbonyl (Figure 2–4). The possibility arises that such an oxygen is hybridized [p, p, sp, sp]. In this case, one lone pair would occupy an sp orbital in line but opposite to the σ bond and the other lone pair would occupy a p orbital normal to both the π bond and the axis of the two sp orbitals (Figure 2–4). Indeed, there is evidence from ultraviolet spectra of carbonyl compounds that this occurs. The alternative possibility is that oxygen is hybridized [p, sp^2, sp^2, sp^2] and that both lone pairs are in sp^2 orbitals.

The decision between these two alternatives is not an insignificant one, for oxygens that have contributed one p orbital and one valence electron to a π molecular orbital system and participate in only one σ bond are by far the majority of the oxygen atoms in a molecule of protein. In a hydrogen-bonding environment, such as the water in which all biochemistry occurs, it appears that these oxygens place their two lone pairs in two sp^2 orbitals. This follows from the fact that in crystallographic structures of small molecules in which an N–H forms a hydrogen bond with such a carbonyl or acyl oxygen, the nitrogen–hydrogen σ bond points to the location where an sp^2 lone pair of electrons would be

A   B

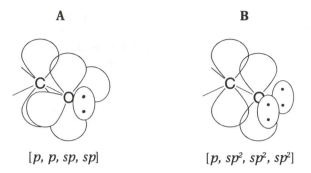

[p, p, sp, sp]   [p, sp², sp², sp²]

Figure 2–4: Two alternative hybridizations for an oxygen in a carbonyl.

located.[1] On the basis of this observation, it will be assumed that carbonyl or acyl oxygens are hybridized [p, sp², sp², sp²], and their two lone pairs will both be designated as sp² by enclosing them in circles (Figure 2–1B). These are σ lone pairs of electrons, they lie within a plane shared with the carbon–oxygen σ bond and normal to the plane of the carbon–oxygen π bond, and they point outward at 120° angles to the carbon–oxygen bond (Figure 2–4B).

The σ structure of a molecule is the basic skeleton producing the σ bonds, the bond angles, and the fixed positions of the localized σ lone pairs. The π electrons are spread over this skeleton above and below the atoms contributing the p orbitals. Neither the bond angles of the molecule, which are defined by hybridization, nor the positions of σ lone pairs, which are localized, can be affected by resonance. It necessarily follows that when one draws two or more resonance structures, one must make certain that the same σ structure is present in each resonance structure and that only the disposition of π electrons differs among them. The best way to ensure this is to draw a σ structure for the second resonance structure identical to the σ structure of the first resonance structure before putting in the π electrons, and to draw an identical σ structure for each of the successive resonance structures before putting in the π electrons. Always include all σ lone pairs oriented as the hybridization of each atom requires. After the set of valid resonance structures has been exhausted, look closely at any lone pair that did not participate and decide if it might not be a σ lone pair. If it is not completing an aromatic complement or being withdrawn by an adjacent π bond, it is probably a σ lone pair in a σ orbital confined only to the one atom.

When the atoms contributing the p orbitals to a system of π molecular orbitals form an unbroken ring rather than being linearly arrayed, the possibility of **aromaticity** arises. In a ring of p orbitals of any size, the energy levels of the individual π molecular orbitals are arrayed in a peculiar pattern. The π molecular orbital with the lowest energy is always the completely overlapping ring with no nodes other than the one at the nuclear plane. This π molecular orbital is occupied by two electrons. The other bonding π molecular orbitals in a ring always come in pairs that have identical energies. Because of Hund's rule, no degenerate pair of orbitals can be filled with electrons to form a stable closed shell until four electrons have been provided simultaneously. These two properties, the one continuous ring occupied by a pair of π electrons and the pairs of orbitals of higher energy occupied by quartets of π electrons, define an aromatic π molecular orbital system. It is an unbroken ring of parallel p orbitals occupied by 2, 6, 10, 14, or 18 π electrons.

From these rules it is clear that a phenyl ring is aromatic, but it is the **nitrogen heterocycles** that are more interesting examples. Pyridine is a neutral, six-membered ring with one nitrogen. Each carbon contributes one valence electron to the π system, so nitrogen can contribute only one to complete the obligatory sextet. This leaves a neutral nitrogen with two remaining valence electrons that end up as a lone pair in the σ structure confined to an sp² orbital. Pyrrole, however, is a five-membered ring. Each carbon gives only one valence electron to the π molecular orbital system, and nitrogen provides the two required to complete the obligatory sextet.

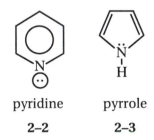

pyridine   pyrrole

2–2   2–3

Nitrogen is left with one valence electron and is forced to form a covalent N–H bond to finish the neutral molecule. Pyridine and pyrrole nitrogens appear throughout aromatic heterocycles and can be recognized readily by applying the rules of aromaticity and discovering what is left over.

An interesting heterocycle that serves as an example of the application of these considerations is porphine

2–4

This is the simplest porphyrin, one of the components of the heme-containing coenzymes. Hidden within this molecule is an unbroken ring of 18 atoms, each contributing a *p* orbital and creating a π molecular orbital system containing 18 π electrons. Therefore, porphine is aromatic. For this to be possible, two of the nitrogens must be pyridine nitrogens and each contribute a *p* orbital and one π electron, and two of the nitrogens must not enter the aromatic ring. The nitrogens that do not enter the aromatic ring nevertheless reside immediately adjacent to it and on the inside of it. They each retain a lone pair of electrons that resembles the π lone pair on the nitrogen of aniline but located endocyclically rather than exocyclically. This requires that each of them, as with the nitrogen in aniline, have three covalent σ bonds, hence the two central hydrogens. The peripheral double bonds in the two five-membered rings with pyridine nitrogens also are bypassed by the aromatic ring and resemble olefins.

Behind the distinctions among π molecular orbital systems, σ bonds, and σ lone pairs is the concept of **orthogonality**. Each π molecular orbital system, each σ bond, and each σ lone pair of electrons is orthogonal to every other π molecular orbital system, σ bond, and σ lone pair in the molecule. As such, each is an independent moiety that does not share electrons with the others. It is this compartmental quality of bonding that has significant impact on the chemistry in which the molecule can engage. It is this fact, rather than a desire to categorize, that renders these distinctions important. They must be clearly made in any drawing of the molecule.

There is also a good reason for making clear distinctions between σ and π lone pairs of electrons. The chemical properties of these two types of lone pairs of electrons are remarkably different. This difference is most clearly expressed in their behavior as bases, and it is the basicity of a lone pair of electrons that, in questionable cases, indicates whether it is a σ or π lone pair of electrons. When the basicity of the lone pair is relied upon as a criterion, a proton is being used to probe its availability. Lone pairs of electrons in π systems are far less basic than those in σ orbitals because σ lone pairs of electrons are localized and directionally oriented by the atomic orbital in which they are confined, while π lone pairs of electrons are delocalized and immersed with the π molecular orbital system.

Problem 2–1: Draw σ–π stereochemical structures as in Figure 2–1B for the *N*-acetyl α-amides of aspartate, asparagine, glutamine, proline, methionine, tyrosine, tryptophan, phenylalanine, histidine, and arginine.

Problem 2–2: The following skeleton structures are various heterocycles. None is intended to be a radical, all have pairwise filled molecular and atomic orbitals, and none has a total of more than one positive or negative charge. No π electrons are shown, and all atoms are shown. All of the compounds are aromatic.

(A) Decide how many π electrons each heterocycle contains to make it aromatic.

(B) Draw resonance forms that indicate the distribution of these π electrons. There may be only one.

(C) Complete the octet for every non-hydrogen atom by adding lone pairs.

(D) Assign formal charges in each resonance form.

(E) Draw the σ structure of each heterocycle, including all lone pairs, and assign hybridization to each atom.

Acids and Bases

The quantitative measure of the basicity of a lone pair of electrons is the microscopic acid dissociation constant of its conjugate acid. In this way all lone pairs of electrons are related to the lone pair on a water molecule. The reaction that defines the microscopic acid dissociation is

(2–1)

where the lone pair on the conjugate base is usually localized (as represented in Reaction 2–1) when X is oxygen, nitrogen, or sulfur but usually delocalized when X is carbon. In this definition, an **acid** is a position within a molecule from which a proton can dissociate to produce a lone pair of electrons, and a **base** is a lone pair of electrons with which a proton can associate, and the reaction occurs in aqueous solution. The two electrons in the σ bond between the central atom (X in Reaction 2–1) and the hydrogen are the same two electrons that form the lone pair in the base, and a bare proton is transferred between the lone pair of the base and a lone pair on the molecule of water and back again. Every acid is always present in solution with its conjugate base, and every base is always present with its conjugate acid.

The equilibrium constant for Reaction 2–1 is

$$K_{eq} = \frac{[H_3O^+][X\odot]}{[H_2O][HX]} \qquad (2\text{–}2)$$

Because $[H_2O] = 55$ M at all times, this term is passed to the left, and for convenience $[H_3O^+]$ is written as $[H^+]$.* These substitutions produce the definition of the microscopic acid dissociation constant

$$K_a \equiv \frac{[H^+][X\odot]}{[HX]} \qquad (2\text{–}3)$$

A **microscopic acid dissociation** constant is the acid dissociation constant of a particular proton in a polyprotic acid. An acid dissociation constant is usually presented as pK_a, where $pK_a = -\log K_a$, solely for convenience. A theoretical justification of this practice is that the pK_a is directly proportional to the change in free energy for Reaction 2–1. The larger the pK_a, the less likely is Reaction 2–1, in the direction it is written and the more likely it is in the opposite direction. Because water is the same in all acid dissociation reactions, the difference in pK_a between two acids is proportional to the free energy for transferring a proton from the one acid to the conjugate base of the other. The smaller the pK_a, the more acidic is the conjugate acid and the less basic, or less available, is the lone pair of electrons on its conjugate base, and vice versa. There are several properties of the position from which the proton dissociates that affect the value of the microscopic pK_a.

The nature of the **central atom** from which the proton dissociates and on which the lone pair remains has a profound effect (Table 2–1). Within the same period of the periodic table, as **electronegativity** increases, the central atom is more capable of supporting the lone pair, and the acidity increases. Atoms in lower periods hold a lone pair of electrons in a larger atomic orbital, making it easier to support, and thus sulfur is much more acidic than carbon even though its electronegativity is the same as that of carbon. Because a localized σ lone pair of electrons on carbon is such a strong base, the only time that there is a lone pair associated with carbon in biochemical situations is when it is a delocalized π lone pair of electrons. Because nitrogen and oxygen are more electronegative elements, delocalized π lone pairs of electrons associated with these elements are rarely bases in biochemical situations, and bases on these atoms are almost always localized σ lone pairs of electrons.

Whether **electronic charge** is created on transfer of a proton to water or merely transferred with the proton also has a large effect on the value of the pK_a (Table 2–1). If it is created, as in the reaction

$$H_2NH + H_2O \rightleftharpoons H_2N\odot^- + H_3O^+ \qquad (2\text{–}4)$$
$$pK_a = 38$$

the dissociation is far less favorable than if it is merely transferred, as in the reaction

$$H_3N^+H + H_2O \rightleftharpoons H_3N\odot + H_3O^+ \qquad (2\text{–}5)$$
$$pK_a = 9.2$$

The successive creation of negative charge on the same polyprotic acid causes each dissociation of a proton to be more difficult than the previous one, and the successive creation of positive charge on the same polybasic molecule causes each association of a proton to be more difficult than the previous one. The larger the molecule over which the accumulating charge can be dispersed, however, the narrower are the increments in pK_a.

The **hybridization** of the central atom affects its pK_a considerably (Table 2–1). All localized σ lone pairs of electrons are in hybrid orbitals formed by mixing one s orbital with one, two, or three p orbitals, as indicated by the designations sp, sp^2, and sp^3, respectively. The fewer the number of p orbitals in the mixture, the greater the fraction of the s orbital distributed into each hybrid and the more s character the hybrid will have. The more s character there is to the orbital, the closer the lone pair of electrons is held next to the nucleus, the less extension along any particular axis will be displayed, and the less basic the orbital will be.

Neighboring electronegative or electropositive atoms also have a significant but less remarkable effect on acidity (Table 2–1). These withdraw or donate electrons by **induction** through σ bonds and decrease or increase the basicity of the lone pair of electrons accordingly.

Because the conjugate acid has a single, σ covalent bond between the central atom and a hydrogen, the pair of electrons that has been protonated in its creation must be or must become σ electrons. If they were a σ lone pair of electrons before the proton was added, **rehybridization** of the central atom is not involved in the protonation. If, however, they were a π lone pair of electrons prior to the protonation, the favorable delocalization energy that they gained within the π system is eliminated as the σ bond is formed. The greater this delocalization energy, the more free energy will be required to protonate the

---

*In fact, the designation H_3O^+ is as misleading as H^+ because the dissociated proton in water is shared either by four molecules of water[2] as the cation $H_9O_4^+$ or by 21 molecules of water[3,4] as the cation $H_{43}O_{21}^+$.

Table 2–1: Electronic Properties Affecting Values of the Acid Dissociation Constant

effect of identity of central atom on acidity[5]

CH_4	$<$	NH_3	$<$	OH_2	$<$	SH_2
$pK_a = 48$		$pK_a = 38$		$pK_a = 15.7$		$pK_a = 7.0$

effect of creation of charge on acidity[6]

OH_3^+	$>$	OH_2
$pK_a = -1.7$		$pK_a = 15.7$

PO_4H_3	$>$	$PO_4H_2^-$	$>$	PO_4H^{2-}
$pK_a = 2.1$		$pK_a = 7.2$		$pK_a = 12.7$

effect of hybridization of the central atom on acidity[5,6]

$HC≡CH$	$>$	$H_2C=CH_2$	$>$	H_3CCH_3
$pK_a = 25$		$pK_a = 44$		$pK_a = 50$

$HC≡NH^+$	$>$	pyridine	$>$	$H_3CNH_3^+$
$pK_a = -10$		$pK_a = 5.2$		$pK_a = 10.6$

$CH_3HC=OH^+$	$>$	$CH_3CH_2OH_2^+$
$pK_a = -10$		$pK_a = -2$

effect of induction on acidity[6]

CF_3CH_2OH	$>$	CHF_2CH_2OH	$>$	CH_2FCH_2OH	$>$	CH_3CH_2OH
$pK_a = 12.4$		$pK_a = 13.1$		$pK_a = 14.2$		$pK_a = 16.0$

$H_5C_2OOCCH_2NH_3^+$	$>$	$H_5C_2OOCC_2H_4NH_3^+$	$>$	$H_5C_2OOCC_3H_6NH_3^+$
$pK_a = 7.7$		$pK_a = 9.1$		$pK_a = 9.7$

lone pair of electrons and the smaller will be the pK_a. This consideration is most clearly exemplified in the protonation of aromatic compounds where the lone pair of electrons is delocalized over the aromatic ring. For example, cyclopentadiene is a strong carbon acid[5]

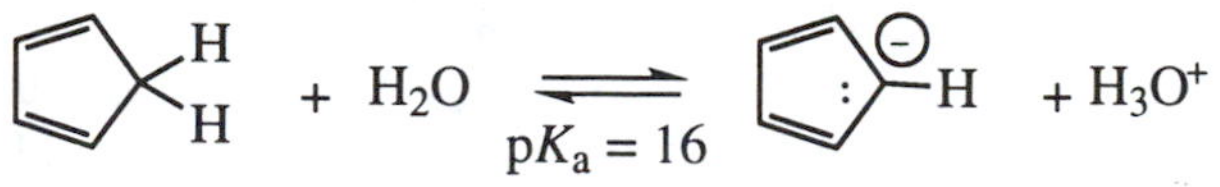

$$(2–6)$$

compared to propane ($pK_a = 50$)[5] because protonation of the lone pair of electrons on the conjugate base of the former destroys the aromaticity of the anion. Likewise, N-methylpyrrole is a very weak base[6]

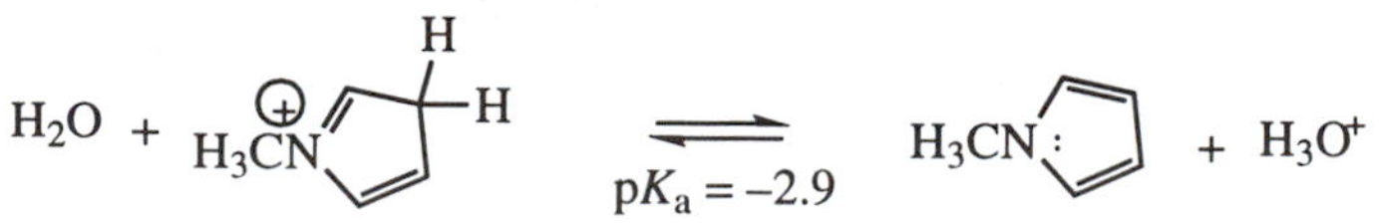

$$(2–7)$$

compared to N-methylazacyclopentane[6]

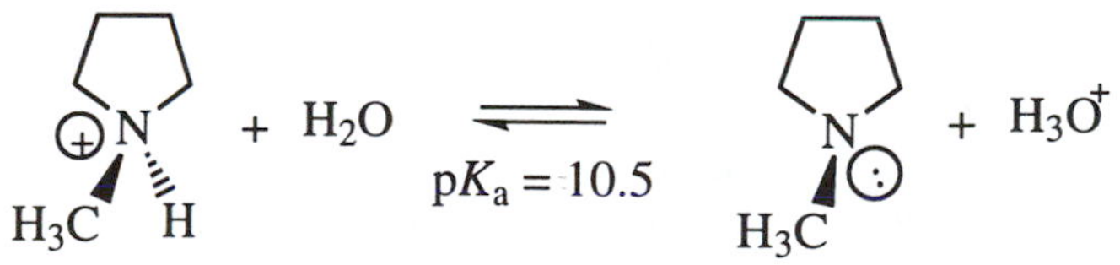

$$(2–8)$$

because protonation of the former destroys its aromaticity. In this case, the comparison is a minimum estimate of the difference in pK_a resulting from aromatic delocalization because pyrrole protonates on carbon rather than on nitrogen. The pK_a for the conjugate acid protonated on nitrogen must be much lower than -2.9.

There are lone pairs of electrons in nonaromatic configurations, the acid–base behavior of which reveals the degree of their π character. The pK_a associated with the lone pair of electrons on aniline ($pK_a = 4.6$)[6] can be compared with the pK_a associated with the lone pair of electrons on cyclohexylamine ($pK_a = 10.6$).[6] The significant difference in basicity demonstrates that the lone pair in aniline is a π lone pair of electrons conjugated to the neighboring π system of the phenyl ring. The pK_a associated with the lone pair of electrons on the nitrogen in an amide ($pK_a = -6$)[7] is even lower than that of the lone pair on aniline and indicates that it is even more delocalized. This is not surprising since the phenyl ring of aniline is otherwise involved in its own aromaticity and the oxygen of the amide has a strong coulomb effect in the highest occupied molecular orbital (Figure 2–3).

Other examples of the use of a proton to evaluate the π character of a lone pair of electrons occur in carbon acids. The pK_a of a methyl group in propene ($pK_a = 43$)[5] is much lower than that in propane ($pK_a = 50$)[5] because the lone pair produced upon the dissociation of a proton from propene conjugates with the neighboring π system of the alkene. The analogous lone pair on carbon in the conjugate base of acetaldehyde ($pK_a = 17.6$)[8] is even less basic because, as occurs with an amide, an oxygen is located two atoms away and exerts a strong coulomb effect on the highest occupied molecular orbital (Figure 2–3). When the π system of the conjugate base is extended from three to five atoms in length, as in the conjugate base of 2,4-dioxopentane ($pK_a = 9$), the lone pair of electrons in the conjugate base becomes even more delocalized and less accessible to protonation.

In making such comparisons, care must be taken to avoid confounding the reasons for the changes in pK_a. The most common confusion is between the effects of hybridization and conjugation. This can be illustrated with the acetate anion. The main reason that the acetate anion, the conjugate base of acetic acid ($pK_a = 4.75$),[6] is a weak base compared to the ethoxide anion, the conjugate base of ethanol ($pK_a = 16$),[6] is that the lone pairs of electrons in the acetate anion are hybridized sp^2 (Figure 2–5) rather than sp^3. The π molecular orbital system of the acetate anion, composed of 4 π electrons in a three-atom system (Figure 2–3), does not provide a pair of electrons to be protonated, notwithstanding any drawing suggesting this to be the case. The protonation occurs on the σ lone pairs orthogonal to the π molecular orbital system, and acetate anion cannot be used as an example of the decrease in basicity that results when a lone pair of electrons is conjugated to a π system.

There is, however, an indirect effect of conjugation on the acidity of acetic acid. When one of the σ lone pairs on the acetate anion is protonated or alkylated, the functional group is no longer symmetric, and the oxygen that has been so modified becomes more electronegative. This change withdraws more electron density onto the protonated or alkylated oxygen, as indicated by the resonance structures

$$\text{(2–9)}$$

The structure on the right indicates that the lone pair of electrons on the alkylated oxygen is less delocalized than

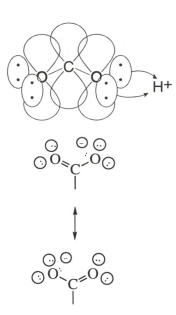

Figure 2–5: Resonance structures and molecular orbital representation of the acetate anion. The resonance structures indicate that one pair of electrons from each oxygen is delocalized and that the π molecular orbital system results from the overlap of three p orbitals, one from the central carbon and one from each of the two oxygens. Each of these three atoms is hybridized [p, sp^2, sp^2, sp^2].

the π lone pair of electrons on an unalkylated oxygen in the carboxylate anion because separation of charge occurs upon delocalization. Nevertheless, the bond between the protonated or alkylated oxygen and the acyl carbon retains double-bond character. This is indicated by an almost 120° angle (116.5°) between an alkyl carbon and an acyl carbon at the oxygen of an ester and a shortening of the bond between the oxygen of an ester and the acyl carbon by 0.09 nm relative to carbon oxygen bonds between sp^2 carbons and oxygens in aryl and vinyl compounds.[9] Therefore, an ester or the conjugate acid of a carboxylic acid retains the overlap of the π molecular orbital system, but the overlap is considerably weakened relative to the unalkylated or unprotonated anion. During protonation of a carboxylate anion, the delocalization in the orthogonal π system is considerably diminished, and this effect destabilizes the conjugate acid and lowers the pK_a. A similar but less pronounced effect of a decrease in delocalization upon protonation occurs with phenol. In this instance, the conjugation in the anion is weaker because negative charge is distributed over three carbons. Consequently, the effect of diminishing this conjugation upon protonation is less, and phenol is a weaker acid than acetic acid.

The acetate anion illustrates another property of a π molecular orbital system—its ability to **redistribute charge**. The excess negative charge in the acetate anion is shared between the two oxygens because the π molecular orbital system is spread over all three atoms. The two electrons in the highest occupied molecular orbital, which account for the negative charge, can reside only over the two oxygens, as there is a node over carbon (Figure 2–3). When one of the oxygens becomes protonated, the π system redistributes and more π electron density is shifted over the oxygen that has become protonated because its coulomb effect has increased. This shift in distribution of charge is reflected in the resonance structure chosen for portraying the conjugate acid

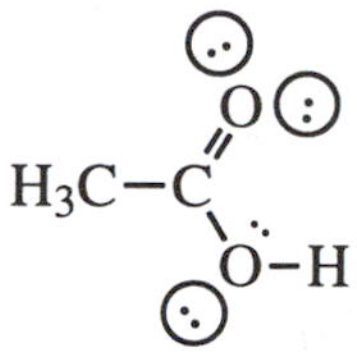

2–5

A similar ability of a π system to redistribute charge is reflected in the lower acidity of *p*-methoxypyridinium (pK_a = 6.67) compared to pyridinium (pK_a = 5.17). This difference results from the ability of the excess electron density of the methoxy substituent to push into the π system through the conjugation represented by the resonance structure

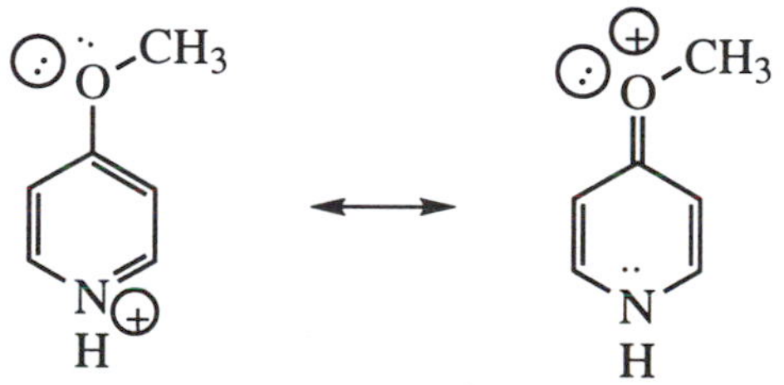

$$(2–10)$$

Contrariwise, an example of a shift of electron density away from the central atom occurs in the *p*-nitrophenolate anion, the associated pK_a of which is 7.2, compared to the phenolate anion, the pK_a of which is 10.0. This can be explained by the resonance structure

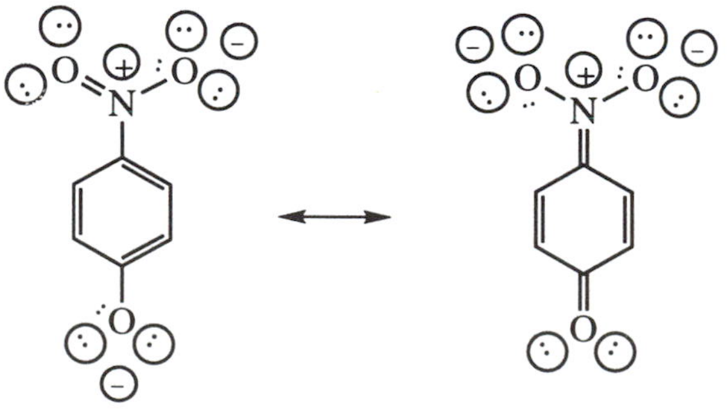

$$(2–11)$$

In each of these examples, the redistribution of charge between electronegative atoms is accomplished by the highest occupied molecular orbital of the π system, which is spread over the whole molecule.

The pK_a of an acid–base is determined by the combination of all of these properties: the electronegativity and hybridization of the central atom, creation of charge, inductive affect, delocalization of the lone pair of electrons, and redistribution of charge.

There are two types of **calculations** performed with acid–bases. The pH of a solution to which a weak acid or weak base has been added can be calculated, or the ratio of the molar concentrations of conjugate acid and conjugate base in a solution of a given pH can be calculated.

The calculation of the pH of a solution upon the addition of an acid–base is an exercise in simultaneous equations. The problem takes the form "Calculate the pH of a solution to which 0.1 mol of sodium acetate has been added for every liter." The equations always used are the **conservation of mass**

$$[HOAc]+[^-OAc]= 0.1\,M \qquad (2–12)$$

where HOAc is acetic acid and $^-$OAc is acetate anion, the **conservation of charge**

$$[^-OAc]+[^-OH]= [Na^+]+[H^+] \qquad (2–13)$$

where $[Na^+]$ = 0.1 M; the **acid dissociation constant** or constants

$$K_a = \frac{[^-OAc][H^+]}{[HOAc]}\,M \qquad (2–14)$$

where pK_a = 4.75 and K_a = 1.78 × 10^{-5} M; and the **water constant**

$$[H^+][^-OH]= 10^{-14}\,M^2 \qquad (2–15)$$

These comprise four, or more, independent simultaneous equations with four, or more, unknowns, such as $[H^+]$, $[^-OH]$, $[^-OAc]$, and $[HOAc]$. These equations can be readily solved for H^+ (1.33 × 10^{-9} M) if the assumption is made that H^+ in Equation 2–13 is negligible relative to the other terms.

The value of this exercise is that the creation of the simultaneous equations and the requirement to cancel certain terms to avoid a cubic or quadratic equation requires an understanding of the acid–base chemistry that is occurring in the solution. For example, one is required to know that the only ions that can be present are H_3O^+, Na^+, ^-OAc, and ^-OH and that sodium acetate is a base so the concentration of protons in the final solution will be very small.

The calculation of the concentrations of a conjugate acid and its conjugate base at a given pH fulfills one or the

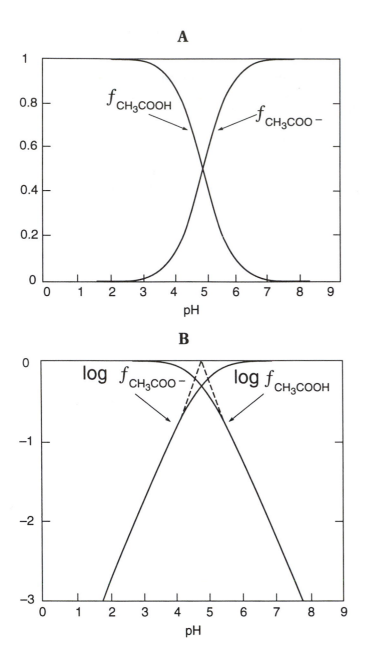

A

B

Figure 2–6: Titration curves for acetic acid. From the acid dissociation constant for acetic acid (pK_a = 4.75), the fraction of the acid–base present in solution as the conjugate acid (f_{CH_3COOH}) or as the conjugate base ($f_{CH_3COO^-}$) as a function of pH can be determined. These values can be plotted directly (A) as a function of pH, or the logarithms of these values (B) can be plotted as a function of pH.

The quantitative behavior of the concentrations of the conjugate acid and conjugate base of each acid–base is described by a titration curve (Figure 2–6) that relates the fraction of the acid–base in the form of the conjugate acid or in the form of the conjugate base to the pH of the solution. This can be presented as the fraction itself (Figure 2–6A), as is usually done, but this presentation leaves the erroneous impression that the fraction of acid goes to zero about 2 pH units above the pK_a and the fraction of base goes to zero about 2 pH units below the pK_a. This misimpression is corrected by examining the logarithms of the fractions as a function of pH (Figure 2–6B). It can be seen that finite fractions of both acid and base are still present at high and low pH, respectively. At a distance of 2 pH units above the pK_a, 1% of the acid–base is in the form of the acid, and this percentage drops off by a factor of 10 for every rise of 1 pH unit but never reaches zero. The importance of this point is that often only one species of the acid–base participates in a chemical reaction, yet the reaction will occur quite well at a pH where the reactive species is present at only 1% or 0.1% or 0.01% or less of the total acid–base. Protonation and deprotonation are extremely rapid reactions; and, as the minor species is consumed in the reaction, it is continuously replaced.

Problem 2–3: Complete the following acid–base equilibria. Draw the structures of the conjugate base and the conjugate acid in proper σ–π stereochemical representation.

(A) $CH_3C\overset{O}{\underset{H}{\big\langle}}$ + H_3O^+ ⇌

(B) $CH_3—\overset{OH}{\underset{OH}{\big\langle}}OCH_3$ + H_3O^+ ⇌

(C) $CH_3C\overset{O}{\underset{OH}{\big\langle}}$ + H_3O^+ ⇌

(D) CH_3CH_2OH + H_3O^+ ⇌

other of two purposes. First, if the solution contains an acid–base of experimental interest, such as an acid–base in a molecule of protein, this calculation will provide the molar concentrations of the conjugate acid and the conjugate base of that acid–base. Second, if a particular buffer is used to stabilize the pH at a particular value, this calculation can be used to determine the concentrations of conjugate acid and conjugate base required for the buffer.

The problem can be stated "Calculate the number of moles of acetic acid and sodium acetate that are present in 2.00 L of an 0.1 M solution of total acetate and acetic acid at pH 5.5." This problem requires only two simultaneous equations, Equations 2–12 and 2–14. Because [H$^+$] is given as 3.16×10^{-6} M there are only two unknowns, [HOAc] and [$^-$OAc], which are 0.015 and 0.085 M, respectively. The answer is 0.03 mol of acetic acid and 0.17 mol of sodium acetate.

(E) 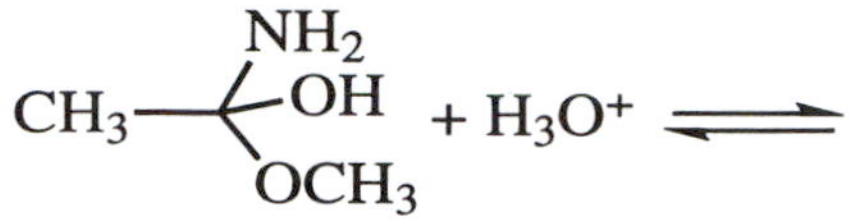 $CH_3-\overset{NH_2}{\underset{OCH_3}{\overset{|}{\underset{|}{C}}}}OH + H_3O^+ \rightleftharpoons$

(F) 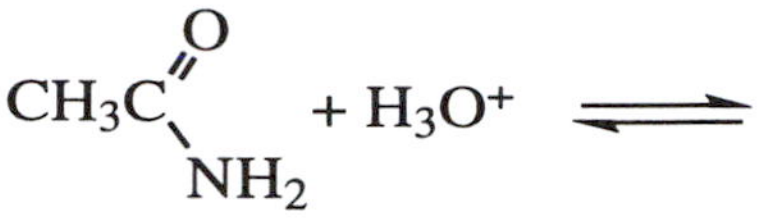 $CH_3\overset{O}{\overset{\|}{C}}\underset{NH_2}{} + H_3O^+ \rightleftharpoons$

Problem 2–4: Write the acid–base equilibrium to which each of the following values of pK_a refer. Write them as chemical reactions, and draw out the structures of the conjugate acid and the conjugate base in σ–π stereochemical representation. In tables[6] of values for pK_a, the name of the compound and the value of the pK_a are all that one is given.

compound	pK_a	
1-amino-2-bromoethane	8.49	
1-aminohexane	10.56	
2,2,2-trifluoroethanol	12.43	
ethanethiol	10.50	
3-hydroxypropyne	13.55	
diethylamine	10.98	
ethanol	−2, 16	
2-hydroxyethanethiol	9.5	
2-aminoethanethiol	8.6, 10.75	
2-chloroethanol	14.31	
morpholine	8.36	
2,2-dichloroethanol	12.89	
diallylmethylamine	8.79	
diethyl ether	−3.5	
1-aminobutane	10.59	
2-hydroxyethanamine	9.50	
allylmethylamine	10.11	
pyrrolidine	11.27	
piperidine	11.22	
piperazine	5.68, 9.82	
pyridine	5.14	
imidazole	7.05, 14.52	
pyrimidine	1.10	
isoquinoline	5.14	
pyrazole	2.48	
aniline	4.62	
o-chloroaniline	2.62	
m-chloroaniline	3.32	
p-chloroaniline	3.81	
p-methylaniline	5.07	
p-methoxyaniline	5.29	
p-nitroaniline	1.02	
phenol	9.95	
p-(trimethylammonium)phenol	8.0	
o-chlorophenol	8.48	
m-chlorophenol	9.02	
p-chlorophenol	9.38	
p-methylphenol	10.19	
p-methoxyphenol	10.20	
p-nitrophenol	7.14	
2-aminobutanoic acid	2.27, 9.68	
N-ethylmorpholine	7.70	
1-aminonaphthalene	3.40	
2-thioethanesulfonate	7.5	
ethyl acetate	25	
1-chloro-2-propanone	16.5	
$CH_3COCH(C_2H_5)CO_2C_2H_5$	12.7	

nicotine	3.13, 8.02
p-hydroxyaniline	5.50, 10.30
1-amino-2,2,2-trifluoroethane	5.7
$CH_3C(NH)NH_2$	12.52
trichloroacetic acid	0.65
fumaric acid	3.03, 4.52
thiazole	2.44
methoxyacetic acid	3.53
thiourea	−0.96

Problem 2–5: From the following list, select the reason for the difference in pK_a between the two molecules in each pair presented below.

(A) hybridization
(B) electronegativity
(C) π donation
(D) σ donation

(E) π withdrawal
(F) σ withdrawal–induction
(G) aromaticity

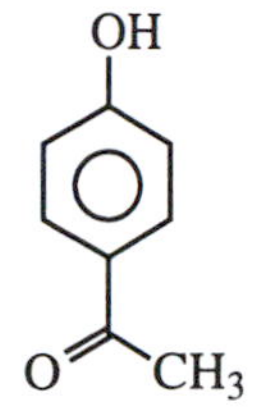

$pK_a = 8.05$ $pK_a = 9.19$

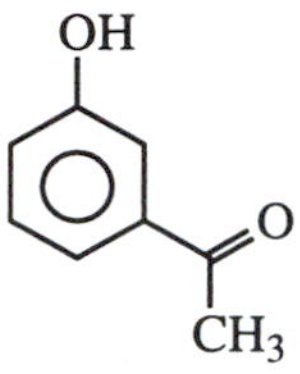

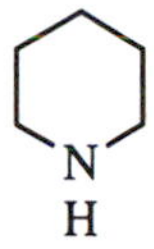

$pK_a = 11.2$ $pK_a = 5.2$

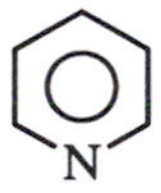

$(CF_3)_3COH$ $(CF_3)_2HCOH$
$pK_a = 5.4$ $pK_a = 9.3$

$pK_a = 7.05$ $pK_a = 8.0$

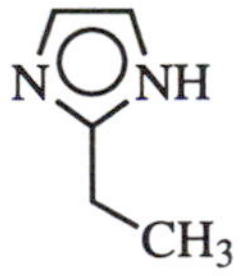

C_2H_5OH $(CH_3)_3\overset{\oplus}{N}C_2H_4OH$
$pK_a = 16.0$ $pK_a = 13.9$

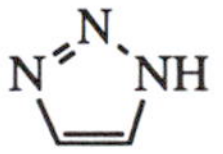

$pK_a = 9.51$ $pK_a = 14.5$

m-nitroaniline   p-nitroaniline
$pK_a = 4.88$ $pK_a = 6.16$

CH_3NO_2 CH_4
$pK_a = 10.3$ $pK_a = 48$

Problem 2–6: What are the exact pHs of

(A) 10^{-2} M acetic acid

(B) 10^{-2} M imidazole acetate

(C) 5×10^{-2} M sodium

(D) 5×10^{-2} M aniline

(E) 10^{-3} M pyridinium chloride

(F) 10^{-2} M *p*-nitroanilinium chloride

(G) 10^{-2} M morpholine

(H) 5×10^{-2} M sodium 2,2-dichloroethoxide

Problem 2–7: Calculate the concentration of anionic imidazole in a 0.1 M solution of imidazole at pH 9.52.

Problem 2–8: Determine the molar concentrations of each species of the weak acids and weak bases in the following solutions.

solute and concentration	pH
0.4 M 1-aminobutane	6.5
0.2 M 1-aminobutane	11.0
0.05 M *p*-chlorophenol	12.0
0.01 M *p*-chlorophenol	7.3
0.01 M *p*-methylaniline	5.0
0.001 M *p*-methylaniline	2.0
0.03 M 2-aminoethanethiol	9.2
0.08 M 2-aminoethanethiol	5.0
0.05 M morpholine	3.5
0.002 M piperazine	7.5
0.03 M ethanol	6.4
0.03 M diethyl ether	8.0
0.03 M 3-hydroxypropyne	4.0

Problem 2–9: From the following information calculate the pH of the final solutions.

buffer species	concentration of buffer (M)	initial pH	amount of NaOH added (mol L^{-1})
(A) imidazole	0.1	6.70	0.02
(B) imidazole	0.03	6.50	0.02
(C) phosphate	0.01	6.80	0.005
(D) phosphate	1.0	6.35	0.1
(E) borate	0.2	9.50	2×10^{-3}
(F) borate	0.15	8.40	0.05
(G) imidazole	0.1	6.50	0.01
(H) imidazole	0.05	7.00	0.02
(I) phosphate	0.2	7.20	0.05
(J) phosphate	0.3	6.20	0.15
(K) borate	0.05	9.40	10^{-3}
(L) borate	0.02	8.60	0.01

Tautomers

A set of tautomers is a set of molecules that differ from each other only in the disposition of their acidic hydrogens. An example is the tautomers of uridine:

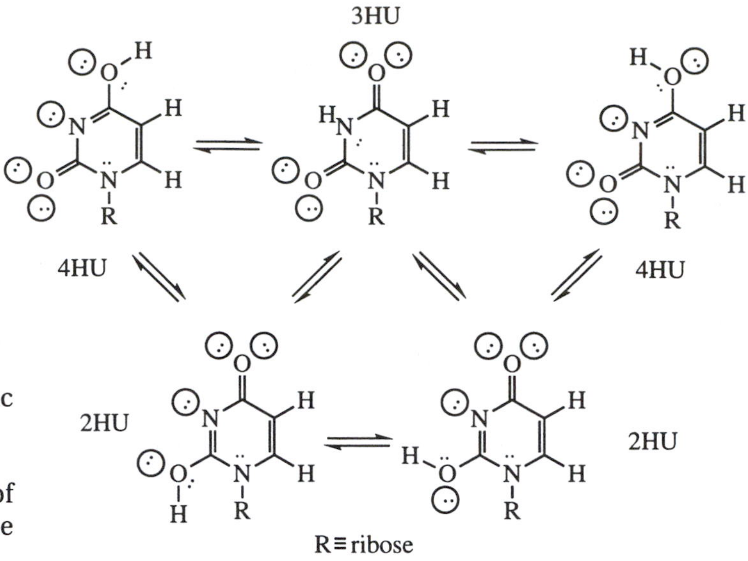

(2–16)

Each of the five members of the set is a distinct molecule with distinct chemical properties. It can be converted to another member of the set by the removal of its acidic hydrogen by a base in solution, almost always a molecule of water (Reaction 2–1), and the readdition of a proton to another lone pair of electrons. None of the lone pairs of electrons in the molecules are disposed properly for the direct intramolecular transfer of a proton. Because these interconversions are acid–base reactions and the intermediate anion

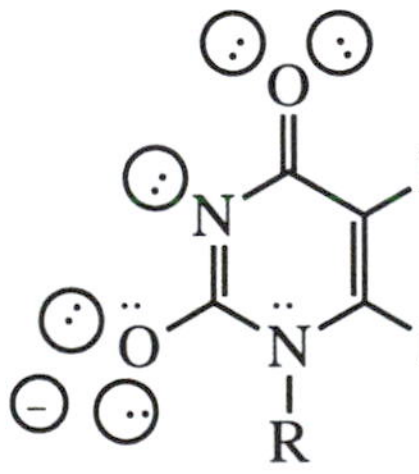

2-6

is a stable molecule, these interconversions are extremely rapid.

At first glance, it would appear that two of the structures in Equation 2–16 are merely rotational isomers of two others and not distinct molecules. The carbon–oxygen bonds of these tautomers, however, should have significant double-bond character, and rotation about them should be hindered, perhaps sufficiently to prevent rotation during the lifetime of the tautomer. This would depend on the rate of dissociation of the proton and the rate of rotation. Therefore, these rotational isomers are formally distinct until demonstrated to be interconvertible significantly by rotation.

In the case of uridine, as the protons are shuffled, the σ structure of the molecule remains constant, and a proton is simply found on a different σ lone pair of anion **2–6**. As there are five σ lone pairs on the anion, there are five tautomers.

In some sets of tautomers, unlike that for uridine, rehybridization of the atoms in the acid–base does occur during tautomerization. Such rehybridization is required to take place when one of the lone pairs that is protonated is a π lone pair in the intermediate base. The usually cited example of this is the keto and enol tautomerization of a carbonyl compound such as acetaldehyde

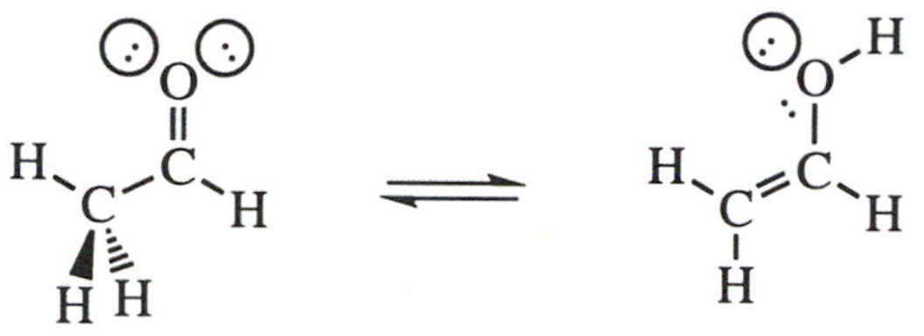

(2–17)

The intermediate in this tautomerization is the enolate anion

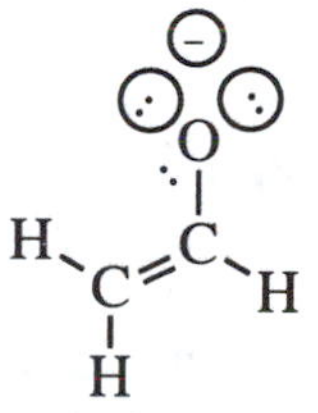

2–7

which has a three-atom π molecular orbital system, and all of the main atoms of the molecule are hybridized [p, sp^2, sp^2, sp^2]. Two of the four π electrons of the anion, however, must be protonated at carbon to form the keto tautomer, an event requiring rehybridization at that carbon. When one of the σ lone pairs on the oxygen is protonated, electronegativities probably dictate that oxygen also rehybridize in the enol tautomer.

There are three aspects of the entire situation that must be clearly distinguished from each other. One is the **set of tautomers** itself. The second is the **resonance structures** that can be written for each member of the set of tautomers. The third is the **acid dissociations** of the parent molecule.

Each of the tautomers in the set can be drawn as a subset of resonance structures. For example, one of the tautomers in Equation 2–16 can be examined in this way

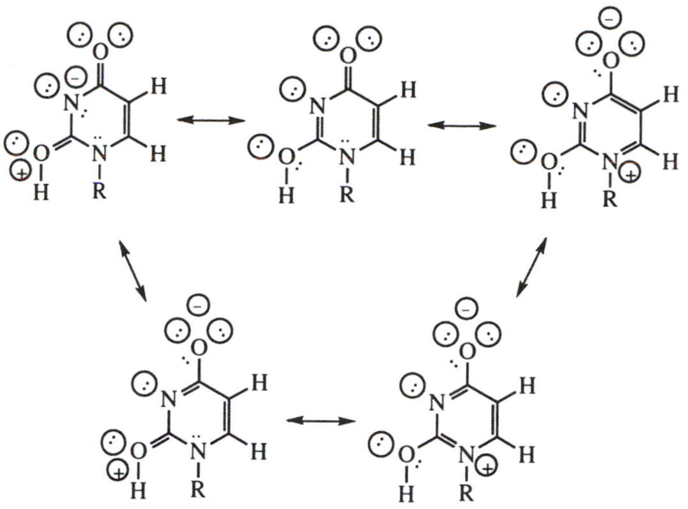

(2–18)

These resonance structures, as distinct from the tautomers themselves, do not have independent existences. In such a subset of resonance structures, as is always required, no σ bond or σ lone pair has engaged in the exercise because they are all orthogonal to the π electrons that are shifted. The resonance structures designate which electrons are the π electrons and which central atoms, namely all of them, are contributing p orbitals to the π molecular orbital system. Each of the five tautomers of uridine can be submitted to this treatment to generate five subsets of resonance structures. It becomes clear that if the hierarchy of set and subset is not always clearly recognized, significant confusion ensues.

Uridine displays an observed macroscopic pK_a^{urid} of 9.2, which reflects the loss of one proton from each of the tautomers to produce anion **2–6**, the sole conjugate base of all five. The ratio among the concentrations of each of the five tautomers is independent of pH because a proton appears on neither side of any chemical equation interconverting any two of them. As the pH increases, the molar concentration of anion **2–6** increases according to the function displayed in Figure 2–6, but with a pK_a of 9.2 instead of 4.75. The sum of the molar concentrations of all of the neutral tautomers decreases accordingly, but the ratio among their concentrations remains unaltered. Again it is important to recall that each tautomer is a distinct molecule present at any pH at a definite molar concentration.

To analyze this situation quantitatively, the two pairs of tautomers that could be rotational isomers of each other will be treated as if they were the same molecule, and the three tautomers of uridine will be abbreviated as 2HU, 3HU, and 4HU. The designations refer to the position assumed by the proton (Equation 2–16). Let U⁻ be anion **2–6**. The equilibria in solution are

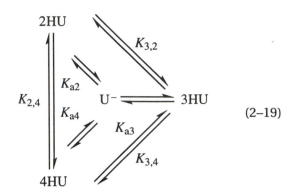

$$(2\text{--}19)$$

where

$$K_{2,4} = \frac{[4HU]}{[2HU]} \qquad K_{3,2} = \frac{[2HU]}{[3HU]} \qquad K_{3,4} = \frac{[4HU]}{[3HU]} \qquad (2\text{--}20)$$

and

$$K_{a2} = \frac{[U^-][H^+]}{[2HU]} \qquad K_{a3} = \frac{[U^-][H^+]}{[3HU]} \qquad K_{a4} = \frac{[U^-][H^+]}{[4HU]} \qquad (2\text{--}21)$$

The three acid dissociation constants K_{a2}, K_{a3}, and K_{a4} are the microscopic acid dissociation constants for the dissociation of the proton from the respective heteroatoms in the given tautomers. The macroscopic acid dissociation constant for uridine is defined by

$$K_a^{\text{urid}} = 6.3 \times 10^{-10} \text{ M} = \frac{[U^-][H^+]}{[2HU]+[3HU]+[4HU]} \qquad (2\text{--}22)$$

A **macroscopic acid dissociation constant** is an acid dissociation constant in which all tautomers with the same number of protons are considered to be indistinguishable. As a result, the molar concentrations of all tautomers with the same number of protons must be summed and only those undivided sums can appear in the expression defining the macroscopic acid dissociation constant. Were there more than one tautomer of the uridine anion, the molar concentrations of all of these tautomers would be summed and that sum would be multiplied by [H⁺] in the numerator. The values measured in an experimental titration are for macroscopic acid dissociation constants.

By simple manipulation it can be shown that

$$K_a^{\text{urid}} = \frac{K_{a2}K_{a3}K_{a4}}{K_{a3}K_{a4} + K_{a2}K_{a4} + K_{a2}K_{a3}} \qquad (2\text{--}23)$$

and

$$K_a^{\text{urid}} = \frac{K_{a3}}{1 + K_{3,2} + K_{3,4}} \qquad (2\text{--}24)$$

Relationships very similar to Equation 2–24 can be obtained by expressing K_a^{urid} in terms of K_{a2} or K_{a4} as well, because all of the equilibrium constants are dependent upon each other, or linked. The linkage is reflected in the relationships

$$K_{3,2} = \frac{K_{a3}}{K_{a2}} \qquad K_{3,4} = \frac{K_{a3}}{K_{a4}} \qquad K_{2,4} = \frac{K_{a2}}{K_{a4}} \qquad (2\text{--}25)$$

The equalities of Equations 2–25 simply state that the ratio among the concentrations of the various tautomers is equal to the inverse of the ratio of their respective microscopic acid dissociation constants, which makes chemical sense. The stronger the bond between the heteroatom and the proton, the smaller will be its intrinsic acid dissociation constant but the greater its relative concentration. It should be noted that the macroscopic acid dissociation constant is always smaller than any of the intrinsic dissociation constants because acids weaker than the strongest acid are always necessarily present.

As far as the effect of a tautomeric acid on the pH of a solution is concerned, it behaves as if it were a simple acid with the pK_a^{urid}. The total concentrations of conjugate base and conjugate acid are used as if they were the concentrations of a simple base and a simple acid.

The biological molecules that illustrate most extensively the various aspects of bonding, acid–bases, and tautomers discussed so far are the bases of the nucleosides. The nucleoside bases are hybrid structures of aromatic heterocycles and carboxylic acid derivatives. The most aromatic base is adenine, the base in adenosine. It is susceptible to electrophilic aromatic substitution at carbon 2 and carbon 8 but is also susceptible to substitution at carbon 6 in reactions that seem to resemble acyl exchange. The most carboxylic base is uracil, the base in uridine. It is unambiguously an *N*-alkyl-*N′*-acyl-*N*-ribosylurea. The carbon–carbon double bond in uracil has almost olefinic character. It is susceptible to addition reactions, unlike the π molecular orbital system in an aromatic compound, which would be susceptible only to substitution.

The nucleoside bases in adenosine, guanosine, and cytidine have exocyclic nitrogens resembling the nitrogen in aniline. The lone pair of electrons on these nitrogens is even more delocalized than that on the nitrogen in aniline (pK_a = 4.6) because the pK_a for the conjugate acid of each of these nitrogens[10] is less than or equal to –2, similar to that for *N*-protonated urea (pK_a ≤ –4). Therefore, each of these exocyclic nitrogens is planar and trigonal, as is always depicted in drawings of the base pairs.

Adenosine has two macroscopic acid dissociation constants (pK_{a1} = 3.6; pK_{a2} = 12.5).[6] These convert the monocation to the neutral molecule and the neutral molecule to the monoanion (Figure 2–7). If all of the

Figure 2–7: Tautomeric forms of adenosine and its cationic conjugate acid and anionic conjugate base.

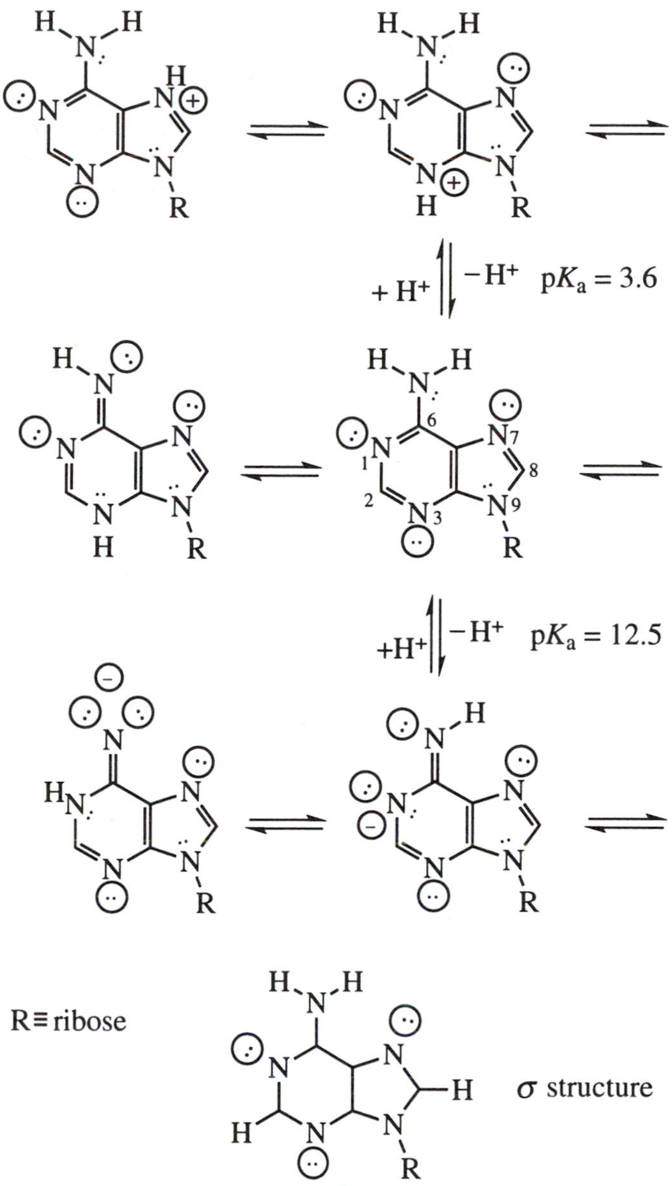

$$+ H^+ \left\Vert\; -H^+ \quad pK_a = 3.6 \right.$$

$$+H^+ \left\Vert\; -H^+ \quad pK_a = 12.5 \right.$$

R ≡ ribose

σ structure

acidic protons were removed from the adenine, it would be a dianion with five σ lone pairs of electrons. These are the five locations that are occupied by the one proton in the monoanion, the two protons in the neutral molecule, and the three protons in the monocation. Only three of the five tautomers of the monoanion, three of the 10 tautomers of the neutral species, and three of the 10 tautomers of the monocation are presented in the figure. Several of the unrepresented tautomers display separation of formal charge and for that reason would be insignificant.

In all of the acid–base reactions and all of the tautomerizations of adenosine, only σ lone pairs are protonated and unprotonated. The number of π electrons remains the same at all times. Adenosine has 12 π electrons. The resonance structures indicate that they redis-

tribute fluidly over the molecule as the σ lone pairs are protonated and unprotonated. In so doing, they are responding to changes in the coulomb effects resulting from these shifts in the locations of the protons.

Problem 2–10:

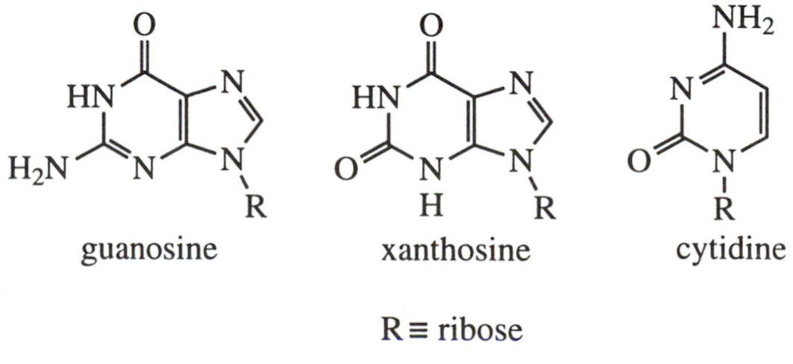

guanosine   xanthosine   cytidine

R ≡ ribose

(A) Draw complete σ structures for the above hetero-cycles in the above tautomeric forms including all σ lone pairs. Draw them with proper bond angles. Abbreviate the ribose as R.

(B) Indicate which protons are involved in tautomeric shifts between which lone pairs. Draw some of the tautomeric forms of these neutral molecules.

(C) How many π electrons are there in each compound?

(D) The macroscopic pK_a values for guanosine are 1.6, 9.2, and 12.5; those for xanthosine are 0.0, 5.5, and 13.0; and those for cytidine are 4.2 and 12.2. Draw a diagram, resembling the one in Figure 2–6, for each of these three heterocycles. Include at least three of the tautomers at each protonation level, if there are three. There are four protonation levels for guanosine and xanthosine and three for cytidine.

(E) How many of the tautomers at each protonation level are insignificant because they require separation of charge?

(F) Draw the σ structure of a tautomer of xanthine which could substitute for adenine in the A-T base pair.

Problem 2–11: Draw in all σ lone pairs but not π electrons, label all acidic hydrogens with an asterisk, and determine the number of π electrons in each of the following nucleoside bases. Draw a neutral resonance form, one with no formal charges.

Problem 2–12: Write the structure of a tautomer of the neutral form of cytidine with no acidic protons on the exocyclic nitrogen and show by drawing all resonance structures that separation of charge cannot be avoided. Write the structure of a tautomer with two protons on the exocyclic oxygen and show that separation of charge cannot be avoided.

Amino Acids

The fundamental, covalent element of a molecule of protein is the polypeptide (Structure **2–8**, below). A **polypeptide** is a long (50–5000 amino acids) linear polymer, the monomers of which are L-α-amino acids. The covalent bonds that link the monomers together to form a polypeptide are amides, which are referred to as **peptide bonds**. Because a molecule of water is lost between two amino acids when a peptide bond is formed, the amino acids, when they are incorporated into a polypeptide, should be referred to as **amino acid residues**. Every polypeptide has the same backbone with an end at which an unbonded primary amine is usually located, the **amino terminus**, and an end at which an unbonded carboxylic acid usually is located, the **carboxy terminus**. At the values of pH usually encountered in living organisms (pH 7–8) the amino terminus ($pK_a = 8$) should be partially protonated and cationic and the carboxy terminus ($pK_a = 3$) should be unprotonated and anionic.

In a polypeptide, the amino acid residues each contribute a **side chain** (the R_i in Structure **2–8**) to the covalent structure. It is the order in which these side chains appear along the polymer that defines the protein. Conceptually, the contribution of an amino acid residue to the structure of the protein can be divided into its α-imido nitrogen, its α-carbon, and its α-acyl carbon and oxygen on the one hand and its side chain on the other. The former always provide the same six atoms (with the minor exception of a carbon for an α-hydrogen in the case of proline) to the backbone of the polypeptide. The structure of this backbone can be treated as if it were a separate molecule, albeit a long tortuous polymer, and the six atoms contributed to it by each amino acid as if they were separate from each side chain. The side chains themselves can be treated as separate entities by detaching them, in the imagination, from their respective α-carbons and replacing the α-carbon with a hydrogen. In this way, a model compound for the amino acid side chain is created.[11] A **model compound** for a particular amino acid residue in a polypeptide would be a small molecule that incorporates the structure of the side chain and any additional structural elements necessary to duplicate the properties of the amino acid residue that are of interest. The model compounds in which the α-carbon is replaced only by a hydrogen are simple, readily available chemicals. For example, in this set, the model compound for glutamic acid (Figure 2–1) would be propionic acid. Another set of model compounds that has been

2–8

used is the *N*-acetyl α-amides of the amino acids.[12] *N*-Acetylaspartic acid α-amide and *N*-acetylglutamic acid α-amide (Figure 2–1) are members of this set.

One of the more important properties of an amino acid residue incorporated in a polypeptide is the acid dissociation constant of any acid–base it contains. The amino acid side chains that contain acid–bases are aspartate, asparagine, serine, threonine, glutamate, glutamine, cysteine, tyrosine, histidine, lysine, tryptophan, and arginine (Table 2–2).[12] The *N*-acetyl α-amides of glutamate and aspartate have been useful in examining the electronic effects of the peptide bonds that surround the α-carbon on the acid dissociation constants of the amino acid side chains in a polypeptide.[12] The α-carbon in an *N*-acetyl α-amide, which is transmitting inductively the electron-withdrawing capacity of both the carboxamide ($\sigma_m = 0.35$)[13] and the acetylimide ($\sigma_m = 0.21$)[13] that are attached to it, seems to have about the same electron-withdrawing capacity as a hydroxyl ($\sigma_m = 0.12$),[13] a cyanomethyl ($\sigma_m = 0.16$),[13] a chloromethyl, a bromomethyl, or a carboxyethyl group.* This conclusion follows from the fact that replacing the α-carbon of *N*-acetylaspartic acid α-amide or *N*-acetylglutamic acid α-amide with any of these functionalities produces little change in the pK_a of the respective carboxylic acid, but the completely aliphatic model compounds, acetic acid and propionic acid, respectively, are significantly less acidic than the respective *N*-acetyl α-amides. These electron-withdrawing substituents can be used to estimate the inductive effect of the polypeptide on the acid dissociations of the various acid–bases on other amino acids (Table 2–2).

These values for the pK_a of the side chains of the amino acids have been shown to be accurate when the amino acid is in a polypeptide if the polypeptide is in the form of a structureless random coil,[16] but when it folds to form a globular protein, significant shifts in the values of these pK_as occur. Neighboring charged functional groups affect the pK_a of a particular acid–base. An adjacent anion makes it harder to remove a proton from an acid and raises its pK_a. An adjacent positive charge makes it easier to remove a proton from an acid and lowers its pK_a. If the acid–base is in an aprotic environment, secluded from water, the more charged form of the acid–base is less stable relative to the less charged form than it is in water. This shifts the pK_a in the direction favoring the less charged form of the acid–base. For example, the pK_a of acetic acid in dimethyl sulfoxide, a relatively polar but aprotic solvent, is 12.9 rather than 4.75, because the anionic conjugate base is poorly solvated by the dimethyl sulfoxide relative to the solvation provided by water. For all of these reasons, when the polypeptide folds to form the native structure, the values for the pK_a of the various amino acids shift away from their ideal values.

* The parameter σ_m is the change in the pK_a of benzoic acid caused by the substituent in the *meta* position. It is used as a quantitative measure of an inductive effect.

Table 2–2: Acid Dissociation Constants for Model Compounds for the Amino Acid Residues[a]

amino acid residues	model compound	pK_a
aspartic acid	in polypeptide (estimate)[b]	**4.0**
	N-acetylaspartic acid α-amide[12]	4.0
	Gly-Gly-Asp-Gly-Gly[14]	3.9
	3-chloropropionic acid[6]	4.1
	hydroxyacetic acid[6]	3.8
	3-bromopropionic acid[4]	4.0
	3-cyanopropionic acid[6]	4.0
	ethyl hydrogen malonate[6]	3.4
	acetic acid[6]	4.75
glutamic acid	in polypeptide (estimate)	**4.3**
	N-acetylglutamic acid α-amide[12]	4.3
	Gly-Gly-Glu-Gly-Gly[14]	4.1
	glutamic acid (microscopic pK_a neutral)[12]	4.5
	4-chlorobutyric acid[6]	4.5
	3-hydroxypropionic acid[6]	4.5
	4-bromobutyric acid[6]	4.6
	4-cyanobutyric acid[6]	4.4
	ethyl hydrogen succinate[6]	4.5
	propionic acid[6]	4.9
serine	in polypeptide (estimate)	**−3, 14.2**
	2-chloroethanol[6]	14.3
	2-bromoethanol[6]	14.4
	2-cyanoethanol[6]	14.0
	ethylglycolate[6]	14.2
	ethanol[5,6]	−2, 16.0
	methanol[6]	15.5
threonine	in polypeptide (estimate)	**−3, 15**
cysteine	in polypeptide (estimate)	**8.7**
	glutathione[15]	8.7
	cysteine (microscopic pK_a neutral)[16]	9.1
	ethanethiol[6]	10.5
	2-mercaptoethanol[6]	9.5
	ethyl mercaptoacetate[6]	8.0
tyrosine	in polypeptide (estimate)	**9.8**
	polytyrosine[17]	9.5
	tyrosine (microscopic pK_a neutral)[18]	9.8
	4-(hydroxymethyl)phenol[6]	9.8
	phenol[6]	10.0
	4-methylphenol[6]	10.2
histidine	in polypeptide (estimate)	**6.4, 14**
	Pro-His-glycinamide[16]	6.4
	histidine hydantoin[19]	6.4
	histidine (microscopic pK_a neutral)[20]	6.0
	imidazole[6]	7.1, 14.5
	4-methylimidazole[20]	7.5
lysine	in polypeptide (estimate)	**10.5**
	1-amino-5-hydroxypentane[6]	10.5
	Ala-Lys-Ala$_n$ (n = 1, 3)[6]	10.5
	1-aminopentane[6]	10.6
glutamine	in polypeptide (estimate)	**−0.5, 17**
	acetamide[5,6]	−0.5, 17
asparagine	in polypeptide (estimate)	**−1, 16**
arginine	in polypeptide (estimate)	**13**
	N-methylguanidine[6]	13.4
tryptophan	in polypeptide (estimate)	**17**
	indole[6]	16.9

[a]The values for the pK_as of the model compounds are from the noted sources for 25 °C. [b]The estimate for each amino acid is based entirely on the values tabulated for the model compounds.

Unfortunately, the free amino acids themselves are very poor model compounds for the amino acids residues in a polypeptide. This arises from the fact that they are either zwitterionic or bear a net charge at all values of pH. Their solubilities, acid–base behavior, and ability to participate in noncovalent interactions are dominated by the carboxylate anion and ammonium cation they contain. Contrary to original expectations, an understanding of the properties of proteins depends very little on an understanding of the properties of the amino acids themselves, while an examination of the structures of the amino acid side chains and the behavior of uncharged model compounds for them does provide essential information.

Alanine, valine, leucine, and **isoleucine** have alkanes as side chains (Figure 2–8A). All of their carbons are hybridized sp^3. Because alkyl groups are sterically more bulky than functional groups containing atoms hybridized $[p, sp^2, sp^2, sp^2]$, steric considerations are more important

in examining the structures of these four amino acids than with most of the others. The view down every carbon–carbon bond should be staggered, and methyl or other alkyl groups should be *trans* to each other in the most stable conformers. The view down the bond between the α-carbon and the β-carbon of a valine residue in a polypeptide (Figure 2–8B) emphasizes the steric interactions between the peptide backbone and the carbons in the two γ positions of this amino acid and, by extension, the steric interactions for each of the others. In every amino acid except alanine and glycine, one or two substituents are attached to the β-carbon, and the β-carbon is hybridized sp^3. This places one or two substituents at either or both of the positions of the two methyl groups of valine.

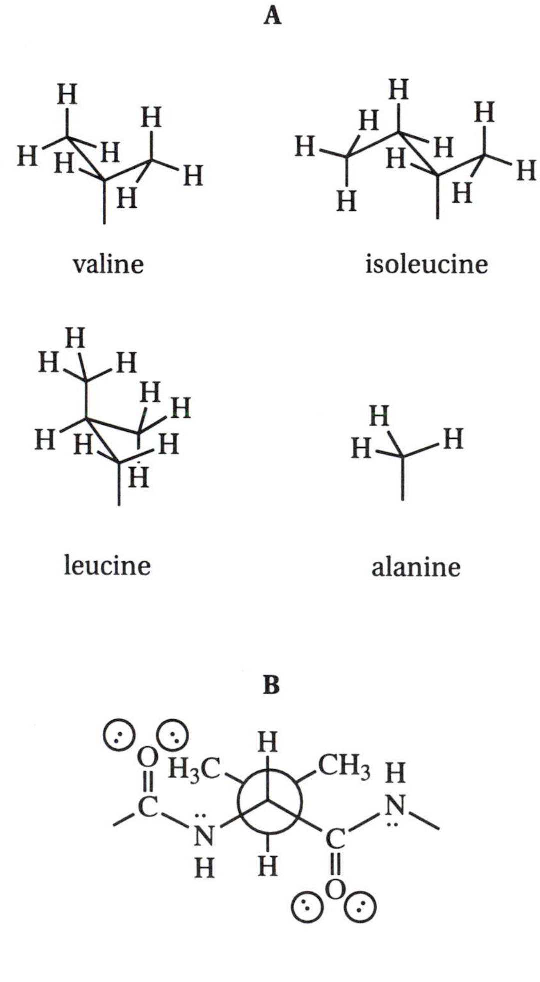

Proline and **glycine** have similar roles as amino acid residues whose effect on the polypeptide is almost entirely steric. Glycine has no side chain at all, merely a hydrogen, and as such can occupy positions in the native structure of a protein that are very cramped. As a result it is one of the most highly conserved amino acids during evolution. A proline, because it is a ring, forces the polypeptide to assume very particular orientations (Figure 2–9). The peptide bond involving the α-nitrogen is planar and rigid but can assume two configurations, *trans* and *cis*. The π system of the amide must be broken for conversion to occur between these two configurations. This requires breaking of partial double bonds, and it is a slow process ($t_{1/2} = 100$ s at pH 4 and 10° C).[21]

Phenylalanine is an aromatic amino acid residue by virtue of its phenyl ring (Figure 2–10). The six π electrons are delocalized in three bonding molecular orbitals over the six carbons that contribute the six p orbitals. This renders the σ structure of the ring planar, and it is sandwiched between two circular clouds of π electrons. The view down the bond between the β-carbon and the γ-carbon of a phenylalanine residue (Figure 2–10) demonstrates that there is a favored orientation that places the polypeptide most distant from the two o-hydrogens of the ring and also avoids eclipse. A phenylalanine residue absorbs ultraviolet light ($\lambda_{max} = 253$ nm, $\varepsilon = 1550$ $M^{-1}cm^{-1}$), and its spectrum displays the usual fine structure seen in unadorned alkylbenzenes.

The side chain of **tryptophan** is indole, which is a benzopyrrole (Figure 2–11). The indole is entirely aromatic, consisting of an unbroken ring of nine atoms each contributing a p orbital, and the aromatic π molecular orbital system contains 10 π electrons. The hydrogen on the pyrrole nitrogen of indole (p$K_a = 17.0$)[6] is slightly less acidic than the hydrogen on a molecule of water (p$K_a = 15.7$). When it departs as a proton, the lone pair left behind is an sp^2 lone pair, and the negative formal charge can be distributed by the π molecular orbital system over all nine atoms in the ring. This delocalization is greater in extent than the delocalization available to pyrrole, which is somewhat less acidic (p$K_a = 17.5$)[6] than indole. Indole is planar with hydrogens directed outward

A

valine

isoleucine

leucine

alanine

B

Figure 2–8: Stereochemical representations of the side chains of valine, isoleucine, leucine, and alanine (A) and the stereochemistry along the bond between the α-carbon and the β-carbon of valine in a polypeptide (B).

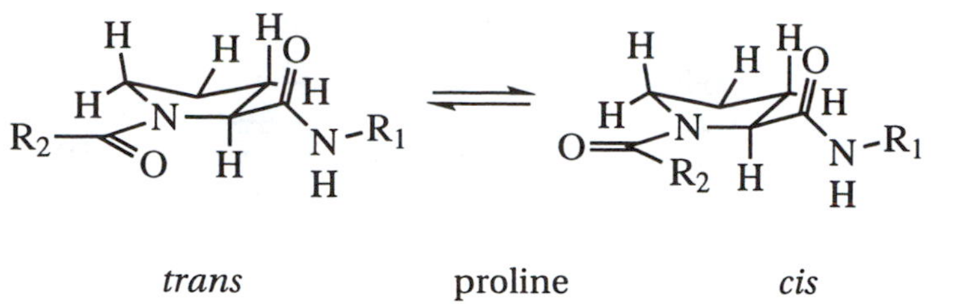

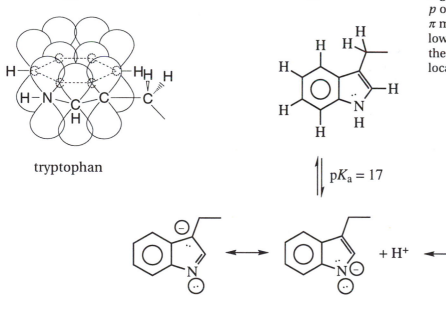

Figure 2–9: Stereochemical representation of two geometric isomers of proline. The amide is planar and occupies a position in the puckered cyclopentyl ring at which eclipse would occur if it were occupied by any of the four carbons. In the *trans* conformation, the two strands of polypeptide (R$_1$ and R$_2$, respectively) emerge in opposite directions. In the *cis* conformation, they emerge in the same direction.

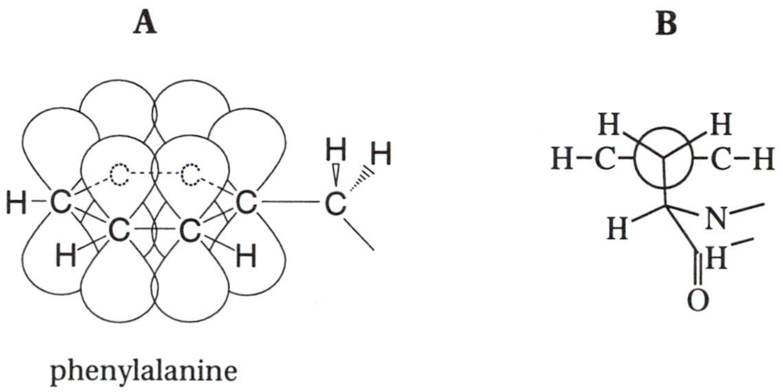

A **B**

phenylalanine

Figure 2–10: Side chain of phenylalanine. (A) The six *p* orbitals that overlap to form the three occupied π molecular orbitals are represented above and below the planar ring. (B) The view down the bond between the β-carbon and the γ-carbon of phenylalanine.

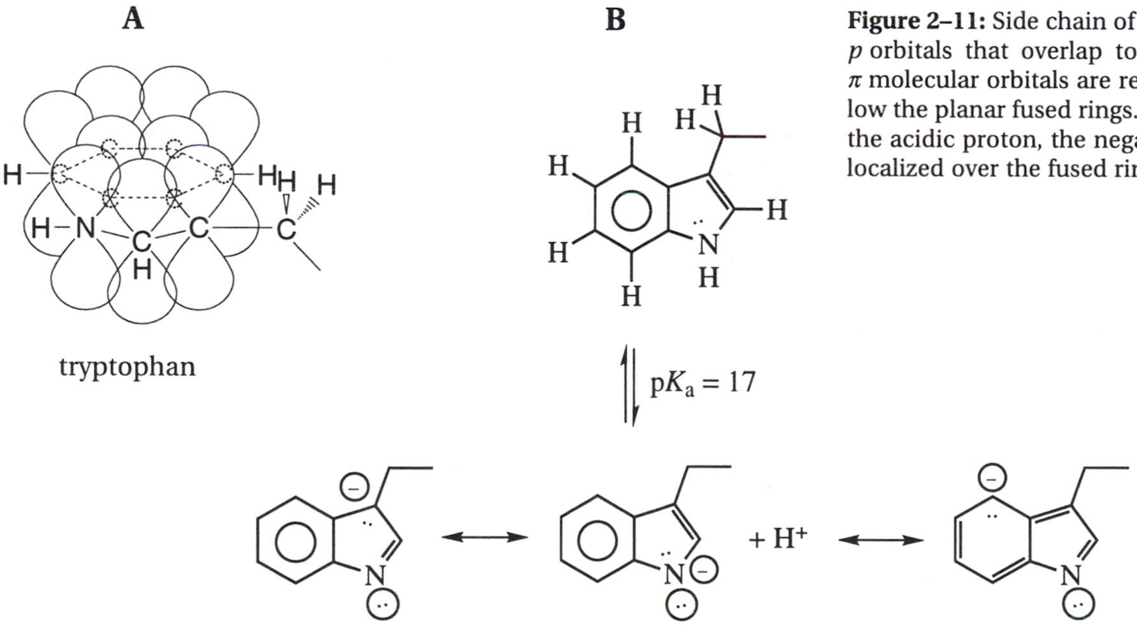

A **B**

tryptophan

pK_a = 17

Figure 2–11: Side chain of tryptophan. (A) The nine *p* orbitals that overlap to form the five occupied π molecular orbitals are represented above and below the planar fused rings. (B) Upon dissociation of the acidic proton, the negative formal charge is delocalized over the fused rings.

+ H$^+$

along its edge and clouds of π electrons above and below the σ plane. The indole ring of a tryptophan residue has the strongest ultraviolet absorption of any functional group in an amino acid (λ_{max} = 281 nm, ε = 5690 M^{-1}cm^{-1}).[22,23]

The side chains of **serine** and **threonine** are primary and secondary aliphatic alcohols resembling ethanol and 2-propanol, respectively, except that they

are more acidic because of the electron withdrawal of the immediately adjacent polypeptide. Their oxygens are hybridized sp^3 and have two σ lone pairs that can act as bases as well as an acidic hydrogen (Figure 2–12). The values of the pK_a for these acid–base reactions can be estimated (Table 2–2) from a series of alcohols containing appropriate electron-withdrawing substituents.

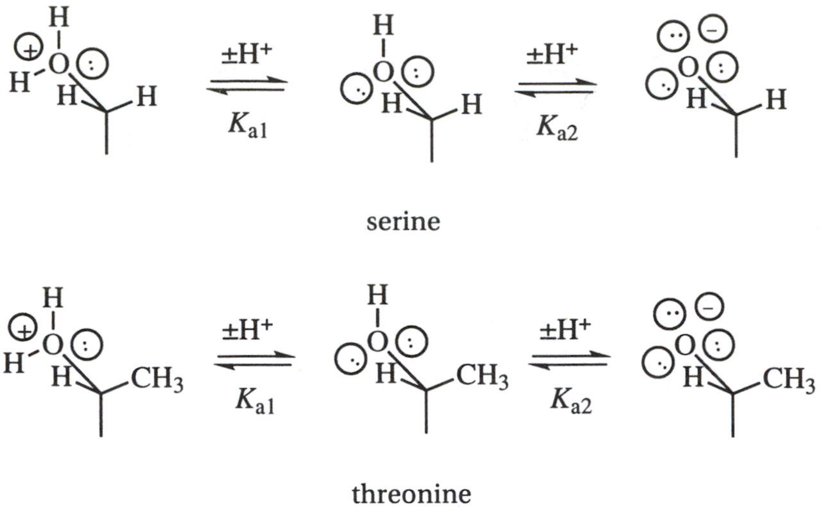

Figure 2–12: Acid–base chemistry of serine and threonine. The neutral side chains have both an acidic proton that can dissociate to give the alkoxide anion and basic lone pairs that can protonate to give the cation.

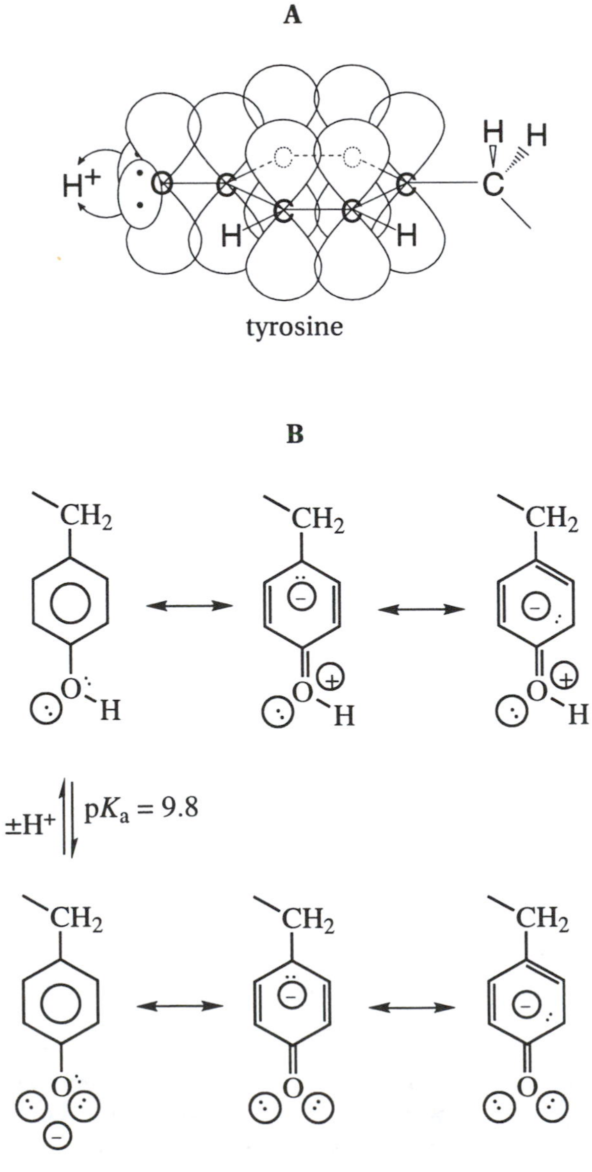

serine

threonine

Figure 2–13: (A) Side chain of tyrosine as the unprotonated anion. The six p orbitals from the ring and the one p orbital from the exocyclic oxygen that overlap in the anion are represented above and below the plane of the ring. The two σ lone pairs on the oxygen are in the plane of the ring at angles of 120° to the carbon–oxygen bond and are the only bases on the anion. (B) Resonance structures of the phenol as the neutral acid and as the tyrosinate anion. The conjugate acid and conjugate base are connected by the acid dissociation.

A

tyrosine

B

$\pm H^+$ ⫫ $pK_a = 9.8$

Tyrosine resembles phenylalanine because it is aromatic and serine because it has a hydroxyl group. As a phenol, however, its properties are distinct from either. Tyrosine (pK_a = 9.8) is more acidic than serine (pK_a = 14.2) because of the ability of the neighboring π system to delocalize the excess electron density of the anion (Figure 2–13). To the extent that one of the lone pairs of electrons on the conjugate acid is delocalized (Figure 2–13B), the lowered pK_a of the hydroxyl would reflect the lowered pK_a of an sp^2 oxygen–hydrogen bond. To the extent that a lone pair is not delocalized in the conjugate acid, the lowered pK_a reflects the stability of the anion relative to the neutral acid. The stability of the anion results from the ability of the π molecular orbital system to spread the excess electron density of the anion over several atoms, in this case one oxygen and three carbons. The ability of a π molecular orbital system to redistribute charge in the absence of rehybridization, as in the phenolate anion, should again be formally distinguished from the transformation of a π lone pair into a σ bond, as in the protonation of the carbon of an enolate. In the case of the phenolate anion, protonation occurs at one of the two σ lone pairs of electrons on oxygen, not on the π lone pair of electrons. A significant change in the ultraviolet spectrum of tyrosine residue occurs when the conjugate acid (λ_{max} = 275 nm, ε = 1410 M^{-1}cm^{-1}) becomes the

A

6 π electrons

B

histidine

K_{a1}^{H3} K_{a2}^{H1}

1-H

H_2^+ K_{a1} K_{a2} H^-

3-H

K_{a1}^{H1} K_{a2}^{H3}

C

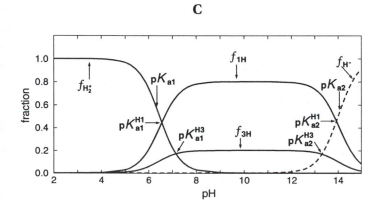

fraction

$f_{H_2^+}$ pK_{a1} f_{1H} f_{H^-} pK_{a2}

pK_{a1}^{H1} pK_{a1}^{H3} f_{3H} pK_{a2}^{H1} pK_{a2}^{H3}

pH

Figure 2–14: (A) Side chain of histidine as the unprotonated imidazolate anion. The five p orbitals from the five atoms in the ring that overlap to form the aromatic heterocycle are represented above and below the plane of the ring. Six electrons occupy the aromatic π molecular orbital system. The two vacant σ lone pairs of electrons at the two nitrogens lie in the plane of the ring. These are the bases that combine with either one proton ($pK_{a1} = 14$) or with a second proton ($pK_{a2} = 6.4$) to produce the neutral forms or the cation, respectively. (B) The cation (H_2^+), the two neutral tautomers (1-H and 3-H), and the anion (H^-) of the side chain of histidine and the associated resonance structures. The macroscopic dissociation constants for each ionization (K_{a1} and K_{a2}) should be distinguished from the microscopic dissociation constants (K_{a1}^{H1}, K_{a1}^{H3}, K_{a2}^{H1}, and K_{a2}^{H3}). (C) Titration curves for the four species: H_2^+, 1-H, 3-H, and H^-. The fraction of the total histidine present as each species is plotted as a function of the pH.

conjugate base (λ_{max} = 293 nm, ε = 2380 M^{-1}cm^{-1}) upon acid dissociation.[22]

The side chain of **histidine** is also an aromatic acid–base (Figure 2–14). Unlike the phenol in tyrosine, the imidazole in a histidine residue has no exocyclic heteroatom whose *p* orbitals may or may not mix with the π molecular orbital system. Neutral imidazole is an aromatic heterocycle containing one pyrrole nitrogen and one pyridine nitrogen, which together contribute three valence electrons to an aromatic ring formed from five *p* orbitals. Six π electrons are located in three bonding molecular orbitals. The six π electrons remain in this aromatic π molecular orbital system at all times but fluidly redistribute in response to changes in coulomb effects as the nitrogens gain or lose protons at their σ lone pairs.

The anion of imidazole (Figure 2–14) is the best place to begin. All atoms are hybridized [p, sp^2, sp^2, sp^2], the two nitrogens each have a σ lone pair of electrons, and both are electronically equivalent. The excess electron density associated with the formal negative charge is distributed by the π system over both nitrogens, and resonance structures can be drawn to show this (Figure 2–14B). The first proton adds to one of the two σ lone pairs in the imidazolate anion to form an sp^2 covalent bond to the hydrogen in an acid–base reaction with a macroscopic pK_{a2} = 14.5. Thus, the imidazolate anion is less basic than the pyrrolate anion (pK_a = 17.5)[6] because its π molecular orbital system can spread the excess electron density over two nitrogens, but the adenosinate anion (pK_a = 12.5) is less basic than the imidazolate anion because its π molecular orbital system can spread the excess electron density over three nitrogens (Figure 2–7).

In the neutral imidazole, the two nitrogens are necessarily nonequivalent because one has a proton attached to it. The proton renders its nitrogen more electronegative, and the lobes of the π molecular orbitals at this location swell accordingly. This is represented in the resonance structures by placing a π lone pair of electrons over this nitrogen, but such formalism is not meant to imply that this becomes a basic position or that this nitrogen rehybridizes. The base is the σ lone pair on the other nitrogen, and it gains a proton in an acid–base reaction with a macroscopic pK_{a1} = 6.4 when the base is histidine in a polypeptide (Table 2–2) or pK_{a1} = 7.1 when it is imidazole itself. This is the acid–base reaction that makes histidine so valuable. The imidazolium cation (pK_{a1} = 7.1) is less acidic than the pyridinium cation (pK_a = 5.2) because its π molecular orbital system can spread the electron deficiency over two nitrogens.

The two ring nitrogens in a histidine residue, unlike the two in imidazole, are not stereochemically equivalent to each other because of the substitution at C4. The behavior, as a function of pH, of the chemical shifts of the nuclear magnetic resonances of the various carbon-13 nuclei of the imidazole ring in histidine, has been compared to their behavior in 1-methylhistidine and 3-methylhistidine. It was concluded from these observations[24] that in aqueous solution the ratio between the two neutral tautomers (Figure 2–14), 1-protiohistidine (1-H) and 3-protiohistidine (3-H), is 4:1. Let H$_2^+$ be the cation. If

$$K_{3,1} = \frac{[\text{1-H}]}{[\text{3-H}]} = 4 \tag{2–26}$$

and

$$K_{a1}^{H3} = \frac{[\text{1-H}][\text{H}^+]}{[\text{H}_2^+]} \tag{2–27}$$

and

$$K_{a1}^{H1} = \frac{[\text{3-H}][\text{H}^+]}{[\text{H}_2^+]} \tag{2–28}$$

and the macroscopic dissociation constant

$$K_{a1} = \frac{([\text{1-H}] + [\text{3-H}])[\text{H}^+]}{[\text{H}_2^+]} = 10^{-6.4} \tag{2–29}$$

then

$$K_{a1}^{H3} = \tfrac{4}{5} K_{a1} = 10^{-6.5} \tag{2–30}$$

and

$$K_{a1}^{H1} = \tfrac{1}{5} K_{a1} = 10^{-7.1} \tag{2–31}$$

Because either one of two protons can dissociate, the imidazolium cation of a histidine residue in a polypeptide has two microscopic acid dissociation constants, pK_{a1}^{H3} = 6.5 and pK_{a1}^{H1} = 7.1.

A word of caution is in order. It is important to distinguish carefully between the use of the macroscopic and microscopic acid dissociation constants. If the imidazole of a histidine residue is on the surface of a molecule of protein and free to rotate around the carbon–carbon σ bonds connecting it to the polypeptide so that both nitrogens can participate in acid dissociations, the macroscopic pK_a will dictate its acid–base behavior as it did in the model compounds. If the imidazole were held in place by the amino acids that surround it in such a way that the surroundings had the same polarity as water and that one of its acidic hydrogens were always engaged in a hydrogen bond with an acceptor that resembled water very closely, the single remaining site available for acid dissociation would display its microscopic acid dissociation constant. A decision to use the microscopic or mac-

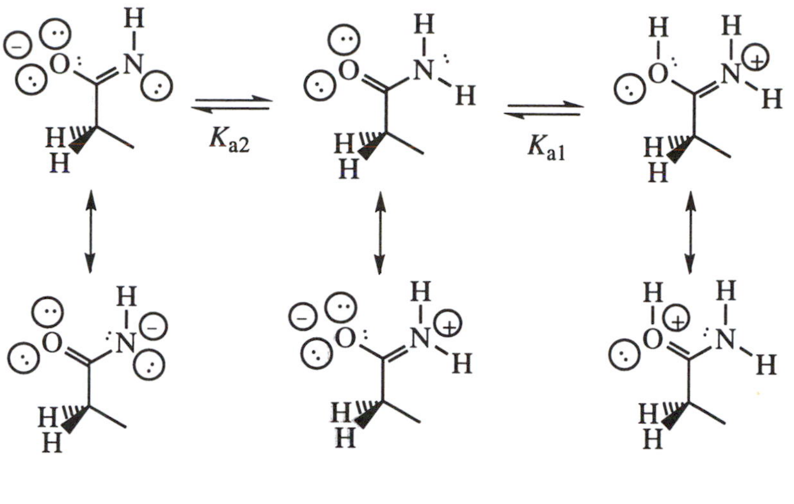

Figure 2–15: Acid–base reactions and resonance structures of a primary amide such as those in asparagine (pictured) and glutamine. The more reasonable of each of the two resonance structures is on the top line.

asparagine

roscopic pK_a implies that the respective situation has been assumed to occur.

The side chains of **aspartic acid** and **glutamic acid** are simple carboxylic acids whose structures and acid–base behavior were used earlier to introduce these properties (Figure 2–5 and Table 2–2). The side chains of **asparagine** and **glutamine** are the primary amides of these two carboxylic acids. The structure of the amide group has also been discussed earlier. Primary amides participate in two acid–base reactions (Figure 2–15). As is the case with imidazole, two protons are removed successively from two heteroatoms separated by one carbon and connected to each other by a π molecular orbital system (compare Figures 2–14B and 2–15). The acid dissociations also proceed from a cation through a neutral compound to an anion. The values for the two acid dissociation constants, however, are more widely separated from each other ($pK_{a1} = -0.5$ and $pK_{a2} = 17$) than the two for imidazole ($pK_{a1} = 7.1$ and $pK_{a2} = 14.5$). Indeed, they are as widely separated as those for water ($pK_{a1} = -1.7$ and $pK_{a2} = 15.7$), which supports the charges on only one oxygen atom.

The side chain of **arginine** (Figure 2–16) contains a cation that slightly resembles the protonated amides of glutamine and asparagine but has a π molecular orbital system larger by one atom. The functional group in arginine is a guanidinium, and it is composed from four atoms, three nitrogens and a central carbon, in the shape of a Y. Each atom contributes one p orbital, and the four mix to produce the one bonding and two nonbonding molecular orbitals shown in Figure 2–16, as well as a fourth antibonding orbital not shown. The guanidinium cation has six π electrons distributed in pairs among the three molecular orbitals presented. The two highest occupied nonbonding molecular orbitals are degener-

ate in energy. They are responsible for distributing two pairs of electrons evenly over the three nitrogen atoms as is described by the resonance structures. This causes the one positive formal charge to be divided evenly among the three nitrogens. An arginine residue ($pK_a = 13$) is less acidic than a histidinium residue ($pK_a = 6.4$) because the positive charge is distributed by the π molecular orbital system over three nitrogens rather than over two.

The guanidinium of an arginine residue defines a plane in which its central carbon, three nitrogens, five hydrogens, and the δ-carbon all reside (Figure 2–16B). The hydrogens bristle from the three nitrogens at 120° angles around the periphery and the flat clouds of π electrons sandwich the σ structure from above and below. The entire structure bears a net positive charge that is neutralized by removing a proton.

The side chain of **lysine**, the other strongly basic amino acid residue ($pK_a = 10.5$), is a simple primary ammonium cation at neutral pH

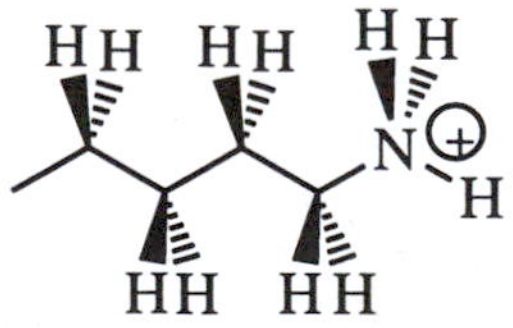

lysine

2–9

With four carbons, it has the longest linear alkane chain among the amino acids. The conformation of lowest free energy should be all *anti* as shown. The introduction of a *gauche* conformation at any of the carbon–carbon bonds should require about 4 kJ mol^{-1} free energy.

A

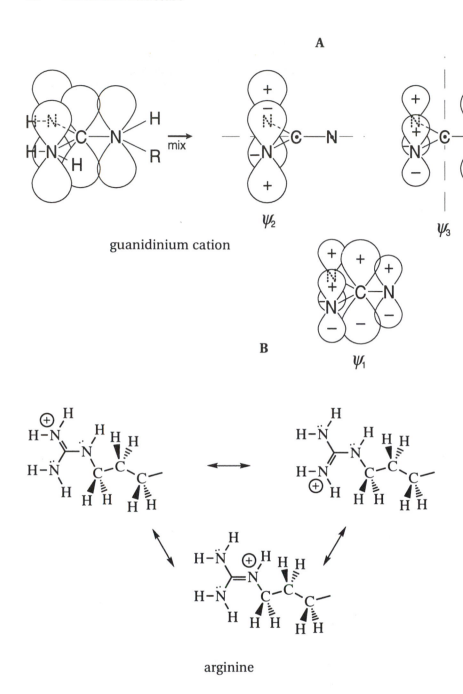

guanidinium cation

ψ_2

ψ_3

B

ψ_1

arginine

Figure 2–16: (A) Occupied molecular orbitals of the guanidinium cation and the resonance structures of the side chain of arginine. The four p orbitals that form the π molecular orbital system of the guanidinium cation mix and form four π molecular orbitals, only three of which are pictured. The two nonbonding π molecular orbitals, ψ_2 and ψ_3, are degenerate in energy and the bonding π molecular orbital, ψ_1, is lowest in energy. The anti-bonding π molecular orbital, ψ_2, the highest in energy, is not represented in the drawing. Six π electrons fill the three π molecular orbitals of lowest energy. (B) The resonance structures of the guanidinium cation in an arginine residue are an alternative way of describing the three occupied molecular orbitals.

The side chain of **cysteine** is fairly acidic (pK_a = 8.7). The thiolate anion that results from the acid dissociation (Figure 2–17) is a strong nucleophile because sulfur is an element of the third row. This is an important point. If the nucleophilicities of a thiolate and an alkoxide were directly proportional to their basicities, at a pH below the pK_a of a cysteine, serine and cysteine would display equivalent nucleophilicities. This would be due to the fact that the difference in the fraction of each that was ionized (Figure 2–6) would be compensated by the difference in their nucleophilicities. The thiolate anion is so much more nucleophilic, however, that at neutral pH cysteine always reacts more rapidly with alkylating electrophiles than any other amino acid in a protein.

The side chain of **methionine**, although it contains a thioether, resembles in its properties those of the amino acids that are purely alkanes, but it is linear rather than branched. The sulfur is large and electron-rich but not very basic; the pK_a for its conjugate acid[5] is about –9.

$$\text{(2-32)}$$

methionine

Methionine is, however, nucleophilic. At low pH, only

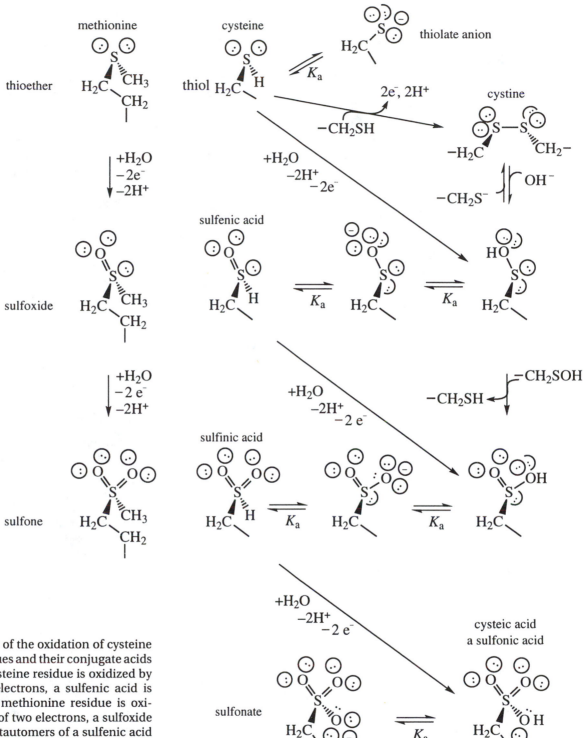

Figure 2–17: Products of the oxidation of cysteine and methionine residues and their conjugate acids and bases. When a cysteine residue is oxidized by the removal of two electrons, a sulfenic acid is formed, and when a methionine residue is oxidized by the removal of two electrons, a sulfoxide is formed. One of the tautomers of a sulfenic acid is the homologue of a sulfoxide. Cystine is formed either by the direct oxidation of two cysteines or by the reaction of the sulfenic acid of cysteine with another cysteine. When a cysteine residue is further oxidized by the removal of two more electrons, a sulfinic acid is formed; and, when a methionine residue is further oxidized by the removal of two more electrons, a sulfone is formed. One of the tautomers of a sulfinic acid is the homologue of a sulfone. Cysteine can be further oxidized by the removal of two more electrons to produce a sulfonic acid.

methionine and cysteine react with alkylating electrophiles.[25]

A drawback of the amino acid residues containing sulfur, methionine and cysteine, is their susceptibility to oxidation by reaction with oxygen, peroxides, or other oxidants. These reactions produce, in addition to disulfides, various oxides of sulfur (Figure 2–17). These are sulfoxides and sulfenic acids, sulfones and sulfinic acids, and sulfonic acids. To understand the bonding in these various products, the best way to begin is to examine sulfate

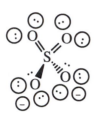

2–10

The sulfate dianion is perfectly tetrahedral, and every sulfur–oxygen bond is the same length. Therefore, sulfur is hybridized [sp^3, sp^3, sp^3, sp^3] and each oxygen must be electronically the same. The sulfur–oxygen bonds are quite short, and this requires that there be double-bond character. This is usually ascribed to mixing of a p orbital on each oxygen with hybrid d orbitals on sulfur in $dp\,\pi$ overlaps to create a tetrahedral $\delta\pi$ molecular orbital system. Because sulfur is an element of the third period, it has vacant, accessible $3d$ orbitals that can be involved in overlap with adjacent $2p$ orbitals to form $dp\,\pi$ bonds. The lobes on a $3d$ orbital are of the proper size to accomplish such an overlap

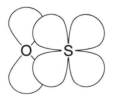

2–11

Each oxygen in sulfate, therefore, is hybridized [p, sp^2, sp^2, sp^2] with two σ lone pairs and a σ bond to sulfur. The five-atom $\delta\pi$ molecular orbital system would contain eight $\delta\pi$ electrons and the excess two negative charges carried within it would be distributed over the four oxygen atoms equally. This distribution can be demonstrated by drawing resonance structures. The double bonds in these resonance structures are $dp\,\pi$ overlaps and not $pp\,\pi$ bonds as those between elements of the second period, and because they are $dp\,\pi$ overlaps, the octet rule can be violated at sulfur.

In all of the oxides of sulfur (Figure 2–17), between the tetrahedral sulfur and the various oxygens, there are σ bonds and $\delta\pi$ bonds or $\delta\pi$ molecular orbital systems through which $\delta\pi$ lone pairs or sets of $\delta\pi$ lone pairs are delocalized. Because sulfur is less electronegative than oxygen (2.4 against 3.5, respectively), the $\delta\pi$ mixing is often quite weak or nonexistent and several of the lone pairs on oxygen designated as $\delta\pi$ lone pairs in Figure 2–17, especially in the compounds with electron-rich sulfurs, probably are not $\delta\pi$ lone pairs.

Sulfenic acids, the first oxidation level of thiols, would be the monothio analogues of peroxides just as disulfides are the dithio analogues of peroxides. One of the tautomers of a sulfenic acid would be the hydrogen analogue of a sulfoxide. Sulfoxides are known compounds, but sulfenic acids have not been isolated. They have been postulated to exist as intermediates in the disproportionation of disulfides, produced by hydroxide in the presence of catalytic amounts of metal ions

$$2RSSR + 4OH^- \xrightarrow[Me^{2+}]{} 3RS^- + RSO_2^- + 2H_2O \quad (2\text{–}33)$$

It has been proposed[1,3] that sulfenic acids themselves would immediately disproportionate

$$2RSOH \rightarrow RSH + RSO_2H \quad (2\text{–}34)$$

in a reaction homologous to but much more rapid than the disproportionation of peroxides. This reaction, however, requires the collision of two sulfenic acids. A sulfenic acid at a cysteine in a molecule of a native protein could be sterically prevented from such a collision and might exist for a significant time. It would, however, be both a strong reductant and a strong oxidant and susceptible to further oxidation and reduction.

Sulfinic acids are stable compounds that can be isolated. One of the tautomers of a sulfinic acid is the hydrogen analogue of a sulfone. Sulfones are stable oxides of sulfur. The sulfonate is the last oxidation state available to an alkylthiol. Sulfonates are also quite stable. Methionine and cysteine are often purposely converted to methionine sulfone and cysteic acid to render them stable to further oxidation.[26]

Oxidations such as those outlined in Figure 2–17 often occur adventitiously and can introduce charge heterogeneity into a protein or peptide owing to the formation of cysteic acid. Such oxidations can also cause functional damage to a protein. It is the adventitious oxidation of a methionine in α_1-antitrypsin, caused by cigarette smoke, that destroys the function of this protein and produces emphysema.[27]

Phosphoserine, phosphothreonine, and phosphotyrosine are formed by posttranslational modification:

serine phosphate

2–12

The phosphate is attached as a monoester of phosphoric acid. When the H in monoester **2–12** is replaced with another alkyl group, a diester of phosphoric acid such as is observed in nucleic acids is produced.

The bonding in phosphate is similar to that in sulfate with $\delta\pi$ molecular orbitals, formed by the overlap of d orbitals on phosphorus and p orbitals on oxygen, which delocalizes $\delta\pi$ lone pairs over two or more oxygen atoms. Phosphorus, however, is even less electronegative than sulfur and less readily shares electrons with oxygen, so the $\delta\pi$ bonding is less extensive in phosphate than sulfate but is certainly involved whenever oxygen is left with three lone pairs. Again, the delocalization in $\delta\pi$ molecular orbitals can be represented by drawing resonance structures.

The acid–base properties of phosphate and esters of phosphoric acid reflect this ability of the $\delta\pi$ molecular orbital system to spread negative charge over two or more oxygens because the acid dissociation constants (Table 2–1) are much closer together than one might expect for a series of steps that each increase the negative charge by 1 unit. The acid dissociation constants for an alkyl monoester of phosphate ($pK_{a1} = 1.7$ and $pK_{a2} = 6.7$)[6] and for a dialkyl diester of phosphate ($pK_a = 1.5$)[6] are close to those of phosphoric acid itself, but sugar phosphates, and presumably also serine phosphate and threonine phosphate, are more acidic ($pK_{a1} = 0.9$ and $pK_{a2} = 6.1$)[6] due to inductive electron withdrawal.

Problem 2–13: 1-Aminopentane is a model compound for the ε-amino group of a lysine residue in a protein.

(A) Draw the structure of a lysine residue incorporated into a polypeptide and the structure of 1-amino-pentane.

(B) The pK_a of a 1-aminopentane is 10.63. What is the pK_a of lysine incorporated into a polypeptide?

(C) Write the acid dissociation of 1-aminopentane.

(D) At pH 7.0, what fraction of the lysine in the peptide Gly-Pro-Lys-Ala-Thr would be in the neutral nucleophilic form? What fraction at pH 12?

(E) The ε-amino group of lysine in a polypeptide reacts readily with acetic anhydride. Write a mechanism for this reaction.

(F) At pH 12, 10 °C, 0.1 M KCl, the lysine in the above pentapeptide would react with acetic anhydride at a rate of 1.3×10^5 M^{-1} min^{-1} (k_N). Write a kinetic mechanism for this reaction at any pH which involves only this rate constant and the acid dissociation constant K_a and solve it for the initial velocity (v_i) of the reaction between lysine and acetic anhydride. Assume that the acid dissociation equilibrium is very rapid compared to k_N.

Problem 2–14: In the peptide CH₃CO-Gly-Glu-Gly-His-NH₂, which acid–bases would be titrating in the region between pH 2 and 11? What are the approximate values for each pK_a? Plot as a function of pH the fraction of each of the three major ionic forms of the peptide present in solution.

Problem 2–15: Two compounds (A and B) have been isolated from a protein by enzymatic hydrolysis. Both have the composition $C_5H_{10}N_2O_3$. The titration behavior of the compounds is the following:

compound A		compound B	
pK_{a1}	3.85	pK_{a1}	2.148
pK_{a2}	8.25	pK_{a2}	9.19

After acid hydrolysis for 20 h in 6 M HCl, both compounds have the composition $C_5H_9NO_4$ and the following titration behavior:

compound A′		compound B′	
pK_{a1}	2.155	pK_{a1}	2.155
pK_{a2}	4.32	pK_{a2}	4.32
pK_{a3}	9.95	pK_{a3}	9.95

(A) What are compounds A and B?

(B) Explain their behavior on titration.

References

1. Taylor, R., Kennard, O., & Versichel, W. (1983) *J. Am. Chem. Soc. 105*, 5761–5766.
2. Eigen, M. (1964) *Angew. Chem., Int. Ed. Engl. 3*, 1–19.
3. Yang, X., & Castleman, A.W. (1989) *J. Am. Chem. Soc. 111*, 6845–6846.
4. Wei, S., Shi, Z., & Castleman, A.W. (1991) *J. Chem. Phys. 94*, 3268–3278.
5. March, J. (1985) *Advanced Organic Chemistry: Reactions, Mechanisms and Structure*, 3rd ed., pp 220–223, McGraw-Hill, New York.
6. Jencks, W.P., & Regenstein, J. (1976) in *Handbook of Biochemistry and Molecular Biology, 3rd Edition; Physical and Chemical Data* (Fasman, G.D., Ed.) Vol. I, pp 305–351, CRC Press, Cleveland, OH.
7. Fersht, A.R. (1971) *J. Am. Chem. Soc. 93*, 3504–3515.
8. Capon, B., & Zucco, C. (1982) *J. Am. Chem. Soc. 104*, 7567–7572.

9. Zacharias, D.E., Murray-Rust, P., Preston, R.M., & Glusker, J.P. (1983) *Arch. Biochem. Biophys. 222*, 22–34.

10. Abrams, W.R., & Kallen, R.G. (1976) *J. Am. Chem. Soc. 98*, 7789–7792.

11. Wolfenden, R.V., Cullis, P.M., & Southgate, C.C.F. (1979) *Science 206*, 575–577.

12. Nozaki, Y., & Tanford, C. (1967) *J. Biol. Chem. 242*, 4731–4735.

13. Hansch, C., Leo, A., Unger, S.H., Kim, K.H., Nikaitani, D., & Lien, E.J. (1973) *J. Med. Chem. 16*, 1207–1216.

14. Keim, P., Vigna, R.A., Nigen, A.M., Morrow, J.S., & Gurd, F.R.N. (1974) *J. Biol. Chem. 249*, 4149–4156.

15. Calvin, M. (1954) in *Glutathione* (Colowick, S., Lazarow, A., Racker, E., Schwartz, D.R., Stadtman, E., & Waelsch, H., Eds.) p 9, Academic Press, New York.

16. Tanford, C. (1962) *Adv. Protein Chem. 17*, 69–165.

17. Tanford, C. (1968) *Adv. Prot. Chem. 23*, 121–283.

18. Martin, R.B., Edsall, J.T., Wetlaufer, D.B., & Hollingsworth, B.R. (1958) *J. Biol. Chem. 233*, 1429–1435.

19. Lennette, E.P., & Plapp, B.V. (1979) *Biochemistry 18*, 3933–3938.

20. Edsall, J.T., & Wyman, J. (1958) *Biophysical Chemistry*, Vol. I, Academic Press, New York.

21. Lin, L., & Brandts, J.F. (1983) *Biochemistry 22*, 559–563.

22. Gratzer, W.B., & Minalyi, E. (1976) in *Handbook of Biochemistry and Molecular Biology, 3rd Edition; Proteins* (Fasman, G.D., Ed.) Vol. I, pp 186–191, CRC Press, Cleveland, OH.

23. Edelhoch, H. (1967) *Biochemistry 6*, 1948–1954.

24. Reynolds, W.F., Peat, I.R., Freedman, M.H., & Lyerla, J.R. (1973) *J. Am. Chem. Soc. 95*, 328–331.

25. Gundlach, H.G., Moore, S., & Stein, W.H. (1959) *J. Biol. Chem. 234*, 1761–1764.

26. Hirs, C.H.W. (1967) *Methods Enzymol. 11*, 59–62.

27. Johnson, D., & Travis, J. (1979) *J. Biol. Chem. 254*, 4022–4026.

CHAPTER 3

Sequences of Polymers

By direct chemical analysis of purified proteins, it has been shown that they are composed of linear polymers of amino acids, referred to as polypeptides. These polymers are formed by a ribosome that reads the messenger RNA and converts the sequence of codons into a sequence of amino acids coupled covalently together in the dictated order. Every polypeptide begins its existence as a single polymer of amino acids of a precise length coupled in a precise order. By and large, this polymer of amino acids is conserved in the mature protein. On its way to maturity, however, various alterations occur. Short segments of amino acids are often removed from the amino- or carboxy-terminal ends of the protein or cut out of the middle, leaving a broken chain. If such an alteration occurs, it causes the actual amino acid sequence of the polypeptide in a mature protein to differ from the sequence encoded in the messenger RNA. The sequence of the amino acids in a mature polypeptide can be determined directly, but this is rarely done anymore. It is far easier to sequence the messenger RNA and translate the sequence of nucleotides into a sequence of amino acids. Correction must be made for the missing segments removed during the maturation of the protein, but this is relatively easy. The lion's share of the original amino acid sequence usually remains in the polypeptides forming the mature protein.

As part of the process that produces the mature protein, various additions are made to the constituent polypeptides. These additions are either chemical modifications of the amino acids themselves or attachment of other compounds to the amino acids. For the most part, these posttranslational modifications are unpredictable, and each presents a challenge in analytical chemistry. There are a series of modifications, however, that consist of the addition of oligosaccharides to particular amino acids. These result in the attachment of a polymer of sugars to the polymer of amino acids, and these modifications are defined by the sequences in which the sugars are attached in these oligomers.

With the exception of the unexpected posttranslational modifications, which are relatively infrequent, defining the covalent structure of a mature polypeptide is an exercise in the sequencing of polypeptides, nucleic acids, and oligosaccharides.

Sequencing of Polypeptides

Each polypeptide (Structure 2–8) has its own length and its own amino acid sequence. The **amino acid sequence** is the order in which the side chains of the amino acids (R_i

in Structure 2–8) are arranged along the polymer. The contiguous lengths of the polypeptides found in molecules of protein, and hence the lengths of their unique sequences, vary between about 50 and 5000 amino acids. The amino acid sequence of a given polypeptide is written as a word, each of whose letters stands for an amino acid. The word begins at the amino terminus and ends at the carboxy terminus.

The amino acid sequence of a polypeptide determines which protein it will become. Bovine pancreatic ribonuclease can be defined as the protein produced in the pancreas of a cow that can cleave ribonucleic acid at random along its length in a reaction that leaves the phosphate on the 2′ and 3′ positions of the products, or it can be defined as a folded polypeptide, 124 amino acids long, with the amino acid sequence KETAAAKFER-QHMDSSTSAASSSNYCNQMMKSRNLTKDRCKPVN-TFVHESLADVQAVCSQKNVACKNGQTNCYQSYSTM-SITDCRETGSSKYPNCAYKTTQANKHIIVACEGNPYVP-VHFDASV. That the sequence is sufficient to define the protein has been demonstrated by total synthesis.[1]

The sequences of polypeptides were, in the past, determined directly. The amino acids in a polypeptide can be removed in steps from the amino-terminal end by the **Edman degradation** (Figure 3–1).[2] The strategy of the Edman degradation relies upon the separation of the chemistry into two discrete steps (labeled 1 and 2 in Figure 3–1) which permits the removal of one amino acid at a time from the polypeptide as the **phenylthiohydantoin**. The phenylthiohydantoins from each step can be positively identified by high-pressure adsorption chromatography.[3]

Only in fortuitous circumstances can the Edman degradation be run for more than 20 or 30 cycles. The necessity for two steps in each cycle and the separation of the shortened polypeptide from the thiazolinone, none of which can be performed in 100% yield, causes the cumulative yield of phenylthiohydantoin to decrease rapidly and noise to increase apace. Side reactions such as random hydrolysis of the polypeptide and cyclization of amino-terminal glutamines to pyrrolidones[4] also increase noise and lower yield, respectively.

Because polypeptides cannot be sequenced in their entirety by the Edman degradation, they are cleaved into pieces, or **peptides**, that can be. The cleavage is performed with reagents that act at the locations of specific amino acid residues in the sequence (Figure 3–2). In this way, high yields of peptides with a particular amino acid sequence can be obtained.

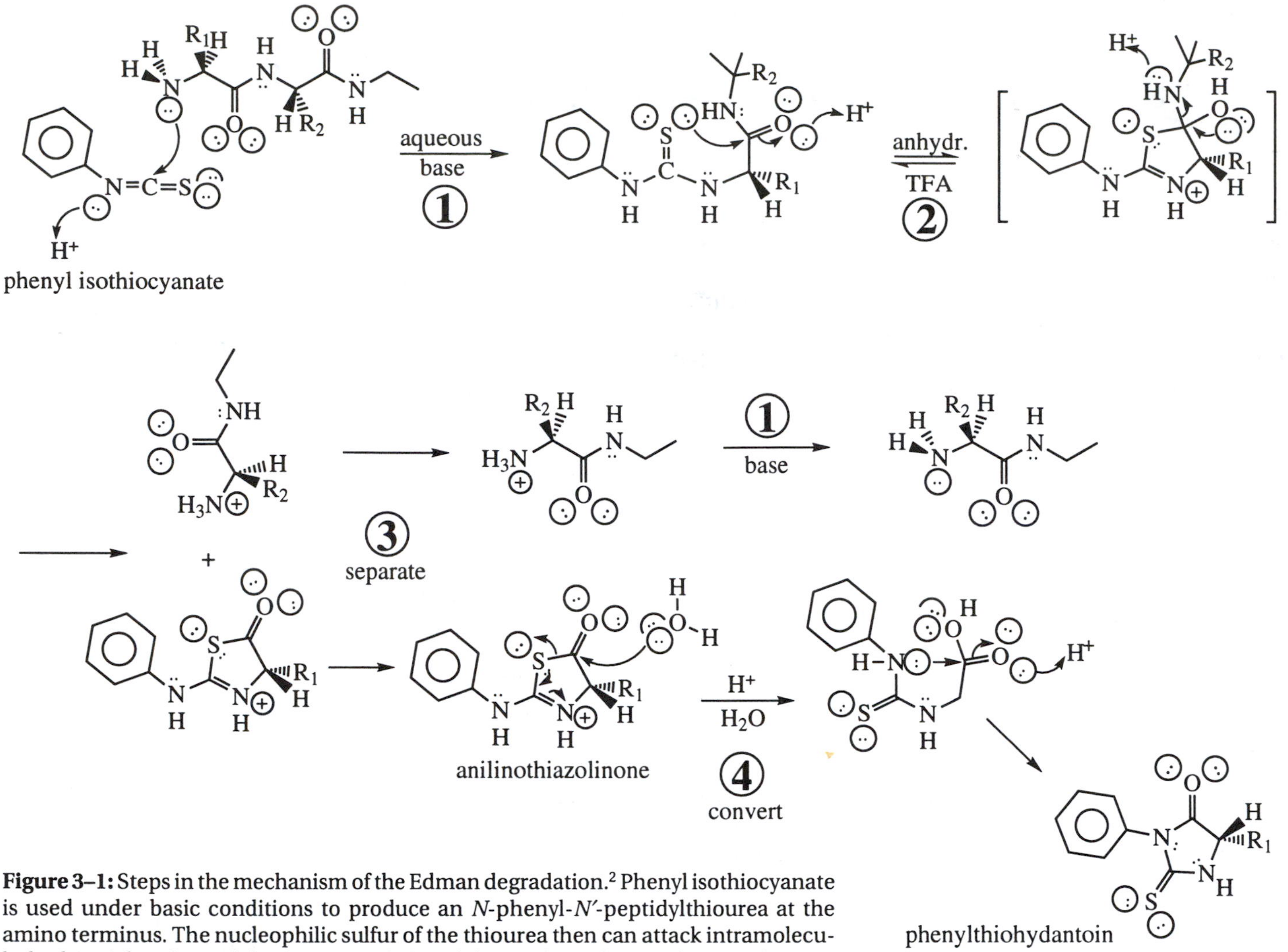

Figure 3–1: Steps in the mechanism of the Edman degradation.[2] Phenyl isothiocyanate is used under basic conditions to produce an *N*-phenyl-*N*′-peptidylthiourea at the amino terminus. The nucleophilic sulfur of the thiourea then can attack intramolecularly the acyl carbon of the first peptide bond, but only under conditions of strong general acid catalysis, which promote protonation of the acyl oxygen. Anhydrous trifluoroacetic acid is used to prevent any unwanted hydrolytic side reactions at this step. The shortened polypeptide that leaves is unreactive at its amino terminus under these conditions owing to protonation. The anilinothiazolinone and the shortened polypeptide are then separated from each other. The shortened polypeptide is recycled through coupling and cleavage. The anilinothiazolinone is opened and recyclized in aqueous acid to produce the phenylthiohydantoin of the first amino acid.

One group of reagents used to perform such specific cleavages are proteolytic enzymes. If polypeptides are to be cleaved by proteolytic enzymes, they must be unfolded or denatured. The folded compact molecule of protein is usually highly resistant to proteolytic digestion for steric reasons. Denaturing the protein that is to be cleaved without simultaneously denaturing the proteolytic enzyme, which is itself a protein, requires some strategy. Usually the chemical modification of one type of amino acid in the polypeptide while it is unfolded in a denaturant such as urea is sufficient to prevent it from refolding after the denaturant is removed. The carboxymethylation of cysteines with 2-iodoacetate after the cystine residues in the protein have been reduced[11]

and the maleylation of lysines[6,12] are examples of this strategy. When proteins that are normally embedded in biological membranes are removed from the membrane, their polypeptides often remain soluble and unfolded and can be cleaved with proteolytic enzymes.[13] Some proteolytic enzymes are themselves quite stable and will function in solutions of denaturants sufficient to unfold the protein to be cleaved.

At times it is useful to cleave a polypeptide at only one or two specific locations in its sequence so that very long fragments can be isolated from it. The most common way that this is done is to take advantage of the resistance of the native, properly folded protein to proteolytic digestion. The consequence of this resistance is that when a

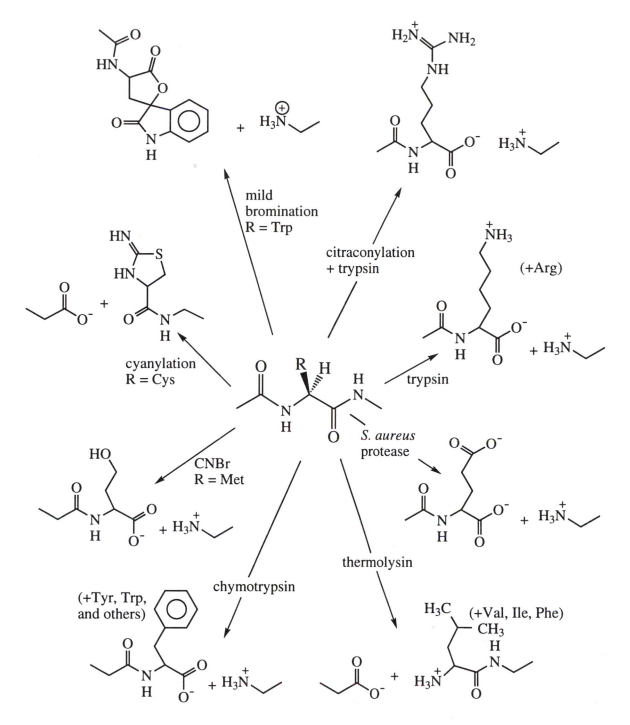

Figure 3–2: Specific cleavage of a polypeptide with enzymes or chemical reagents. Trypsin hydrolyzes the peptide bonds on the carboxy-terminal sides of lysine and arginine residues with high specificity to produce a series of peptides. Each of these peptides has the respective lysine or arginine at its carboxy terminus.[5] The lysine residues can be rendered incapable of being recognized by trypsin by modification with succinic anhydride, maleic anhydride,[6] or citraconic anhydride.[7] The latter two modifications are reversible, and the lysines can be regenerated, after cleavage with trypsin, to yield a series of unmodified peptides whose carboxy-terminal residues are the respective arginines. A proteolytic enzyme from the bacterium *Staphylococcus aureus*, strain V8, hydrolyzes polypeptides with high specificity at the peptide bonds on the carboxy-terminal sides of glutamic acid residues.[8] Under the proper conditions, the same enzyme also can be made to hydrolyze the bonds on the carboxy-terminal side of aspartic acid residues. Thermolysin, a proteolytic enzyme from the bacterium *Bacillus thermoproteolyticus*, hydrolyzes polypeptides at peptide bonds on the amino-terminal sides of leucine, isoleucine, valine, phenylalanine, methionine, and occasionally alanine and tyrosine.[9] Chymotrypsin usually catalyzes the hydrolysis of the amide bonds on the carboxy-terminal sides of phenylalanine, tyrosine, and tryptophan.[10] Chemical cleavage can be performed at methionine residues with cyanogen bromide, at cysteine residues by cyanylation, and at tryptophan residues by mild bromination.

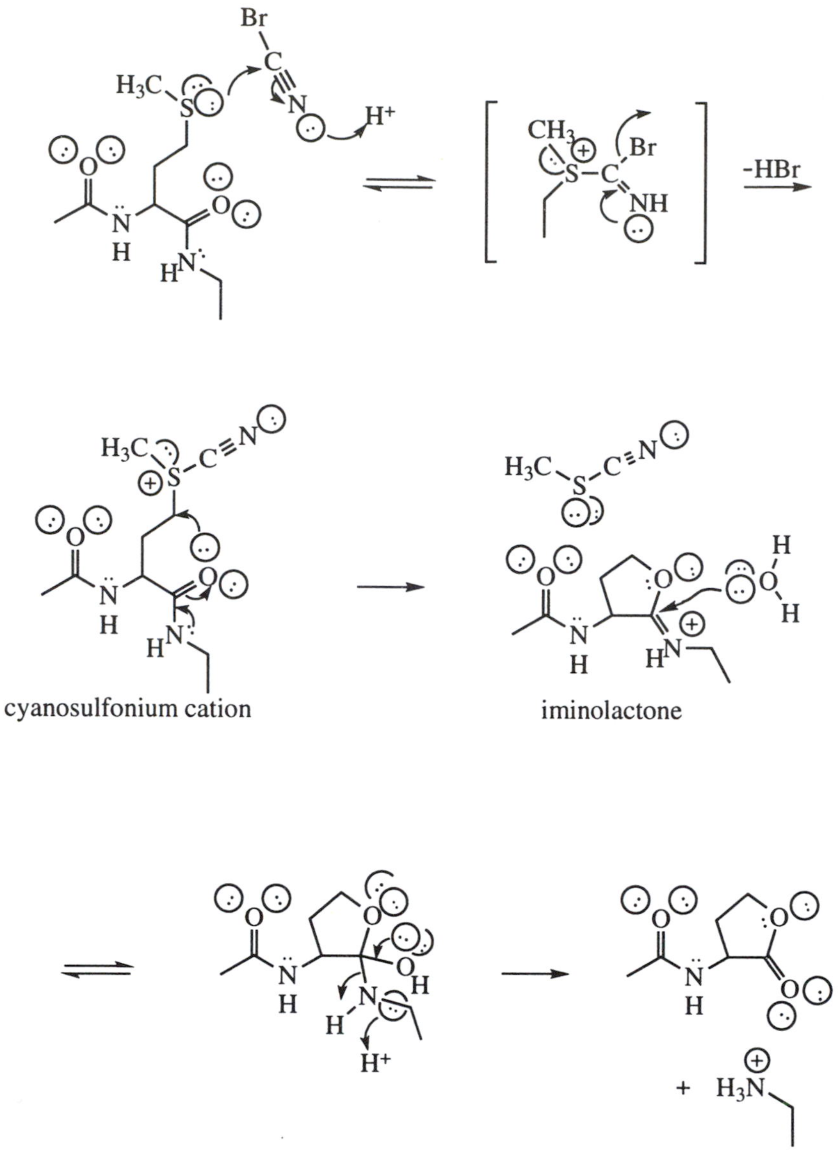

Figure 3–3: Mechanism of cyanogen bromide cleavage of a polypeptide on the carboxy-terminal side of a methionine. At acidic pH, a methionine residue, because it is not protonated, remains nucleophilic enough to react in an acyl exchange reaction with cyanogen bromide to produce a cyanosulfonium cation. This cationic center causes the adjacent carbon to be electrophilic. This electrophile is five atoms away from the weakly nucleophilic acyl oxygen of the same amino acid, and an intramolecular, nucleophilic substitution ensues. The conjugate acid of the iminolactone formed in this nucleophilic substitution is susceptible to hydrolysis under the acidic conditions. This hydrolysis produces a mixture of the lactone and the open γ-hydroxycarboxylic acid of homoserine at the carboxy terminus of the resulting peptide.

cyanosulfonium cation

iminolactone

properly folded protein is treated with trypsin or chymotrypsin, often only one or two of its peptide bonds are exclusively hydrolyzed, and this produces the long fragments desired. Because this is completely the result of steric effects, no control over the location of the sites of cleavage, other than that exerted by the intrinsic specificity of the proteolytic enzyme, can be exercised.

Polypeptides can also be cleaved chemically. The paradigm of chemical cleavages is that produced by cyanogen bromide (Figure 3–3)[14]. Several other chemical cleavages of limited usefulness have been developed. 2-Nitro-5-thiocyanatobenzoate induces cleavage on the amino-terminal side of cysteine residues (Figure 3–4), but the yield is less than quantitative and the amino terminus of the product is blocked.[15] The amino terminus can be regenerated as an amino-terminal alanine by treatment with Raney nickel.[16] Cleavage at tryptophan residues can be performed chemically with brominating agents under heterolytic conditions[17]

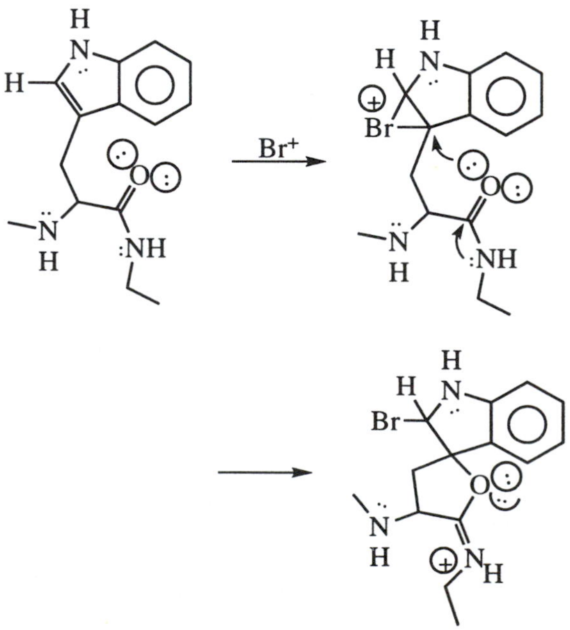

(3–1)

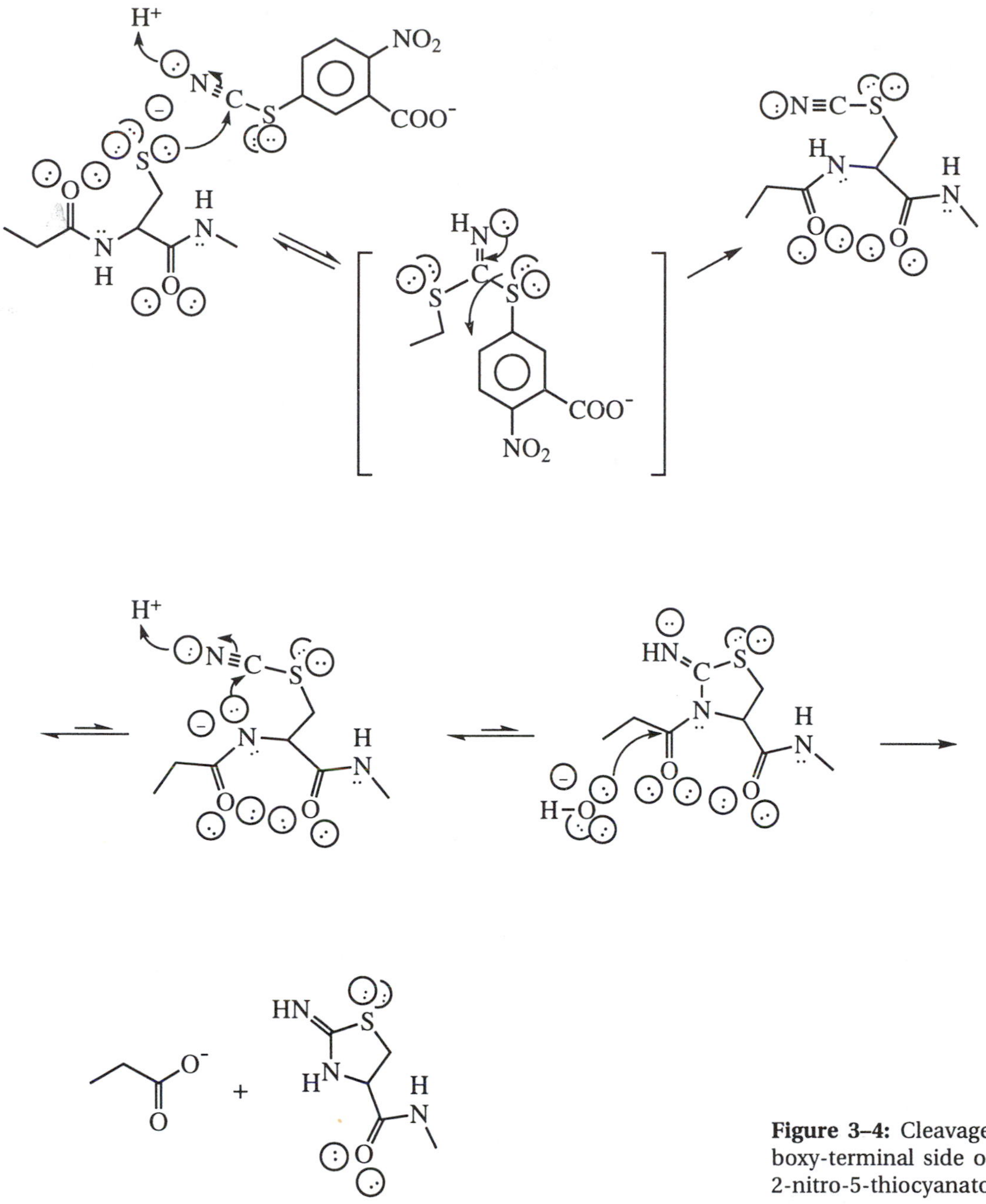

Figure 3–4: Cleavage of a polypeptide to the carboxy-terminal side of cysteine by cyanylation with 2-nitro-5-thiocyanatobenzoate.

This reaction proceeds through a bromonium cation that results from insertion of Br⁺ into the olefin between carbons 2 and 3 of the indole to create an electrophilic center. A nucleophilic attack of the acyl oxygen five atoms away occurs as in the cyanogen bromide cleavage. The resulting iminolactone then hydrolyzes to produce cleavage of the amide to the carboxy-terminal side of the tryptophan and to release a fragment with a free amino terminus. The olefin between carbons 2 and 3 in indole is an easily brominated position, and the mildest brominating agent capable of reacting at this location should be used under the mildest conditions to avoid widespread bromination of the polypeptide elsewhere.[18]

A chemical cleavage that can produce very large fragments from a polypeptide is the cleavage that occurs preferentially at the peptide bond between an aspartate and a proline under mildly acidic conditions (Figure 3–5).[19] This reaction results from intramolecular attack of the carboxylate anion of the 3-carboxy group of the aspartate on its own acyl carbon, the acyl oxygen of which has been protonated, to produce, upon departure of the amide nitrogen of the proline, an anhydride, which is subsequently hydrolyzed.[20] The cleavage occurs preferentially at proline because the amine in the tetravalent intermediate is by far the poorer leaving group but proline, because it is a hindered secondary amine, is the best leaving group of all of the amino acids.

Figure 3–5: Cleavage of a polypeptide at the peptide bond between an aspartate and a proline under acidic conditions.

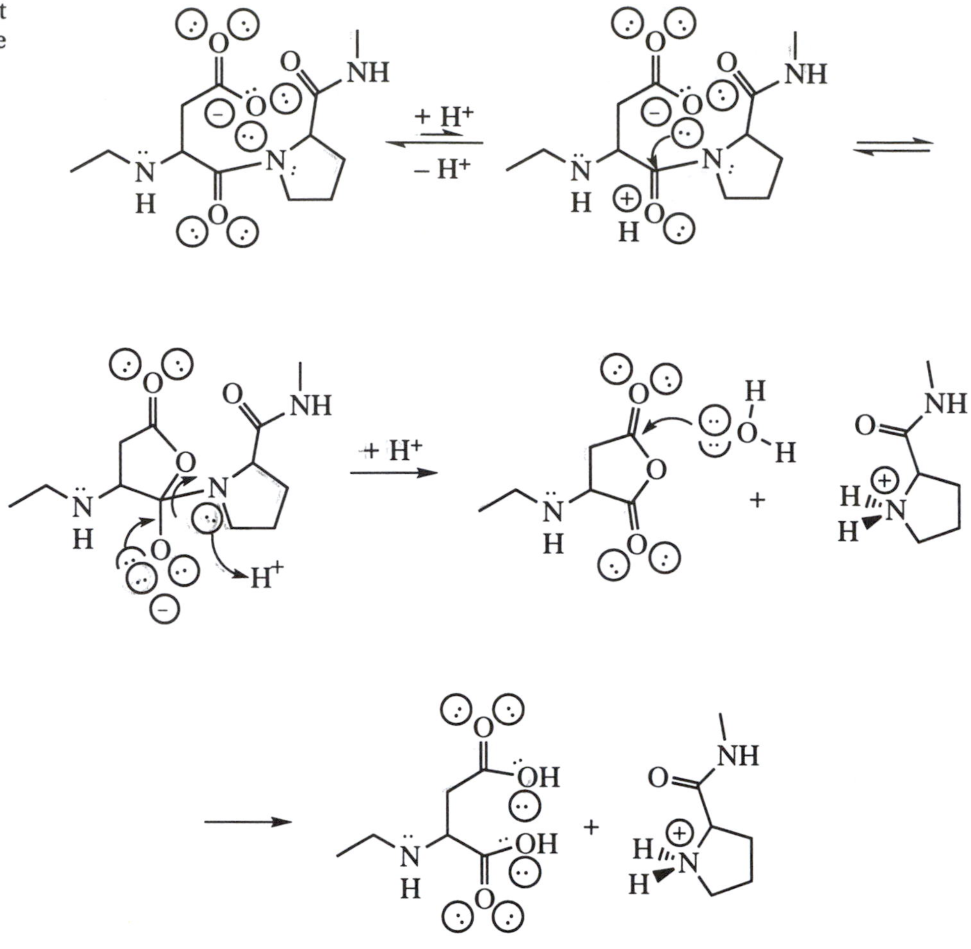

Each of these treatments produces a different set of peptides from a given polypeptide, and the complex mixtures that result must be separated chromatographically or electrophoretically. Electrophoresis on paper soaked in an aqueous buffer of a given pH can be used to separate short peptides.[21] They can also be separated by descending chromatography on paper in mixed organic and aqueous solvents. These two procedures can be combined by running electrophoresis in one dimension followed by chromatography at right angles to spread the peptides over a rectangular field known as a **peptide map** (Figure 3–6).[22] When paper is used as a medium, the regions in which the peptides are located can be cut out, and the peptides can be eluted from them.

The peptides that result from a specific cleavage can also be separated by column chromatography. Molecular exclusion chromatography is used to separate the mixture into groups of peptides of different lengths (Figure 3–7).[23] Because the larger peptides often aggregate or precipitate, these columns are generally run in solutions of denaturants such as urea or under conditions such as extremes of pH, either basic[23] or acidic.[24] At high or low pH, the net charges on all of the peptides are negative or positive, respectively, and aggregation is prevented by mutual electrostatic repulsion. Large peptides can also be rendered more soluble by modification of all of the

lysine residues with citraconic anhydride to increase their net negative charge at neutral and basic pH.[24]

The smaller peptides, either those isolated by molecular exclusion chromatography or those from the whole digest, can be separated by cation-exchange chromatography on sulfonated polystyrene (Figure 3–8A)[25,27] or adsorption chromatography at high pressure under acidic conditions on alkylated silica gel (Figure 3–8B)[28]. The latter method can also be used with large peptides such as cyanogen bromide fragments.[28] The resolution obtained with either cation-exchange chromatography or high-pressure adsorption chromatography (Figure 3–8) are very similar, but the latter has become the method of choice because of its rapidity, the continuous spectrophotometric monitoring it permits, and its adaptability to samples containing very small quantities of peptide. Large peptides are often separated by ion-exchange chromatography on appropriately modified matrices of cellulose or dextran.[29] In all cases, the art of the chromatography lies in choosing solvents and buffers that will dissolve the peptides, meet the demands of the chromatographic process chosen for the separation, and be easily removed from the peptides after they have been separated.

Once the peptides have been purified, their amino acid composition is usually determined by hydrolysis

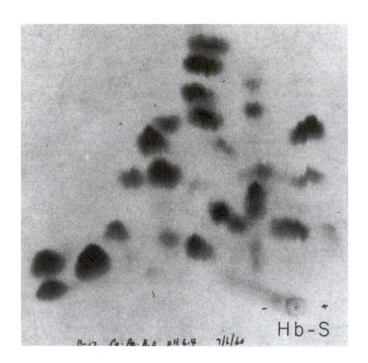

Figure 3–6: Separation of the peptides in a tryptic digest of human hemoglobin S on paper.[22] The mixture of peptides was applied to the origin in the lower, right-hand corner. Electrophoresis (negative pole to left) was performed at pH 6.4 for 2.5 h at 1 kV followed by chromatography (from bottom to top) with a solvent of isoamyl alcohol–pyridine–water (35:35:30). Reprinted with permission from ref. 22. Copyright 1961 Elsevier Science Publishers B.V.

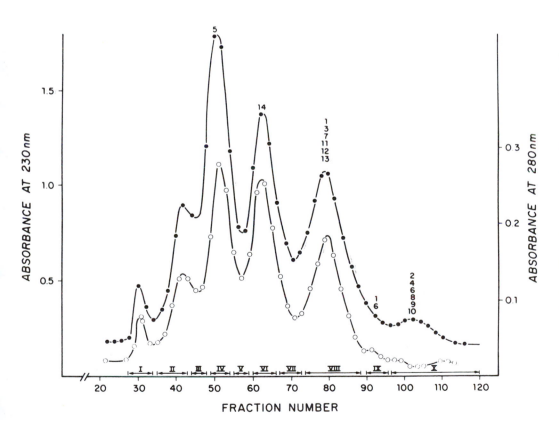

Figure 3–7: Separation of peptides produced by cyanogen bromide cleavage of *S*-carboxymethylated human phosphoglycerate kinase.[23] The cyanogen bromide peptides (50 mg) were applied to a column (1.9 cm × 150 cm) of Sephadex G-75 run in 0.2 M ammonium bicarbonate. The effluent fractions were monitored by absorbance at 230 nm (●) and 280 nm (○). Pools (I–X) were made as indicated. The numbers indicate which fragments, identified later in other separations, were in each pool. Reprinted with permission from ref 23. Copyright 1980 *Journal of Biological Chemistry.*

(Figure 3–9), performed under vacuum in 6 M HCl, followed by quantitative cation-exchange chromatography on sulfonated polystyrene (Figure 1–4). In this way, if the peptide is pure and not too long, the amount of each amino acid in the peptide can be determined. Often, however, the peptides are sequenced directly because procedures for sequencing by automated Edman degradation[3] have become more sensitive than procedures for amino acid analysis.

In its present applications, the Edman degradation is

A

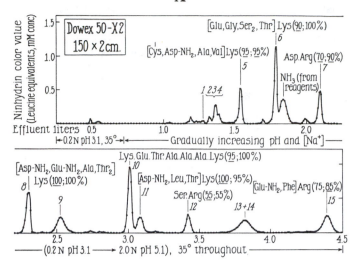

Figure 3–8: Separation of small peptides by ion-exchange chromatography or adsorption chromatography. (A) A tryptic digest of oxidized bovine ribonuclease (200 mg) was submitted to chromatography on a column (1.8 cm × 150 cm) of sulfonated polystyrene.[25] The peptides were eluted with an irregularly but monotonically increasing gradient of sodium ion and pH. The peptides were located by colorimetric assay of small samples from fractions. Concentrations of amino acids produced by alkaline hydrolysis were determined colorimetrically by reaction with ninhydrin. The colorimetric assay was for primary amino groups and relied on the absorbance at 570 nm produced by ninhydrin (ninhydrin color value) following hydrolysis of each sample in base. The absorbances are expressed as concentrations (millimolar) by using the extinction coefficient of leucine in the assay (leucine equivalents). The composition of each peptide, determined by amino acid analysis, is noted in brackets. The yield of each peptide is in parentheses. Reprinted with permission from ref 25. Copyright 1956 *Journal of Biological Chemistry*. (B) A tryptic digest of *S*-carboxymethylated human factor VIII (0.8 nmol) was submitted to high-pressure liquid chromatography on a column (0.46 cm × 25 cm) of octadecylated silica equilibrated with 0.1% trifluoroacetic acid. The peptides were eluted with a linear gradient (0–70%) of acetonitrile in 0.1% trifluoroacetic acid.[26] Peptides were detected by their absorbance at 210 nm (solid line) and 280 nm (dashed line). Absorbance (A_{210} or A_{280}, respectively) is presented as a function of retention time (minutes). Reprinted with permission from ref 26. Copyright 1984 *Nature*.

B

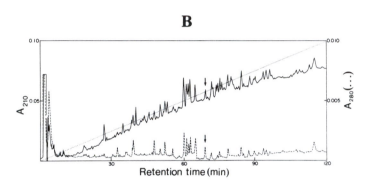

Figure 3–9: Hydrolysis of a polypeptide to its constituent amino acids under acidic conditions.

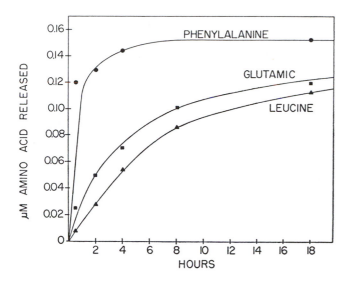

Figure 3–10: Rate of release of amino acids by carboxypeptidase from α-corticotropin, a peptide 39 amino acids in length, the carboxy-terminal sequence of which is –AFPLEF. Carboxypeptidase (at a mass ratio of 1:100 relative to α-corticotropin) was added to a 0.66 mg mL^{-1} solution of α-corticotropin at pH 8.5 and 40 °C. Samples (1mL) were removed at noted times (hours) to determine the micromoles of each amino acid released.[47] Reprinted with permission from ref 47. Copyright 1955 *Journal of Biological Chemistry.*

performed on peptides or polypeptides noncovalently[30] or covalently[31,32] attached to thin membranes of glass fiber[3] or poly(vinylidene difluoride).[33] Because the peptide remains attached to a solid phase, the reagents, in solution, can be sequentially applied to and removed from the peptide efficiently. It is also possible to transfer polypeptides that have been separated by electrophoresis onto these supports and then submit them to sequencing.[33–35]

One of the problems with relying on a peptide that is being sequenced to remain adsorbed to the glass fiber or the poly(vinylidene difluoride) is that very hydrophobic peptides often rinse away in the washes with organic solvents that are part of the sequencing cycle. One way to avoid these problems during the sequencing of such hydrophobic peptides is by performing mass spectroscopy of small fragments of the larger peptides.[36] The small fragments are produced by partial acid hydrolysis, separated by gas chromatography, and identified by mass spectroscopy. It is also possible to confirm the amino acid composition of peptides isolated during a sequencing study by mass spectroscopy.[37]

The amino-terminal amino acid of a peptide can be identified by reacting the peptide with 5-(N,N-dimethylamino)naphthylenesulfonyl chloride (dansyl chloride).[38] This sulfonylating reagent reacts with the amino-terminal amine and the primary amines of lysines and with the phenolic oxygens of tyrosines to produce highly fluorescent sulfonamides (dansyl amines) and sulfonic phenyl esters, respectively. The fluorescent peptide is then hydrolyzed in strong acid, and the products of hydrolysis are separated by thin-layer chromatography in two dimensions. The only amino acid whose α-amino group has been modified by the reaction with the sulfonyl chloride is identified as a fluorescent spot at a characteristic location on the chromatographic field.

Occasionally, proteolytic enzymes known as exopeptidases, such as carboxypeptidases A, B, or Y, or leucine aminopeptidase, can be used to assist in determining or confirming the sequence of a peptide. These enzymes remove amino acids one at a time from one of the two ends of the peptide. Because the shortened peptide released as a product by one of these enzymes immediately becomes a reactant for another cleavage, these digestions do not release the amino acids from the respective end in a stepwise fashion but by a progressive process (Figure 3–10), and absolute information about sequence beyond three or four residues from the end is rarely obtained with one of these enzymes alone.

The grand strategy for determining the sequence of a polypeptide directly is to separate and sequence all of the peptides from one particular cleavage, to cleave the protein at a set of different locations, to identify all of the peptides in this second set that contain the points of cleavage for the first set, and to sequence these overlapping peptides to learn the order in which the first peptides are arranged in the intact polypeptide. The dramatic epics[17,39–46], in each of which this strategy was applied to another protein and its sequence was revealed, are no longer produced. The expectation and excitement surrounding each of them is only dimly remembered. In their place are myriads of short essays that present the sequences of often very long polypeptides. This flood of information has been possible because the sequences of polypeptides are now determined by sequencing DNA complementary to the messenger RNA that encodes them.

Suggested Reading

Suzuki, N., & Wood, W.A. (1980) Complete Primary Structure of 2-Keto-3-deoxy-6-phosphogluconate Aldolase, *J. Biol. Chem.* 255, 3427–3435.

Problem 3–1: Write a complete mechanism for this chemical reaction. Draw in important lone pairs and indicate the combination of nucleophiles and electrophiles with arrows. Use protons where appropriate. For what purpose is this chemical reaction used? Write the step-by-step cycle for using this reaction to accomplish this purpose.

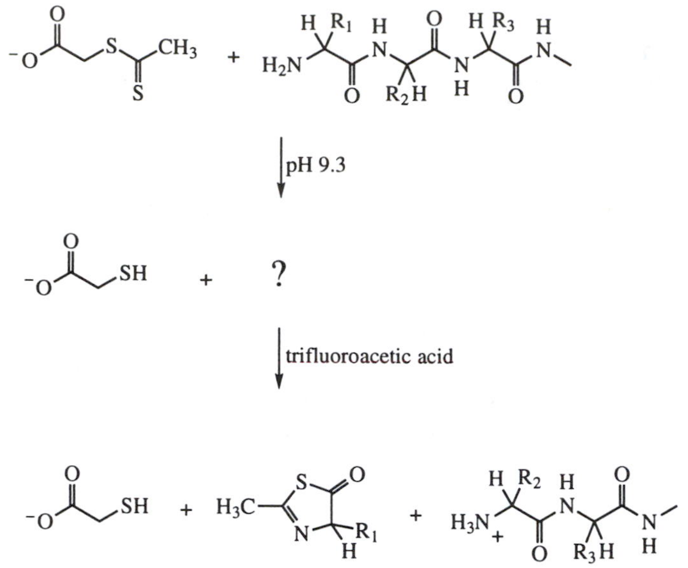

Problem 3–2: A cyanogen bromide fragment has been purified from a digest of certain protein. Consider the following information. The compositions shown in parentheses are those obtained following complete acid hydrolysis in 6 M HCl, 110 °C, for 24 h.

(A) Complete acid hydrolysis
  (1) (Arg, Glu, 2 Gly, homoserine (Hse), Leu, Lys, Phe, Ser, Val)

(B) Amino terminus
  (2) (Val)

(C) Amino acid composition of peptides from tryptic digest
  (3) (Arg, Glu, Gly, Val)
  (4) (Gly, Lys, Phe, Ser)
  (5) (Hse, Leu)

(D) Edman degradation

peptide	cycle	
	1	2
(3)	Val	Glu
(4)	Phe	Ser

(E) Reaction with 2,3-butanedione, followed by tryptic digest
  (6) (Arg, Glu, 2 Gly, Lys, Phe, Ser, Val)
  (7) (Hse, Leu)

What is the sequence of the fragment? With which amino acid side chain does 2,3-butanedione react?

Problem 3–3: Deduce the sequence of an unknown peptide from the following information.

(A) Amino acid composition of intact peptide
  (Ala, Arg, 2 Glu, Gly, Leu, Lys, 2 Ser, Thr)

(B) Tryptic peptides
  (1) (Glu, Thr)
  (2) (Gly, Lys, Ser)
  (3) (Ala, Arg, Glu, Leu, Ser)

(C) Trypsin followed by 5-(N,N-dimethylamino)-napthylenesulfonyl chloride (dansyl chloride) yields dansyl-Ser, dansyl-Ala, and dansyl-Thr.

(D) Peptides produced by digestion with thermolysin
  (4) (Ala, Gly, Lys, 2 Ser)
  (5) (Arg, 2 Glu, Leu, Thr)

(E) At pH 8.0, tryptic peptide 3 moved on electrophoresis with a positive charge.

Problem 3–4: Deduce the sequence of a peptide from the following information.

(A) Tryptic peptides
  (1) (Ala, Glu, Phe)
  (2) (Arg, Gln, Ser, Val)
  (3) (His, Lys, Val)

(B) Carboxypeptidase A
  (4) Ala then Phe and Glu

(C) Modification with methyl acetimidate followed by trypsin
  (5) (Arg, Gln, His, Lys, Ser, 2 Val)
  (6) (Ala, Glu, Phe)

(D) Amino-terminal amino acids
  peptide 1, Phe
  peptide 2, Val
  peptide 3, His

(E) Edman degradation of peptide 2
  Val, Ser, Gln

Sequencing of Nucleic Acids

Nucleic acids are linear polymers whose monomers are **nucleoside 5′-monophosphates**

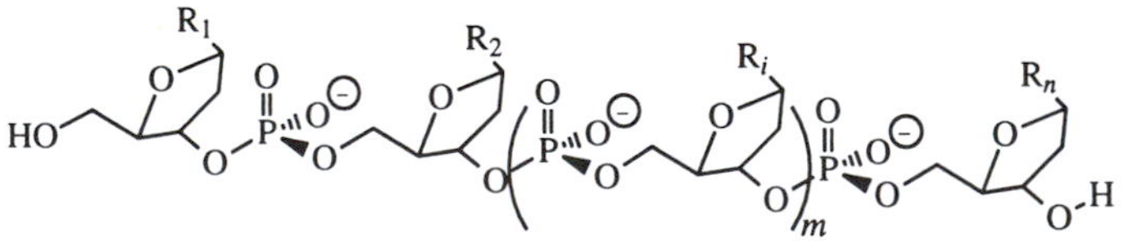

3–1

The covalent bonds that link the monomers together to form the polymer are diesters of phosphoric acid that connect the 3′-hydroxyl of one nucleoside and the 5′-hydroxyl of the next. Nucleic acids are divided structurally and biologically into ribonucleic acids (RNA), which have a 2′-hydroxyl on each of their furanosyl rings, and deoxyribonucleic acids (DNA), which are unsubstituted at the 2′ position of their furanosyl rings. Aside from this distinction, every nucleic acid has the same polymer backbone with one end at which a 5′-hydroxyl group that is not incorporated into the polymer is located. At the other end of the polymer, there is an unincorporated 3′-hydroxyl group. These hydroxyls marking the respective 5′ and 3′ ends are often modified as phosphate monoesters. At the pH usually encountered in living organisms (pH 7–8), the oxygens on each of the phosphate diesters are unprotonated and each monomer bears a full negative charge. Each nucleic acid has its own length and its own sequence in which the nucleotide bases, R_i in **3–1**, are arranged. A nucleic acid sequence is written as a word, each of whose letters stands for a nucleotide base. Unless otherwise noted, the word begins at the 5' end and ends at the 3' end.

There are four nucleoside phosphates incorporated into a nucleic acid as it is synthesized biologically. These four respective nucleoside phosphates are distinguished by the heterocyclic bases they contain (R_i in **3–1**):

pyrimidines   purines

3–2	**3–3**	**3–4**	**3–5**
uracil	cytosine	adenine	guanine
U	C	A	G

These bases are flat, planar, cyclic hybrids of aromatic heterocycles, such as pyrimidine and imidazole, and acyl functional groups, such as amidines, ureas, and amides. The nucleoside bases are divided into the categories of purines and pyrimidines, the larger word designating the smaller rings. Uracil is the base in the nucleoside uridine, cytosine is the base in the nucleoside cytidine, adenine is the base in the nucleoside adenosine, and guanine is the base in the nucleoside guanosine. Uracil is incorporated into DNA as the nucleoside phosphate of its 5-methyl derivative, which is called thymine (T), so that in sequences of DNA the letter U is replaced by T. Thymine is the base in the nucleoside thymidine. Within each nucleoside, the base is attached to its respective pentose in an azaacetal (or *N*-glycosidic) linkage between a pyridine nitrogen or an imidazole nitrogen of the pyrimidine or purine, respectively, and the aldehydic carbon of the furanose. Nucleoside phosphates are referred to as **nucleotides**. The names of the corresponding nucleotides are uridylic acid, cytidylic acid, adenylic acid, guanylic acid, and thymidylic acid.

Deoxyribonucleic acid usually occurs as a double-helical complex of two of these polymers running in opposite directions paired through their bases, so that the sequence of one strand read 5' → 3' **complements** the sequence of the other strand read 3' → 5'. The positions in the sequence of the one strand occupied by adenosine, guanosine, thymidine, and cytidine are paired with positions in the sequence of the other occupied by thymidine, cytidine, adenosine, and guanosine, respectively.

Polypeptides are synthesized biologically by ribosomes that translate a sequence of nucleosides in single-stranded messenger RNA into a sequence of amino acids in a polypeptide. The two corresponding words written in the respective sequences are in the same language, the language of the structure of the protein, and they have the same spelling, but the alphabets are different. The alphabet of the polypeptide sequence consists of the 20 amino acids; the alphabet of the messenger RNA consists of triplets of nucleosides known as the genetic code. Each triplet specifies a particular amino acid, and the triplets are sequentially arranged in the same order as the amino acids of the protein encoded by the message. Because the sequence of the nucleosides, however, is continuous and does not indicate how they are grouped as triplets, there are three ways to divide a sequence of nucleosides into triplets, or three distinct reading frames, only one of which encodes the sequence of the protein. If the sequence and the correct reading frame of a messenger RNA have been determined, it can be immediately translated on paper into the sequence of the polypeptide for which it is responsible.

Almost every protein molecule present at a particular time in a living organism is being continuously produced by ribosomes from messenger RNA molecules, and it follows that if a protein is found in a tissue, the messenger RNA encoding it should be there as well. Messenger RNA can be isolated as a complex mixture of all of the messages normally being expressed in a particular tissue, but there are no satisfactory methods available for isolating individual messenger RNAs from that mixture.

The stratagem devised to obtain the nucleic acid sequence of a particular single-stranded messenger RNA is to transcribe all of the single-stranded messenger RNAs in the mixture into double-helical DNAs and separate them biologically. Deoxyribonucleic acid that has the same sequence in one of its two complementary strands as the sequence of messenger RNA is referred to as complementary DNA (cDNA). Messenger RNA is transcribed into cDNA in the laboratory by first synthesizing a double helix containing the messenger RNA as one strand and DNA as the other. This is performed by the enzyme RNA-directed DNA polymerase. The complement to the strand of DNA is then synthesized with the enzyme DNA-directed DNA polymerase, which produces the double-helical DNA required. This double-helical DNA contains as the sequence of one of its strands the original sequence of the messenger RNA encoding the amino acid sequence of a particular protein.

This complex mixture of cDNAs is incorporated into the DNA of a bacteriophage or a plasmid. In this form the cDNA can be biologically replicated at will, to provide as much as is necessary for sequencing. In this form it can also be sorted. When a population of bacteria is infected under the proper conditions with the bacteriophages or transfected with the plasmids containing these cDNAs, each bacterium that is infected or transfected picks up only one bacteriophage or one plasmid. Because the plasmid or the DNA in the bacteriophage usually contains a gene conferring resistance to an antibiotic, the bacteria that have been infected or transfected can be selected by growing the population in the presence of that antibiotic. The plasmid or the bacteriophage can then be replicated as the host bacterium is replicated. When a suitably diluted suspension of these bacteria is wiped over a field of nutrient agar and allowed to grow, individual colonies, or **clones**, each arising from one bacterium and each containing copies of only one of the original cDNAs, appear upon the field.

The trick is to discover which of these little clones, clearly visible to the naked eye but numbering in the thousands to hundreds of thousands, happens to contain the cDNA whose nucleic acid sequence encodes the protein of interest. **Screening** is the process of discovering the desired clone. Although several methods have been devised for screening, the most rapid and unambiguous method is to synthesize chemically a short fragment of radioactive single-stranded DNA, referred to as a **probe**, whose sequence is a short segment of the amino acid sequence of the protein written in the genetic code. When the DNA of a clone containing that particular short nucleic acid sequence is denatured, the sequence from the opposite strand of the double helix, the strand that is complementary to the nucleic acid sequence of the probe, will also be present. The complementary nucleic acid sequence on the denatured DNA will combine with the probe to form a short segment of double-helical DNA, and in this way the probe is captured. This renders the colony containing the desired cDNA radioactive.

An example[26,48] will illustrate this screening procedure. The protein factor VIII is one of the proteins that are together responsible for the cascade of events leading to the clotting of the plasma of mammalian blood. Factor VIII was digested with trypsin, and the peptides that resulted from the digestion were separated by adsorption chromatography at high pressure (Figure 3–8B).[26] One of these peptides happened to be resolved quite cleanly from its neighbors (arrow in Figure 3–8B), and it was submitted to Edman degradation. The amino acid sequence determined for this peptide was AWAYFSDVDLEK. A segment of radioactive, single-stranded DNA with the nucleic acid sequence CTTTTCCAGGTCAAGTCGGA-GAAATAAGCCCAAGC was synthesized chemically. This sequence is the complement of a spelling of the amino acid sequence of this peptide written in the genetic code. This probe was captured by the denatured, single-stranded DNA in 15 clones out of the 500,000 screened for DNA containing this nucleic acid sequence.[48] Bacteria from each of these 15 clones were separately grown on a large scale, the inserted DNA was cut out of the DNA of the bacteriophage that had been carrying it with a site-specific deoxyribonuclease, and this cDNA was sequenced. The nucleic acid sequences obtained, when translated into amino acid sequences, were overlapping

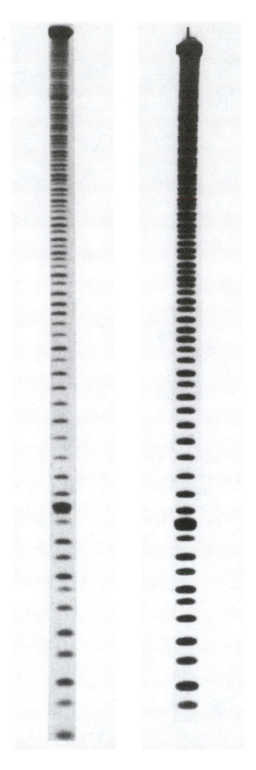

Figure 3–11: Separation of fragments of end-labeled RNA by electrophoresis on polyacrylamide gels.[49] 4.5S Ribonucleic acid, isolated from spinach chloroplasts, was labeled at its 3' end with [5'-^{32}P]cytidine 3',5'-bisphosphate and T_4-polyribonucleotide synthase (ATP). The end-labeled RNA was partially digested under alkaline conditions and then submitted to electrophoresis on slabs of 12% polyacrylamide cast in a buffered solution of 7 M urea. The two lanes in the figure were loaded with different amounts of sample and were run for different lengths of time. Reprinted with permission from ref 49. Copyright 1982 *Journal of Biological Chemistry.*

regions of the same overall amino acid sequence and each contained the sequence AWAYFSDVDLEK. Four other fragments of factor VIII were also submitted to Edman degradation and their amino acid sequences could also be found in the amino acid sequence translated from the nucleic acid sequences of the 15 clones. Comparisons like these between directly determined amino acid sequences of a particular protein and the amino acid sequence translated from the cDNA are often used to substantiate the identification of the cDNA as that of the particular protein.

Deoxyribonucleic acids, such as the 15 cDNAs isolated in the experiment just described, can be rapidly sequenced because denatured, single-stranded nucleic acids, both RNA and DNA, behave with extraordinary regularity upon electrophoresis. When a ribonucleic acid, such as the 107-residue 4.5S ribosomal RNA from the chloroplasts of spinach,[49] is enzymatically elongated by one nucleotide at its free 3′ hydroxyl with [5′-^{32}P]cytidine 3′,5′-bisphosphate and then submitted to partial alkaline hydrolysis, a random mixture of fragments of all possible lengths and all possible beginning and ending points within the sequence is produced. Only those fragments that begin at the original 3′ end, however, are radioactive. In the case of the 4.5S rRNA, these formed a set of 108 unique fragments that were of all possible lengths between one and 108 nucleotides. When this mixture was submitted to electrophoresis under denaturing conditions on a gel cast from 12% acrylamide and the radioactive components were located by placing the polyacrylamide gel on a photographic film, a regular array of bands, referred to as a ladder, could be observed (Figure 3–11).[49] Each of these bands, with one interesting exception that will be discussed later, represents a single-stranded RNA that begins at the labeled 3′ end of the original 4.5S rRNA, because it is radioactive, and is one nucleotide longer than the nucleic acid in the band below it in the figure.

This ability of electrophoresis on polyacrylamide gels to separate nucleic acids only on the basis of their length arises from the properties of these polymers and the nature of the electrophoresis. The free electrophoretic mobility of denatured single-stranded DNA at $(\Gamma/2) = 0.01$ M, pH 7.5, and 0 °C is $(1.82 \pm 0.02) \times 10^{-4}$ cm^2 V^{-1} s^{-1} and does not vary[50] with its length over a range of molar masses between at least 0.26×10^6 and 130×10^6 g mol^{-1}. The free electrophoretic mobility of denatured RNA under the same conditions is $(1.77 \pm 0.05) \times 10^{-4}$ cm^2 V^{-1} s^{-1}, and it also shows no tendency to vary with length.[50] The electrophoretic mobilities of single-stranded DNA and RNA also satisfy the relationship $u_i = u_i^\circ e^{-iKRTa}$ (Equation 1–73),[51] and the free mobilities extrapolated from their behavior on polyacrylamide gels are in reasonable agreement with those measured directly.[52] Because their free mobilities are all the same, it is only the resistance posed by the polyacrylamide, e^{-iKRTa}, that separates the nucleic acids of the various lengths. It is not surprising that this sieving, accomplished at the

molecular level by the strands of polyacrylamide, should be a regular, continuous, monotonic function of the lengths of the nucleic acids (Figure 3–11).

Suppose a single-stranded deoxyribonucleic acid, labeled at its 5′ end by phosphorylation with [^{32}P]phosphate, has been cleaved in a low yield and randomly on the 5′ side of each of the guanylates in its sequence. This will have produced a series of radioactive fragments of different length, each of which ends at a nucleotide whose only distinction is that it preceded a guanylate in the original sequence. When the products of this partial cleavage are submitted to electrophoresis, a series of radioactive bands will appear whose mobilities correspond to only those bands in the ladder whose 3′-terminal nucleotide precedes a guanylate. The knowledge that the cleavage occurred only at guanylates and the position of the products in the ladder identifies the position of every guanylate in the original sequence.

Suppose further that four samples have been prepared from the original single-stranded deoxyribonucleic acid such that they contain radioactive fragments, all of which begin at the original 5′ end, because they were rendered radioactive by phosphorylating that location, but end at every nucleotide preceding an adenylate, a guanylate, a cytidylate, or a thymidylate, respectively. When these samples are submitted to electrophoresis, side by side, every band in the ladder will be represented in the four lanes, but each band in the ladder will be found in only one of the lanes. As one scanned the pattern, from the bands of greatest mobility to the bands of least mobility, one would encounter each band of the ladder in its proper succession. The lane in which each successive band was found would have been determined by the identity of the nucleotide that follows its actual 3′-terminal nucleotide in the complete sequence of the original DNA. By starting with the band of greatest mobility and noting its lane and the lane in which each successive band of lower mobility occurs, one would be reading the sequence of the DNA in the direction from 5′ to 3′.

The strategy for sequencing DNA illustrated by this simplified situation requires that a set of fragments of radioactive, single-stranded DNA be produced. Each of these fragments must have as its 5′ terminus the same nucleotide in the complete nucleic acid sequence to be determined, but this does not have to be the actual 5′ terminus of the original piece of DNA. For example, this result could be achieved by cleaving all of the molecules of the original DNA at the same nucleotide with a specific endonuclease. Every position in the portion of the complete nucleic acid sequence to be determined on a particular polyacrylamide gel must be represented by a radioactive fragment that ends at this position and that has been produced in sufficient yield to be seen as a dark band on a sheet of photographic film exposed to the polyacrylamide gel. The observer must have enough information about each fragment to associate its 3′ terminus with a particular nucleotide, adenylate, guan-

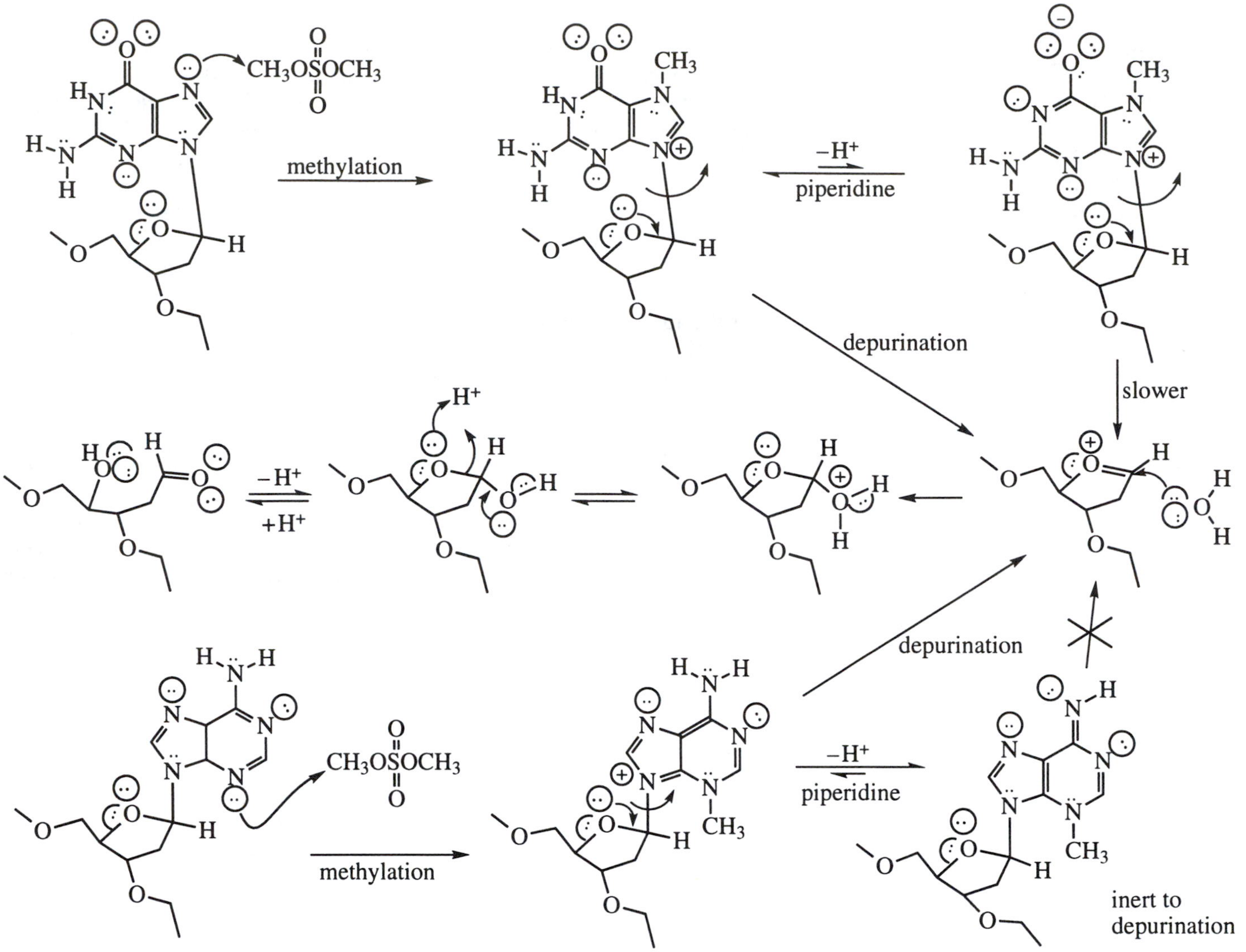

Figure 3–12: Depurination and opening of the furanose at methylated adenines and guanines in intact DNA. Guanine methylates more rapidly than adenine; and, because of its higher pK_a, methylated guanine depurinates more readily in the presence of piperidine than does methylated adenine. These two features permit almost exclusive depurination at guanine. Following depurination, the polyribose has a labile acetal at the depurinated position and the furanose can open.

ylate, cytidylate, or thymidylate. In practice, this information is either that the 3′ terminus of a particular fragment precedes a particular nucleotide in the complete nucleic acid sequence or that its 3′ terminus is a particular nucleotide. There are two methods, chemical and enzymatic, for producing such a set of fragments. Neither corresponds exactly to the simplified illustration just de-scribed, but both satisfy the requirements of the strategy.

In the chemical method of Maxam and Gilbert,[53] reagents that take advantage of the hybrid nature of the nucleotide bases, which are partly aromatic heterocycles and partly acyl derivatives, are used to cleave the single-stranded DNA, labeled at its 5′ terminus, at locations occupied by a particular base. The chem-ical cleavages used are based on reactions previously developed to remove selectively either purine bases or pyrimidine bases from DNA. Such reactions are referred to as depurinations[54] or depyrimidinations,[55] respectively.

When single-stranded DNA is methylated with dimethylsulfate, the lone pairs on the N7 imidazole nitrogens of the guanines and the N3 pyridine nitrogens of the adenines display the greatest nucleophilicity.[56] The pyrimidines are relatively inert because their nitrogens are the nitrogens of acyl functionalities. The methylation of either of these two nitrogens turns their respective purine bases into good leaving groups and hence weakens the *N*-glycosidic linkage (Figure 3–12). Depurination from these methylated positions occurs at neutral pH.[56] A discrimination between adenine and guanine can be made because guanine happens to methylate 5–10 times more rapidly than adenine,[56] and when only partial methylation is performed, as is always the case in sequencing reactions, the depurination at each of the positions occupied by a guanine is 5–10-fold greater than the depurination at a position occupied by an adenine (G > A conditions). The depurination can be made even more specific for guanine by raising the pH (G conditions),[57] because 3-methyladenine readily loses a proton, after which it is no longer a leaving group, while 7-methylguanine remains a leaving group even at the higher pH (Figure 3–12). Depurination of both unmethylated adenines and unmethylated guanines occurs at acidic pH,[54] but depurination at adenines predominates (G < A conditions).[53] The protonations of adenine and guanine that occur under these conditions are formally equivalent to the methylations but cannot be so readily controlled. Positions at which guanine is located can be recognized because depurination predominates there under the G conditions or the G > A conditions. Positions at which adenine is located can be recognized because depurination predominates there under G < A conditions, decreases under G > A conditions, and does not occur under G conditions.

Positions at which cytosine and thymine occur in the sequence of a single-stranded DNA can be partially depyrimidinated by mild treatment with aqueous hydrazine.[54,55,58,59] Hydrazine attacks thymine and cytosine specifically because it is a reagent directed against electrophilic acyl carbons, and the pyrimidines are more like acyl derivatives in their behavior than are the purines, which are more aromatic. Hydrazinolysis of cytidylates and thymidylates has been shown to proceed through intermediate 5-ureido-3-amino-3-pyrazolines and 5-ureido-3-hydroxy-3-pyrazolines, respectively, to produce 1-ureidonucleotides. These react further with the hydrazine to yield hydrazones of deoxyribose[59] at the depyrimidinated positions (Figure 3–13). Fortuitously, this reaction is suppressed at thymine, but not cytosine, by high concentrations of sodium chloride.[53] Positions at which either cytosine or thymine are located are depyrimidinated in the absence of sodium chloride (C + T conditions) and positions at which cytosine is located are depyrimidinated even in the presence of sodium chloride (C conditions).

Ribonucleic acid differs from DNA in its susceptibility to hydrolysis at its phosphate diesters under mildly basic conditions (Figure 3–11). This results from intramolecular nucleophilic catalysis of the hydrolysis by the 2′-hydroxyl next to every phosphate diester on a 3′-hydroxyl. Deoxyribonucleic acid, because it lacks any free hydroxyls, is stable to mild base. Upon depurination (Figure 3–12) or depyrimidination (Figure 3–13), however, a free 4-hydroxyl is liberated adjacent to both the 3-phosphate diester and the 5-phosphate diester on the depurinated or depyrimidinated nucleotide. It is this

4-hydroxyl, acting as an intramolecular nucleophilic catalyst

that is presumably[60] responsible for the lability of depurinated[60] and depyrimidinated[55] positions in DNA to cleavage under mildly basic conditions. In terms of sequencing DNA, the DNA that has been partially depurinated or depyrimidinated, respectively, at locations whose identity has been controlled by the conditions is hydrolyzed at each of these locations by treatment with base (0.1 M NaOH or 0.5 M piperidine) to produce fragments that have as their 3′ terminus (Reaction 3–2) a nucleotide that preceded, in the original nucleic acid sequence, a target for the depurination or depyrimidination.

In the enzymatic method of Sanger, Nicklen, and Coulson,[61] the fragments required for the electrophoresis are made by synthesizing complementary strands of DNA using the single-stranded DNA to be sequenced as templates for DNA-directed DNA polymerase (Figure 3–14).

The enzyme currently in widest use is the DNA polymerase from bacteriophage T7 that has been covalently modified to eliminate its 3′ → 5′ exonuclease activity.[62] This enzyme synthesizes a single strand of DNA complementary to a single-stranded template by consecutively attaching the nucleotide complementary to the next nucleotide on the template to the growing 3′ end of the synthetic nucleic acid. The monomers are present in solution as their activated 5′-triphosphates. The newly synthesized polymer of DNA is rendered radioactive by including [α-^{32}P]dATP in the synthetic mixture. The successive fragments that have at their 3′ end only a particular nucleotide are produced by including a small

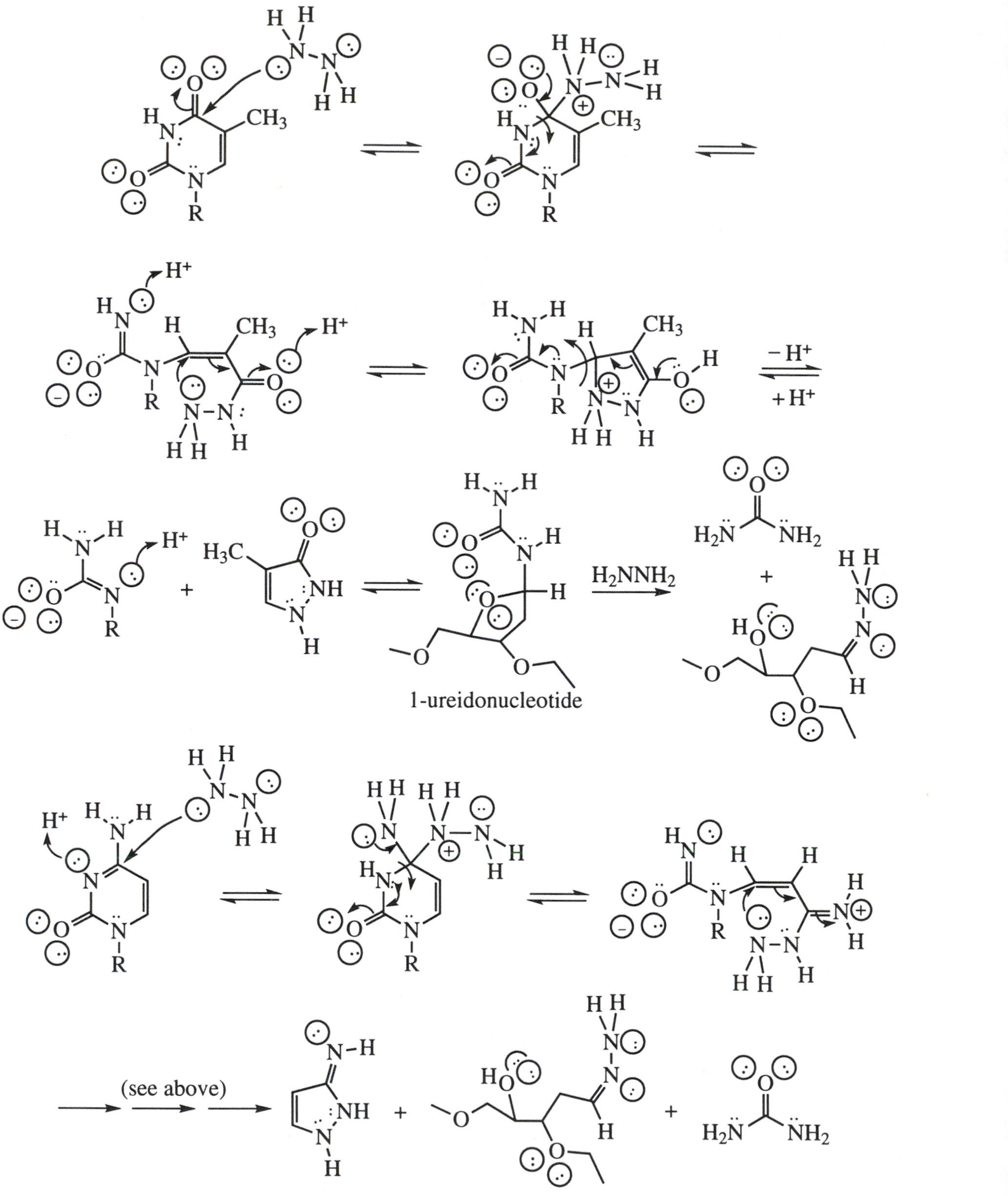

Figure 3–13: Depyrimidination and opening of the furanose at cytosines and thymines in intact DNA. Hydrazine attacks nucleophilically at the acyl carbons in thymine and cytosine, respectively, initiating formation of an *N*-deoxyribosylurea (a 1-ureidonucleotide). The 1-ureidonucleotide decomposes and the hydrazone of the opened furanose is the eventual product.

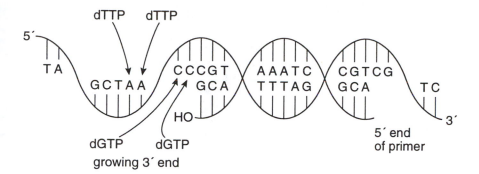

Figure 3–14: Elongation of a strand of DNA by DNA-directed DNA polymerase. The enzyme adds deoxyribonucleotides sequentially to the growing 3′ end of a single-stranded DNA, the primer, held in a double helix with a longer complementary strand of deoxyribonucleic acid, by using the template of the longer strand. Each nucleotide in the template is paired with its complementary nucleotide as the shorter strand elongates.

amount of 2′,3′-dideoxythymidine triphosphate, 2′,3′-dideoxycytosine triphosphate, 2′,3′-dideoxyguanosine triphosphate, or 2′,3′-dideoxyadenosine triphosphate, respectively, in each of four synthetic reactions along with the 2′-deoxythymidine triphosphate, 2′-deoxycytosine triphosphate, 2′-deoxyguanosine triphosphate, and 2′-deoxyadenosine triphosphate present in all of them. Occasionally, a 2′,3′-dideoxynucleotide is incorporated into one of the growing polymers, and its incorporation terminates polymerization because that polymer then lacks the 3′-hydroxyl necessary for elongation. In this way fragments satisfying two of the requirements for electrophoretic sequencing are produced.

The last requirement, that every fragment have as its 5′-terminus the same position in the complete sequence is satisfied by taking advantage of the requirement of DNA polymerase for a short, preformed segment of a double-helical DNA to provide a 3′-OH from which the new strand can be elongated. To initiate the reaction a short segment of single-stranded DNA, known as a **primer**, which is complementary to a segment of the DNA to be sequenced, is annealed to the template to provide the necessary 3′-hydroxyl. Because the DNA polymerase starts at the primer when it synthesizes a complementary, radioactive single strand of DNA, the sequence of the primer can be chosen so that the newly synthesized DNA will begin at a particular point in the sequence of the template. The complementary sequence to which the primer is annealed can be a short piece of DNA of known sequence that has been deliberately attached to the 3′ end of the DNA to be sequenced,[63] or it can be any internal sequence for which a complementary fragment of single-stranded DNA happens to be available.[61] Often this is a probe that had been made for purposes of screening. It is also possible to use an oligonucleotide that has the same sequence as a segment near the 3′ end of the longest single-stranded fragment that provided readable nucleotide sequence in the last set of polyacrylamide gels, to extend the sequencing of the template further to its 5′ end. In this way, one can walk along a very long template.

The polyacrylamide gels that result from the application of these two methods, the chemical and the enzymatic, are very similar in appearance (Figure 3–15). Sequence is read from the bottom (shortest fragments) to the top (longest fragments), 5′ to 3′. In the chemical method the sequence of the original single-stranded DNA is being read. In the enzymatic method the sequence of the complement of the original single-stranded DNA is being read. Since DNA is normally double-helical and normally contains the two complementary sequences running in opposite directions anyway, either sequence is formally the sequence of the DNA, as long as the correct direction (5′ → 3′) is assigned to the sequence by the observer.

These original methods were both based on using fragments of nucleic acid made radioactive by incorporating [32P]phosphate. In this way, the bands can be detected by autoradiography. More recently, reagents have been developed to label the various bands with fluorescent labels. In one method, based on the enzymatic strategy, each dideoxynucleotide is tagged with a fluorescent functional group that has a different emission maximum so that sequence can be read from a single lane of a polyacrylamide gel.[64]

Usually, DNA in a quantity sufficient for sequencing is prepared by cloning a bacterium containing a plasmid bearing the DNA of interest and isolating the plasmids from a large batch of cells. Recently, the polymerase chain reaction[65] has been developed as an alternative method for preparing significant quantities of a particular segment of DNA from a minute quantity of any DNA. All that is required is that the segment of DNA to be amplified is flanked on either side by known nucleotide sequences. Two primers are made, one complementary to the flanking sequence at one end and the other complementary to the flanking sequence at the other. These two primers, however, must be complementary to opposite strands of the initial double-stranded DNA. The initial DNA is melted and the primers are annealed. Deoxyribonucleic acid polymerase is then used to elongate from the 3′ end of each primer (Figure 3–14). This produces two copies of duplex DNA over the segment of interest. The new DNA is melted and reannealed with the same two primers and elongation is performed again to produce four copies of duplex DNA for the segment of

A

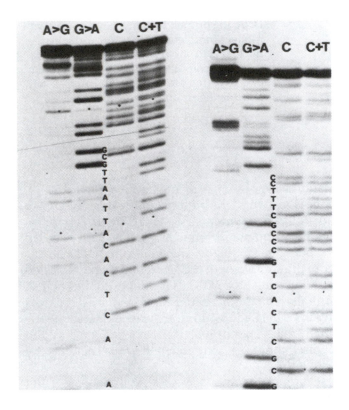

B

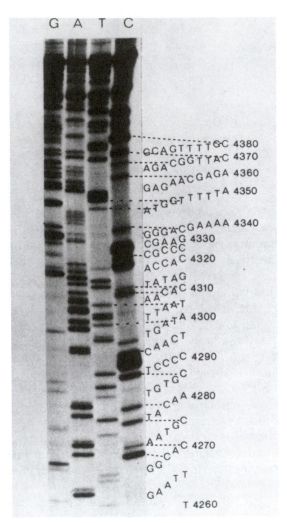

Figure 3–15: Sequencing of DNA. (A) A fragment of single-stranded DNA, 64 bases long, was phosphorylated at its free 5′ end with [γ-³²P]MgATP and polynucleotide 5′-hydroxyl kinase. It was divided into four aliquots that were then submitted to depurination either in 0.5 M hydrochloric acid (A > G) or following methylation at neutral pH (G > A) or to depyrimidination with hydrazine in the absence (C + T) or presence (C) of 2 M NaCl. The depurinated or depyrimidinated DNA was subsequently cleaved with strong base, 0.1 M NaOH or 0.5 M piperidine, and the end-labeled fragments were separated on a gel of 20% polyacrylamide cast in a buffered solution of 7 M urea.[53] Reprinted with permission from ref 53. Copyright 1977 National Academy of Sciences. (B) A fragment of DNA complementary to a short segment of the nucleic acid sequence of the single-stranded DNA from the bacteriophage φX174 was annealed to the template as a primer. The initiation complex was divided into four aliquots. Elongation of the primer in these separate samples was performed with DNA-directed DNA polymerase in the presence of 2′,3′-dideoxyguanosine triphosphate (lane G), 2′,3′-dideoxyadenosine triphosphate (lane A), 2′,3′-dideoxythymidine triphosphate (lane T), and 2′,3′-dideoxycytosine triphosphate (lane C). Each sample contained a small amount of [α-³²P]MgATP to render the newly synthesized DNA radioactive. Following these respective reactions each sample was digested with the *Hae*III restriction endonuclease, which cut the DNA within the primer, so that all of the newly synthesized DNA would start with the same nucleotide at the 5′ end. The single-stranded, radioactive fragments of DNA in each sample were separated by electrophoresis on a slab of 12% polyacrylamide.[61] Reprinted with permission from ref 61. Copyright 1977 National Academy of Sciences.

interest. By repeated cycles of melting and annealing, the amount of the desired DNA increases exponentially. If the heat-stable DNA polymerase from *Thermus aquaticus* is used for the elongation, new polymerase does not have to be added after each melting cycle.[66]

The DNAs normally sequenced are quite long, from hundreds to tens of thousands of nucleotides. A particular sequence of DNA can be read only to a certain length (300–400 nucleotides) before the bands corresponding to the longer and longer rungs of the ladder can no longer be separated unambiguously by electrophoresis. To sequence long DNAs efficiently, they must be cleaved into fragments, just as polypeptides have to be cleaved into peptides. The fragments are obtained, just as peptides are obtained during the sequencing of proteins, by using enzymes to cleave the larger polymer into smaller fragments. To produce discrete fragments, as in the case of the peptides used in sequencing proteins, the positions in the larger polymer at which cleavage occurs must be particular.

Cleavage is confined to particular sites in long pieces

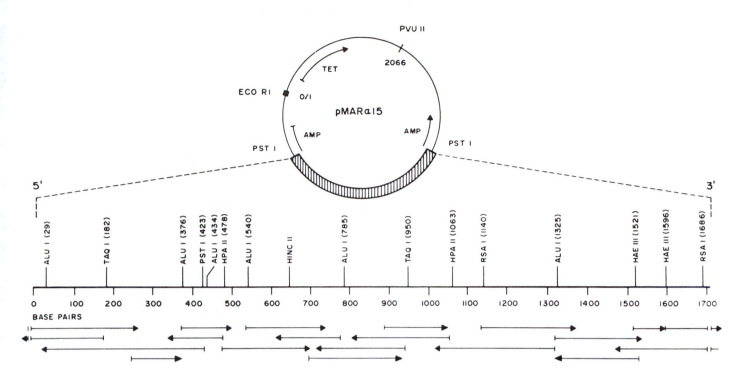

Figure 3–16: Restriction map of a large fragment of DNA cut out of a plasmid.[68] A large fragment of cDNA (17 kb) was removed with the site-specific deoxyribonuclease *Pst*I from the circular plasmid pMAR α 15, which had been originally constructed from the circular plasmid pBR322. The plasmid pMAR α 15 was isolated during a screening procedure for cDNA encoding the α polypeptide of the murine acetylcholine receptor. The fragment of cDNA was purified by electrophoresis and submitted to a series of digestions with the noted site-specific deoxyribonucleases. The patterns of fragments established the restriction map displayed. The arrows below the restriction map indicate which restriction fragments were submitted to sequencing from which 5′ end. The positions in the nucleic acid sequence cleaved by each restriction endonuclease are identified by numbers in parentheses. Reprinted with permission from ref 68. Copyright 1985 Oxford University Press.

of double-helical DNA by using **site-specific deoxyribonucleases**, sometimes referred to as restriction endonucleases. As trypsin and thermolysin catalyzed hydrolysis of amide bonds in a protein only next to particular amino acids to produce specific peptides, so site-specific deoxyribonucleases catalyze the hydrolysis of double-helical DNA only at phosphate diesters within particular sequences of nucleotides. These hydrolyses produce specific **restriction fragments**. Unlike the situation in protein sequencing, however, a large number of site-specific deoxyribonucleases[67] are available, whose specificities vary in their complexity. The particular sequence recognized by a given site-specific deoxyribonuclease can be four, five, or six nucleotides long. The longer the sequence recognized, the less frequently it will occur in the DNA, and the longer will be the restriction fragments. By trial and error, a pattern of fragments ideally suited to the demands of sequencing can be prepared.

The double-helical restriction fragments produced from double-helical DNA are rapidly separated by preparative electrophoresis on polyacrylamide gels. They are usually visualized by using fluorescent dyes. Their length can be estimated from their electrophoretic mobilities. The order in which a given set of fragments is arranged is determined by restriction mapping. To produce a **restriction map** of a large piece of DNA, it is cleaved separately with several site-specific deoxyribonucleases. The fragments produced in each separate digestion are isolated and assigned a length by electrophoresis. Each of these fragments of DNA is then submitted to digestion by the other site-specific deoxyribonucleases, and the shorter fragments that result are isolated and assigned a length. This dissection is contin-

ued until the products observed, which are designated by their length and the pedigree of the cleavages that produced them, are consistent with only one distribution of cleavage sites through the original DNA. This unique distribution, the restriction map, orders the different fragments that have been obtained relative to the complete sequence.

An example will serve to illustrate the complete process.[68] A clone containing the cDNA encoding the α polypeptide of the murine nicotinic acetylcholine receptor was identified by screening. The cloned cDNA was isolated as an intact double-helical polymer from the plasmid that had been carrying it. It was digested with the following site-specific deoxyribonucleases: *Alu*I, which cleaves at the nucleic acid sequence AG ↓ CT; *Taq*I, which cleaves at T ↓ CGA; *Pst*I, which cleaves at CTGCA ↓ G;

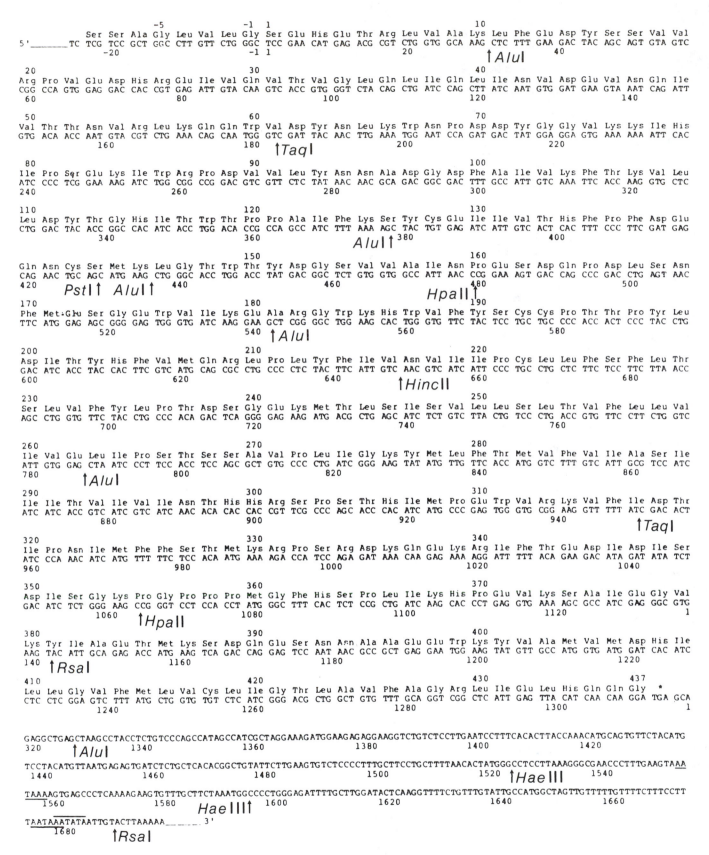

Figure 3–17: Nucleic acid sequence and deduced amino acid sequence for the α polypeptide of murine nicotinic acetylcholine receptor. The nucleotides are presented in the 5' to 3' direction for the coding strand. Both sequences are numbered starting with the first amino acid in the mature protein. The first eight amino acids in the presented sequence are removed posttranslationally. The initiation codon for translation was not on the cloned piece of cDNA. The asterisk marks the codon at which translation is terminated. The restriction sites that produced the restriction map (Figure 3–16) are identified in the nucleic acid sequence. Reprinted with permission from ref 68. Copyright 1985 Oxford University Press.

*Hpa*II, which cleaves at C ↓ CGG; *Hae*III, which cleaves at GG ↓ CC; *Rsa*I, which cleaves at GT ↓ AC; and *Hinc*II, which cleaves at GTPy ↓ PuAC. The pattern of fragments obtained when these enzymes were used in various combinations was only consistent with one restriction map (Figure 3–16).[68] For example, the *Hpa*II fragment between positions 478 and 1063 would give three fragments about 60, 240, and 280 base pairs in length upon digestion with site-specific deoxyribonuclease *Alu*I. The order in which they occur in the *Hpa*II fragment could be determined by gathering the following observations. Deoxyribonuclease *Taq*I would cut only the *Alu*I fragment that is about 280 base pairs in length to yield the same fragment, about 120 base pairs long, that it would produce from one end of the *Hpa*II fragment. Deoxyribonuclease *Hinc*II would cut only the *Alu*I fragment that is about 240 base pairs in length to give a fragment about 140 base pairs in length. This fragment, together with the *Alu*I fragment about 60 base pairs in length, would form the fragment about 200 base pairs in length produced during the digestion of the *Hpa*II fragment with deoxyribonuclease *Hinc*II alone.

When restriction fragments of a convenient size had been produced from this cDNA encoding the α-polypeptide of the murine acetylcholine receptor, a selection of them were sequenced (arrows in Figure 3–16). Each sequence began at the 5′ end of one of the two complementary strands of a double-helical fragments and was read as far as was possible. Together, all of these individual sequences produced the complete sequence of the cDNA (Figure 3–17). Of the six reading frames in the completely sequenced double-helical cDNA, the one coding for the sequence of the α polypeptide of murine nicotinic acetylcholine receptor was easily identified by locating the sequence that encodes the amino acid sequences on which the probes used to screen the clones were based.

Suggested Reading

Valenzuela, P., Quiroga, M., Zaldivar, J., Rutter, W.J., Kirschner, M.W., & Cleveland, D.W. (1981) Nucleotide and corresponding amino acid sequences encoded by α- and β-tubulin mRNAs, *Nature 289*, 650–655.

Problem 3–5: A segment of double-stranded DNA, 12 base pairs in length, was cut from a chromosome by the site-specific deoxyribonuclease *Alu*I. After dephosphorylation, polynucleotide 5′-hydroxyl kinase and [γ-^{32}P]MgATP were used to label the two 5′ ends of this fragment with ^{32}P. The two strands (A and B) of the double helix were separated by electrophoresis. They were then individually subjected to hydrazinolysis, in the presence or absence of 2 M NaCl, followed by treatment with 0.5 M piperidine to produce partial cleavage at cytosines only or at both cytosines and thymines, respectively. The partial cleavage mixtures from each strand were submitted to electrophoresis separately, and the

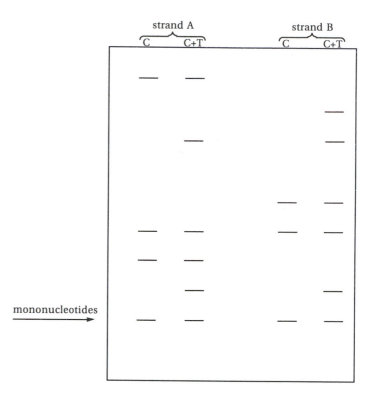

polyacrylamide gels were subjected to autoradiography to locate the radioactive fragments. The pattern of fragments is presented in the following diagram.

What is the sequence of the original piece of double-stranded DNA? Indicate 5′ and 3′ ends of each strand in your answer as well as base pairing.

Problem 3–6: A fragment of single-stranded RNA, 488 nucleotides long, was obtained from one of the ribosomal RNAs of rat liver, 28S rRNA, by treatment with α-sarcosin. It was treated with alkaline phosphatase to remove any phosphate from its 5′ end, and then with [γ-^{32}P]MgATP and T4 polynucleotide 5′-hydroxyl kinase to attach a radioactive phosphate to its 5′ end. The sample was then split into five separate portions. They were treated with the following reagents, respectively:

(I)   NaOH

(II) ribonuclease T_1, which cleaves on the 3′ side of G

(III) ribonuclease U_2, which cleaves on the 3′ side of A

(IV) ribonuclease PhyM, which cleaves on the 3′ sides of A and U

(V)   ribonuclease BC, which cleaves on the 3′ sides of U and C

The alkaline hydrolysis (I) and the enzymatic digestions (II–V) were carefully controlled so that only a small amount of cleavage occurred at each sensitive position. Each of the five mixtures was then placed in an adjacent lane on a polyacrylamide gel and submitted to electrophoresis followed by autoradiography. A tracing of that autoradiogram is presented at right, above. An autoradiogram only registers radioactive fragments.

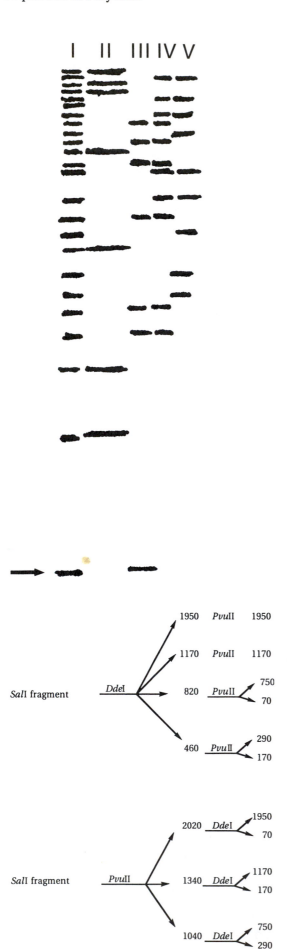

Each lane is labeled with the appropriate roman numeral. The arrow marks the position of the mononucleotides.

(A)   Starting with the nucleotide on the 5′ end, write the sequence of the α-sarcosin fragment covered by the gel. Indicate clearly 5′ → 3′ polarity.

(B)   Look carefully at the gel and then give a reason for including the digest in lane I.

Problem 3–7: A piece of double-stranded DNA about 4360 base pairs in length was produced by the site-specific deoxyribonuclease *Sal*I. When this was digested with the site-specific deoxyribonucleases *Dde*I and *Pvu*II the fragments described in the diagram to the left were obtained. The numbers are the approximate lengths of the fragments.
Construct a restriction map.

Problem 3–8: A piece of double-stranded DNA about 5300 base pairs in length has been produced by the action of the site-specific deoxyribonuclease *Eco*R1. When this fragment was digested with the site-specific deoxyribonucleases *Hind*III, *Kpn*I, and *Bam*HI the following results were obtained. The numbers are the approximate lengths of the fragments.

Construct a restriction map.

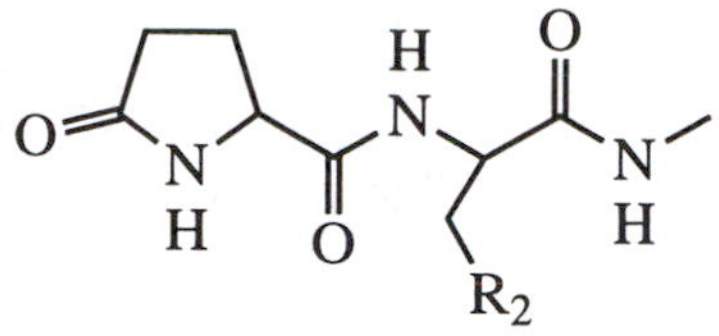

pyroglutamate
at amino terminus

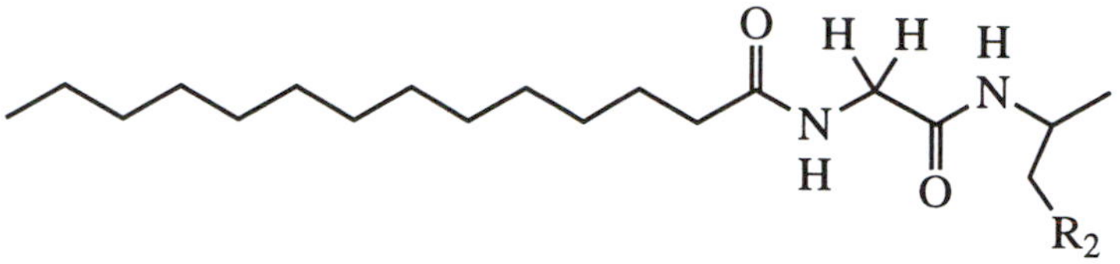

N-tetradecanoylglycine
at amino terminus

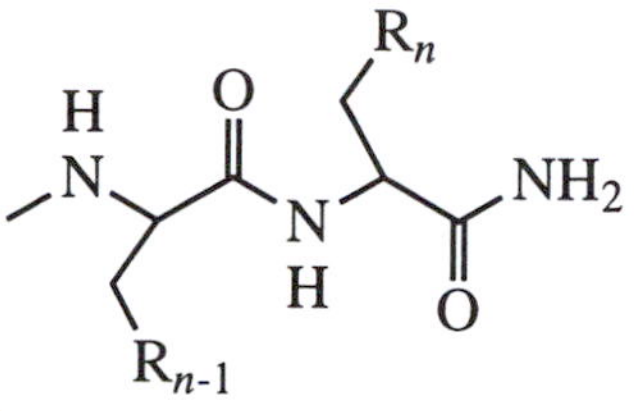

primary amide
at carboxy terminus

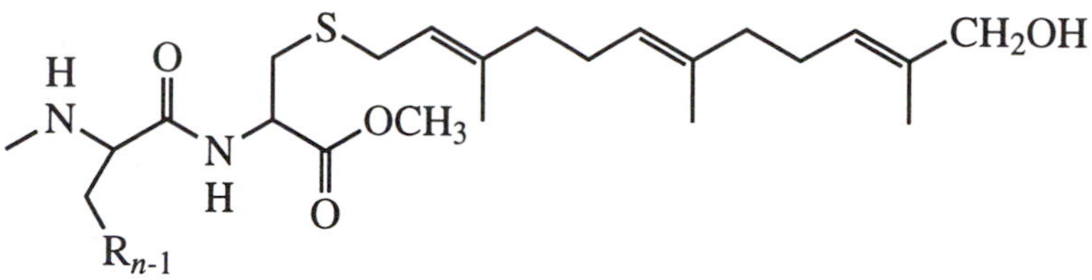

S-farnesyl-*O*-methylcysteine
at carboxy terminus

Figure 3–18: Posttranslational modifications that occur at the termini of a polypeptide.

Posttranslational Modification[69]

With the exception of the evanescent N^{α}-formyl group on its amino terminus and perhaps the 21st primary amino acid, selenocysteine,[70] the infant polypeptide as it emerges from the peptidyltransferase site on the ribosome is a polymer containing only the 20 amino acids encoded by the particular messenger RNA, each coupled to its neighbors by the amides of the peptide backbone. It is this covalent structure and only this covalent structure that can be read by the investigator in the sequence of the messenger RNA if she knows which methionine codon is the initiation site. The covalent structures of many proteins, however, do not remain in this untouched state but are biologically modified. A **posttranslational modification** is any change in the covalent structure of a polypeptide that occurs after its emergence from the ribosome.

Some of these modifications of the original covalent structure are performed by proteolytic enzymes. These normal modifications must be distinguished from artifactual modifications that occur, for example, during the purification of a protein. In the course of normal, natural modification, the polypeptides of certain proteins are cleaved internally either for architectural purposes or as a mechanism for controlling their enzymatic activities. An example of the former is the trimming of folded proinsulin to produce insulin. An example of the latter is the activation of proteolytic enzymes in the pancreatic secretions or the serum by internal proteolytic cleavages. Almost all of the proteins of animals are posttranslationally shortened by the removal of one or more of the amino acids from their amino terminus, but some proteins have particular segments removed from their amino termini as they are passed from one compartment in the cell to another compartment. These amino-terminal signal sequences[71] address the proteins to the proper locations, and their removal is presumably involved in keeping them there. These successive removals of portions of the amino-terminal sequence have led to the terms preproprotein and proprotein. Peptide hormones are often synthesized within much larger polypeptides that are then specifically cleaved into the biologically active pieces.[72] In all of these instances new amino termini or carboxy termini appear, and the situation can usually be sorted out by separating the pieces and sequencing them by Edman degradation and by digestion with carboxypeptidases because the complete amino acid sequence of the precursor polypeptide is known from the cDNA. In contrast to simple enzymatically catalyzed hydrolysis of the polypeptide at specific locations, covalent modification of the individual amino acids yields products[69] that cannot be characterized simply by sequencing.

The amino terminus of a polypeptide can be *N*-methylated[73,74] or *N*-acylated, either intramolecularly, as in pyroglutamate (Figure 3–18), or externally, as when it is formylated,[75] acetylated,[76] glucuronylated,[77] or *n*-tetradecanoylated (Figure 3–18).[78] *n*-Tetradecanoyl amides of amino termini were first found on proteins known as kinases. The existence of these fatty acylated amino termini was established by isolating an amino-terminal peptide, $CH_3(CH_2)_{12}COHN$Gly-Asn-Ala, from cAMP-dependent protein kinase and confirming its structure by chemical degradation and by mass spectroscopy with fast atom bombardment.[79] Such a chemical demonstration of a modification at the amino terminus of a polypeptide should be distinguished from an unsupported conjecture that the amino terminus is blocked when the Edman degradation fails. The example of the *n*-tetradecanoyl amide also illustrates the point that the elucidation of the structure of a novel covalent modification that

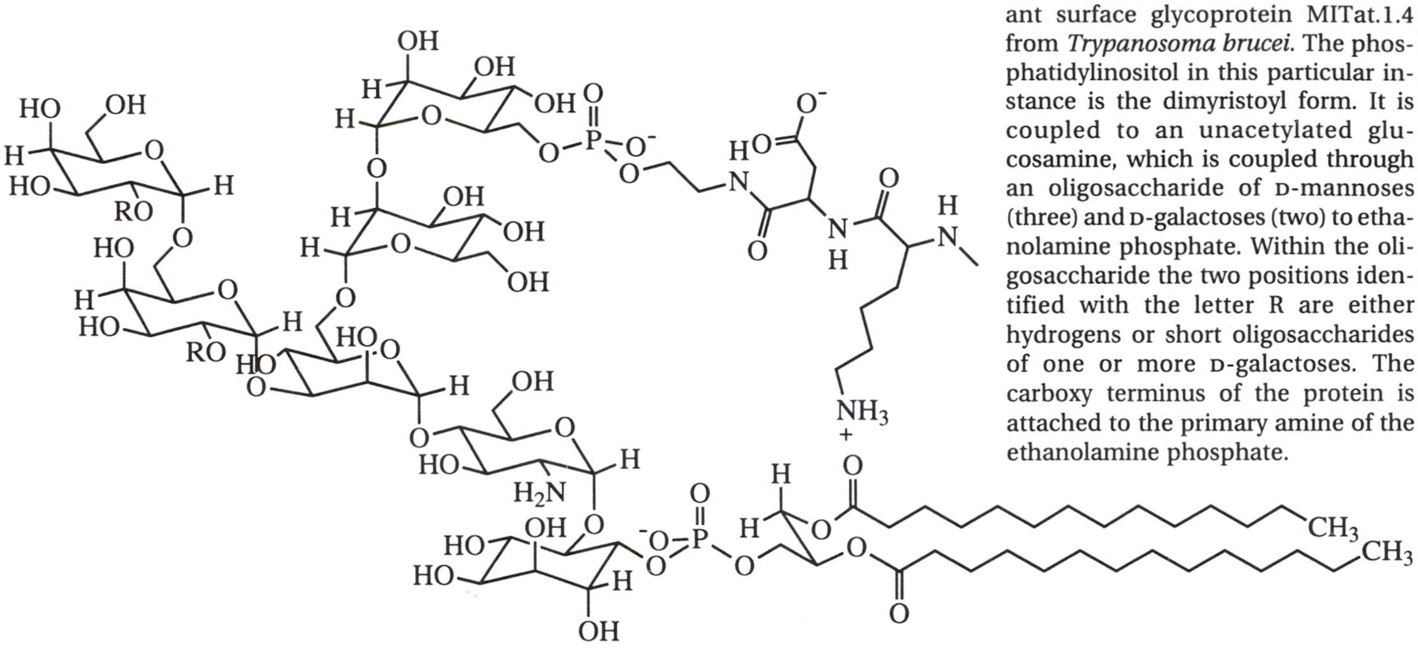

Figure 3–19: Structure of the linkage between phosphatidylinositol and the carboxy terminus of a polypeptide in a phosphatidylinositol- (PI-) linked protein.[86] The structure shown is that for the linkage at the end of the variant surface glycoprotein MITat.1.4 from *Trypanosoma brucei*. The phosphatidylinositol in this particular instance is the dimyristoyl form. It is coupled to an unacetylated glucosamine, which is coupled through an oligosaccharide of D-mannoses (three) and D-galactoses (two) to ethanolamine phosphate. Within the oligosaccharide the two positions identified with the letter R are either hydrogens or short oligosaccharides of one or more D-galactoses. The carboxy terminus of the protein is attached to the primary amine of the ethanolamine phosphate.

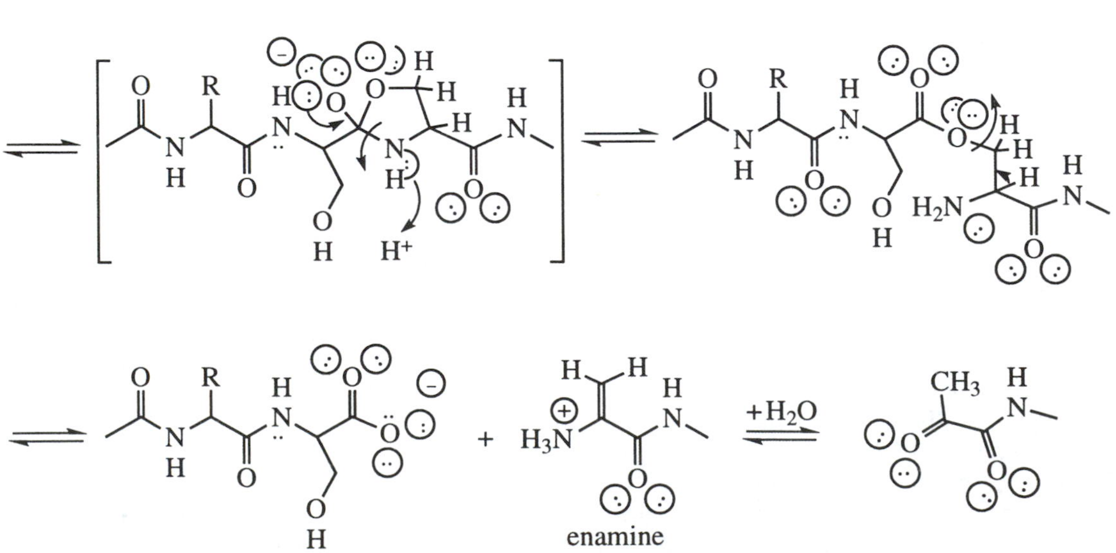

Figure 3–20: Formation of an N^{α}-pyruvyl amino terminus from a serine in the middle of a polypeptide.

has been discovered on a polypeptide is a problem in analytical organic chemistry.

The carboxy terminus of a polypeptide can also be modified, for example, as the primary amide (Figure 3–18). A very unusual posttranslational modification of the carboxy terminus of a number of proteins involved in regulatory functions has recently been observed.[80] The polypeptide as it is synthesized from the messenger RNA[81] has the carboxy-terminal sequence -Cys-X-X-X, where X represents one of several possible amino acids. During the posttranslational modification of the polypeptide, the last three amino acids are removed and the cysteine is doubly modified (Figure 3–18). A farnesyl group or a geranylgeranyl group[82] is added to the sulfur of the cysteine in an allylthioether, and the new carboxy terminus is methylated to form the methyl ester.[83,84] It is thought that this makes the carboxy terminus sufficiently hydrophobic to bind to biological membranes.[80] A similar posttranslational modification, but at the amino terminus, has been observed in a murein lipoprotein from bacterial outer membrane in which an amino-terminal cysteine is acylated by a fatty acid at the α-amino group and the sulfur forms a thioether with carbon 3 of a 1,2-diacylglycerol, the acyl groups of which are also fatty acids.[85]

An extensive posttranslational modification of the carboxy terminus occurs in certain proteins that are bound tightly to the extracellular surface of protozoal and animal cells. It has the effect of connecting covalently the carboxy terminus of the protein to a phospholipid, phosphatidylinositol, within the bilayer of the plasma membrane. The carboxy terminus is linked directly through an amide to an ethanolamine, which is linked through a phosphate diester to the mannose of an oligosaccharide, which, in turn, is linked by a glycosidic linkage to phosphatidylinositol, a phospholipid (Figure 3–19).[86]

An interesting example in which internal proteolytic cleavage is followed by covalent modification of the newly produced amino terminus is offered by the posttranslational modification of the protein histidine decarboxylase.[87] Following its synthesis, this protein is cleaved internally, presumably in an intramolecularly catalyzed reaction, at the amide between two adjacent serine residues in its sequence. During the cleavage, the serine to the carboxy-terminal side of the cleaved amide is converted to a pyruvate that remains acylated to the α-amino group of the next amino acid in the sequence (Figure 3–20). During the reaction, the oxygen atom on the original serine residue is transferred to the new carboxy terminus,[88] an observation consistent with the mechanism presented in the figure. An α-ketobutyryl group[89] is found as an acyl substituent at the amino terminus of one of the two polypeptides composing threonine dehydratase, and it presumably arises by a similar mechanism from a threonine in the protein.

In addition to the amino acid residues at the amino and carboxy termini, amino acid residues in the interior of a polypeptide are often modified on their side chains. When this occurs, the derivative remains an L-α-amino acid residue because its carboxyl group and its α-amino group are protected by the amides of the backbone. It is often possible to hydrolyze the polypeptide, liberate the modified amino acid, and demonstrate its peculiarity by its unusual mobility upon cation-exchange chromatography (Figure 1–4). Usually the hydrolysis is performed enzymatically to avoid destruction of the modified amino acid that might occur in strong acid or strong base. Once its presence is recognized, enough of the peculiar amino acid must be purified to perform a proof of its structure by chemical analysis.

Recently, electrospray mass spectroscopy[90] has been applied to the analysis of posttranslational modification. In this procedure, an entire protein molecule or a peptide can be vaporized from a small droplet in one of its positively charged protonation states and its mass can be precisely determined. In this way, it was shown that the amino-terminal tryptic peptide from the β polypeptide of the A_{1b} variant of hemoglobin differed from the same peptide from hemoglobin A by the mass of one pyruvyl group, presumably attached as the ketimine to its amino terminus.[91] This technique was also used to show that each of the three repeating segments of the dihydrolipoyl acetyltransferase of pyruvate dehydrogenase had a molecular mass (9764, 10,149, and 10,210 Da, respectively) equal to that calculated form its known amino acid sequence plus the mass of a covalently attached dihydrolipoyl group (188 Da).

There are many posttranslationally modified amino acids that have been isolated from naturally occurring polypeptides (Table 3–1, Figure 3–21).[69] The length of Table 3–1 gives the erroneous impression that posttranslational modifications are common. Aside from glycosylation and phosphorylation, the incidence of any of these modifications is quite limited, often being confined to only one protein or one small family of proteins. For example, two of the earliest recognized posttranslational modifications were the 5-hydroxylysine and 4-hydroxyproline (Figure 3–21) that are formed in the posttranslational **monooxygenation** of the protein collagen. A tyrosine residue in the enzyme amine oxidase (copper-containing) is monooxygenated twice to form 3,6-dihydroxytyrosine.[146] The *p*-quinone of this post-translationally modified amino acid is believed to function as an electrophile by forming an imine with the amine that is oxidized in the enzymatic reaction. Many amino acid side chains become **methylated** post-translationally. Examples are N,N,N-trimethyllysine, N^ω,N^ω-dimethylarginine, and *O*-methylaspartate (Figure 3–21). *O*-Methylaspartates have turned out to be, at least in part, the modified side chains of D-aspartates in the polypeptide. This has led to speculation that this methylation may be involved in the repair of locations in the polypeptide in which an L-aspartate residue has nonenzymatically and

Table 3–1: Posttranslational Modifications of Amino Acid Residues in Proteins[69]

type of modification	derivative of amino acid residue found[a]
phosphorylation	O-phosphonoserine,[92–94] O-phosphonothreonine,[93–95] O-phosphonotyrosine,[94,96,97] O-phosphonoglutamate[98] N-phosphonolysine,[99] N-phosphonohistidine,[99–101] N-phosphonoarginine,[101] S-phosphonocysteine[102]
sulfation	O-sulfotyrosine[103]
carboxylation	4-carboxyglutamate,[104,105] 3-carboxyaspartate[106]
aromatic substitution	3,5-diiodotyrosine,[107–109] 3-iodotyrosine,[107,109,110] 3,5-dibromotyrosine,[111] 3-bromotyrosine,[110] 3-bromo-5-chlorotyrosine,[112] 3-chlorotyrosine,[107] 3,5-dichlorotyrosine,[107] 4-iodohistidine,[109] O-(3,5-diiodo-4-hydroxyphenyl)-3,5-diiodotyrosine (thyroxine, T_4),[113–115] O-(3-iodo-4-hydroxyphenyl)-3,5-diiodotyrosine (triiodothyronine, T_3),[116,117] 2-[3-carboxamido-3-(trimethylammonio)propyl]histidine (diphthamide)[118,119]
methylation	N-methyllysine,[120–122] N,N-dimethyllysine,[121–123] N,N,N-trimethyllysine,[121,122,124] N^ω-methylarginine,[125,126] $N^\omega,N^{\omega'}$–dimethylarginine,[126,127] N^ω,N^ω-dimethylarginine,[126] N^3-methylhistidine,[126,128] O–methylisoaspartate[129] S-methylmethionine,[130] N-methylasparagine,[131,132] N-methylglutamine[133]
acylation	N-acetyllysine,[134,135] O-palmitoylthreonine,[136] S-palmitoylcysteine[137–139]
monooxygenation	5-hydroxylysine,[140] N,N,N-trimethyl-5-hydroxylysine,[141] N,N,N-trimethyl-O-phosphono-5-hydroxylysine,[141] 4-hydroxyproline,[142–144] 3-hydroxyproline,[145] 2,5-dihydroxytyrosine[146]
oxidation	6-deamino-6-oxolysine (allysine),[140,147] 6-deamino-5-hydroxy-6-oxolysine[140]
ADP-ribosylation	N^ω-(ADP-ribosyl)arginine,[148,149] N-(ADP-ribosyl)asparagine,[150] N-(ADP-ribosyl)lysine[151] O-(ADP-ribosyl)-glutamate,[152] 1-[N-(ADP-ribosyl)]-2-[3-carboxamido-3-(trimethylammonio)propyl]histidine[153,154]
nucleotidylation	O-(5′-adenylyl)tyrosine,[155] O-(5′-uridylyl)tyrosine[156]
hydrolysis	citrulline from arginine,[157] ornithine from arginine,[158] aspartate from asparagine[159,160] glutamate from glutamine[160]
glycosylation	O-(polymannosyl)serine,[161] O-(polymannosyl)threonine,[161] O-[oligo(α1,2)galactosyl]serine,[162] O-[3-O-(β-glucosyl)-α-fucosyl]threonine,[163] O-[2-O-(α-glucosyl)-β-galactosyl]-5-hydroxylysine,[164] O-(β-xylosyl)serine,[165,166] O-[4-O-(β-galactosyl)-β-xylosyl]serine[165,166] S-digalactosylcysteine,[167] S-triglucosylcysteine,[168] O-(glucosylarabinosyl)hydroxyproline,[169] O-(N-acetylglucosaminyl)serine[170]
cross-links between side chains of two amino acids	lysine in amide linkage with aspartate,[171] lysine in amide linkage with glutamate,[172,173] cysteine in thioether to histidine,[174] cysteine in thioether to tyrosine,[175] cystine
covalently bound coenzymes	heme in thieoether linkages to two cysteines,[176] S-phycoerythrobilinylcysteine,[177] 8a-(S-cysteinyl)-8a-hydroxyflavin adenine dinucleotide,[178] 8a-S-cysteinylflavin mononucleotide,[179] N-(8a-flavinyl)histidine,[179] 8a-(O-tyrosyl)flavin adenine dinucleotide,[180] N-biotinyllysine,[181] N-lipoyllysine,[182] N-retinyllysine,[183] O-(4′-phosphonopantetheinyl)serine[184]

[a]Only posttranslational modifications of the side chains of amino acids that are neither the amino terminus nor the carboxy terminus of a polypeptide in a protein are tabulated.

unfortunately been converted into a D-aspartate residue.[185] Fatty acids have also been found to be attached, in at least one well-documented case,[136] to the hydroxyls of threonine residues (Figure 3–21). 4-Carboxyglutamate (Figure 3–21),[186,187] a posttranslational derivative of a glutamate residue,[188] is found in some of the proteins that bind calcium strongly or that are involved in calcium metabolism.[189] Tyrosine residues can be **halogenated**,[107] for example, with iodine, by electrophilic aromatic substitution at the positions ortho to the hydroxyl (Figure 3–21). The side chains of threonine, serine, and tyrosine[190] are sometimes **phosphorylated** on their oxygens (Figure 3–21). A tyrosine residue can also be adenylated through a phosphodiester to its phenolic oxygen (Figure 3–21).[155] As with phosphorylation, this modification regulates an enzymatic reaction. In another instance, an adenylate is added to an arginine residue through an N-glycosidic linkage to a ribose in a process referred to as ADP-ribosylation (Figure 3–21). There also are two rather strange posttranslational modifications, those producing thyroxine[113–115] and diphthamide,[118,119] that seem to be

the result of electrophilic aromatic substitution.

One way in which two or more of the amino acid residues in a polypeptide can be modified coincidentally is during the formation of a **covalent cross-link** between them or among them. The cross-link can be intramolecular, connecting two or more amino acid residues in the same polypeptide, or intermolecular, connecting two or more amino acid residues in different polypeptides. There is no practical distinction between these two outcomes because the linkage is invariably made after the polypeptides have folded into their native structure and, subsequently, formed specific intermolecular complexes among themselves. This folding and intermolecular assembly is what brings the two or more amino acid side chains that will be cross-linked into atomic contact with each other. Therefore, it is irrelevant whether the amino acid residues started out on the same polypeptide or different polypeptides or whether they are at positions within the amino acid sequence of the same polypeptide that are close to or distant from each other. The only deciding factor is that they are immediately adjacent to each other

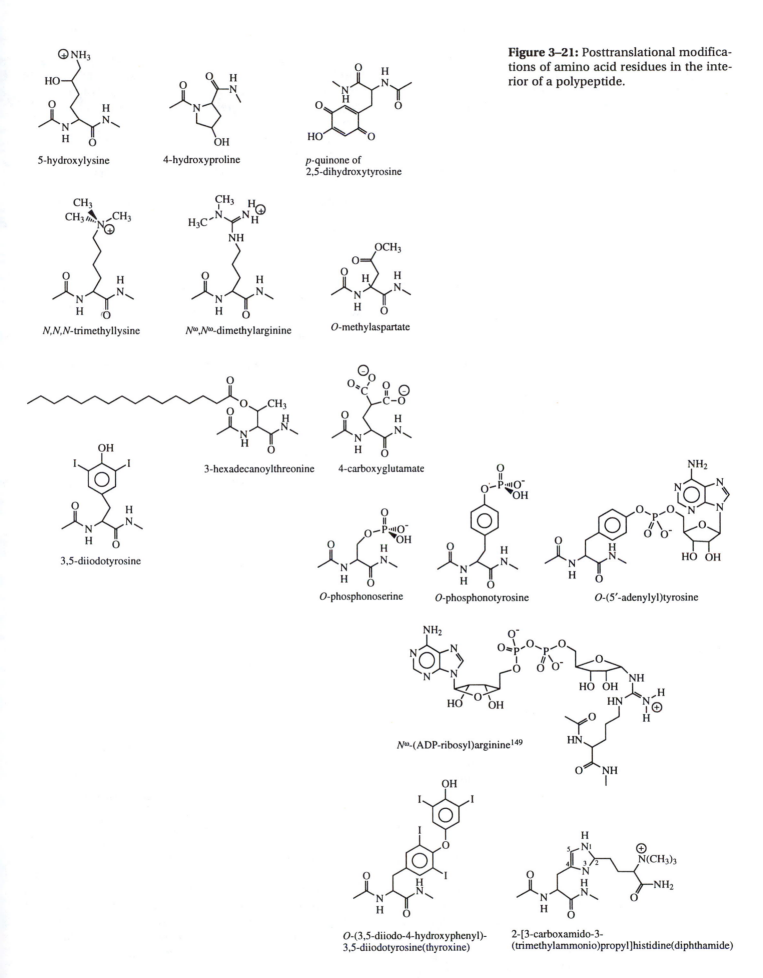

Figure 3–21: Posttranslational modifications of amino acid residues in the interior of a polypeptide.

5-hydroxylysine

4-hydroxyproline

p-quinone of 2,5-dihydroxytyrosine

N,N,N-trimethyllysine

*N*ω,*N*ω-dimethylarginine

O-methylaspartate

3-hexadecanoylthreonine

4-carboxyglutamate

3,5-diiodotyrosine

O-phosphonoserine

O-phosphonotyrosine

O-(5′-adenylyl)tyrosine

*N*ω-(ADP-ribosyl)arginine[149]

O-(3,5-diiodo-4-hydroxyphenyl)-3,5-diiodotyrosine(thyroxine)

2-[3-carboxamido-3-(trimethylammonio)propyl]histidine(diphthamide)

in the final structure of the mature protein. A simple example of a covalent cross-link is an amide between a lysine residue and a glutamate residue. Such a cross-link is formed from a glutamine residue and a lysine residue, both within a folded polypeptide, by the enzyme protein-glutamine γ-glutamyltransferase.[191]

Thioethers between a cysteine residue and the aromatic rings of either a histidine residue (3–6)[174] or a tyrosine residue (3–7)[175]

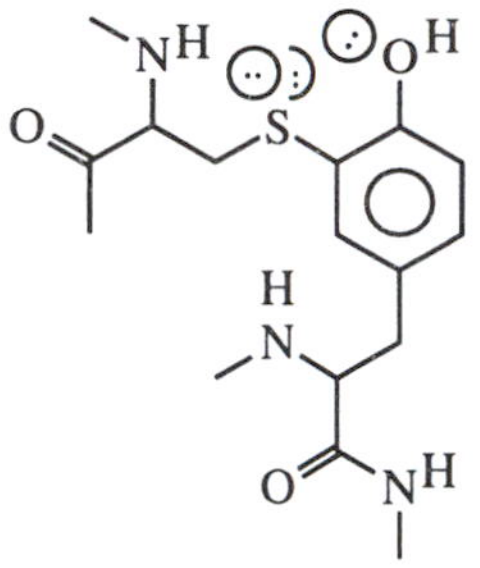

3–6 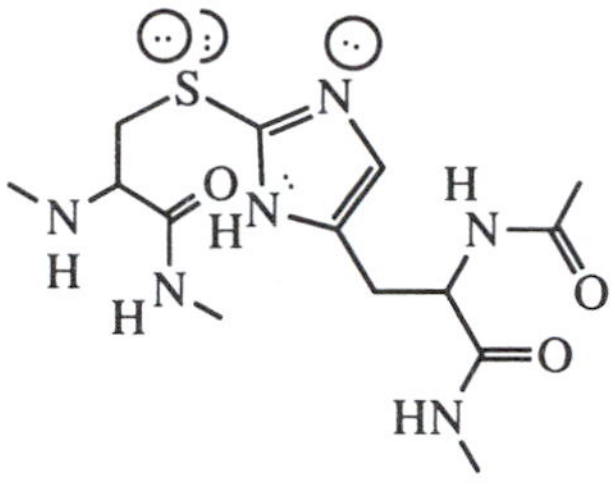   **3–7**

have been identified in the copper monooxygenases, tyrosinase and galactose oxidase, respectively. These cross-linked amino acid residues would arise from electrophilic aromatic substitution. They are thought to be able to exist as stable neutral radicals complexed with cupric cation and thereby form a species capable of participating in the two-electron transfers performed by these enzymes

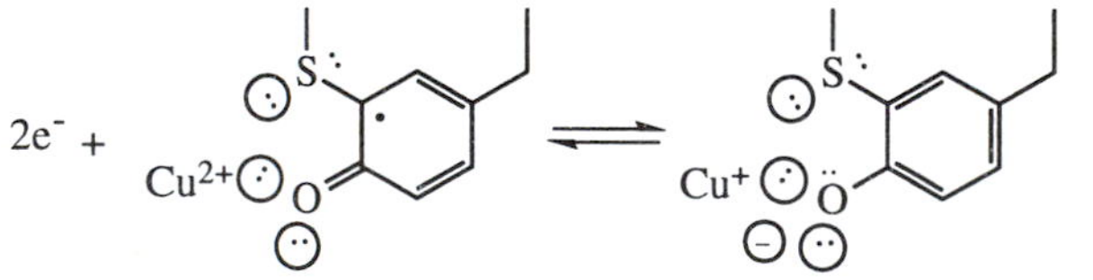

(3–3)

A large number of cross-links form in both collagen and elastin as the direct result of the posttranslational oxidation of lysine residues in these proteins to the corresponding aliphatic aldehydes in a reaction performed by the enzyme protein-lysine 6-oxidase.[192] These aldehydes are formed in the vicinity of each other as well as in the vicinity of 5-hydroxylysines, lysines, and hydroxyaldehydes, all also derived from lysine residues. A dazzling array of aliphatic carbonyl chemistry is initiated by the formation of the reactive aldehydes in this environment, including aldol condensations, imine formations, dehydrations of β-hydroxyaldehydes, dehydrations of aliphatic alcohols, Michael additions, and oxidations (Figure 3–22). Only four out of the more than 25 cross-links that result from this uncontrolled flurry of reactions[140] are displayed in the figure.

One of the most common covalent cross-links is the **disulfide** that forms when two cysteine residues are oxidatively coupled to form one cystine residue (Figure 3–23). While cysteine is unstable under the conditions necessary to hydrolyze proteins in strong acid, cystine is stable, and its appearance on an ion-exchange chromatogram between alanine and valine (Figure 1–4) establishes the presence of a cystine residues in a protein. The ultimate oxidant in the production of a cystine residue from two cysteine residues is usually molecular oxygen.

The interior of most cells has a high concentration of a small, free thiol, such as glutathione, that reduces back to cysteine any cystine that forms in the cytoplasm, in a reaction known as **disulfide interchange** (Figure 3–23). The net effect of disulfide interchange is to set the cystine residues in a protein in equilibrium with the disulfide of the small thiol such that

$$K_{eq} = \frac{[RS\text{–}SR][Prot(SH)_2]}{[RSH]^2\,[Prot(S\text{–}S)]} \qquad (3\text{–}4)$$

where RSH is the small thiol and RS–SR its disulfide. This expression can be rearranged

$$\frac{K_{eq}[RSH]^2}{[RS\text{–}SR]} = \frac{[Prot(SH)_2]}{[Prot(S\text{–}S)]} \qquad (3\text{–}5)$$

where $[Prot(SH)_2]$ is the molar concentration of the reduced pair of adjacent cysteine residues in the folded polypeptide and $[Prot(S\text{–}S)]$ is the concentration of cystine residues, also in the folded polypeptide. The reaction as written is first-order in reduced protein because the oxidation of the reduced protein is usually intramolecular, but $[Prot(SH)_2]$ could be replaced with $[ProtSH]^2$ in intermolecular or formally intermolecular instances. The point made is the same, namely, that the greater the ratio $[RSH]^2[RS\text{–}SR]^{-1}$, the less cystine will be found in proteins.

In the cytoplasm, [RSSR] is kept at a low level enzymatically. For example, when RSH is glutathione, the enzyme glutathione reductase accomplishes this

$$H^+ + GSSG + NADPH \underset{\substack{\text{glutathione}\\\text{reductase}}}{\rightleftharpoons} 2GSH + NADP^+ \qquad (3\text{–}6)$$

The result of all of these considerations is that proteins confined to the cytoplasm do not contain cystine, while proteins removed from the cytoplasm often do contain cystine.

A protein is usually prepared for sequencing by reducing its cystine residues with a small thiol such as 2-mercaptoethanol and then alkylating all of the resulting cysteines with a reagent such as iodoacetic acid to prevent their reoxidation (Figure 3–23). This creates the stable amino acid residue S-(carboxymethyl)cysteine in place of the unstable cysteine. In such experimental situations where cystines must be reduced, a problem arises. If the protein remains folded, its cystine residues are generally more stable, relative to the adjacent cysteine residues that would be produced upon reduction, than is the disulfide of a small reducing agent such as 2-

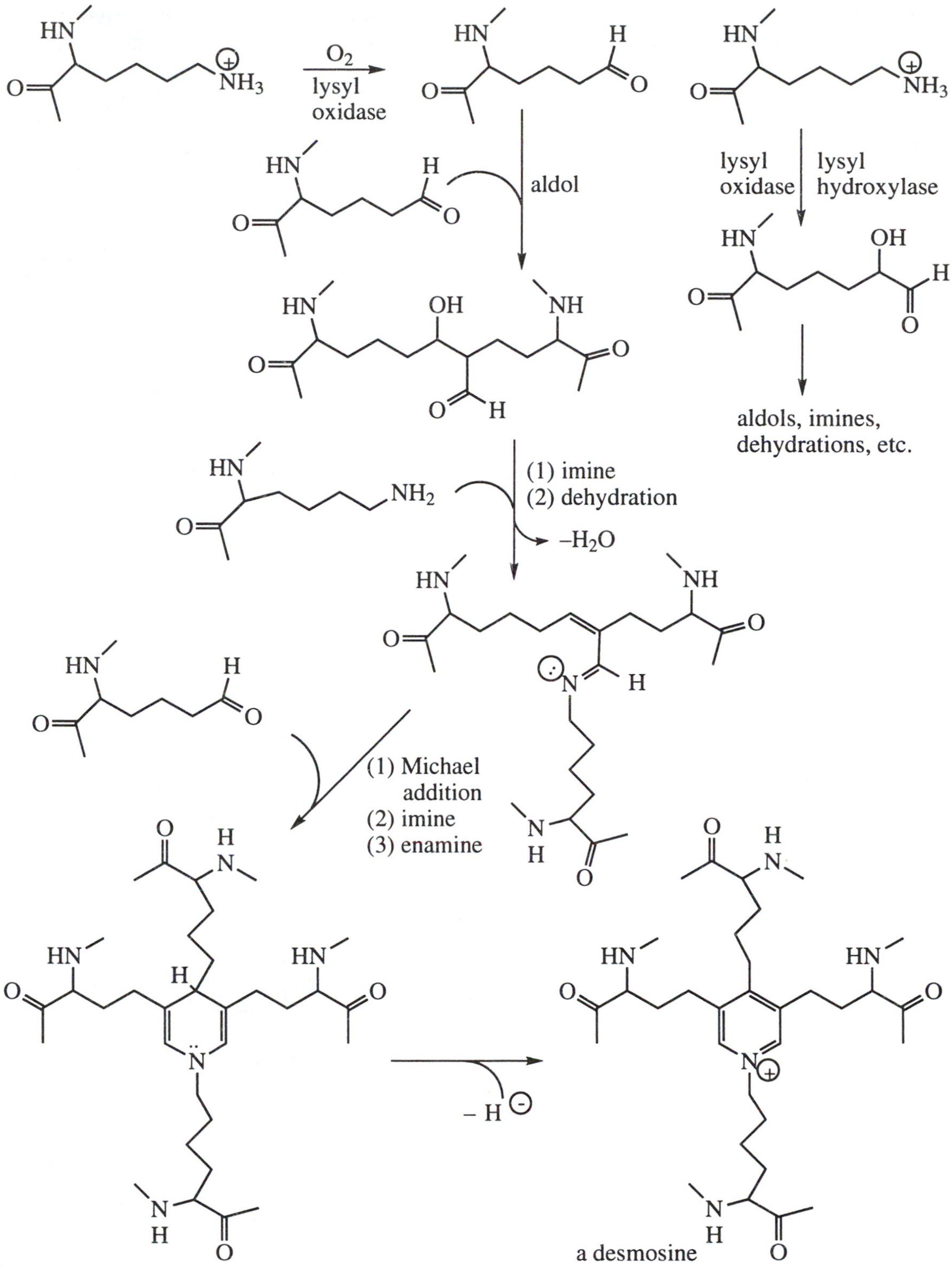

Figure 3–22: Examples of the formation of four of the more than 25 cross-links initiated by the formation of 6-deamino-6-oxolysine in collagen by protein-lysine 6-oxidase (lysyl oxidase). Shown is an aldol condensation to produce the first cross-link. The β-hydroxyaldehyde can then form an imine with a lysine that dehydrates to an α,β-unsaturated imine, a product that cross-links three amino acids. The enol of another aldehyde can add to the α,β-unsaturated imine, and the initial enamine can condense to an imine with the carbonyl of the aldehyde. This forms a dihydropyridine that cross-links four amino acid residues. The pyridinium cation formed upon oxidation of the dihydropyridine is a desmosine linking the four amino acid residues. Upper right corner: The 6-deamino-5-hydroxy-6-oxolysine formed by the combined action of lysyl oxidase and protein-lysine 5-monooxygenase (lysyl hydroxylase) produces an α-hydroxyaldehyde, which is susceptible to an even more complicated set of modifications.

mercaptoethanol, relative to its thiol form. This results from the fact that the reaction in the protein is intramolecular in the direction of oxidation but the reaction of the 2-mercaptoethanol is intermolecular in the direction of oxidation (Equation 3–4). The problem is solved either by using a dithiol such as 2,3-dihydroxy-1,4-dithiobutane

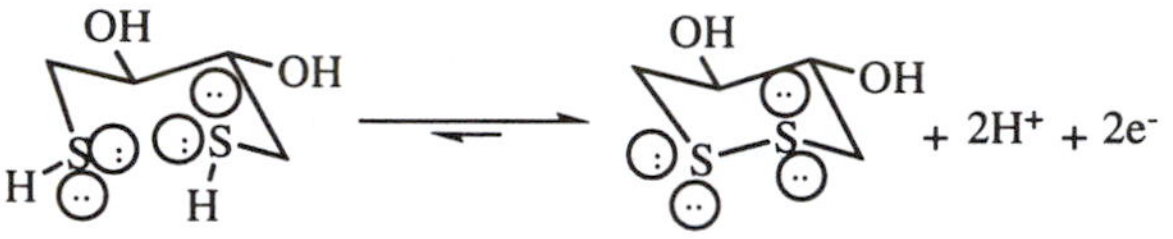

$$+ 2H^+ + 2e^- \quad (3\text{--}7)$$

which causes the reaction of the reducing reagent to be intramolecular and produces a very stable disulfide as a product or by unfolding the protein with urea or guanidinium chloride, which renders the reaction of the cysteine residues in the protein formally intermolecular in the direction of oxidation.

In extracytoplasmic proteins that do contain cystine residues as a posttranslational modification, the two cysteine residues that are connected to each other are almost always[193] distant from each other in the amino acid sequence. For example, in ribonuclease, a protein formed from a polypeptide 124 amino acids long, the cystines are formed from Cysteine 96 and Cysteine 40, Cysteine 58 and Cysteine 110, Cysteine 26 and Cysteine 84, and Cysteine 65 and Cysteine 72. When the polypeptide folds to form the native structure of the protein, cysteine residues that were distant from each other in the unfolded polypeptide are juxtaposed, and they are oxidized to cystine

$$R_1CH_2SH + HSH_2CR_2 + \tfrac{1}{2}O_2 \rightleftharpoons R_1CH_2SSH_2CR_2 + H_2O \quad (3\text{--}8)$$

The identification of the two cysteine residues that are connected in a particular cystine in a native protein requires the isolation of a peptide containing only those

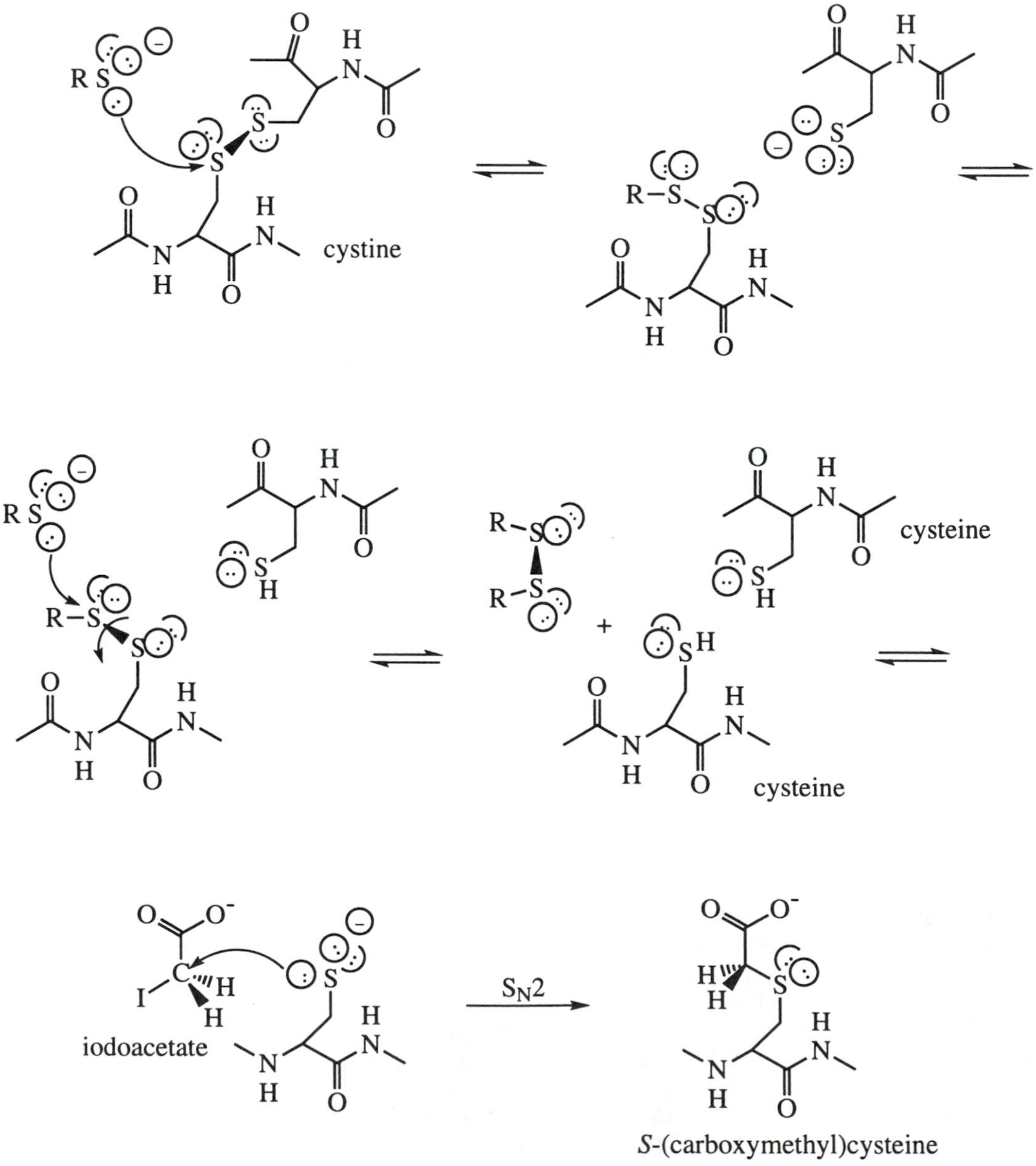

Figure 3–23: Reduction of a cystine residue by disulfide interchange and carboxymethylation of the resulting cysteine. The cystine connecting two segments of polypeptide is exposed to an external thiolate (RS⁻) that displaces the cysteinyl anion by nucleophilic substitution and is in turn removed by another thiolate. The thiolate anions of the released cysteine residues react in a bimolecular nucleophilic substitution with iodoacetate.

two cysteines still joined as the cystine.[194] Such peptides with intact cystine residues are almost always two shorter peptides held together by the cystine itself because the two cysteines participating in a cystine are almost always distant from each other in the sequence while the peptide containing the cystine residue must be small enough to be isolated chromatographically or electrophoretically. For example, one of the peptides from ribonuclease isolated from a digest of the protein performed with pepsin, trypsin, and chymotrypsin was composed of the two smaller peptides NGQTNCYH and NVACK, covalently coupled by a cystine between the two respective cysteine residues.[194] From this result it could be concluded that in native ribonuclease, Cysteine 65 is coupled by a disulfide to Cysteine 72. A sensitive method has been described for monitoring electrochemically chromatograms of such proteolytic digests to detect peptides containing intact cystine residues.[193]

An ingenious method for identifying the pairs of cysteine residues coupled as cystine was developed by Brown and Hartley[195] and is classified as a diagonal method. A digest of the unreduced protein is separated by electrophoresis along one dimension. This array of peptides, usually each unresolved from its neighbors, is exposed to performic acid vapors that cleave all of the cystines by oxidation

$$RSSR + 5HCO_3H + H_2O \rightarrow 2H^+ + 2RSO_3^- + 5HCO_2H \quad (3-9)$$

The linear array of peptides is then submitted to electrophoresis in a second dimension perpendicular to the first. All peptides lacking disulfides and unaffected by the performic acid move the same distance that they did before and end upon the diagonal of the two dimensions. Each peptide containing a disulfide and split by the oxidation yields two new components that end up off the diagonal because they now have different electrophoretic mobilities than they had originally when they were attached to each other. The connections that existed before the oxidation can be identified because the two products from the larger peptide had the same mobility in the first dimension and remain aligned with each other during the second dimension.

The logic underlying the diagonal method has wider applications. If any mixture is separated in one dimension, the array of components is exposed to a selective chemical reaction, and the array is then separated in a second dimension identical to the first but orthogonal to it, any components of the original mixture altered by that chemical reaction will be readily isolated as components that run off of the diagonal. For example, such a strategy has been used to isolate selectively peptides containing methionine by diagonal column chromatography.[196,197]

Suggested Reading

Carr, S.A., Biemann, K., Shoji, S., Parmelee, D.C., & Titani, K. (1982) *n*-Tetradecanoyl is the NH$_2$-terminal blocking group of the catalytic subunit of cyclic AMP-dependent protein kinase from bovine cardiac muscle, *Proc. Natl. Acad. Sci. U.S.A. 79*, 6128–6131.

Brown, J.R., & Hartley, B.S. (1966) Location of disulphide bridges by diagonal paper electrophoresis, *Biochem. J. 101*, 214–228.

Problem 3–9: Draw the structure of a polypeptide with an amino-terminal cysteine residue the α-amine of which is acylated with palmitate and the sulfur of which forms a thioether with C3 of a 1,2-dipalmitoyl-3-deoxyglycerol.

Problem 3–10: A remarkable feature of the enzyme glutamate–ammonia ligase from *Escherichia coli* is that its catalytic properties depend on the conditions of growth of the *E. coli* from which it is purified. Thus the enzyme purified from *E. coli* grown on NH$_4$Cl and glucose (Type I) is less sensitive to inhibition by AMP than is the enzyme purified from *E. coli* grown on glutamate and glycerol (Type II). The Type II enzyme can be converted into Type I enzyme if it is treated with snake venom phosphodiesterase.

When Type II enzyme was digested with snake venom phosphodiesterase and subsequently precipitated out of solution with trichloroacetic acid, the supernatant solution contained material having a maximum absorbance at 260 nm.

Hydrolysis of the Type I and Type II enzymes was performed by a series of proteolytic enzymes. Both enzymes (Type I and Type II) were split into the same number of peptides, but one decapeptide from the Type II enzyme differed in chromatographic behavior from the similar decapeptide from the Type I enzyme. The single different peptide isolated from the Type II enzyme had the following composition after acid hydrolysis:

Asp	2
Glu	2
Pro	3
Gly	1
Leu	1
Tyr	1
adenine	1
D-ribose	1
phosphate	1

This peptide was not the amino-terminal peptide of the polypeptide chain. From an acid–base titration, the following pK_a values were deduced for the decapeptide isolated from Type II enzyme before and after treatment with snake venom phophodiesterase:

	pK_a value	
number of groups	before	after
1	3.4	3.4
1	3.7	3.7
4	4–4.5	4–4.5
1	7.8	7.8
1		6.0
1		10.0

How does the covalent structure of the Type II enzyme differ from that of the Type I enzyme? Explain each result described above on the basis of the proposed structure.

Problem 3–11: Write a step-by-step series of reactions that show how the cross-link dehydromerohistidine would form in collagen from three lysine residues and a histidine residue.

Oligosaccharides of Glycoproteins

Living organisms are formed from three types of covalent polymers: proteins, nucleic acids, and polysaccharides. Polysaccharides occur biologically as unmixed polymers such as agarose (Figure 1–7), cellulose (Figure 1–3), hyaluronic acid, chitin, glycogen, and starch, but polysaccharides are also attached to infant polypeptides as posttranslational modifications, usually of serines, threonines, or asparagines. Such posttranslational modifications produce **glycoproteins**. To define the complete covalent structure of a glycoprotein, not only must the amino acid sequence of the protein be established, but the monosaccharide sequence of the attached polysaccharides must be as well.

The polysaccharides found as posttranslational modifications of proteins are referred to as **oligosaccharides**. Some of the rarely occurring oligosaccharides[163] and their sites of attachment have been listed along with the other posttranslational modifications in Table 3–1. There

is also a small group of glycoproteins that are posttranslationally modified at their carboxy termini with covalently attached glycophospholipids.[80] The most commonly encountered oligosaccharides in glycoproteins from animals, however, are branched polymers attached through *N*-acetylglucosamine to asparagine residues or through *N*-acetylgalactosamine to serine and threonine residues (Figures 3–24, 3–25, and 3–26).

The monomers from which these polymers are constructed are monosaccharides. The seven major monosaccharides that are found in the oligosaccharides of animal glycoproteins are D-glucose, *N*-acetyl-D-glucosamine, D-mannose, D-galactose, *N*-acetyl-D-galactosamine, L-fucose, and the sialic acids (Figure 3–27). Unlike the situation with the other six monosaccharides, a variety of different sialic acids are known (greater than 25) that are derivatives of the same compound, neuraminic acid, which is modified variously by acylation. Several other monosaccharides, such as D-xylose and D-glucosamine, are found occasionally in oligosaccharides.

The covalent bonds that link the monosaccharides together are acetals, and occasionally ketals. These acetals and ketals are referred to as **glycosidic linkages**. Glycosidic linkages are formed between the only carbonyl carbon on each monosaccharide, enclosed within a pyranose ring as a hemiacetal, and one of the hydroxyls of the preceding monosaccharide in the polymer. A glycosidic linkage is formed between the oxonium cation of the pyranose and a lone pair of electrons on a nitrogen or an oxygen (see 3–10 at bottom of page).

Branching of the polymer occurs whenever two or more of the hydroxyls on one of the monosaccharides participate in glycosidic linkages. Each of the oligosaccharides on a glycoprotein can be thought of as beginning at the monosaccharide that is attached to the polypeptide. The point of attachment is either an *O*-glycosidic linkage formed between the carbonyl carbon of this initial monosaccharide and the hydroxyl of a serine or threonine residue or an *N*-glycosidic linkage formed between the initial monosaccharide and the amide nitrogen of an asparagine residue. The first monosaccharide in an oligosaccharide attached to a serine or a threonine is usually *N*-acetylgalactosamine; the first sugar in an oligosaccharide attached to asparagine is always *N*-

(3–10)

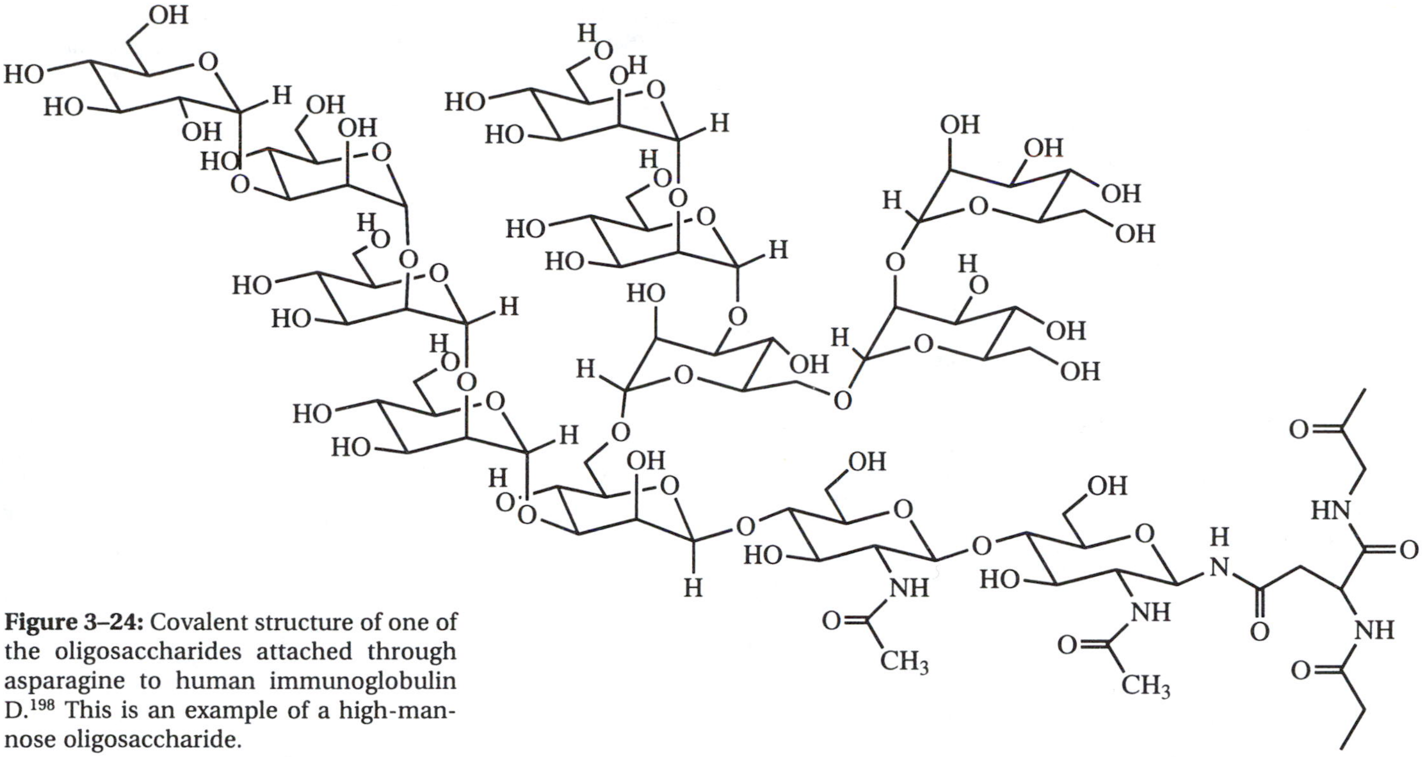

Figure 3–24: Covalent structure of one of the oligosaccharides attached through asparagine to human immunoglobulin D.[198] This is an example of a high-mannose oligosaccharide.

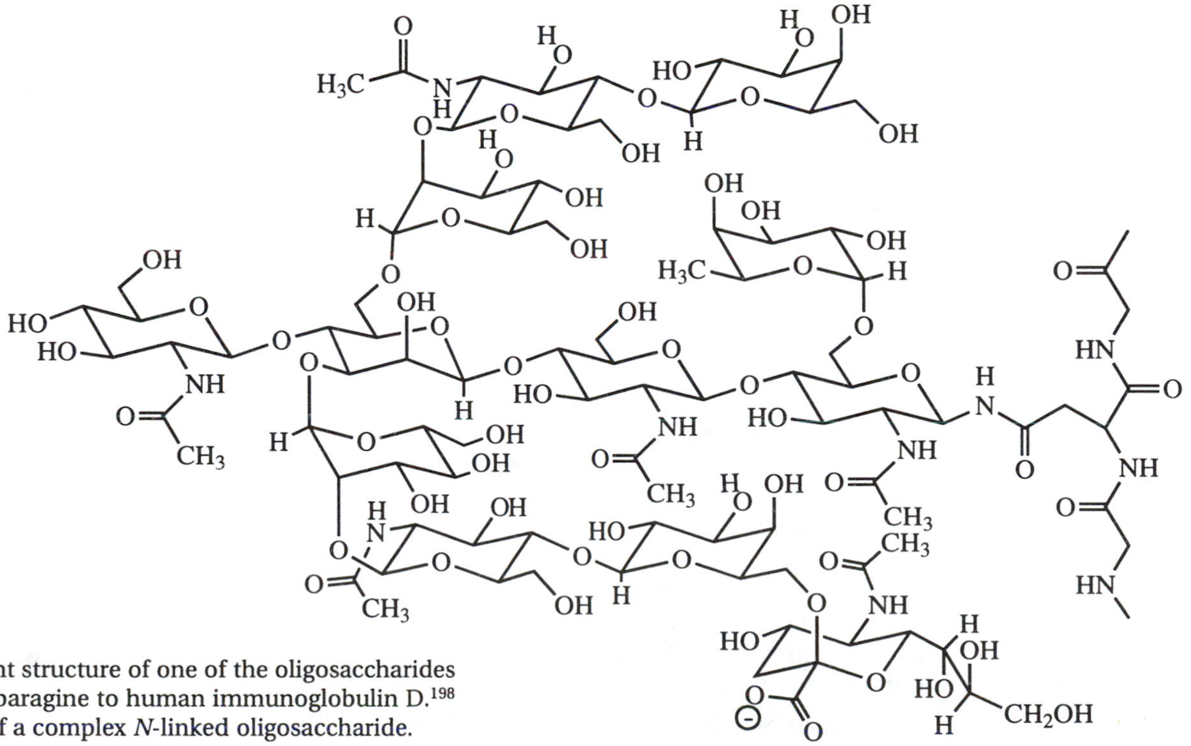

Figure 3–25: Covalent structure of one of the oligosaccharides attached through asparagine to human immunoglobulin D.[198] This is an example of a complex N-linked oligosaccharide.

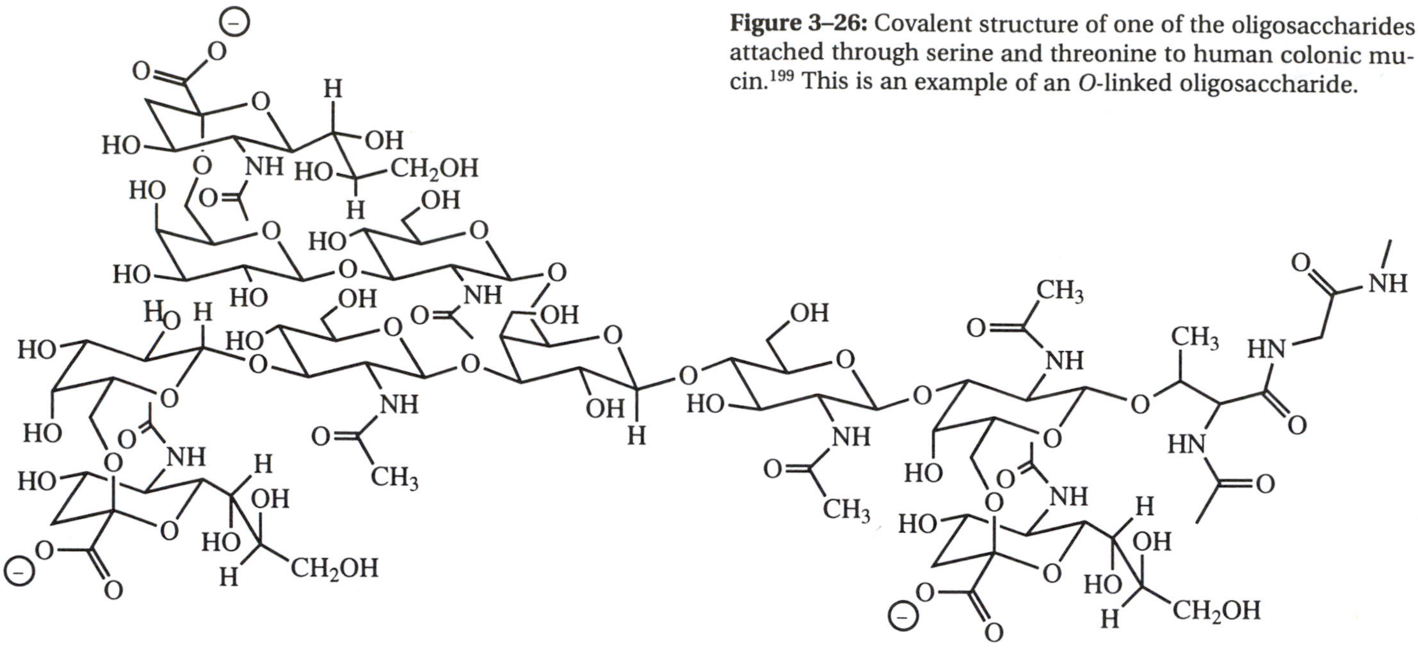

Figure 3–26: Covalent structure of one of the oligosaccharides attached through serine and threonine to human colonic mucin.[199] This is an example of an *O*-linked oligosaccharide.

acetylglucosamine. Peripheral to this initial monomer, the polymer will be found to branch at several points and end at each of several unsubstituted monosaccharides that occupy the last positions on the branches. There are usually 2–8 monosaccharides counting from the initial monosaccharide to the end of a branch and 1–10 branches in a typical oligosaccharide on a glycoprotein. Often a branch has only one monosaccharide.

Writing the sequences of these oligosaccharides is complicated by the requirements that each monosaccharide must be noted, the particular hydroxyl in the preceding monosaccharide to which it is attached through its carbonyl carbon must be noted, and the anomeric state of the carbonyl carbon must be noted. It is usually assumed that unless otherwise stated the monosaccharides in the oligosaccharide are of the D stereochemistry, where, for example, L-fucose would be an exception to be noted, and in the pyranose form, where, for example, galactofuranose would be an exception to be noted. To confuse matters further, the sequences of oligosaccharides are usually written from right to left beginning with the monosaccharide attached directly to the protein. The anomeric state of the carbonyl carbon and the hydroxyl to which it is attached are noted to the right of the name of each monosaccharide. For example, GlcNAc(β1,4) states that an *N*-acetylglucosamine is attached through its carbonyl carbon, carbon 1, to the 4-hydroxyl of its immediate predecessor to the right in the written sequence, by a β-anomeric acetal. In addition, any modifications of the monosaccharides, such as *O*-acetylation, *O*-sulfation, *O*-phosphorylation, de-*N*-acetylation, or *O*-methylation, must be noted. Such derivatives are encountered occasionally.

There is yet a further peculiarity of oligosaccharides, that of **microheterogeneity**. Because serious biological problems usually do not arise when unfinished oligosaccharides are produced, in contrast to the devastation that would occur if unfinished proteins and nucleic acids were produced, natural selection has not enforced uniformity on the synthesis of oligosaccharides. In fact, although some finished glycoproteins are homogeneous, in most instances the synthesis of the oligosaccharides is a rather haphazard affair, and each oligosaccharide that ends up attached at a particular protein is usually unfinished. Each, however, is unfinished in a different way. Each is missing a different set of monosaccharides. As a result, 10 or 15 different oligosaccharides may be found on the amino acid residue at the same position of the amino acid sequence in different molecules of the same protein. Once they have been separately identified and individually sequenced, each of these oligosaccharides can be recognized as a different, incomplete realization of only one complete sequence. This prototypical sequence is often longer than any one of the sequences of the actual oligosaccharides, but each of the sequences of the actual oligosaccharides is a piece of the prototype and every sugar in the prototype is represented in one of the actual oligosaccharides. Whether the prototype including all of the actual sequences is the most complete sequence that could have been produced or is itself only an incomplete realization of a longer sequence can never be decided unequivocally.

Another view of microheterogeneity, in opposition to the view that it is haphazard and purposeless, is that it has a role in producing many different **glycoforms** of the same protein. This would increase the functional range

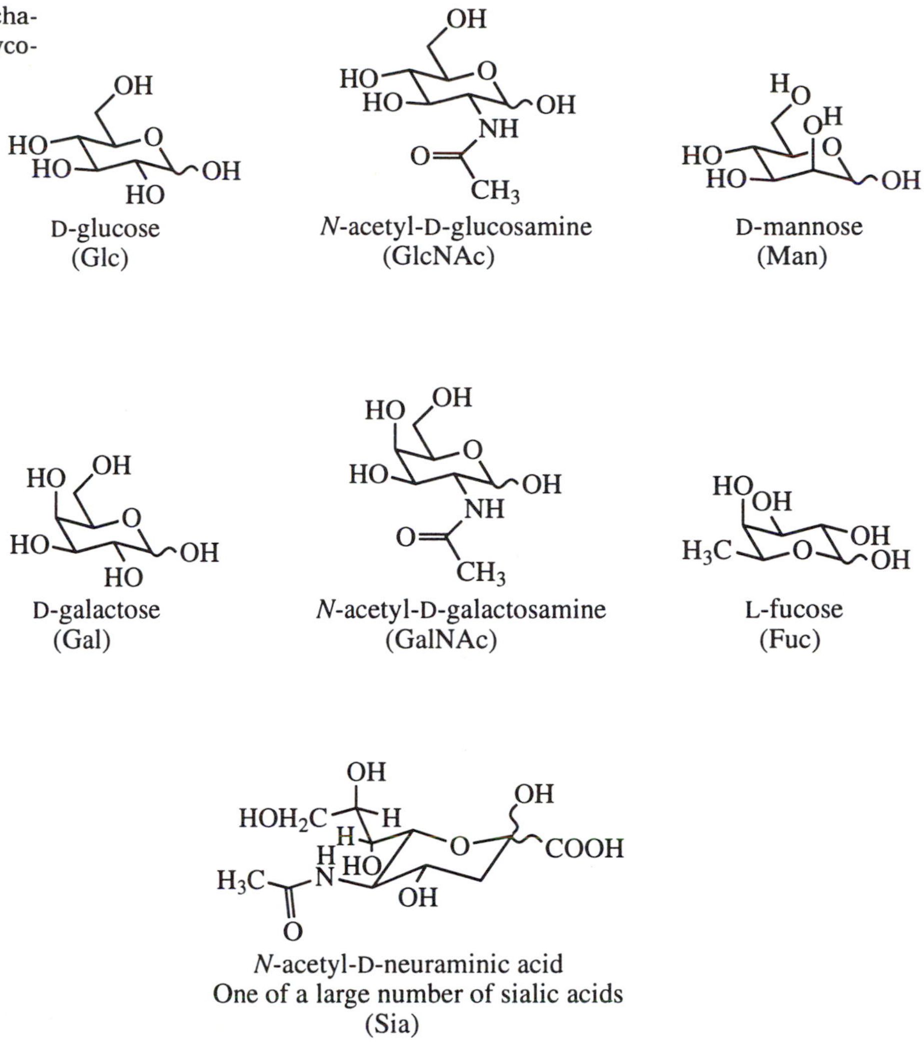

Figure 3–27: Seven most common monosaccharides composing the oligosaccharides of glycoproteins.

D-glucose
(Glc)

N-acetyl-D-glucosamine
(GlcNAc)

D-mannose
(Man)

D-galactose
(Gal)

N-acetyl-D-galactosamine
(GalNAc)

L-fucose
(Fuc)

N-acetyl-D-neuraminic acid
One of a large number of sialic acids
(Sia)

of these proteins and would be advantageous in particular situations. For example, the microheterogeneity observed in a partial set of the oligosaccharides isolated from human colonic mucin (Table 3–2) may permit the oligosaccharides on this glycoprotein to ensnare many different species of bacteria, each of which binds specifically to only one or a few oligosaccharide sequences. All told, 13 oligosaccharides were isolated from this protein and sequenced. The largest (entry 7 in Table 3–2) is displayed in Figure 3–26. All of the other 12, including the first six listed in Table 3–2, were incomplete realizations of the largest.

The sequence of the monosaccharides in an oligosaccharide is established by chemical analysis. The starting material in this analysis is a purified preparation of the glycoprotein itself. Often the glycoprotein has been made radioactive to facilitate the analysis. If the glycoprotein is produced in cells that are grown in the presence of one or two radioactive monosaccharides, for example, [3H]mannose and [14C]glucosamine, the glycoprotein will be radioactive at each of the mannoses and each of the N-acetylglucosamines in its oligosaccharide, and these can be quantified separately in any prepara-

tion. Chemical or enzymatic degradations can be performed and the consequent successive losses of radioactivity provide information about the sequence of the oligosaccharide.[200] By controlling the exposure of the cells to the radioactive precursor, different steps in the synthesis of the glycoprotein can also be followed.

The oligosaccharides attached to the glycoprotein, whether they are radioactive or not, are isolated by digesting the protein, purifying **glycopeptides**, and releasing the oligosaccharides from these glycopeptides by chemical or enzymatic cleavage. The glycoproteins are digested with proteolytic enzymes, and the peptides produced are separated chromatographically. The chromatography is usually performed by a procedure, such as adsorption chromatography on reverse-phase chromatographic media, that separates the peptides on the basis of only their amino acid sequence so that all oligosaccharides attached to a particular amino acid residue in the sequence of a glycoprotein are isolated together. From an examination of the amino acid sequences of a large number of these glycopeptides, it has been concluded that N-linked oligosaccharides are always attached to asparagines that have either a serine or a threonine two

Table 3–2: Some of the Oligosaccharides Isolated from Human Colonic Mucin[199]

(1)   Sia(α2,6)GalNAc[a]

(2)   GlcNAc(β1,3)GalNAc

(3)   GlcNAc(β1,3)GalNAc
        /
   Sia(α2,6)

(4)   Gal(β1,4)GlcNAc(β1,3)GalNAc

(5)   Gal(β1,4)GlcNAc(β1,3) GalNAc
                          /
              Sia(α2,6)

(6)   GlcNAcβ(1,3)Gal(β1,4)GlcNAc(β1,3)GalNAc

(7)   Sia(α2,6)Gal(β1,3)GlcNAc(β1,3)
                                              \
   Sia(α2,6)Gal(β1,3)GlcNAc(β1,6) Gal(β1,4)GlcNAc(β1,3)GalNAc
                                              /
                      Sia(α2,6)

[a]For abbreviations see Figure 3–27.

amino acids further on in the amino acid sequence (Asn-X-Ser/Thr).[201] The serines and threonines to which *O*-linked oligosaccharides are attached are evidently not designated by any pattern in the surrounding sequence of amino acids[202] but tend to be clustered in regions of the polypeptide rich in serines, threonines, and prolines.

The chemical or enzymatic cleavage used to release the several microscopically heterogeneous oligosaccharides from a particular glycopeptide depends upon the glycosidic linkage. For *N*-linked oligosaccharides, endoglycosidases specific for cleavage within the common segment GlcNAc(β1,4)GlcNAcAsn are usually used. For example, mannosyl-glycoprotein endo-β-*N*-acetyl-glucosaminidase (endoglycosidase F) catalyzes the hydrolysis of the glycosidic linkage between the two *N*-acetylglucosamines and releases the oligosaccharide missing its initial monosaccharide. Oligosaccharides in *N*-glycosidic linkage to asparagine can also be released from the glycopeptide by hydrazinolysis[203] and reacetylated with acetic anhydride, and the aldehyde at C1 of the initial *N*-acetylglucosamine can then be reduced to the primary alcohol with Na[^{3}H]BH$_4$.[198] This reduction eliminates the aldehyde, simplifies the subsequent chemistry, and renders the oligosaccharide radioactive, if it is not so already. Oligosaccharides in *O*-glycosidic linkage to serine and threonine are usually released from the glycopeptides by treatment with base, which promotes β-elimination (Figure 3–28). The treatment with base is performed in the presence of Na[^{3}H]BH$_4$ to prevent, by reduction of the aldehyde at C1, the destruction of the oligosaccharide from its reducing end and to make the released oligosaccharide radioactive.

It is at this point that the technical consequences of microheterogeneity are experienced. Instead of one pure oligosaccharide released in a quantity equimolar to the amount of glycopeptide, a mixture of many oligosaccharides is produced, each present in a correspondingly small quantity. This mixture is first separated into neutral and anionic oligosaccharides chromatographically. Chromatographic systems that separate the oligosaccharides on the basis of their molar mass are then used to perform further separations. At this stage, an indication of the size of each oligosaccharide in the set is obtained. After the sialic acids have been removed from the anionic oligosaccharides by hydrolysis in mild acid and separately analyzed, the composition of each oligosaccharide is determined. This can be done by methanolysis under acidic conditions to cleave the acetals and coincidentally form methyl glycosides (Figure 3–29) that are identified by their mobilities on gas chromatography. It is also possible to analyze the composition of the oligosaccharide directly by submitting it to hydrolysis in acid and separating the resulting monosaccharides and deacetylated amino sugars by anion-exchange chromatography (Figure 3–30).[204]

The sequence of an oligosaccharide is determined by indirection. A series of chemical and enzymatic reactions is performed on the oligosaccharide and the outcome of each of these reactions is assessed directly or by determining the change in composition of the oligosaccharide that occurs. The results of these various reactions are gathered until in their entirety they are only consistent with one of the many possible structures for the oligosaccharide. This one structure is then considered to be the actual structure. The reactions used in this process are periodate oxidation, Smith degradation, treatment with glycosidases, and methylation analysis. These results are often supplemented with nuclear magnetic resonance spectroscopy and mass spectroscopy.

When sodium metaperiodate (NaIO$_4$) is dissolved in

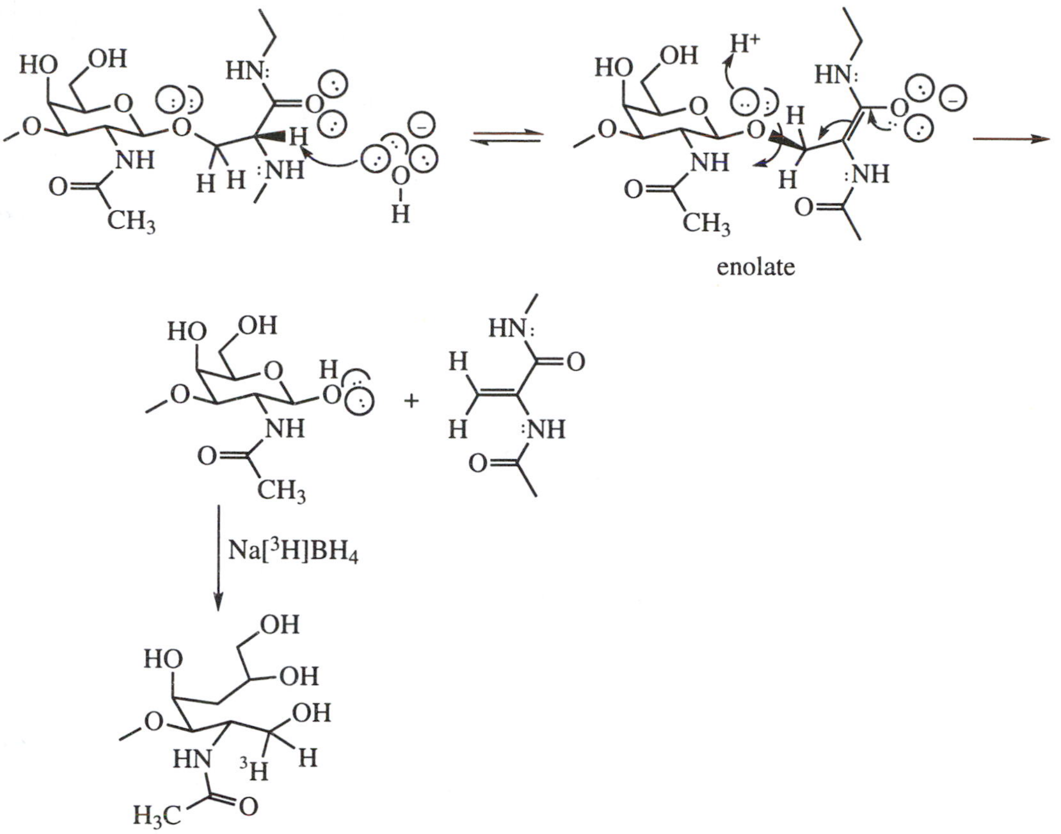

Figure 3–28: β-Elimination of an *N*-acetylgalactosamino oligosaccharide from serine by strong base. The initial step in the reaction is the removal of the proton α to the amide to produce the 1,2-diamino-enolate, which in turn ejects the alcohol to form a dehydroalanine. The reducing end of the oligosaccharide is then reductively labeled with Na[³H]BH₄.

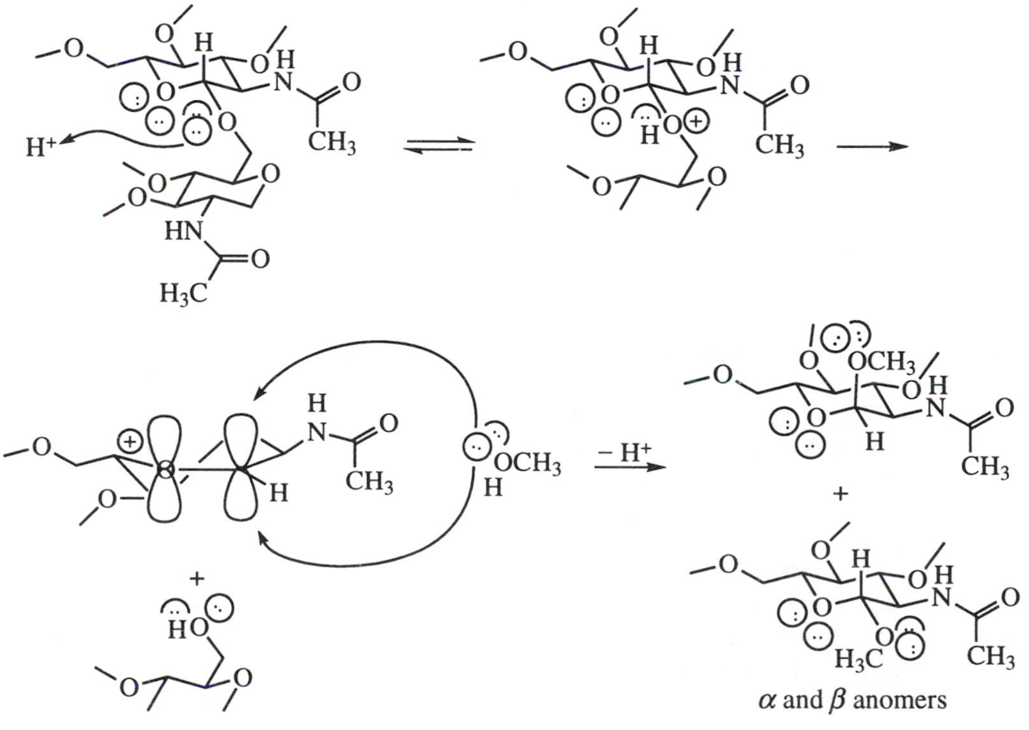

Figure 3–29: Acidic methanolysis of oligosaccharides. Protonation of the acetal oxygen produces a leaving group, the departure of which gives the planar oxonium cation. Addition of methanol to either face of the oxonium cation produces a mixture of the α and β anomers of the methyl glycoside, respectively.

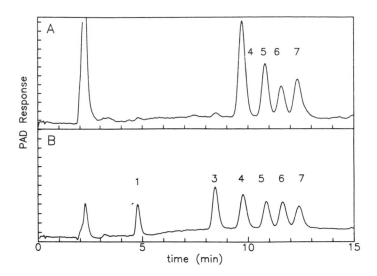

Figure 3–30: Chromatographic analysis of the hydrolysate of a glycopeptide to quantify its composition of monosaccharides.[204] A sample (300 pmol) of a purified, homogeneous glycopeptide (with the composition $Gal_3GlcNAc_5Man_3$) was hydrolyzed for 4 h at 100 °C in 4 M trifluoroacetic acid. The acid was removed by evaporation, and the hydrolysate was submitted to chromatography (panel A) on a column (0.46 cm × 25 cm) of anion-exchange resin equilibrated and eluted with 22 mM NaOH. The strong base renders the sugars sufficiently anionic to be separated by the chromatographic medium. The concentration of monosaccharide was monitored continuously with a pulsed amerpometric detector (PAD). The detector responds to the current resulting from the uptake of electrons at a gold electrode (PAD response). The electrode is poised at +50 mV, which is a sufficiently positive potential to oxidize the polyols of the monosaccharides. It is this oxidation at the surface of the electrode that produces the current. Standards (25 pmol) were run (panel B) under the same conditions and separately identified as the monosaccharides: 1, fucose; 3, galactosamine; 4, glucosamine; 5, galactose; 6, glucose; 7, mannose. From the areas of the peaks of the standards it could be calculated that the original hydrolysate contained 1.1 nmol of glucosamine, 0.79 nmol of galactose, and 0.75 nmol of mannose. The glucose observed was a contaminant. Reprinted with permission from ref 204. Copyright 1988 Academic Press, Inc.

water at acidic pH (pH 3–6) it forms a mixture of acidic hydrates referred to as periodic acid (HIO_4). Periodic acid cleaves polyalcohols such as monosaccharides at the carbon–carbon bonds between vicinal diols and produces two carbonyls from the two hydroxyls (Figure 3–31). Both of the hydroxyls in the vicinal diol must be free for periodic acid to cleave it. The disappearance of a monosaccharide during treatment with periodic acid demonstrates that, in the intact oligosaccharide, the sugar that disappeared had at least two adjacent hydroxyls unbonded in glycosidic linkages. Oxidation by periodic acid can be performed sequentially by the Smith degradation (Figure 3–31). This series of reactions takes advantage of the lability to acid of a glycosidic linkage at carbon 1 of a sugar that has been cleaved by periodic acid and whose resulting aldehydes have been reduced with sodium borohydride. In theory this reaction should be able to cleave sugars sequentially from the ends of the branches inward, but in practice only one cycle is usually successful because the selectivity for acyclic acetals is not great.

A more informative sequence of cleavages can often be performed with **exoglycosidases**. These are enzymes that remove particular sugars from the ends of branches. They are highly specific for the sugar removed, the anomeric state of the glycosidic linkage, and sometimes the location of the hydroxyl from which the bond has been formed. Examples of such exoglycosidases would be β-galactosidase, α-L-fucosidase, β-N-acetylglucosamidase, and α-N-acetylneuraminidase. An example of the specificity for the hydroxyl would be the α-2,3-neuraminidase from Newcastle disease virus. The disappearance of a monosaccharide after exposure of the oligosaccharide to an exoglycosidase is evidence that that monosaccharide was at the end of a branch and attached to it by a glycosidic linkage of the designated anomeric stereochemistry. Several specific endoglycosidases, which cleave an oligosaccharide internally at particular bonds with high specificity, are also available. Examples of these enzymes are endo-1,4-β-galactosidase and endoneuraminidase. Many of these exoglycosidases and endoglycosidases are now available from expression in bacteria containing DNA encoding these proteins. The cDNA isolated from the original sources of these glycosidases has been transferred to bacteria by recombinant techniques, and the proteins can be produced in high yield by these bacteria. One advantage is that these purified enzymes are uncontaminated by other glycosidases.

An oligosaccharide can be chemically methylated on all of its free hydroxyls. This is done by forming the sodium alkoxides of the hydroxyls in a solution of dimethyl sulfoxide; the sodium salt of the dimethyl sulfoxidate anion is used as the base. The alkoxides are then methylated with methyl iodide.[205] The methylated product is hydrolyzed at its glycosidic linkages, and the various methylated monosaccharides that result are identified. In this way, the hydroxyls at which the various monosaccharides were bonded in glycosidic linkages can be determined because they are unmethylated in the products. For example, the oligosaccharide drawn in Figure 3–24 yielded 2,3,4,6-tetramethylmannose; 3,4,6-trimethyl-mannose; 2,4-dimethylmannose; and smaller amounts of 2,4,6-trimethylmannose and 3,6-dimethyl-N-acetyl-glucosamine.[198] The appearance of each of these products is consistent with the structure ultimately proposed for the intact oligosaccharide.

Although it is limited by the amounts (750 nmol) of

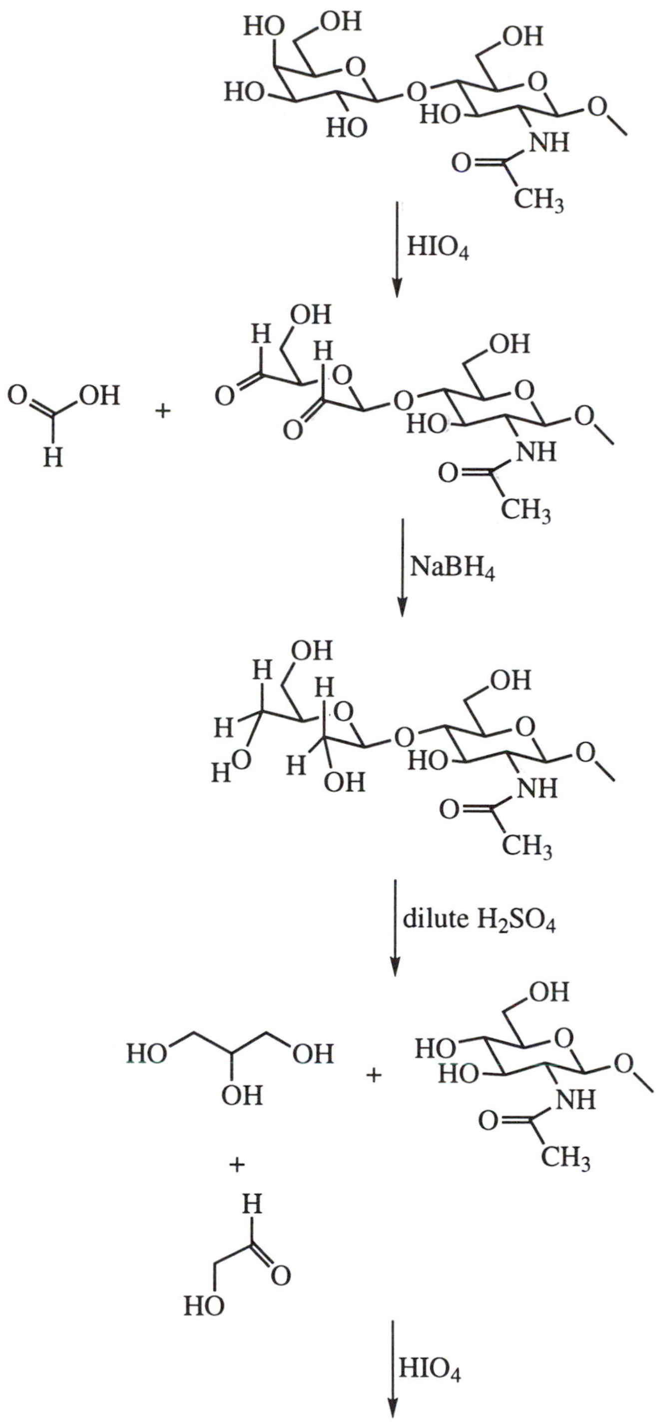

Figure 3–31: Periodate oxidation and Smith degradation. Periodate oxidation (HIO_4) cleaves the carbon–carbon bond between any two unbonded vicinal hydroxyls. The products are most readily understood by first putting a hydroxyl at each position on a carbon that was involved in the carbon–carbon bond and then turning hydrates to aldehydes and orthoacids to acids. Following periodate oxidation, the aldehydes are reduced with borohydride anion, and the acetals containing the degraded monosaccharides are cleaved selectively in weak acid. This frees hydroxyls that were previously in glycosidic linkages and makes certain monosaccharides, which were resistant before, now susceptible to periodate oxidation.

oligosaccharide needed, nuclear magnetic resonance spectroscopy has been applied to solving the sequence of oligosaccharides. The analysis has been most successful in cases where the oligosaccharide is a member of a class of oligosaccharides, such as those with high mannose content, the structures of which are predictable and from which many well-characterized standards are available to assist in the assignments of the various resonances.[206–208] In one instance, the structures of a nested set of oligosaccharides of increasing length, from a less well-characterized class of oligosaccharides, were determined entirely by nuclear magnetic resonance spectroscopy.[209]

Mass spectroscopy has also been applied to structural studies of glycopeptides.[210] It is of limited usefulness because it can only provide information about the number of hexoses, N-(acetylamino) hexoses, and sialic acids, respectively, present in a given glycopeptide. This is because the molecular mass of a polysaccharide is the same regardless of how the monosaccharides are connected and which epimers are present. For example, no distinction can be made by mass spectrometry between mannose and galactose. When combined with chemical modifications such as methylation, however, more information can be gathered by the use of this technique to analyze the products of the reactions. The one significant advantage of mass spectroscopy is that mixtures of glycopeptides or oligosaccharides can be analyzed because each component in the mixture produces a different molecular ion.[211]

From an examination of the sequences of the oligosaccharides attached to glycoproteins from animal sources, several generalizations can be drawn. The most common of these oligosaccharides can be divided into three classes. The **high-mannose oligosaccharides** begin with two N-acetylglucosamines linked (β1,4) to each other, the first attached to asparagine, and contain 5–9 mannoses (Figure 3–24). The **complex N-linked oligosaccharides**, because they are biosynthetically derived from the high-mannose oligosaccharides, also begin with two N-acetylglucosamines linked (β1,4) to each other, the first attached to asparagine, followed by three branched mannose residues. Beyond this structural core, variable amounts of N-acetylglucosamine, galactose, fucose, various sialic acids, and occasionally N-acetylgalactosamine are attached (Figure 3–25). The **O-linked oligosaccharides** begin with an N-acetylgalactosamine linked to serine or threonine and contain variable amounts of N-acetylglucosamine, galactose, N-acetylgalactosamine, fucose, and various sialic acids (Figure 3–26).

Most, if not all,[206] of the high-mannose oligosaccharides from the proteins of animal tissues are incomplete realizations of one complete, basic structure (Table 3–3, entry 1).[212] This uniformity results from the fact that this unit is transferred in its entirety to the targeted asparagine residue on the glycoprotein.[212] Then, in a specific sequence of steps, catalyzed by three exomannosidases, it is shortened until all of the mannoses in (α1,2) linkage have been removed. When only five mannoses remain, the oligosaccharide is then elongated in a highly specific sequence of steps by specific glycosyltransferases to produce complex N-linked oligosaccharides. After the last step of this elongation, another mannosidase removes two more mannoses to leave the three found in the mature complex N-linked oligosaccharide. At the end of this process, most of these complex N-linked oligosaccharides are also incomplete realizations of one basic structure (Table 3–3, entry 2).[208] Fucoses are found attached to some of the complex N-linked oligosaccharides in (α1,6) linkage to one of the internal N-acetyl-glucosamines. It seems that, within the same protein, the oligosaccharides on certain asparagine residues will remain as high-mannose oligosaccharides exclusively, while the oligosaccharides on other asparagine residues are completely converted to complex N-linked oligosaccharides.[198] At each location, the oligosaccharides will display microheterogeneity but only within a given class.

The O-linked oligosaccharides display much less uniformity than the N-linked. This may result from the fact that they are built up one sugar at a time rather than as intact units.[214] The two O-linked oligosaccharides drawn in Figure 3–26 and presented in Table 3–3, respectively, include some of the common structural features of this class. The branches are constructed from the basic repeating unit, Gal(β1,3 or 4)GlcNAc(β1,3 or 4 or 6). Branching usually occurs at galactose or the initial N-acetylgalactosamine, rarely if ever at N-acetylglucosamine, with the exception that fucose is found in (α1,4) linkage to penultimate N-acetylglucosamines in addition to (α1,2) linkage to peripheral galactoses. N-Acetylglucosamine is usually attached in (β1,3 or 4 or 6) linkage to galactose or the initial N-acetylgalactosamine. The basic repeating unit can begin at the initial N-acetylgalactosamine (Figure 3–26), or one intervening galactose from which the basic repeating unit then begins (Table 3–3) will be found attached to the initial N-acetylgalactosamine. The ends of branches are often capped by galactose or N-acetylgalactosamine in (α1,3) linkage or N-acetylgalactosamine in (α1,4) linkage. Sialic acids are found in (α2,6) or (α2,3) linkage to galactoses or the initial N-acetylgalactosamine. Many variations on these patterns are observed.[199,202,209,213,215–217] Often O-linked oligosaccharides are very short. An example would be NeuNAc(α2,3)- Gal(β1,3)[NeuNAc(α2,6)]GalNAc.[202,215] All of these regularities seem to result from the fact that the sugars are added one at a time from the initial N-acetylgalactosamine outward by a limited set of glycosyltransferases. These enzymes are specific for particular sugars and attach them only to particular hydroxyls on particular sugars within the growing oligosaccharide.

Suggested Reading

Baenziger, J.U., & Fiete, D. (1979) Structure of the Complex Oligosaccharides of Fetuin, *J. Biol. Chem. 254*, 789–795.

Table 3–3: Representatives[208,212,213] of the Three Main Classes of Oligosaccharides on Animal Glycoproteins

(1) High-Mannose Oligosaccharide

Man(α1,2)Man(α1,6)

Man(α1,2)Man(α1,3)Man(α1,6)Man(β1,4)GlcNAc(β1,4)GlcNAcβAsn[a]

Glu(α1,2)Glu(α1,3)Glu(α1,3)Man(α1,2)Man(α1,3)

(2) Complex N-Linked Oligosaccharide

Sia(α2,6)Gal(β1,4)GlcNAc(β1,6)

Sia(α2,6)Gal(β1,4)GlcNAc(β1,2)Man(α1,6)

GlcNAc(β1,4)Man(β1,4)GlcNAc(β1,4)GlcNAcβAsn[b]

Sia(α2,6)Gal(β1,4)GlcNAc(β1,2)Man(α1,3)

Sia(α2,6)Gal(β1,4)GlcNAc(β1,4)

(3) O-Linked Oligosaccharide

Gal(β1,4)GlcNAc(β1,6)

Fuc(α1,4)

Gal(β1,4)GlcNAc(β1,3)Gal(β1,3)GalNAcSer/Thr[c]

Gal(β1,4)GlcNAc(β1,6)

---

[a]From Chinese hamster ovary cells. [b]From human plasma α_1-acid glycoprotein. [c]From blood group A active glycoprotein in human ovarian cyst fluid.

Problem 3–12: Complete the following reactions:

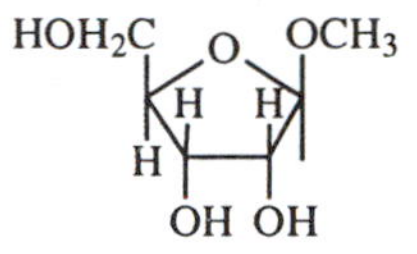

+ nHIO$_4$ ⟶

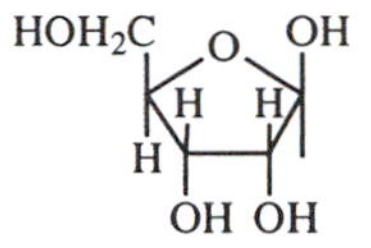

+ nHIO$_4$ ⟶

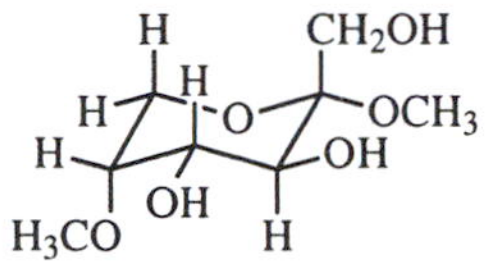

+ nHIO$_4$ ⟶

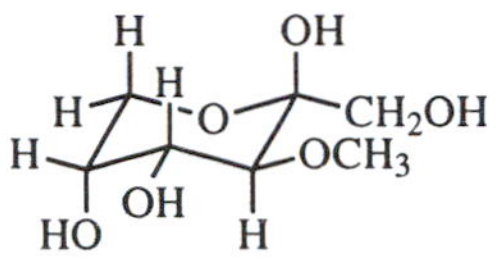

+ nHIO$_4$ ⟶

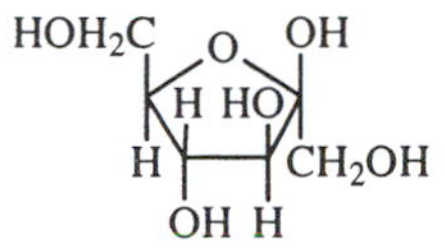

+ nHIO$_4$ ⟶

Problem 3–13: A polysaccharide, which is a polymer of glucose only, has been isolated. It is treated in the following way:

$$\text{polyglucose} + \text{NaBH}_4 \longrightarrow X$$

$$X + \text{CH}_3\text{I} \longrightarrow Y$$

$$\text{H}_2\text{O} + Y \xrightarrow[60\ ^\circ\text{C}]{\text{HCl}} Z$$

Z is a mixture containing the following distribution of methylated sugars.

monosaccharide	mole percent
2,3,6-trimethylglucose	81.9
2,3-dimethylglucose	9.0
2,3,4,6-tetramethylglucose	8.8
1,2,3,5,6-pentamethylglucitol	0.2

(A) On the average, how many sugar residues does each polysaccharide molecule contain, how many branch points are there, and how many nonreducing ends are there?

(B) Draw structures of the linkages in the main linear polymer and the structure of a branch point.

(C) If the polysaccharide were treated with periodic acid, what percentage of the glucose would be destroyed?

Problem 3–14: Immunoglobulin E is a protein isolated from human serum. This protein was digested extensively with pronase, and a glycopeptide was isolated that had the following composition.

monosaccharide	mole ratio
mannose	6.0
N-acetylglucosamine	2.0
Pro	2.5
Gly	1.6
Thr	1.4
Lys	1.3
Asn	1.1
Ser	0.8

This glycopeptide was digested first with α-mannosidase and then with β-mannosidase, enzymes that successively remove mannose residues in α and β linkage, respectively. A core glycopeptide remained that contained two N-acetylglucosamines and one mannose plus all the above amino acids. This core peptide was exhaustively methylated and hydrolyzed with 4 M HCl at 100 °C for 4 h. The result was 3,4,6-trimethyl-N-acetylglucosamine, 3,6-dimethyl-N-acetylglucosamine, and 2,4,6-trimethyl-mannose in roughly equimolar amounts.

(A) What is the linkage between the protein and the oligosaccharide?

(B) Which oxygen is never methylated on aldopyranoses such as mannose and N-acetylglucosamine?

(C) What is the sequence of the monosaccharides and the linkages in the core glycopeptide?

Problem 3–15: Phytohemagglutinin is a glycoprotein from lima beans. A glycopeptide has been isolated from this protein following exhaustive pronase digestion. Determine its structure from the following information. The compositions of single, intact, homogeneous glycopeptides or oligosaccharides are enclosed within parentheses.

(A) Composition
(mannose$_4$, N-acetylglucosamine$_2$, Asp)

(B) Exhaustive methylation, acid hydrolysis

methylated sugar	mol (mol of 2,4-dimethylmannose)$^{-1}$
2,3,4,6-tetramethylmannose	1.95
3,4,6-tetramethylmannose	1.10
2,4-dimethylmannose	1.00
3,6-dimethylglucosamine	1.90

(C) Periodate oxidation followed by mild acid hydrolysis yields a smaller glycopeptide and no free sugar of any kind. The composition of the smaller glycopeptide is

(mannose$_1$, *N*-acetylglucosamine$_2$, Asp)

(D) Mannosidase treatments of initial glycopeptide produced

mannosidase	mol of mannose released (mol glycopeptide)$^{-1}$
α-mannosidase (*Arthrobacter* GJM-1)	0.9
(α1,2)mannosidase (*Aspergillus niger*)	1.1
α-mannosidase (jack bean)	3.0
α-mannosidase (jack bean) followed by β-mannosidase (*A. niger*)	3.8
β-mannosidase (*A. niger*) alone	< 0.2

(E) Glycopeptide core remaining after digestion with α-mannosidase (jack bean) and β-mannosidase (*A. niger*) was

GlcNAc(β1,4)GlcNAc-Asn

(F) The glycopeptide remaining after α-mannosidase (*Arthrobacter* GJM-1) digestion was isolated and exhaustively methylated. 2,3,4-Trimethylmannose, 3,4,6-trimethylmannose, and 2,3,4,6-tetramethylmannose were obtained in approximately equal amounts.

Draw a structure for this glycopeptide that is consistent with all of these observations.

Problem 3–16: A glycopeptide has been purified from a pronase digest of thyroglobulin. From the following information determine the complete structure of the oligosaccharide. Draw the linkage to the amino acid side chain in the peptide portion. The compositions of single, intact, homogeneous glycopeptides or oligosaccharides are enclosed within parentheses.

(A) Composition
(Asp, Gly, Val, mannose$_3$, acetate$_2$, glucosamine$_2$)

(B) Pronase + α-mannosidase
Gly
Val
3 mannose
(Asp, glucosamine$_2$, acetate$_2$)

(C) Chick oviduct extract
(mannose$_3$, glucosamine$_2$, acetate$_2$)
(Asp, Gly, Val)

(D) Exhaustive methylation and acid hydrolysis
(amounts not determined)
3,6-dimethylglucosamine
2,3,4,6-tetramethylmannose
2,X-dimethylmannose (X = 3 or 4)

(E) Periodate oxidation
(Asp, Gly, Val, mannose$_1$, glucosamine$_2$, acetate$_2$)

(F) Reduction and acid hydrolysis of oligosaccharide from (C)
3 mannose
1 glucosamine
1 glucosaminitol

(G) α-Mannosidase treatment of oligosaccharide from (C)
3 mannose
[di-*N*-acetylglucosaminyl(β1,4)glucosamine]*

References

1. Gutte, B., & Merrifield, R.B. (1971) *J. Biol. Chem.* 246, 1922–1941.
2. Edman, P. (1953) *Acta Chem. Scand.* 7, 700–701.
3. Hewick, R.M., Hunkapillar, M.W., Hood. L.E., & Dreyer, W.J. (1981) *J. Biol. Chem.* 256, 7990–7997.
4. Smyth, D.G., Stein, W.H., & Moore, S. (1962) *J. Biol. Chem.* 237, 1845–1850.
5. Hirs, C.H.W., Moore, S., & Stein, W.H. (1956) *J. Biol. Chem.* 219, 623–642.
6. Butler, P.J.G., Harris, J.I., Hartley, B.S., & Leberman, R. (1969) *Biochem. J. 112*, 679–689.
7. Dixon, H.B.F., & Perham, R.N. (1968) *Biochem. J. 109*, 312–314.
8. Houmard, J., & Drapeau, G.R. (1972) *Proc. Natl. Acad. Sci. U.S.A. 69*, 3506–3509.
9. Heinrikson, R.L. (1977) *Methods Enzymol. 47*, 175–189.
10. Smyth, D.G. (1967) *Methods Enzymol. 11*, 214–231.
11. Sela, M., White, F.H., & Anfinsen, C.B. (1959) *Biochim. Biophys. Acta 31*, 417–426.
12. Brattin, W.J., & Smith, E.L. (1971) *J. Biol. Chem. 246*, 2400–2418.
13. Nicholas, R.A. (1984) *Biochemistry 23*, 888–898.
14. Gross, E., & Witkop, B. (1962) *J. Biol. Chem. 237*, 1856–1860.
15. Jacobson, G.R., Schaffer, M.H., Stark, G.R., & Vanaman, T.C. (1973) *J. Biol. Chem. 248*, 6583–6591.
16. Otieno, S. (1978) *Biochemistry 17*, 5468–5474.
17. Witkop, B. (1961) *Adv. Prot. Chem. 16*, 221–321.
18. Burstein, Y., & Patchornik, A. (1972) *Biochemistry 11*, 4641–4650.
19. Landon, M. (1977) *Methods Enzymol. 47*, 145–149.
20. Piszkiewicz, D., Landon, M., & Smith, E.L. (1970) *Biochem. Biophys. Res. Commun. 40*, 1173–1178.
21. Ryle, A.P., Sanger, F., Smith, L.F., & Kitai, R. (1955) *Biochem. J. 60*, 541–556.
22. Baglioni, C. (1961) *Biochim. Biophys. Acta 48*, 392–396.
23. Huang, I., Welch, C.D., & Yoshida, A. (1980) *J. Biol. Chem. 255*, 6412–6420.
24. Shoji, S., Parmelee, D.C., Wade, R.D., Kumar, S., Ericsson, L.H., Walsh, K.A., Neurath, H., Long, G.L., Demaille, J., Fischer, E.H., & Titani, K. (1981) *Proc. Natl. Acad. Sci. U.S.A. 78*, 848–851.
25. Hirs, C.H.W., Moore, S., & Stein, W.H. (1956) *J. Biol. Chem. 219*, 623–642.

---

*Identified by comparison to authentic compound.

26. Vehar, G.A., Keyt, B., Eaton, D., Rodriguez, H., O'Brien, D.P., Rotblat, F., Oppermann, H., Keck, R., Wood, W.I., Harkins, R.N., Tuddenham, E.G.D., Lawn, R.M., & Capon, D.J. (1984) *Nature 312*, 337–347.

27. Bradshaw, R.A., Garner, W.H., & Gurd, F.R.N. (1969) *J. Biol. Chem. 244*, 2149–2158.

28. Mahoney, W.C., & Hermodson, M.A. (1980) *J. Biol. Chem. 255*, 11199–11203.

29. Koide, A., Titani, K., Ericsson, L.H., Kumar, S., Neurath, H., & Walsh, K.A. (1978) *Biochemistry 17*, 5657–5672.

30. Tarr, G.E., Beecher, J.F., Bell, M., & McKean, D.J. (1978) *Anal. Biochem. 84*, 622–627.

31. Pappin, D.J.C., Coull, J.M., & Köster, H. (1990) *Anal. Biochem. 187*, 10–19.

32. Laursen, R.A. (1971) *Eur. J. Biochem. 20*, 89–102.

33. Matsudaira, P. (1987) *J. Biol. Chem. 262*, 10035–10038.

34. Vandekerchove, J., Bauw, G., Puype, M., VanDamme, J., & VanMontague, M. (1985) *Eur. J. Biochem. 152*, 9–19.

35. Moos, M., Nguyen, N.Y., & Liu, T. (1988) *J. Biol. Chem. 263*, 6005–6008.

36. Gerber, G.E., Anderegg, R.J., Herlihy, W.C., Gray, C.P., Biemann, K., & Khorana, H.G. (1979) *Proc. Natl. Acad. Sci. U.S.A. 76*, 227–231.

37. Svennson, B., Svendsen, I., Hojrup, P., Roepstorff, P., Ludvigsen, S., & Poulsen, F.M. (1992) *Biochemistry 31*, 8767–8770.

38. Hartley, B.S. (1970) *Biochem. J. 119*, 805–822.

39. Hirs, C.H.W., Moore, S., & Stein, W.H. (1960) *J. Biol. Chem. 235*, 633–647.

40. Canfield, R.E. (1963) *J. Biol. Chem. 238*, 2698–2707.

41. Hartley, B.S. (1964) *Nature 201*, 1284–1291.

42. Edmundson, A.B. (1965) *Nature 205*, 883–887.

43. Edelman, G.M., Cunningham, B.A., Gall, W.E., Gottlieb, P.D., Rutishauser, U., & Waxdal, M.J. (1969) *Proc. Natl. Acad. Sci. U.S.A. 63*, 78–85.

44. Titani, K., Koide, A., Ericsson, L.H. Kumar, S., Hermann, J., Wade, R.D., Walsh, K.A., Neurath, H., & Fischer, E.H. (1978) *Biochemistry 17*, 5680–5693.

45. Fowler, A., & Zabin, I. (1978) *J. Biol. Chem. 253*, 5521–5525.

46. Watt, K.W.K., Cottrell, B.A., Strong, D.D., & Doolittle, R.F. (1979) *Biochemistry 18*, 5410–5416.

47. Harris, J.I., & Li, C.H. (1955) *J. Biol. Chem. 213*, 499–507.

48. Wood, W.I., Capon, D.J., Simonsen, C.C., Eaton, D.L., Gitschier, J., Keyt, B., Seeburg, P.H., Smith, D.H., Hollingshead, P., Wion, K.L., Delwart, E., Tuddenham, E.G.D., Vehar, G., & Lawn, R.M. (1984) *Nature 312*, 330–337.

49. Kumagai, I., Pieler, T., Subramanian, A.R., & Erdmann, V.A. (1982) *J. Biol. Chem. 257*, 12924–12928.

50. Olivera, B.M., Baine, P., & Davidson, N. (1964) *Biopolymers 2*, 245–257.

51. Fisher, M.P., & Dingman, C.W. (1971) *Biochemistry 10*, 1895–1899.

52. Richards, E.G., & Lecanidou, R. (1971) *Anal. Biochem. 40*, 43–71.

53. Maxam, A.M., & Gilbert, W. (1977) *Proc. Natl. Acad. Sci. U.S.A. 74*, 560–564.

54. Tamm, C., Hodes, M.E., & Chargaff, E. (1952) *J. Biol. Chem. 195*, 49–63.

55. Chargaff, E., Rust, P., Temperli, A., Morisawa, S., & Danon, A. (1963) *Biochim. Biophys. Acta 76*, 149–151.

56. Lawley, P.D., & Brooks, P. (1963) *Biochem. J. 89*, 127–128.

57. Maxam, A., & Gilbert, W. (1980) *Methods Enzymol. 65*, 499–560.

58. Temperli, A., Turler, H., Rust, P., Danon, A., & Chargaff, E. (1964) *Biochim. Biophys. Acta 91*, 462–476.

59. Hayes, D.H., & Hayes-Baron, F. (1967) *J. Chem. Soc. C*, 1528–1533.

60. Tamm, C., Shapiro, H.S., Lipshitz, R., & Chargaff, E. (1953) *J. Biol. Chem. 203*, 673–688.

61. Sanger, F., Nicklen, S., & Coulson, A.R. (1977) *Proc. Natl. Acad. Sci. U.S.A. 74*, 5463–5467.

62. Tabor, S., & Richardson, C.C. (1987) *Proc. Natl. Acad. Sci. U.S.A. 84*, 4767–4771.

63. Messing, J., Crea, R., & Seeburg, R. (1981) *Nucleic Acids Res. 9*, 309–321.

64. Prober, J.M., Trainor, G.L., Dam, R.J., Hobbs, F.W., Robertson, C.W., Zagursky, R.J., Cocuzza, A.J., Jensen, M.A., & Baumeister, K. (1987) *Science 238*, 336–341.

65. Mullis, K., Faloona, F., Scharf, S., Saiki, R., Horn, G., & Erlich, H. (1986) *Cold Spring Harbor Symp. Quant. Biol. 51*, 263–273.

66. Saiki, R.K., Gelfand, D.H., Stoffel, S., Scharf, S.J., Higuchi, R., Horn, G.T., Mullis, K.B., & Erlich, H.A. (1988) *Science 239*, 487–491.

67. Robert, R.J. (1983) *Nucleic Acids Res. 11*, r135–r167.

68. Boulter, J., Luyten, W., Evans, K., Mason, P., Ballivet, M., Goldman, D., Stengelin, S., Martin, G., Heinemann, S., & Patrick, J. (1985) *J. Neurosci. 5*, 2545–2552.

69. Uy, R., & Wold, F. (1977) *Science 198*, 890–896.

70. Zinoni, F., Birkmann, A., Stadtman, T.C., & Bock, A. (1986) *Proc. Natl. Acad. Sci. U.S.A. 83*, 4650–4654.

71. Blobel, G., & Dobberstein, B. (1975) *J. Cell Biol. 67*, 852–862.

72. Keutmann, H.T., Lampman, G.W., Mains, R.E., & Eipper, B.A. (1981) *Biochemistry 20*, 4148–4155.

73. Chang, C.N., Schwartz, M., & Chang, F.N. (1976) *Biochem. Biophys. Res. Commun. 73*, 233–239.

74. Stock, A., Clarke, S., Clarke, C., & Stock, J. (1987) *FEBS Lett. 220*, 8–14.

75. Milligan, D.L., & Koshland, D.E. (1990) *J. Biol. Chem. 265*, 4455–4460.

76. Persson, B., Flinta, C., vonHeijne, G., & Jornvall, H. (1985) *Eur. J. Biochem. 152*, 523–527.

77. Lin, T., & Kolattukudy, P.E. (1980) *Eur. J. Biochem. 106*, 341–351.

78. Towler, D.A., Eubanks, S.R., Towery, D.S., Adams, S.P., & Glaser, L. (1987) *J. Biol. Chem. 262*, 1030–1036.

79. Carr, S.A., Biemann, K., Shoji, S., Parmelee, D.C., & Titani, K. (1982) *Proc. Natl. Acad. Sci. U.S.A. 79*, 6128–6131.

80. Hancock, J.F., Magee, A.I., Childs, J.E., & Marshall, C.J. (1989) *Cell 57*, 1167–1177.

81. Brake, A.J., Brenner, C., Najarian, R., Laybourn, P., & Merryweather, J. (1985) in *Protein Transport and Secretion* (Gething, M., Ed.) pp 103–108, Cold Spring Harbor Laboratory, Cold Spring Harbor, NY.

82. Yamane, H.K., Farnsworth, C.C., Xie, H., Howald, W., Fung, B.K., Clarke, S., Gelb, M.H., & Glomset, J.A. (1990) *Proc. Natl. Acad. Sci. U.S.A. 87*, 5868–5872.

83. Ishibashi, Y., Sakagami, Y., Isogai, A., & Suzuki, A. (1984) *Biochemistry 23*, 1399–1404.

84. Stimmel, J.B., Deschenes, R.J., Volker, C., Stock, J., & Clarke, S. (1990) *Biochemistry 29*, 9651–9659.

85. Hantke, K., & Braun, V. (1973) *Eur. J. Biochem. 34*, 284–296.

86. Ferguson, M.A.J., & Williams, A.F. (1988) *Annu. Rev. Biochem. 57*, 285–320.

87. Recsei, P.A., & Snell, E.E. (1973) *Biochemistry 12*, 365–371.

88. Recsei, P.A., Huynh, Q.K., & Snell, E.E. (1983) *Proc. Natl. Acad. Sci. U.S.A. 80*, 973–977.

89. Kapke, G., & Davis, L. (1975) *Biochemistry 14*, 4273–4276.

90. Fenn, J.B., Mann, M., Meng, C.K., Wong, S.F., & Whitehouse, C.M. (1989) *Science 246*, 64–71.

91. Prome, D., Blouquit, Y., Ponthus, C., Prome, J., & Rosa, J. (1991) *J. Biol. Chem. 266*, 13050–13054.

92. Lipmann, F. (1933) *Biochem. Z. 262*, 3–8.

93. Taborsky, G. (1974) *Adv. Protein Chem. 28*, 1–210.

94. Hunter, T. (1987) *Cell 50*, 823–829.

95. deVerdier, C. (1952) *Nature 170*, 804–805.

96. Eckhart, W., Hutchinson, M.A., & Hunter, T. (1979) *Cell 18*, 925–933.

97. Hunter, T., & Cooper, J.A. (1985) *Annu. Rev. Biochem. 54*, 897–930.

98. Cohen-Solal, L., Cohen-Solal, M., & Glimcher, M.J. (1979) *Proc. Natl. Acad. Sci. U.S.A. 76*, 4327–4330.

99. Chen, C.C., Bruegger, B.B., Kern, C.W., Lin, Y.C., Halpern, R.M., & Smith, R.A. (1977) *Biochemistry 16*, 4852–4855.

100. DeLuca, M., Ebner, K.E., Hultquist, D.E., Kreil, G., Peter, J.B., Moyer, R.W., & Boyer, P.D. (1963) *Biochem. Z. 338*, 512–525.

101. Smith, L.S., Kern, C.W., Halpern, R.M., & Smith, R.A. (1976) *Biochem. Biophys. Res. Commun. 71*, 459–465.

102. Pigiet, V., & Conley, R.P. (1978) *J. Biol. Chem. 253*, 1910–1920.

103. Huttner, W.B. (1982) *Nature 299*, 273–276.

104. Nelsestuen, G.L., Zytkovicz, T.H., & Howard, J.B. (1974) *J. Biol. Chem. 249*, 6347–6350.

105. Stenflo, J., Fernlund, P., Egan, W., & Roepstorff, P. (1974) *Proc. Natl. Acad. Sci. U.S.A. 71*, 2730–2733.

106. McTigue, J.J., Dhaon, M.K., Rich, D.H., & Suttie, J.W. (1984) *J. Biol. Chem. 259*, 4272–4278.

107. Welinder, B.S. (1972) *Biochim. Biophys. Acta 279*, 491–497.

108. Henze, M. (1907) *Hoppe-Seyler's Z. Physiol. Chem. 51*, 64.

109. Wolf, J., & Covelli, I. (1969) *Eur. J. Biochem. 9*, 371–377.

110. Roche, J. (1952) *Experientia 8*, 45–84.

111. Ackermann, D., & Müller, E. (1941) *Hoppe-Seyler's Z. Physiol. Chem. 269*, 146–157.

112. Hunt, S., & Breuer, S.W. (1971) *Biochim. Biophys. Acta 252*, 401–404.

113. Kendall, E.C. (1919) *J. Biol. Chem. 39*, 125–147.

114. Harington, C.R. (1944) *Proc. R. Soc. London B 132*, 223–238.

115. McQuillan, M.T., & Trikojus, V.M. (1972) in *Glycoproteins: Their Composition, Structure, and Function* (Gottschalk, A., Ed.) 2nd ed., pp 926–963, Elsevier, Amsterdam.

116. Gross, J., & Pitt-Rivers, R. (1953) *Biochem. J. 53*, 645–650.

117. Roche, J., Michel, R., & Tata, J. (1953) *Biochim. Biophys. Acta 11*, 543–547.

118. VanNess, B.G., Howard, J.B., & Bodley, J.W. (1980) *J. Biol. Chem. 255*, 10710–10716.

119. Chen, J.C., & Bodley, J.W. (1988) *J. Biol. Chem. 263*, 11692–11696.

120. Ambler, R.P., & Rees, M.W. (1959) *Nature 184*, 56–57.

121. Paik, W.K., & Kim, S. (1971) *Science 174*, 114–119.

122. Paik, W.K., & Kim, S. (1980) *Protein Methylation*, John Wiley, New York.

123. Paik, W.K., & Kim, S. (1967) *Biochem. Biophys. Res. Commun. 27*, 479–483.

124. Hempel, K., Lange, H.W., & Birkofer, L. (1968) *Naturwissenschaften 55*, 37.

125. Paik, W.K., & Kim, S. (1970) *J. Biol. Chem. 245*, 88–92.

126. Ghosh, S.K., Paik, W.K., & Kim, S. (1988) *J. Biol. Chem. 263*, 19024–19033.

127. Baldwin, G.S., & Carnegie, P.R. (1971) *Science 171*, 579–581.

128. Vijoyasora, C., & Rao, B.S.N (1987) *Biochim. Biophys. Acta 923*, 156–165.

129. Lowenson, J.D., & Clarke, S. (1990) *J. Biol. Chem. 265*, 3106–3110.

130. Forooqui, J.Z., Tuck, M., & Paik, W.K. (1985) *J. Biol. Chem. 260*, 537–545.

131. Swanson, R.V., & Glazer, A.N. (1990) *J. Mol. Biol. 214*, 787–796.

132. Klotz, A.V., Leary, J.A., & Glazer, A.N. (1986) *J. Biol. Chem. 262*, 17350–17355.

133. Lhoest, J., & Colson, C. (1977) *Mol. Gen. Genet., 154*, 175–180.

134. Gershey, E.L., Vidali, G., & Allfrey, V.G. (1968) *J. Biol. Chem. 243*, 5018–5022.

135. DeLange, R.J., Smith, E.L., Fambrough, D.M., & Bonner, J. (1968) *Proc. Natl. Acad. Sci. U.S.A. 61*, 1145–1146.

136. Stoffel, W., Hillen, H., Schröder, W., & Deutzmann, R. (1983) *Hoppe-Seyler's Z. Physiol. Chem. 364*, 1455–1466.

137. Jing, S., & Trowbridge, I.S. (1987) *EMBO J. 3*, 2581–2585.

138. Schmidt, M., Schmidt, M.F.G., & Rott, R. (1988) *J. Biol. Chem. 263*, 18635–18639.

139. Schmidt, M.F.G. (1989) *Biochim. Biophys. Acta 988*, 411–426.

140. Gallop, P.M., Blumenfeld, O.O., & Seifter, S. (1972) *Annu. Rev. Biochem. 41*, 617–672.

141. Nakajima, T., & Volcani, B.E. (1970) *Biochem. Biophys. Res. Commun. 39*, 28–33.

142. Udenfriend, S. (1966) *Science 152*, 1335–1340.

143. Bornstein, P. (1974) *Annu. Rev. Biochem. 43*, 567–603.

144. Berg, R.A., & Prockop, D.J. (1973) *J. Biol. Chem. 248*, 1175–1185.

145. Ogle, J.D., Arlinghaus, R.B., & Logan, M.A. (1962) *J. Biol. Chem. 237*, 3667–3673.

146. Janes, S.M., Mu, D., Wemmer, D., Smith, A.J., Kaur, S., Maltby, D., Burlingame, A.L., & Klinman, J. (1990) *Science 248*, 981–987.

147. Stassen, F.L.H. (1976) *Biochim. Biophys. Acta 438*, 49–60.

148. Moss, J., & Vaughn, M. (1977) *J. Biol. Chem. 252*, 2455–2457.

149. Oppenheimer, N.J. (1978) *J. Biol. Chem. 253*, 4907–4910.

150. Manning, D.R., Fraser, B.A., Kahn, R.A., & Gilman, A.G. (1984) *J. Biol. Chem. 259*, 749–756.

151. Ueda, K., & Hayaishi, O. (1985) *Annu. Rev. Biochem. 54*, 73–100.

152. Ogata, N., Ueda, K., & Hayaishi, O. (1980) *J. Biol. Chem. 255*, 7610–7615.

153. Oppenheimer, N.J., & Bodley, J.W. (1981) *J. Biol. Chem. 256*, 8579–8581.

154. Pappenheimer, A.M. (1977) *Annu. Rev. Biochem. 46*, 69–94.

155. Shapiro, B.M., & Stadtman, E.R. (1968) *J. Biol. Chem. 243*, 3769–3771.

156. Adler, S.P., Purich, D., & Stadtman, E.R. (1975) *J. Biol. Chem. 250*, 6264–6272.

157. Harding, H.W.J., & Rogers, G.E. (1976) *Biochim. Biophys. Acta 427*, 315–324.

158. Slettin, K., Aakesson, I., & Alvsaker, J.O. (1971) *Nature New Biol. 231*, 118–119.

159. Midelfort, C.F., & Mehler, A.H. (1972) *Proc. Natl. Acad. Sci. U.S.A. 69*, 1816–1819.

160. Robinson, A.B., Scotchler, J.W., & McKerrow, J.H. (1973) *J. Am. Chem. Soc. 95*, 8156–8159.

161. Nakajima, T., & Ballou, C.E. (1974) *J. Biol. Chem. 249*, 7685–7694.

162. Muir, L., & Lee, Y.C. (1969) *J. Biol. Chem. 244*, 2343–2349.

163. Hallgren, P., Lunblad, A., & Svensson, S. (1975) *J. Biol. Chem. 250*, 5312–5314.

164. Spiro, R.G. (1967) *J. Biol. Chem. 242*, 4813–4823.

165. Lindahl, V., & Róden, L. (1966) *J. Biol. Chem. 241*, 2113–2119.

166. Lindahl, V., & Róden, L. (1972) in *Glycoproteins: Their Composition, Structure, and Function.* (Gottschalk, A., Ed.) 2nd Ed., pp 491–517, Elsevier, Amsterdam.

167. Lote, C.J., & Weiss, J.B. (1971) *FEBS Lett. 16*, 81–85.

168. Weiss, J.B., Lote, C.J., & Bobinski, H. (1971) *Nature New Biol. 234*, 25–26.

169. Miller, D.H., Lamport, D.T.A., & Miller, M. (1972) *Science 176*, 918–920.

170. Torres, C., & Hart, G.W. (1984) *J. Biol. Chem. 259*, 3308–3317.

171. Klostermeyer, H., Rabbel, K., & Reimerdes, E. (1976) *Hoppe-Seyler's Z. Physiol. Chem. 357*, 1197–1199.

172. Pisano, J.J., Finlayson, J.S., & Peyton, M.P. (1969) *Biochemistry 8*, 871–876.

173. Harding, H.W.J., & Rogers, G.E. (1971) *Biochemistry 10*, 624–630.

174. Lerch, K. (1982) *J. Biol. Chem. 257*, 6414–6419.

175. Ito, N., Phillips, S.E.V., Stevens, C., Ogel, Z.B., McPherson, M.J., Keen, J.N., Yadav, K.D.S., & Knowles, P.F. (1991) *Nature 350*, 87–90.

176. Margoliash, E., & Schejter, A. (1966) *Adv. Protein Chem. 21*, 113–286.

177. Williams, V.P., & Glazer, A.N. (1978) *J. Biol. Chem. 253*, 202–211.

178. Walker, W.H., Kenney, W.C., Edmondson, D.E., & Singer, T.P. (1974) *Eur. J. Biochem. 48*, 439–448.

179. Edmondson, D.E., & Singer, T.P. (1976) *FEBS Lett. 64*, 255–265.

180. McIntire, W., Edmondson, D.E., Singer, T.P., & Hopper, D.J. (1980) *J. Biol. Chem. 255*, 6553–6555.

181. Maloy, W.L., Bowien, B.U., Zwolinski, G.K., Kumar, K.G., Wood, H.G., Ericsson, L.H., & Walsh, K.A. (1979) *J. Biol. Chem. 254*, 11615–11622.

182. Hale, G., & Perham, R.N. (1980) *Biochem. J. 187*, 905–908.

183. Tanase, S., Kojima, H., & Morino, Y. (1979) *Biochemistry 18*, 3002–3007.

184. Vanaman, T.C., Wakil, S.J., & Hill, R.L. (1968) *J. Biol. Chem. 243*, 6420–6431.

185. Barber, J.R., & Clarke, S. (1985) *Biochemistry 24*, 4867–4871.

186. Stenflo, J., Fernlund, P., Egan, W., & Roepstorff, P. (1974) *Proc. Natl. Acad. Sci. U.S.A. 71*, 2730–2733.

187. Nelsestuen, G.L., Zytkovicz, T.H., & Howard, J.B. (1974) *J. Biol. Chem. 249*, 6347–6350.

188. Esmon, C.T., Sadowski, J.A., & Suttie, J.W. (1975) *J. Biol. Chem. 250*, 4744–4748.

189. Price, P.A., Poser, J.W., & Raman, N. (1976) *Proc. Natl. Acad. Sci. U.S.A. 73*, 3374–3375.

190. Eckhart, W., Hutchinson, M.A., & Hunter, T. (1979) *Cell 18*, 925–933.

191. Folk, J.E., & Finlayson, J.S. (1977) *Adv. Protein Chem. 31*, 1–133.

192. Williamson, P.R., & Kagan, H.M. (1986) *J. Biol. Chem. 261*, 9477–9482.

193. Kellaris, K.V., & Ware, D.K. (1989) *Biochemistry 28*, 3469–3482.

194. Spackman, D.H., Stein, W.H., & Moore, S. (1960) *J. Biol. Chem. 235*, 648–659.

195. Brown, J.R., & Hartley, B.S. (1966) *Biochem. J. 101*, 214–228.

196. Hartley, B.S. (1970) *Biochem. J. 119*, 805–822.

197. Degen, J., & Kyte, J. (1978) *Anal. Biochem. 89*, 529–539.

198. Mellis, S.J., & Baenziger, J.U. (1983) *J. Biol. Chem. 258*, 11546–11556.

199. Podolsky, D.K. (1985) *J. Biol. Chem. 260*, 15510–15515.

200. Kornfeld, S., Li, E., & Tabas, I. (1978) *J. Biol. Chem. 253*, 7771–7778.

201. Spiro, R.G. (1970) *Annu. Rev. Biochem. 39*, 599–638.

202. Mellis, S.J., & Baenziger, J.U. (1983) *J. Biol. Chem. 258*, 11557–11563.

203. Takasaki, S., Mizuochi, T., & Kobata, A. (1982) *Methods Enzymol. 83*, 263–268.

204. Hardy, M.R., Townsend, R.R., & Lee, Y.C. (1988) *Anal. Biochem. 170*, 54–62.

205. Hakomori, S. (1964) *J. Biochem. (Tokyo) 55*, 205–208.

206. Van Kuik, J.A., VanHalbeek, H., Kamerling, J.P., & Vliegenthart, J.F.G. (1986) *Eur. J. Biochem. 159*, 297–301.

207. Nomoto, H., Takahashi, N., Nagaki, Y., Endo, S., Arata, Y., & Hayashi, K. (1986) *Eur. J. Biochem. 157*, 233–242.

208. Fournet, B., Montreuil, J., Strecker, G., Dorland, L., Haverkamp, J., Vliegenthart, J.F.C., Binette, J.P., & Schmid, K. (1978) *Biochemistry 17*, 5206–5214.

209. Dua, V.K., Rao, B.N.N., Wu, S., Dube, V.E., & Bush, C.A. (1986) *J. Biol. Chem. 261*, 1599–1608.

210. Sasaki, H., Ochi, N., Dell, A., & Fukuda, M. (1988) *Biochemistry 27*, 8618–8626.

211. Sasaki, H., Bothner, B., Dell, A., & Fukuda, M. (1987) *J. Biol. Chem. 262*, 12059–12076.

212. Li, E., Tabas, I., & Kornfeld, S. (1978) *J. Biol. Chem. 253*, 7762–7770.

213. Wu, A.M., Kabat, E.A., Nilsson, B., Zopf, D.A., Gruezo, F.G., & Liao, J. (1984) *J. Biol. Chem. 259*, 7178–7186.

214. Strous, G.J.A. (1979) *Proc. Natl. Acad. Sci. U.S.A. 76*, 2694–2698.

215. Thomas, D.B., & Winzler, R.J. (1969) *J. Biol. Chem. 244*, 5943–5946.

216. Slomiany, B.L., Murty, V.L.N., & Slomiany, A. (1980) *J. Biol. Chem. 255*, 9719–9723.

217. Adamany, A.M., Blumenfeld, O.O., Sabo, B., & McCreary, J. (1983) *J. Biol. Chem. 258*, 11537–11545.

CHAPTER 4

Crystallographic Molecular Models

To this point, it has been established that proteins are composed of long polymers of amino acids and that these polymers are posttranslationally modified by processes that alter the polypeptide or the side chains of the amino acids or that add oligosaccharides to the polypeptides. All of the specific covalent bonds connecting all of the atoms in each of the posttranslationally altered polypeptides composing a particular protein can be defined by chemical analysis. The bond lengths and fixed bond angles of the monomers and of the bonds coupling the monomers into polymers are known precisely. With these values, every bond length, the hybridization of each of the atoms, and every fixed bond angle in each complete, posttranslationally modified polypeptide can be assigned unambiguously. From this information a long flexible molecular model of a particular posttranslationally modified polypeptide can be constructed with high precision.

The problem with defining the complete structure of any polymer, polypeptides included, is the rotational degrees of freedom about the large number of exocyclic, unconjugated single bonds that are present in the polymer. In a finished polypeptide there are from 500 to 10,000 of these single bonds. In a commercial polymer, such as polystyrene, rotation about its many single bonds causes each molecule of the polymer, even though it may be covalently identical to other molecules of the polymer in the sample, to assume a different three-dimensional structure; and, if the polymer is in solution, the structure of each molecule usually changes constantly and randomly with time. The polypeptides in a protein, however, assume only one unchanging structure, or a small number of interchanging structures, unique to the amino acid sequences of those polypeptides. Each molecule of the same protein assumes the same or one of a small number of three-dimensional structures. This structure or these few structures are unique and defined because almost all of the exocyclic single bonds composing the backbones of the polymers and most of the exocyclic single bonds of the side chains in the protein are confined to particular dihedral angles. It is crystals of proteins that have provided both this insight and the opportunity to observe these structures.

The existence of a crystal of any protein permits certain conclusions to be drawn about that protein. As in organic chemistry, it can be concluded that the molecules in the crystal are all covalently identical to each other. Furthermore, if a crystal exists, the covalently identical molecules can be present only in a small number of specific three-dimensional conformations. In the case of proteins, all of the molecules in the crystal usually have the same structure or one of a small number of almost identical structures. It is now also known that the structure of a molecule of protein in a crystal is essentially identical to its only structure or one of its few structures when it is free in solution. When the crystal is submitted to X-ray crystallography, that unique structure can be observed.

Maps of Electron Density[1]

Suppose that one could see X-radiation. If one were to pick up a crystal of purified protein and tumble it in his hand under a beam of X-radiation of one wavelength, it would glitter as does a jewel in a ray of sunlight. There would be, however, a peculiarity to this glitter. A jewel glitters because its facets reflect the sunlight as small individual mirrors. This means that if one follows a facet carefully as the jewel turns, one would see that it is always reflecting the sunlight and realize that the glittering sensation only arises because the eye is at rest with respect to the moving reflected beam. The glitter from a crystal of protein, however, arises because its facets produce flashes, and these flashes occur only when each facet is aligned in one precise direction relative to the direction of the incident beam of X-radiation. The reason for this is that the flashes are produced by the summation of the **reflections** from a stack of evenly spaced, parallel mirrors, and such an arrangement produces diffraction.

If one played with the crystal of protein long enough, it would become clear that there were axes running through it. Rotation about any one of these axes would produce flashes that were regularly arrayed. This regular array of flashes would be reminiscent of the array of reflections that emanates from one of the rotating mirrored spheres in a ballroom. These spheres are made by attaching small mirrors to the surface of a sphere in a regular pattern. A set of these mirrors is usually placed in succession around the equator of the sphere and then other sets are placed in regular rows following lines of latitude above and below the equator. If the rows of mirrors were placed above and below the equator on lines of latitude that had values of

$$\theta = \sin^{-1}(n\beta) \qquad (4\text{–}1)$$

where β is any constant, then the mirrored sphere, rotating in the beam of a single spotlight, would flash in a way reminiscent of the way flashes that emanate from a crystal of protein rotated about one of its axes. One

A

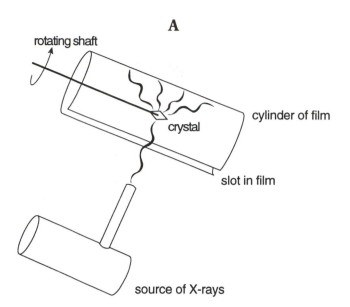

rotating shaft

crystal

cylinder of film

slot in film

source of X-rays

B

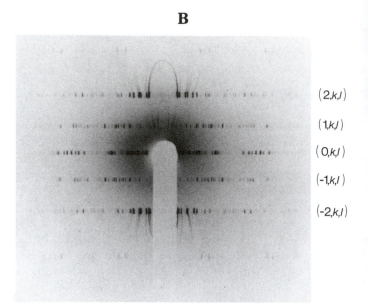

$(2,k,l)$

$(1,k,l)$

$(0,k,l)$

$(-1,k,l)$

$(-2,k,l)$

Figure 4–1: (A) Schematic drawing of a camera used to take rotation photographs of crystals turning about one of the crystallographic axes. (B) A photograph from such a camera.[1] The axis of rotation was aligned vertically with respect to the film as it is displayed. The white shadow in the center of the photograph is of a beam stop used to protect the film from the majority of the X-rays, which pass through and around the crystal. The beam was pointed at the circular top of the beam stop and would have burned a hole in the film in the center of the middle layer line. The five layer lines are labeled as if the rotation had occurred around the **a** axis of the crystal. The middle layer line $(0,k,l)$ is the equator. Reprinted with permission from ref 1. Copyright 1968 Macmillan.

difference, however, would be that while each mirror on the sphere continuously reflects the spotlight when it is on the illuminated side, as can be discerned by following the reflected beams on the walls, each of the mirrors in the crystal of protein reflects only when it passes through certain precise orientations, as could also be discerned by watching the patterns of the flashes on the walls. In addition, the mirrors in the rotating crystal, referred to as the **reflecting planes**, reflect onto the walls behind the crystal as well as in front of the crystal because the crystal is not opaque to X-radiation, as is the ballroom sphere to light, and both sides of each mirror can reflect.

The easiest way to verify this behavior is photographically. A crystal is placed on the axis of a cylinder of photographic film in an orientation such that one of its principal axes is parallel to the axis of the cylinder. It is mounted on the end of a shaft whose axis is coincident with the axis of the cylinder of film and which can be rotated. The cylinder of film has a slot so that a beam of X-radiation perpendicular to the axis of rotation can be directed upon the crystal (Figure 4–1A). After an appropriate exposure, the film is developed. If the crystal contains an orthorhombic lattice, the image observed is that of reflected flashes arrayed on lines of latitude (Figure 4–1B). Each line of latitude, referred to as a **layer line**, arises from all of the mirrors that are tilted at the same angle with respect to the beam of incident X-radiation. Because the spots on the film produced by the flashes occur along layer lines, the tilt of the mirrors relative to the axis of the crystal must only be able to assume certain values. Because the layer lines are made up of discrete spots, each the result of one flash, each mirror must reflect only when the angle between its face and the incident beam assumes unique values.

The profound insight into this curious phenomenon was the realization that the remarkable variations in the intensities of the flashes (Figure 4–1B) contained information and that, from the information they contained, the atomic structure of the molecules from which the crystal was formed could be deduced. With the promise that this is the reward, one can now ask, what are these mirrors, why do they flash, and why does each one flash with a different intensity?

A crystal of protein is a solution to a warehousing problem. It is a solid object formed from a huge number of the same protein molecules, neatly stacked as the boxes or barrels in a warehouse, with the vacancies between the molecules of the protein filled with solvent. It is, for all intents and purposes, an infinite, three-dimensional array of identical enantiomeric objects. It can be shown that there are only 71 ways to arrange enantiomeric objects to form an infinite array. Each crystal represents one of these 71 solutions to the problem.

Every crystal can be divided into a set of boxes, each of which is identical in its size, shape, contents, and the arrangement of its contents to every other one. These boxes are always parallelepipeds, and they are referred to as unit cells. A **unit cell** is the smallest parallelepiped of matter that, by only simple translational movements

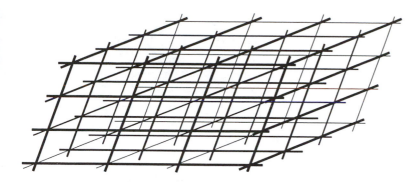

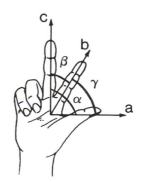

Figure 4–2: A **triclinic** lattice. This lattice is the most general because none of the sides of the unit cells (*a*, *b*, or *c*) is the same length and none of the three angles (α, β, or γ) is 90° or 120°. If $\alpha = \gamma = 90°$, the lattice would be **monoclinic**. If $\alpha = \beta = \gamma = 90°$, the lattice would be **orthorhombic**. If $a = b$ and $\alpha = \beta = \gamma = 90°$, the lattice would be **tetragonal**. If $a = b = c$ and $\alpha = \beta = \gamma \neq 90°$, the lattice would be rhombohedral. If $a = b$, $\alpha = \beta = 90°$, and $\gamma = 120°$, the lattice would be **hexagonal**. If $a = b = c$ and $\alpha = \beta = \gamma = 90°$, the lattice would be **cubic**. The axes are defined by the right-hand rule. Hand reprinted with permission from ref 1. Copyright 1968 Macmillan.

along the three axes of the crystal, can be stacked to create and fill completely the whole crystal. Keep in mind that each of these parallelepipeds is filled with molecules of protein, and surrounding solvent, that are necessarily arranged in space in certain positions and orientations.

In any crystal, three sets of planes can be constructed. Within each of these three sets, all constituent planes must be parallel to each other. Every one of the planes in each of the three sets must intersect planes from both of the other two sets, and every one of the parallelepipeds that results from these intersections must be the same unit cell. The partition of space accomplished by three sets of planes so defined is accompanied by the creation of a network of lines, each of which is the intersection of two of these planes. This network of lines is a **lattice** (Figure 4–2) encaging the unit cells.

Unfortunately, each crystal can be divided into several unique lattices, each of which satisfies the definition, and any translational movement, no matter how small, of any one of these lattices produces another equally satisfactory lattice. The lattice chosen by a crystallographer as the fundamental lattice for a given crystal depends entirely on its suitability for the calculations that she must perform. One lattice is chosen, however, during a proce-

dure known as indexing, and this choice determines the fundamental unit cell (Figure 4–3). The three **axes** of the fundamental unit cell are arbitrarily designated **a**, **b**, and **c** by the right-hand rule (Figure 4–4). The length of the unit cell along each axis is designated *a*, *b*, or *c*, respectively.

There are other ways to divide the space occupied by the crystal into different sets of unit cells using sets of parallel planes. This can be most easily seen by reference to a two-dimensional picture (Figure 4–4). This drawing can be thought of as one of the lattice planes in a monoclinic crystal. In this case, the view presented in the figure is up the **a** axis, and each line can be considered to be the intersection of a plane perpendicular to the page. Each set of the parallel planes depicted in the figure is constructed so that its members pass through the origins of the fundamental unit cells, and the origin of each fundamental unit cell is contained in one of the planes of each set. Each set creates a new array of unit cells (Figure 4–5). Most of these arrays do not produce lattices because their unit cells are not formed from three intersecting sets of parallel planes, but they are arrays of genuine unit cells nevertheless. By extension it is clear that there is an infinite number of ways to divide the lattice into a set of

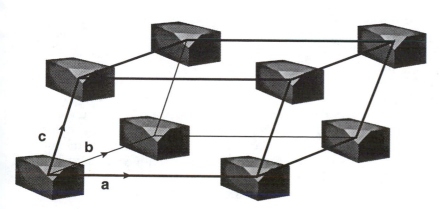

Figure 4–3: The unit cell in a triclinic lattice showing the relationship between the distribution of matter and the boundaries of the unit cell. Even when the unit cell is not chosen to enclose one of the repeating objects, it contains a total of one complete object.

Figure 4–4: Sets of parallel lines, defining sets of unit cells, that pass through a two-dimensional lattice. The sets of parallel lines with indices (1,1), (2,1), and (3,1) and (–1,1), (–2,1), and (–3,1) are presented.

unit cells that are bounded by at least one set of parallel planes (Figure 4–5). Each of these sets of planes can be distinguished from the others by giving it an index, *hkl*, where *h*, *k*, and *l* are integers. X-radiation is reflected and diffracted by every set of parallel planes for which the spacing between the planes is larger than $\lambda/2$, where λ is the wavelength of the X-radiation. Each of the layers of unit cells sandwiched between two of the parallel planes in one of these sets is one of the mirrors that reflects the X-radiation. Because these layers of unit cells are stacked regularly and there are a large number of layers, X-rays are emitted only from the crystal when diffraction occurs.

Electrons **scatter** X-radiation, and molecules are clouds of electrons confined within atomic and molecular orbitals. The molecule or molecules of protein and the molecules of solvent distributed through any unit cell in a crystal are also clouds of electrons, and they will scatter X-radiation. Electrons scatter X-radiation by being excited to vibrate by the oscillating electric field of the incident beam and then radiating X-radiation of the same wavelength in all directions.

Consider a set of parallel planar tiles of thickness *ds*. The tiles in this set have been chosen so that the upper surface of each of them is one of the planes forming the boundary of a layer of unit cells in a crystal of protein molecules. Each member of this set of tiles is the top tile of a laminated unit cell such as the one displayed in Figure 4–6. Each unit cell is one of the set of unit cells defined by one of the sets of parallel planes (Figures 4–4 and 4–5) that divide the crystal. Every unit cell within this

set of planes contributes a top tile, all of these top tiles are parallel to each other, and all are identical to each other in their distribution of electrons. Each top tile will scatter the incident X-radiation, and the scattered waves from the whole set of these top tiles will sum to give the emitted radiation from that set. It can be shown[1] that when the angle of the incident beam of X-radiation assumes a particular value, θ_{hkl}, with respect to the set of planes that defines this set of unit cells, the reflections from all of the top tiles in all of the unit cells will add in phase to produce a burst of flash or X-radiation by diffraction. The value of θ_{hkl} at which this happens is defined by Bragg's law

$$\theta_{hkl} = \sin^{-1}\left(\frac{n\lambda}{2d_{hkl}}\right) \qquad (4\text{–}2)$$

where *n* is any integer, λ is the wavelength of the incident X-radiation, and d_{hkl} is the perpendicular distance, or **Bragg spacing**, between the planes on which the tiles are located.

The reason that diffraction occurs even though the distribution of electrons is not uniform throughout each top tile and even though each top tile is almost never directly above or directly below its identical neighbors (Figure 4–6) is that, through the process of diffraction, the scattering has become optically equivalent to a reflection of the incident X-radiation. One property of reflected electromagnetic radiation is that it automatically sums in

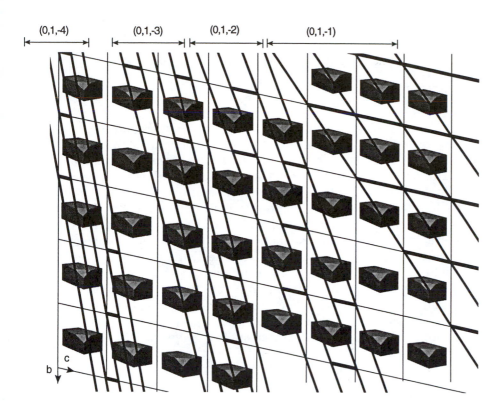

(0,1,-4)   (0,1,-3)   (0,1,-2)   (0,1,-1)

Figure 4–5: Sets of unit cells created by sets of parallel planes. Assume this to be a monoclinic lattice and each line the intersection of a perpendicular plane $(0, k, l)$ with the page. Each new set of parallel planes produces a new set of unit cells. The index of each set is given at the top. Each type of unit cell cuts the repeating object into different segments. In each unit cell, however, the segments, if put together, form one complete object.

phase regardless of the positions in the plane of reflection at which the individual events that cause the reflection occur (Figure 4–7). As a result, the reflecting elements can be anywhere in the planar tiles and the regularly arrayed, identical planar tiles can be translated with respect to each other along axes parallel to the planes, without affecting the angle θ_{hkl} or the intensity of the diffracted reflection. It is this insensitivity of reflected electromagnetic radiation to translation that creates the requirement that a unit cell be only a translational repeating

unit. Since reflection is insensitive to translation within the reflecting plane, the angle at which the diffraction occurs is only dictated by the perpendicular distance between the planes, d_{hkl} (Equation 4–2). The reflections emanating at these angles from the crystal are referred to as Bragg reflections.

This phenomenon of reflection and diffraction explains the flashes emitted by the crystal as it is rotated about one of its principal axes (Figure 4–1B). They arise as each set, hkl, of reflecting planes rotates into the respective angle, θ_{hkl}, with the incident beam, diffraction occurs, and a flash of radiation is reflected outward at an angle, θ_{hkl}, to the planes. By examination of Figure 4–5, it is clear that each set of parallel planes creates the same number of unit cells from a given volume of the crystal, and the same number of unit cells must be contributing to each flash in Figure 4–1B, yet each flash has a different amplitude. These differences in amplitude can be explained by examining the tiles of the unit cell below the top one that was examined before.

As one progresses through the unit cell in a direction normal to the defining plane (Figure 4–6), a series of parallel tiles, each of width ds, are encountered, and each has a different distribution of electrons than the ones above it or below it. This results from the fact that the distribution of electrons is irregular within a unit cell (Figure 4–5). By the definition of a unit cell, however, each successive tile is a member of its own set of identical, parallel tiles, each of which is a distance d_{hkl} above or below identical neighbors in vertically adjacent unit cells, and each of these sets of tiles must also diffract at the

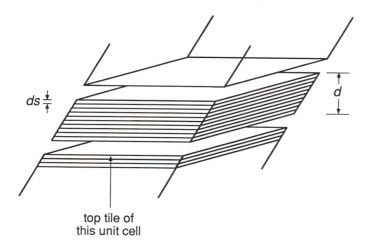

Figure 4–6: Unit cells divided into tiles of thickness ds. By this division each unit cell becomes a stack of tiles. Each tile in the stack has a different two-dimensional distribution of electron density because each tile contains a different section through the unit cell.

Figure 4–7: Incident electromagnetic radiation at an angle ϕ to a plane of reflection emerges from the reflection in phase regardless of the locations of the points on the plane at which reflection occurs.

same θ_{hkl} as the set immediately above and the set immediately below. Its emitted radiation, however, will be of a different amplitude from that of the sets of tiles above it and below it because it contains a different section through the distribution of electrons in the unit cell, and hence a different number of electrons. Its reflection will also be out of phase with the reflections from the other sets because it comes from a different level in the stack. It is the summation of the diffracted reflections from each of the consecutive sets of identical tiles that together fill the unit cells that determines the amplitude and phase of the X-ray emitted by the mirror produced by a particular set of planes. It is the dependence of the amplitude and phase on the number of electrons that happens to fall into each tile that imparts the information to the diffracted reflection.

The reflection from the set of planes *hkl* can be represented as a vector, $\mathbf{F}_{hkl}$, because it can be assigned both a magnitude and a direction relative to a set of two-dimensional coordinates. The magnitude is defined by amplitude of the reflection, $|F_{hkl}|$, and the direction in the plane is defined by its phase. The **phase** of the reflected X-ray, α_{hkl}, is the distance between a crest of the emitted wave and a point of reference common to all of the emitted X-rays. It is expressed in units of wavelength so that, were its value an integer, there would be an integral number of wavelengths between the point of reference and the crest. Because the wave is periodic, the phase is usually expressed, for convenience, as a number between 0 and 1. It is also possible to express the phase in units of radians, in which case its magnitude is $2\pi\alpha_{hkl}$. The vector $\mathbf{F}_{hkl}$ is assigned a direction, defined by the angle $2\pi\alpha_{hkl}$ (in radians) with respect to the coordinates. In this way, if α_{hkl} were an integer, the angle of the vector $\mathbf{F}_{hkl}$ within the two-dimensional coordinates would be 0. With these definitions, the vector $\mathbf{F}_{hkl}$ can be represented in complex coordinates as

$$\mathbf{F}_{hkl} = \left|F_{hkl}\right|e^{2\pi i\alpha_{hkl}} \tag{4–3}$$

where i designates $\sqrt{-1}$ and

$$e^{2\pi i\alpha} = \cos 2\pi\alpha + i \sin 2\pi\alpha \tag{4–4}$$

Such a vector is called a **structure factor**, and each structure factor is identified by its index *hkl*.

At this point it becomes clear that just as the choice of parallelepiped used as the fundamental unit cell was arbitrary, so the choice of the boundaries of a unit cell is arbitrary. Because every tile in the stack that forms the

unit cell (Figure 4–6) is no better than any other, the plane chosen as the upper boundary of the unit cell could be anywhere so long as the planes above and below it respectively occur at the levels where the pattern repeats. Crystals are seldom cooperative and the molecules packed within them do not fit as intact entities into a neat box. But this is irrelevant because the solution to a crystallographic calculation gives the distribution of electrons in an arbitrary unit cell. The distribution of electrons in any number of adjacent unit cells can be constructed by simply stacking unit cells next to each other. If a large enough pile is made, a complete molecule will be found somewhere in the pile.

Aside from crystallizing the protein to begin with, which is often done by other investigators anyway, the two problems facing a crystallographer are the problem of indexing and the problem of phasing. The problem of indexing has a geometric solution, but the problem of phasing has no exact solution.

The problem of indexing is a game of mirrors. As with all games, it is captivating and takes on a life of its own. The crystallographer plays the game in **reciprocal space**; and, as such, learning to live in reciprocal space is a rite of passage. But it is not necessary to live in reciprocal space to understand the problem of indexing.

It takes little imagination to look at the sets of planes in Figure 4–4, to realize that the angular disposition of the sets of planes can be regularly but infinitely varied, to realize that the third dimension, forced to remain parallel to the page in Figure 4–4, could also be tilted through an infinity of specific angles, to realize that each set of planes produces two reflections, and to come to the conclusion that a large number of reflections will have to be kept track of, and the number is purposely kept as large as possible.

Each set of planes is identified by giving it an **index**, *hkl*, sometimes referred to as a Miller index. The index is referred to the axes of the fundamental unit cell. From an examination of Figure 4–5, it can be seen that the parallel planes that define a given set of unit cells always intersect the axes of the fundamental unit cell at intervals that are the quotient of the length of the fundamental unit cell along that axis (*a*, *b*, or *c*, respectively) and an integer. As the tilt of the planes relative to that axis increases, so does the magnitude of this integer, monotonically and continuously. A given set of parallel planes (Figure 4–8) is assigned three integers, *h*, *k*, and *l*. The magnitude of the integer *h* is the number of times the planes intersect the length of the fundamental unit cell along its **a** axis, where each pair of intersections that necessarily occur at the two exact ends of this length are considered as one

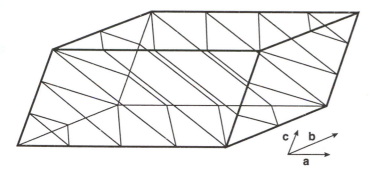

Figure 4–8: Assignment of an index to a set of reflecting planes. The index h, k, or l relative to a given axis, **a**, **b**, or **c**, respectively, is the number of times the respective axis is intersected by the set of planes over the length of the unit cell.

intersection. The magnitude of the integer k is the number of times the planes in the set intersect the length of the unit cell along the **b** axis; and the magnitude of the integer l, the **c** axis. When the set of planes is parallel to one of the axes of the fundamental unit cell, it is assigned 0 for the respective index. The signs of the integers h, k, and l are designated relative to the source of the X-radiation. Each plane has two faces, and either can reflect. The signs of the integers depend on which of these faces has reflected the X-radiation. This in turn is determined by the position of the source because the direction of the reflection is the same as the direction from which the radiation impinges on the set. If movement normal to the planes and toward the source is in the direction of the **a** axis, h is given a positive sign; if movement normal to the planes and toward the source is in the direction opposite that of the **a** axis, h is given a negative sign. The same rule is applied to the **b** axis and its integer k and to the **c** axis and its integer l.

The assignment can be illustrated by referring to Figure 4–8. The magnitudes of h, k, and l for the set of planes illustrated in the figure are 4, 2, and 3, respectively, and they all have the same sign. If the source of X-radiation were to the right, above, or behind the view presented, these would be the (4,2,3) set of planes; if the source of X-radiation were to the left, below, or in front of the view presented, these would be the (−4,−2,−3) set of planes. Notice that as the crystal is rotated, both faces, the positive face and the negative face, eventually diffract. This exercise illustrates the point that when the crystal is rotated in the beam of X-radiation (Figure 4–1), the two reflections originating from the two faces of each set of planes have indices of opposite sign. These pairs of reflections of opposite sign are referred to as Friedel pairs.[1]

Each reflection in Figure 4–1B originated from one of the sets of parallel planes, and the index of that set must be assigned to that reflection. Rather than discuss this process at length, it is enough to note that this can be done with certainty. The concept of reciprocal space simplifies this process of assignment and also permits an engineer to design a camera that can display the reflections on a sheet of photographic film in the order of their index number. A precession photograph (Figure 4–9) is the product of such a camera. The blackness of a spot on this film, or the frequency at which quanta pass through this region of space, is a measurement of the intensity of the reflection, I_{hkl}. Since the intensity is the flux of quanta through an area, the intensity of a reflection, I_{hkl} is equal to the square of the amplitude of the wave, $(F_{hkl})^2$.

The intensities, and hence the amplitudes, of all of the reflections whose values of h, k, and l fall within chosen limits are measured and indexed. Together they produce a data set. A **data set** is a three-dimensional matrix centered on an origin (0,0,0) in which is entered the amplitudes of all the reflections that have been measured from a given crystal. Each amplitude is entered at the location in the matrix that has an index identical to the index of the reflection. The amplitude entered at position hkl in this matrix is referred to as F_{hkl}.

Several days are required to collect a data set with a precession camera, because each reflection must be collected individually. The time can be shortened by using a multiwire area detector[3] that collects amplitudes from a large array of reflections simultaneously. It is also possible to use the high fluxes of X-rays of a broad spectrum of wavelengths that are emitted by synchrotrons to shorten the time required to obtain a data set to several seconds.[4]

The size and shape of the fundamental unit cell can be calculated from the angles at which the reflections emerge from the crystal and hence the spacings on the photographs (Figures 4–1B and 4–9). The size and shape of the fundamental unit cell and the data set itself are the only directly measurable quantities available to the crystallographer, and they are ultimately the information used to calculate the distribution of electrons in the crystal of a protein. The unobservable phases, whose values are inescapably required for the calculation, must be ascertained by comparing several data sets, each obtained from a different altered form of the original crystal.

At this point it is possible to explain the pattern of reflections seen in the rotation photograph in Figure 4–1B taken of a crystal with an orthorhombic lattice. The central layer line, which intersects at its midpoint the axis of the collimated beam of X-rays in the camera, is referred to as the **equator**. Assume that the crystal is being rotated about the **a** axis and that the axis of the beam of X-rays is perpendicular to the axis about which the crystal is rotated. Define the angle v, which is the angle between the axis of rotation and the beam of reflected X-rays (Figure 4–10). The sets of planes with an index h of 0, referred to as the (0, k, l) sets of planes, contain only planes parallel to the axis of rotation. As the rotation of the crystal brings each set of these (0, k, l) planes into the proper angle θ_{0kl} with respect to the beam of X-rays, reflection occurs. Because each set of these planes is parallel to the axis about which the crystal is rotated and because the angle of reflection equals the angle of incidence, each flash emerges at an angle $v = 90°$. The reflections from the $(0, k, l)$ sets of planes are the layer line of reflections on the equator. There is, however, no easily discerned pattern in which the reflections with particular values for k and l are distributed along the equator. Now consider all the sets of planes with an index h of 1, referred to as the $(1, k, l)$ sets of planes. Each of the planes in these sets makes an angle with the axis of rotation such that all values of v for these sets are the same and all the reflections lie on the first layer line. The same argument can be made for the other layer lines. Therefore, all of the reflections with the same first index lie on the same layer line, and each successive layer line out from the equator is of successively higher or successively lower first index. In each layer line, however, the pattern in which the reflections with successive second or with successive third indices occur is complex. A precession camera is able to pick out these reflections from a given layer line systematically and display them as an array (Figure 4–9). Each photograph from a precession camera is only the reflections with a fixed value of one index but all values of

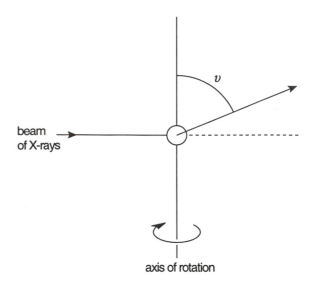

Figure 4–10: Definition of the angle v, the angle between the axis of rotation and the beam of reflected X-radiation.

the other two indexes within the range of resolution chosen. Therefore, a data set is compiled from many precession photographs, each providing the data for a different value of the fixed index.

Any piece of matter, at a given instant ($< 10^{-12}$ s), has a particular distribution of molecules, and this distribution of molecules causes the matter to have a **distribution of electron density**, $\rho(x,y,z)$. If the matter is a gas or a liquid, the rapid redistribution of the molecules causes $\rho(x,y,z)$ to change with time. The collection and measurement of the reflections necessary to perform a determination of molecular structure usually takes hours, and the distribution of electron density of a liquid or a gas averaged over this period of time is absolutely uniform. The liquid regions of aqueous solvent within a crystal of protein, which account for 40–75% of its volume,[5] are, as a result, featureless. To the extent that matter is a solid, its molecules remain fixed in space, and solids have well-defined distributions of electron density. The protein molecules in the crystal remain fixed, except for thermal vibrations, and the distribution of electron density in the regions of the crystal that contain the protein is the average over these vibrational displacements. Within these limits, it is featured.

The distribution of electron density in a crystal is a periodic function, by definition, and it is this periodicity that leads to the reflections. It can be shown[1] that for a crystal of protein

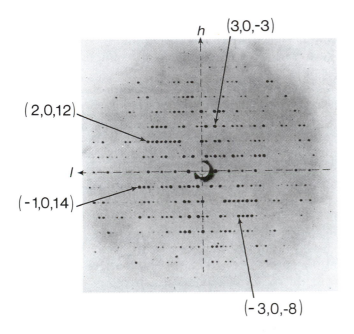

Figure 4–9: X-ray precession photograph[2] of reflections from a crystal of adenylate kinase (unit cell: $a = b = 4.85$ nm; $c = 14.1$ nm). Only reflections with an index k of 0 were arrayed by the camera in this particular experiment. The h and l axes of the array are indicated. The index of each reflection can be assigned by inspection from its position in the array. Several reflections are indexed as examples. One can consider this photograph as an array of all of the reflections in an equatorial layer line from a photograph such as the one in Figure 4–1B laid out by the precession camera systematically upon the field. The spacing between the lines of reflections is directly proportional to the reciprocals of the a and c dimensions of the unit cell. Reprinted with permission from ref 2. Copyright 1977 Academic Press.

$$\rho(x, y, z) = \frac{1}{V} \sum_{h=-\infty}^{\infty} \sum_{k=-\infty}^{\infty} \sum_{l=-\infty}^{\infty} \mathbf{F}_{hkl} e^{-2\pi i(hx+ky+lz)} \quad (4–5)$$

where V is the volume of the unit cell and the $\mathbf{F}_{hkl}$ are all of the structure factors from the crystal properly indexed. The coordinates x, y, and z in Equation 4–5 are referred to

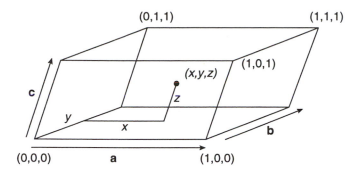

Figure 4–11: Assignment of coordinates x, y, and z to a point within a unit cell. The coordinate axes for the distances x, y, and z are coincident with the crystallographic axes. The distances in each direction are measured along these axes regardless of the angles between them.

the major axes **a**, **b**, and **c**, respectively, of the fundamental unit cell (Figure 4–11). This usually produces a coordinate system that is not orthogonal. Three crystallographic vectors, **a**, **b**, and **c**, are defined such that their respective directions are coincident with the major axes. The lengths, a, b, and c, of these crystallographic vectors are equal to the three dimensions of the unit cell along the three major axes expressed in absolute length (nanometers). The three distances x, y, and z that are the coordinates of a point within the unit cell are measured along the three major axes (Figure 4–11). The units in which these three distances are expressed in Equation 4–5, and the equations that follow, are fractional distances along these axes, where $x = Xa^{-1}$, $y = Yb^{-1}$, and $z = Zc^{-1}$, and X, Y, and Z are the absolute distances (nanometers) along each axis. When Equation 4–5 is combined with Equation 4–3

$$\rho(x,y,z)=\frac{1}{V}\sum_h \sum_k \sum_l |F_{hkl}|e^{-2\pi i(hx+ky+lz-\alpha_{hkl})} \quad (4\text{–}6)$$

This equation can be simplified by noting that

$$e^{iw} = \cos w + i \sin w \quad (4\text{–}7)$$

and that, because a complete data set is a complete set of Friedel pairs, all terms in $i \sin w$ cancel in pairs.[1] As a result

$$\rho(x,y,z)=\frac{1}{V}\sum_h \sum_k \sum_l |F_{hkl}| \cos [2\pi(hx+ky+lz-\alpha_{hkl})]$$

$$(4\text{–}8)$$

The value of examining Equation 4–8 is that it demonstrates that the electron density at any point in the unit cell can be calculated explicitly by inserting the amplitudes of the reflections, properly indexed, and their respective phases. The way that the calculation of a map of

electron density is performed is to divide the fundamental unit cell into a large number of points with coordinates x_p, y_q, z_r. Insertion of all available values for $|F_{hkl}|$, α_{hkl}, h, k, l, and the coordinates of the point x_p, y_q, z_r into Equation 4–8 produces the value of $\rho(x, y, z)$ at the point x_p, y_q, z_r. All points in three-dimensional space with similar values of $\rho(x_p, y_q, z_r)$ are connected by contoured surfaces, or all points in sections through the unit cell with similar values of $\rho(x_p, y_q, z_r)$ are connected by contour lines, and this produces a **map of electron density** (Figure 4–12).

The summations of Equation 4–8 are theoretically infinite, but any finite number of reflections will give an approximate solution. The effect of summing over only a finite number of reflections in Equation 4–8 is to blur $\rho(x, y, z)$. The fewer the reflections used (assuming all reflections indexed at less than the maximum values of h, k, and l that are chosen are included), the more blurred $\rho(x, y, z)$ becomes.

Associated with each reflection, F_{hkl}, is a distance d_{hkl}, which is the Bragg spacing of its respective set of reflecting planes. As h, k, and l increase, d_{hkl} decreases (Figures 4–4 and 4–5). It is customary to note the d_{hkl} for the reflections of the highest index numbers included in the data set and refer to this number as the **resolution** of the calculated map of electron density. Usually 10,000–100,000 reflections are measured and indexed to calculate a map of electron density. If the unit cell has dimensions of 10 nm or less, the nominal resolution of this data set usually will be 0.2–0.4 nm, and the individual values of h, k, and l will lie between −15 to −25 and 15 to 25, respectively. Thirty to 50 unique precession photographs (Figure 4–9) could contain all of the reflections with these indices.

The choice of "resolution" as the word used to define the smallest spacing between the planes, or the largest angle θ, of the reflections included in the data set is unfortunate. This choice implies that the property being noted is the same property as the resolution defined in optics and leaves the erroneous impression that it states something about the quality of the final map of electron density. It does not. The difficulty is not collecting and indexing reflections, which ultimately determines the minimum d_{hkl}, but establishing accurate phases for each reflection.[5] In practice, at a given resolution it is the quality of the phases that defines the quality of the map of electron density. It has been shown that if all of the correct amplitudes are used but the phases are set at an arbitrary number, the map of electron density calculated is meaningless. If, however, the correct phases are used and the amplitudes are all set at the same value, a fairly good map of electron density can be calculated.[7] Unfortunately, the quality of the phases cannot be defined by a simple number.

The use of Equation 4–6 or 4–8 requires that the phase, $2\pi\alpha_{hkl}$, of each reflection in a data set be measured, and this is generally accomplished for crystals of protein by

Figure 4–12: Map of electron density calculated from a crystal of deoxyribonuclease I. Portions of six contiguous sections (0.08 nm section^{-1}) through the unit cell (monoclinic; a = 13.2 nm, b = 5.5 nm, and c = 3.8 nm; β = 91.4°) perpendicular to the **y** axis were stacked together.[6] The contours represent three levels of electron density and were drawn through points on each section whose electron densities were 10%, 20%, and 30%, respectively, of the the maximum electron density observed. Reprinted with permission from ref 6. Copyright 1984 IRL Press.

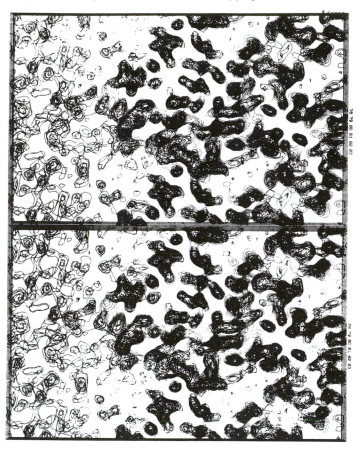

multiple isomorphous replacement. Suppose there are two crystals of protein, alike in almost every way and hence isomorphous. At one or a few specific locations on each of the molecules of protein in one of the crystals, an atom or several atoms that have a large number of electrons has been attached; the other crystal is of the unadorned protein. One of the requirements placed on the bound atoms is that they have high electron density, in other words, a large number of electrons in a small volume. For this reason, an atom of a heavy metal such as mercury, gold, or uranium is chosen. Another requirement is that these atoms of heavy metal occupy specific points in the unit cell. This requirement is fulfilled if the atoms of heavy metal are bound to particular amino acids or clusters of amino acids on the surface of each molecule of protein. The reflections from these two crystals will have the same values for all θ_{hkl} and hence the same display of reflections, but the amplitudes of the reflections will differ (Figure 4–13) as well as the phases.

The magnitude and phase of a given reflection from a

crystal can be calculated if the distribution of atoms in a unit cell of that crystal is known

$$\mathbf{F}_{hkl} = \sum_j f_j e^{2\pi i(hx_j + ky_j + lz_j)} \qquad (4\text{–}9)$$

where f_j is the scattering factor for atom j and x_j, y_j, and z_j are its positions. The scattering factor is determined by the number of electrons in atom j and, unfortunately, by θ_{hkl} as well, but f_j values have been tabulated for all atoms and systematic values of θ.

It can be seen that, since Equation 4–9 is a summation

$$\mathbf{F}_{hkl}^{\mathrm{H+P}} = \mathbf{F}_{hkl}^{\mathrm{P}} + \mathbf{F}_{hkl}^{\mathrm{H}} \qquad (4\text{–}10)$$

where $\mathbf{F}_{hkl}^{\mathrm{H+P}}$ is the value for a given reflection from the crystal containing the heavy metal atom, $\mathbf{F}_{hkl}^{\mathrm{P}}$ is the same reflection from the unadorned crystal, and $\mathbf{F}_{hkl}^{\mathrm{H}}$ would be the reflection from a crystal in which only the heavy metal atoms were present and were at their locations in the existing isomorph. This summation can be presented geometrically (Figure 4–14A). If one knew where the heavy metals were located in the unit cell, the vectors $\mathbf{F}_{hkl}^{\mathrm{H}}$, both amplitudes and phases, could be calculated. Discovering the locations of the heavy metal atoms is an art, the description of which is dramatic but not germane to this discussion. Their locations are eventually determined, and these locations are used to calculate each of the values of $\mathbf{F}_{hkl}^{\mathrm{H}}$.

Unless the metal chosen displays strong anomalous scattering, at least two isomorphous crystals, each substituted with a heavy metal in a different way are required, in theory, for a unique determination of the phases. The data that are available are $|F_{hkl}^{\mathrm{P}}|$, $|F_{hkl}^{\mathrm{H+P}}|$, and $\mathbf{F}_{hkl}^{\mathrm{H}i}$ where i refers to each of the isomorphous replacements from the crystals of which reflections have been gathered. From Equations 4–3 and 4–10, these data provide a set of i simultaneous vector equations for each reflection. In theory, any two of these vector equations can be solved for the phase, $2\pi\alpha_{hkl}^{\mathrm{P}}$, of reflection F_{hkl}^{P}.

This solution is most readily understood geometrically. The amplitude of the vector $\mathbf{F}_{hkl}^{\mathrm{P}}$ is known from the data set, but not its phase, $2\pi\alpha_{hkl}^{\mathrm{P}}$. Therefore, what is known about $\mathbf{F}_{hkl}^{\mathrm{P}}$ defines a circle of radius F_{hkl}^{P} with its center at origin O (Figure 4–14B). Both the amplitude and the phase of a given $\mathbf{F}_{hkl}^{\mathrm{H}}$ are known, and this vector can be placed so that its head is at the origin. Its tail defines a new origin A which defines the position of the tail of vector $\mathbf{F}_{hkl}^{\mathrm{H+P}}$ in the vector sum (Figure 4–14A). The phase of vector $\mathbf{F}_{hkl}^{\mathrm{H+P}}$ is unknown; and, therefore, the known amplitude of the vector $\mathbf{F}_{hkl}^{\mathrm{H+P}}$ defines a second circle. Because the vector sum must balance (Equation 4–10), the two points where the two circles intersect must represent two possibilities for the one actual vector sum. In theory, the correct one of the two possibilities can be determined by going through the same steps with the

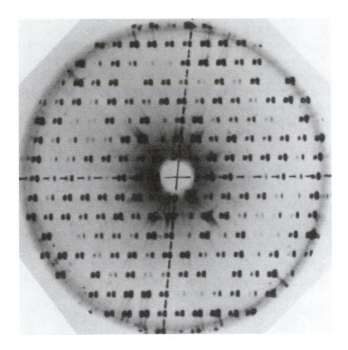

Figure 4–13: Diffraction of X-rays by a crystal of triclinic egg-white lysozyme.[8] This is a section through the full three-dimensional pattern, taken with a Buerger precession camera. Two photographs are superimposed slightly out of horizontal register to show changes in intensities produced by isomorphous introduction of heavy atoms into the crystal. Left spot of each pair, native lysozyme; right spot, crystal after $HgBr_4$ has diffused in.[9] This is a photograph of the $0kl$ set of reflections with the l axis horizontal and k axis nearly vertical. The photograph contains all reflections needed to compute a projection of the structure down the **a** axis to Bragg spacings of 0.4 nm. Reprinted with permission from ref 8. Copyright 1964 Academic Press.

scattering of some heavy metals to the reflection F_{hkl}^{H+P} is different in amplitude from its contribution to $F_{-h,-k,-l}^{H+P}$ and this difference can be predicted. Because of this predictable difference, each isomorphous replacement becomes two isomorphous replacements, each unique to one half of the data set.

Some examples of isomorphous replacements that have been used in crystallographic investigations should make these considerations less abstract. Each isomorphous replacement is a different crystal, usually obtained by soaking a crystal of the unmodified protein in a solution of the compound containing the heavy metal. These compounds can be simple ions, such as WO_4^{2-}, $Pt(CN)_4^{2-}$, $Au(CN)_2^-$, Hg^{2+}, or $Pt(NH_3)_2^{4+}$. Such ions are chelated at certain specific locations by functional groups on the surface of the protein. Some organomercuric compounds, such as ethyl mercurithiosalicylate or diphenylmercury, are bound at specific locations on the protein while other organomercuric compounds, such as mersalyl, o-mercuriphenol, or p-mercuribenzoate, react covalently with thiols on the protein. More complicated organomercury compounds such as 5-mercuride-oxyuridine monophosphate,[12] 3-acetoxymercuri-4-aminobenzenesulfonamide,[13] or ethylmercuriphosphate,[14] have been designed as analogues of ligands specific to the protein.

At least two and perhaps as many as six (Figure 4–14C) or seven isomorphous replacements are made. To obtain a map of electron density for L-arabinose binding protein, isomorphous replacements were made by soaking crystals in CdI_2, p-mercuribenzoate, 2-mercuri-4-nitrophenol, and $Pt(NH_3)_2Cl_4$.[15] In the case of catalase, six separate isomorphous replacements were prepared with $CH_3HgOCOCH_3$, K_2PtCl_4, $Pb(CH_3COO)_2$, K_2PtCl_6, $K_2Pt(NO_2)_4$, and $K_2UO_2F_5$, respectively.[16] In the case of apoferritin, only two isomorphous replacements, made with p-mercuribenzoate and $K_2UO_2F_5$, respectively, were used to determine the phases.[17]

Once the positions of two or more separate sets of heavy metal atoms are known within the unit cell, the reagents can be used in pairs to generate additional unique isomorphous replacements. The advantage is that because the positions in each of the original isomorphous replacements are already available, the positions in the combined isomorphous replacement can be readily established. Isomorphous replacements were made from crystals of alcohol dehydrogenase with $K_2Pt(CN)_4$ and $KAu(CN)_2$, and the positions of the platinum and gold, respectively, in the resulting unit cells were determined. In combination, these two anions produced a third isomorphous replacement.[18] From crystals of deoxyribonuclease I, it was possible to make three isomorphous replacements, one each with $TbCl_3$, K_2PtCl_4, and $Pb(NO_3)_2$, which could then be used in the three possible combinations to generate three additional, unique isomorphous replacements.[6]

Because the quality of the final map of electron den-

data from a second isomorphous replacement because the phase of F_{hkl}^P must be the same in both, and only one of the two possibilities for F_{hkl}^P, namely, the actual vector sum, should be the same in both. A particularly gratifying example of this way of choosing the correct point defining the head of vector F_{hkl}^P was the definition of the phase of reflection $F_{9,1,-2}^P$ for a crystal of hemoglobin using six different isomorphous replacements (Figure 4–14C). All seven circles intersect at the same point and define the phase $\alpha_{9,1,-2}$. This is, of course, the best example from the thousands of reflections for hemoglobin; and, in practice, the circles almost never intersect in the same spot or even near the same spot. The phase of each F_{hkl}^P must be estimated by taking a statistical average of all of the points of intersection.[5] The uncertainty in this average value for each phase is then used to weight the contribution of the respective reflection to the summations of Equation 4–5 or 4–8.

The phenomenon of anomalous scattering can also be used to assist in assignment of phases.[11] This approach takes advantage of the fact that the contribution of the

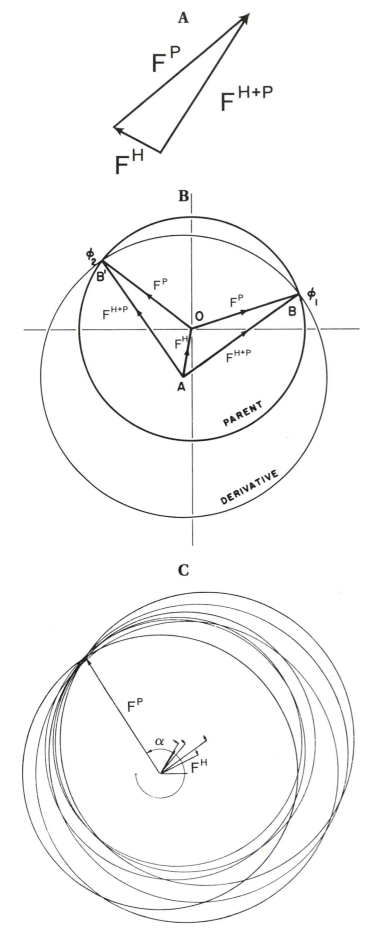

Figure 4–14: Assignment of phases by isomorphous replacement. (A) The vector equation that must define the actual relationship between the three actual vectors $\mathbf{F^P}$, $\mathbf{F^H}$, and $\mathbf{F^{H+P}}$. (B) The amplitudes $|F^P|$ (parent) and $|F^{H+P}|$ (derivative) define two circles.[5] The centers of these two circles must be at the head and tail, respectively, of $\mathbf{F^H}$. The two points of intersection (ϕ_1 and ϕ_2) are possible locations for the head of $\mathbf{F^P}$ in the actual vector equation. Reprinted with permission from ref 5. Copyright 1977 Academic Press. (C) Seven circles, the one defined by $|F^P|$ and the six defined by $|F^{Hi+P}|$ from six isomorphous derivatives, for the reflection $\mathbf{F}_{9,1,-2}$ from a crystal of hemoglobin.[10] The origins of each of the six circles for the isomorphous replacements are displaced from the origin of the circle for the native protein by the respective vector $\mathbf{F^{H_i}}$ calculated from the particular distribution of heavy metals in the unit cell. Reprinted with permission from ref 10. Copyright 1961 Royal Society.

sity (Figure 4–12) depends so heavily on the quality of the phases, the uninvolved observer can evaluate the results only if he is informed. It is important to learn how many isomorphous replacements were made and which were used to calculate the phases. It is also essential to see at least a portion of the calculated map of electron density (Figure 4–12) to get a feeling for its quality. It must be emphasized that the calculation of the initial map of electron density from the phases derived from isomorphous replacement is an unavoidable step in crystallography, and the quality of this map can affect significantly the remainder of the process. The work involved in obtaining this initial map is extensive, and many crystallographic experiments are designed specifically to avoid this work.

When the map of electron density within the unit cell or from several neighboring unit cells is examined, the electron density that corresponds to the intact molecule of protein can be discerned. Since a large fraction of the crystal is liquid water, which is featureless, the protein, which is fixed and highly featured, stands out (Figure 4–12). The compact globule of electron density eventually assigned to an individual molecule of protein usually has an overall size and shape consistent with its frictional coefficient and other molecular parameters. Within this globular solid, features can be seen in relatively sharp detail, but only rarely at atomic resolution.

Suggested Reading

Stout, G.H., & Jensen, L.H. (1968) X-ray Structure Determination, Macmillan, New York.

Problem 4–1: The amplitude of a particular reflection from a crystal of protein, $|F_p|$, is 22.2. The amplitude of the reflection with the same index from a crystal of the first isomorphous replacement, $|F_1|$, is 24.2. The reflection with the same index calculated from the established positions of the heavy metal ions in the unit cell of the first isomorphous replacement has an amplitude $|F_{M1}|$ of 5.4 and a phase of 110°. The amplitude of the reflection with the same index from a crystal of the second isomorphous replacement $|F_2|$ is 21.0. The reflection with the same index calculated from the established positions of the heavy metal ions in the unit cell of the second isomorphous replacement has an amplitude $|F_{M2}|$ of 8.9 and a phase of 65°. Estimate graphically the phase for the reflection of this index from the crystal of protein alone.

The Molecular Model

A continuous tube of electron density can be observed to meander through and account for the globule of featured electron density assigned to the intact molecule of protein in the map of electron density. Sections of one such continuous tube can be seen embedded in the flat slice of electron density presented in Figure 4–12. This tube is the polypeptide of the protein (Structure **2–8**) that has folded to assume the native structure of the molecule. It is into this tube that a molecular model of the known covalent structure of the polypeptide is inserted by the crystallographer.

Once the covalent sequences of the polypeptides, the covalent sequences and points of attachment of any covalently bound oligosaccharides, and the identity and points of attachment of any other posttranslational modifications have been established, molecular models of the fully modified and glycosylated polypeptides known to constitute the molecule of protein whose map of electron density has been determined can be constructed. These models incorporate bond lengths and bond angles that have been measured with high precision during crystallographic studies of small molecules. These small molecules used as standards are molecules whose covalent structures are identical to segments of polypeptide, the side chains of the amino acids, segments of oligosaccharide, or the monosaccharides in the oligosaccharide. The complete, atomically accurate models of the modified polypeptides can be made from pieces of brass wire held together by cylindrical clips, or they can be made from lines displayed on the monitor of a computer (Figure 4–15). As with any molecular model of such a size and complexity, the one of a polypeptide is a flexible, protean object that assumes a new shape each time one of the clips loosens or each time rotation around one of its acyclic single bonds is elicited on the screen.

The model of the polypeptide must be inserted into the continuous tube that runs through the map of electron density. This insertion is performed visually by the crystallographer.[19] There are no rules capable of directing this process that are as successful as the intuition and accumulated knowledge of the crystallographer. There is no machine so accurate and reliable as his eye. The complete amino acid sequence of the protein is essential if this insertion procedure is to be successful because the correspondence between the sequence in which amino acids of different sizes are known to occur along the polypeptide and the sequence in which protrusions of different size occur at regular intervals along the tube of electron density (Figure 4–15) ensures that the insertion is progressing properly.

The 20 different amino acids are, in order of increasing electron density (Figure 4–16), glycine, alanine, serine, proline, cysteine, valine, threonine, aspartate, asparagine, leucine, isoleucine, glutamate, glutamine, methionine, lysine, histidine, phenylalanine, arginine, tyrosine, and tryptophan. In terms of electron density, many of them are indistinguishable, for example, cysteine, valine, and threonine, or aspartate, asparagine, leucine, and isoleucine, and only a few of them, for example, tryptophan, tyrosine, and phenylalanine, are of sufficient size and peculiar enough shape to be identified provisionally with the protrusions jutting out from the continuous tube in the map of electron density (Figure 4–15). Together, however, the sequence in which the amino acids are

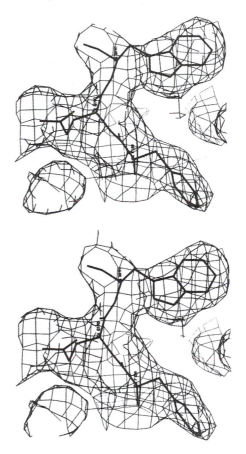

Figure 4–15: Map of electron density[6] for deoxyribonuclease in the region assigned to the sequence FQW which occupies positions 189, 190, and 191 in the amino acid sequence of the protein. A segment of the tube of continuous electron density is shown. The tube enters at the left of the figure and departs at the right of the figure. Protrusions for the side chains of phenylalanine and tryptophan are significant features. A skeletal model of the primary structure has been placed within the tube of electron density. Reprinted with permission from ref 6. Copyright 1984 IRL Press.

Figure 4–16: Silhouettes of the side chains of the amino acids. Space-filling models of the side chains of the amino acids were constructed with the program Chem 3D Plus™. In each model the α carbon, the α carboxylate, and the α amino group of the amino acid were replaced by a single hydrogen. Each of the models was rotated to produce the largest silhouette while in all cases keeping the bond between the β carbon and the hydrogen replacing the α carbon vertical and in the plane of the page. These views were chosen for all of the side chains except the aromatic amino acids. In the case of the aromatic amino acids, each of these views placed the aromatic ring perpendicular to the page, as in the right-hand view for phenylalanine. For each of the aromatic side chains, therefore, a view was chosen in which the plane of the ring was in the plane of the page so that the silhouette was as large as possible. In this way, each of the two-dimensional silhouettes represents the relative three-dimensional bulk of each side chain. To produce the silhouettes, the hydrogens were erased from the models, and all of the atoms were turned black. In all of the silhouettes, except those of the aromatic side chains, the α carbon would occupy the position of the label. In each of the silhouettes of the side chains of the aromatic amino acids, the β carbon is directly above the label. The number of electrons is indicated in parentheses.

arranged in a given protein and their relative sizes usually provide a sufficient amount of information to fit a molecular model of the polypeptide into the map unambiguously.

The reassurance provided by the agreement between the known amino acid sequence of the polypeptide and the sequence of the sizes of the protrusions along the continuous tube of electron density is not inconsequential. Usually, segments of the tube in the map of electron density are blurred or missing because they are too flexible in the actual molecule of protein in the crystal, or one segment will cross another segment too closely to follow each tube confidently through the intersection. Such ambiguities have led to serious errors in the insertion of polypeptides into maps of electron density,[20] and these errors are often corrected by paying close attention to the patterns of the protrusions along the tube and correlating them with the sequence.[21]

Such errors in tracing the polypeptide occur frequently enough that any crystallographic molecular model should be considered provisional until it has been shown to agree with other independent observations. This is an important point because crystallography has often assumed the mantle of infallibility. Even incorrect crystallographic molecular models look completely convincing. It should be pointed out, however, that they are convincing not because of the way the polypeptide is folded but because they are constructed with covalent bonds of the correct lengths and angles. These latter features are not indicative of the reliability of the crystallographic molecular model because they were incorporated automatically.

As the polypeptide is inserted into the map of electron density, certain regular patterns appear in the arrangement of the polyamide backbone. They are referred to as the **secondary structure** of the protein. The regular patterns that have been seen in the structures of proteins are α **helices** (c2 Kin.1),* β **structures,** and β **turns**. α Helices and β structures were first observed in hypothetical models built by Pauling and his collaborators (Figure 4–17).[22,23] After the models had been built, it was found that certain of their dimensions were consistent with molecular dimensions that had been observed in X-ray diffraction patterns from oriented fibers of protein such as hair and silk,[23,24] but it was not until much later that these structures were found in electron density maps of protein molecules. Several strands of β structure are often joined together to form **pleated sheets** (Figure 4–17). Such sheets can be formed from strands all running **parallel** to each other or alternating in their orientation and thus each **antiparallel** to its neighbors or from a mixture of these two arrangements (c2 Kin.3). Tight turns (Figure 4–17D)

were first noticed by Venkatachalan in the crystallographic molecular models of a cyclic hexapeptide, a short tetrapep-tide, and the protein lysozyme.[25]

Aside from the covalent bonds of the polyamide backbone, the main structural element responsible for these secondary structures is the hydrogen bond, which is a noncovalent interaction that forms between the dipole of the nitrogen–hydrogen bond of one amide and one or two of the lone pairs of electrons on the acyl oxygen of another amide. These hydrogen bonds are indicated in Figure 4–17 by dashed lines. A hydrogen bond connects the amide oxygen of each amino acid in a sequence coiled into an α helix with the amide nitrogen–hydrogen of the amino acid four positions farther along (Figure 4–17A). In pleated sheets of parallel β structure (Figure 4–17B), the amide nitrogen–hydrogen and the amide oxygen from an amino acid in one of the polypeptides are connected by hydrogen bonds to the amide oxygen and amide nitrogen–hydrogen, respectively, of amino acids two positions apart from each other in the sequence of a neighboring polypeptide to form a ring containing 12 atoms. In pleated sheets of antiparallel β structure (Figure 4–17C), hydrogen-bonded rings of 14 atoms and 10 atoms alternate along a ladderlike structure. In tight turns (Figure 4–17D) the only structural element that defines the configuration is a hydrogen bond between the amide oxygen of the first amino acid in the turn and the amide nitrogen–hydrogen of the fourth and last amino acid in the turn.

These secondary structures enforce particular geometries on the configuration of the polypeptide. The tight turn causes the polypeptide to double back on itself, often to form a hairpin whose two tines are cross-connected in antiparallel β structure. An α helix has a right-handed pitch, and the absolute stereochemistry of the L-amino acids causes each side chain, the R groups in Figure 4–17A, to cant toward the amino terminus of the helix. The side chains protrude from the helical core at intervals of about 100°. β Structure is pleated when viewed from the side (Figure 4–17B, C) owing to unavoidable steric requirements resulting from the angles of the covalent bonds along the polypeptide. In pleated sheets of β structure, the side chains of the amino acids in the sequence of each strand alternately protrude to one side and then the other of the surface in which the strands of polypeptide lie.

When the wire model of the polypeptide has been inserted into the tube of electron density, the final structure of its folded conformation represents a skeleton of the actual molecule of protein. This **crystallographic molecular model** (c2 Kin.4) is the product of inserting atomically accurate molecular models of known covalent structures into a map of electron density. The crystallographic molecular model should never be confused with the actual structure of the molecule itself. The crystallographic molecular model is really no more than the coordinates of the centers of all of the atoms in the macroscopic, mechanical model that was inserted by the crystallographer into the map of electron density. It lacks

*Cross-references in the format (c*n* Kin.*m*) are to the Kinemage Supplement; the integer *n* refers to the number of the chapter and the integer *m* to the number of the image in that chapter of the Kinemage Supplement. See the "Preface" on page vii of this volume for more information.

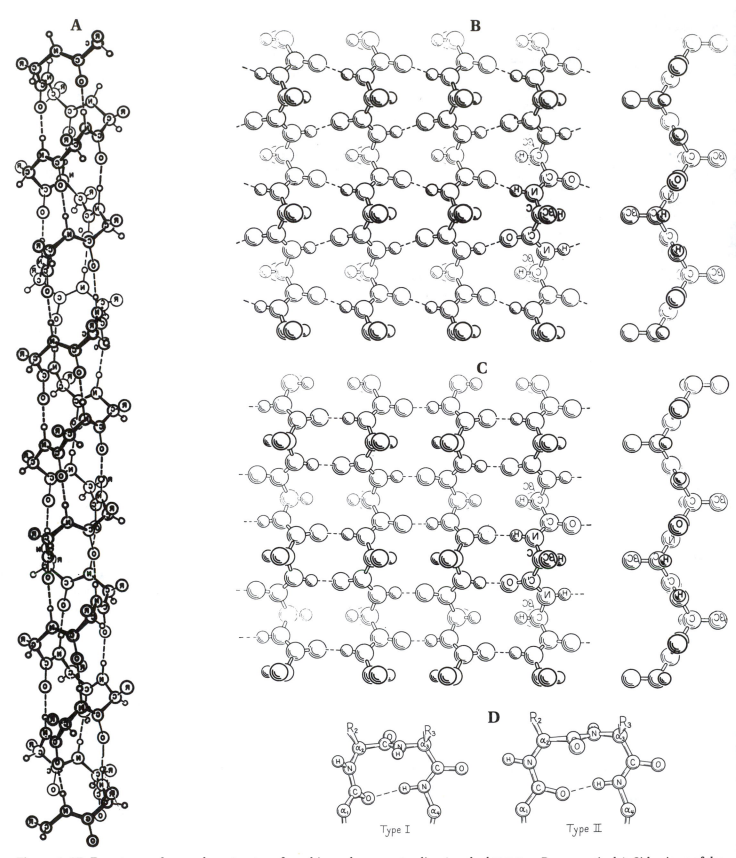

Figure 4–17: Four types of secondary structure found in molecules of protein: (A) α helix,[22] (B) parallel β sheet,[23] (C) antiparallel β sheet,[23] and (D) two types of β turn (Type I and Type II).[26] The polyamide backbone can be traced by the pattern $\cdots$N, Cα, CO, N, Cα, CO, N, Cα,CO$\cdots$ and the side chains are the groups protruding (marked Я, Ɔᵦ, or R$_b$, respectively). Side views of the β sheets are shown to the right of each overhead view to demonstrate the pleats. Reprinted with permission from refs 22, 23, and 26. Copyright 1951 National Academy of Sciences and 1981 Academic Press.

Figure 4–18: Crystallographic molecular model of penicillopepsin, a protein from the mold *Penicillium anthinellum*.[27] In the first skeletal model (A), both the peptide backbone and the side chains of the amino acids are displayed, and no potential hydrogen bonds are indicated. In the second skeletal model (B), the side chains are left out and the crystallographer has assigned intuitively the locations of hydrogen bonds. The detailed skeletal model can be represented diagramatically (C). Reprinted with permission from ref 27. Copyright 1983 Academic Press.

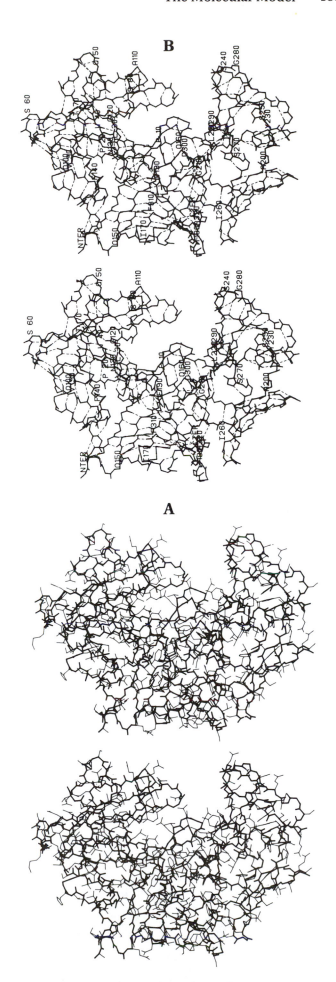

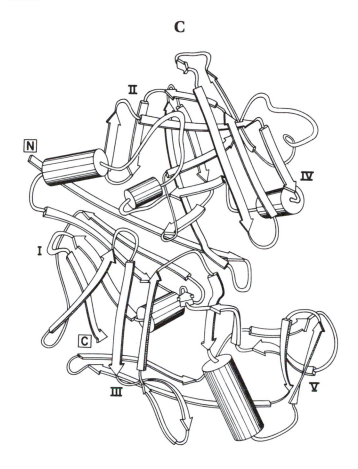

most of the important information about the actual molecule, in particular, the location and distribution of the outer-shell electrons that are responsible for all of the chemistry in which the actual molecule participates. Another important omission is hydrogen atoms. Because hydrogen atoms have no inner-shell electrons, they are never seen in maps of electron density. Any hydrogens in an X-ray crystallographic molecular model are there only because the crystallographer knows that they must be, even though they were not observed. An example of a crystallographic molecular model is the one constructed for the protein penicillopepsin (Figure 4–18A).[27] To obtain a full understanding of this molecular model, it must be viewed stereoscopically. The two related panels of Figure 4–18 show the same views of the model; but, in Figure 4–18B, the side chains of the amino acids have been removed to focus attention on only the polyamide

A

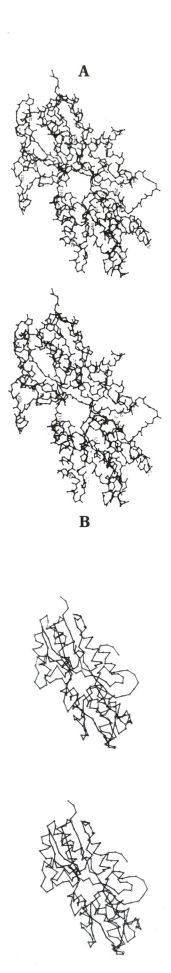

B

C

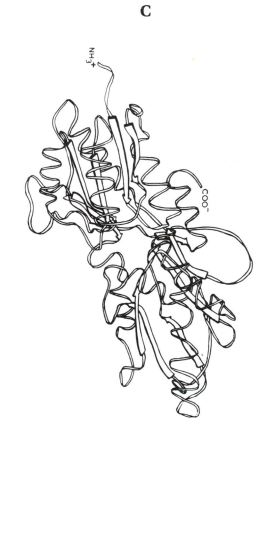

Figure 4–19: Crystallographic molecular model[28,29] of the L-arabinose binding protein from the bacterium *Escherichia coli*. In the skeletal model (A), all of the atoms along the backbone of the polypeptide [⋯N, Cα, CO, N, Cα, CO, N, Cα, CO,⋯] are presented. In the α carbon diagram (B), the positions of the α carbons of the amino acids are designated by points and the points are joined by line segments. This presents a clearer picture of the patterns of secondary and tertiary structure. These patterns can also be represented diagramatically (C). Reprinted with permission from refs 28 and 29. Copyright 1981 Academic Press and 1977 *Journal of Biological Chemistry*.

backbone of the polypeptide (c2 Kin.4). In both panels, the amino terminus is on the upper right at about 10 o'clock and the carboxy terminus is to the back at about 8 o'clock. You should follow the polypeptide through the whole model in Figure 4–18B. Note the α helices, β structures, and tight turns. Compare what you see to the models presented by Pauling (Figure 4–17A, B, and C). Note that α helices are rigid tubes while β structures are sinuous and flexible. Distinguish between sheets of β structure formed from three or more strands and ribbons of β structure formed from only two strands. Now follow the polypeptide through the crystallographic molecular model in Figure 4–18A. Note the disposition of the side chains along secondary structures, and try to identify some of the amino acids. This figure emphasizes the fact that some amino acids are isosteric and indistinguishable crystallographically.

The arrangement of secondary structures in three dimensions produces the tertiary structure of a folded polypeptide. The **tertiary structure** of a polypeptide is the complete configuration into which its peptide backbone is folded in a native protein. The tertiary structure observed in a crystallographic molecular model is often presented diagramatically (Figure 4–18C) in a drawing where flat arrows are used to represent strands of polypeptide in β structure, with the head of the arrow at the carboxy terminus of the strand to provide the direction in which the chain is oriented, and cylinders are used to represent α helices. The tertiary structure of penicillopepsin, which you have explored in detail in Figure 4–18B, is represented, in the same orientation, by the diagram in Figure 4–18C. Follow the polypeptide through Figure 4–18, panels B and C, simultaneously.

Penicillopepsin (Figure 4–18) contains mostly antiparallel β structure where adjoining strands are often connected at one end by tight turns. Parallel β structure can be produced when the carboxy terminus of one strand in a sheet of parallel strands is connected to the amino terminus of another strand in the same sheet by an α helix that runs across one or the other face of the sheet. Examples of such an arrangement are found in the crystallographic molecular model of the L-arabinose binding protein (Figure 4–19).[28,29] There are two regions in this protein that contain parallel β structures in the form of twisted sheets four to six strands in width flanked by α helices. Follow the polypeptide through Figure 4–19A to obtain a feeling for this arrangement. A diagram representing the folding of the polypeptide viewed from the same angle (Figure 4–19C) may assist you.

Another common way of presenting the complete structure of a folded polypeptide is to display the position of only the α carbon atoms of the polypeptide backbone, each connected to its immediate neighbors in the sequence by a line segment. A representation of the L-arabinose binding protein in this format (Figure 4–19B) illustrates its clarity.

The representations of the structure of a protein molecule that have been presented so far are either skeletons or diagrams of the crystallographic molecular model. The advantage of the skeletons is that the whole molecule can be examined simultaneously even in its interior. As with all molecules, flesh resides upon the bones in the form of the electron clouds that produced the map of electron density in the first place. It is possible to construct a model of a molecule of protein from space-filling units of the kind developed by Pauling and Corey (CPK models) after the coordinates of the individual atoms in three dimensions have been gathered from the skeletal model. A three-dimensional photograph of such a model of the protein lysozyme can be seen in Volume 243 of the *Journal of Biological Chemistry*.[30] This photograph produces a reliable mental image of the molecular structure of a properly folded polypeptide and is worth studying. Computers are also able to produce stereoscopic views of the space-filling structure (Figure 4–20).

After insertion of the model of the polypeptide into the map of electron density, any oligosaccharides attached to the protein must also be accounted for. From the co-

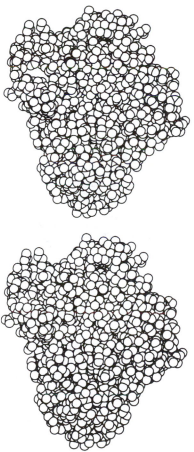

Figure 4–20: Stereo, space-filling diagram of the crystallographic molecular model of one of the subunits of citrate (*si*) synthase from porcine heart.[32] The extension at the top of the structure is normally buried in the other subunit of the dimer, which is not displayed in this figure. Reprinted with permission from ref 32. Copyright 1982 Academic Press.

Figure 4–21: Insertion of oligosaccharide into assigned electron density.[31] After the polypeptide had been inserted into the map of electron density for the Fc fragment of human immunoglobulin G, a large feature of electron density adjacent to Asparagine 297, an asparagine known to carry a complex *N*-linked oligosaccharide, was observed. A 1.0-nm section through this feature is presented and a skeletal model of the partial sequence of the oligosaccharide, GlcNAc(β1,2)Man(α1,6)[Man (α1,3)]Man, has been inserted into the envelope of electron density. Reprinted with permission from ref 31. Copyright 1976 Walter de Gruyter.

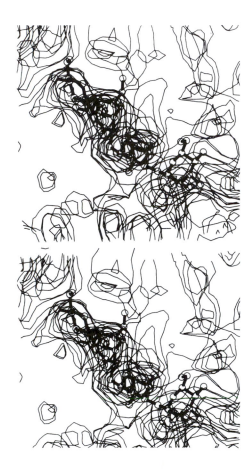

fully solvated, flexible, and structureless. This absence of a fixed structure is carried into the crystal, and the region within the unit cell occupied by the oligosaccharide is often featureless (Figure 4–12). Attempts to assign an atomic structure to such regions are probably irrelevant to an understanding of the behavior of an oligosaccharide in a biological situation where it will have no defined structure anyway.

Aside from the polypeptide and its posttranslational modifications, there is an additional group of compounds that are found attached to molecules of protein. These are referred to as coenzymes. Their electron density must be located and accounted for in the map of electron density assigned to the molecule of protein. Coenzymes are associated with proteins to provide chemical capabilities that cannot be provided by the side chains of the amino acids alone. Examples of coenzymes are pyridoxal phosphate, thiamin pyrophosphate, flavin adenine dinucleotide, biotin, lipoic acid, heme, chlorophyll, and ubiquinone. Coenzymes are either covalently bonded to the polypeptide as additional examples of posttranslational modifications (Table 3–1) or enclosed within it so tightly that they form an integral structural component. In either case, the coenzyme never leaves the protein and is incorporated with the protein into a crystal. At the moment, these molecules will be considered to be merely small clouds of electrons that have interesting shapes. Among this group of compounds are inorganic cations, such as Fe^{2+}, Zn^{2+}, Mg^{2+}, Mn^{2+}, Ca^{2+}, and Cu^{2+}, and anions, such as $HOPO_3^{2-}$) and S^{2-}, that are found tightly bound to certain proteins.

The electron density contributed by coenzymes known to be associated with a protein is always clearly featured because these molecules are enclosed within the protein and precisely aligned for functional purposes. The shapes of most coenzymes are unique, and they can usually be inserted unambiguously into one of the envelopes of electron density unfilled by the polypeptide. One of the most exciting examples of the arrangement of coenzymes within a map of electron density is found in the crystallographic structure (c13 Kin.2) of the photosynthetic reaction center,[33,34] the protein responsible for converting sunlight into electrochemical energy. This protein and the cytochrome that cocrystallizes with it together contain no fewer than 13 coenzymes: four hemes, four bacterio-chlorophylls *b*, two bacteriopheophytins *b*, two quinones, one atom of iron, and several carotenoids. Bacteriochlorophyll *b*, with its characteristic queue, could be inserted into several of the envelopes of electron density in the map (Figure 4–23), and they could be distinguished from the envelopes associated with the very similar bacteriopheophytins by the bulge of electron density due to the magnesium ions present in the former but missing from the latter. The molecular model for the cytochrome associated with the reaction center (Figure 4–24) displays the characteristic, intimate association between coenzymes and the polypeptides that enfold them. In this case

valent structure of the protein, it is known where the oligosaccharide should be located in the map because the asparagines, serines, or threonines to which it is attached have been identified. For example, the envelope of electron density attributable to the oligosaccharide attached to Asparagine 297 in the polypeptide of the Fc fragment of human immunoglobulin G could be readily located once the sequence had been inserted into the map.[31] The covalent sequence of the sugars known to comprise the oligosaccharide then could be inserted into this envelope (Figure 4–21). In this particular case, the oligosaccharide was surrounded by the protein and forced to assume a somewhat defined configuration. Usually, oligosaccharides are located on the outer surface of a protein and protrude into the aqueous phase surrounding it (Figure 4–22). Under these circumstances, they are

the four hemes of the cytochrome are embraced by α helices arranged to compose the entire structure.

If one assumes for the moment that a particular crystallographic molecular model has been constructed correctly and represents accurately the molecules of protein as they are packed in the crystal, what relationship does this structure have to the molecules of protein when they are in solution in the cytoplasm?

The existence of a map of electron density requires that every unit cell in the crystal contain the same distribution of matter. If the unit cell contains only one molecule of protein, then every molecule of protein in the crystal must have exactly the same structure. It has always been observed that in unit cells containing more than one molecule of protein, the several maps of electron density for the several molecules are very similar and differ from each other only at the surfaces of the molecules where differences in crystal packing have caused flexible side chains or backbone to assume slightly different orientations. Within the limits of resolution, however, the rest of the structures of the several molecules in the same unit cell are always identical. It follows that all of the molecules of protein within a given crystal have essentially the same structure.

There is little doubt that this one structure present in the crystal is the same as the unique structure, or one of a limited number of unique structures, assumed by the protein in free solution and referred to as its **native structure** or native structures. First, the crystal is 40–70% water.[5] This water usually surrounds each molecule of protein almost entirely, and the contacts between molecules of protein in the crystal are not extensive. Consequently, the molecule of protein is still dissolved in the same aqueous solution from which it crystallized. Second, there are several instances in which the same protein has been crystallized under two different conditions, and it was found to be incorporated into the two resulting unit cells with completely different orientations, yet the two respective maps of electron density were almost indistinguishable from each other and could be superposed.[5] Third, crystals of a protein usually retain its enzymatic activity,[5] albeit sometimes at a lower rate, and this also indicates that the structure of the protein has not changed during its crystallization. Fourth, Raman spectroscopy can be performed on solids as well as liquids, and when the Raman spectrum of ribonuclease in solution was compared to its Raman spectrum in the crystal, the two were virtually identical in the region of the amide III vibrations, a region which would be sensitive to any changes in the structure of the polypeptide chain that might have occurred during crystallization.[36]

There are proteins that have been shown to be able to assume two stable structures in rapid equilibrium with each other in solution, and, in some cases, two different crystals can be made, each exclusively incorporating one of these respective structures. The crystallization and

Figure 4–22: Crystallographic molecular model of the glycoprotein bovine deoxyribonuclease I.[35] Structured electron density could be observed in the vicinity of Asparagine 18 in the crystallographic model, but it was only large enough to contain the first two N-acetylglucosamines and one of the mannoses of the high-mannose oligosaccharide (GlcNAc$_2$Man$_5$) known to be attached at this asparagine. The other four mannoses were arbitrarily positioned in the crystallographic molecular model. This model provides another opportunity to trace the polypeptide through the structure. Reprinted with permission from ref 35. Copyright 1986 Academic Press.

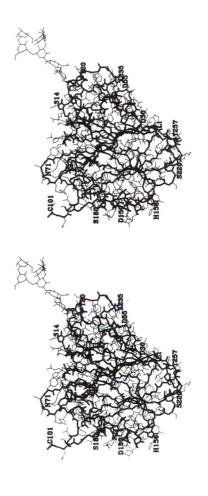

elucidation of the structures of deoxyhemoglobin and oxyhemoglobin provide an example.[37] When such crystals are exposed to a ligand that binds to the protein they contain and coincidentally elicits the change in the structure of the protein, the crystals will often shatter[38] as the protein assumes the new structure, incompatible with the former crystal lattice. In addition to presenting another observation consistent with the conclusions that the molecules of protein in the crystal retain the potentialities that they assume in solution, this observation suggests why some crystals are not enzymatically active. If expression of enzymatic activity requires that the pro-

Figure 4–23: Electron density assigned to a bacteriochlorophyll b,[34] one of the coenzymes in the reaction center from *Rhodopseudomonas viridis*. A skeletal model of the known atomic structure of the coenzyme has been placed within the envelope of electron density. Note the bulge of electron density in the center of the coenzyme that results from the magnesium cation. Reprinted with permission from ref 34. Copyright 1984 Academic Press.

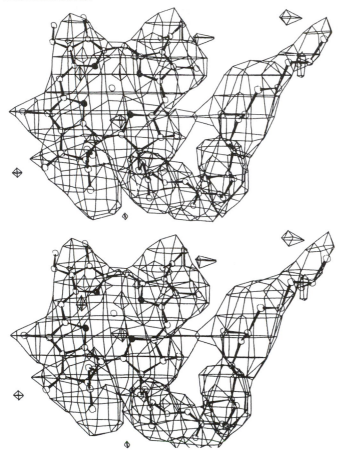

tein change its shape slightly and reversibly each time it catalyzes the reaction and that change in shape is sterically hindered by the lattice of the crystal, the protein would not be able to display activity.

Crystals of citrate (*si*) synthase provide an example of such a situation.[32] This protein can be crystallized under different sets of conditions that yield two different types of crystals containing different conformations of the protein. From a careful examination of the maps of electron density for these two conformations, it became clear that each time the enzyme in free solution converts acetyl-SCoA and oxaloacetate into citrate and CoASH, it passes back and forth between these two conformations. Neither crystal is enzymatically active, but upon dissolving either, full activity is restored. The conclusion drawn was that the packing of the molecules of protein in the crystal sterically prevented the movement between the

two conformations necessary for enzymatic activity, not that either crystallographic molecular model was unrepresentative of the enzyme.

The most compelling argument for the identity of the structure seen in the crystal and the structure assumed by the protein in solution is that the structure seen in the crystal makes sense. Over the two decades that crystallographic molecular models of high resolution have been available for examination, what has been seen has consistently provided reasonable explanations for the behavior of the respective proteins in solution. These explanations have stimulated experiments to test those explanations that have usually yielded informative results. Often an experiment will rule out a hypothesis based on an examination of the structure, but the more informed reexamination of the structure that then occurs usually turns up the original error of judgment. The fact that a crystallographic molecular model makes sense is an unambiguous verification that it represents the actual structure of the molecule of protein even when it is in the crystal, let alone in solution.

Suggested Reading

Wyckoff, H.W., Tsernoglou, D., Hanson, A.W., Know, J.R., Lee, B., & Richards, F.M. (1970) The Three-Dimensional Structure of Ribonuclease S: Interpretation of an Electron Density Map at a Nominal Resolution of 2 Å, *J. Biol. Chem. 245*, 305–328.

Problem 4–2: Pig heart citrate (*si*) synthase crystallizes from solution at pH 7.4. The crystals are tetragonal, which means that the unit cell is a rectangular parallelepiped with all angles 90°. The dimensions of the unit cell are $a = b = 7.74$ nm and $c = 19.64$ nm. A crystal was submitted to diffraction with X-radiation generated from a rotating anode of copper. The Kα emission of the copper ($\lambda = 0.154$ nm) was selected for the source of the X-rays. On graph paper draw a view of the unit cell looking down the c axis with the set of (2,–4,0) planes intersecting it. At what angle θ to the incident beam of X-radiation will the reflection from the (2,–4,0) set of planes emerge from the crystal?

Problem 4–3: The structure at the top right of the next page was drawn from a crystallographic molecular model of a particular protein.[39] It depicts only a small portion of the entire molecule. Trace the polypeptide backbone through the structure.

(A) How many lengths of polymer enter the figure?
(B) Identify as many of the amino acids as you can along the polymer, and write out its sequence or sequences.
(C) Identify the symbols for the individual atoms. Which atoms are not depicted? Why?

Refinement

Every crystallographic molecular model starts out as a molecular model of the polypeptide that has been manually inserted into the initial map of electron density. The initial map of electron density for a protein of unknown structure is always calculated (Equation 4–6) from the amplitudes of the properly indexed reflections, which are determined by direct measurement of their intensities, and the phases of these reflections, which are estimated by the procedure of multiple isomorphous replacement. The molecular model of the polypeptide is a skeletal model reproducing the known amino acid sequence and built with ideal bond lengths and bond angles that have been measured precisely in small molecules. Because the initial map of electron density is never of high enough quality to pick out individual atoms, this molecular model of the polypeptide can only be inserted so that it fits into the tube of electron density as successfully as possible. The acyl oxygens of the peptide bonds are inserted into the small bulges emerging regularly along the tube and the side chains of the amino acids are positioned as efficiently as possible in the larger bulges. The pattern in which the larger bulges vary in size should match the variation in size of the side chains in the amino acid sequence.

At this stage, the accuracy of the crystallographic molecular model is sufficient to define the patterns in which secondary structures are arranged. Because the individual atoms did not appear in the initial map of electron density and because the molecular model has been inserted in the absence of this information, the initial crystallographic molecular model does not have sufficient accuracy to establish atomic details. These are of importance in their own right as well as being critical in understanding most of the biological functions of proteins.

Until 1975, almost all of the crystallographic molecular models that were available had been built directly from initial maps of electron density calculated from observed amplitudes and observed phases. Recently, the practice of refining molecular models has been adopted to increase their accuracy. Although the fold of the polypeptide chain and the general positions of the side chains of the individual amino acids usually do not

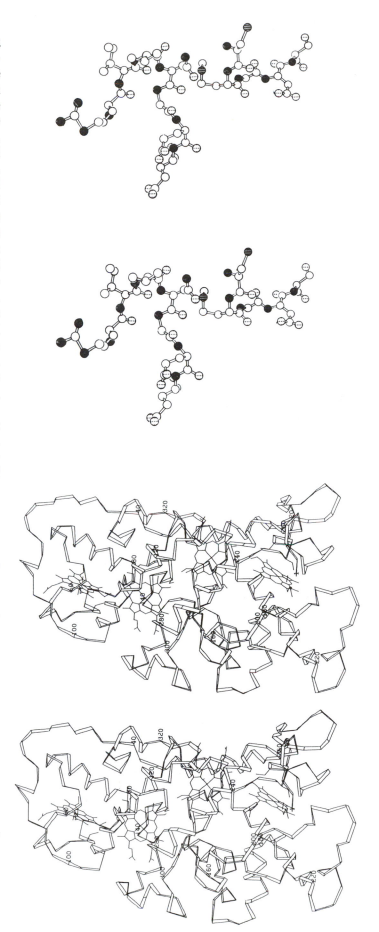

Figure 4–24: Crystallographic molecular model of the cytochrome associated with the reaction center from *R. viridis.*[33] The ribbon diagram has a crease located at the position of each α carbon in the folded polypeptide. Positioned at the proper locations within this array of α carbons are skeletal models of the four hemes identified in the map of electron density. Each of the cysteines responsible for covalently connecting each heme to the polypeptide is displayed at its respective position in the sequence. Reprinted with permission from ref 33. Copyright 1985 *Nature.*

change significantly upon refinement, the atomic details of both the polypeptide and the side chains almost always change dramatically. If the dramatic changes that occur during refinement do bring the molecular model closer to reality, then atomic details in initial, unrefined molecular models are best ignored until the refinement has validated their existence. Whether or not the atomic details currently described in refined crystallographic molecular models will be revised in the future as they have been in the past is unknown.

The **refinement** of a crystallographic molecular model is the systematic adjustment of the positions of its atoms and the addition to it of molecules of solutes and water so that the amplitudes of the set of reflections calculated from the model reproduces the amplitudes of the observed reflections as closely as possible. The first step in a refinement is to calculate the amplitudes of the reflections that the initial molecular model itself would produce, so that these amplitudes can be compared to the amplitudes of the observed reflections. Once the initial molecular model has been built to the satisfaction of the crystallographer, the coordinates of each atom other than the hydrogens within the unit cell can be determined by direct measurements of the model. A set of theoretical reflections can be calculated from these coordinates and the scattering functions for each atom by Fourier transformation of the theoretical electron density of the molecular model (Equation 4–9). The amplitudes of this set of calculated reflections are referred to as the **calculated amplitudes**, and the set of these amplitudes is designated $|F_c|$. The set of simultaneously **calculated phases** is designated as α_c. The amplitudes of the original experimental data set or any subset thereof are referred to as the **observed amplitudes** and designated $|F_o|$. The set of phases estimated by isomorphous replacement are euphemistically referred to as the **observed phases** and the set containing their values is designated as α_o. All of these designations, $|F_c|$, α_c, $|F_o|$, and α_o, refer to three-dimensional matrices, each containing 5000–100,000 elements, all individually indexed as either $|F_{hkl}|$ or α_{hkl}, respectively.

The only directly observed quantities are the observed amplitudes $|F_o|$, and they are the only parameters against which the success of the construction of any molecular model can be judged. If the molecular model were an exact representation of the molecules of protein, small solutes, and water within the crystal and there were no systematic errors in the observed data set, the calculated amplitudes $|F_c|$ would be identical to the observed amplitudes $|F_o|$. It is traditional to quantify the degree of this correspondence with a **crystallographic R-factor**

$$R = \frac{\sum_i \left| |F_o|_i - |F_c|_i \right|}{\sum_i |F_o|_i} \qquad (4-11)$$

where $|F_o|_i$ and $|F_c|_i$ are the amplitudes of two respective reflections i of the same index, or

$$R = \left[\frac{\sum_i \left(|F_o|_i - |F_c|_i \right)^2}{\sum_i |F_o|_i^2} \right]^{1/2} \qquad (4-12)$$

This use of two expressions for the same parameter is somewhat confusing, but the two different parameters both decrease as the calculated reflections become closer in value to the measured reflections. The summation is performed over all available pairs of corresponding observed and calculated reflections or some subset of the available pairs. Once the resolution of the data set is less than about 0.5 nm, so that the electron density of the solvent can be properly reproduced, the differences between the initial molecular model and the real structure usually become more significant the higher the resolution, and the value of the R-factor has a tendency to increase in magnitude as the data set is expanded to include reflections of higher and higher resolution. Therefore the resolution of the data set must be known to assess the significance of the value of the R-factor.

The value of the R-factor is often presented as a measure of the validity of a particular crystallographic molecular model. Such claims should be ignored. Aside from the significant differences in the definition of the R-factor which make it difficult to compare its values from one crystallographic molecular model to the next, an incorrect crystallographic molecular model can, nevertheless, give a reasonable R-factor. For example, an incorrect crystallographic molecular model[40] for the ferredoxin from *Azotobacter vinelandii* to Bragg spacings of 0.2 nm had an R-factor of 0.24, while the later, presumably correct, crystallographic molecular model[41] had an R-factor of 0.21 to Bragg spacings of 0.2 nm. An incorrect crystallographic molecular model[42] for the *ras* protein to Bragg spacings of 0.3 nm had an R-factor of 0.29, while the later, presumably correct, crystallographic molecular model[43] had an R-factor of 0.23 to Bragg spacings of 0.26 nm. It is not the value of the R-factor that should be used as a validation of the model but the agreement of the model with independent chemical observations. In the case of the ferredoxin from *A. vinelandii*, it was disagreements between the earlier crystallographic molecular model and several direct chemical observations of the protein that prompted a reevaluation.[41] Both of these examples were situations in which the chains were incorrectly traced in the original maps of electron density, and this produced very large errors in the molecular model. Smaller errors may often go undetected.

Usually, the initial molecular model yields an R-factor of 0.30–0.60. This means that the amplitudes calculated from the model differ on the average by 30–60% from the observed amplitudes. At first glance this seems alarming because a completely random acentric structure would give an *R*-factor of 0.59. It is not so disturbing, however, because it is obvious from direct observation (Figure 4–12) that a unique structure has been defined by the map of electron density. Nevertheless, such a large value of the *R*-factor indicates that the molecular model does not duplicate the structure of the actual molecule and suggests that there is room for improvement. The improvements made in the structure after the initial model has been constructed are known as **refinements**. The goal of all refinement is to produce a molecular model whose calculated reflections have amplitudes as close as possible to the respective observed amplitudes. This is accomplished by adjusting the positions of each of the atoms in the model in such a way that the *R*-factor decreases in magnitude. Only when it is realized that models of molecules of protein have 500–10,000 atoms that are not hydrogen and that the movement of any one of these atoms in the model affects the amplitudes of all of the reflections in the set $|F_c|$ is the task of refinement placed in a proper perspective.

The most easily understood way to perform a refinement proceeds by calculating **difference maps of electron density**. When two sets of crystallographic amplitudes are available for the same structure or for two structures so similar that the same set of phases can be used for both, a difference map of electron density, $\Delta\rho(x,y,z)$, can be calculated from

$$\Delta\rho(x,y,z) = \frac{1}{V} \sum_h \sum_k \sum_l \left(|F_{hkl}| - |F'_{hkl}|\right)e^{-2\pi i(hx+ky+lz-\alpha_{hkl})}$$

(4–13)

where $|F|$ and $|F'|$ refer to the two available sets of amplitudes. Equation 4–13 produces a map that has positive electron density wherever $\rho(x,y,z)$ is greater than $\rho'(x,y,z)$ and negative electron density wherever $\rho(x,y,z)$ is less than $\rho'(x,y,z)$, where $\rho(x,y,z)$ and $\rho'(x,y,z)$ are the two maps of electron density that would be calculated directly from the respective amplitudes and phases.

Difference maps of electron density have many more uses than in refinement; but, in this particular instance, $|F|$ is the set $|F_o|$ and $|F'|$ is the set $|F_c|$, and α_{hkl} can be either observed phases or calculated phases. The difference map of electron density indicates where the molecular model differs from the actual molecule. Adjustments can then be made in the model at the locations where significant differences occur. Unfortunately, as adjustments are made at one point in the molecular model, unavoidable shifts occur elsewhere, and the new *R*-factor calculated with the adjusted model usually does not change dramatically and often increases. A new difference map must be calculated to locate the new problems and the process repeated. This approach is an

example of tuning, and it is slow and ultimately unsatisfactory.

There are several approaches to refinement that are designed to discover the optimal shifts of all of the atoms simultaneously by computation rather than manual manipulation. These techniques are all based on the minimization of a particular multivariate function by solving large sets of simultaneous differential equations through matrix methods. Suppose that there is a multivariate function θ where

$$\theta = f(x_1, x_2, x_3, \ldots x_n)$$

(4–14)

that the values for the variables x_j have been assigned initial magnitudes, and that one wishes to discover the individual shifts, Δx_j, in the magnitudes of the assigned values of each of the variables that will produce a minimum numerical value for θ. It can be shown[44] that

$$\mathbf{A}_{i,j} \times \mathbf{h}_j = \mathbf{k}_i$$

(4–15)

Equation 4–15 defines a linear system of n simultaneous differential equations where $\mathbf{A}_{i,j}$ is the square matrix $(n \times n)$ whose elements are

$$a_{ij} = \frac{\partial^2\theta}{\partial x_i \partial x_j}$$

(4–16)

$\mathbf{h}$ is the vector whose elements are the individual shifts, Δx_j, in each x_j required to minimize θ, and $\mathbf{k}$ is the vector whose elements are $\partial\theta/\partial x_i$. Equation 4–15 is solved[45] for $\mathbf{h}$, and its solution defines the shifts, Δx_j, in each variable Δx_i, required to produce a minimum value for θ.

Suppose the variables x_j are the positions of the atoms in the molecular model of a protein, and

$$\theta = \sum_p w_p \left(|F_o|_p - |F_c|_p\right)^2$$

(4–17)

where $|F_o|_p$ is the observed amplitude of a particular reflection p, $|F_c|_p$ is the calculated amplitude of the same reflection p, and w_p is a weight assigned to a given reflection. The magnitude of w_p varies with the certainty of the value of each observed amplitude. The values of $|F_o|$ are fixed quantities, but the vales of $|F_c|$ are direct functions of the positions of the atoms in the unit cell of the molecular model. These positions are the variables x_j. The solution of Equation 4–15, the vector $\mathbf{h}$, would then be a list of the shifts in the positions of each atom of the molecular model that would produce a minimum value of θ and, presumably, a minimum value of *R*-factor. This differs from manual tuning in that the effects of all shifts are considered simultaneously.

This simple but computationally complex approach suffers from the drawback that the shifts of the atoms are unconstrained. In other words, every atom in the molecular model would be allowed to shift regardless of the shifts imposed upon its neighbors. If the data set extended to high enough resolution, this would not be a problem because the map of electron density itself would confine each atom to the vicinity of its proper location. But because the data set does not extend to atomic resolution, some other means must be used to confine the individual atoms of the molecular model. As one is dealing with the covalent structure of a polypeptide rather than a distribution of unconnected atoms, the bond lengths and bond angles of the covalent bonds connecting the atoms can be used to correlate their motions. For example, when any one of the carbon atoms in the phenyl ring of a phenylalanine is shifted, all the others must also be shifted accordingly because they are all covalently attached to each other. To accomplish this, constraints are added to the minimization to force the motions of the bonded atoms to be correlated.[46] The definition of θ is changed so that

$$\theta = \sum_p w_p \left(|F_o|_p - |F_c|_p \right)^2 + \sum_q w_q \left(d_q'^2 - d_{c,q}^2 \right)^2 \quad (4\text{--}18)$$

where d_q' is the ideal distance between any two atoms that are rigidly connected by the covalent structure, $d_{c,q}$ is the distance between them in the final, refined molecular model, and w_q is a weight whose magnitude is chosen on the basis of how constrained the particular distance must be. If the two atoms whose positions are x_i and x_j, respectively, are directly attached to each other, w_q is large. If there are three or four covalent bonds between them, w_q is small. By adding the second term in Equation 4–18, bond distances and any rigid bond angles, such as those in a phenyl ring, are retained during the minimization.

This approach has been further improved[47] by allowing for the fact that neighboring atoms in a real molecule are not held rigidly with respect to each other but are subject to vibrational modes. This improvement replaces the distances in Equation 4–18 with variances of interatomic distances. In this way the displacements resulting from the vibrations can be explicitly estimated during the refinement as well as the mean positions of the atoms.

Alternatively, it has been proposed[48] that θ can be written as

$$\theta = \sum_p w_p \left(|F_o|_p - |F_c|_p \right)^2 + w_e E \quad (4\text{--}19)$$

where E is a theoretically calculated value of the potential energy for the molecular model and w_e is a weight given

to this term. The weight w_e is arbitrarily adjusted to make it more or less important during the refinement. In this approach, covalent bonds between atoms remain because their distortion would produce a major increase in E. This approach has an advantage over the consideration of only interatomic distances (Equation 4–18) because any shift in an atom in the molecular model causing it to overlap another atom automatically causes E to increase dramatically. The disadvantage of using E is that, once overlaps are eliminated and covalent bonds retained, the refinement is influenced by a large number of electrostatic, noncovalent forces imposed by the theoretical function and these may or may not be realistic. These biases influence the shifts of the atoms dictated by $\mathbf{h}$.

In practice, refinement always proceeds pragmatically. For example, Equation 4–15 is used to calculate shifts in the atoms that cause θ to assume a minimum value. These shifts are used to calculate a new set of atomic positions x_j and another cycle of refinement is performed with these new positions as initial values for the x_j. In each of the first few of these cycles the value of R drops (Figure 4–25), but the decreases become more modest at each cycle until no further progress can be made. The reason for this is that the refinement process has become trapped within a local minimum of the function θ because refinement performed in this way will only progress as long as θ is decreasing. Manual tuning, however, is one way to escape from the local minimum in which the process is trapped. At some point (designated ΔF in Figure 4–25), the decision is made to calculate a difference map of electron density. Adjustments of the current molecular model are made by manual tuning, and this allows the minimization to enter a new trajectory. After this trajectory reaches a new local minimum, more tuning is performed. As the process advances, molecules of **water** (oxygen atoms) and anions and cations from the crystallization solution that are bound at specific locations on the surface or in the interior of the molecules of protein begin to appear in the difference maps and they become sharp and reproducible features. They appear in the difference maps of electron density because they are fixed at certain locations in the real unit cell by their specific noncovalent interactions with the molecules of protein but have not yet been incorporated into the molecular model. When the identity and location of these molecules of water and these ions become sufficiently unambiguous, they are incorporated as atoms in the molecular model (this was done at cycle 98 in Figure 4–25). Their inclusion causes a significant decrease in the R-factor because they are as real a feature of the actual crystallographic unit cell as the individual amino acids in the polypeptide, and they contribute accordingly to $|F_o|$. For example, in the refinement for deoxyribonuclease I described in Figure 4–25, the inclusion of the water molecules observed in difference maps rendered the molecular model much more realistic and permitted the refinement to produce a significantly lower minimum of the *R*-factor than it had before they were included. This reasonable consequence suggests that the refinement is registering reality.

The process just described requires a significant amount of time. Whenever the refinement reaches a plateau and no further progress is evident (Figure 4–25), the molecular model must be examined in detail and manually adjusted with the assistance of a map of difference electron density before a new trajectory can be initiated. The reason for this is that the refinement is performed in such a way that the *R*-factor must decrease at each step, and this causes the refinement to become trapped in a local minimum for the *R*-factor rather than searching out the global minimum that would represent the best fit of the molecular model to the experimental data set.

It has been found that one way to avoid the time-consuming manual adjustments required during the refinement is to combine molecular dynamics and refinement.[49] In a molecular dynamics simulation, atoms are positioned in space and a potential energy function E_i, incorporating the potential energies of the covalent bonds and the potential energies of nonbonded interactions, is calculated. The atoms are then given kinetic energies appropriate to a certain temperature and allowed to move within this potential energy function according to classical laws of motion for a short interval (less than 1 fs). The new positions in turn create a new potential energy function and the atoms are allowed to move again in response to these new potential energies, and so forth.

When molecular dynamics is used in crystallographic refinement, the empirical potential energy function, E_i, for each step i in the usual molecular dynamics calculation is augmented by an effective potential energy

$$E_x = w_x \sum_p w_p \left(\left| F_o \right|_p - \left| F_c \right|_p \right)^2 \qquad (4\text{–}20)$$

where w_x is a weighting factor chosen so that E_x has the same magnitude as E_e and $|F_c|$ is the set of reflections calculated for the instantaneous distribution of atoms after each step in the molecular dynamics calculation. This effective potential energy, E_x, constrains the atoms during the molecular dynamic trajectory to the vicinity they occupied in the original molecular model, but if a high enough kinetic energy is applied, the atoms can move as much as 0.3 nm from their initial positions.[50] This is what allows the structure to break out of local minima of the function θ.

The process proceeds in several steps referred to as simulated annealing.[51] Initially a high kinetic energy is applied to the atoms (high temperature), and then the kinetic energy is decreased to finish within a minimum of potential energy. It is while the kinetic energy is high that

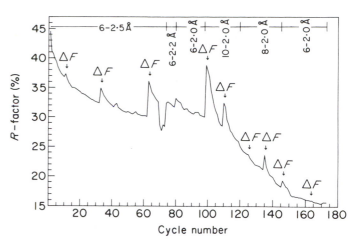

Figure 4–25: Progress of a refinement of the crystallographic molecular model of deoxyribonuclease I.[35] An initial molecular model was constructed by inserting a molecular model of the polypeptide into the initial map of electron density. The *R*-factor of the initial map was 0.45 (expressed in the figure as percent). The *R*-factor is presented as a function of the number of cycles of least-squares refinement performed. At the cycles indicated by ΔF, difference maps were constructed from $|F_o|$ and $|F_c|$ using the calculated phases of the molecular model to that point. These difference maps were used to rebuild the model manually and establish a new trajectory for the refinement. At cycle 98, molecules of water were added to the molecular model at locations identified in the maps of difference electron density. The resolution range included in the data sets at each cycle is indicated at the top of the figure. The final *R*-factor of the refined molecular model was 0.16. Reprinted with permission from ref 35. Copyright 1986 Academic Press.

local minima of potential energy, and hence local minima of the function θ, can be passed through. It has been shown that in this way the *R*-factor can be minimized with much less need for manual adjustment of the molecular model.[52]

It should be noted, however, that the molecular dynamics procedure, used for the purpose of pushing the refinement out of local minima, includes coincidentally a large number of empirical and not necessarily realistic constraints, and the final structure will incorporate these constraints. In addition, ideal bond lengths and bond angles have been enforced upon the crystallographic molecular model because if they were not, the refinement could not have been performed at all. Therefore, if a refinement were performed entirely by the computer, the final molecular model would be confined by all of these implicit and often unsubstantiated constraints. To verify that these have not biased the final structure, careful inspections of difference maps of electron density are always required to identify locations where the actual structure of the protein deviates from these simple expectations.

This is routinely done by using **omit maps** of difference electron density. The first 10 amino acids in the final refined crystallographic molecular model are omitted and the truncated model that results is used to calculate $|F_c|^{omit}$ and α_c^{omit}. The observed data set $|F_o|$ and $|F_c|^{omit}$ and α_c^{omit} are used to calculate (Equation 4–13) a difference map of electron density. In this difference map the omitted segment appears as positive electron density. This positive electron density has the advantage that its details are defined only by the observed data set because nothing is present at this location in the truncated molecular model. The 10 amino acids in the refined molecular model in this region are adjusted, if necessary, to fit within this electron density. Then the next 10 amino acids are omitted and so forth down the whole polypeptide, to incorporate into the refined molecular model the ways in which the actual structure of the protein deviates from the ideal structure dictated by ideal bond lengths and bond angles and empirical functions of potential energy. It should be stressed at this point that the goal of all refinement is to produce a crystallographic molecular model that reproduces as accurately as possible the actual structure of the molecule of protein, including all of its perversities, rather than some ideal structure consistent with a set of theoretical potential energy functions.

Inherent in the process of refinement is the ability to produce a crystallographic molecular model by **molecular replacement**. Many proteins of interest are closely related to other proteins for which a crystallographic molecular model is already available. This relationship can be established by aligning the amino acid sequences of the two proteins. For example, the amino acid sequence of the aspartyl proteolytic enzyme from Rous sarcoma virus ($n_{aa} = 124$) can be aligned with the amino acid sequence of the aspartyl proteolytic enzyme from human immunodeficiency virus, type 1 ($n_{aa} = 99$), so that there are 30 identical residues and four gaps, all in the shorter protein.[53] It necessarily follows that the structures of these two proteins are superposable. The three long gaps in the shorter amino acid sequence (10, 5, and 6 amino acids, respectively) can be assumed to represent loops in the larger protein missing from the smaller. A crystallographic molecular model, produced by the normal method of multiple isomorphous replacement, was available for the larger of the two proteins, that from Rous sarcoma virus.[54] Crystals of the protein from the human immunodeficiency virus were produced, and a data set was collected from them. The side chains of the amino acids in the crystallographic molecular model of the protein from Rous sarcoma virus were replaced with the corresponding side chains in the aligned amino acid sequence of the protein from the human immunodeficiency virus. The loops corresponding to the gaps in the alignments were removed from the model to produce a preliminary molecular model for the protein from the human immunodeficiency virus. This model was aligned in the unit cell defined by the data set collected from crystals of this protein. This preliminary model was then submitted to refinement to produce a final structure with an R-factor of 0.18.[55] As this example illustrates, the method of molecular replacement avoids the difficulties of multiple isomorphous replacement.

How much more reliable is a refined crystallographic molecular model than an initial model built into the map of electron density produced by the phases determined experimentally by multiple isomorphous replacement? It is true that the R-factor is much smaller, but this is not surprising because the decrease occurred automatically. It is gratifying, however, because the refined molecular model is usually not remarkably different from the original molecular model but has an electron density that produces reflections whose amplitudes are much closer to the observed amplitudes, which are the only directly observed quantities.

Often the success of the refinement is touted by showing that the map of electron density calculated from $|F_o|$ and the α_c of the final molecular model has features that very closely resemble phenyl rings or other equally characteristic side chains, but this is illusory. A map of electron density constructed from $|F_c|$ and α_c would have to have features precisely resembling these side chains because $|F_c|$ and α_c are calculated from the molecular model itself, which always has ideal bond angles and bond lengths for the entire polypeptide. The minimization has automatically caused $|F_c|$ to be very similar to $|F_o|$. Therefore, it is not surprising that the details explicitly put into the model by the crystallographer should reappear when the map of electron density is reconstructed from $|F_o|$ and α_c.

In fact, an important characteristic of crystallography is hidden in this particular illusion. The amplitude of a reflection can only assume a value greater than 0. The phase of a reflection, operating through the cosine (Equation 4–8), can have little effect or it can erase the amplitude or it can invert the sign of the amplitude or it can do anything in between. This causes the phases to have more impact on the final map of electron density than the amplitudes. A set of unitary amplitudes combined with the correct phases will always give a more accurate map of electron density than a set of correct amplitudes combined with a set of phases that are all forced to be identical to each other. Yet it is only amplitudes that can be observed. The phases obtained from isomorphous replacement experiments are unreliable, and phases calculated from a molecular model bias the final map toward resembling the molecular model. It is always informative to keep track of what phases are being used by a crystallographer in the calculation of a particular map of electron density. When there is a decision to be made, α_c is usually chosen because it produces a more gratifying map of electron density.

A similar difficulty in evaluating the success with which a particular refinement has reproduced the real structure of the molecule of protein is the nature of the constraints applied. To progress efficiently, a refinement must be forced to retain bond lengths and bond angles or be at the minimum of an empirical function for potential energy; however, every constraint enforced upon the refinement is automatically incorporated into the final structure regardless of its reality. For example, if one of the constraints in Equation 4–18 is that every peptide bond shall be planar, the planarity of the peptide bonds in the final structure cannot be cited as a measure of the success of the refinement. If one of the constraints inadvertently introduced by Equation 4–19 is that the conformation along no carbon–carbon bond can eclipse vicinal methyls or methylenes, the absence of such eclipsed conformations cannot be cited as a measure of the success of the refinement. These are not minor criticisms. The structure of a map of electron density improves so breathtakingly upon refinement that those changes that were enforced by the crystallographer must be clearly separated from those changes that arise only from the real molecular structure in the crystal. It is only these unconstrained and often unexpected features of the refined map of electron density which clearly state that it is an improvement.[56]

Suggested Reading

Oefner, C., & Suck, D. (1986) Crystallographic Refinement and Structure of DNase I at 2-Å Resolution, *J. Mol. Biol. 192*, 605–632.

References

1. Stout, G.H., & Jensen, L.H. (1968) *X-ray Structure Determination*, Macmillan, New York.
2. Sachsenheimer, W., & Schulz, G.E. (1977) *J. Mol. Biol. 114*, 23–36.
3. Xuong, N., Sullivan, D., Nielsen, C., & Hamlin, R. (1985) *Acta Crystallogr. B 41*, 267–269.
4. Hajdu, J., Machin, P.A., Campbell, J.W., Greenhough, T.J., Clifton, I.J., Zurek, S., Gover, S., Johnson, L.N., & Elder, M. (1987) *Nature 329*, 178–181.
5. Matthews, B.W. (1977) in *The Proteins* (Neurath, H., & Hill, R.L., Eds.) 3rd ed., Vol. III, pp 404–590, Academic Press, New York.
6. Suck, D., Oefner, C., & Kabsch, W. (1984) *EMBO J. 3*, 2423–2430.
7. Fraser, R.D.B., & MacRae, T.P. (1969) in *Physical Principles and Techniques in Protein Chemistry* (Leach, S.J., Ed.) Part A, pp 59–100, Academic Press, New York.
8. Dickerson, R.E. (1964) in *The Proteins* (Neurath, H., Ed.) 2nd ed., Vol. II, pp 603–778, Academic Press, New York.
9. Harris, J.I., & Li, C.H. (1955) *J. Biol. Chem. 213*, 499–507.
10. Cullis, A.F., Muirhead, H., Perutz, M.F., Rossmann, M.G., & North, A.C.T. (1961) *Proc. R. Soc. London, A 265*, 15–38.
11. Blow, D.M. (1958) *Proc. R. Soc. London, A 247*, 302–336.
12. Ollis, D.L., Brick, P., Hamlin, R., Xuong, N.G., & Steitz, T.A. (1985) *Nature 313*, 762–767.
13. Liljas, A., Kannan, K.K., Bergsten, P., Waara, I., Fridborg, K., Strandberg, B., Carlbom, U., Jarup, L., Lövgren, S., & Petef, M. (1972) *Nature New Biol. 235*, 131–137.
14. Black, C.C.F., & Evans, P.R. (1974) *J. Mol. Biol. 84*, 585–601.
15. Gilliland, G.L., & Quiocho, F. (1981) *J. Mol. Biol. 146*, 341–362.
16. Vainshetein, B.K., Melik-Adamyan, W.R., Barynin, V.V., Vagin, A.A., Grebenko, A.I., Borisov, V.V., Bartels, K.S., Fita, I., & Rossman, M.G. (1986) *J. Mol. Biol. 188*, 49–61.
17. Banyard, S.H., Stammers, D.K., & Harrison, P.M. (1978) *Nature 271*, 282–284.
18. Branden, C.-I., Eklund, H., Nordstrom, B., Boiwe, T., Soderlund, G., Zeppezauer, E., Ohlsson, I., & Akeson, A. (1973) *Proc. Natl. Acad. Sci. U.S.A. 70*, 2439–2442.
19. Wyckoff, H.W., Tsernoglou, D., Hanson, A.W., Knox, J.R., Lee, B., & Richards, F.M. (1970) *J. Biol. Chem. 245*, 305–328.
20. Brändén, C., Eklund, H., Nordström, B., Boiwe, T., Söderlund, G., Zeppezauer, E., Ohlsson, I., & Åkeson, Å. (1973) *Proc. Natl. Acad. Sci. U.S.A. 70*, 2439–2442.
21. Eklund, H., Nordström, B., Zeppezauer, E., Söderlund, G., Ohlsson, I., Boiwe, T., Söderberg, B., Tapia, O., & Brändén, C. (1976) *J. Mol. Biol. 102*, 27–59.
22. Pauling, L., Corey, R.B., & Branson, H.R. (1951) *Proc. Natl. Acad. Sci. U.S.A. 37*, 205–211.
23. Pauling, L., & Corey, R.B. (1951) *Proc. Natl. Acad. Sci. U.S.A. 37*, 729–740.
24. Perutz, M.F. (1951) *Nature 167*, 1053–1054.
25. Venkatachalam, C.M. (1968) *Biopolymers 6*, 1425–1436.
26. Richardson, J.S. (1981) *Adv. Protein Chem. 34*, 167–338.
27. James, M.N.G., & Sieleki, A.R. (1983) *J. Mol. Biol. 163*, 299–361.
28. Gilliland, G.L., & Quiocho, F.A. (1981) *J. Mol. Biol. 146*, 341–362.
29. Quiocho, F.A., Gilliland, G.L., & Phillips, G.N. (1977) *J. Biol. Chem. 252*, 5142–5149.
30. Harte, R.A., & Rupley, J.A. (1968) *J. Biol. Chem. 243*, 1663–1669.
31. Deisenhofer, J., Colman, P.M., Epp, O., & Huber, R. (1976) *Hoppe-Seyler's Z. Physiol. Chem. 357*, 1421–1434.
32. Remington, S., Wiegand, G., & Huber, R. (1982) *J. Mol. Biol. 158*, 111–152.
33. Deisenhofer, J., Epp, O., Miki, K., Huber, R., & Michel, H. (1985) *Nature 318*, 618–623.
34. Deisenhofer, J., Epp, O., Miki, K., Huber, R., & Michel, H. (1984) *J. Mol. Biol. 180*, 385–398.
35. Oefner, C., & Suck, D. (1986) *J. Mol. Biol. 192*, 605–632.
36. Yu, N., & Jo, B.H. (1973) *J. Am. Chem. Soc. 95*, 5033–5037.
37. Shaanan, B. (1983) *J. Mol. Biol. 171*, 31–59.
38. Haurowitz, F. (1938) *Hoppe-Seyler's Z. Physiol. Chem. 254*, 266–274.
39. Freer, S.T., Kraut, J., Robertus, J.D., Wright, H.T., & Xuong, N.H. (1970) *Biochemistry 9*, 1997–2009.
40. Ghosh, D., O'Donnell, S., Furey, W., Robbins, A.H., & Stout, C.D. (1982) *J. Mol. Biol. 158*, 73–109.
41. Stout, C.D. (1989) *J. Mol. Biol. 205*, 545–555.
42. DeVos, A.M., Tong, L., Milburn, M.V., Matias, P.M., Jancarik, J., Noguchi, S., Nishimura, S., Miura, K., Ohtsuka, E., & Kim, S. (1988) *Science 239*, 888–893.
43. Pai, F.F., Kabsch, W., Krengel, U., Holmes, K.C., John, J., & Wittinghofer, A. (1989) *Nature 341*, 209–214.

44. Waser, J. (1963) *Acta Crystallogr. A 16*, 1091–1094.

45. Hetenes, M.R., & Stiefel, E. (1952) *J. Natl. Bur. Stand. 49*, 409–436.

46. Konnert, J.H. (1976) *Acta Crystallogr. A 32*, 614–617.

47. Konnert, J.H., & Hendrickson, W.A. (1980) *Acta Crystallogr. A 36*, 344–350.

48. Jack, A., & Levitt, M. (1978) *Acta Crystallogr. A 34*, 931–935.

49. Brunger, A.T., Karplus, M., & Petsko, G.A. (1989) *Acta Crystallogr. A 45*, 50–61.

50. Brunger, A.T., Kuriyan, J., & Karplus, M. (1987) *Science 235*, 458–460.

51. Kirkpatrick, S., Gelatt, C.D., & Vecchi, M.P. (1983) *Science 220*, 671–680.

52. Brunger, A.T. (1988) *J. Mol. Biol. 203*, 803–816.

53. Weber, I.T., Miller, M., Jaskolski, M., Leis, J., Skalka, A.M., & Wlodawer, A. (1989) *Science 243*, 928–931.

54. Miller, M., Jaskolski, M., Rao, J.K.M., Leis, J., & Wlodawer, A. (1989) *Nature 337*, 576–579.

55. Wlodawer, A., Miller, M., Jaskolski, M., Sathyanarayana, B.K., Baldwin, E., Weber, I.T., Selk, L.M., Clawon, L., Schneider, J., & Kent, S.B.H. (1989) *Science 245*, 616–621.

56. Artymiuk, P.J., & Blake, C.C.F. (1981) *J. Mol. Biol. 152*, 763–782.

CHAPTER 5

Noncovalent Forces

Crystallographic studies have demonstrated that a molecule of protein, dissolved in aqueous solution, is composed of folded polypeptides, each of which is folded into a structure identical or almost identical to the structure of other polypeptides of the same amino acid sequence. A polypeptide, as it emerges from the ribosome, is a fluid polymer of undefined structure. Each newly synthesized polypeptide then folds spontaneously to assume its unique secondary and tertiary structure.

The folding of polypeptides to form the native structure of a protein, the association of folded polypeptides to form multimeric proteins, and the binding of substrates, coenzymes, or other molecules to proteins proceed without the formation of covalent bonds and are controlled by noncovalent forces. It appears that four noncovalent forces are involved in these chemical reactions: ionic interactions, hydrogen bonds, the hydrophobic effect, and van der Waals forces. In the refined crystallographic molecular model of a folded protein, the consequences of these noncovalent forces are evident. The chemical and physical properties of these interactions, as they occur in aqueous solution, must be understood before their consequences can be appreciated. Consequently, a discussion of these interactions must precede a detailed description of the atomic details of refined crystallographic molecular models. None of the four categories of noncovalent forces—ionic interactions, hydrogen bonds, the hydrophobic effect, and van der Waals forces—can be completely separated from all of the others. Van der Waals forces must play a part in each of the other three phenomena, hydrogen bonds can be considered to be special cases of ionic interactions, almost all ionic interactions in biochemical situations involve hydrogen bonds, and the hydrophobic effect is to a large degree the reflection of hydrogen bonding in the solvent. It is informative, however, to discuss each of these categories separately to focus on their unique properties.

Each of these types of interactions can be considered to be a special case of a noncovalent association between two molecules, A and B, or two segments, A and B, of the same molecule. For the situations under discussion, our attention will be directed to such a reaction as it would occur in aqueous solution. A general chemical equation for these associations is

$$A(H_2O)_x + B(H_2O)_y \rightleftharpoons A \cdot B(H_2O)_z + (x+y-z)H_2O$$

$$(5-1)$$

The species $A(H_2O)_x$ and $B(H_2O)_y$ are the separated solutes dissolved in water and surrounded on all sides by water. Presumably, there are a certain number of water molecules, x and y, respectively, that are significantly affected by the presence of A or B. The effects of the solute on the surrounding molecules of water and the effects of the surrounding molecules of water on the solute are referred to as **hydration**. At a particular molecule of solute at a particular instant, a particular number of water molecules are affected significantly by the presence of the solute. This number fluctuates with time, and the coefficients x, y, and z are intended to represent averages over a range of possible configurations of hydration. When A and B associate to form the noncovalent complex, that complex will also be surrounded by the solvent, and there will be a number of water molecules, z, that are significantly influenced by the complex. As A·B always has a smaller surface area than the sum of the surface areas of A and B, z should be less than $x + y$, and $(x + y - z)$ molecules of water will return to the bulk phase of the solvent when the complex is formed.

The change in standard free energy for the overall reaction can be expressed as

$$\Delta G^\circ = \Delta G^\circ_{A \cdot B} + \Delta G^\circ_{HYD(A \cdot B)} + \Delta G^\circ_{rH_2O} - \Delta G^\circ_{HYD(A)} - \Delta G^\circ_{HYD(B)}$$

$$(5-2)$$

where $\Delta G^\circ_{HYD(i)}$ refers to the standard free energy of hydration between each of the solutes and its surrounding waters of hydration, $\Delta G^\circ_{A \cdot B}$ refers to the direct standard free energy of interaction between A and B, and $\Delta G^\circ_{rH_2O}$ is the change in standard free energy experienced by the $x + y - z$ molecules of water as they leave shells of hydration and return to the bulk aqueous phase. It will become apparent that all of the terms on the right-hand side of Equation 5–2 except sometimes the first are remarkably influenced by the fact that this reaction occurs in water as a solvent. No noncovalent interaction has the same outcome in any other solvent. For example, hydrogen bonds and ionic interactions are stable in almost any other solvent but are dissociated by water, while the hydrophobic effect is observed only when the solvent is water. To appreciate fully these influences of water on the outcome of noncovalent associations, the structure of liquid water must be understood.

Water

The properties of liquid water, when considered in their entirety, are unlike those of any other liquid. For example, the surface tension of water at 20 °C is 73 dyne cm^{-1}, while those of most other liquids are between 20 and 40 dyne cm^{-1}. The dielectric constant of water is 80, while the dielectric constants of other liquids, with few exceptions, are less than 30. The high melting point and boiling point of water, for a molecule of its size and composition, are well-publicized anomalies. Not only are the numerical values of the physical constants anomalous, but the qualitative behaviors of the thermodynamic properties of the liquid, when it is exposed to variations of physical forces such as pressure, temperature, electric field, and electromagnetic energy, are unique. The details of these peculiarities provide an intuitive picture of the structure of liquid water that can serve as a basis for understanding the behavior of solutes such as polypeptides in this solvent. Unfortunately, there is no adequate molecular model for the structure of liquid water, and an informed intuitive picture is the closest approach to reality currently available.

A water molecule in the dilute, ideal **vapor** is an oxygen atom bonded covalently to two hydrogen atoms. Quantum mechanical calculations[1,2] of the isolated molecule in the vacuum seem[3] to support the conventional orbital picture of an oxygen hybridized sp^3 with two covalent bonds to two hydrogens and two σ lone pairs of electrons; these four substituents are oriented tetrahedrally around the oxygen. The HOH bond angle[4] is 104.5°, distorted from 109.5° by the electron repulsion of the lone pairs or by a rehybridization, driven by energy of promotion, that gives the oxygen–hydrogen σ bonds more p character. The oxygen–hydrogen bond lengths are 0.096 nm.

In more concentrated vapor, **dimers of water** form (Figure 5–1). From results of molecular beam microwave spectroscopy, the mean structure of the dimer can be calculated.[5] The two oxygens are separated by a distance of 0.298 nm. One of the four hydrogens lies on the line of centers between the two oxygens, and it is covalently bonded to one of them, which is referred to as the proton donor. The other oxygen, which is referred to as the proton acceptor, also has two hydrogens covalently bonded to it. The plane defined by the two hydrogens and the oxygen of the proton acceptor is inclined at an angle of 60° to the line of centers between the two oxygen

atoms. This means that the four substituents around the acceptor—the two hydrogens, the shared hydrogen, and the lone pair of electrons—are tetrahedrally arrayed in the dimer. This arrangement suggests that the oxygen–hydrogen bond on the donor points directly at one of the two σ lone pairs of electrons on the acceptor oxygen. The axis of the sp^3 orbital in which that σ lone pair resides should be congruent with the line of centers. The interaction between the hydrogen–oxygen σ bond on the donor molecule of water and the σ lone pair of electrons on the acceptor is an unhindered example of a hydrogen bond. The formation of dimers and higher oligomers in steam contributes significantly to its nonideal behavior at higher concentrations of water in the gas phase.

The ice that is in equilibrium with liquid water at atmospheric pressure and 0 °C is known as ice Ih. **Ice Ih** is a tetrahedral diamond lattice of oxygen atoms (Figure 5–2A), each 0.276 nm from its nearest neighbor.[6] The oxygens are held in the lattice by hydrogen bonds to each of their four nearest neighbors. Between any oxygen atom and each of its four nearest neighbors in the lattice is one hydrogen atom. At any instant, each hydrogen is covalently bound to one of the two oxygens between which it is found, and every oxygen has only two hydrogens covalently bound to it. These two requirements create a situation in which only a predictable number of arrangements for these hydrogens can occur, and this number of arrangements can explain almost exactly the observed residual entropy of ice Ih at 0 K.[4] There is a significant amount of empty space in ice Ih (Figure 5–2B), and this is one of the properties permitting it to be less dense than the liquid with which it can be in equilibrium.

The structure and properties of ice Ih and water vapor have been exhaustively investigated and unambiguously established. At atmospheric pressure, **liquid water** lies between these two extremes on the phase diagram, and its properties can be compared with them. It is in the transitions between solid and liquid and between liquid and vapor that insight into the structure of the liquid can be gained.

When ice melts, the reaction involves an **enthalpy of fusion**; and, when the liquid vaporizes, the reaction involves an **enthalpy of vaporization**. The enthalpy of water at atmospheric pressure can be plotted as a function of absolute temperature (Figure 5–3). On this plot, the discontinuities at the melting point and boiling point are the enthalpy of fusion and the enthalpy of vaporization, respectively. At 25 °C, the enthalpy of fusion is 8.0 kJ mol^{-1}, and the enthalpy of vaporization is 44 kJ mol^{-1}.

The high isopiestic **heat capacity**, C_P, of liquid water (Figure 5–3) also indicates that it is highly structured. Calculations of the heat capacity of both ice Ih and water vapor, from the known vibrational energy levels of these two substances, agree quite closely with observed values of their respective heat capacities.[4] The observed value of the heat capacity of liquid water, however, is almost twice that calculated from its estimated vibrational energy

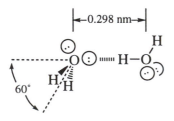

Figure 5–1: Qualitative picture of the dimer of two molecules of water in a dimer in the gas phase. The distances and angles were obtained by microwave spectroscopy.[5]

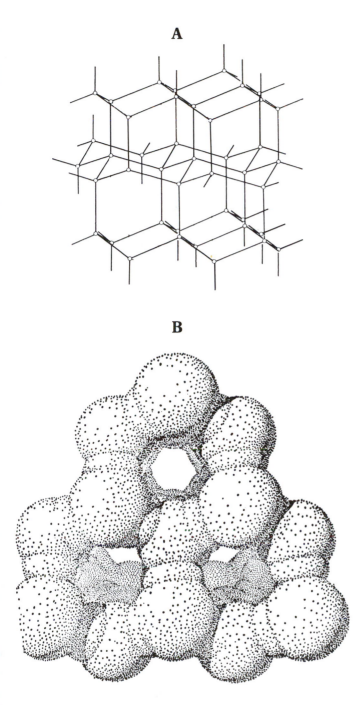

A

B

Figure 5–2: Structure of ice Ih. (A) Tetrahedral lattice.[4] In a tetrahedral lattice each atom or molecule occupies a position indicated by the small open circles. It forms equivalent connections with its four nearest neighbors, and because every connection is equivalent and every nearest neighbor is the same, the nearest neighbors are arrayed tetrahedrally and at equal distance. (B) Representation of space-filling models of molecules of water arrayed on the tetrahedral lattice of ice Ih. The atomic radii of the oxygen atoms are their tabulated van der Waals radii (0.14 nm), the bond lengths are the bond lengths in ice Ih (0.28 nm), and the hydrogens are sandwiched between the oxygens. Reprinted with permission from ref 4. Copyright 1969 Clarendon Press.

(Figure 5–4). This excess or configurational heat capacity can be explained by postulating that much of the hydrogen-bonded structure of ice remains in the liquid and its gradual deterioration as the temperature is raised is responsible for the anomalous absorption of heat. The high and relatively constant value for the heat capacity throughout the range of temperature between 0 and 100 °C suggests that the hydrogen-bonded network in the liquid is gradually and constantly deteriorating as the temperature is rising.

Another indication of the extensive hydrogen-bonded structure in liquid water is its high static **dielectric constant** ($\varepsilon = 88$ at 0 °C), which is almost equivalent to that of ice Ih ($\varepsilon = 100$ at 0 °C).[4] The large value for the dielectric constant of ice Ih is usually explained semi-quantitatively[4] as a result of the high correlation among the orientations of the individual dipole moments of the water molecules caused by their rigid arrangement in the hydrogen-bonded lattice. When an electric field is applied, the dipole moments reorient cooperatively, producing the large dielectric constant. The fact that the liquid has almost the same dielectric constant as the solid indicates that much of the lattice remains.

The molar volume of ice (19.6 cm^3 mol^{-1}) is somewhat greater than that of liquid water (18.0 cm^3 mol^{-1}) at 0 °C and much greater than the molar volume that would be expected if spheres the radius of molecules of water (0.14 nm) were randomly packed in an unstructured, disordered array (10 cm^3 mol^{-1}).[4] The large molar volume of ice Ih is due to the vacant space created by the fact that oxygens are held in a tetrahedral array by the hydrogen-bonded network (Figure 5–2B). When ice melts, the molecules of water are allowed to occupy some of the vacant space in the hydrogen-bonded lattice and the density increases. A related fact is that the molar volume of liquid water increases as temperature is decreased below 4 °C, presumably because the expansion caused by the strengthening of the hydrogen-bonded lattice is greater than the usual contraction resulting from the decrease in thermal energy. It is only above 4 °C that the latter effect becomes dominant. The contraction of water upon melting and the expansion of the liquid upon cooling below 4 °C are almost unprecedented. Diamond, silicon, and germanium are tetrahedal solids that also float upon their melts, as ice floats upon water.

Aside from these peculiar features, the molar volumes of ice and liquid water at 0 °C are both large and not that different from each other, and this suggests that much of the vacant space created by the hydrogen-bonded lattice in the solid remains in the liquid. This possibility could also explain the unique decrease in isothermal compressibility that occurs in liquid water as temperature is raised from 0 to 50 °C (Figure 5–5).[4] The **coefficient of isothermal compressibility** of a liquid, γ_T, is the fractional decrease in volume produced by the application of pressure at constant temperature

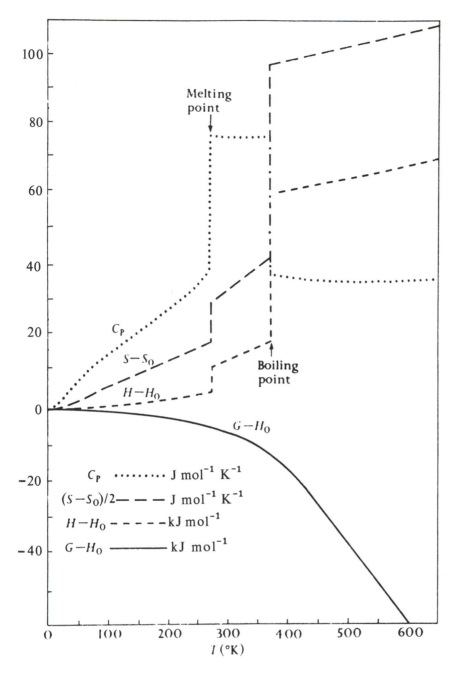

Figure 5–3: Enthalpy $(H - H_0)$, entropy $(S - S_0)$, free energy $(G - H_0)$, and isopiestic heat capacity (C_P) as a function of temperature for water at unit atmosphere pressure.[4] The quantities of H_0 and S_0 are the enthalpy and entropy, respectively, at 0 K. Enthalpy and free energy are in units of kilojoules mole^{-1}, heat capacity is in units of joule mole^{-1} Kelvin^{-1} and the values of entropy in joule mole^{-1} Kelvin^{-1} are arbitrarily multiplied by 2 to avoid coincidence between the curves for enthalpy and entropy. This enthalpy of vaporization is more than twice that of a liquid such as chloro–methane ($\Delta H_{vap} = 19$ kJ mol^{-1} at 25 °C), which is a polar solvent ($\varepsilon = 13$) containing molecules incapable of forming hydrogen bonds but much larger than water (50.5 g mol^{-1}). This comparison illustrates the fact that the enthalpy of vaporization of water is anomalously large. In the sum of the two reactions, fusion and vaporization, all of the hydrogen bonds in ice Ih are lost. The fact that most of the enthalpy change occurs upon vaporization and the fact that the heat of vaporization is anomalously large suggest that liquid water retains most of the hydrogen bonds present in ice Ih. Adapted with permission from ref 4. Copyright 1969 Clarendon Press.

$$\gamma_T = -\frac{1}{V}\left(\frac{\partial V}{\partial P}\right)_T \qquad (5\text{–}3)$$

in units of reciprocal pressure (bar^{-1}). In almost every other liquid, isothermal compressibility increases monotonically with temperature. In liquid water at low temperatures, most of the structured vacant space of ice Ih remains when the transition from solid to liquid occurs, and this structured vacant space is gradually replaced with randomly distributed, unstructured vacant space, similar to that in other liquids as the temperature is raised. The high compressibility at low temperatures results from the ability of the lattice to decrease its volume upon the application of pressure at the expense of the significant vacant space among the oxygen atoms.

The idea that liquid water at low temperature retains a structure similar to that of ice Ih is also supported by the very small **coefficient of thermal expansion** of liquid water. Upon heating at atmospheric pressure between temperatures of 20 and 30 °C, other liquids expand about 3 times more rapidly than does water (Figure 5–6).[7] As pressure is applied, however, the coefficient of thermal expansion for water increases while the coefficients of thermal expansion for other liquids decrease. At high pressures both water and other liquids have about the same coefficient of thermal expansion. If liquid water at atmospheric pressure is extensively hydrogen-bonded with an expanded structure similar to that of ice Ih (Figure 5–2B), then as the temperature is raised, the decrease in structured empty volume due to the deterioration of this hydrogen-bonded lattice could almost can-

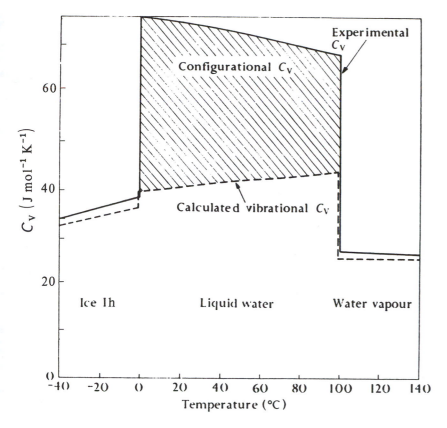

Figure 5–4: Separation of the experimental heat capacity of water (solid line) into vibrational (dashed line) and configurational (shaded difference) components.[4] The vibrational heat capacity of ice Ih was calculated from the two vibrational absorption bands of lowest energy ($v = 840$ cm^{-1} and $v = 230$ cm^{-1}), the "vibrational" heat capacity of water vapor was calculated from the vibrational, rotational, and translational energies of the water molecules, and the "vibrational" heat capacity of the liquid was calculated on the assumption that each molecule in the liquid has three hindered degrees of translation and three hindered librations. Adapted with permission from ref 4. Copyright 1969 Clarendon Press.

cel the increase in unstructured volume due to increased thermal motion. As pressure is applied, however, it could also cause the hydrogen-bonded structure to deteriorate or restructure and cause the liquid to have a more normal coefficient of thermal expansion. In this view, the ability of pressure to change the structure of the liquid is due to the fact that liquid water is in an extensively hydrogen-bonded form at normal pressures but not at higher pressures. A similar process is operating when ice melts under the blade of an ice skate.

Such a transition between an ordered and a less ordered state caused by an increase in pressure would also explain why the application of pressure decreases the **viscosity** of liquid water rather than increasing it as it does the viscosities of other liquids.[7] The viscosity of water is anomalously large in the first place ($\eta = 1.00 \times 10^{-2}$ g cm^{-1} s^{-1} at 20 °C) compared to the viscosity of liquids such as acetonitrile ($\eta = 0.36 \times 10^{-2}$ g cm^{-1} s^{-1} at 20 °C), pentane ($\eta = 0.24 \times 10^{-2}$ g cm^{-1} s^{-1} at 20 °C), and carbon disulfide ($\eta = 0.36 \times 10^{-2}$ g cm^{-1} s^{-1} at 20 °C).

Additional evidence for the retention of a significant fraction of the hydrogen-bonded lattice in liquid water is provided by scattering of X-rays. When a beam of X-rays is passed through a liquid, the X-rays are scattered by the electrons of the molecules in the liquid. The intensity of the scattered X-rays varies as a function of the angle between the incident beam and the direction at which they emerge from the solution. This angular dependence of the intensity can be used to calculate a **radial molecular correlation function**, $G_M(r)$. This function is an approximation[8] of the variation of electron density as a

function of the radial distance from any one molecule in the liquid. The actual variation of electron density is distinguished from its approximation by designating it as $g(r)$. The function $G_M(r)$ registers any local variations in the electron density of the liquid, relative to the mean electron density of the liquid, that are maintained around any one of the molecules. Because it is a relative quantity, the value of $G_M(r)$ is unity when the electron density is equal to the mean electron density. Any variations in density that are observed are assumed to be permanent features of the structure of the liquid. As $G_M(r)$ or $g(r)$ is proportional to the electron density as a function of radial distance from a central molecule, they can be used to calculate the total number of molecules of solvent in a spherical shell between r_1 and r_2 by the integral[8]

$$n_s = \frac{p_o}{\gamma} \int_{r_1}^{r_2} 4\pi r^2 g(r)\, dr \qquad (5\text{–}4)$$

where n_s is the number of molecules of solvent in that shell, p_o is the bulk electron density of the liquid, and γ is the number of electrons in each molecule of solvent (10 in the case of water).

The radial molecular correlation function for liquid water has been determined over a range of temperatures from 4 to 200 °C (Figure 5–7).[8] At fairly long distances from any one molecule of water (> 0.8 nm) the function becomes unity and does not vary noticeably. Therefore, beyond 0.8 nm from any given molecule the liquid is, on the average, homogeneous. A significant peak of density

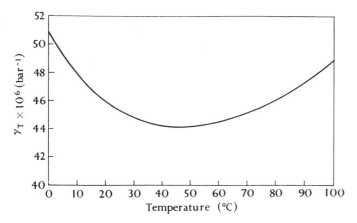

Figure 5–5: Coefficient of isothermal compressibility, γ_T, for liquid water at unit atmosphere, presented as a function of temperature.[4] Reprinted with permission from ref 4. Copyright 1969 Clarendon Press.

are four nearest neighbors at 0.276 nm, 12 next neighbors at 0.45 nm, and 12 farther neighbors at 0.52 nm (Figure 5–2). A distribution of liquid molecules of water confined to the diamond lattice of ice Ih would produce a radial molecular correlation function with a distinct minimum between the first four neighbors and the next group of 24 (Figure 5–7).[10]

When the radial molecular correlation functions of liquid water at 4 °C and liquid molecules of water confined to the lattice of ice Ih are compared, several differences are noted. Although still a prominent feature, the maximum in ice Ih centered at around 0.5 nm is considerably broadened in the liquid. This indicates that the hydrogen-bonded network has become considerably more elastic in water than in ice, permitting the second and third groups of neighbors to approach the molecule at the origin much more closely, rather than being held at a distance by a rigid lattice. There also seems to be too much electron density in the actual liquid between the first maximum and the second maximum.[10] This has been interpreted to mean that molecules are able to break out of the lattice and become interstitial molecules of water, transiently occupying the vacant spaces (Figure 5–2B).[11]

So far the discussion has emphasized similarities between ice Ih at 0 °C and liquid water at low temperatures. There are, of course, remarkable differences. The most obvious is the fact that ice is a solid and water is a liquid. To be a liquid, layers of molecules must be able to slide past layers of other molecules above and below them. In the case of water this requires extensive and simultaneous disruption of continuous layers of hydrogen bonds

occurs, however, at 0.28 nm. Integration of this peak[9] for the curve at 4 °C, upon the assumption that it is a Gaussian function, indicates that it is produced by about four nearest neighbors. In ice there are four nearest neighbors to each water molecule and they are held at a distance of 0.276 nm. It can be assumed that these are retained in the liquid. That the peak is centered at a distance so close to the hydrogen-bonded distance in ice has been interpreted to mean that each water molecule in the liquid has about four hydrogen-bonded nearest neighbors.

A radial molecular correlation function can be calculated[9] for liquid molecules of water confined to the tetrahedral lattice of ice Ih (Figure 5–7). In ice Ih, there

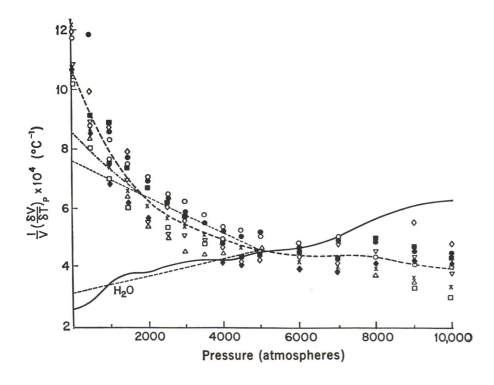

Figure 5–6: Coefficients of thermal expansion for several liquids at 25 °C as a function of applied pressure. The liquids are (×) PCl_3, (○) CH_3OH, (◊) CS_2, (●) C_2H_5Cl, (■) C_2H_4I, (△) C_2H_5OH, (▽) C_3H_9OH, (♦) isobutyl alcohol, and (□) n-$C_5H_{11}OH$. The solid curve is the coefficient of thermal expansion for liquid water at 25 °C as a function or pressure. Reprinted with permission from ref 7. Copyright 1970 American Association for the Advancement of Science.

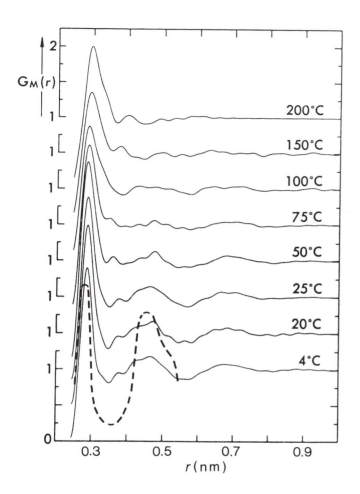

Figure 5–7: Molecular correlation functions for liquid water[8] and ice Ih.[10] The molecular correlation functions for liquid water at several temperatures (solid lines) were calculated from the angular dependence of the intensity of the scattered X-rays from samples of pure water through which a collimated beam of X-rays was passed. A molecular correlation function for liquid molecules arranged on the lattice of ice Ih (dashed line) was calculated from the length of the hydrogen bonds in ice Ih (0.276 nm) and the fact that the oxygen atoms lie upon a tetrahedral diamond lattice. The calculation was performed assuming that the distributions of electron density around the maxima defined by the lattice could be approximated by error functions. The width of the first error function was made the same as the width of the first maximum in liquid water at 4 °C and the widths of the two subsequent error functions were made proportional to the square of their distances from the origin.[6] Adapted with permission from ref 6, originally from ref 8. Copyright 1971 American Institute of Physics.

in the liquid as it flows. This capacity to flow is far more evident in water than in ice.

When individual molecules in water change their relative positions, hydrogen bonds must be broken and reformed elsewhere. The capacity to change positions is reflected in the process of **self-diffusion**, a measure of the rate at which the average molecule of water diffuses through a condensed phase of water molecules. The self-diffusion coefficient for ice Ih is about 10^{-10} cm^2 s^{-1} at 0 °C, and for liquid water it is 1.4×10^{-5} cm^2 s^{-1} at 5 °C.[4] This difference of 10^5 demonstrates that water molecules can exchange their hydrogen-bonded neighbors far more rapidly in the liquid than in the solid. To the extent that this exchange involves breaking and making of hydrogen bonds, the hydrogen bonds in the liquid are weaker than those in the solid.

An even more easily understood measurement of the rate at which a molecule of water can detach itself from the hydrogen-bonded network in the liquid in order to reorient itself is the **dielectric relaxation** of liquid water. The dielectric constant of a chemical substance is a function of the frequency of the alternating electric field used to measure it. Tabulated dielectric constants are usually static dielectric constants that are measured with an alternating electric field with a frequency of alternation so low that the measured values may be confidently

extrapolated to zero frequency. The low frequency of alternation allows the molecules in the substance more than ample time to align themselves, as far as they are able, with the electric field while the measurement is made. If the frequency of the applied field, however, is gradually increased, at some point the molecules in the substance will be unable to invert their alignments at rates sufficient to keep up with the alternations of the applied field. Their inability to keep up results from intermolecular forces that hinder their rotation. The dielectric relaxation time is the time that an applied field must be in operation before half of the increase in dielectric constant due to the rotation of the molecules aligning them with the field has occurred. The dielectric relaxation time of ice Ih at 0 °C is 2×10^{-5} s, that of liquid water at 0 °C is 2×10^{-11} s, and that of a water molecule in a dilute solution of water in benzene is 1×10^{-12} s.[4] Although a water molecule in liquid water is constrained so that it rotates 20 times more slowly than it does in a condensed phase lacking hydrogen bonds, it rotates 10^6 times faster in liquid water than in ice. This again demonstrates that the hydrogen bonds between water molecules in liquid water are weaker than those in ice.

This weakening of the hydrogen bonds in the liquid is also reflected in a shift that occurs in the frequency of the maximum **infrared absorption** of the oxygen–hydrogen stretching vibration of water when it melts.[4] The frequency at which a covalent bond absorbs infrared electromagnetic energy is correlated with its bond energy. The greater the bond energy, the higher the frequency of the light required to excite its vibration. In the case of the stretching frequency of the oxygen–hydrogen bond in water, the stronger the hydrogen bond in which it participates, the weaker will be the covalent oxygen–hydrogen bond itself and the lower the frequency of its absorption. The stretching frequency of the oxygen–hydrogen bond in ice Ih is 3220 cm^{-1}, in liquid water it is 3490 cm^{-1}, and in the vapor it is 3700 cm^{-1}. In the vapor, no hydrogen

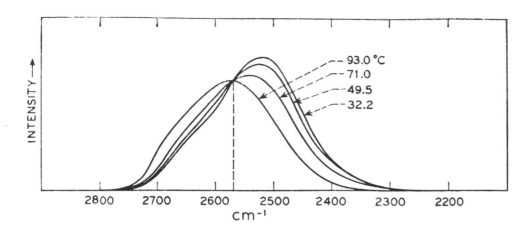

Figure 5–8: Raman spectrum of the oxygen–deuterium stretching vibration of DOH in dilute solution in H_2O taken at several temperatures.[12] The 435.8-nm line from a mercury lamp was used to excite a 6.2 M solution of D_2O in H_2O to obtain a Raman spectrum. The intensity scale is arbitrary. The vertical line at 2570 cm^{-1} marks the position of an isosbestic point.[7] Reprinted with permission from ref 7, originally from ref 12. Copyright 1968 American Institute of Physics.

bond weakens the oxygen–hydrogen bond. In ice Ih, a strong fixed hydrogen bond weakens the oxygen–hydrogen bond significantly. In liquid water, less than half the decrease in frequency between the vapor and ice Ih occurs, presumably because the hydrogen bonds formed when the vapor is converted to the liquid are much weaker than the hydrogen bonds formed when the vapor is converted to the solid.

The infrared stretching frequency of an isolated oxygen–hydrogen or oxygen–deuterium bond, uncoupled to the other oxygen–hydrogen or oxygen–deuterium bonds in the liquid, can be observed by making a dilute solution of water in deuterium oxide or a dilute solution of deuterium oxide in water, respectively. In such situations, the width of the uncoupled oxygen–hydrogen or oxygen–deuterium stretching vibrations are much sharper than they are in spectra of either of the pure liquids. They can be sharpened further by recording Raman spectra. In the case of the oxygen–deuterium stretching vibration in a dilute solution of D_2O in H_2O, the Raman spectrum[12] appears to show two types of oxygen–deuterium bonds that change in relative proportion as the temperature is changed between 30 and 90 °C (Figure 5–8). The isosbestic point in the collection of spectra suggests that there are only two distinct species present. This was originally proposed as evidence for two classes of water in the liquid, one less hydrogen-bonded than the other.[12]

It has long been thought that the properties of liquid water could be explained if it were a mixture of two classes of molecules, a highly structured, hydrogen-bonded class and a more normal liquid class of molecules.[13] It has already been noted that the existence of unbonded, interstitial water could explain the lack in liquid water of the minimum at 0.35 nm in the radial molecular correlation function of ice Ih.[11] The lack of this minimum, however, could also be explained by postulating that the radial molecular correlation function was actually the weighted average of a hydrogen-bonded phase and a disordered liquid phase coexisting in the same sample.

There are, however, several results that are inconsistent with the possibility that liquid water is such a mixture. The isosbestic point in Figure 5–8 disappears when the measurements are made over a larger range of temperatures.[7] Moreover, no evidence for two components is found in the infrared spectrum of a dilute solution of water in deuterium oxide.[7] Further, no simple binary mixture that varies in its composition with temperature and pressure can explain both the coefficient of thermal expansion and the temperature coefficient of the isothermal compressibility of liquid water simultaneously.[6] Finally, small-angle neutron scattering results indicate that no inhomogeneities larger than a few molecular dimensions can exist in liquid water[9] for longer than 10^{-11} s.[4]

This short description of the debate over the mixture model could be extended to a lengthy discussion of models that have been proposed for liquid water over the last 100 years. At the moment, nothing can be gained from this scholasticism, because no model has been able to explain all of the properties of liquid water. The mental picture of liquid water that forms intuitively as its peculiarities are described is presently more adequate than any sophisticated physical description of its structure.

Suggested Reading

Eisenberg, D., Kauzmann, W. (1969) *The Structure and Properties of Water*, Clarendon Press, Oxford, England.

Problem 5–1: The isopiestic heat capacity of a substance, C_p, is defined as the amount of heat required to raise one mole of the substance one degree in temperature at constant pressure. The units of this quantity are joules degree^{-1} mole^{-1}.

A substance has a certain intrinsic enthalpy at 0 K, and this intrinsic enthalpy increases as the temperature increases and the substance absorbs heat

$$H_T - H_0 = \int_0^T C_p \, dT + \Delta H_{pc}$$

where $H_T \equiv$ intrinsic enthalpy at $T = T$, $H_0 \equiv$ intrinsic enthalpy at $T = 0$ K, and $\Delta H_{pc} \equiv$ enthalpy changes for all phase transitions between 0 K and T. The heat capacity of H_2O is the following function of temperature

$$C_p = (0.172)T \text{ J deg}^{-1} \text{ mol}^{-1} \quad [T = 0-60 \text{ K}]$$

$$C_p = 2.47 + (0.129)T \text{ J deg}^{-1} \text{ mol}^{-1} \quad [T = 60-273 \text{ K}]$$

$$C_p = 77.5 \text{ J deg}^{-1} \text{ mol}^{-1} \quad [T = 273-373 \text{ K}]$$

and

$$\Delta H_{\text{fusion}} = 6.0 \text{ kJ mol}^{-1} \text{ at } 0 \text{ °C}$$

$$\Delta H_{\text{vap}} = 40.7 \text{ kJ mol}^{-1} \text{ at } 100 \text{ °C}$$

Use this experimental data to draw a graph of $H_T - H_0$ as a function of temperature from 0 to 375 K.

The intrinsic entropy of a substance is related to the heat capacity by the following equation

$$S_T - S_0 = \int_0^T \frac{C_p}{T} dT + \Delta S_{\text{pc}}$$

where S_T = intrinsic entropy at $T = T$, S_0 = intrinsic entropy at $T = 0$, and ΔS_{pc} = entropy changes at phase transitions.

When phase transitions occur, $\Delta G° = 0$ at the transition temperature. Use $\Delta H°_{\text{fusion}}$ and $\Delta H°_{\text{vap}}$ to calculate $\Delta S°_{\text{fusion}}$ and $\Delta S°_{\text{vap}}$. Plot $S_T - S_0$ as a function of T. If $G_T - G_0 = H_T - H_0 - S_T T + S_0 T$, are changes in G, as the temperature changes, greater than or less than changes in H in the case of H_2O?

Ionic Interactions

The possibility that a positively charged cation might interact favorably with a negatively charged anion and bring two molecules or two segments of the same polymer together has a lasting appeal. Such an association seems plausible because, as everyone knows, unlike charges attract each other. When a positive ion in a solution encounters a negative ion and a complex between these two ions is formed, it is referred to as an **ion pair**. In terms of Reaction 5–1, a hydrated ion pair forms whenever a hydrated anion associates with a hydrated cation. In this reaction, the various changes of standard free energy identified in Equation 5–2 can be separately considered by writing the following thermodynamic cycle

A⁺ (g) + B⁻ (g) →(ΔG°_{A⁺·B⁻}) A⁺·B⁻ (g)
+ + +
xH₂O yH₂O zH₂O
↓ΔG°_{HYD(A⁺)} ↓ΔG°_{HYD(B⁻)} ↓ΔG°_{HYD(A⁺·B⁻)}
A⁺(H₂O)_x + B⁻(H₂O)_y →(ΔG°_{IP}) A⁺·B⁻(H₂O)_z + (x + y − z) H₂O

(5–5)

It is easiest to consider the changes of standard enthalpy first because they constitute the main contribution to the changes in standard free energy and they have been measured least ambiguously. The standard enthalpy change when a positive ion and a negative ion associate in the gas phase, $\Delta H°_{A⁺·B⁻}$, is governed by electrostatics. To the extent that the Born–Haber cycle is able to provide accurate estimations of crystal lattice energies only from simple electrostatic theory,[14] the standard enthalpy of formation of an ion pair in the gas phase from the two separated ions, A⁺ and B⁻, if electron repulsion is ignored, should be

$$\Delta H°_{A⁻B⁺} = (Z_{A⁺})(Z_{B⁻})e_a^2 \left(\frac{1}{a_{A⁺} + a_{B⁻}} \right) N_{Av} \quad (5–6)$$

where Z_i is the charge on the ion, e_a is the elementary charge (4.80×10^{-10} esu), and $a_{A⁺}$ and $a_{B⁻}$ are the radii of the two ions. Values for the ionic radii in crystalline lattices, based on crystallographic studies of salts,[15] are usually used for $a_{A⁺}$ and $a_{B⁻}$.[6,16] The standard enthalpy of formation defined by Equation 5–6 can be presented for monovalent ions ($Z_{A⁺} = -Z_{B⁻} = 1$) as a function of the sum of the two ionic radii (Figure 5–9).

When an ion is transferred from the gas phase to water, there is a large release of heat. This large negative change in standard enthalpy is referred to as the **standard enthalpy of hydration**, $H°_{HYD(I)}$. Measurements of these standard enthalpies of hydration have been tabulated[6] for a number of monovalent, divalent, and trivalent spherical ions. The values for the spherical monovalent cations and anions can be presented as a function of their ionic radii (Figure 5–9).

The large negative standard enthalpies of hydration for ions are generally explained as the result of the ability of the fixed charge on an ion to gather around itself a layer of tightly held molecules of water that are oriented either with the positive ends of their dipoles, their hydrogens, toward an anion or the negative ends of their dipoles, their lone pairs, directed toward a cation (Figure 5–10). From measurements of the standard enthalpy of formation for complexes between monovalent cations in the gas phase and 1–7 molecules of water, it has been concluded[17] that about four molecules of water are sufficient to hydrate a cation such as NH_4^+, H_3O^+, H_2COH^+, Li^+, or Na^+. This result suggests that the innermost shell of the layer of hydration around an ion is not very large. When the standard enthalpy changes for the formation of 1:1 complexes between a molecule of water and various cations and anions in polar nonaqueous solvents were determined, however, the values observed were quite small ($0 > \Delta H° > -3.0$ kcal mol⁻¹).[18] This result suggests that the large standard enthalpies of hydration observed for ions arise as much, if not more, from the influence exerted by the ion over a significant region of the water surrounding it as from the specific, intimate noncovalent contacts between the ion and its immediate neighbors. If this is the case, the bulk dielectric of the water, which is a measure of the macroscopic response of the solvent to a fixed electrostatic charge, can be used to explain standard enthalpy of hydration.

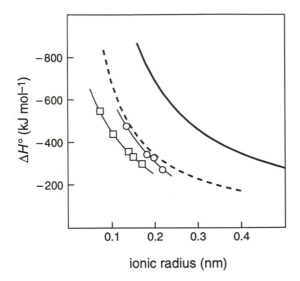

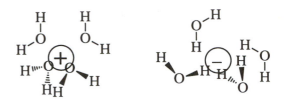

Figure 5–10: Schematic drawing of molecules of water with the negative ends of their dipoles directed toward a cation and the positive ends of their dipoles directed toward an anion.

Figure 5–9: Electrostatic enthalpies and standard enthalpies of hydration. The standard enthalpy change for bringing together a positive monovalent cation and a negative monovalent cation in the vacuum is presented as a function of the sum of the two respective ionic radii (solid dark line), as calculated from Equation 5–6. The standard enthalpies of hydration[6] for monovalent cation (open squares) are presented as a function of their ionic radii. The ions are, in order of increasing radius, Li^+, Na^+, K^+, Rb^+, and Cs^+. The line connecting the points is drawn by hand. The standard enthalpies of hydration[6] for monovalent anions (open circles) are presented as a function of their ionic radii. The ions are, in order of increasing ionic radius, F^-, Cl^-, Br^-, and I^-. The line connecting the points is drawn by hand. The standard enthalpy change for the hydration of a monovalent ion of either charge, based on the assumption that the standard enthalpy of hydration is due only to the difference in self-charging energies in the vacuum and in water, is presented as a function of ionic radius (dark dashed line). The line was calculated from Equation 5–8. All enthalpies are presented in kilojoules mole^{-1} for a standard temperature of 25 °C.

In electrostatic theory there is a quantity known as the self-charging energy, which is the energy required to charge a sphere of a given radius, a_i, in a medium of dielectric constant ε.[19] The self-charging energy, E_{sc}, for placing the charge, Z_i, on an ion i of radius a_i would be

$$E_{sc} = \frac{Z_i^2 e_a^2}{2a_i \varepsilon} N_{Av} \qquad (5–7)$$

where e_a is the elementary charge (4.80×10^{-10} esu). The standard enthalpy change $\Delta H_{sc}°$ associated with the electro-static energy required to move an ion from the vacuum ($\varepsilon = 1$) to water ($\varepsilon = 78$) would be the difference in the two self-charging energies

$$\Delta H_{sc}° = \frac{Z_i^2 e_a^2}{2a_i}\left[\frac{1}{\varepsilon_{H_2O}} - 1\right] N_{Av} \qquad (5–8)$$

This standard enthalpy change for a monovalent ion can be presented as a function of ionic radius (Figure 5–9). It can be seen that the standard enthalpies of hydration for spherical anions are very close to the values of the difference in self-charging energy, between the vacuum and a medium whose dielectric constant is that of water, for a sphere of unit charge with the same radius as the anion. The observed standard enthalpies of hydration for the cations, however, are less than the values expected from differences in self-charging energy. The small differences between $\Delta H_{sc}°$ and $\Delta H_{HYD(B^-)}°$ and the large differences between $\Delta H_{sc}°$ and $\Delta H_{HYD(A^+)}°$ have been explained as being due either to increases in the "effective radius"[20] of the ions or to decreases in the "effective dielectric constant around the ion."[21] Which of these conjectures is more realistic is unknown. Nevertheless, self-charging can explain the shapes and slopes of the curves connecting the experimental values for standard enthalpies of hydration.

The standard enthalpy of hydration for a monovalent ion pair, $\Delta H_{HYD(A^+ \cdot B^-)}°$, can be estimated from electrostatic theory just as standard enthalpies of hydration were estimated. It should be equal to the difference between the sum of the standard enthalpies of charging the two ions separately and the standard enthalpy of bringing them to within a certain distance of each other

$$\Delta H_{HYD(A^+ \cdot B^-)}° = \frac{e_a^2}{2}\left(\frac{1}{a_{A^+}} + \frac{1}{a_{B^-}} - \frac{2}{d_{A^+ \cdot B^-}}\right)\left[\frac{1}{\varepsilon_{H_2O}} - 1\right] N_{Av} \qquad (5–9)$$

where $d_{A^+ \cdot B^-}$ is the distance between the monovalent ions in the ion pair; when the two ions are as close together as possible, $d_{A^+ \cdot B^-}$ equals $a_{A^+} + a_{B^-}$. It can be shown that

$$\frac{0.828}{a_0} < \left(\frac{1}{a_{A^+}} + \frac{1}{a_{B^-}} - \frac{2}{a_{A^+} + a_{B^-}}\right) < \frac{1}{a_0} \qquad (5–10)$$

where a_0 is the radius of the smaller of the two monovalent ions. It follows that the standard enthalpy of hydration for a monovalent ion pair is slightly less than the standard enthalpy of hydration for the smaller of the two ions alone.[19]

The overall change in standard enthalpy for the formation of an ion pair in aqueous solution should be

$$\Delta H^\circ_{IP} = \Delta H^\circ_{A^+ \cdot B^-} + \Delta H^\circ_{HYD(A^+ \cdot B^-)} - \Delta H^\circ_{HYD(A^+)} - \Delta H^\circ_{HYD(B^-)}$$

$$(5\text{--}11)$$

From an examination of Figure 5–9, it becomes clear that for monovalent ions this change in standard enthalpy is a small difference between several very large numbers, and its value could be either positive or negative. This conclusion, that the standard enthalpy change has a small value, is supported by estimating the electrostatic energy involved in bringing two monovalent ions of opposite charge together in a medium with a uniform dielectric constant equal to that of water

$$\Delta H^\circ_{IP} \cong \frac{e_a^2}{\varepsilon_{H_2O}} \left(\frac{1}{a_{A^+} + a_{B^-}} \right) N_{Av} \qquad (5\text{--}12)$$

For $a_{A^+} \geq 0.1$ nm and $a_{B^-} \geq 0.1$ nm, -9 kJ mol^{-1} $< \Delta H^\circ_{IP} < 0$ kJ mol^{-1}.

These changes in standard enthalpy do demonstrate quite clearly why an ion pair sequestered in the middle of a folded polypeptide would be unstable relative to the separated hydrated ions in solution. The only reason an ion pair is almost stable in aqueous solution is that there is considerable standard enthalpy of hydration for the ion pair itself, $\Delta H^\circ_{HYD(A^+ \cdot B^-)}$. In the center of a protein, this standard enthalpy of hydration would not be exerted and the ion pair would be much less stable. There will never be sufficient electrostatic energy in the ion pair alone to overcome the large negative standard enthalpies of hydration that are lost when the separated ions are removed from water during the folding of the protein.

The **standard entropies of hydration**,[22] in marked contrast to the standard enthalpies of hydration, are small. Values of the entropies of hydration for a series of small monovalent ions of either charge lie between -67 and $+21$ J K^{-1} mol^{-1} when the two standard states are chosen as the molten salt at one mole fraction in the ion and the ideal solution at one mole fraction in the ion.[22] At 298 K, these standard entropies of hydration would cause the standard free energies of hydration to differ from enthalpies of hydration by less than 4%, certainly less than the error in the estimation of enthal-pies of hydration from experimental data.[6]

The very small standard entropies of hydration seem at first glance to be inconsistent with the formation of a firmly held layer of oriented water around an ion, which is one explanation for the large standard enthalpies of hydration. The apparent inconsistency is usually explained by noting that a sphere of hydration around either an anion or a cation cannot merge flawlessly with the hydrogen-bonded lattice of the bulk water. Therefore, there must be an outer sphere of disorder between the inner sphere of order and the order of the lattice. The negative standard entropy change of forming the sphere of hydration should be canceled by the positive standard entropy change of forming this outer zone of transition.[6]

At least one experimental observation indicates that ion pairs between simple monovalent anions and cations are not stable in water and will not display a negative standard free energy of formation. Two of the more common ionic groups on a molecule of protein are the carboxylates of its glutamates and aspartates and the primary ammonium cations of its lysines. The dielectric increment is the change in the dielectric constant of a solution with the concentration of an added solute. The dielectric increments for a series of zwitterionic amino acids containing an ammonium and a carboxylate, namely, glycine, 3-aminopropionate, 4-aminobutyrate, 5-aminopentanoate, and 6-aminohexanoate, have been measured. The values display a monotonic increase with the number of methylenes between the positively charged ammonium cation and the negatively charged carboxylate. The values observed are in agreement with theoretical calculations of their magnitude from a simple model in which the distance between the positive charge and the negative charge is only determined by free rotation around the carbon–carbon bonds connecting them.[23] Were the formation of an ion pair between an ammonium cation and a carboxylate anion a favorable interaction in aqueous solution, this regularity could not have occurred. In glycine, an intramolecular ion pair cannot form. In 3-aminopro-pionate and 4-aminobutanoate, excellent intramolecular ion pairs, forming rings five and six atoms in size, should form even more readily than a similar intermolecular ion pair. If these intramolecular ion pairs were able to form, however, the dielectric increments of these two amino acids should both be less than that of glycine, yet no anomaly is observed in their dielectric increments compared to the other compounds within the complete series.

There is considerable evidence from observations such as this one that in aqueous solution, ion pairs between monovalent anions and cations are unstable relative to the separate ions. This, however, merely reiterates the common observation that simple monovalent salts are usually soluble in water. As an ionic lattice is more stable than an ion pair, ion pairs must be unstable because monovalent salts dissolve.

Although ion pairs do not have net favorable standard free energies of formation in aqueous solution and do not contribute to the stability of a folded polypeptide, the repulsion of amino acids of like charge can destabilize a particular structure. This distinction is illustrated by the effect of ionic strength on the stability of coiled coils of α helices.[24] A coiled coil of α helices is a stable structure that forms when two α helices coil around each other in a supercoil. This supercoil stabilizes the two α helices sufficiently that they can form in water. Very few isolated α helices are stable in aqueous solution, but a coiled coil is one way to circumvent this problem. A series of peptides designed to form coiled coils were synthesized chemically. One of the peptides (n_{aa} = 30) had glutamates at the positions flanking its hydrophobic core; the other (n_{aa} = 30) had arginines flanking the core. The stability of the heterodimeric coiled coil formed from a positively charged peptide and a negatively charged peptide was not affected by changing the ionic strength, and this result indicates that electrostatic interactions such as ion pairing were not contributing to the standard free energy of formation of that coiled coil. The stabilities of homodimers formed from either two of the positively charged peptides or two of the negatively charged peptides, however, decreased significantly as the ionic strength of the solution was lowered, and this result demonstrates that these complexes were destabilized by charge repulsion.

The reason that ion pairs do not form even under ideal circumstances is that water hydrates the separated charges so strongly. This hydration is a specific example of a general phenomenon known as solvation, which is a feature of any solution. **Solvation** is the interaction between any solute and the solvent that surrounds it. In all cases, solvation plays a major role in the character of the chemistry that can occur in any solution. This is particularly true of the mechanisms of chemical reactions because the solvation of transition states and intermediates of high energy is a major factor in their stability. There are two features of water, its ability to hydrate isolated formal charge and its ability to provide protons, that have a profound effect on its ability to promote particular types of reactions.

Because water can hydrate charges so effectively, the predominant and characteristic chemistry of an aqueous solution is heterolytic; when covalent bonds are broken in aqueous solution, the two electrons usually remain together on one or the other of the two atoms. Heterolytic reactions generate charged species from neutral species, and water is the most effective solvent, to a significant degree, at solvating these charged intermediates. There is, however, a type of free radical chemistry that also involves charged intermediates. Certain reactions proceed by single electron transfer between donor and acceptor, and when electron transfer occurs between neutral molecules, radical anions or radical cations, or both, are created as intermediates. These charged radical species can also be solvated by water.

A **protic solvent** is a solvent that can act as an acid by providing protons. In aqueous solutions there are both hydronium cations (pK_a = –1.7) and water molecules (pK_a = 15.7) that can act as Bronsted acids. In addition, the oxygen–hydrogen bond is polar, so that a neutral molecule of water can protically hydrate a negative charge

either directly (Figure 5–10) or through a hydrogen bond. The fact that water is a protic solvent also promotes heterolysis and electron transfer, because protonation often precedes or coincides with these processes. Water is the fundamental solvent for acid–base reactions.

The chemistry of biological systems seems to have arisen almost entirely by improvements in the catalysis of reactions that began as uncatalyzed processes in aqueous solution. The peculiar properties of water promote heterolytic mechanisms and electron transfer because of the solvation that is unique to water among all the other solvents available to a chemist. It is not surprising to learn that biochemistry is dominated by heterolytic chemistry and that the occasional free radicals that are encountered usually result from electron transfer.

Proteins are generally dissolved in aqueous solutions whose **ionic strength** is 0.1–0.3 M, and the effect of such ionic strengths on ionic interactions, although small, should be noted. The activity coefficients of electrolytes decline sharply from a value of unity at very low concentrations to minimum values at about 0.3 M, and above that they range from 0.05 to 0.8.[25] When activity coefficients of solutes are less than unity, it means that the solute is behaving with a chemical potential less than it would have if it were an ideal solute at the same concentration, and its tendency to leave the solution is less than it should be. At low concentrations, the activity coefficients of ions are near 1. This means that as long as they are far enough apart their activities increase in proportion to their concentration as expected for any solute. As their concentrations become high enough that each ion begins to experience the presence of the others, that presence decreases the tendency of that ion to leave the solution. This decrease arises from the tendency of the ions in the solution to depart from a random distribution in such a way that a region enriched in the ions of opposite charge forms around each individual ion, similar to the ionic layer that forms beside a layer of fixed charge in an ionic double layer (Equation 1–62).[25] These ionic layers make dissolved ions even more stable in the aqueous solution than they would be even if they were ideal solutes, and this is what causes the decrease in their activity coefficients. This effect of ionic strength makes the formation of an ion pair even less likely than it would be in the absence of added salt, because the formation of the ion pair would involve the loss of a considerable fraction of the ionic layers around each separated ion. The presence of the ionic layer also makes it more difficult to remove an ionic functional group from a solution of moderate ionic strength than it would be to remove it from pure water. The activity coefficients for most ionic solutes are between 0.2 and 0.8 at the ionic strengths encountered in biochemical situations, and these values should lead to decreases in the standard free energies of hydration of –4 kJ mol^{-1} and –0.4 kJ mol^{-1}, respectively.[25]

Although ion pairs between simple monovalent cations and anions usually have positive standard free energies of formation, there are two situations in which ionic interactions are favorable. Ion pairs involving divalent ions often have negative standard free energies of formation. For example, ion pairs of $Ca^{2+} \cdot SO_4^{2-}$ and $Mg^{2+} \cdot SO_4^{2-}$ are present in solutions of these salts, and ion pairs between divalent cations such as Ba^{2+}, Ca^{2+}, and Mg^{2+} and hydroxide ion in aqueous solution show appreciable stabilities.[16]

The other situation in which ionic interactions become favorable is encountered when chelation can occur. Chelation is the binding of an ion to another molecule, the chelating agent. The chelating agent contains two or more functional groups of opposite charge to the bound ion that can simultaneously associate with it, or it contains two or more dipoles that simultaneously can be favorably directed toward the bound ion, or it contains some combination of such charges and dipoles. It has already been mentioned that the binding of monovalent cations and anions by proteins is thought to involve particular binding sites that have advantageous dispositions of functional groups with charge opposite the charge on the bound ion. Chelation, however, assumes a preexisting arrangement of two or more charged groups or dipoles that create a pocket within which an ion can be held, and this arrangement does not exist in an unfolded polypeptide or in isolated anions and cations in solution. Chelation could be important in forming an interface between two folded polypeptides or binding a charged substrate to an enzyme, but there is no way of estimating the magnitude of its effect even in specific instances.

Suggested Reading

Parsegian, A. (1969) Energy of an Ion Crossing a Low Dielectric Membrane: Solutions to Four Relevant Electrostatic Problems, *Nature* **221**, 844–846.

Problem 5–2: Calculate the standard enthalpy of hydration for an ion pair between a sodium ion (a_{Na^+} = 0.097 nm) and a chloride ion (a_{Cl^-} = 0.181 nm) from the difference in the self-charging energy of the ion pair between the vacuum and a medium of dielectric constant equal to that of water ($\varepsilon_{H_2O} = 78$).

Problem 5–3: Consider the series of six compounds $H_3N^+(CH_2)_nCOO^-$, where $n = 1$–6. Within each of these molecules there is a carboxy group and an amino group, which bear opposite charges at pH 7. The negative charge on the carboxy group is located between the two oxygens.

(A) For each compound, construct with molecular models the conformation of the molecule in which the positively charged nitrogen is positioned as close as possible to one of the negatively charged oxygens.

(B) In which of the molecules is it possible to form an ion pair that juxtaposes NH_3^+ and O^-?

The dipole moment (μ) of a particular molecular structure which contains fixed charges is equal to the product of the magnitude of the charges Q and the distance, r, that separates them: $\mu = Qr$. In each structure you have made, Q is the same but r changes.

(C) Examine the structures you have drawn and rank them in order of increasing dipole moment. Indicate ranking using the symbols < and =.

(D) The observed dipole moments for these molecules dissolved in water at pH 7.0 are

n	1	2	3	4	5	6
μ	12 D	15 D	18 D	20 D	22 D	24 D

Explain why the actual dipole moments for these molecules fail to agree with the theoretical predictions that you made in part C.

The Hydrogen Bond

The hydrogen bond is a noncovalent force that arises between an acid, known as the **donor**, A–H, and a base, known as the **acceptor**, ⊙B. It is an intermediate on the trajectory of an acid–base reaction (see Mechanism 5–13, below.[26,27]

The two central complexes in Mechanism 5–13 are each held together by a hydrogen bond, but the overall reaction is the transfer of a proton between two lone pairs of electrons. An example of this relationship between a hydrogen bond and an acid–base reaction is the self-dissociation of water. In the liquid most of the water molecules participate in hydrogen bonds, and these hydrogen bonds are the intermediate steps in the production of a hydroxide anion and a hydronium cation. The anion and the cation are produced when the proton in a hydrogen bond between two molecules of water moves momentarily from donor to acceptor and the hydrogen bond between the new donor, the hydronium cation, and the new acceptor, the hydroxide anion, dissociates to yield the free species.

(5–13)

Before the complications of a more detailed description of the bond itself are discussed, it is important to accept that the hydrogen bond exists. Its **manifestations** are the alterations it effects in the physical and chemical properties of liquids, gases, and solids. The nonideal behavior of certain gases can be explained by the existence of hydrogen-bonded oligomers of the molecules composing the gas. The water dimer (Figure 5–1) is an example of such a situation; its existence lowers the pressure of water vapor. Abnormally high values for the enthalpy of vaporization or abnormally low values for the standard enthalpy of mixing can often be explained as the result of either the breaking of hydrogen bonds as the molecules depart the liquid or the formation of hydrogen bonds as a donor and acceptor are mixed, respectively. When an acceptor is added to a solution of a donor, the infrared spectrum of a solution often displays a new absorption band, at a lower frequency and adjacent to the absorption of the A–H stretching vibration observed with a solution of the donor alone. This new absorption increases in magnitude in proportion to the amount of acceptor added while the amplitude of the absorption of the unshifted stretching vibration of the A–H bond of the donor decreases in proportion. The new stretching vibration is assigned to that of the A–H covalent bond within a hydrogen bond between the donor and the added acceptor. A similar observation is made in nuclear magnetic resonance spectra of mixtures of donors and acceptors. In this case, two separate absorptions are not observed because the rates at which the hydrogen bonds are interchanged among the molecules in the solution are faster than the time resolution of the method, but the chemical shift of the hydrogen participating in the A–H bond moves downfield until it reaches a maximum value, associated with the chemical shift of the hydrogen within the hydrogen bond.

Taken together, these commonly encountered observations demonstrate three features of a hydrogen bond. First, a hydrogen bond causes two molecules to associate with each other and form a complex that prevents them from changing their relative positions as readily as they would otherwise; in other words, it correlates their movements. Second, there is a release of heat associated with the formation of this complex. Third, the hydrogen in the A–H covalent bond of the donor experiences a change in its environment during the formation of this complex. The results of both infrared and nuclear magnetic resonance spectroscopy are consistent with a weakening of the covalent bond between A and H concomitant with either a movement of the hydrogen away from the electrons of the σ bond or a movement of the electrons of the σ bond away from the hydrogen, respectively.

The arrangement of the atoms in crystallographic molecular models of small molecules that display these physical manifestations of hydrogen bonding usually displays a pattern that can be assigned to the hydrogen bond itself. The positions in the unit cell of the atoms of the second and third period of the periodic table, for example, carbon, nitrogen, oxygen, and sulfur, are determined by X-ray crystallography, and the arrangement of the hydrogens, often as deuterium atoms, can be determined by neutron diffraction. While hydrogen has very little ability to scatter X-rays, inasmuch as it has no core electrons, deuterium scatters neutrons as readily as carbon or oxygen;[28] and deuterium atoms are prominent features in maps of neutron scattering density, as opposed to hydrogens in maps of electron density. Furthermore, hydrogen has a negative scattering amplitude for neutrons while deuterium has a positive scattering amplitude. This causes difference maps of neutron scattering density for deuterated against protonated molecules to display sharp maxima where the hydrogens are normally located in the molecules.

In crystallographic structures of molecules known to be hydrogen bonded, the bond is recognized as an enforced orientation of the donor and acceptor (Figure 5–11). Associated with this orientation are certain **bond lengths** and **bond angles**.[28] It is these bond lengths and bond angles that are the most important property of a hydrogen bond as far as the structures of properly folded proteins are concerned. The hydrogen bond provides no net standard free energy to the process of folding, but hydrogen bonds are responsible for aligning atoms and holding them at precise distances and angles to each other in the folded structure. The A–H σ bond of the donor in a hydrogen bond is pointed at the basic atom, $\odot$B, of the acceptor. The hydrogen within the bond is usually on the line of centers between A and B. In an asymmetrical hydrogen bond, where the hydrogen is located preferentially on the donor, AH, the distance between A and H, when A is either nitrogen or oxygen, is 0.10 nm. In rare instances, however, the hydrogen is found halfway between donor and acceptor, and a symmetrical hydrogen bond results.[16] The distance, d, between A and B is always less than it would be if the hydrogen on the donor atom and the atom acting as the acceptor were simply in van der Waals contact. For example, in a hydrogen bond of the type O–H$\odot$N (Mechanism 5–13), the distance between oxygen and nitrogen is 0.28 ± 0.01 nm,[28] while the distance between carbon and nitrogen in a van der Waals contact of the type C–H$\odot$N would be 0.35 nm. It is this shortened distance between donor and acceptor that reflects the bonding. The bond lengths, d, for most types of hydrogen bonds (Table 5–1) lie between 0.25 and 0.30 nm, but the bond angles of the various hydrogen bonds are more variable. A general rule is that angle a between the axis of the bond and a σ covalent bond to the donor, AH (Figure 5–11) will reflect closely the hybridization of atom A, while angle b between the axis of the bond and a σ covalent bond to the acceptor, B, although much more flexible than angle a, will tend to reflect the hybridization of the lone pair of electrons on atom B.

The type of hydrogen bond that accounts for the

Table 5–1: Length of Hydrogen Bonds[a]

A–H···B	compounds	average bond length[b] (nm)
OH⊙O	carboxylic acids	0.26 ± 0.01[c]
OH⊙O	phenols	0.27 ± 0.01[c]
OH⊙O	alcohols	0.27 ± 0.01[c]
OH⊙N	all O–H	0.28 ± 0.01[c]
NH⊙O	ammoniums	0.29 ± 0.01[c]
NH⊙O	amides	0.29 ± 0.01[c]
NH⊙O	amines	0.30 ± 0.01[c]
NH⊙N	all N–H	0.31 ± 0.01[c]

[a]The values in this table are reproduced directly from tables in ref 29. [b]These are the distances between the heteroatoms, nitrogens or oxygens. [c]These standard deviations may be standard deviations of actual lengths or standard deviations of the measurement or both.

majority of those in biological macromolecules, both proteins and nucleic acids, is the hydrogen bond between the sp^2 lone pair on an acyl oxygen as an acceptor and the nitrogen–hydrogen bond of an acyl derivative as a donor (Figure 5–11B). The geometries of a series of hydrogen bonds between either a carbonyl oxygen or an acyl oxygen and a nitrogen–hydrogen bond were systematically examined. From the crystallographic molecular models of 900 such intermolecular hydrogen bonds, the bond lengths and bond angles were compiled.[29] For the collection of all such hydrogen bonds examined, the mean nitrogen–oxygen distance, d, was 0.297 nm. If it is assumed that the acyl or carbonyl carbon and its three σ bonds define a plane and that the line of centers between the nitrogen and oxygen defines a line, two angles define the hydrogen bond: angle b, the angle in the plane between the projection upon the plane of the line and the carbon–oxygen bond (Figure 5–11B), and angle c, the angle that the line of centers between the nitrogen and oxygen makes with the plane itself (Figure 5–11C). Angle c determines how far the nitrogen is above or below the plane, and $d \sin c$ is the actual distance the nitrogen is above or below the plane. In the hydrogen bonds examined, the nitrogen atom was usually within 0.1 nm ($\sin c = 0.33$) of the plane defined by the acceptor (Figure 5–12). If the carbonyl oxygen or acyl oxygen participates as the acceptor in two hydrogen bonds, there will be a strong tendency for angle b to be 120° (Figure 5–12B), the angle expected from an sp^2 hybridization of its two lone pairs. If the carbonyl oxygen or acyl oxygen, however,

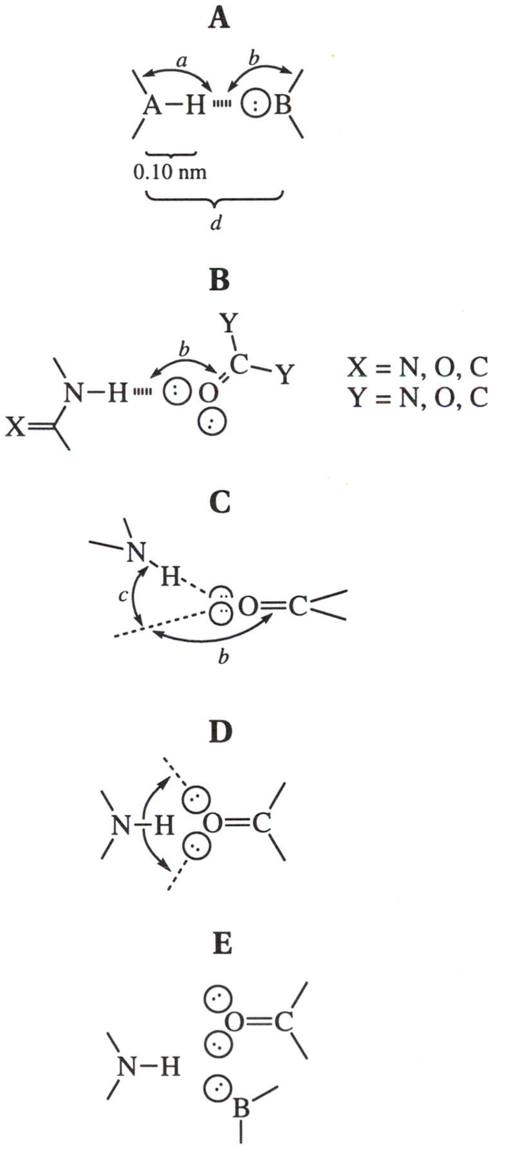

Figure 5–11: Relationships defining bond angles for hydrogen bonds. (A) In a simple hydrogen bond between AH and ⊙B, the line of center between A and B creates an axis, the axis of the hydrogen bond. The angle a is the angle between that axis and a σ covalent bond to the atom A. The angle b is the angle between that axis and a σ covalent bond to the atom B. The distance d is the length of the hydrogen bond. (B) The bond angle b for a hydrogen bond between an amido nitrogen and a carbonyl oxygen or an acyl oxygen is the angle between the projection of the axis of the hydrogen bond on the plane defined by the carbonyl group or the acyl group and the σ bond connecting the carbon to the oxygen. (C) The bond angle c for a hydrogen bond between an amido nitrogen and a carbonyl oxygen or acyl oxygen is the angle between the axis of the hydrogen bond and the plane defined by the carbonyl group or the acyl group. (D) The bond angle b between the axis of a hydrogen bond between a nitrogen–hydrogen donor and the σ covalent bond of a carbonyl oxygen and acyl oxygen can vary over a range bounded by the lone pairs of electrons on the oxygen if the oxygen is otherwise unoccupied. (E) In several instances, the axis of the σ bond between the central atom of the donor and the hydrogen lies between two lone pairs of electrons from two different atoms and one donor interacts with two acceptors.

participates as an acceptor in only one hydrogen bond, angle *b* will still show a slight preference for 120° but the angle can also assume other values between 120° and 180° with almost equal facility (Figure 5–12A). It is as though the nitrogen–hydrogen bond can pivot over the electron cloud formed by both lone pairs when one of them is empty (Figure 5–11D), and the location it eventually assumes in the crystal is determined by forces other than the hydrogen bond itself. This apparent ability of a nitrogen donor to associate with two lone pairs on the same atom may be related to its ability to participate in a bifurcated hydrogen bond[30] in which it associates with two lone pairs from separate atoms (Figure 5–11E). The fact that the nitrogen is not confined strictly to the plane of the acyl group (Figure 5–12), although it has a strong preference for that plane, demonstrates that the nitrogen–hydrogen dipole can also pivot up or down out of the plane about a single lone pair (Figure 5–12B) or about two lone pairs (Figure 5–12A). All of the observations presented in Figure 5–12 are for acyl oxygens that are not in carboxylates. In the case of the oxygens in carboxylates, the tendency for the nitrogen to reside in the plane of the carboxylate is lessened and the tendency for angle *b* to assume 120° is increased.[29]

There has been some disagreement over the ability of sulfur to participate as an acceptor in a hydrogen bond because of the poor overlap between its atomic orbitals and those of nitrogen or oxygen. In a survey of crystallographic structures for a number of compounds in which nitrogen donors and sulfur acceptors both appear,[31] juxtapositions were frequently observed and these were consistent with hydrogen bonds of the type NH☉S. The most telling observation in favor of the existence of such hydrogen bonds was the fact that the nitrogen–sulfur distances (0.33–0.35 nm) were shorter than the distance expected from purely van der Waals contact.

The **strength** of a hydrogen bond is expressed in thermodynamic parameters. The **standard enthalpy of formation**, or the heat released when the bond forms, is a measure of the electronic strength of the bond. It is usually the property that is referred to when the strength of the bond is discussed indiscriminately. The **standard free energy of formation** determines the degree to which the hydrogen bond will be favored over the unbonded reactants. Its magnitude is complicated by the fact that it is a function of both the standard enthalpy of formation, the electronic term, and the **standard entropy of formation**, the quantitative measure of the change in disorder occurring during the reaction. The standard entropy of formation is always a negative term because order increases when hydrogen bonds are formed. Unfortunately, the standard entropy of formation also depends on the choice of units for concentration. The **association equilibrium constant**, which is usually the quantity that is directly measured, is connected directly to the standard free energy of formation, not to the standard enthalpy of formation.

Ordinarily, the values of these thermodynamic properties are obtained systematically.[28] A method of measurement is chosen that can provide values for the molar concentration of free donor, [HA], the molar concentration of free acceptor, [B☉], and the molar concentration of hydrogen bonds, [B☉HA], in a solution. The total concentrations of donor and acceptor are systematically varied at a given temperature, and the association equilibrium constants are measured for each set of concentrations

$$K_{AHB} = \frac{[B\ominus HA]}{[B\ominus][HA]} \qquad (5\text{–}14)$$

From the association equilibrium constant at a particular temperature, the standard free energy of formation of the hydrogen bond can be calculated

$$\Delta G^\circ_{AHB} = -RT \ln K_{AHB} \qquad (5\text{–}15)$$

The variation of the equilibrium constant with temperature is determined experimentally; and, from these observations, the standard enthalpy of formation of the hydrogen bond can be calculated

$$\left(\frac{\partial \ln K}{\partial T^{-1}} \right)_P = -\frac{\Delta H^\circ_{AHB}}{R} \qquad (5\text{–}16)$$

Finally, the standard entropy of formation is calculated from the experimental results by the relationship

$$\Delta S^\circ = \frac{\Delta H^\circ - \Delta G^\circ}{T} \qquad (5\text{–}17)$$

The standard enthalpies of formation for hydrogen bonds of biological interest lie between 12 and 24 kJ (mol · bond)$^{-1}$ (Table 5–2), when the donor and acceptor are dissolved in organic solvents such as CCl_4 or benzene. In spite of these favorable standard enthalpies of formation, the equilibrium constants for the formation of hydrogen bonds, even when they are measured in organic solvents, are quite small (Table 5–3). The small magnitude of these values results from the fact that the negative standard enthalpy of formation is canceled to a considerable degree by a negative standard entropy of formation (Equation 5–17). Therefore, even in the best of circumstances, a hydrogen bond is a weak interaction.

The standard enthalpy of formation and the standard free energy of formation of a hydrogen bond are functions of the difference in pK_a between the donor and the acceptor.[16] For example, a reference donor, *p*-fluorophenol, was chosen, and a series of acceptors were used to form hydrogen bonds with this donor in CCl_4 at 25 °C[32,33]

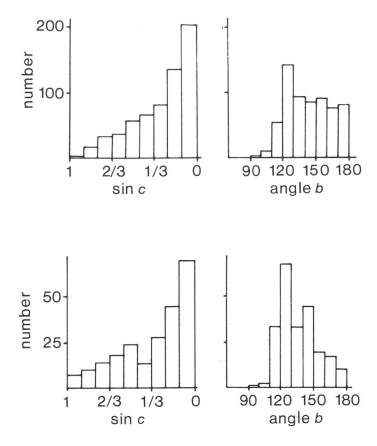

Figure 5–12: Distribution of values for the angle b (Figure 5–11B) and the angle c (Figure 5–11C) over a population of hydrogen bonds between nitrogen–hydrogen donors and carbonyl oxygen acceptors or acyl oxygen acceptors observed in crystallographic molecular models of small molecules.[29] (A) Hydrogen bonds involving a carbonyl oxygen or acyl oxygen in which the oxygen atom accepts no other hydrogen bonds. (B) Hydrogen bonds involving a carbonyl oxygen or acyl oxygen in which the oxygen atom accepts one other hydrogen bond. In each of the four panels, the number of bonds falling within a range of values is plotted as the value of the ordinate. In the two left panels, the values on the abscissa defining the ranges are values of the sine of angle c (sin c). In the two right panels, the values on the abscissa defining the ranges are values of the angle b in degrees. Adapted with permission from ref 29. Copyright 1983 American Chemical Society.

$$B\ominus + HOC_6H_4F \underset{K_{AHB}}{\rightleftharpoons} B\ominus HOC_6H_4F \qquad (5\text{–}18)$$

The correlation between standard enthalpy of formation and pK_a for these hydrogen bonds was[33]

$$\Delta H^\circ_{AHB} = [-1.3\,\text{kJ mol}^{-1}]pK_a + b_j \qquad (5\text{–}19)$$

where b_j(kilojoules mole^{-1}) assumes different values for each category of base examined; for example, b_j has different values for carbonyl oxygens, pyridines, and primary amines. Within a particular category, however, the standard enthalpies of formation are linearly related by Equation 5–19. This correlation states that as the acceptor becomes a stronger base (as the pK_a of its conjugate acid becomes larger), the hydrogen bond becomes stronger (the standard enthalpy of formation becomes more negative). The correlation between standard free energy of formation and pK_a for hydrogen bonds between p-fluorophenol and various bases of different pK_a is[32]

$$\Delta G^\circ_{AHB} = [-1.3\,\text{kJ mol}^{-1}]pK_a + c_j \qquad (5\text{–}20)$$

where c_j (kilojoules mole^{-1}) assumes a different value for each category of bases. The similarity of the slopes of these two correlations (Equations 5–19 and 5–20)[33] demonstrates that the variation in standard free energy of formation arises entirely from a variation in standard enthalpy of formation. The standard enthalpy of formation[33–35] and the standard free energy of formation[36] of a hydrogen bond were also linearly related, respectively, to the pK_a of the donor in similar experiments in which a common acceptor and a systematic series of donors were used.

Therefore, the standard free energy of formation of a hydrogen bond in an organic solvent increases as the acidity of the donor increases and as the basicity of the acceptor increases. This increase in the standard free

Table 5–2: Standard Enthalpies of Formation for a Series of Biochemically Important Hydrogen Bonds in Organic Solvents[a]

hydrogen bond	solvent	$\Delta H°$ (kJ mol^{-1})
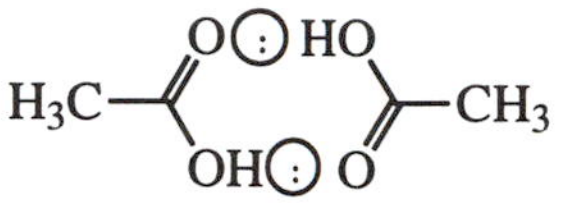	CCl_4	-45[b]
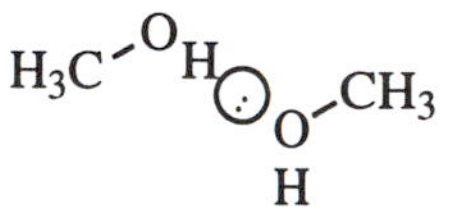	CCl_4	-20
	CCl_4	-13
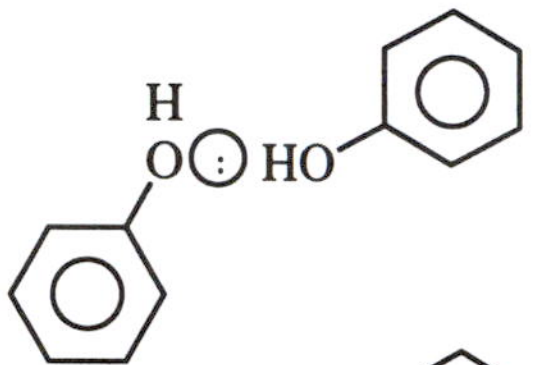	CCl_4	-18
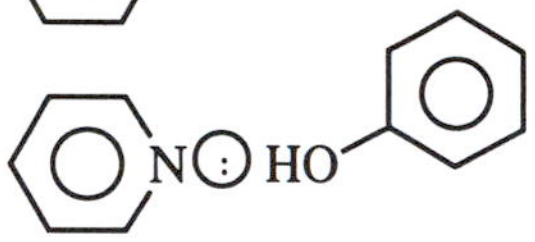	CCl_4	-21
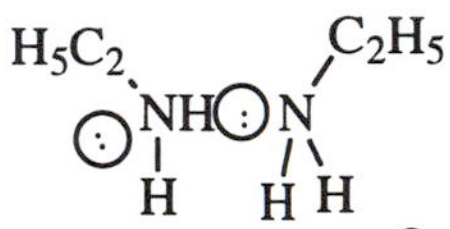	neat	-15
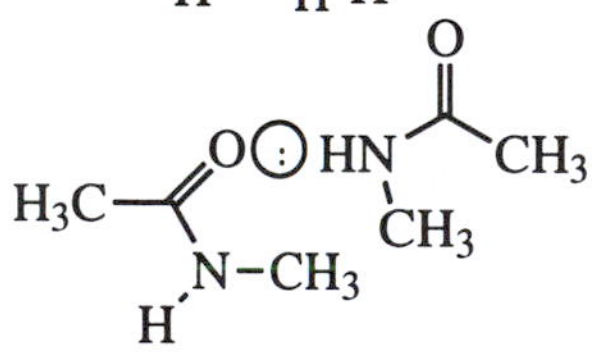	benzene	-15
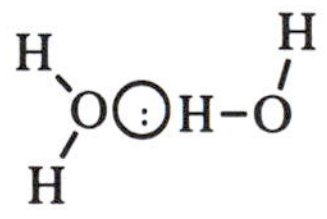	neat	-14

[a]Values copied directly from tables in ref 28. [b]Value for formation of the two hydrogen bonds of the dimer.

energy of formation results from an equivalent increase in the standard enthalpy of formation for the hydrogen bond. If one considers the situation of a hydrogen bond in which the acceptor remains the same and a series of donors of increasing acidity are examined, the standard enthalpy of formation of the hydrogen bond will decrease as the pK_a of the donor decreases until the pK_a of the donor is equivalent to the pK_a of the conjugate acid of the acceptor. If the acidity of the donor is increased further, the proton will be transferred between donor and acceptor and the conjugate acid of the former acceptor becomes the new donor and the conjugate base of the former donor becomes the new acceptor. If the pK_a of the former donor is decreased below the pK_a of the conjugate acid of the former acceptor, the standard enthalpy of formation for the hydrogen bond, because donor and acceptor have switched roles, will begin to increase. A corresponding argument could be made for the situation in which the donor remains the same and a series of acceptors of increasing pK_a is examined.

From these considerations it follows that the strength of the hydrogen bond is determined by the difference in pK_a between donor and the conjugate acid of the acceptor; the smaller the difference, the stronger the bond. If this is the case, it must follow that the strongest possible hydrogen bond in a given series is the one in which the pK_a of the donor is equal to the pK_a of the conjugate acid of the acceptor. This is why a hydrogen bond between an acid and its conjugate base is usually the strongest hydrogen bond in which that acid or that base can participate.

The frequency shift of the infrared absorption of an oxygen–hydrogen stretching vibration has been used to examine the effect of stereochemistry on the strength of intramolecular hydrogen bonds.[37] A series of pyridines substituted at the o-position by $-CH_2OH$, $-CH_2CH_2OH$, or $-CH_2CH_2CH_2OH$ were examined. These functional groups can form hydrogen-bonded rings with the pyridine nitrogen that are four, five, or six atoms in size, respectively, when the hydrogen is not counted. All three compounds displayed an absorption that could be assigned to a shifted OH stretching vibration. The hydrogen bonds increased in standard enthalpy of formation ($\Delta v = 192$, 203, and 357 cm^{-1}, respectively) as the ring became larger. The ring with six atoms, the only one large enough to permit the hydrogen bond to be linear, displayed the largest frequency shift, an observation consistent with its being the strongest. This result suggests that in a cyclic hydrogen-bonded structure, the hydrogen bond should be considered as a somewhat longer and more flexible covalent bond between the two heteroatoms, and the hydrogen should not be counted as one of the atoms in the ring.

With a sense of the geometry of a hydrogen bond and an understanding of the strength of a hydrogen bond in hand, it is possible to appreciate the debate surrounding the **electronic structure** of a hydrogen bond. It is as yet unknown whether a hydrogen bond is ionic or covalent. If it is ionic, the force between donor and acceptor is the simple electrostatic force of their fixed dipoles. The A–H bond of the donor is a dipole, electronegative on the heteroatom, A, and electropositive on the hydrogen. The σ orbital of the acceptor, ⊙B, is electronegative on the lone pair and electropositive on the heteroatom, B. These two dipoles should attract each other electrostatically when oriented in the same direction.

It is possible, however, that the hydrogen bond is **covalent** and thereby involves both the overlap of atomic orbitals to form molecular orbitals and the delocalization of valence electrons over the three participating atoms, A,

Table 5–3: Association Constants of Hydrogen Bonds[a]

hydrogen bond	solvent	temperature (°C)	K_{AHB} (M^{-1})
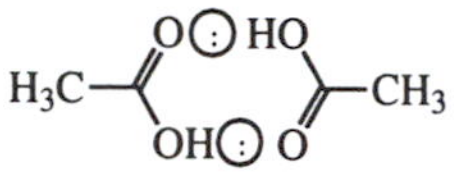 H_3C—C(OH·O)(O·HO)—CH_3	benzene	30	130
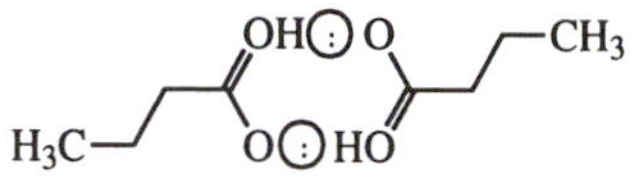 H_3C—...(OH·O)(O·HO)...—CH_3	benzene	30	430
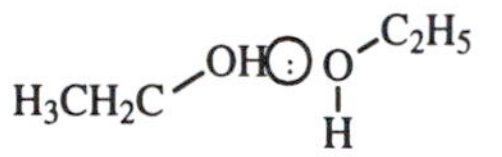 H_3CH_2C—OH·O—C_2H_5	CCl_4	25	0.64
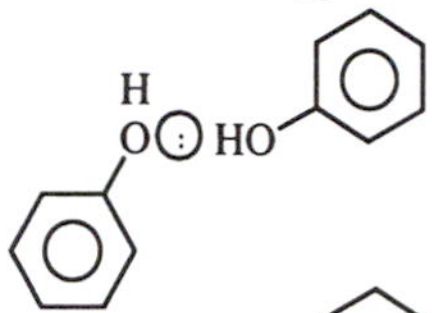 phenol···phenol	CCl_4	21	2.3
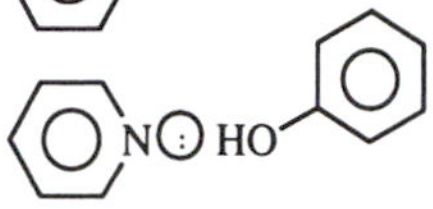 pyridine N·HO—phenol	CCl_4	20	55
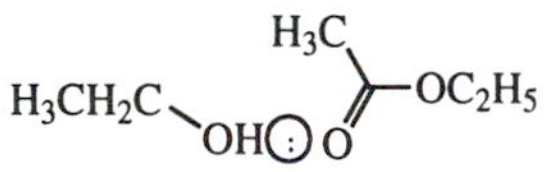 H_3CH_2C—OH·O—C(CH_3)—OC_2H_5	CCl_4	25	1.7
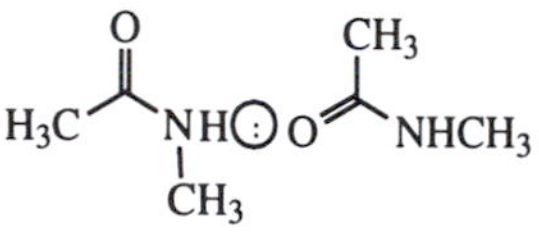 H_3C—C(O)—N(CH_3)H·O—C(CH_3)—$NHCH_3$	benzene	25	6.2

[a]Values copied directly from tables in ref 28.

H, and B. Such a covalent bond would result from the mixing of the sp^2 or sp^3 orbital on the central atom, A, of the donor, the $1s$ orbital of the hydrogen, and the sp^2 or sp^3 orbital of the lone pair of electrons of the acceptor, B, to form a **molecular orbital system** with three molecular orbitals (Figure 5–13A). A covalent hydrogen bond can also be represented by writing two resonance structures (Figure 5–13B). As always, the resonance structures represent the molecular orbitals. The σ bond that switches between the two resonance structures represents the bonding σ molecular orbital spread over the three atoms. The lone pair that switches between the two structures represents the nonbonding molecular orbital occupied by two electrons. The advantage of the resonance structures is that they give a clearer picture of the asymmetry of the bond. In an asymmetric bond, one of the two resonance structures will be less favorable than the other.

The distinction between a dipolar, electrostatic interaction and a covalent bond is also relevant to an evaluation of the properties of an ionized hydrogen bond. An **ionized hydrogen bond** is a hydrogen bond between a cationic donor and an anionic acceptor. An example of

such an ionized hydrogen bond would be the hydrogen bond between a carboxylate anion as the acceptor and an ammonium cation as the donor

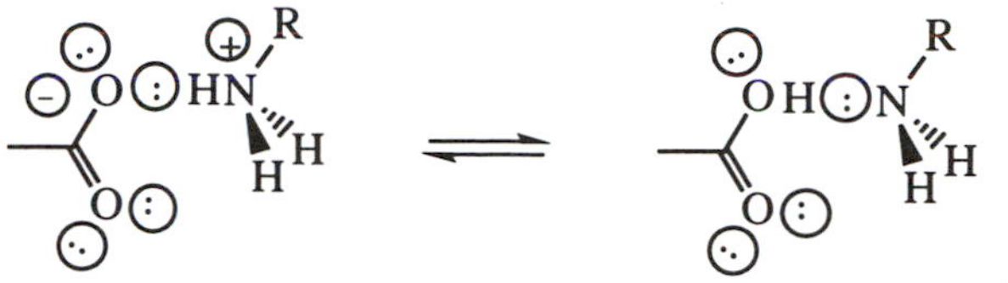

(5–21)

It is ionized hydrogen bonds that are often designated as ion pairs or salt bridges in crystallographic molecular models of proteins. In almost every instance in which such a hydrogen bond is found in these molecular models, it is unknown on which atom the proton resides. If it remains on the stronger base, as is likely, then the hydrogen bond should be an ion pair. Certainly, in the majority of crystallographic studies of small molecules in which the location of the hydrogen has been observed, it usually appears to form a normal σ bond with the stronger base. When solvated with water, the separation of charge necessitated by this arrangement should easily be supported for reasons already discussed. Whether the hydrogen bond is ionized or not, however, depends formally on whether or not the proton is on the stronger base, in which case the bond would be ionized, or on the weaker base, in which case it would be neutral. When an ionized hydrogen bond is transferred from surroundings of high dielectric such as water to surroundings of low dielectric such as the center of a protein, it is possible that the proton shifts its location to the weaker base so that separation of charge, which would be incompatible with the new environment, is eliminated.

An estimate of the effect of dielectric constant on the location of the proton in an ionized hydrogen bond has been formulated,[38] and it was concluded that the dielectric constant of the surroundings would have to be less than about 4 before the shift of the proton in the ionized hydrogen bond represented in Equation 5–21 would be favored. If this were the case, most ionized hydrogen bonds would be ionized even when surrounded by protein.

These calculations, however, were based on electrostatics and assumed that unit formal charges were concentrated on oxygen and nitrogen, respectively. This assumption requires that the relationship between the structures on either side of Equation 5–21 be a tautomeric equilibrium between two discrete states as it is represented. If, however, a hydrogen bond is a covalent interaction and electrons are delocalized over the three atoms (Figure 5–13), then the distribution of charge in the system is far more complex and it cannot be assumed that two states, one in which oxygen bears unit negative charge and nitrogen bears unit positive charge and one in which these two atoms are neutral, describe the system. Rather, charge is distributed by the delocalized electrons

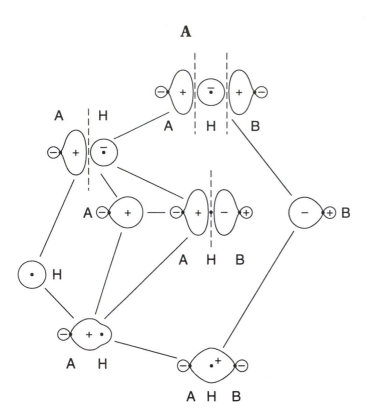

Figure 5–13: Molecular orbitals (panel A) and resonance structures (panel B) for a covalent hydrogen bond. The molecular orbitals are for a symmetric hydrogen bond in which the donor and the conjugate acid of the acceptor are equivalent. The covalent molecular orbital system is formed from two sp^2 or sp^3 orbitals, one from atom A and one from atom B, respectively, and the s orbital on hydrogen. These three atomic orbitals combine to form the three molecular orbitals—bonding, nonbonding, and antibonding—shown in the middle of the diagram. The final molecular orbital system is constructed in steps by first mixing the atomic orbitals of atom A and the hydrogen to form the two molecular orbitals—bonding and antibonding—of the A–H covalent bond and then mixing the A–H molecular orbital system with the atomic orbital on atom B containing the lone pair of electrons. The resonance structures (panel B) are written for the unsymmetrical hydrogen bond between the nitrogen–hydrogen bond of an amide and the lone pair on an acyl oxygen. The σ bond that shifts in the two structures reflects the overlap in the bonding molecular orbital. The σ lone pair of electrons that shifts in the two structures reflects the two electrons in the central nonbonding molecular orbital.

of the molecular orbital system and imbalances of charge in that molecular orbital system would respond continuously to a continuous change in the dielectric constant of the surroundings. In this case, the proper relationship between the structures on either side of Equation 5–21 is one of resonance. Because of the ambiguity in this type of noncovalent interaction, it is more informative to refer to it as an ionized hydrogen bond than as a salt bridge or an ion pair to indicate that separation of unit charge may be only a formal possibility rather than an actual fact.

There are experimental results suggesting that the degree of ionization of a hydrogen bond between a charged donor and a charged acceptor decreases in environments with low dielectric constants. For example, dry solids formed between polylysine and a number of phenols exhibited hydrogen bonds that were less than 50% ionic even when the phenol was 2 units in pK_a more acidic than the conjugate acid of the lysine.[39] The degree

of ionization, however, increased as the solids were hydrated. Similar results were observed for hydrogen bonds between polyhistidine and various carboxylic acids.[40] Hydrogen bonds between benzoic acid (pK_a = 4.2) and triethylamine (pK_a = 9.8), two compounds whose values of pK_a differ by as much as those of the side chains of glutamic acid and lysine, form hydrogen bonds in solvents such as carbon tetrachloride (ε = 2.3), cyclohexane (ε = 2.0), and carbon disulfide (ε = 2.6) that exhibit little or no ionization.[41]

To this point, the hydrogen bonds that have been discussed are those formed in aprotic organic solvents such as carbon tetrachloride or benzene. The situation changes dramatically when the donor and acceptor are dissolved in water because of the competition of the donors and acceptors of the water molecules themselves. In retrospect, the fact that the concentrations of donors and acceptors in water are both 110 M is a sufficient observation in itself to lead to the conclusion that the hydrogen bond between a solute, A–H, and a solute, ⊙B, would be extremely unlikely to form.

The hydrogen bond between the nitrogen–hydrogen bond of an amide and the lone pair of electrons on the acyl oxygen of another amide can be used again as an example of the majority of the hydrogen bonds in biological macromolecules. When *N*-methylacetamide is dissolved in carbon tetrachloride or dioxane, an absorption appears in the infrared spectra of the various solutions that can be assigned[41a] to the N–H stretching vibration in a hydrogen bond of the structure

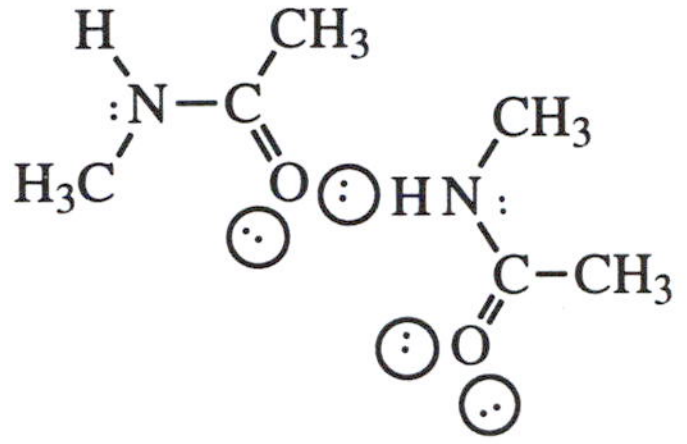

5–1

This is the hydrogen bond holding α helices and β structure together in a molecule of protein and the base pairs together in DNA. From the magnitude of the absorption as a function of concentration, it could be calculated that the standard free energy of formation of this hydrogen bond at 25 °C in CCl$_4$ is -3.8 kJ mol^{-1}, when an infinitely dilute solution of *N*-methylacetamide is defined as the standard state and the association equilibrium constant is expressed in units of reciprocal molarity (Table 5–3). When the *N*-methylacetamide was dissolved in water, the infrared absorption arising from hydrogen bond **5–1** could be barely detected even at a concentration of 12.5 M. From what was observed, the standard free energy of formation of the hydrogen bond in aqueous solution was judged[41a] to be about $+13$ kJ mol^{-1} at 25 °C.

A more complete picture of the situation was gained by measuring the standard free energy of transfer of *N*-methylacetamide between H$_2$O and CCl$_4$ at 25 °C.[42] It was found to be $+17$ kJ mol^{-1}. The complete standard free energy diagram for the hydrogen bond between two amides (Figure 5–14) suggests that the intrinsic standard free energy of the hydrogen-bonded complex is almost identical in the two solvents, and its transfer between them should be an isoergonic reaction. The difference between CCl$_4$ and H$_2$O, however, is due mainly to the higher stability of the separated donors and acceptors in the H$_2$O. From this observation, it can be concluded that the large unfavorable standard free energy of transfer for *N*-methylacetamide between H$_2$O and CCl$_4$ reflects the necessity to break hydrogen bonds between the water and the *N*-methylacetamide before the transfer can occur.

If this is the case, the formation of the hydrogen bond in water can be written, in analogy to Reaction 5–1, as

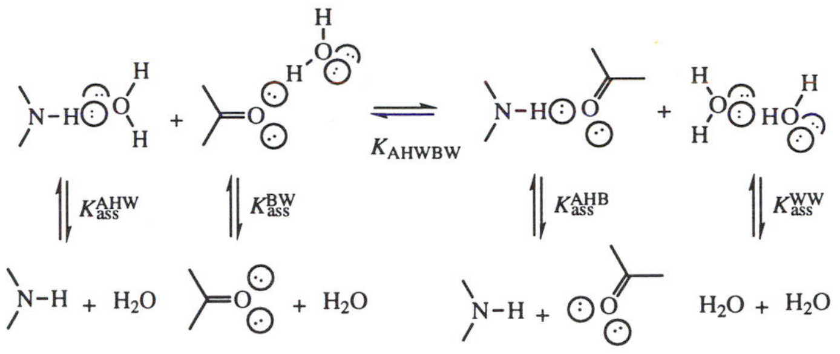

$$(5\text{--}22)$$

where the association constant for any of the complexes, K_{ass}^{XY} is

$$K_{ass}^{XY} = \frac{[X \odot Y]}{[X][Y]} \qquad (5\text{--}23)$$

It is entirely possible[26] that the concentrations of unbonded donors and unbonded acceptors in aqueous solutions are negligible and that only the upper part of Reaction 5–22 is thermodynamically relevant.

The equilibrium constant for the upper part of Reaction 5–22, K_{AHWBW}, is defined by

$$K_{AHWBW} = \frac{[B \odot HA][H_2O \odot H_2O]}{[B \odot H_2O][H_2O \odot HA]} \qquad (5\text{--}24)$$

where, in the particular case of *N*-methylacetamide, AH is the amide nitrogen–hydrogen bond and B⊙ is a lone pair of electrons on the acyl oxygen. The equilibrium constant actually observed, K_{app}^{AHB}, is

$$K_{app}^{AHB} = \frac{[B \odot HA]}{([B \odot H_2O] + [B \odot])([H_2O \odot HA] + [HA])} \qquad (5\text{--}25)$$

from which it follows that

$$K_{app}^{AHB} = \frac{K_{ass}^{AHB}}{\left(1 + \frac{K_{ass}^{AHW}}{(K_{ass}^{WW})^{1/2}}[H_2O \odot H_2O]^{1/2}\right)\left(1 + \frac{K_{ass}^{BW}}{(K_{ass}^{WW})^{1/2}}[H_2O \odot H_2O]^{1/2}\right)} \qquad (5\text{--}26)$$

If, as is reasonable, $K_{ass}^{AHW} > (K_{ass}^{WW})^{1/2}$, $K_{ass}^{BW} > (K_{ass}^{WW})^{1/2}$, and $[H_2O \odot H_2O] > 1$ M, it follows that

$$K_{app}^{AHB} \cong \frac{K_{AHWBW}}{[H_2O \odot H_2O]} \qquad (5\text{--}27)$$

where $[H_2O \odot H_2O]$ is the molar concentration of hydrogen bonds in the water.

The difference in pK_a between the nitrogen–hydrogen bond in *N*-methylacetamide as a donor (pK_a = 16) and the oxygen–hydrogen bond in water as a donor (pK_a = 15.7) should be negligible, as should be the difference between the lone pair of electrons on the acyl oxygen of *N*-methylacetamide (pK_a = –1.5) and the lone pair of electrons on water (pK_a = –1.7) as acceptors. Therefore, the standard enthalpy of formation for the following four hydrogen bonds should be very similar

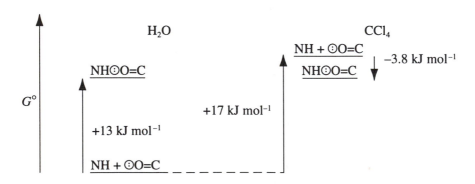

and the standard enthalpy change for the upper part of Reaction 5–22 should be zero. If the upper part of Reaction 5–22 were isoentropic as well as isoenthalpic, so that $K_{AHWBW} \cong 1$ (Equation 5–24), then K_{app}^{AHB} would be equal to $[H_2O\odot H_2O]^{-1}$. As the observed standard free energy of formation for hydrogen bond **5–1** in aqueous solution is about +13 kJ mol^{-1}, $K_{app}^{AHB} \cong$ (190 M)$^{-1}$, which is in the range expected for the reciprocal of the concentration of hydrogen bonds in pure water, $[H_2O\odot H_2O] \leq 110$ M. The conclusion to be drawn from these considerations is that

a hydrogen bond in aqueous solution will always have a small apparent association equilibrium constant and a large apparent standard free energy of formation because the concentration of hydrogen bonds between water molecules in the solution is a hidden and very significant term in the apparent association equilibrium constant (Equation 5–27).

The hydrogen bond represented by that of *N*-methylacetamide accounts for the majority of those found in proteins, and yet its standard free energy of formation is positive by a considerable degree. From this it follows that each hydrogen bond of this type in a protein represents an energetic liability rather than an asset. It is possible, however, that some other combination of donor and acceptor might produce a hydrogen bond strong enough to overcome the competition of the water and provide a negative standard free energy of formation. In assessing this possibility, it would be useful to have an equation that could be used to estimate the apparent equilibrium constant, K_{app}^{AHB}, for the formation of any hydrogen bond in aqueous solution. Such an equation has been derived[26] and has been demonstrated to be reliable.[43]

Consider the hydrogen bond

$$B[\odot H\!-\!]A$$

5–6

and focus in turn on the portions within the brackets and without the brackets. The portion within the brackets is

Figure

Figure 5–14: Standard free energy diagram for a hydrogen bond between the amide nitrogen (NH) and acyl oxygen ($\odot$O=C) of *N*-methylacetamide.[42] The standard free energies of association for the hydrogen bond in water (+13 kJ mol^{-1}) and carbon tetrachloride (–3.8 kJ mol^{-1}) are positioned in relation to each other on the diagram by the standard free energy of transfer (+17 kJ mol^{-1}) of unbonded *N*-methylacetamide between the two solvents. The standard free energy of transfer was calculated from the distribution coefficient of *N*-methylacetamide between water and carbon tetrachloride, extrapolated to infinite dilution, with the concentrations of solute in each solvent expressed in units of molarity.

the same for all hydrogen bonds. The structures without the brackets on either side affect the intrinsic enthalpy of the hydrogen bond by donating or withdrawing electrons from this central structure.[44] The net result of their action is an intrinsic enthalpy, H_{AHB}^{int}, that is proportional to the product of the two respective σ constants[44]

$$H_{AHB}^{int} = v^{\odot H}\sigma_A \sigma_B \qquad (5\text{–}28)$$

where $v^{\odot H}$ is a constant of proportionality. These σ constants are the same terms used in physical organic chemistry to provide a quantitative assessment of the ability of any group to withdraw or donate electrons and cause changes in standard enthalpy in any similar situation.

For the upper part of Reaction 5–22, when the specific example of *N*-methylacetamide is replaced by the general reaction between A–H and $\odot$B, the standard enthalpy change for the reaction should be

$$\Delta H_{AHWBW}^{\circ} = H_{AHB}^{int} + H_{WW}^{int} - H_{AHW}^{int} - H_{BW}^{int} \qquad (5\text{–}29)$$

or

$$\Delta H_{AHWBW}^{\circ} = v^{\odot H}(\sigma_A - \sigma_{OH})(\sigma_B - \sigma_{H_2O}) \qquad (5\text{–}30)$$

where σ_{OH} is the σ constant for OH taking the place of A in **5–6** and σ_{H_2O} is the σ constant for H$_2$O taking the place of B in **5–6**. As the values of the σ constants are proportional to the values of pK_a for the appropriate acids

$$\Delta H_{AHWBW}^{\circ} = 2.303RT\tau(\text{p}K_a^{HA} - \text{p}K_a^{HOH})(\text{p}K_a^{HB} - \text{p}K_a^{H_3O^+}) \qquad (5\text{–}31)$$

where τ incorporates the constants of proportionality. If it is assumed that the standard entropy change in the upper part of Reaction 5–22 is usually zero and that differences in standard enthalpy are the only significant determinants of relative strengths, then

$$-\log K_{ass}^{AHWBW} = \tau(\text{p}K_a^{HA} - \text{p}K_a^{HOH})(\text{p}K_a^{HB} - \text{p}K_a^{H_3O^+}) \qquad (5\text{–}32)$$

and[26]

$$\log K_{app}^{AHB} = \tau(\text{p}K_a^{HA} - \text{p}K_a^{HOH})(\text{p}K_a^{H_3O^+} - \text{p}K_a^{HB}) - \log[\text{H}_2\text{O}\odot\text{H}_2\text{O}] \qquad (5\text{–}33)$$

where the values of the measured pK_a have been corrected statistically for the number of protons, *p*, on the conjugate acid and the number of lone pairs, *q*, on the conjugate base

$$K_a^{corr} = \frac{q}{p}K_a^{meas} \qquad (5\text{–}34)$$

Equation 5–33 can be used to calculate the association equilibrium constant, and hence the standard free energy of formation, of any hydrogen bond in aqueous solution.

This relationship was validated[43] by an examination of the formation of hydrogen bonds between a series of phenolate anions as acceptors and ammonium cations as donors at 25 °C

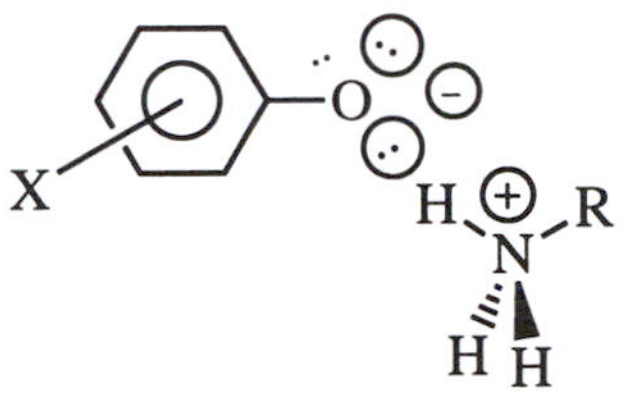

5–7

where the substituents X and R were various electron-donating and electron-withdrawing groups chosen to vary the pK_a values of the donor and acceptor. The logarithms of the association equilibrium constants for the formation of these hydrogen bonds varied with the pK_a of either the donor (Figure 5–15) or the acceptor as predicted by Equation 5–33. Extrapolating the relationships to either pK_a^{HA} = pK_a^{HOH} or pK_a^{HB} = p$K_a^{H_3O^+}$ gave the same value, 2.0, for log [H$_2$O$\odot$H$_2$O]. This numerical value is a reasonable estimate for the logarithm of the concentration of hydrogen bonds in liquid water, where [H$_2$O$\odot$H$_2$O] $\leq$ 110 M, and it is in reasonable agreement with the results gathered independently with *N*-methylacetamide.

The value of τ (Equations 5–31 and 5–33) at 25 °C and 2 M ionic strength was found to be 0.013, from which it follows that even the strongest possible hydrogen bond, where the pK_a of the donor equals the pK_a of the acceptor, would have an association equilibrium constant in aqueous solution of less than 1 M^{-1}. For example, the hydrogen bond between the lone pair of electrons on imidazole (pK_a^{HB} = 6.4) and the nitrogen–hydrogen bond of the imidazolium cation (pK_a^{HA} = 6.4) would have an apparent association equilibrium constant of only 0.040 M^{-1} at 25 °C. At pH 6.4, a 2 M solution of imidazole would only have a concentration of hydrogen bonds **5–8** equal to 0.04 M.

5–8

If the donor and acceptor for a given hydrogen bond are hydrated solutes in aqueous solution before the hydrogen bond is formed, Equation 5–31 states that the

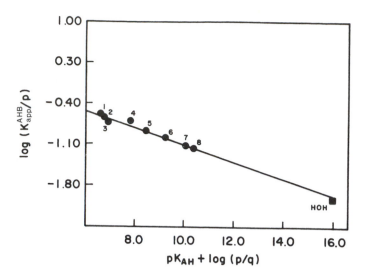

Figure 5–15: Relationship between the apparent association equilibrium constants (K_{app}^{AHB}) for a series of hydrogen bonds between the phenolate anion and a series of aliphatic ammonium cations in aqueous solution and the acid dissociation constants K_a^{AH} for those ammonium cations.[43] The associations between the phenolate anion and the ammonium cations were followed spectrophotometrically by changes in absorbance at 300 nm as ammonium cation was added to an aqueous solution of phenolate anion at 2 M ionic strength and 25 °C. The values of the apparent association equilibrium constants were divided by the number of protons (p) on the respective ammonium cation to correct the molar concentration of the free cation to the molar concentration of donors. The acid dissociation constants were also statistically corrected by multiplying the observed acid dissociation constants by the number of lone pairs on the conjugate base (q) and dividing by the number of protons on the conjugate acid (p) so that the corrected values reflect the molar concentrations of the protons on the respective conjugate acid and lone pairs of electrons on the respective conjugate base. The logarithms of the apparent association equilibrium constants are linearly correlated with the logarithms of the corrected acid dissociation constants by a line with slope 0.15. The value of log ($K_{app}^{AHB}p^{-1}$) calculated by Equation 5–33 for an acid with a pK_a equal to that of water is indicated by a filled square. The ammonium cations used were (1) hydroxylammonium cation, (2) piperazine dication, (3) *sym*-tetramethylethylenediammonium dication, (4) *N,N,N*-trimethyl-ethylenediammonium dication, (5) ethylenediammonium dication, (6) 2-hydroxy-1,3-diaminopropane dication, (7) 1,3-diaminopropane dication, and (8) (2-hydroxyethyl)ammonium cation. Adapted with permission from ref 43. Copyright 1986 American Chemical Society.

standard enthalpy of formation for the bond will always be a small negative number. Equation 5–33 states that the standard free energy of formation of that hydrogen bond will be increased by about +11 kJ mol^{-1} because of the competition by the high concentration of water. This competition is an entropic effect. It follows that a hydrogen bond in aqueous solution cannot be strengthened significantly by any alteration of the acid dissociation constants of donor and acceptor. There is a way, however, to decrease its standard free energy of formation by increasing its standard entropy of formation.

Suggested Reading

Jencks, W.P. (1987) Hydrogen Bonds, in *Catalysis in Chemistry and Enzymology*, Chapter 6, pp 323–350, Dover, New York.

Klotz, I.M., & Farnham, S.B. (1968) Stability of an Amide Hydrogen Bond in an Apolar Environment, *Biochemistry 7*, 3879–3882.

Problem 5–4: Draw structures that represent all of the possible hydrogen bonds that can form between the following pairs of molecules when they are dissolved in a nonpolar solvent. Draw the structures with proper geometry and include all lone pairs of electrons.

(A) two molecules of *N*-methylacetamide

(B) *N*-methylacetamide and ethanol

(C) ethanol and water

(D) *N*-methylacetamide and water

(E) urea and *N*-methylacetamide

(F) acetic acid and *N*-methylacetamide

(G) 4-methylimidazole and *N*-methylacetamide

(H) 4-methylimidazole and ethanol

(I) acetic acid and ethanol

Problem 5–5: Draw the precise structure of each of the following hydrogen bonds in the most stable geometry (bond angles and distances). Include all lone pairs of electrons.

I. 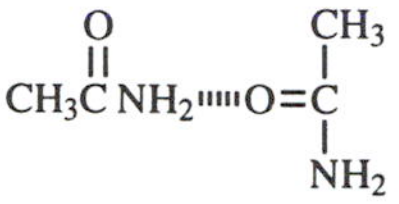 $CH_3\overset{O}{\overset{\|}{C}}NH_2\text{''''}O\overset{CH_3}{\underset{NH_2}{=C}}$

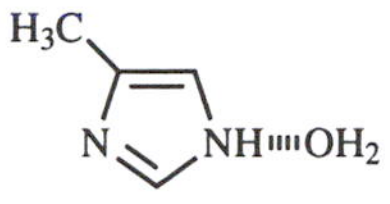

II. 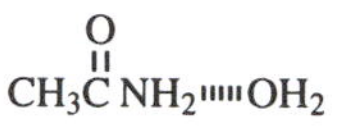 $CH_3\overset{O}{\overset{\|}{C}}NH_2\text{''''}OH_2$

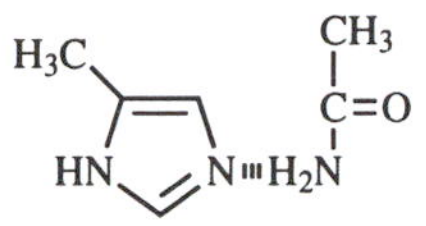

III. 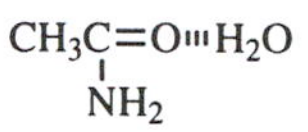 $CH_3\overset{}{\underset{NH_2}{C}}=O\text{'''}H_2O$

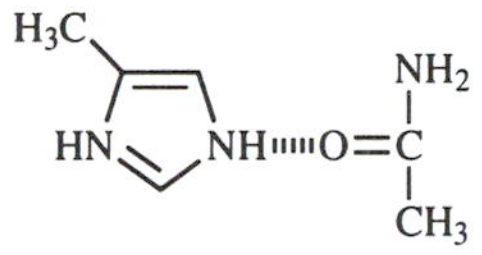

IV. 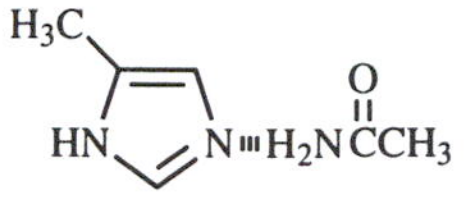 H_3C — HN⎯N''' $H_2N\overset{O}{\overset{\|}{C}}CH_3$

V. H_3C — N⎯NH''' OH_2

VI. H_3C — HN⎯N''' $H_2N\overset{CH_3}{\underset{}{C}}{=}O$

VII. H_3C — HN⎯NH''' $O{=}\overset{NH_2}{\underset{CH_3}{C}}$

VIII. $H_2O\text{''''}H_2O$

Write the acid dissociations to which the following apparent pK_as refer, and correct them for number of protons and number of lone pairs.

	pK_{a1}	pK_{a2}
$CH_3\overset{O}{\overset{\|}{C}}NH_2$	-0.51	15.70
H_2O	-1.75	15.75
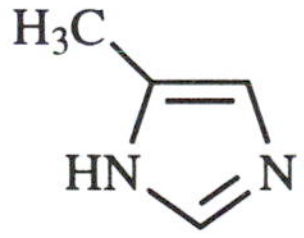 H_3C—HN⎯N	7.51	15.10

Rank the eight hydrogen bonds in order of strength.

Problem 5–6: When hydrogen bond donors and acceptors are dragged into the center of a molecule of protein as it folds, they search for partners. Hydrogen bonds between different regions of secondary structure together in the interior of a protein are hydrogen bonds between donors and acceptors on amino acid side chains. Draw the structure of only the most stable hydrogen bond that forms between side chains of the following pairs of amino acid side chains. Include all relevant angles and distances around the various hydrogen bonds.

(A) glutamic acid and histidine

(B) serine and tyrosine

(C) glutamine and histidine

Problem 5–7: The panel below is an infrared spectrum of propionamide in carbon tetrachloride. The bands marked M_1 and M_2 are the absorptions of the monomeric propionamide and the bands marked P_1, P_2, P_3, P_4, and P_5 are absorptions of hydrogen-bonded species. As the concentration of propionamide is increased, within these two sets the amplitudes remain in constant ratio to each other and must be different absorptions of the same species.

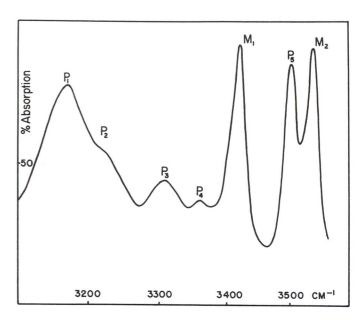

Below is a table of amplitudes from the infrared spectra for various total concentrations of propionamide at 298 K in CCl_4.

[propionamide]$_{TOT}$ (M)	A_{M1}	A_{P1}
1.71×10^{-3}	0.415	0.055
2.15×10^{-3}	0.509	0.083
2.43×10^{-3}	0.566	0.102
4.72×10^{-3}	0.981	0.308
6.90×10^{-3}	1.32	0.558
10.40×10^{-3}	1.79	1.020

Recall that, by Beer's law, [monomer] = $(\varepsilon_{M1})^{-1}A_{M1}$ and [polymeric species] = $(\varepsilon_{P1})^{-1}A_{P1}$. If the hydrogen bonding is a dimerization, it should be described by the following equation:

$$2\text{ propionamide} \rightleftharpoons \text{propionamide}_2$$

$$K_{eq} = \frac{[\text{propionamide}_2]}{[\text{propionamide}]^2}$$

(A) Show that the data are consistent with a dimerization.

(B) Use the data to determine K_{eq} (298 K). (Hint: [propionamide]$_{TOT}$ = [propionamide] + 2[propionamide$_2$].)

Values of K_{eq} were determined at a number of different temperatures.

temp (K)	K_{eq} (M^{-1})
303	35.5
313	24.6
329	13.3
298	_____ (your number)

(C) What is $\Delta H°$ for the dimerization?

(D) The appearance of the species P_1–P_5 in the infrared spectrum can only be adequately explained if propionamide has two hydrogen bonds that form a cyclic structure. Draw that structure.

(E) What is $\Delta H°$ for each mole of hydrogen bond?

Problem 5–8: Poly [d(AT) · d(AT)] melts at 67 °C and poly [d(GC) · d(GC)] melts at 102 °C. The usual explanation for this observation is that there is one more hydrogen bond in a G · C pair than in an A · T pair. Consider, however, this explanation in terms of Figure 5–14. Assume that when the heterocyclic bases in single strands of randomly coiled DNA form a double helix, the interior σ faces of the bases are transferred from water to a completely nonpolar environment and that the proper tautomers for base pairing are present at the pH of the experiment.

(A) Write a complete equation for the formation of an A · T pair and a G · C pair as the double helix forms, drawing in all hydrogen–bonded waters to all donors and acceptors on each side of the equation and the hydrogen bonds that form between the residual waters. Assume that hydrogen–bond donors and acceptors accessible to water in the major and minor grooves retain their hydrogen bonds with H$_2$O.

(B) Why is poly [d(GC) · d(GC)] more stable than poly [d(AT) · d(AT)]?

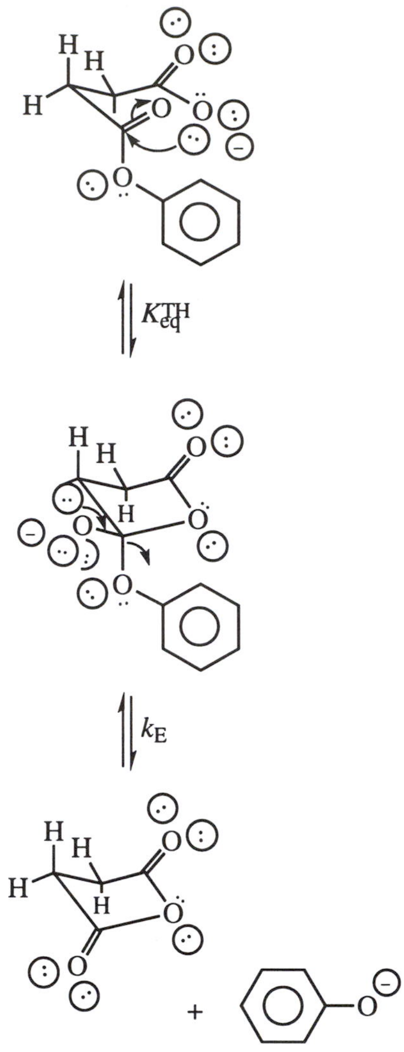

Intramolecular and Intermolecular Processes: Molecularity and Approximation

Intramolecular chemical reactions often occur at rates much faster than equivalent **intermolecular** reactions, and intramolecular associations often occur with association equilibrium constants much larger than those of equivalent intermolecular associations. A particularly informative series illustrating such effects can be gathered[45] from among the reactions involving intramolecular nucleophilic catalysis of the hydrolysis of phenyl esters by the carboxylate anion. The mechanism for this nucleophilic catalysis has been shown to involve the formation of an intermediate anhydride which in the intramolecular examples would be cyclic (Figure 5–16). These reactions should involve a preequilibrium in which

Figure 5–16: Mechanism for the formation of the intramolecular anhydride that is the intermediate in the hydrolysis of phenyl succinate by intramolecular nucleophilic catalysis. The tetrahedral intermediate that leads to the anhydride has both a phenolate and a carboxylate as potential leaving groups; and, because the latter is the better, it should decompose to reactant much more frequently than to anhydride.

the tetrahedral intermediate is formed from which the phenolate is eventually ejected. It is the ejection of the phenolate that is monitored as the reaction progresses. The rate constant, in units of reciprocal seconds, for the appearance of phenolate would be equal to $K_{eq}^{TH} k_E$, where K_{eq}^{TH} is the equilibrium constant for the formation of the tetrahedral intermediate. If it is assumed that for all compounds in the series the rate constant k_E, which in all cases is for a chemically equivalent first-order relaxation, has the same value, then the differences in observed rates result from differences in K_{eq}^{TH}, an equilibrium constant.

A comparison of the first-order rate constants of phenolate release for a series of intramolecular reactions[45] to the estimated pseudo first-order rate constant for the intermolecular reaction between phenyl acetate and acetate anion, when the catalysis by acetate anion proceeds through acetic anhydride as an intermediate,[45a,46] indicates how large these intramolecular accelerations of rate and increases in association can be (Table 5–4). The increase in K_{eq}^{TH} of 2.2×10^5 in the case of phenyl succinate compared to that for phenyl acetate and 1 M acetate is similar in magnitude to the 3×10^5 increase in the association equilibrium constant for the intramolecular formation of succinic anhydride compared to the intermolecular formation of acetic anhydride at 1 M acetic acid,[47] and this suggests that the changes seen are representative. An enhancement of 5×10^5, when compared to nucleophilic attack by 1 M hydroxide, was seen in the intramolecular attack of alkoxide in the intramolecular S_N2 reaction[48]

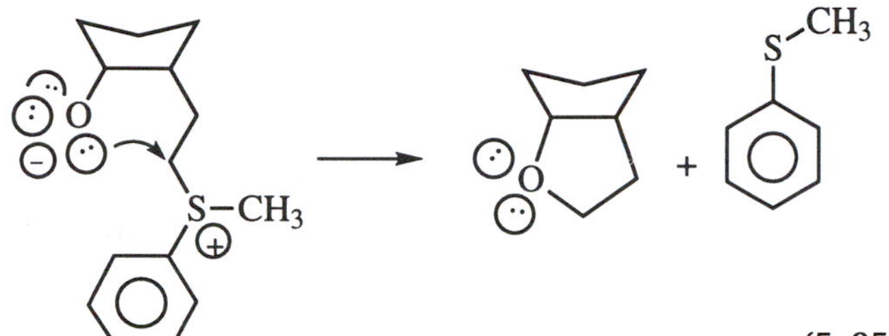

$$(5\text{–}35)$$

Again, this can be compared with the enhancement of 2×10^5 for the intramolecular nucleophilic enhancement in phenyl succinate (Table 5–4), also a situation in which a five-membered ring is produced.

Many other examples of intramolecular accelerations of rates or increases in association equilibrium constants have been reported,[16,46] but the accelerations presented in Table 5–4 are among the largest that have been noted for each particular size of ring formed during each of these intramolecular reactions. In some of the other instances, electronic effects are difficult to separate from the effects of approximation. For example, in the intramolecular reaction

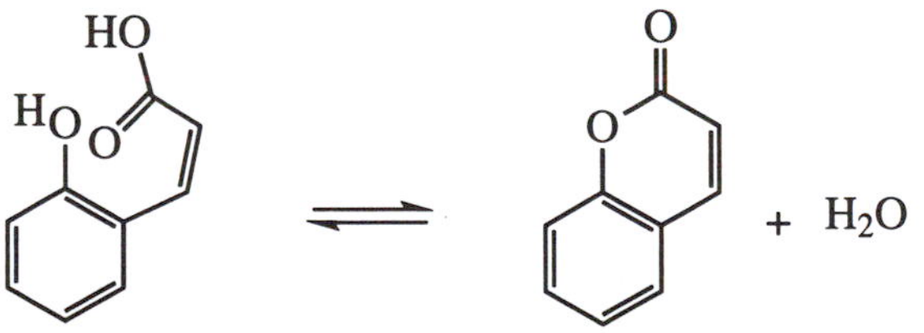

$$(5\text{–}36)$$

the two reactants are connected electronically by the π system in addition to being juxtaposed by the σ system.

At one time it was fashionable to refer to these accelerations in rate or increases in association as due to an increase in the **effective molarity** of one of the reactants brought about by attaching it covalently to the other. As more exaggerated examples of this phenomenon were reported, however, the unreality of discussing concen-

Table 5–4: Relative Rate Constants[45,46] for Anhydride Formation[a]

phenyl ester	relative rate of anhydride formation[b]	$T\Delta S^{\circ\ddagger}_{approx}$[c] (kJ mol^{-1})
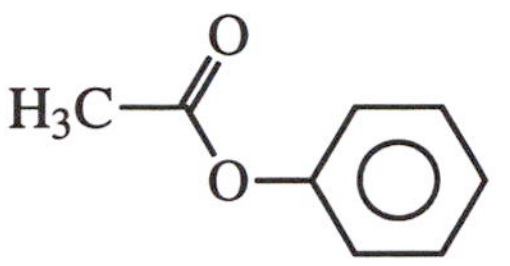	1.0 (1 M CH$_3$COO$^-$)	
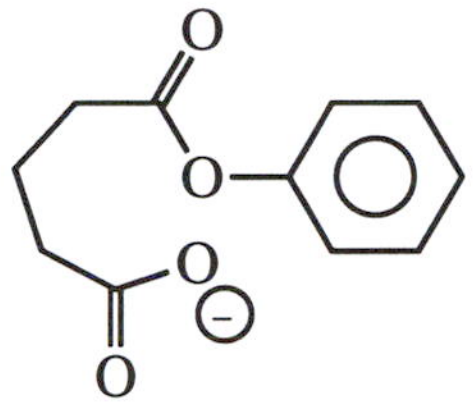	1×10^3	< -17
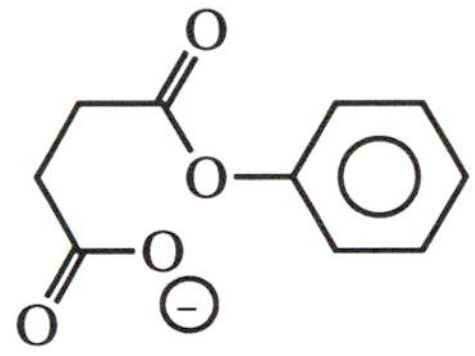	2×10^5	< -30
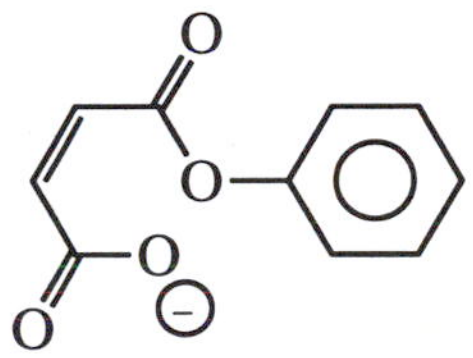	1×10^7	< -40
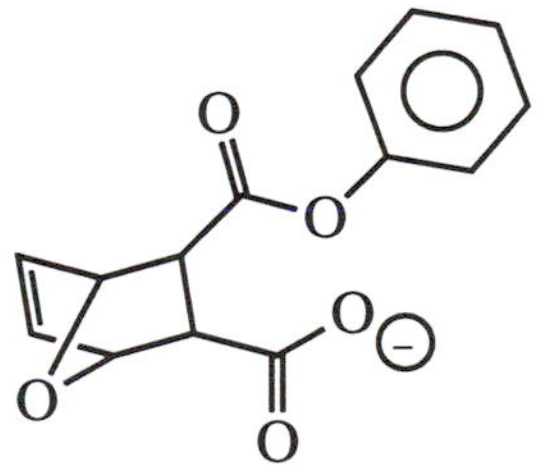	5×10^7	< -44

[a]The first-order rate constants for the hydrolysis of the various phenyl monoesters of dicarboxylic acids were determined[45] as a function of pH in the range pH 4–8. From the pH–rate behavior of each of these rate constants, the first-order rate constant for intramolecular nucleophilic catalysis of the respective hydrolysis by the appended carboxylate could be calculated. From these values, the first-order rate constants for intramolecular anhydride formation could be calculated. These rate constants were determined for each phenyl monoester at 25, 30, or 35 °C. The values of these rate constants were adjusted to the same temperature and presented relative to the first-order rate constant for intramolecular anhydride formation from phenyl glutarate.[45] These first-order rate constants could be related to the pseudo first-order rate constant for the intermolecular formation of acetic anhydride from excess acetate anion and phenyl acetate during the intermolecular nucleophilic catalysis of the hydrolysis of phenyl acetate by acetate anion.[46] [b]First-order rate constants for anhydride formation are presented relative to the calculated[46] pseudo first-order rate constant for the formation of acetic anhydride from phenyl acetate and 1.0 M acetate anion, which is arbitrarily assigned the value of 1.0. [c]The standard entropy of approximation calculated by Equation 5–44 assuming $\Delta\Delta H^{\ast\ddagger} \cong 0$ was multiplied by 298 K. The standard state was 1 M for acetate anion.

trations of 5×10^7 M became apparent,[47] and a more sophisticated view of the situation was required.

In all instances in which an intramolecular reaction is compared to an intermolecular reaction, the difference observed in the two rate constants k_{intra} and k_{inter}, or the two equilibrium constants K_{eq}^{intra} and K_{eq}^{inter}, is due in large part to an increase in the standard entropy of activation or the change in standard entropy, respectively, caused simply by the fact that a unimolecular reaction is being compared to a multimolecular reaction. This increase in the standard entropy of activation or in the standard entropy change for the reaction results from the fact that the standard entropy of approximation is missing from the standard free energy change for the intramolecular reaction. The **standard entropy of approximation**, $\Delta S_{approx}^{\circ\ddagger}$ or $\Delta S_{approx}^{\circ}$, is the change in standard entropy due solely to bringing the separate reactants together into the same molecule or into the same complex, respectively, prior to the reaction. It is a negative number because the intrinsic entropy of two separate reactants relative to the transition state or to the unimolecular product of a reaction is greater than the intrinsic entropy of one molecule containing the two reactants or of a complex into which the two reactants have been assembled relative to the transition state or the product. Because the standard entropy of approximation is missing from the standard entropy of activation for the intramolecular reaction, owing to the fact that approximation has already been accomplished synthetically, the standard entropy of activation or standard entropy change for the intramolecular reaction is greater than the standard entropy of activation or standard entropy change for the intermolecular reaction.

These relationships can be expressed in equations. The standard entropy of activation, $\Delta S_{intra}^{\circ\ddagger}$, or standard entropy change of the intramolecular reaction, ΔS_{intra}°, should be related to the standard entropy of activation, $\Delta S_{inter}^{\circ\ddagger}$, or standard entropy change, ΔS_{inter}°, of the intermolecular reaction by

$$\Delta S_{inter}^{\circ\ddagger} = \Delta S_{intra}^{\circ\ddagger} + \Delta S_{approx}^{\circ\ddagger} \qquad (5\text{--}37)$$

or

$$\Delta S_{inter}^{\circ} = \Delta S_{intra}^{\circ} + \Delta S_{approx}^{\circ} \qquad (5\text{--}38)$$

if the same reaction with the same change in standard enthalpy is occurring once the reactants have been approximated. If this were an adequate description of the situation, then

$$R \ln \frac{k_{intra}}{k_{inter}} = \Delta S_{intra}^{\circ\ddagger} - \Delta S_{inter}^{\circ\ddagger} = -\Delta S_{approx}^{\circ\ddagger} \qquad (5\text{--}39)$$

or

$$R \ln \frac{K_{eq}^{intra}}{K_{eq}^{inter}} = \Delta S_{intra}^{\circ} - \Delta S_{inter}^{\circ} = -\Delta S_{approx}^{\circ} \qquad (5\text{--}40)$$

Because $\Delta S_{approx}^{\circ\ddagger} < 0$ or $\Delta S_{approx}^{\circ} < 0$, $k_{intra} > k_{inter}$ or $K_{eq}^{intra} > K_{eq}^{inter}$, respectively.

The magnitude of the standard entropy of approximation is determined by the difference between two other standard entropy changes, the standard entropy of molecularity and the standard entropy of rotational restraint.[47] The formation of a transition state or a unimolecular product during an intermolecular reaction requires that two or more independent molecules become one molecule, and this involves a considerable decrease in standard entropy. The standard entropy change responsible for this decrease, the **standard entropy of molecularity**, $\Delta S_{molec}^{\circ\ddagger}$ or ΔS_{molec}°, respectively, has a negative value and is a major, unavoidable, unfavorable term in the standard free energy of activation or change in standard free energy in any intermolecular reaction. In an intramolecular reaction, however, the decrease in standard entropy due to the standard entropy of molecularity does not occur, because reactants are already on only one molecule, and this has the effect of increasing dramatically the standard entropy of activation or change in standard entropy for the intramolecular reaction relative to the intermolecular reaction and hence increasing its rate or yield of product.

There is affiliated with an intramolecular reaction, however, a **standard entropy of rotational restraint**, which, conversely, is irrelevant to an intermolecular reaction. It arises from the fact that the formation of the transition state or product during an intramolecular reaction requires that a portion of the rotational entropy in the molecule be eliminated because only a fraction of the accessible rotational isomers can participate in the reaction productively. The standard entropy change accompanying this decrease in the number of rotational isomers, the standard entropy of rotational restraint, $\Delta S_{rot}^{\circ\ddagger}$ or ΔS_{rot}°, has a negative value and its inclusion causes the standard entropy of activation or standard entropy change for the reaction to be smaller than it would be if no rotation occurred and only a productive rotational isomer were present.

The relationships between the standard entropy of approximation and the standard entropy of molecularity and standard entropy of rotational restraint are

$$\Delta S_{approx}^{\circ\ddagger} = \Delta S_{molec}^{\circ\ddagger} - \Delta S_{rot}^{\circ\ddagger} \qquad (5\text{--}41)$$

or

$$\Delta S_{approx}^{\circ} = \Delta S_{molec}^{\circ} - \Delta S_{rot}^{\circ} \qquad (5\text{--}42)$$

The magnitudes of each of these terms can be discussed in turn.

The standard entropy of molecularity is the decrease in standard entropy that should accompany the change of an intermolecular reaction to a rigidly oriented intramolecular reaction.[47] In the specific case of a bimolecular reaction, the two independent reactants have six translational and six rotational degrees of freedom, but the one transition state or the one molecule, formed by the association of the two others, should have only three translational and three rotational degrees of freedom. The standard entropy change associated with the loss of the three translational and three rotational degrees of freedom during this inescapable association, calculated for the situation in which the two reactants and the transition state or product are dissolved in a solution at 25 °C with a standard state of 1 M in solutes, has been estimated[47] to be between −190 and −210 J K^{-1} mol^{-1}. This estimate can be compared to the standard entropy change observed for a simple bimolecular reaction, such as the dimerization of cyclopentadiene in the liquid phase, during which the standard entropy change is −130 to −170 J K^{-1} mol^{-1} with the same choice of standard state. The difference between the calculated standard entropy change and the observed standard entropy change in the particular instance of cyclopentadiene can be completely accounted for by the presence of low frequency vibrations in the dimer that could not be present in the two monomers because of their smaller size. It has been concluded[47] that −190 to −210 J K^{-1} mol^{-1} is an adequate estimate for $\Delta S^{\circ\ddagger}_{molec}$, the maximum decrease in standard entropy change expected from converting a bimolecular reaction into a unimolecular reaction, when molarities are used as units of concentration for standard states. If mole fractions are used as units of concentration for standard states and the concentration of solvent is about 50 M, which is about the concentration of water, the same range for expected standard entropy of approximation for a bimolecular reaction would be −150 to −180 J K^{-1} mol^{-1}.

As the example of the dimerization of cyclopentadiene illustrates, an intramolecular reaction, because it usually involves a larger and more flexible molecule than any of the reactants in an intermolecular reaction, can never realize all of this favorable standard entropy of molecularity. Major factors in decreasing the portion of the standard entropy of molecularity that an intramolecular reaction will enjoy are the internal rotations within the intramolecular reactant itself. These rotations decrease the probability that the necessary juxtaposition of reactants will occur. For example, in the case of phenyl succinate (Figure 5–16) the rotational orientation around three carbon–carbon bonds must be appropriate if the carboxyl oxygen is to be placed adjacent to the acyl carbon. It has been estimated[47] from the results of thermodynamic and kinetic measurements from a number of intramolecular reactions that the standard entropy of rotational restraint decreases by about 20 J K^{-1} mol^{-1} for every bond that lies between the two atoms participating directly in the reaction and about which free rotation can occur.

When two similar intramolecular reactions are compared, for which it is assumed that differences between their standard enthalpies of formation are negligible[45]

$$\Delta\Delta S^{\circ} = R \ln\left(\frac{K_{eq1}}{K_{eq2}}\right) \tag{5–43}$$

where $\Delta\Delta S^{\circ}$ is the difference between their standard entropy changes and K_{eq1} and K_{eq2} are the respective association equilibrium constants. The change in relative rate of phenoxide release in going from phenyl glutarate to phenyl succinate (Table 5–4) is 230-fold and in going from phenyl succinate to the phenyl ester of the fused ring is also 230-fold. In each comparison, one less carbon–carbon bond around which free rotation is allowed is found in the more constrained member of the pair. The changes in rate, presumably reflecting differences in K^{TH}_{eq}, are equivalent in each comparison to 45 J K^{-1} mol^{-1}. It has been noted, however, that the case of the nucleophilic catalysis of the hydrolysis of phenyl esters provides the largest standard entropy of rotational restraint (carbon–carbon bond)$^{-1}$ yet observed.[47] One explanation for the unusually large change in the relative rates as axes of rotation are eliminated in this particular case would be that the differences are not entirely entropic in origin but also involve favorable decreases in the standard enthalpy of formation of the tetravalent intermediate.

With small molecules, the effect of approximation on the rate constant or equilibrium constant is usually significant only when the two central atoms that participate in the reaction become involved in a five-membered or six-membered ring in the transition state or product. A four-membered ring is usually too strained, because of the normal bond angles of commonly encountered molecules, to provide any favorable approximation. A seven-membered ring, if there is free rotation about every bond, has too small a value of ΔS°_{rot} to exhibit a $\Delta S^{\circ}_{approx}$ small enough to have a noticeable effect on rates or equilibrium. For example, even in the intramolecular nucleophilic catalysis of phenyl ester hydrolysis, a series of reactions unusually prone to intramolecular catalysis, phenyl adipate would show a rate of phenolate release due to nucleophilic catalysis only 4-fold greater than that for the same reaction of phenyl acetate in 1.0 M sodium acetate. Large, rigid molecules in which the two atoms that must react are more than six atoms apart yet close enough to collide are difficult to construct, with the relevant exception of a molecule of protein or nucleic acid.

The magnitude of the actual difference in standard entropy of activation or change in standard entropy between a given intramolecular reaction and the corresponding intermolecular reaction will be less than the magnitude of the standard entropy of approximation because vibrational degrees of freedom, unavailable to the reactants in the intermolecular reaction, are available to the necessarily larger reactant in the intramolecular reaction and because steric effects that do not apply to the intermolecular reaction are often unavoidable consequences of designing the intramolecular reactant. If the magnitude of $\Delta\Delta S^{\circ\ddagger}$ or $\Delta\Delta S^{\circ}$, the actual difference between the standard entropies of activation or standard entropy changes in the reactions, must be less than the magnitude of $\Delta S^{\circ\ddagger}_{approx}$ or $\Delta S^{\circ}_{approx}$, respectively, then

$$T\Delta S^{\circ\ddagger}_{approx} < -RT \ln\left(\frac{k_{intra}}{k_{inter}}\right) - \Delta\Delta H^{\circ\ddagger} \qquad (5\text{--}44)$$

where k_{intra} and k_{inter} are the intramolecular and intermolecular rate constants, or

$$T\Delta S^{\circ}_{approx} < -RT \ln\left(\frac{K^{intra}_{eq}}{K^{inter}_{eq}}\right) + \Delta\Delta H^{\circ} \qquad (5\text{--}45)$$

where K^{intra}_{eq} and K^{inter}_{eq} are the intramolecular and intermolecular association equilibrium constants. If the differences in the actual standard enthalpies of activation, $\Delta\Delta H^{\circ\ddagger}$, or the actual standard enthalpies of formation, $\Delta\Delta H^{\circ}$, are known, the estimates of the upper limits for $\Delta S^{\circ\ddagger}_{approx}$ or $\Delta S^{\circ}_{approx}$ can incorporate them. If they are unknown, they can be assumed to be zero, for the sake of argument, but such an assumption can be misleading.

In several reactions displaying large increases in the rate of the reaction or the yield of the product due to covalent approximation of the reactants, the major effect of the approximation is on the standard enthalpy change of the reaction.[16] For example,[30,49] 2,2,3,3-tetramethylsuccinanilide displays a rate of aniline release at pH 5, 1200 times greater than that of succinanilide itself. This increase in rate, however, which is equivalent to a change in the standard free energy of activation of −18 kJ mol^{-1}, is accompanied by a change in the standard enthalpy of activation, $\Delta\Delta H^{\circ\ddagger}$, of −25 kJ mol^{-1}. Therefore, in this case the standard entropy of activation actually decreases as the rate of the reaction is enhanced by approximation. An explanation for this unusual behavior may be the compression that must occur in either the transition state or the tetrahedral intermediate of the tetramethyl analogue because of the vicinal dimethyls on the other side of the ring

5–9

The steric forces on the back of the ring, because they involve van der Waals repulsions among the methyl groups, would be enthalpic effects. They will be the forces that either compress the attacking oxygen against the acyl carbon and lower the standard enthalpy of the

transition state or lock the tetravalent intermediate into the ring and thereby enhance exocyclic ejection of aniline.

A similar enthalpic effect could explain the enhancement in the rate of acid-catalyzed lactonization of 3-(o-hydroxyphenyl)propionate when three methyls are incorporated into its structure to produce 3-[o-hydroxy-o'-methylphenyl]-2,2-dimethylpropionate[50]

$$(5\text{--}46)$$

where $R_1 = R_2 = R_3$ and all are either hydrogen or methyl. The observed difference in the rates of acid-catalyzed lactonization for the trimethyl compound relative to the trihydro compound of 5×10^{10} could arise in large part from steric effects that decrease the standard enthalpy of activation rather than increasing the standard entropy of activation. Unfortunately, no standard enthalpies of activation were measured in this case.

When the upper limits of $T\Delta S^{\circ}_{approx}$ are calculated from the relative rates of the intramolecular and biomolecular nucleophilic catalysis of the hydrolysis of phenyl esters, on the assumption that $\Delta\Delta H^{\circ}$ is equal to zero,[45] they are all equal to or greater than −44 kJ mol^{-1} (Table 5–4). If it is assumed that the fused ring retains two rotational axes, which is a generous assumption, $\Delta S^{\circ}_{approx}$ should be about −150 J K^{-1} mol^{-1} and $T\Delta S^{\circ}_{approx}$ about −44 kJ mol^{-1}. The agreement between experiment and theory is certainly fortuitous.

At least three points are illustrated by this exercise. First, the largest intramolecular increases in rate or degree of association yet measured, with the educational exceptions of the cases involving severe compressive steric effects in the transition states or products, are of a magnitude less than that expected for the transformation of an intermolecular reaction into a fully constrained intramolecular reaction. Second, the maximum decrease in the standard free energy of activation or standard free energy of association to be expected when a bimolecular reaction is turned into an intramolecular reaction is −60 kJ mol^{-1}, which would produce a rate enhancement of 3×10^{10} M. Third, the larger the number of bonds about which rotation can occur between the atoms that must collide during the reaction, the smaller will be the decrease in standard free energy or standard free energy of activation to be expected when an intermolecular reaction becomes an intramolecular reaction.

With these points in mind, the issue of intramolecular hydrogen bonds in aqueous solution can be readdressed. The standard enthalpy change for a hydrogen bond forming in water should be quite small, but possibly of a negative value (Equation 5–31). The competition of water molecules for donor and acceptor seems to contribute an entropic effect whose magnitude is -38 J K^{-1} mol^{-1}, which is $R \ln 100$, when concentrations of donor and acceptor are expressed in molarity. In an intramolecular association, the consequent elimination of the standard entropy of approximation should compensate for the entropic deficit caused by the presence of the water.

Evidence for the existence of intramolecular hydrogen bonds within solutes dissolved in aqueous solution has been reported. The extensive lore surrounding involvement of hydrogen bonds in the equilibrium acid–base behavior of the monoanions of dicarboxylic acids is by and large equivocal,[16] but it has been noted[27] that decreases in the rates of the reactions of the acidic hydrogens in the monoanions of salicylates (**5–10**) and severely constrained dicarboxylic acids such as di-(*n*-propyl)malonic acid (**5–11**) with hydroxide anion

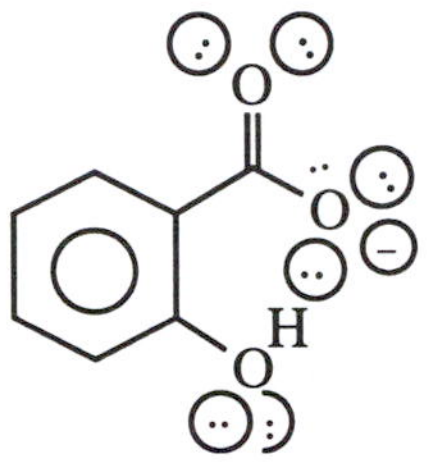

5–10 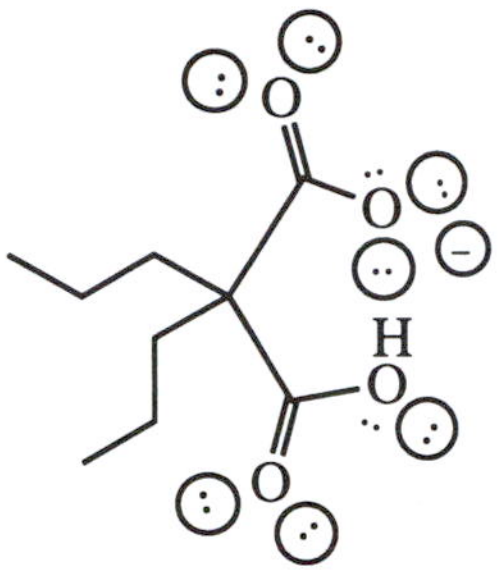   **5–11**

suggest that they contain intramolecular hydrogen bonds whose standard free energy of formation is around -15 kJ mol^{-1}. In the hydrogen bond of dicarboxylic acid **5–11**, the standard enthalpy of formation would be quite favorable because the pK_a of the donor is quite close to the pK_a of the conjugate acid of the acceptor (Equation 5–31), and this might explain the negative value of the standard free energy of formation. A series of compounds capable of forming intramolecular hydrogen bonds either between the pyrrole nitrogen–hydrogen bond on imidazole as donor (pK_a = 15) and a carboxylate as acceptor (pK_a = 5) or between the pyridine lone pair on imidazole (pK_a = 7.5) as acceptor and the nitrogen–hydrogen bond on an ammonium cation as donor (pK_a = 10) has been described (Table 5–5).[51] As the standard entropy of approximation was decreased by confining the juxtaposed donor and acceptor more severely, or the difference in pK_a was decreased, the equilibrium constant for the intramolecular hydrogen bond

$$K_{AHB} = \frac{[\mathrm{B} \ominus \mathrm{HA}]}{[\mathrm{B} \ominus + \mathrm{HA}]} \qquad (5\text{–}47)$$

increased in magnitude (Table 5–5).

The question that these results beg is whether or not the hydrogen bond in a conformation such as an α helix or a hairpin of β structure could ever be rendered favorable by standard entropy of approximation (Figure 5–17). If it is assumed that one or the other of these configura-

Table 5–5: Intramolecular Hydrogen Bonds in Water[a]

hydrogen bond	ΔpK_a[b]	$K_{AHB}^{intra c}$
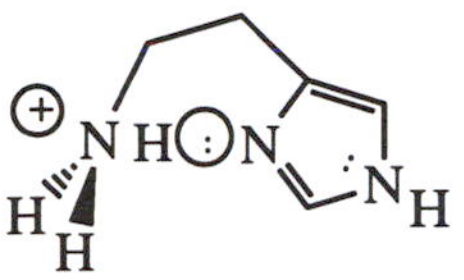	3	2
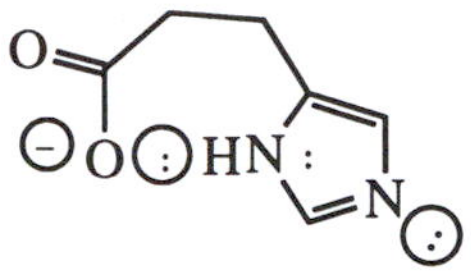	10	< 1
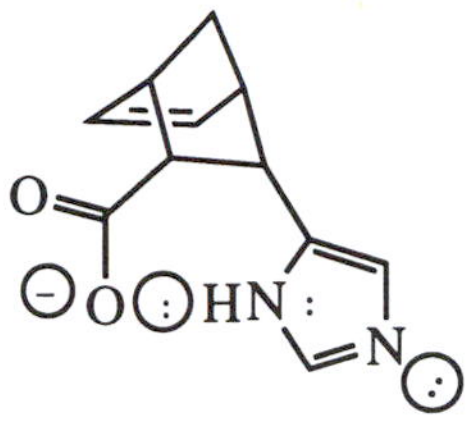	10	1.5
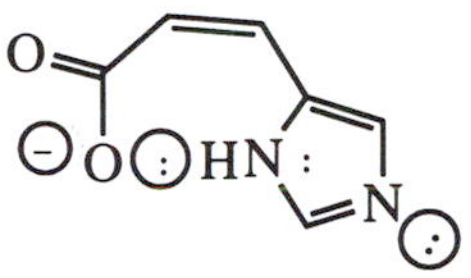	13	13

[a]The equilibrium constants for the formation of the noted intramolecular hydrogen bonds in aqueous solution were estimated from values of the tautomeric equilibrium constants for the respective 4-substituted imidazoles determined by ^{15}N nuclear magnetic resonance.[51] It was assumed that a difference between the value of the tautomeric equilibrium constant for a 4-substituted imidazole in which a hydrogen bond can form and the value of the tautomeric equilibrium constant for a similar compound in which a hydrogen bonding could not form was due to the formation of the noted hydrogen bond. It was also assumed that the value for the tautomeric equilibrium constant for the form of the hydrogen-bonding species in which the bond was not formed was equal to that of the reference compound and that the observed excess of one of the two tautomers represented entirely the hydrogen-bonded form. [b]Difference in pK_a between donor and acceptor. [c]Value for K_{AHB}^{intra} is for the ratio of the concentration of the hydrogen-bonded species (see column 1) to the concentration of the same tautomer not hydrogen bonded.

tions has already been initiated, could the standard free energy of formation for the next hydrogen bond during the propagation have a negative value? In Figure 5–17, the next donor and acceptor in each structure are marked with asterisks. Because the reaction occurs in aqueous solution, the standard enthalpy of formation for this hydrogen bond is zero (Equation 5–31). The value for the standard free energy of formation for this hydrogen bond will be determined in part by the difference between the unfavorable competition of the water and the favorable elimination of standard entropy of approximation that structures provide. For the α helix, there are two bonds about which rotation can occur between donor and acceptor;[52] for the hairpin of β structure, there are four. It

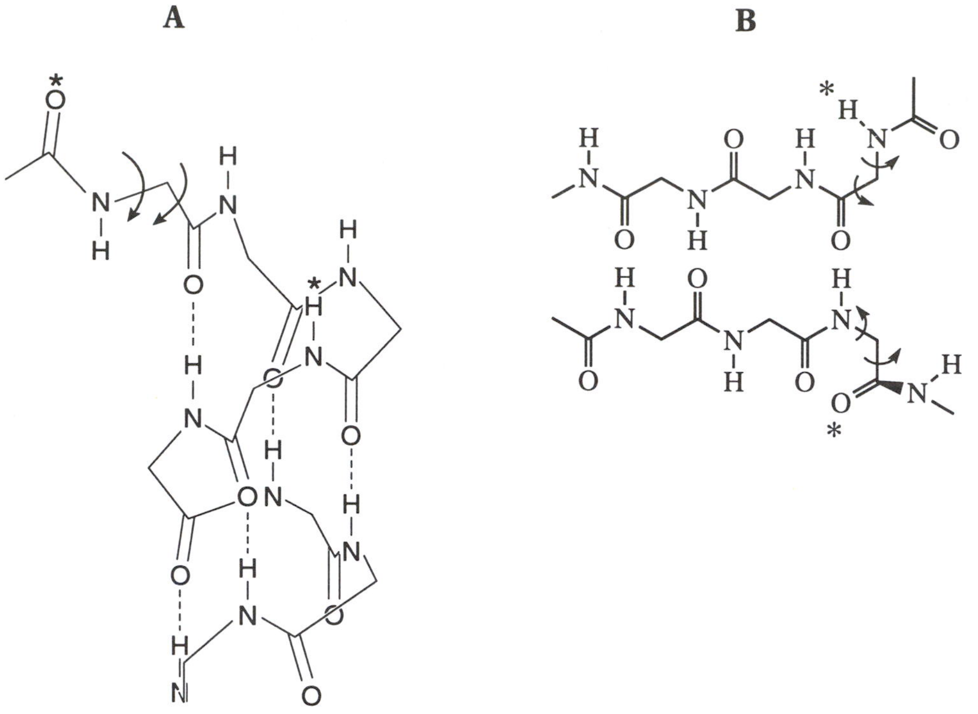

A

B

Figure 5–17: Intramolecular formation of a hydrogen bond to elongate an α helix or a β hairpin. (A) To add the next hydrogen bond in an elongating α helix, the acyl oxygen bearing the asterisk must combine with the amide N–H bearing the other asterisk. The two bonds about which rotation can occur between the last acyl group fixed in the α helix and the acyl oxygen in question have been highlighted with arrows. (B) To add the next hydrogen bond in an elongating β hairpin, the acyl oxygen bearing the asterisk must combine with the amide N–H bearing the other asterisk. In this case there are four bonds about which rotation must be restricted before the hydrogen bond can form.

should, however, be remembered that when the problem is stated in these terms, the difficulties involved in the initiation of either of these structures are ignored, and these are even more formidable.[52]

Almost all small, linear peptides fail to form α helices when they are dissolved at room temperature in water at neutral pH,[53] even if they have the same amino acid sequence as an α helix in a crystallographic molecular model. It was noted, however, that a short peptide with the amino acid sequence KETAAAKFERQHHse, where Hse is homoserine, can fold to produce, in low yield, an α helix at 1 °C in 33 mM Na_2SO_4. This peptide is the cyanogen bromide fragment (Figure 3–3) containing the amino-terminal 12 amino acids of ribonuclease, which form an α helix in the crystallographic molecular model of this protein. At equilibrium at 3 °C, about 15% of this peptide is in α-helical structure between pH 4 and pH 5,[53] but the percentage of α helix decreases to less than 3% as the temperature is raised to 20 °C. This indicates that this α-helical conformation is only marginally stable.

The effect of pH on the amount of α helix in the peptide KETAAAKFERQHHse led to the conclusion that this structure was stabilized by a hydrogen bond between the carboxylate of Glutamate 9 and the imidazolium of Histidine 12.[53] These two amino acids would be adjacent to each other in the α helix (Figure 4–17). This consideration led to the design of a peptide with the sequence AEAAAKEAAAKEAAAKE, which displayed 80% α-helical structure at 0 °C at pH 7.[54] By design, the amino acid sequence of this peptide should permit three hydrogen bonds to form, between glutamates and lysines at posi-

tions 2 and 6, 7 and 11, and 12 and 16, respectively, when the peptide assumes an α-helical conformation, and it was concluded that these three hydrogen bonds were responsible for the high stability of this particular α helix. The two additional hydrogen bonds added sufficient stability that at 20 °C, where no α-helical structure remained in the peptide with only one potential external hydrogen bond, the peptide with three potential external hydrogen bonds still displayed 60% α-helical content.

These results suggest that in the absence of any additional interactions, the standard free energy of formation of an α helix from a linear peptide will always be positive at neutral pH and temperatures above 0 °C. The formation of several hydrogen bonds between several side chains coincident with the formation of an α helix can provide the additional favorable free energy necessary to stabilize the α helix. This stabilization must mean that these external hydrogen bonds have negative standard free energies of formation, presumably because a significant fraction of the standard entropy of approximation has been eliminated by juxtaposing donor and acceptor. By extension, this effect implies that the formation of the next hydrogen bond in an elongating α helix is slightly exergonic and that the reason α helices almost never form from short peptides is that initiation is significantly endergonic, as has been predicted theoretically.[52] It has already been noted that the unfavorable standard free energy of formation for an α helix can be overcome by designing peptides that have amino acid sequences that promote the formation of α-helical coiled coils.[55] Such coiled coils are stabilized by the hydrophobic effect.

An interesting example of the cooperative formation of hydrogen bonds in aqueous solution between two molecules of biological interest occurs in the case of the deoxydinucleotide, 5′-phosphodeoxyguanylyl(3′-5′)deoxycytidine (pdG-dC). A hydrogen-bonded dimer of this self-complementary dinucleotide forms with an association equilibrium constant of 8 M^{-1} at 2 °C and pH 7.5.[56] If it has the common base pairing, six hydrogen bonds hold the two halves together. The rings of the bases are rigid; and, once two of the hydrogen bonds in each base pair form, the standard free energy of formation of the third must incorporate the undiluted standard entropy of molecularity. Furthermore, the nucleotide bases on the two monomers are probably already stacked one on top of the other,[57] an association that at least brings all six donors and acceptors to the same side of the monomer if not in proper alignment. This stacking of the bases results from the hydrophobic effect. It must contribute significantly to this association by roughly aligning the bases before the dimerization occurs.

Suggested Reading

Page, M.I., & Jencks, W. (1971) Entropic Contributions to Rate Accelerations in Enzymic and Intramolecular Reactions and the Chelated Effect, *Proc. Natl. Acad. Sci. U.S.A.* 68, 1678–1683.

Problem 5–9: The rate constant for the reaction between trimethylamine and phenyl acetate to produce the *N,N,N*-trimethylacetamide cation [$(CH_3)_3{}^+NCOCH_3$] and phenoxide anion is 8×10^{-3} M^{-1} min^{-1} at 20 °C. The rate constant for the intramolecular aminolysis of phenyl 4-(*N,N*-dimethylamino)butyrate to yield the *N,N*-dimethyl-2-oxopyrrolidinium cation and phenoxide anion is 10 min^{-1} at 20 °C.

(A) Calculate the "effective molarity" of trimethylamine that would be required for the rate of the intermolecular reaction to equal the rate of the intramolecular reaction.

(B) Calculate an upper limit for $\Delta S^{\circ\ddagger}_{approx}$ for the intramolecular reaction on the assumptions that the solvent for both of these reactions is water and that the standard enthalpies of activation are the same for the two reactions. Use mole fraction as the units for the second-order rate constant.

Problem 5–10: Why would the peptide with the amino acid sequence SEEEKKKKEEEEKKKKF display 60% α helix at pH 8.3 and 4 °C?

The Hydrophobic Effect[58]

The hydrophobic effect is exemplified by the fact that oil and water do not mix. The reason is that water is more stable when the oil is not dissolved in it than when the oil is. This failure of oil and water to mix is only the most extreme manifestation of the tendency of liquid water to expel solutes that are not ions and that do not have significant numbers of donors and acceptors of hydrogen bonds. An ionic solute is held in water by large, negative standard enthalpies of hydration. Solutes that have donors and acceptors of hydrogen bonds are held in water by the hydrogen bonds they form with it. Solutes that neither are ions nor have donors and acceptors of hydrogen bonds are expelled from liquid water. This expulsion is referred to as the hydrophobic effect.

The nature of the hydrophobic effect was succinctly described by G.S. Hartley in 1936:[59]

> The antipathy of the paraffin chain for water is, however, frequently misunderstood. There is no question of actual repulsion between individual water molecules and paraffin chains, nor is there any very strong attraction of paraffin chains for one another. There is, however, a very strong attraction of water molecules for one another in comparison with which the paraffin–paraffin or paraffin–water attractions are very slight.

As is clear from this description, the term hydrophobic is misleading, if its etymology is examined closely.[60] The oil does not dislike the water. In fact, measurements of interfacial energies suggest that the oil prefers the water to itself.[61] Rather, water ejects the oil because water molecules have a powerful like for other water molecules.

In its most general case, the hydrophobic effect upon a hydrophobic solute A can be formally represented by the chemical equation[62,63]

$$A(H_2O) \rightleftharpoons A(\text{solvent } i) \qquad (5–48)$$

In this reaction, a molecule of the solute A is transferred from water to some other solvent. In the case of the failure of oil and water to mix, A is a molecule of oil and the other solvent is the liquid oil itself. Because, in general, pure phases of the different solutes A in a comparison may in themselves have unique peculiarities, Reaction 5–48 is often studied in a systematic fashion by choosing a common solvent for all of the transfers.

The hydrophobic effect can be quantified[62,63] by measuring a standard free energy of transfer. The **standard free energy of transfer** of solute A between water and solvent i, $\Delta G^{\circ}_{A,H2O \rightarrow \text{solvent } i}$, is the change in standard free energy when solute A, dissolved in water at standard state, is transferred from the water into another solvent at standard state. An aqueous phase and a phase of another immiscible solvent are placed in contact with each other directly or indirectly. The solute of interest, for example, hexane or benzene or cyclopentane, is added to the system at low concentration, and its partition between the two phases is allowed to reach equilibrium. The concentration of the solute in each phase is measured, and the partition coefficient, ${}^A K_p$, is calculated

$$^A K_p = \frac{X_A^{\text{solvent } i}}{X_A^{H_2O}} \qquad (5–49)$$

where $X_A^{\text{solvent } i}$ and X_A^{H2O} are the concentrations of solute A in the phase of the other solvent, solvent i, and the aqueous phase, respectively, in units of mole fraction (N). The standard free energy of transfer is[62,63]

$$\Delta G_{A,H_2O \to \text{solvent } i}^{\circ} = \lim_{X_A \to 0} (-RT \ln {}^A K_p) \qquad (5\text{--}50)$$

The standard free energy of transfer is equal to the difference in standard free energy between solute A solvated with solvent i and solute A solvated with water. This difference in **standard free energy of solvation** is the difference in chemical potential between a mole of solute A dissolved in solvent i at standard state and a mole of solute A dissolved in water at standard state. In each case, in order to focus only on standard free energies of solvation, the conditions must be such that solute A is surrounded entirely by either water molecules or solvent molecules, and no molecule of either solute A or the surrounding molecules of water or solvent is affected in its behavior by the presence of another molecule of solute A. This is the reason for the limit in Equation 5–50, which defines the standard state of infinite dilution. In this way, the only contributions to the difference in standard free energy of solvation are the specific interactions between the molecule of solute A and the solvent i or the water.

The chemical potential, μ_A, of solute A dissolved in a particular solution is

$$\mu_A = \mu_A^{\circ} + RT \ln \gamma_A X_A \qquad (5\text{--}51)$$

where γ_A is the activity coefficient for solute A. The concentration-dependent term $RT \ln \gamma_A X_A$ is a direct measure of the increase in standard free energy that occurs as the number of molecules of solute A increases relative to the other molecules in the solution. This increase in standard free energy directly increases the tendency of A to leave the solution and hence raises its chemical potential. If solute A were an ideal solute, this term $RT \ln \gamma_A X_A$ would be entirely due to and directly proportional to the decrease in standard entropy caused by the increase in the number of solute molecules relative to the other molecules in the solution, the chemical potential of solute A would increase only as a logarithmic function of its concentration, and γ_A would be unity at all values of X_A.

Solutes, however, are never ideal. As the concentration of solute A increases, individual molecules of solute A collide more frequently and can interact with each other. If two molecules of solute A attract each other more than one molecule of solute A and one molecule of the solvent do, it will seem, when the concentration of solute A is increased, as though there were fewer molecules of solute A in the solution than there actually are, and γ_A will be less than 1. If two molecules of solute A repel each other more than one molecule of solute A and one molecule of the solvent do, it will seem as though there were more molecules of solute A in solution and γ_A will be greater than 1.

As the concentration of solute A increases in the solution, the increases in the interactions among the molecules of solute A cause the exponent of its chemical potential to deviate from linear behavior with respect to concentration (Figure 5–18). At sufficiently low concentration, however, it is always the case that

$$\mu_A = \mu_A^{\circ} + RT \ln X_A \qquad (5\text{--}52)$$

which is a mathematical statement of Henry's law. At equilibrium, $\mu_{A,H_2O} = \mu_{A,\text{solvent } i}$ and

$$\Delta G_{A,H_2O \to \text{solvent } i}^{\circ} = \mu_{A,\text{solvent } i}^{\circ} - \mu_{A,H_2O}^{\circ} \qquad (5\text{--}53)$$

If a molecule of solute A is transferred from 1 L of water to 1 L of another solvent that has fewer molecules of solvent for every liter than does water, the chemical potential of solute A will increase simply because it is diluted by fewer molecules of solvent. Therefore, the larger the molar volume of the solvent, the larger will be the mole fraction of solute at a given molar concentration. Associated with such a transfer is a change in the

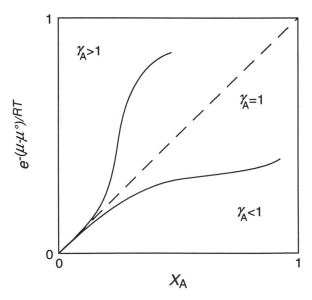

Figure 5–18: Hypothetical behavior of the chemical potential of real solutes as their mole fraction is increased. The exponential of the chemical potential is plotted as a function of the mole fraction. An ideal solute, with an activity coefficient of unity (represented by the dashed line), shows a direct proportionality between mole fraction and $\exp[-(\mu - \mu^{\circ})(RT)^{-1}]$ which is the consequence of Henry's law. Nonideal solutes with activity coefficients greater than or less than unity (represented by the solid lines) deviate from ideal behavior. At low enough concentrations, however, where molecules of solute can no longer interact, ideal behavior is observed with nonideal solutes.

Table 5–6: Thermodynamic Properties of Transfer from Water to Another Solvent[a]

solute	solvent	$\Delta G^\circ_{H_2O \to solv}$ (kJ mol^{-1})	$\Delta H^\circ_{H_2O \to solv}$ (kJ mol^{-1})	$T\Delta S^\circ_{H_2O \to solv}$ (kJ mol^{-1})	$\Delta C^\circ_{P(H_2O \to solv)}$ (J K^{-1} mol^{-1})	ref
CH_4	benzene	−11	+12	+22		6, 63
CH_4	CCl_4	−12	+10	+24		6, 63
C_2H_6	benzene	−16	+9	+25	−250	6, 63
C_2H_6	CCl_4	−15	+7	+22		6, 63
C_2H_6	C_2H_6	−16	+10	+26	−280	58
C_3H_8	C_3H_8	−21	+8	+28	−290	6, 63, 64
n-C_4H_{10}	n-C_4H_{10}	−25	+4	+28	−290	6, 63, 64
C_6H_6	benzene	−19	−2	+17	−430	6, 58, 64
$C_6H_5CH_3$	$C_6H_5CH_3$	−22	−3	+20	−450	6, 58, 64
C_2H_5OH	C_2H_5OH	−3.3	+10	+13	−150	6, 64, 65
n-C_3H_7OH	n-C_3H_7OH	−6.6	+10	+17	−220	6, 64, 65
i-C_3H_7OH	i-C_3H_7OH	−5.1	+13	+18	−220	6, 64, 65
n-C_4H_9OH	n-C_4H_9OH	−9.3	+10	+19	−280	6, 64, 65
n-$C_5H_{11}OH$	n-$C_5H_{11}OH$	−13	+8	+22	−350	58, 65

[a]Values calculated from the thermodynamic behavior of the partition coefficient (Equation 5–50), expressed in units of mole fraction. All values are for a temperature of 25 °C.

standard entropy of mixing which is irrelevant to the solvation itself. This change of entropy should not be included inadvertently in the standard free energy of transfer, hence the choice of mole fraction in Equation 5–49.[62,63]

The standard free energies of transfer of hydrophobic solutes between water and an organic solvent such as benzene, carbon tetrachloride, or the liquid solute itself are negative (Table 5–6). It is this negative change in standard free energy that produces the hydrophobic effect and permits noncovalent forces based on it to be favorable in aqueous solution. The hydrophobic effect is the only noncovalent force in aqueous solution that proceeds with a net negative change in standard free energy, and it is thought to provide all of the driving force for the folding of polypeptides and the formation of interfaces between subunits in oligomeric proteins.

The explicit reason for the negative change in standard free energy at physiological temperatures is that the changes in standard entropy are larger than the changes in standard enthalpy (Table 5–6). At 25 °C, the change in standard enthalpy, which in most cases is positive or unfavorable, is overcome by a much larger positive favorable change in standard entropy. This peculiarity has led to the maxim that the hydrophobic effect is entropy-driven, but this is a misleading view. It has been pointed out[66] that, because the magnitude of the change in standard heat capacity is large[66a] and because the change in standard heat capacity is negative (Table 5–6), the standard enthalpy change and standard entropy change for the hydrophobic effect are required to vary dramatically with temperature (Figure 5–19). As the tem-

perature increases, the standard enthalpy change for transfer becomes more and more exothermic. At high enough temperatures, however, the standard entropy change passes through zero and becomes endergonic. As a result, at intermediate temperatures the reaction changes from an entropically driven process to an enthalpically driven process; and, at high temperatures, the change in standard entropy is actually unfavorable even thought the transfer itself remains favorable because the change in standard free energy does not vary significantly with temperature. This result illustrates the fact that the most characteristic feature of the hydrophobic effect is not its change in standard entropy but its change in standard heat capacity.[66a]

The changes in standard thermodynamic properties, such as standard entropy, standard enthalpy, and standard heat capacity, that are associated with the hydrophobic effect (Table 5–6) have been interpreted as changes in the thermodynamic properties of the water surrounding the solute as it leaves the aqueous phase and changes in the thermodynamic properties of the nonpolar solvent as it enters. The reason that these changes of the thermodynamic variables are assigned to the respective solvents rather than the solutes is that the solutes in most cases are too small, as in the case of methane, ethane, or propane, or too rigid, as in the case of benzene, to account internally for the large changes that are observed.

There are several observations which suggest that a more rigid hydrogen-bonded lattice, similar to the lattice in ice Ih, surrounds dissolved hydrophobic solutes.[67] Macroscopic solids known as clathrates form spontaneously when hydrophobic solutes are mixed with pure

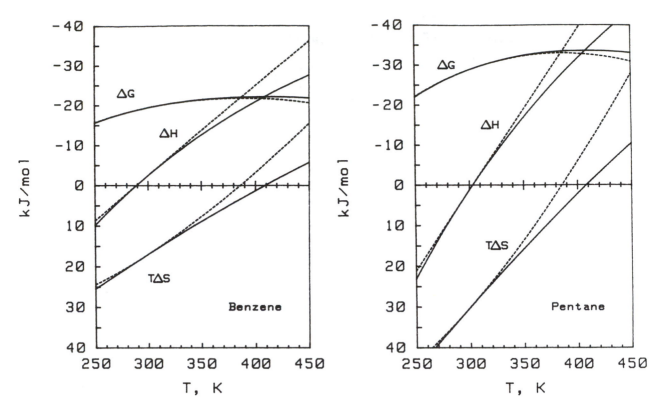

Figure 5–19: Temperature dependence for the changes in standard free energy (ΔG), standard enthalpy (ΔH), and standard entropy, multiplied by temperature ($T\Delta S$), for the transfer of benzene between water and benzene (left panel) and the transfer of pentane between water and pentane (right panel) as a function of temperature.[66] The lines were calculated from the values of the thermodynamic parameters measured in the range from 0 to 40 °C and either the assumption that the observed behavior of ΔC_P° can be fit by an analytic function of temperature and that analytic function can be extrapo- lated beyond the range of measurement (solid lines) or the assumption that ΔC_P° is independent of temperature and has a value that is the mean of the observed values (dashed lines). The values of ΔC_P° over the range of measurements seem to decrease somewhat with increasing temperature and this deviation is the basis for adjusting the value of ΔC_P° for changes in temperature. The trends are the same regardless of the assumption. Reprinted with permission from ref 66. Copyright 1988 Academic Press.

water at proper molar ratios. These solids are crystalline hydrates that are composed of isolated individual molecules of solute encased in rigid hydrogen-bonded networks of water molecules. The thermodynamic parameters associated with such clathrates are of a magnitude sufficient to lead to the conclusion that very similar rigid networks of water molecules should surround hydrophobic solutes when they are present at dilute concentration as well.[68] Whether or not clathrates are of relevance to the hydrophobic effect, their very existence has subconsciously influenced our views of the process. The unusually large changes in standard heat capacity (Table 5–6)[66a] associated with the hydrophobic effect have also been interpreted to result from an increase in the order of the water surrounding the solute. It should be recalled that it is the gradual melting of the hydrogen-bonded lattice that is supposed to be responsible for the anomalously high heat capacity of liquid water itself, and increasing the amount or degree of structure of this lattice should produce even greater capacity for melting and, hence, greater heat capacities. The partial molar volume of a hydrophobic solute in water is usually about 13 cm³ mol⁻¹ less than that of the same solute in other solvents,[69] and this has been thought to result from the efficient packing of the solute within a cage of hydrogen-bonded waters that resembles the networks in clathrates.[70] It has already been noted that ice Ih contains a large amount of vacant space which a solute could occupy (Figure 5–2). Both the increase in the dielectric relaxation time[16,71] and the anomalously large increase in the viscosity[16] observed when a hydrophobic solute is added to water also indicate that a more rigid shell of water forms around the solute than the water in the bulk solvent.

If it is the case, however, that the water surrounding a hydrophobic solute is more structured and held within a more rigid hydrogen-bonded lattice than water in the bulk phase, there should be compensatory thermodynamic changes[16] associated with this increase in structure. Specifically, the enthalpy of the aqueous solution should be less than the enthalpy of liquid water because the hydrogen bonds in this more structured cage should be stronger. If a noncovalent chemical transformation occurs in any solution, the change in standard entropy observed is usually compensated by a change in standard enthalpy. This observation can be stated mathematically[6] as

$$\Delta H^\circ(\alpha) = \Delta H' + T^c \Delta S^\circ(\alpha) \tag{5-54}$$

where α refers to any noncovalent process and $\Delta H'$ and T_c are parameters peculiar to that process. Many noncovalent processes occurring in water[72] satisfy this relationship with $T_c = 280 \pm 10$ K. Parenthetically, this means that values of changes in standard enthalpy or changes in standard entropy for a noncovalent process occurring entirely in aqueous solution are monotonously uninformative because they register only compensatory changes in the structure of the solvent. In the particular case of the hydrophobic effect, however, it has been noted[16] that the large decrease in standard entropy presumably associated with the formation of a rigid cage of hydration as the solute enters water is not accompanied by the large decrease in standard enthalpy to be expected from compensation (Table 5–6). The argument is that the missing enthalpy in this reaction is the enthalpy that was required to crack open the lattice of the liquid water to create a **cavity** for the hydrophobic solute. Because standard entropy and standard enthalpy should compensate in the formation of the shell of hydration and have little effect on the overall reaction due to this cancellation, it is this positive enthalpy of opening the lattice, or conversely, the negative enthalpy realized upon collapsing the cavity, that produces the hydrophobic effect. The positive enthalpy required to open the lattice results from the simple fact that some of the hydrogen bonds of the fluid lattice within liquid water must be broken irretrievably when a cavity is formed for hydrophobic solute. The empty donors and acceptors of such broken hydrogen bonds can be observed in molecular dynamics simulations of aqueous solutions of hydrophobic solutes.[73] In this view, the expulsion of hydrophobic solutes from aqueous solution results from the fact that liquid water is a hydrogen-bonded fluid.

Either an emphasis on the increase in standard enthalpy required to form a cavity or an emphasis on the decrease in standard entropy associated with the formation of the cage of hydration that is the wall of the cavity suggests that the size of the cavity occupied by a hydrophobic solute should be the dominant molecular factor in the magnitude of the hydrophobic effect. The **accessible surface area** of a molecule of solute can be calculated theoretically[76] by asking a digital computer to roll a sphere the size of a molecule of solvent over the surface of a molecular model of the solute.[75] The center of the sphere will trace a surface (Figure 5–20), and the area of this surface is referred to as the accessible surface

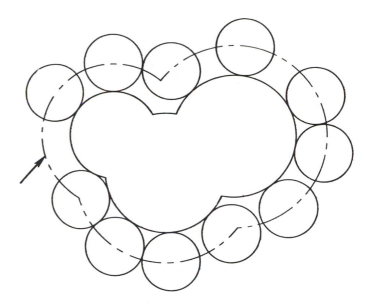

Figure 5–20: Definition of cavity surface area[74] or accessible surface area.[75] The surface traced by the center of a sphere the radius of a water molecule (0.15 nm) as it is rolled over the surface of a molecular model of the solute is the accessible surface area and is denoted in the figure by the broken line. Adapted with permission from ref 74. Copyright 1972 American Chemical Society.

area,[75] or cavity surface area.[74] The relative values of accessible surface area are not significantly affected[74] by the size of sphere chosen ($r = 0.15 – 0.35$ nm), so the set of accessible surface areas, as long as they have all been calculated with same sphere, are reliable and characteristic properties of a set of molecules. If the sphere is chosen to have the radius of a molecule of water (0.15 nm), the accessible surface area of a hydrophobic solute should be the surface area of the cavity that it occupies in an aqueous solution. It has been shown[74] that within a series of saturated alkanes, the logarithm of the solubility of the pure liquid alkane in water (moles liter^{-1}) is correlated with the accessible surface area of the solute (Figure 5–21A). This solubility is for the transfer reaction

solute A (liquid, unit mole fraction) $\longrightarrow$ solute A (aqueous, unit molarity)

$$\tag{5-55}$$

If the solute contains a phenyl group, the standard free energy of transfer between its own liquid and water is +6 kJ mol^{-1} more favorable (Figure 5–21B).[74] This increased solubility of arenes in water can be explained as the result of the basicity of their π electrons.

The correlation between accessible surface area and the hydrophobic effect among small alkanes and arenes has been extended to the conclusion that the hydrophobic effect in any process will be directly proportional to the accessible surface area removed from contact with water. It should be recognized, however, that the relationships

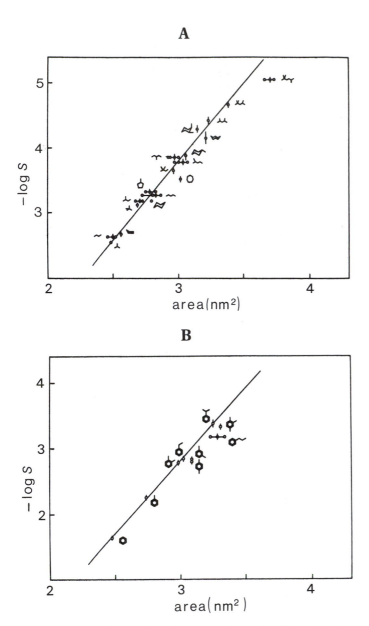

Figure 5–21: Relationships between the solubility of hydrocarbons in water and their cavity surface areas.[74] The negative logarithms of the solubilities (–log *S*) of the liquid hydrocarbons in water[77] in moles liter⁻¹ were used directly and were plotted as a function of cavity surface area (Figure 5–20). (A) Aliphatic hydrocarbons; (B) aromatic hydrocarbons. Reprinted with permission from ref 74. Copyright 1972 American Chemical Society.

arguments based on surface area are made, the molecular details of the process are not known with sufficient precision to make this distinction. This uncertainty is most acute in the simple association of two hydrophobic solutes in water because in this process one cavity disappears.

The noncovalent interaction between two hydrophobic solutes in water can be dissected by referring directly to Reaction 5–1 and Equation 5–2. The term $\Delta G^{\circ}_{\text{A·B}}$ in Equation 5–2 should be the standard free energy of attraction between the two solutes after they have associated with each other. For hydrophobic solutes, this is the van der Waals force arising from the attraction between transient dipoles in one molecule and induced dipoles in the other. These van der Waals forces between the two solutes have negative standard enthalpies of formation and should be favorable contributions to the hydrophobic effect. It has been argued[16] that these van der Waals forces can be ignored because they will be of approximately the same magnitude as the van der Waals forces, lost during the reaction, between molecules of water and the separated solutes. The issue of van der Waals forces between water and the separated solutes, however, presents its own problems, and it is best to note the existence of $\Delta G^{\circ}_{\text{A·B}}$ explicitly. This term will become more important as the argument proceeds.

The standard free energies of hydration, $\Delta G^{\circ}_{\text{HYD(A)}}$, $\Delta G^{\circ}_{\text{HYD(B)}}$, and $\Delta G^{\circ}_{\text{HYD(A·B)}}$, are the standard free energies of formation for the shells of hydration surrounding the isolated solutes and the intermolecular complex, respectively. To the extent that their standard enthalpies of formation compensate their standard entropies of formation (Equation 5–54), these standard free energy changes should be insignificant; to the extent that they do not, the fusion of the two cavities to produce one cavity of less total surface area should proceed with net negative change in standard free energy. The standard free energy change experienced by the water left over after the cavities have fused, $\Delta G^{\circ}_{\text{r H2O}}$, should incorporate the negative change in standard enthalpy that occurs when the number of hydrogen bonds between molecules of water increases as the total cavity surface area decreases. Because two cavities have become one cavity and the net change has been the loss of one cavity, it is not clear that the overall standard free energy change of Reaction 5–1 will be directly proportional to the accessible surface area lost during the juxtaposition of solute A and solute B. In an intramolecular isomerization such as the folding of a polypeptide, however, one could argue[78] that both the product, the folded polypeptide, and the reactant, the unfolded polypeptide, occupy only one cavity that merely decreases in accessible surface area as the reaction proceeds, so that the decrease in standard free energy should be proportional to the decrease in accessible surface area.

It has been pointed out that examining the hydrophobic effect by constructing a thermodynamic cycle

illustrated in Figure 5–21 do not intersect the origin. Therefore, the practice of relating the hydrophobic effect to total surface area must be engaged in with care because standard free energy of transfer is not directly proportional to surface area; it is only correlated with surface area. When a cavity disappears completely, rather than simply decreasing in surface area, the change in standard free energy will not be proportional to the change in surface area. In many of the instances in which

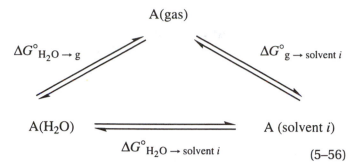

$$\Delta G^{\circ}_{H_2O \rightarrow solvent\ i} \qquad (5\text{–}56)$$

provides further insight into the phenomenon.[66,79,80] The two standard free energies of transfer between the gas phase and the two liquid phases have the usual function of measuring standard free energies of solvation. In a simple Born–Haber cycle, for example, standard enthalpies of solvation are being calculated, and the choice of standard state is not so critical. In the present situation, standard free energies of solvation are being calculated, and the choice of standard states is quite important.

The usual choice of standard state for solvation is the solute at infinite dilution in the solvent so that the solute is fully solvated and no interactions occur among the molecules of solute. Under these circumstances Henry's law is obeyed, and fugacity is proprotional to mole fraction. Because of Henry's law, the correct unit of concentration, when solutes dissolved in several different solvents are being compared, is the mole fraction; and the activity of the solute at infinite dilution is its mole fraction. The volume of a mole of solute at this standard state is the partial molar volume (cubic centimeters mole^{-1}) of the solute A in the solvent i

$$\bar{V}^{solvent\ i}_{A} = \lim_{n_A \to 0} \left(\frac{\partial V}{\partial n_A} \right)_{T,P} \qquad (5\text{–}57)$$

where V is the volume of the solution and n_A is the moles of solute A added to pure solvent. The standard state of solvated solute is chosen to be infinite dilution; and, if the mole fraction of the dissolved solute were extrapolated to unity, the volume of the solution would be the partial molar volume of the solute.

The usual choice of standard state for a gas is the real gas extrapolated to zero pressure. This eliminates the nonideal behavior of the real gas. The behavior of a real gas extrapolated to zero pressure becomes that of an ideal gas, for which fugacity is proportional to pressure. Because of this proportionality, the units of concentration for a gas are usually pressure, and the activity of the gas is defined as its pressure in atmospheres. At standard state, the mole fraction of the gas is unity.

To avoid both the standard entropy of mixing and changes in volume at constant pressure during the transfer of solute from the gas phase to the solution, the volume occupied by a mole of the solute in the gas phase must be equal to the volume occupied by a mole of the solvated solute in the solution. Consider a large volume of solution at standard state in contact with a large volume of the gaseous solute. If the pressure of the gas is such that 1 mol of gaseous solute has the same volume as the partial molar volume of the solute in the solution, when 1 mol of solute is transferred from the gas to the solution at constant pressure the volume of the system does not change. Under these circumstances the transfer is both isochoric and isobaric. As a result, the standard entropy of mixing is 0, no work is performed by the system, and the transfer occurs at constant pressure.

To achieve this condition the gas must be compressed to a volume equal to the partial molar volume of the solute in the solution. At 1 atm pressure the volume occupied by an ideal gas at a concentration of unit atmosphere and a temperature of 298.15 K is 24,500 cm^3 mol^{-1}. The standard free energy change for the compression of that mole of gas to a volume equal to the partial molar volume of the solute is

$$\Delta G^{\circ}_{compression} = RT \ln \left(\frac{24,500\ cm^3 mol^{-1}}{\bar{V}^{solvent\ i}_{A}} \right)_{T,P} \qquad (5\text{–}58)$$

The standard free energy of solvation is the standard free energy of transfer between an ideal gas of the solute, compressed mathematically to the partial molar volume of the solute when it is dissolved in the solvent, and a solution at standard state when the concentration of solute is unit mole fraction. Therefore, the standard free energy of transfer is the standard free energy for the reaction

$$solute\ A\left(g,\ V = \bar{V}^{solvent\ i}_{A}\right) \longrightarrow solute\ A\ (solution,\ unit\ mole\ fraction) \qquad (5\text{–}59)$$

This quantity can be obtained from the usually tabulated standard free energy of transfer for the reaction[80]

$$solute\ A\ (g,\ 1\ atm) \longrightarrow solute\ A\ (solution,\ unit\ mole\ fraction) \qquad (5\text{–}60)$$

by adding $\Delta G^{\circ}_{compression}$. The standard free energy of transfer described by Equation 5–59 reflects most clearly the change in standard free energy between unsolvated solute and solvated solute.

The standard free energies of transfer of a number of hydrophobic solutes between the gas and various solvents have been collected,[80] and extracts from these tabulations, corrected for $\Delta G^{\circ}_{compression}$, are presented in Table 5–7. The first point to be noted is that the solvation of most hydrophobic solutes by most solvents proceeds with a negative change in standard free energy. This is to be expected because the van der Waals forces that arise as the solute is surrounded by solvent have negative standard enthalpies of formation. Water is the clear exception among the solvents examined. The published

Table 5–7: Standard Free Energy of Transfer[a] from the Condensed Gas ($V_{gas} = \bar{v}_A$) to Solvent[b]

									iso-butane	cyclo-pentane	cyclo-hexane		
solvent	CH_4	C_2H_6	C_3H_8	$n\text{-}C_4H_{10}$	$n\text{-}C_5H_{12}$	$n\text{-}C_6H_{14}$	$n\text{-}C_7H_{16}$	$n\text{-}C_8H_{18}$	iso-butane	cyclo-pentane	cyclo-hexane	$(CH_3)_4C$	$(C_2H_5)_4C$
decane	−1.8	−6.2	−9.1	−12	−14	−17	−19.5	−23					
benzene	+0.1	−4.1	−6.7	−9.6	−12	−15	−17	−21	−8.3	−14	−17	−9.6	−22
methanol	+2.3	−0.8	−2.6	−4.9	−5.9	−8.6	−11	−12	−3.7	−8.3	−11	−4.7	−14
dimethyl sulfoxide	+4.1	−0.3	−1.0	−3.1	−3.8	−5.8	−7.5	−9.6					
water[c]	+10.3	+10.6	+12	+13	+14	+15.5	+16	+17.5	+14	+9.4	+9.9	+15	+15
water[d]	+10.9	+12	+15	+18	+21	+24	+27	+29					

[a]Values are in kilojoules mole^{-1} at 25 °C. [b]Values[80] for the change in standard free energy for Reaction 5–60 using mole fraction for units of concentration in the solution were corrected by compressing the volume of the gas to the partial molar volume of the solute. Partial molar volumes were calculated[24] for each solute from the formulas developed by Traube.[69] [c]Values obtained using mole fraction as units of concentration in the solution. [d]Values obtained[81] using molarity as units of concentration in the solution and only correcting for the expansion of the solution caused by dissolving the solute with Equation 5–61.

tabulation[80] presents standard free energies of transfer into 23 solvents other than water. With the exception of ethylene glycol, which is also strongly hydrogen-bonded, the nonaqueous solvent showing the least negative standard free energies of transfer was dimethyl sulfoxide, which has been included in Table 5–7. Even dimethyl sulfoxide, however, fails to demonstrate the extreme behavior of water.

The considerations leading to the values for the standard free energies of transfer in Table 5–7 have been formulated on the assumption, based on Henry's law, that the concentration in the condensed phase determining the chemical potential for the solute in that phase is its mole fraction. It has been argued, however, that the more fundamental and thermodynamically relevant concentration of the solute in the condensed phase is its molarity rather than its mole fraction.[81] If this were the case, the fundamental variable would be the moles of solute in a liter of solution rather than the moles of solute relative to a mole of solvent. In this other approach, molarity is used for the concentration of solute in both the vapor and the condensed phase, and this convention avoids the need to compress the gas to the partial molar volume of the solute to perform an isochoric, isobaric transfer. A correction to the standard free energy of transfer must be made, however, for the expansion of the volume of the solution produced by the introduction of the solute. This term, which is considered as an standard entropy of mixing, has in the limit of infinite dilution of the solute in the solvent the magnitude[81]

$$-T\Delta S_{mix,molarity} = RT(1 - r_{vol}) \qquad (5\text{–}61)$$

where r_{vol} is the ratio between the partial molar volume of the solute and the molar volume of the solvent. When this correction is applied to the standard free energies of transfer between the vapor and water (Table 5–7), calculated with the molarities of the solutes in the solutions rather than the mole fractions, the standard free energies of transfer for the alkanes between the vapor and water are significantly more positive (last line, Table 5–7). It is also argued[81] that the same correction should be applied to standard free energies of transfer between water and liquid alkane (Table 5–6), and this would make these values more negative. For example, the standard free energy of transfer for ethane between water and ethane would be −18 kJ mol^{-1}, that for propane between water and propane would be −23 kJ mol^{-1}, and that for n-butane between butane and water would be −29 kJ mol^{-1}. In both cases, the differences resulting from the choice of units for concentration are significant, but it is unclear which is the correct approach.

The values for the standard free energies of transfer, based on concentrations of mole fraction (Table 5–7, except last line), can be presented graphically by plotting the standard free energy of transfer for a given solute into a given solvent as a function of the standard free energy of transfer for that solute into benzene, which can be used arbitrarily as a reference solvent (Figure 5–22).[80] The solvents hexane, cyclohexane, carbon tetrachloride, toluene, phenyl bromide, and phenyl iodide all yield standard free energies of transfer between those of benzene and decane; the solvents phenyl chloride, N-methylpyrrole, 1-octanol, 1-butanol, 1-propanol, and ethanol all give standard free energies of transfer intermediate between those of benzene and methanol; and the solvents acetonitrile, propylene carbonate, and nitromethane all display free energies of transfer between those of methanol and dimethyl sulfoxide. All of these solvents show clear decreases in standard free energy of transfer as the size of the solute is increased (Table 5–7), while water shows an increase in the standard free energy of transfer. The difference in behavior between water and the other solvents is the hydrophobic effect: the exclusion of hydrophobic solutes from aqueous solution.

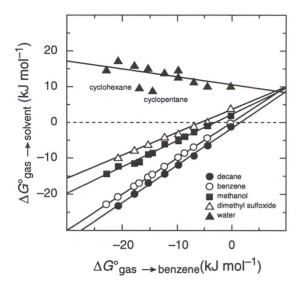

Figure 5–22: Standard free energy of transfer for a given solute from the gas phase into a given solvent plotted as a function of the standard free energy of transfer for that solute from the gas phase into benzene. The values plotted are taken directly from Table 5–7. Benzene was chosen as the reference solvent because more values for standard free energy of transfer into benzene were available and benzene has behavior similar to decane.

It is likely that the common intersection of the five lines in Figure 5–22 indicates the true level at which the standard free energy of transfer equals zero. This intersection should occur at the position occupied by a solute so small and inert that it can exhibit neither van der Waals forces nor the hydrophobic effect and, therefore, has a standard free energy of transfer of 0 into all solvents. The value on the abscissa for such a solute on the plot in Figure 5–22 should lie to the right of methane, the smallest solute in the series. The common intersection of the lines in Figure 5–22 occurs very near the value for the standard free energy of transfer of helium, a solute that displays almost the same standard free energy of transfer into all of the nonaqueous solvents (+4.2 kJ mol^{-1} into decane, +7 kJ mol^{-1} into benzene, +8 kJ mol^{-1} into methanol, and +10 kJ mol^{-1} into di-methyl sulfoxide)[80] and which is the least polarizable of all solutes because it has only two electrons.* It should therefore be the smallest and most inert of the existing solutes. That the true zero of the standard free energy of transfer should be greater than the calculated zero makes sense when the remaining difference between the two standard states is considered. The gas should have more translational and rotational entropy than the liquid solvated solute because translation and rotation are less obstructed in the gas than the liquid. It has been estimated[47] that the loss of translational and rotational entropy in going from the gas

to the liquid for molecules of this size should be between 0 and 13 kJ mol^{-1} at 298 K. This range would also be consistent with the level of standard free energy at which the lines intersect. If the standard free energies of transfer must still be corrected for the differences in entropy of translation and rotation, the magnitude of the correction should be correlated with the entropies of vaporization, which in turn are correlated with the boiling points of the solutes.[47] Therefore, as one moves from right to left in Figure 5–22, the decrease in each individual standard free energy of transfer due to this correction should become greater, causing the slopes of all of the lines to become more positive as well as shifting the zero point upward.

If water participates in van der Waals interactions with hydrophobic solutes just as the other solvents do, the unfavorable hydrophobic effect is greater in magnitude than these favorable van der Waals forces and more than overcomes them. Several attempts have been made to estimate experimentally the magnitude of the van der Waals forces between water and solutes so that the full magnitude of the hydrophobic effect could be calculated by difference. In one attempt,[80] it was proposed that solvation of noble gases by water might proceed without a hydrophobic effect and that their behavior could be used to estimate van der Waals forces, but peculiarities in their interaction with water, in particular their rather large negative standard entropies of solvation,[82] suggest that they also experience a hydrophobic effect.

As the polarities of the solvents increase, the slopes of the lines in Figure 5–22 become less negative. This effect can be explained by noting that as the solvents become more polar, more standard free energy is required to form a cavity within them, and this change is deducted from the favorable standard free energy of interaction between solute and solvent. In this view, water is simply the extreme example of the difficulty in forming a cavity.

When examined from the viewpoint of standard free energies of solvation, the hydrophobic effect, originally defined as the tendency of hydrophobic solutes to leave water and enter another solvent, is the difference between $\Delta G^{\circ}_{g \to \text{solvent } i}$ and $\Delta G^{\circ}_{g \to H_2O}$, or graphically, the distance between a point for water in Figure 5–22 and the respective point for solvent i in Figure 5–22. As water excludes all hydrophobic solutes to about the same extent, the decrease in standard free energy of transfer between water and a particular solvent among a series of different solutes of increasing size could be considered to be due mainly to the decreases in standard free energy of solvation manifested by the nonaqueous solvent. This standard free energy of solvation should result almost entirely from van der Waals forces.

Within a limited set of solutes, such as acyclic alkanes (Table 5–7), the standard free energies of transfer between the gas and the various solvents are correlated with the accessible surface areas of the solutes (Figure 5–23). When the standard free energies of transfer are presented in this way, it becomes clear that the majority

---

*Helium, however, is significantly less soluble in water ($\Delta G^{\circ}_{gas \to H_2O}$ = +13 kJ mol^{-1}) than in other solvents.

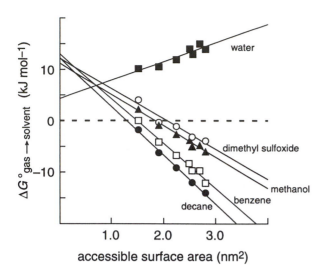

Figure 5–23: Correlation between accessible surface area and standard free energy of transfer. The standard free energies of transfer into the noted solvents were from Table 5–7. The accessible surface areas were from published tables.[74] The compounds chosen to display the correlation were the acyclic aliphatic hydrocarbons (from left to right) methane, ethane, propane, isobutane, *n*-butane, neopentane, and *n*-pentane.

of the correlation between accessible surface area and standard free energies of transfer between water and liquid alkane arises from the correlation between accessible surface area and solvation of the solute by the alkane itself. This conclusion suggests that the widely relied upon correlation between the hydrophobic effect and accessible surface area may be only the adventitious consequence of the fact that transfer between water and some other isotropic solvent was used as the basis for its definition. What was observed in this correlation was mainly the correlation between normal solvation and accessible surface area rather than any correlation between the hydrophobic effect exerted by the water itself and accessible surface area. This consideration would throw into question any estimate of the magnitude of a hydrophobic effect that was based solely on changes of accessible surface area when properties other than transfer into another isotropic solvent are being estimated.

This practice of relying only on accessible surface area is particularly disturbing when it is used to estimate the magnitude of the change in standard free energy due to the hydrophobic effect manifested during the folding of a polypeptide, the formation of a multimeric protein, or the binding of a ligand to a protein. The transfers that occur during these processes are not between liquid water and another solvent but between liquid water and the interior of a molecule of protein. There is no reason to assume that the interior of a molecule of protein behaves as if it were an isotropic solvent.

These considerations bring the argument back to the van der Waals forces between the molecules brought together by the hydrophobic effect, back to $\Delta G°_{A \cdot B}$ in

Equation 5–2. Regardless of whether or not water engages in van der Waals interactions of the same magnitude as those between the two molecules themselves, the results in Figure 5–22 suggest that water *behaves* as though it does not engage in van der Waals interactions. If so, regardless of what type of cancellation is occurring, once the solutes have come together the standard free energy of their association is significantly affected by the standard free energy of the van der Waals forces between themselves. As a result, and ironically, it is the van der Waals forces between the associated solutes that determine the magnitude of the hydrophobic effect in any particular circumstance, and the magnitude of the van der Waals force felt by a functional group in the interior of a molecule of protein may be very different from the magnitude of the van der Waals force it would experience in any particular isotropic solvent.

In conclusion, it is hard to argue that much has been learned by dissecting the hydrophobic effect beyond the observation that oil and water do not mix, and the hydrophobic effect remains somewhat mysterious. Nevertheless, when a polypeptide is present in water in its unfolded state, the hydrophobic surfaces scattered along its length are unstable relative to any state in which they are in contact with each other and out of contact with water. This is the hydrophobic effect. Because ions and donors and acceptors of hydrogen bonds are more stable in the hydrated state than in any state in which they are isolated from water, even if they are fully paired in ion pairs and hydrogen bonds, they cannot provide net favorable standard free energy to the process. The hydrophobic effect is the only noncovalent force that provides net favorable standard free energy to drive the folding of a polypeptide.

Suggested Reading

Abraham, M.H. (1979) Free Energies of Solution of Rare Gases and Alkanes in Water and Nonaqueous Solvents: A Quantitative Assessment of the Hydrophobic Effect, *J. Am. Chem. Soc. 101*, 5477–5484.

Problem 5–11: The anomalously high heat capacity of liquid water is presumably due to the fact that as the liquid is heated, a certain amount of its hydrogen-bonded structure is lost. This extra heat capacity (beyond that calculated from vibrations and translations of the molecules) is called the configurational heat capacity ($C_{P,cf}$). For water, $C_{P,cf} = 30$ J K^{-1} mol^{-1} (T = 273–373 K).

(A)   Calculate ΔS_{cf} and ΔH_{cf}, the configurational standard entropy and standard enthalpy changes, which are associated with heating water from 293 to 313 K.

It is difficult to say with certainty what fraction of the hydrogen-bond character is lost over this temperature range, but these numbers give a rough estimate of the ratio of the changes of standard enthalpy and standard entropy ($\Delta H/\Delta S$) to be expected when the structure of water increases or decreases in the liquid. Consider the following phase transfer reaction

$$C_3H_8(H_2O) \rightleftharpoons C_3H_8(\text{pure phase})$$

(B) If all the change in standard entropy (Table 5–6) is due to a decrease in water structure as the "icelike" regions which surrounded the alkane melt, what standard enthalpy change must accompany this decrease in water structure?

(C) What is the contribution to the change in standard free energy ($\Delta G°$) in the above reaction that results from the melting of structured water around the alkane?

(D) What other effects may account for the remainder of the $\Delta G°$?

(E) Which of these processes is the most important in determining the $\Delta G°$ for the separation of propane and H_2O?

Problem 5–12: Consider the following transfer reaction

$$RCH_2OH(H_2O) \rightleftharpoons RCH_2OH \text{ (pure)}$$

This chemical equation describes the transfer of an alcohol from water to a pure phase. As such it describes the tendency of the alcohol to remove itself from water, a hydrophobic effect. For any substance A under any circumstances there is associated an intrinsic standard free energy, or chemical potential (μ)

$$\mu_A = \mu°_A + RT \ln X_A$$

where μ_A is the chemical potential of A under the experimental circumstances, $\mu°_A$ is the chemical potential of A at standard state and unit concentration, and X_A is the concentration of A in units of mole fraction. The standard free energy change for the transfer reaction is

$$\Delta G_{alc,H_2O} = \mu_{alc} - \mu_{alc,H_2O} = \mu°_{alc} - \mu°_{alc,H_2O} - RT \ln X_{alc,H_2O}$$

where μ_{alc} is the chemical potential of pure alcohol, μ_{alc,H_2O} is the chemical potential of the alcohol dissolved in water at a concentration of X_{alc,H_2O}, where X_{alc,H_2O} is the concentration of the alcohol in the water phase in units of mole fraction. At equilibrium, when alcohol saturates the water phase, $\Delta G = 0$ and

$$\mu°_{alc} - \mu°_{alc,H_2O} = \Delta G°_{alc,H_2O \to alc} = RT \ln X^{sat}_{alc,H_2O}$$

It is easy to measure the [alcohol] at saturation when H_2O and pure alcohol are shaken in a two-phase system (separatory funnel) and the aqueous phase is removed. The following results were obtained at 25 °C

alcohol	[alcohol]$_{sat}$ (mol L^{-1})
n-butanol	0.97
n-pentanol	0.25
n-hexanol	0.059
n-heptanol	0.0146

(A) Change these numbers into X^{sat}_{alc,H_2O}, making the assumption that

$$\frac{\text{moles of alcohol} + \text{moles of } H_2O}{\text{liter}} \cong 55 \text{ mol L}^{-1}$$

(B) Calculate $\Delta G°_{alc,H_2O \to alc}$ for each case. The units are kilojoules mole^{-1}. Plot $\Delta G°_{alc,H_2O \to alc}$ as a function of the number of carbon atoms (n).

(C) Determine the slope of your plot, $\Delta G°_{CH2,H_2O \to alc}$

$$\Delta G°_{alc,H_2O \to alc} = \Delta G°_{alc} + n\Delta G°_{CH2,H_2O \to alc}$$

(D) The term $\Delta G°_{CH2,H_2O \to alc}$ is the standard free energy change associated with the transfer of a methylene group from H_2O to pure hydrocarbon. Is this transfer favored or unfavored?

(E) The term $\Delta G°_{alc}$ is simply the standard free energy change due to the fact that the molecule you are using is an alcohol. Extrapolate to determine $\Delta G°_{alc,H_2O \to alc}$ for *n*-propanol.

(F) Determine $\Delta S°_{CH2}$ in the equation $\Delta S° = \Delta S°_{alc} + n\Delta S°_{CH2}$ and $\Delta H°_{CH2}$ in the equation $\Delta H° = \Delta H°_{alc} + n\Delta H°_{CH2}$ from the values in Table 5–6.

(G) Is $\Delta H°$ or $\Delta S°$ the major contributor to the hydrophobic effect on a CH_2 group at 25 °C?

Hydropathy

As ionic interactions, hydrogen bonds, and the hydrophobic effect are considered in turn, it becomes clear that the functional groups participating in each of these processes—ionic groups, donors and acceptors of hydrogen bonds, and hydrophobic surfaces—have strong favorable and unfavorable interactions with water. Ionic solutes display large negative standard enthalpies of hydration that dominate their behavior in water. When these ions are withdrawn from water, large investments of free energy must be made to strip the shells of hydration from them. Solutes with donors and acceptors of hydrogen bonds form hydrogen bonds with water molecules that have significant negative standard free energies of formation because of the high molar concentration of the water in the solution. When donors and acceptors of hydrogen bonds are withdrawn from water, significant investments must be made to counter these standard free energies. Hydrophobic solutes leave aqueous solution with a strong preference for almost any other condensed phase because when they are dissolved in water, they behave as if they cannot form any net favorable interactions with it (Figure 5–22). When they are withdrawn from water into another condensed phase, significant favorable changes in standard free energy are realized. Each of these outcomes arises from the respective changes in the structure of water that accompany the particular transfer of the functional groups between water and a nonaqueous phase.

Viewed in this light, few solutes elicit indifferent responses from water. Solutes are either hydrophilic, demonstrating an affinity for water, or hydrophobic, demonstrating an aversion to water. These strong feelings about water are referred to as **hydropathy**. Hydropathy defines a continuous spectrum, at one end of which is **hydrophilicity** and at the other **hydrophobicity**.

It was suggested by Hine and Mookerjee[83] that the hydrophilic or hydrophobic character of a solute could be judged from its standard free energy of transfer between water and the gas. Assembling values from the tables of Hine and Mookerjee[83] and providing several previously unmeasured values, Wolfenden, Andersson, Cullis, and Southgate[84,85] have presented the standard free energies of transfer between water and the gas for model compounds of the side chains of the amino acids in which the α carbon has been replaced by hydrogen (Table 5–8). These values reflect the magnitudes of the standard free energies required to remove the various functional groups present in the amino acids from water at pH 7. As previously noted, the alkanes are expelled spontaneously from water with standard free energies of

transfer between −10 and −15 kJ mol^{-1}. Toluene, as are most arenes (Figure 5–21), is less hydrophobic because of the presence of the basic π cloud of the phenyl ring. The hydrogen-bond donors and acceptors on methanethiol, ethanol, and methanol increase their standard free energies of transfer by either +15 kJ mol^{-1} in the case of the thiol or +30 kJ mol^{-1} in the case of the alcohols compared to alkanes of the same size. Presumably, this is due to the requirement for breaking hydrogen bonds between water and these donors and acceptors to remove them into the gas phase. Both 3-methylindole and 4-cresol have standard free energies of transfer +20 kJ mol^{-1} greater than toluene, and this must also arise from the presence of either the weak pyrrole donor or the hydroxyl, respectively, that form hydrogen bonds with the donors and acceptors on water.

In the case of the amino acids that are charged at pH 7, such as glutamic acid, aspartic acid, histidine, lysine, and arginine, the partition coefficients measured for transfer of the model compounds between water and the gas are for acidic solutions or basic solutions in which they are dissolved entirely as the neutral conjugate acids or neutral conjugate bases, respectively. At pH 7 only a fraction of the actual amino acid will be present as the neutral species, and this will decrease the value of the partition coefficient for transfer from water to the gas. This decrease in the partition coefficient can be incorporated into the standard free energy of transfer with the formula[84]

$$\Delta G^{\circ}_{A,H_2O \to g} = \lim_{X_A \to 0} (-RT \ln \alpha^A K_p) \qquad (5\text{–}62)$$

Table 5–8: Standard Free Energies of Transfer of Model Compounds for the Amino Acids Between Water and the Gas Phase at 25 °C and pH 7[a]

amino acid	model compound	$\Delta G^{\circ}_{H_2O \to g}$ (kJ mol^{-1})	amino acid	model compound	$\Delta G^{\circ}_{H_2O \to g}$ (kJ mol^{-1})	amino acid	model compound	$\Delta G^{\circ}_{H_2O \to g}$ (kJ mol^{-1})
L	isobutane	−14	C	methanethiol	+3	K	butylamine[b]	+34
I	butane	−13	T	ethanol	+18	Q	propionamide	+36
V	propane	−12	S	methanol	+19	N	acetamide	+38
A	methane	−10	W	3-methylindole	+20	E	propionic acid[b]	+38
F	toluene	−1	Y	4-cresol	+21	H	4-methylimidazole[b]	+40
M	ethylmethylsulfide	+2				D	acetic acid[b]	+42
						R	methylguanidine[b]	+79

peptide bond N-methylacetamide +39 kJ mol^{-1}

[a]The values for the standard free energies of transfer from water to the gas phase for the various model compounds were obtained from several tables.[83–85] They were usually presented as the transfer of the compound between the standard state of the real gas at infinitely low partial pressure with concentration expressed in atmospheres and the standard state of the infinitely dilute solution with concentration expressed in molarity. The units of concentration in the dilute solution were changed to mole fraction by dividing molar concentrations of solute by the molar concentration of water. The partial molar volumes of the solutes[86] were calculated[24] for each solute from the formulas developed by Traube.[69] Free energies of compression (Equation 5–58) of the gas to a volume equal to the partial molar volume of the solute were then calculated. These values were subtracted from the standard free energies of transfer corrected to units of mole fraction to obtain the final values for the standard free energy of isochoric transfer from water to the compressed gas. [b]Values for the pK_a of the various amino acids in a polypeptide (Table 2–2) were used to correct the standard free energies of transfer of the neutral compounds[84,85] for the standard free energy of neutralization required at pH 7 (Equation 5–62).

where ^AK_p is the partition coefficient for the model compound and α is the fraction that is un-ionized at pH 7. The values for α were calculated with the values of the pK_a for the amino acids in a polypeptide (Table 2–2) rather than with the values for the pK_a of the model compounds themselves. The fraction of un-ionized species, α, varies from 0.8 for histidine to 10^{-6} for arginine. Even the standard free energy necessary to neutralize N-methylguanidinium and transfer the neutral compound into the gas ($+79$ kJ mol^{-1}) is still less than the standard free energy that would be required to transfer it into the gas as a cation (Figure 5–9). Therefore, all of the tabulated values should refer to the most likely reactions, namely, neutralization followed by transfer. For example, the only reasonable reaction for the transfer of n-butylamine, dissolved in H_2O at pH 7, to the gas would be

$$(5\text{–}63)$$

One of the most striking, and perhaps informative, facts about the charged amino acids in a polypeptide is that all of them can be neutralized in simple acid–base reactions. This is a common property of almost all organic cations and anions, with sulfonates and tetraalkylammonium cations among the notable exceptions. Nevertheless, it could be argued that these latter two functionalities, which do appear frequently in biological settings, have been purposely excluded from among the amino acids by natural selection because they cannot be neutralized. When charged amino acid side chains are transferred to a region with a low dielectric constant, as occurs during the folding of a polypeptide, they will enter as the uncharged conjugate acids or conjugate bases. The standard free energy required to neutralize the charges of glutamic acid, aspartic acid, and lysine and then remove their hydrogen-bond donors and acceptors from water is greater by about 50 kJ mol^{-1} than the standard free energy of transfer for an alkane of the same size.

The amides and 4-methylimidazole have remarkably large positive standard free energies of transfer. In the case of the amides, it might have been argued that when they were removed from water, two acceptors and two donors were left unoccupied for a net loss of two hydrogen bonds; however, the standard free energy of transfer for N-methylacetamide, the model for the peptide bond, is identical to that for acetamide even though a net loss of only one hydrogen bond occurs when it is removed from water. It has been suggested that in the case of the amides the acceptors may be more important than the donors,[87] even though they are very weak bases. That the standard free energy of transfer for acetamide or propionamide is about $+12$ kJ mol^{-1} greater than that for neutral acetic acid or propionic acid, respectively, suggests that amides

form stronger hydrogen bonds with water than might be expected from their acid dissociation constants. The large positive standard free energy of transfer for 4-methylimidazole is presumably due to the two facts that it is aromatic and that it has a strong donor and a strong acceptor.

The point emphasized by these experimental results is that many of the side chains of the amino acids have a very high affinity for water, and a good deal of standard free energy must be spent whenever they are removed from it. This is the complement of the fact that the only favorable standard free energy gained from noncovalent forces in aqueous solution arises from the removal of the hydrophobic amino acids from contact with water. At the beginning, every amino acid within a newly synthesized, unfolded polypeptide is completely solvated by water. During the folding of the polypeptide, the formation of an interface between subunits, or the binding of a ligand, there is a net transfer of individual amino acids, segments of polypeptide backbone, and small solutes from water to the interior of a native protein. In analogy with Equation 5–56, the standard free energies of transfer from the aqueous solution should be the standard free energies of transfer between water and the gas (Table 5–8). Those solutes removed from water are transferred into a new environment, the interior of the protein. The standard free energies of transfer into this new environment are the second half of the reaction, but the standard free energies of transfer into the interior of a protein from the gas cannot be predicted with any certainty.

Presumably, noncovalent forces with negative standard free energies of formation arise as the amino acids and the polypeptide backbone are packed into the interior. The standard free energy of transfer for the removal of hydrophobic amino acids from the aqueous phase is a negative change in standard free energy, but a small one (Table 5–8). The hydrophobic amino acids, however, are being transferred into a new environment that should not have the aversion displayed by liquid water but should respond with favorable van der Waals forces, perhaps resembling in their magnitude the favorable negative standard free energies of solvation for nonpolar solutes displayed by all solvents other than water (Figure 5–22). The standard free energy of transfer for the removal of the hydrogen-bond donors and acceptors from the aqueous phase, especially those of the polypeptide backbone itself, are large positive standard free energies of transfer (Table 5–8). In the interior of the protein, however, these donors and acceptors participate in hydrogen bonds whose standard free energy of formation may be rendered even more negative because standard entropies of approximation are not required for the association (Figure 5–17). Standard entropies of approximation are at a minimum whenever a set of hydrogen bonds forms cooperatively.

The interior of a properly folded molecule of protein is tightly packed in a defined conformation. It thus resembles closely the interior of a solid rather than a liquid, and the van der Waals forces arising when amino acids or segments of polypeptide backbone are transferred into it are those that would arise in a solid rather than in a liquid. This is a significant difference because in solids dipoles and polarizable regions remain more fixed in their relative orientations rather than being averaged over all orientations as they are in a liquid. Even more troublesome, when the question of reproducing the behavior of this solid is considered, is the fact that the interior of a protein is not a systematic solid, such as a crystalline or microcrystalline mineral or an amorphous glass. Although it is a highly integrated system, shaped by natural selection for the performance of a definite function, very little uniformity can be found in the interior of a particular molecule of protein. Because it is the product of evolution by natural selection, the interior has all of the haphazard character of an acre of woodland. It is unlikely that any solvent could reproduce its properties.

Eventually it may be possible to calculate the standard free energy of transfer into the interior of a particular molecule of protein from the respective positions and orientations of each of the constituents in its crystallographic molecular model, but at the moment the standard free energies of transfer between the gas and another solvent are the only reliable observations available. It has been proposed that the standard free energy of transfer of an amino acid, a segment of polypeptide, or a small solute from water to the interior of a protein should be similar to the standard free energy of transfer for a model compound of that amino acid or segment of polypeptide between water and a solvent whose properties resemble those of the interior of a protein. Scales of hydropathy based on this intuition have been presented. They differ in the personal preferences of their proponents for the type of model compounds and the particular solvent chosen as the basis of the scale.

The first such scale was the hydrophobicity scale proposed by Nozaki and Tanford,[88] which, as its name implies, was confined to only one end of the spectrum. It was based on the solubilities of the zwitterionic amino acids in ethanol that had been previously tabulated by Cohn and Edsall.[23] By subtracting the standard free energy of transfer for glycine between water and ethanol from the standard free energies of transfer for hydrophobic amino acids between water and ethanol, they estimated the standard free energies of transfer for the side chains alone between water and ethanol. The implication in formulating a scale of this type is that the interior of a protein resembled ethanol in its interaction with hydrophobic amino acids, and this may not be far from the truth because all nonaqueous solvents display similar standard free energies of solvation for hydrophobic solutes (Figure 5–22).

Since this first scale was proposed, many others have appeared, and they have usually been expanded spectra including all 20 of the amino acids, hydrophilic as well as hydrophobic. They have been based on several different standard free energies of transfer. The original description of the hydrophobic effect was based on observations of abnormal decreases in the surface tension of water that result when hydrophobic solutes display a preference for the surface of an aqueous solution rather than its interior,[6,89] and a scale of hydropathy based on the change in the surface tension of an aqueous solution with the change in concentration of the different amino acids has been presented.[90] The hydrophobicity scale based on the solubilities of the amino acids in ethanol has been expanded to include uncharged, hydrophilic amino acids.[91] The standard free energies of transfer of the model compounds for the amino acids between water and the gas (Table 5–8) have also been used to create a scale of hydropathies.[84,85] The standard free energies of transfer of various solutes between water and 1-octanol have been proposed as the parameters for a general scale for the hydrophobic effect.[92] The standard free energies of transfer of the N-acetyl α-amides of each of the amino acids (Figure 2–1) between water and 1-octanol have been determined, and they have been used to construct a scale of hydropathies.[93] It has been proposed that N-cyclohexyl-2-pyrrolidone would be a better solvent to use as reference for standard free energies of transfer into the interior of a protein,[94] and standard free energies of transfer for the amino acids between water and N-cyclohexyl-2-pyrrolidone have been measured and used to construct a scale of hydropathies.

In competition with these scales based on standard free energies of transfer are scales derived from the locations of the various amino acid side chains in crystallographic molecular models of native proteins. The logic in this case is that the purpose of all of these scales is to estimate contributions due to changes in solvation during the folding of a polypeptide and the degree with which particular amino acids are buried in the interior or exposed to the solvent should directly indicate how hydrophobic or hydrophilic, respectively, they are. In these computations, the accessible surface area (Figure 5–20) of each amino acid in a set of crystallographic molecular models is individually determined. These individual accessibilities are then grouped by amino acid, and average accessibilities for each amino acid are calculated. The uncertainty in these calculations is in the calculation of these averages, and the three scales of hydropathy based on the accessible surface area in molecular models of folded polypeptides[78,95,96] are not equivalent, even though they are based on very similar raw data.

Finally, there are the scales of hydropathies for the amino acids that are based on mixtures of the pure scales discussed so far. In one case,[97] a scale based on the accessible surface area of amino acids in crystallographic models was modified by a theoretical calculation of the standard free energy required to break hydrogen bonds and neutralize charge. In another,[86] the standard free energies of transfer between water and the gas and a

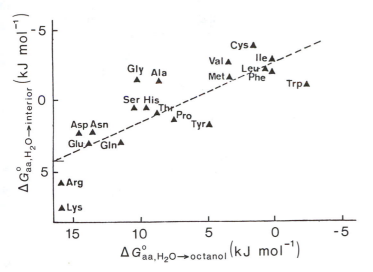

Figure 5–24: Correlation between the standard free energies of transfer of *N*-acetyl α-amides of the amino acids between octanol and water with the degree to which the amino acids are buried in the interior of a molecule of protein.[93] The partition coefficients for the distribution of the *N*-acetyl α-amides of the 20 amino acids between water at pH 7 and octanol at room temperature were measured (concentrations in molarity) and standard free energies of transfer, $\Delta G^\circ_{aa,H_2O \to octanol}$, were calculated (Equation 5–50). Each of the 5220 amino acids in the crystallographic molecular models of 22 proteins was identified as either buried (less than 0.2 nm² of accessible surface area) or accessible to water (greater than 0.2 nm² of accessible surface area).[100] For each type of amino acid a partition ratio [number buried (number accessible)⁻¹] was calculated, and from this partition ratio, a standard free energy of transfer, $\Delta G^\circ_{aa,H_2O \to interior}$ was calculated (Equation 5–50). Adapted with permission from ref 93. Copyright 1983 Elsevier.

tabulation of accessible surface areas were combined with personal preference to produce a scale of hydropathies. In a third case,[98] a consensus scale of hydropathies was inferred from two scales based on standard free energy of transfer and three scales based on accessible surface area in folded polypeptides. In a fourth case,[99] the correlation between surface area and the hydrophobic effect (Figure 5–21), the standard free energy required to neutralize charged amino acids (Equation 5–62), and estimates of the standard free energy for withdrawing each individual hydrogen-bond donor and acceptor from water were combined to obtain a scale of estimated standard free energies of transfer for each of the amino acids, when located in an α helix, from water to a phase of hydrocarbon.

At low resolution all of these scales are similar to each other. The amino acids whose side chains are alkanes, namely, leucine, isoleucine, and valine, are the most hydrophobic amino acids; the charged amino acids whose pK_a is farthest from pH 7, arginine, lysine, glutamate, and aspartate, are the most hydrophilic; and neutral but polar amino acids such as serine and threonine reside in the middle; but the details of the ranking and the relative magnitudes of the parameters are quite different. At the moment, each of these attempts to estimate the standard free energy of transfer for each of the amino acids between water and the interior of a protein has its particular proponents, some more forceful than others, and there is no unambiguous way to choose among them or assess whether any of them is realistic or unrealistic.

The usual criterion for the reliability of each scale is to demonstrate either that it correlates with the distribution of amino acids between the surface and the interior of a protein, if it is based on standard free energy of transfer (Figure 5–24),[73,85] or that it correlates with standard free energies of transfer, if it is based on the distribution of amino acids between the surface and the interior.[95] None

of these correlations suggest that any one of the scales is more realistic than any of the others.

Suggested Reading

Kyte, J., & Doolittle, R.F. (1982) A Simple Method for Displaying the Hydropathic Character of a Protein, *J. Mol. Biol.* **157**, 105–132.

Problem 5–13: Consider a saturated solution of the substance A. In this case, the solution of A is in equilibrium with solid A and

$$\mu_{A,\text{sat solution}} = \mu_{A,\text{solid}}$$

$$\mu^\circ_{A,\text{sat solution}} + RT \ln X_{A,\text{sat}} = \mu^\circ_{A,\text{solid}}$$

since $X_{A,\text{solid}} = 1$. If you wish to compare two different solvents and their effects on A

$$\mu^\circ_{A,\text{solution 1}} + RT \ln X_{A,\text{sat 1}} = \mu^\circ_{A,\text{solid}}$$

$$\mu^\circ_{A,\text{solution 2}} + RT \ln X_{A,\text{sat 2}} = \mu^\circ_{A,\text{solid}}$$

$$\mu^\circ_{A,\text{solution 2}} - \mu^\circ_{A,\text{solution 1}} = \Delta G^\circ_{A,1 \to 2}$$

$$\Delta G^\circ_{A,1 \to 2} = RT \ln \left(\frac{X_{A,\text{sat 1}}}{X_{A,\text{sat 2}}} \right)$$

where $\mu^\circ_{A,\text{solution 1}}$ is the chemical potential of A in solvent 1 at a concentration of 1 mole fraction, $X_{A,\text{sat}}$ is the concentration of A in a saturated solution in mole fraction units, and $\Delta G^\circ_{A,1 \to 2}$ is the standard free energy change when A is transferred from solvent 1 to solvent 2. The quantity $\Delta G^\circ_{A,1 \to 2}$ is a measure of the change in standard free energy for the following type of reaction

$$\text{amino acid (H}_2\text{O)} \rightleftharpoons \text{amino acid (EtOH)}$$

where ethanol serves as a model for the interior of a protein.

Given the following data

amino acid	concentration at saturation [g (100 g of solvent)$^{-1}$]	
	H$_2$O	EtOH
glycine	25.16	0.00382
leucine	2.17	0.0196

(A)   Calculate $X_{\text{amino acid,sat}}$ for these four situations.

(B)   Calculate $\Delta G^\circ_{\text{A,H2O}\rightarrow\text{EtOH}}$ for glycine and leucine.

(C)   Use the value you have for glycine to subtract away the contribution of $^-$OOCCH$_2$NH$^+_3$ to the solubility of leucine. The remainder is an estimate of the standard free energy of transfer of the leucine side chain from H$_2$O to ethanol.

Problem 5–14:

(A)   Draw the structure of the glutamine side chain and divide it into hydrophobic or hydrogen-bonding regions. Label each region on your drawing and indicate all hydrogen-bond donors and acceptors with D or A, respectively.

(B)   Estimate the $\Delta G^\circ_{\text{glutamine,H2O}\rightarrow\text{ethanol}}$ contributed only by the hydrophobic regions of the side chain.

(C)   The $\Delta G^\circ_{\text{transfer,H2O}\rightarrow\text{ethanol}}$ for the peptide unit (–CH$_2$CONH–) is +4.8 kJ mol^{-1}. Estimate the $\Delta G^\circ_{\text{glutamine,H2O}\rightarrow\text{ethanol}}$ for the whole glutamine side chain.

(D)   The solubility of glutamine (concentration at saturation) in water is 4.6 g (100 g of H$_2$O)$^{-1}$. The solubility of glutamine in ethanol is 4.59×10^{-4} g (100 g ethanol)$^{-1}$. Calculate $\Delta G^\circ_{\text{transfer,H2O}\rightarrow\text{ethanol}}$ of the glutamine side chain.

Problem 5–15: Consider the following table:

A	0.25	G	0.16	P	−0.07
R	−1.76	H	−0.40	S	−0.26
N	−0.64	I	0.73	T	−0.18
D	−0.72	L	0.53	W	0.37
C	0.04	K	−1.10	Y	0.02
E	−0.62	M	0.26	V	0.54
Q	−0.69	F	0.61		

(A)   What are the letters and what is the intention of assigning these numbers to these letters?

(B)   What common property is shared by the letters with the positive numbers?

(C)   What common property is shared by the letters with the negative numbers?

(D)   Divide the letters with the positive numbers into two groups based on differences in chemical properties. Why are the numbers in one of these groups less positive than the numbers in the other?

(E)   Divide the letters with negative numbers into two groups on the basis of differences in chemical properties. Why are the numbers in one of these groups more negative than the numbers in the other?

(F)   On what types of measurements could the numbers assigned to the letters be based?

(G)   What is the most important factor determining the sign and magnitude of the numbers assigned to the letters?

(H)   Draw the interactions with water that are one of the reasons that R has a value of −1.76. There are two reasons that R has such a low value: the interactions you have just drawn and another of its properties. What are these two reasons?

References

1. Del Bene, J., & Pople, J.A. (1970) *J. Chem. Phys. 52*, 4858–4866.
2. Hankins, D., Moscowitz, J.W., & Stillinger, F.H. (1970) *J. Chem. Phys. 53*, 4544–4554.
3. Symons, M.C.R. (1972) *Nature 239*, 257–259.
4. Eisenberg, D., & Kauzmann, W. (1969) *The Structure and Properties of Water*, Clarendon Press, Oxford, England.
5. Dyke, T.R., & Muenter, J.S. (1973) *J. Chem. Phys. 59*, 3125–3127.
6. Edsall, J.T., & McKenzie, H.A. (1978) *Adv. Biophys. 10*, 137–207.
7. Frank, H.S. (1970) *Science 169*, 635–641.
8. Narten, A.H., & Levy, H.A. (1971) *J. Chem. Phys. 55*, 2263–2269.
9. Narten, A.H., & Levy, H.A. (1969) *Science 165*, 447–454.
10. Morgan, J., & Warren, B.E. (1938) *J. Chem. Phys. 6*, 666–673.
11. Narten, A.H., Danford, M.D., & Levy, H.A. (1967) *Discussions Faraday Soc. 43*, 97–107.
12. Walrafen, G.E. (1968) *J. Chem. Phys. 48*, 244–251.
13. Stillinger, F.H., & Rahman, A. (1974) *J. Chem. Phys. 60*, 1545–1557.
14. Moore, W.J. (1972) *Physical Chemistry*, 4th ed., pp 890–894, Prentice-Hall, New York.
15. Pauling, L. (1960) *The Nature of the Chemical Bond*, 3rd ed., p 449, Cornell University Press, Ithaca, NY.
16. Jencks, W.P. (1969) *Catalysis in Chemistry and Enzymology*, McGraw-Hill, New York.
17. Meof-Ner, M. (1984) *J. Am. Chem. Soc. 106*, 1265–1272.
18. Benoit, R.L., & Lam, S.Y. (1974) *J. Am. Chem. Soc. 96*, 7385–7390.
19. Parsegian, A. (1969) *Nature 221*, 844–846.
20. Stokes, R.H. (1964) *J. Am. Chem. Soc. 86*, 979–982.
21. Noyes, R.M. (1962) *J. Am. Chem. Soc. 84*, 513–522.
22. Cox, B.G., & Parker, A.J. (1973) *J. Am. Chem. Soc. 95*, 6879–6804.
23. Cohn, E.J., & Edsall, J.T. (1943) *Proteins, Amino Acids, and Peptides*, Reinhold, New York.

24. O'Shea, E.K., Lumb, K.J., & Kim, P.S. (1994) *Curr. Biol.*, in press.

25. Moore, W.J. (1972) *Physical Chemistry*, 4th ed., pp 420–476, Prentice-Hall, New York.

26. Hine, J. (1972) *J. Am. Chem. Soc. 94*, 5766–5771.

27. Eigen, M. (1964) *Angew. Chem., Int. Ed. Engl. 3*, 1–19.

28. Pimentel, G.C., & McClelland, A.L. (1960) *The Hydrogen Bond*, Freeman, San Francisco, CA.

29. Taylor, R., Kennard, O., & Versichel, W. (1983) *J. Am. Chem. Soc. 105*, 5761–5766.

30. Taylor, R., Kennard, O., & Versichel, W. (1984) *J. Am. Chem. Soc. 106*, 244–248.

31. Donohue, J. (1969) *J. Mol. Biol. 45*, 231–235.

32. Taft, R.W., Gurka, D., Joris, L., Schleyer, P., & Rakshys, J.W. (1969) *J. Am. Chem. Soc. 91*, 4801–4809.

33. Arnett, E.M., Mitchell, E.J., & Murty, T.S.S.R. (1974) *J. Am. Chem. Soc. 96*, 3875–3891.

34. Stymne, B., Stymne, H., & Wettermark, G. (1973) *J. Am. Chem. Soc. 95*, 3490–3494.

35. Arnett, E.M. (1963) *Prog. Phys. Org. Chem. 1*, 223–403.

36. Rubin, J., Senkowski, B.Z., & Panson, G.S. (1964) *J. Phys. Chem. 68*, 1601–1602.

37. Kuhn, L.P., Wires, R.A., Ruoff, W., & Kwart, H. (1969) *J. Am. Chem. Soc. 91*, 4790–4793.

38. Honig, B.H., & Hubbell, W.L. (1984) *Proc. Natl. Acad. Sci. U.S.A. 81*, 5412–5416.

39. Kristof, W., & Zundel, G. (1980) *Biophys. Struct. Mech. 6*, 209–225.

40. Lindeman, R., & Zundel, G. (1978) *Biopolymers 17*, 1285–1304.

41. DeTar, D.F., & Noval, R.W. (1970) *J. Am. Chem. Soc. 92*, 1361–1365.

41a. Klotz, I.M., & Franzen, J.S. (1962) *J. Am. Chem. Soc. 84*, 3461–3466.

42. Klotz, I.M., & Farnham, S.B. (1968) *Biochemistry 7*, 3879–3882.

43. Stahl, N., & Jencks, W.P. (1986) *J. Am. Chem. Soc. 108*, 4196–4205.

44. Hine, J. (1962) *Physical Organic Chemistry*, 2nd ed., pp 81–103, McGraw-Hill, New York.

45. Bruice, T.C., & Pandit, U.K. (1960) *J. Am. Chem. Soc. 82*, 5858–5865.

45a. Bruice, T.C., & Turner, A. (1970) *J. Am. Chem. Soc. 92*, 3422–3428.

46. Bruice, T.C. (1970) in *The Enzymes: Kinetics and Mechanism* (Boyer, P.D., Ed.) 3rd ed., Vol. II, pp 217–279, Academic Press, New York.

47. Page, M.I., & Jencks, W.P. (1971) *Proc. Natl. Acad. Sci. U.S.A. 68*, 1678–1683.

48. Coward, J.K., Lok, R., & Takagi, O. (1976) *J. Am. Chem. Soc. 98*, 1057–1058.

49. Higuchi, T., Eberson, L., & Herd, A.K. (1966) *J. Am. Chem. Soc. 88*, 3805–3808.

50. Milstein, S., & Cohen, L.A. (1972) *J. Am. Chem. Soc. 94*, 9158–9165.

51. Roberts, J.D., Yu, C., Flanagan, C., & Birdseye, T.R. (1982) *J. Am. Chem. Soc. 104*, 3945–3949.

52. Zimm, B.H., & Bragg, J.K. (1959) *J. Chem. Phys. 31*, 526–535.

53. Bierzynski, A., Kim, P.S., & Baldwin, R.L. (1982) *Proc. Natl. Acad. Sci. U.S.A. 79*, 2470–2474.

54. Marqusee, S., & Baldwin, R.L. (1987) *Proc. Natl. Acad. Sci. U.S.A. 84*, 8898–8902.

55. O'Neil, K.T., & DeGrado, W.F. (1990) *Science 250*, 646–651.

56. Krugh, T.R., & Young, M. (1975) *Biochem. Biophys. Res. Commun. 62*, 1025–1031.

57. Ogasawara, N., & Inoue, Y. (1976) *J. Am. Chem. Soc. 98*, 7054–7060.

58. Tanford, C. (1980) *The Hydrophobic Effect: Formation of Micelles and Biological Membranes,* John Wiley and Sons, New York.

59. Hartley, G.S. (1936) *Aqueous Solutions of Paraffin-Chain Salts*, Hermann & Cie., Paris, as quoted in Tanford, C. (1973) *The Hydrophobic Effect: Formation of Micelles and Biological Membranes*, p viii, John Wiley & Sons, New York.

60. Hildebrand, J.H. (1979) *Proc. Natl. Acad. Sci. U.S.A. 76*, 194.

61. Tanford, C. (1979) *Proc. Natl. Acad. Sci. U.S.A. 76*, 4175–4176.

62. Kauzmann, W. (1954) in *The Mechanism of Enzyme Action* (McElroy, W.D., & Glass, B., Eds.) pp 70–120, Johns Hopkins Press, Baltimore, MD.

63. Kauzmann, W. (1959) *Adv. Protein Chem. 14*, 1–63.

64. Franks, F., & Reid, D.S. (1973) in *Water: A Comprehensive Treatise, Water in Crystalline Hydrates: Volume 2, Aqueous Solutions of Simple Non-electrolytes* (Franks, F., Ed.) pp 323–380, Plenum Press, New York.

65. Arnett, E.M., Kover, W.B., & Carter, J.V. (1969) *J. Am. Chem. Soc. 91*, 4028–4034.

66. Privalov, P.L., & Gill, S.J. (1988) *Adv. Protein Chem. 39*, 191–234.

66a. Edsall, J.T. (1935) *J. Am. Chem. Soc. 57*, 1506–1507.

67. Frank, H.S., & Evans, M.W. (1945) *J. Chem. Phys. 13*, 507–532.

68. Hafemann, D.R., & Miller, S.L. (1969) *J. Phys. Chem. 73*, 1392–1397.

69. Traube, J. (1899) *Samml. Chem. Chem. Tech. Vortr. 4*, 255–332.

70. Nemethy, G., & Scheraga, H.A. (1962) *J. Chem. Phys. 36*, 3382–3400.

71. Haggis, G.H., Hasted, J.B., & Buchanan, T.J. (1952) *J. Chem. Phys. 20*, 1452–1465.

72. Lumry, R., & Biltonen, R. (1969) in *The Structure and Stability of Biological Macromolecules* (Timasheff, S.N., & Fasman, G.D., Eds.) pp 65–212, Marcel Dekker, New York.

73. Ravishankar, G., Mehotra, P.K., Mezei, M., & Beveridge, D.L. (1984) *J. Am. Chem. Soc. 106*, 4102–4108.

74. Hermann, R.B. (1972) *J. Phys. Chem. 76*, 2754–2759.

75. Lee, B., & Richards, F.M. (1971) *J. Mol. Biol. 55*, 379–400.

76. Lee, B.K., & Richards, F.M. (1971) *J. Mol. Biol. 55*, 379–400.

77. McAuliffe, C. (1966) *J. Phys. Chem. 70*, 1267–1275.

78. Chothia, C. (1976) *J. Mol. Biol. 105*, 1–14.

79. Cramer, R.D. (1977) *J. Am. Chem. Soc. 99*, 5408–5412.

80. Abraham, M.H. (1979) *J. Am. Chem. Soc. 101*, 5477–5484.

81. Sharp, K.A., Nicholls, A., Friedman, R., & Honig, B. (1991) *Biochemistry 30*, 9686–9697.

82. Abraham, M.H. (1980) *J. Am. Chem. Soc. 102*, 5912–5913.

83. Hine, J., & Mookerjee, P.K. (1975) *J. Org. Chem. 40*, 292–297.

84. Wolfenden, R., Cullis, P.M., & Southgate, C.C.F. (1979) *Science 206*, 575–577.

85. Wolfenden, R., Andersson, L., Cullis, P.M., & Southgate, C.C.B. (1981) *Biochemistry 20*, 849–855.

86. Kyte, J., & Doolittle, R.F. (1982) *J. Mol. Biol. 157*, 105–132.

87. Wolfenden, R. (1978) *Biochemistry 17*, 201–204.

88. Nozaki, Y., & Tanford, C. (1971) *J. Biol. Chem. 246*, 2211–2217.

89. Traube, J. (1891) *Ann. Chem. 265*, 27.

90. Bull, H.B., & Breese, K. (1974) *Arch. Biochem. Biophys. 161*, 665–670.

91. Segrest, J.P., & Feldman, R.J. (1974) *J. Mol. Biol. 87*, 853–858.

92. Hansch, C., & Leo, A. (1979) *Substituent Constants for Correlation Analysis in Chemistry and Biology*, Wiley and Sons, New York.

93. Fauchére, J.L., & Pliska, V. (1983) *Eur. J. Med. Chem., Chim. Ther. 18*, 369–375.

94. Lawson, E., Sadler, A.J., Harmatz, D., Brandau, D.T., Micanovic, R., MacElroy, R.D., & Middaugh, C.R. (1984) *J. Biol. Chem. 259*, 2910–2912.

95. Janin, J. (1979) *Nature 277*, 491–492.

96. Guy, H.R. (1985) *Biophys. J. 47*, 61–70.

97. von Heijne, G., & Blomberg, C. (1979) *Eur. J. Biochem. 97*, 175–181.

98. Eisenberg, D., Weiss, R.M., Terwilliger, T.C., & Wilcox, W. (1982) *Faraday Symp.Chem. Soc. 17*, 109–120.

99. Engelman, D.M., Steitz, T.A., & Goldman, A. (1986) *Annu. Rev. Biophys. Biophys. Chem. 15*, 321–353.

100. Janin, J. (1979) *Nature 277*, 491–492.

CHAPTER 6

Atomic Details

It is within the refined crystallographic molecular model of a protein that the consequences of the noncovalent forces just described can be viewed. As the polypeptide folded to form the native structure of the protein represented by the model, a significant fraction of its backbone had to be withdrawn from water. Secondary structure formed not because it was inherently beautiful but because it offered an efficient way to maintain the total number of hydrogen bonds in the solution. As the polypeptide folded, the donors and acceptors of hydrogen bonds in the side chains of the amino acids, as well as any ionized side chains, remained, by and large, on the surface of the structure to maintain their hydration. The networks of hydrogen-bonded water molecules hydrating the surface of the folded structure are prominent features of the refined crystallographic molecular model. The core of the model is formed mostly from hydrophobic amino acids, the exclusion of which from the solvent drove the folding process.

As remarkable as these consequences are in the refined crystallographic molecular model, it has become clear upon close inspection that they do not, except indirectly, produce the final structure. The most important determinant of the final native structure is the steric effect. In retrospect, this should not be so surprising. Steric effects are always the most overwhelming among the different forces influencing the outcome of a chemical reaction. They are rarely discussed at great length because they are so easy to understand. No two fragments of matter may occupy the same place at the same time. The folding of a polypeptide, however, is a steric nightmare. Not only must all of the functional groups fit together in a confined space without overlapping, but all of the functional groups are connected together by the polypeptide. Although the outcome of any one of these games of packing atoms cannot be predicted, the ultimate solutions to the vast array of steric problems encountered during each game can be appreciated by examining the final native structure. Both steric effects operating along the polypeptide backbone and those engendered between the side chains as the elements of secondary structure attempt to fit together to produce the final native structure of the protein are represented by the crystallographic molecular model. These steric effects and the noncovalent forces are the players in each of the games.

Secondary Structure of the Polypeptide Backbone

As has already been noted, because of its π molecular orbital system (Figure 2–3), the amide of the **peptide bond** is planar. The dihedral angle, ω, assigned to this amide is defined by looking down the carbon–nitrogen bond from carbon to nitrogen

trans 6–1   6–2   cis 6–3

The sign of the angle is determined by the right-hand rule. In *trans* peptide bonds, the angle ω is 180°, and the two strands of polypeptide leaving the amide in the two direction depart diagonally. In *cis* peptide bonds, the angle ω is 0°, and the two α carbons of the two departing strands are eclipsed. In a protein, at any particular location in the amino acid sequence, the peptide bond is either *trans* in every molecule of that particular protein or *cis* in every molecule of that particular protein. The reason for this is that the stereochemical difference between these two geometric isomers is substantial, and only one will be compatible with the structure at that location in the folded polypeptide. In the available crystallographic molecular models, almost all of the locations where a *cis* peptide bond is found have proline as the amino acid on the carboxy-terminal side. In these peptide bonds, the proline provides the amide nitrogen, and the amide formed is a secondary amide. In this secondary amide, there is little preference for the *trans* stereochemistry because both positions on the amide nitrogen are sterically equivalent, and the equilibrium constant between *cis* and *trans* peptide bonds involving proline is between 0.1 and 1.[1] In proteins, about 25% of the peptide bonds in which proline provides the amide nitrogen are *cis* peptide bonds.[2] The other amino acids form primary amides. Because the hydrogen of a primary amide is much smaller than the alkyl substitutent, the equilibrium constant heavily favors the *trans* form. *Cis*-peptide bonds in which an amino acid other than proline provides the amide nitrogen are very rare. A few, however, such as those between Serine 197 and Tyrosine 198, Proline 205 and Tryptophan 206, and Arginine 272 and Aspartate 273

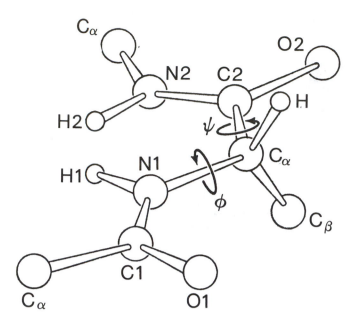

Figure 6–1: Designation of the two dihedral angles ϕ and ψ to the amino- and carboxy-terminal sides, respectively, of the α carbon (C_α) of an amino acid in a polypeptide.[6] The first carbon of a side chain (C_β) corresponds to that of amino acids in the L-configuration. Adapted with permission from ref 6. Copyright 1963 Academic Press.

in bovine carboxypeptidase A[3] and between Glycine 95 and Glycine 96 in *Escherichia coli* dihydrofolate reductase,[4] have been seen in refined molecular models. These are probably locations where a *cis* peptide bond is unavoidable, but evolution by natural selection has not yet replaced the carboxy-terminal amino acid with a proline. Other than the occasional situations in which *cis* peptide bonds occur, all of the peptide bonds in refined molecular models are *trans* with values of angle ω equal to $180° \pm 3°$, although deviations from $180°$ as large as $10°$ have been observed.[5]

It was noted by Ramachandran, Ramakrishnan, and Sasisekharan,[6] before crystallographic structures of proteins became available, that there are severe steric effects hindering rotation about the single bonds of a polypeptide. The dihedral angle of the bond between an α carbon and the adjacent amide nitrogen is designated as ϕ; and that of the bond between an α carbon and the adjacent amide acyl carbon, as ψ (Figures 6–1 and 6–2). Every amino acid in the protein has a **dihedral angle ϕ** and a **dihedral angle ψ** associated with its α carbon (c1 Kin.1). The signs of these dihedral angles are determined by the right-hand rule (Figure 6–2). To follow what is about to be described, you should build a model of the structure shown in Figure 6–1.

When the view from an α carbon down the bond to an amide nitrogen is observed, it can be seen that in every *trans* peptide bond the acyl oxygen, O1, of the previous amino acid in the polypeptide leans forward (Figure 6–

2B, C). When the dihedral angle ϕ is greater than $+310°$ ($-50°$), the amide, N2–C2–O2, collides with acyl oxygen O1; and, when the dihedral angle ϕ is less than $+180°$, the side chain, H–C_β–C_γ, collides with acyl oxygen O1. Therefore, values of dihedral angle ϕ greater than $+310°$ ($-50°$) and less than $+180°$, with the exception of a small gap at ϕ of $+60°$ (Figure 6–2C), are forbidden.

When the view from an α carbon down the bond to an acyl carbon is observed, it is clear that in all *trans* peptide bonds the hydrogen H1 on the amide nitrogen of the next amino acid leans forward (Figure 6–2D, E). When the dihedral angle ψ is greater than $+200°$ ($-160°$), the side chain, H–C_β–C_γ, collides with this hydrogen; and, when dihedral angle ψ is less than $+30°$, the amide, N1–C1–O1, collides with this hydrogen (Figure 6–2D). The latter collision is not a serious one so long as the value of dihedral angle ϕ remains around $+90°$ so the amide N1–C1–O1 can squeeze past the hydrogen H2 sideways (Figure 6–2F). When dihedral angle ψ is between $+290°$ ($-70°$) and $+320°$ ($-40°$), hydrogen H2 on the amide nitrogen N2 can fit between amide nitrogen N1 and the side chain with little difficulty (Figure 6–2E). All of these steric effects can be summarized[7] in a Ramachandran plot (Figure 6–3).

Refinements of crystallographic molecular models using Equation 4–18 for calculation of the function θ usually do not constrain the values for dihedral angles ϕ and ψ. Even though they are not constrained, however, their values converge upon the allowed regions in a Ramachandran plot during the refinement. For example, although many of the values for dihedral angles ϕ and ψ for the various amino acids along the polypeptide in the initial molecular model of deoxyribonuclease I were scattered beyond the allowed regions in a Ramachandran plot before refinement was performed (Figure 6–4A), they clustered within the enclosures after the refinement had been completed (Figure 6–4B). Because this convergence was not enforced by the choice of $(d_q'^2 - d_{c,q}^2)$ in Equation 4–18, its occurrence can be used as evidence that the refined structure is closer to reality than the unrefined.

Almost all of the amino acids found outside of the enclosures in Figure 6–4B are glycines. In addition to being able to reside in cramped locations, glycine lacks a β carbon, and all of the steric collisions involving the β carbon (Figures 6–2 and 6–3) are irrelevant. Therefore, the dihedral angles around a glycine have a larger compass. In particular, the regions of the Ramachandran plot in which dihedral angle ϕ lies between $+60°$ and $+180°$ or dihedral angle ψ lies between $-70°$ and $-180°$ represent areas where either O1 or H2, respectively, clash with the side chain of any amino acid other than glycine.

The regions of the Ramachandran plot in which both dihedral angles ϕ and ψ are positive is one of the most poorly occupied (Figure 6–4B). In addition to the glycines that would be expected to fall in this region, however, asparagines are sometimes found.[8] When the amino

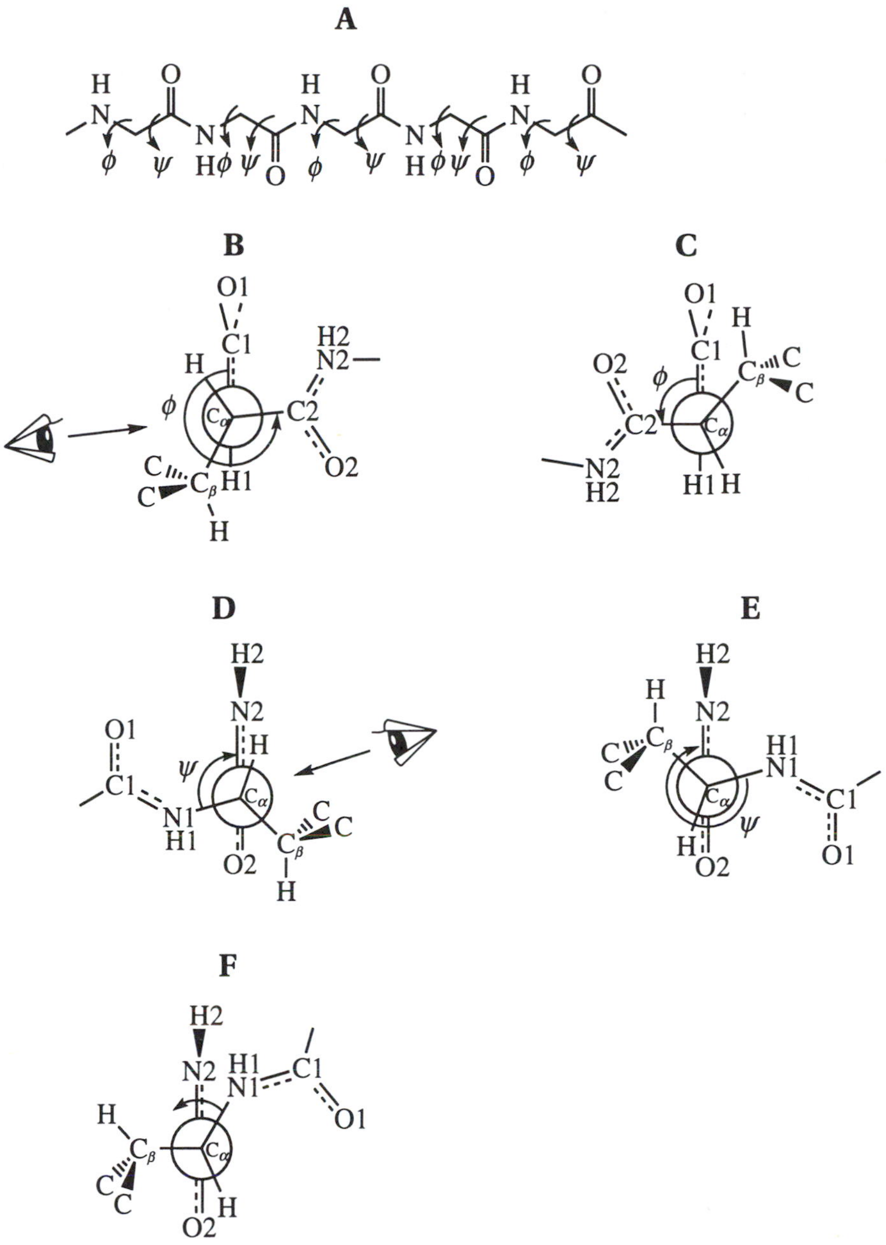

A

Figure 6–2: Definitions of the dihedral angles ϕ and ψ and the steric effects of rotation. (A) Pattern in which the bonds associated with the dihedral angles ϕ and ψ are distributed along a polypeptide. (B) View down the bond between C_α and the amide nitrogen N1 that precedes it along the polypeptide. The dihedral angle ϕ is defined as that between the bond connecting N1 and C1 and the bond connecting C_α and C2 (Figure 6–1). Its sign is determined by the right-hand rule. The configuration shown ($\phi = +270°$ or $\phi = -90°$) is in the most sterically free range for ϕ ($-45°$ to $-180°$) because in this range the hydrogen on C_α can slip under the acyl oxygen O1. (C) View down the same bond as in (B), but with ϕ at $+90°$, produced by rotation about only the bond on the axis of the view. When ϕ is $+60°$, the acyl oxygen, O1, sits between the carbon of the next peptide bond, C2, and the first carbon of the side chain, C_β. This would be the value of ϕ in a left-handed α helix. (D) The same configuration presented in (B) is viewed along the bond between C_α and the acyl carbon of the carboxyl group, C2. The eyes indicate the views interconverting (B) and (D). The dihedral angle ψ is defined as that between the bond connecting C_α and N1 and the bond connecting C2 and N2 (Figure 6–1). Its sign is determined by the right-hand rule. The configuration shown ($\psi = +100°$) is in the most sterically free range for ψ ($+15°$ to $+190°$) because the hydrogen on C_α can slip below H2. (E) View down the same bond as in (D) but with ψ at $+285°$ ($-75°$). When ψ is $+300°$ ($-60°$), H2 lies between the first carbon of the side chain, C_β, and the nitrogen of the amino-terminal peptide bond, N1. This is the value of ψ in a right-handed α helix (Figure 4–17). (F) Steric effect between H2 and N1 that occurs when ψ is $0°$.

B   **C**

D   **E**

F

acids, both of whose dihedral angles are positive, were listed for a set of refined molecular models, 60% were glycines, 12% were asparagines, and each of the other amino acids had a percentage less than 5% (C. Chothia and A.M. Lesk, personal communication). The fact, however, that 28% of the amino acids that had positive values for both of their dihedral angles were amino acids other than glycine and asparagine demonstrates that the steric effects governing the Ramachandran plot are not inflexible. Relatively minor widenings of the various bond angles in the structure (Figure 6–1) can permit dihedral angles that are forbidden to the rigid plastic models with which such steric effects are explained, but each of these distortions exacts a price in free energy.

An interesting feature of these particular results with deoxyribonuclease I, which is also universally observed, is that quite a few amino acids have values for dihedral angle ϕ between $-120°$ and $-60°$ when dihedral angle ψ is between $-30°$ and $+30°$ (Figure 6–4). This is the region in which the amide, H1–N1–C1–O1, is squeezed against the amide nitrogen and hydrogen, N2 and H2, of the next amino acid (Figure 6–2F). This tight fit can be accommodated only if the tetrahedral angle, N1–C_α–C2 (Figure 6–1), is wider than the usual $109.5°$ of a carbon hydridized sp^3. In refined crystallographic molecular models this angle is observed to be wider[9] than expected, with a mean of $112°$ and deviations up to $120°$. This widening of the tetrahedral angle may be a response to the steric crowding on this side of each α carbon along the whole polypeptide as well as to the particular situation displayed in Figure 6–2F.

In refined molecular models,[5,10] the amino acids within **right-handed α helices** have dihedral angles of $\phi = -64°$ $\pm 6°$ and $\psi = -40° \pm 8°$ (Figure 6–2E), and these values fall

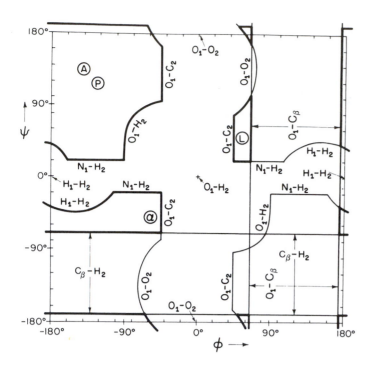

Figure 6–3: Diagram illustrating the steric effects producing the Ramachandran plot.[7] The two dimensions of the plot are the dihedral angles ψ and ϕ (Figure 6–2). Boundaries are drawn between allowed and forbidden regions obtained from a molecular model in which each atom was a hard sphere of the appropriate atomic radius. The clashing atoms are identified on the forbidden side of the boundary with the same numbering system as in Figures 6–1 and 6–2. There are only four allowed regions: the large region including the values for parallel (Ⓟ) and antiparallel (Ⓐ) β sheet, the region including the values for right-handed α helix (Ⓐ) and right-handed 3_{10} helix, the region including the values for left-handed α helix (Ⓛ), and an insignificant triangle at $\phi = +60°$ and $\psi = +180°$. The clashes can be understood by referring to Figure 6–2. For example, if $\phi = -90°$ and $\psi = +100°$ (Figure 6–2B,D) and ϕ is increased to $-45°$, O1 clashes with H2; if ψ is decreased to $+20°$, N1 clashes with H2. If $\phi = -60°$ and $\psi = -60°$ (Figure 6–2E) and ϕ is decreased to $-185°$, O1 runs into C_β; if ψ is decreased to $-70°$, H2 runs into C_β. Adapted with permission from ref 7. Copyright 1977 *Journal of Biological Chemistry.*

A

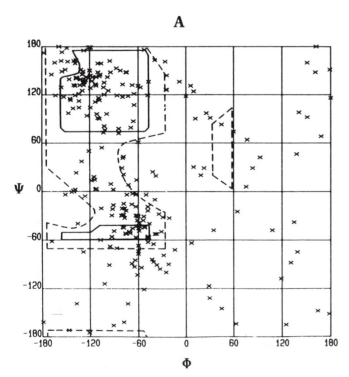

B

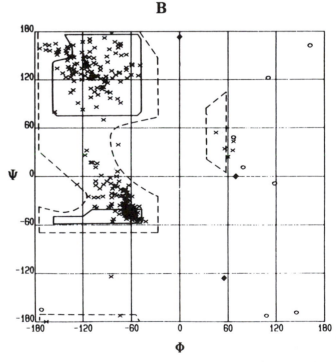

Figure 6–4: Effect of refinement on the values of ϕ and ψ for the amino acids in deoxyribonuclease I.[5] Each $\times$ in one of the diagrams represents the values for the dihedral angles ϕ and ψ for one of the amino acids in a crystallographic molecular model of deoxyribonuclease I. The boundaries in the Ramachandran plot are defined by the steric effects represented diagrammatically in Figure 6–3. Unbroken lines surround regions of no hindrance; and broken lines, regions of little hindrance. It has been decided that the clash between N1

or H1 with H2 around $\phi = -90°$ and $\psi = -30°$ (Figure 6–2F) is slight but that the open region around $\phi = +60°$ and $\psi = +60°$ (Figure 6–2C) is not entirely unhindered; these account for the major differences between the boundaries in Figure 6–3 and those shown in these plots. (A) Unrefined, initial molecular model. (B) Refined, final molecular model. Glycines are denoted by open circles, cystines by filled squares, and all other amino acids by $\times$. Reprinted with permission from ref 5. Copyright 1986 Academic Press.

within one of the enclosures in a Ramachandran plot (Figure 6–4B). This suggests that there are no severe steric problems along the polypeptide in a right-handed α helix. A left-handed α helix of L-amino acids would have dihedral angles of $\phi = +60°$ and $\psi = +40°$, also within one of the enclosures (Figure 6–3). This latter enclosure, however, is the small one arising from the conformation in which the acyl oxygen, O1, fits between the next acyl group, C2 and O2, and the side chain (Figure 6–2C). In refined structures, about 2% of the amino acids usually have dihedral angles of ϕ and ψ clustered around this enclosure (Figure 6–4B). This fact demonstrates that these are accessible rotational isomers, yet left-handed α helices are almost[11] never seen. In theoretical calculations of potential energy, this location on the Ramachandran plot is not a significant minimum;[9,12] and it may be that an extended sequence of such rotational isomers, which would be required to form a left-handed α helix, in contrast to the few isolated examples that are observed, would be sterically unstable.

The original right-handed α helix (Figure 4–17), built before crystallographic structures were available, was constructed with 3.69 amino acids for every turn, which would have produced a rotational angle for each amino acid of 98° and a rise of 0.147 nm for each amino acid.[13] In refined crystallographic molecular models of proteins,[5,10] in which these dimensions were not constrained, right-handed α helices display rotational angles for each amino acid of 99° ± 7° and a rise for each amino acid of 0.15 ± 0.02 nm. The agreement is uncanny.

The original molecular model of the α helix and most subsequently constructed molecular models of proteins have been built so that the acyl oxygens along the α helix participate as an acceptor to only one hydrogen bond, namely, the one in which the nitrogen–hydrogen bond from the appropriate amide was the donor (Figure 4–17). The tacit assumption was that the hydrogen bond would be one in which the nitrogen–hydrogen bond would pivot on the lone pairs of the acyl oxygen (Figure 5–11C). As should have been expected, however, it has been observed in refined molecular models that the acyl carbon–oxygen bonds often tilt away[14] from the axis of an α helix so that the acyl oxygens can form a second hydrogen bond with a molecule of water (Figure 6–5A).[4] Usually this occurs only on the side of an α helix facing the solvent.

It has also been observed that when an α helix contains a serine or threonine, the hydroxyl on the side chain has a tendency to be located in a position similar to that occupied by one of the waters in Figure 6–5A with respect to the acyl oxygens on the amino acid three or four positions ahead of that serine in the α helix (Figure 6–5B).[16] The hydroxyl of the serine is presumed to act as a donor in a second hydrogen bond to one of these acyl oxygens just as a molecule of water does in the other situation. For example, if Alanine 29 in Figure 6–5A were a serine, its hydroxyl would take the place of Water 632 or

Water 657 in the respective hydrogen bonds to acyl oxygens. Such intramolecular hydrogen bonds are quite frequently encountered in α helices. In the refined molecular model of myoglobin, a protein with a large amount of α helix, six out of the 11 serines and threonines in the protein form hydrogen bonds with the acyl oxygens on amino acids three or four positions ahead of them in the sequence of the protein.[15]

It has been noted that the acyl oxygens along an α helix can be divided into two groups, those participating in two hydrogen bonds and those in only one.[17] Those participating in two hydrogen bonds show an amide nitrogen–acyl oxygen–acyl carbon bond angle of about 150°, while those with only one show an angle of 160°. These differences are very slight, but it seems to be the case that if all of the doubly hydrogen-bonded acyl oxygens are on one side of the helix, namely, the side facing the solvent, this small bias can cause the helix to bend.[17] A proline in the middle of an α helix can also put a bend into it. For example, Proline 183 in the middle of an α helix 30 amino acids long in citrate synthetase[18] causes the α helix to bend abruptly by 40°.

Because an α helix is formed by hydrogen bonds and because hydrogen bonds are perhaps only aligned dipoles, the α helix may be a collection of molecular dipoles all aligned in the same direction. The dipole moment of an isolated peptide bond is estimated to be 3.5 D,[19] and the peptide bonds in an α helix are held with their dipoles parallel to the axis so that the positive poles point to the amino-terminal end of the α helix and the negative poles to the carboxy-terminal end (Figure 6–6). Such an arrangement of dipoles is thought to create an electrostatic field of the respective polarity, whose magnitude is 1 V at 0.3 nm and 0.5 V at 0.5 nm, from each end of the α helix, if the α helix is greater than 10 amino acids long.[19] Although these are less than twice the voltages that would be felt at the same distances from two adjacent, isolated peptide bonds, it is thought that this amplification produced by aligning the peptide bonds in an α helix is significant. A great deal of circumstantial evidence has been gathered to suggest that such an electrostatic field, equal to about 50 kJ mol^{-1} for a univalent ion at 0.5 nm from the end, is exerted at each end of an α helix.

Unfortunately, the original calculations of the magnitude of the electric field assumed that the α helix existed in a uniform dielectric of 2, and no account of the dielectric constant of the medium surrounding it was taken. Later calculations[20] have incorporated this contribution, and it was found that if the protein was approximated by a solid sphere of dielectric 3.5 in a solvent of dielectric 80 (water), even when the α helix was completely within the sphere of low dielectric, the electric field around the α helix was dramatically less than the electric field in a uniform dielectric of 3.5. Furthermore, if the ends of the α helix were at the surface of the sphere, in contact with the solvent, the electric field decreased

Figure 6–5: Occupation of the second hydrogen-bond acceptor on the acyl oxygen of a peptide bond in an α helix by water (A) or the hydroxyl of a serine (B). (A) α Helix between Proline 25 and Arginine 33 in the crystallographic molecular model[4] of dihydrofolate reductase from *Escherichia coli*. The numbered circles are fixed water molecules in the vicinity of the α helix. Waters 609, 632, and 657 occupy locations consistent with the formation of hydrogen bonds to the adjacent acyl oxygens, in addition to the hydrogen bonds those oxygens accept from the appropriate amide nitrogens. Reprinted with permission from ref 4. Copyright 1982 *Journal of Biological Chemistry*. (B) Molecular model of an α helix in which the hydroxyl of the side chain of a serine is shown acting as a donor of a hydrogen bond to the acyl oxygen of the amino acid either three or four positions ahead of it in the amino acid sequence.[16] The serine hydroxyl group can swing to complete a hydrogen bond to either acyl oxygen. Reprinted with permission from ref 16. Copyright 1984 Academic Press.

A

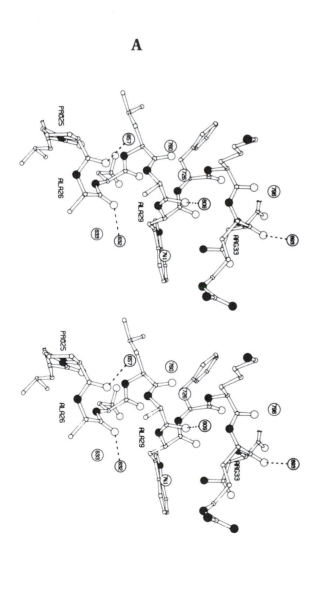

B

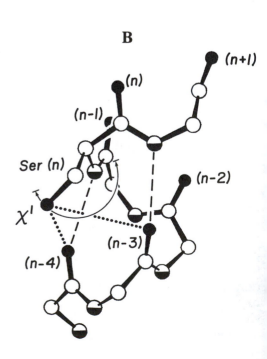

A

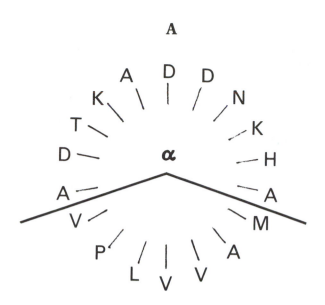

B

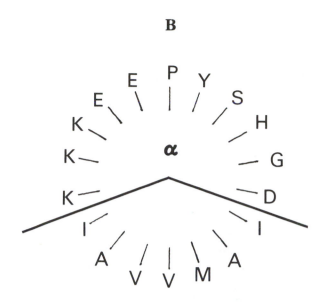

Figure 6–6: Two segments of amino acid sequence displayed on helical wheels. (A) Sequence from the α polypeptide of human hemoglobin ($_{60}$KKVADALTNAVAHVDDMP$_{77}$). The amino terminus is the lysine at 10:30 and the sequence is read at 100° intervals. This sequence is an α helix running across the surface of the crystallographic molecular model of hemoglobin (c3 Kin.3). (B) Amphipathic helical sequence from the α polypeptide of acetylcholine receptor ($_{455}$PDVKSAIEGVKYIAEHMK$_{472}$).

even further to negligible levels. Moreover, if the dielectric constant of the interior of a protein is greater than 3.5, the magnitude of the electric field would decrease accordingly in inverse proportion. Finally, the solution around a molecule of protein always contains electrolytes that would further diminish the electric field. For all of these reasons, it is unknown whether or not electrostatic free energies of significant magnitude are exerted by an α helix within a protein, although the possibility is often discussed.

When an α helix traverses the surface of a protein as a straight continuous rod, its surfaces directed toward the protein are hydrophobic and those directed toward the solution are hydrophilic (c2 Kin.1). This asymmetry of hydropathy is sometimes reflected in the amino acid sequence of the protein and can be identified by constructing a helical wheel.[21] Around a circle, successive amino acids in the sequence are placed at 100° intervals (Figure 6–6). This represents the view down an α helix much as a Newman projection represents the view down a carbon–carbon bond. The asymmetry in the distribution of hydropathy is easily observed. If a segment of amino acid sequence in a polypeptide, when placed upon a helical wheel, reveals such a pattern, this is evidence that that segment is an α helix in the folded polypeptide. Such α helices are referred to as **amphipathic α helices**.

The values for the dihedral angles ϕ and ψ found in **β structure** ($\phi = -130°$, $\psi = +120°$) lie within the largest allowed region of the Ramachandran plot (Ⓐ and Ⓟ in Figure 6–3). These dihedral angles place the hydrogen on the α carbon under the preceding acyl oxygen, O1 (Figure 6–2D) and under the next amide hydrogen, H2 (Figure 6–2D), respectively. These are the least hindered of all the rotational isomers, and β structure experiences no serious steric problems around its α carbons.

It is obvious from an examination of the polypeptide backbones of the proteins presented in Chapter 4 that, because of this rotational freedom, β structure is far more pliant and unpredictable than an α helix, and efforts to define regular patterns have been less informative than time spent looking at different refined molecular models. The original β-pleated sheets (Figure 4–17) have turned out to be highly idealized. There are, however, at least two notable structural features of β structure.

When a number of strands of β structure do form a sheet, the sheet has a **twist** (c2 Kin.2, c2 Kin.3, and c4 Kin.1) to its surface (Figure 6–7).[22] This is supposed to arise from the fact that the enclosure on the Ramachandran plot in which the dihedral angles ϕ and ψ for parallel and antiparallel β sheets reside has more open area for smaller values of dihedral angle ϕ and larger values of dihedral angle ψ than the values that would give a flat sheet. Deviations tend to be biased toward these smaller values of dihedral angle ϕ and larger values of dihedral angle ψ, and this bias creates the twist in the sheet.[23] It may simply be the case, however, that twisted β sheets have surfaces against which other segments of secondary structure, such as α helices, can be more efficiently packed and that packing efficiency dictates the hand and magnitude of the twist.

Another feature of β structure is the **β bulge**.[24] In this instance one of the amino acids is skipped in the regular pattern of hydrogen bonding between two antiparallel strands. The hydrogen bond that would have incorporated the nitrogen–hydrogen bond of the skipped amide incorporates the nitrogen–hydrogen bond of the next amide instead. This causes the β structure to bulge at the location of the skipped amino acid (Figure 6–8), and the bulge is located where the strands change direction. This change in direction can take two forms. If the β structure remains as a sheet in roughly the same plane, the β bulge puts a bend in the structure. A β bulge, however, also can occur at a location where a large sheet of β structure folds over upon itself to form a sandwich of two opposed β sheets.

A third regular structure, in addition to α helices and β structure, found in the crystallographic molecular models of proteins is the β turn. A **β turn** is any structure that has a hydrogen bond between the acyl oxygen of the first amino acid in the turn and the nitrogen–hydrogen bond of the amide of the fourth amino acid in the turn (Figure 6–9). Usually such a hydrogen bond causes the polypeptide to reverse its direction (Figure 4–17). The original description of β turns by Venkatachalam[25] was based on structures built with molecular models. He proposed that

there would be six types of β turns. The fundamental distinction is between type I and type II. A β turn of type I is represented by the β turn in Figure 4–17 with its second acyl oxygen down, and a β turn of type II is represented by the β turn in Figure 4–17 with its second oxygen up. Each of these two fundamental types could also be built with molecular models in such a way that each of their four dihedral angles, ϕ_2, ψ_2, ϕ_3, and ψ_3, had the opposite sign from that in the original two types. These were called type I' and type II', respectively. In each of these two types, the polypeptide backbone is the mirror image of the polypeptide backbone in the corresponding β turns of type I and type II in Figure 4–17, but the amino acids must remain L-amino acids. Finally, two other types of β turns, of historical significance, were defined as type III and type III'.

The prototype for β turns of type III is the 3_{10} helix (Figure 6–10) originally proposed by Taylor[26] and included by Bragg, Kendrew, and Perutz[27] in their catalogue of all helices that would have rotational angles for each residue that were integral quotients of 360°. This is a helix that has hydrogen bonds between the acyl oxygen of amino acid i and the nitrogen–hydrogen bond of amino acid $(i + 3)$, the pattern that is the primary definition of the β turn. This type of hydrogen bond forms a ring of 10 atoms, if the hydrogen of the amide is counted as an atom, hence the subscript on 3_{10}. A 3_{10} helix has three amino acids for each turn, hence the 3 in 3_{10}. This smaller repeat makes it a tighter helix than the α helix. Short segments of 3_{10} helix are occasionally seen in refined molecular models, but they are never more than five or six amino acids in length. These short segments of 3_{10} helix have distorted bond angles along the polypeptide, and this observation suggests that the strain in such a tight helix is considerable.[2] This would explain their rarity. The most frequent location for a short segment of 3_{10} helix is at the end of an α helix. A turn of 3_{10} helix in the middle of an α helix can put an elbow into it. For example, a turn of 3_{10} helix at Serine 143 and Leucine 144 in the center of an α helix in deoxyribonuclease I[5] causes an abrupt bend of 22°. Four amino acids of 3_{10} helix with one hydrogen bond is very similar to a β turn of type I and can perform the same role. It is designated as a β turn of type III. A turn in which the polypeptide backbone assumes a configuration that is the mirror image of a β turn of type III would be a β turn of type III'.

In refined crystallographic molecular models, β turns are designated both by the existence of a hydrogen bond between the acyl oxygen on the first amino acid and the nitrogen–hydrogen bond on the fourth amino acid and by the proximity of the α carbons of the first and fourth amino acids. In general, these two α carbons are 0.5–0.6 nm apart.[28] Those configurations designated by these rules as β turns can be grouped into the six categories proposed by Venkatachalan (Table 6–1). It was only after refined molecular models became available that the clear tendency of these structures to fall into these six catego-

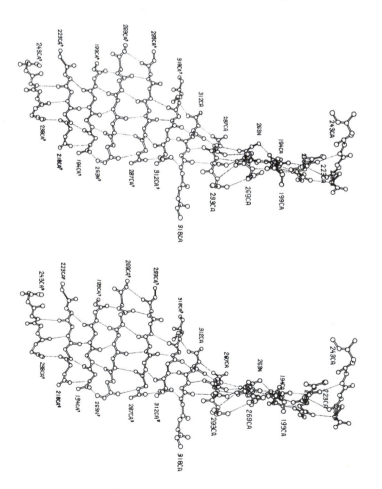

Figure 6–7: Parallel β-pleated sheet within the crystallographic molecular model (c2 Kin.1) of alcohol dehydrogenase.[22] The 12-stranded β sheet is composed of six parallel strands from each of the subunits of the dimer joined in an antiparallel orientation. Reprinted with permission from ref 22. Copyright 1976 Academic Press.

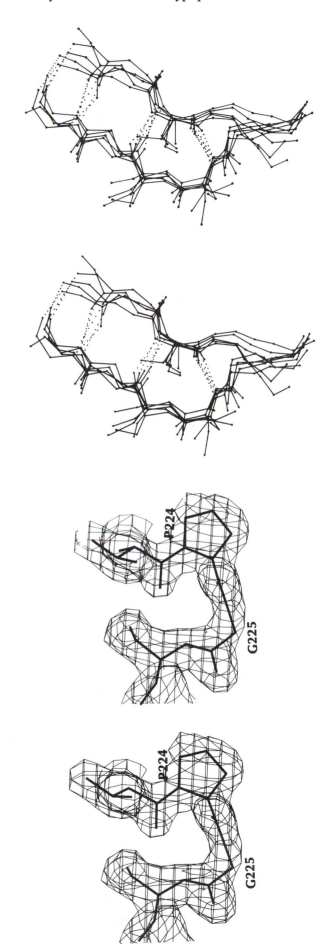

Figure 6–8: Five examples of β bulges from the crystallographic molecular models of various proteins, superposed upon themselves to indicate their uniformity.[24] The skipped amide NH in each case is in the center left of the structure. Hydrogen bonds are indicated by dotted lines. The five β bulges are formed by F41, C42, and L33 of chymotrypsin (c15 Kin.1); A86, K87, and K107 of chymotrypsin; L107, S108, and A196 of concanavalin A; I90, Q91, and V120 of carbonic anhydrase C; and I15, K16, and K24 of micrococcal nuclease, where the first two amino acids listed flank the vacant amide NH and the third provides the amide NH and acyl oxygen on the other strand. Only the side chains of the three central amino acids are included in the figure. Reprinted with permission from ref 24. Copyright 1978 National Academy of Sciences.

ries became apparent, because in unrefined structures the orientation of the polypeptide backbone could not be defined with sufficient accuracy (Figure 6–4A).

The distinction between a β turn of type I and a β turn of type III appears to be arbitrary because most of the standard deviations of their dihedral angles overlap. In practice, type III structures are usually designated as such because they occur either at the ends of α helices or as short 3_{10} helices of five or six amino acids in length. When it is realized that β turns of type I and type III are essentially indistinguishable, they can be considered as a single group that accounts for 70% of all β turns. The dihedral angles at both of the α carbons in these two types of β turns (Table 6–1) fall in the region on the Ramachandran plot between dihedral angles of $\phi = -60°$ and $-120°$ and $\psi = 0°$ and $-30°$ (Figure 6–4B). This is the region in which the two successive amides are squeezed against each other (Figure 6–2F). Presumably the return on the investment of energy necessary to squeeze them against each other and widen the tetrahedral bond at the α carbon is the efficient reversal of the direction of the polypeptide. It is probably the case that β turns of type I and type III account for most of the amino acids that fall in this region of the Ramachandran plot.

About half of the examples of the remaining four types of β turn are of type II (Table 6–1). The values for the dihedral angles ϕ_2 and ψ_2 in these β turns fall in the largest enclosure on the Ramachandran plot, but those for the dihedral angles ϕ_3 and ψ_3 fall in a region that can be occupied only by an amino acid without a β carbon (Table 6–1, Figure 6–3). Therefore, only glycine should

Figure 6–9: Type II β turn in the crystallographic molecular model of deoxyribonuclease I.[5] The β turn contains a hydrogen bond between the acyl oxygen of Valine 223 and the amide nitrogen of Serine 226. Proline and glycine occupy the two central positions. The map of difference electron density is for $(2|F_o| - |F_c|)$ with α_c, where $|F_c|$ and α_c are for the refined molecular model. The molecular skeleton is the refined molecular model. Reprinted with permission from ref 5. Copyright 1986 Academic Press.

Table 6–1: Frequency and Dihedral Angles of the Various Types of β Turns

type	number[a]	dihedral angles[b] (deg)			
		ϕ_2	ψ_2	ϕ_3	ψ_3
I	24	−64 ± 8	−19 ± 8	−90 ± 9	−2 ± 11
I′	4	52 ± 3	41 ± 4	87 ± 6	−11 ± 14
II	12	−61 ± 6	132 ± 1	82 ± 12	3 ± 15
II′	4	63 ± 5	−126 ± 5	−80 ± 9	−11 ± 10
III	36	−58 ± 11	−33 ± 15	−64 ± 7	−26 ± 15
III′	3	60 ± 3	24 ± 13	65 ± 3	34 ± 4

[a]These are the number of each type found in the refined molecular models of lysozyme,[10] α-lytic protease,[29] deoxyribonuclease I,[5] and penicillopepsin.[30] [b]Mean and standard deviations of the dihedral angles for amino acids $i + 1$ and $i + 2$ in the β turns from the same four proteins.

occupy the third position in a β turn of type II. Although this is usually the case, there are exceptions, such as Asparagine 69 in α-lytic protease.[29]

The mirror image configurations, in which the polypeptide backbone mirrors the respective basic β turn but the amino acids remain, of necessity, L-amino acids, are quite rare. The third amino acid in a β turn of type I′ and the second amino acid in a β turn of type II′ should be glycine residues, but again a few exceptions have been noted, such as Cysteine 170 in deoxyribonuclease I.[5] It has been noted[31,32] that when an antiparallel β hairpin reverses itself in the tightest possible β turn, where the hydrogen bond of the β turn is also the last hydrogen bond between the tines of the hairpin, the β turn is usually type I′ or type II′.

Aside from the requirement that glycine occupy certain positions of a β turn for steric reasons, there are some clear preferences[2,33] for other amino acids. Because β turns are almost always at the surface of a protein, they contain hydrophilic amino acids more frequently than hydrophobic amino acids. About one-third of all β turns have proline at their second position (Figure 6–9), and 60% of all β turns have either aspartic acid, asparagine, or glycine at their third position (Figure 6–9).[33]

In every protein there are also segments of polypeptide that do not assume the configuration of an α helix, β structure, a 3_{10} helix, or a β turn. These segments of **random meander** pass about the protein as would an

α helix or a strand of β structure. They are usually found on the surface of the molecule, and occasionally one of them will loop out quite a ways from the core of the structure. Even though there is no regular pattern to random meander, the values of the dihedral angles φ and ψ are still confined to the minima in the Ramachandran plot because these minima are defined by local steric effects. This places the angles for random meander within the same regions occupied by the dihedral angles for β structure or α helix, respectively. The distinction between α helix, β structure, and random meander cannot be made by comparing single values of the dihedral angles φ and ψ but only by identifying repeating patterns that extend over several amino acids or several strands of polypeptide. In random meander, no such pattern is evident.

All of the regular structures in which the polypeptide participates are given their regularity by **hydrogen bonds** between the nitrogen–hydrogen bonds of the amides and the acyl oxygens of the amides. These hydrogen bonds can be easily identified in refined molecular models, and it can be assumed that they exist. The bond length for such unambiguous hydrogen bonds, expressed as the distance between nitrogen and oxygen, is 0.29 ± 0.015 nm.[11,29,30,34] The small standard deviation is notable. The angular dependence of these bonds can be expressed either in reference to the nitrogen–hydrogen bond of the one amide (Figure 6–11A) or the carbon–

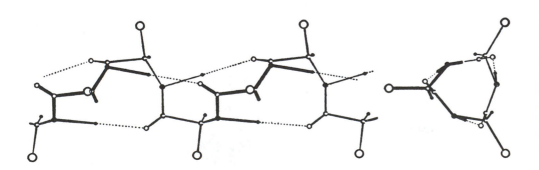

Figure 6–10: The 3_{10} helix.[27] Drawings of a side view of the 3_{10} helix (left) and a top view (right) are presented. The structure has an amino acid at every 120° along the helix and a rise of 0.17 nm for each amino acid. This gives the structure an exact repeat every three amino acids. Reprinted with permission from ref 27. Copyright 1950 Royal Society (London).

A

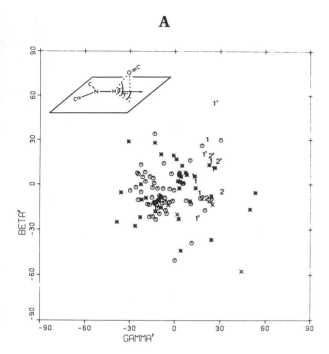

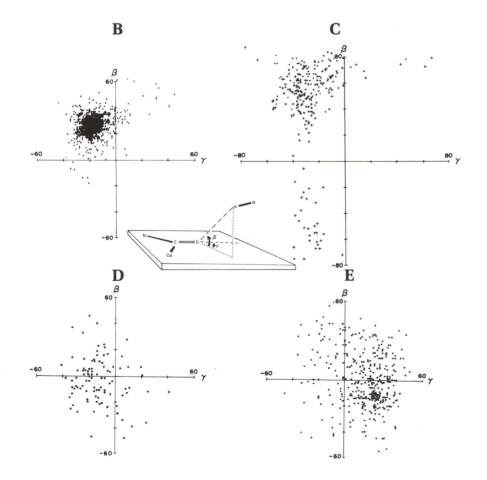

Figure 6–11: Bond angles for the hydrogen bonds in regular structures formed by a folded polypeptide. (A) Bond angles at the amide NH. Two angles are defined (inset), the angle γ' within the plane of the amide and the angle β out of the plane of the amide.[10] When γ' is 0°, the acyl oxygen is in the plane containing the NH bond normal to the plane of the amide. When β is 0°, the acyl oxygen is in the plane of the amide. The plot is for all of these bond angles for the hydrogen bonds between the peptide bonds in the refined crystallographic molecular model of α-lytic protease.[29] Symbols are (O) β structure, (×) α helix, and (*) random meander. The bond angles for hydrogen bonds in β turns are designated by the number corresponding to the turn type. Reprinted with permission from ref 29. Copyright 1985 Academic Press. (B–E) Bond angles at the carbon–oxygen bond of the amide. Two angles are defined (inset), the angle γ within the plane of the amide and the angle β out of the plane of the amide. Angles are defined relative to the axis of the carbon–oxygen bond and the plane of the amide. These angles at each hydrogen bond involving the peptide bonds in the refined crystallographic molecular models of 15 proteins[34] are plotted for hydrogen bonds in α helices (B), β turns (C), parallel β structure (D), and antiparallel β structure (E). Each mark is for the angles β and γ of one of the hydrogen bonds included in the set. Reprinted with permission from ref 34. Copyright 1984 Pergamon Press.

oxygen bond of the other amide (Figure 6–11B–E).[34] As expected, the angles around the nitrogen–hydrogen bond of the donor are much more confined than those around the lone pairs on the acyl oxygen of the acceptor. A deviation from 0° of either of the angles around the nitrogen–hydrogen bond (Figure 6–11A) places the hydrogen off the line of centers between the nitrogen and the oxygen and bends the bond. The angles of the hydrogen bond relative to the carbon–oxygen bond of the acceptor (Figure 6–11B–E) vary over a greater latitude. In keeping with the rigidity of α helices and flexibility of β structure, the angles around the acyl oxygens in β structure are much more variable (Figure 6–11D, E) than those around the acyl oxygens in α helices (Figure 6–11B). In none of these regular structures, however, is there any tendency for the values of the angles around the acyl oxygen to cluster at $\beta = 0°$ and $\gamma = \pm 60°$, the positions at which the nitrogen–hydrogen bond would point directly at one or the other of the lone pairs on the acyl oxygen (Figure 5–11). It has, however, already been noted that, even in crystallographic molecular models of small, unconstrained molecules, the preference for these angles is not remarkable (Figure 5–12), and there seems to be little energetic cost in pivoting the donor over the surface of the acyl oxygen distal to the acyl carbon (Figure 5–11D). Therefore, in regular structures such as an α helix or β structure, it is the steric requirements of these structures that easily take precedence.

Even in refined molecular models, the identification of a hydrogen bond is subjective. It is often based on the fact that two heteroatoms are simply within a certain distance of each other. The dimensions of unquestionable hydrogen bonds in regular structures, however, suggest that a more objective definition of a hydrogen bond[10] would be a situation where the heteroatoms of the donor and the acceptor are less than 0.34 nm from each other; the angle A–H–B, where A is the donor and B the acceptor, is between 150° and 180° (Figure 6–11A); and the distance between the theoretical position of the hydrogen and the heteroatom of the acceptor is less than 0.24 nm. When these definitions are applied to the rather

featureless distribution of distances between nitrogens and oxygens in a refined molecular model, that distribution can be divided into hydrogen bonds and non-hydrogen bonds (Figure 6–12).[10]

It is likely that the nitrogen–hydrogen bond of every peptide bond in a folded polypeptide participates as a donor in a hydrogen bond, either with water, with an acyl oxygen of another peptide bond, or with a lone pair of electrons on a side chain.[34] Almost every one of these hydrogen bonds should have the same enthalpy of formation as the hydrogen bonds between water and these same nitrogen–hydrogen bonds in the unfolded polypeptide because the majority[35] of the acceptors (80%) within a polypeptide have acid dissociation constants associated with their σ lone pairs of electrons that are almost the same as that of the lone pairs of electrons on water ($pK_a = -1.7$). In a solution of protein, for example, the cytoplasm of a cell, the concentration of acceptors of hydrogen bonds exceeds the concentration of donors because there are 305 acceptors but only 185 donors for every 100 amino acids in a protein.[35] The presence of nucleic acid and carbohydrate only increases this disparity. Therefore, every time the nitrogen–hydrogen bond of the amide in a peptide bond is transferred from water into the interior of a protein, where it invariably participates as a donor in a hydrogen bond, the total concentration of hydrogen bonds in the solution does not change and the enthalpy of formation of that hydrogen bond must be zero. As both the competition of the water for this donor (Equation 5–33) and the entropy of approximation involved in forming the regular structures of the polypeptide backbone (Figure 5–17) are entropic terms, they can be combined in the larger question of the change in entropy accompanying the folding of the polypeptide.

When viewed with this perspective, the regular structures assumed by the polypeptide serve the purpose of maintaining the total concentration of hydrogen bonds in the solution. It seems highly unlikely that a protein could fold without withdrawing considerable numbers of its peptide bonds from contact with water. Were the donors on these peptide bonds withdrawn from the

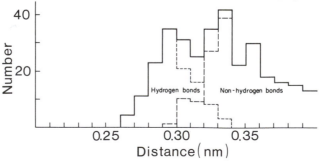

Figure 6–12: Histogram of all of the contact distances (nanometers) between nitrogen atoms and oxygen atoms in the refined crystallographic molecular model of human lysozyme.[10] The number of pairs of nitrogens and oxygens that are a given distance apart in the molecular model is plotted as a function of those distances. If a hydrogen bond is defined as a nitrogen and an oxygen less than 0.34 nm apart, hydrogen and oxygen less than 0.24 nm apart, and the angle N–H–O between 150° and 180°, the histogram can be divided into nitrogen–oxygen contacts that are hydrogen bonds and contacts that are not hydrogen bonds. Reprinted with permission from ref 10. Copyright 1981 Academic Press.

solvent without subsequently participating in hydrogen bonds in the interior of the protein, there would be an unavoidable deficit of enthalpy (+15 kJ mol^{-1}, Table 5–2). This way of viewing the situation also suggests that the fact that only a small fraction of the acyl oxygens end up with two donors (Figure 6–5) is inconsequential because many acceptors were vacant before folding occurred anyway.

There may also be a favorable hydrophobic effect accompanying the removal of the backbone from water during the folding of a polypeptide.[36] None of the available experimental observations, however, demonstrate that this is the case. For example, the transfer of the hydrogen-bonded dimer of N-methylacetamide between water and carbon tetrachloride is isergonic (Figure 5–14).[37] Were this the isosteric dimer of isobutane, a large hydrophobic effect would have been evident (Table 5–6). It could be argued that the hydrophobic effect is exactly canceled by the necessity of removing a donor and an acceptor from water during this transfer. This removal does decrease the total number of hydrogen bonds in the system, a problem not encountered when the backbone of the polypeptide is removed from water during the folding of a protein, as long as all of its donors eventually find acceptors.

Suggested Reading

James, M.N.G., & Sielecki, A.R. (1983) Structure and Refinement of Penicillopepsin at 1.8-Å Resolution, *J. Mol. Biol. 163*, 299–361.

Prediction of Secondary Structure

It has been proposed that certain amino acids or short sequences of amino acids may impose upon a folding polypeptide biases toward the formation of particular secondary structures at locations where they reside in the native structure of a protein. One hears terms such as "helix-forming" or "helix-breaking" amino acids.[38] Originally, these distinctions were based on the observed preferences of homopolymers of the various amino acids to assume α helices or sheets of β structure or to remain structureless at various temperatures, ionic strengths, concentrations of cosolvents, and values of pH.[39] If the biases observed in these studies of homopolymers were strong ones and integral to the particular amino acid in any setting, it would be possible to discover parameters that could be assigned to each of the amino acids or short linear combinations of amino acids and that would permit the secondary structure assumed by particular sequences in a properly folded polypeptide to be predicted accurately from the amino acid sequence alone.

Several algorithms have been developed to perform such predictions. The method of Chou and Fasman[38] is based on a set of protein conformational parameters derived from the observed frequencies with which each of the amino acids occurs in α helices, β structure, β turns, or random meander in 29 proteins of known structure. The method of Lim is based on an intricate but entirely theoretical set of stereochemical rules[40] derived from considerations of steric effects and distribution of polar and nonpolar amino acids in α helices and β sheets.[41] The method of Garnier, Osguthorpe, and Robson[42] is based on the frequency with which each of the 20 amino acids occurs at positions in the sequence of a polypeptide known from crystallographic molecular models to be in α helices, β structure, tight turns, or random meander and the frequency with which each of the 20 amino acids occurs at each of the eight positions preceding and each of the eight positions following a position in the sequence known to be in an α helix, β structure, a β turn, or a random meander. This last method was based on crystallographic molecular models of 25 proteins of known sequence and conformation.

None of these methods provides reliable predictions of the distribution of secondary structure in the native protein formed from a polypeptide of known sequence for which a crystallographic molecular model is unavailable. This has been demonstrated in two unrelated and independent experiments by applying these algorithms to the sequences of 38 or 24 proteins, respectively, whose crystallographic molecular models became available after the parameters upon which the algorithms were based had been derived.[43,44] There was remarkable agreement between the results of these two tests. When the algorithms were required to designate each position in the amino acid sequences of these proteins as located in one of the three categories, α helix, β structure, or β turn/random meander, only 50% of the designations agreed with the crystallographic molecular models, regardless of which of the three methods was used. When all three were combined the results were exactly the same.[44] When tight turns and random meanders were considered as separate categories, no improvement in the reliability of the predictions was noted.[44] The various predictive schemes were not noticeably more successful at predicting the participation of a position in the sequence in any particular one of the three categories.[43]

In both of these tests, it was noted that when the methods were applied to the sequences of the proteins that served as the data sets in deriving the respective parameters, the success of the predictions increased significantly.[43,44] This result illustrates an important consideration in evaluating statistical methods based on large numbers of parameters. The larger the number of parameters in any numerical analysis, the greater will be its success in reproducing the data on which it was based. No evaluation of the success of a predictive method that relies on its ability to reproduce the data should ever be taken seriously in any situation.

To the extent that these predictive methods have all failed, it can be concluded that particular amino acids or particular short linear combinations of different amino acids do not exert any strong influence on the ultimate secondary structure assumed by the segment of polypeptide in which they are found. To the extent that they have succeeded—to the extent that a frequency of correct choices between three possibilities of 50% can be shown to be statistically significant—they may have revealed some intrinsic preferences among the amino acids for certain secondary structures. This in itself, however, may be illusory because β sheets tend to reside in the center of crystallographic molecular models of proteins and α helices on the periphery. Thus β sheets would have a different composition of amino acids than α helices for reasons other than any intrinsic preferences of the amino acids themselves for particular secondary structures. Regardless of whether these predictions mean anything or not, they should not be taken seriously.

Suggested Reading

Kabsch, W., & Sander, C. (1983) *Fed. Eur. Biochem. Soc. Lett.* *155*, 179–182.

Stereochemistry of the Side Chains

It was pointed out in a discussion of the crystallographic molecular model of chymotrypsin, which was one of the first refined molecular models,[9] that certain rotational isomers of the side chains of the amino acids seemed to be preferred. As more highly refined molecular models of proteins have become available, a number of the stereochemical features of these conformations have been even more clearly defined. As would be expected, all rotational isomers are staggered rather than eclipsed, a fact that is reflected in the strong tendency for dihedral angles along carbon–carbon bonds between atoms that are hybridized sp^3 to assume values near 60°, 180°, and 300° (Figure 6–13).[30] In an extensive tabulation[45] of the values for the dihedral angles of all of the side chains in 19 highly refined crystallographic molecular models, almost all of the mean values for the dihedral angles of carbon–carbon bonds connecting atoms that are hybridized sp^3 are within 10° of one of these three values.

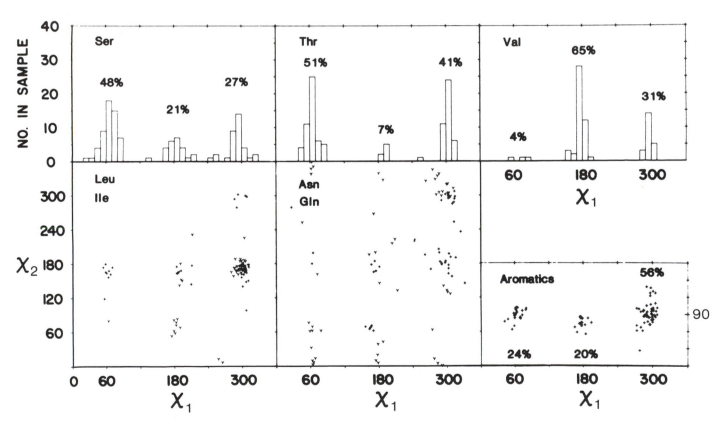

Figure 6–13: Histograms of the distributions of the values for the dihedral angles χ_1 and χ_2 for the first two carbon–carbon bonds of the side chains of amino acids in the refined crystallographic molecular models of five proteins: penicillopepsin, *Streptomyces griseus* serine proteinase A, *S. griseus* serine proteinase B, the third domain of the ovomucoid inhibitor, and α-lytic proteinase.[30] The abbreviation of each side chain appears in the upper left-hand corner of the panel. Serine, threonine, and valine have no observable dihedral angles for the second carbon–carbon bond (χ_2), so in these instances frequency is plotted as a function of the only value of the dihedral angle χ_1. Leucine, isoleucine, asparagine, and glutamine have observable dihedral angles for both χ_1 and χ_2, and in these cases each mark represents the value of these two angles for one of these side chains in these molecular models. Because of symmetry, the values for χ_2 for the aromatic amino acids tyrosine, phenylalanine, and tryptophan fall only between 0° and 180°. Reprinted with permission from ref 30. Copyright 1983 Academic Press.

The dihedral angle along the bond between C_α and C_β in an amino acid is designated χ_1. It is the dihedral angle between the bond to the amide nitrogen, the heaviest of the three atoms around C_α, and the bond to the atom attached to C_β that has the highest priority in the Cahn–Ingold–Prelog system (Figure 6–14). The sign on this dihedral angle is determined by the right-hand rule.

The stereochemistry about this bond between C_α and C_β is dominated by the polypeptide rather than the rest of the side chain. Valine is the logical place to begin the discussion of this sterochemistry because its two methyl groups can assess the steric bulk of the three substituents on the α carbon. This is due to the fact that, in each of the three staggered conformations (two of which are displayed in Figure 6–14), one of these three substituents must reside between the two methyl groups, the most hindered locations. Because the smallest functional group should occupy this position most frequently, the distribution of the dihedral angles χ_1 of the valines in refined molecular models (Figure 6–13) states that the hydrogen on the α carbon is smaller (56%) than the nitrogen of the preceding amide (31%), which is much smaller than the acyl carbon of the following amide (4%). This behavior is completely consistent with the assessment of steric bulk based on preferences of various substituents on cyclohexane for equatorial over axial locations. The increase in free energy[46] for placing an acetoxy group in an axial location rather than an equatorial location is 2.9 kJ mol^{-1}, but the increase in free energy for placing a methoxycarbonyl group in an axial location rather than an equatorial location is 5.4 kJ mol^{-1}. Isoleucine (Figure 6–14) reinforces these preferences by showing a distribution[47] of analogous stereochemical conformations similar to that for valine (66%, 17%, and 17%, respectively).[45]

Threonine is isosteric with valine, but the designation of the dihedral angle χ_1 of threonine is 240° out of phase with that of the dihedral angle χ_1 of valine because of the precedence of the (S)-oxygen (Figure 6–14). The rotational isomer of threonine ($\chi_1 = 60°$, 48% of all threonines)[45] in which the two substituents on C_β surround the nitrogen of the preceding amide (Figure 6–14) is more than twice as frequent as the analogous isomer ($\chi_1 = 300°$, 26% of all valines)[45] of valine relative to the respective isomers in which hydrogen is surrounded. The most likely explanation for this difference is the fact that a hydroxyl group is significantly smaller than a methyl group. Another possibility, however, is that a dihedral angle $\chi_1 = 60°$ places the hydroxyl of threonine in the proper position to form either a hydrogen-bonded ring with its own acyl oxygen (Figure 6–15A) or a hydrogen-bonded ring with the acyl oxygen of the amino acid that precedes it in the sequence (Figure 6–15C). Because the hydrogen is within the hydrogen bond, the former ring can be considered to be five-membered and the latter, six-membered. The five-membered ring requires that dihedral angle ψ for the threonine be 90°, and the six-membered ring requires that dihedral angle ϕ for the

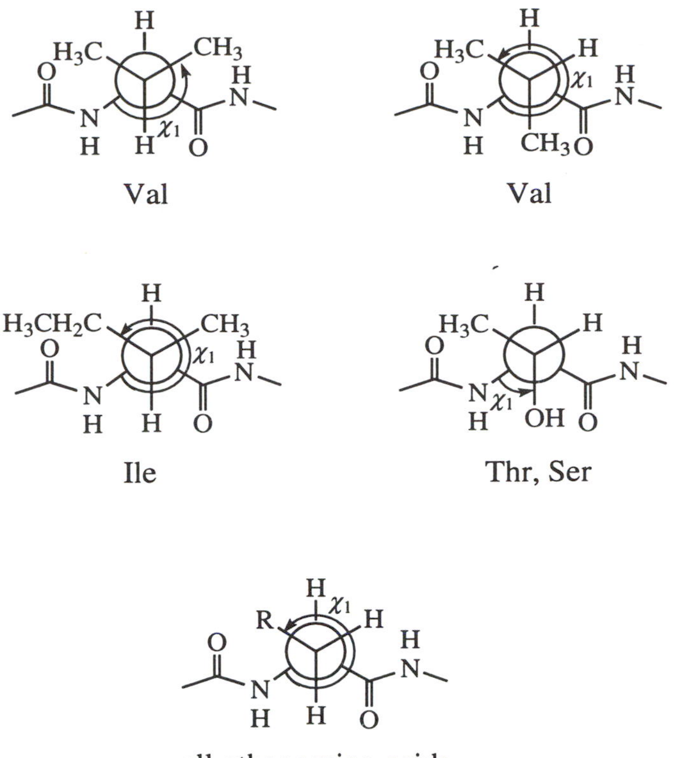

Figure 6–14: Definition of the dihedral angle χ_1 for the carbon–carbon bond between the α carbon and the β carbon of an amino acid in a polypeptide. All dihedral angles follow the right-hand rule. For valine, χ_1 is the dihedral angle between the carbon–nitrogen bond and the bond to the *pro-R* methyl group. For isoleucine, threonine, and serine, χ_1 is the dihedral angle between the carbon–nitrogen bond and the bond to the substituent of higher priority, namely, the ethyl group or the hydroxyl group, respectively. For all other relevant amino acids, the dihedral angle χ_1 is the angle between the carbon–nitrogen bond and the bond to the remainder of the side chain.

threonine be 60°. The former dihedral angle is quite common (Figure 6–4B), and this suggests that such a structure (Figure 6–15A) is easily formed. The latter dihedral angle ($\phi = 60°$), however, is seldom found (Figure 6–4B), and this structure (Figure 6–15C) would be infrequent. The configurational isomer of the six-membered ring (Figure 6–15D), however, would have a dihedral angle $\chi_1 = 300°$ and a dihedral angle $\phi = 180°$. This latter dihedral angle is also rarely seen (Figure 6–4B); however, by increasing angle ϕ to greater than 180° a six-membered hydrogen-bonded ring can still be formed. An example of such a six-membered ring is found at Threonine 19 in penicillopepsin.[30] The other configurational isomer of the five-membered ring (Figure 6–15B) would have a dihedral angle $\chi_1 = 180°$, which would place the acyl carbon of the threonine between the methyl and hydroxyl groups, a χ_1 angle only rarely observed (Figure 6–13). This latter configuration of the five-membered ring should be available to serine, however, which lacks

A

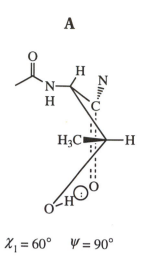

$\chi_1 = 60°$ $\psi = 90°$

B

$\chi_1 = 180°$ $\psi = 30°$

Figure 6–15: Some cyclic hydrogen-bonded structures in which a threonine (or serine) would form a five-membered ring with its own acyl oxygen (A and B) or a six-membered ring with the acyl oxygen of the preceding amino acid (C and D). The hydrogen is not counted because it is within the hydrogen bond.

C

$\chi_1 = 60°$ $\phi = 60°$

D

$\chi_1 = 300°$ $\phi = 180°$

the methyl group; and 23% of all serines have dihedral angles $\chi_1 = 180°$ (Figure 6–13),[45] perhaps, in part, reflecting the presence of such cyclic structures. Five-membered hydrogen-bonded rings of the type shown in Figure 6–15A would explain why serines show such a high percentage of dihedral angles $\chi_1 = 60°$ (48%),[45] in contrast to all other amino acids with only one substituent on the β carbon, which have a low percentage of dihedral angles $\chi_1 = 60°$ (11% of all side chains with one and only one substituent on C_β).[45,47] Side chains of aspartates and asparagines can participate as acceptors in similar five-membered and six-membered hydrogen-bonded rings with the amide protons of the polypeptide to their amino-terminal and carboxy-terminal sides, respectively, and asparagines can form six-membered hydrogen-bonded rings with the acyl oxygen to their carboxy-terminal sides.

The amino acids, other than serine, with one and only one substituent on C_β have a preference (55%)[45,47] for dihedral angles $\chi_1 = 300°$ (Figures 6–13 and 6–16). This preference for 300° among these amino acids is understandable because it places the single substituent on the β carbon between the two least bulky substituents around the α carbon (Figure 6–14).

The dihedral angle χ_2 is assigned to the bond between C_β and C_γ in an amino acid. In linear amino acids such as glutamine, glutamic acid, lysine, arginine, and methionine, 70% of the dihedral angles χ_2 are 180°, which is the angle expected for an *anti* conformation at this carbon-carbon bond.[45,47] The remainder are split equally between the two *gauche* conformations at dihedral angles χ_2 of 60° and 300°. In the aromatic amino acids phenylalanine, tyrosine, and tryptophan the conformation whose di-hedral angle χ_2 is designated as 90°, in which the plane of the ring is perpendicular to the bond between the α carbon and the β carbon (Figure 2–10), is heavily preferred (96% of these residues).[45] It has also been observed in refined maps of electron density from crystallography by neutron diffraction that the hydrogens on all methyl groups are staggered,[48] but no other result could have been observed because methyl groups in proteins should be free to rotate and populate only energy minima significantly.

Cystine is an amino acid under peculiar steric constraints (Figure 6–17). The distribution of the two dihedral angles χ_1 of cystine[16] shows the same sequence and relative magnitude of preferences for 300°, 180°, and 60° as those of any other amino acid with only one uncompli-

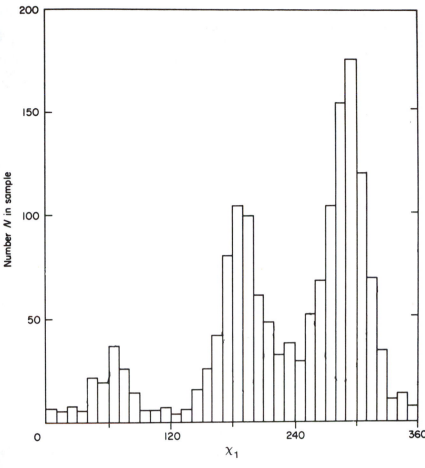

200
150
100
50
0
Number N in sample
0
120
240
360
χ1

Figure 6–16: Histogram of the distribution of the angle χ_1 for all of the amino acids with one and only one substituent on the β carbon.[47] The values for χ_1 (Figure 6–14) for all of the tryptophans, tyrosines, phenylalanines, histidines, methionines, leucines, aspartates, asparagines, glutamates, glutamines, lysines, and arginines in 19 crystallographic molecular models were tabulated. The frequency with which each value of the dihedral angle χ_1 occurred is plotted as a function of the value of that angle (in degrees). Adapted with permission from ref 47. Copyright 1978 Academic Press.

cated substituent on the β carbon and is indistinguishable from that in Figure 6–16. The disulfide itself, because it is a dithioperoxide, is electronically required to have a dihedral angle along the sulfur–sulfur bond (c2 Kin.5) similar to the dihedral angle of hydrogen peroxide, which is 94°. If the angle were 90°, the four lone pairs, two on each sulfur, would be as far from being parallel to each other as is possible, and this orientation would be the most stable electronically[49]

Cβ
S
Cβ

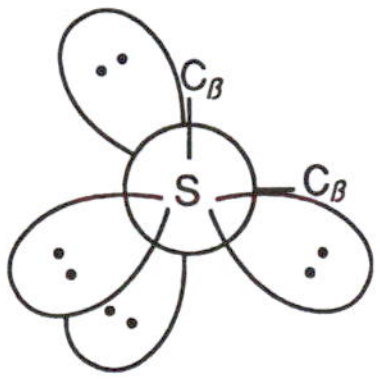

6–4

The steric interactions between the β carbons, however, widen the angle somewhat from 90°. The dihedral angle along the sulfur–sulfur bond in a disulfide[2] is always within 10° of +95° or –95°.

The bonds between the β carbons and the sulfurs in a cystine should have a preference for dihedral angles χ_2 equal to 180° as with most other amino acids, but in refined molecular models, values around broad maxima of 60° and 300° are heavily preferred[2]

χ2
Cα
S
H
Cβ
H

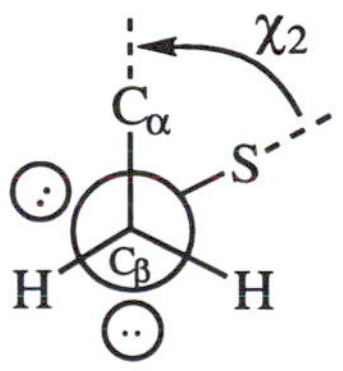

6–5

Each of the two bonds between the β carbons and the sulfurs should be fairly long and not severely confined by the adjacent sulfur or the lone pairs on the immediate sulfur itself, and they are probably the most compliant of the bonds in the disulfide. Because the disulfide connects two strands of polypeptide engaged in many other interactions, it probably sustains considerable torque. Presumably it is the dihedral angles χ_2 that accommodate this torque. Because the dihedral angle of the disulfide itself must always be +95° or –95°, it is only the two dihedral angles χ_2 that can adjust to fit the distance between the two α carbons of a cystine in a protein. If the dihedral angles χ_2 were required to be 180°, the two α carbons would have to be about 0.9 nm apart, a rather distant connection. Because the distances between the two α carbons of cystines in native proteins always fall between 0.45 and 0.7 nm,[2] the dihedral angles χ_2 assume many other values, and rarely 180°.

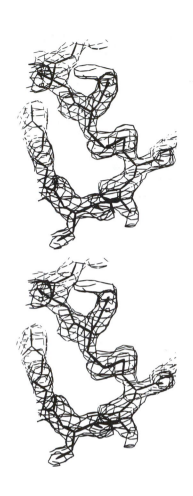

Figure 6–17: Cystine between Cysteine 55 and Cysteine 144 in the crystallographic molecular model (c5 Kin.1) of copper/zinc superoxide dismutase.[50] The map of difference electron density is for $(2|F_o| - |F_c|)$ where $|F_c|$ is calculated from the refined molecular model. The molecular skeleton is the refined molecular model itself. The two sulfur atoms of the disulfide are clearly resolved and greater in electron density than the atoms surrounding them. The dihedral angle at the disulfide is –95° as expected. Reprinted with permission from ref 50. Copyright 1982 Academic Press.

protein, which remains in contact with the water, and its interior, which is more or less withdrawn from the water. This bias reflects the hydropathy of the individual amino acids.

The accessible surface area of a properly folded protein is determined by rolling a sphere the size of a molecule of water over the surface of a refined crystallographic molecular model.[51] The surface of a molecule of protein is not smooth but highly irregular, covered with cracks, crevasses, and ridges (Figures 6–18 and 4–20). It has a much larger accessible surface area than a sphere of the same size. A large fraction of the amino acids in a protein make contact with the water, and this contact varies in its extent from one amino acid to the next. Therefore, the definition of accessible and buried amino acids is complicated.[36,52]

Usually a particular amino acid in a crystallographic molecular model is designated as buried if less than a certain amount of its surface area is accessible. It has been shown[52] (Figure 6–18) that

$$n^{1/3} - n_B^{1/3} = \kappa \qquad (6-1)$$

where n is the total number of amino acids in a protein, n_B is the number designated as buried by some rule, and κ is a constant. What this equation states is that the buried amino acids are found within a roughly spherical solid that is smaller than the roughly spherical solid containing all of the amino acids (inset, Figure 6–18). The spherical shell of width d in the inset to Figure 6–18 contains all of the amino acids that are accessible by the rule that has been chosen. This shell can be considered to be the depth of penetration of water into the interior of the protein owing to its irregular surface. If buried residues are defined as only those amino acids completely inaccessible to water, d is fairly large (1.0 nm) and the buried residues are deep in the interior. If buried residues, however, are defined as any residues having accessible surface area of less than 0.2 nm², then d is only 0.5 nm and far more amino acids are considered to be buried. The number of buried amino acids, defined by this latter rule, as a function of the total number of amino acids in the molecular models of 22 proteins is presented graphically in Figure 6–18. The fraction of the total number of amino acids that are defined as buried increases regularly as the size of the protein increases.

All of the stereochemical observations discussed so far, because they are either consistent with the behavior of small molecules or otherwise make sense, are further evidence that refined molecular models of proteins are more realistic than unrefined molecular models.[30] They also leave the impression that there is far less flexibility involved in the folding of a protein than there seems to be at first glance. Conformations that demand dihedral angles to assume values other than those of lowest energy, which are normally the conformations most heavily populated in small molecules, require that extra energy be spent to occupy them. It turns out that there is not much extra energy to go around.

Suggested Reading

Janin, J., Wodak, S., Levitt, M., & Maigret, B. (1978) Conformation of Amino Acid Side Chains in Proteins, *J. Mol. Biol. 125*, 357–386.

Hydropathy of the Side Chains

There is an obvious bias in the distribution of the constituent amino acids between the surface of a molecule of

Accessible surface areas of the amino acids in molecular models of proteins have been calculated, and the fractions of each type of amino acid scored as buried have been tabulated in terms of at least three different definitions (Table 6–2). When the stringent definition is used, namely, that a buried amino acid in the molecular model of the protein must have no accessible surface area, the frequencies with which most of the amino acids are buried is very low and the statistics become unreliable. When the definition is relaxed, more amino acids are scored as buried, and discriminations become more dependable.

Half of the amino acids have total accessible surface areas between 1.5 and 2.0 nm^2 and therefore are of similar size. Small amino acids such as alanine are probably buried more often simply because they are easier to surround, and large amino acids such as tryptophan are harder to surround completely and bury, especially in the smaller proteins. These stereochemical problems must contribute to the observed distributions.

Nevertheless, it has already been noted that the frequencies with which the various amino acids are buried are correlated[55] with the free energies of transfer for their model compounds from water to the gas phase (Table 5–8) and also with many of the other scales of hydropathy (Figure 5–24). This correlation is strongest when the amino acids are distributed into three main groups, hydrophobic, apathetic, and hydrophilic. Within each of these groups, however, there is no significant correlation between extent of burial and any scale of hydropathy derived from free energies of transfer. Presumably, the reason for the lack of correlation within the main groups is that protein folding is not a transfer between solvents. With this in mind, it still can be stated that if an amino acid is hydrophobic, it is more likely to be buried; and, if it is hydrophilic, it is more likely to remain in contact with the water in the folded polypeptide.

One of the most hydrophobic amino acids is cystine. Unfortunately a clear distinction between cysteine and cystine is rarely made in these tabulations. The few results that are available suggest that cystine is as often buried as isoleucine, valine, or leucine.

The aromatic amino acids have several interesting peculiarities. Phenylalanine, because it is aromatic, should be more hydrophilic than isoleucine, valine, and leucine. For example, in the refined molecular model of chymotrypsin, one face of the π molecular orbital system of every phenylalanine is in contact with the aqueous phase.[9] Nevertheless, phenylalanine is buried more frequently than its free energy of transfer between water and the gas would predict. The frequency with which histidine is buried is also greater than its hydrophilicity would suggest.

It is curious that tyrosine and tryptophan, also two large aromatic side chains, should be so hydrophilic in the locations they occupy in folded polypeptides. In fact, free energies of transfer for model compounds of tryp-

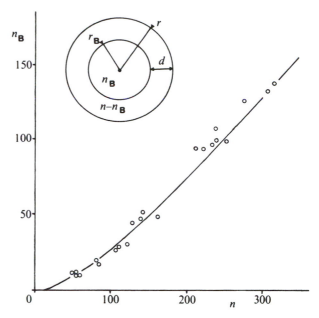

Figure 6–18: Relationship between the number of buried amino acids and the total number of amino acids in a series of crystallographic molecular models of globular proteins.[52] Twenty-two crystallographic molecular models of globular proteins were examined, each represented by a point on the graph. The accessible surface area of each amino acid in each molecular model was determined.[51] Every amino acid with an accessible surface area less than 0.2 nm^2 was designated as buried. The number of buried amino acids (n_B) was then plotted as a function of the total number (n) of amino acids in each protein. The inset illustrates a simplified model representing a molecular model as a sphere and assuming water can penetrate an irregular surface to a depth d. If this approximates the situation, Equation 6–1 should describe the distribution of n_B and κ should be proportional to d. The line drawn in the figure was for Equation 6–1 with $\kappa = 1.62$, which gives a value of d of 0.5 nm. Adapted with permission from ref 52. Copyright 1979 *Nature*.

tophan or tyrosine between water and a solvent such as 1-octanol (Figure 5–24)[56] or ethanol,[57] rather than free energies of transfer between water and the gas phase, have always suggested that tryptophan and tyrosine are more hydrophobic than they seem to be when they are found in a protein.[57] The explanation for this is probably that solvents such as ethanol and 1-octanol are able to form hydrogen bonds with the one donor on the indole and the donor and acceptors on the phenol, rendering them more soluble in these solvents than they would be in a hydrocarbon. This would be consistent with the observation that the hydroxyl on tyrosine usually remains in contact with the water in refined crystallographic molecular models.[9]

The necessity of maintaining hydrogen bonds is a particularly severe problem with the indole of tryptophan. Solvents such as ethanol and 1-octanol, in which tryptophan is remarkably soluble,[57] have twice as many acceptors of hydrogen bonds as donors. As the indole

Table 6–2: Removal of Amino Acids from Water in Molecular Models of Proteins

amino acid	accessible surface area[a] (nm^2)	fraction buried[b]			
		buried 100%[c]	buried 95%[c]	less than 0.2 nm^2 accessible[d]	mean of fraction surface buried[c]
Ile	1.80	0.18	0.60	0.76	0.90
Val	1.60	0.18	0.54	0.74	0.88
Phe	2.20	0.14	0.50	0.69	0.89
Leu	1.80	0.16	0.45	0.71	0.87
Met	2.05	0.11	0.40	0.66	0.83
Ala	1.15	0.20	0.38	0.63	0.78
Trp	2.60	0.04	0.27	0.62	0.88
Ser	1.20	0.08	0.22	0.46	0.62
Thr	1.45	0.08	0.23	0.41	0.60
His	1.95	0.02	0.17	0.44	0.78
Tyr	2.30	0.03	0.15	0.34	0.74
Glu	1.85	0.03	0.18	0.24	0.74
Asp	1.50	0.04	0.15	0.27	0.67
Asn	1.60	0.03	0.12	0.30	0.61
Gln	1.90	0.01	0.07	0.23	0.61
Lys	2.10	0	0.03	0.05	0.60
Arg	2.40	0	0.01	0.10	0.51

[a]For entire amino acid, both its side chain and its contribution to the backbone, in the tripeptide Gly-X-Gly.[53,54] [b]Fraction of the total number of that amino acid in a series of crystallographic molecular models that are buried by the noted criterion. [c]Ref 36. [d]Ref 52.

contains only a donor, a net of one hydrogen bond is created every time an indole is transferred from water to an alcohol because a solution of indole in water has more donors than acceptors and an alcohol has more acceptors than donors. When indole is transferred, empty donors disappear from water and empty acceptors disappear in the alcohol. This precaution of tallying the net change in hydrogen bonding in any reaction by counting donors and acceptors is often overlooked. Presumably such imbalances have a major effect on the distribution of amino acids between the surface and the interior of a molecule of protein or the coupling of the donors and acceptors of hydrogen bonds withdrawn from water during the process of folding the polypeptide. Because a solution of protein always has more acceptors than donors, it will always be more difficult to remove donors from the aqueous phase than acceptors; hence tryptophan should be more hydrophilic than it appears.

An extreme example of the strong tendency of tryptophan to retain its hydrogen bond with water occurs in the structure (c12 Kin.1) of the Bence-Jones protein Rhe. A tryptophan in the center of the crystallographic molecular model of this protein, though completely buried, is engaged in a hydrogen bond with a molecule of water sitting next to its indole nitrogen (Figure 6–19). This molecule of water is trapped in the interior during the folding of the polypeptide. In γ-II crystallin, two of the tryptophans are also hydrogen-bonded to buried molecules of water.[58] Usually, however, tryptophan retains the hydrogen bond to the nitrogen–hydrogen bond of its indole in less dramatic ways. For example, all of the donors in the indoles of the tryptophans of chymotrypsin (c15 Kin.1) retain hydrogen bonds with the solvent or another acceptor in the interior.[9] The other two tryptophans in γ-II crystallin (c5 Kin.3) form hydrogen bonds with acyl oxygens.[58] In deoxyribonuclease I, all of the tryptophans, though mostly buried, retain contact with the solvent at their nitrogen–hydrogen bonds.[5]

Glutamine and asparagine are more straightforward examples of the effect of the hydrogen bonds with water in the unfolded polypeptide on the location of that amino acid in the folded polypeptide. Complete withdrawal of glutamine or asparagine from water during folding would result in a net disappearance of two hydrogen bonds from the solution. The difficulty of simultaneously regaining both of these lost hydrogen bonds in the interior of the protein seems to be great enough that almost every glutamine and asparagine in a protein ends up in the folded polypeptide fully exposed to the aqueous phase.[47]

One of the most notable features of the accessibilities of the amino acids in the native structure of a protein is that in the folded polypeptide, the accessible surface areas of all types are less than they are in the unfolded polypeptide (Table 6–2). The accessible surface areas tabulated are for the entire amino acid in a polypeptide, both side chain and backbone segments. Usually, the backbone is buried before the side chain, so a significant portion of the mean fraction of surface buried for each type of amino acid is accounted for by the fragment common to all of the amino acids. This backbone portion, however, cannot account for greater than 0.6–0.7 nm² of buried surface area because the accessible surface area of glycine in a polypeptide[36] is only 0.75 nm². Therefore even the most accessible side chains, arginine, lysine, and glutamine (with a mean buried surface of 1.2 nm²), have more than 0.5, 0.55, and 0.45 nm² of the surface area of their side chains buried, respectively. The regions of each side chain that are buried are usually the carbon–hydrogen bonds or the electron clouds of the π molecular orbital systems. It is in the last column of Table 6–2 that the hydrophobicity of tryptophan is manifested, because 90% of its accessible surface area is usually buried. Presumably, there is a hydrophobic effect associated with the partial removal of these hydrophobic portions of the side chains from contact with the water as well as the removal at other locations of entire hydrophobic amino acids from water.

Because the accessible surface area of a folded polypeptide in its native configuration is so large, most of the amino acids in the final structure reside at the surface (Figure 6–18) and are only partially buried. To assess the hydrophobic effect that was contributed by burying these portions of each amino acid, the contribution of each atom in an amino acid to its overall hydropathy should be extracted. From these atomic parameters, the free energy of transfer for only those portions of each amino acid that are actually buried could be calculated. It has already been noted[60] that free energies of transfer for individual solutes between water and the gas could be dissected into the individual contributions of each covalent bond that they contained. A similar dissection has since been performed upon the set of free energies of transfer for the N-acetyl α-amides of the amino acids between water and 1-octanol.[61] A series of parameters based on individual atoms, rather than covalent bonds, has been presented that can reproduce the original scale of hydropathy with acceptable precision. Presumably every scale of hydropathy presently in use can be so dissected.

As is the case with the backbone of the polypeptide, when donors of hydrogen bonds on the side chains of the amino acids are removed from the water and stripped of their hydrogen bonds with the solvent, there would be a considerable loss of enthalpy if they did not find new partners in the interior of the protein. The hydrogen-bonded rings in which serine and threonine may participate (Figure 6–15) would be consequences of this need, but they would not be structurally significant, except where they cause local changes in the path of the polypeptide. It might be supposed that buried hydrogen bonds between side chains on different segments of secondary structure would be important factors because these would be capable of organizing significant regions of the pro-

Figure 6–19: Structures surrounding Tryptophan 36 in the refined crystallographic molecular model (c12 Kin.1) of Bence-Jones protein Rhe.[59] This protein is a compact globular structure formed from a folded polypeptide 114 amino acids in length. Tryptophan 36 lies in the center of the molecule sandwiched between two antiparallel β-pleated sheets comprising the entire first and second half of the polypeptide, respectively. A water molecule (O118) is buried with the tryptophan and forms a hydrogen bond to the NH of the indole. Reprinted with permission from ref 59. Copyright 1983 Academic Press.

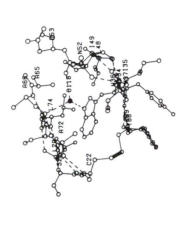

tein.[36] Of the buried hydrogen bonds between side chains, however, only about 20% are the type that connect different segments of secondary structure; the other 80% connect donors and acceptors within the same α helix, 3_{10} helix, β sheet, or β turn.[36] It is probably the case that the steric requirements of packing the secondary structures efficiently and avoiding empty space are more important than hydrogen bonds in organizing the structure and the few buried hydrogen bonds that do occur between segments of secondary structure are adventitious.

A

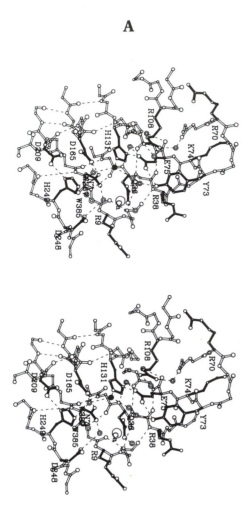

Figure 6–20: Examples of clusters of hydrogen bonds among side chains. (A) A large cluster of hydrogen bonds in the active site of the crystallographic molecular model of deoxyribonuclease I.[5] Reprinted with permission from ref 5. Copyright 1986 Academic Press. (B) Diagrammatic representation of a cluster of hydrogen bonds among some of the side chains in the crystallographic molecular model of human lysozyme.[10] Reprinted with permission from ref 10. Copyright 1981 Academic Press.

One of the remarkable features of partially buried hydrogen bonds is that they tend to be clustered. For example, of the 54 side chains in myoglobin that form hydrogen bonds with atoms in the protein other than bound water, 16 participate in eight closed pairs, nine participate in three closed triplets, but 29 participate in larger clusters.[15] These clusters often incorporate buried water. Examples of such clusters occur in deoxyribonuclease I (Figure 6–20A)[5] and in lysozyme (Figure 6–20B).[10] In these clusters, charged amino acids participate as donors and acceptors of hydrogen bonds as readily as uncharged amino acids and there is no obvious balancing between positive and negative charges (Figure 6–20A). It could be the case that these clusters, by permitting hydrogen bonds to form cooperatively, exploit entropy of approximation sufficiently to allow a net free energy of formation that is negative. If this is the case, their formation contributes favorably to the free energy of folding the polypeptide.

Clusters of hydrogen bonds serve to orient functionally important amino acids. For example, a "complex network of hydrogen bonds" serves to orient the six histidines (c5 Kin.1) responsible for chelating the copper and the zinc in superoxide dismutase.[50] Histidine 57 in the active site of chymotrypsin (c15 Kin.1) is oriented by a hydrogen bond to Aspartate 102, which in turn is oriented by three other hydrogen bonds, one to each of its three remaining acceptors.[9] Histidine 31 in deoxyribonuclease I is functionally important and is held in position by the cluster in which it participates (Figure 6–20A). It is interesting that the hydrogen bond in this last case forces the dihedral angle χ_2 of Histidine 31 to assume an unfavorable value when it is positioned properly. Carboxylic acids, histidines, and arginines are most susceptible to such pinning because they have donors and acceptors at two or more separate locations on their side chains, and they are rigid structures because of their π molecular orbital systems. These features make them easily immobilized.

Whether hydrogen-bond donors and acceptors form clusters and exclude hydrophobic amino acids or hydrophobic amino acids themselves have an affinity for each other, clusters of hydrophobic amino acids also occur in the regions of the interior of a molecule of protein most distant from the surface. Clusters of aro-

B

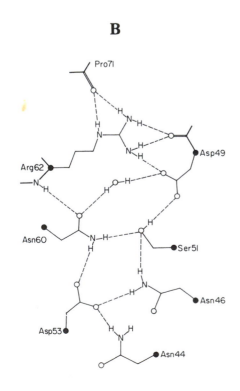

Figure 6–21: Examples of clusters of hydrophobic amino acids in the interior of a protein. (A) Cluster of nine aromatic side chains from Tyrosine 165, Phenylalanine 32, Phenylalanine 167, Phenylalanine 140, Phenylalanine 153, Phenylalanine 141, Phenylalanine 310, Tryptophan 191, and Phenylalanine 126 is found in the crystallographic molecular model of penicillopepsin.[30] The map of difference electron density is for $(2|F_o| - |F_c|)$ with α_c, and the solid lines are the refined crystallographic molecular model. Reprinted with permission from ref 30. Copyright 1983 Academic Press. (B) Hydrophobic cluster in the crystallographic molecular model (c3 Kin.3) of myoglobin.[15] The cluster is composed of Leucine 135 (12H), Isoleucine 111 (12G), Isoleucine 28 (9B), Methionine 131 (9H), Leucine 72 (15E), Leucine 69 (12E), Valine 114 (15G), Histidine 24 (5B), Leucine 76 (19E), Tryptophan 14 (12A), and Valine 13 (11A). The top panel is a ball-and-stick presentation of the molecular model, and the bottom panel is the same view of the same amino acids presented in a space-filling format at the same magnification. Reprinted with permission from ref 15. Copyright 1977 Academic Press.

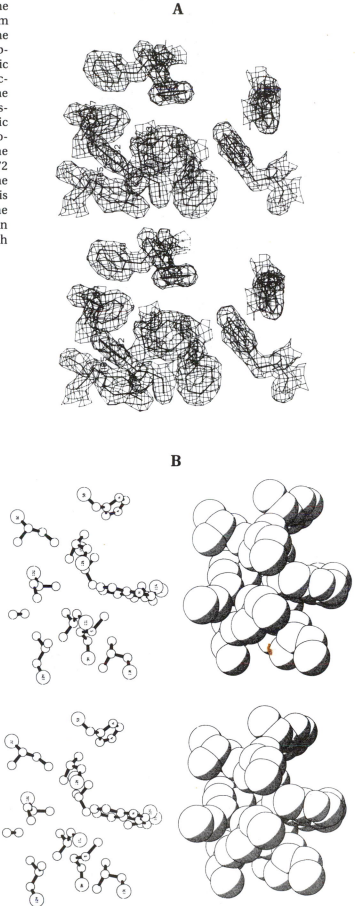

A

B

matic amino acids, such as the one in penicillopepsin (Figure 6–21A), are fairly common features. The tendency for aromatic amino acids, cystines, and arginines to cluster has been explained in terms of favorable overlaps of their π molecular orbitals,[58] but just as often aromatic and aliphatic amino acids intermingle. A characteristic example of this (c3 Kin.3) occurs in one of the regions of myoglobin (Figure 6–21B).[15] It is the interdigitation of the hydrophobic residues within such clusters in the center of the protein that orients the β sheets and α helices in the native structure. One interesting aspect of the cluster of aromatic amino acids in penicillopepsin (Figure 6–21) is that it illustrates the tendency of two phenyl rings, in isolation from water, to form a complex in which the planes of the two rings are at 90° to each other.[62] This orientation is commonly observed in the crystallographic molecular models of proteins, and theoretical calculations for benzene dimers in the gas phase suggest that it is the energetically favored arrangement.[63]

Suggested Reading

Chothia, C. (1976) The Nature of the Accessible and Buried Surfaces in Proteins, *J. Mol. Biol. 105*, 1–14.

Problem 6–1: The amino acid sequence of the chymotrypsin inhibitor I from Russet Burbank potatoes is

```
                5                    10                   15
Pro Ile Cys Thr Asn Cys Cys Ala Gly Tyr Lys Gly Cys Asn Tyr
                20                   25                   30
Tyr Ser Ala Asn Gly Ala Phe Ile Cys Glu Gly Gln Ser Asp Pro
                35                   40                   45
Lys Lys Pro Lys Ala Cys Pro Leu Asn Cys Asp Pro His Ile Ala
                50
Tyr Ser Lys Cys Pro Arg Ser
```

On the next page is its crystallographic molecular model (reprinted with permission from ref 64; copyright 1989 Academic Press):

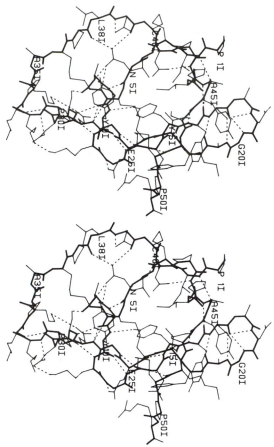

(A) Make a list of as many of the hydrogen bonds as you can identify. Use the notation Y15OH–Y10pepNH, where Y15 is the amino acid Tyrosine 15, OH refers to its phenolic OH, Y10 is Tyrosine 10, pep refers to the atoms of Tyrosine 10 in the polypeptide backbone, and NH refers to the amide nitrogen of Tyrosine 10 in the polypeptide.

(B) Make a list of the four cystines in the molecule. Use the notation Cys85–Cys95 to identify the pairs. Notice the expected dihedral angles of 90°.

(C) Identify the participants in a small hydrophobic cluster.

(D) Make a list of the secondary structure. Identify each element of secondary structure by its name and the letters and numbers of the participating amino acids.

Packing of the Side Chains

In the refined molecular model of a protein, the space between the α helices, β sheets, and random meander is filled necessarily with the side chains of the amino acids that protrude from the polypeptide at each α carbon. It is the specific interactions among these side chains that dictate the relative orientations assumed by these secondary structures.

The structure of a molecule of protein is organized in such a way that the amino acids in its interior are packed with admirable efficiency, as if they were pieces in a three-dimensional puzzle. One of the most striking consequences of this efficiency is that the alkyl groups in the center of a molecule of protein are confined to specific conformations rather than being free to rotate as they would be in a liquid. Another consequence of the economy in arranging the side chains of the amino acids is that the volume of a molecule of protein is quite small. The volume occupied by the atoms in a molecule of protein can be calculated by summing all of its individual atomic volumes, and the actual volume of the molecule can be calculated from its partial specific volume and its molecular weight. From these two volumes, it can be calculated that 75% of the volume of a molecule of protein is occupied by atoms.[65,66] In most organic liquids only about 45% of the volume is occupied by atoms, in water only 36%, and in a solid of hexagonally packed spheres only 75% of the volume would be filled by atoms.[65] In crystals of small organic molecules, 70–80% of the volume is filled by atoms.[66]

At first glance, these results seem incompatible with the observation that the **partial specific volume** of a protein (usually 0.72–0.75 mL g^{-1}) can be calculated quite accurately from the molar volumes of its constituents.[67,68] This latter calculation, however, does not treat the constituents as independent solutes in free solution. In fact, if each side chain were an independent solute, each of their partial molar volumes would include a covolume,[69] which is a volume that arises simply because a particular constellation of atoms is an independent molecule dissolved in a given solvent. These covolumes are substantial. For water[68] the covolume of a solute is 14 cm^3 mol^{-1}, and for organic solvents[69] it is 25 cm^3 mol^{-1}. Therefore, the sum of the partial molar volumes of the components of a protein, were they each separate molecules in solution, would be significantly greater than its actual partial molar volume. To the extent that its covolume arises from the fact that a solute is in free solution in a given solvent, the fact that the partial molar volume of a protein is the sum of the atomic volumes of its substituents with no added covolume states that those substituents are not in free solution. This of course is true; they are economically packed into a solid.

The partial molar volume of a protein is observed to be the sum of the atomic volumes of its constituents, but each protein has a unique structure. This suggests that each structure, although it is unique, incorporates the requirement that its volume be as small as possible. The minimization of molecular volume is an important noncovalent force in the folding of a molecule of protein, and it dictates many of the features of the structure, in particular the alignment of the elements of secondary structure. This noncovalent force minimizing the empty space within a molecule of protein can be considered to be a consequence of the hydrophobic effect, if the hydrophobic effect is defined as the tendency of water to minimize the volume of the cavity occupied by any solute.

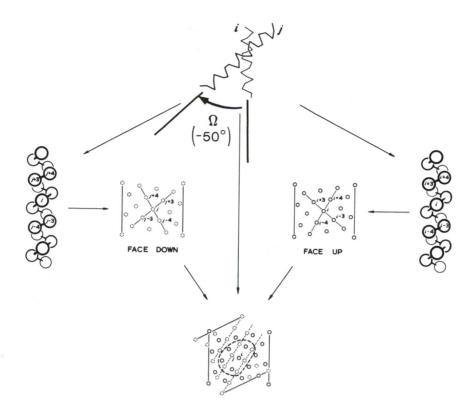

Figure 6–22: Use of superimposed helical nets to describe the contacts at an interface between two α helices. (Top) The angle between two adjacent α helices, i and j, is defined as the angle Ω; its sign is determined by the right-hand rule. (Right) α Helix i in a vertical orientation is numbered out from its center. The central amino acid is given the designation i, those below are designated by negative integers, and those above, by positive integers. The relative orientations in which amino acids $i - 7$, $i - 4$, $i - 3$, i, $i + 3$, $i + 4$, and $i + 7$ are distributed can be projected onto a plane tangential to the position of amino acid i. These seven projected points define a unique lattice, or helical net. For α helix i the lattice is face up, in the orientation of the original α helix. (Left) α Helix j also defines the same lattice as that defined by α helix i, but, because α helix j is to be opposed to α helix i, face to face, the helical net for α helix j is flipped over, face down. (Bottom) The two helical nets, the one for α helix i face up and the one for α helix j face down, are then opposed and rotated with respect to each other until maximum interdigitation of the lattice points is achieved. The angles at which maximum interdigitation occurs in the helical nets will be the angles at which maximum interdigitation occurs between the amino acids of the two α helices. Adapted with permission from ref 71. Copyright 1981 Academic Press.

The details of three types of regular packing have been examined: α helices packed with α helices, β sheets packed with β sheets, and α helices packed upon β sheets. In all of these situations the structures take up orientations with respect to each other that permit the side chains that protrude from each of them to interdigitate. This **interdigitation** is the reason that there is very little vacant space in the interior of a molecule of a protein. If it can be assumed that the configuration of minimum volume is the preferred configuration in the condensed phase, then these interdigitations are favorable arrangements. In order to form as many interdigitations as possible, the individual segments of secondary structure are forced to assume preferred orientations with respect to each other. Viewed in this perspective, packing is a structural force just as the formation of hydrogen bonds between buried donors and acceptors is a structural force, but packing is probably far more important than hydrogen bonds.

The orientation between two α helices, two sheets of β structure, or an α helix and a sheet of β structure can be assigned an angle Ω.[70] The Ω between two α helices is the angle between their two axes (Figure 6–22). The sign on Ω is given by the right-hand rule. Consequently, the angle Ω in Figure 6–22 has a negative sign. Because the pattern in which the amino acids protrude from an α helix has a 2-fold rotational axis of pseudosymmetry at each position in the α helix (focus on position i at the

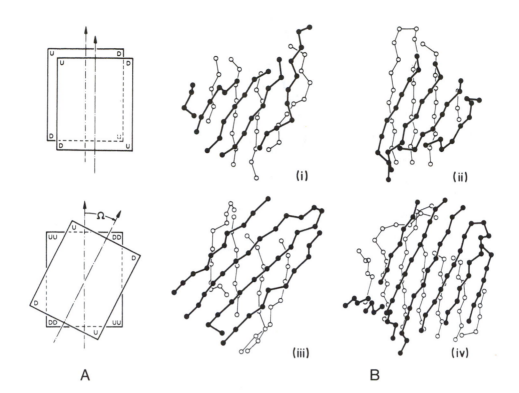

Figure 6–23: Opposition of two sheets of β structure.[70] (A) The angle Ω between two opposed sheets of β structure is the angle between the two vectors in the two respective sheets that are parallel to the strands of polypeptide composing the sheets. (B) Examples of opposed sheets of β structure from crystallographic molecular models. The positions of the α carbons in the molecular model of the individual strands of polypeptide are designated by closed circles for the upper sheet and open circles for the lower sheet. The α carbons for a given strand are connected by line segments. The crystallographic molecular models from which these four examples were drawn are (i) immunoglobulin fragment V_{REI} (c12 Kin.1), (ii) prealbumin, (iii) superoxide dismutase (c5 Kin.1), and (iv) concanavalin A. Reprinted with permission from ref 70. Copyright 1977 National Academy of Sciences.

right of Figure 6–22), the axis has no direction associated with it and a value of Ω = –50° is topologically equivalent to a value of Ω = +130°. The angle Ω between two sheets of β structure is the angle between the direction of the parallel or antiparallel strands in one sheet and the direction of the strands in the other sheet (Figure 6–23). The right-hand rule determines the sign of Ω, and the angle Ω in Figure 6–23 is, therefore, negative. No distinction is made between parallel and antiparallel sheets (c2 Kin.8 and c5 Kin.1) or the amino- and carboxy-terminal ends of a given strand of β structure because all combinations of these distinct stereochemistries produce almost the same pattern in which the side chains are distributed across the face of the sheet (Figure 4–17). The angle Ω between an α helix and a sheet of β structure is the angle between the axis of the helix and the direction of the strands of β structure (Figure 6–24).

Two α helices may pack next to each other in several ways that are compatible with the interdigitation (c3 Kin.4) of their side chains.[71] In a coiled coil (c8 Kin.5), two strands of α helix twist around each other to permit the

side chains to interdigitate at the interface between them.[72] The same type of interdigitation can be achieved with two straight α helices if they are inclined at +27° [(2 × 360°) – (7 × 99°)] to each other (c3 Kin.1 and c13 Kin.2). Examples of helices inclined with angles between 0° and +40° and interdigitating in this way are common.[71]

The most frequently observed angle Ω between two α helices,[71] however, is around –50° (c3 Kin.3), the angle used to construct Figure 6–22. One-third of all adjacent α helices in molecular models of globular proteins are inclined with respect to each other by an angle Ω between –60° and –40°. When an α helix is observed from one side (Figure 6–25), it can be seen that the side chains are arranged in sets of parallel ridges and grooves. For example, residues 8 and 4; 19, 15, and 11; and 22 and 18 in Figure 6–25 form a set of ridges and grooves, but so do residues 4 and 1, 11 and 8, 18 and 15, and 22 and 19. If one α helix is opposed to another, the first set of these ridges will fit into the second set of these grooves, and conversely when the angle between the two helices is –50° (Figure 6–22). An example of the interdigitation that

Figure 6–24: Packing of an α helix on a β-pleated sheet.[70] (Top right) The angle Ω between the α helix and the β-pleated sheet is the angle between the axis of the α helix and the vector parallel to the strands of polypeptide in the β-pleated sheet. (A) View down an α helix of the orientations of the amino acids on one face. (B) The usually observed twist of a β-pleated sheet (Figure 6–7) with the corners designated as up or down. (C) An α helix sitting on the surface of a twisted β-pleated sheet. The amino acids on the lower face of the α helix are in black and designated by number. The locations of the four corners of the twisted β-pleated sheet are designated by letters. (D) The three relationships that are possible between the straight axis of the α helix and the various curvatures in a twisted β-pleated sheet. The curvature encountered by the α helix is determined by the value of angle Ω. Reprinted with permission from ref 70. Copyright 1977 National Academy of Sciences.

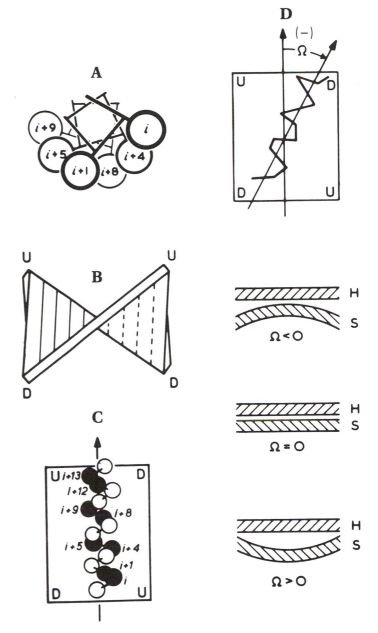

occurs in such situations is found between two adjacent α helices in the molecular model of carboxypeptidase (Figure 6–26).[71]

There are a number of other values for the angle Ω between two α helices that promote less favorable interdigitations of the side chains, and examples of all of them have been observed.[71] Because several possibilities exist and because α helices can tighten or loosen to accommodate different angles close to the ideal values, the distribution of angle Ω between −90° and +90° is fairly uniform[71] with the exception of the striking and sharp peak at −50°. There is also a suggestion of a broad maximum at +20°, which is the arrangement analogous to the one in a coiled coil that, in theory, should produce efficient packing.

The broad range of configurations seen with two α helices is in contrast to the packing of **two sheets of β structure** against each other. In this case, only two orientations are observed with significant frequency. In a sheet of β structure, the side chains of the amino acids along a given strand alternate in protruding above the sheet and below the sheet (Figure 4–17). On a given side of the sheet, the protrusions form approximately a square array aligned with the strands of β structure (Figure 6–27). There are two ways for two square arrays to interdigitate if they are flat, one with Ω = 0°(c5 Kin.4), the other with Ω = 90°. Because the sheets twist, however (Figure 6–7), and because the array is not exactly square (Figure 6–27), real sheets of β structure pack at angles Ω of either −30° ± 15° (Figure 6–23)[70,73,74] or 90° (Figure 6–28).[75] Regardless of whether the β sheets pack at 90° (c5 Kin.2) or −30° (c5 Kin.1), however, the side chains fit tightly together at the interface.

Were two sheets of β structure to be packed parallel to each other with angle Ω equal to 0°, their two twists would match. The twist, however, causes the residues protruding from the bottom of the top sheet to lean one way and those protruding from the top of the bottom sheet to lean the other,[73,74] and when the side chains interdigitate the sheets cannot be parallel to each other

but are forced to assume an angle Ω equal to −30°. The associations between the two parallel β sheets in prealbumin illustrates the stereochemistry at such a parallel interface between two twisted sheets (Figures 6–23 and 6–29). In a β barrel, where eight strands of parallel β structure form a cylindrical structure, the cylinder is usually flattened along one of its axes to produce an interface equivalent to that between two parallel β sheets (c2 Kin.8). Otherwise vacant space would be left in the middle of the structure. The same is true of an antiparallel β barrel (c5 Kin.1).

When two sheets of β structure are aligned orthogonally to each other (c5 Kin.2), the twists oppose one another and one pair of diagonal corners are closer together than the other pair of diagonal corners (Figure 6–30). It is usually at the close corners that the polypep-

A   **B**

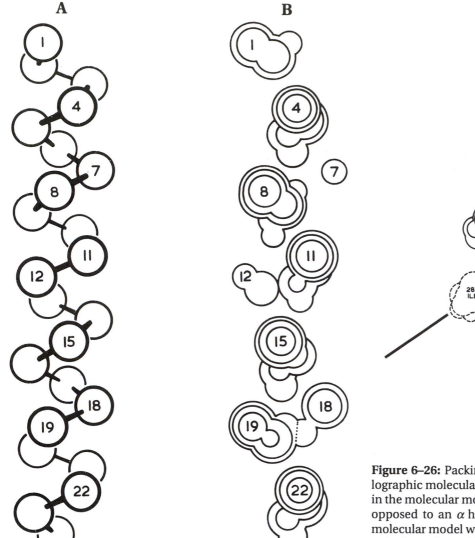

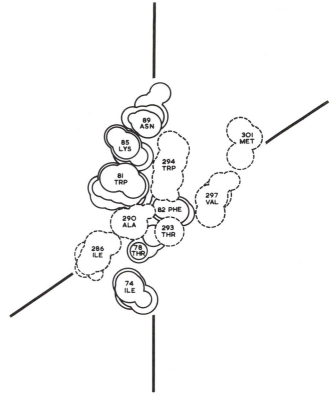

Figure 6–25: Orientation of the side chains in an α helix. (A) The amino acids in an α helix are numbered consecutively from top to bottom with the front face accentuated. (B) A topographic map of the front face of an α helix of a polyalanine. The contours are at intervals of 0.05 nm and the topmost contour, that surrounding the numbers 4, 15, and 22, is at 0.475 nm from the axis of the α helix. The numbers are on the methyl carbons of the appropriate alanine. Reprinted with permission from ref 71. Copyright 1981 Academic Press.

Figure 6–26: Packing of two opposed α helices in the crystallographic molecular model of carboxypeptidase A.[71] An α helix in the molecular model comprising amino acids 72–90 is tightly opposed to an α helix comprising amino acids 285–307. The molecular model was converted into a space-filling format and only the amino acid side chains along these two α helices were displayed. A set of three parallel planes was cut through the interface between the two helices and superposed to create a topographic map through the interface. The planes were approximately parallel to the two helical axes and at intervals of 0.1 nm. The contours of the side chains in the lower α helix are solid, those in the upper are broken. Each amino acid in the map is designated by its number in the amino acid sequence, and lines designating the axes of the two α helices are included. The angle Ω between these two helices[71] is −56°. Reprinted with permission from ref 71. Copyright 1981 Academic Press.

tide connects one sheet to the other.[75] The interdigitations of the amino acids in the three layered orthogonal β sheets of penicillopepsin illustrates the packing observed between such orthogonal β sheets (Figure 6–28).

An **α helix lying upon a sheet of β structure** (c2 Kin.7) usually has its axis almost parallel to the strands of the sheet because the α helix is straight, the sheet is twisted, and a straight rod can contact a twisted surface only when it is either parallel or perpendicular to the axis of the twist (Figure 6–24).[70] The angle Ω observed[76] between α helices and adjacent sheets of β structure is

usually around 0°, and almost all values fall between −20° and +10°. The interface (Figure 6–31)[76] between three of the α helices and one of the β sheets in lactate dehydrogenase illustrates the fit between these segments of secondary structure. Note that the side chains from the α helices lie upon the gaps between the residues in the sheet of β structure. Because a sheet of β structure twists appropriately, the α helices lying across its surface are aligned next to each other with angles Ω of about −40° between adjacent pairs (c2 Kin.2, c4 Kin.1, and c4 Kin.5) even when they cleave tightly to the

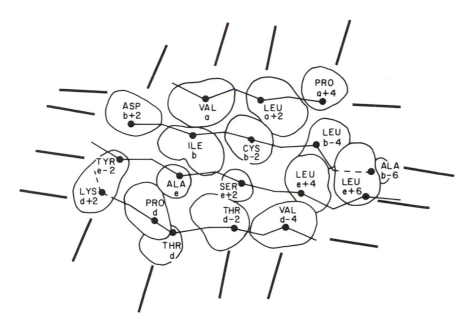

Figure 6–27: Distribution of side chains on one side of a sheet of β structure.[74] The representation is that of a four-stranded antiparallel sheet of β structure in the crystallographic molecular model of (c12 Kin.1) the light constant domain from γ-globulin (λ) New. The strands of polypeptide of the sheet of β structure in the crystallographic molecular model were placed in the plane of the page and each α carbon was connected by a line segment. The α carbon of each amino acid the side chain of which protruded below the plane of the page was marked with a dark dot. The side chains protruding below the page from these positions were displayed in space-filling format, and their projections on the plane of the page were traced. Each projection is labeled with the name of the particular amino acid. The amino acids in one column, valine, isoleucine, alanine, and proline, were assigned letters determined by the order in which their strands occur in the amino acid sequence. The other amino acids were numbered in the direction in which the particular strand ran across the page. The view is from above and of the amino acids hanging down from the sheet. Lines indicate a lattice on which the α carbons of these amino acids lie. Reprinted with permission from ref 74. Copyright 1981 Academic Press.

surface of the sheet.[76] This value for angle Ω is very close to the $-50°$ that produces the most frequently observed type of interdigitation between two α helices. Therefore, both the interfaces between the α helices and the sheet of β structure and the interfaces among the α helices themselves can exist simultaneously in optimum orientations. It is also possible, however, that the twist of such a sheet of β structure arises from the requirement that the α helices upon it be positioned at the proper angles to maximize the interdigitation of their amino acids.

It is the efficiency in packing the elements of secondary structure together that creates the almost solid state of the interior of a molecule of protein in which rotational motion is severely constrained and each side chain assumes only one of the possible rotational isomers. Even upon the surface of a molecule of protein, however, the side chains are usually confined to only one conformation. The conformations of the side chains of

the amino acids in a refined molecular model can be divided into three categories: discrete, discrete but disordered, and disordered. A side chain with a discrete conformation populates only one rotational isomer, within the resolution of the map of electron density. A side chain with a discrete but disordered conformation spends its time in only two or three rotational isomers so that its electron density becomes a discernable hybrid of these few isomers. A side chain with a disordered conformation is flexible enough to populate so many rotational isomers that its electron density fades into the noise. As the data set is increased to higher resolution and the refinement is improved, disordered residues decrease in number as discrete but disordered and discrete residues become clearer in outline. In the four highly refined and highly resolved molecular models of lamprey hemoglobin, myohemerythrin, erabutoxin, and crambin, only 3% of the side chains are disordered (11 out of 375 amino

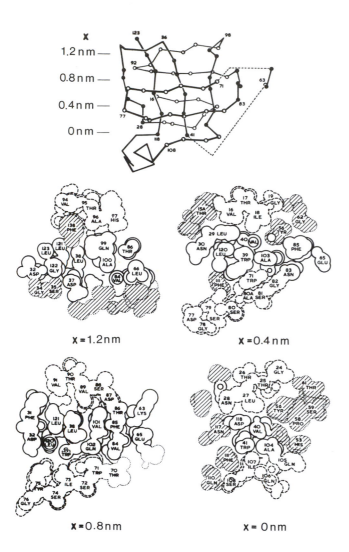

Figure 6–28: Packing of amino acids at the interface between three alternately perpendicular sheets of β structure in the crystallographic molecular model of penicillopepsin.[75] (Top) The α carbon atoms of the molecular model between amino acids 16 and 123 are connected by line segments to provide a tracing of the polypeptide. The three-layered sandwich is viewed from above and consists of a three-stranded sheet of β structure (parallel, parallel, antiparallel) on top of a four-stranded sheet of β structure (parallel, parallel, antiparallel, antiparallel) on top of a second four-stranded sheet of β structure (parallel, antiparallel, parallel, antiparallel). Sections 0.4 nm apart were cut through the three-layered sandwich in a space-filling molecular model. The planes of the sections were horizontal and normal to the page as indicated. The packing between the sheets can be viewed in cross section. Amino acids are numbered, and all amino acids in a given pleated sheet are either in solid outline or broken outline. Amino acids in hatched outline are not in the sheets of β structure. In the cross sections the strands of the sheets at the top and bottom run parallel to the page while the sheet in the middle runs perpendicular to the page. Reprinted with permission from ref 75. Copyright 1982 American Chemical Society.

Figure 6–29: Packing of the amino acids at the interface between two opposed sheets of β structure in the crystallographic molecular model of prealbumin. A tracing of the α carbons of the strands in this structure is presented in Figure 6–24. This is represented diagrammatically in the upper inset, where the locations of the horizontal sections through the structure are designated. The sections are normal to the sheets, and the strands run normal to the planes of the sections. Three sections through a space-filling model of this structure are presented spaced at 0.8 and 0.7 nm. The amino acids in the upper sheet (in which the four strands run parallel, antiparallel, parallel, antiparallel) are enclosed in solid lines; those in the lower sheet (in which the four strands run parallel, antiparallel, antiparallel, parallel) are enclosed in broken lines. The straight lines indicate the orientation of the interface which twists in a right-handed sense as the sections proceed through the structure. Reprinted with permission from ref 73. Copyright 1981 National Academy of Sciences.

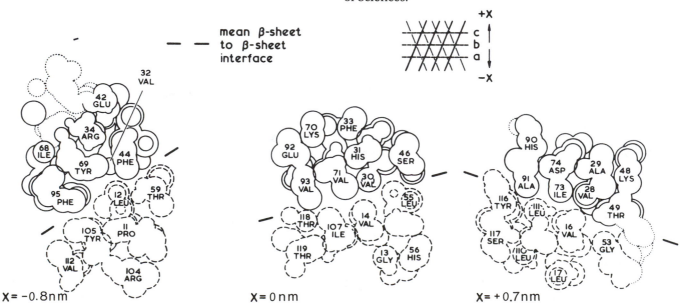

Figure 6–30: Abstract representation of the packing of two orthogonal sheets of β structure against each other (c5 Kin.2) that incorporates the right-handed twist usually associated with such structures. The front view illustrates the orthogonal disposition of the strands of the two respective sheets. The top view illustrates the twist, and the fact that the strands can join at the two corners (A and B) that are opposed by the separate twists, and the fact that two of the corners are splayed by the twist. The view of the opened interior demonstrates how the structure can be produced by folding over two coplanar sheets of β structure joined at one corner by a continuous strand shared by both sheets. The bottom two sheets of β structure in the three-layered sandwich of penicillopepsin (Figure 6–28) have this topology. They are connected at two diagonal corners and splayed at the other two. Reprinted with permission from ref 75. Copyright 1982 American Chemical Society.

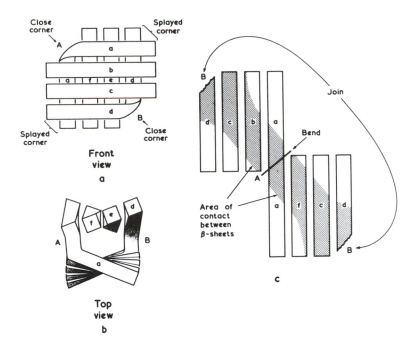

Figure 6–31: Schematic formation of the interface between three α helices and a sheet of β structure found in the crystallographic molecular model (c4 Kin.1) of lactate dehydrogenase.[76] (A) The sheet of β structure is presented in the top part of the figure with the strands running vertically in the x direction and the axis of the right-handed twist parallel to the y axis. Side chains of the amino acids on the upper face of the sheet are identified and numbered by the amino acid sequence of the protein. The three α helices that will form the interface with the sheet of β structure are presented in the lower part of the figure with the face that will participate in the final interface directed upward. The axes of the three helices (αG3, αB, and αC) are almost parallel to the vertical x axis. Side chains that will participate in the interface are identified and numbered. (B) When the three α helices are rotated 180° around the y axis and placed upon the sheet of β structure as they are in the molecular model, the interface is produced by the interdigitation of the highlighted amino acids from the sheet and the highlighted amino acids from three α helices, respectively. It is these interdigitations that position the three α helices upon the sheet of β structure. Adapted with permission from ref 76. Copyright 1980 Academic Press.

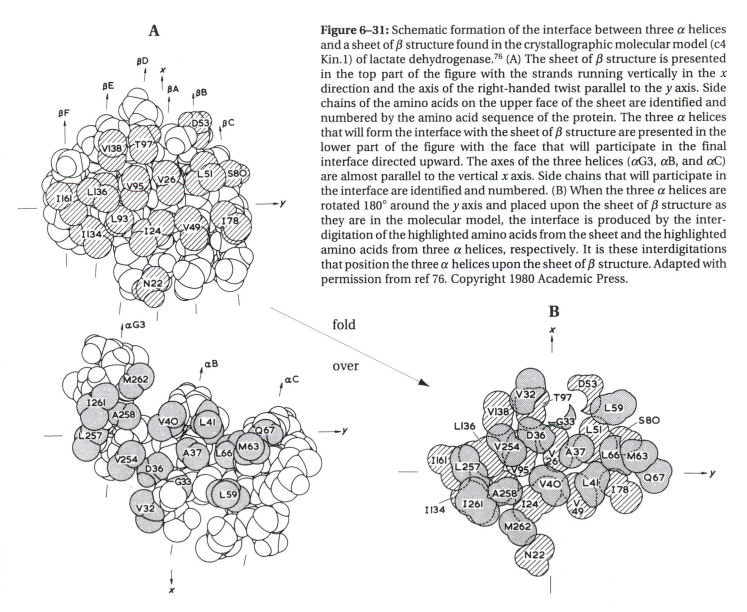

acids), and only 8% are discrete but disordered (32 out of 375 amino acids).[77] The rest of the side chains, including most of the side chains on the surfaces of the proteins, are in one discrete conformation. It is possible that this will be true for all highly refined and highly resolved molecular models.

When pairs of refined molecular models of proteins that were either homologous in sequence or identical but in different crystallographic unit cells were compared, 70% of the conserved residues and 86% of the identical amino acids were present in the same rotational isomer.[78] This observation also suggests that even on the surface of a molecule of protein there are significant constraints on the rotational freedom of the side chains built into a particular structure. These constraints could be steric, or they could involve hydrogen bonds among the side chains or between the side chains and the water of hydration.

Suggested Reading

Chothia, C., Levitt, M., & Richardson, D. (1977) Structure of Proteins: Packing of α-Helices and Pleated Sheets, *Proc. Natl. Acad. Sci. U.S.A. 74*, 4130–4134.

Water

About 40–70% of the volume of a crystal of protein is occupied by water.[79] It fills the large vacant spaces among the folded polypeptides. The majority of the molecules of water in a crystal of protein are liquid and disordered over the time required to collect a data set. Regardless of the degree of refinement, the regions containing this disordered water remain featureless and have a mean electron density similar to that of liquid water.[80] These regions of the unit cell can be treated as solids of uniform electron density that have the shape of the disordered regions peculiar to the particular crystal. These solids are incorporated as such into the molecular model of the unit cell from the first cycle of the refinement because when these water-filled lacunas are added explicitly to the molecular model, the *R*-factor decreases significantly.[80]

During the refinement, maps of difference electron density are frequently calculated. In addition to the large spaces filled with disordered water, small discrete peaks of positive electron density become regularly recurring features of these difference maps. Because no reasonable rearrangement of the atoms of the molecular model of the protein is able to erase these features and because they are unaccompanied by adjacent peaks of negative electron density indicative of a misalignment of the molecular model, these peaks are assumed to represent either individual molecules of water or individual molecules of solutes from the solution in which the crystals were formed, usually a concentrated solution of ammonium sulfate. Sulfate is an ion with a large number of electrons, and any peaks of electron density representing sulfate can usually be recognized with little difficulty. For example, in the maps of difference electron density prepared during the refinement of myoglobin, two of the positive peaks of electron density could be unambiguously recognized as sulfate anions bound to the molecule of protein.[15] Often proteins are crystallized from glycols or other polyols, and molecules of these solutes are also easily recognized. The ammonium cation is indistinguishable in its electron density from a molecule of water, but proteins at neutral pH rarely bind cations so it is usually assumed that the smaller peaks of electron density are molecules of water. A molecule of water found in the refined molecular model of a protein can be defined as a positive peak of electron density that persists

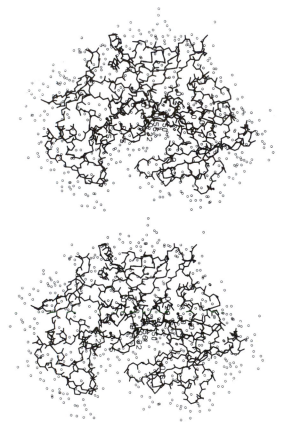

Figure 6–32: Locations of the ordered molecules of water in the crystallographic molecular model of penicillopepsin.[30] At various cycles after cycle 20 of the refinement of this molecular model, it was decided that certain members of the array of as yet unassigned peaks of positive density in the map of difference electron density were molecules of water, and a molecule of water was placed at each of these positions. The 319 unique water molecules so positioned are located in the figure relative to the polypeptide backbone without the side chains. Molecules of water that maintain contact with the bulk solvent either directly or through other molecules of water are represented by open circles. The 13 water molecules in the interior of the molecular model that do not maintain contact with the bulk solvent are represented by filled circles and labeled with the letter O and a number. Reprinted with permission from ref 30. Copyright 1983 Academic Press.

Figure 6–33: Molecules of water in the interior of crystallographic molecular models of proteins. (A) A single peak of positive electron density (designated by the cross) assigned to a molecule of water, unconnected to any other peak of positive density, within the interior of the molecular model of deoxyribonuclease I.[5] The map of difference electron density is for $(2|F_o| - |F_c|)$ with α_c, so the peak is prominent. The molecular skeleton is the refined molecular model itself. The molecule of water is hydrogen-bonded to the amide nitrogen–hydrogen bonds of Isoleucine 193 and Isoleucine 211 and the acyl oxygens of the same two amino acids. Reprinted with permission from ref 5. Copyright 1986 Academic Press. (B) A cluster of four peaks of positive electron density (O50, O56, O32, and O46) assigned to four molecules of water in the interior of the crystallographic molecular model of penicillopepsin.[30] These four molecules of water are held in an array of hydrogen bonds with donors and acceptors in the interior of the model, but Water 46 forms one hydrogen bond with Water 174 that is in a poorly occupied location in contact with the bulk solvent. As such, this structure is a finger of water passing into the interior. Reprinted with permission from ref 30. Copyright 1983 Academic Press.

A

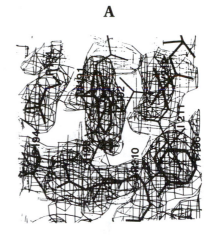

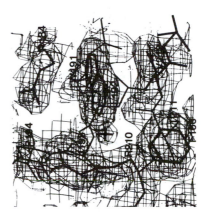

B

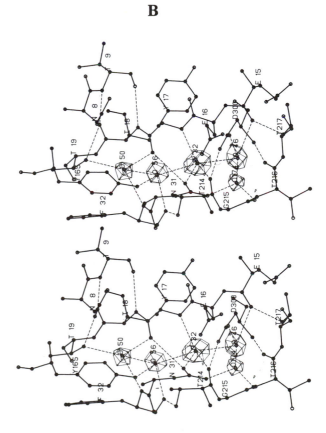

in its location in maps of difference electron density as the refinement progresses and that has a magnitude 0.2–1.0 times the magnitude expected for a stationary molecule of water. These sometimes vague peaks of electron density present in a difference map at the moment when the decision is made by the crystallographer that they are molecules of water should be distinguished from the more solid peaks of electron density that appear at the same positions in the map of electron density calculated with F_o and α_c after the contributions of these molecules of water have been included in α_c.

The molecules of **ordered water** in the final refined molecular model surround the molecule of protein, fill deep clefts in its surface, and are found in its interior (Figure 6–32). They represent locations in the actual molecule of protein that are consistently occupied by one or another molecule of water. With the exception of the molecules of water buried deeply in the molecule of protein, the occupants of these locations are constantly changing, probably almost as rapidly as the molecules of water change positions in the hydrogen-bonded lattice of liquid water itself; it is the locations that remain fixed relative to the molecule of protein.

One of the more unexpected observations has been the discovery of molecules of water buried in the interior of proteins with no direct contact with the solvent. These occur as single forlorn molecules (Figures 6–19 and 6–33A) or as small clusters of two or more molecules (Figure 6–33B) surrounded by donors and acceptors of hydrogen bonds from the protein itself. In the crystallographic molecular model of penicillopepsin ($n_{aa} = 323$), 10 molecules of water making no contact with the solvent have been located in the interior of the protein,[30] three as a triplet and seven as singlets; in the model of α-lytic protease ($n_{aa} = 198$), nine water molecules making no contact with the solvent have been located, three as a triplet, four as two doublets, and two as singlets.[29] In the molecular model of deoxyribonuclease I ($n_{aa} = 257$), five

Figure 6–34: Molecules of water in a deep fissure in the surface of the crystallographic molecular model of deoxyribonuclease I.[5] The map of difference electron density is for $(2|F_o| - |F_c|)$ with α_c. The peaks of positive electron density assigned to molecules of water are labeled with asterisks. The fissure is formed from several loops of the polypeptide in which the amino acids are numbered by the sequence of the protein. The molecular skeleton is the refined molecular model itself. Because the fissure is wide and in contact with the solvent over much of its boundary, most of the hydrogen bonds are between molecules of water, but the entire cluster is held in place by hydrogen bonds to donors and acceptors on the protein that surrounds it. Reprinted with permission from ref 5. Copyright 1986 Academic Press.

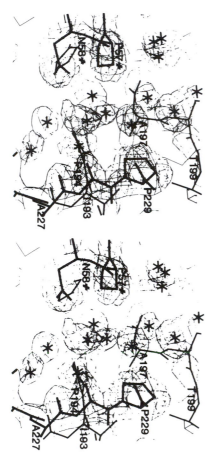

molecules of water are buried out of contact with the solvent.[5]

Water molecules are also buried in cracks and fissures connected to the solvent. For example, in penicillopepsin a finger of five water molecules extends from the surface into the interior (Figure 6–33B). A more common situation is for a fairly broad depression to be filled with an extended cluster of water (Figure 6–34).

The ordered water found covering the open surface of the refined molecular model (Figure 6–32) is held in its discrete locations by hydrogen bonds to donors and acceptors on the surface. For an ordered molecule of water to appear on a map of difference electron density, real molecules of water must sit at exactly the same location in a significant fraction of the 10^{14} asymmetric

units in a crystal. This pinning is accomplished by particular asparagines, glutamines, lysines, aspartates, arginines, and glutamates distributed over the surface of the individual molecules of protein. These donors and acceptors hold the molecules of water in extended, fixed hydrogen-bonded networks. It is the donors and acceptors on the protein that anchor these networks in the unit cell. At their peripheries the networks contain water molecules attached only to other water molecules, and at the far edges the networks fade into the disorder of the bulk solvent in the crystal. The number of water molecules that are included as individuals in the refined molecular model is probably a function more of the resolution of the data set, the peculiarities of the refinement, and the subjective decisions of the crystallographer than of any real distinction between ordered and disordered water in the actual crystal, if any such distinction can ever be made.

Networks resembling clathrates are rarely[81] found around hydrophobic functional groups on the surface of refined molecular models.[4] If they are present in the crystal but are not pinned to the same locations in each unit cell or if they are rearranging continuously while the data set is being gathered, they would not be seen in the refined molecular model. If the charged functionalities on the surface of a protein are surrounded by spheres or semispheres of hydration, as the paradigm associated with the hydration of spherical ions suggests (Figure 5–10), the molecules of water in these shells of hydration are not pinned, because no indication of their presence is seen in the maps of difference electron density. These points reemphasize the fact that only specific locations occupied by molecules of water for long periods of time appear as distinct features in maps of electron density. In a typical liquid, locations shift and rearrange at least 10^{10} times faster than they should to appear stationary in crystallography.

Because the number of water molecules included in a refined crystallographic molecular model is arbitrary, only those attached directly to the molecule of protein should display characteristic properties. There are several features of such molecules of water. About two-thirds make only one hydrogen bond to the protein[80,82] and one-third make two or more hydrogen bonds. The mean number of hydrogen bonds between these molecules of water and the protein is 1.7.[30] The average distance between water and a donor nitrogen is 0.29 ± 0.02 nm and between water and an acceptor oxygen is 0.29 ± 0.02 nm.[30,80] Of the water molecules directly bound to protein, 42% act as donors to acyl oxygens of the polypeptide backbone, 16% act as acceptors from nitrogen–hydrogen bonds of the polypeptide backbone, and 42% are in hydrogen bonds to donors and acceptors on the side chains.[5,30,80] In the refined molecular model of lysozyme,[80] of the waters bound to side chains on the surface of the protein, 24% were donors to carboxylates, 13% were donors to primary amides, 13% were acceptors from primary amides, 14% were acceptors from primary alkyl ammoniums, 14% were acceptors from guanidiniums, 14% were hydrogen-bonded to alkyl hydroxyls, and 6% were hydrogen-bonded to phenolic hydroxyls.

Side chains that are disordered in a crystallographic molecular model are usually at its surface in the most accessible locations. Because they are at the surface, they are also hydrophilic and are probably even more hydrated than the ordered side chains. Any molecules of water bound to these disordered side chains are never seen in the refined molecular model. Therefore, the mean number of waters bound to each side chain is probably an underestimate of the actual values. Nevertheless, of the side chains to which bound water can be assigned in lysozyme,[80] aspartic acids have a mean of 2.0 waters; lysines, a mean of 1.8 waters; asparagines, a mean of 1.6 waters; glutamic acids, a mean of 1.5 waters; threonines, a mean of 1.2 waters; tyrosines, a mean of 1.0 waters; serines, a mean of 0.7 waters; and glutamines, a mean of 0.7 waters bound to their side chains. More than 50% of the arginines are disordered, but those that can be observed have 1.5 waters bound.

There are a number of physical measurements which register the fact that each molecule of protein in solution has water bound to it in an irregular meshwork of hydration. The molecules of water in this meshwork of hydration differ from the molecules of water in the bulk of the solution away from the molecule of protein in several of their physical properties. While examining these properties, it should be kept in mind that each of the molecules of water surrounding a molecule of protein is in a different situation (Figure 6–32) and that the relationship of

each one of these molecules of water to the molecule of the protein will be different depending on its situation. The contribution of each of these molecules to the statistical behavior that produces the value of the physical property being measured, namely, **hydration**, will be a unique function of its situation, and the physical measurement is only an average over all of these contributions.

Mathematically, this heterogeneity can be expressed as a weighted mean

$$\delta_{H_2O} = \frac{M_{H_2O}}{M_p} \sum_{i=1}^{n} w_i \qquad (6-2)$$

where δ_{H_2O} is the grams of water of hydration for every gram of protein, M_{H_2O} is the molar mass of water (18 g mol^{-1}), M_p is the molar mass of the protein, and the sum of a set of statistical weights w_i, assigned individually to each molecule of water i, is taken over all n water molecules in the vicinity of the protein. The statistical weight w_i expresses the degree of influence the molecule of protein has over the behavior of water molecule i. When $w_i = 1$, the water molecule i is fixed, as if covalently, to the molecule of protein. When $w_i = 0$, the water molecule i is in the bulk solvent uninfluenced in its behavior by the presence of the molecule of protein.

In all cases, the measurements yield a simple number, δ_{H_2O}, the grams of water bound for every gram of protein (Table 6–3). It is not surprising that this number varies with the method used to obtain it, as more or less of the molecules of water surrounding the protein differ more or less from the water in the bulk solvent in the particular

Table 6–3: Hydration of Proteins[a]

protein	self-diffusion[b] of H$_2$O^{18}	dielectric relaxation[c]	solid at RH = 90%[d]	excluded volume sugar[e]	excluded volume (NH$_4$)$_2$SO$_4$[f]	NMR frozen solution[g]	frictional coefficient diffusion[h]	frictional coefficient viscosity[h]
ribonuclease (c2 Kin.4)			0.35	0.18 0.46				
lysozyme			0.25	0.45			< 0.89	
myoglobin (c3 Kin.3)		0.25	0.42		0.46		< 0.35	< 0.59
chymotrypsinogen (c15 Kin.1)			0.29	0.31 0.50	0.26	0.37	< 0.52	
β-lactoglobulin			0.32 0.34	0.56	0.29		< 0.72	< 0.61
ovalbumin	0.18		0.29			0.31	< 0.45	
hemoglobin			0.37 0.30		0.31	0.45	< 0.36	< 0.69
serum albumin			0.32	0.31		0.43	< 1.07	< 0.75

[a]All units are grams of water (gram of protein)$^{-1}$. [b]Ref 83. [c]Ref 84. [d]Refs 85 and 86; RH is relative humidity. [e]Refs 87 and 88. [f]Ref 85. [g]Ref 89. [h]Ref 90.

behavior measured by the particular procedure. Under no circumstances should the layer of hydration surrounding a molecule of protein be pictured as a uniform layer clearly distinguished from the water in the bulk of the solution by some discontinuous boundary.

The hydration of molecules of protein has been measured in several ways (Table 6–3). The self-diffusion of water decreases when protein is added to the solution,[91] and this decrease can be explained if it is assumed that the water of hydration is fixed and does not participate significantly in self-diffusion. With this assumption, the amount of water bound to the protein can be calculated. The dielectric constant of a solution of protein decreases discontinuously as the frequency of the alternating electric field used to measure that dielectric constant becomes greater than the ability of the molecules of protein to reorient in response to its alterations[92] and another discontinuous decrease is observed as the frequency becomes greater than the ability of the water in the bulk solvent to reorient. Between these two extremes, there is a third dielectric relaxation, which is assigned[84] to the waters of hydration bound to the protein. These waters have dielectric relaxations 10–100-fold slower than the waters in the bulk solution. From the spectrum of these dielectric relaxations, the concentration of these relatively immobilized molecules of water and the amount of water bound to the protein can be calculated. These molecules of water, however, are not fixed to the protein, or they would be required to rotate with it, and their dielectric relaxation would be indistinguishable from that of the protein itself.

Solid powders of dry protein always have water incorporated in them and the amount of this water of hydration can be chemically determined. A more systematic approach is to equilibrate the dry powder, either as a precipitate, as a microcrystalline solid, or as visible crystals, with air of a certain relative humidity. It has been proposed that air at 90% relative humidity is the appropriate choice.[86] Below this value the powders tend to become glasses,[86] and above this value they become hygroscopic. The amount of water bound by a solid powder of a given protein at 90% relative humidity can be taken as its water of hydration.

The water of hydration surrounding a protein seems to be unable to dissolve certain solutes that are otherwise freely soluble in water. This failure to dissolve a solute can be expressed[87] as the partial derivative $(\partial g_1 / \partial g_2)_{T, \mu_1, \mu_3}$. This is the change in the grams of water in a solution, ∂g_1, when the grams of protein in the solution are changed, ∂g_2, at constant chemical potential of water, μ_1, and constant chemical potential of the excluded solute, μ_3. If $(\partial g_1 / \partial g_2)_{T, \mu_1, \mu_3}$ is a positive number, it means that water must be added to the solution in addition to the protein to maintain constant chemical potential of water and solute. The extra water is assumed to be water of hydration. To perform these measurements a solution of the protein is usually brought into equilibrium by dialysis

with a solution containing only water and the other solute.[87] This is the procedure adopted when glucose,[87] lactose,[87] and sucrose[88] are used as excluded solutes (Table 6–3). It turns out that ammonium sulfate is also one of these excluded solutes. Equilibrium can be achieved in this case by adding crystals of protein to a solution of ammonium sulfate in which they will not dissolve and allowing the solution of ammonium sulfate to equilibrate with the free water within the crystals.[85] The water in the crystals that excludes ammonium sulfate can be determined by difference.

When a solution of protein is frozen, the water of hydration freezes below the freezing point of the water in the bulk solution. For example, at –3 °C, 0.51 g of water (g of protein)$^{-1}$; at –5 °C, 0.46 g of water (g of protein)$^{-1}$; and at –7 °C, 0.41 g of water (g of protein)$^{-1}$ remained unfrozen in a solution of ovalbumin.[93] Unfrozen water is more mobile than frozen water, and the two can be distinguished by nuclear magnetic resonance.[89] The amount of unfrozen water in a frozen solution of protein at –35 °C has been designated as water of hydration.

Upper limits on the amount of water that migrates with a molecule of protein through the solution can be calculated from the frictional coefficient.[90] The radius of a hard sphere the same volume as a molecule of protein can be calculated from its molar mass and partial specific volume, and the radius of the sphere that would have the same frictional coefficient as the molecule of protein can be calculated with Equation 1–57. The latter sphere is always larger than the former. If it is assumed that the entire difference in volume is water forced to move with the molecule of protein, an upper limit to the amount of bound water can be calculated (Table 6–3). It is an upper limit because molecules of protein are not spheres and a particle with the same volume as a given sphere but a different shape will always have a larger frictional coefficient than that sphere. How much of the difference between the two radii is due to hydration and how much to differences in shape has never been ascertained unambiguously for any protein. The numbers tabulated are not intended to be estimates of hydration, but upper limits of the hydration.

There are several remarkable features of this tabulation (Table 6–3). The values for bound water are all similar, and each technique produces values that, although they do not agree, are in the same range (0.2–0.4 g g^{-1}), which is about 2 mol of water (mol of amino acids)$^{-1}$. There seems to be no significant difference in the amount of bound water for every gram of protein over a 5-fold range in size of the proteins, between ribonuclease ($n_{aa} = 124$) and serum albumin ($n_{aa} = 581$). For a small protein such as lysozyme ($n_{aa} = 129$) or dihydrofolate reductase ($n_{aa} = 162$), these results indicate that there should be 200–300 molecules of bound water for every molecule of protein. In a crystal of lysozyme, 140 molecules of water had locations that were sufficiently distinct to be incorporated into the refined molecular model.[80] In a crystal of dihydrofolate reductase, 264 molecules of ordered water had sufficiently distinct locations to be incorporated in the refined molecular model.[4] Whether these ordered molecules of water bear any relation to the bound water detected by the physical measurements is uncertain.

There is a highly significant correlation between the accessible surface area of a crystallographic molecular model of a protein and the total number of amino acids it contains, regardless of whether it is a monomer or an oligomer.[94] As a result of this correlation, the mean accessible surface area for each amino acid falls gradually and monotonically from 0.53 nm^2 (amino acid)$^{-1}$ when the protein contains 100 amino acids to 0.30 nm^2 (amino acid)$^{-1}$ when the protein contains 2000 amino acids. From the definition of accessible surface area (Figure 5–20) it follows that a molecule of water, held by hydrogen bonds at 0.28 nm from its nearest neighbors, can cover about 0.07 nm^2 of accessible surface area if waters are assumed to pack in hexagonal array or 0.09 nm^2 if they are in a tetrahedral lattice (Figure 5–2). This means that there are about 7 waters (amino acid)$^{-1}$ immediately adjacent to the surface of a protein containing 100 amino acids and 4 waters (amino acid)$^{-1}$ immediately adjacent to the surface of a protein containing 2000 amino acids. These limits would be equivalent to 1.2 and 0.7 g of water (g of protein)$^{-1}$, respectively. For the proteins gathered in Table 6–3, which all contain less than 600 amino acids, the lower limit on the amount of immediately adjacent water would be 0.9 g of water (g of protein)$^{-1}$. Therefore, the water of hydration determined by physical measurements is considerably less than the amount of water required to cover the surface of a molecule of protein with a continuous layer of water.

This fact is entirely consistent with the heterogeneity that must exist among the waters of hydration. Some waters at the surface are held tightly ($w_i \cong 1.0$), but most are only loosely influenced by the protein ($w_i < 1$) and contribute only partially to the weighted mean (Equation 6–2). Therefore it is not surprising that the weighted mean is less than the limit calculated by simply counting every immediately adjacent molecule of water and presuming it to be fixed to the protein. Also, the range over which the amount of immediately adjacent molecules of water (0.9–1.2 g g^{-1}) varies among the proteins of the size of those contributing to Table 6–3 is narrow; this explains why all of the proteins seem to have about the same degree of hydration, within the variation of the measurements.

Suggested Reading

Blake, C.C.F., Pulford, W.C.A., & Artymiuk, P.J. (1983) X-ray Studies of Water in Crystals of Lysozyme, *J. Mol. Biol. 167*, 693–723.

Problem 6–2: Assign the hydrogens to donors and acceptors in Figure 6–33B.

Ionic Interactions and Hydrogen Bonds

Because of the strong hydration of ions, or because of the high dielectric constant of liquid water, or because of both of these factors, it is unlikely that ion pairs between monovalent anions such as formates or acetates and monovalent cations such as alkyl ammoniums, imidazoliums, or guanidiniums have negative free energies of formation in aqueous solution. No evidence for the existence of significant concentrations of such ion pairs between monovalent anions and cations of simple solutes has been presented. This observation means that during the folding of a protein, the formation of an interface, or the binding of a small ligand, the formation of ion pairs cannot provide a favorable contribution to the change in free energy of the reaction. Nevertheless, hydrogen bonds between aspartates or glutamates and lysines or arginines are quite common in refined molecular models of proteins. In keeping with the extreme hydrophilicity of these four amino acids (Table 6–2), these hydrogen bonds are almost always found on the surface of the protein, near or in contact with the water and presumably in a region of high dielectric constant. Therefore, they should be ionized (Equation 5–21). Presumably, this could be verified by neutron diffraction.[95]

In refined molecular models of proteins, the association between oppositely charged amino acids usually assumes the structure of a hydrogen bond. For example, one of the more common ionized hydrogen bonds is that between a glutamate or an aspartate and an arginine. Occasionally this will be a complex held together by two complementary hydrogen bonds[15,29,50]

6–6

Often the σ planes of the carboxylate and the guanidinium are inclined at an angle with respect to each other in such a way that the intersection of these two planes is a line passing through the two lone pairs on the oxygens. In the complex (c15 Kin.3) between trypsinogen and the Arginine 15 analogue of pancreatic trypsin inhibitor,[96] Arginine 15 of the inhibitor is situated adjacent to Aspartate 189 of the trypsin. The plane of the arginine is inclined at an angle of about 50° to the plane of the aspartate (Figure 6–35). The two nitrogen–hydrogen bonds of the arginine point to the positions in space where the complementary lone pairs on the oxygens would be located (Structure **6–6**), rather than at the oxygens themselves. Both nitrogens are within 0.32 nm of the respective oxygens.

The frequency with which ionized hydrogen bonds are observed in refined molecular models (Table 6–4)[5,10,29,30] is no greater than the probability that they would occur at random. Only hydrogen bonds between

Table 6–4: Frequency of Hydrogen Bonds Between Side Chains

hydrogen bond	number[a]	probability[b] (%)	hydrogen bond	number	probability[b] (%)
(guanidinium···amide)	2	5	(amide···amide)	4	3
(guanidinium···carboxylate)	10	13	(amide···hydroxyl)	9	6
(guanidinium···hydroxyl)	6	9	(amide···carboxyl)	7	6
(ammonium···amide)	0	4	(hydroxyl···hydroxyl)	15	5
(ammonium···carboxylate)	5	11	(imidazole HN···N HX)	3	2
(ammonium···hydroxyl)	0	8	(imidazole N···NH X)	4	2
(carboxylate···amide)	12	8	(carboxyl···hydroxyl/amide)	1	2
(carboxylate···carboxyl)	3	4			
(carboxylate···hydroxyl)	17	13			

[a]From tables in refs 5, 10, 29, and 30. [b]Probability that the hydrogen bond would occur at random calculated only from frequencies of functional groups[35] in proteins and their respective number of donors or acceptors, assuming no preferences for type of hydrogen bond.

two side chains are considered in the tabulation, and the probability that a certain hydrogen bond will form at random is calculated from the frequency with which the amino acids occur in the usual protein[35] and the number of equivalent donors or equivalent acceptors present on each amino acid. If anything, ionized hydrogen bonds are observed less frequently than predicted by this calculation of probability. This may be due to the fact that both charged donors and charged acceptors will tend to be more exposed to the solvent and less likely to form hydrogen bonds. A reciprocal argument could be invoked to explain the fact that hydrogen bonds between hydroxyl groups are more frequent than expected (Table 6–4), because amino acids bearing hydroxyl groups are often buried (Table 6–2). Nevertheless, with few exceptions, the frequencies with which each of the particular hydrogen bonds are observed are about those expected from the probability that the respective donor and acceptor would encounter each other at random, regardless of charge. An example of the interchangeability of charged and uncharged donors and acceptors of hydrogen bonds occurs in phycobiliproteins where a hydrogen bond between arginine and aspartic acid in one species is replaced by an isosteric hydrogen bond between two glutamines in another[97]

$$(6\text{--}3)$$

This, however, may not be very common because the amino acids surrounding a charged hydrogen bond may have been selected for their ability to solvate the charges and such a tailored environment might resist the neutralization of the bond.[98]

There is one serious uncertainty about the hydrogen bonds between a donor and an acceptor, both of which are on side chains. It is the designation of such hydrogen bonds in refined molecular models that forms the basis for the frequencies listed in Table 6–4. The problem is that these designations change significantly as the resolution of the data set increases or the refinement of the molecular model progresses because small changes in the dihedral angles within the side chains can move potential donors and acceptors within or beyond the distances designated as the only definition of a hydrogen bond. For example, a set of 12 hydrogen bonds on the surface of myoglobin between pairs of amino acid side chains in which both of the partners have been conserved by natural selection throughout all myoglobin sequences had been identified in a refined molecular model of the protein.[15] When the resolution of the data set was increased and the refinement significantly improved,[99] seven of these hydrogen bonds, four of which had been between oppositely charged side chains, were

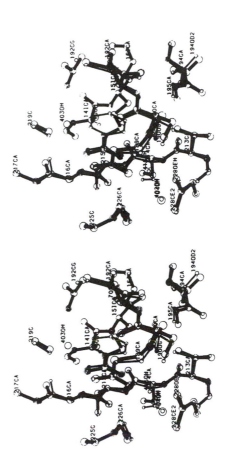

Figure 6–35: Hydrogen bond between Aspartate 189 of trypsinogen and Arginine 15 of a derivative of pancreatic trypsin inhibitor in which arginine has been incorporated at position 15 in the amino acid sequence in place of lysine. The drawing is of a crystallographic molecular model (c15 Kin.3) of the tight complex between these two molecules of protein.[96] The entire side chains of Aspartate 189, Serine 190, Cysteine 191, Aspartate 194, Valine 213, and Serine 214 are presented, as well as fragments of the side chains of Glutamine 192, Tryptophan 214, and Tyrosine 228 of trypsinogen. The letters CA, CB, CG, OG, CD, OD, CE, and OE refer to carbons or oxygens in the α, β, γ, δ, or ε position on the side chains. The peptide backbone for positions 189–192, 194, 195, and 213–217 of trypsinogen is also displayed. Arginine 15 and a portion of the surrounding peptide backbone for the porcine trypsin inhibitor is also presented with the letter I used to designate this segment. The molecular model of the complex between the porcine trypsin inhibitor and trypsinogen (bold lines) is superposed on the crystallographic molecular model of the complex between benzamidine and trypsinogen (thin lines) to show that the hydrogen bond to Aspartate 189 is identical in the two complexes. This identity validates the designation of this interaction as a hydrogen bond. Reprinted with permission from ref 96. Copyright 1984 Springer-Verlag.

Figure 6–36: Hydrogen-bonded network involving Arginine 138 and Aspartate 194 in the crystallographic molecular model of α-lytic protease.[29] Two strands of polypeptide from the interior of the protein are displayed as well as a buried lone internal molecule of water (O2). The interaction between the two oppositely charged residues, the arginine and the aspartate, is mediated by hydrogen bonds to Threonine 143. Reprinted with permission from ref 29. Copyright 1985 Academic Press.

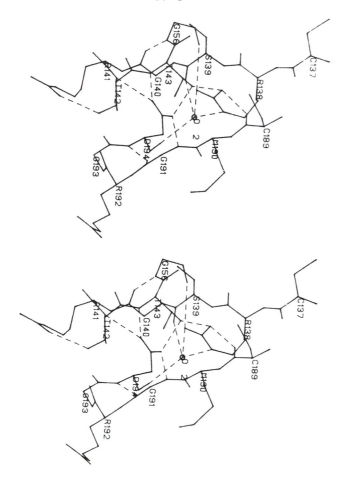

cient. For example, a positive amino acid and a negative amino acid will not be directly hydrogen-bonded to each other but linked through an intermediate, as are Arginine 138 and Aspartate 194 in α-lytic protease (Figure 6–36).[29] As if to illustrate the irrelevance of placing positive next to negative, Arginine 8P and Arginine 308 in pepsinogen, albeit each hydrogen-bonded to carboxylates, are nevertheless stacked on top of each other, their π molecular orbital systems parallel to each other, within a buried hydrogen-bonded cluster.[100]

Ribonuclease offers an example of a situation in which an ionized hydrogen bond thought to contribute favorably to the stability of a folded polypeptide turned out to be irrelevant. When six out of the first 15 amino acids in ribonuclease S (c2 Kin.4) were replaced by alanines, the polypeptide could still adopt its normal structure, and the folded polypeptide displayed enzymatic activity whose rate was 35% that of the unaltered protein. The ionized hydrogen bond between Glutamate 2 and Arginine 10, seen in the crystallographic molecular model of the normal protein, was not present in the crystallographic molecular model of the altered protein because Glutamate 2 had swung away from Arginine 10.[101]

One of the unexpected observations resulting from an examination of refined molecular models is the high frequency with which hydrogen bonds between donors and acceptors, each from the protein itself, occur on the surface of the folded polypeptide.[58] It had been thought that the competition from water would prevent such intramolecular hydrogen bonds from forming. It is possible that the donors and acceptors are confined sufficiently that the expenditure of entropy of approximation overcomes competition from water. It is also possible that these hydrogen bonds are simply the random result of the participation of all of the donors and acceptors on the surface of the molecule of protein in the hydrogen-bonded network of the water surrounding it (Figure 6–32). In this case, these hydrogen bonds would be the only fortuitous outcome of the fact that the positions of these donors and acceptors in the larger lattice happen to be adjacent to each other. This hydrogen-bonded network of waters and donors and acceptors from the protein itself should be a rather fluid structure. The refined molecular model probably represents only the structure of lowest energy in a constantly fluctuating environment.

Regardless of the reason for their existence, the formation of these hydrogen bonds on the surface of the protein is subsequent to and dependent upon the folding of the polypeptide, and their formation probably does not provide significant net negative free energy to this process. It has been proposed, however, that the formation of these networks of hydrogen bonds on the surface could decrease the rate at which the protein unfolds and add to its kinetic thermal stability.[58]

Most of the amino acids the side chains of which contain the acid–bases in a molecule of protein are found at its outer surface in contact with the aqueous phase, as

no longer present in the crystallographic molecular model (C. Chothia and A.M. Lesk, personal communication). This result suggests that all assignments of hydrogen bonds between two amino acid side chains on the surface of a protein should be regarded with skepticism.

One of the arguments in favor of ion pairs is that they conveniently neutralize charge by placing negative next to positive. Unfortunately, molecules of protein are the product of evolution by natural selection; and, because ion pairs provide no stability, convenience is of little consequence. Ironically, ion pairs seem to be more frequent on the surface of a protein than in its interior, where neutralization would be more important. Buried, charged amino acid side chains usually end up in hydrogen-bonded clusters (Figure 6–20), and the neutralization of charge in such confined regions, if it occurs at all, always seems to be opportunistic rather than effi-

can be demonstrated by measuring their accessible surface area.[102] For the major acid–bases in a protein whose dissociation constants are in the accessible range of pH—the glutamic acids, aspartic acids, histidines, tyrosines, and lysines—the value for the solvent accessibility of their heteroatoms is usually greater than 0.3.[103–105] These acid–bases in contact with the solvent participate in acid dissociations.

The acid–base titration curve of a native protein (Figure 1–11) must be the summation of the individual titrations of the accessible acid–bases on its surface. For every 100 amino acids, a normal protein contains about five aspartic acids, six glutamic acids, two histidines, three tyrosines, and six lysines.[35] The carboxylic acids of the aspartic acids (pK_a = 4.0) and glutamic acids (pK_a = 4.4) account for most of the dissociation of protons between pH 2 and 5.5. The lysines (pK_a = 10.4) and tyrosines (pK_a = 9.8) account for most of the dissociation of protons between pH 8 and 12. These are the two major features of the titration curve of a protein because these two groups of amino acids account for the majority (90%) of the acid–bases present in the protein. The histidines account for most of the small amount of dissociation that occurs between pH 5.5 and 8.

As the pH is decreased below the isoelectric point, the protein gains net positive charge as each proton associates, and as the pH is increased above the isoelectric point, the protein gains net negative charge as each proton dissociates. This change of net charge with decreases and increases of pH causes the addition of each successive proton or the removal of each successive proton, respectively, to be more difficult. The reason for this is that the gathering of net charge on a molecule, even one as large as a protein, is an unfavorable reaction relative to dispersing those charges evenly throughout the solution.

That the electrostatic work of creating this charge influences the observed titration curve is easily demonstrated by changing the ionic strength (Figure 1–11). An increase in ionic strength shrinks the counterionic boundary layer around each individual, charged amino acid in the protein (Equation 1–62) and causes them to exert a decreased effective electrostatic charge in their influence on neighboring acid–bases undergoing titration. This in turn decreases the electrostatic work that must be done to create charge on the neighboring acid–bases and shifts the titration curve closer to the curve that would have been seen if each acid–base titrated only according to its intrinsic pK_a. This electrostatic shielding due to increased ionic strength produces a steepening of the titration curve for the protein both below and above its isoelectric point (Figure 1–11).

It is possible to correct roughly[106] for the electrostatic work involved in creating charge on the molecule by assuming that in a given region of the titration curve, for example, between pH 2 and 5.5, only one type of acid–base is titrating and all of the members of this set have the

same intrinsic pK_a, pK_a^{int}. Then it is assumed that the charge on the molecule of protein, Q_i, is proportional to the mean proton charge, $^H\bar{Z}_i$, and that the intrinsic pK_a^{int}, which is proportional to a free energy, is shifted arithmetically by the electrostatic work, which is a free energy and which should be proportional to $^H\bar{Z}_i$. In short, it should be the case that

$$pK_a^{obs} = pK_a^{int} + w^H\bar{Z}_i \qquad (6\text{–}4)$$

where pK_a^{obs} is the pK_a displayed by the protein in this region and w is simply an empirical constant of proportionality.[24] At any point on a titration curve

$$pK_a^{obs} = -\log\frac{f_A[H^+]}{f_{HA}} \qquad (6\text{–}5)$$

where f_A and f_{HA} are the fractions of the acid–base that are in the form of the conjugate base and the conjugate acid, respectively. After the combination of Equations 6–4 and 6–5

$$pH - \log\frac{f_A}{f_{HA}} = pK_a^{int} + w^H\bar{Z}_i \qquad (6\text{–}6)$$

As f_A and f_{HA} can be read directly from any point on the titration curve and $^H\bar{Z}_i$ can be directly calculated for the same point on the titration curve, an analytical plot of Equation 6–6 provides an estimate of pK_a^{int} for any region of the titration curve (Figure 6–37). Only w should be a function of the ionic strength as it changes the widths of the boundary layers around the various charges. It follows that the lines defined by Equation 6–6 for titration curves measured at different ionic strengths should intersect at $^H\bar{Z}_i = 0$, and they often do (Figure 6–37).

The values of the intrinsic acid dissociation constants obtained by these corrections for electrostatic work agree with expectation (Table 2–2) to a certain extent. The value for pK_a^{int} of the carboxyl groups in several proteins whose titration curves between pH 2 and 5.5 have been analyzed in this way[106] are between 4.0 and 4.8. The titration of tyrosine residues in a native protein can be followed independently by using the large difference in ultraviolet absorbance between the phenol and the phenolate anion to calculate f_A and f_{HA}.[107] The values of pK_a^{int} corrected for electrostatic work, for tyrosines in several proteins[7] are between 9.4 and 10.8. The contribution of tyrosine to the titration curve between pH 8 and 12 can then be deducted from the overall curve, and values for pK_a^{int} for the lysines in these same proteins can be calculated. They lie between 9.8 and 10.4.

The titration curves of proteins usually fail to meet expectations in one key aspect. There are usually too few acid–bases contributing to the titration.[106] The deficit is most easily noticed in the case of histidine and tyrosine.

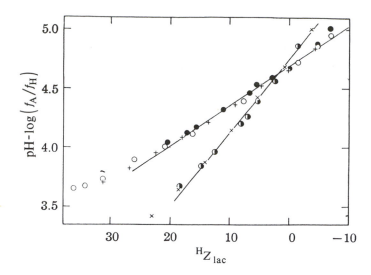

Figure 6–37: Data from an acid–base titration of β-lactoglobulin[108] presented in the form of Equation 6–6. A deionized solution of β-lactoglobulin was adjusted to the desired ionic strength by adding solid KCl. The titration curve of the protein was then determined by following the pH of the solution as a function of added H^+ or OH^-. From these observations, $^H\bar{Z}_{lac}$ as a function of pH was calculated. The function $pH - \log(f_A/f_{HA})$ could be calculated by treating the titration curve observed between pH 1 and 7 as only the titration of the 51 carboxyl groups of β-lactoglobulin. The change in $^H\bar{Z}_{lac}$ in this range was 50 mol (mol of protein)$^{-1}$, as expected. From the value at any point on the curve, a value for f_A/f_{HA} could be calculated. Values of $pH - \log(f_A/f_{HA})$ could then be plotted as a function of $^H\bar{Z}_{lac}$. Reprinted with permission from ref 108. Copyright 1959 American Chemical Society.

The number of moles of protons dissociating from a mole of protein between pH 5.5 and 8 is often less than the moles of histidine in a mole of that protein. This deficit can be explained by assuming that the acid dissociation constants of one or more histidines have been shifted to higher values and their titrations have become buried in those of the large number of carboxylates. The moles of tyrosine whose ultraviolet absorption displays the expected shift between pH 9 and 11 upon formation of the phenolate anion is often less than the total moles of tyrosine present in a mole of the protein. For example, only four of the six tyrosines in ribonuclease can be titrated[106,109] and only two of the four tyrosine residues of chymotrypsinogen can be titrated[106] within accessible ranges of pH. With most proteins, values of pH greater than 11 are inaccessible, so it can be said only that each of these tyrosines has a $pK_a > 11$.

Both the decreases in the pK_a of histidines and the increases in the pK_a of tyrosines implied or demonstrated by these results are reasonable. If these shifts in pK_a are due to burying the residues in the interior of the protein, even though they remain accessible to the solvent and capable of acid–base reactions, their neutral forms should become more stable relative to their charged forms. In most cases, the missing acid–bases in a titration curve are assumed to be buried in the interior of the folded polypeptide. Such buried acid–bases can be seen in the crystallographic molecular models of proteins. For example, in the crystallographic molecular model (c2 Kin.4) of ribonuclease, Tyrosine 25 is almost completely buried (the solvent accessibility of its phenolic oxygen is only 0.02) and Tyrosine 97 is completely buried.[103] It has been assumed that these are the two tyrosines in the native protein that do not participate in acid–base titrations.

The pK_a of a particular amino acid on the surface of a particular protein will be determined by the local electrostatic potential. Now that refined molecular models of proteins have become available, attempts have been made to calculate the electrostatic potential on the sur-

face and within the interior of the actual molecule of protein. The original approach to this problem was to treat the protein as a sphere of uniform dielectric and to distribute charged residues over its surface with a spacing dictated by the molecular model of the protein.[110] The electrostatic potential at any specific position on the surface was determined by summing the electrostatic potentials exerted at that point by each of the fixed positive and negative charges.

The original purpose of this calculation was to assess the effect of adjacent charged amino acids on the acid–base titration of a given amino acid. At a given pH the amount of charge on any acid–base in a protein will be determined by its intrinsic acid dissociation constant (Table 2–2), the electrostatic potential in its vicinity, and the effective dielectric constant in its vicinity. The electrostatic potential in its vicinity partly determines the electrostatic work required to charge an acid–base or discharge it during its acid–base reaction. For example, if the amino acid in question were glutamic acid and the electrostatic potential in its vicinity were positive because it happened to be surrounded by two or three lysines or arginines, the electrostatic work required to remove a proton form this glutamic acid to produce the carboxylate anion would be less than if the electrostatic potential were zero, and its apparent pK_a would be lower than its intrinsic pK_a.

The observed pK_a for an acid–base is determined by the electrostatic potential, but the electrostatic potential is determined by the degree of ionization of every other acid–base in the protein. As the pH is changed to titrate the acid–base of interest, the degrees of ionization of other charged residues also change. As all of the ionized acid–bases mutually interact, apparent acid dissociation constants are calculated by an interactive procedure repeated until a self-consistent set of theoretical values is obtained.[110]

When the model of a sphere of uniform dielectric constant having fixed charges placed at the locations occupied by the acid–bases in a molecular model was used to calculate titration curves, it was found that shifts in the values of the acid dissociation constants from their intrinsic values large enough to duplicate the observed values for the acid dissociation constants could be obtained only when the charges were moved into the sphere, away from its surface,[111] and the dielectric constant of the sphere was adjusted arbitrarily.[110] Because the sphere has a low dielectric constant and the surrounding solvent a high dielectric constant, the deeper into the sphere the various charges are placed, the greater will the electrostatic forces between them become, and the greater will be the mutual effects on their separate acid–base behaviors. The arbitrary nature of these adjustments caused concern.[110]

It has also been proposed that, rather than lowering every acid–base the same arbitrary depth into the interior of the sphere of uniform dielectric constant, the experimentally observed accessible surface area of each particular acid–base in the actual protein could be used as a direct measure of its degree of burial.[103] The effect of these individual burials is applied by arbitrarily multiplying the electrostatic interaction, W_{ij}, between the amino acid of interest, amino acid i, and any other amino acid j by the factor $(1 - \overline{SA}_{ij})$, where $\overline{SA}_{ij}$ is the average accessible surface area of amino acids i and j. The electrostatic interaction, W_{ij}, is the electrostatic work required to charge amino acid i in the presence of charged amino acid j. It is the electrostatic interactions, W_{ij}, that are summed over all j to determine the electrostatic potential at amino acid i. The effect of the term $(1 - \overline{SA}_{ij})$ is to render the electrostatic interaction negligible when both amino acids, i and j, are completely surrounded by water and at full force when both are completely buried in the protein. The acid dissociation constants for histidines in hemoglobin,[104] myoglobin (c3 Kin.3),[105] and ribonuclease (c2 Kin.4)[103] have been calculated in this way and shown to agree satisfactorily with the observed values for their individual acid dissociation constants. Complete acid–base titration curves have also been calculated for hemoglobin[104] and myoglobin[105] in this way and shown to agree with the observed acid–base titration curves.

An alternative treatment of this problem[112,113] is to create a solid that has the same irregular shape as the crystallographic molecular model of a protein. The dielectric constant within this solid is given a low value ($\varepsilon < 10$) and the dielectric outside the solid a high value ($\varepsilon = 80$). The Poisson–Boltzmann equation can be used to account for the mobile ions in the surrounding solvent and thus provides the ability to calculate electrostatic potentials at any ionic strength. Appropriate unit charges are placed where formally charged atoms occur in the crystallographic molecular model. From this model, the electrostatic potential at any point can be calculated. This approach has been used to reproduce the effect of the site-directed mutation of a particular charged amino

acid on the pK_a of another amino acid in the same protein. For example, when Aspartate 99 in subtilisin BPN (c3 Kin.4) is mutated to serine, the pK_a of Histidine 64 shifts 0.40 unit downfield when it is measured at an ionic strength of 0.010 M. The shift calculated with the model just described is 0.31 unit. In all the examples of site-directed mutation for which this model was applied, however, the mutations were usually of fully exposed residues, so the majority of the electrostatic effect was carried through the aqueous phase. For this reason, the variation of the dielectric constant chosen for the interior of the protein, and probably the variation of the shape of the protein as well, had little effect on the shifts of pK_a calculated. It has also been pointed out that it is much easier to calculate differences in electrostatic field resulting from the elimination of a single charge than it is to calculate the absolute value of the electrostatic potential produced by a constellation of charges in a nonuniform dielectric.[98]

One of the more difficult problems that remains unsolved is that of the dielectric constant that should be assigned to the interior of a molecule of protein.[114] In trypsin (c15 Kin.2), Histidine 57 is hydrogen bonded to Aspartate 102.[115] Aspartate 102 is completely buried and Histidine 57 is buried by greater than 50%.[9] The half of Histidine 57 that is buried is the half that is hydrogen-bonded to Aspartate 102. It has been demonstrated by neutron diffraction that Histidine 57 is the imidazolium and Aspartate 102 is the carboxylate, and the hydrogen bond is ionic. This observation is mitigated somewhat by the fact that Histidine 57 is also hydrogen-bonded to another anion, but it suggests that the local dielectric constant experienced by these particular buried groups is significantly greater than 10. This possibility is not unreasonable for a region within a polar solid in the immediate vicinity of the bulk aqueous phase. It is quite likely, however, that the dielectric constant fluctuates significantly throughout the interior of a molecule of protein, making a homogeneous treatment unrealistic.[114]

A different approach to the estimation of electrostatic potential is to calculate its magnitude at a specific point in a molecule of protein or in its surroundings by a complete electrostatic treatment.[116] Mulliken atomic charges are assigned to every atom in a refined molecular model of the protein. The sum of the electrostatic potentials of all of these atomic charges exerted at the point of interest can be performed with a digital computer. Three-dimensional electrostatic fields can be calculated in this way. Whether or not these electric fields are consistent with any observed property of the real protein in a real solution is not yet known. As in all calculations of this type, the electrostatic potential assigned to a particular point in space is a small difference between many large numbers, none of whose magnitudes are known with remarkable precision.

There are several reasons for wishing to know the actual distribution of electrostatic potential within a molecule of protein, upon its surface, and in its vicinity. The acid dissociation constants of particular amino acids could be estimated.[111] The favorable or unfavorable effect of electrostatic potential on the binding of small, charged solutes could be ascertained.[103] The electrostatic forces exerted on other molecules, proteins or solutes, in the solution surrounding the particular molecule of protein could be assessed.[116] In the few imperfect calculations made so far, however, electrostatic forces have turned out to be quite small, usually less than 10 kJ mol^{-1}, whenever the fact that a real molecule of protein is dissolved in an aqueous solution of ionic strength 0.1 M is taken into consideration. An aqueous solution of 0.1 M ionic strength dissipates an electrostatic potential with remarkable efficiency.

Suggested Reading

Takano, T. (1977) Structure of Myoglobin Refined at 2.0 Å Resolution. II Structure of Deoxymyoglobin from Sperm Whale, *J. Mol. Biol. 110*, 537–568.

References

1. Brandts, J.F., Halvorson, H.R., & Brennan, M. (1975) *Biochemistry 14*, 4953–4963.
2. Richardson, J.S. (1981) *Adv. Protein Chem. 34*, 167–339.
3. Rees, D.C., Lewis, M., Honzatko, R.B., Lipscomb, W.N., & Hardman, K.D. (1981) *Proc. Natl. Acad. Sci. U.S.A. 78*, 3408–3412.
4. Bolin, J.T., Filman, D.J., Matthews, D.A., Hamlin, R.C., & Kraut, J. (1982) *J. Biol. Chem. 257*, 13650–13662.
5. Oefner, C., & Suck, D. (1986) *J. Mol. Biol. 192*, 605–632.
6. Ramachandran, G.N., Ramakrisnan, C., & Sasisekharan, V. (1963) *J. Mol. Biol. 7*, 95–99.
7. Mandel, N., Mandel, G., Trus, B., Rosenberg, J., Carlson, G., & Dickerson, R.E. (1977) *J. Biol. Chem. 252*, 4619–4636.
8. Ravichandran, V., & Subramanian, E. (1981) *Int. J. Pept. Protein Res. 1*, 121–186.
9. Birktoft, J.J., & Blow, D.M. (1972) *J. Mol. Biol. 68*, 187–240.
10. Artymiuk, P.J., & Blake, C.C.F. (1981) *J. Mol. Biol. 152*, 763–782.
11. Matthews, B.W., Weaver, L.H., & Kester, W.R. (1974) *J. Biol. Chem. 249*, 8030–8044.
12. Maigret, B., Pullmann, B., & Perahia, D. (1971) *J. Theor. Biol. 31*, 269–285.
13. Pauling, L., Corey, R.B., & Branson, H.R. (1951) *Proc. Natl. Acad. Sci. U.S.A. 37*, 205–211.
14. Watson, H.C. (1969) *Prog. Stereochem. 4*, 229–361.
15. Takano, T. (1977) *J. Mol. Biol. 110*, 537–568.
16. Gray, T.M., & Matthews, B.W. (1984) *J. Mol. Biol. 175*, 75–81.
17. Blundell, T., Barlow, D., Borkakoti, N., & Thornton, J. (1983) *Nature 306*, 281–283.
18. Remington, S., Wiegand, G., & Huber, R. (1982) *J. Mol. Biol. 158*, 111–152.
19. Hol, W.G.J., van Duijnen, P.T., & Berendsen, H.J.C. (1978) *Nature 273*, 443–446.
20. Rogers, N.K., & Sternberg, M.J.E. (1984) *J. Mol. Biol. 174*, 527–542.
21. MacLachlin, A.D., & Stewart, M. (1975) *J. Mol. Biol. 98*, 293–304.
22. Eklund, H., Nordstrom, B., Zeppezauer, E., Söderlund, G., Ohlsson, I., Boiwe, T., Söderberg, B.O., Tapia, O., Brändén, C.I., & Åkeson, Å. (1976) *J. Mol. Biol. 102*, 27–59.
23. Chothia, C. (1973) *J. Mol. Biol. 75*, 295–302.
24. Richardson, J.S., Getzoff, E.D., & Richardson, D.C. (1978) *Proc. Natl. Acad. Sci. U.S.A. 75*, 2574–2578.
25. Venkatachalam, C.M. (1968) *Biopolymers 6*, 1425–1436.
26. Taylor, H.S. (1941) *Proc. Am. Philos. Soc. 85*, 1–12.
27. Bragg, L., Kendrew, J.C., & Perutz, M.F. (1950) *Proc. R. Soc. London, A 203*, 321–357.
28. Dijkstra, B.W., Kalk, K.H., Hol, W.G.J., & Drenth, J. (1981) *J. Mol. Biol. 147*, 97–123.
29. Fujinaga, M., Delbaere, L.T.J., Brayer, G.D., & James, M.N.G. (1985) *J. Mol. Biol. 183*, 479–502.
30. James, M.N.G., & Sielecki, A.R. (1983) *J. Mol. Biol. 163*, 299–361.
31. Sibanda, B.L., & Thornton, J.M. (1985) *Nature 316*, 170–174.
32. Efimov, A. (1986) *Mol. Biol. (USSR) 20*, 250–260.
33. Chou, P.Y., & Fasman, G.D. (1977) *J. Mol. Biol. 115*, 135–175.
34. Baker, E.N., & Hubbard, R.E. (1984) *Prog. Biophys. Mol. Biol. 44*, 97–179.
35. Doolittle, R.F. (1981) *Science 214*, 149–159.
36. Chothia, C. (1976) *J. Mol. Biol. 105*, 1–14.
37. Klotz, I.M., & Farnham, S.B. (1968) *Biochemistry 7*, 3879–3882.
38. Chou, P.Y., & Fasman, G.D. (1978) *Adv. Enzymol. Relat. Areas Mol. Biol. 47*, 45–148.
39. Fasman, G.D. (1967) *Poly-α-Amino Acids: Protein Models for Conformational Studies*, Marcel Dekker, New York.
40. Lim, V.I. (1974) *J. Mol. Biol. 88*, 873–894.
41. Lim, V.I. (1974) *J. Mol. Biol. 88*, 857–872.
42. Garnier, J., Osguthorpe, D.J., & Robson, B. (1978) *J. Mol. Biol. 120*, 97–120.
43. Kabsch, W., & Sander, C. (1983) *FEBS Lett. 155*, 179–182.
44. Nishikawa, K. (1983) *Biochim. Biophys. Acta 748*, 285–289.
45. Ponder, J.W., & Richards, F.M. (1987) *J. Mol. Biol. 193*, 775–791.
46. Roberts, J.D., & Caserio, M.C. (1977) *Basic Principles of Organic Chemistry*, 2nd ed., p 457, Benjamin, Menlo Park, CA.
47. Janin, J., Wodak, S., Levitt, M., & Maigret, B. (1978) *J. Mol. Biol. 125*, 357–386.
48. Kossiakoff, A.A., & Shteyn, S. (1984) *Nature 311*, 582–583.
49. Nelsen, S.F., Teasley, M.F., Bloodworth, A.J., & Eggelte, (1985) *J. Org. Chem. 50*, 3299–3302.
50. Tainer, J.A., Getzoff, E.D., Beem, K.M., Richardson, J.S., & Richardson, D.C. (1982) *J. Mol. Biol. 160*, 181–217.
51. Lee, B.K., & Richards, F.M. (1971) *J. Mol. Biol. 55*, 379–400.
52. Janin, J. (1979) *Nature 277*, 491–492.
53. Miller, S., Janin, J., Lesk, A., & Chothia, C. (1987) *J. Mol. Biol. 196*, 641–656.
54. Chothia, C. (1974) *Nature 248*, 338–339.
55. Wolfenden, R.V., Cullis, P.M., & Southgate, C.C.F. (1979) *Science 206*, 575–577.
56. Fauchére, J.L., & Pliska, V. (1983) *Eur. J. Med. Chem. Chim. Ther. 18*, 369–375.
57. Nozaki, Y., & Tanford, C. (1971) *J. Biol. Chem. 246*, 2211–2217.
58. Wistow, G., Turnell, B., Summers, L., Slingsby, C., Moss, D., Miller, L., Lindley, P., & Blundell, T. (1983) *J. Mol. Biol. 170*, 172–202.

59. Furey, W., Wang, B.C., Yoo, C.S., & Sax, M. (1983) *J. Mol. Biol. 167*, 661–692.
60. Hine, J., & Mookerjee, P.K. (1975) *J. Org. Chem. 40*, 292–297.
61. Eisenberg, D., & McLachlin, A.D. (1986) *Nature 319*, 199–203.
62. Burley, S.K., & Petsko, G.A. (1985) *Science 229*, 23–28.
63. Jorgensen, W.L., & Severance, D.L. (1990) *J. Am. Chem. Soc. 112*, 4768–4774.
64. Greenblatt, H.M., Ryan, C.A., & James, M.N.G. (1989) *J. Mol. Biol. 205*, 201–228.
65. Klapper, M.H. (1971) *Biochim. Biophys. Acta 229*, 557–566.
66. Richards, F.M. (1974) *J. Mol. Biol. 82*, 1–14.
67. Cohn, E.J., McMeekin, T.L., Edsall, J.T., & Blanchard, M.H. (1934) *J. Am. Chem. Soc. 56*, 784–794.
68. Cohn, E.J., & Edsall, J.T. (1943) *Proteins, Amino Acids, and Peptides*, Reinhold, New York.
69. Traube, J. (1899) *Sammlung Chem. Chem. Tech. Vort. 4*, 255–331.
70. Chothia, C., Levitt, M., & Richardson, D. (1977) *Proc. Natl. Acad. Sci. U.S.A. 74*, 4130–4134.
71. Chothia, C., Levitt, M., & Richardson, D. (1981) *J. Mol. Biol. 145*, 215–250.
72. Crick, F.H.C. (1953) *Acta Crystallogr. 6*, 689–697.
73. Chothia, C., & Janin, J. (1981) *Proc. Natl. Acad. Sci. U.S.A. 78*, 4146–4150.
74. Cohen, F.E., Sternberg, M.J.E., & Taylor, W.R. (1981) *J. Mol. Biol. 148*, 253–272.
75. Chothia, C., & Janin, J. (1982) *Biochemistry 21*, 3955–3965.
76. Janin, J., & Chothia, C. (1980) *J. Mol. Biol. 143*, 95–128.
77. Smith, J.L., Hendrickson, W.A., Honzatko, R.B., & Sheriff, S. (1986) *Biochemistry 25*, 5018–5027.
78. Lesk, A.M., & Chothia, C.H. (1986) *Philos. Trans. R. Soc. London, A 317*, 345–356.
79. Matthews, B.W. (1977) in *The Proteins* (Neurath, H., & Hill, R.L., Eds.) 3rd ed., Vol. III, pp 404–590, Academic Press, New York.
80. Blake, C.C.F., Pulford, W.C.A., & Artymiuk, P.J. (1983) *J. Mol. Biol. 167*, 693–723.
81. Teeter, M.M. (1984) *Proc. Natl. Acad. Sci. U.S.A. 81*, 6014–6018.
82. Watenpaugh, K.D., Margulis, T.N., Sieker, L.C., & Jensen, L.H. (1978) *J. Mol. Biol. 122*, 175–190.
83. Fisher, H.F. (1965) *Biochim. Biophys. Acta 109*, 544–550.
84. Grant, E.H., Mitton, B.G.R., South, G.P., & Sheppard, R.J. (1974) *Biochem. J. 139*, 375–380.
85. McMeekin, T.L., Groves, M.L., & Hipp, N.J. (1954) *J. Am. Chem. Soc. 76*, 407–413.
86. Bull, H.R., & Breese, K. (1968) *Arch. Biochem. Biophys. 128*, 488–496.
87. Arakawa, T., & Timasheff, S.N. (1982) *Biochemietry 21*, 6536–6544.
88. Lee, J.C., & Timasheff, S.N. (1981) *J. Biol. Chem. 256*, 7193–7201.
89. Kuntz, I.D., Brassfield, T.S., Law, G.D., & Purcell, G.V. (1969) *Science 163*, 1329–1331.
90. Tanford, C. (1961) *The Physical Chemistry of Macromolecules*, John Wiley, New York.
91. Wang, J.H. (1954) *J. Am. Chem. Soc. 76*, 4755–4763.
92. Oncley, J.L. (1943) in *Proteins, Amino Acids, and Peptides* (Cohn, E.J., & Edsall, J.T., Eds.) pp 543–568, Reinhold, New York.
93. Bull, H.B., & Breese, K. (1968) *Arch. Bioch. Biophys. 128*, 497–502.
94. Miller, S., Lesk, A.M., Janin, J., & Chothia, C. (1987) *Nature 328*, 834–836.
95. Kossiakoff, A.A. (1982) *Nature 296*, 713–721.
96. Bode, W., Walter, J., Huber, R., Wenzel, H.R., & Tscheshe, H. (1984) *Eur. J. Biochem. 144*, 185–190.
97. Schirmer, T., Huber, R., Schneider, M., Bode, W., Miller, M., & Hackert, M.L. (1986) *J. Mol. Biol. 188*, 651–676.
98. Warshel, A. (1987) *Nature 330*, 15–16.
99. Phillips, S.E.V. (1980) *J. Mol. Biol. 142*, 531–554.
100. James, M.N.G., & Sielecki, A.R. (1986) *Nature 319*, 33–38.
101. Taylor, H.C., Komoriya, A., & Chaiken, I.M. (1985) *Proc. Natl. Acad. Sci. U.S.A. 82*, 6423–6426.
102. Lee, B.K., & Richards, F.M. (1971) *J. Mol. Biol. 55*, 379–400.
103. Matthew, J.B., & Richards, F.M. (1982) *Biochemistry 21*, 4989–4999.
104. Matthew, J.B., Hananaia, G.I.H., & Gurd, F.R.N. (1979) *Biochemistry 18*, 1919–1928.
105. Botelho, L.H., Friend, S.H., Matthews, J.B., Lehman, L.D., Hananaia, G.I.H., & Gurd, F.R.N. (1978) *Biochemistry 17*, 5197–5205.
106. Tanford, C. (1962) *Adv. Protein Chem. 17*, 69–165.
107. Crammer, J.L., & Neuberger, A. (1943) *Biochem. J. 37*, 302–310.
108. Nozaki, Y., Bunville, L.G., & Tanford, C. (1959) *J. Am. Chem. Soc. 81*, 5523–5529.
109. Lenstra, J.A., Bolscher, G.J.M., Stob, S., Beintema, J.J., & Kaptein, R. (1979) *Eur. J. Biochem. 98*, 385–397.
110. Tanford, C., & Roxby, R. (1972) *Biochemistry 11*, 2192–2198.
111. Tanford, C., & Kirkwood, J.G. (1957) *J. Am. Chem. Soc. 79*, 5333–5339.
112. Gilson, M.K., & Honig, B.H. (1987) *Nature 330*, 84–86.
113. Sternberg, M.J.E., Hayes, F.R.F., Russell, A.J., Thomas, P.G., & Fersht, A.R. (1987) *Nature 330*, 86–88.
114. Gilson, M.K., Rashin, A., Fine, R., & Honig, B. (1985) *J. Mol. Biol. 183*, 503–516.
115. Kossiakoff, A.A., & Spencer, S.A. (1981) *Biochemistry 20*, 6462–6474.
116. Getzoff, E.D., Tainer, J.SA., Weiner, P.K., Kollman, P.A., Richardson, J.S., & Richardson, D.C. (1983) *Nature 306*, 287–290.

CHAPTER 7

Evolution

The proteins that exist today are the products of evolution by natural selection. The detailed history of this evolution is both a history of change along a particular lineage and a history of the creation of new lineages. Two new species arise from one ancestral species as soon as subpopulations of that ancestral species become so different from each other that two individuals, one from each of the subpopulations, are no longer able to breed successfully. When two species diverge from one, a new lineage arises. Even when two closely related species that have only recently diverged from their common ancestor are compared, the amino acid sequences of the same respective proteins from each species will often differ at one or more positions. For example, myoglobin from domestic sheep differs in amino acid sequence at three of its 143 positions from myoglobin of domestic goats.

The reason for this divergence is that once speciation has occurred and interbreeding becomes impossible, two versions of the same protein are established. These two versions begin to evolve in isolation from each other. Mutations occur at random in the respective genes encoding each version, once in a while one of these mutations is fixed by genetic drift or natural selection independent of any fixation occurring in the same gene from the other species, and slowly the amino acid sequences of the encoded proteins become different, one position at a time. Because the geologic instant at which the two species were established from one common ancestor coincides with the instant at which the two versions of the same protein began to evolve separately, amino acid sequences retain the history of speciation. This history can be reconstructed by comparing the amino acid sequences of the same protein from an array of different species.

By comparing amino acid sequences, connections can usually be made only as far back as the common ancestors of prokaryotes and eukaryotes. What has been found, however, is that the tertiary structure of a particular protein from distantly related species, when viewed in crystallographic molecular models, changes less rapidly than its amino acid sequence during evolution by natural selection. Because of this, comparisons of crystallographic molecular models permit us to look back in evolutionary history to the time at which the individual proteins themselves were diverging from common ancestors: to the time, for example, when L-lactate dehydrogenase and glyceraldehyde-3-phosphate dehydrogenase diverged from their common ancestor. Through such comparisons, the speciation of proteins can be traced. Because amino acid sequences change more rapidly than tertiary structures, only a few of the pedigrees of proteins, those that diverged recently in geologic time, can be traced by comparing amino acid sequences. Most of our insight into the speciation of proteins has come from comparisons of tertiary structures.

From the comparisons that can be made among the limited set of tertiary structures that are now available, it has become clear that the larger proteins often, if not always, arose during evolution by the chance fusion of two genes encoding smaller polypeptides, each of which could fold independently and each of which usually had an independent function prior to the fusion. In this way, larger and larger proteins appeared. If a particular fusion produced a protein that was not impaired functionally, the new gene for the larger protein may have been fixed in the population by genetic drift; or, if the fusion produced a protein with advantageous features, the new gene for the larger protein may have been fixed in the population by natural selection.

The history of these fusions can be observed in the domains from which these larger proteins are constructed. The domains of a protein are discrete regions in the tertiary structure of that protein which arose from separate, previously independent proteins that were fused together, one after the other, to produce the present protein. Because a polypeptide shorter than about 70 amino acids usually cannot fold spontaneously to form a tertiary structure, domains, when defined in this way, are usually larger than this. They appear in the crystallographic molecular model as independently folded regions. Because they are the fundamental units in the evolution of proteins, domains must be identified by a set of conservative, objective criteria, if our description of the history of a set of proteins is to be accurate.

It may be possible, by examining crystallographic molecular models, to trace the ancestry of the proteins that presently exist, in a sense to derive a molecular phylogeny of the proteins. Because most of the existing proteins were produced by fusion of smaller units, this molecular phylogeny of the proteins must be based on a reconstruction of two processes. First, the family trees of the individual, ancestrally related domains from different proteins must be reconstructed. In almost every instance these family trees must be based on patterns in which the secondary structures are arranged to form the tertiary structures of the domains being compared because homology in amino acid sequence has been completely lost. Second, the separate events that produced each of the fusions of the independent domains to produce the larger chimeric proteins must also be reconstructed.

Although the most interesting question may be how the large array of existing proteins arose from a much smaller array of smaller proteins present in the distant past, it should be stressed that new proteins are continuously being made by this process of fusion of different pieces. Evolution by natural selection has not ceased. We know this because, in some instances, domains that have homologous amino acid sequences can be found in otherwise completely different proteins. Because homology in amino acid sequence disappears quite rapidly over geologic time, these domains must have been separately incorporated into their respective proteins fairly recently; the greater the homology of amino acid sequence among them, the more recently the separate fusions must have occurred.

Molecular Phylogeny from Amino Acid Sequence

The amino acid sequences of a set of related polypeptides retain a record of the history of their evolution by natural selection. That record provides information about the speciation of organisms and the specialization of tissues. The evolutionary history of the species that these polypeptides represent is read from aligned amino acid sequences.

As the amino acid sequences of the same protein from different species became available, it was usually found that they were similar enough to be readily aligned with each other. An **alignment** of two or more amino acid sequences is a display in which positions that are thought to be directly related to each other from the respective sequences are aligned directly above and below each other. The decision that the aligned positions are related is based on the fact that they are occupied by the same amino acid or the fact that they are surrounded by similar sequences of amino acids. An example of such an alignment of five amino acid sequences is presented in Figure 7–1. If you run your eye along the aligned sequences for the cytochrome c_{550} from the bacterium *Paracoccus denitrificans* and the cytochrome c_2 from the bacterium *Rhodospirillum rubrum*, it is quite clear that they are related to each other and that the relationships are not the result of random coincidences. As aligned, identical amino acids occupy the same positions in these two sequences 42% of the time, which is considerably more frequent than such coincidences would occur by chance.

It is often the case, especially as the relationship becomes more distant, that gaps have to be introduced into one or more of the sequences to achieve the proper alignment. A **gap** is a series of blank spaces inserted into an amino acid sequence that is missing some amino acids present in a related amino acid sequence. The two gaps that must be introduced into the sequence of the cytochrome c_2 from *R. rubrum*, at the positions of IQAPDGT and VAEK in the sequence of *P. denitrificans*, are clearly required to produce the obvious alignments that occur around them. The actual sequence of the cytochrome *c* from *R. rubrum* is HTFDQG and TEMKAK

in these two regions and, obviously, contains no gaps. From a comparison of only these two sequences, it cannot be ascertained whether the protein from *P. denitrificans* was elongated in such a location or the protein from *R. rubrum* was shortened in such a location during evolution. A decision, however, can often be made on this point when the sequences of the same protein from a number of species are compared. For example, four amino acids, DAVG, are missing from the fibrinopeptide A of the cape buffalo. These four amino acids, or their homologues, are present in the fibronopeptides A from all other artiodactyls (17 examples).[4] It can be concluded that these four amino acids were deleted during the very recent evolution that produced the cape buffalo.[4]

The conclusion that is drawn from an alignment, such as the one between the cytochromes *c* of *P. denitrificans* and *R. rubrum*, is that the two species which carry these two proteins both evolved from the same ancestral species or, in other words, that these two species **share a common ancestor**. In the distant past, when only the common ancestral species was present, all of the individuals in the population of that ancestral species contained the cytochrome *c* whose amino acid sequence was unique to that species. As natural selection operated upon the genetic variation present within the population of that ancestral species, varieties arose that were adapted to different ecological niches. These varieties eventually diverged sufficiently to become separate species. At that point, the genes for cytochrome *c* in these two new species, the ones that were to be the ancestors of *P. denitrificans* and *R. rubrum*, became disconnected, and their amino acid sequences from that time forth were altered independently and continuously by mutation and natural selection. The differences and similarities between these two present sequences are the accumulated result of the individual steps in this alteration.

Evolution by natural selection is usually viewed from its optimistic side. Natural selection operates on the variation inherent in any large population of a given species to shift the distribution of its abilities gradually in a direction that makes that species or its descendants more successful. Beneficial traits are patiently nurtured and multiplied. A major portion, if not all, of the variation upon which natural selection operates to achieve this progress is variation in the sequences of the proteins within the population of a given species.

It is unlikely, however, that more than a small number of the differences seen when two aligned amino acid sequences are compared (Figure 7–1) reflect improvements in the ability of the individuals of that species to survive relative to that of individuals of other species or their common ancestors. There is no evidence that the cytochrome *c* from either *P. denitrificans* or *R. rubrum* is an improved version of the cytochrome *c* that was used by their common ancestor or that any of the proteins whose amino acid sequences are being

```
                                    G                            G
                                    F                            D
    ESR        D          G         Y                 D          Q
    PTA        EG         A    G    A    Q            A     K    R
    SDT  RD KS      E   L     E  T TI   S VV  T      P     E    K
   SIES ENL TTQ   EA   E   GKNLG GQ    N FTN QA SVV  A    A   AA
   NAKN ATI IMR   SL   GI   DAAAPQSTA A H IYS HS TSQKFT SNGMIN
   GDVAKGKKTFVQK—CAQCHTV—————ENGGKHKVGPNLWGLFGRKTGQAEGYSYTDANKS—————
        10          20              30        40        50

   NEGDAAKGEKEF——NKCKACHMIQAPDGTDIKGGKTGPNLYGVVGRKIASEEGFKYGEGILEVAEKN
   EGDAAAGEK——VSKKCLACHTF—————DQGGANKVGPNLFGVFENTAAHKDNYAYSESYTEM——KA
   GDVAKGKKTFVQK—CAQCHTV—————ENGGKHKVGPNLWGLFGRKTGQAEGYSYTDANKS—————
    EDPEVLFKNKGCVACHAI————————DTKMVGPAYKDVAAKFAGQA———————————————
   YDAAAGKATYDAS—CAMCH——————————KTGMMGAPKVGDKAAWAPHI———————————————


                                        Q
                                        A    N
   Q T      S                           K    G               N
    E D Q    Y                      D   T    K            TV D
    L  EPPH PK                   P      T    S  V         QLKE
  S  N QQKV HD    K            S G      P    E  T         EETSK
  AN T KEEN RV  L       V        A V    P A  D  A    S    VDKAN
  MAVI GDNDMFIF T   S FM         V T  LSAE   ND  ENIITFMLKSCA
  KGIVWNNDTLMEYLENPKKYI————————PGTKMIFAGIKKK———GERQDLVAYLKSATS
       60        70              80        90        100

  PDLTWTEANLIEYVTDPKPLVKKMTDDKGAKTKMTFKMGK——————NQADVVAFLAQBBPB
  KGLTWTEANLAAYVKDPKAFVLEKSGDPKAKSKMTFKLTK————DDEIENVIAYLKTKL
  KGIVWNNDTLMEYLENPKKYI————————PGTKMIFAGIKKK———GERQDLVAYLKSATS
  ————GAEAELAQRIKNGSQGV———————WGPIPMPPNAVS————DDEAQTLAKWVLSQK
  —————AKGMNVMVANSIKGYK———————GTKGMMPAKGGNPKLTDAGVGNAVAYMVGQSK
```

Figure 7–1: Alignment of the amino acid sequences of cytochromes *c* and mutational change observed at each of the positions in the common sequence. The five amino acid sequences below the numerical scale are the aligned amino acid sequences of the cytochromes *c* from *Paracoccus denitrificans*[1] (NEGDAAK⋯), *Rhodospirillum rubrum*[1] (EGDAAA⋯), tuna[1] (GDVAK⋯), *Pseudomonas aeruginosa*[2] (EDPEV⋯), and *Chlorobium thiosulfatophilum*[3] (YDAAA⋯). The dashes represent gaps that must be made in one of the amino acid sequences to align it reasonably with the other sequences. When five sequences are aligned in this way, the size of each gap is arbitrarily determined by the number of extra amino acids in the one or more sequences that do not have a gap. The amino acid sequence of cytochrome *c* from the tuna is repeated immediately above the numerical scale, which is based on this sequence from the tuna. Above each position in this top sequence is a list of the other amino acids found in this position in a collection of 40 cytochromes *c* from various eukaryotes. Letters below the horizontal lines in each of the columns are variations found among cytochromes *c* of animals, and letters above the horizontal lines are the additional variations found in fungi and plants, more distantly related eurkaryotes.

presently compared are improved versions. It is believed[5] that the majority of the differences that accumulate in the sequences of the same protein in two lineages, following their divergence from their common ancestor, are neutral replacements. A **neutral replacement** is a change of one amino acid for another that is harmless enough that the biological function of the protein does not deteriorate sufficiently to cause the elimination of the replacement by natural selection. These replacements arise from mutations in the DNA encoding the protein. Each replacement that is now in existence has spread through the population of a species, or become fixed, by genetic drift. When we view aligned sequences of the same protein from different species, we are examining the record of this gradual increase in entropy.

This increase in entropy, however, is heavily biased. From examining aligned amino acid sequences of the same protein from many species, it is clear that each position in the underlying sequence that gives the protein its unique character is under a different degree of negative selective pressure. Mutations can occur with equal frequency at any position in the sequence of the DNA encoding for the sequence of a protein. Each of these individual mutations is assessed by natural selection, and the majority[5] disappear almost immediately because they are deleterious. A deleterious mutation is one that adversely affects the function of the protein. For example, in the human population, there are mutant forms of hemoglobin that bind oxygen improperly or are unstable proteins.[6] These represent deleterious mutations that survive for a limited time before disappearing from the population. These mutant forms can be contrasted with fetal hemoglobin that has been fixed in the human population because it is a stable protein and has beneficial properties. The most deleterious mutation is one that kills the individual in which it arises before that individual has had an opportunity to mate or otherwise reproduce after the mutation occurs. The more critical a particular amino acid in the sequence of the protein is to its function, the less prone will that position be to substitution over time. For this reason, the aligned amino acid sequences of the same protein from an array of species present a record of the scope of the intolerance to variation expressed at each position in the sequence of the protein.

This record can be read from examining consecutively each position in the aligned sequences of a large collection of the same proteins from different species. Above the numerical scale in Figure 7–1, the sequence of cytochrome c from tuna is presented, and above each of its positions in a column of letters are tallied the amino acids found there in the cytochromes c from 40 other eukaryotes.[7,8] The horizontal lines in each of these columns of letters separate amino acids found in the sequences of cytochromes c from animals, of which far more are available, from amino acids found in the sequences of cytochromes c from fungi and plants, which represent more distant relationships. Aligned below this array are the sequences of the cytochromes c from P. denitrificans,[1] R. rubrum,[1] tuna,[1] Pseudomonas aeruginosa,[2] and Chlorobium thiosulfatophilum.[3] Four of the five lower sequences come from prokaryotic sources and represent the most distant relationships, both between each prokaryote and the tuna and between each pair of prokaryotes.

The information contained in the amino acid sequences of the cytochromes c is the most detailed available about the progress of evolution by natural selection as it fashions the sequence of a protein. A few of the positions in these aligned sequences have remained absolutely **invariant**, for example, Cysteine 14, Cysteine 17, Histidine 18, and Methionine 80; but it is known that these are functionally irreplaceable. Some positions such as Glycine 6, Glycine 41, and Proline 71 would have been judged invariant if only a smaller group of sequences were available. It should be clear that the widely used designation of invariant is always provisional.

Some of the changes observed seem to be conservative replacements. A **conservative replacement** is a replacement at a position in which only similar amino acids, either in size or chemical properties, can be tolerated. For example, only valine, isoleucine, and leucine, each of whose side chains is an alkane, seem to occur in position 57 of cytochrome c (Figure 7–1). Either glycine, alanine, or proline, each of whose side chains is small, seems to be necessary in position 41. Either serine, threonine, glutamine, or histidine, each of whose side chains is polar, but uncharged, seems to be necessary in position 42. In addition, at positions 42 and 43, at least among the eukaryotes, tandem replacements occur; when Glutamine 42 changes to threonine or serine, Alanine 43 changes to serine or valine, respectively. These changes could reflect compensating adjustments in size. The earlier belief, however, that each position could be assigned unambiguously to one of a few categories, for example, invariant, conservative, physicochemically constant, and variable,[9] was probably an optimistic one arising from the small number of sequences available at the time. The majority of the changes observed cannot be easily explained, and this suggests that even the specific explanations just presented may themselves be rationalizations of more subtle processes that are not understood. A close examination of the actual results, however, sequence position by sequence position, does produce an intuitive feeling for the play of evolution.

In addition to the capacity for a particular amino acid to be tolerated at a particular position in the sequence of a protein, the nature of the genetic code itself also affects the patterns in which replacements in the sequence occur during evolution. Because there are three bases coding for each amino acid and mutation occurs at one base at a time, certain replacements should be more common than others.[5] There are, however, some interesting apparent exceptions to this generalization which occur in the comparisons of the various eukaryotic cytochromes *c* (Figure 7–1). For example, at position 31, only asparagine and alanine are found; at position 72, only lysine and serine; at position 45, only lysine and glycine; and at position 19, only glycine and threonine. Each of these four replacements would require that two bases of the respective codon mutate simultaneously. Although these are unlikely events, the constraints on the occupation of these positions seem to have been severe enough to confine the replacements to these choices, at least among the eukaryotes.

These, however, may only be apparent anomalies. When a detailed history of the mutational events in a family of proteins has been assembled, such as in the history of the evolution of artiodactyl fibrinopeptides,[5] it has been observed that replacements requiring the mutation of only one base are far more frequent than those requiring two simultaneous mutations. It is possible that in circumstances where two simultaneous mutations seem to have occurred, the amino acid sequence of the protein displaying the intermediate single mutation, although it exists, has not yet been determined.

There have been several attempts to analyze the frequency with which the various replacements occur at positions in aligned sequences. If a number of sequences, all known to be related, are mutually aligned, the frequency at which any pair of the 20 amino acids appears at the same position can be determined and normalized by the frequency at which that pair would have appeared by chance. Such a distribution of **normalized frequencies of mutual occupancy** has been calculated from 17 such multiple alignments,[10] and the 20 most frequent pairs and 20 least frequent pairs out of the 190 possible pairs are presented in Table 7–1. Pairs whose interchange would require the mutation of only one base in a codon seem to occupy the same position more frequently than those requiring two or three. Only about 15 pairs have a normalized frequency more than 2 times that expected by chance, and only about 15 have a frequency less than one-third that expected by chance. This means that about 85% of the possible pairs occur at frequencies between 0.4 and 2, a result consistent with the observation that the actual distributions at particular positions in a sequence are more diverse than might be expected. The tolerance or intolerance of natural selection that is reflected in the pairs with the highest and lowest frequencies, respectively, taking into account the base changes required, seems to reflect the size and chemical properties of the respective amino acids (Table 7–1).

Another approach to examining the progress of natural selection has been to calculate a mutation probability for every pair of possible replacements.[7] This was accomplished by reconstructing a probable sequence of events in the evolution of 10 groups of very closely related sequences. Sequences for common ancestors were predicted and all of the replacements that should have occurred following the divergence of the progeny from that ancestor were tabulated to provide the basis for the calculation of probabilities. The results of this study were presented as mutation probabilities. A **mutation probability** is defined as the probability that a certain replacement will occur during a time long enough for a total of two replacements to accumulate for every 100 amino acids of the sequence. The top 20 values of these mutation probabilities are also presented in Table 7–1. Almost all of the 20 replacements with the lowest frequencies of mutual occupancy were found not to have occurred in this latter study and therefore had formal mutation probabilities of zero.

Although there is considerable overlap, the frequencies of mutual occupancy and the mutation probabilities do not coincide even though the same families of proteins were used in the two calculations. This is partly due to differences in the procedures used to make these two calculations, but the major difference seems to be one of time. The goal of the calculation of mutation probability was to focus on the shortest possible period of time to encompass each step in evolution. It was intended to be a measure of the probability at any instant that a given replacement will be accepted and fixed. The frequency of mutual occupancy measures how tolerant natural selection is to a given replacement over a longer period of time. If pairs with high values of normalized frequency of mutual occupancy but low values of mutational probability are singled out, it seems that differences in size, shape, and chemical properties are more important in the short term than the long term. Another peculiar result is that replacements involving alanine seem much more frequent in the short term than the long term. This may be due to the fact that alanine, because it is the β carbon contained in all of the other amino acids in isolation, can act as intermediate replacement in a stepwise transition between two other side chains. This reinforces the impression that over the long run many more replacements are tolerated than in the short run because compensating replacements, which permit more unusual replacements to occur, have had the time to happen elsewhere in the protein.

As more and more time has passed following the divergence of the amino acid sequences of two proteins from that of their common ancestor, the similarities become weaker and the number of insertions and deletions increases until it is difficult to align them. This has not prevented people from trying to do so. Digital computers are used to align such distantly related sequences or to search large collections of amino acid sequences for any pairs that might be related to each other.[11]

The problem of aligning two distantly related amino acid sequences is usually solved by constructing a matrix.[12] If one sequence A has p amino acids, arranged in the order $a_1a_2a_3...a_p$, and the other sequence B has q amino acids, arranged in the order $b_1b_2b_3...b_q$, the product of these two vectors is a matrix C whose coefficients, c_{ij}, are equal to $a_i \times b_j$, where a_i and b_j are particular amino acids. For example, $a_i \times b_j$ would be Asn $\times$ Thr when asparagine appeared at position i in sequence A and threonine appeared at position j in sequence B. The numerical value assigned to the particular $a_i \times b_j$ depends on the various schemes devised to weight the comparisons.

The simplest scheme[11] is to decide that when $a_i = b_j$, $a_i \times b_j = 1$, and when $a_i \neq b_j$, $a_i \times b_j = 0$. This produces a matrix whose coefficients, c_{ij}, are either 1 or 0. When the amino acid in position a_i in the first sequence is the same as the amino acid in position b_j in the second sequence,

$c_{ij} = 1$; when they are different, regardless of the difference, $c_{ij} = 0$. Such a matrix, spread upon a two-dimensional field, can be represented diagrammatically by placing a dot on every position with a score of 1 (Figure 7–2).[13] In such a **dot matrix**, the alignment is represented by diagonal strings of dots. In the matrices comparing human, monkey, and fish, the diagonals are obvious and unbroken. In the matrix comparing human and bacterium, the alignment is a set of at least three diagonal segments that can be picked out by eye if the figure is tilted and viewed along the diagonal direction. The skips

Table 7–1: Normalized Frequency of Mutual Occupancy and Mutation Probability for Various Pairs of Amino Acid Replacements

pair[a]	normalized frequency of mutual occupancy[b]	base change[c]	mutation probability[d] (%)	pair[e]	normalized frequency of mutual occupancy	base change
Y/F	5.4	1	1.3	W/P	0.0	2
W/Y	5.1	2	0.2	Q/F	0.1	3
W/F	3.3	1	0.4	G/F	0.1	2
L/M	3.2	1	1.1	D/I	0.2	2
F/I	2.8	1	2.3	N/F	0.2	2
D/E	2.7	1	1.9	P/Y	0.2	2
E/Q	2.7	1	0.6	E/F	0.3	3
R/K	2.6	1	0.9	K/F	0.3	2
I/L	2.4	1	0.7	Y/G	0.3	2
L/F	2.3	1	0.3	Q/I	0.3	2
H/R	2.3	1	0.4	G/M	0.3	2
T/S	2.1	1	1.5	D/W	0.3	3
R/Q	2.1	1	0.3	F/D	0.3	2
S/N	1.9	1	1.2	P/M	0.3	2
N/D	1.8	1	0.8	K/I	0.4	1
F/M	1.8	2	0.2	N/P	0.4	2
M/I	1.8	2	0.3	N/I	0.4	1
L/V	1.7	1	0.5	I/G	0.4	2
M/V	1.7	1	0.9	Y/Q	0.4	2
H/N	1.6	1	0.4	L/E	0.4	2
___[f]						
S/A	1.5	1	1.6			
G/A	1.2	1	0.8			
P/A	1.4	1	0.8			
T/A	1.3	1	0.8			
V/A	1.1	1	0.6			
E/A	1.5	1	0.6			
K/N	1.3	2	0.6			
S/D	1.2	2	0.5			

[a]Interchange of one amino acid with the other observed at particular positions in aligned amino acid sequences. [b]Number of times that the two members of the pair have been found at the same position in a set of aligned amino acid sequences divided by the number of times that the two members would have been found at the same position by chance.[10] A value of 1.0 would indicate that the two members of the pair are randomly interchanged; a value greater than 1.0 indicates that there is a positive bias for replacing the one with the other. [c]Minimum number of base changes required to change a codon for one member of the pair into a codon for the second. [d]Probability that one member of the pair would be replaced by the other member of the pair as the result of fixation of that particular mutation in the population.[7] The probability is for the replacement to occur during a time long enough for only two mutations to be fixed for every 100 amino acids in the sequence of the protein. [e]All pairs whose values (≤ 0.4) of normalized frequency of mutual occupancy are among the lowest 20. Each of these pairs has a value of mutation probability equal to zero. [f]Pairs listed above this line have the 20 highest values (≥ 1.6) of normalized frequency of mutual occupancy. They are listed in order of this parameter. The pairs below this line are all pairs whose normalized frequencies of mutual occupancy are not among the highest 20 but whose values (≥ 0.5) of mutation probability are among the highest 20.

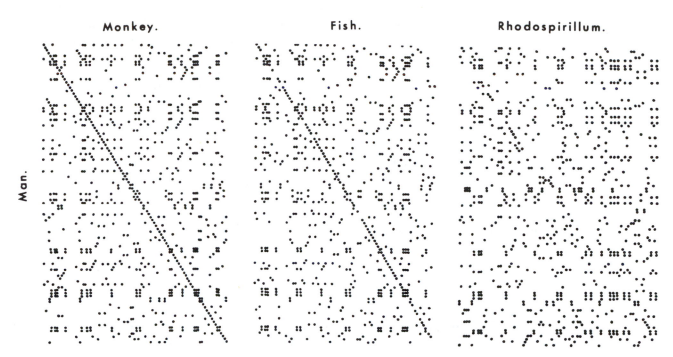

Figure 7–2: Dot matrices[13] for the amino acid sequences of the cytochromes *c* from the rhesus monkey, the tuna, and the bacterium *Rhodospirillum rubrum*, each compared to the amino acid sequence of human cytochrome *c*. The sequence of human cytochrome *c* is the vertical vector (top to bottom, amino to carboxy terminus), and the respective sequences with which it is compared are the horizontal vectors (left to right, amino to carboxy terminus). A dot is placed in the matrix when the amino acids at those two positions, horizontal and vertical, are the same. Reprinted with permission from ref 13. Copyright 1970 Springer-Verlag.

between segments are the gaps in the alignments (Figure 7–1).

There are, however, 20 different amino acids and 190 different outcomes for $a_i \times b_j$ when $a_i \neq b_j$, if one assumes symmetry, namely, that Asn × Thr = Thr × Asn. It has always seemed that some of these outcomes are more probable and that recognition of this probability might enhance the ability to align distantly related sequences. The ultimate goal in aligning two amino acid sequences is to decide whether position i in sequence A and position j in sequence B arose from the same position in the sequence of a common ancestor or position i in sequence A and position j in sequence B are unrelated to each other either because protein A and protein B do not share a sufficiently recent common ancestor or because the two sequences are misaligned. If the two amino acid sequences are unrelated, $a_i \times b_j$ is governed solely by chance. If they are descended from the same position in an ancestral sequence, then $a_i \times b_j$ should retain some of the biases enforced by natural selection, and these biases, if they can be quantified, should be considered while reaching a decision. If a particular replacement has a higher probability of occurring as a result of evolutionary change than it does of occurring as a result of random change, then whenever that particular replacement is encountered, those two positions have a higher prob-

ability of being evolutionarily related than of being unrelated. For example, every $a_i \times b_j$ where a_i and b_j are interconvertible by only one base change should have a higher probability of being evolutionarily related than those $a_i \times b_j$ where a_i and b_j are interconvertible only by two or three base changes.[14] Every $a_i \times b_j$ where a_i and b_j are similar in size or chemical properties should have a higher probability of being evolutionarily related than those in which they are dissimilar.[10] The normalized frequency with which the replacement $a_i \times b_j$ occurs in other sequences that have been unambiguously aligned (Table 7–1) can be used directly to weight $a_i \times b_j$. The mutation probability[7] can be used to weight the entries. The net effect of each of these various weighting schemes, or some combination of them, is to assign a number, other than 0, to every coefficient c_{ij} of the matrix where $a_i \neq b_j$. The magnitude of this number is thought to be directly proportional either to the effect of natural selection relative to chance on the particular replacement $a_i \times b_j$ or to the probability that the particular replacement $a_i \times b_j$ would arise as the result of evolution rather than chance. If the weighting scheme is based on probability, logarithms of the probabilities are used as entries in the matrix so that the summation to be performed will represent products of probabilities.[7]

When the matrix has been constructed to the taste of

the practitioner, the alignment can be performed.[12] Any alignment of the two sequences A and B can be represented as a set of diagonal segments running through the matrix such that each a_i and each b_j is represented by only one point in these segments or by none of the points in these segments. Two such diagonal segments can be picked out in the dot matrix comparing the cytochromes c of humans and *Rhodospirillum* (Figure 7–2). The end of one of these diagonal segments must be either above or to the left of the beginning of the next diagonal segment when the diagonals run from top left to bottom right as in Figure 7–2. Each discontinuity associated with the end of one diagonal segment and the beginning of the next is a horizontal shift or a vertical shift representing a gap in one of the sequences being aligned. Associated with each individual alignment is an alignment score

$$AS = \sum c_{ij} - Pg \qquad (7\text{–}1)$$

where the sum is over all c_{ij} intersected by the diagonal segments, P is a penalty assessed for creating a gap, and g is the number of gaps in the particular alignment. The penalty assessed is judged to be the probability of a gap appearing during evolution by natural selection on the same numerical scale used to assign values to c_{ij}. For example, it could be decided that when $a_i = b_j$, $a_i \times b_j = 1$; when $a_i \neq b_j$, $a_i \times b_j = 0$; and when a gap is introduced, $P = 2.5$.[11] This choice states that a gap, on the average, is 2.5 times more unlikely than a replacement. A computer can be programmed to find the path of diagonal segments through the matrix that has the largest alignment score.[12]

The most important responsibility of an investigator who performs such a computation and produces an alignment with the maximum alignment score is also to provide a reliable assessment of its **statistical significance**. The accepted criterion for this assessment is a statistical evaluation of combinations of alignments produced from randomly jumbled sequences of the same length and amino acid composition as the actual amino acid sequences.[7] First the two actual sequences are aligned, and a maximum alignment score for the optimum alignment is calculated. Then each of the two actual sequences is randomly jumbled a number of times to produce for each a set of nonsense sequences that have the same amino acid composition and length as the actual amino acid sequence from which they were generated. This produces two sets of jumbled amino acid sequences, one derived from each actual sequence. All of the many combinations of one jumbled sequence from one of these two sets and one jumbled sequence from the other set are aligned by the same algorithm that was used to align the two actual, unjumbled sequences, and a large collection of maximum alignment scores for the nonsense sequences is gathered in this way. The mean and standard deviation of the alignment scores of this collection of jumbled sequences are calculated by the usual

statistical formulas. The number of standard deviations that the alignment score for the two actual amino acid sequences lies above the mean for the maximum alignment scores for the jumbled sequences is a measure of the confidence that can be assigned to the decision that the two actual sequences share a common ancestor and to the decision that the alignment has juxtaposed positions in the sequence that have evolved independently from the same position in the ancestral sequence.

There is a frequently encountered sleight of hand that is practiced in the alignment of amino acid sequences and that violates the rules of statistics. This trick is to align two sequences and then select only the regions in which there is a higher frequency of coincidence for the statistical test. Because the sample has been preselected, it usually shows a higher frequency of coincidences than occurs when jumbled sequences of the same small regions are compared. Ordinarily, statistical evaluation of an alignment of two amino acid sequences shorter than those of complete, naturally occurring, and logically defensible domains within the native protein should not be accepted without the closest scrutiny.

As the procedures for aligning sequences have been applied to the rapidly expanding library, many statistically significant relationships have been discovered (Table 7–2). Most of these make sense from a functional standpoint, but unsuspected connections have often been discovered. Examples from Table 7–2 would be the alignments of ovalbumin and antithrombin III and chymotrypsinogen and haptoglobin.

Several practical conclusions have been reached. The various schemes that assign different weights to the various $a_i \times b_j$ in the matrix are no more reliable at detecting distant relationships than a simple system of assigning 1 to identities and 0 to nonidentities.[14] It is possible that the rules governing the frequency of mutual occupancy among distantly related proteins, which is where the assignments are so interesting, are quite different from those governing it among the closely related proteins used to construct the tables. If the relevant rules for distantly related proteins could be discovered, their incorporation might improve the present algorithms. Only the distance in standard deviations from the mean of the jumbles, however, can be used to decide if a particular algorithm is an improvement or not. One feature of the alignment of two related sequences is that identities cluster (Figures 7–1 and 7–2). This tendency might be used to increase the capacity to detect evolutionary relationships.

At the present time statistically significant alignments can be made only on two amino acid sequences that have identical amino acids in 15% or greater of their aligned positions.[14] If a set of three or four amino acid sequences can be assembled, however, that are from a set of proteins that share some structural or functional feature, it is often possible to demonstrate with high statistical confidence that the members of this set all share the same

Table 7–2: Examples of Pairs of Proteins Thought to Share a Common Ancestor on the Basis of an Alignment of the Two Amino Acid Sequences[a]

protein I[b]	protein II[b]	% identity[c]	gaps[d]	gaps/100a a	standard deviations[e]
chymotrypsinogen a, bovine (245)	chymotrypsinogen b, bovine (245)	79.2	0	0	65.8
hemoglobin β, human (146)	hemoglobin γ, human (146)	73.3	0	0	40.7
carbonic anhydrase b, human (260)	carbonic anhydrase c, human (259)	60.6	1	0.4	56.8
chymotrypsinogen a, bovine (245)	trypsinogen, bovine (229)	46.1	6	2.6	22.6
lysozyme (egg white), chicken (129)	lactalbumin (milk), human (123)	38.2	3	2.4	10.7
viral coat protein, PF1 (46)	viral coat protein, Xf (44)	40.5	1	2.3	5.0
hemoglobin α, human (141)	myoglobin, human (153)	27.0	1	0.7	9.3
ovalbumin, chicken (386)	antithrombin III, human (423)	28.1	6	1.6	14.3
parvalbumin, carp (108)	troponin c, bovine (161)	27.0	2	2.0	6.1
cytochrome c, pig (104)	cytochrome f, Spirulina maxima (89)	27.0	3	2.9	7.2
β2-microglobulin, human (100)	immunoglobulin, κ–constant region, human (102)	18.8	1	1.0	3.5
plastocyanin, spinach (99)	azurin, Pseudomonas (128)	27.3	4	4.0	2.9
histocompatibility antigen, mouse (173)	immunoglobulin SH λ chain[f] (183)	16.7	3	1.7	4.1
leghemoglobin, yellow lupin (153)	invertebrate hemoglobin, midge (151)	22.0	3	2.0	2.6
chymotrypsinogen b, bovine (245)	haptoglobin β, human (245)	19.4	4	1.6	5.4

[a]Alignment was performed on a matrix where $a_i \times b_j = 1$ when $a_i = b_j$ and $a_i \times b_j = 0$ when $a_i \neq b_j$ and the gap penalty was 2.5. When $a_i = b_j = $ cysteine, $a_i \times b_j = 2.0$. Reproduced from ref 11. [b]Two proteins the amino acid sequences of which were aligned. Number of amino acids in each protein is shown in parentheses. [c]Percentage of the positions in the aligned sequences at which the same amino acid was found in both sequences. [d]Total number of gaps that had to be introduced to get the best alignment. [e]Distance in standard deviations that the alignment score for the actual sequences was above the mean of the alignment scores of 36 comparisons of jumbled sequences. [f]Only represents a portion of the entire sequence.

common ancestor even when pairwise comparisons between the members of the set fail to demonstrate convincing homology.[15,16] In these instances, the statistical significance only becomes convincing when the whole set is aligned together.

The aligned amino acid sequences of polypeptides have been used to provide information about **speciation**. The sequences of the same protein from a collection of different species, for example, the sequences of the cytochromes c from different eukaryotic species,[17] serve as the data on which such studies are based. Each pair of aligned amino acid sequences is compared position by position and the minimum number of mutations required to accomplish each replacement is noted. These are added to obtain the total number required to convert either of the two into the other. Each of these totals for every possible pair of sequences is tabulated. These numbers are referred to as the minimal mutational distances.[18] The **minimal mutational distance** is the smallest number of base changes that could change the amino acid sequence of one protein into that of another. Each pair of sequences in the collection being compared has a

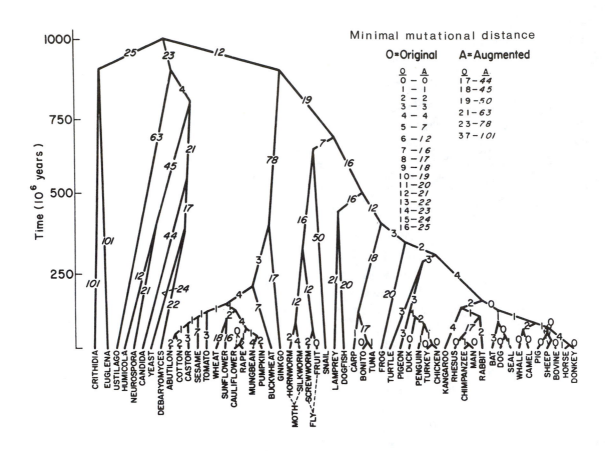

Figure 7–3: Phylogenetic tree[17] for the cytochromes *c* from 53 species of eukaryotes. Each of the possible 52! pairs of sequences was aligned and the minimal mutational distance between each pair was tabulated. These numerical values were then adjusted statistically for mutations that would have not left a trace. The magnitudes of these corrections are indicated in the inset. The augmented mutational distances were used to construct the phylogenetic tree. If one passes along the branches of the tree between any two species, the numbers on the branches that are passed through sum to give the augmented mutational distance. For example, the augmented mutational distance between carp and dogfish is 67. The noted length of the branches in mutational distance and the branch points are determined by this requirement. In this figure, each branch point is positioned on the vertical axis at the geologic time (in units of 10⁶ years) that has passed since the species connected by that branch point shared a common ancestor, as determined by the fossil record. Adapted with permission from ref 17. Copyright 1976 Academic Press.

minimal mutational distance. It should be emphasized that the actual number of mutations that were fixed in the two lineages represented by each of the comparisons between two species is almost always greater than the number calculated, often by a large factor, because neither multiple mutations at the same site in either of the two sequences nor mutations producing the same amino acid at the same site in both of the two sequences have been accounted for. The tabulated values of minimal mutational distances are then used to construct a tree whose branches connect the species being compared (Figure 7–3).[17] The tree is arranged so that the connections made produce the most parsimonious sequence of mutational events that can reproduce the observed minimal mutational distances.

The major contributors to the minimal mutational distances calculated for each pair of aligned amino acid

sequences are the regions of the protein that have experienced the greatest change over time. Unfortunately, these are also the most difficult segments of the amino acid sequences to align convincingly. As a result, the choice of the method used to align the sequences can have a significant effect on the structure of the final tree. With this in mind, a method of progressive alignment of amino acid sequences has been developed to provide the most suitable and internally consistent alignments of a large collection of sequences from different species.[19] The basis of this method is assumption from the beginning that all of the amino acid sequences to be aligned share a common ancestor and arose through a process of speciation. The most closely related sequences are aligned first, and the gaps in these more certain alignments are retained as the more distant alignments are made. This is advantageous because it is the uncertainty in the precise

```
                                                    *     *
hghu                GHFTEEDKATI  TSLW   GKV  NVEDAGGETLGRLLVVYPWTQRFFDSFGNLSSASAIMGNPK  VKAHGKKVLTSLG
hbhu                VHLTPEEKSAV  TALW   GKV  NVDEVGGEALGRLLVVYPWTQRFFESFGDLSTPDAVMGNPK  VKAHGKKVLGAFS
hahu                VLSPADKTNV   KAAW   GKVGAHAGEYGAEALERMFLSFPTTKTYFPHF  DLSH      GSAQ  VKGHGKKVADALT
heha   PITDHGQPPTLSEGDKKAI       RESW   PQIYKNFEQNSLAVLLEFLKKFPKAQDSFPKFSAKKS  HLEQDPA  VKLQAEVIINAVN
hbrl   PIVDSGSVAPLSAAEKTKI       RSAW   APVYSNYETSGVDILVKFFTSTPAAQEFFPKFKGMTSADQLKKSAD  VRWHAERIINAVN
myhu                GLSDGEWQLV   LNVW   GKVEADIPGHGQEVLIRLFKGHPETLEKFDKFKHLKSEDEMKASED  LKKHGATVLTALG
mycr                SLQPASKSAL   ASSWKTLAKDAATIQNNGATLFSLLFKQFPDTRNYFTHFGNM  SDAEMKTTGV  GKAHSMAVFAGIG
haew            KKQCGVLEGLKVKSEWGRAYGSGHDREAFSQAIWRATFAQVPESRSLFKR       VHGDHTSDPA  FIAHAERVLGGLD
hety             TDCGILQRILVLQQWAQVYSVGESRTDFAIDVFNNFFRTNPD  RSLFNR       VNGDNVYSPE  FKAHMVRVFAGFD
gpfb               GAFTEKQEALVNSSW  EAFK   GNIPQYSVVFYTSILEKAPAAKNLFSF    LANGVDPTNPK  LTAHAESLFGLVR
hbvs                MLDQQTINIIKATV  PVLK   EHGVTITTTFYKNLFAKHPEVRPLFD      MGRQESLEQPKALAMTVLAAAQNIE


hghu   DAIKHLD    DLKGTFAQLSELHCDKLHVDPENFKLLGNVLVTVLAIHFGKEFTPEVQASWQKMV   TGVASALSSRYH
hbhu   DGLAHLD    NLKGTFATLSELHCDKLHVDPENFRLLGNVLVCVLAHHFGKEFTPPVQAAYQKVV   AGVANALAHKYH
hahu   NAVAHVD    DMPNALSALSDLHAHKLRVDPVNFKLLSHCLLVTLAAHLPAEFTPAVHASLDKFL   ASVSTVLTSKYR
heha   HTIGLMDKEAAMKKYLKDLSTKHSTEFQVNPDMFKELSAVFVSTM      GGKAAYEKLF   SIIATLLRSTYDA
hbrl   DAVASMDDTEKMSMKFRDLSGKHAKSFQVDPQYFKVLAAVIADTV      AAGDAGFEKLM   SMICILLRSAY
myhu   GILKKKGHHE    AEIKPLAQSHATKHKIPVKYLEFISECIIQVLQSKHPGDFGADAQGAMNKAL   ELFRKDMASNYKE  LGFQG
mycr   SMIDSMDDADCMNGLALKLSRNHIQRKIGASRFGE  MRQVFPNFLDEALGGGASCDVKGAWDALL   AYLQDNKQA   QA  L
haew   IAISTLDQPATLKEELDHLQVQHEGRKIPDNYFDA  FKTAILHVVAAQLGERCYSNNEEIHDAIACDGFARVLPQVLERG  IKGHH
hety   ILISVLDDKPVLDQALAHYAAFH   LQFGTIPFKA  FGQTMFQTIAEHI      HGADIGAWRAC   YA    EQIVT  G  ITA
gpfb   DSAAQLRANGAVVAD  AALGSIHSQKGVSNDQFLV  VKEALLKTLKQAV     GDKWTDQLSTALELA   YDELAAAI  KKAYA
hbvs   NLPAILPAVKKIAVKHCQAGVAAAHYPIVGQELLGAIKEVLGDAATDDI      LDAWGKAYGVIADV   FIQVEADLYAQAVE
```

Figure 7–4: Multiple alignment of 11 globins by the progressive method.[19] The amino acid sequences aligned were γ polypeptide of human hemoglobin (hghu), β polypeptide of human hemoglobin (hbhu), α polypeptide of human hemoglobin (hahu), hagfish hemoglobin (heha), lamprey hemoglobin (hbrl), human myoglobin (myhu), gastropod myoglobin (mycr), hemoglobin of the earthworm *Lumbricus* (haew), hemoglobin of the earthworm *Tylorrhynchus* (hety), leghemoglobin from kidney bean (gpfb), and hemoglobin from the bacterium *Vitreoscilla* (hbvs). Reprinted with permission from ref 19. Copyright 1987 Springer-Verlag.

location of the gaps that must be inserted to align distantly related sequences that creates the greatest uncertainty in the final value for the minimal mutational distance. An example of the product of this method is the progressive alignment of the amino acid sequences of 11 globins (Figure 7–4). The important feature of these alignments is that the gaps are confined to specific locations rather than being more randomly distributed as would result from simple pairwise alignments.

There are two reasons for arranging the minimal mutational distances as a tree, and they often operate at cross purposes. These are the obligation to present a tree reflecting the actual mutational distances and the desire to present a tree that duplicates the events of evolution by natural selection. Ideally, the tree should be an economical way of presenting the $(n - 1)!$ values for the minimal mutational distances between all of the pairs of sequences that are being compared. These are the only real data. The length of the limbs should represent the minimal mutational distance between each existing sequence and the unknown sequence inferred for the common ancestor represented by one of the forks. The tendency, however, to correct the numbers for the mutations that could not be scored is often too strong, and the lengths of the limbs are usually derived numbers based on the preference of the presenter for a stochastic method[17] of correcting the values, a logarithmic method

of correcting the values,[14] or some other method. In the example presented in Figure 7–3, the original minimal mutational distances were adjusted by a stochastic method, but the original values can be reconstructed from the scale included.

The desire to correct for the unobserved mutations stems from the other purpose behind such a presentation, that these trees may eventually present an accurate phylogenetic history of the species. To assess their success in this role, one must distinguish two pieces of information these trees convey. The branching order in such trees conveys a historical sequence of the relationships among the species represented, and these historical sequences seem to be reasonable, based on the fossil record and anatomical resemblances. The phylogenetic tree also, however, conveys estimates of the actual distances from existing species to common ancestors, and it is here that one must view these trees with caution.

The problem is time. In Figure 7–3, the time elapsed since divergence from a common ancestor, as determined from the fossil record, is presented as the ordinate. It is clear that neither the original values of the minimum mutational distances nor the corrected values correspond very satisfactorily to these times. The problems involved in associating a time with the length of a limb on the minimal mutational tree, aside from the correction for unscored mutations, seem to be at least the following.[5] First, the rate at which replacements occur in a sequence and the range of acceptable replacements is remarkably different for each position in the sequence, and this property has defied attempts to derive a simple equation relating evolutionary time and minimal mutational distance. Second, there are examples of accelerated changes occurring along only one branch of a tree containing species that seem indistinguishable from each other. For example, rat ribonuclease seems to have accumulated replacements at 4 times the rate of its close relatives the ribonucleases from mice, muskrats, and hamsters.[20] Third, the size of the populations of a given species or its generation time may affect the rate at which mutations become fixed. This may explain why the sequences of the cytochromes *c* from three closely related strains of the bacterial genus *Pseudomonas* show as much variation in their sequences (differences of 22–39%)[21] as is shown between mammals and amphibians (18%) or mammals and insects (33%). These and other problems have not been solved, and trees directly or indirectly presenting minimal mutational distances should not yet be confused with phylogenetic trees constructed from a complete fossil record, regardless of their similarity in appearance. The most that can be said at the moment is that trees constructed from minimal mutational distances can be produced that are consistent with accepted phylogeny. Attempts to use such trees of minimal mutational distances to rearrange phylogenetic trees[22] seem premature.

At the point at which the lineages of two presently existing species diverged from their common ancestor, the gene for a particular protein carried by the common ancestor had become two separate and disconnected genes, one carried by each of the new, independent ancestral species. At that time, natural selection began to operate on these two genes independently, and the differences now observed in the sequences of the same protein from the two existing species began to accumulate. A similar disconnection of two genes for the same protein can occur within a single genome by gene duplication.[5] As a result of a mistake in replication, the DNA in an organism suddenly contains two copies of the same gene. If this duplex arrangement spreads through the population by genetic drift and becomes fixed, the genome of the affected species will now contain two copies of the same gene. Both copies will often continue to produce their respective proteins; but, because of the disconnection, the amino acid sequences of these two proteins have become capable of independent variation. What results are isoforms of a given protein, or isoenzymes of a given enzyme. The advantage to this ancestral species and those that diverge from it is that the two isoforms can gradually specialize to meet separate demands. These demands are often expressed at the level of individual tissues within the organism, and the sequences of the two isoenzymes or isoforms may have diverged in

part to satisfy the particular demands of two sets of tissues. If this has indeed occurred, it would be an example of positive selection, the process by which advantageous mutations are preferentially fixed in the populations at the expense of the parental types, and it is in such an adaptation that positive selection of amino acid sequences will perhaps be most readily detected. Positive selection would perpetuate those changes at particular positions in the sequence that in turn produce changes in biological properties rendering the protein more suitable to a particular set of tissues.

An example of the evolution of a family of isoenzymes occurs among the lactate dehydrogenases. Among the tissues of mammals and birds, at least three isoenzymes of lactate dehydrogenase have been identified. From this observation, it may be inferred that an individual mammal or bird contains within its genome three discrete genes encoding three discrete lactate dehydrogenases. Complete sequences are available for several of these proteins.[23] They have been aligned and a tree of minimal mutational distances was constructed from these alignments (Figure 7–5).[24] The tree suggests that the three isoenzymatic lactate dehydrogenases diverged from their common ancestor before the appearance of the vertebrates. This conclusion has been supported in a more recent phylogenetic tree constructed from an even larger collection of amino acid sequences.[23] Because each of these isoenzymes, or appropriate mixtures of them, are found in different tissues, it can be concluded that the natural selection which has produced them in their present guise has operated at the level of the tissue rather than that of the the whole organism. To the extent that different tissues are constructed from different isoforms of the same proteins, they can be considered to have evolutionary histories that are independent of the histories of the organisms carrying them.

As the number of available amino acid sequences has increased, the number of statistically significant alignments has also increased. This has allowed some sets of amino acid sequences to be grouped into superfamilies. A **superfamily** is any one of the complete sets formed from a group of amino acid sequences all of which can be shown to share a common ancestor by the existence of statistically significant alignments connecting all of the members of the set. Examples of these superfamilies are the globin superfamily (Figure 7–4), the immunoglobulin superfamily, the cytochrome superfamily, the tyrosine kinase superfamily, and the serine proteinase superfamily. In most cases, all of the proteins in any one of the superfamilies, as the names of the superfamilies themselves indicate, have similar functions, and it would have been reasonable to assume that they were related even if the amino acid sequences were not available. To demonstrate that two proteins that have very different functions nevertheless share a common ancestor, it is usually necessary to examine crystallographic molecular models.

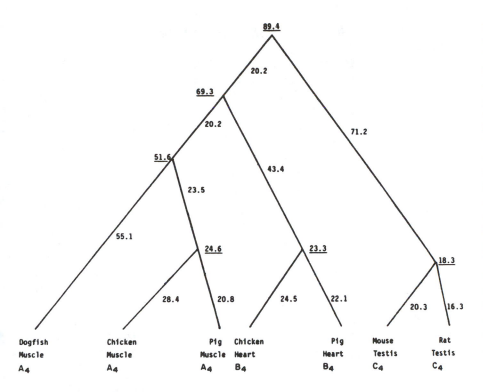

Figure 7–5: Phylogenetic tree of seven vertebrate lactate dehydrogenase isozymes.[24] The phylogenetic relationship among seven lactate dehydrogenase isozymes, namely, dogfish A_4 (muscle), chicken A_4 (muscle), pig A_4 (muscle), chicken B_4 (heart), pig B_4 (heart), mouse C_4 (testis), and rat C_4 (testis), is indicated by the most parsimonious tree. The number on each leg is the minimal mutational distance required to account for the descent from the ancestor, and the number underlined is the average of its immediate descendents. The minimal mutational distance in any one interval is not an integer because of averaging over all equally most parsimonious solutions for that topology. The total number of nucleotide substitutions is 366. The count does not include insertions or deletions. The root is arbitrarily placed halfway between the two most distantly related groups. Reprinted with permission from ref 24. Copyright 1983 *Journal of Biological Chemistry*.

Suggested Reading

Feng, D.F., Johnson, M.S., & Doolittle, R.F. (1985) Aligning Amino Acid Sequences: Comparison of Commonly Used Methods, *J. Mol. Evol. 21*, 112–125.

Problem 7–1: Calculate alignment scores (Equation 7–1) for the five cytochromes *c* aligned in Figure 7–1 based on the rules that $a_i = b_j$, $a_i \times b_j = 1$; $a_i \neq b_j$, $a_i \times b_j = 0$; $P = 2.5$.

Problem 7–2: This exercise will illustrate the method for assessing the validity of a particular alignment of two sequences. Pick a number between 1 and 80 at random and write it on a piece of paper. Turn to Figure 7–1 and the alignment of the two amino acid sequences of the cytochromes *c* from *P. denitrificans* and *R. rubrum*, respectively. Start at the amino acid in the sequence of the cytochrome *c* from *P. denitrificans* corresponding to the number you picked at random and write the next 20 amino acids in that sequence across the page. Below this sequence write the corresponding amino acids of the aligned sequence from the cytochrome *c* of *R. rubrum*, as in the figure. Calculate an alignment score (Equation 7–1) for these segments of aligned sequences. Take 20 playing cards and to each of them assign one of the 20 amino acids in the amino acid sequence from the cytochrome *c* of the amino acid sequence without a gap. Shuffle the cards well and deal them into a row. Copy out the jumbled sequence dictated by this shuffle on a piece of paper. Align the jumbled sequence with the segment of real amino acid sequence from the cytochrome *c* of the other bacterium by shifting and gapping until you think the alignment will give the highest alignment score. Record that score. Repeat the process five times. How do the alignment scores of the jumbled sequences compare to the alignment score of the one unjumbled sequence?

Problem 7–3: On the basis of their locations in the crystallographic molecular models of proteins, their structural roles, and their chemical properties, the amino acids can be divided into three categories: hydrophobic, neutral, and hydrophilic. The hydrophobic amino acids are isoleucine (I) valine (V), leucine (L), phenylalanine (F), cystine, methionine (M), and alanine (A). The neutral amino acids are glycine (G), cysteine (C), threonine (T), tryptophan (W), serine (S), tyrosine (Y), and proline (P). The hydrophilic amino acids are histidine (H), glutamate (E), glutamine (Q), aspartate (D), asparagine (N), lysine (K), and arginine (R).

The following alignment is from Figure 7–1. The alignment is based on the crystallographic molecular models of the two cytochromes *c* in which these amino acid sequences occur.

```
Q K _ C A Q C H T V E N G G K H
N K G C V A C H A I _ _ _ D T K
```

(A) Construct a 15 × 13 matrix on a sheet of graph paper for the two segments of sequence involved in this alignment using the following rules:

(1) $a_i \times b_j = 1$ for an identity
(2) $a_i \times b_j = 0.6$ for hydrophobic × hydrophobic
(3) $a_i \times b_j = 0.6$ for neutral × neutral
(4) $a_i \times b_j = 0.6$ for hydrophilic × hydrophilic
(5) $a_i \times b_j = 0.2$ for hydrophobic × neutral

(6) $a_i \times b_j = 0.2$ for neutral × hydrophilic
(7) $a_i \times b_j = 0.0$ for hydrophobic × hydrophilic

(B)   Trace the alignment presented above through the matrix.

(C)   Calculate an alignment score for that trajectory if the gap penalty is 2.5.

(D)   What is the most serious difficulty with the rules?

Problem 7–4: From the genetic code, calculate the minimum number of base changes between the amino acid sequences of the γ and β chains of human hemoglobin as they are aligned in Figure 7–4. To do this, make a table containing all of the replacements between the two sequences, find the minimum number of base changes required for each, and add up the individual minimum base changes to obtain the total.

Problem 7–5: The sequences of the fibrinopeptides A and B from a series of primates are given in the table below. Construct a tree of minimal mutational distances. Treat a gap as if it were 2 base changes.

Molecular Phylogeny from Tertiary Structure

Just as the amino acid sequences either of the same protein from two different species or of two different proteins from two different species or the same species can be aligned, so can their tertiary structures be superposed.[25] The crystallographic molecular models of two proteins that have tertiary structures so similar that they are thought to share a common ancestor are chosen for comparison. Those pairs of atoms that unambiguously occupy equivalent positions in equivalent strands of secondary structure in the two crystallographic models are identified. To **superpose** these two crystallographic molecular models is to translate and rotate one of them relative to the other until the sum of the squares of the distances between these pairs of equivalenced atoms is minimized. Because two different crystallographic models are being superposed, the structures never coincide exactly. An example of a superposition (Figure 7–6) is that between the crystallographic molecular model of the thiol proteolytic enzyme present in the fruit of the papaya

primate	fibrinopeptide A	fibrinopeptide B
green monkey	ADTGEGDFLAEGGGVR	PCA[a]-GVNGNEEGLFGGR
human	ADSGEGDFLAEGGGVR	PCA-GVNDNEEGFFSAR
drill	ADTGDGDFITEGGGVR	PCA-GVNGNEEGLFGGR
macaque	ADTGEGDFLAEGGGVR	NEESPFSGR
chimpanzee	ADSGEGDFLAEGGGVR	PCA-GVNDNEEGFFSAR

[a] Pyrrolidone-5-carboxylic acid, a cyclized form of glutamine.

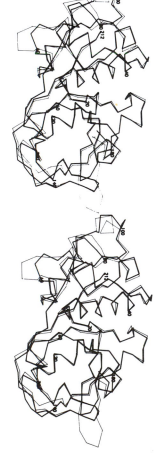

Figure 7–6: Superposition of the α-carbon backbones of the crystallographic molecular models of papain (heavy lines) and actinidin (light lines).[26] All of the pairs of atoms in the two crystallographic molecular models of these two proteins that could be identified as occupying equivalent positions (888 atoms) were chosen for the initial superposition. After the sum of the squares of the distances between these pairs of atoms was minimized by rotating and translating one of the models relative to the other, all pairs of atoms the square of whose distances were greater than 3 standard deviations from the mean of the square were eliminated. Several cycles of successive superposition and elimination were performed until 73% of the original 888 atoms were retained as the reference point for the final least-squares superposition. The α-carbon atoms in the final superposed molecular models were joined by line segments to generate the figure. Reprinted with permission from ref 26. Copyright 1985 Academic Press.

Figure 7–7: Relationship between superposition of crystallographic molecular models and the alignment of amino acid sequence.[27] Thirty-two pairs of homologous proteins were chosen for which a crystallographic molecular model of each member of the pair was available. Structural cores were defined for each superposed pair of molecular models by including all equivalenced atoms in only the polypeptide backbone the distance between which was less than 0.3 nm. The cores for the pairs that were most distantly related included only 50% of the amino acids in the sequence. The root mean square distance between pairs of conjoined atoms of the cores were calculated and plotted against the frequency at which the same amino acid was found at the equivalent positions in the two cores, expressed as a percentage. By this procedure, the percent identity is considerably higher than it would be if the entire sequences had been aligned. Reprinted with permission from ref 27. Copyright 1986 IRL Press.

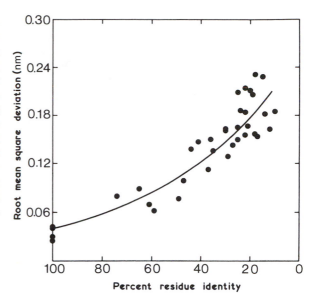

(*Carica papaya*) and that of the same enzyme from the fruit of the kiwi (*Actinidia chinensis*).[26] The best alignment of the sequences of these two proteins that could be made was one in which only 48% on the positions contained identical amino acids, yet the superposition is very close.

The degree to which two superposed structures coincide is usually quantified by the square root of the mean of the values for the squares of the distances between pairs of equivalent α-carbon atoms in the two sequences. This is referred to as the **root mean square deviation.** Unfortunately, there is a habit of discarding arbitrarily the distances between any pairs that are widely separated and including only values for the **cores** of the two molecules,[27] a practice that obscures the meaning of the criterion. In the case of the two thiol proteolytic enzymes superposed in Figure 7–6, the root mean square deviation between equivalent members in the two subsets of α-carbons composing the 73% of the protein judged to be the core was 0.04 nm,[26] which was considerably smaller than the nominal resolution (0.17 nm) of the two structures.

Whenever the crystallographic molecular models of two proteins whose amino acid sequences can be aligned with statistical significance have been compared, they could be readily superposed.[28–33] This universal observation can be presented quantitatively by relating percent identity between the aligned amino acid sequences and root mean square deviation of superposed crystallographic molecular models (Figure 7–7).[27] When the percent residue identity in the core reached around 15%, a point well below that at which statistically meaningful alignments of the complete amino acid sequences could no longer be made, the root mean square deviation of the α-carbons of the core residues (50% or greater of the total α-carbons) was only 0.2 nm, and the similarity in topology among the structures being compared was still unmistakable.

From these considerations, it can be concluded that whenever two proteins have amino acid sequences that can be aligned with statistical significance, they will also have superposable tertiary structures. For example, the fact that the amino acid sequences of the three enzymes Ca^{2+}-transporting ATPase, Na^+/K^+-transporting ATPase, and H^+-transporting ATPase can be aligned[34] demonstrates that their tertiary structures are superposable.[35] This rule is important because far more sequences are available than tertiary structures. If the amino acid sequence of a protein the crystallographic molecular model of which is unavailable can be related to a protein for which a crystallographic molecular model is available through a valid alignment of the two amino acid sequences, reliable conclusions can be drawn about the unknown tertiary structure by comparison with the known structure.

The process by which evolution by natural selection proceeds in the shaping of a protein is evident when the crystallographic molecular models of various cytochromes *c* are compared (Figure 7–8). The eukaryotic cytochromes *c* are indistinguishable from each other in tertiary structure,[36] if those from rice and tuna are assumed to represent the evolutionary extremes, and the eukaryotes are represented by the structure of the protein from the tuna (structure C in Figure 7–8). The other four structures in Figure 7–8 are the four bacterial cytochromes *c* whose sequences are aligned in Figure 7–1. They are from *C. thiosulfatophilum* (Figure 7–8A), *P. aeruginosa* (Figure 7–8B), *R. rubrum* (Figure 7–8D), and *P. denitrificans* (Figure 7–8E). The alignments presented in Figure 7–1 were based in part on superpositions[1,37] of these crystallographic molecular models. It has been argued that when they are available, such structurally derived alignments can be more reliable than statistically based alignments. This is an interesting argument because it assumes that no relative movement of β strands parallel to

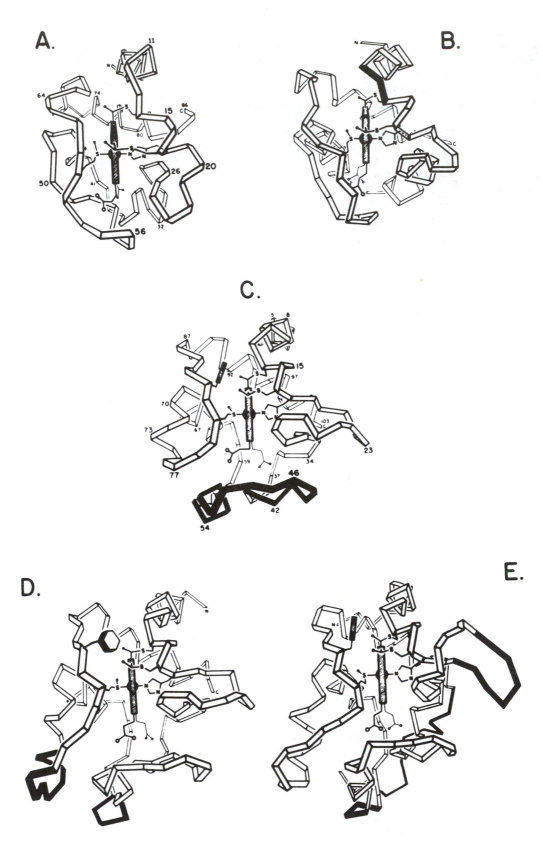

Figure 7–8: Tertiary structures of five cytochromes *c*.[2,39] Ribbon diagrams were made from the crystallographic molecular models of (A) cytochrome *c*-555 from *C. thiosulfatophilum*, (B) cytochrome *c*-51 of *P. aeruginosa*, (C) cytochrome *c* of tuna mitochondria, (D) cytochrome *c*₂ of *R. rubrum*, and (E) cytochrome *c*-550 of *P. denitrificans*. The sequences of these five cytochromes are aligned in Figure 7–1. Shaded portions of the ribbon diagrams highlight loops in which a high variability of the basic structure is encountered. The central structure for tuna cytochrome *c* is numbered with the same numbers as those used in Figure 7–1. Reprinted with permission from ref 2. Copyright 1982 Academic Press.

each other or advancement of the screw of an α helix has occurred during evolution. There is some support for this assumption.[38]

When the crystallographic molecular models of the cytochromes *c* (Figure 7–8) are compared, the reason for the gaps in the aligned sequences is immediately apparent. They represent loops in the longer protein that are missing from the shorter protein. These loops (darkened in the figure) can appear or disappear because sequence positions at their bases are near enough to each other to be connected without disrupting the structure in a major way. It has usually been observed that the insertions and deletions that are found in proteins whose tertiary structures can be superposed and that are responsible for the gaps in the aligned amino acid sequences occur in regions such as these that are peripheral to the central elements of the structure. An interesting spreading apart of two flexible flaps to accommodate the large loop present in the cytochrome *c* from tuna but missing in the two smaller bacterial cytochromes is also apparent.

The details of the variations that occur in the tertiary structure of a protein as amino acids are slowly replaced at the toleration of natural selection have been examined by superposing the nine available crystallographic molecular models of different globins.[38] The globins are a family of proteins for which many sequences are available. They include myoglobins, hemoglobins, erythrocruorins, and leghemoglobins. The crystallographic molecular model of each of the nine globins for which models are available could be superposed on the other eight models, and their sequences have been aligned from these superpositions.[38] Each globin is formed from eight α helices stacked one upon the others as a bundle of sticks would be in a fire. As the sequence of the protein has varied, the interdigitation of the amino acid side chains situated between the α helices has adjusted to accommodate changes in their size, and this has caused the helices to shift as rigid bodies with respect to each other. These adjustments are necessary because the amino acid side chains between the α helices are tightly packed together and many atomic contacts occur. As the shifting proceeds, accommodating changes in size, the individual pairs of atomic contacts persist between two amino acids at different positions in the amino acid sequence but next to each other in the tertiary structure even though the identities of the amino acids themselves change.

The buried amino acids in these crystallographic models of the globins can be distinguished from the amino acids on the surfaces. In practice this was done by rolling a sphere whose diameter was that of a molecule of water over the surface of space-filling crystallographic molecular models of each globin. All amino acid side chains with less than 0.1 nm^2 of surface area that could be touched by such a sphere were judged to be buried, and all other amino acid side chains were judged to be on the surface. In pairwise comparisons, the frequency at which identical residues were found at the same buried positions was

twice that at which identical residues were found at the same surface positions (see legend to Figure 7–7). It could be concluded that this behavior of natural selection reflected the more severe steric restrictions in the tightly packed interior.

About 60 amino acids out of the 140 in the polypeptide of a typical globin remain in very similar locations in the superposed crystallographic molecular models and account for the core of the native structure. Only half of these are buried; the ones on the surface remain fixed because they are within α helices which are themselves rigid structures. The regions in which the greatest variation in sequence and tertiary structure occurs are in the seven loops connecting the eight helices. This is due to their almost exclusive location at the surfaces of the molecular models but may also reflect the changes in the end to end distances between the α helices that were required to accommodate the slow shifts of the helices relative to each other as the packing among them was altered by mutation.

These observations suggest that the degree of conservation that is displayed by a position in the sequence of a protein may provide an indication of its location in the tertiary structure. Positions showing the least tolerance to replacement are often located on the interior of the protein and those displaying the greatest tolerance tend to be located on flexible surface loops, but the tendency is not overwhelming.

One common way in which a hypothesis about the role of a particular amino acid in the structure of a molecule of protein is tested is to examine how its location is occupied in the sequences of a family of homologous proteins. If the occupation of that location is confined to only those amino acids suitable for a particular role, this reinforces the importance of that role to the common identity of the proteins in that family. One protein from one species is never anything other than a representative of all of the proteins that are homologous to it in all species. No one particular protein is any more a paradigm of this family than any one individual is a paradigm of the population of a biological species. Therefore, characteristic and important positions in the structure of one protein from the family must also be characteristic and important in all the others. Conversely, any position in the common sequence of the family that shows unconstrained variation in its occupation is irrelevant to the fundamental identity of the protein that defines the family.

A refined molecular model of myoglobin from the sperm whale has been prepared,[28] and the structural roles of those 82 amino acids that were invariant among the 24 myoglobins that had been sequenced at the time were tabulated (Table 7–3).[28] This list represents a combination that has been retained since the time that all of these myoglobins shared a common ancestor. Myoglobins and hemoglobins diverged before the divergence of the myoglobins themselves. Positions marked (Hb) in

Table 7–3: Role and Location of Invariant Residues in Myoglobin[a,b]

structural location[c]	amino acid[d]	location[e]	role[f]
2NA	LEU	I	in contact with helix H: L[g]
1A	SER	E	hydrogen bond to GLU 4A NH to start forming helix A: TS
4A	GLU	S	hydrogen bond to SER 1A NH to stabilize LEU 2NA: DQE
5A	TRP	S	between LEU 2NA and LYS 2EF: KWRIA
8A	VAL	I	in contact with helix H; in bottom hydrophobic cluster: IV
9A	LEU	E	: KTLRAE
12A	TRP	S	hydrogen bond to GLU 16A, which in turn is bound to LYS 20E to hold helices A and E together (Hb)[h]: WF
14A	LYS	E	hydrogen bond to asp (glu, gln, asn) 4GH to stabilize GH-corner (Hb): KPDE
15A	VAL	I	in bottom hydrophobic cluster: VIF
16A	GLU	E	hydrogen bond to TRP 12A: G–EYAKN
1B	ASP	E	hydrogen bond to gly 4B NH to stabilize AB-corner: HND
5B	HIS	I	hydrogen bond to HIS 1GH to stabilize GH-corner: YVHSD
6B	GLY	I	in close contact with GLY 8E: GPT
10B	LEU	I	in hydrophobic cluster on HIS 7E side (Hb): LF
11B	ILE	S	blocking an opening between helices B and D: EGIVY
13B	LEU	I	in hydrophobic cluster on HIS 7E side: MLFHV
14B	PHE	I	in hydrophobic cluster on HIS 7E side: FL
1C	HIS	S	in close contact with phe (leu) 7G: FYHTDA
2C	PRO	E	sharp turn from helix B to helix C (Hb): P
3C	GLU	E	hydrogen bond to GLU 3C NH: TWEAS
4C	THR	I	van der Waals contact with heme, hydrogen bond to HIS 1C CO to start forming helix C (Hb): TAMI
5C	LEU	S	blocking an opening formed by helix C and CD-corner: KQLEAMK
6C	GLU	E	in contact with LYS 8CD through a water molecule: TAERD
1CD	PHE	I	van der Waals contact with heme parallel to heme plane; in HIS 7E side hydrophobic cluster (Hb): F
2CD	ASP	E	hydrogen bond to LYS 5CD to stabilize CD-corner: PEDGTS
4CD	PHE	I	in hydrophobic cluster on HIS 7E side of heme (Hb): F
5CD	LYS	E	hydrogen bond to ASP 2CD to stabilize CD-corner: G–KSAL
6CD	HIS	E	in contact with ASP 2CD CO through a water molecule to stabilize CD-corner: DHGK
7CD	LEU	S	in hydrophobic cluster on HIS 7E side (Hb): L–G
8CD	LYS	E	in contact with ASP 2CD CO through a water molecule to stabilize CD-corner: SKTG
5D	MET	I	blocking an opening formed by CD-corner and helix D: VMLIP
1E	SER	S	hydrogen bond to LEU 4E CO to start forming helix E: NSDT
2E	GLU	E	in contact with lys (arg) 5E through a water molecule: APE
4E	LEU	I	in hydrophobic cluster on HIS 7E side: VLF
6E	LYS	E	hydrogen bond to neighboring molecule: KRAEQ
7E	HIS	I	hydrogen bond to the sixth ligand molecule; van der Waals contact with heme (Hb): HL
8E	GLY	I	in close contact with GLY 6B (Hb): GA
11E	VAL	I	van der Waals contact with heme; in hydrophobic cluster on HIS 7E side: VI
12E	LEU	I	in bottom hydrophobic cluster: ALGIVF
14E	ALA	S	van der Waals contact with heme: ASEFL
15E	LEU	I	in bottom hydrophobic cluster. Van der Waals contact with heme vinyl group: LFVI
16E	GLY	S	in contact with TRP 12A: TSGDY
18E	ILE	I	in hydrophobic cluster on HIS 8F side: AGI
19E	LEU	I	in bottom hydrophobic cluster: VLIA
20E	LYS	E	hydrogen bond to GLU 16A to keep helices A and E stable: AGHKSI
1EF	LYS	E	hydrogen bond to neighboring molecule: HKSEQ
2EF	LYS	E	hydrogen bond to glu (asp) 2A to stabilize amino terminus of helix A: DKG–
3EF	GLY	E	: –TGV
5EF	HIS	S	hydrogen bond to ASP 18H to stabilize EF-corner and helix H: MLHKIS
7EF	ALA	E	: NGAS
4F	LEU	I	in hydrophobic cluster on HIS 8F side; in contact with heme (Hb): LVF
5F	ALA	S	in close contact with helix H: SAGV
7F	SER	S	van der Waals contact with heme; hydrogen bond to pro (his) 3F CO or LEU 4F CO or HIS 8F N(δ): LSKRV
8F	HIS	I	the fifth ligand to heme iron (Hb): H
9F	ALA	S	in close contact with helix H: AKV
1FG	LYS	E	(invisible) (Hb): KSYR
2FG	HIS	S	van der Waals contact with heme; hydrogen bond to propionic acid residue to stabilize heme and FG-corner: LHFG
3FG	LYS	E	(invisible): RHKEV
1G	PRO	E	sharp turn from FG-corner to helix G: DPKTA
5G	LEU	I	van der Waals contact with heme; in hydrophobic cluster on HIS 8F side: FL
6G	GLU	E	hydrogen bond to ARG 16H to stabilize helices G and H: KRENP
8G	ILE	I	van der Waals contact with heme: LIFV
9G	SER	S	hydrogen bond to LEU 5G CO: SGARK
12G	ILE	I	in bottom hydrophobic cluster: LIF
1GH	HIS	S	hydrogen bond to HIS 5B to stabilize GH-corner: LFHITV
5GH	PHE	I	in bottom hydrophobic cluster: FMW
3H	ASP	E	hydrogen bond to ala (val) 4H NH: APED
7H	ALA	S	in close contact with helix A: SAG
8H	MET	I	in bottom hydrophobic cluster: LYMFW
10H	LYS	E	hydrogen bond to GLU 4A to stabilize amino-terminal end of helix A (Hb): KAI
11H	ALA	I	in hydrophobic cluster: FVALT
12H	LEU	I	in bottom hydrophobic cluster: LVY
13H	GLU	E	in contact with asn 9H through a water molecule: ASERD
14H	LEU	E	: SGLMDTE
15H	PHE	I	van der Waals contact with heme; in hydrophobic cluster on HIS 8F side: VFIL
16H	ARG	S	hydrogen bond to GLU 6G to stabilize helices G and H: SARF
18H	ASP	E	hydrogen bond to HIS 5EF to stabilize EF-corner and helix H: VADFM
20H	ALA	E	: TARIFK
23H	TYR	I	hydrogen bond to ile (val) 4FG CO to stabilize C-end of helix H (Hb): YLM
24H	LYS	E	hydrogen bond to the carboxy terminus: RHKED
1HC	GLY	E	
4HC	GLY	E	

[a]Adapted from ref 28. [b]The amino acids listed are those that are invariant in all myoglobins, and the structural roles assigned are those in the crystallographic molecular model of myoglobin. [c]Position in the common crystallographic molecular model of the globin superfamily. Capital letters (A–H) indicate which α helix, from amino- to carboxy-terminal; and the numbers, the position in the α helix. Double letters refer to turns between the respective helices. The globins are all bundles of eight α helices. [d]Amino acids that are invariant over all myoglobins. [e]Location in the crystallographic molecular model of myoglobin: I, internal; E, external; S, surface crevice. [f]Three-letter amino acid abbreviations given in uppercase letters represent invariant residues in myoglobin; those given in lowercase letters are not invariant. [g]Amino acids appearing at each of these positions in nine superposed globins are noted. Dash indicates deletion. [h]Amino acids noted with (Hb) are invariant in all mammalian hemoglobins and common in all myoglobins.

Table 7–3 are those residues invariant among hemoglobins and myoglobins, and they represent amino acids that have been retained for an even longer period of time. Finally, only the amino acids that appear at these 82 positions in the nine globins aligned by superposition have been entered into the tabulation. An examination of Table 7–3 reinforces several features of the atomic structure of molecules of protein in general.

Positions in the sequence that are buried in hydrophobic clusters are the most invariant. Usually three or four members of the group, isoleucine, valine, phenylalanine, leucine, methionine, and alanine, will substitute among themselves in this role, but occasionally only one or two are suitable. For example, only leucine is found at position 2NA and only valine or isoleucine at position 11E in the globins. These two preferences may reflect the constraints of the intricate, interlocking stereochemistry in the interior. In two locations, positions 1CD and 4CD, only phenylalanine is found among all the globins, and presumably in this location the flat disk of the phenyl ring is essential to maintain the structure.

There are usually a number of locations in the structure of a protein where difficulties resulting from the packing of the backbone of the polypeptide arise. At position 2C in the globins, a proline seems essential to enforce a sharp turn. When two strands of polypeptide are forced too closely together, these tight locations, such as positions 6B, 8E, 5F, and 7H in the globins, are occupied by glycine, proline, alanine, serine, or threonine. Both serine and threonine, by forming cyclic hydrogen bonds (Figure 6–15), hug the polypeptide. These tight fits can also result from the juxtaposition of a large and bulky amino acid. The amino acid at position 16E in the globins is crowded by the tryptophan at position 12A in both hemoglobin and myoglobin.

Occasionally, an amino acid will form a hydrogen bond to a donor or an acceptor on the backbone at the beginning or end of an α helix. An example would be Threonine 4C in myoglobin and hemoglobin. It is often stated (Table 7–3) that this arrangement has the effect of initiating the α helix. The fact that at this location other globins lack an amino acid capable of forming a hydrogen bond and still contain the helix suggests that the assignment of such a purpose is an overstatement, at least in this case. Remarkably, four pairs of participants in ionized hydrogen bonds between side chains on the surface of myoglobin are invariant in the short term. When these particular interactions are examined, however, over all nine of the globins, which represent a much longer history of evolution, all of these hydrogen bonds are found to be dispensable (Table 7–4).

A deeply buried position in the sequence of a folded polypeptide remains invariably hydrophobic, but a buried location near the surface will occasionally erupt toward the water. For example, at position 65 in cytochrome c (Figure 7–1) an arginine appears at a location usually occupied by hydrophobic amino acids. Presumably the alkane portion of the arginine traverses the hydrophobic region and the guanidinium can push through the surface into the solvent.

Often a hydrophilic location on the exterior becomes occupied by a hydrophobic amino acid. For example, position 9A in the globins (Table 7–3) is on the exterior of the protein and is usually occupied by hydrophilic amino acids, but in the myoglobins it is occupied by leucine. Such a substitution has no effect on the free energy of folding for myoglobin compared to the other globins because the leucine is solvated equivalently in both the unfolded and folded polypeptide. This may explain why such exchanges are so common during evolution. There is, however, a price to be paid for such an exchange because a hydrophobic amino acid that replaces a hydrophilic amino acid on the surface of a protein renders it less soluble. The helical polymers formed by hemoglobin S, in which a glutamate on the surface has been replaced by a valine, are an example of such a problem. That they are helical polymers comes as no surprise because the interface created by this mutation was accidental. Similar mutations of hydrophilic amino acids into hydrophobic amino acids presumably have created the interfaces within multimeric proteins.

Table 7–4: Evolutionary Variation of Ionized Hydrogen Bonds[a]

structural location[b]	pairs of amino acids								structural location	pairs of amino acids							
2CD	P	E	P	D	**D**	P	G	T	S[c]	1B	H	N	H	N	**D**	D	N	–	N
5CD	–	G	–	G	**K**	K	S	A	L	19G	H	H	H	H	**R**	G	R	H	V
5EF	M	L	L	L	**H**	K	M	I	S	2EF	D	D	D	D	**K**	D	D	–	G
18H	V	A	V	A	**D**	L	A	M	V	2A	P	P	A	G	**E**	A	A	A	E

[a]Four invariant ionized hydrogen bonds that were present in the earlier refined crystallographic molecular model (Table 7–3)[40] and the latest refined crystallographic molecular model[41] of myoglobin were chosen for examination. The two amino acids forming each of these four hydrogen bonds were conserved among all of the myoglobins. The amino acids occupying these positions in the nine different globins are listed. [b]Positions in the common crystallographic molecular model of the globin superfamily. [c]The amino acid in each of the globins is paired above and below with the other amino acid from that same globin. The pairs for myoglobin are in **boldface**.

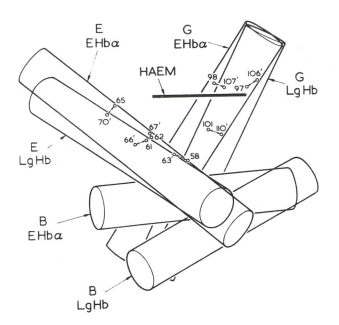

Figure 7–9: Arrangement of the α helices and contact residues that form part of the heme pocket in the α subunit of horse hemoglobin (EHbα) and in lupin leghemoglobin (LgHb).[43] The hemes in the two proteins are superposed. The three α helices are designated as B, E, and G, in order of their arrangement in the globin molecule. The positions of homologous pairs of residues that are in contact with the heme are indicated by open circles joined by arrows. The coupling of the shifts at the E–B and B–G helix interfaces keeps the residues that form the heme pocket in the same relative positions. Reprinted with permission from ref 43, originally from ref 38. Copyright 1980 Academic Press.

The fact that the mutation of a hydrophilic amino acid on the surface of a molecule of protein into a hydrophobic amino acid is energetically neutral may explain the observation that the surface of a molecule of protein usually contains a significant number of hydrophobic amino acids. For example, in the refined molecular model of chymotrypsin,[42] 20% of the total number of leucines, valines, and isoleucines are completely accessible to the aqueous phase and 30% are still partially accessible even though they are more than 50% buried, yet chymotrypsin does not aggregate catastrophically when it is in solution. These amino acids can be thought of as the product of neutral mutations, and a mutation of any one of them that puts a hydrophilic amino acid in place of the hydrophobic amino acid would also be a neutral mutation.

The globins also provide a particularly informative example of the focused constraints that natural selection places on the gradual shifts in position among segments of secondary structure during evolution. The invariant feature of both the structure and the function of a globin is the heme. Through all of the alterations encountered during evolution, the amino acids responsible for surrounding the heme and supporting it within the protein were required by natural selection to fulfill these roles. The record of this series of accommodations can be inferred from superposing molecular models of present globins such that their hemes are made to coincide. The situation is most graphically illustrated when the α subunit from equine hemoglobin is superposed in this way on leghemoglobin from lupin (Figure 7–9).[43] Over this long period of evolution, the amino acids supporting the heme have shifted their positions relative to it by only small distances. At the same time, however, the ends most distant from the heme of the two α helices in which these functionally critical amino acids reside (E and G in Figure 7–9) have shifted significantly in their position and

another α helix that provides no amino acids in contact with the heme (B in Figure 7–9) has shifted even more.

In any protein, a few amino acids that are critical for its function can be identified. Over evolution, natural selection maintains the relative separations and orientations among these amino acids because if it did not, the protein could no longer be what it is. The more distant a location within the protein is from these invariant points of reference, however, the more likely its position will drift as mutations accumulate that shift the orientations of the segments of secondary structure within the overall molecular structure of the protein.

A considerably more drastic evolutionary rearrangement of the tertiary structure of a protein can be seen in a comparison of the crystallographic molecular models of the phospholipase A2 from cobra venom and the phospholipase from bee venom.[44] From aligned amino acid sequences and superposed crystallographic molecular models, there is no doubt that these two proteins share a common ancestor. The face of the protein from cobra venom that interacts with the surface of a biological membrane, in which the reactants for the enzyme are found, is formed by the first 20 amino acids of the polypeptide. When the amino acid sequence of the protein from bee venom, however, was aligned, with the help of the crystallographic models, with that from cobra venom, it could be seen that these first 20 amino acids, present in the protein from the cobra, were missing from the protein from the bee. The missing tertiary structure is supplied but from the carboxy-terminal half of the polypeptide of the protein from the bee. In the protein from the cobra, the last 16 amino acids (amino acids 111–126) following the last central α helix of the tertiary structure meander off in one direction. In the protein from the bee, there are another 58 amino acids (amino acids 77–134) following the same central helix; and, after emerging from that α helix, the polypeptide folds over to take a completely different path. On this path it fills in the vacancy caused by the missing amino terminus before meandering off to finish at the opposite side of the molecule from that at which the polypeptide from the cobra ends. In filling in the vacancy, an α helix from the carboxy-terminal end of the polypeptide from the bee takes the place of an α helix from the amino-terminal end of the polypeptide from the cobra but runs in the opposite direction. In the same comparison, it can also be seen that a loop in the middle of the polypeptide from the cobra (amino acids 71–91) has been transferred by a tandem internal gene fusion to the carboxy terminus (amino acids 95–128) of the polypeptide from the bee. From the locations of the various rearrangements, it is clear that the transfer of this loop occurred after the carboxy terminus reoriented and filled in the vacancy left by the loss of the amino terminal.

As the number of crystallographic molecular models has increased, instances have become more common in which two proteins that display no statistically significant homology in amino acid sequence nevertheless have segments of tertiary structure that can be superposed. An example of such a recurring structure is found in the crystallographic molecular models of L-lactate dehydrogenase,[25] alcohol dehydrogenase,[45] phosphoglycerate kinase,[40] and phosphorylase.[46] This common segment is 140–200 amino acids in length and occurs at different locations in the overall sequences of these proteins. It is formed from the amino acids in the sequence between Asparagine 21 and Glycine 162 in dogfish L-lactate dehydrogenase, Phenylalanine 207 and Serine 392 in equine phosphoglycerate kinase, Serine 193 and Phenylalanine 319 in equine liver alcohol dehydrogenase, and Asparagine 559 and Arginine 713 in rabbit muscle phosphorylase. All four structures can be superposed.[25,46] The superposition of these regions from phosphorylase and lactate dehydrogenase is presented in Figure 7–10A.

This particular topological pattern of secondary structures can be defined as a doubly wound, parallel β sheet. It is a six-stranded sheet of parallel β structure flanked on both sides by α helices. The basic rhythm of the recurring theme is β structure, α helix, β structure, α helix, β structure, random meander, β structure, α helix,

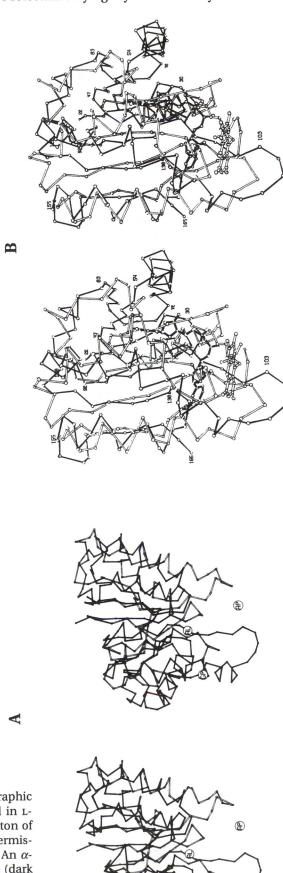

Figure 7–10: Doubly wound, parallel β sheets found in two crystallographic molecular models. (A) An α-carbon skeleton of this structure found in L-lactate dehydrogenase (dark lines) superposed on the α-carbon skeleton of this structure found in phosphorylase (open lines).[46] Reprinted with permission from ref 46. Copyright 1976 *Journal of Biological Chemistry*. (B) An α-carbon skeleton of this structure found in L-lactate dehydrogenase (dark lines) superposed on the α-carbon skeleton of this structure found in flavodoxin (open lines).[25] The bound NAD+ in the crystallographic molecular model of lactate dehydrogenase has been included. Only the skeleton of L-lactate dehydrogenase is numbered. Reprinted with permission from ref 25. Copyright 1974 *Nature*.

β structure, α helix, β structure. You should trace this pattern in Figure 7–10A. The six strands of β structure all run parallel to each other to form a pleated sheet, and the helices arch above or below the sheet to connect the end of one strand of β structure to the next. The complete and concise theme is developed in L-lactate dehydrogenase (Figure 7–10), and there are variations on this theme in the other proteins. For example, in phosphoglycerate kinase, there are two α helices after the second strand of β structure and a long additional loop after the third strand of β structure, and in alcohol dehydrogenase the last α helix is replaced by an additional strand of antiparallel β structure. An interesting variation occurs after the first strand of β structure in the structure from phosphorylase, where a large bulge has appeared that pushes up the loop between the third and fourth strands of β structure (Figure 7–10A).

Flavodoxin also seems to share this pattern but with a more significant variation. In this protein, the second α helix and the third strand of β structure have been deleted (Figure 7–10B). This deletion seems to have been very similar to those seen in cytochrome c (Figure 7–8) in that the loop containing the α helix and the strand of β structure has simply been pinched off from the open end of the common structure. Nevertheless, the superposition of flavodoxin upon the corresponding region from L-lactate dehydrogenase is quite close, even though

the sequences of these two superposed polypeptides appear to be completely unrelated. When the sequences are aligned, even with the assistance of the superposition, they have identical amino acids in only 9% of their aligned positions.

The conclusion that has been drawn from these superpositions is that all of these regions from these very different proteins together share a common evolutionary ancestor. As these structures represent only a portion of each of the presently existing proteins, and as the rest of the proteins bear no resemblance, this common ancestor must have been a small primordial protein that was combined covalently with other small primordial proteins by gene fusion to produce larger chimeric proteins. **Gene fusion** is an infrequent process in which genomic DNA is recombined incorrectly so that segments of different genes become fused together rather than as in the usual process of recombination in which allelic segments of the same gene are interchanged in precise alignment. The evidence of these early gene fusions can still be observed in the progeny. The reason that each resides at a different location in the sequences of the present proteins is that each was combined with different proteins in different orders during these gene fusions. This description of evolutionary history requires that at one time in the distant past each of the segments of these polypeptides now folding to produce each of these superposable regions was not attached to the remainder of the polypeptide to which it is now joined. If this is the case, then each of the doubly wound β-pleated sheets and the other regions of each of these proteins were at one time separate proteins, and the evolution of proteins is a history of the joining together of smaller domains to produce ever larger proteins.

Suggested Reading

Rossmann, M.G., Moras, D., & Olsen, K.W. (1974) Chemical and Biological Evolution of a Nucleotide-Binding Protein, *Nature 250*, 194–199.

Domains

Alcohol dehydrogenase from equine liver is a protein composed of a polypeptide 374 amino acids in length. As has been noted, the polypeptide between Serine 193 and Phenylalanine 319 assumes a doubly wound, parallel β sheet in the native enzyme, a structure seen in the upper right of Figure 7–11 that recurs in other proteins. The amino-terminal half of the protein, however, is composed of polypeptide folded into a very different conformation, and this conformation occurs in none of the proteins in which the carboxy-terminal structure is found.[47] When the tracing of the α-carbon skeleton (Figure 7–11) is viewed from the proper angle, the region between Serine 193 and Phenylalanine 319 seems to be folded independently of the rest of the protein.

Figure 7–11: α-Carbon skeleton of the crystallographic molecular model of alcohol dehydrogenase from equine liver.[48] The polypeptide is numbered from amino terminus to carboxy terminus. The domain from Serine 193 to Phenylalanine 319 is a doubly wound, parallel β sheet. The amino-terminal half of the protein (before Serine 193) is folded separately. The last 55 amino acids adhere to the amino-terminal domain. Reprinted with permission from ref 48. Copyright 1976 Academic Press.

A number of observations, among them the ones just described, have led to the conclusion that the native structures of many folded polypeptides can be divided into independent domains. A **domain** is any region within the native tertiary structure of a protein for which evidence can be provided of an existence independent of the rest of the polypeptide. There are several types of independent existence that qualify a region within the native structure as a domain.

The most obvious evidence that two regions of the same protein are domains is that a limited cleavage of the polypeptide releases independent fragments of the protein which retain their respective native structures. In such instances the two or more separated fragments would represent **detachable domains**. The paradigm of a protein with detachable domains is immunoglobulin G, a circulating antibody responsible for binding to foreign proteins or antigens. Thirty years ago, Porter[49] demonstrated that intact immunoglobulin G could be cleaved by the thiol proteolytic enzyme papain into three pieces of almost equal sizes. Two of these detachable domains, the Fab fragments, retained the ability to bind antigens, and the third, the Fc fragment, was stable enough that it crystallized spontaneously during its isolation. The Fab fragments could be readily separated by cation-exchange chromatography from the Fc fragments without any loss of biological activity. Crystallographic studies of Fab fragments, Fc fragments, and the full immunoglobulin G (Figure 7–12) have shown that the detached and separated domains retain the respective structures that they had in the intact molecule before it was cleaved.[50] In Figure 7–12, the Fab portions are the two vertical wings and the Fc portion is the horizontal trunk, and cleavage occurs at the two joints.

It was unfortunate that the association between limited proteolytic cleavage and domains was established with these elegant experiments. The more important fact that Porter separated the detached domains from each other and demonstrated that each was still structurally intact is often forgotten. It has already been noted that a protein must be prepared for complete proteolytic digestion by unfolding it. This is because most of the peptide bonds susceptible to proteolytic digestion in the unfolded polypeptide are not susceptible to digestion in the native folded polypeptide. The cleavage of a native protein by a proteolytic enzyme usually occurs at only one or two locations. The sites at which proteolytic cleavage of a native protein can occur are exposed loops of polypeptide on its surface. Such loops are rarely situated between domains, and domains are rarely connected by such loops. If a polypeptide is cut at only one or two positions by a proteolytic enzyme when it is folded in its native conformation, this is not evidence that the fragments observed compose separate domains in that native conformation.

If, however, the protein can be digested and the pieces that result can be separated as biologically active

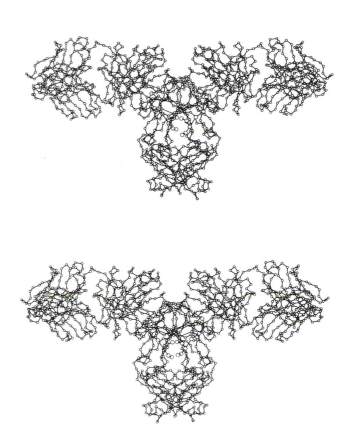

Figure 7–12: α-Carbon skeleton of the crystallographic molecular model of a human immunoglobulin G designated Dob.[50] The smaller circles are the α-carbons, and the larger circles represent hexoses. The two Fab domains are aligned vertically, and a horizontal 2-fold rotational axis of symmetry bisects the molecule through the Fc domain. Reprinted with permission from ref 50. Copyright 1977 National Academy of Sciences.

or structurally intact moieties, they are detachable domains. Few examples of this have been reported, and among those are the following. The enzyme sulfite oxidase can be cleaved with either trypsin, chymotrypsin, or papain to produce two detached domains that can be separated from each other by molecular exclusion chromatography.[51] One retains the ability to transfer electrons from sulfite to $Fe(CN)_6^{3-}$; the other retains the spectrum characteristic of the heme in its native environment. The transfer of electrons from sulfite all the way to the ultimate oxidant, cytochrome *c*, can no longer occur because the domain containing the heme is no longer attached to the domain at which the sulfite is oxidized. Anion carrier is a protein in the plasma membrane of erythrocytes and is responsible for anion transport. It can be cleaved with chymotrypsin to produce a water-soluble domain that can be readily separated from the other domain, which remains in the membrane.[52] Between them the two detached domains retain the biological functions that are displayed by the intact protein, and both seem to retain the structures they had in the intact protein.

There is a protein in the bacterium *Escherichia coli* that is responsible for the two enzymatic activities of aspartate kinase and homoserine dehydrogenase. It is composed of one polypeptide 820 amino acids in length. When this protein is digested with a proteolytic enzyme from *Streptomyces griseus*, the polypeptide is cut into two pieces 520 and 300 amino acid residues in length.[53] The longer piece comprises the carboxy-terminal 65% of the intact molecule; the shorter, the amino-terminal 35%.[54] The carboxy-terminal piece retains the homoserine dehydrogenase activity. The amino-terminal piece, however, shows only 1% or less of the aspartate kinase activity originally present, and even this small amount of activity could result from an undetected contaminant. A piece retaining normal levels of aspartate kinase activity, however, can be prepared genetically by deleting the carboxy-terminal 35% of the polypeptide.[55] It seems clear that the amino-terminal half of the polypeptide folds to form aspartate kinase; and the carboxy-terminal half, homoserine dehydrogenase. The fact that aspartate kinase is entirely lost upon digestion, which is limited to a region less than 10 amino acids in length even when other proteolytic enzymes are used,[54] suggests that the digestion occurs well to the amino-terminal side of the boundary between the enzymatic domains. If this is the case, the cleavage within the domain responsible for aspartate kinase causes it to unfold and the unfolded amino-terminal fragment to separate from the domain responsible for homoserine dehydrogenase, which remains folded. If the point of proteolytic cleavage were a true boundary between two detachable domains, rather than an adventitious loop of polypeptide on the surface of the aspartate kinase, the aspartate kinase activity, which can readily be expressed by the genetically deleted protein, should have been unaffected. Aspartate kinase–homoserine dehydrogenase, then, is an example of a protein that has domains that cannot be detached from each other, at least not by this proteolytic cleavage.

Numerous examples of proteins such as aspartate kinase–homoserine dehydrogenase, which contain single polypeptides folded to produce two or more distinct enzymatic activities, are known. They belong among a more general class of proteins known as multienzyme complexes. A **multienzyme complex** is a protein that, although it is a single, discrete macromolecule, is able to catalyze two or more enzymatic activities. It is generally assumed that each of the enzymatic activities in one of these multienzyme complexes is expressed by its own unique domain in the folded polypeptide or polypeptides. Such an **enzymatic domain** within a larger protein is a domain that is by itself independently responsible for a particular enzymatic activity.

Two paradigms of a multienzyme complex would be the fatty acid synthases found in fungi and animals, respectively. In prokaryotes and plants,[56,57] the seven enzymes and the acyl carrier protein responsible for the synthesis of fatty acids from acetyl-SCoA and malonyl-SCoA are individual proteins that can be separated and individually purified. In fungi and animals, however, all of these activities are expressed by a single protein. In fungi, the protein is constructed from two polypeptides that are encoded by different genes and are completely different in sequence from each other.[58,59] Their lengths are 1890 and 1980 amino acids, respectively.[58,60] The fatty acid synthase from animals, however, is constructed from only one polypeptide, 2440 amino acids in length.[61] All seven of the enzymatic activities and the acyl carrier protein are located on the single polypeptide comprising the animal enzyme, and the domains responsible for each have been identified in the amino acid sequence.[61] The order in which these enzymatic domains occur on the single polypeptide of the animal fatty acid synthase appears to be unrelated to the orders in which they appear on the two unique polypeptides from fungi.[62] On the basis of this fact, it has been concluded that all or most of the gene fusions that produced the animal protein and the fungal protein must have occurred as independent events after the lineages of these two kingdoms diverged from their common ancestor. These separate processes would be ones in which each enzymatic domain has been shuffled into a larger protein as a modular unit of evolution. Nevertheless, the individual domains, even though fused in different orders, still retain significant homologies in amino acid sequence when the one responsible for a given activity in the fungal enzyme is aligned with the one responsible for the same activity in the animal enzyme.[62] It follows that the tertiary structure of each pair of corresponding domains in the enzymes from fungi and animals would be superposable even though they occur in different orders along the different polypeptides, and it can be assumed that each enzyme in each chimeric protein retains the compact, integral tertiary structure that it had before the gene fusions occurred. By this chain of logic, the fact that several enzymatic activities are expressed by several individual proteins in some species yet all of them are expressed by only one protein formed from one folded polypeptide in other species is sufficient evidence of an independent existence to conclude that the latter protein is constructed from enzymatic domains.

Animal fatty acid synthase has been submitted to digestion by proteolytic enzymes. It is not surprising that a polypeptide of this length ($n_{aa} = 2440$) shows several locations in its native configuration at which it is susceptible to proteolytic cleavage,[63] but whether or not these coincide with the boundaries between domains has not been settled. Upon limited digestion with trypsin, however, the enzymatic domain responsible for the thioesterase activity can be detached and purified[64] without affecting any of the remaining activities.[65] In fact, it can be interchanged with a free thioesterase from a different tissue to alter the distribution of products from the complete enzyme.[66]

There are other examples of proteins containing enzymatic domains. In *E. coli* there is a protein responsible for both the chorismate mutase reaction and the prephenate dehydratase reaction. The bifunctional protein is constructed from a single polypeptide 360 amino acids in length.[67] In several other species of bacteria these two respective reactions are catalyzed by two different monofunctional proteins.[68] In fungi, three enzymes involved in the biosynthesis of histidine[69,70] are enzymatic domains of the same folded polypeptide, 850 amino acids in length, and five enzymes involved in the biosynthesis of the aromatic amino acids are enzymatic domains of the same folded polypeptide, 1350 amino acids in length.[71]

In animal tissues, there is a single protein responsible for the reactions catalyzed by carbamoyl-phosphate synthase (glutamine-hydrolyzing), aspartate transcarbamylase, and dihydroorotase,[72] reactions catalyzed by discrete proteins in bacteria. This trifunctional protein is composed of one polypeptide, 1800 amino acids in length. It can be dissected with the proteolytic enzyme elastase through a series of cleavages that eventually produce two smaller proteins, 400 and 360 amino acids in length.[73] The larger retains all of the dihydroorotase activity; and the smaller, all of the aspartate transcarbamylase activity originally present in the undigested protein. The activity of carbamoyl-phosphate synthase (glutamine-hydrolyzing), however, is lost as the digestion proceeds and can be associated with none of the fragments smaller than 1700 amino acids in length. In situations such as this, digestion of one domain, for example, the carbamoyl-phosphate synthase domain, at some point on its surface could cause it to unfold and render the polypeptide much more susceptible to proteolytic cleavage in a region forming the boundary between that unfolded domain and a neighboring domain. An example of such a pruning of an unfolded segment of polypeptide from a properly and compactly folded protein by proteolytic digestion occurred during the production of hybrids of different portions of staphylococcal nuclease.[74]

Proteins constructed from enzymatic domains presumably arose as the result of the fusion of the individual genes that encoded their smaller ancestors. There is some suggestion that the likelihood of this fusion was enhanced when the ancestral genes were gathered together in the same operon and consequently placed adjacent to each other in the DNA.[70] Many artificial fusions of two genes to produce chimeric proteins have recently been performed and the products that result from these artificial fusions seem to be little affected functionally.[75]

When two or more regions of a protein unfold independently of each other, they can be considered to be **separately melting domains**. Although such domains can be detected by studying the behavior of the protein during its denaturation in solutions of agents such as urea and guanidinium chloride,[76] a more sensitive method is calorimetry. Fibrinogen is a protein constructed from two copies of each of three polypeptides that are combined in such a way that the intact protein contains a central detachable domain, domain E, and two identical peripheral, detachable domains, domains D. The two domains D are attached to domain E by ropes constructed from three strands of α helix,[77,78] and domains D and E can be detached from each other by cleaving the ropes with proteolytic enzymes. When fibrinogen is submitted to differential scanning calorimetry,* two clearly separated transitions can be observed (Figure 7–13). These have been assigned to the melting of domains D and E, respectively.[79]

The melting, or unfolding, of the separately melting domains of fibrinogen is an irreversible process under the conditions chosen,[79] but unfolding and refolding back to the native structure are often reversible processes, even in a calorimeter. Domains that by themselves reversibly unfold and refold would be **independently folding domains**. They should be distinguished from domains that can only unfold separately. Plasminogen is a protein composed of at least seven domains. These are five homologous sequences about 90 amino acids in length that repeat consecutively within the entire sequence and two additional segments of polypeptide on each side of this pentuplication. Several of these domains or combinations of these domains can be detached and isolated separately. The reversible unfolding and refolding of five of these detached pieces could be followed by differential scanning calorimetry. These individual measurements could be combined to show that the rather complex, fully reversible calorimetric curve obtained with the intact protein was actually the sum of seven independent transitions.[41]

In many instances where independently folding domains are found in a protein, it is unclear whether or not they could fold properly if detached from their neighbors.

---

*A differential scanning calorimeter is used to measure the difference in the absorption of heat, as the temperature is raised, between a solution containing a protein and an identical solution lacking the protein. Two cells, sample and reference, contain precisely matched coils that introduce identical quantities of heat into each of them and establish a constant rate of temperature increase. The sample cell has an auxiliary coil that provides the additional heat necessary to keep its temperature exactly the same as that of the reference cell. The power supplied to the auxiliary heater is a measure of the excess heat absorbed by the sample, the endothermic heat flow. A protein unfolds, or melts, as the temperature rises, and this transition proceeds with the absorption of heat. This absorption of heat is a convenient way to follow the progress of the unfolding.

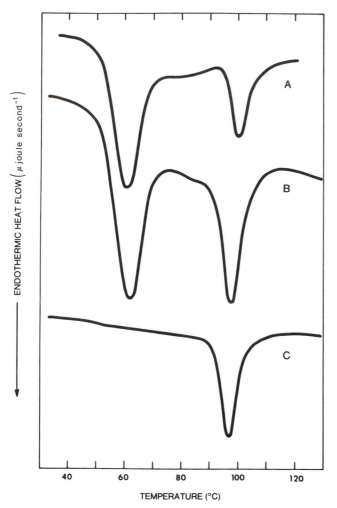

Figure 7–13: Thermal melting of domains D and E of bovine fibrinogen.[79] (A) Native intact fibrinogen. A solution (26 μL) of 88 mg mL^{-1} native intact fibrinogen was introduced into the sample chamber of the differential scanning calorimeter, and endothermic heat flow (microjoule second^{-1}) into the sample in excess of the flow into an identical solution lacking the protein was recorded as a function of temperature (°C). (B) A solution (25 μL) of a 2:1 molar mixture of the proteolytically detached and chromatographically purified domains D and E was used as the sample at a final concentration of 101 mg mL^{-1}. (C) A solution (22 μL) containing only proteolytically detached and chromatographically purified domain E was used as the sample at a final concentration of 47 mg mL^{-1}. Upper calibration mark is for traces A and B, and lower calibration mark is for trace C. Heating rate for all traces was 10 °C min^{-1}. Adapted with permission from ref 79. Copyright 1974 National Academy of Sciences.

There are reports that fragments, produced by the cleavage of certain polypeptides and purified chromatographically in their unfolded state, can refold to assume the conformation they had in the native protein from which they were derived and therefore must represent independently folding domains. For example, a fragment of the polypeptide comprising the neutral proteinase thermolysin from *Bacillus thermoproteolyticus* can be pro-

duced by cleavage with cyanogen bromide and purified by molecular exclusion chromatography in its unfolded state. This fragment contains the last 111 amino acids of the intact protein (sequence positions 206–316). It can be induced to refold. The solution that results contains protein that has an α-helical content, determined spectroscopically, close to that predicted from the crystallographic molecular model of thermolysin.[80] The protein in this solution melts, but at a temperature 20 °C below that at which native thermolysin, cut between residues 225 and 226, melts. The refolded protein unfolds in solutions of the denaturant guanidinium chloride, but at concentrations of guanidinium half those at which the nicked thermolysin unfolds. If the shorter polypeptide were folding to assume the same structure it had in native thermolysin, that structure is now much less stable. But the results are also consistent with it assuming a completely different conformation. There is no evidence, such as analyses by molecular exclusion chromatography or disc gel electrophoresis, that this refolded fragment is unaggregated, monodisperse, and homogeneous. It is possible that this solution contains aggregated protein folded in many different conformations whose average α-helical content resembles that expected.

A claim that an independently folding domain has been produced requires an unambiguous demonstration that the domain can refold into the conformation it assumes in the parent protein. For example, one of the detachable domains of plasminogen, known as Kringle 4, is able to bind lysine even after it has been separated from the remainder of the protein. This fragment of the protein can be unfolded and its three cystines cleaved by disulfide interchange with dithiothreitol. It will then refold to form its original native structure, as demonstrated by the ability of this refolded structure to bind lysine and to enforce the formation of only the properly paired disulfides upon its exposure to oxygen.[81] In this case, the proper pairing of the cysteines, located at distant positions in the amino acid sequence, is the result of their proper juxtaposition in a properly folded polypeptide.

There are examples of proteins in which two domains in the same folded polypeptide, while rigidly retaining within themselves their separate conformations, can move with respect to each other. Usually they are attached to each other by a much narrower region in the protein that acts as a hinge to permit this movement. Domains related in this way can be referred to as **hinged domains**. Such rearrangements have been demonstrated by obtaining maps of electron density from different crystals of the same protein in two different conformations, referred to as open and closed (Figure 7–14).[82] These maps of electron density are static images, and the kinetics of the transition between these two conformations cannot be studied in this way. They usually represent two independent crystallographic experiments because the two crystals of the protein in its two respective conformations are so different as to be not isomorphous.

Crystallographic evidence for hinged domains has been provided for hexokinase,[82] citrate synthase,[83] alcohol dehydrogenase,[84] and viral coat proteins, such as that from the tomato bushy stunt virus.[85] In all of these instances, there are two domains that swing on the hinge. Either one of the two domains in the open conformation can be superposed upon the same domain in the closed conformation, and often only a few changes in the structure of either domain occur upon closing the protein. In alcohol dehydrogenase, a loop of polypeptide in a region of one of the domains farthest from the hinge, where the largest movements occur, has to be pushed aside to permit closure,[84] but no changes seem to occur in the core regions of the two domains in this protein as they move. The movements of all of these pairs of hinged domains have been considered to be rigid-body rearrangements of the separate domains.[82–85]

The regions of these proteins that act as the hinges must adjust to absorb the torque generated by the movements of the domains. In alcohol dehydrogenase this involves a relatively minor rearrangement, the sliding of two helices in this region across one another.[86] In citrate synthetase, however, the distortions arising from the movements seem to spread evenly over almost the entire structure of the smaller of the two domains,[86] and this belies the description of the movements in this protein as rigid-body rearrangements.[83] It does provide, however, an interesting example of the plasticity available to proteins the size of this small domain, which contains 160 amino acids folded into five stacked α helices. This smaller domain elastically expands outward in an arclike movement during the closure.[86]

Because there is such a variation in both the extent of the protein distorted by the closing of the hinge and the number of strands composing the hinge, there is some hesitation in assigning an independent evolutionary existence to all hinged domains, and hence even designating them as domains. In proteins where only one or two strands of polypeptide pass through the hinge and the two domains swing as intact rigid bodies, the case for their independent evolutionary existence and the case for the protein as a product of gene fusion is quite strong. If several strands pass through the hinge and the conformational change closing the structure spreads over a large portion of one domain, as in citrate (*si*) synthase, so that it does not appear to respond as a rigid body, the possibility that the two domains represent separate, fused ancestors seems less likely.

In alcohol dehydrogenase, however, the two hinged domains do seem to represent independent ancestral proteins. The doubly wound, parallel β sheet between Serine 193 and Phenylalanine 319 (Figure 7–11) is one of the two hinged domains that moves during the closing of alcohol dehydrogenase. It has already been noted that it is a type of domain that recurs in several other proteins, in association with various other domains that do not resemble its particular partner in alcohol dehy-

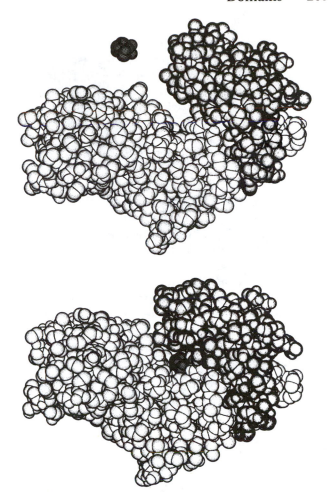

Figure 7–14: Space-filling representations of two crystallographic molecular models of hexokinase. Hexokinase was crystallized in the absence of substrates and the molecules were incorporated into the crystals (unit cell; a = 16.65 nm, b = 5.92 nm, c = 5.85 nm) in the open conformation.[82] These crystals produced the top crystallographic model. Hexokinase was also crystallized in the presence of glucose and $HOPO_3^{2-}$ and the molecules were incorporated into the crystals (unit cell; a = 14.48 nm, b = 7.88 nm, c = 6.19 nm) in the conformation in which the protein had closed around the glucose. These latter crystals produced the bottom crystallographic molecular model. The two domains are highlighted by using different shades of gray. Glucose is the stippled molecular model. The domains undergo a rigid-body reorientation of 12° about the hinge. Reprinted with permission from ref 82. Copyright 1980 Academic Press.

drogenase.[87] A **recurring domain**, such as this one, is a domain that is folded with a tertiary structure that can be superposed on the tertiary structures of other domains in other proteins of otherwise entirely different structure. A recurring domain is a compact module used in several distinct situations. Because of this recurrence in different situations, there is little doubt that a domain of this type has had an independent existence.

Pyruvate kinase is one of the more interesting examples of a protein built from recurring domains.[88] Its central domain is superposable on the entire folded polypeptide of triosephosphate isomerase. To one side of this central domain is another domain folded with the same topology as one of the 12 superposable domains in immunoglobulin (Figure 7–12).[47] To the other side of the central domain is a domain that can be superposed on half of the doubly wound, parallel β sheet found in alcohol dehydrogenase and lactate dehydrogenase.[88]

Recurring domains sometimes appear to be associated with a particular functional role. For example, p-hydroxybenzoate hydroxylase and glutathione reductase both contain a recurring domain about 160 amino acids in length.[89] In both of these enzymes, the domain serves to bind tightly an integral flavin coenzyme, and it has been referred to as the "FAD-binding domain."[89]

Internally repeating domains are members of a set of consecutive segments within the same polypeptide, each homologous in amino acid sequence to the other members of the set or each folded in a tertiary structure superposable on the tertiary structures of the other members of the set. An example of such internally repeating domains are the 12 domains, four in each of the two long

polypeptides and two in each of the two short polypeptides, that compose immunoglobulin G (Figure 7–12). Each of these internally repeating domains is a seven-stranded barrel of antiparallel strands of β structure. Each is superposable on all the others and shares significant homologies in amino acid sequence with some of the others. Such internally repeating domains are assumed to arise from internal duplication or the consecutive internal multiplication of a gene encoding a smaller protein. This implies that the single, unrepeated amino acid sequence or tertiary structure of a single one of these domains existed on its own at some time in the past, before the multiplication occurred. As such, each of these domains in the enlarged protein has had an independent existence.

Quite a few proteins have **internal duplications**. For example, the two halves of the doubly wound, parallel β sheet as it presently occurs in L-lactate dehydrogenase can be superposed upon each other (Figure 7–15). It has been proposed that the entire structure arose itself from a gene duplication in which the two segments of polypeptides encoded by the duplicated gene remained consecutively attached to each other and then began to evolve

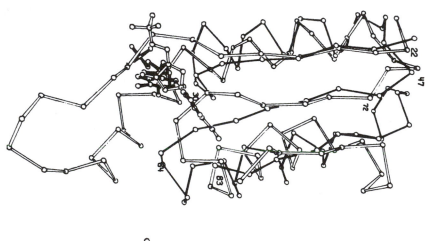

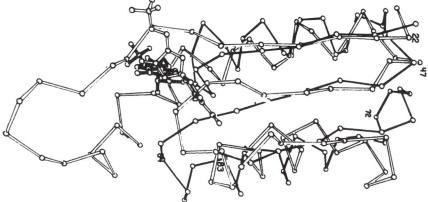

Figure 7–15: One half of the doubly wound, parallel β sheet found in lactate dehydrogenase (open line), as an α-carbon skeleton, superposed on the other half of the same doubly wound, parallel β sheet (dark line).[25] Only the first half is numbered. Adenosine monophosphate (open lines) and nicotinamide mononucleotide (dark lines) are included as they are bound to the two halves of this structure in the crystallographic molecular model. Reprinted with permission from ref 25. Copyright 1974 *Nature.*

α
β

Figure 7–16: Hypothetical model of the protein spectrin from erythrocytes.[95] A repeating pattern has been detected in the amino acid sequences of the two different polypeptides (n_{aa}^{α} = 2430, n_{aa}^{β} = 2200) composing this protein. The repeating pattern is 106 amino acids in length and occurs 22 times in the α polypeptide[44] and about 18–20 times in the β polypeptide.[95] From numerous physical measurements, it has been concluded that each repeating domain is a bundle of three α helices and that the bundles are strung together as shown. Reprinted with permission from ref 95. Copyright 1984 *Nature.*

independently but within the same protein. That this did happen is supported by the fact that domains are found in the crystallographic molecular models of pyruvate kinase[88] and phosphoglycerate kinase that are long enough to superpose only on half of the entire structure from L-lactate dehydrogenase.[90] The lineages leading to these two shorter regions of polypeptide could have diverged before the gene duplication produced the common ancestor of the larger. The serine proteolytic enzymes,[91] thiosulfate sulfur transferase,[92] and hexokinase[93] are other proteins that seem to contain internal duplications. Serum albumin is an example of a protein in which an internal triplication has been identified; the demonstration of this triplication came about by aligning repeating sequences within the same polypeptide.[94]

A few proteins have many consecutive internally repeating domains. Examples would be the five independently melting domains of plasminogen that arose from a quintuplication or the 38 internally repeating domains consecutively occurring in the two unique folded polypeptides of the protein spectrin.[95] In the latter case, the folded domains are thought to sit as pearls on a string and create a long flexible protein (Figure 7–16). This protein offers an excellent example of the absence of a correlation between locations at which proteolytic cleavages occur upon the surface of a native protein and the boundaries between its domains. Of the 15 cleavages of native spectrin produced by trypsin,[96] only four occur that are even near the boundaries of the domains.[95]

A fragment containing the covalently attached lipoyl coenzymes from the intact protein dihydrolipoyl transacetylase has been cleaved from the rest of the protein with trypsin and isolated by molecular exclusion chromatography.[97] Although it was designated a detachable domain on the basis that its lipoyl functionalities could be enzymatically acetylated, the designation was questionable. Lipoamide unattached to any protein can also be enzymatically acetylated, and thus this reaction is not indicative of the native structure. The protein isolated also had a larger frictional coefficient than it would have had if it were compactly folded, and this also suggested that it had been altered after it became detached. Although no unequivocal evidence was presented that this fragment in its isolated form retained the structure it had in the intact protein, there is independent evidence

that it actually is a domain of the transacetylase polypeptide. The region of the amino acid sequence of the transacetylase contained in the fragment is composed of three consecutive internally repeating domains, and the last of these three ends very near the point at which the fragment was cleaved from the rest of the protein.[20]

In the amino acid sequences of a number of vertebrate proteins, segments have been observed that can be aligned with other segments within the same protein as well as segments in other proteins.[98] These recurring homologous segments are often 40–80 residues in length and can appear many times in the same protein, either as internally repeating domains[99] or in combination with other different recurring domains. Their boundaries seem to coincide with exons in the DNA, and they may be distributed by **exon shuffling**. If this is the case, they are distributed among the proteins in which they occur as modules. A **modular domain** is a domain encoded by an exon that appears in the genomic DNA of two or more proteins that otherwise are unrelated to each other. An example of several modular domains would be the one fibronectin type I modular domain, the one epidermal growth factor modular domain, and the two protease Kringle modular domains all within the much longer amino acid sequence of tissue plasminogen activator. Modular domains have been noted in the epidermal growth factor precursor, the low-density lipoprotein receptor, fibronectin, complement component 9, and urokinase. The patterns in which they occur suggest that their insertion is a random event.

When the tertiary structure of alcohol dehydrogenase is examined (Figure 7–11) it seems possible to divide it reasonably into two domains. It is now known that one of these is a recurring domain and that both of them together are hinged domains. Even if this were not known, however, it has been argued that a judicious decision that these were distinct domains could still have been made by inspection of the crystallographic molecular model alone. In this sense, these are two structural domains. A **structural domain** has been defined as a "section of peptide chain that can be enclosed in a compact volume . . . by a closed surface . . . , and is characterized by possession of two terminal points."[100] These two terminal points are the point at which the polypeptide enters the compact volume enclosed by the surface and the point at which it exits. This definition does not possess a requirement for evidence of the independent existence of the domain. It could be argued that evidence will eventually be gathered for the independent existence of each of the structural domains now designated. For example, phosphoglycerate kinase is constructed from two structural domains, each formed from one continuous length of polypeptide possessing two terminal points and clearly capable of being enclosed by continuous surfaces surrounding compact volumes,[40] and there is some equivocal evidence suggesting that they are hinged.[101]

The difficulty with the definition of structural domains is that it is subjective. Even though the closed surfaces chosen are the ones that seem to be reasonable, in most cases, other choices, which usually produce a greater number of smaller domains, could be made that would satisfy the same definition. In fact, when this basic definition was used to derive a set of objective rules to divide any given crystallographic molecular model into domains, the tertiary structures of the 22 proteins examined could be divided unambiguously into from as few as two to as many as 10 structural domains. The number of domains so defined increased monotonically with the lengths of the respective polypeptides, which varied from 58 to 450 amino acids.[102] The mean length of the polypeptide in these structural domains was about 50 amino acids, which seems too small to be an evolutionarily significant unit. Most of these irreducible units could be combined with one or more of their neighbors to produce larger structural domains. The relevance of these small segments either to the evolution of one of these proteins or to its structure is not apparent.

The intuitive impression persists, nevertheless, that the tertiary structures of many proteins, as revealed in their crystallographic molecular models, can be divided into two or more autonomous structural domains. It seems to be the case that anyone examining these structures would make the same decision, but there is no way to verify this surmise. Some examples of proteins that are thought to contain structural domains are DNA polymerase I[103] and the hemaglutinin glycoprotein of influenza virus.[104] An interesting example of a structural domain for which there is independent evidence of its existence occurs in catalase from *Penicillium vitale*. This protein has a structural domain formed from the carboxy-terminal 160 amino acids in its sequence (amino acids 510–670)[105] that is missing entirely from bovine liver catalase,[106] even though the two proteins are superposable throughout the other structural domain.

It is unclear whether or not these different words (detachable, enzymatic, separately melting, independently folding, hinged, recurring, internally repeating, modular, and structural) distinguishing the term domain would be acceptable or are desirable. It is possible that each property defines essentially the same thing.[47] As a domain, by definition, is a structure that may be now or

has been in the past an independent entity, the various categories could be simply different ways of identifying members of a large group of **fundamental units of protein structure**. This group contains all the of the smaller units from which the larger proteins that now exist were constructed, and the various domains that now exist in any one protein were at one time unique, unattached, stable, folded polypeptides that were the ancestors of those portions of the entire polypeptide now containing them. These primodial proteins were then individually fused together during evolution by natural selection.

It is this role as a fundamental unit of evolution that lends luster to the title of domain and elicits the desire to grant it. But the term domain should remain an operational designation, closely tied to the particular evidence presented in each case. Problems can arise when it is applied indiscriminately. In particular, it often happens that when the term is used to describe a region of a protein for a very specific reason, all of the connotations associated with it have a way of attaching themselves to that region. For example, a structural domain subliminally gains the status of an independently folding domain, or an enzymatic domain is assumed to be also a detachable domain. Such confusion should be avoided.

Suggested Reading

Ploegman, J.H., Drenth, G., Kalk, K.H., & Hol, W.G.J. (1978) Structure of Bovine Liver Rhodanese I. Structure Determination at 2.5-Å Resolution and a Comparison of the Conformation and Sequence of Its Two Domains, *J. Mol. Biol. 123*, 557–594.

Porter, R.R. (1959) The Hydrolysis of Rabbit γ-Globulin and Antibodies with Crystalline Papain, *Biochem. J. 73*, 119–126.

Problem 7–6: There is an enzyme in vertebrate liver responsible for three enzymatic activities: phosphoribosylamine–glycine ligase, phosphoribosylglycinamide formyltransferase, and aminoimidazole ribonucleotide synthase. It is composed of a single polypeptide 1000 amino acids in length. When the protein was digested with chymotrypsin, two products were produced that could be separated from each other. They were composed of polypeptides 450 and 550 amino acids in length. The larger retained the phosphoribosylamine–glycine ligase; the smaller, the phosphoribosylglycineamide formyltransferase activity. The aminoimidazole ribonucleotide synthetase activity was lost. In *E. coli*, the aminoimidazole ribonucleotide synthetase reaction is catalyzed by a monofunctional protein composed of a polypeptide 330 amino acids in length. Discuss and explain these observations in terms of detachable domains and enzymatic domains.

Problem 7–7: What conclusion concerning pantetheine phosphate adenyltransferase and dephospho-CoA kinase can be drawn from this table?

Purification of Pantetheine Phosphate Adenylyltransferase and Dephospho-CoA Kinase from Porcine Liver (600 g)

purification step	volume (mL)	protein (mg mL^{-1})	transferase activity (units mL^{-1})	transferase specific activity (units mg^{-1})	kinase activity (units mL^{-1})	kinase specific activity (units mg^{-1})	transferase/ kinase ratio	yield (%)	purifi- cation
1700g supernatant	1650	83			0.0165	1.98×10^{-4}		100	1
protamine sulfate supernatant	1680	32			0.0156	4.87×10^{-4}		96	2.5
(NH$_4$)$_2$SO$_4$ fraction + Sephadex G-25	605	30			0.0395	1.32×10^{-3}		87	6.7
DEAE–cellulose chromatography	485	6.7	0.095	0.014	0.045	6.7×10^{-3}	2.1	80	33
procion Red-Sepharose eluate	380	0.208	0.113	0.543	0.055	0.264	2.05	76	1330
blue Sepharose CoA eluate	104	0.041	0.305	7.4	0.15	3.64	2.03	57	18300
Sephadex G-150	75	0.028	0.213	7.61	0.103	3.68	2.07	29	18500

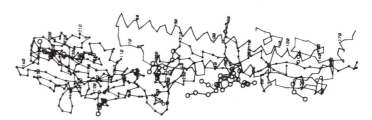

Problem 7–8: Which of these three tertiary structures have structural domains? How many are there in each? (Reprinted with permission from refs 87, 103, and 104. Copyright 1975, 1981, and 1985 *Nature*.)

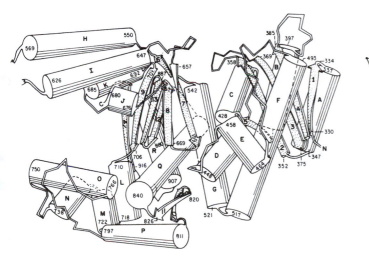

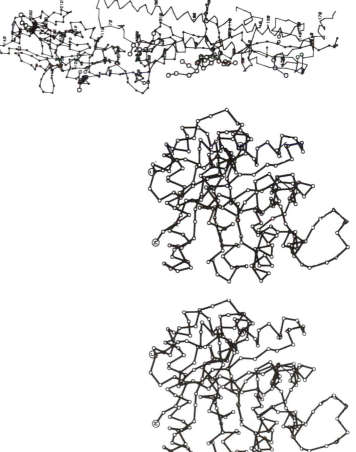

Molecular Taxonomy

As the purview of the study of the evolution of proteins has expanded, the language originally applied to the phylogeny of the species has been subtly expropriated. The implication is that individual proteins can be treated as if they were individual species and that a phylogeny of the proteins might eventually be formulated. It is believed that the proteins observed today evolved from a much smaller group of less elaborate primordial proteins, just as the species observed today are thought to have evolved from a much smaller group of less elaborate primordial species. These primordial proteins are now represented by the domains of presently existing proteins. Establishing the evolutionary relationships among the proteins, however, may be far more difficult to do than establishing the evolutionary relationships among the species. Unfortunately, there is no fossil record of proteins. It is also quite clear that the evolutionary divergence that produced most of the proteins that are universally distributed among present living organisms, for example, the metabolic enzymes, occurred before the divergence of the organisms themselves. This follows from the observation that the proteins from all living organisms responsible for one particular biological function are usually superposable, but proteins from the same organism responsible for two different functions are usually impossible to relate to each other. Thus the lineages of these fundamental proteins may have remained almost unbranched since the evolution of the earliest organisms, and the radiation producing these lineages must have occurred before that time. It is also clear, however, from examining amino acid sequences and crystallographic molecular models that more specialized proteins have been arising continuously throughout evolution and are still arising today. These newer proteins are usually members of classes peculiar to a particular kingdom or phylum, and one of the challenges is to identify their ancestral relationships to the more universally distributed proteins.

It is hoped that, as the number of tertiary structures elucidated by crystallography grows, an anatomical collection of the proteins large enough to form the basis for a comprehensive taxonomy can be assembled. When Linnaeus developed his taxonomy of the species, it is unlikely that he realized the reason for its existence. It is only when taxonomy was connected to the theory of evolution through natural selection that an exercise in cataloguing became something more profound. At the present time, taxonomy in biology is one of the methods by which evolutionary relationships are established. The desire to establish the evolutionary history of proteins has led, in an interesting inversion of history, to the conclusion that a taxonomic system of the proteins might form the basis for an evolutionary explanation.

The most extensive taxonomic system for proteins proposed so far is that of Richardson.[47] It is based on categories that distinguish the tertiary structures of crystallographic molecular models by the topological relationships of the α helices and β structure they contain. It relies on the assumption, for which there is observational support,[37,38] that absolute orientations among secondary structures shift significantly and randomly as evolution proceeds but that the positions of the secondary structures themselves in the amino acid sequence and their topological relationships are more stable. The tertiary structures that are classified in this system are those of individual domains because domains rather than complete proteins are considered to be the descendants of the smaller proteins that were combined by gene fusion to produce the larger ones we now observe, and because it is assumed that, following these gene fusions, no drastic changes occurred in the topologies of the constituent domains, it seems reasonable to focus on the domains in isolation rather than on the randomly conjoined chimeras containing them.

Five of the major categories into which the tertiary structures of domains are sorted in this taxonomic system can be derived from one topological operation (Figure 7–17).[47] In this way a simple primordial motif, the flattened superhelical coil of a long hairpin, with a small number of alterations, can give rise to these major classes. Consider a polar curve that doubles back upon itself to form a hairpin. Twist the hairpin thus formed so that it folds into two turns of a right-handed superhelix (Figure 7–17A). Compress this superhelix until its neighboring segments in front and behind come in contact and then incorporate the segments into the surface of a flattened cylinder (Figure 7–17B). This produces a flattened barrel with eight staves whose polarity alternates as one proceeds around it. This flattened cylinder can be rolled as the tread on a tank to produce eight different barrels that resemble each other but that vary in the juxtapositions of the staves across the center. The connections between the segments, which define the topology of the line, remain unaltered during such rolling. If the cylinder in any of its guises is cut between the first and last segments in the hairpin (segments 1 and 8 in Figure 7–17) and spread upon a plane, a jelly roll[47] (Figure 7–17C) is produced. Shorten the hairpin by removing the two most peripheral segments (cuts numbered 1 in Figure 7–17 to remove segments 1 and 8). A new flattened barrel is created (dotted lines in Figure 7–17B) with six staves that alternate in polarity. If this smaller flattened cylinder is cut between the first and last segments in the hairpin (segments 2 and 7 in Figure 7–17) and spread upon a plane, a Greek key[47] (Figure 7–17D) is produced. Shorten the hairpin by removing the two most peripheral segments (cuts numbered 2 in Figure 7–17 to remove segments 2 and 7). A new flattened barrel is created with four staves that alternate in polarity. If this cylinder is cut between the first and last segments in the hairpin (segments 3 and 6 in Figure 7–17) and flattened upon a plane, an up-down-up-down[47] pattern is produced.

The polar curve in the topological exercise can be substituted with a polypeptide either in an α helix or as a strand of β structure, and the staves of the flattened barrel will be either α helices or strands of β structure, respectively. If they are strands of β structure, they are gathered as antiparallel sheets (Figure 4–17). These topological alternatives produce five of the categories of the present taxonomic system: up-down-up-down, antiparallel α-helical bundles; Greek key, antiparallel α-helical bundles; up-down-up-down, antiparallel β barrels; Greek key, antiparallel β barrels; and jelly roll, antiparallel β barrels. These five categories can be represented respectively by myohemerythrin (Figure 7–18A), the β subunit of hemoglobin (Figure 7–18B), the second domain of papain (Figure 7–18C), the second domain of pyruvate kinase (Figure 7–18D), and the third domain of the coat protein of tomato bushy stunt virus (Figure 7–18E). From examining this figure closely, it is obvious that if the topological scheme displayed in Figure 7–17 was

the initial mechanism of folding, individual staves of the barrels have drifted significantly from their original positions, and more recent secondary structures have arisen at the ends of the barrels and in the loops connecting the staves.

The last three major categories in the taxonomic system are quite different from the first five and from each other. The sixth category, represented by the first domain in L-lactate dehydrogenase (Figure 7–18F), is that of doubly α-helical-wound, parallel β sheets. The last two categories are those of singly α-helical-wound, parallel β barrels and open-faced, antiparallel β sheets. These are represented by triosephosphate isomerase (Figure 7–18G) and glutathione reductase domain 3 (Figure 7–18H), respectively.

Each of these categories has many members and the membership grows as more and more crystallographic molecular models become available. For example, members of the class of α-helical-wound, parallel β barrels (Figure 7–18G) are encountered frequently. One-tenth of the present crystallographic molecular models are of this type. More importantly, they resemble each other sufficiently that it has been concluded that they all arose from one common ancestor.[107] The purpose of constructing each of these taxonomic categories is to imply that all of the members of a particular category share a common ancestor.

The majority of the domains whose tertiary structures have been determined crystallographically can be forced into these eight categories. The ones now occupying the miscellaneous category may eventually be included in additional categories whose definitions become apparent as more structures are solved. The broad range of structures presently subsumed under several of these categories can be appreciated by examining a diagram of the topologies of a collection of the tertiary structures

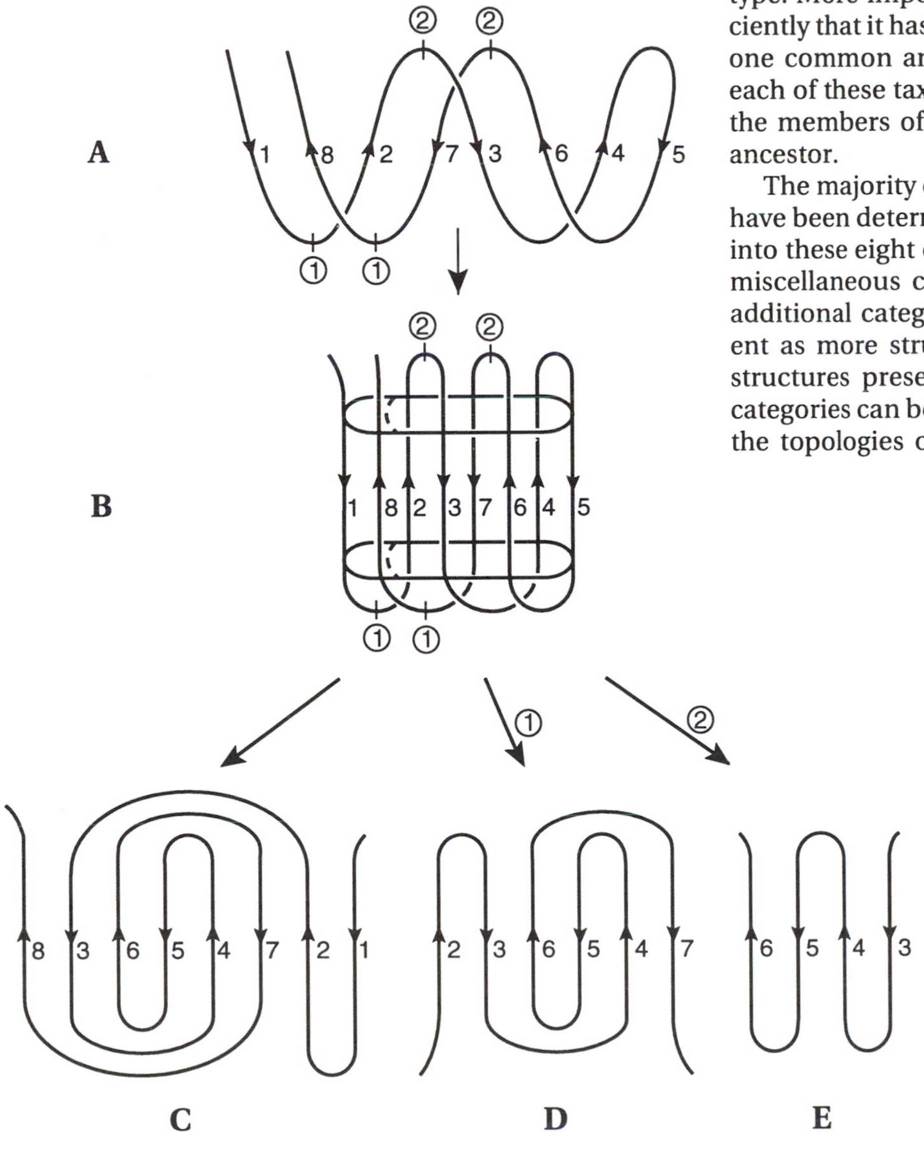

Figure 7–17: Topological explanation of how the patterns of the folded polypeptides defining five of the main categories of domains could have arisen by a common mechanism.[47] (A) A long hairpin of β structure or α helix is twisted into a superhelical coil. (B) That superhelical coil is flattened into a barrel. (C) The order in which the strands occur around the flattened barrel (8, 3, 6, 5, 4, 7, 2, 1) is that of a jelly roll. (D) If strands 1 and 8 are cut away, or were never there to begin with, the order in which the strands occur around the smaller flattened barrel (2, 3, 6, 5, 4, 7) is that of a Greek key. (E) If strands 1 and 2 and strands 7 and 8 are cut away, or were never there to begin with, the order in which the strands occur around the smaller flattened barrel (6, 5, 4, 3) is up-down-up-down.

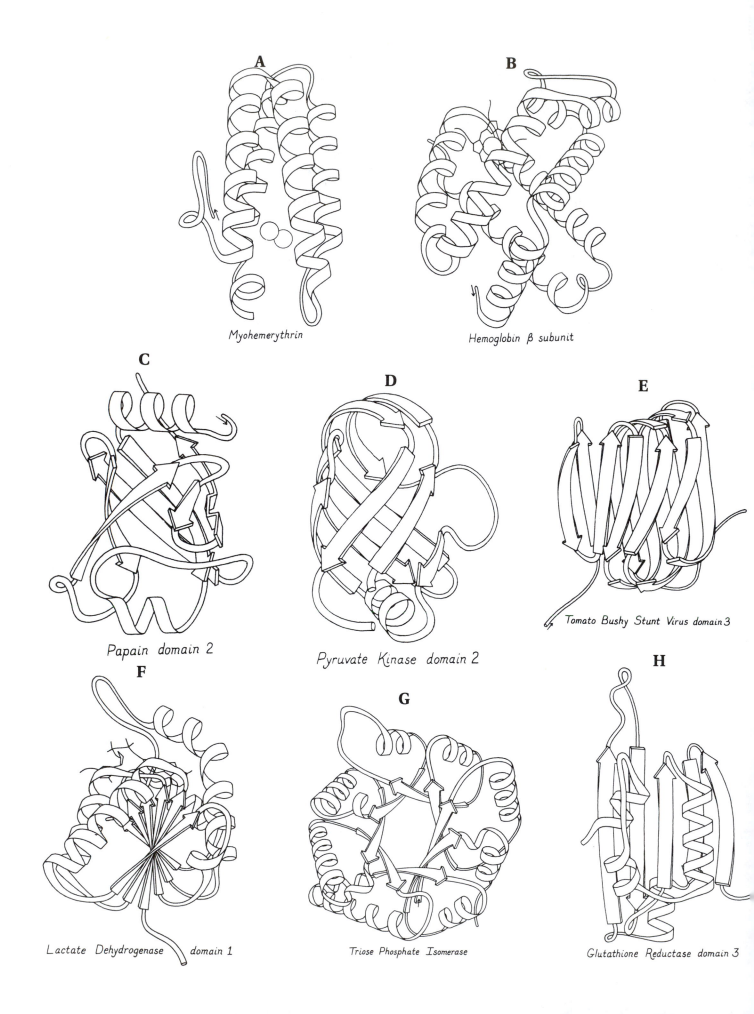

A

Myohemerythrin

B

Hemoglobin β subunit

C

Papain domain 2

D

Pyruvate Kinase domain 2

E

Tomato Bushy Stunt Virus domain 3

F

Lactate Dehydrogenase domain 1

G

Triose Phosphate Isomerase

H

Glutathione Reductase domain 3

Figure 7–18: Representative tertiary structures of the eight major categories of a taxonomic system for classifying domains.[47] Each of these ribbon diagrams has been drawn from the respective crystallographic molecular model. (A) Myohemerythrin is an example of an up-down-up-down, antiparallel α-helical bundle. (B) The β subunit of hemoglobin is an example of a Greek key, antiparallel α-helical bundle. This can be seen if, from the amino terminus indicated by the arrow, the first long, bent α helix, the second short α helix, and the fourth through the seventh long α helices are numbered 1 through 6, respectively, and it is assumed that α helix 1 has drifted 90° away from being parallel to α helix 6. (C) Domain 2 of papain is an example of an up-down-up-down β barrel if the last two short and bent β strands are ignored. (D) Domain 2 of pyruvate kinase is an example of a Greek key, antiparallel β barrel if the short strand of β structure, β strand 3, between β strands 2 and 4 is ignored. (E) Domain 3 from tomato bushy stunt virus is an example of a jelly roll, antiparallel β barrel if the first two, amino-terminal β strands are ignored. (F) Domain 1 of L-lactate dehydrogenase is an example of a doubly wound, parallel β sheet. (G) Triosephosphate isomerase is an example of a singly α-helical-wound, parallel β barrel. (H) Domain 3 of glutathione reductase is an example of an open-faced β sandwich. Adapted with permission from ref 47. Copyright 1981 Academic Press.

Figure 7–19: Topological arrangements of the central β sheets in several representatives of doubly α-helical-wound, parallel β sheets.[47] The β sheets seen in each crystallographic molecular model were flattened upon a plane. The order in which they occurred in the amino acid sequence of the polypeptide was noted as well as whether the connecting loops, usually α helices, were above or below the plane. The dark arrows represent each strand in the order in which they appear across the sheet. The open lines represent connections above the plane, and the thin lines represent connections below the plane. Domain 1 of L-lactate dehydrogenase (Figure 7–10), domain 2 of alcohol dehydrogenase (Figure 7–11), domain 2 of phosphoglycerate kinase, domain 2 of the catalytic subunit of aspartate transcarbomylase, and domain 2 of phosphorylase (Figure 7–10A) serve as paradigms for this category. Their topologies are all the same. Reprinted with permission from ref 47. Copyright 1981 Academic Press.

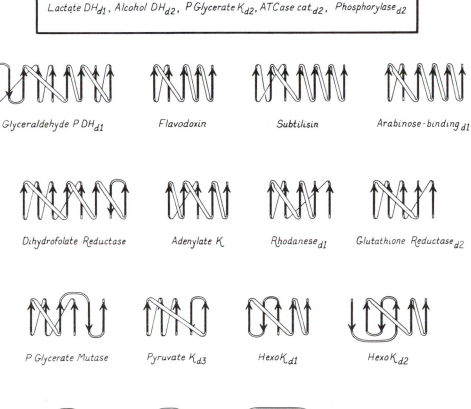

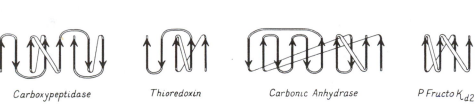

included in the one category of doubly α-helical-wound, parallel β sheets (Figure 7–19). The five structures serving as the paradigms for this category are topologically equivalent, but the other structures represent variations of greater and greater discord on the main theme. For example, the first domain of glyceraldehyde-3-phosphate dehydrogenase is an embellished variation, flavodoxin is a shortened variation, and adenylate kinase is a rearranged variation of flavodoxin. To the extent that the relative magnitude of these variations reflects the degree to which these domains are evolutionarily related to each other, this taxonomic system of proteins resembles the taxonomic system for species. To the extent that these categories reflect only some very favorable solution to the problem of folding a linear polymer, they may be the product of convergent evolutionary processes and contain no phylogenetic information.

Suggested Reading

Richardson, J.S. (1981) Protein Anatomy, *Adv. Protein Chem. 34*, 167–339.

Problem 7–9: Construct a phylogenetic tree from Figure 7–19.

References

1. Timkovich, R., & Dickerson, R.E. (1976) *J. Biol. Chem. 251*, 4033–4046.
2. Meyer, T.E., & Kamen, M.D. (1982) *Adv. Protein Chem. 35*, 105–212.
3. VanBeeuman, J., Ambler, R.P., Meyer, T.E., Kamen, M.D., Olson, J.M., & Saw, E.K. (1976) *Biochem. J. 159*, 757–774.
4. Mross, G.A., & Doolittle, R.F. (1967) *Arch. Biochem. Biophys. 122*, 674–684.
5. Doolittle, R.F. (1979) in *The Proteins* (Neurath, H., & Hill, R.L., Eds.) 3rd ed., Vol. IV, pp 1–118, Academic Press, New York.
6. Perutz, M.F., & Lehmann, H. (1968) *Nature 219*, 902–909.
7. Dayhoff, M.O. (1972) *Atlas of Protein Sequence and Structure*, Vol. 5, National Biomedical Research Foundation, Washington, DC.
8. Dayhoff, M.O. (1978) *Atlas of Protein Sequence and Structure*, Vol. 5, Suppl. 3, National Biomedical Research Foundation, Washington, DC.
9. Margoliash, E., & Schejter, A. (1967) *Adv. Protein Chem. 21*, 113–286.
10. McLachlan, A.D. (1972) *J. Mol. Biol. 64*, 417–437.
11. Doolittle, R.F. (1981) *Science 214*, 149–159.
12. Needleman, S.B., & Wunsch, C.D. (1970) *J. Mol. Biol. 48*, 443–453.
13. Gibbs, A.J., & McIntyre, G.A. (1970) *Eur. J. Biochem. 16*, 1–11.
14. Feng, D.F., Johnson, M.S., & Doolittle, R.F. (1985) *J. Mol. Evol. 21*, 112–125.
15. Jue, R.A., Woodbury, N.W., & Doolittle, R.F. (1980) *J. Mol. Evol. 15*, 129–148.
16. Johnson, M.S., & Doolittle, R.F. (1986) *J. Mol. Evol. 23*, 267–278.
17. Moore, G.W., Goodman, M., Callahan, C., Holmquist, R., & Moise, H. (1976) *J. Mol. Biol. 105*, 15–37.
18. Fitch, W.M., & Margoliash, E. (1967) *Science 155*, 279–284.
19. Feng, D., & Doolittle, R.F. (1987) *J. Mol. Evol. 25*, 351–360.
20. Stephens, P.E., Darlison, M.G., Lewis, H.M., & Guest, J.R. (1983) *Eur. J. Biochem. 133*, 481–489.
21. Ambler, R.P., & Wynn, M. (1973) *Biochem. J. 131*, 485–498.
22. Dene, H., Sazy, J., Goodman, M., & Romero-Herrera, A.E. (1980) *Biochim. Biophys. Acta 624*, 397–408.
23. Crawford, D.L., Constantino, H.R., & Powers, D.A. (1989) *Mol. Biol. Evol. 6*, 369–383.
24. Li, S.S., Fitch, W.M., Pan, Y.E., & Sharief, F.S. (1983) *J. Biol. Chem. 258*, 7029–7032.
25. Rossmann, M.G., Moras, D., & Olsen, K. (1974) *Nature 250*, 194–199.
26. Kamphuis, I.G., Drenth, J., & Baker, E.N. (1985) *J. Mol. Biol. 182*, 317–329.
27. Chothia, C., & Lesk, A.M. (1986) *EMBO J. 5*, 823–826.
28. Takano, T. (1977) *J. Mol. Biol. 110*, 537–568.
29. Poljak, R.J. (1978) *CRC Crit. Rev. Biochem. 5*, 45–84.
30. Biesecker, G., Harris, J.I., Thierry, J.C., Walker, J.E., & Wonacott, A.J. (1977) *Nature 266*, 328–333.
31. Kossiakoff, A.A., Chambers, J.L., Kay, L.M., & Stroud, R.M. (1977) *Biochemistry 16*, 654–664.
32. Tang, J., James, M.N.G., Hsu, I.N., Jenkins, J.A., & Blundell, T.L. (1978) *Nature 271*, 618–621.
33. Bolin, J.T., Filman, D.J., Matthews, D.A., Hamlin, R.C., & Kraut, J. (1982) *J. Biol. Chem. 257*, 13650–13662.
34. Serrano, R.S., Kielland-Brandt, M.C., & Fink, G.R. (1986) *Nature 319*, 689–693.
35. Kyte, J. (1981) *Nature 292*, 201–204.
36. Ochi, H., Hata, Y., Tanaka, N., Kakudo, M., Sakurai, T., Aihara, S., & Morita, Y. (1983) *J. Mol. Biol. 166*, 407–418.
37. Chothia, C., & Lesk, A.M. (1985) *J. Mol. Biol. 182*, 151–158.
38. Lesk, A.M., & Chothia, C. (1980) *J. Mol. Biol. 136*, 225–270.
39. Almassy, R.J., & Dickerson, R.E. (1978) *Proc. Natl. Acad. Sci. U.S.A. 75*, 2674–2678.
40. Banks, R.D., Blake, C.C.F., Evans, P.R., Haser, R., Rice, D.W., Hardy, G.W., Merrett, M., & Phillips, A.W. (1979) *Nature 279*, 773–777.
41. Novokhatny, V.V., Kudinov, S.A., & Privalov, P.L. (1984) *J. Mol. Biol. 179*, 215–232.
42. Birktoft, J.J., & Blow, D.M. (1972) *J. Mol. Biol. 68*, 187–240.
43. Chothia, C., & Lesk, A.M. (1987) *Cold Spring Harbor Symp. Quant. Biol. 52*, 399–405.
44. Sahr, K.E., Laurila, P., Kotula, L., Scarpa, A.L., Coupal, E., Leto, T.L., Linnenbach, A.J., Winkelmann, J.C., Speicher, D.W., Marchesi, V.T., Curtis, P.J., & Forget, B.G. (1990) *J. Biol. Chem. 265*, 4434–4443.
45. Ohlsson, I., Nordström, B., & Brändén, C. (1974) *J. Mol. Biol. 89*, 339–354.
46. Fletterick, R.J., Syqush, J., Semple, H., & Madsen, N.B. (1976) *J. Biol. Chem. 251*, 6142–6146.
47. Richardson, J.S. (1981) *Adv. Protein Chem. 34*, 167–339.
48. Eklund, H., Nordström, B., Zeppezauer, E., Söderlund, G., Ohlsson, I., Boiwe, T., Söderberg, B., Tapia, O., Brändén, C., & Åkeson, Å. (1976) *J. Mol. Biol. 102*, 27–59.
49. Porter, R.R. (1959) *Biochem. J. 73*, 119–126.
50. Silverton, E.W., Navia, M.A., & Davies, D.R. (1977) *Proc. Natl. Acad. Sci. U.S.A. 74*, 5140–5144.
51. Sutherland, W.M., Winge, D.R., Rajagopalan, K.V., & Wiley, R.D. (1978) *J. Biol. Chem. 253*, 8747–8752.

52. Appell, K.C., & Low, P.S. (1981) *J. Biol. Chem. 256*, 11104–11111.
53. Faxel, A., Müller, K., LeBras, G., Garel, J., Véron, M., & Cohen, G.N. (1983) *Biochemistry 22*, 158–165.
54. Sibilli, L., LeBras, G., LeBras, G., & Cohen, G.N. (1981) *J. Biol. Chem. 256*, 10228–10230.
55. Véron, M. Falcoz-Kelly, F., & Cohen, G.N. (1972) *Eur. J. Biochem. 28*, 520–527.
56. Caughey, I., & Kekwick, R.G.O. (1982) *Eur. J. Biochem. 123*, 553–561.
57. Shimakata, T., & Stumpf, P. (1982) *Arch. Biochem. Biophys. 218*, 77–91.
58. Mohamed, A.H., Chirala, S.S., Mody, N.H., Huang, W., & Wakil, S.J. (1988) *J. Biol. Chem. 263*, 12315–12325.
59. Chirala, S.S., Kuziora, M.H., Spector, D.M., & Wakil, S.J. (1987) *J. Biol. Chem. 262*, 4231–4240.
60. Chirala, S.S., Kuziora, M.A., Spector, D.M., & Wakil, S.J. (1987) *J. Biol. Chem. 262*, 4231–4240.
61. Holzer, K.P., Liu, W., & Hammes, G.G. (1989) *Proc. Natl. Acad. Sci. U.S.A. 86*, 4387–4391.
62. Chang, S., & Hammes, G.G. (1989) *Proc. Natl. Acad. Sci. U.S.A. 86*, 8373–8376.
63. Mattick, J.S., Tsukamoto, Y., Nickless, J., & Wakil, S.J. (1983) *J. Biol. Chem. 258*, 15291–15299.
64. Lin, C.Y., & Smith, S. (1978) *J. Biol. Chem. 253*, 1954–1962.
65. Smith, S., Agradi, E., Libertini, L., & Dilpeenan, K.N. (1976) *Proc. Natl. Acad. Sci. U.S.A. 73*, 1184–1188.
66. Libertini, L.J., & Smith, S. (1979) *Arch. Biochem. Biophys. 192*, 47–60.
67. Gething, M.H., & Davidson, B.E. (1976) *Eur. J. Biochem. 71*, 327–336.
68. Duggleby, R.G., Sneddon, M.K., & Morrison, J.F. (1978) *Biochemistry 17*, 1548–1554.
69. Minson, A.C., & Creaser, E.H. (1969) *Biochem. J. 114*, 49–56.
70. Keesey, J.K., Bigelis, R., & Fink, G.R. (1979) *J. Biol. Chem. 254*, 7427–7433.
71. Gaertner, F.H., & Cole, K.W. (1977) *Biochem. Biophys. Res. Commun. 75*, 259–264.
72. Coleman, P.F., Suttle, D.P., & Stark, G.R. (1977) *J. Biol. Chem. 252*, 6379–6385.
73. Mally, M.I., Grayson, D.R., & Evans, D.R. (1981) *Proc. Natl. Acad. Sci. U.S.A. 78*, 6647–6651.
74. Taniuchi, H., & Anfinsen, C.B. (1971) *J. Biol. Chem. 246*, 2291–2301.
75. Rechler, M.M., & Bruni, C.B. (1971) *J. Biol. Chem. 246*, 1806–1813.
76. Betton, J., Desmadril, M., Mitraki, A., & Yon, J.M. (1984) *Biochemistry 23*, 6654–6661.
77. Doolittle, R.F., Goldbaum, D.M., & Doolittle, L.R. (1978) *J. Mol. Biol. 120*, 311–325.
78. Williams, R.C. (1981) *J. Mol. Biol. 150*, 399–408.
79. Donovon, J.W., & Mihalyi, E. (1974) *Proc. Natl. Acad. Sci. U.S.A. 71*, 4125–4128.
80. Vita, C., Fontana, A., & Chaiken, I.M. (1985) *Eur. J. Biochem. 151*, 191–196.
81. Trexler, M., & Patthy, L. (1983) *Proc. Natl. Acad. Sci. U.S.A. 80*, 2457–2461.
82. Bennett, W.S., & Steitz, T.A. (1980) *J. Mol. Biol. 140*, 211–230.
83. Remington, S., Wiegand, G., & Huber, R. (1982) *J. Mol. Biol. 158*, 111–152.
84. Eklund, H., Samama, J., Wallén, L., Brändén, C.-I, Åkeson, Å., & Jones, T.A. (1981) *J. Mol. Biol. 146*, 561–587.
85. Harrison, S.C., Olson, A.J., Schuff, C.E., & Winkler, F.K. (1978) *Nature 276*, 368–373.
86. Lesk, A.M., & Chothia, C. (1984) *J. Mol. Biol. 174*, 175–191.
87. Banner, D.W., Bloomer, A.C., Petsko, G.A., Phillips, D.C., Pogson, C.I., Wilson, I.A., Corran, P.H., Furth, A.J., Milman, J.D., Offord, R.E., Priddle, J.D., & Waley, S.G. (1975) *Nature 255*, 609–614.
88. Levine, M., Muirhead, H., Stammers, D.K., & Stuart, D.I. (1978) *Nature 271*, 626–630.
89. Wierenga, R.K., & Drenth, J. (1983) *J. Mol. Biol. 167*, 725–739.
90. Levine, M., Muirhead, H., Stammers, D.K., & Stuart, D.I. (1978) *Nature 271*, 626–630.
91. McLachlan, A.D. (1979) *J. Mol. Biol. 128*, 49–79.
92. Ploegman, J.H., Drent, G., Kalk, K.H., & Hol, W.G.J. (1978) *J. Mol. Biol. 123*, 557–594.
93. McLachlan, A.D. (1979) *Eur. J. Biochem. 100*, 181–187.
94. McLachlan, A.D., & Walker, J.E. (1977) *J. Mol. Biol. 112*, 543–558.
95. Speicher, D.W., & Marchesi, V.T. (1984) *Nature 311*, 177–180.
96. Speicher, D.W., Morrow, J.S., Knowles, W.J., & Marchesi, V.T. (1982) *J. Biol. Chem. 257*, 9093–9101.
97. Bliele, D.M., Hackert, M.L., Pettit, F.H., & Reed, L.J. (1981) *J. Biol. Chem. 256*, 514–519.
98. Doolittle, R.F. (1985) *Trends Biochem. Sci. 10*, 233–237.
99. Doolittle, R.F. (1989) in *Prediction of Protein Structure and the Principles of Protein Conformation* (Fasman, G.D., Ed.) pp 599–623, Plenum, New York.
100. Wetlaufer, D.B. (1973) *Proc. Natl. Acad. Sci. U.S.A. 70*, 697–701.
101. Pickover, C.A., McKay, D.B., Engleman, D.M., & Steitz, T.A. (1979) *J. Biol. Chem. 254*, 11323–11329.
102. Rose, G.D. (1979) *J. Mol. Biol. 134*, 447–470.
103. Ollis, D.L., Brick, P., Hamlin, R., Xuong, N.G., & Steitz, T.A. (1985) *Nature 313*, 762–766.
104. Wilson, I.A., Skehel, J.J., & Wiley, D.C. (1981) *Nature 289*, 366–373.
105. Vainshtein, B.K., Melik-Adamyan, W.R., Barynin, V.V., Vagin, A.A., Grebenko, A.I., Borisov, V.V., Bartels, K.S., Fita, I., & Rossmann, M.G. (1986) *J. Mol. Biol. 188*, 49–61.
106. Murthy, M.R.N., Reid, T.J., Sicignano, A., Tanaka, N., & Rossmann, M.G. (1981) *J. Mol. Biol. 152*, 465–499.
107. Farber, G.K., & Petsko, G.A. (1990) *Trends Biochem. Sci. 15*, 228–234.

CHAPTER 8

Counting Polypeptides

Almost all of the proteins found in a living organism are discrete macromolecules. Their component parts were originally synthesized by ribosomes from messenger RNAs, each of which encoded a polypeptide of a precise, finite length. These polypeptides folded into defined conformations and were posttranslationally modified. Only a specific and well-defined number of these folded polypeptides gathered together to form the macromolecular complex that is the finished, existing molecule of the protein. In a few instances, such as the proteins actin, keratin, and collagen, a large and undefined number of polypeptides combine to form a **polymeric protein**, which in theory could continue to polymerize indefinitely. Such proteins, however, are the exceptions. The molecules of most proteins are specific, clearly defined structures.

The **subunit stoichiometry** of a protein is the number of each type of folded polypeptide that have combined to produce the specific structure. At this level of definition, each of the polypeptides is identified only by its length. The **length of a polypeptide** is the number of amino acids it contains, n_{aa}, an integer that is either dimensionless or has the units of amino acids (molecule of polypeptide)$^{-1}$ or mole of amino acids (mole of polypeptide)$^{-1}$. The length of a polypeptide is usually a precisely known quantity because its amino acid sequence is usually available.

A great deal of effort has been expended in discovering the subunit stoichiometries of proteins. The original approach to this information was to determine the molar mass of the intact protein, to separate the individual polypeptides composing the protein, to quantify the mass ratio among the various polypeptides, and to determine the molar mass of each of the separated polypeptides. The measurement of the molar masses of intact proteins was at one time a major area of biophysical research; and, although this presently attracts almost no attention, a short discussion of the physical properties that provide this information is still considered educational.

The individual folded polypeptides composing an intact native protein are separated and catalogued analytically by electrophoresis on polyacrylamide gels cast in solutions of the detergent sodium dodecyl sulfate. The separation that is effected by these polyacrylamide gels relies on their ability to sieve the unfolded polypeptides. The constituent polypeptides of a protein are separated preparatively by chromatography that depends either on sieving or on ion exchange of the unfolded, dissociated polymers. The separated polypeptides are shown to be homogeneous and unique by peptide mapping.

The major weaknesses of the original approach to defining the subunit stoichiometry of a protein were the extreme care with which the initial measurements of the molar mass of the intact protein had to be performed, as well as the expense involved, and the unreliability of the assessments of the mass ratios among the constituent polypeptides and of their molar masses. The present approach to defining the subunit stoichiometry of a protein avoids these problems. The individual polypeptides composing a protein are still separated and catalogued by electrophoresis and shown to be unique and homogeneous by peptide mapping. The length of each of the constituent polypeptides is assessed either by the electrophoresis itself or, preferably, by sequencing the appropriate cDNA. Any glycosylation is quantified analytically. The number of each polypeptide composing the intact protein is determined by covalently cross-linking the protein to various degrees of completion and identifying the various intermediate covalent complexes and the limit complex. The possibility that the intact molecule of protein may participate in intermolecular dimerization or higher oligomerization rather than being monodisperse in solution is assessed by quantitative cross-linking. When a molecule of a protein contains a fixed, definite number of more than one folded polypeptide, the protein is referred to as an **oligomeric protein** to distinguish it from a polymeric protein. The different polypeptides in an oligomeric protein are defined by their lengths and distinguished by assigning them consecutive letters of the Greek alphabet or, occasionally, by giving them names describing their functions, if they are known. For example, deoxyhemoglobin at its normal concentrations is constructed from two polypeptides, α and β, each present in two copies to produce the complex $(\alpha\beta)_2$. Nicotinic acetylcholine receptor has the composition $\alpha_2\beta\gamma\delta$; L-lactate dehydrogenase, $(\alpha_2)_2$; DNA-directed RNA polymerase from *Escherichia coli*, $\alpha_2\beta\gamma\delta$; and 6-phospho-2-dehydro-3-deoxygluconate aldolase, α_3. An example of designations based on function would be aspartate transcarbamylase. It is composed of two polypeptides each present in six copies. The longer polypeptides form the catalytic subunits that are responsible for the enzymatic activity; the shorter polypeptides form the regulatory subunits that adjust the rate of the enzymatic activity. The polypeptides are thus designated c and r, respectively; and the whole complex, $(c_3)_2(r_2)_3$. The grouping of polypeptides into subsets, for example, the groups of two polypeptides in L-lactate dehydrogenase, arises from the symmetries in which the folded polypeptides are arranged within the intact molecule of an oligomeric protein.

A **subunit** of an oligomeric protein is a portion of the intact molecule of protein that contains only complete, properly folded polypeptides. A subunit is usually only one folded polypeptide, but it can also be several folded polypeptides that together form a recognizable unit. The important distinction is that a subunit is a unit independent of the total structure. For example, aspartate transcarbamylase can be dissociated into a mixture of c_3 trimers, referred to as catalytic subunits, and r_2 dimers, referred to as regulatory subunits. Both types of subunits remain properly folded and assembled in isolation. An intact molecule of aspartate transcarbamylase is a combination of integral numbers of these subunits, two catalytic and three regulatory.

Some oligomeric proteins, when they are dissolved at certain concentrations, are mixtures of two different combinations of subunits in equilibrium with each other. For example, oxygenated hemoglobin is an equilibrium mixture of $\alpha\beta$ dimers and $(\alpha\beta)_2$ tetramers. Most oligomeric proteins, however, have a particular subunit composition, and hence composition of polypeptides, that does not vary unless harsh conditions are applied. The subunit stoichiometry of the protein is the number of each that produces the compact, unique macromolecule that remains dissolved and unassociated with its neighbors. A solution of a pure oligomeric protein will usually be monodisperse. A **monodisperse** solution of a particular solute is a solution in which every molecule of that solute is of the same size and shape.

When the map of electron density from a crystal of an oligomeric protein is examined, the complete molecule can be discerned. It is recognized as a large, independent feature in the map that is formed from several folded tubes of electron density. Although not essential, it is reassuring to know before the map is examined how many polypeptides are combined to produce the protein that has been crystallized. This determination can be made by a combination of sequencing, molecular sieving, and cross-linking. It is now so routine to do this that few oligomeric proteins whose subunit stoichiometry has not already been established are examined crystallographically.

Molar Mass

The only indubitably reliable method for determining the length and amino acid composition of a polypeptide, and, consequently, its molar mass, is to sequence it. At the moment, the sequences of a large array of readily available polypeptides are known, and they form a collection of standards each of whose lengths, n_{aa}, is an exact quantity. It is often, but not always, the case that the amino acid sequence of a newly purified polypeptide is known from the nucleotide sequence of its cDNA before enough of it becomes available to study its physical properties in detail. This situation has inverted the classical strategy of physical measurements in which the molar mass of a protein was one of the ultimate discoveries rather than something precisely known from the beginning. For this reason, the discussion presented in this section is mainly of historical interest and describes measurements now rarely performed, but it provides an explanation of some of the points of view rooted in the earlier techniques for determining molar mass that have attached themselves to the present situation. These must be understood to appreciate the techniques now used to determine the length of a polypeptide and the stoichiometry of the polypeptides in an oligomeric protein.

The **molar mass** of a protein is the number of grams in a mole of that protein. The **molecular mass** of a protein is the mass of a single molecule of that protein expressed in relative units. These relative units are referred to as either atomic mass units (amu) or daltons (Da). Both an atomic mass unit and a dalton (1.6606×10^{-24} g) are $\frac{1}{12}$ the mass of carbon isotope 12. Molar mass is the quantity that is determined by the measurements of physical behavior such as osmotic pressure, sedimentation equilibrium, and light scattering because the units on the final quantity are usually grams mole^{-1}.

Because both sedimentation equilibrium and light scattering are alternative measurements of osmotic pressure, all three techniques measure the same fundamental physical property of the solution of protein, namely, its osmotic pressure. Osmotic pressure is a colligative property of the solution. A **colligative property** of a solute is a physical property that is a function only of the moles of independent particles of that solute in a standard volume of the solution. If the solution were monodisperse, if the only osmotically active particles present were individual molecules of the protein, and if the concentration of the protein in grams centimeter^{-3} were known precisely, this molar concentration could be used to calculate the molar mass of the protein.

Osmotic pressure is the pressure exerted by impermeant solutes when a solution containing those impermeant solutes is separated by a semipermeable membrane from another solution identical in every way to the first except that it lacks the impermeant solutes. A solute is **impermeant** to a particular barrier if it cannot pass through that barrier. A **semipermeable membrane** is a membrane through which all of the components of the two solutions can pass freely except the impermeant solutes. It should be obvious that the chemical and physical properties of the membrane define which solutes are permeant and which are impermeant, which are osmotically silent and which are osmotically active, respectively. In the case of experimental measurements of the osmotic pressure of solutions of proteins, a membrane is chosen that is porous to the small molecules in the solution but the pores of which are too small to pass the molecules of protein. This is usually a sheet formed from a polymeric material that has spaces between the polymers wide enough to pass small molecules and ions but too small to pass macromolecules. The membranes of living cells, however, can be far more subtle in their discrimination among solutes.

It has already been noted that a protein is a polyelectrolyte. The fact that it bears a net charge at a given pH complicates the measurements of osmotic pressure, and also sedimentation equilibrium and light scattering. In the absence of added electrolytes, gradients or discontinuities of electrical potential would form during all three of these procedures as the concentration of the protein was purposely varied. These gradients or discontinuities of electrical potential can be eliminated in the case of most proteins by adding a simple electrolyte such as KCl to the solution to a concentration of around 0.1 M. The resulting increase in ionic strength decreases significantly the thickness of the ionic double layer (Equation 1–62) so that it encloses the molecule of protein tightly, completely neutralizes its charge, and turns it into a neutral macromolecule. The added electrolyte also eliminates any local gradients of electrical potential caused either by separating two phases by a semipermeable membrane,[1] as in measurements of osmotic pressure, or by gravitational forces exerted on ions of unlike mass, as in sedimentation equilibrium.[2] In addition to an electrolyte, a buffer may also be added to the solution to maintain the pH.

Pressure is a force that results from the tendency of molecules to expand the confines in which they are contained and fill a larger volume. When molecules such as impermeant solutes exert an osmotic pressure, they cannot expand their confines, as can gases, by entering vacant space because they are held by intermolecular forces within a condensed phase. The volume in which they are confined, however, can expand if solution from the other side passes across the semipermeable membrane into the solution containing the impermeant solutes. This is the liquid analogy to a balloon expanding to fill more space. Just as the balloon expands until the external pressure is equivalent to the internal pressure of the trapped gas, the solution containing the impermeant molecules expands until an external pressure equal to its osmotic pressure is applied to it. Operationally, the osmotic pressure of a solution is the external pressure that must be applied to the solution containing the impermeant solute to prevent any expansion in its volume by the net movement of fluid through the semipermeable membrane.

It can be shown[3] that the osmotic pressure exerted by a solution of impermeant solutes is formally equivalent to the pressure exerted on a container filled with gases that are impermeant to the walls of that container. The molecules of the impermeant solute are formally equivalent to the molecules of the gas, and the solvent and permeant solutes are as physically silent as the vacuum in which the gas is suspended. For an ideal gas

$$P = \frac{n_\mathrm{m} RT}{V} \qquad (8\text{--}1)$$

where P is the pressure (dynes centimeter^{-2}) exerted by the gas, R is the gas constant (8.315×10^7 dyne cm K^{-1} mol^{-1}), T is the temperature (kelvins), n_m is the number of moles of the gas, and V is the volume (centimeters^3) in which it is confined. The pressure exerted by a nonideal gas, however, is

$$P = RT \left[\frac{n_\mathrm{m}}{V} + B\left(\frac{n_\mathrm{m}}{V}\right)^2 + C\left(\frac{n_\mathrm{m}}{V}\right)^3 \cdots \right] \qquad (8\text{--}2)$$

where $n_m V^{-1}$ is the concentration (moles centimeter^{-3}) of the gas and the coefficients B, C, and so forth are referred to, respectively, as the second virial coefficient, the third virial coefficient, and so forth. The virial coefficients can be shown to correct the behavior of the nonideal gas for the specific properties that make it nonideal, such as the finite dimensions of the molecules that fill, or exclude, some of the volume of the container, the intermolecular forces between the molecules of the gas, and the tendency of molecules of the gas to dimerize or polymerize. For all of the same reasons,[3] the osmotic pressure, Π (dynes centimeter^{-2}), exerted by a nonideal, impermeant solute S is

$$\Pi = RT \left([S] + B[S]^2 + C[S]^3 \cdots \right) \qquad (8\text{--}3)$$

where $[S]$ is the molar concentration of the solute. If the impermeant solute is protein i

$$\lim_{[\text{protein}] \to 0} \Pi = RT[\text{protein } i] \qquad (8\text{--}4)$$

Equation 8–4 states that at low enough concentrations of protein, the osmotic pressure observed will be directly proportional to the molar concentration of the protein.

The molar concentration of the protein in the solution cannot be known if the molar mass of the protein is unknown, but the concentration of the protein in the solution must be known in some type of units. From the ultraviolet spectrum of the solution, the concentration (moles centimeter^{-3}) of the tryptophan in the protein might be known.[4] From a colorimetric assay, the concentration (moles centimeter^{-3}) of peptide bonds in the solution might be known.[5] From total amino acid analysis,[6] the concentration (moles centimeter^{-3}) of amino acids in the solution might be known. From a dry weight measurement of the protein, the grams of dry weight in a milliliter of solution (grams centimeter^{-3}) might be known. Regardless of the units, the value for the concentration of protein in the units for the quantity measured can be designated as C_p (units centimeter^{-3}). It follows that

$$W[\text{protein } i] = C_\mathrm{p} \qquad (8\text{--}5)$$

where W is a constant of proportionality. The units on W are moles of tryptophan (mole of protein)$^{-1}$, moles of peptide bonds (mole of protein)$^{-1}$, moles of amino acids (mole of protein)$^{-1}$, or grams of dry weight (mole of protein)$^{-1}$, respectively. It is only coincidental that the last is usually chosen. It is this exercise that defines what a colligative property is and illustrates that osmotic pressure does not measure molar mass directly. The final result can be always traced back to an independent measurement of the concentration of the protein. When Equation 8–5 is incorporated into Equation 8–4

$$\lim_{C_p \to 0} \frac{\Pi}{C_p} = \frac{RT}{W} \qquad (8\text{–}6)$$

and the intercept of $\Pi(C_p)^{-1}$ as C_p is decreased to zero provides the value of W (Figure 8–1). If the units of concentration, C_p, were grams of protein centimeter^{-3} and the only protein present were the one of interest, W would be the molar mass of the protein.

Although the most useful units, for purposes of calculation, in which to express osmotic pressure are dynes centimeter^{-2}, it is often tabulated in units of centimeters. These are the centimeters that the level of the solution containing the protein can rise above the level of the solution lacking the protein as a result of the expansion of the former at the expense of the latter. This additional layer of solution exerts a pressure because of the force of gravity. The units of centimeters are converted into dynes centimeter^{-2} by multiplying by the density of the solution lacking the protein (grams centimeter^{-3}) and the gravitational acceleration (980.6 cm s^{-2} at sea level, 45° latitude) felt by the excess fluid on top of the solution of protein.

The complete equation describing the actual behavior of the osmotic pressure, at low concentrations of protein, is

$$\Pi = RT\left(\frac{C_p}{W} + BC_p^2 + CC_p^3 \ldots\right) \qquad (8\text{–}7)$$

where B, C, . . . are the virial coefficients expressed in appropriate units.

In Figure 8–1, the chosen concentrations of the protein, bovine serum albumin, are in the range where only the first virial coefficient, B, is significant, and the results are presented as if the equation were

$$\frac{\Pi}{C_p} = RT\left(\frac{1}{W} + BC_p\right) \qquad (8\text{–}8)$$

It can be seen that when the charge on the protein was changed by changing the pH, the value of the second virial coefficient, B, and hence the slope of the line, changed in response to changes in the various parameters of the solution, but in agreement with Equation 8–6, the intercept seems to have remained the same.

The fact that the slopes of the two lines in Figure 8–1 are different arises from the **Donnan effect**. Because a protein bears a net charge at a given pH and because the two solutions on either side of the semipermeable membrane must be electroneutral, the concentrations of electrolytes cannot be the same in the two solutions. Suppose that the only electrolyte in the solution is NaCl, the solution containing the protein is designated α, and the solution in the other compartment is designated β. To preserve electroneutrality

$$[Na^+]_\beta = [Cl^-]_\beta \qquad (8\text{–}9)$$

and

$$Q_i[\text{protein } i] + [Na^+]_\alpha = [Cl^-]_\alpha \qquad (8\text{–}10)$$

where Q_i is the charge on protein i. If activity coefficients are ignored for the moment,

$$\mu_{NaCl} = \mu°_{NaCl} + RT \ln[Na^+][Cl^-] \qquad (8\text{–}11)$$

where μ_{NaCl} is the chemical potential of the NaCl. The chemical potential of the NaCl must be the same on both sides of the membrane, and $\mu°_{NaCl}$, the standard chemical potential of NaCl, is always the same, so

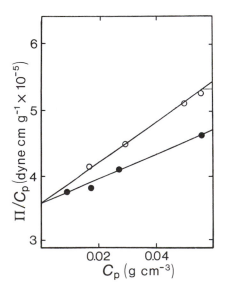

Figure 8–1: Osmotic pressure of solutions of bovine serum albumin.[7] The incremental osmotic pressure [$\Pi(C_p)^{-1} \times 10^{-5}$], where Π is in dynes centimeter^{-2} and C_p is in grams of protein centimeters^{-3}, is plotted as a function of C_p. The apparatus was at 25 °C, and within the apparatus the solution of protein was separated from the solution lacking the protein by a membrane of nitrocellulose polymer. The osmotic pressure of the solution was the external pressure that had to be applied to the solution of protein, in excess of the atmospheric pressure on the solution lacking the protein, to prevent its expansion. The excess pressure was measured with a toluene manometer, and millimeters of toluene was converted to dynes centimeter^{-2}. The concentration of protein, following the measurement, in each of the solutions containing protein was determined by a dry weight analysis that was corrected for the weight of other solutes present. The solutions contained 0.15 M NaCl as supporting electrolyte. The behaviors of the osmotic pressure at pH 7.0 (○) and pH 5.37 (●) are shown. Adapted with permission from ref 7. Copyright 1946 American Chemical Society.

$$[Na^+]_\alpha \, [Cl^-]_\alpha = [Na^+]_\beta \, [Cl^-]_\beta \qquad (8\text{--}12)$$

at equilibrium. Combining these equations

$$[Na^+]_\beta = [Na^+]_\alpha \left(1 + \frac{Q_i[\text{protein } i]}{[Na^+]_\alpha}\right)^{\frac{1}{2}} \qquad (8\text{--}13)$$

$$[Cl^-]_\beta = [Cl^-]_\alpha \left(1 + \frac{Q_i[\text{protein } i]}{[Cl^-]_\alpha}\right)^{\frac{1}{2}} \qquad (8\text{--}14)$$

Equations 8–13 and 8–14 state that if the protein is positively charged, there will be more chloride and less sodium in compartment α than in compartment β, or that if the protein is negatively charged, there will be more sodium and less chloride in compartment α than in compartment β. These imbalances of molecular species will affect the osmotic pressure because the electrolytes cannot distribute freely across the semipermeable membrane and they become osmotically active. Although there is no exact solution to these equations, at low concentrations of protein and at small values of Q_i[8]

$$\lim_{[\text{protein}]\to 0} \Pi = RT\,[\text{protein } i]\left(1 + \frac{Q_i^2[\text{protein } i]}{4[\text{NaCl}]}\right) \quad (8\text{--}15)$$

From Equation 8–15 it can be seen that the Donnan effect is expressed in the second virial coefficient, B, as demonstrated in Figure 8–1. For molecules with large values of Q_i, such as nucleic acids, this approximation fails badly. From inspection of Equation 8–15 it can be seen that when $[\text{NaCl}] > 5Q_i[\text{protein } i]$, the Donnan effect causes less than a 5% error in osmotic pressure and that as $[\text{protein } i]$ approaches zero, the Donnan effect also approaches zero.

When a solution of protein is submitted to **sedimentation equilibrium**, it is placed in the strong centrifugal field created by a spinning rotor, and the distribution of solutes, particularly that of the protein, is allowed to reach equilibrium. One way of viewing this is to imagine the protein redistributing until the centrifugal force upon it is equal but opposite in sign to the force of diffusion that arises from the gradient of its concentration. If negligible gradients of electrical potential form because sufficient electrolyte has been added to the solution, the solutes and solvent will redistribute until their chemical potential at each position in the chamber balances the centrifugal potential that they experience. From the equality produced by this balance, the following exact relationship can be derived[2]

$$\frac{d\ln C_p}{dr^2} = \left(\frac{\omega^2}{2}\right)\left(\frac{\partial\rho}{\partial C_p}\right)_\mu \left(\frac{d\Pi}{dC_p}\right)^{-1} \quad (8\text{--}16)$$

where C_p is the concentration of protein expressed in any

units, r is the distance (centimeters) of any point in the chamber from the center of the rotor, ω is the rate of revolution (radians second^{-1}) of the rotor, ρ is the density (grams centimeter^{-3}) of the solution of protein, and $(\partial\rho/\partial C_p)_\mu$ is the change in density of the solution as a function of the concentration of protein at constant chemical potential of solvent and all other solutes other than the protein. Because the differential on the left in Equation 8–16 is that of the natural logarithm of the concentration of protein, it will have the same numerical value regardless of the units chosen to express the concentration of the protein. If it is recognized that

$$\frac{d\Pi}{dC_p} = RT\left(\frac{1}{W} + 2BC_p + 3CC_p^2\ldots\right) \qquad (8\text{--}17)$$

and

$$\lim_{C_p\to 0}\left(\frac{1}{W} + BC_p + CC_p^2\ldots\right) = \frac{1}{W} \qquad (8\text{--}18)$$

Equation 8–16 can be rearranged to give[2]

$$\lim_{C_p\to 0}\frac{d\ln C_p}{dr^2} = \frac{\omega^2}{2RT}\left(\frac{\partial\rho}{\partial C_p}\right)_\mu W \qquad (8\text{--}19)$$

As predicted by Equation 8–19, when a solution of protein is submitted to centrifugation and the distribution of that protein is allowed to reach equilibrium, the gradient of concentration that forms is such that a plot of ln C_p against r^2 is a straight line (Figure 8–2). From the slope of this plot and a value for $(\partial\rho/\partial C_p)_\mu$, the value of W can be calculated.

There is some confusion between the use in the present instance of a centrifugal field to measure the molar concentration of a solution of protein and its use to measure the sedimentation coefficient, a hydrodynamic property of an individual molecule of the protein. In sedimentation equilibrium, the centrifugal field is used as a convenient device to create a gradient of concentration. Because the centrifugal potential can be calculated directly, the chemical potential of the solute, and hence its molar concentration, can be calculated. This is a measurement at equilibrium. A measurement of a hydrodynamic property of a molecule of protein, however, is, as the name implies, a kinetic measurement, not a measurement at equilibrium. In such a kinetic measurement, the rate of movement of the molecule of protein under an applied force is measured. Free electrophoretic mobility is an example of such a hydrodynamic property. The confusion arises because centrifugal force can also be used to promote the movement of a molecule of protein through a solution. This use of centrifugal force is unrelated to the use of centrifugal force to create a gradient of concentration. The confusion also arises because the same instrument, an analytical ultracentrifuge, is used to make each of the measurements, even though they are unrelated to each other.

The term $(\partial \rho / \partial C_p)_\mu$, the change in the density of the solution as protein is added to it at constant chemical potential of the other solutes, is determined in a separate experiment. Its units, as is the case with the magnitude of $\Pi(C_p)^{-1}$ found in Equation 8–6, are determined by the units chosen for C_p, the concentration of protein. By chance, it happens that if C_p is in grams centimeter^{-3} and ρ is expressed in grams centimeter^{-3}, $(\partial \rho / \partial C_p)_\mu$ is dimensionless. In theory, the measurement of this differential requires the tabulation of the macroscopic density of a series of solutions containing increasing, precisely known concentrations of protein and brought to equilibrium, at the appropriate osmotic pressure, by dialysis against a solution identical to the solution used to prepare the samples for centrifugation but lacking the protein. In practice, the term $(\partial \rho / \partial C_p)_\mu$ is approximated by the value of $(1 - \bar{v}_{prot} \rho_{sol})$, where $\bar{v}_{prot}$ is the partial specific volume (centimeters^3 gram^{-1}) of the protein in the particular solution chosen and ρ_{sol} is the density (grams centimeter^{-3}) of the solution in the absence of the protein. This commonly used approximation assumes that C_p can be expressed in grams centimeter^{-3} and implicitly dictates a choice of unit for the entire equation. This assumption remains hidden in the definition of the partial specific volume

$$\bar{v}_{prot} = \left(\frac{\partial V}{\partial m_{prot}} \right)_{T, P} \qquad (8\text{–}20)$$

where m_{prot} is the mass (grams) of protein added to a solution and V is the resulting volume of the solution (centimeters^3). The accuracy of this measurement is no greater than the accuracy with which the concentration of protein in grams centimeter^{-3} can be known. Nor is this requirement avoided by the use of values of $\bar{v}_{prot}$ calculated from the amino acid composition. In effect, this latter approximation simply relies on the care with

which protein concentrations in grams centimeter^{-3} were determined in the earlier experiments demonstrating that such a calculation was reliable.[10]

Equation 8–19 represents an extrapolation to C_p equal to zero. The purpose of the extrapolation is to eliminate the effect of the virial coefficients (Equation 8–17) on $d\Pi/dC_p$. The same protein, bovine serum albumin, under similar conditions ($\Gamma/2 = 0.1 - 0.15$M), was used in the experiments described in Figures 8–1 and 8–2. From the values of the second virial coefficients, B, of Figure 8–1, it can be calculated that at the actual concentrations examined in Figure 8–2, the nonideality of the solution should only have affected the molar mass determined from the slope of the line by less than 0.1% of its value. Uncoiled polypeptides or highly charged macromolecules such as DNA would have had much larger virial coefficients, and the measurements would have been significantly affected. The molar mass calculated[9] from Figure 8–2, 64,500 g mol^{-1}, agrees quite closely with the actual value of the molar mass of bovine serum albumin, 66,300 g mol^{-1}, a value that was subsequently established by its amino acid sequence.[11]

Light scattering is a property of any solution. It arises because a solution is a collection of molecules undergoing random movements rather than an instantaneously uniform continuum of electrons. Scattered light emerges from a solution at all angles to the incident direction of a beam of light passing through the solution. The sources of the scattered light are electrons in the sample that oscillate in response to the alternating electric field of the light and in turn emit light. The light is emitted in directions other than that of the exciting light. The susceptibility of electrons in molecules to this phenomenon arises from their respective polarizabilities. If the solution were a uniform, unvarying distribution of electrons, the emitted light would always be canceled by interference and no scattering would result. Therefore, scattering from a solution arises from regional fluctuations in polarizability on a scale smaller than the wavelength of

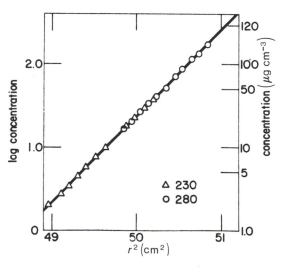

Figure 8–2: Sedimentation equilibrium of bovine serum albumin.[9] A solution of bovine serum albumin was placed in an optical cell in a rotor. The rotor was placed in an ultracentrifuge and spun at 24,630 rpm. The distribution of protein in the optical cell was followed by scanning the absorbance of the solution along the radial axis of the cell. After 18 h, the distribution of protein was no longer changing and equilibrium had been reached. The absorbance as a function of radial distance from the center of rotation of the rotor was measured at wavelengths of 280 nm (○) and 230 nm (△). Absorbance was converted to concentration of protein (micrograms centimeter^{-3}) and the logarithm (log concentration) was taken. Radial distance, r, was measured in centimeters and the respective values were squared (r^2). Adapted with permission from ref 9. Copyright 1966 American Chemical Society.

light. These fluctuations are related to local fluctuations in the concentrations of the components of the solution. The majority of the electrons in the solution are on molecules of water, and it is fluctuations in the local concentrations of water that are the major contributors to the scattered light in the absence of the protein. When protein is present, the scattering arising from the molecules of protein in the solution is in addition to this background scattering.

The scattered light from a source of collimated, unpolarized light is measured by placing a detector at an angle θ to the beam of unscattered light passing through the sample (Figure 8–3) and at a distance r (centimeters) from the sample. The difference in intensity, i_θ, between light scattered by a unit volume of the solution of protein [photons second^{-1} (centimeter^3 of solution)$^{-1}$] and that scattered by an identical solution (also in photons second^{-1} centimeter^{-3}) not containing protein is measured and reported relative to the intensity of the incident light, I_0 (photons second^{-1}). It can be shown that[12]

$$\lim_{\theta \to 0} \frac{r^2 i_\theta}{I_0(1+\cos^2\theta)} = RTKC_p \left(\frac{\partial \tilde{n}}{\partial C_p}\right)_{P,\mu}^2 \left(\frac{d\Pi}{dC_p}\right)^{-1} \quad (8\text{–}21)$$

where $\tilde{n}$ is the refractive index (dimensionless) of the solution of protein and the optical constant K (moles centimeter^{-4}) is defined by

$$K \equiv \frac{2\pi^2 \tilde{n}_0^2}{\lambda^4 N_{Av}} \quad (8\text{–}22)$$

where $\tilde{n}_0$ is the refractive index of the solution in the absence of the protein, λ is the wavelength (centimeters)

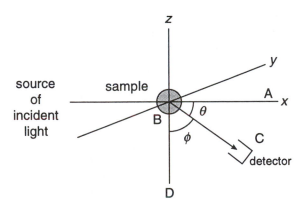

Figure 8–3: Angular dependence of scattered light. The angular dependence of scattered light is related to a coordinate system in which the x axis is the beam of incident light and the origin is the sample. The angle θ is the angle ABC where A is a point on the x-axis beyond the sample, AB is along the x axis, B is the position of the sample, and C is the position of the detector. The angle ϕ is the angle DBC where D is any point on the z axis, which is defined as an axis parallel to the electric vector of the light from a polarized source, B is the position of the sample, and C is the position of the detector.

of the light, and N_{Av} is Avogadro's number (6.022×10^{23} mol^{-1}). It should be noted that the units on C_p, the concentration of protein, cancel in Equation 8–21. This again illustrates that the concentration of protein can be expressed in any units. Equation 8–21 also illustrates that light scattering is a measurement of the osmotic pressure of the solution of protein. The limit in Equation 8–21 is taken to eliminate any optical interference that might arise if the dimensions of the protein are close to the magnitude of the wavelength of the light.

At the present time, the light source usually used for measurements of light scattering is a laser; and, if it is not already polarized, the light is usually passed through a polarizer so that polarized light is scattered by the sample. The intensity of the scattered light from a source of polarized light has a different angular dependence than that from a source of unpolarized light. The oscillating electric vector of the polarized light defines the z axis of a coordinate system with the sample at the origin (Figure 8–3). As defined, the x, y plane is normal to the oscillating electric vector. If ϕ is the angle between the z axis and the ray of scattered light entering the detector, then

$$\lim_{\theta \to 0} \frac{r^2 i_\theta}{I_0(2\sin^2\phi)} = RTKC_p \left(\frac{\partial \tilde{n}}{\partial C_p}\right)_{P,\mu}^2 \left(\frac{d\Pi}{dC_p}\right)^{-1} \quad (8\text{–}23)$$

where θ is still the angle between the beam of unscattered light emerging from the sample and the ray of scattered light entering the detector. If the detector is confined to the x, y plane (Figure 8–3), the angle ϕ is always 90° and $\sin^2\phi = 1$. In this configuration, to perform the necessary extrapolation to $\theta = 0$, the angle θ can be varied over all values without changing the angle ϕ. When unpolarized light is used, $\cos^2\phi$ changes continuously as the limit of $\phi \to 0$ is taken, and this complicates the extrapolation.

It is convenient to define a quantity, R_θ, known as Rayleigh's ratio (centimeters^{-1}) to eliminate the dimensions of the apparatus from the calculation. For unpolarized light

$$R_\theta \equiv \frac{r^2 i_\theta}{I_0(1+\cos^2\theta)} \quad (8\text{–}24)$$

and for polarized light when $\phi = 90°$

$$R_\theta \equiv \frac{r^2 i_\theta}{I_0} \quad (8\text{–}25)$$

When Equation 8–21 or 8–23 is combined with Equation 8–24 or 8–25

$$\lim_{\substack{\theta \to 0 \\ C_p \to 0}} R_\theta = KC_p \left(\frac{\partial \tilde{n}}{\partial C_p}\right)_{P,\mu}^2 W \quad (8\text{–}26)$$

The double limit in Equation 8–26 is often taken by a procedure known as the Zimm plot,[12] but with proteins

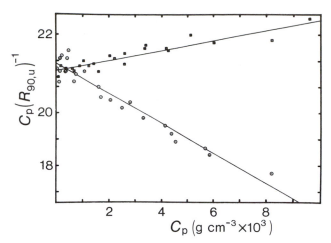

Figure 8–4: Light scattering by solutions of bovine serum albumin.[13] The Rayleigh ratio $R_{90,u}$ (centimeters^{-1}) was determined for each of a series of solutions of bovine serum albumin, each at a different concentration, C_p (grams centimeter^{-3}), by measuring the scattering of unpolarized light at an angle θ of 90°. The quotient $C_p(R_{90,u})^{-1}$ was calculated for each measurement and plotted as a function of $C_p \times 10^3$. Measurements were made in 0.15 M sodium chloride ($\square$,$\blacksquare$) or in water ($\bigcirc$) of solutions prepared by diluting an isoionic solution of albumin into the appropriate solutions. Adapted with permission from ref 13. Copyright 1954 American Chemical Society.

of normal dimensions, the variation of R_θ with θ is small and inconsequential, and the extrapolation to $\theta = 0$ is a minor correction and often ignored. The extrapolation to $C_p = 0$ is always required because the virial coefficients (Equation 8–7) can be appreciable (Figure 8–4). To perform this extrapolation a rearrangement of Equations 8–23 and 8–26, incorporating the virial coefficients explicitly, is performed:

$$\lim_{\theta \to 0} \frac{C_p}{R_\theta} = \frac{1}{K}\left(\frac{\partial \tilde{n}}{\partial C_p}\right)_{P,\mu}^{-2}\left(\frac{1}{W} + 2BC_p + 3CC_p^2 \dots\right) \quad (8\text{–}27)$$

In Figure 8–4, two experiments with bovine serum albumin under different conditions, with and without added electrolyte, are presented. As with the measurements shown in Figure 8–1, it can be seen that the virial coefficient, B, changes appreciably with changes in conditions; in this case it even inverts in sign, probably because the protein is participating in a concentration-dependent oligomerization in the absence of electrolyte. The intercepts, however, again remain the same. It has been shown that under the same conditions with the same protein, the same values of the second virial coefficient are obtained by measurements of either osmotic pressure or light scattering.[14] This result demonstrates that the virial coefficient is a property of the solution itself rather than the method of measurement, and it provides further evidence that these techniques are both measuring the same property of the solution.

The value of $(\partial \tilde{n}/\partial C_p)_{P,\mu}$ is the change in the refractive index of the solution as the concentration of protein is increased, at constant chemical potential of the other solutes such as electrolytes and buffers. Each of the solutions of protein used to make the determination of $(\partial \tilde{n}/\partial C_p)_{P,\mu}$, as well as the solution used in the determination of the light scattering itself, should be equilibrated by dialysis against a solution identical except for the protein to obtain a constant chemical potential of the other solutes throughout.

From an examination of Equation 8–26, it is clear again that the unit chosen for C_p, the concentration of protein (units centimeter^{-3}), determine the units (units mole^{-1}) of the parameter W. If the concentration of protein is known in grams centimeter^{-3}, the units on W will be grams mole^{-1}, or molar mass. The determination of the molar mass, however, will only be as accurate as the measurement of the concentration of protein.

Recently, electrospray mass spectroscopy[15] has been applied to determining the molecular mass of proteins. Individual molecules of protein are atomized in individual droplets and the solvent is evaporated to produce vaporized, ionized molecules of the protein, each in one of its ionization states, and the vaporized, ionized molecules are submitted to mass spectroscopy. In this way precise masses of the molecule of protein can be determined. For example, it was possible to show that the molecular mass of the blue copper protein rusticyanin from *Thiobacillus ferrooxidans* was 16,552 Da, which is within 1 Da of the mass calculated from its amino acid sequence.[16] Because amino acid sequences of proteins are usually known before enough is available to submit to analysis, the major application of this technique is to assess posttranslational modification. For example, in the case of rusticyanin the conclusion drawn from the mass spectral analysis was not that the molecular mass of the protein was 16,552 Da but that the protein lacked posttranslational modifications.

Now that the sequences of so many proteins are known, as well as their subunit stoichiometries, precise values of molar mass can be calculated for a large array of proteins from their atomic compositions. These can be compared with values determined, before this information was available, by osmotic pressure, sedimentation equilibrium, and light scattering (Table 8–1). By and large, the agreement between the actual values of molar mass and the measured values is quite close, and this in itself validates the methods.

The problem with molar mass, no matter how accurately it can be determined, is that it means very little to most people. Once it was clear that proteins were polymers of amino acids, sometimes posttranslationally modified, the reason behind all determinations of molar mass has been to learn how many amino acids are contained in a given polypeptide and how many polypeptides are contained in a given protein. These are quantities that anyone can understand. Unfortunately, the results have

Table 8–1: Comparison of the Actual Molar Masses of Selected Proteins and the Molar Masses Determined by Light Scattering, Osmotic Pressure, and Sedimentation Equilibrium[a]

protein	subunit stoichiometry	molar mass (g mol^{-1} × 10^{-3})				references
		actual	light scattering	osmotic pressure	sedimentation equilibrium	
human serum albumin	α	66.5	70	69	68	17, 18
bovine pancreatic ribonuclease	α	13.69			13.7, 13.0	17, 19, 20
bovine β-lactoglobulin	α_2	36.6	36	35, 39	38	17, 19, 20
chicken lysozyme	α	14.32	14.8	17.5, 16.6		19, 20
porcine L-lactate dehydrogenase	α_4	145.9	143	146	141	18, 21
rabbit phosphorylase *b*	α_2	194.6			185	22, 23
mammalian glyceraldehyde-3-phosphate dehydrogenase	α_4	143	145			20, 24
bovine catalase	α_4	233			240	25–27
rabbit muscle fructose-bisphosphate aldolase	α_4	156.9			142, 158	28, 29
mammalian apoferritin	$\alpha_n\beta_m$ (m + n = 24)	510	430			30, 31
E. coli aspartate transcarbamylase	$\alpha_6\beta_6$	310			310	20, 32, 33
bovine chymotrypsinogen	α	25.7		36		19, 20
porcine pepsin A	α	39.5		36	39	19
Pseudomonas putida 6-phospho-2-dehydro-3-deoxygluconate aldolase	α_3	71.8			73	34
E. coli aspartate kinase I–homoserine dehydrogenase I	α_4	357	358		360	35, 36
bovine glutamate dehydrogenase	α_6	333	313		320	37–39

[a]The actual molar mass of each protein was calculated from the amino acid sequences of its constituent polypeptides and their stoichiometry in the complex.

seldom been presented in these terms even though they always could have been. As it happens, the mean molar mass of an amino acid in most proteins is a reasonably constant number, 111 ± 4 g mol^{-1} (Table 8–2). It follows that the number of amino acids in most proteins can be estimated by simply dividing a measurement of its molar mass by 111 g mol^{-1}, after accounting for glycosylation and other posttranslational modifications.

It should not be imagined that the molar mass is a more fundamental number while the number of amino acids in a protein is somehow an approximation. It has already been pointed out that expressing C_p in the units of grams centimeter^{-3} during determinations of molar mass is an arbitrary choice. Were C_p expressed in terms of moles of amino acids centimeter^{-3}, each of the procedures would have necessarily provided as accurate a value of W in terms of moles of amino acids (mole of protein)$^{-1}$ rather than grams (mole of protein)$^{-1}$. The ability to determine molar masses by physical measurements has always relied ultimately upon the ability of the investigator to make an accurate measurement of dry weight. All values of molar mass can be traced back to such a determination. The difficulties involved in measurements of dry weight have been noted,[10] and more than anything else, the accuracy of the values in Table 8–1 are a testimony to the careful measurements of these quantities. Accurate dry weight measurement, however,

requires more protein than is usually available, and other less reliable measures of protein concentration have necessarily supplanted it.

Part of the description of a particular protein is an enumeration of the length of each of the polypeptides from which it is composed and the number of each subunit that it contains. At one time this information could be most conveniently learned by ascertaining both the molar mass of the entire protein and the molar mass of the isolated individual polypeptides. The history of this quest is interesting but beyond the scope of the present discussion. In two celebrated instances, that of aspartate transcarbamylase and that of fructose-bisphosphate aldolase,[46] disagreements arose over the results from such measurements. These particular disagreements coincided with the development of the two techniques that have supplanted almost entirely the earlier methods of determining molar mass that were just described. These newer procedures are both based on the electrophoresis of complexes between polypeptides and dodecyl sulfate upon gels of polyacrylamide. In one procedure, sieving is used to display the different types of polypeptides in the protein and provide accurate estimates of the length of each. In the other procedure, patterns of covalently cross-linked polypeptides separated by electrophoresis are used to count the number of each polypeptide present in the whole protein.

Table 8–2: A Tabulation of the Mean Grams (Mole of Amino Acid)$^{-1}$ in a Set of Proteins

polypeptide[a]	type of protein	length[b]	total number of amino acids in protein	polypeptide stoichiometry	grams (mole of amino acids)$^{-1}$ [c]	references
cod parvalbumin	cytoplasmic	113	113	α	107.1	18
chicken lysozyme	extracytoplasmic, enzymatic	129	129	α	110.9	20
R17 coat protein	virus coat	129			106.2	20
human hemoglobin ($\alpha + \beta$)	cytoplasmic	141 + 146	574	$(\alpha\beta)_2$	108.9	20
bovine chymotrypsinogen	extracytoplasmic, enzymatic	245	245	α	104.7	20
Halobacterium halobium bacteriorhodopsin	membrane-spanning	248		α_3	107.3	40
dogfish L-lactate dehydrogenase	cytoplasmic, enzymatic	329	1316	$(\alpha_2)_2$	110.7	18
human immunoglobulin G, Eu ($\alpha + \beta$)	extracytoplasmic	446 + 214	1320	$(\alpha\beta)_2$	108.9	20
human fibrinogen ($\beta + \gamma$)	extracytoplasmic, fibrous	415 + 411	2964	$(\alpha\beta\gamma)_2$	117.3	18, 41
human serum albumin	extracytoplasmic	585	585	α	113.6	18
rabbit phosphorylase *a*	cytoplasmic, enzymatic	841	3364	$(\alpha_2)_2$	115.8	22
murine anion carrier	membrane-spanning	929	1858	α_2	110.8	42
ovine Na$^+$/K$^+$-transporting ATPase (α subunit)	membrane-spanning	1016	1318	$\alpha\beta$	110.5	43, 44
rat myosin (α subunit)	cytoplasmic, fibrous	1939	4560	$\alpha\beta\gamma\alpha\beta\sigma$	115.5	18, 45
					111±4	

[a] Proteins composed of one or more polypeptides, the sequences of which were available, were chosen as examples. An attempt was made to include examples of all types of proteins, but extremely unusual proteins such as collagen were avoided. The constituent polypeptides whose compositions were used are indicated. [b] The lengths of the polypeptides chosen for analysis are presented in numbers of amino acids. [c] Calculated by dividing the molar mass of the protein portion of the polypeptide or polypeptides by the length or combined lengths, respectively.

Suggested Reading

Schachman, H.K., & Edelstein, S.J. (1966) Ultracentrifuge Studies with Absorption Optics. IV. Molecular Weight Determinations at the Microgram Level, *Biochemistry* 5, 2681–2705.

Problem 8–1: Calculate the molar mass of ribonuclease from its sequence.

Problem 8–2: Calculate the molar masses of the proteins that have the following incremental osmotic pressures at 25 °C.

protein	$\lim_{C_p \to 0} \Pi$
L-lactate dehydrogenase	173 dyne cm^{-2}(g of protein)$^{-1}$ L
β-lactoglobulin	0.722 cm H$_2$O (g of protein)$^{-1}$ L
bovine serum albumin	3.57 × 10^5 dyne cm (g of protein)$^{-1}$

Problem 8–3: Calculate the molar mass of aspartate kinase I–homoserine dehydrogenase I from the data in this figure (adapted with permission from ref 47; copyright 1968 Springer-Verlag).

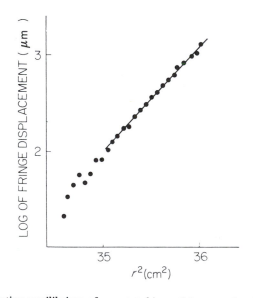

Sedimentation equilibrium of aspartate kinase I–homoserine dehydrogenase I from *E. coli*. Interference optics were used. Protein concentration: 0.6 mg mL^{-1}. The logarithm of the distance in micrometers between blank corrected fringes and zero level concentration is plotted against r^2, the distance from the center of the rotor to the point at which the measurement was made. The fringe displacement (in micrometers) is directly proportional to the concentration of protein in grams centimeter^{-3}. The concentration near the meniscus being zero, the left-most points are scattered; those with a fringe displacement less than 100 μm were not used for computations.

Use the approximation

$$\left(\frac{\partial \rho}{\partial C_p}\right)_\mu = 1 - \bar{v}_{prot}\,\rho_{sol}$$

and assume $\rho_{sol} = \rho_{H_2O}$. The partial specific volume of the protein is 0.737 cm^3 g^{-1}, the rotor was spinning at 11,272 revolutions min^{-1} (1 revolution is 2π radians), and the temperature was 23 °C. Fringes are optical features that are directly proportional to the concentration of protein. Remember that the concentration of protein can be expressed in any units.

Problem 8–4: Human hemoglobin is a protein formed from two α polypeptides, two β polypeptides, and four hemes for a total molar mass of 64,452 g mol^{-1}. It is referred to as an $(\alpha\beta)_2$ tetramer to indicate that it is a heterotetramer formed from two $\alpha\beta$ heterodimers. The osmotic pressure at 20 °C of a solution of hemoglobin, at a concentration of 3 g L^{-1}, that had been flushed exhaustively with N_2 gas was 1750 dyne cm^{-2}. When the solution was flushed with O_2 gas, the osmotic pressure increased to 2500 dyne cm^{-2} even though the concentration of the hemoglobin was unchanged.

(A) Assume ideal behavior and that the virial coefficients are zero and calculate the number average molecular weight for each circumstance.

(B) Explain the values that you obtain.

Problem 8–5: The enzyme glutamine synthase, which catalyzes the ATP-dependent condensation of glutamate and ammonia, was isolated from *E. coli*. It was purified by $(NH_4)_2SO_4$ fractionation followed by ion-exchange chromatography. The final preparation of the enzyme was considered to be a pure protein because a single peak was obtained on repeated ion-exchange chromatography. The protein is a complex of 12 identical polypeptides each 468 amino acids in length. This has been ascertained by sequencing its genomic DNA.

Glutamine synthase was dissolved in 6 M guanidinium chloride, and the osmotic pressure of this solution was determined with a high-speed osmometer at 20 °C. The following results were obtained

protein concentration g L^{-1}	pressure (mmHg)
2	0.69
4	1.40
8	2.87
10	3.63
15	4.73
20	7.96

(A) What is the molar mass of the protein in 6 M guanidinium chloride?

(B) What has this solvent done to the protein?

Problem 8–6: The light scattering from a series of solutions of ovalbumin of different concentrations was measured at an angle θ of 90°. The apparatus sampled the light scattered from a volume of the solution of 1.8 cm^3 with a photomultiplier tube at 10 cm from the center of the sample. The differences between the scattering of the buffer alone and the scattering of each solution were used to calculate the Rayleigh ratio, R_{90}, for each solution. The intensity of the incremental scattered light was less than 0.001% of the intensity of the incident light. The Rayleigh ratios for light of wavelength 327 nm were as follows

C_p (g cm^{-3})	R_{90} (cm^{-1})
4.3×10^{-3}	3.6×10^{-4}
5.8×10^{-3}	4.8×10^{-4}
8.6×10^{-3}	7.0×10^{-4}
9.7×10^{-3}	8.0×10^{-4}
13.7×10^{-3}	11.6×10^{-4}

The refractive index $\bar{n}_0$ of the solution without the protein was 1.333, the increment of the refractive index $[(\partial\bar{n}/\partial C_p)_{P,\mu}]$ for the protein at $\lambda = 327$ nm was 0.1883 cm^3 g^{-1}, and the temperature was 25 °C.

(A) Determine graphically the limit

$$\lim_{C_p \to 0} \frac{C_p}{R_{90}}$$

(B) Assume that the Rayleigh ratio, R_θ, does not vary significantly with variation in θ for this protein at this wavelength and that

$$\lim_{\substack{\theta \to 0 \\ C_p \to 0}} \frac{C_p}{R_\theta} = \lim_{C_p \to 0} \frac{C_p}{R_{90}}$$

and use the value for this limit to calculate the molar mass of ovalbumin.

Problem 8–7: The isoelectric pH of bovine serum albumin in a solution of 0.15 M NaCl is 5.37. From the data in Figure 8–1, calculate the net charge Q_{SA} on the serum albumin at pH 7.0. Assume that the difference in slope between the two lines is due entirely to the Donnan effect.

Electrophoresis on Gels of Polyacrylamide Cast in Solutions of Dodecyl Sulfate

The sodium salt of dodecyl sulfate ($H_{25}C_{12}OSO_3^-$) is a detergent widely used commercially to dissolve nonpolar substances in water. It accomplishes this purpose by forming micelles of dodecyl sulfate in aqueous solution. A micelle of dodecyl sulfate at moderate ionic strength (0.2 M) contains about 100 of the anions in an oblate ellipsoid that is 3 nm across at its minor axis.[15] All of the anionic sulfonates are at the surface of the ellipsoid and the hydrocarbon is in the center. It is the hydrocarbon core of the micelle that dissolves individual molecules of a nonpolar substance, producing the detergent properties. Although dodecyl sulfate must be in the form of micelles to interact with proteins, the complexes that result between the anions of dodecyl sulfate and polypeptides do not seem to involve discrete micelles.

When sodium dodecyl sulfate[48] is added to a solution of protein at a concentration greater than its critical micelle concentration and at ratio greater than 2 g of dodecyl sulfate (g of protein)$^{-1}$, all of the polypeptides present in the solution unfold, separate from each other, and become coated with the dodecyl sulfate. The amount of dodecyl sulfate coating the unfolded, separated polypeptides at saturation is usually a function of only their total length. Pitt-Rivers and Impiombato[49] observed that within a series of globular, water-soluble proteins, each polypeptide would bind 0.54 ± 0.01 molecule of dodecyl sulfate for every amino acid in its sequence. The important point, however, is not the numerical value of this ratio but the fact that it is very constant (less than a 2% variation) regardless of the protein examined, as long as it is of the usual water-soluble, globular variety. This rather unexpected behavior presumably results from the fact that the amino acid compositions of all such proteins are very similar to each other. This regularity, however, is observed only when all of the cystines in the proteins, if there were any, have been cleaved.[49] Usually this is done by disulfide interchange with a small thiol (Figure 3–23).

The complexes that are formed between dodecyl sulfate and polypeptides are extended structures and have been variously described as tubular rods whose length is directly proportional to the length of the polypeptide[50] or micellar pearls of dodecyl sulfate on a string of the flexible polypeptide.[51] No definitive description of their structure is available, but there is no evidence that the dodecyl sulfate in these complexes is present in discrete packets of 100 molecules of detergent, as would be expected if the micelles present in the absence of the protein were simply incorporated intact into a long string upon the unfolded polypeptide.

As with nucleic acids and presumably for the same reasons, the complexes between dodecyl sulfate and those polypeptides that bind a constant ratio of the strongly anionic detergent all display the same free electrophoretic mobility, $-2.62 \pm 0.04 \times 10^{-4}$ cm^2 s^{-1} V^{-1}, regardless of the length of the polypeptide.[51] In the case of nucleic acids, the invariance of the free electrophoretic mobility with length results from the uniform distribution of negative charge along the regular polymer, and presumably this is also a necessary condition met by the complexes between dodecyl sulfate and polypeptides. With nucleic acids, however, this is a covalently conferred, intrinsic property of the phosphodiesters of the backbone rather than the fortuitous and less reliable inclination of the polymer to bind a charge-conferring species uniformly along its length. As such, any polypeptide that binds dodecyl sulfate abnormally should have a different free electrophoretic mobility. When the amount of dodecyl sulfate bound to a series of polypeptides was purposely decreased, the magnitudes of their free electrophoretic mobilities also decreased.[51]

As is the case with nucleic acids,[53] native proteins (Figure 1–17), and other macromolecules submitted to electrophoresis on gels of polyacrylamide or other polymeric supports, the electrophoretic mobilities of complexes between dodecyl sulfate and polypeptides (Figure 8–5) follow the relationship

$$u_i = u_i^{\circ} e^{-iK_r T_a} \qquad (8\text{–}28)$$

where T_a is the concentration of acrylamide (in percent)

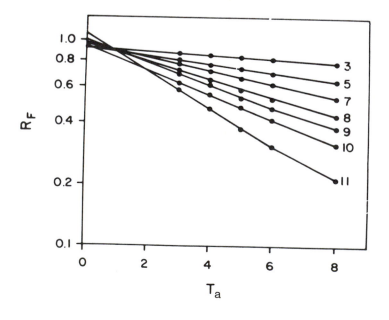

Figure 8–5: Relative electrophoretic mobilities, R_f, of complexes of polypeptides and dodecyl sulfate.[52] Various proteins—myoglobin (3), chymotrypsinogen (5), L-lactate dehydrogenase (7), ovalbumin (8), glutamate dehydrogenase (9), bovine serum albumin (10), and phosphorylase (11)—were dissolved separately in a solution containing a concentration of dodecyl sulfate sufficient to saturate the polypeptides. Each was then submitted to electrophoresis on gels of polyacrylamide cast in a solution of 1% sodium dodecyl sulfate. A series of gels were used for each protein that differed in the percent acrylamide (T_a) from which they were cast. The gels were stained for protein, and the distance migrated by the protein was divided by the distance migrated by a dye, Pyronine-Y, of low molecular weight to obtain the relative electrophoretic mobility (R_f) of each polypeptide at each percent acrylamide. The assumption made was that the mobility of the Pyronine-Y, because it is such a small molecule, would be unaffected by the percent acrylamide. Reprinted with permission from ref 52. Copyright 1972 *Journal of Biological Chemistry*.

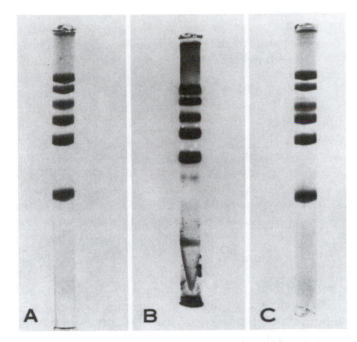

Figure 8–6: Separation of polypeptides by electrophoresis on gels of polyacrylamide cast in a solution of 0.2% sodium dodecyl sulfate.[54] Proteins containing the polypeptides were dissolved in solutions of sodium dodecyl sulfate sufficient to saturate them. They were submitted to electrophoresis on tubular gels (0.6 cm × 10 cm) cast from 10% acrylamide in 0.2% sodium dodecyl sulfate. Following electrophoresis, the gels were stained for protein. The polypeptides run on gel A were those composing catalase (n_{aa} = 506), porcine cardiac fumarate hydratase (n_{aa} = 466), rabbit muscle fructose-bisphosphate aldolase (n_{aa} = 361), rabbit muscle glyceraldehyde-3-phosphate dehydrogenase (n_{aa} = 332), human carbonic anhydrase B (n_{aa} = 260), and horse cardiac myoglobin (n_{aa} = 153). The polypeptides run on gel B were the same as on gel A, but the myoglobin was omitted. The polypeptides run on gel C were catalase, fumarate hydratase, horse liver alcohol dehydrogenase (n_{aa} = 374), glyceraldehyde-3-phosphate dehydrogenase, carbonic anhydrase, and myoglobin. Reprinted with permission from ref 54. Copyright 1969 *Journal of Biological Chemistry.*

from which the gel was cast and ^iK_r is a constant unique to the partic-ular polypeptide referred to as its **retardation coefficient**. Because $u_i°$ is the free electrophoretic mobility of the complex between dodecyl sulfate and polypeptide i and $u°$ is the same for all complexes between dodecyl sulfate and well-behaved polypeptides, this relationship predicts that the lines in Figure 8–5 should intersect at the axis of the ordinate when T_a is equal to zero, which is almost the case. Because each complex has a unique retardation coefficient, electrophoresis on gels of polyacrylamide can be used to separate these complexes one from the other (Figure 8–6).

As with nucleic acids, the electrophoretic mobilities of complexes between dodecyl sulfate and polypeptides on gels of polyacrylamide are a regular function of the length of the polypeptides. To understand this property of the electrophoretic separations, the process known as sieving must be understood.

Suggested Reading

Weber, K., & Osborne, M. (1969) The Reliability of Molecular Weight Determinations by Dodecyl Sulfate–Polyacrylamide Gel Electrophoresis, *J. Biol. Chem. 244*, 4406–4412.

Sieving

Sieving of macromolecules, for example, proteins, nucleic acids, or complexes between dodecyl sulfate and polypeptides, occurs during molecular exclusion chromatography and electrophoresis on polymeric supports. **Sieving** is the discrimination between macromolecules on the basis of size that is accomplished by a random network of linear polymers. In molecular exclusion chromatography, the network forms the beads among which the mobile phase percolates and into which the macromolecules diffuse to enter the stationary phase. In electrophoresis, the network forms an obstacle course through which the macromolecule must pass as it moves in the direction of the electric field.

Consider a geometric solid of any shape within a network of lines thrown at random through a volume of space completely containing the solid. An equation[55] for the probability that none of these lines intersects the solid, $P(ni)$, was derived during the solution of an unrelated topological problem,[56] and

$$P(\text{ni}) = e^{-lS/4} \tag{8–29}$$

where l is the density of the lines (cm cm^{-3}) and S is the surface area of the solid (cm^2). Assume that a macromolecule is a geometric solid and a network of chemical polymers is a network of lines. The fraction of the total volume available to a macromolecule, K_{av}, in a stationary phase of randomly arranged linear polymers should be that fraction of the total volume whose occupation by the macromolecule does not cause any polymer to intersect the macromolecule. In this case,

$$^iK_{av} = e^{-bT_P S_i^{app}} \tag{8–30}$$

where $^iK_{av}$ is the distribution coefficient of macromolecule i defined by Equation 1–15, T_P is the concentration of polymer in percent [grams (100 cubic centimeters)$^{-1}$], S_i^{app} is the apparent surface area (centimeters^2) of the macromolecule i, and b is a proportionality constant to convert, among its other roles, the concentration of polymer [grams (100 cubic centimeters)$^{-1}$] into its linear density [centimeters (cubic centimeter)$^{-1}$].

Because the polymers are not lines but solids themselves, the apparent surface area of the macromolecule is not its real surface area, whatever that would be. The apparent surface of the macromolecule lies outside its actual surface by a distance equal to the sum of the widths of any tight shells of hydration around either the macromolecule or the polymer and the width of the polymer itself. All of the dimensions that cause the poly-

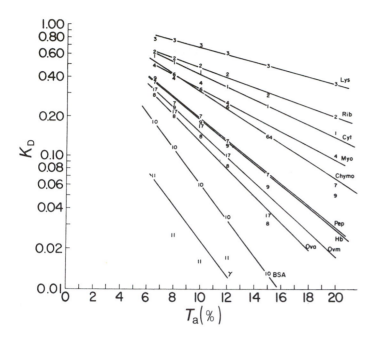

Figure 8–7: Distribution coefficients for a series of globular proteins submitted to molecular exclusion chromatography on polyacrylamide gels of varying composition.[55,57] A series of gels cast from different concentrations, T_a(percent), of acrylamide were separately fragmented to form suspensions of polyacrylamide granules of different porosities. Columns were made from these chromatographic media, and a set of standard globular proteins were submitted to molecular exclusion chromatography on these columns and their respective elution volumes were used to calculate the respective distribution coefficients, K_D (Equation 1–16). The distribution coefficient K_D is always directly proportional to the distribution coefficient K_{av}. The natural logarithms of the values of K_D are plotted as a function of T_a. The proteins were (1) cytochrome c, (2) ribonuclease (c2 Kin.4), (3) lysozyme, (4) myoglobin (c3 Kin.3), (6) chymotrypsinogen (c15 Kin.1), (7) pepsin, (8) ovalbumin, (9) hemoglobin, (10) serum albumin, (11) immunoglobulin G, and (17) ovomucoid. Adapted with permission from ref 55. Copyright 1970 National Academy of Sciences.

mer not to be a line and the macromolecule not to be a dry smooth solid object are incorporated into the dimensions of an apparent macromolecule that is larger than the actual macromolecule. When the actual macromolecule collides with the actual polymer, the apparent macromolecule collides with a line in the center of the polymer.

This model predicts that if a series of beaded stationary phases of increasing concentration of polymer is used to separate the same set of standard macromolecules by molecular exclusion chromatography, then

$$^iK_{av} = e^{-{^iK_r}T_P} \qquad (8–31)$$

If this is so, then $\ln\,^iK_{av}$ should be a linear function of T_P. Such behavior has been observed (Figure 8–7).[55,57] It has been demonstrated that the relationship of Equation 8–31 describes the behavior of both proteins[57,58] and

polysaccharides[58] during molecular exclusion chromatography on gels of both polyacrylamide[57,59] and linear dextrans.[58]

If Equation 8–30 also describes behavior during molecular exclusion chromatography, then when T_P is fixed, $-\ln\,^iK_{av}$ should be a linear function of S_i^{app}. It has been shown by Ogston[59] that if a set of macromolecules were all spheres of radius R_i and the polymers of the network were infinitely long right cylinders of radius r, then

$$S_i^{app} = 4\pi(r+R_i)^2 \qquad (8–32)$$

The partial molar volume of a molecule of protein can be calculated from its amino acid composition.[61,62] As the amino acid compositions of most proteins are very similar, their partial molar volumes should be directly proportional to the number of amino acids each contains. To the extent that a molecule of protein is a sphere, the number of amino acids it contains, $^in_{aa}$, should determine its radius by the relationship

$$R_i^\circ = \left[\frac{3}{4\pi}\left(\frac{^in_{aa}\overline{V}_{aa}}{N_{Av}}\right)\right]^{1/3} \qquad (8–33)$$

where $\overline{V}_{aa}$ is the mean partial molar volume of the amino acids in the usual protein (82 cm^3 mol^{-1}) and the superscript has been added to R_i to indicate that this is a sphere equivalent in volume to the volume of the protein, which is never exactly a sphere. When Equations 8–30, 8–32, and 8–33 are combined

$$\left(\frac{-\ln\,^iK_{av}}{4\pi bT_P}\right)^{1/2} = r + \left[\frac{3}{4\pi}\left(\frac{^in_{aa}\overline{V}_{aa}}{N_{Av}}\right)\right]^{1/3} \qquad (8–34)$$

and a plot of $(-\ln\,^iK_{av})^{1/2}$ against $(^in_{aa})^{1/3}$ should be a linear relationship. When the data of Andrews,[63] for a series of proteins submitted to molecular exclusion chromatography upon cross-linked dextran, are displayed in this fashion,[55] they are linearly related (Figure 8–8B). The intercept with the abscissa is at a negative value as predicted by Equation 8–34, and this intercept yields a value of r, the mean radius of the polymers of dextran, of 0.35 nm, which is not unreasonable.

Several of the points in Figure 8–8B deviate from the line that was drawn, most notably that for immunoglobulin G (γ-globulin). The shape of this molecule is quite different from a sphere (Figure 7–12). One way to quantify its deviation from spherical behavior is to define a **frictional ratio**, $f(f_0)^{-1}$, which is simply the quotient between the measured frictional coefficient of the protein and the frictional coefficient it would have if its mass were distributed to form a hard sphere of the same partial specific volume. The measured frictional coefficient, f, is calculated from the diffusion coefficient (Equation 1–54) and the ideal frictional coefficient is calculated from

$$f_0 = 6\pi\eta R^\circ \qquad (8–35)$$

Figure 8–8: Sieving of globular proteins by molecular exclusion chromatography.[63] The proteins, dissolved in 0.1 M KCl at pH 7.5, were submitted to molecular exclusion chromatography on a column (2.5 cm × 50 cm) of Sephadex G-200. The volumes at which the several proteins eluted from the column were tabulated. The distribution coefficient K_{av} for each was calculated (Equations 1–14 and 1–15) from the elution volume, V_e, the void volume, V_0, and the included volume, V_i, as determined by the volume at which sucrose eluted. It was assumed that $V_i = V_{H_2O}$ and that $V_T = (1 - f_{poly})^{-1} V_{H_2O}$ and that $W_r = 20$ mL g^{-1}. The quantity $(-\ln K_{av})^{1/2}$ is plotted (A) as a function of the cubic root of the product of the number of amino acids in each protein, n_{aa}, and the intrinsic viscosity of each protein, $[\eta]$, (B) as a function of the cubic root of the number of amino acids in each protein, and (C) as a function of the Stokes radius, a, of each protein calculated from its diffusion coefficient by Equation 1–57. The proteins in panel A were, in order of increasing total number of amino acids n_{aa}, cytochrome c (n_{aa} = 104), myoglobin (n_{aa} = 153, c3 Kin.3), chymotrypsinogen (n_{aa} = 245, c15 Kin.1), ovalbumin (n_{aa} = 385), serum albumin (n_{aa} = 581), malate dehydrogenase (n_{aa} = 628), transferrin (n_{aa} = 680), immunoglobulin G (γ-globulin, n_{aa} = 1324), glyceraldehyde-3-phosphate dehydrogenase (n_{aa} = 1328), L-lactate dehydrogenase (n_{aa} = 1332), fructose-bisphosphate aldolase (n_{aa} = 1444), fumarate hydratase (n_{aa} = 1864), catalase (n_{aa} = 2020), β-galactosidase (n_{aa} = 4084), and apoferritin (n_{aa} = 4368).

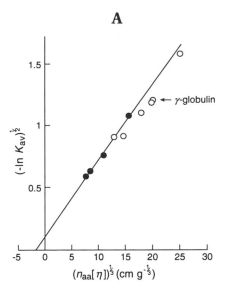

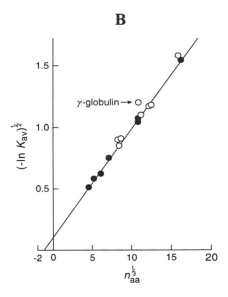

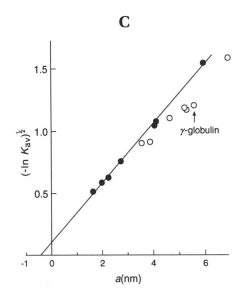

where $R°$ is defined by Equation 8–33. For γ-globulin with a diffusion coefficient[19] of 4.0×10^{-7} cm^2 s^{-1}, $f(f_0)^{-1}$ is 1.58; for the most spherical of proteins, values of 1.1–1.2 are observed.[19] The values of the frictional ratio are always greater than 1 because the water bound to a protein increases its actual frictional coefficient.

The filled points in Figure 8–8B are those for all of the proteins chosen by Andrews which happen to have frictional ratios less than 1.2. They are, in ascending order of n_{aa} (with the frictional ratios in parentheses), cytochrome c (1.09), myoglobin (1.16, c3 Kin.3), chymotrypsinogen (1.12, c15 Kin.1), ovalbumin (1.18), glyceraldehyde-3-phosphate dehydrogenase (1.16), L-lactate dehydrogenase (1.17), and apoferritin (1.15).* The line in Figure 8–8B was drawn through these points because they should be the most ideal examples. It can be seen that several of the points for proteins with larger frictional ratios indicate that they are behaving as if they were larger than they are, which makes sense if their behavior is only a function of their surface area.

It has been argued that rather than using n_{aa}, the number of amino acids in a protein, as the fundamental variable in describing its behavior when it is submitted to sieving on molecular exclusion chromatography, the effective radius, or Stokes radius, a, calculated from its diffusion coefficient (Equation 1–57) should be used.[64,65]

---

*Values of frictional ratios calculated from diffusion coefficients cited by Andrews[63] and actual number of amino acids in each protein.

When the same data displayed in Figure 8–8B are replotted against the effective radii for the various proteins, no significant improvement is seen (Figure 8–8C). Something can be learned, however, when the line is again drawn through the points for the seven most globular proteins listed above, whose properties should be least affected by the change to a from $n_{aa}^{1/3}$. It can be seen that the use of the effective radius significantly overcompensates for the deviations from the linear behavior displayed in Figure 8–8B for some of the proteins that have large frictional ratios.

It has also been argued that rather than using n_{aa} as the variable, the product between the molar mass, M, of the protein and the intrinsic viscosity, $[\eta]$, of the protein should be used.[66] It has been shown that this product correlates more accurately than molar mass alone with the mobilities on molecular exclusion chromatography of the individual members of a large collection of different random polymers with a wide range of shapes and sizes. This proposal, unlike that for the effective radius, has a firm theoretical basis because the product, $[\eta]M$, is directly proportional to the hydrodynamic volume of a particle. When some of the same data displayed in Figure 8–8B are replotted against $(n_{aa}[\eta])^{1/3}$ the deviation of immunoglobulin G (γ-globulin) decreases, but the overall scatter increases (Figure 8–8A). This may be due to the difficulty in measuring accurately the relatively small intrinsic viscosities of globular proteins. Although the correlation with hydrodynamic volume may be theoretically more defensible, it is clear that the correlation between $(-\ln K_{av})^{1/2}$ and $n_{aa}^{1/3}$ is the most accurate, useful, uncomplicated, and reasonable approach, at least for globular proteins. For example, it is hard to explain why proteins with irregular shapes would behave as if they were smaller than spherical ones, as is the case in both panels A and C of Figure 8–8.

It is the correlation between $^{i}K_{av}$ and $^{i}n_{aa}$ that is exploited when data from molecular exclusion chromatography are used to estimate the number of amino acids contained within a protein of interest.[63] This estimation requires that the distribution coefficients, $^{i}K_{av}$, for a series of uncomplicated standard proteins of known $^{i}n_{aa}$ be used to define the line for the column chosen for the particular experiment. The estimate for the number of amino acids in the protein of interest is interpolated from the known values for the standards. A standard line for a particular chromatographic column must be established by running standards on that column, because the properties of each commercial batch of chromatographic medium are unique.[58]

As a macromolecule moves through a polymeric network during electrophoresis, it is also being sieved. In this case, it must travel through the network in a kinetic process, rather than equilibrating with the internal volume of a bead, but it appears that this distinction is inconsequential. It has been argued[55] that one can view the solid matrix of the polymerized gel as an array of screens through which the macromolecule must travel. A random cross section through a random three-dimensional network of lines will provide a distribution of points. The probability that none of these points lies within the randomly placed, random cross section of a geometric solid of any shape is still described by Equation 8–29.[55] If those points represent one of the screens in the gel, if the macromolecule can pass only through openings in that screen large enough so that no point forming that screen is found within the cross section of the macromolecule, and if the rate of its movement through the screen is proportional to the probability that openings of the proper size or larger will be encountered, then the mobility of a macromolecule through a gel during electrophoresis should be described by

$$u_i = u_i^{\circ} e^{-bT_P S_i^{app}} \qquad (8\text{–}36)$$

It has already been noted that the electrophoretic mobilities of proteins (Figure 1–17), nucleic acids,[53] and complexes between dodecyl sulfate and polypeptides satisfy Equation 8–28, and therefore their behavior is also consistent with Equation 8–36. It should also be the case that

$$^{i}K_r = b S_i^{app} \qquad (8\text{–}37)$$

If it is assumed that a series of proteins chosen to test this relationship resembles a series of spheres and that Equations 8–32 and 8–33 are still valid approximations, then

$$\left(\frac{^{i}K_r}{4\pi b}\right)^{1/2} = r + R_i^{\circ} \qquad (8\text{–}38)$$

and a plot of $(^{i}K_r)^{1/2}$ against $(^{i}n_{aa})^{1/3}$ for this series should yield a linear relationship.[55] When the data of Hedrick and Smith[67] and Bryan[68] for the retardation coefficients, $^{i}K_r$, of a series of proteins sieved by electrophoresis through gels cast from increasing concentrations of acrylamide, T_a, are plotted in this fashion, they do display linear behavior (Figure 8–9).[55,68] Again, the intercept with the abscissa is at a negative value as predicted by Equation 8–38, and this intercept yields a value of r, the mean radius of the polyacrylamide, of 0.7 nm.[55] It has been suggested by Hedrick and Smith[67] that this linear correlation permits the number of amino acids in a protein of unknown size to be estimated.

Two types of macromolecules that are of interest in biochemistry are globular macromolecules, such as proteins in their native state, and **extended polymers**, such as RNA, single-stranded DNA, and unfolded or denatured polypeptides. Extended polymers, whose shapes are unable to be approximated as spheres, nevertheless display regular behavior when they are submitted to sieving. The apparent surface areas, S_i^{app}, of extended, flexible polymers, such as unfolded polypeptides or single-stranded nucleic acids, should increase linearly with their lengths, $^{i}n_{aa}$, because as each monomer is added, it increases S_i^{app}

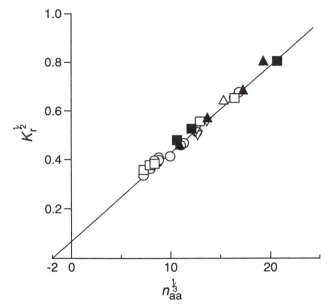

Figure 8–9: Relationship between retardation coefficient K_r and number of amino acids n_{aa} for the electrophoresis of globular proteins.[67,68] A series of globular proteins were submitted to electrophoresis, each protein on a series of gels cast from solutions of increasing concentrations of acrylamide. The slopes of the lines of the logarithm of the relative mobility plotted against percent of acrylamide were tabulated as K_r. The values of $K_r^{1/2}$ were then plotted as a function of $(n_{aa})^{1/3}$ for proteins used by Hedrick and Smith[67] for which amino acid sequences were available ($\square$); proteins used by Hedrick and Smith for which only physical measurements of molar mass were available ($\triangledown$); proteins used by Bryan[68] for which amino acid sequences were available ($\bigcirc$); and proteins used by Bryan for which only physical measurements of molar mass were available ($\triangle$). Darkened symbols are for adventitious oligomers of a particular protein.

by the same increment, once the polymer is beyond a certain length. In this case,

$$S_i^{app} = c + d\left({}^i n_{aa} \right) \tag{8–39}$$

where c incorporates all of the properties of homologous short polymers only a few segments in length.

When polypeptides are dissolved in solutions of guanidinium chloride, they unfold and become separated, individual random coils.[70] A series of these random coils, whose lengths are precisely known, has been submitted to molecular exclusion chromatography on beaded agarose, and the values of K_D (Equation 1–16) have been reported.[69] Combining Equations 1–15, 1–16, 8–30, and 8–39

$$-\ln {}^i K_D = \ln K_{av}^{STD} + bT_P[c + d({}^i n_{aa})] \tag{8–40}$$

where K_{av}^{STD} is the distribution coefficient of the small molecule used to determine the apparent internal volume. This equation predicts that a plot of $\ln {}^i K_D$ against ${}^i n_{aa}$ should be linear, and it is (Figure 8–10). It has been

proposed that the length of a polypeptide, the sequence of which was unavailable, could be estimated from its distribution coefficient by using such a standard curve.[69]

The regular behavior of single-stranded nucleic acids upon electrophoresis is crucial to the strategies for determining their sequences. The relative electrophoretic mobilities of the components in the ladder of single-stranded RNA displayed in Figure 3–11 can be measured from the photograph. Each relative mobility can in turn be related to the relative mobility of one chosen as a standard, for example, the mobility of the component containing 30 nucleotides.[71] If Equations 8–28, 8–37, and 8–39 are combined, and if it is remembered that u_i° for all single-stranded nucleic acids is the same, then

$$-\ln\left(\frac{u_i}{u_{30}}\right) = bT_a d({}^i n_b - 30) \tag{8–41}$$

This predicts that a plot of $\ln [u_i(u_{30})^{-1}]$ against ${}^i n_b$ should be linear, and it is (Figure 8–11) with an interesting exception. In the separation displayed in Figure 3–11 it could be ascertained, because sequencing was being performed,[71] that the bands for the components in the ladder representing single-stranded ribonucleic acids of lengths 24–27 had all overlapped, causing a commonly encountered artifact referred to as a compression. It is usually assumed that a compression results from the ability of the 3′ end of the nucleic acid to double back upon itself and form a double-stranded hairpin as soon as the length of the nucleic acid becomes greater than a critical value in the expanding series. This compression is the discontinuity in Figure 8–11. As the series approaches the discontinuity, it behaves regularly, because no hairpin is imminent. At the discontinuity and beyond it, the hairpin is present in each component, but it is eventually found far enough in the interior for the series to resume its linear behavior with the same slope it had previously but with a slight displacement. The displacement indicates that the polymer is behaving as if it were smaller than it actually is, presumably because the surface area of the double-helical hairpin in its interior is smaller than the surface area of the same number of nucleotides in a single-stranded state.

Denatured polypeptides and single-stranded nucleic acids are examples of well-defined extended polymers. Complexes between dodecyl sulfate and polypeptides, because they are not chemically defined covalent polymers, are not so well understood. Nevertheless, both the behavior of polypeptides dissolved in solutions of guanidinium chloride (Figure 8–10) and the behavior of single-stranded nucleic acids (Figure 8–11) when they are respectively submitted to sieving suggest that the extended, unfolded complexes that form between dodecyl sulfate and polypeptides, which resemble the former in their unfolded state and the latter in both their distribution of negative charge and unfolded state, should display electrophoretic mobilities correlated with the length

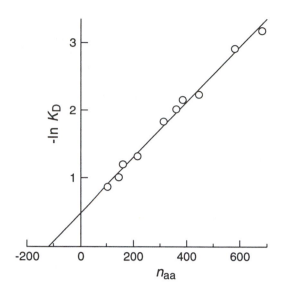

Figure 8–10: Sieving of unfolded, randomly coiled polypeptides on molecular exclusion chromatography.[69] Each of a series of proteins was dissolved in 6 M guanidinium chloride and 0.1 M 2-mercaptoethanol and submitted to chromatography on a column (90 cm × 1.5 cm) of beaded 6% agarose. The elution volume of each protein was used to calculate its distribution coefficient K_D. The negative natural logarithm of K_D is plotted as a function of n_{aa}. The polypeptides chosen were those composing cytochrome c ($n_{aa} = 104$), hemoglobin ($n_{aa} = 144$), β-lactoglobulin ($n_{aa} = 162$), immunoglobulin G light chain ($n_{aa} = 220$), malate dehydrogenase ($n_{aa} = 314$), fructose-bisphosphate aldolase ($n_{aa} = 362$), ovalbumin ($n_{aa} = 385$), immunoglobulin G heavy chain ($n_{aa} = 450$), bovine serum albumin ($n_{aa} = 581$), and transferrin ($n_{aa} = 679$).

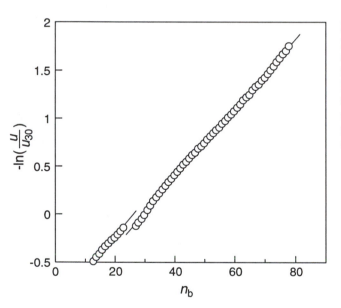

Figure 8–11: Sieving of single-stranded ribonucleic acid on electrophoresis in a gel of polyacrylamide. The distance between the origin and the final position of each band on the gel in Figure 3–11 was measured. These distances were each divided by the distance for the band corresponding to the ribonucleic acid 30 bases in length to obtain mobilities relative to this internal standard $[u(u_{30})^{-1}]$. The negative natural logarithms of these mobilities are plotted as a function of the lengths, in bases, n_b, of each single-stranded ribonucleotide.

of the polypeptides. In fact, it was noted by Shapiro, Viñuela, and Maizel[72] that this is the case. The electrophoretic mobility of the complex between dodecyl sulfate and the polypeptide i is generally reported as a relative mobility

$$^iR_f = \frac{u_i}{u_{STD}} \qquad (8\text{–}42)$$

where u_{STD} is the mobility of a standard, either a small dye that can be readily followed visually or one of the boundaries on a discontinuous gel. The advantage of the latter point of reference is that its absolute mobility can be calculated. Equations 8–28 and 8–42 can be combined, and

$$\ln(^iR_f) = \ln\left(\frac{u^\circ_{STD}}{u^\circ_i}\right) - K_r^{STD}T_a + T_a{}^iK_r \qquad (8\text{–}43)$$

This can be combined with Equations 8–37 and 8–39, and

$$\ln(^iR_f) = \ln\left(\frac{u^\circ_{STD}}{u^\circ_i}\right) - K_r^{STD}T_a + bT_a[c + d(^in_{aa})] \qquad (8\text{–}44)$$

Because u_i° should be the same for all complexes between well-behaved polypeptides and dodecyl sulfate[51] and u°_{STD}, K_r^{STD}, and T_a are all constant, this equation predicts that a plot of $-\ln R_f$ against n_{aa} should be linear. When the natural logarithms of the relative mobilities measured by Weber and Osborn[54] are plotted as a function of n_{aa}, they conform to this expectation (Figure 8–12).

At the present time, the method almost universally used to estimate the length of a polypeptide, the sequence of which is not yet known, is to determine the mobility of its complex with dodecyl sulfate upon electrophoresis on polyacrylamide gels. The mobility of the unknown is compared to the mobilities of complexes between dodecyl sulfate and standard polypeptides of known length, usually by plotting the data as in Figure 8–12. It should be realized, however, that the widespread reliance on this method is based on the assumption that the polypeptide of interest binds the same amount of dodecyl sulfate (amino acid)$^{-1}$ as the standards used. A comparison of Figure 8–11, which describes the behavior of a series of polymers in which the uniformity of the charge distribution is covalently dictated, with Figure 8–12, which describes the behavior of a series of polymers in which the uniformity of charge distribution depends only on a fortuitous consistency in its composition producing a fortuitous consistency in its ability to bind a small electrolyte, emphasizes the weakness of this assumption. It has also been observed that the complexes between dodecyl sulfate and proteins that are themselves highly charged, such as histones, or that are significantly glycosylated do not have mobilities on polyacrylamide gels that reflect only their length, and this procedure is unreliable in these instances.

A further problem that also should be realized is that beyond the ranges of relative mobilities displayed in Figure 8–12, the linear behavior of complexes between dodecyl sulfate and polypeptides often fails. This was originally pointed out by Weber and Osborn,[54] and it manifests itself in the tendency of complexes between dodecyl sulfate and very long polypeptides to travel faster than their lengths should permit. Because it is impossible to predict in what range of polymer lengths nonlinear behavior will become significant, a large collection of standard polypeptides whose mobilities are close to and on either side of the mobility of the unknown should be chosen and the length of the unknown should be estimated by interpolation.

The failure of complexes between dodecyl sulfate and long polypeptides to behave as if they were geometric solids may be due to a change in the mechanism of sieving. It has been proposed that when the long dimension of a severely elongated macromolecule is significantly greater than the mean spacing between the fibers of polymer in a sieve, it will be hindered from reorienting significantly about any axes normal to that long dimension by the network itself. If so, it will be forced to move through the network as a worm in a randomly meandering burrow.[73] At very low electric field gradients, the mobility of such a wormlike molecule should be proportional to $(^i n_{aa})^{-1}$ rather than $e^{-i n_{aa}}$ (Equation 8–44). At field gradients in the range normally used for electrophoresis the burrow has a strong tendency to become aligned with the field, causing the mobility of the worm to become almost independent of $^i n_{aa}$.[74] It is possible that the deviation of the retardation coefficients of complexes between dodecyl sulfate and longer polypeptides from linear behavior on gels of one polyacrylamide concentration[54,75] results from a change in the mechanism of sieving that occurs as the longer dimension of the polymer becomes so long that it is forced to travel through the network as a worm rather

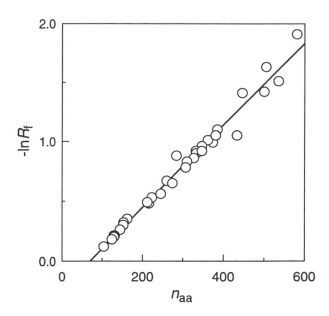

Figure 8–12: Sieving of complexes between dodecyl sulfate and unfolded polypeptides on electrophoresis in gels of polyacrylamide cast in solutions of sodium dodecyl sulfate.[54] A series of proteins were submitted to electrophoresis as described in Figure 8–6. A marker dye was included in each sample, the mobility of which served as an internal standard. Its position was marked on each gel at the end of each run and the gels were stained. The distance migrated by each polypeptide was divided by the distance migrated by the marker dye to obtain a relative mobility R_f. The negative natural logarithm of R_f is plotted as a function of the number of amino acids, n_{aa}, in the polypeptide.

than as a randomly reorienting, flexible geometric solid. The point at which this change in mechanism would occur would be a function not only of the length of the polymer but also of the spacing of the fibers in the network. If this is the case, it would explain the common observation that the behaviors of complexes between dodecyl sulfate and long polypeptides often become linear again when the concentration of polymer in the network is decreased.[75]

Suggested Reading

Rodbard, D., & Chrambach, A. (1970) Unified Theory for Gel Electrophoresis and Gel Filtration, *Proc. Natl. Acad. Sci. U.S.A.* 65, 970–977.

Problem 8–8: A series of standards was run to calibrate a column of Sephadex G-75 so that it could be used to estimate the number of amino acids in α-lactalbumin. The void volume of the column, V_0, was 71 mL, and the total volume of the bed, V_T, was 226 mL. The following elution volumes were observed

protein	n_{aa}	elution volume
cytochrome c	104	138 mL
myoglobin	153	127 mL
ribonuclease	124	136 mL
chymotrypsinogen	245	113 mL
α-lactalbumin		131 mL

(A) Calculate $K_{av} = (V_e - V_0)/(V_T - V_0)$ for every protein.

(B) Estimate the number of amino acids in α-lactalbumin.

Problem 8–9: Phosphorylase was dissolved in a solution of sodium dodecyl sulfate sufficient to saturate the protein and submitted to electrophoresis on a polyacrylamide gel cast in a solution of sodium dodecyl sulfate. The relative mobilities of the phosphorylase and several standard proteins were measured.

protein	length of polypeptide (amino acids)	mobility relative to marker dye
myosin	1939	0.10
β-galactosidase	1021	0.16
serum albumin	581	0.33
catalase	505	0.37
glutamate dehydrogenase	500	0.43
fumarase	466	0.47
aldolase	362	0.56
phosphorylase		0.23

Estimate the length of the polypeptide composing phosphorylase.

Problem 8–10: Estimate the length of the polypeptide that composes porcine gastric pepsin from the following relative mobilities of complexes between the polypeptides and sodium dodecyl sulfate on electrophoresis on polyacrylamide gels.

polypeptide	n_{aa}	distance migrated (cm)
serum albumin	581	1.78
immunoglobulin G heavy chain	450	2.92
glyceraldehyde-3-phosphate dehydrogenase	332	4.80
carbonic anhydrase B	260	6.12
aspartate transcarbamylase catalytic polypeptide	310	5.22
carboxypeptidase A	307	5.48
D-amino acid oxidase	347	4.80
pepsin		5.06

Cataloguing Polypeptides

Rather than the intact oligomeric complexes of folded polypeptides separated during the electrophoresis of native proteins, the components separated when a mixture of different proteins is submitted to electrophoresis in the presence of dodecyl sulfate represent individual, unfolded, unassociated polypeptides. The only difference is the addition of the dodecyl sulfate. These electrophoretic separations, therefore, catalogue the polypeptides present in a sample[54] rather than the proteins. A graphic example of such a catalogue can be seen in Figure 1–21. When a purified protein whose homogeneity has been verified by electrophoresis in its native state is submitted to electrophoresis in the presence of dodecyl sulfate, the pattern observed gives a catalogue of the different polypeptides of which that one protein is composed.

These catalogues are reliable provided their shortcomings are recognized and eliminated. It has already been noted that on discontinuous electrophoresis, components of high mobility often fail to escape the descending boundary. Complexes between dodecyl sulfate and very short polypeptides are often trapped in this way and are unresolved. It is also the case that for reasons not well understood, all complexes between dodecyl sulfate and polypeptides less than 100 amino acids in length seem to have the same electrophoretic mobility,[76] regardless of the concentration of polyacrylamide. This lower limit below which resolution fails can be lowered to about 25 amino acids in length by adding 8 M urea to the polyacrylamide gel.[77,78] Because the cleavage of one peptide bond out of the hundreds present in an intact polypeptide always produces two new polypeptides that will be separated from each other and from their parent by electrophoresis in the presence of dodecyl sulfate, any proteolytic degradation of the native protein during or before its purification can artifactually multiply the apparent number of polypeptides without significantly altering the native protein. Proteolytic enzymes often unfold more slowly than other proteins upon exposure to dodecyl sulfate, and they cleave their unfolded neighbors before they in turn die. If the purified protein is contaminated even with minute amounts of the proteolytic enzymes, which are always present in a homogenate, they can degrade the polypeptides during the preparation of the sample. Because these cleavages of the denatured polypeptides are produced at random, as opposed to the unique cleavages usually produced during the proteolytic degradation of a native protein, they cause polypeptides in the sample to disappear into hundreds of fragments smeared over the field, each present in very low yield. Such an apparent disappearance of a polypeptide or polypeptides, however, can also occur during a purification rather than during preparation of the sample. For example, it was once thought that the subunit stoichiometry of nicotinic acetylcholine receptor from the electric eel was simpler than that from the electric ray until it was demonstrated that the missing polypeptides appeared when indigenous proteolytic enzymes, unavoidably present during the purification, were intentionally inactivated.[79] Yet the acetycholine receptor originally purified, even though it had been cut up, was still biologically active. If disulfides are not fully cleaved, cross-linked and un-cross-linked forms of the same polypeptide will appear as separate components. These and other artifacts must be recognized and eliminated[80] before the patterns observed give a reliable assessment of the different polypeptides present in a protein.

The various components separated by electrophoresis in the presence of dodecyl sulfate may or may not represent single, unique sequences of amino acids. The majority of the time they do, but it is possible that one of the components could represent two different polypeptides whose lengths are so close to each other that they cannot be resolved or that one or more of the components represent fragments of a larger component, also seen on the same gel. To resolve these ambiguities, peptide maps are performed.

A **peptide map** is a display of the peptides produced when a polypeptide is digested with a specific proteolytic enzyme. The display is usually performed by chromatographically separating the digest in two dimensions (Figure 3–6) to produce a characteristic pattern or map. Peptide maps are extremely sensitive methods for assessing the similarity of two polypeptides, demonstrating that one polypeptide is a fragment of another, or revealing that one of the components on a polyacrylamide gel represents two polypeptides that fortuitously have the same electrophoretic mobility. The most reliable maps are obtained from tryptic digests of a polypeptide because trypsin is the most specific and dependable of the proteolytic enzymes. Polypeptides usually contain about 5 mol % arginine and 7 mol % lysine,[81] and for every 100 residues in length, about 12 tryptic peptides should be present in the digest. Each spot on the map represents a different tryptic peptide.

When the polypeptide is a long one, there are so many spots on a tryptic peptide map that they begin to overlap. This problem is solved by modifying the tyrosine residues in the protein with radioactive iodine by electrophilic aromatic substitution. Because there are only 3–4 tyrosines for every 100 amino acids in a typical protein,[81] only about one-quarter of the tryptic peptides are rendered radioactive and an autoradiogram of the map is less cluttered than the entire map itself, but just as unique to the particular polypeptide. Such a map of tyrosine-containing chymotryptic peptides was used to show that the α polypeptides of Na$^+$/K$^+$-transporting ATPase, both polypeptides about 1000 amino acids in length, with 24 mol of tyrosine (mol of polypeptide)$^{-1}$, were very similar if not identical to each other (Figure 8–13).[82]

Peptide mapping is sometimes performed by adsorption chromatography at high pressures because these separations are more rapidly accomplished,[83] but this procedure is not so informative because it involves a separation in only one dimension. If two short peptides from two different proteins have the same relative mobility on both electrophoresis and chromatography, this observation is a stronger indication that they have the

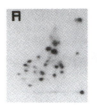 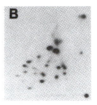

Figure 8–13: Peptide maps of tyrosine-containing chymotryptic peptides from the α polypeptide of Na$^+$/K$^+$-transporting ATPase.[82] After Na$^+$/K$^+$-transporting ATPase was purified from rat liver or rat kidney by immunoadsorption, its α polypeptide was isolated by electrophoresis on polyacrylamide gels in solutions of sodium dodecyl sulfate. Each of the respective polypeptides was chemically modified at its tyrosines by electrophilic aromatic substitution with ^{125}I. The radioactive polypeptides were then digested separately with chymotrypsin, and the digests were separated in two dimensions on thin layers of cellulose. Electrophoresis was performed from right to left followed by ascending chromatography with butanol/pyridine/water/acetic acid (65:50:40:10 v/v/v/v) from bottom to top. Peptides containing o-[^{125}I]iodotyrosine were identified by placing photographic film over the chromatogram. (A) Map from α polypeptide of kidney; (B) map from α polypeptide of liver. Adapted with permission from ref 82. Copyright 1986 American Chemical Society.

same sequence than if they both have the same distribution coefficient between a mobile and a stationary phase.

If two polypeptides yield peptide maps similar enough that they are judged to be related, the homology between their two sequences will be very high; if they yield peptide maps that cannot be regarded as similar, they may still have clearly homologous sequences. In two-dimensional mapping either of all the peptides in a proteolytic digest or of just the tyrosine-containing peptides, or in peptide mapping by adsorption chromatography, the evaluation of the similarity of two polypeptides is based on comparison of the two patterns in which the peptides are displayed (Figure 8–13). Only if a significant fraction of the peptides on the two maps have the same relative positions on the field and produce a pattern that can be recognized is the judgment made that the two polypeptides are similar to each other. This implicit criterion of similarity requires that a significant fraction of the respective peptides be identical to each other in sequence for the decision to be made that the two polypeptides are related. One difference in the sequence of two otherwise identical peptides is usually sufficient to cause them to have different mobilities. For example, because the mean tryptic peptide is eight amino acids in length, differences between the sequences of two polypeptides at more than 20% of the positions will cause the two maps to be completely different, yet the two polypeptides may still be similar enough to be judged homologous in sequence. Each of the four polypeptides of nicotinic acetylcholine receptor, although they are all homologous in sequence to each other (averaging 40% identity), yields a completely different peptide map.[84] The order in which the three ways of detecting homologies among proteins fail as the homologies become more distant is peptide mapping before alignment of amino acid sequences and alignment of amino acid sequences before superposition of tertiary structures.

The initial triumph of analytical peptide mapping was in the examination of a mutant hemoglobin. It had been proposed that the difference between normal hemoglobin, referred to as hemoglobin A, and hemoglobin S, a hemoglobin producing pathological distortions in erythrocytes, was due to a small difference in the amino acid sequence of the two proteins.[85] It was then shown that one and only one of the tryptic peptides on the two respective peptide maps of the two respective proteins displayed an altered mobility (Figure 8–14).[86] It was concluded that all of the peptides whose mobilities were the same between the two maps had identical sequences and the same relative locations in the two intact polypeptides, but that the one peptide whose mobility was different had a sequence that differed between the two proteins by at least one amino acid. Because it is unlikely that the only two or three changes in the sequence of a polypeptide would occur in the same tryptic peptide, this result alone was substantial evidence that the two proteins differed from each other at only one location in their respective sequences. This was soon shown to be true by complete amino acid sequencing.

A similar strategy was used to evaluate the differences in the amino acid sequences of actin from several tissues.[87,88] This protein occurs as isoforms in different tissues. Each member of a set of actins chosen for the experiment, each of which had been isolated from a different species or a different tissue—a total of eight in all—was digested with trypsin, and each peptide map was compared to the peptide map of actin from rabbit skeletal muscle, the complete amino acid sequence of which was known. In all cases, the majority of the tryptic peptides were distributed over the map in the same pattern as the corresponding peptides on the map from the standard, and this permitted the various maps to be aligned with that of the standard. The peptides occupying the same positions in a pair of maps were assumed to be identical to each other in amino acid sequence. Amino acid analysis was used to verify these identities. Each unique peptide on the maps of the various unknowns was eluted and sequenced. Each of these amino acid sequences could be aligned with one of the tryptic peptides in the sequence of actin from rabbit muscle, and in this way the amino acid replacements in the sequences of the other actins could be readily established. This set of experiments relied on the fact that, aside from the first six amino acids in each sequence, all of the actins compared show about 95% identity when their amino acid sequences are aligned. This level of identity was what produced the underlying pattern that permitted the maps to be aligned and, in turn, permitted the ready identification of the peculiar peptides.

Whenever two or more polypeptides appear upon electrophoresis of a purified protein in the presence of dodecyl sulfate, the possibility that the smaller polypeptide or polypeptides are fragments of the larger should be examined. Such a relationship can be established by peptide mapping. The protein ankyrin from human erythrocytes is present in the cell under physiological conditions as the complete polypeptide and three progressively smaller fragments. That these four polypeptides represent such a nested set derived by proteolysis from the largest could be demonstrated by producing peptide

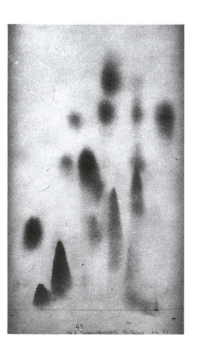
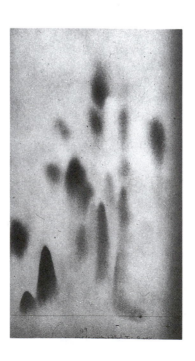

Figure 8–14: Comparison of tryptic peptide maps of hemoglobin A (left panel) and hemoglobin S (right panel).[86] The respective hemoglobins were denatured at 90 °C for 4 min and the denatured proteins were digested with trypsin (1:50 trypsin/hemoglobin). The digests were spotted on sheets of chromatographic paper. The peptides were separated by electrophoresis at pH 6.4 (negative pole to the left) and ascending chromatography in 1-butanol/acetic acid/water (3:1:1). The peptides were visualized with ninhydrin. Reprinted with permission from ref 86. Copyright 1958 Elsevier Science Publishers.

maps of each.[89] This was done by separating these polypeptides on polyacrylamide gels in solutions of sodium dodecyl sulfate, iodinating their tyrosines, digesting them with trypsin, and producing tryptic peptide maps. These peptide maps all displayed the same pattern, but the maps from the smaller polypeptides lacked one or two of the peptides present in those from the next larger one. When samples containing ankyrin in its native state were deliberately digested with chymotrypsin, two fragments each about half the apparent length of the entire polypeptide of ankyrin appeared as the ankyrin disappeared. It could be shown that these fragments were derived from the two halves of the ankyrin polypeptide because the two respective peptide maps of the two fragments contained only peptides present in the map of ankyrin but neither of the two maps contained any of the peptides present in the other.

Peptide mapping can be used to determine whether an apparently unique component resolved by electrophoresis in the presence of dodecyl sulfate represents only one polypeptide or two or more polypeptides that fortuitously have the same electrophoretic mobility. The electrophoretic mobility of the complex between dodecyl sulfate and the polypeptide or polypeptides in question can be used to estimate the length of the polypeptide or polypeptides. The mole percent of lysine and arginine in the protein can be either determined directly by total amino acid analysis or estimated from the fact that this number is usually about 12 mol %. The mole percent of lysine and arginine and the length of the polypeptide or polypeptides can be used to estimate the number of tryptic peptides that should be produced if the component observed on the gel does represent only one unique polypeptide. If the number of peptides observed on the map agrees with this expectation, the component probably represents only one unique polypeptide. If there are about twice as many spots as expected, it must represent two different polypeptides. If the component does represent two or more polypeptides, it should be possible to separate them chromatographically. The two or more separated polypeptides should each give unique peptide maps whose sum should be the peptide map of the original mixture.

When phosphoglycerate dehydrogenase was saturated with dodecyl sulfate and submitted to electrophoresis, one component was observed whose electrophoretic mobility was that of a polypeptide 360 amino acids in length. The content of the lysine plus arginine in the protein was determined to be 9.6 mol %. A tryptic digest of the protein was separated by cation-exchange chromatography, and each of the pools from this first dimension was submitted to electrophoresis on paper. This two-dimensional peptide map displayed 39–40 major peptides. The content of tryptophan in the protein was 1.0 mol %, and four of the peptides gave a positive test for tryptophan. If all of the polypeptides in this protein are identical, there should have been 36 tryptic peptides,

four of which should have contained tryptophan. The agreement between the observed numbers and the expected numbers led to the conclusion that phosphoglycerate dehydrogenase was composed of identical polypeptides.[90]

Citrate (si) synthase was shown to be composed from polypeptides estimated to be 410 amino acids in length. Its content of lysine plus arginine was determined to be 9.3 mol %. Thirty-nine tryptic peptides should have been seen on the peptide map if the enzyme is constructed from identical copies of only one polypeptide; 41 peptides were observed.[91]

Glutamate-tRNA ligase was submitted to electrophoresis in the presence of dodecyl sulfate, and a single component was observed whose mobility was that of a polypeptide 500 amino acids in length. The protein had a content of lysine plus arginine of 12 mol %; tryptophan, 1.0 mol %; arginine, 6.3 mol %; and cysteine, 1.0 mol %. It could be concluded[91a] that the component observed upon electrophoresis represented only one polypeptide because the tryptic peptide map of the protein displayed 55 peptides, 30 of which gave a positive test for arginine, five of which gave a positive test for tryptophan, and five of which were rendered radioactive after the protein was reduced and carboxymethylated with [^{14}C]iodoacetic acid (Figure 3–23).

D-Serine dehydratase is composed of polypeptides whose apparent length upon electrophoresis in the presence of dodecyl sulfate was 430 amino acids. Its content of lysine, arginine, histidine, tryptophan, and pyridoximine phosphate were determined to be 4.5, 4.0, 2.5, 0.9, and 0.19 mol %, respectively. The tryptic map displayed 35 peptides, 17 positive for arginine, eight positive for histidine, four positive for tryptophan, and one positive for pyridoximine.[92]

The use of peptide mapping to provide evidence for the homogeneity of the polypeptides in a protein relies on the assumption that the trypsin has digested the polypeptide completely. This should be independently demonstrated. For example, initial tryptic digests of the polypeptides composing glucose-6-phosphate isomerase produced only two-thirds to three-fourths as many peptides as had been expected from the assumption that they were all identical. It was found that less base was consumed during the digestion than should have been, and this suggested that the digestion had been incomplete. When the protein was carbamylated on all of its lysines and then digested with trypsin, the quantity of base consumed during the digestion and the number of peptides observed on the map were those expected theoretically.[93] A more sensitive measure of complete tryptic digestion is to compare the total content of lysine and arginine in the digest to the amount of lysine and arginine released from the peptides in the digest when a sample is in turn digested with carboxypeptidase.

Peptide mapping has also been used to demonstrate that what seemed to be a single component, or two or more closely related components, upon electrophoresis in the presence of dodecyl sulfate actually represented two unique polypeptides. Upon electrophoresis in dodecyl sulfate, the molybdenum–iron protein of nitrogenase gave two components of very similar and often indistinguishable electrophoretic mobility, whose apparent lengths were 540 amino acids. When the protein was reduced, carboxymethylated with [^{14}C]iodoacetic acid, and submitted to amino acid analysis, its content of ([^{14}C]carboxymethyl)cysteine was 1.7 mol %. Eleven of the tryptic peptides on a peptide map of the reduced and carboxymethylated protein were radioactive when nine were expected. Instead of passing this off as the result of incomplete digestion or inaccurate values for content of cysteine, the investigators proceeded to show that when the protein was dissolved in urea, to unfold its polypeptides, two polypeptides could be isolated by ion-exchange chromatography on (carboxymethyl)cellulose. Both were submitted to reduction, carboxymethylation, and peptide mapping. Four of the radioactive, cysteine-containing peptides from the map of the total protein were found on the map of one of the polypeptides, and the other seven radioactive peptides from the map of the total protein were found on the map of the other polypeptide, and no overlaps occurred between the two maps. It could be concluded that the molybdenum–iron protein had the subunit stoichiometry $\alpha\beta$.[94]

A similar situation arose with transcarboxylase. When this protein was submitted to electrophoresis in the presence of dodecyl sulfate, a component was present whose apparent length was 550 amino acids, but under some conditions it would split into two bands of equal intensity. It was found that the native enzyme could be dissociated into two proteins at pH 9.0 that could be separated from each other by molecular exclusion chromatography. Each of these proteins was composed of polypeptides whose apparent lengths were 550 amino acids.[95] Although they were almost indistinguishable in length, the polypeptides in these separated proteins produced completely different tryptic peptide maps.[96]

The electron transfer flavoprotein is another protein that is composed of two polypeptides whose lengths are very similar. It was originally believed to be a dimer of two identical polypeptides.[97] Under certain circumstances, however, two narrowly separated components would appear upon electrophoresis in the presence of dodecyl sulfate, and these were different enough to be separated by preparative electrophoresis. Each was cleaved separately with cyanogen bromide, and the fragments produced were in turn saturated with dodecyl sulfate and separated in one dimension by electrophoresis in 8 M urea (Figure 8–15).[98] The maps produced in this way from each separated polypeptide were unique, and their sum was equal to the map of the intact protein. The fact that the polypeptide stoichiometry of the electron transfer flavoprotein is $\alpha\beta$ explains the observation that there is only 1 mol of flavin (1.8 mol of polypeptide)$^{-1}$.

Often there is a functional basis for the fact that two or more different polypeptides are present in the same protein. Many **multienzyme complexes**, rather than being composed of a string of enzymatic domains each formed from a different region of the same polypeptide, are constructed from individual subunits gathered to-

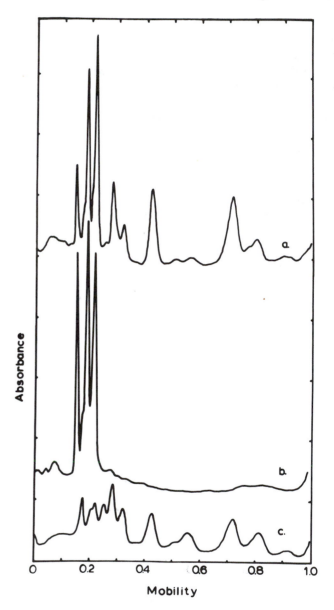

Figure 8–15: Electrophoresis on a polyacrylamide gel of peptides derived from cleavage of electron transfer flavoprotein from porcine liver.[98] All proteins were reduced and alkylated with iodoacetamide before cleavage. Intact electron transfer flavoprotein (a), its α polypeptide (b), or its β polypeptide (c) were digested with cyanogen bromide (25 mM) in 88% formic acid and 0.03% sodium dodecyl sulfate for 24 h at 25 °C. The fragments produced were separated on gels cast from 13.4% acrylamide in 0.1% sodium dodecyl sulfate and 8 M urea and stained for protein. The gels were then scanned for absorbance as a function of length, and length was converted to mobility relative to the mobility of a marker dye. Reprinted with permission from ref 98. Copyright 1983 *Journal of Biological Chemistry.*

gether in a larger complex. Functionally there is no distinction between these two types of multienzyme complexes because the important feature is that the different enzymes are gathered together, whether they are gathered as domains on the same polypeptide or as different subunits.

Methylmalonyl-CoA carboxytransferase is an example of a multienzyme complex formed from different subunits. One of the two polypeptides described earlier is folded to produce a protein that in isolation[99] can catalyze the reaction

$$\text{methylmalonyl-SCoA} + \text{biotin} \rightleftharpoons$$
$$\text{propionyl-SCoA} + \text{carboxybiotin} \quad (8\text{–}45)$$

and the other, a protein that can catalyze[99] the reaction

$$\text{carboxybiotin} + \text{pyruvate} \rightleftharpoons \text{oxaloacetate} + \text{biotin}$$
$$(8\text{–}46)$$

The entire enzyme is composed of these two subunits and a third polypeptide bearing covalently attached biotin as a posttranslational modification of one of its lysines. The intact multienzyme complex catalyzes the overall reaction

$$\text{methylmalonyl-SCoA} + \text{pyruvate} \rightleftharpoons$$
$$\text{propionyl-SCoA} + \text{oxaloacetate} \quad (8\text{–}47)$$

In the case of the molybdenum–iron protein of nitrogenase, which is constructed from two polypeptides, one of the two subunits, formed from one of the folded polypeptides, contains the molybdenum and some of the iron while the other subunit contains iron–sulfur clusters of a ferredoxin type. This assigns different functional roles to each of the subunits and explains the stoichiometries of the molybdenum and iron found in the intact protein.[94]

It is likely that at least some of the proteins thought to be composed of two different types of subunits, because they are isolated as complexes of two different polypeptides, are actually the products of a posttranslational modification involving the intentional or artifactual cleavage of what was initially a single polypeptide. The internal modification of histidine decarboxylase producing the *N*-pyruvyl amino terminus (Figure 3–20) coincidentally produces two shorter polypeptides of length 81 and 250 amino acids from the originally intact precursor.[100] Before the modification occurs, the protein is constructed from six copies, each folded as an independent subunit, of a single polypeptide.

It was once believed that vertebrate acetyl-CoA carboxylase was constructed from two or three different polypeptides, each present in one or two copies.[101] This belief was reinforced by the fact that the enzyme from *E. coli* is a multienzyme complex constructed from three different subunits,[102] albeit from polypeptides of lengths much shorter than the polypeptides found in the enzyme from vertebrates. When the vertebrate enzyme was purified by a rapid procedure employing affinity adsorption, it was found to be composed from only one polypeptide, the length of which (2100 amino acids) was more than twice the individual lengths of the separate polypeptides seen previously.[103] Only two copies of this longer polypeptide are present in the native protein. It was concluded that the smaller polypeptides seen previously were the products of artifactual proteolytic digestion.

The distinction between a protein that contains two different polypeptides because of a posttranslational cleavage and one that has been assembled from two separately translated polypeptides is significant. The posttranslational cleavage of a protein usually occurs after the entire polypeptide has folded because it is only in the folded polypeptide that the cleavage can be directed to a precise location. Although significant changes may occur in the vicinity of the cleavage, the overall structure of the protein remains what it was before the cleavage. For all intents and purposes, a polypeptide cleaved either naturally or artifactually after it has folded remains structurally a folded single polypeptide. On the other hand, when a protein is assembled from separately translated polypeptides, they first must fold independently before they can recognize each other and join together. Each subunit begins as and remains as a discrete entity and appears as such when a crystallographic molecular model is viewed. Yet when the polypeptides are resolved on a polyacrylamide gel these two very different situations cannot be distinguished, and, as with the confusion surrounding the connotations of the term domain, the urge arises to confer the structural attributes of separate subunits to a protein that may contain two polypeptides only by virtue of posttranslational modification.

It is possible that the number of proteins assembled from two or more different subunits containing polypeptides with different sequences and different lengths has been overestimated. Even if this is not so, such heterooligomers represent only a small minority (15%) of the oligomeric proteins.[104] Most oligomeric proteins are constructed from only one polypeptide that is present in two or more identical copies in the complete protein. The length of that polypeptide is estimated by electrophoresis in the presence of dodecyl sulfate. The number of copies present in the intact protein is counted.

Suggested Reading

Dowhan, W., & Snell, E.E. (1970) D-Serine Dehydratase from *Escherichia coli*. II. Analytical Studies and Subunit Structure, *J. Biol. Chem. 245*, 4618–4628.

Problem 8–11: Glucose-6-phosphate isomerase catalyzes the conversion

$$\text{D-glucose-6-phosphate} \rightleftharpoons \text{D-fructose-6-phosphate}$$

The enzyme has been purified from rabbit muscle. When it was dissolved in a solution of sodium dodecyl sulfate and submitted to electrophoresis, the following results were obtained

protein	n_{aa}	R_f
ovalbumin	385	0.42
catalase	505	0.32
serum albumin	581	0.26
glucose-6-phosphate isomerase		0.30

The enzyme was denatured in 6 M guanidinium chloride and modified with potassium cyanate ($K^+O=C=N^-$), the reagents were removed by dialysis, and the protein was digested with trypsin. The following two maps were made of this tryptic digest. The only difference in these two maps is the pH at which the electrophoresis was performed.

The amino acid composition of the protein has been determined

amino acid	moles (100,000 g of protein)$^{-1}$	amino acid	moles (100,000 g of protein)$^{-1}$
K	61.8	I	53.4
H	35.9	L	87.2
R	32.2	Y	18.5
D	87.7	F	45.9
T	62.0	G	66.6
S	58.0	A	63.9
E	96.2	V	51.5
P	37.9	M	21.6

(A) What is the length of the polypeptides composing glucose-6-phosphate isomerase?

(B) How many different polypeptides does the protein contain?

(C) Which amino acid side chain are you certain the cyanate modified? Why?

(D) What conclusions did you draw from the peptide maps?

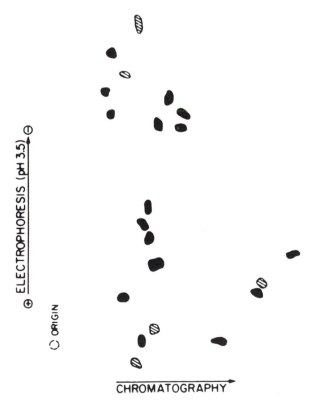

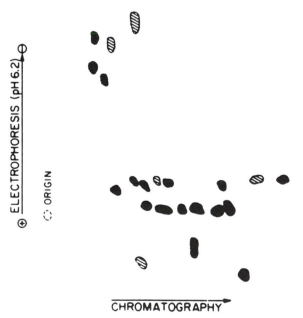

Tryptic peptide maps of carbamylated rabbit muscle glucose-6-phosphate isomerase.[105] Electrophoresis: right map, pyridine/acetic acid/water (520:1.4:1000 by volume), pH 6.2; left map, pyridine/acetic acid/water (7:66:1927 by volume), pH 3.5. Chromatography: descending chromatography with butanol/acetic acid/pyridine/water (15:10:3:12 by volume). Maps were performed on sheets (46 cm × 57 cm) of chromatographic paper and peptides were located by ninhydrin.

Problem 8–12: The molar mass of phospho-2-dehydro-3-deoxygluconate aldolase from *Pseudomonas putida* has been estimated as 73,000 g mol^{-1} by sedimentation equilibrium. The enzyme was dissolved in a solution of sodium dodecyl sulfate and submitted to electrophoresis. Its mobility on the gel as well as the mobilities of several standards are tabulated

protein	n_{aa}	R_f
carbonic anhydrase	260	0.62
chymotrypsinogen	245	0.74
trypsin	223	0.76
myoglobin	153	1.00
PDD aldolase		0.75

The protein was reduced, carboxymethylated, and digested with trypsin. There were 24 spots observed on a two-dimensional map of this tryptic digest; three were positive for tyrosine, as detected by Pauli stain, and 13 were positive for arginine.

The amino acid composition of the protein was determined (see table at bottom of page).

The protein was dissolved in 8 M urea, 2-mercaptoethanol was added, and the mixture was incubated for 4 h. The reduced protein was then alkylated with [^{14}C]iodoacetate, dialyzed, and digested with trypsin. The tryptic digest was run on an ion-exchange column, and four peaks of radioactivity were observed. Each radioactive peak was further purified to homogeneity. The four were shown by composition to be unique and each contained one (carboxymethyl)cysteine. Their compositions were $C_1D_2S_1E_3A_6I_2L_2$, $C_1D_1E_1A_1I_2K1$, $C_1D_2T_1G_2A_1V_2F_1R_1$, and $C_1D_1T_1E_1G_1A_1V_1L_1R_1$.

(A) What is(are) the length(s) of the polypeptide(s) composing this enzyme?

(B) How many different polypeptides are there in the protein?

(C) What conclusions can you draw from the tryptic peptide map?

(D) What conclusions can you draw from the tryptic peptides containing ([^{14}C]carboxymethyl)cysteine?

Problem 8–13: A protein has been purified to homogeneity. It has the following properties.

(1) When the protein is reduced with 2-mercaptoethanol and run on a sodium dodecyl sulfate gel in the presence of standards, the following results are obtained

protein	n_{aa}	mobility
β-lactoglobulin	162	0.70
myoglobin	153	0.73
lysozyme	129	0.81
ribonuclease	124	0.82
cytochrome *c*	104	0.87
protein X		0.77

(2) The amino acid composition of the protein is as follows

amino acid	mol (100 mol)$^{-1}$	amino acid	mol (100 mol)$^{-1}$
G	6.9	Y	2.0
A	12.5	W	0.9
S	5.4	C	1.0
T	5.4	M	1.0
P	5.0	D	8.6
V	10.7	E	5.9
I	0.0	R	2.9
L	12.6	H	6.5
F	5.2	K	7.6

(3) The protein was digested with trypsin, and a peptide map was prepared. It contained 26 well-defined peptides.

(4) The peptide map was stained for various amino acid side chains: five peptides were positive for arg-inine, 13 peptides were positive for histidine, four peptides were positive for methionine, three peptides were positive for tryptophan, and seven peptides were positive for tyrosine.

(5) The protein was carboxymethylated with [^{14}C]iodoacetic acid and digested with trypsin. Three tryptic pep-

Table for Problem 8–12

amino acid	mol (100 mol)$^{-1}$	amino acid	mol (100 mol)$^{-1}$	amino acid	mol (100 mol)$^{-1}$
C[a]	1.7	G	9.2	Y	1.3
D+N	7.4	A	13.6	F	3.2
T[b]	4.5	V	6.5	K	3.1
S[b]	3.6	M	3.0	H	0.5
E+Q	8.9	I	8.2	R	6.6
P	6.9	L	9.5	W[c]	1.6

[a]As (carboxymethyl)cysteine at 24 h. [b]Extrapolated to zero hour. [c]Determined spectroscopically.

tides containing radioactive (carboxy-methyl)cysteine were isolated by ion-exchange chromatography, and each of these was shown to have a unique amino acid composition and to contain 1 mol of Cys (mol of peptide)$^{-1}$.

(A)   What is the length of a polypeptide composing protein X?

(B)   How many different polypeptides compose protein X?

(C)   Explain the peptide maps.

Cross-Linking

There arose a disagreement over the number of polypeptides contained in fructose-bisphosphate aldolase. The physical methods for estimating the molar mass of the native protein and the molar mass of its constituent polypeptides were unable to decide between three and four. It should be noted that everyone had an equal chance of being correct, so the point is not who turned out to be right, but that the question could not be resolved simply by arguing over the numbers. What was needed instead was a different kind of experiment, and it was provided.

When the fructose-bisphosphate aldolase in a homogenate from rat brain was submitted to electrophoresis in its native state, five evenly spaced components displaying enzymatic activity are observed (Figure 8–16). Penhoet, Kochman, Valentine, and Rutter[106] decided that this must be due to the fact that in the brain, two isoenzymatic polypeptides designated α and β are translated from two different messenger RNAs continuously and coincidentally. These polypeptides fold separately to form monomeric subunits that then combine at random with subunits of their own kind or of the other isoenzymatic type to produce hybrids of the stoichiometries α_4, $\alpha_3\beta$, $\alpha_2\beta_2$, $\alpha\beta_3$, and β_4. The two different subunits, α and β, differ slightly in the sequences of their polypeptides and hence in their charge. Each hybrid in turn has a different electrophoretic mobility because each has a different total charge, Q. The hybrids are capable of forming in the first place because the two different polypeptides are homologous in their sequences, have superposable tertiary structures in their folded state, share a common ancestor, and have not diverged sufficiently from that common ancestor to have lost the ability to combine with each other in the same way that they are required to do with subunits identical to themselves. If this explanation is correct, the number of polypeptides in any molecule of aldolase can be determined by simply counting the components on the electrophoretic separation. There must be four.

These investigators proved that the components observed upon electrophoresis were the hybrids that they proposed by isolating each of them, dissociating each to subunits, reassociating the subunits, and demonstrating

that each of these reassociated samples contained a new set of hybrids consistent with the stoichiometry assigned to the parent (Figure 8–16). For example, α_4 gave back only α_4 and β_4 gave back only β_4, but $\alpha_2\beta_2$ gave all five hybrids in approximately binomial ratios. It was also shown that an equal mixture of α_4 and β_4 when dissociated and reassociated reproduced all five hybrids.

Although it is possible to make electrophoretic variants by simply succinylating any native protein of interest,[107] rather than relying on isoforms coincidently provided by nature, the number of polypeptides in a native protein is rarely counted in the manner just described. The point, however, is not that hybridization solves the problem of determining the stoichiometries of polypeptides but that an experiment can be designed so that the number of polypeptides can be counted directly instead of calculated from physical measurements. This new way of looking at the problem stimulated the development of a technique that could be used to count the number of polypeptides in almost any soluble protein. This technique relies upon cross-linking.

A reagent capable of covalently cross-linking two polypeptides is a chemical compound that contains at least two electrophilic functional groups attached to each other by the remainder of the molecule. Aromatic diisocyanates, such as m-xylene diisocyanate[108] (Figure 8–17), were among the first such reagents to be used for this purpose, but they suffer from problems of low solubility. A widely employed cross-linking agent, and the one originally used to count polypeptides,[109] is dimethyl suberimidate (Figure 8–17). By far the most prevalent and accessible nucleophiles on the surface of a molecule of protein are the primary amines of its lysines, and the majority of the reagents, such as the diisocyanates and the diimidoesters, are directed to lysines.

An imidoester reacts with a primary amine to form an amidine (Figure 8–18). When a molecule of dimethyl suberimidate happens to react at one of its ends with a lysine on one polypeptide and at its other end with a lysine from another polypeptide, those two polypeptides become covalently cross-linked and migrate together upon electrophoresis in the presence of dodecyl sulfate with the mobility of a polypeptide whose length is equal to the sum of the lengths of the two polypeptides so joined.

At concentrations of protein below 10 μM, no significant intermolecular cross-linking between separate molecules of protein occurs when a solution of protein is mixed with a cross-linking agent such as dimethyl suberimidate. Instead, what is observed are the products that result from intramolecular cross-linking among the fixed number of polypeptides of which the protein is composed.[109] When the products of such intramolecular cross-linking are separated on a polyacrylamide gel, a ladder of bands, vaguely reminiscent of the ladders seen with randomly cleaved nucleic acids, is observed (Figure 8–19).[110] These ladders, however, end abruptly because

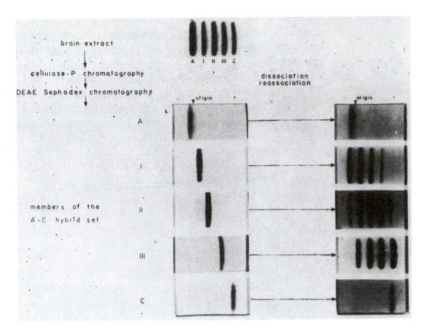

Figure 8–16: Dissociation and reassociation of oligomeric hybrids of fructose-bisphosphate aldolase. A clarified homogenate from rabbit brain was submitted to electrophoresis on cellulose acetate in its native state at pH 8.6 and the strip was then stained for the activity of fructose-bisphosphate aldolase. Five evenly spaced bands of enzymatic activity were observed (top pattern). The homogenate was then submitted to substrate elution from cellulose phosphate followed by anion-exchange chromatography on (diethylaminoethyl)cellulose (Figure 1–3). The five components that differed in electrophoretic mobility could be identified on these chromatograms by their enzymatic activity and could be cleanly separated from each other in this way. Each was submitted to electrophoresis separately at pH 8.6 and only one component with enzymatic activity was found in each, respectively (the five patterns to the left). Each of these single components was then exposed to 0.33 M H_3PO_4 at pH 2.0 for 30 min at 0 °C and then brought back to pH 7.5. The low pH served to dissociate the subunits of each hybrid and the neutralization reassociated them but at random. Each dissociated and reassociated hybrid was then resubmitted to electrophoresis, and the strips of cellulose acetate were stained for enzymatic activity (the five patterns to right). Reprinted with permission from ref 106. Copyright 1967 American Chemical Society.

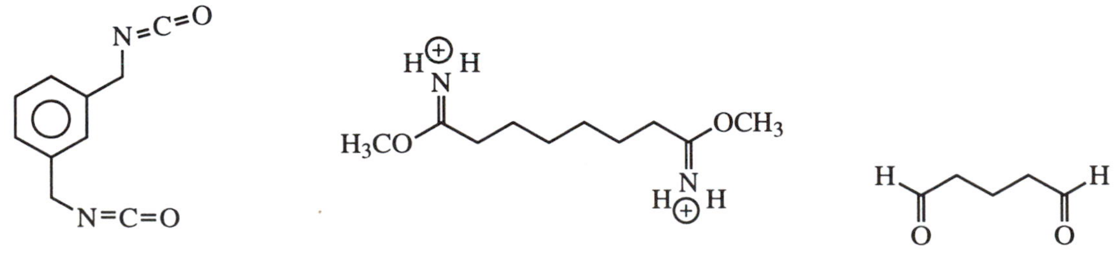

m-xylylene diisocyanate

dimethyl suberimidate

glutaraldehyde

Figure 8–17: Reagents for cross-linking proteins.

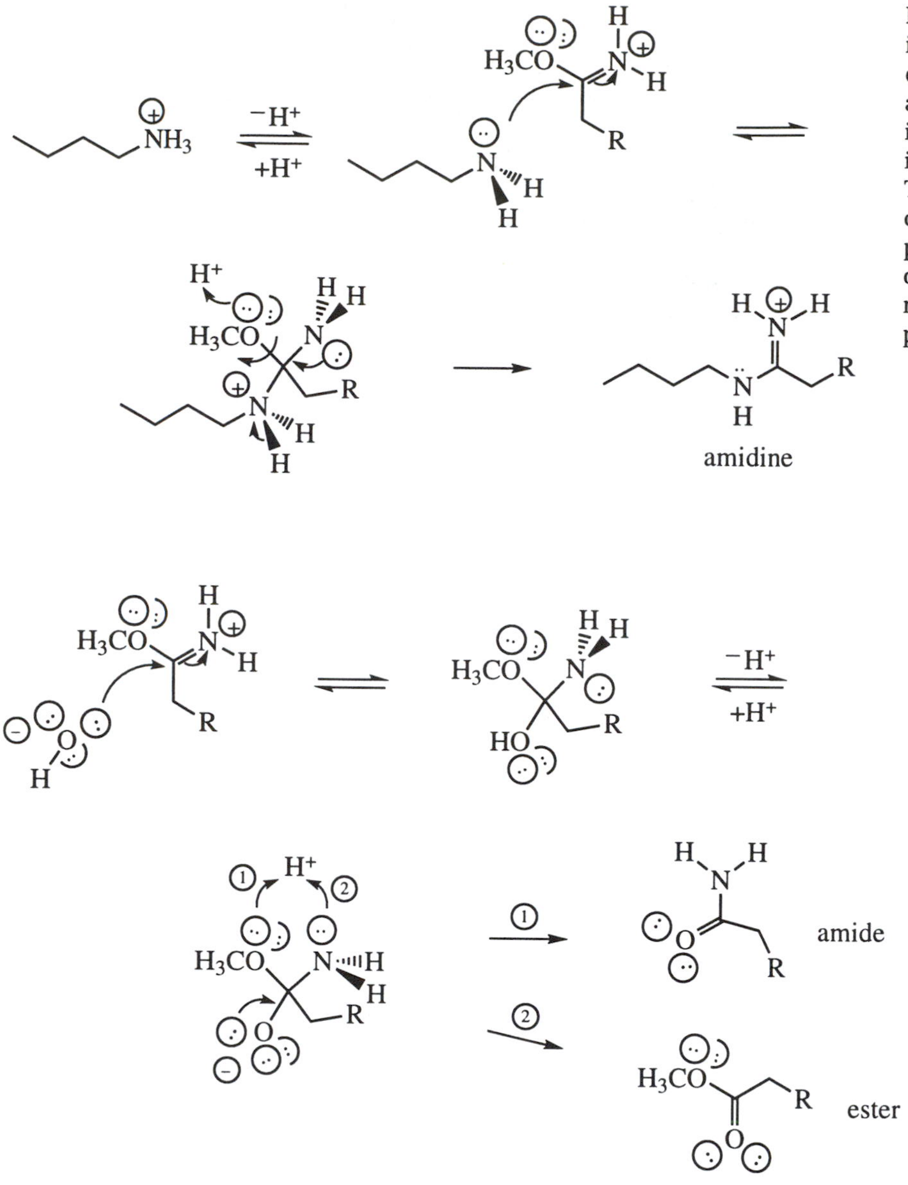

Figure 8–18: Mechanism of the reaction of an imidoester with a primary amine such as lysine or with hydroxide anion. The free base of the amine combines with the acyl carbon of the imidate to produce a tetravalent intermediate in which methanol is the best leaving group. The product is an amidine. Hydroxide anion combines with the acyl carbon of the imidate to produce a tetravalent intermediate with an oxyanion strong enough to push out either methanol to produce the amide or ammonia to produce the ester.

no more polypeptides can be cross-linked than are present in the complete protein. A count of the rungs in the ladder provides the number of polypeptides in the protein. In the case of glycerol kinase, the protein used for the analyses presented in Figure 8–19, it could be concluded that it is composed of four and only four polypeptides.[110] Peptide maps showed that all four of these polypeptides are identical to each other.[110]

There are two reassurances that should be provided. The relative mobility of each of the bands in the ladder should be shown to be a regular function of its number (graph in Figure 8–19). This is a reassurance that one of the members of the set was not missing from the pattern. The cross-linking reaction should be forced to completion so that only the highest oligomer is seen (gel C, inset to Figure 8–19). This result provides the reassurance that this oligomer does represent a true limit to the reaction and that the solution contains only one unique multimer of the polypeptides rather than a mixture of multimers

each containing a different number of polypeptides. It also rules out the possibility that any intermolecular cross-linking is occurring. Random intermolecular cross-linking yields a distribution that gradually declines in its amplitude with band number rather than displaying an abrupt limit at a certain unique polymer size. It is this discontinuous behavior that is the logical basis for believing the results, other than the fact that they give the correct answer.[109]

It is in fulfilling the requirement that the reaction be forced to completion that dimethyl suberimidate usually fails. The reason is simple to understand. The reaction of an imidoester with a primary amine in aqueous solution is a competition between formation of the desired amidine and hydrolysis of the imidoester to amide and ester (Figure 8–18). There is no way to avoid this competition by some informed adjustment of the conditions. It complicates the reaction because when one end of the diimidoester has attached to a lysine there is a significant probability that its other end has either already been hydrolyzed or will hydrolyze before it can react with another lysine. A high percentage of the lysines end up with defunct reagent after all of the lysines have become amidines and no further reaction can occur regardless of how much reagent is added. Because of this, it is often the case that not all of the polypeptides can be cross-linked among themselves. The logical chemical solution to this problem would be to use an electrophile at the two ends of the cross-linking reagent that is more selective for a primary amine relative to a hydroxide anion, but this has not been explored systematically. Rather, it has been inadvertently accomplished.

The most versatile cross-linking agent is glutaraldehyde. Unfortunately, the chemistry of its reactions with proteins has never been elucidated. Presumably as an aliphatic aldehyde it engages in all of the same chemical reactions with lysine that occur after aliphatic aldehydes are produced in collagen by lysyl oxidase (Figure 3–22). It is part of the lore surrounding glutaraldehyde that the freshly distilled reagent is far less active; this is meant to suggest that compounds derived from glutaraldehyde itself, such as its aldol and dehydrated aldol, are important to the cross-linking. Be that as it may, glutaraldehyde produces cross-linked products in the highest yield of any known bifunctional reagent. In situations when the cross-linking reaction with dimethyl suberimidate cannot be forced to completion, cross-linking with glutaraldehyde will usually produce the fully cross-linked protein in high yield. The efficiency of glutaraldehyde has permitted it to be used to provide a quantitative catalogue of the various oligomeric complexes present in a heterodisperse solution of a single, pure protein.

Quantitative cross-linking is cross-linking carried to an extent sufficient to connect covalently every polypeptide in a macromolecular complex to every other polypeptide in the same macromolecular complex, either directly or indirectly, but not to any polypeptide in another macromolecular complex. In order for glutaraldehyde to perform quantitative cross-linking, the formation of intermolecular covalent connections between independent, unassociated oligomeric complexes in the solution must be negligible because every covalent complex must represent only the product of intramolecular cross-linking, and every multimeric complex in the solution must be completely cross-linked within itself so that every one of its constituent polypeptides is covalently attached to all of the others. Cross-linking with appropriately high concentrations of glutaraldehyde satisfies these requirements.[111-113]

This can be illustrated with a simple observation. A monodisperse solution containing a homogeneous population of one oligomeric protein, such as L-lactate dehydrogenase,[111] a protein known to be composed of four identical polypeptides, was mixed with glutaraldehyde. The reaction was permitted to proceed a short period of time, and the products were examined by electrophoresis on polyacrylamide gels in the presence of dodecyl sulfate. At high enough concentrations of glutaraldehyde, the only component that is observed contains the number of polypeptides, four, known to be present in the protein, all covalently attached to each other (Figure 8–20). No larger products, which would have resulted from intermolecular cross-linking, and no smaller products,

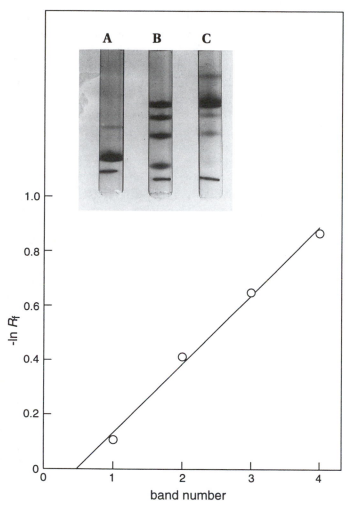

Figure 8–19: Definition of the number of subunits in glycerol kinase.[110] Three aliquots of a solution of glycerol kinase were mixed with final concentrations of 10 mg mL^{-1} of dimethyl suberimidate (inset, gel C), 0.25 mg mL^{-1} of dimethyl suberimidate (inset, gel B), or no additions (inset, gel A). After 4 h at pH 8.5, the samples were dissolved in a solution of sodium dodecyl sulfate and subjected to electrophoresis on polyacrylamide gels cast in 0.1% sodium dodecyl sulfate. The gels were stained for protein. The irregular line at the bottom of each gel is a line of India ink used to mark the position of the marker dye at the end of the electrophoresis. The thick band of stain just above this mark is the un-cross-linked polypeptide of glycerol kinase. The mobilities of each component on gel B relative to the marker dye were calculated from the photograph, and the negative natural logarithm of each is plotted against a scale of successive integers. Adapted with permission from ref 110. Copyright 1971 *Journal of Biological Chemistry.*

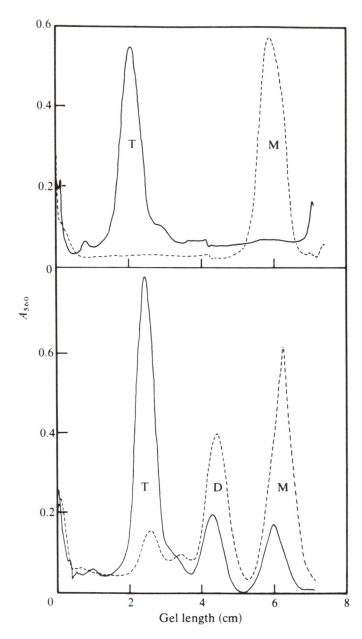

Figure 8–20: Cross-linking of L-lactate dehydrogenase.[111] Samples of L-lactate dehydrogenase were submitted to cross-linking for 2 min with 0.4 M glutaraldehyde at 20 °C and the reaction was terminated by adding sodium dodecyl sulfate to 20 μM. The samples were then submitted to electrophoresis on polyacrylamide gels, the gels were stained for protein, and the absorbance at 560 nm as a function of the distance migrated is presented. (A) Cross-linking in a solution of native tetramers (20 nM). Equivalent samples were either cross-linked (solid line) or not cross-linked (dashed line). In the former, only fully cross-linked tetramers (T) are observed; in the latter, only un-cross-linked monomers (M). (B) Cross-linking of reassembling L-lactate dehydrogenase after it was dissociated into subunits. L-Lactate dehydrogenase was dissociated into subunits at pH 2.3 and then diluted into a solution at pH 7.6 (final concentration of subunits 340 nM). After 210 s (dashed line) and 1 h (solid line), samples were removed and complexes (M, monomer; D, dimer; T, tetramer) were catalogued by cross-linking. Reprinted with permission from ref 111. Copyright 1979 *Nature*.

which would have resulted from incomplete intramolecular cross-linking, are observed.

While this example illustrates how glutaraldehyde can be used to define the oligomer present in a monodisperse solution of a protein, it can also be used to provide a catalogue of the oligomers in a heterodisperse solution. When Na^+/K^+-transporting ATPase is dissolved in a solution of nonionic detergent, a mixture is formed of various oligomers, $(\alpha\beta)_n$, which are combinations of the two different polypeptides, α and β, from which this protein is composed. The fraction of the total protein engaged in each of these oligomers, respectively, could be rapidly and accurately determined by quantitative cross-linking (Figure 8–21).[113] From the pattern displayed it could be calculated that before the glutaraldehyde had been added to the solution, the fractions of the protein present as the $(\alpha\beta)_4$ tetramer, the $(\alpha\beta)_3$ trimer, the $(\alpha\beta)_2$ dimer, and the $\alpha\beta$ monomer were 0.25, 0.19, 0.28, and 0.31, respectively. When L-lactate dehydrogenase is transferred to a solution at pH 2.3, it dissociates into monomers consisting of single folded polypeptides, and when it is transferred back to pH 7.6 it slowly (over an hour) reassociates to its normal tetrameric state. The concentration of monomer, dimer, and tetramer can be ascertained at any given minute by removing a sample and cross-linking it with glutaraldehyde (Figure 8–20).[111] An extensive study of the kinetics of this stepwise association could be performed in this way.[112] The rate and mechanism of the association of the subunits of the catalytic trimer of aspartate transcarbamylase could also be monitored by quantitative cross-linking.[114]

Cross-linking has also been used to determine the stoichiometric ratio between two dissimilar subunits. For example, the fact that the α and β polypeptides of succinate–CoA ligase (ADP-forming) disappear in concert almost entirely during the formation of a covalent $\alpha\beta$ heterodimer[116] means that they must be present in equimolar ratio in the protein (Figure 8–22). A similar result was observed with the two dissimilar polypeptides composing Na^+/K^+-transporting ATPase.[117] In either of these cases, if the polypeptides had not been present in equimolar ratio, either one polypeptide would have disappeared from its position on the polyacrylamide gel more rapidly than the other during the formation of a covalent $\alpha\beta$ heterodimer or significant amounts of covalent products of the form $\alpha_2\beta$ or $\alpha\beta_2$ would have appeared in addition to the covalent $\alpha\beta$ heterodimer.

Instead of focusing on either the unique, completely cross-linked complex that defines the polypeptide stoichiometry of a pure protein in monodisperse solution (Figure 8–20) or the pattern of completely cross-linked complexes that provides a catalogue of the oligomers in a heterogeneous mixture (Figure 8–21), the patterns seen in the ladders from partially cross-linked samples of a protein (Figure 8–19) can be examined. These patterns can provide information about the arrangement of the

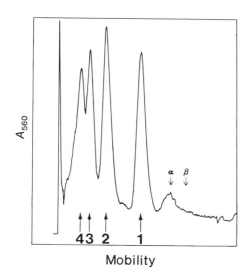

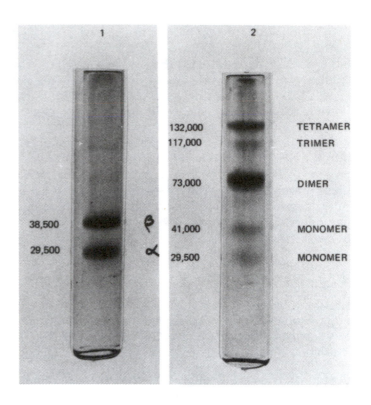

Figure 8–21: Cross-linking of the oligomers in a heterodisperse solution of Na$^+$/K$^+$-transporting ATPase.[113] A suspension of biological membranes containing only Na$^+$/K$^+$-transporting ATPase was dissolved in a 5 mM solution of the nonionic detergent octaethyleneglycol dodecyl ether. After 45 min at 22 °C, sodium dodecyl sulfate was added and the sample was submitted to electrophoresis on a gel cast from 3.6% acrylamide. After the gel was stained, it was scanned for absorbance at 550 nm as a function of the distance migrated. By their mobilities the components were identified as covalently cross-linked $\alpha\beta$ monomer, $(\alpha\beta)_2$ dimer, $(\alpha\beta)_3$ trimer, and $(\alpha\beta)_4$ tetramer of the enzyme, which in its simplest native form is a monomer ($\alpha\beta$) of one α subunit and one β subunit in noncovalent association.[115] Reprinted with permission from ref 113. Copyright 1982 American Chemical Society.

Figure 8–22: Cross-linking of succinate–CoA ligase (ADP-forming).[116] A solution of succinate–CoA ligase (1.0 mg mL^{-1}) was cross-linked with dimethyl suberimidate (2.0 mg mL^{-1}) for 30 min. The resulting covalent complexes were dissolved in a solution of sodium dodecyl sulfate and submitted to electrophoresis. Gel 1: Un-cross-linked control showing the α and β polypeptides from which the enzyme is composed. Gel 2: Cross-linked product. The components observed were assigned as the covalent $\alpha\beta$ heterodimer, the covalent heterotrimer, and the covalent $(\alpha\beta)_2$ heterotetramer on the basis of their apparent molecular masses (numbers to the right of each gel) determined by their mobilities on electrophoresis relative to a set of polypeptides of known molecular mass. Reprinted with permission from ref 116. Copyright 1975 *Journal of Biological Chemistry.*

subunits in the oligomer. Succinate–CoA ligase (ADP-forming) is a protein composed of two different polypeptides, 350 and 260 amino acids in length, each present in two copies. These polypeptides could have been arranged in at least two ways to produce stoichiometries of either $(\alpha\beta)_2$ or $\alpha_2\beta_2$. The former designation implies that the association between an α polypeptide and a β polypeptide is more intimate than that between either two α polypeptides or two β polypeptides; the latter implies the reverse. Examples illustrating this distinction would be either a dimer of two identical polypeptides, each of which was posttranslationally cleaved to produce a pair of subunits, each an entwined α polypeptide and β polypeptide, or a heterotetramer assembled from a fully folded α_2 dimer and a fully folded β_2 dimer, respectively. Succinate–CoA ligase (ADP-forming) was reacted with enough dimethyl suberimidate to cross-link completely the α and β polypeptides among themselves but not enough to produce a high yield of the completely cross-linked product.[116] The covalent $\alpha\beta$ heterodimer was by far the major product of this incomplete reaction. A reasonable yield of the covalent heterotetramer, $(\alpha\beta)_2$ was also produced during the reac-

tion, but little or no covalent trimer, covalent $\alpha\alpha$ homodimer, or covalent $\beta\beta$ homodimer was seen (Figure 8–22). This result demonstrates that the formation of a cross-link between one α polypeptide and one β polypeptide in native succinate–CoA ligase (ADP-forming) is a far more likely event than the formation of a cross-link between either two α polypeptides or two β polypeptides, and it suggests that the proper designation for the arrangement of the subunits in the native, un-cross-linked protein is $(\alpha\beta)_2$.

A similar disparity among the products of a partial cross-linking reaction was seen when several tetrameric proteins, each composed of four identical polypeptides, were examined.[118] L-Lactate dehydrogenase, pyruvate kinase, fructose-bisphosphate aldolase, fumarate hydratase, and catalase were each partially cross-linked with various dimethyl diimidates. In each case, the yield of the covalent trimer was 2–6-fold lower than that of either the covalent dimer or the covalent tetramer. This result is reminiscent of the one seen with succinate–CoA ligase (ADP-forming), and its explanation is the same. Within each of these molecules there must be associations between some pairs of α polypeptides that are more intimate than the associations between other pairs of α polypeptides. The results suggest that the proper designation for the arrangement of the subunits in each of these proteins is $(\alpha_2)_2$. To understand why this is so, the rules governing the evolution of oligomeric proteins must be understood. These rules are based on rotational axes of symmetry.

Suggested Reading

Davies, G.E., & Stark, G.R. (1970) Use of Dimethyl Suberimidate, a Cross-Linking Agent, in Studying the Subunit Structure of Oligomeric Proteins, *Proc. Natl. Acad. Sci. U.S.A. 66*, 651–656.

Problem 8–14: Assume that each pure hybrid of fructose-bisphosphate aldolase shown in Figure 8–16 was dissociated completely and reassociated at random during the experiment described in the figure. Refer to the two types of subunits as α and γ.

(A) What is the respective subunit structure of each of the five purified hybrids?

(B) Assume the subunits reassemble at random by the binomial formula $a^4 + 4a^3c + 6a^2c^2 + 4ac^3 + c^4$, where a is the fraction of the dissociated subunits that are α and c is the fraction that are γ and predict the ratio of components expected from the reassociation of each of the five dissociated hybrids.

References

1. Tanford, C. (1961) *Physical Chemistry of Macromolecules*, John Wiley & Sons, New York.
2. Eisenberg, H. (1981) *Q. Rev. Biophys. 14*, 141–172.
3. McMillan, W.G., & Mayer, J.E. (1945) *J. Chem. Phys. 13*, 276–305.
4. Edelhoch, H. (1967) *Biochemistry 6*, 1948–1954.
5. Gornall, A.G., Bardawill, C.J., & David, M.M. (1949) *J. Biol. Chem. 177*, 751–766.
6. Moczydlowski, E.G., & Fortes, P.A.G. (1981) *J. Biol. Chem. 256*, 2346–2356.
7. Scatchard, G., Batchelder, A.C., & Brown, A. (1946) *J. Am. Chem. Soc. 68*, 2320–2331.
8. vanHolde, K.E. (1985) *Physical Biochemistry*, 2nd ed., Prentice-Hall, Englewood Cliffs, NJ.
9. Schachman, H.K., & Edelstein, S.J. (1966) *Biochemistry 5*, 2681–2705.
10. Cohn, E.J., & Edsall, J.T. (1943) *Proteins, Amino Acids, and Peptides*, Reinhold, New York.
11. Brown, J.R. (1976) *Fed. Proc. 35*, 2141–2144.
12. Zimm, B.H. (1948) *J. Chem. Phys. 16*, 1099–1116.
13. Dandliker, W.B. (1954) *J. Am. Chem. Soc. 76*, 6036–6039.
14. Edsall, J.T., Edelhoch, H., Lontie, R., & Morrison, P.R. (1950) *J. Am. Chem. Soc. 72*, 4641–4656.
15. Fenn, J.B., Mann, M., Meng, C.K., Wong, S.F., & Whitehouse, C.M. (1989) *Science 246*, 64–71.
16. Ronk, M., Shively, J.E., Shute, E.A., & Blake, R.C. (1991) *Biochemistry 30*, 9435–9442.
17. Tanford, C. (1961) *Physical Chemistry of Macromolecules*, John Wiley & Sons, New York.
18. Dayhoff, M.O. (1978) in *Atlas of Protein Sequence and Structure*, Vol. 5, Suppl. 3, National Biomedical Research Foundation, Washington, DC.
19. Edsall, J.T. (1953) in *The Proteins, Chemistry, Biological Activity, and Methods* (Neurath, H., & Bailey, K.E., Eds.) Vol. I, Part B, pp 549–726, Academic Press, New York.
20. Dayhoff, M.O. (1972) *Atlas of Protein Sequence and Structure*, Vol. 5, National Biomedical Research Foundation, Washington, DC.
21. Jaenicke, R., & Knof, S. (1968) *Eur. J. Biochem. 4*, 157–163.
22. Titani, K., Koide, A., Ericsson, L.H., Kumar, S., Hermann, J., Wade, R.D., Walsh, K.A., Neurath, H., & Fischer, E. (1978) *Biochemistry 17*, 5680–5693.
23. Seery, V.L., Fischer, E.H., & Teller, D.C. (1967) *Biochemistry 6*, 3315–3327.
24. Dandliker, W.B., & Fox, J.B. (1955) *J. Biol. Chem. 214*, 275–283.
25. Schroeder, W.A., Shelton, J.R., Shelton, J.B., Robberson, B., & Apell, G. (1969) *Arch. Biochem. Biophys. 131*, 652–658.
26. Samejima, T., & Yang, J.T. (1963) *J. Biol. Chem. 238*, 3256–3261.
27. Murthy, M.R.N., Reid, T.J., Sicignano, A., Tanaka, N., & Rossmann, M.G. (1981) *J. Mol. Biol. 152*, 465–499.
28. Stellwagen, E., & Schachman, H.K. (1962) *Biochemistry 1*, 1056–1068.
29. Kawahara, K., & Tanford, C. (1966) *Biochemistry 5*, 1578–1591.
30. Richter, G.W., & Walker, G.F. (1967) *Biochemistry 6*, 2871–2880.
31. Boyd, D., Vecoli, C., Belcher, D.M., Jain, S.K., & Drysdale, J.W. (1985) *J. Biol. Chem. 260*, 11755–11761.
32. Gerhart, J.C., & Schachman, H.K. (1965) *Biochemistry 4*, 1054–1062.
33. Hoover, T.A., Roof, W.D., Folterman, K.F., O'Donovan, G.A., Bencini, D.A., & Wild, J.R. (1983) *Proc. Natl. Acad. Sci. U.S.A. 80*, 2462–2466.
34. Hammerstedt, R.H., Möhler, H., Decker, K.A., & Wood, W.A. (1971) *J. Biol. Chem. 246*, 2069–2074.
35. Falcoz-Kelly, F., Janin, J., Saari, J.C., Veron, M., Truffa-Bachi, P., & Cohen, G.N. (1972) *Eur. J. Biochem. 28*, 507–519.
36. Katinka, M., Cossart, P., Sibilli, L., Saint-Girons, I., Chalvignac, M.A., LeBras, G., Cohen, G.N., & Yaniv, M. (1980) *Proc. Natl. Acad. Sci. U.S.A. 77*, 5730–5733.
37. Cassman, M., & Schachman, H.K. (1971) *Biochemistry 10*, 1015–1024.
38. Moon, K., & Smith, E.L. (1973) *J. Biol. Chem. 248*, 3082–3088.
39. Eisenberg, H., & Tomkins, G.M. (1968) *J. Mol. Biol. 31*, 37–49.
40. Ovchinnikov, Y.A., Abdulaev, N.G., Feignina, M.Y., Kiselev, A.V., & Lobanov, N.A. (1979) *FEBS Lett. 100*, 219–224.

41. Watt, K.W.K., Cottrell, B.A., Strong, D.D., & Doolittle, R.F. (1979) *Biochemistry 18*, 5410–5416.

42. Kopito, R.R., & Lodish, H.F. (1985) *Nature 316*, 234–238.

43. Shull, G.E., Schwartz, A., & Lingrel, J.B. (1985) *Nature 316*, 691–695.

44. Shull, G.E., Lane, L.K., & Lingrel, J.B. (1986) *Nature 321*, 429–431.

45. Strehler, E.E., Strehler-Page, M., Perriard, J., Periasamy, M., & Nadal-Ginard, B. (1987) *J. Mol. Biol. 190*, 291–317.

46. Rosenbusch, J.P., & Weber, K. (1971) *J. Biol. Chem. 246*, 1644–1657.

47. Truffa-Bachi, P., vanRapenbusch, R., Janin, J., Gros, C., & Cohen, G.N. (1968) *Eur. J. Biochem. 5*, 73–80.

48. Burgess, R.R. (1969) *J. Biol. Chem. 244*, 6168–6176.

49. Pitt-Rivers, R., & Impiombato, F.S.A (1968) *Biochem. J. 109*, 825–330.

50. Reynolds, J.A., & Tanford, C. (1970) *J. Biol. Chem. 245*, 5161–5165.

51. Shirahama, K., Tsuji, J., & Takagi, T. (1974) *J. Biochem. (Tokyo) 75*, 309–319.

52. Banker, G.A., & Cotman, C.W. (1972) *J. Biol. Chem. 247*, 5856–5861.

53. Fisher, M.P., & Dingman, C.W. (1971) *Biochemistry 10*, 1895–1899.

54. Weber, K., & Osborn, M. (1969) *J. Biol. Chem. 244*, 4406–4412.

55. Rodbard, D., & Chrambach, A. (1970) *Proc. Natl. Acad. Sci. U.S.A. 65*, 970–977.

56. Cornfield, J., & Chalkley, H.W. (1951) *J. Washington Acad. Sci. 41*, 226–229.

57. Fawcett, J.S., & Morris, C.J.O.R. (1966) *Sep. Sci. 1*, 9–26.

58. Laurent, T., & Killander, J. (1964) *J. Chromatogr. 14*, 317–330.

59. Morris, C.J.O.R. (1966) *Protides Biol. Fluids 14*, 543–551.

60. Ogston, A.G. (1958) *Trans. Faraday Soc. 54*, 1754–1757.

61. Cohn, E.J., McMeekin, T.L., Edsall, J.T., & Blanchard, M.H. (1934) *J. Am. Chem. Soc. 56*, 784–794.

62. Cohn, E.J., & Edsall, J.T. (1943) *Proteins, Amino Acids, and Peptides as Ions and Dipolar Ions*, Reinhold, New York.

63. Andrews, P. (1965) *Biochem. J. 96*, 595–606.

64. Ackers, G.K. (1964) *Biochemistry 3*, 723–730.

65. Siegel, L.M., & Monty, K.J. (1966) *Biochim. Biophys. Acta 112*, 346–362.

66. Grubisic, Z., Rempp, R., & Benoit, H. (1967) *J. Polym. Sci., Polym. Lett. 5*, 753–759.

67. Hedrick, J.L., & Smith, A.J. (1968) *Arch. Biochem. Biophys. 126*, 155–164.

68. Bryan, J.K. (1977) *Anal. Biochem. 78*, 513–519.

69. Fish, W.W., Mann, K.G., & Tanford, C. (1969) *J. Biol. Chem. 244*, 4989–4994.

70. Tanford, C. (1968) *Adv. Protein Chem. 23*, 121–282.

71. Kumagai, I., Pieler, T., Subramanian, A.R., & Erdmann, V.A. (1982) *J. Biol. Chem. 257*, 12924–12928.

72. Shapiro, A.L. Viñuela, E., & Maizel, J.V. (1967) *Biochem. Biophys. Res. Commun. 28*, 815–820.

73. Lumpkin, O.J., & Zimm, B.H. (1982) *Biopolymers 21*, 2315–2316.

74. Lumpkin, O.J., Dejardin, P., & Zimm, B.H. (1985) *Biopolymers 24*, 1573–1593.

75. Peterson, G.L., & Hokin, L.E. (1981) *J. Biol. Chem. 256*, 3751–3761.

76. Williams, J.G., & Gratzer, W.B. (1971) *J. Chromatogr. 57*, 121–125.

77. Swank, R.T., & Munkries, K.K. (1971) *Anal. Biochem. 39*, 462–477.

78. Kyte, J., & Rodriguez, H. (1983) *Anal. Biochem. 133*, 515–522.

79. Lindstrom, J., Cooper, J., & Tzartos, S. (1980) *Biochemistry 19*, 1454–1458.

80. Weber, K., Pringle, J.R., & Osborn, M. (1972) *Methods Enzymol. 26*, 3–27.

81. Doolittle, R.F. (1981) *Science 214*, 149–159.

82. Hubert, J.J., Schenk, D.B., Skelly, H., & Leffert, H.L. (1986) *Biochemistry 25*, 4156–4163.

83. Fullmer, C.S., & Wasserman, R.H. (1979) *J. Biol. Chem. 254*, 7208–7212.

84. Lindstrom, J., Merlie, J., & Yogeeswaran, G. (1979) *Biochemistry 18*, 4465–4470.

85. Pauling, L., Itano, H.A., Singer, S.J., & Wells, I.C. (1949) *Science 110*, 543–548.

86. Ingram, V.M. (1958) *Biochim. Biophys. Acta 28*, 539–545.

87. Vandekerckhove, J., & Weber, K. (1978) *Proc. Natl. Acad. Sci. U.S.A. 75*, 1106–1110.

88. Vandekerckhove, J., & Weber, K. (1978) *Eur. J. Biochem. 90*, 451–462.

89. Luna, E.J., Kidd, G.H., & Branton, D. (1979) *J. Biol. Chem. 254*, 2526–2532.

90. Grant, G.A., & Bradshaw, R.A. (1978) *J. Biol. Chem. 253*, 2727–2731.

91. Singh, M., Brooks, G.C., & Srere, P.A. (1970) *J. Biol. Chem. 245*, 4636–4640.

91a. Kern, D., Potier, S., Boulanger, Y., & Lapointe, J. (1979) *J. Biol. Chem. 254*, 518–524.

92. Dowhan, W., & Snell, E.E. (1970) *J. Biol. Chem. 245*, 4618–4628.

93. James, G.T., & Noltmann, E.A. (1973) *J. Biol. Chem. 248*, 730–737.

94. Lundell, D.J., & Howard, J.B. (1978) *J. Biol. Chem. 253*, 3422–3426.

95. Green, N.M., Valentine, R.C., Wrigley, N.G., Ahmad, F., Jacobson, B., & Wood, H.G. (1972) *J. Biol. Chem. 247*, 6284–6298.

96. Zwolinski, G.K., Bowien, B.U., Harmon, F., & Wood, H.G. (1977) *Biochemistry 16*, 4627–4637.

97. Hall, C.L., & Kamin, H. (1975) *J. Biol. Chem. 250*, 3476–3486.

98. McKean, M.C., Beckman, J.D., & Frerman, F.E. (1983) *J. Biol. Chem. 258*, 1866–1870.

99. Chuang, M., Ahmad, F., Jacobson, B.E., & Wood, H.G. (1975) *Biochemistry 14*, 1611–1619.

100. Recesi, P.A., Huynh, Q.K., & Snell, E.E (1983) *Proc. Natl. Acad. Sci. U.S.A. 80*, 973–977.

101. Guchhait, R.B., Zwergel, E.E., & Lane, M.D. (1974) *J. Biol. Chem. 249*, 4776–4780.

102. Guchhait, R.B., Polakis, E., Dimroth, P., Stoll, E., Moss, J., & Lane, M.D. (1974) *J. Biol. Chem. 249*, 6633–6645.

103. Song, C.S., & Kim, K. (1981) *J. Biol. Chem. 256*, 7786–7788.

104. Darnell, D.W., & Klotz, I.M. (1975) *Arch. Biochem. Biophys. 166*, 651–682.

105. James, G.T., & Noltmann, E.A. (1973) *J. Biol. Chem. 248*, 730–737.

106. Penhoet, E., Kochman, M., Valentine, R., & Rutter, W.J. (1967) *Biochemistry 6*, 2940–2949.

107. Meighan, E.A., & Schachman, H.K. (1970) *Biochemistry 9*, 1163–1176.
108. Singer, S.J. (1959) *Nature 183*, 1523–1524.
109. Davies, G.E., & Stark, G.R. (1970) *Proc. Natl. Acad. Sci. U.S.A. 66*, 651–656.
110. Thorner, J.W., & Paulus, H. (1971) *J. Biol. Chem. 246*, 3885–3894.
111. Hermann, R., Rudolf, R., & Jaenicke, R. (1979) *Nature 277*, 243–245.
112. Hermann, R., Jaenicke, R., & Rudolph, R. (1981) *Biochemistry 20*, 5195–5201.
113. Craig, W.S. (1982) *Biochemistry 21*, 2667–2674.
114. Burns, D.L., & Schachman, H.K. (1982) *J. Biol. Chem. 257*, 8638–8647.
115. Craig, W.S. (1982) *Biochemistry 21*, 5707–5717.
116. Teherani, J.A., & Nishimura, J.S. (1975) *J. Biol. Chem. 250*, 3883–3890.
117. Craig, W.S., & Kyte, J. (1980) *J. Biol. Chem. 255*, 6262–6269.
118. Hucho, F., Müllner, H., & Sund, H. (1975) *Eur. J. Biochem. 59*, 79–87.

CHAPTER 9

Symmetry

Most of the proteins in a living organism are multimeric. A **multimeric protein** is a protein composed of more than one folded polypeptide. Each of the polypeptides composing such a protein folds to form a tertiary structure unique to the amino acid sequence of that polypeptide. These folded polypeptides then associate with each other to form the multimeric protein. The arrangement in space of these folded polypeptides in the final protein is its **quaternary structure**. Most multimeric proteins are composed of multiple copies of only one particular folded polypeptide. These folded structures, each necessarily identical to the others when free in solution, combine together to form the final molecule of protein. In these instances, each of the folded polypeptides in the crystallographic molecular model can be formally designated a **protomer**[1] of the final overall structure.

Some multimeric proteins, like the isoenzymatic hybrids of fructose-bisphosphate aldolase, are composed of two or more distinct folded polypeptides that are, nevertheless, each the offspring of the same common ancestor. Although different in amino acid sequence and in the atomic details of their tertiary structure, these folded polypeptides are still related closely enough to participate together to form the complete macromolecule much as identical protomers would participate in a multimeric protein. In such instances, each of the individual folded polypeptides, although actually different, can be considered to be one of the indistinguishable protomers of the overall structure. In contrast to the hybrids formed by the isoenzymatic protomers of fructose-bisphosphate aldolase, other proteins built from such similar but distinct protomers usually incorporate those protomers in unvarying ratios. For example, hemoglobin is always an $(\alpha\beta)_2$ tetramer and acetylcholine receptor is always an $\alpha_2\beta\gamma\delta$ pentamer, even though the α and β protomers of hemoglobin or the α, β, γ, and δ protomers of acetylcholine receptor are homologous in amino acid sequence and necessarily superposable. These exclusive stoichiometries are established by the distinct atomic interactions between the protomers that take place as the multimer assembles.

Some multimeric proteins contain two or more dissimilar, unrelated polypeptides. For example, aspartate transcarbamylase contains both folded catalytic polypeptides and folded regulatory polypeptides. Even though the polypeptides composing such a protein are so different, the final structure produced, when observed as a crystallographic molecular model, can usually be divided formally into identical protomers, each containing one folded copy of each of the unique polypeptides. A particular number of these protomers arranged in an array unique to that protein makes up its final structure. Because every protomer is identical to all the others, the arrangement of the protomers of the protein is formally equivalent to the assembly of a multimeric protein composed only of identical subunits.

From these considerations it follows that, with a few peculiar exceptions, the crystallographic molecular model of a multimeric protein can be divided formally into a set of identical or almost identical protomers. The rules that govern the way these protomers are arranged in space to produce the complete molecular structure of the entire protein seem to be the same whether the protomers are identical or are merely homologous. In a multimeric protein with a fixed number of identical or homologous protomers, those protomers are arrayed around rotational axes of symmetry or rotational axes of pseudosymmetry, respectively. In a multimeric protein with an indefinite number of identical or homologous protomers, those protomers are arrayed around a screw axis of symmetry or a screw axis of pseudosymmetry, respectively.

Rotational and Screw Axes of Symmetry

The fundamental symmetry operations that are available to asymmetric objects like proteins when they assemble into multimeric structures are rotational and screw axes of symmetry. If a protein is constructed with rotational symmetry, only a finite number of protomers produce the final structure. If a protein is assembled with screw symmetry, a potentially infinite number of protomers can usually combine to produce a polymer of indefinite length. In one isolated instance observed so far, a protein, hexokinase, although it is assembled with screw symmetry, is nevertheless forced to have only two protomers in the final structure. Such finite structures assembled with screw symmetry are thought to be very rare.

Consider two proteins that illustrate the observation that multimeric proteins constructed from identical protomers are assembled around rotational or screw axes of symmetry. The proteins are triose-phosphate isomerase (Figure 9–1A), a dimer built from two identical protomers (c2 Kin.8) that are folded polypeptides 247 amino acids in length,[2] and actin (Figure 9–1B,C), a protein that forms helical polymeric fibers of indefinite length with each fiber built from many identical protomers that are folded polypeptides 375 amino acids in length.[3]

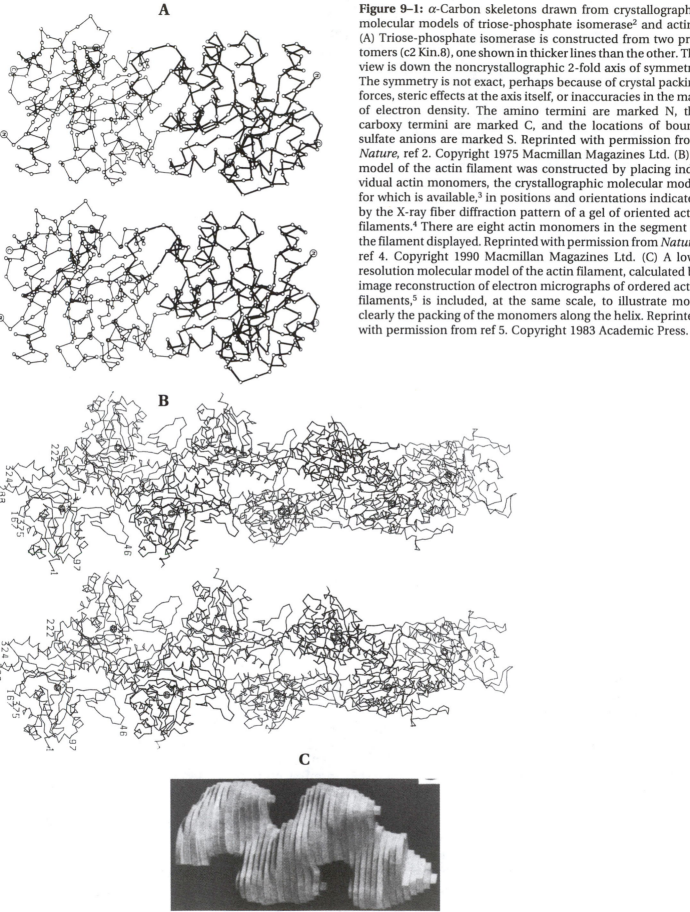

A

B

C

Figure 9–1: α-Carbon skeletons drawn from crystallographic molecular models of triose-phosphate isomerase[2] and actin.[4] (A) Triose-phosphate isomerase is constructed from two protomers (c2 Kin.8), one shown in thicker lines than the other. The view is down the noncrystallographic 2-fold axis of symmetry. The symmetry is not exact, perhaps because of crystal packing forces, steric effects at the axis itself, or inaccuracies in the map of electron density. The amino termini are marked N, the carboxy termini are marked C, and the locations of bound sulfate anions are marked S. Reprinted with permission from *Nature*, ref 2. Copyright 1975 Macmillan Magazines Ltd. (B) A model of the actin filament was constructed by placing individual actin monomers, the crystallographic molecular model for which is available,[3] in positions and orientations indicated by the X-ray fiber diffraction pattern of a gel of oriented actin filaments.[4] There are eight actin monomers in the segment of the filament displayed. Reprinted with permission from *Nature*, ref 4. Copyright 1990 Macmillan Magazines Ltd. (C) A low-resolution molecular model of the actin filament, calculated by image reconstruction of electron micrographs of ordered actin filaments,[5] is included, at the same scale, to illustrate more clearly the packing of the monomers along the helix. Reprinted with permission from ref 5. Copyright 1983 Academic Press.

An almost exactly 2-fold rotational axis of symmetry runs through the center of the α_2 dimer of triose-phosphate isomerase. A **rotational axis of symmetry** is a line about which a structure can be rotated by $360°/n$, where n is an integer larger than 1, to superpose upon itself. The value of n determines the fold of the symmetry. The 2-fold rotational axis of symmetry in the crystallographic molecular model of triose-phosphate isomerase is a line perpendicular to the plane of the page in Figure 9–1A. If the image of the upper protomer in the figure is rotated 180° about this axis, it superposes almost exactly on the lower protomer. Because of this rotational axis of symmetry, the two protomers in the protein are almost indistinguishable.

Through the center of the indefinitely long helical polymer of actin, of which only a segment is presented in Figure 9–1B, runs a screw axis of symmetry. A **screw axis of symmetry** is a line, passing through a structure, about which the structure can be rotated by an angle less than 360° and along which the structure can be simultaneously translated to superpose upon itself. In the crystallographic molecular model of the actin filament, the screw axis of symmetry is a vertical line parallel to the plane of the page. If the image of any protomer is transposed in a left-handed sense by being rotated 166° in a clockwise direction around this axis while it is lifted 2.8 nm in a direction parallel to it, it superposes on the next protomer in the fiber. Because this operation can be repeated infinitely, all of the protomers are indistinguishable from each other.

A rotational axis of symmetry (Figure 9–1A) is necessarily 2-fold, 3-fold, 4-fold, and so forth. This requirement arises from the fact that as one of the images of a protomer is being rotated to superpose it on its neighbor, all of the images of its neighbors are also simultaneously being rotated (Figure 9–1A). When the rotation is completed, each of the images must superpose on its respective partner. This can be accomplished only if the protomers are arrayed about the axis at angles to each other that are integral quotients of 360° ($360/n$). The integer in the denominator defines the number of times the rotation can be accomplished before returning to the beginning. The rotational axis of symmetry within triose-phosphate isomerase is a 2-fold rotational axis of symmetry. At the end of two superpositions, the original locations are regained. For the rotation to superpose the images of all protomers simultaneously each time, no translation along the axis can occur.

Screw axes of symmetry are defined by a rotation and a translation, and they are designated as left-handed or right-handed. If the translation is not zero, a screw axis of symmetry produces a helical array. A spiral staircase is an example of a helical array. A designation of left-handed means that the helix produced by the screw axis of symmetry progresses clockwise as it rises, and a designation of right-handed means that it progresses counterclockwise as it rises. By the principle that the majority

rules, right-handed screws are given a positive sign and left-handed screws are given a negative sign. For example, one of the screw axes of symmetry in the actin polymer, the left-handed one, has a rotational angle of −166°. It is also possible to generate an actin polymer with a right-handed screw axis of +194° or two coaxial right-handed screw axes of +28°. A rotational axis of symmetry is simply a special case of a screw axis of symmetry where the translation is zero and the rotational angle is required to assume values that are integral quotients of 360°.

A screw axis of symmetry, other than the special case of a rotational axis of symmetry, in which translation cannot occur, does not require the angular steps to be integral quotients of 360°. As the image of one of the protomers is rotating and rising, the point of superposition need not occur at any particular angular disposition along the helix produced by the screw axis. It was the realization that a screw axis of symmetry could take on any arbitrary rotational angle that allowed Pauling to visualize the α helix while everyone else was trapped in the illusion[6,7] that a helix of amino acids would have to incorporate a rotational angle that was an integral quotient of 360°. The α helix, as it occurs in proteins,[8,9] has a screw axis of pseudosymmetry defined by a rotation of 99°.

There is, however, one requirement that limits the angles and translations permitted a screw axis of symmetry. This requirement is a corollary of the law that no two pieces of matter can occupy the same space at the same time; its operation is most readily observed by comparing the crystallographic molecular models of hexokinase (Figure 9–2) and actin (Figure 9–1). Through the center of the α_2 dimer of hexokinase (Figure 9–2) there runs a screw axis. It is a vertical line parallel to the plane of the page. If the image of the lower protomer (c10 Kin.2) is transposed on a right-handed screw by being rotated +156° around this axis as it is simultaneously lifted by 1.4 nm in a direction parallel to the axis, it superposes on the upper protomer.[10] Because of the translation and the rotation of less than 180°, the two subunits can always be distinguished from each other; one is the "down" protomer and one is the "up" protomer.

A helical polymer of actin can be constructed by placing one protomer upon an origin, properly oriented with respect to the axis of symmetry; transposing its image around the axis −166° and along the axis 2.8 nm; placing another protomer at this next location; and repeating this process infinitely. Suppose this were attempted with the subunit for hexokinase. Place one protomer upon an origin, properly oriented with respect to the axis. Move the image of this subunit 156° around the axis and 1.4 nm along the axis. Place another protomer at this next location. Move the image again 156° around the axis and 1.4 nm along the axis and try to place another protomer at this third location. This third subunit would have to overlap the first, and the principle of the impenetrability of matter would be violated. The problem here is that the translation of the screw axis in hexokinase is insufficient to move the image of the upper protomer far enough along to clear the protomer below it as the helical path completes the turn. Therefore, two protomers can be combined to form a dimer, but three or more cannot be combined to form a polymer. The parameters of the helix generating actin do not cause such collisions. Actin forms a helical polymer of indefinite length while hexokinase is limited to a dimer.

In theory, the same requirement restricting hexokinase to be a dimer also restricts triose-phosphate isomerase to being a dimer. Because no translation occurs along the axis in triose-phosphate isomerase, the third

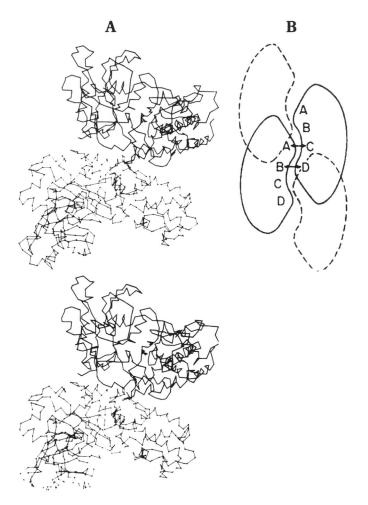

A   **B**

Figure 9–2: Crystallographic molecular model of hexokinase.[10] (A) Stereo drawing of the α-carbon backbone of the yeast hexokinase dimer. The thickness of the line segments connecting the positions of the α-carbons in the crystallographic molecular model (c10 Kin.2) differs between the two protomers. (B) Schematic arrangement of the array of intersubunit interactions in a closed structure built around a screw axis of symmetry where there is translation. The dashed outlines indicate where adjacent protomers would add but for the steric effect. Reprinted with permission from ref 10. Copyright 1976 Academic Press.

protomer would completely intersect the first if the game just played with hexokinase were repeated with triose-phosphate isomerase. It is the inescapable and obvious rules of this game that cause hexokinase and triose-phosphate isomerase to be closed structures[1] and actin to be an open structure. A **closed structure** is a structure built upon a rotational axis of symmetry or a screw axis of symmetry in which the addition of further protomers by that symmetry operation is precluded. An **open structure** is a structure built upon a screw axis of symmetry to which protomers can be infinitely added by that symmetry operation. If a multimeric protein has a closed structure, it is referred to as **oligomeric**; if it has an indefinite length, it is referred to as **polymeric**.

Each multimeric protein composed of identical protomers, even a fiber of indefinite length, can be considered to be the manifestation of a set of **interfaces** between its protomers. Each of these interfaces includes all of the points of contact that lie between two protomers in the structure, and each is formed by the association of two complementary **faces**, one from each of the two protomers. These faces are particular regions on the respective surfaces of the two associating protomers. When the structures of the protomers are identical to each other, each necessarily possesses, on its own surface, all of the unique faces forming the interfaces found in the complete molecule.

Following its biosynthesis, a polypeptide folds to form a structure capable of recognizing and being recognized by other folded polypeptides. If it is to combine with its twins to form a multimeric protein constructed from identical protomers, it must do so in a series of individual steps, and each step must involve the formation of an interface. The atomic contacts within that interface are as specific as the atomic contacts throughout the protein because the structure of that interface is locally indistinguishable from the structure of any other region within either of the subunits (Figure 9–1A). For the same reasons that a folded polypeptide assumes a precise and unique atomic structure, the interface between two subunits has a precise and unique atomic structure.

Any association between any two identical asymmetric objects always defines one and only one screw axis, angle of rotation, and translation that can superpose the image of one of the objects onto the other object. It follows that the precise atomic structure of any interface formed between two identical folded polypeptides automatically defines three parameters: a unique screw axis, an angle of rotation about that screw axis, and a translation along that screw axis that will superpose the image of one of the folded polypeptides upon the other. Either these three parameters are consistent with a fiber or they are consistent with a closed structure. If they are consistent with a fiber, the very fact that the one interface can form means that many others will subsequently form. A series of such interfaces is a fiber.

The interfaces, which are repeated throughout a

multimer composed of identical protomers, are the origin of the geometry of the final structure, and they in turn are created by evolution through natural selection. As such, their creation is determined by a completely random process. As time passes, variation in the identity of the amino acids on the surface of a given monomeric protein occurs. At some point, in some organism, in some species, a constellation of amino acids appears that permits a stable interface to form between two of these monomers. Natural selection operates at this point. It takes little imagination to realize that if the vast majority of multimeric proteins were not closed structures, the cell would rapidly fill with fibers and become a solid, inflexible object incapable of the pliability essential to life. The difficulties encountered with helical polymers of hemoglobin S in sickled erythrocytes dramatically illustrate this problem. If a monomeric protein were to sustain a series of mutations dictating that it combine with its twins in such a way that a fiber necessarily results, this set of mutations would probably be eliminated by natural selection. Mutations leading to closed structures, however, may be neutral initially, but oligomeric proteins have potentials denied to monomeric proteins, and the appearance of an oligomeric protein during evolution is the first step in the eventual exploitation of these potentials. Nevertheless, if the interface is compatible with a closed structure, it can initially be fixed by genetic drift as a neutral variation. With its fixation within that species, the protein has become an oligomer of identical protomers.

There remains one perplexing fact. As in the case of triose-phosphate isomerase (Figure 9–1A), the vast majority of multimeric proteins that have been examined conclusively are built around rotational axes of symmetry. Often these rotational axes of symmetry can be proven to be exact; they always seem to be exact. In fact, hexokinase (Figure 9–2) seems to be one of the few exceptions to this rule. If rotational axes of symmetry are severely restricted cases of screw axes of symmetry, and if a screw axis of symmetry is compatible with a closed structure, why are multimeric proteins always rotationally symmetric?

The main difference between a dimer like triose-phosphate isomerase and a dimer like hexokinase lies in the respective interfaces defining these structures. In a rotationally symmetric dimer such as triose-phosphate isomerase, individual interactions between the two protomers come in sets of identical pairs. The best way to see this is to consider the two loops on either side of the rotational axis of symmetry (Figure 9–1A). These loops interdigitate with portions of the opposite protomers. Consider a specific position in the segment of the sequence of the upper protomer within this loop. The amino acid at this location is making several contacts with amino acids from the lower protomer. Suppose a mutation occurred here that strengthened these interactions by a certain increment. Because the lower protomer was read from the same gene, at the same sequence

position in its loop the same favorable change would occur automatically so the increment for the whole interface would be twice that of each individual increment. The same argument could be made for each location in the interface.

In a protein built on a screw axis of symmetry, however, the amino acids at the same sequence positions in the two protomers never interact with the same amino acids from the other protomer across the interface (Figure 9–2B). A mutation, occurring anywhere in the interface, that adds an increment of stability to the dimer is not duplicated automatically across the axis of symmetry. It follows that as variation proceeds during evolution, the incremental changes that occur within an interface built around a 2-fold rotational axis of symmetry are amplified 2-fold relative to those that occur within each interface around a screw axis. This conclusion is valid whether one of these interfaces has already appeared during evolution or is merely incipient.

The formation of an interface between two identical monomeric proteins, which is the evolutionary event that precedes the appearance of a multimeric protein, is not an all or none phenomenon. The chemical reaction in question is

$$2\alpha \rightleftharpoons \alpha_2 \qquad (9\text{–}1)$$

Associated with this reaction is a change in standard free energy, and it is this change in standard free energy that determines the extent of the reaction. The numerical value of this change in free energy is dictated by the particular interactions that occur among the amino acids within the interface. The particular interactions that occur are the product of evolution by natural selection. Each explicit variation in one of these interactions adds or subtracts an increment of free energy to the overall change. If the increments are automatically doubled, overall decreases in the standard free energy change for the reaction proceed more rapidly over evolutionary time.

Incremental increases in the standard free energy change, however, are also doubled. Although rotationally symmetric dimers should appear more frequently, they should also disappear more frequently, unless they represent advantageous variations. Improvements in a certain protein are retained by natural selection, and their retention is unaffected by the frequency with which retrograde changes arise. Mutations turning the dimer back into a monomer would be eliminated if they were disadvantageous. It is possible that oligomerization of a protein has an immediate advantageous effect. If it did, the fact that most oligomeric proteins are built around 2-fold rotational axes of symmetry would be a reflection of the fact that such oligomers arise with a high frequency and of the fact that they are fixed by natural selection because oligomers are advantageous.

Because events were discussed in the opposite order of the normal progression, a summary of the historical and logical sequence seems appropriate. As a result of genetic variation among the individuals in a given species, a constellation of amino acids appears on the surface of a monomeric protein within one of those individuals. The constellation causes molecules of that previously monomeric protein to associate with each other. This association necessarily creates an interface. This interface necessarily forces the two protomers it brings together to be related to each other by a particular and unique screw axis of symmetry. The association between the two or more protomers created by this screw axis of symmetry is tested by natural selection. Occasionally, a fiber, which necessarily results from a screw axis that is not closed, is advantageous and is retained. Most of the time, however, the survivors of natural selection are closed oligomeric proteins whose interfaces dictate rotational axes of symmetry.

Suggested Reading

Monod, J., Wyman, J., & Changeux, J.P. (1965) On the Nature of Allosteric Transitions: A Plausible Model, *J. Mol. Biol. 12*, 88–118.

Problem 9–1: Using as your three examples triose-phosphate isomerase, dimeric yeast hexokinase, and filamentous actin, discuss the topics of rotational axes of symmetry, screw axes, open structures, closed structures, interfaces, and helical polymers.

Space Groups

In addition to the translational operations among the unit cells, screw and rotational axes of symmetry also occur within crystals of proteins. They are the fundamental operations that define the space groups. A **space group** of identical enantiomeric objects is a potentially infinite array of those objects, the positions and orientations of which are related to each other by screw and rotational axes of symmetry and translational operations. For reasons mainly associated with the phenomenon of diffraction, a unit cell has been defined solely in terms of translation. If molecules of protein were always stacked in crystals so that each of them had exactly the same rotational orientation, all unit cells would be of the same type and each would contain only one molecule of protein. Packing the same molecules in different rotational orientations, however, is not forbidden, and strangely shaped enantiomeric objects, such as proteins, usually are packed with greater efficiency by placement in different rotational orientations (Figure 9–3). If these rotational orientations are to be compatible with the infinite regular array that is a crystal, they must be related by particular symmetry operations. Dismissing mirror symmetry, which is irrelevant to enantiomeric objects, one is left with axial symmetry.

In the simple *P*2 lattice represented in Figure 9–3, the array of unit cells portrayed represents one of the layers in a three-dimensional crystal. Within each unit cell, the two identical enantiomeric objects are related by a 2-fold rotational axis of symmetry perpendicular to the page. Because of this type of rotational disposition, which arises only because of the increased efficiency of packing, the unit cell ends up containing two of the enantiomeric objects rather than one. Each of the 2-fold rotational axes in the center of each of these unit cells is itself a rotational axis of symmetry for the entire array if it is assumed that the array is infinitely propagated in three dimensions. There are also three sets of 2-fold rotational axes of symmetry between the unit cells. It is a feature of axes of symmetry in crystals that more than one set appears at a time.

In crystals, as opposed to individual oligomers and polymers, screw axes of symmetry, as well as rotational axes of symmetry, are required to have rotational angles that are integral quotients of 360°. This arises from the fact that as the image of one of the asymmetric objects from which the crystal is composed is rotating and rising around the screw axis, the images of every other asymmetric object in the lattice are rotating and rising around the same axis. At the completion of the operation all of the images in the lattice must superpose on identical partners. This can only occur if the rotations are 2-fold, 3-fold, 4-fold, or 6-fold. No other rotational multiplicities are compatible with an infinite array of asymmetric objects. Space groups never have 5-fold rotational or screw axes of symmetry because an infinite repetitive array of pentagonal objects cannot be formed. Any translational distance, compatible with an unclosed screw axis of symmetry, is compatible with an infinite array. Techni-

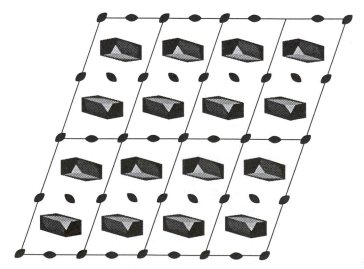

Figure 9–3: Packing of asymmetric objects in an array in which they alternately assume two different rotational orientations. The symbol ● indicates a 2-fold rotational axis of symmetry perpendicular to the page.

cally no crystal is infinite, but it always has the potential to be so, and this potential is all that matters.

The unit cell is the smallest unit that, by only simple translational movements, can be stacked to create and completely fill the crystal. The **crystallographic asymmetric unit** is the smallest unit that can be distributed both by translation and by the axes of symmetry to create and completely fill the whole crystal. The difference is that while the unit cells are related by only translational operations, crystallographic asymmetric units are related by rotational or screw axes of symmetry or both, as well as by translation. The lattice in Figure 9–3 is created by distributing the single asymmetric unit about rotational axes of symmetry and by translational operations.

To this point, either ribbon diagrams or the α-carbon backbones of actual molecules of protein have been sufficient to illustrate the role of a crystallographic asymmetric unit. Using such complicated structures in the following discussion to illustrate properties of symmetry would be confusing and hard to follow. Instead, it is easier to use an asymmetric unit familiar to all chemists, as well as one that is easy to draw, namely, a small enantiomeric molecule. Lactic acid is one of the smallest enantiomeric molecules, and it has been chosen to illustrate the symmetry relationships.

Although there are four sets of rotational axes in the array of Figure 9–3, it is considered to be a simple 2-fold array, designated by the single integer 2, because all of its axes are parallel to each other and no set can exist without all of the others. The presence of a rotational axis of symmetry in a space group is indicated by an unadorned integer: 2, 3, 4, or 6. The presence of a screw axis of symmetry is indicated by an integer, 2, 3, 4, or 6, followed by another integer in subscript. The main integer, n, is the integer by which 360° is divided to obtain the rotational angle of the steps. The integer in subscript, m, determines the fraction, m/n, of the unit cell over which the translation occurs with each rotational step. The translation is always right-handed to the rotation. You should convince yourself that a 3_1 screw is right-handed and a 3_2 screw is left-handed.

The designation of the space group in a crystalline array takes the form of a capital letter[*] followed by several numbers. An example would be $P2_12_12_1$, which would mean a primitive lattice made of parallelepipeds with sets of orthogonal 2-fold screw axes of symmetry (Figure 9–4). The arrangements of these axes, in the absence of a diagram like that of Figure 9–4, can be learned only by consulting a table.[11]

The crystallographic asymmetric unit has a volume that is always an integral quotient of the unit cell and is thus equal to or smaller than the unit cell. An integral number of asymmetric units, but not necessarily of intact asymmetric units, composes a unit cell. For example, one whole, two halves, four fourths, and eight eighths gathered from 15 different asymmetric units can together create a unit cell containing a total of four asymmetric units (Figure 9–4).

The space group imposes certain constraints on the structure of the unit cell. For example, in the $P2$ space group that produces the lattice of Figure 9–3, the 2-fold rotational axes of symmetry must be normal of the plane of Figure 9–3 or the superposition cannot occur. Therefore, each unpictured asymmetric unit above and below the plane of the page in the lattice must be perpendicularly aligned with one of the asymmetric units in the plane of the page. This requires that the two angles of the fundamental unit cell aligning the axis out of the page be precisely 90°. A lattice where two of the angles must be 90° is referred to as monoclinic. In the $P2_12_12_1$ space group displayed in Figure 9–4, the necessarily orthogonal screw axes force the unit cell to be a rectangular parallelepiped and the lattice to be orthorhombic. Each space group other than the most primitive, $P1$, which lacks axial symmetry entirely, enforces one or more of the angles of the unit cell to be 90° or 120° or enforces one or more of the axes of the unit cell to be the same length or enforces requirements on both angles and lengths. These are not coincidental identities but required identities. They are dictated by the symmetry operations and are thus exact quantities. If the angle in the monoclinic crystal of Figure 9–2 were not exactly 90°, the crystal would be filled with fractures and would not be a crystal.

Reality seems to take place in Cartesian space, and there are only 71 ways that an infinite array of identical crystallographic asymmetric units, each containing protein, can be arranged in Cartesian space to produce a crystal. Each of these is a different space group. Every crystal of protein has its crystallographic asymmetric units arrayed in one of these space groups. The space group is established as the crystal nucleates and grows in the dish; it cannot be dictated by the investigator. He or she can only try to change the conditions of crystallization in the hope that another space group will be generated by the process. This is often attempted because the identity of the space group determines how difficult it will be to calculate a map of electron density.

Three of the most common space groups in which proteins crystallize and from which maps of electron density can be determined with little difficulty are $P2_12_12_1$ (Figure 9–4), $P3_121$ (Figure 9–5), and $C2$ (Figure 9–6). There are no rotational axes of symmetry, only screw axes of symmetry, among the asymmetric units in the $P2_12_12_1$ space group. In the $C2$ and $P3_121$ space groups, pairs of asymmetric units are disposed around 2-fold rotational axes that are inherent in the space group. In both of these

---

[*]The capital letter refers to the particular relationship between the underlying lattice and the unit cell for the space group of interest. These relationships are primitive (P), C-face center (C), A-face centered (A), B-face centered (B), all-face centered (F), body centered (R), or hexagonally centered (H).

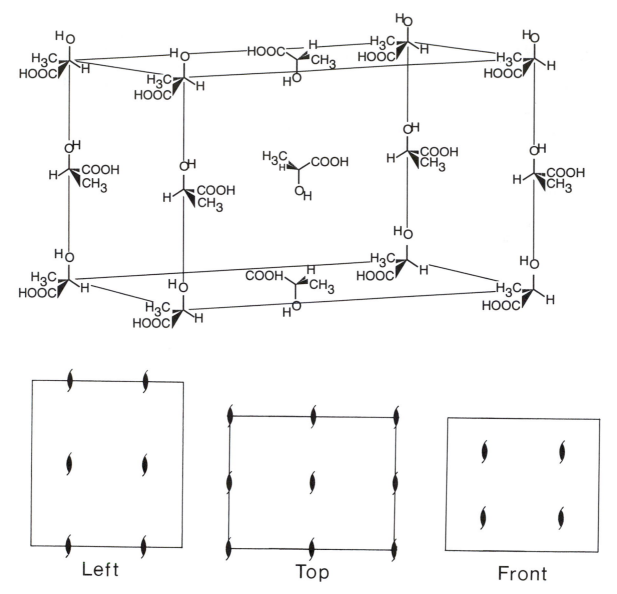

Figure 9–4: Molecules of lactic acid arranged in a $P2_12_12_1$ lattice. The $P2_12_12_1$ lattice is orthorhombic with all three crystallographic angles equal to 90°, and the unit cell is rectangular. This results from the fact that there are three sets of 2-fold screw axes, each set is necessarily orthogonal to the others, and each set of screw axes is parallel to one of the crystallographic axes. There are four screw axes for each unit cell in each of the three sets. The molecule of lactic acid is the crystallographic asymmetric unit, and from any one molecule of lactic acid the entire lattice can be created by performing the operations of 2-fold screw symmetry in the three orthogonal directions. Vertical 2-fold screw axes of symmetry coincide with each vertical column of lactic acids, of which there are two for each unit cell. The other two vertical screw axes of symmetry for each unit cell coincide with vertical lines in the center of each vertical face. When rotated 180° about any one of these vertical axes while simultaneously rising half of a unit cell, the lattice superposes on itself. The four parallel horizontal screw axes of symmetry passing from back right to front left for each unit cell are in the top face, bottom face, and halfway between the top face and bottom face, one-quarter and three-quarters of the way across the unit cell. The four parallel horizontal screw axes of symmetry passing from back left to front right through each unit cell are at one-quarter and three-quarters of the distance between top and bottom face and one-quarter and three-quarters of the distance between the side faces. These dispositions of the screw axes of symmetry are presented diagrammatically below the figure. The symbols in these three diagrams, which are of the left side, the top, and the right front of the unit cell, respectively, are the 2-fold screw axes of symmetry seen end-on.

A

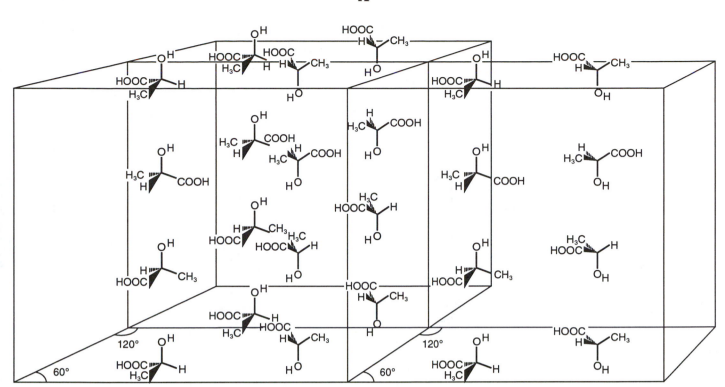

B

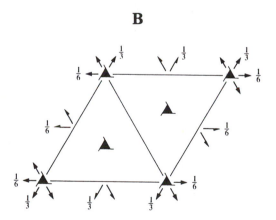

Figure 9–5: Molecules of lactic acid arranged in a $P3_121$ lattice. (A) The $P3_121$ lattice is trigonal with one crystallographic angle of 120°. The angle of 120° is in the bottom faces of the three unit cells that are drawn. The $P3_121$ lattice has a set of right-handed 3-fold screw axes of symmetry. In the drawing, two of the 3-fold screw axes of symmetry of each unit cell coincide with the vertical columns of lactic acid molecules. The third 3-fold screw axis of symmetry of each unit cell coincides with the vertical edge of each unit cell. Because the crystallographic angle is 120°, counterclockwise rotation of 120° about this latter axis while simultaneously rising one-third of a unit cell causes the three upward columns of lactic acid around this axis to superpose on themselves and the three downward columns of lactic acid around this axis to superpose on themselves. Normal to each of the 3-fold screw axes of symmetry running along each vertical edge of the unit cells and intersecting perpendicularly these vertical axes are 2-fold rotational axes of symmetry arrayed at 60° angles to each other. In the bottom face of the unit cell the first 2-fold rotational axis of symmetry is the diagonal bisecting the 120° angle. Each successive 2-fold rotational axis of symmetry is 60° counterclockwise to the one below it and one-sixth of the distance up the axis of the unit cell. (B) This diagram[11] indicates the arrangement of all of the axes of symmetry in the $P3_121$ space group. The view is looking down onto the face of the unit cell containing the 120° angle. The flared triangles denote 3-fold screw axes normal to the page. The double-headed arrows are 2-fold rotational axes of symmetry, and the single-headed arrows are 2-fold screw axes of symmetry parallel to the plane of the page. The fractions indicate how far up the vertical edge or vertical face of the unit cell the axes are found. Reprinted with permission from ref 11. Copyright 1983 D. Reidel.

latter cases, each and every asymmetric unit is related to at least one of its adjacent twins by a particular 2-fold rotational axis of symmetry. This unique relationship establishes a particular pair of twins. This pair is exceptional because the whole lattice can be divided into an array of these pairs, and in each of these pairs the orientation of the two twins to each other is the same. In the $C2$ space group pictured in Figure 9–6A, such a pair of twins includes any two lactic acid molecules that have their carboxylic acid functionalities opposite the hydroxyls of their neighbors. Every lactic acid molecule in the crystal participates in one and only one such symmetric relationship. In the $P3_121$ lattice, the pairs of twins are established by 2-fold rotational axes of symmetry

A

B

C

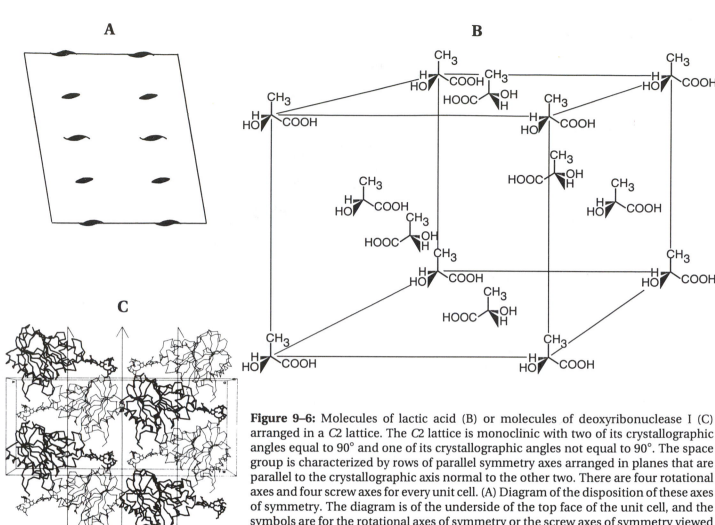

Figure 9–6: Molecules of lactic acid (B) or molecules of deoxyribonuclease I (C) arranged in a *C*2 lattice. The *C*2 lattice is monoclinic with two of its crystallographic angles equal to 90° and one of its crystallographic angles not equal to 90°. The space group is characterized by rows of parallel symmetry axes arranged in planes that are parallel to the crystallographic axis normal to the other two. There are four rotational axes and four screw axes for every unit cell. (A) Diagram of the disposition of these axes of symmetry. The diagram is of the underside of the top face of the unit cell, and the symbols are for the rotational axes of symmetry or the screw axes of symmetry viewed end-on. (B) Each molecule of lactic acid is the crystallographic asymmetric unit. The crystallographic angle that is not equal to 90° is in the bottom face of the unit cell and the axes of symmetry are all vertical. The planes containing the alternating screw and rotational axes of symmetry are parallel to the side faces of the unit cell and pass through the front face at intervals of one-quarter and three-quarters of the unit cell. The screw axes are in the front face, the back face, and halfway between the front and the back face. The rotational axes of symmetry are at one-quarter and three-quarters of the distance between the front and back face. (C) In this drawing of deoxyribonuclease I packed in its crystal,[8] all of the axes of symmetry run horizontally, rather than vertically, and the unit cell has been shifted so that the planes containing the axes of symmetry, which are parallel to the plane of the page, now lie in the front face, the back face, and halfway between the front and the back face of the unit cell. The half arrows indicate 2-fold screw axes, and the full arrow indicates a 2-fold rotational axis. The three marked axes are in the plane passing through the center of the unit cell. The crystallographic angle in the side faces is 91.4°, and the other two crystallographic angles are 90°. Reprinted with permission from ref 8. Copyright 1986 Academic Press.

relating lactic acid molecules that contact each other through the faces defined by the hydroxyl, the carboxyl, and the hydrogen.

The particular 2-fold rotational axes of symmetry connecting these pairs of rotationally symmetric twins are crystallographic axes. A **crystallographic axis of symmetry** is one of the axes of symmetry that defines the space group of the crystal. It exists only when the protein is in

a crystal. The 2-fold rotational axis of symmetry running through the center of triose-phosphate isomerase and connecting its twin subunits is a **molecular axis of symmetry**. It exists whether the protein is in solution or in a crystal. Crystallographic axes and molecular axes arise under different circumstances. The molecular axes of symmetry are created as the oligomeric protein assembles in the cell, and the crystallographic axes of symmetry are created as the crystal grows in a dish. The two sets of axes of symmetry are independent properties. They can, but they are never required to, coincide. If a molecular rotational axis of symmetry coincides with a crystallographic rotational axis of symmetry, it can be stated unequivocally that the molecular axis is an exact rotational axis of symmetry because a crystallographic axis of symmetry is necessarily an exact rotational axis of symmetry. If it were not exact, crystal growth could not continue because the small equal deviations between the actual rotational operation and an exact rotational operation would add up across the crystal and eventually produce an interruption in the lattice.

The results from several crystallographic experiments serve to illustrate the distinction between crystal symmetry and molecular symmetry and the consequences of their coincidence. Triose-phosphate isomerase crystallizes in the space group $P2_12_12_1$. The crystallographic asymmetric unit is the α_2 dimer, and the 2-fold molecular rotational axis of symmetry within the dimer (Figure 9–1A) cannot coincide with a crystallographic rotational axis of symmetry because there is none.[2] Glyceraldehyde-3-phosphate dehydrogenase, a tetramer composed of four identical polypeptides, crystallizes in the space group $P2_12_12_1$, and again the asymmetric unit is necessarily the entire tetramer; however, the spacing of the heavy metal atoms in the isomorphous replacement[12] and the final map of electron density[13] were both consistent with the tetramer being a dimer of dimers constructed around three apparently exact 2-fold molecular rotational axes of symmetry orthogonal to each other. Glutathione peroxidase, a tetramer composed of four identical polypeptides, crystallizes in the space group $C2$, and the asymmetric unit is the α_2 dimer.[14] This means that one of the molecular axes of symmetry coincides with a crystallographic axis of symmetry and that the two dimers composing the tetramer must be related to each other by an exact 2-fold molecular rotational axis of symmetry. L-Lactate dehydrogenase, a tetramer composed of four identical polypeptides, crystallizes in the space group $F422$, and one single folded polypeptide is the asymmetric unit.[15] This means that the tetramer is a dimer of dimers constructed around three exact 2-fold molecular rotational axes of symmetry that coincide with the three precisely orthogonal crystallographic rotational axes of symmetry. Phosphorylase b, a dimer composed of two identical polypeptides, crystallizes in the space group $P4_32_12_1$, and the asymmetric unit is one folded polypeptide.[16] This means that the dimer is constructed upon an exact 2-fold molecular rotational axis of symmetry. Alcohol dehydrogenase, a dimer composed of two identical polypeptide chains, crystallizes in the space group $C22_12_1$, and the molecular 2-fold rotational axis of symmetry coincides with one of the crystallographic 2-fold rotational axes of symmetry.[17]

Each of the last three proteins has all of its subunits arranged with a perfect rotational symmetry that can be conclusively proven. The coincidences of the molecular axes of symmetry and the crystallographic axes of symmetry that permitted these proofs, however, were by chance, and it seems reasonable to assume that the first three proteins, that just happened to crystallize in space groups incompatible with one or more of their molecular rotational axes of symmetry, are no less symmetric than the last three. While it is sometimes possible to deduce the whole symmetry of the protein from the crystallographic symmetry, the absence of the appropriate coincidences permitting this deduction does not mean that the molecule of protein lacks symmetry.

Suggested Reading

Hahn, T., Ed. (1983) *International Tables for Crystallography, Vol. A: Space-Group Symmetry*, D. Reidel, Dordrecht, The Netherlands.

Problem 9–2: In the $P3_121$ space group portrayed in Figure 9–4, every lactic acid molecule is related to one of its neighbors by the same 2-fold rotational axis of symmetry. Draw two lactic acid molecules arranged around that specific rotational axis of symmetry.

Oligomeric Proteins

It has already been noted that the great majority (85%)[18] of oligomeric proteins are constructed from the same folded polypeptide present in multiple copies. The frequencies with which these various homooligomers occur (Table 9–1) can be rationalized by examining each of the operations of rotational symmetry that govern their construction.

Half of all homooligomers are **dimers** (Table 9–1). As far as can be determined from the crystallographic structures currently available, all of them, with the educational exception of hexokinase, are rotationally symmetric (c7 Kin.5). In addition to triose-phosphate isomerase (Figure 9–1A), citrate (*si*) synthase serves as an example of

Table 9–1: Frequency of Homooligomers

no. of polypeptides	percent observed[a]
2	50
4	35
3	5
6	10
8	3
10	1
12	2

[a]The table of oligomeric stoichiometries published by Darnell and Klotz[18] was used to calculate these frequencies. Because this is a selected and incomplete list, some numbers were rounded to the nearest 5%.

such an arrangement (Figure 9–7).[19] It crystallizes in the space group $P4_12_12$, and in the crystal the molecular 2-fold rotational axis of symmetry coincides with a crystallographic 2-fold rotational axis of symmetry. This means that the dimer is precisely symmetric.[19] This can be verified by a close examination of the figure. Note the particularly intimate contact along the interface as the secondary structures from each protomer mimic each other across the axis. Note that the detailed contacts in the interface are no different from those in any other region in one of the protomers except for this mimicry.

The two lactic acid molecules illustrate the features of a 2-fold rotational axis of symmetry. The methyl–hydroxyl contact is duplicated identically on either side of the axis. Both protomers direct the same face to the viewer. The structure is unquestionably closed. Every position in the common sequence of the two polypeptides whose amino acid is enclosed within the interface from one of the protomers is also enclosed from the other.

In many proteins, the polypeptides of which contain internal duplications, the two superposable domains are related by a 2-fold rotational axis of pseudosymmetry (c5 Kin.3). Thiosulfate sulfurtransferase is an example of such an arrangement (Figure 9–8) as are the arabinose binding protein (Figure 4–19) and chymotrypsinogen.[20] Presumably, these are the remains of the 2-fold rotational axes of symmetry in the dimers of two identical protomers that were the ancestors of each of these proteins before the gene duplication occurred. The duplicated polypeptide incorporated the original rotational axis of symmetry, but following the duplication the two halves began to evolve separately and diverge.

Not all folded polypeptides with internally repeating domains display such rotational axes of pseudosymmetry, and there is no direct correspondence between internally repeating domains and rotational axes of symmetry. Immunoglobulin G (Figure 7–12) contains two different polypeptides, one long and one short, that are both composed of internally repeating domains. Within neither of the polypeptides are any of the adjacent internally repeating domains related by a rotational axis of symmetry.

One-third of all soluble homooligomers are **tetramers** (Table 9–1). Those freely soluble tetramers whose structures have been solved by crystallography are dimers of dimers built around three orthogonal 2-fold rotational axes of symmetry. This is the molecular symmetry 222. An example of this arrangement occurs in phosphoglycerate mutase (Figure 9–9).[22] The protein crystallizes in the space group $C2$.[23] One of the molecular 2-fold rotational axes coincides with a crystallographic 2-fold rotational axis of symmetry and is exact. The dimer found within the crystallographic asymmetric unit appears also to be precisely symmetric, at least at the resolution of the map of electron density.[22]

A **molecular asymmetric unit** is that component of the structure of an oligomeric molecule which, when submitted to the appropriate symmetry operations, creates the entire structure. In the $(\alpha_2)_2$ tetramer of phosphoglycerate mutase (Figure 9–9), the molecular asymmetric unit is, by inspection, the single folded polypeptide. If the one polypeptide is positioned in space, its image is rotated 180° around one of the 2-fold rotational axes of symmetry, and another identical folded polypeptide is placed where the image of the first has thus been positioned, one of the three dimers in the tetramer is created. If the image of this dimer is then rotated about another of the 2-fold rotational axes of symmetry and another iden-

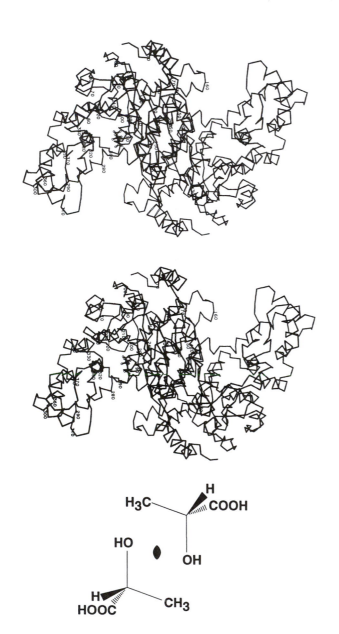

Figure 9–7: Two-fold rotational symmetry. The α-carbon skeleton of citrate (*si*) synthase,[19] drawn from the crystallographic molecular model of this dimeric protein, is presented in the upper panel. Two molecules of lactic acid, arranged with the same symmetry as the protomers of the protein, are drawn in the lower panel. Upper panel reprinted with permission from ref 19. Copyright 1982 Academic Press.

tical dimer is placed where the image of the first has been positioned, then the entire tetramer is created.

Technically, a crystal contains all of the molecular axes of symmetry even if they do not coincide with the crystallographic axes of symmetry, but those molecular axes that do not coincide with the crystallographic axes cannot be proven to exist. If by inspection it is decided that they do exist, then the molecular asymmetric unit is also the crystallographic asymmetric unit and can be operated upon to create the entire crystal. The first symmetry operations performed upon the molecular asymmetric unit to create the crystal are the rotations about the noncrystallographic molecular axes of symmetry, which result in the creation of the crystallographic asymmetric unit from the smaller molecular asymmetric unit. For example, with phosphoglycerate mutase, rotation about one of the two noncrystallographic molecular rotational axes of symmetry creates the crystallographic asymmetric unit of the $C2$ space group.

The molecular asymmetric unit in a molecule of protein is enantiomeric. The protomers of a protein, a domain within the protomer of a protein, or any other structural entity of a protein should not be represented by any object that is achiral. The habit of using circles, spheres, squares, cubes, or triangles for this purpose obscures the relationships that are required to occur among chiral objects and implies that relationships forbidden to chiral objects can occur.

The fact that the symmetry illustrated by the lactic acid molecules is also the symmetry of phosphoglycerate mutase can be verified by rotating and superposing each of the protomers and pairs of protomers in the protein with the same symmetry operations noted for the arrangement of lactic acid molecules. The same points noted in the description of the evolution of a rotationally symmetric dimer from an ancestral monomeric protein are of equal validity in describing the evolution of a rotationally symmetric dimer of dimers from an ancestral dimeric protein. Also, for the same reasons, a dimer of dimers is a closed structure.

In a dimer of dimers, three different pairs of interfaces can be formed. The methyl–hydroxyl interface above the plane of the page in Figure 9–9 is repeated below the plane of the page. The carboxylic–methyl interface at the top of the figure is repeated at the bottom, and the carboxyl–hydroxyl interface on the left is repeated on the right of the figure. Each of these different pairs of interfaces can be found in the structure of the protein as well. Each of the three pairs of interfaces contains within itself one of the rotational axes of symmetry.

Because each of the three pairs of interfaces is completely different, they each have different strengths, reflected in their free energies of association. In most dimers of dimers, because of steric effects, one of these pairs of interfaces is not formed or is only slightly formed. In the case of phosphoglycerate mutase, it is the methyl–hydroxyl interface, where only the tips of the subunits

touch. This interface is obviously the weakest. One of the other two pairs of interfaces must be the strongest, and if there is any dissociation of the tetramer into dimers, it is this strongest pair of interfaces that will be retained in the two dimers. Assume that the two hydroxyl–carboxyl interfaces in phosphoglycerate mutase are the strongest, and imagine the tetramer splitting at the methyl–carboxyl interfaces to form two dimers.

Hemoglobin is an honorary homotetramer. It is built from two different polypeptides, α and β, each present in two copies to provide the four protomers. The two polypeptides are homologous in sequence[24] and their tertiary structures (c3 Kin.3) are superposable.[25] In the heterotetramer, the four protomers of hemoglobin occupy the same arrangement as the four protomers of phosphoglycerate mutase. The α–α interface and the β–β interface are the rudimentary pair, so the structure can be represented as $(\alpha\beta)_2$. Designate each of the protomers

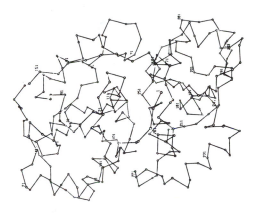

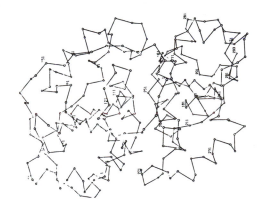

Figure 9–8: α-Carbon skeleton drawn from the crystallographic molecular model of thiosulfate sulfurtransferase.[21] The amino acid sequence of the protein does not contain significant internal homology (< 12% identity when the two halves are aligned) but the internal duplication is manifest in the crystallographic molecular model of the folded polypeptide. The two domains are related to each other by a 2-fold rotational axis of pseudosymmetry normal to the plane of the page running through the center of the macromolecule. Reprinted with permission from ref 21. Copyright 1978 Academic Press.

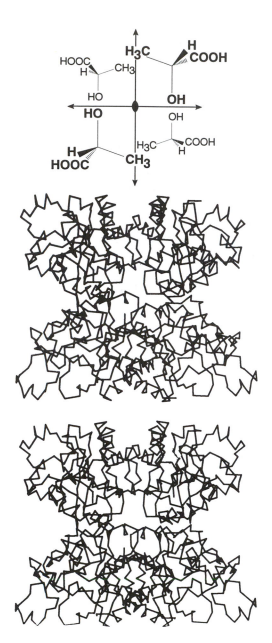

Figure 9–9: A dimer of dimers with 222 molecular symmetry. The α-carbon skeleton of tetrameric phosphoglycerate mutase, drawn from the crystallographic molecular model,[22] is presented in the lower panel. Four molecules of lactic acid, arranged with the same symmetry as the protomers of the protein, are presented in the upper panel. The double-headed arrow designates a 2-fold rotational axis of symmetry in the plane of the page. The symbol ● designates a 2-fold rotational axis of symmetry normal to the plane of the page. The lactic acid molecules in the upper right and lower left are above the plane of the page, and those at the upper left and lower right are below the plane of the page. Lower panel reprinted with permission from *Nature*, ref 22. Copyright 1974 Macmillan Magazines Ltd.

two species, dog and human, respectively.[27] In this experiment, hemoglobins from two species were used to permit electrophoretic separation of the hybrids. It was shown that the hybrids $\alpha'\beta$ and $\alpha\beta'$ could be artificially formed and easily separated electrophoretically, from each other and from the parent dimers, but these hybrids also could not be shuffled by the reaction of Equation 9–2. Therefore one of the two different pairs of interfaces between α and β subunits in the hemoglobin tetramer[25] must be much stronger than the other pair of interfaces. This series of observations results from the fact that each of the three different pairs of interfaces in a tetrameric protein is unique.

Probably the most peculiar of the tetramers yet solved crystallographically is catalase (Figure 9–10).[28] In this structure the amino termini of the constituent polypeptides are threaded through eyes formed by loops of polypeptide in the middle of the sequences of symmetrically related neighbors. These rotationally symmetric knots tie pairs of protomers together as intimate dimers. Two of these knotted dimers associate back-to-back to form the tetramer.

Trimers and **hexamers** are both oligomers built around 3-fold rotational axes of symmetry. The frequency with which they arise during evolution is 6 times lower than the frequency with which dimers and tetramers, oligomers built around 2-fold rotational axes of symmetry, arise (Table 9–1). In fact, at one time it was thought, erroneously, that 3-fold rotational axes of symmetry did not occur in proteins. The bacteriochlorophyll α-protein is a trimer (Figure 9–11).[29] It crystallizes in the $P6_3$ space group with only one folded polypeptide in the asymmetric unit, and the 3-fold rotational axis of symmetry found in this protein must be exact. Each interface of the three that produce the trimer is the same.

It is by considering the problem of assembling a trimer that the reason for the scarcity of trimers becomes apparent. It should be recalled that the interface is the feature that evolves and not the oligomer. A dimer built around a 2-fold rotational axis of symmetry is held together by an interface centered on the rotational axis of symmetry. The axis divides the complete interface into two identical halves (Figure 9–1A). Each half is the formal equivalent to

in Figure 9–9 as either α or β and convince yourself that there are two different kinds of interfaces between α and β protomers in hemoglobin.

The oxygenated form of hemoglobin in solution participates in the dissociation reaction

$$(\alpha\beta)_2 \overset{K_d}{\rightleftharpoons} 2\alpha\beta \qquad (9\text{–}2)$$

and the value of the dissociation constant K_d, where

$$K_d = \frac{[\alpha\beta]^2}{[(\alpha\beta)_2]} \qquad (9\text{–}3)$$

is 1.0×10^{-6} M.[26] In this reaction, the same interface always remains in the dimer. This last conclusion arises from the fact that no hybrid dimers of the type $\alpha'\beta$ or $\alpha\beta'$ are formed under conditions where the reaction of Equation 9–2 is rapidly interconverting tetramers and dimers in a mixture of two hemoglobins, $(\alpha\beta)_2$ and $(\alpha'\beta')_2$, from

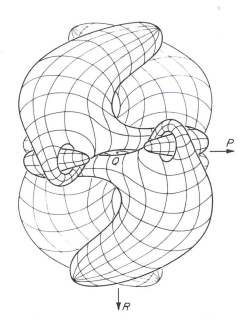

Figure 9–10: Diagrammatic representation of the knotlike structure joining a pair of the four protomers in catalase. The diagram[28] was made from the crystallographic molecular model of this protein by drawing solids, with smooth surfaces, roughly the shape of each subunit. Reprinted with permission from ref 28. Copyright 1986 Academic Press.

one of the three identical interfaces distributed around the 3-fold rotational axis of symmetry in a trimer (Figure 9–11). Because termolecular collisions rarely occur, the assembly of an oligomeric protein must proceed through a series of bimolecular steps. A bimolecular collision producing a dimer automatically involves the formation of the two halves of its interface simultaneously and incorporates the free energies of formation of both halves into the immediate product. The first step in the assembly of a trimer, however, is the collision of two monomers to form only one of its three interfaces. This first interface, standing alone, must exist long enough or form often enough for the third protomer to complete the ring. The incremental decreases of free energy associated with favorable mutations are not automatically doubled during the evolution of this initial, and intermediate, interface in the assembly of the trimer as they are in the evolution of the initial, and final, interface of the dimer. Therefore, trimers should appear less frequently than dimers during evolution.

In the trimer's favor, however, is the fact that, as with the two halves of the interface in a dimer, its three identical interfaces can evolve simultaneously, a fact that increases the incremental decrease in free energy change in the overall formation of the complete oligomer for each favorable mutation. If the 3-fold axis of symmetry in the bacteriochlorophyll α-protein were not a rotational axis of symmetry but a closed screw axis, one of the interfaces could not be equivalent to the other two because the ring could not be completed. It is most likely that the peculiar interface of the three would not fit

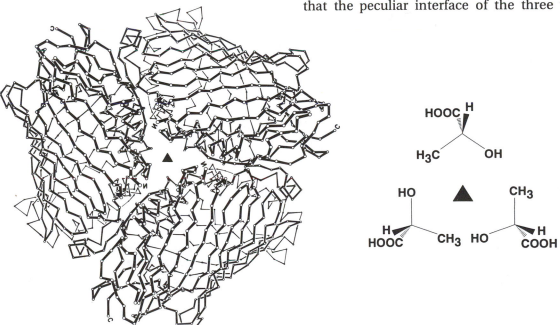

Figure 9–11: Three-fold rotational symmetry. The α-carbon skeleton of bacteriochlorophyll α-protein, drawn from the crystallographic molecular model of this trimeric protein,[29] is presented in the left panel. Three lactic acid molecules, arranged with the same symmetry, are presented in the right panel. The solid triangle in the center of each arrangement is the symbol used to indicate a 3-fold rotational axis of symmetry normal to the plane of the page. Left panel reprinted with permission from ref 29. Copyright 1979 Academic Press.

A

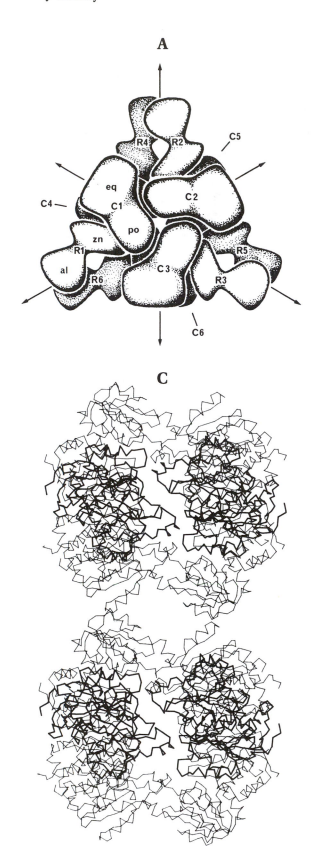

B

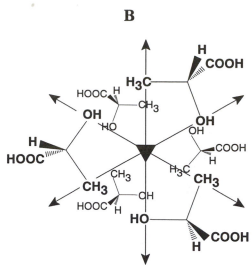

Figure 9–12: A dimer of trimers with 32 molecular symmetry. (A) The drawing of the protomers of aspartate transcarbamylase arranged around the molecular axes of symmetry was made from the crystallographic molecular model of this hexameric protein.[30] The six folded c polypeptides are designated by capital letters C, and the six folded r polypeptides, by capital letters R. The folded c polypeptides can be divided into two structural domains designated eq and po, and the folded r polypeptides can be divided into two structural domains designated al and zn. The view is down the 3-fold rotational axis of symmetry (white triangle in the center of the diagram), and the three 2-fold rotational axes of symmetry are indicated by double-headed ar-rows. (B) Six molecules of lactic acid arranged with the same 32 symmetry as the drawing in (A). The molecules of lactic acid at 1, 5, and 9 o'clock are above the lane of the page; and those at 3, 7, and 11 o'clock, below the plane of the page. The three 2-fold rotational axes are in the plane of the page. (C) α-Carbon skeletons of only the six folded catalytic polypeptides in the crystallographic molecular model, viewed down one of the 2-fold rotational axes of symmetry. Panels A and C reprinted with permission from ref 30. Copyright 1985 National Academy of Sciences.

C

together properly because it would be formed from the same two faces required to associate in a different way from the way they associated at the other two interfaces. Whether a screw trimer, analogous to the screw dimer of hexokinase, does exist is not known, but none has been observed. Certainly it would be a weaker structure than a rotationally symmetric trimer because of the one misaligned interface.

The only hexameric homooligomers that have been observed are dimers of trimers. The structure that illustrates this arrangement most clearly is aspartate transcarbamylase.[30] This protein is a homooligomer even though it is made from two different polypeptides. The protomer of the homooligomer is formed from one folded c polypeptide and one folded r polypeptide, and aspartate transcarbamylase is a hexamer (Figure 9–12). The protein crystallizes in the space group *P*321, and the crystallographic 3-fold rotational axis of symmetry coincides with the molecular 3-fold rotational axis of symmetry.[30]

A dimer of trimers contains a central 3-fold rotational axis of symmetry and three 2-fold rotational axes of symmetry perpendicular to it and at 60° to each other. The view down one of the 2-fold rotational axes of symmetry (Figure 9–12C) illustrates the superpositions

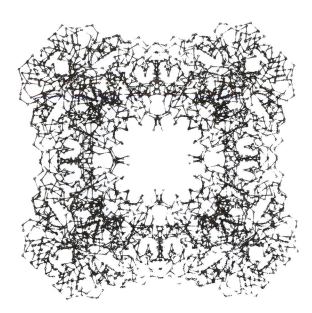

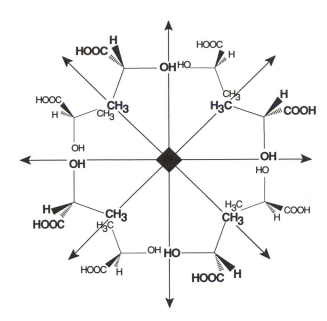

Figure 9–13: A dimer of tetramers with 42 molecular symmetry. The α-carbon skeleton of the octamer of (*S*)-2-hydroxy-acid oxidase, drawn from the crystallographic molecular model of the enzyme,[33] is presented in the left panel. The structure is viewed along the molecular 4-fold rotational axis of symmetry with two of the four 2-fold rotational axes of symmetry in vertical and horizontal orientations, respectively. A drawing of eight molecules of lactic acid arrayed with the same symmetry is presented in the right panel. The molecules at 2, 5, 8, and 11 o'clock are above the plane of the page, and the other four are below the plane of the page. The four 2-fold rotational axes of symmetry are in the plane of the page, and the filled square denotes a 4-fold rotational axis of symmetry normal to the plane of the page. Left panel reprinted with permission from ref 33. Copyright 1985 National Academy of Sciences.

that are intrinsic to this arrangement and makes the point that every folded polypeptide is equivalent to and indistinguishable from every other in the hexamer.

In a dimer of trimers there are three different sets of interfaces. First, there are the six identical interfaces, arrayed around the 3-fold axis of symmetry, that create the trimers. In aspartate transcarbamylase, these are between folded c polypeptides. Second, there are the three identical interfaces at one end of each 2-fold rotational axis of symmetry. In aspartate transcarbamylase, these are between two folded r polypeptides, at 12, 4, and 8 o'clock. Third, there are the three identical interfaces at the other end of each 2-fold rotational axis of symmetry. In aspartate transcarbamylase, these are between two folded c polypeptides, at 2, 6, and 10 o'clock.

Almost no freely soluble oligomeric proteins that are tetramers, pentamers, or hexamers containing 4-, 5-, or 6-fold rotational axes of symmetry, respectively, have ever been observed. All freely soluble tetramers seem to be dimers of dimers. Immunoglobulin M remains unique as a pentamer,[31] inasmuch as the only other freely soluble protein reported as a pentamer of identical polypeptides[18] turned out to be a hexameric dimer of trimers.[32] All hexamers seem to be dimers of trimers. There are, however, octamers, decamers, and dodecamers, containing 4-, 5-, and 6-fold rotational axes of symmetry and constructed as dimers of tetramers, pentamers, and hexamers, respectively. These arrangements are il-

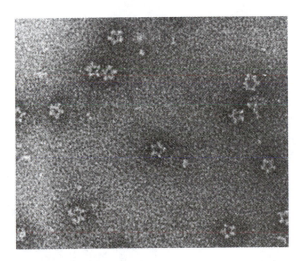

Figure 9–14: A dimer of pentamers with 52 molecular symmetry. Molecules of arginine decarboxylase[34] were attached to a carbon film on an electron microscopic grid and negatively stained with sodium phosphotungstate. They were viewed in an electron microscope at a magnification of 240000×. The 5-fold rotational axis of symmetry produces star-shaped images. Reprinted with permission from ref 34. Copyright 1968 *Journal of Biological Chemistry.*

lustrated by (*S*)-2-hydroxy-acid oxidase (Figure 9–13),[33] arginine decarboxylase (Figure 9–14),[34] and glutamate–ammonia ligase (Figure 9–15).[35] Each of these structures is a simple extension of the dimer of trimers and requires no further discussion of its symmetry. (*S*)-2-Hydroxy-acid oxidase crystallizes in the space group *I*422. As a result, all of the molecular rotational axes of symmetry coincide with crystallographic rotational axes of symmetry, and the molecular symmetry must be exact.

The reason that dimers of each of these larger rings are seen when undimerized rings are not may be due to the inability of a single ring to withstand twisting forces. A dimer, containing two rings, should be more rigid than one ring alone. In any case octamers, decamers, and dodecamers are very rare (Table 9–1).

Suggested Reading

Buehner, M., Ford, G.C., Moras, D., Olsen, K.W., & Rossmann, M.G. (1974) Three-Dimensional Structure of D-Glyceralde-hyde-3-phosphate Dehydrogenase, *J. Mol. Biol. 90*, 25–49.

Problem 9–3: Make a xerographic copy of this figure. Reprinted with permission from ref 14. Copyright 1983 *European Journal of Biochemistry.*

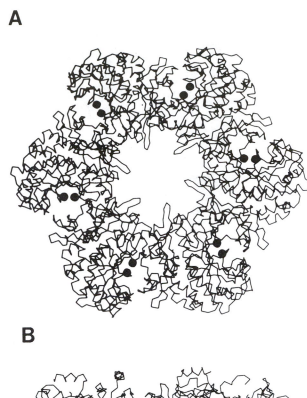

A

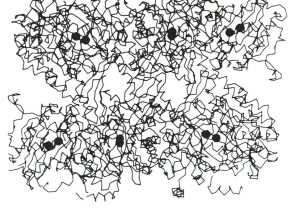

B

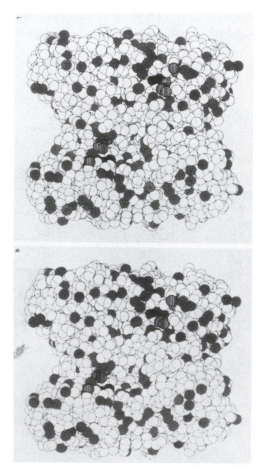

Figure 9–15: A dimer of hexamers with 62 molecular symmetry. (A) The α-carbon skeletons of the six protomers of glutamate-ammonia ligase that form one of the two hexameric rings of the dodecameric enzyme,[35] viewed down the 6-fold rotational axis of symmetry. (B) The α-carbon skeletons of the six protomers in the front of the dodecamer when it is viewed down one of its six 2-fold rotational axes of symmetry normal to the 6-fold rotational axis of symmetry. Each of these partial views was drawn from the crystallographic molecular model of the entire dodecamer.[36] Reprinted with permission from ref 36. Copyright 1989 *Journal of Biological Chemistry.*

Using a ruler where necessary, draw all of the rotational axes of symmetry on one of the two members of the stereo pairs. Use the abbreviations for rotational axes found in the *International Tables for Crystallography.*[11]

Isometric Oligomeric Proteins

There are a number of oligomeric proteins in which more than 12 protomers assemble together to produce a structure that, rather than being a circular disc (Figure 9–15), is an isometric solid, roughly spherical or cubic in shape. Crick and Watson[37] pointed out that there are only three ways to arrange identical asymmetric objects, such as folded polypeptides, to form an isometric object (Table 9–2). Both octahedral and icosahedral oligomeric proteins have been observed; whether or not the tetrahedral dodecameric proteins predicted in this list actually exist is not yet known. The **icosahedral oligomeric proteins** are represented by viral protein coats, and several of these structures have been solved crystallographically.

The **protein coat** of a virus is a thick layer of protein that serves the purpose of enclosing the viral nucleic acid. This nucleic acid encodes the genetic information necessary for the virus to control the host parasitically and divert the purpose of the host from its own growth and replication to the growth and replication of the virus. These requirements can be satisfied only by a fairly large molecule of nucleic acid, and it all must fit within the coat so that it can be protected from the environment. A viral protein coat is made from a large number of identical or homologous protomers. In a spherical virus, these protomers are arranged about the icosahedral rotational axes of symmetry to produce spherical shells that can enclose the nucleic acid. Because protein molecules are usually constructed of polypeptides 100–800 amino acids in length, only an icosahedral arrangement of subunits can produce a shell almost large enough to enclose sufficient nucleic acid to support viral infection, but it is not quite large enough.

In a strict icosahedral arrangement, 60 identical asymmetric objects must be arranged about 31 rotational axes of symmetry (Table 9–2) to produce the shell. Each rotational axis of symmetry passes through the center of the structure. In every icosahedral arrangement, the relative angular dispositions of these axes is always the same, and any of the icosahedrally symmetric structures can be used to define these dispositions. The distribution of these rotational axes can be appreciated by examining regular icosahedral polygons (c11 Kin.1) such as the icosahedron, the dodecahedron, and the rhombic triacontahedron (Figure 9–16). The icosahedron and the dodecahedron are constructed from plane figures, 20 triangles or 12 pentagons, respectively, that each have a central, precise rotational axis of symmetry and cannot themselves be asymmetric objects as are proteins. Both the intersection of these two figures (Figure 9–16C) and the rhombic triacontahedron (Figure 9–16D) can be constructed from identical asymmetric units, 60 puckered quadrilaterals or 60 triangles, respectively, that do not have any internal rotational axes of symmetry, are formally asymmetric objects, and could be molecules of protein.

Each of these latter two polygons can be divided into

Table 9–2: Isometric Arrangements of Identical Asymmetric Objects[37]

symmetry	rotational axes of symmetry[a]	number of asymmetric units	platonic solid with these symmetry elements
tetrahedral	three 2-fold	12	tetrahedron
	four 3-fold		
octahedral	six 2-fold	24	cube
	four 3-fold		octahedron
	three 4-fold		
icosahedral	15 2-fold	60	dodecahedron
	10 3-fold		icosahedron
	six 5-fold		

[a]A rotational axis of symmetry is a line passing through the center of the oligomeric structure. Because it is a line, it extends in both directions from the center. As a result, each axis of symmetry passes out of the oligomeric structure at two opposite points, and at each of these two points there is a symmetric arrangement of asymmetric objects on the surface of the structure.

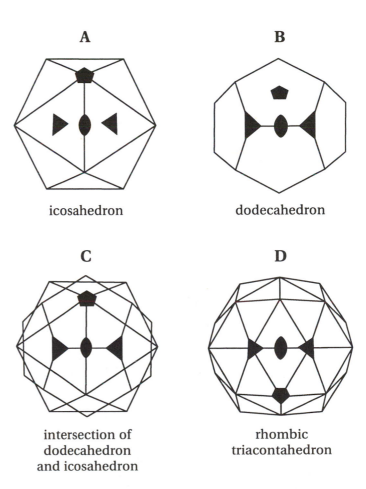

A icosahedron

B dodecahedron

C intersection of dodecahedron and icosahedron

D rhombic triacontahedron

Figure 9–16: Several regular solids that have icosahedral symmetry. (A) The icosahedron itself, (B) the dodecahedron, (C) the solid formed by the intersection of an icosahedron and a dodecahedron, and (D) the rhombic triacontahedron.

60 equivalent radial sectors by connecting the center of the solid with a series of lines to each of its vertices. In each case, another strictly similar polygon can be drawn farther out on each of these axially directed lines, and the spaces between the inner and outer polygons would define 60 identical tiles. If there were a protein the shape of such a tile, 60 copies of it could be assembled to form an icosahedral shell. Proteins, however, are not tiles and do not obediently and exclusively occupy such a sector in a biological object such as a virus.

It is not any regular polygon that defines a viral protein coat built from 60 protomers but the rotational axes of symmetry. Regardless of how intertwined or encroaching the protomers in such a viral protein coat become, the

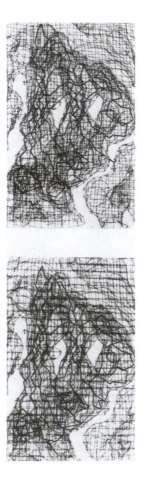

Figure 9–17: Protomer of an icosahedral viral coat. The map of electron density for the coat of satellite tobacco necrosis virus[39] is presented in a view from the outside of the virus looking in. The area included contains one complete protomer and segments of protomers surrounding it. The α-carbon skeleton of the crystallographic molecular model of the central protomer is placed within its electron density. The exact 5-, 3-, and 2-fold rotational axes of symmetry are indicated by the three respective integers (c11 Kin.3). Reprinted with permission from ref 39. Copyright 1988 Academic Press.

number and relative angular dispositions of the 31 rotational axes of symmetry that dictate the positions of those protomers are permanent features of the structure. If 60 identical folded polypeptides are arranged by 31 such exact rotational axes of symmetry and their shapes are so constructed as to mesh symmetrically at each of the boundaries among themselves, they will necessarily form a tight icosahedral shell.

Viewed in this light, the interfaces among the protomers are again the fundamental determinants of the multimeric structure. First, at one vertex of the individual protomer there must be a 5-fold rotational axis of symmetry created by five interfaces, each formed from two complementary faces that on the same subunit are at a 72° angle to each other. This interface creating the 5-fold rotational axis of symmetry is repeated five times around a pentameric ring of protomers (Figure 9–16C or 9–16D). At each of the five equivalent edges of this pentagon, which do not have to coincide with the boundary of only one protomer (Figure 9–16C), there must be an interface connecting the pentagon to another identical pentagon in the usual way that two identical proteins are associated, which is around a 2-fold axis of symmetry (Figure 9–7). This particular 2-fold axis of symmetry, however, must incline the two pentagons with respect to each other so that their respective 5-fold rotational axes of symmetry both intersect the 2-fold axis of symmetry and form the angle required to exist between the 5-fold axes in an icosahedron, which is 63° 32′. If two such interfaces, the one defining the pentamer and the other this angle of 63° 32′, are built into a protomer of the protein, 60 such protomers will automatically assemble into an icosahedral shell (c11 Kin.2).

The structures of a number of icosahedral viruses have been solved crystallographically at a resolution high enough to trace the folded polypeptide chains of the individual protomers accurately. At this resolution the arrangement of the protomers in the crystallographic molecular model is unambiguous, and the icosahedral symmetry, with its three precisely arranged sets of rotational axes of symmetry, is universally observed. All of these viruses are spherical. The spherical shells are fairly thick and resemble heavy but hollow rubber balls. The nucleic acid in the interior is featureless and no information is available concerning its structure.

The smallest virus whose structure has been solved is satellite tobacco necrosis virus. Its protein coat is composed of 60 identical protomers, each one a folded polypeptide (c11 Kin.6) 195 amino acids in length. In the low-resolution model of a complete protomer within the coat (Figure 9–17)[38,39] the shape that permits it to fit into the shell is most clearly seen (c11 Kin.3). One of the 5-fold rotational axes of symmetry of the complete viral coat (see Figure 9–16D) is at the left vertex of the protomer as it is displayed in Figure 9–17; one of the 3-fold axes is at the right vertex; and one of the 2-fold axes is about two-thirds of the way from the 3-fold axis to the top vertex of

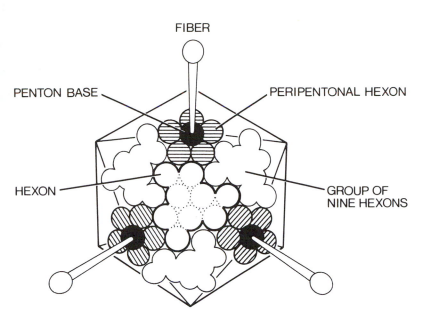

FIBER

PENTON BASE

PERIPENTONAL HEXON

HEXON

GROUP OF
NINE HEXONS

Figure 9–18: A schematic view of the adenovirus capsid showing the major external proteins arranged on the surface of an icosahedron.[40,42] The structure residing at the 12 intersections of the six 5-fold rotational axes with the coat are referred to as pentons. The individual protomers filling in the space between the pentons are referred to as hexons. They can be distinguished as peripentonal hexons or group-of-nine hexons on the basis of their location in the structure. All of the icosahedral axes of symmetry are retained in the structure. The view presented is down one of the 3-fold axes of symmetry and each 2-fold axis of symmetry is halfway between two 5-fold axes and halfway between two 3-fold axes of symmetry. Reprinted with permission from ref 40. Copyright 1984 John Wiley and Sons.

the protomer. The axes are at icosahedral angles to each other. In the assembled shell, the interfaces among the subunits are all tight, and the 60 protomers together completely enclose the interior.

The problem with satellite tobacco necrosis virus is that it is too small. The name gives it away. It is a satellite virus, which means it must infect the host at the same time that a larger virus does because it does not have enough nucleic acid to produce an infection independently. In order to be large enough to include sufficient DNA to produce an infection unaided, the size of the viral protein coat must be made larger. There are two ways viral protein coats are made larger, neither of which is to make a larger protomer.

One way is to slide the 12 pentamers outward on their 5-fold rotational axes of symmetry and fill in the empty space with the same or another protein. This is the strategy employed to construct the coat of adenovirus (Figure 9–18).[40] The protein covering the vacated space is a sheet of nine protomers known as hexons. Each sheet is centered on one of the icosahedral 3-fold rotational axes of symmetry. A different, but similar, strategy is used in the viral protein coats of the polyoma viruses.[41] As with the adenovirus, each 5-fold rotational axis of symmetry is occupied by a pentameric ring of identical protomers, and these rings occupy positions on the 5-fold rotational axes of symmetry so as to leave about as much vacant space among them as that among the pentons of an adenovirus. The vacant space, however, is paved with homopentamers identical to the pentamers on the 5-fold rotational axes. For each pentamer on a 5-fold rotational axis of symmetry, there are five that fill in the vacant space. These interstitial pentamers are arranged precisely around the 3-fold rotational axes of symmetry and the 2-fold rotational axes of icosahedral symmetry, but the strong interfaces among these interstitial pentamers

occur at only the 2-fold rotational axis of symmetry and local 2-fold and 3-fold rotational axes of pseudosymmetry within the intervening space. Using this strategy of sliding the pentamers out along the 5-fold axis of symmetry and filling in the intervening space, viral protein coats can, in theory, be made any size, consistent with the 2-fold and 3-fold icosahedral rotational axes of symmetry, but once the structures in the intervening space become too large, serious problems of assembly should arise.

The other way to make a larger icosahedral virus is to take advantage of **pseudosymmetry**. It was noted by Caspar and Klug[43] that the triangle from which the rhombic triacontahedron (Figure 9–16D) is constructed, although formally an asymmetric object because two vertices lie at 3-fold rotational axes of symmetry, one vertex lies at a 5-fold rotational axis of symmetry, and only one of its three edges lies on a 2-fold rotational axis of symmetry, was nevertheless an equilateral triangle and locally symmetric. They proposed that if three identical folded polypeptides, related to each other by a local 3-fold rotational axis of pseudosymmetry, could fill this triangle, it would have three times more area than if only one folded polypeptide filled it, and that the shell could then contain 3.5 times more nucleic acid. This would only require (Figure 9–16D) that the folded polypeptide at one vertex of the triangle be arrayed around a 5-fold rotational axis of symmetry (72° for each step) while the folded polypeptides at the other two vertices be arrayed around 6-fold rotational axes of pseudosymmetry (60° for each step). Tomato bushy stunt virus,[44] the structure of which has been determined crystallographically, has such an arrangement of polypeptides (Figure 9–19). Each of its 60 identical protomers is formed from three folded polypeptides, subunits A, B, and C, each of an identical sequence 386 amino acids in length, the tertiary structures of which, when they are in the viral protein coat, are

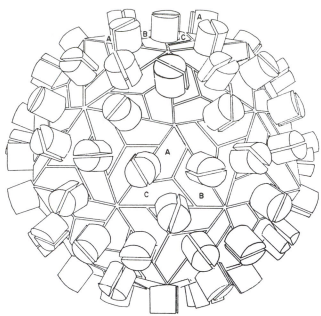

Figure 9–19: Arrangement of the 180 subunits in the coat of the tomato bushy stunt virus.[44] Each tile is a single folded polypeptide, and all of the polypeptides are identical to each other in amino acid sequence. Three folded polypeptides, designated A, B, and C, are arrayed around a local 3-fold axis of pseudo-symmetry (c11 Kin.5) to produce the trimeric protomer of the icosahedral array. The vertices occupied by the A subunit are the icosahedral 5-fold rotational axes of symmetry and the axes occupied by the B and C subunits are the icosahedral 3-fold axes of symmetry but also local 6-fold axes of pseudosymmetry. Two-fold rotational axes of symmetry relate C subunits, and local 2-fold rotational axes of pseudosymmetry relate A and B subunits. Each subunit has a protrusion that runs up its associated 2-fold rotational axis. The diagram was adapted from the crystallographic molecular model of this viral coat.[44] Reprinted with permission from ref 44. Copyright 1983 Academic Press.

very similar and superposable, although not exactly. The small differences in their respective conformations (c11 Kin.4) permit each of them to play its required role. For example, only subunit C is adjacent to the true 2-fold rotational axes of symmetry. Subunits B and C must alternate around the exact 3-fold and pseudo-6-fold rotational axes of symmetry, while subunit A is at the exact 5-fold rotational axes of symmetry. These different roles produce alterations in the structures of these folded polypeptides that are most obvious at the interfaces among the homotrimers; it is here that the strain of requiring the same protein to adopt different orientations is the strongest.

The packing at these interfaces has been described in detail for the viral protein coat of southern bean mosaic virus,[45] which is closely related to tomato bushy stunt virus. The three identical folded polypeptides, each 260 amino acids in length, are arranged around a 3-fold rotational axis of pseudosymmetry (Figure 9–20A) to create the homotrimeric protomer (c11 Kin.5). The A subunits are arranged around the exact icosahedral 5-fold rotational axes of symmetry (Figures 9–16, 9–19, and 19–20B). Each of the B and C subunits uses the same vertex to form the 6-fold rotational axis of pseudosymmetry (Figures 9–19 and 9–20C) as that used by the A subunit to form the 5-fold rotational axis of symmetry. Careful inspection of Figure 9–20B,C shows that the two unique defining interfaces are similar but significantly adjusted to accommodate the differences in the angular requirements around the two axes.

A similar but even more extreme adjustment must be made at the interfaces among the homopentameric protomers of polyoma viruses that fill in the space between the homopentamers on the 5-fold icosahedral axes of symmetry.[41] Significantly different interfaces, on the same surface of each unique folded polypeptide, must be

made around the icosahedral 2-fold rotational axis of symmetry and the local 2-fold and 3-fold rotational axes of pseudosymmetry in the space between the homopentameric subunits on the icosahedral 5-fold rotational axes of symmetry. This requires that three different interfaces be accommodated by the same surface of a folded polypeptide in the different distinct locations, even though all of the polypeptides are identical in sequence. This is accomplished by forming the interfaces with flexible strands of polypeptide rather than interdigitated secondary structure.

The next step in complexity in the pseudosymmetric icosahedral viral coats is an obvious one. As the three folded polypeptides A, B, and C are not in identical environments, they need not be identical in amino acid sequence. In some icosahedral viruses, gene triplication of the nucleic acid encoding the amino acid sequence of the coat protein has occurred to produce three genes. The comoviruses, of which cowpea mosaic virus and beanpod mottle virus are examples, are an interesting intermediate case in this process.[46] In these viruses, two of the subunits in the heterotrimeric protomer are internally repeating domains on the same polypeptide. This suggests that the sequence of events was a gene duplication that gave rise to two genes producing two separate polypeptides followed by a gene duplication of one of these genes producing a single polypeptide with two internally repeating domains. In the viruses with three separate proteins, the next step in the evolutionary process was to split the latter duplicated gene so that it then produced two smaller polypeptides, each containing one domain of the ancestral polypeptide. Following the triplication, each of the three genes evolved independently to produce three polypeptides, or one polypeptide and two domains, each of different sequence and each presumably incorporating changes rendering it more successful at occupying its respective position in the viral protein coat. Four icosahedral viruses that have accomplished the complete evolutionary transition, poliovirus,[47] rhinovirus,[48] Mengo virus,[49] and foot-and-mouth disease virus[50] have had their structures solved crystallographically. These viruses are spheres (Figure 9–21A,B) whose surfaces are paved with heterotrimeric protomers in icosahedral array (Figure 9–21C). That the three different folded polypeptides forming these latter four viral protein coats have arisen from a gene triplication follows from the fact that, in each case, the three polypeptides folded in their native conformations are superposable.[47-50] If this is a definitive correspondence, it demonstrates that the ancestors of each of these viruses were constructed from 180 polypeptides of identical sequence and had the pseudosymmetry of tomato bushy stunt virus or southern bean mosaic virus.

It has been shown that the single folded polypeptide composing one protomer of satellite tobacco necrosis virus is superposable on any one of the folded polypeptides forming the trimeric protomer of either tomato bushy stunt virus or southern bean mosaic virus.[51] It has also been pointed out that the packing of the folded polypeptides of either southern bean mosaic virus or tomato bushy stunt virus is similar to the packing of the protomers of satellite tobacco necrosis virus. When the 3-fold rotational axis of symmetry in satellite tobacco necrosis virus is aligned with the local 3-fold rotational axis of pseudosymmetry within one of the trimeric protomers of one of the other viruses (Figure 9–22),[52] the three 5-fold rotational axes of symmetry in satellite tobacco necrosis virus coincide with one of the 5-fold rotational axes of symmetry and two of the 6-fold rotational axes of pseudosymmetry in the other virus. It has also been observed[53] that when the first 61 amino acids of the coat protein of southern bean mosaic virus are removed, the remainder of the folded polypeptide can assemble to produce a hollow icosahedral shell containing only 60 folded polypeptides instead of the usual 180. In this new oligomeric protein, the original arrangements around the 5-fold rotational axes of symmetry have been retained and the 3-fold rotational axes of pseudosymmetry in the centers of the protomers of the orignal structure have become the true 3-fold rotational axes of symmetry in the new structure (as depicted in Figure 9–22). This latter result demonstrates that these two icosahedral struc-

Figure 9–20: α-Carbon skeletons of the folded polypeptides composing the coat of southern bean mosaic virus drawn from the crystallographic molecular model of the entire viral coat.[45] The entire crystallographic molecular model has 180 identical polypeptides all folded into the same tertiary structure. There are three different environments, however, in which the folded polypeptides are found that can be designated A, B, and C (Figure 9–19). (A) Two folded polypeptides arrayed about the 3-fold rotational axis of pseudosymmetry at position A (at 12 o'clock) and position B (at 4 o'clock), respectively. (B) Two folded polypeptides arrayed about the 5-fold rotational axis of symmetry both in positions A. (C, next page) Two folded polypeptides arrayed about the 3-fold rotational axis of symmetry or local 6-fold rotational axis of pseudosymmetry at position B (2 o'clock) and position C (at 4 o'clock). Reprinted with permission from ref 45. Copyright 1983 Academic Press.

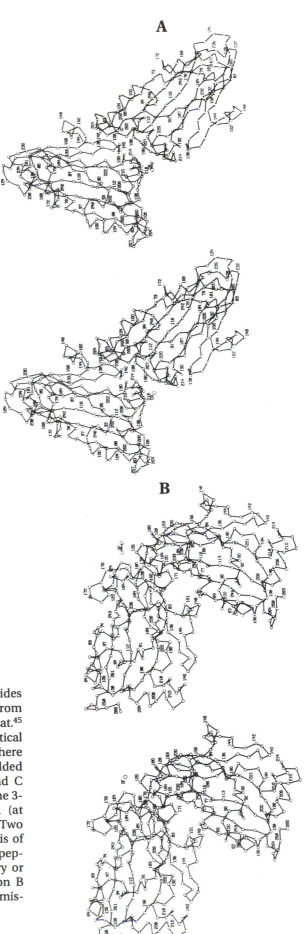

Figure 9–20C

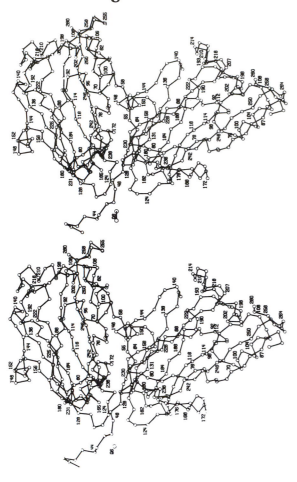

Figure 9–21

B   A

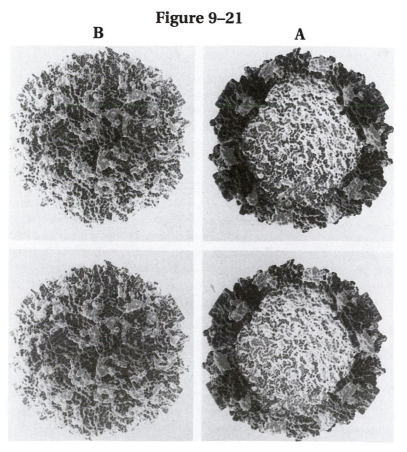

Figure 9–21: Space-filling models of the folded polypeptides assembled into the icosahedral coat of poliovirus as drawn from the crystallographic molecular model of this icosahedral oligomeric protein.[47] (A) A view into the central cavity of the viral coat in which the viral RNA is packed. (B) A view of the surface of the viral coat in which the atoms contributed by each of the three different types of folded polypeptides, VP1, VP2, and VP3, have been given different shades of grey. Panels A and B reprinted with permission from ref 47. Copyright 1985 American Association for the Advancement of Science. (C) A diagrammatic representation of the surface of an icosahedral capsid[49] in the same orientation as B to illustrate the distribution of the various folded polypeptides around the icosahedral axes of symmetry. Panel C reprinted with permission from ref 49. Copyright 1987 American Association for the Advancement of Science.

C

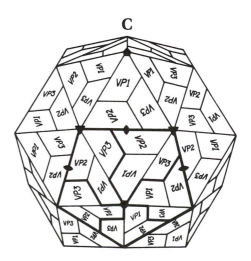

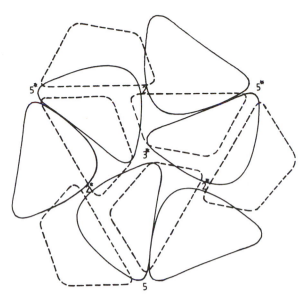

Figure 9–22: Comparison of the packing of the respective folded polypeptides in the coat of satellite tobacco necrosis virus (unbroken lines) and the coat of tomato bushy stunt virus or southern bean mosaic virus (broken lines).[52] The former viral coat has 60 folded polypeptides in exact icosahedral symmetry; the latter has 180 folded polypeptides in icosahedral pseudo-symmetry. The exact 3-fold rotational axis of symmetry of satellite tobacco necrosis virus is aligned with the 3-fold rotational axis of pseudosymmetry in the center of the protomer of the other two. This alignment causes the exact 5-fold rotational axes of symmetry at the bottom of the diagram to coincide and concurrently causes the two other exact 5-fold rotational axes of symmetry at the other two vertices in satellite tobacco necrosis virus to coincide with the two 6-fold rotational axes of pseudosymmetry at the other two vertices of the protomer from tomato bushy stunt virus or southern bean mosaic virus. Reprinted with permission from ref 52. Copyright 1982 Academic Press.

tures, the pseudosymmetric with 180 polypeptides and the symmetric with 60 polypeptides, are readily interconverted. All of these results taken together clearly indicate that these three viruses, tomato bushy stunt, southern bean mosaic, and satellite tobacco necrosis, share a common ancestor.[51] The fact that satellite tobacco necrosis virus is a parasite on another virus and the fact that a viral protein coat of its size cannot carry enough nucleic acid suggests that its ancestor originally had a larger protein coat built from 60 homotrimeric protomers.

The symmetric and pseudosymmetric icosahedral viral coats discussed so far, those of satellite tobacco necrosis, tomato bushy stunt, southern bean mosaic, cowpea mosaic, beanpod mottle, Mengo, and foot-and-mouth disease viruses, poliovirus, and rhinovirus, are all those of viruses carrying single-stranded RNA that has positive copies of the viral messenger RNAs. These viruses infect eukaryotic cells, both animals and plants. From the descriptions that have just been given of the crystallographic molecular models of the protein coats of these viruses, it follows that the ancestors of all of these eukary-

otic positive-strand RNA viruses had icosahedral protein coats constructed from 60 homotrimeric protomers, each of which was constructed from three polypeptides of identical sequence arranged about a local 3-fold rotational axis of pseudosymmetry. The most remarkable discovery, however, is that all of the folded polypeptides (c11 Kin.6) comprising all of these eukaryotic positive-strand RNA viruses, whether their hosts are plants or animals, are superposable.[47–49,51] This suggests that all of these viruses share one common ancestor. This remarkable possibility, if it is true, would not be hard to explain. Three unique faces, each creating an independent set of repeating interfaces among the 180 identical folded polypeptides, would have had to evolve on the surface of the same monomer. The three unique interfaces produced by these three unique faces are the interface responsible for the 3-fold rotational axis of pseudosymmetry within the protomer (Figures 9–11 and 9–20A), the interface responsible for the 5-fold rotational axis of symmetry at one vertex of each subunit, and the interface responsible for the 2-fold rotational axis of symmetry that orients the two pentagons it connects. Five-fold rotational axes of symmetry are rare products of evolution (Table 9–1) and the 2-fold rotational axis of symmetry, though common, would be required to be located at a particular disposition relative to the 5-fold axis of symmetry. These constraints are probably not so rigid as they seem. If the angles are close enough, the significant stability of such an edifice, each of whose associations strengthens all of the others, could force the interfaces to rearrange sufficiently to accommodate the icosahedral arrangement. This cooperativity in the construction of the shell, which should resemble the cooperativity among the supports of a building, also permits the interfaces to be weaker than interfaces that must stand alone. Nevertheless, that such a monomer, with such faces, has arisen rarely during evolution would not be a surprising fact.

The eukaryotic positive-strand RNA viruses, however, are not the only family of icosahedral viruses. Adenoviruses and polyoma viruses represent other families, unrelated to the eukaryotic positive-strand RNA viruses, but they are constructed with quite different strategies. There is, however, a positive-strand RNA virus, MS2, that infects bacterial hosts. It also has a pseudosymmetric icosahedral protein coat composed of 60 homotrimers, but the folded polypeptides do not appear to be related to the folded polypeptides of the eukaryotic positive-strand RNA viruses.[54]

There is an oligomeric enzyme, heavy riboflavin synthase,[55] that has 60 protomers, each a single polypeptide 145 amino acids in length, that are arranged with icosahedral symmetry. This protein resembles a possible intermediate species in the evolution of icosahedral viral protein coats, and its existence suggests that viral protein coats could have evolved from an ancestral protein that originally had a completely different function.

There is one enzyme that serves as an example of an

octahedral oligomeric protein containing 24 subunits, each a single folded polypeptide and each occupying an equivalent position in the oligomer. This enzyme now appears biologically as three independent species: dihydrolipoamide acetyltransferase, dihydrolipoamide acyltransferase,[56] and dihydrolipoamide succinyltransferase. The polypeptides composing dihydrolipoamide acetyltransferase and dihydrolipoamide succinyltransferase, respectively, are homologous in amino acid sequence (30% identity between the aligned sequences of the polypeptides of the two enzymes in the bacterium *Escherichia coli*),[57] and both consist of a carboxy-terminal domain 300 amino acids in length, showing the highest degree of similarity (35% identity), and one to three internally repeating and detachable amino-terminal domains bearing lipoic acid. Therefore, these two enzymes share a common ancestor as well as a similar enzymatic reaction and a common multimeric structure. The similarities between dihydrolipoamide succinyltransferase or dihydrolipoamide acetyltransferase and the less well characterized dihydrolipoamide acyltransferase[56] leave little doubt that this third enzyme also shares the same common ancestor.

The X-ray diffraction pattern produced by crystals of dihydrolipoamide acetyltransferase is that of the space group *F*432, and the unit cell is only large enough to contain one molecule of enzyme.[58] Therefore, the three 4-fold, four 3-fold, and six 2-fold crystallographic rotational axes of symmetry of this octahedral space group must coincide with three 4-fold, four 3-fold, and six 2-fold molecular rotational axes of symmetry in the octahedral protein, and these axes of symmetry must be exact.

The geometry of this arrangement can be appreciated by considering the intersection of a cube and an octahedron (Figure 9–23A). While both a cube and an octahedron have faces that have internal rotational axes of symmetry and that could not be occupied by a single folded polypeptide, the 24 puckered quadrilaterals formed by their intersection are rotationally asymmetric structures within each of whose boundaries a single folded polypeptide could reside. Again, however, it should be emphasized that it is only the arrangement of the 24 identical protomers around the 13 rotational axes of symmetry that is the essential aspect of the structure. In the case of dihydrolipoamide succinyltransferase,[58] dihydrolipoamide acetyltransferase from *E. coli*,[59] and dihydrolipoamide acyltransferase from animals,[56] the 3-fold rotational axes of symmetry must be the axes around which most of the protein is concentrated because in electron micrographs these molecules have the appearance of squares with vacant centers (Figure 9–23B). It is through these vacant centers that the 4-fold rotational axes of symmetry must pass.

Dihydrolipoamide acetyltransferase from animals is an interesting exception in this family. It is formed from 60 identical folded polypeptides arranged with icosahedral symmetry.[60] In its structure, the empty spaces among the protomers appear to lie along the icosahedral 5-fold rotational axes of symmetry. This observation suggests that the strong interfaces among the protomers around the octahedral 3-fold rotational axes of symmetry of the other proteins have become the strong interfaces among

A

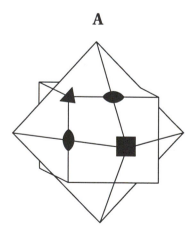

B

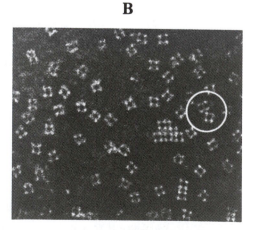

Figure 9–23: Octahedral symmetry. (A) The intersection of a cube and an octahedron is a structure that is assembled from 24 identical, rotationally asymmetric units arranged around the 4-fold, 3-fold, and 2-fold rotational axes of octahedral symmetry. (B) Electron micrograph of dihydrolipoamide succinyltransferase, a protein with 24 identical folded polypeptides arranged with octahedral symmetry.[58] A solution of the protein in 0.25% potassium phosphotungstate (the negative stain) was sprayed onto a film of carbon supported on an electron microscopic grid by a network of polymeric strands. After drying, to permit the negative stain to form an electron-dense glass embedding the molecules of protein, the specimen was viewed in the electron microscope at a magnification of 250000×. Panel B reprinted with permission from ref 58. Copyright 1971 National Academy of Sciences.

the protomers around the icosahedral 3-fold rotational axes of symmetry in animal dihydrolipoamide acetyltransferase. The significant vacancies along the 5-fold rotational axes of symmetry cause animal dihydrolipoamide acetyltransferase to differ from the icosahedral viral protein coats or heavy riboflavin synthetase in that it is not a tight shell but an open spherical net, unsuitable for enclosing other macromolecules.

Dihydrolipoamide acetyltransferase, dihydrolipoamide succinyltransferase, and dihydrolipoamide acyltransferase each forms the central octahedral core of a respective multienzyme complex that is responsible for the oxidative decarboxylation of different 2-oxoacids: pyruvate, 2-oxoglutarate, and either 2-oxoisovalerate, 2-oxoisocaproate, or 2-oxo-3-methylvalerate, respectively. Each of these multienzyme complexes is constructed from three different types of folded polypeptides. Each type of folded polypeptide catalyzes one of the three reactions (Reactions 9–4 through 9–6) whose sum is an oxidative decarboxylation in which NAD^+ is the formal oxidant (Reaction 9–7)

$$RC\overset{O}{\overset{\|}{C}}COOH + \overset{S}{\underset{S}{|}}{>} lipoamide \rightleftharpoons RC\overset{O}{\overset{\|}{C}}S\text{-}lipoamide\text{-}SH + CO_2$$

$$(9\text{–}4)$$

$$RC\overset{O}{\overset{\|}{C}}S\text{-}lipoamide\text{-}SH + HSCoA \rightleftharpoons RC\overset{O}{\overset{\|}{C}}SCoA + HS\text{-}lipoamide\text{-}SH$$

$$(9\text{–}5)$$

$$HS\text{-}lipoamide\text{-}SH + NAD^+ \rightleftharpoons \overset{S}{\underset{S}{|}}{>} lipoamide + NADH + H^+$$

$$(9\text{–}6)$$

$$RC\overset{O}{\overset{\|}{C}}COO^- + NAD^+ + HSCoA \rightleftharpoons RC\overset{O}{\overset{\|}{C}}SCoA + NADH + CO_2$$

$$(9\text{–}7)$$

where

$$\overset{S}{\underset{S}{|}}{>} lipoamide$$

is oxidized lipoamide and HS-lipoamide-SH is reduced lipoamide. The three acyltransferases, forming respectively the octahedral core of each complex, catalyze Reaction 9–5 and also contain covalently attached lipoamide. The other two enzymatic activities, Reactions 9–4 and 9–6, are catalyzed by the appropriate 2-oxoacid decarboxylase and lipoamide dehydrogenase, respectively.

These latter enzymes, the 2-oxoacid decarboxylases and lipoamide dehydrogenase, are simple dimers composed of two polypeptide chains, and these dimers adorn the outer surface of the central core, which is a molecule of the respective acyltransferase. They must each be attached to the octahedral core through interfaces formed between a face on the core and a face on the respective dimer. Because a dimer is built around a 2-fold rotational

axis of symmetry, every dimer necessarily has two identical faces, each complementary to its respective face on the octahedral core. Because the core is octahedral, it necessarily has 24 identical faces, each complementary to a face on a dimer of the respective 2-oxoacid carboxylase, and 24 other identical faces, each complementary to a face on a dimer of lipoamide dehydrogenase. The two identical faces on each dimer are arrayed about its 2-fold rotational axis of symmetry; the two different sets of 24 faces on the octahedral core are each arrayed about its 13 exact rotational axes of symmetry.

As the dispositions of the two sets of faces on the core must display octahedral symmetry, one might expect the arrangements of the dimers of the respective 2-oxoacid decarboxylase and the lipoamide dehydrogenase over the exterior surface of the core to reflect this symmetry; however, this is not the case. Careful examination of electron micrographs of complexes containing one dihydrolipoamide succinyltransferase and only two dimers of 2-oxoglutarate decarboxylase[61] showed that a dimer of 2-oxoglutarate decarboxylase is bound to a site on the core midway between a 2-fold rotational axis of symmetry and a 4-fold rotational axis of symmetry (Figure 9–23A). Therefore only one interface, formed from one face on the dimer and one face on the core, must attach the 2-oxoglutarate decarboxylase to the core because if the two identical faces on the dimer were both occupied with complementary faces on the same core, the dimer would have to sit with its 2-fold rotational axis of symmetry coinciding with one of the 2-fold rotational axes of the core. Only in this situation would the two pairs of complementary faces on the dimer be able to associate. The symmetry-related face on an attached dimer of 2-oxoglutarate decarboxylase finds itself empty because it is sterically inaccessible to another molecule of dihydrolipoamide succinyltransferase and consequently is unable to reach a symmetrically arrayed, complementary face on its own core. If the other face on the dimer were accessible to another dihydrolipoamide succinyltransferase, a polymer would result. In retrospect, if both a dimer of 2-oxoglutarate decarboxylase and a dimer of lipoamide dehydrogenase had occupied each of the 2-fold rotational axes of symmetry at the 12 locations at which they emerge from the core, the resulting stack of dimers along each of these axes would be an awkward arrangement.

The more interesting and unexpected conclusion, however, is that for every one of the 24 equivalent faces on the core that is occupied with a dimer of 2-oxoglutarate decarboxylase, the three other identical faces arrayed around the respective 4-fold rotational axis of symmetry (Figure 9–23A) remain empty.[61] This follows from the observations that one dihydrolipoamide succinyltransferase has only six dimers of 2-oxoglutarate decarboxylase attached to it in the 2-oxoglutarate decarboxylase complex, that two dimers of 2-oxoglutarate decarboxylase were almost never found on the same cubic face of the core in electron micrographs of the 2:1 complexes of decarboxylase and succinyltransferase, and that the distribution of pairs of dimers around the core in these 2:1 complexes is consistent with the distribution expected if only one dimer could reside around each of the 4-fold axes of symmetry emerging at the six locations around the core.[61] The complementary face on the dihydrolipoamide succinyltransferase to which a face on the dimer of 2-oxoglutarate decarboxylase attaches must be too close to the 4-fold rotational axis of symmetry, and the dimer must be too large, to permit more than one dimer to bind to the four identical faces around this axis. In the saturated complex between 2-oxoglutarate decarboxylase and dihydrolipoamide succinyltransferase there are 12 folded polypeptides of the former and 24 folded polypeptides of the latter, but this ratio is not dictated by symmetry so much as by steric effects. The six dimers of 2-oxoglutarate decarboxylase are randomly attached, each at only one of the four possible locations around each of the six locations on the surface of the core at which the 4-fold rotational axes of symmetry emerge. A similar but more extreme example of such substoichiometry is observed in the attachment of the symmetrical dimer of dihydrolipoamide dehydrogenase to the acetyltransferase core of mammalian pyruvate dehydrogenase. To this icosahedral core, formed from 60 folded polypeptides, only six dimers of the lipoamide dehydrogenase will bind at saturation.[62]

Another multienzyme complex that displays an unequal ratio between two constituent polypeptides, for similar reasons, is methylmalonyl-CoA carboxyltransferase.[63] The core of this complex, which catalyzes Reaction 8–45, is a hexamer, presumably built around a 3-fold rotational axis and three 2-fold rotational axes of symmetry (Figure 9–12). The central core seems to have the necessary 32 symmetry.[64] At each of the two trimeric surfaces of the core of methylmalonyl-CoA carboxyltransferase (top and bottom in Figure 9–12), there are three symmetrically displayed faces. Each of these faces can form an interface with a complementary face on the dimer constituting the second enzyme of the complex, which catalyzes Reaction 8–46. The symmetrically related second face of the dimer remains empty because it cannot reach a complementary face on the opposite side of the core, and only a face on the opposite side of the core related by a 2-fold rotational axis of symmetry could be complementary to the other face on the dimer. No other face on the surface of the trimer to which a particular dimer is attached can be related to the face on the trimer at which the dimer is attached by a 2-fold rotational axis of symmetry. Methylmalonyl-CoA carboxyltransferase ends up as an oligomer constructed from the hexameric core and six peripherally attached dimers.

The protein clathrin forms isometric cagelike structures that assemble around small pinocytotic vesicles as they bud inward from the plasma membrane of an animal cell. The polypeptide is 1600 amino acids in length, and when folded it produces a tubular protein, 45 nm in length and 2.5 nm in diameter.[65] The protomer from which the cages are formed is a trimer of these polypeptides, all three joined together at one end around a 3-fold rotational axis of symmetry to produce a triskele with bent arms.[65]

Different numbers of these triskeles can assemble to produce intact cages of various isometric shapes between 70 and 200 nm in diameter.[66] The wires of the mesh forming these cages are presumed to be formed from two or more intertwined arms of the triskeles.[67] Each and every vertex in each and every cage is a junction of three wires, and each vertex must contain the 3-fold rotational axis of symmetry at the nexus of an individual triskele. The mesh itself is always formed of pentagons and hexagons of wire.[66] It is the elongated and flexible nature of the subunit that permits the one protein to generate such a wide variety of isometric oligomeric proteins.

Suggested Reading

Rossmann, M.G., Abad-Zapatero, C., Hermodson, M.A., & Erickson, J.W. (1983) Subunit Interactions in Southern Bean Mosaic Virus, *J. Mol. Biol. 166*, 37–83.

Problem 9–4: Make a xerographic copy of Figure 9–19. On that xerographic copy designate every true 2-, 3-, 5-, and 6-fold rotational axis of symmetry. Use the standard symbols for this designation. In the same figure designate some of the local 2-, 3-, and 6-fold rotational axes of pseudosymmetry by the symbols P_2, P_3, and P_6, respectively.

Problem 9–5: Ferritin is an oligomeric protein involved in the storage of iron in animals. A crystallographic molecular model of the protein is available. The following figure is a schematic representation of the protomers arranged in the oligomer [reprinted with permission from ref 68; copyright 1984 Royal Society (London)].

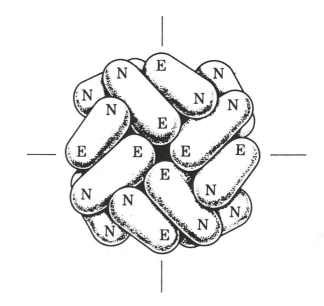

Each sausage-shaped subunit is one protomer. The letters N and E serve to identify the two distinct ends of the protomer: the one end which contains the amino terminus of the polypeptide and the other end in which an α helix designated as the E α helix is located, respectively.

(A) With what type of symmetry are the protomers of ferritin arranged?

(B) Make a xerographic copy of the figure; and, on your copy, identify all of the rotational axes of symmetry defining this arrangement of protomers.

Helical Polymeric Proteins

Helical fibers formed from identical globular monomers of protein have useful properties, and there are several examples of them. Each time an interface is created by evolution between two complementary faces on the surface of the same globular protein, no matter how the two faces are disposed on its surface, a distinct screw axis of symmetry is defined. Most of these screw axes would be open and generate helical polymers of the monomer, so the existing helical polymers must be the very few that have escaped elimination by natural selection. The surprising fact is that there are so few helical polymeric proteins.

A helix can be defined by three parameters: its radius, its handedness, and its pitch, which is the distance it rises for each complete turn. It can also be defined by four parameters: its radius, its handedness, a recurring radial angle dividing the helix into equal segments of arc, and the rise for each equal segment of arc. In a helical polymeric protein the second definition makes more sense because the successive protomers can be considered to be repeating segments of arc.

An interface between two identical molecules of a protein can generate several types of helical polymers. The actin helix (Figure 9–1B,C) is an example of the simplest type in which the protomers ascend one step at a time around the screw axis. In the actin polymer, the screw axis of symmetry passes through a corner of each protomer, the helix is not much wider than the protomer (Figure 9–1C), and the radial angle between successive protomers is fairly large (–166°).

If the generating interface creates a screw axis of symmetry where the radial angle between successive protomers is fairly small, so that there are several protomers in one turn of the helix, and where the rise for each monomer is just enough to cause the protomers in each turn of the helix to lie upon the protomers from the turn below it, then a singly threaded cylinder is formed. If the generating interface also creates a screw axis of symmetry that is separated by some distance from the monomer itself, rather than passing through it, this threaded cylinder will be hollow inside. An example of such a hollow, singly threaded cylinder is tobacco mosaic virus (Figure 9–24).[69] Each of the protomers in tobacco mosaic virus is a single identical folded polypeptide. The angle between successive protomers is 22°, and the rise for each protomer is 0.14 nm,[70] which brings the protomers in the next turn of the helix into contact with the protomers in the preceding turn because the turns are separated by 2.3 nm, the width of a protomer. Between two successive turns there are interfaces among the protomers. Because of the screw symmetry, every protomer provides at its lower surface the upper faces for the interfaces with the protomers below it and at its upper surface the lower faces for the interfaces with the protomers above it. Because every protomer is the same, each of these pairs of interfaces is the same and repeats along the thread every 22°.

The generating helix emphasized in the drawing of tobacco mosaic virus (Figure 9–24) is a right-handed helix of pitch 2.29 nm [(360/22) × 0.14]. If the structure is

A   **B**

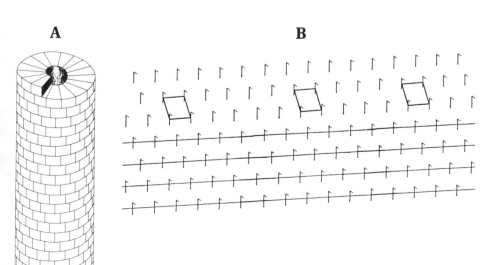

Figure 9–24: Diagrammatic representations of the helical arrangement among the protomers of the coat of tobacco mosaic virus.[69] (A) Individual protomers form a singly threaded helical screw that is a hollow cylinder. Each protomer has the same orientation and the successive protomers are related by a rise of 0.14 nm and a rotation of 22° along the helix. Reprinted with permission from ref 69. Copyright 1972 Federation of American Societies for Experimental Biology. (B) The surface helix of tobacco mosaic virus flattened onto the page. The hollow cylinder in (A) was split along an approximately vertical plane passing through its wall and then flattened onto the page. Reprinted with permission from ref 70. Copyright 1974 Academic Press.

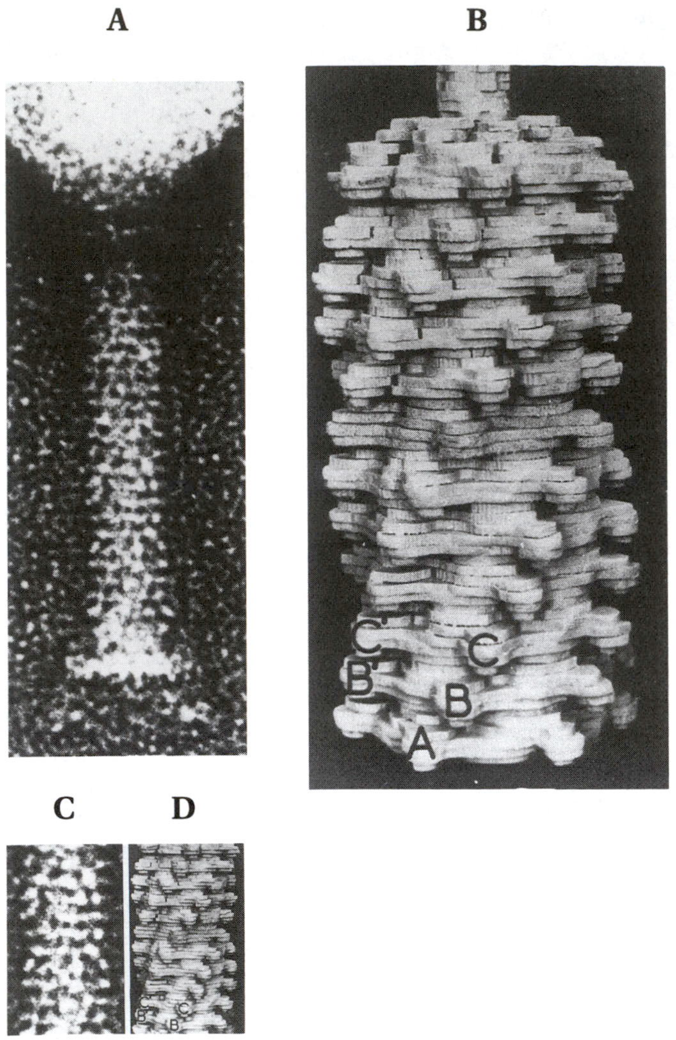

A   B

C   D

Figure 9–25: The helical surface lattice on the extended tail of the bacteriophage T4.[71] (A) Electron micrograph of an extended tail of bacteriophage T4. Bacteriophage were negatively stained with uranyl acetate on electron microscopic grids and examined in the electron microscope at a magnification of 140000×. The head of the bacteriophage is the large round object at the top of the electron micrograph and the tail extends down from the head. The tail is a hextuply threaded helical tube. (B) Three-dimensional reconstruction of the electron micrograph in (A). The electron micrograph was scanned with a densitometer and the Fourier transform of the two-dimensional distribution of density was calculated. This produced an array of peaks that defined the reciprocal lattice of a six-start helical array that repeats almost exactly every seven layers in the vertical direction. On the basis of this assignment of surface lattice, the three-dimensional distribution of density was calculated from the amplitudes and phases in the Fourier transform of the original two-dimensional distribution of density. Because the transform is made from a calculated Fourier transform, both amplitudes and phases are available. (C) A section of the electron micrograph displayed in (A). (D) The three-dimensional reconstruction in (B) at the same magnification as the micrograph in (C). Notice that the two images (C and D) resemble each other quite closely. Reprinted with permission from ref 71. Copyright 1975 Academic Press.

examined closely, however, it can be seen that there is a collection of helices of steeper pitch running through it. The protomers in tobacco mosaic virus are arranged upon a **helical surface lattice** that contains all of these other helices. The helical surface lattice can be displayed in two dimensions by fracturing the cylinder along a line vertical to the central axis and flattening the cylinder upon the page (Figure 9–24B). When the helical surface lattice of tobacco mosaic virus is viewed in this format, it can be seen that, in addition to the single helix of pitch 2.29 nm running through the lattice, there are also sets of 16 parallel right-handed helices and sets of 16 parallel left-handed helices. Any one of these sets of helices can also define the structure. Along the set of 16 right-handed helices of the steepest pitch and the set of 16 left-handed helices of steepest pitch, consecutive interfaces among protomers occur. Within each set of parallel helices, right-handed or left-handed, respectively, all of the interfaces are identical, but the interfaces within the left-handed helices are different from those within the right-handed helices, and both are different from the interfaces within the single shallow right-handed helix.

A helical surface lattice does not have to have a single shallow helix of one or the other hand acting as the generating operation. The extended tail of the T4 bacteriophage is built from rings of protomers. Each ring has six protomers arranged around a 6-fold rotational axis of symmetry. The rings are stacked upon each other by successive interfaces that cause them to be out of alignment in a right-handed sense by 17° (Figure 9–25). This creates a helical surface lattice that is a hextuply threaded cylinder. The lattice contains both a set of six parallel helices of left-handed sense and a set of six parallel helices, of steeper pitch, with a right-handed sense. Two of the steep right-handed helices are designated by capital letters. Because the fact that each of the six helices in one of these sets is formed from protomers of the same folded polypeptide necessitates that the interfaces among the six distinct helices all be the same, each helix assumes the same angular disposition to the one below it, and the structure is completely uniform. In the multiply threaded cylindrical tail of the T4 bacteriophage, the radial angle between successive protomers in one of the six right-handed helices is 17°, the same as the angle offsetting the rings.

It is the interfaces among the protomers in a helical polymer that determine the angle between successive globular protomers in the helix or set of helices. It is these interfaces that produce a helical filament such as actin or a singly or multiply threaded cylinder such as tobacco mosaic virus or the extended tail of the T4 bacteriophage. These interfaces are created by the interdigitations of the amino acids on the surfaces of the adjacent monomers, and the angle between successive monomers that these interfaces dictate can have any numerical value. Technically, this means that the helix or helices never position a monomer exactly above any monomer below it in the helix. Therefore, there is no translationally repeating unit in a rigid, biological helical structure. What this means is that a helical polymeric protein has difficulty crystallizing in three dimensions, and crystals of helical polymers suitable for high-resolution crystallographic studies have never been produced. One approach to determining the structure of a helical polymeric protein is to crystallize the monomer and construct a crystallographic molecular model at atomic resolution and simultaneously to determine the structure of the helical polymer at high enough resolution to position crystallographic models of these monomers in the proper orientation at the locations they occupy in the polymer. Such a strategy has been applied to the polymer of actin (Figure 9–1B).[4] The details of the structure of the intact polymer required by this strategy can be determined by image reconstruction[72] of electron micrographs.

Under appropriate circumstances, an **electron microscope** is capable of magnifying the image of a protein sufficiently that individual protomers can be distinguished (Figures 9–14, 9–23, and 9–25) and their shapes almost distinguished. A beam of collimated electrons is impinged upon the sample, and those that pass through it without being deflected sufficiently to leave the beam are then focused by electromagnetic lenses. The contrast in the image is caused by the distribution within the sample of the ability to deflect electrons from their path. Provided that phase contrast is not occurring, a small region of the sample formed of matter with a high efficiency for deflecting electrons will appear dark in the image; if another small region deflects electrons poorly as they pass through it, it will appear light in the image. The sample placed in an electron microscope usually has to reside in a chamber under very high vacuum at moderate temperatures. This means that the sample has to be a solid with a low vapor pressure. In the case of molecules of protein, this requires that they be encased in a solid matrix, which is always a glass.

To solve both the problems of contrast and solid support simultaneously, the glass is usually formed by drying a solution of a salt containing a heavy metal such as uranium or tungsten. Either uranyl acetate or sodium phosphotungstate, for example, will form an electron-dense glass when a film of a solution containing it is dried. This electron-dense glass will surround, encase, and support a molecule of protein that had been present in the solution from which the film was made. The glass of heavy metal encasing the molecule of protein forms a three-dimensional boundary or mold that has the shape of the molecule of protein. It is the dark image of this mold of electron-dense glass encasing the light image of the electron-translucent molecule of protein that is observed in the micrograph. Helical polymeric proteins, such as the tail of T4 bacteriophage (Figure 9–25), can be observed in this way. This procedure is known as **negative staining**.

It is now possible to insert thin layers of amorphous ice, which is also a glass, into an electron microscope.[73]

When a molecule of protein is embedded in such a glass, the contrast observed results from the fact that the atoms in a molecule of protein are more efficient at deflecting electrons than the molecules of water in amorphous ice. A positive image of the molecule, rather than a negative image, is observed.

The three-dimensional distribution of the ability to deflect electrons within the sample is known as the distribution of scattering density, $\theta(x,y,z)$. If a map of this function can be produced at as high a resolution as possible, the details of the structure of either the mold in which the molecule of protein is encased or the molecule of protein itself can be observed. **Image reconstruction** is a computational method that is used to calculate $\theta(x,y,z)$ from the images of molecules of proteins observed in electron micrographs.[72] In all cases, the electron micrograph, which is a photographic negative, is the experimental data submitted to these calculations, and the photographic negative used in a particular reconstruction must be presented to the reader so that she may appreciate the point of departure (Figure 9–25).

A molecule of protein has a certain distribution of electron density $\rho(x,y,z)$, and when molecules of protein are arrayed in a crystal they create a periodic, three-dimensional distribution of electron density. This periodic array diffracts X-rays to produce a diffraction pattern that is also periodic. The angular dispositions of the reflections in the diffraction pattern are determined by both the angles among the axes of the fundamental unit cell and its dimensions. The dimensions and axial angles of the unit cell can be calculated from these angular dispositions. The diffraction pattern of the crystal is the Fourier transform of the periodic distribution of electron density it contains. The magnitudes and phases of the maxima in the diffraction pattern can be calculated from the distribution of electron density within the unit cell by digital Fourier transformation. Conversely, the distribution of electron density in the unit cell can be calculated from the amplitudes and phases of the diffraction maxima by digital Fourier transformation.

A helical polymer of protein in its mold of negative stain has a certain distribution of scattering density, $\theta(x,y,z)$, which is a periodic function because the helix is periodic. Each protomer is a unit cell in this helical array. The Fourier transform of this periodic array is also a periodic function. From the spatial disposition of its maxima, the angle between successive unit cells in the helix, the rise for each unit cell, and the number of helical threads in this structure can be calculated. From the amplitudes and phases of the maxima of the Fourier transform, the distribution of scattering density within the unit cell and its mold can be calculated. Every helical array of identical asymmetric objects has a Fourier transform; the task is to discover it.

The depth of focus in an electron microscope is larger than the width of the specimen, and all points in the specimen are in focus in the final micrograph. As such, the micrograph represents the projection of the three-dimensional distribution of scattering density onto a two-dimensional surface.[72] Because the micrograph directly detects transmitted electrons rather than deflected electrons, it is not a direct projection of the distribution of scattering density, but a correction can be performed to obtain the projection of the scattering density. The Fourier transform, $F(X,Y,Z)$, of any three-dimensional distribution of scattering density is[74]

$$F(X,Y,Z) = \iiint_{object} \theta(x,y,z) e^{2\pi i(xX+yY+zZ)} dx\, dy\, dz$$

$$(9\text{--}8)$$

When $Z = 0$

$$F(X,Y,0) = \iint \sigma(x,y) e^{2\pi i(xX+yY)} dx\, dy \qquad (9\text{--}9)$$

where

$$\sigma(x,y) = \int \theta(x,y,z)\, dz \qquad (9\text{--}10)$$

The function $\sigma(x,y)$ is the projection of the three-dimensional distribution of scattering density. This set of relationships states that the two-dimensional Fourier transform of the projection of the distribution of scattering density is the central section of the three-dimensional Fourier transform of the unprojected distribution of scattering density. This central section of the three-dimensional Fourier transform is obtained by digitizing the optical density of the micrograph, correcting each point so that it is proportional to scattering density, and calculating a digitized Fourier transform by computer.

In the case of a helical polymeric protein embedded in negative stain, the central section of its three-dimensional Fourier transform systematically intersects all of the maxima in its three-dimensional Fourier transform. If properly indexed, the amplitudes and phases of these intersections can be used to calculate $\theta(x,y,z)$ just as the properly indexed amplitudes and phases of the X-ray diffraction pattern could be used to calculate $\rho(x,y,z)$. Figure 9–25B presents an example of such a reconstruction from the electron micrograph of Figure 9–25A. It should be remembered that this approach succeeds in producing the molecular structure of a helical polymeric protein only when it is able to produce a reconstructed image of sufficient resolution to orient unambiguously an atomic model of the crystallographic structure of the monomer within the polymer[4] and detect any conformational changes that occur when the monomer enters the polymer.

There are also helical polymeric proteins that are constructed from long, flexible strands of polypeptide rather than globular protomers. These proteins resemble the helical cables that are used in the construction of suspension bridges. The smallest structural element in a cable is a **strand**. Two or more strands are twisted around each other to make a **rope**. Two or more ropes are then twisted around each other to make a **cable**. The strands

from which ropes of protein are made are polypeptides read from messenger RNA, and for this reason, each strand must have a discrete length. This in turn means that the cable is built from uniform segments of rope that are overlapped to provide the necessary tensile strength.

The arrangement of the strands in the molecular rope and the ropes in the cable is elucidated by permitting a macroscopic fiber to diffract X-rays. A macroscopic fiber, built from billions of the molecular cables, is placed in a beam of X-rays. The cables are all more or less aligned with the axis of the macroscopic fiber. The helical arrays of the strands in the ropes and the helical arrays of the segments of rope in the cable have certain regularly recurring dimensions associated with them that give rise to diffraction. The dimensions and helical parameters of these arrays can be calculated from the angles at which the reflections emerge from the fiber. For example, a series of reflections are produced by a tendon from the tail of a rat when the tendon is placed in a beam of X-rays. This series arises from a helical array that repeats every 67 nm,[75,76] and this dimension, along with others, must be incorporated into the model for the complete cable of the collagen from which the tendon is formed.

Collagen is the helical polymeric protein from which is formed the tough, flexible material composing tendons, intercellular matrix, the matrix of bone, and many strong, plastic sheets of various shapes and sizes found in animals. The basic structural element in most of these macroscopic structures is the fibril of collagen, which is a cylindrical thread of indefinite length, 200–800 nm wide. This thread in turn is formed from molecular cables of collagen, each probably as long as the fibril, packed side by side in register.

The strand from which a molecular cable of collagen is formed is a polypeptide about 1040 amino acids in length. There are 10 or more isoforms of this polypeptide in a given animal, each encoded by a different gene, but all essentially the same length and all homologous in sequence among themselves. Commencing at some point within 20 amino acids of the amino terminus,[77] the sequence of each adopts a repeating pattern that places a glycine at every third amino acid.[78] This pattern continues uninterrupted for 1014 amino acids.[79,80] The last 10–30 amino acids at the carboxy terminus do not display this Gly-X-Y pattern.[80] About one-third of the amino acids in the central region of collagen, other than the glycines, are prolines.

The central 1014 amino acids of the polypeptide of collagen assume a helical structure that is distinct from the α helix. This helix is a right-handed helix that repeats at slightly less than every third amino acid. It is more extended than an α helix in that it has a rise of 0.29 nm for each amino acid, and this prevents intrastrand hydrogen bonding. Because every third amino acid is glycine and the repeat is slightly less than every third amino acid, three of these helical strands can wrap around each other

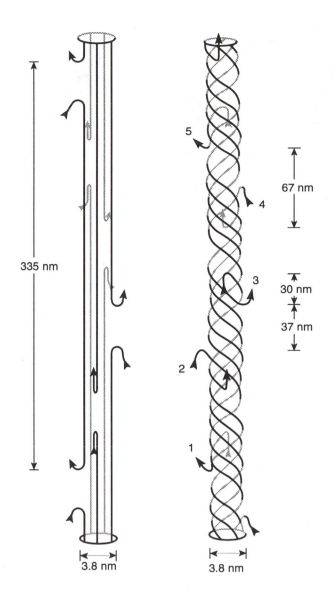

Figure 9–26: Arrangement of the triple-helical ropes of collagen in the cable.[76] The individual segments of rope, each a triple helix formed from three 1014 amino acid segments from the central regions of three collagen polypeptides, are represented by vertical lines with splayed amino- and carboxy-terminal ends. The individual segments of rope are arrayed in a right-handed helical distribution (left panel). This cable is then twisted to the left (right panel) to generate the final cable.

left-handed twist is sufficient to convert the 5-fold right-handed screw axis of symmetry of the original array into a 4-fold right-handed screw axis that generates the final array. The cable is a helical array of five stacked left-handed helices, and the identical interfaces among the adjacent ropes determine the regular, uniform alignment of one rope with its neighbor above and its neighbor below. These interfaces are among amino acids on the surfaces of the segments of rope that make contact as the five helices sit one on top of the other (Figure 9–26). The dimensions of Figure 9–26, as with all diagrams of collagen,[76] are very deceptive. The length, relative to the width, has been decreased by a factor of 8 to be able to illustrate the symmetry. In the actual molecular cable of collagen the twist is 8 times less remarkable.

To produce the collagen fibril or a larger structure, the cables must be able to pack side by side. Their 4-fold right-handed screw axis of symmetry permits them to pack in a square array within a $P4_1$, $I4_1$, or $I4$ space group,[76] and the pattern of reflections from a tendon is consistent with any one of these three tetragonal space groups. The 5-fold screw axis of symmetry of the untwisted cable would be incompatible with extensive side by side packing, but these three space groups, based on a 4-fold screw axis of symmetry, are by definition capable of infinite propagation in all three directions. Presumably this is the reason that sufficient twist is introduced into the cable to give it a 4-fold screw axis of symmetry. The interfaces between the cables of collagen in the three-dimensional array are all identical because as the cable is ascended (Figure 9–26) it displays, at rises of 67 nm, the same face to the front, then to the left side, then to the back, and then to the right side, and these four identical faces can associate in turn with complementary faces on the four neighbors in a square array, whose complementary faces are also repeating four times around their screw axes.

As the cables enter these side by side arrays, they are aligned in register, and the crystallographic asymmetric unit, which is 67 nm of cable, creates a repeating pattern on the surface of the fibril itself. This pattern can be seen in the electron microscope. It appears as alternating thickenings and thinnings along a desiccated fibril that repeat every 67 nm. These thickenings and thinnings are believed to represent the alternation in register of regions within the molecular cables five ropes in thickness and those of four ropes in thickness, respectively (Figure 9–26). Fibrils of collagen also stain positively by binding heavy metals at specific locations on the surfaces of the ropes

with a leisurely right-handed twist to form a segment of rope 300 nm in length. In this segment of rope each glycine is directed into the center of the triple helix. That every third residue is glycine allows the three strands to approach each other intimately enough for interstrand hydrogen bonds to form among the peptide bonds of the three polypeptides and produce a strong rope.

The segments of rope (represented by vertical lines in Figure 9–26) are placed side by side in a right-handed helical array. The radial angle between each successive protomer, which is an individual segment of rope, would be 72° and the helix would repeat every five protomers if the ropes were held perfectly vertical (Figure 9–26). The cable that they form, however, has a gentle left-handed twist that is sufficient to bring the amino terminus of the fourth segment of rope farther along in the cable directly above the amino terminus of any given segment of rope in the cable. If the amino termini (the tails of the arrows found at the levels labeled 1–5 in Figure 9–26) are connected in sequence around the cable they trace a right-handed helix that repeats every fourth step. Therefore the

where there are high, unbalanced constellations of negative charge. Because the segments of rope are placed in register by the side to side packing of the cables, these positions to which heavy metals bind form bands across the fibrils or across sheets of collagen. The pattern of bands is quite reproducible,[81] and it repeats every 67 nm.[82] From an examination of the patterns in which charged amino acids occur in the amino acid sequences of the polypeptides, it can be shown that the pattern in which these bands occur on the ropes is entirely consistent[82] with the triple helical array of strands and pentahelical array of ropes that has been proposed (Figure 9–26) to explain the angular dispositions of the X-ray diffraction pattern from an oriented tendon.[76] This correlation provides independent support for the model of the structure of the cable.

Tropomyosin is a helical protein that does not spontaneously form cables in solution, but it serves as a simple example of a **coiled coil of α helices**.[83–85] It is a protein composed of two identical polypeptides, 284 amino acids in length. Each polypeptide is folded entirely as one α helix. The two identical α-helical strands of tropomyosin twist around each other to form a left-handed segment of rope referred to as a coiled coil.[83] It is the interface between the two α helices in the rope that enforces the left-handed twist.

Suppose that the amino acids emerged from a right-handed α helix at successive angles of precisely 102⁶⁄₇° instead of about 99°. Every seven amino acids (7 × 102⁶⁄₇ = 720) the angular dispositions of the side chains would repeat precisely. Two such tightened α helices could be placed next to each other, with their axes parallel, in such a way that their side chains would interdigitate regularly along the interface (Figure 9–27).

Every seventh side chain in one helix would sit next to every seventh side chain of the other, and every side chain four amino acids to the carboxy-terminal side of every seventh side chain in one polypeptide would sit next to every side chain four residues to the carboxy-terminal side of every seventh side chain in the other. The two helices could comfortably sit side by side for an indefinite length because the topology of the interface would repeat precisely every seven amino acids.

α Helices, however, do not have angles between successive residues of 102⁶⁄₇°, but less than that. Crick[83] pointed out that such an interface, permitting the advantageous interdigitation and repeating every seven residues, nevertheless could be retained if the two helices, instead of being parallel to each other, twisted around each other in a left-handed rope such that the twist of the rope exactly compensated for the difference between the actual angle between successive amino acids in the α helix and 102⁶⁄₇°. If the actual angle between successive amino acids in a right-handed α helix is 99°,[8,9] the two α helices would have to twist around each other in a left-handed sense at −3⁶⁄₇° for every position in one of the sequences. Although the actual twist of the rope should reflect a tradeoff between energy required to tighten the α helix and energy required to bend the α helix into the supercoil, in tropomyosin the twist in the rope is −3.9° for every position in the sequence,[84] almost too close to the expected value.

Crick also calculated the X-ray fiber diffraction pattern expected from such a coiled coil and was able to explain why the meridional reflection in the pattern that would normally arise from the 0.54-nm pitch of an untwisted α helix should shorten to a pitch of 0.51 nm when the α helix becomes twisted into a coiled coil. A prominent meridional reflection, representing a repeat of 0.51 nm,

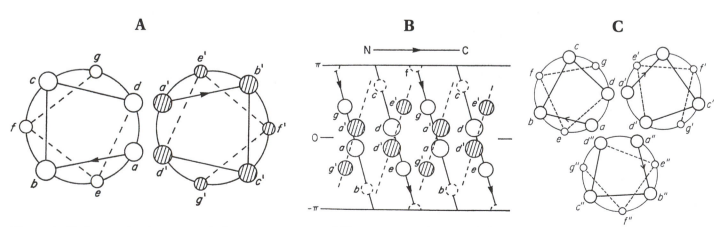

Figure 9–27: Interaction between α helices in a coiled coil.[84,86] (A) Alignment of two tightened α helices with 3.5 amino acids for each turn rather than 3.6. Amino acids from amino- to carboxy-terminal are designated as *a*, *b*, *c*, *d*, *e*, and *f*, and the view is end-on looking from amino- to carboxy-terminal amino acid. Every seven amino acids the orientations would repeat, and this would place the amino acid after amino acid *g* precisely below amino acid *a* and so forth. (B) The two α helices in (A) are cut along a vertical plane passing through amino acids *f* and flattened, one on top of the other. This view illustrates the interdigitations of amino acids *a* and of amino acids *d*. (C) Three tightened α helices running parallel to each other. In this arrangement also, amino acids *a* can interdigitate with amino acids *d*. Reprinted with permission from refs 84 and 86. Copyright 1975 and 1978 Academic Press.

had been observed previously[87] in the diffraction patterns of oriented fibers of keratin, myosin, and fibrinogen, and it is now known that such a reflection is indicative of coiled coils of α helices.

The sequence of tropomyosin can be divided into successive units, or heptads, each seven amino acids in length. The first and fourth amino acid in each heptad are the most deeply buried amino acids in the interface between the two strands in the rope (Figure 9–27). These most deeply buried locations are isolated from the water surrounding the rope. The sequence positions of the side chains sequestered there are usually occupied by nonpolar amino acids such as leucine, valine, isoleucine, alanine, phenylalanine, tyrosine, and methionine.[84]

The synthetic peptide N-acetyl-RMKQLEDKVEELLS KNYHLENEVARLKKLVGER, which represents a portion of the amino acid sequence of the transcription factor GCNR, spontaneously forms a homodimeric coiled coil.[88] The heptad repeat of a coiled coil is readily apparent in its amino acid sequence. A crystallographic molecular model for this coiled coil (c8 Kin.5) has been constructed.[89] As predicted,[88] the hydrophobic amino acids of the heptad repeat interdigitate (Figure 9–27) along the interface between the two α helices to form the hydrophobic core of the structure and to produce the supercoil. They pack so closely together and so efficiently that there is almost no vacant space in the core of the structure. The two identical α helices are parallel to each other and packed in precise register so that each hydrophobic side chain packs against its twin from the other α helix. The asparagines at position 16 pack side by side in the core of the coiled coil, and the two possible hydrogen bonds between the respective amide protons and acyl oxygens are formed. The axes of the two α helices are 0.93 nm apart,

the supercoil is uniform along its length, and the twist of the supercoil is –3.0° for each position in the amino acid sequence, a value very close to that of tropomyosin. The impression made by the crystallographic molecular model is that the coiled coil is a highly regular structure that could be extended to any length.

In a three-stranded coiled coil, the rope still places the first and fourth amino acids of a heptad in its core (Figure 9–27C). An example of a three-stranded coiled coil may occur in fibrinogen.[86] It is a protein constructed from two copies of three different polypeptides to produce the structure $(\alpha\beta\gamma)_2$. The three polypeptides in each molecular asymmetric unit are intramolecularly cross-linked by several cysteines. In each of the three polypeptides, there is a pair of cysteines four amino acids apart in the respective sequences. These three pairs of cysteines, one pair from each polypeptide, are coupled among themselves to connect the three polypeptides in a disulfide ring. One hundred eleven amino acids farther along in the sequence of each polypeptide another disulfide ring is located. An explanation for this curious regularity is that in the regions between the two disulfide rings, the three polypeptides run parallel to each other. Between the two disulfide rings, the sequences of the three polypeptides display a heptad repeat of nonpolar amino acids. When oriented fibrinogen itself or oriented fibrin, a macroscopic polymeric material formed from fibrinogen, is permitted to diffract X-rays, a meridional reflection arising from a repeating dimension of 0.51 nm is observed.[87] Based on these observations, a hypothetical molecular model of a three-stranded coiled coil was built from the amino acid sequences of the three polypeptides in this region of the fibrinogen molecule (Figure 9–28).

Three classes of filaments can be observed within

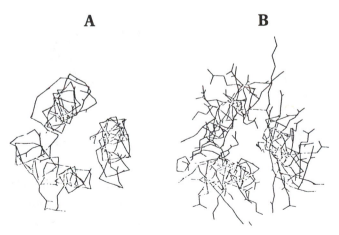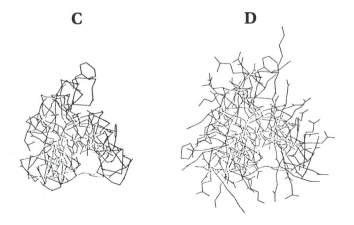

A   **B**   **C**   **D**

Figure 9–28: Different features of a hypothetical molecular model of a segment of the three-stranded coiled coil in fibrinogen.[86] All views are the same orientation looking down a segment containing three α helices each 30 amino acids in length. Each α helix has the amino acid sequence of one of the three aligned amino acid sequences of the α, β, and γ polypeptides of fibrinogen. The model is displayed by computer graph-ics. (A) The polyamide backbone of the three α helices. (B) The polyamide backbone and only the polar amino acid side chains. (C) The polyamide backbone and only the nonpolar side chains. (D) All of the atoms of the hypothetical molecular model. Reprinted with permission from ref 86. Copyright 1978 Academic Press.

A

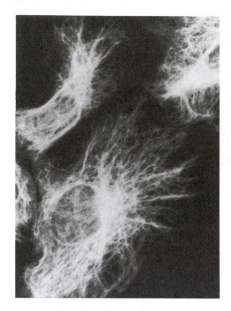

B

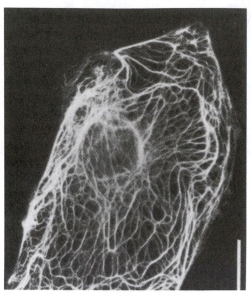

Figure 9–29: Distribution of intermediate filaments of vimentin (A) and keratin (B) in animal cells.[92] Hamster Nil-8 cells (A) or kangaroo rat PtK-2 cells (B) were grown on glass cover slips. The cells were fixed with methanol, rinsed with acetone, and dried in the air. Antiserum raised in guinea pigs to either vimentin (A) or epidermal prekeratin (B) was then applied to the respective cells. After the antiserum was rinsed away, those antibodies bound to the intermediate filaments could be visualized by adding fluorescein-labeled immunoglobulins specific for guinea pig immunoglobulin. The covalently attached fluorescein causes the intermediate filaments to which it is bound to fluoresce. Reprinted with permission from ref 92. Copyright 1978 National Academy of Sciences.

animal cells. Microfilaments, or thin filaments, are filaments of actin whose basic structural element is the actin helix (Figure 9–1C). Tropomyosin lies in the grooves of the actin helix, each dimeric tropomyosin rope spanning seven actins, and the globular protein troponin adorns the microfilament at regular intervals.[90] These microfilaments are about 5 nm in width.[90] Microtubules are triply threaded, hollow, helical cylinders,[91] constructed from the globular protein tubulin. They are 20 nm in width. Intermediate filaments are the third class. They are filaments 6–11 nm in width, intermediate between microfilaments and microtubules.

Intermediate filaments were originally considered to be a heterogeneous class of polymeric proteins grouped together only because they were similar in width. Within this class are tonofilaments, neurofilaments, cellular keratin filaments, desmin filaments, glial filaments, and vimentin filaments. Each of these subclasses occurs in a different set of tissues, and they form intermediate filaments that often seem very different in their distribution through a cell (Figure 9–29).[92] It is now known, however, that all of these filaments are constructed from polypeptides that are homologous in sequence[93] (Figure 9–30) and thus necessarily share a common, superposable structure. That an intermediate filament can be constructed from only one of these polypeptides has been demonstrated by reassembling filaments from a pure homogeneous preparation of a given polypeptide.[94] These polypeptides form helical cables of indefinite length.

One of these intermediate filaments, keratin, composes the fibers in the composite material that forms hair and horn. The quill of a porcupine represents a large array of more or less aligned keratin cables, and X-ray diffraction from such specimens[95] provide dimensions of the helical arrays in these cables.[96] Meridional reflections

representing a repeat of 0.51 nm are strong features of such diffraction patterns, and this demonstrates that these cables are built from α-helical coiled coils.

The strand from which an intermediate filament is constructed is a polypeptide folded into an α helix. Two or three α helices—the number is not yet known[94]—twist around each other in a left-handed coiled coil to produce the rope, as in tropomyosin or fibrinogen. The heptad permitting the formation of this rope can be noticed in those regions of the sequences that are involved in its formation when sequences from several of these polypeptides are aligned (Figure 9–30).[93] The amino acids at the heptad positions are not always nonpolar, but in the few cases that they are not, remarkable conservation is displayed.

As in collagen, the sequences producing the segments of rope are found in the middle portions of the polypeptides. Collagen, however, has only 5–30 amino acids at each end that are not directly incorporated into the ropes. In the various polypeptides of intermediate filaments an amino-terminal segment 100–150 residues in length precedes the region involved in the coiled coil, which is about 190 amino acids in length, and a carboxy-terminal segment 100–1600 amino acids in length follows the coiled coil depending on the type of intermediate filament the polypeptide produces. Presumably these additional segments protrude at regular intervals along the cable, as do the two ends of collagen (Figure 9–26) and it would be these protrusions that cause each type of intermediate filament to have a different width and a different tissue-specific function.

If all 190 amino acids are involved in the coiled coil, two or three central regions of the polypeptide forming an intermediate filament, twisted into a coiled coil, would produce a segment of rope 28 nm long. There is a short

Coil 1b

```
        •       •       •       •       •       •       •       •       •       •       •       •       •       •       •
HE₂  7  F E Q Y I N N L R R Q L D S I V G E R G R L D S E L R G M Q D L V E D F K N K Y E D E I N K R T A A E
7  134  F E G Y I E T L R R E A E C V E A D S G R L S S E L N H V Q E V L E G Y K K K Y E E E V A L R A T A E
8  105  Y F R T I E E L Q Q K I L C A K S E N S R L V I E I D N A K L A S D D F R T K Y E S E R S L R Q L V E
HE₁ 102  Y F K T I E D L R N K I L T A T V D N A N V L L Q I D N A R L A A D D F R T K Y E T E L N R M S V E
D  146  Y E E L R E L R R Q V D A L T G Q R A R V E V E R D N L L D N L Q K L K Q R L Q E E I Q L K Q E A E
V  148  Y E E M R E L R R Q V D Q L T N D K A R V E V E R D N L A E D I M R L R E K L Q E E M L Q R E E A E
G   86  Y Q A E L R E L R L R L D Q L T A N S A R L E V E R D N F A Q D L G T L R Q K L Q D E T N L R L E A E
NF₂ 150  Y D Q E I R E L R A T L E L V N H E K A Q V Q L D S D H L E E D I H R L K E R F E E E A R L R D D T E
NF₃  ————————————————— M R G A V L R L G A A R G Q L R L E Q E H L L E D I A H V R Q R L D D E A R Q R Q E A E
```

```
        •       •       •       •       •       •       •       •       •       •       •       •       •
HE₂  58  N E F V T L K K D V D A A Y M N K V E L Q A K A D T L T D E I N F L R A L Y D A E L S Q M Q T H I S D
7  185  N E F V A L K K D V D C A Y V R K S D L E A N V E A L I Q E I D F L R R L Y E E E I R V L Q A N I S D
8  156  S D I N S L R R I L D E L T L C K S N L E A E V E S L K E E L L C L K Q N H E E E V N T L R S Q L G D
HE₁ 153  A D I N G L R R V L D E L T L A R A D L E M Q I E S L K E E L A Y L K K N H E E E M N A L R G Q V G G
D  197  N N L A A F R A D V D A A T L A R I D L E R R I E S L Q E E I A F L K K V H E E E I R E L Q A Q L Q E
V  199  S T L Q S F R Q D V D N A S L A R L D L E R K V E S L Q E E I A F L K K L H D E E I Q E L Q A Q I Q E
G  137  N N L A A Y R Q E A D E A T L A R V D L E R K V E S L E E E I Q F L R K I Y E E E V R D L R E Q L A Q
NF₂ 201  A A I R A L R K D I E E A S L V K V E L D K K V Q S L Q D E V A F L R S N H E E E V A D L L A Q I Q A
NF₃  A A A R A L A R F A Q E A E A A R V E L Q K K A Q A L Q E E C G Y L R R H H Q E E
```

Figure 9–30: Alignment of portions of the amino acid sequences of the polypeptides composing various intermediate filaments.[93] The aligned segments come from different locations in the amino acid sequences of the various polypeptides. The numbers indicate the sequence positions in the various polypeptides at which the alignment commences. The abbreviations are HE₁, human epidermal keratin (50 kDa); HE₂, human epidermal keratin (56 kDa); 8, sheep wool keratin 8c-1; 7, sheep wool keratin 7c; D, chicken desmin; V, hamster vimentin; G, murine glial fibrillary acidic protein; NF₂, porcine neurofilament protein NF-M; and NF₃, porcine neurofilament protein NF-H.

region in the middle of that 190 amino acids that lacks the heptad repeat, but even if this bulged completely out of the rope it would shorten to only about 24 nm. The cable that is an intermediate filament is formed from a central core of three of these segments of rope around which is wrapped an outer cylinder formed from about nine of these segments of rope.[96] This produces a cable about 8 nm in diameter because each rope is about 2 nm in diameter. The problem is that on the surface of the outer hollow cylinder there is a helix with a pitch of 22 nm, which is the period also seen in electron micrographs of intermediate filaments themselves. It is not clear how the segments of rope 28 nm in length are arranged to produce a surface helix with a pitch of 22 nm.

In all of the cables that have been discussed, the faces and interfaces can be divided into three groups. There is a continuous interface between or among the strands, containing within itself the central axis of the rope and twisting around that axis with the twist of the rope. Between the ropes in the cable there are interfaces, but they are formed from faces on each rope composed of small regions of surface on each strand alternating between or among the strands as the rope is ascended. The cables themselves may have faces on them to promote side to side associations, and these faces are composed of

small regions of surface on strands from the same or different ropes that are encountered in turn as the cable is ascended (Figure 9–26).

Suggested Reading

Amos, L.A., & Klug, A. (1975) Three-Dimensional Image Reconstructions of the Contractile Tail of T4 Bacteriophage, *J. Mol. Biol.* **99**, 51–73.

MacLachlan, A.D., & Stewart, M. (1975) Tropomyosin Coiled-Coil Interactions: Evidence for an Unstaggered Structure, *J. Mol. Biol.* **98**, 293–304.

Problem 9–6: The following is a segment of amino acid sequence from the coil region of human epidermal keratin:

```
T A A E N E F V T L K K D V D A A Y M N K V E L Q
A K A D T L T D E I N F L R A L Y D A E L S Q M Q
T
```

(A)   Write out this sequence in the same format as the following diagram of a portion of the amino acid sequence of α-tropomyosin (reprinted with permission from ref 84; copyright 1975 Academic Press).

```
     Asp     Lys     Asp     Glu     Glu     Lys     Ala     Arg     Glu
  Gln     Ser     Gly     Lys     Asp     Leu     Asp     Ser     Leu
     Glu     Lys     Glu     Ala     Lys     Lys     Asp     Arg     Glu
 -Leu-----Leu-----Thr-----Tyr-----Ala-----Ala-----Ala-----Leu-----Val-
     Leu     Leu     Leu     Leu     Leu     Ala     Val     Ile     Leu
  Glu     Gln     Glu     Ser     Gln     Glu     Glu     Asn     Glu
     Val     Lys     Asp     Lys     Glu     Thr     Ala     Gln     Asp
  Asp     Lys     Asp     Glu     Glu     Lys     Ala     Arg     Glu
```

In your diagram place the appropriate amino acids from the sequence of human epidermal keratin along the center line as was done in the diagram of the sequence of α-tropomyosin.

(B) What is the role of the amino acids placed along the center line?

(C) Circle the two amino acids in your diagram that do not seem to fit this role.

(D) How may they be excused?

References

1. Monod, J., Wyman, J., & Changeux, J.P. (1965) *J. Mol. Biol.* 12, 88–118.

2. Banner, D.W., Bloomer, A.C., Petsko, G.A., Phillips, D.C., Pogson, C.I., Wilson, I.A., Corran, P.H., Further, A.J., Milman, J.D., Offord, R.E., Priddle, J.D, & Waley, S.G. (1975) *Nature* 255, 609–614.

3. Kabsch, W., Mannherz, H.G., Suck, D., Pai, E.F., & Holmes, K.C. (1990) *Nature* 347, 37–44.

4. Holmes, K.C., Popp, D., Gebhard, W., & Kabsch, W. (1990) *Nature* 347, 44–49.

5. Smith, P.R., Fowler, W.E., Pollard, T.D., & Aebi, U. (1983) *J. Mol. Biol.* 167, 641–660.

6. Bragg, L., Kendrew, J.C., & Perutz, M.F. (1950) *Proc. R. Soc. London, A 203,* 321–357.

7. Huggins, M.L. (1943) *Chem. Rev. 32,* 195–218.

8. Oefner, C., & Suck, D. (1986) *J. Mol. Biol. 192,* 605–632.

9. Artymiuk, P.J., & Blake, C.C.F. (1981) *J. Mol. Biol. 152,* 737–762.

10. Steitz, T.A., Fletterick, R.J., Anderson, W.F., & Anderson, C.M. (1976) *J. Mol. Biol. 104,* 197–222.

11. Hahn, T., Ed. (1983) *International Tables for Crystallography, Volume A. Space-Group Symmetry,* D. Reidel, Dordrecht, The Netherlands.

12. Buehner, M., Ford, G.C., Moras, D., Olsen, K.W., & Rossmann, M.G. (1974) *J. Mol. Biol. 82,* 563–585.

13. Buehner, M., Ford, G.C., Moras, D., Olsen, K., & Rossmann, M.G. (1974) *J. Mol. Biol. 90,* 25–49.

14. Epp, O., Ladenstein, R., & Wendl, A. (1983) *Eur. J. Biochem. 133,* 51–69.

15. Adams, M.J., Ford, G.C., Koekoek, R., Lentz, P.J., McPherson, A., Rossmann, M.G., Smiley, I.E., Schevitz, R.W., & Wonacott, A.J. (1970) *Nature 227,* 1098–1103.

16. Sprang, S., & Fletterick, R.J. (1979) *J. Mol. Biol. 131,* 523–551.

17. Eklund, H., Nordstrom, B., Zeppezauer, E., Soderlund, G., Ohlsson, I., Boiwe, T., Soderberg, B., Tapia, O., & Brändén, C. (1976) *J. Mol. Biol. 102,* 27–59.

18. Darnell, D.W., & Klotz, I.M. (1975) *Arch. Biochem. Biophys. 166,* 651–682.

19. Remington, S., Wiegand, G., & Huber, R. (1982) *J. Mol. Biol. 158,* 111–152.

20. MacLachlan, A.D. (1979) *J. Mol. Biol. 128,* 49–79.

21. Ploegman, J.H., Drent, G., Kalk, K.H., & Hol, W.G.J. (1978) *J. Mol. Biol. 123,* 557–594.

22. Campbell, J.W., Watson, H.C., & Hodgson, G.I. (1974) *Nature 250,* 301–303.

23. Campbell, J.W., Hodgson, G.I., Watson, H.C., & Scopes, R.K. (1971) *J. Mol. Biol. 61,* 257–259.

24. Dayhoff, M.O. (1972) *Atlas of Protein Sequence and Structure,* Vol. 5, National Biomedical Research Foundation, Washington, DC.

25. Perutz, M.F., Muirhead, H., Cox, J.M., & Goaman, L.C.G. (1968) *Nature 219,* 131–139.

26. Ip, S.H.C., & Ackers, G.K. (1977) *J. Biol. Chem. 252,* 82–87.

27. Park, C.M. (1970) *J. Biol. Chem. 245,* 5390–5394.

28. Melick-Adamyan, W.R., Barynin, V.V., Vagin, A.A., Borisov, V.V., Vainshtein, B.K., Fita, I., Murthy, M.R.N., & Rossmann, M.G. (1986) *J. Mol. Biol. 188,* 63–72.

29. Matthews, B.W., Fenna, R.E., Bolognesi, M.C., Schmid, M.F., & Olson, J.M. (1979) *J. Mol. Biol. 131,* 259–285.

30. Krause, K.L., Volz, K.W., & Lipscomb, W.N. (1985) *Proc. Natl. Acad. Sci. U.S.A. 82,* 1643–1647.

31. Feinstein, A., & Munn, E.A. (1969) *Nature 224,* 1307–1309.

32. Parks, E.H., Ernst, S.R., Hamlin, R., Xuong, N.H., & Hackert, M.L. (1985) *J. Mol. Biol. 182,* 455–465.

33. Lindqvist, Y., & Brändén, C. (1985) *Proc Natl. Acad. Sci. U.S.A. 82,* 6855–6859.

34. Boeker, E.A., & Snell, E.E. (1968) *J. Biol. Chem. 243,* 1678–1684.

35. Almassy, R.J., Janson, C.A., Hamlin, R., Xuong, N., & Eisenberg, D. (1986) *Nature 323,* 304–309.

36. Yamashita, M.M., Almassy, R.J., Janson, C.A., Cascio, D., & Eisenberg, D. (1989) *J. Biol. Chem. 264,* 17681–17690.

37. Crick, F.H.C., & Watson, J.D. (1956) *Nature 177,* 473–475.

38. Unge, T., Liljas, L., Strandberg, B., Vaara, I., Kannan, K.K., Fridborg, K., Nordman, C.E., & Lentz, P.J. (1980) *Nature 285,* 373–378.

39. Montelius, I., Liljas, L., & Unge, T. (1988) *J. Mol. Biol. 201,* 353–363.

40. Burnett, R.M., Grütter, M.G., & White, J.L. (1984) *J. Mol. Biol. 185,* 105–123.

41. Liddington, R.C., Yan, Y., Moulai, J., Sahli, R., Benjamin, T.L., & Harrison, S.C. (1991) *Nature 354,* 278–284.

42. Burnett, R.M. (1984) in *Biological Macromolecules and Assemblies, Virus Structures: Virus Structures* (McPherson, A., & Jurnak, F.A., Eds.) Vol. 1, pp 337–385, Wiley, New York.

43. Caspar, D.L.D., & Klug, A. (1962) *Cold Spring Harbor Symp. Quant. Biol. 27*, 1–24.

44. Olson, A.J., Bricogne, G., & Harrison, S.C. (1983) *J. Mol. Biol. 171*, 61–63.

45. Rossmann, M.G., Abad-Zapatero, C., Hermodson, M.A., & Erickson, J.W. (1983) *J. Mol. Biol. 166*, 37–83.

46. Chen, Z., Stauffacher, C., Li, Y., Schmidt, T., Bomu, W., Kamer, G., Shanks, M., Lomonosoff, G., & Johnson, J.E. (1989) *Science 245*, 154–159.

47. Hogle, J.M., Chow, M., & Filman, D.J. (1985) *Science 229*, 1358–1365.

48. Rossmann, M.G., Arnold, E., Erickson, J.W., Frankenburger, E.A., Griffith, J.P., Hecht, H., Johnson, J.E., Kamer, G., Luo, M., Mosser, A., Rueckert, R.R., Sherry, B., & Vriend, G. (1985) *Nature 317*, 145–153.

49. Luo, M., Vriend, G., Kamer, G., Minor, I., Arnold, E., Rossmann, M.G., Boege, U., Scraba, D.G., Duke, G.M., & Palmenberg, A.C. (1987) *Science 235*, 182–191.

50. Acharya, R., Fry, E., Stuart, D., Fox, G., Rowlands, D., & Brown, F. (1989) *Nature 337*, 709–716.

51. Rossmann, M.G., Abad-Zapatero, C., Murthy, M.R.N., Liljas, L., Jones, T.A., & Strandberg, B. (1983) *J. Mol. Biol. 165*, 711–736.

52. Liljas, L., Unge, T., Jones, T.A., Fridborg, K., Lovgren, S., Skoglund, U., & Strandberg, B. (1982) *J. Mol. Biol. 159*, 93–108.

53. Erickson, J.W., Silva, A.M., Murthy, M.R.N, Fita, I., & Rossmann, M.G. (1985) *Science 229*, 625–629.

54. Valegard, K., Liljas, L., Fridborg, K.I., & Unge, T. (1990) *Nature 345*, 36–41.

55. Bacher, A., Ludwig, H.C., Schnepple, H., & Ben-Shaul, Y. (1986) *J. Mol. Biol. 187*, 75–86.

56. Chuang, D.T., Hu, C.C., Ku, L.S., Markovitz, P.J., & Cox, R.P. (1985) *J. Biol. Chem. 260*, 13779–13786.

57. Spencer, M.E., Darlinson, M.G., Stephens, P.E., Duckenfield, I.K., & Guest, J.R. (1984) *Eur. J. Biochem. 141*, 361–374.

58. DeRosier, D.J., Oliver, R.M., & Reed, L.J. (1971) *Proc. Natl. Acad. Sci. U.S.A. 68*, 1135–1137.

59. Bleile, D.M., Munk, P., Oliver, R.M., & Reed, L.J. (1979) *Proc. Natl. Acad. Sci. U.S.A. 76*, 4385–4389.

60. Bleile, D.M., Hackert, M.L., Pettit, F.M., & Reed, L.J. (1981) *J. Biol. Chem. 256*, 514–519.

61. Wagenknecht, T., Francis, N., & DeRosier, D.J. (1983) *J. Mol. Biol. 165*, 523–541.

62. Wu, T.L., & Reed, L.J. (1984) *Biochemistry 23*, 221–226.

63. Green, N.M., Valentine, R.C., Wrigley, N.G., Ahmad, F., Jacobson, B., & Wood, H.G. (1972) *J. Biol. Chem. 247*, 6284–6298.

64. Skrzypczak-Jankun, E., Tulinsky, A., Fillers, J.P., Kumar, K.G., & Wood, H.G. (1986) *J. Mol. Biol. 188*, 495–498.

65. Ungewickell, E., & Branton, D. (1981) *Nature 289*, 420–422.

66. Pearse, B.M.F., & Robinson, M.S. (1984) *EMBO J. 3*, 1951–1957.

67. Vigers, G.P.A., Crowther, R.A., & Pearse, B.M.F. (1986) *EMBO J. 5*, 529–534.

68. Ford, G.C., Harrison, P.M., Rice, D.W., Smith, J.M.A., Treffrey, A., White, J.L., & Yariv, J. (1984) *Philos. Trans. R. Soc. London, B 304*, 551–565.

69. Klug, A. (1972) *Fed. Proc. 31*, 30–42.

70. Finch, J.T., & Klug, A. (1974) *J. Mol. Biol. 87*, 633–640.

71. Amos, L.A., & Klug, A. (1975) *J. Mol. Biol. 99*, 51–73.

72. DeRosier, D.J., & Klug, A. (1968) *Nature 217*, 130–134.

73. Dubochet, J., LePault, J., Freeman, R., Berriman, J.A., & Homo, J.C. (1982) *J. Microsc. 128*, 219–237.

74. Crowther, R.A., DeRosier, D.J., & Klug, A. (1970) *Proc. R. Soc. London, A 317*, 319–340.

75. Miller, A., & Wray, J.S. (1971) *Nature 230*, 437–439.

76. Miller, A., & Parry, D.A.D. (1973) *J. Mol. Biol. 75*, 441–447.

77. Dayhoff, M.O. (1978) *Atlas of Protein Sequence and Structure*, Vol. 5, Suppl. 3, National Biomedical Research Foundation, Washington, DC.

78. Hofmann, H., Fietzek, P.P., & Kuhn, K. (1978) *J. Mol. Biol. 125*, 137–165.

79. Highberger, J.H., Corbett, C., Dixit, S.N., Yu, W., Seyer, J.M., Kang, A.H., & Gross, J. (1982) *Biochemistry 21*, 2048–2055.

80. Bernard, M.P., Chu, M., Myers, J.C., Ramirez, F., Eikenberry, E.F., & Prockop, D.J. (1983) *Biochemistry 22*, 5213–5223.

81. Hodge, A.J., & Schmidt, F.O. (1960) *Proc. Natl. Acad. Sci. U.S.A. 46*, 186–206.

82. Meek, K.M., Chapman, J.A., & Hardcastle, R.A. (1979) *J. Biol. Chem. 254*, 10710–10714.

83. Crick, F.H.C. (1953) *Acta Crystallogr. 6*, 689–697.

84. MacLachlan, A.D., & Stewart, M. (1975) *J. Mol. Biol. 98*, 293–304.

85. Pauling, L., & Corey, R.B. (1953) *Nature 171*, 59–66.

86. Doolittle, R.F., Goldbaum, D.M., & Doolittle, L.R. (1978) *J. Mol. Biol. 120*, 311–325.

87. Bailey, K., Astbury, W.T., & Ruddall, K.M. (1943) *Nature 151*, 716–717.

88. O'Shea, E.K., Rutkowski, R., & Kim, P.S. (1989) *Science 243*, 538–542.

89. O'Shea, E.K., Klemm, J.D., Kim, P.S., & Alber, T. (1991) *Science 254*, 539–544.

90. Wakabayashi, T., Huxley, H.E., Amos, L.A., & Klug, A. (1975) *J. Mol. Biol. 93*, 477–497.

91. Amos, L.A., & Klug, A. (1974) *J. Cell Sci. 14*, 523–549.

92. Franke, W.W., Schmid, E., Osborn, M., & Weber, K. (1978) *Proc. Natl. Acad. Sci. U.S.A. 75*, 5034–5038.

93. Geisler, N., Fischer, S., Vandekerckhove, J., VanDamme, J., Plessmann, U., & Weber, K. (1985) *EMBO J. 4*, 57–63.

94. Renner, W., Franke, W.W., Schmid, E., Giesler, N., Weber, K., & Mandelkow, E. (1981) *J. Mol. Biol. 149*, 285–306.

95. Fraser, R.D.B., & MacRae, T.P. (1971) *Nature 233*, 138–140.

96. Fraser, R.D.B., & MacRae, T.P. (1976) *J. Mol. Biol. 108*, 435–452.

CHAPTER 10

Chemical Probes of Structure

Crystallographic molecular models have been obtained for only a small minority of the proteins that have been purified. This is not due simply to the fact that only 30 years have elapsed since the first molecular models were published. There are also a number of technical problems associated with the crystallographic method, and these have proved baffling, if not insurmountable, in many instances. Often a purified protein has not been crystallized. Often the crystals of protein deteriorate too rapidly in the beam of X-rays. Often the space group is too complex to permit a ready solution. Often the crystals of a particular protein are microscopically disordered. Often the isomorphous replacements have not been obtained.

There are indications that many of these problems may be solved eventually. For example, crystals of membrane-spanning proteins, a class of proteins that are extremely difficult to crystallize, have recently been obtained and used successfully for high-resolution crystallography.[1,2] The development of area detectors, which permit a complete data set to be gathered in a very short period of time, have shortened the required length of exposure to the beam.[3] Changing the conditions of crystallization[4] or the species from which the protein has been purified[5] can sometimes give crystals that have a different space group or are more ordered. As more reagents containing heavy metals become available, the odds against finding a suitable set of isomorphous replacements decrease. Yet it still seems likely that the majority of the proteins that have been or will be purified will never yield high-resolution maps of electron density.

In the absence of a map of electron density, the molecular structure of a protein is studied by a diverse collection of techniques. These approaches can be conveniently divided into three classes: the use of chemical probes, the use of immunochemical probes, and the use of physical measurements. The chemical probes most commonly used are electrophilic reagents that modify particular amino acids.

Several types of amino acids in a molecule of protein are susceptible to covalent modification.[6] As all of the reactive functional groups are **nucleophiles**, the reagents used to modify proteins are usually **electrophiles**. Serine and threonine contain nucleophilic oxygens, but they are indistinguishable from those of the water and cannot be modified. The amino acids possessing nucleophilic sites that can be modified are cysteines, methionines, lysines, histidines, tyrosines, glutamates, aspartates, arginines, and tryptophans. The electrophiles modify these amino acids by reactions that couple the electrophile covalently

to them. In the process, the ability of the amino acid to act as an acid–base is often lost because an atom of the electrophile, usually a carbon, forms a covalent bond to the conjugate base of the central heteroatom. The proton that occupies the conjugate acid, which is the smallest possible atom, is replaced with the whole molecular structure of the electrophile, and this also increases the size of the side chain dramatically. After its modification, the amino acid is no longer able to participate in any particular role it might have had in the function of the protein or fit into the same space. Accordingly, the function or the structure of the protein is often disrupted.

Chemical modifications of amino acids in a protein are used for many different purposes. Most of the uses are designed imaginatively to answer a particular question about a particular protein, so it is impossible to give an exhaustive list of the reasons for modifying amino acids in proteins. A few examples, however, will indicate why such experiments are so common.

The most common purpose for using covalent modification is to demonstrate that a particular type of amino acid is involved in the function of the protein. For example, fibrinogen, upon activation, polymerizes to form long helical polymers that produce a clot. The initial polymerization is noncovalent, but the polymer is then strengthened by posttranslational cross-links. When the lysines of fibrinogen were modified by amidination, the initial noncovalent polymer could form normally, but the covalent cross-links could not.[7] This evidence was the basis for the prediction that the posttranslational cross-links were amides between glutamates and lysines. The most common use of covalent modification to study the function of a protein is the observation of the inactivation of an enzyme by covalent modification of amino acids in its active site. A complete description of this strategy will be presented in a later chapter.

There are, however, many other purposes for covalently modifying proteins. Covalent modification can result in the dissociation of the subunits of a protein. For example, succinylation of its lysines caused the hemerythrin of *Goldfingia gouldi* to dissociate into its eight identical subunits.[8] Covalent modification can also be used to change the electrophoretic mobility of a protein by converting, for example, positively charged lysines into negatively charged carboxylates.[9] When this is performed reversibly, the protein will travel with a different mobility before the reaction is reversed than after the reaction is reversed. Covalent modifications can be used to prevent proteolytic enzymes, for example, trypsin, from digesting the protein at particular amino acids, for example, arginine.[10] Covalent modifications are also used to introduce foreign functional groups into proteins. For example, functional groups that absorb visible light[11] or have strong fluorescence[12] may be introduced so that their spectral properties can be used in physical studies.

When a protein is modified covalently, either the modification is performed under conditions that produce a high yield of the desired product, and this is used in further experiments, or the chemical reaction itself between the protein and the reagent is monitored, and its kinetics are used to make arguments about the properties of a particular amino acid in the native protein. For example, the dependence of the rate of the modification on the pH of the solution can be used to obtain the pK_a of the amino acid.[13] Differences in rate constants for the reaction of amino acids of a particular type with the same electrophile can be used to assess differences in their accessibility to the solution in the native structure of the protein.[14]

In all of these experiments, the possibility that the covalent modification itself disrupts the global conformation of the protein must be considered. If this happens, effects of the modification on the function of the protein might be attributed to local changes around the specific amino acid when they actually result from the disruption of the entire structure of the protein. The modified protein should always be examined to rule out this possibility. For example, it could be shown by following electrophoretic mobility, sedimentation rate in the ultracentrifuge, and optical rotation that exhaustive amidination of the lysines in either serum albumin or immunoglobulin G had no measurable effect on the structures of these proteins.[15]

In designing an experiment that involves the covalent modification of a protein, the usual desire is that the reagent chosen react with only one type of amino acid. Because cysteines, methionines, lysines, histidines, and tyrosines are similarly reactive nucleophiles, this is not a simple task.

The issue of specificity is best addressed by examining the reaction of a simple alkylating agent, such as **iodoacetamide**, with the nucleophiles present in a protein. Iodoacetamide has been shown to alkylate cysteines, lysines, histidines, and methionines.[16] The four reactions are

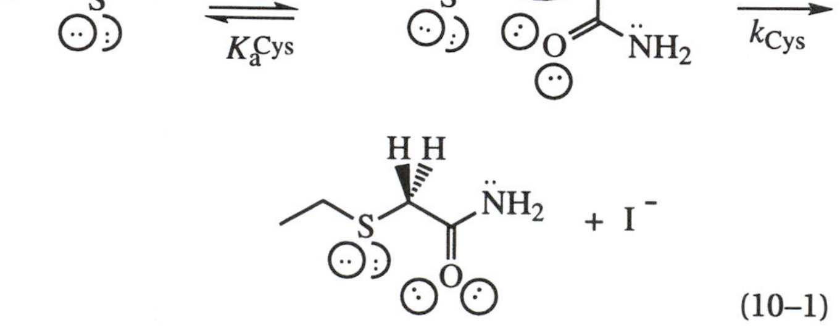

(10–1)

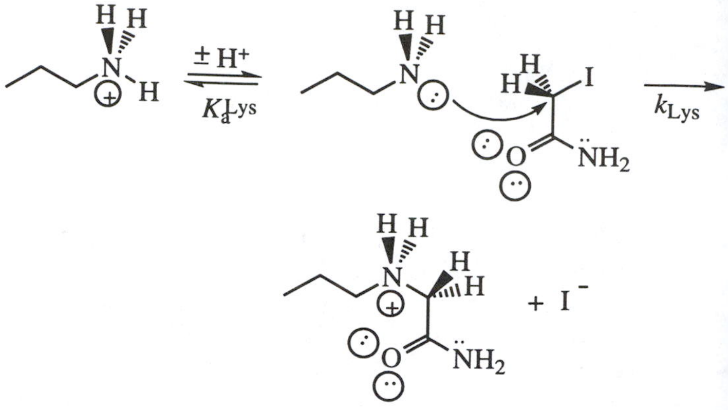

(10–2)

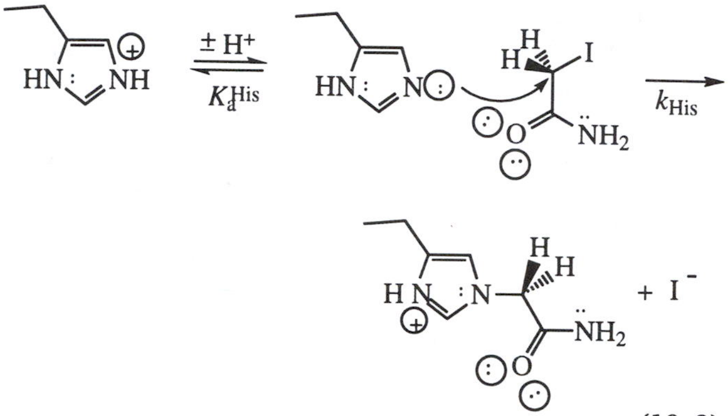

(10–3)

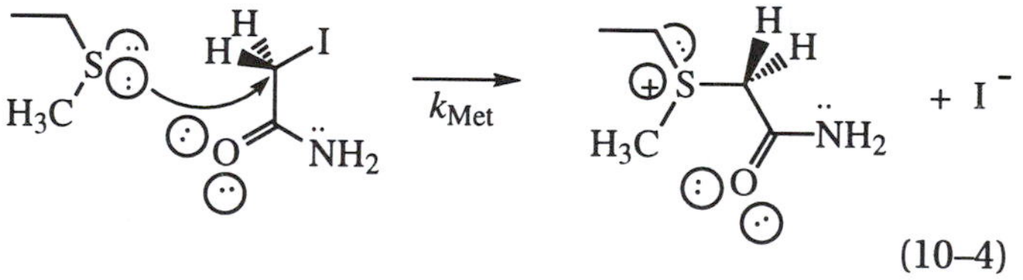

(10–4)

When a protein is exposed to iodoacetamide all four of these amino acids disappear from amino acid analyses of the reaction mixtures, at rates that depend on the pH,[17] and the carboxymethyl products[18,19] appear in concert. The first three reactions require that the amino acid be in the form of its conjugate base.

The rate of the reaction between lysine and iodoacetamide can be described by

$$\frac{d[Lys]_{TOT}}{dt} = -k_{Lys}[RH_2N\odot][\text{iodoacetamide}] \quad (10-5)$$

where $[Lys]_{TOT}$ is the total concentration of unmodified lysine (both protonated, RNH_3^+, and unprotonated, $RH_2N\odot$) at a particular time, t. Substitution of the appropriate terms into Equation 10–5 leads to

$$\frac{d[Lys]_{TOT}}{dt} = -\left(\frac{k_{Lys}K_a^{Lys}}{K_a^{Lys}+[H^+]}\right)[Lys]_{TOT}[\text{iodoacetamide}]$$

(10–6)

If the concentration of iodoacetamide is so large that it remains constant throughout the reaction and the pH does not change, Equation 10–6 describes a pseudo-first-order reaction. The pseudo-first-order rate constant, k'_{Lys} governing the disappearance of lysine with time is

$$k'_{Lys} = -\left(\frac{k_{Lys}K_a^{Lys}}{K_a^{Lys}+[H^+]}\right)[\text{iodoacetamide}] \qquad (10\text{–}7)$$

and

$$\frac{d[\text{Lys}]_{TOT}}{dt} = -k'_{Lys}[\text{Lys}]_{TOT} \qquad (10\text{–}8)$$

When this is rearranged

$$\frac{d[\text{Lys}]_{TOT}}{[\text{Lys}]_{TOT}} = -k'_{Lys}\,dt \qquad (10\text{–}9)$$

and integrated from $t = 0$ to $t = t$

$$\int_{[\text{Lys}]^0_{TOT}}^{[\text{Lys}]_{TOT}} \frac{d[\text{Lys}]_{TOT}}{[\text{Lys}]_{TOT}} = -\int_0^t k'_{Lys}\,dt \qquad (10\text{–}10)$$

where $[\text{Lys}]^0_{TOT}$ is the initial concentration of unmodified lysine

$$\left|\begin{matrix}[\text{Lys}]_{TOT}\\ [\text{Lys}]^0_{TOT}\end{matrix}\right. \ln[\text{Lys}]_{TOT} = -\left|\begin{matrix}t\\0\end{matrix}\right. k'_{Lys} \qquad (10\text{–}11)$$

It follows that

$$\ln[\text{Lys}]_{TOT} = \ln[\text{Lys}]^0_{TOT} - k'_{Lys}t \qquad (10\text{–}12)$$

and

$$[\text{Lys}]_{TOT} = [\text{Lys}]^0_{TOT}\,e^{-k'_{Lys}t} \qquad (10\text{–}13)$$

As in all first-order reactions, the rate constant k'_{Lys} is the slope of the line obtained when $\ln[\text{Lys}]_{TOT}$ is plotted against time.

Each of the first three reactions (Reactions 10–1 through 10–3) has a formally equivalent mechanism, and the pseudo-first-order rate constants k'_i of each of them, where i designates the particular amino acid, are of the same form as k'_{Lys} (Equation 10–7) with the appropriate rate constants, k_{Cys} or k_{His}, and acid dissociation constants, K_a^{Cys} or K_a^{His}, substituted for k_{Lys} and k_a^{Lys}, respectively. The variation in each of these rate constants k'_i can be presented graphically (Figure 10–1).

At values of pH greater than the pK_a^{Lys} of lysine ($[H^+] < K_a^{Lys}$) almost all of the primary amine is the conjugate base, and both k'_{Lys}, which is equal to k_{Lys} (Equation 10–7), and the rate of the reaction of lysine with iodoacetamide are independent of pH. At values of pH below pK_a^{Lys} ($[H^+] > K_a^{Lys}$) the concentration of the unprotonated conjugate base of lysine, and hence both the rate of its reaction with iodoacetamide and k'_{Lys} (Equation 10–7) decreases by 1 logarithmic unit for each decrease of 1 unit in pH (Figure 2–6).

The rate of the reaction of the unprotonated conjugate base of histidine with iodoacetamide, which is governed solely by k_{His} (Equation 10–3), is correlated to the rate of the reaction of the unprotonated conjugate base of lysine with iodoacetamide, k_{Lys} (Equation 10–2), through the Brønsted relationship

$$\log\left(\frac{k_{His}}{k_{Lys}}\right) = -\beta\log\left(\frac{K_a^{His}}{K_a^{Lys}}\right) \qquad (10\text{–}14)$$

Below a certain pH (in this case pH 8) histidine reacts more rapidly than lysine, while above that pH lysine reacts more rapidly than histidine because β is usually between 1.0 and 0. If β were zero, the unprotonated conjugate bases of both histidine and lysine would react at equal rates with iodoacetamide ($k_{His} = k_{Lys}$), and the curves for lysine and histidine would coincide at values of

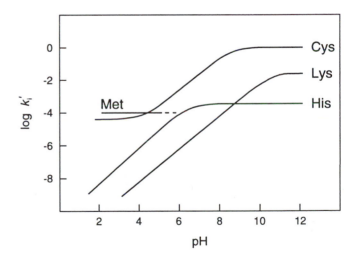

Figure 10–1: Variation with pH of the logarithm of the pseudo-first-order rate constant, k'_i, for the reaction between methionine, cysteine, histidine, or lysine in a protein with iodoacetamide. The first-order rate constants are on a relative scale with the rate constant of cysteinate anion arbitrarily assigned a value of 1.0. The individual lines were drawn according to the respective forms of Equation 10–7 with pK_a values of 6.5 for histidine, 8.5 for cysteine, and 10.5 for lysine. The relative vertical positions of the lines were fixed by assuming[17] that methionine reacts 10 times more slowly than cysteine at pH 5.5, histidine reacts 30 times more slowly than cysteine at pH 5.5, methionine reacts at the same rate at pH 5.5 and 8.5, lysine reacts 2.5 times more quickly than methionine at pH 8.5, and histidine reacts 2 times more slowly than methionine at pH 8.5. The rate of the reaction with cysteine below pH 4 is assumed to be invariant with pH, as is the rate of the reaction with methionine, but with a rate constant less than that of methionine.

pH above the pK_a of lysine. At values of pH below the pK_a of histidine, the pseudo-first-order rate constant, k'_{His}, for its reaction with iodoacetamide decreases by 1 logarithmic unit for each decrease of 1 unit in pH. At low pH, therefore, the rates of the reactions of both lysine and histidine with iodoacetamide decrease in concert and remain in constant ratio to each other. If β were equal to 1.0 (Equation 10–14), the rate constants for the modification of both lysine, k'_{Lys}, and histidine, k'_{His} (Equation 10–7), would be equal at low pH and the lines for histidine and lysine in Figure 10–1 would coincide at values of pH less than pK_a^{His}.

The rate constant, k_{Cys}, for the reaction of the unprotonated conjugate base of cysteine with iodoacetamide is significantly greater than that of even unprotonated lysine. Because sulfur is a third-row element and nitrogen is a second-row element, cysteine is more nucleophilic than lysine, even though the pK_a (8.5) associated with its lone pair of electrons is less than the pK_a (10.5) associated with the lone pair of electrons on lysine. With the appropriate substitutions, Equation 10–7 governs the behavior of the rate of the reaction of cysteine with iodoacetamide as a function of pH at values of pH above and below the pK_a of cysteine. Cysteine (Reaction 10–1), however, unlike lysine (Reaction 10–2) and histidine (Reaction 10–3), retains two nucleophilic lone pairs of electrons after its protonation and can react with iodoacetamide as the conjugate acid in a reaction analogous to that of methionine (Reaction 10–4).

The reaction of methionine with iodoacetamide is invariant with pH because its acid dissociation constant ($pK_a^{Met} = -9$) is below accessible ranges of pH. The rate constant for the reaction of protonated cysteine with iodoacetamide should be lower than the rate constant, k_{Met}, for the reaction of methionine because of hyperconjugation. At low pH the rate of the reaction of cysteine with iodoacetamide should level off at a value lower than k_{Met} and also should become invariant with pH because the concentration of neutral cysteine is invariant with pH.

The individual behavior of each of the reactions determines the specificity of iodoacetamide. At the lowest values of pH, methionine is the most reactive amino acid. As the pH is increased, into the range where the concentration of the thiolate anion becomes sufficiently large, cysteine becomes the most reactive amino acid at all higher values of pH. As the pH is increased, histidine becomes as reactive as methionine because the lone pair of electrons of its conjugate base ($pK_a^{His} = 6.5$) is so much more basic than that of methionine ($pK_a^{Met} = -9$). At even higher values of pH, lysine becomes more reactive than histidine. All of these consequences determine which amino acid reacts most rapidly with iodoacetamide at a particular pH.

The reagent used to modify the amino acids of a protein may itself also be affected by alterations in pH. **Methyl acetimidate** is a reagent that modifies lysines with high specificity (Figure 10–2). The specificity results in part from the fact that while the product of the reaction with lysine is a stable amidine, the analogous products with cysteine, methionine, glutamate, aspartate, tyrosine, and histidine are unstable and they decompose as quickly as they are produced. The effect of pH on the rate of the reactions between amines and imidates has been explained mechanistically (Figure 10–2)[20] with the assumption that the reactive form of the lysine is the free base and the reactive form of the imidate is the cationic conjugate acid. For methyl acetimidate,[21] $pK_a^{AI} = 7.5$, and the concentration of cationic imidate should be decreasing as the pH is raised above pH 7, while that of the free base of lysine should be increasing.

At the highest values of pH, the rate of the reaction between lysine and methyl acetimidate is governed by the equation

$$\frac{d[Lys]_{TOT}}{dt} = -k_{AI}[RH_2N\ominus][R'=NH_2^+] \quad (10\text{–}15)$$

from which it follows that

$$\frac{d[Lys]_{TOT}}{dt} = -\left(\frac{k_{AI}K_a^{Lys}[H^+]}{\left(K_a^{Lys}+[H^+]\right)\left(K_a^{AI}+[H^+]\right)}\right)[Lys]_{TOT}[AI]_{TOT} \quad (10\text{–}16)$$

where $[AI]_{TOT}$ is the total concentration of methyl acetimidate, both conjugate acid, $R'=NH_2^+$, and conjugate base, $R'=NH$.

If $[AI]_{TOT}$ is high and constant throughout the course of the reaction, and the pH does not change, Equation 10–16 describes a pseudo-first-order reaction. The pseudo-first-order rate constant, k'_{AI}, governing the disappearance of lysine with time would be

$$k'_{AI} = \frac{k_{AI}K_a^{Lys}[H^+][AI]_{TOT}}{\left(K_a^{Lys}+[H^+]\right)\left(K_a^{AI}+[H^+]\right)} \quad (10\text{–}17)$$

When $pK_a^{AI} < pK_a^{Lys} < pH$

$$k'_{AI} = \frac{k_{AI}[H^+]}{K_a^{AI}}[AI]_{TOT} \quad (10\text{–}18)$$

and k'_{AI} decreases by 1 logarithmic unit for each increase of 1 unit in pH. Between the two acid dissociation constants, when $pK_a^{AI} < pH < pK_a^{Lys}$, Equation 10–17 would predict that

$$k'_{AI} = \frac{k_{AI}K_a^{Lys}}{K_a^{AI}}[AI]_{TOT} \quad (10\text{–}19)$$

and the rate of the reaction should be almost invariant with pH, and when $pH < pK_a^{AI} < pK_a^{Lys}$

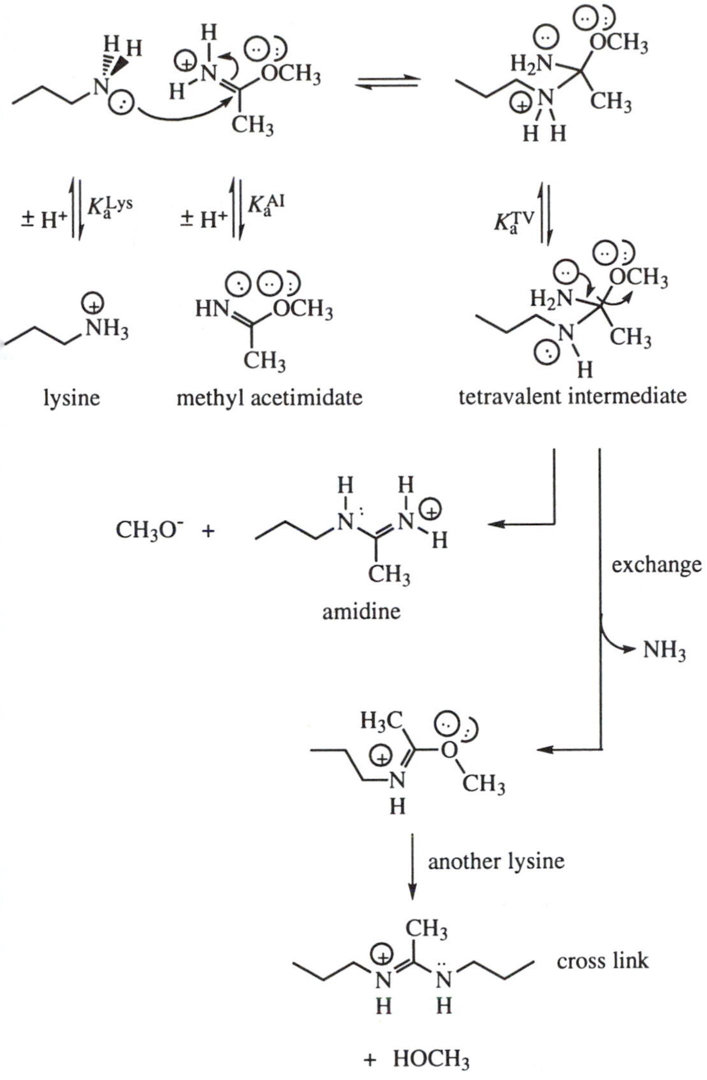

Figure 10–2: Mechanism for the reaction between the free base of lysine and the conjugate acid of methyl acetimidate.[20] The products of the reaction can be either the amidate of lysine and ammonia or the amidate of two lysines. The latter reaction produces cross-links within the protein but is rarer than the former.

$$k'_{AI} = \frac{k_{AI} K_a^{Lys}}{[H^+]} [AI]_{TOT} \qquad (10-20)$$

and k'_{AI} decreases by 1 logarithmic unit for every decrease of 1 unit in pH. While the reaction of amines with ethyl benzimidate is more complicated than this simple picture,[20] in the case of the reaction of methyl acetimidate with the lysines in unfolded aldolase,[21] the plateau between pK_a^{AI} and pK_a^{Lys} is observed as predicted and the decrease in the rate of amidation does not occur until below pH 7.5. Between pH 7.5 and 9 the rate of the modification of lysine in a protein by methyl acetimidate is invariant with pH as predicted by Equation 10–17.

The fact that the modification of a protein is usually performed in aqueous solution limits the reagents that can be used. On the one hand, problems of solubility of the electrophile often arise, thereby restricting its useful range of concentrations. On the other hand, because water is itself a nucleophile, decomposition of the reagent through hydrolysis often occurs. Methyl acetimidate, unlike iodoacetamide, reacts quite readily with water. Between pH 6.8 and 8.4, the half-time for its hydrolysis at 20 °C is about 30 min.[21]

When the reagent chosen for a particular modification decomposes rapidly, measurements of its rate of reaction with the protein are complicated by this decomposition. In the case of methyl acetimidate, the situation can be represented by the kinetic mechanism

$$\text{lysine + methyl acetimidate} \xrightarrow{k_1} \text{amidine}$$

$$\downarrow k_2 \qquad\qquad (10-21)$$

$$\text{products of hydrolysis}$$

In this situation, where hydrolysis is occurring coincidentally with modification, it can be shown[22] that

$$\ln\left(\frac{[Lys]_{TOT}^0}{[Lys]_{TOT}^0 - [amidine]_f}\right) = \frac{k_1}{k_2}[AI]_{TOT}^0 \qquad (10-22)$$

where $[Lys]_{TOT}^0$ and $[AI]_{TOT}^0$ are the initial total molar concentrations of lysine and methyl acetimidate and $[amidine]_f$ is the final concentration of modified lysine when all of the methyl acetimidate has been consumed either by reaction with lysine or by hydrolysis. In effect, this allows the reaction to be studied more leisurely. A series of mixtures containing the protein are prepared with increasing concentrations of methyl acetimidate at constant temperature and pH. The reaction is allowed to reach completion. The yield of amidine is assessed. From these results and Equation 10–22, the ratio $k_1 (k_2)^{-1}$ can be determined. If the rate constant k_2 has been measured in a separate experiment under identical conditions, k_1 can be obtained directly.

Alkyl imidates, such as methyl acetimidate, react specifically and in high yield with **lysine** in proteins. In the case of myoglobin, for example, the two major products of the reaction with methyl acetimidate were proteins amidinated at every primary amine except either Lysine 77 or the amino terminus.[23] An amidine ($pK_a = 12.5$) is positively charged at pH 7; and, as a result, no change in the charge of the protein occurs during the amidination of its lysines.

Another way to direct the modification exclusively to lysines is to take advantage of the fact that primary amines such as lysine are the only functional groups on a protein that react with **aldehydes** to form imines (Figure 10–3). The conjugate acid of the imine can then be reduced with sodium borohydride to produce the secondary amine. Both formaldehyde[24] and pyridoxal 5'-phosphate[25] have been used as the aldehyde. The former is the more reactive aldehyde and produces much higher yields of alkylated lysine; the latter is more selective and under the proper conditions will modify only the most nucleophilic lysines in a protein.

Isothiocyanates are also specific for the primary amino groups of lysines, as well as the amino terminus, of a protein

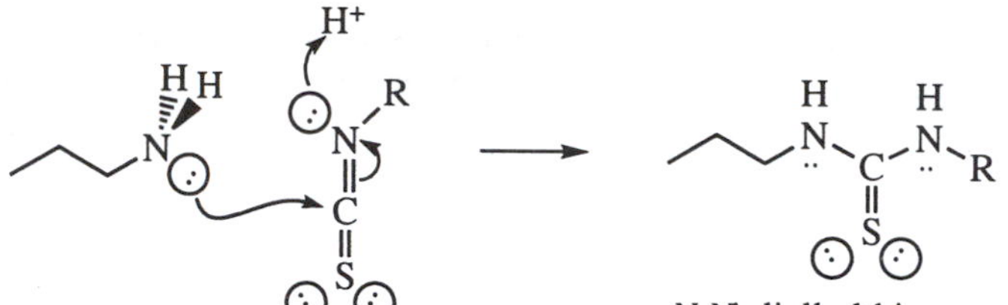

N,N'-dialkylthiourea

$$(10–23)$$

The products of the modification are N,N'-dialkyl-thioureas. Presumably, isothiocyanates are specific for lysine because the products they would form with the other nucleophilic amino acids are unstable under the reaction conditions. A similar reaction occurs with **isocyanates**

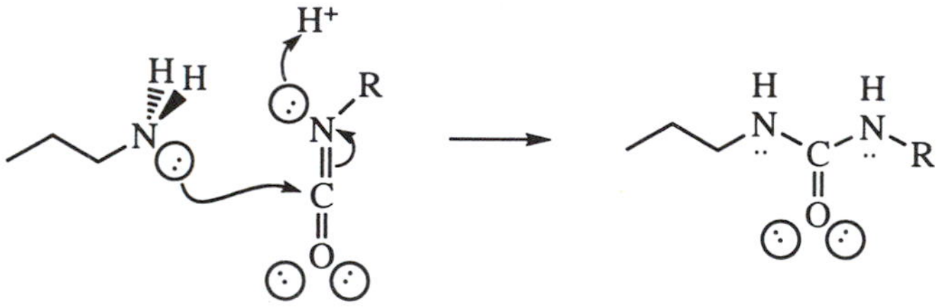

$$(10–24)$$

The products are N,N'-dialkylureas. Alkyl and aryl isocyanates reacted with cysteine and tyrosine as well to produce products that can be hydrolyzed back to the unmodified amino acids under alkaline conditions,[6] to leave only the lysines modified.

A large collection of **acylating agents** react with lysine and acylate other nucleophilic amino acids as well. The general reaction performed by these reagents is acyl transfer (Figure 10–4). As in synthetic organic chemistry, the reagent is chosen for its electrophilicity. The properties of the leaving group X determine both the electrophilicity and the rate at which the reagent will modify the lysines in a protein. Because the leaving group departs from a Lewis acid, the tendency for the leaving group to depart from a proton, another Lewis acid, will reflect its tendency to depart from the tetravalent intermediate (Figure 10–4). Therefore the larger the acid dissociation constant K_a^{LG} of the conjugate acid of the leaving group X, the more reactive will be the reagent. If the leaving group is the carboxylic acid itself, the reagent is an anhydride such as trifluoroacetic anhydride ($pK_a^{LG} = 0.2$) or acetic anhydride[26] ($pK_a^{LG} = 4.8$). Other leaving groups that have been used in the modification of lysine are azide[6] ($pK_a^{LG} = 4.7$), N-hydroxysuccinimide[27] ($pK_a^{LG} = 6.0$),[28] imidazole[26] ($pK_a^{LG} = 7.0$), ethyl carbonate[29] ($pK_a^{LG} = 7$), and ethanethiol[30] ($pK_a^{LG} = 10.5$).

All of these acylating agents react as readily with cysteine, tyrosine, and histidine as they do with lysine to form the respective S-, O-, or N-acyl derivatives. Unlike the S-, O-, or N-alkyl derivatives formed during the reaction with an alkylating reagent such as iodoacetamide (Reactions 10–1 through 10–4), these acyl derivatives of cysteine, tyrosine, and histidine are often unstable and decompose spontaneously or can be decomposed intentionally under conditions that leave the lysines in the

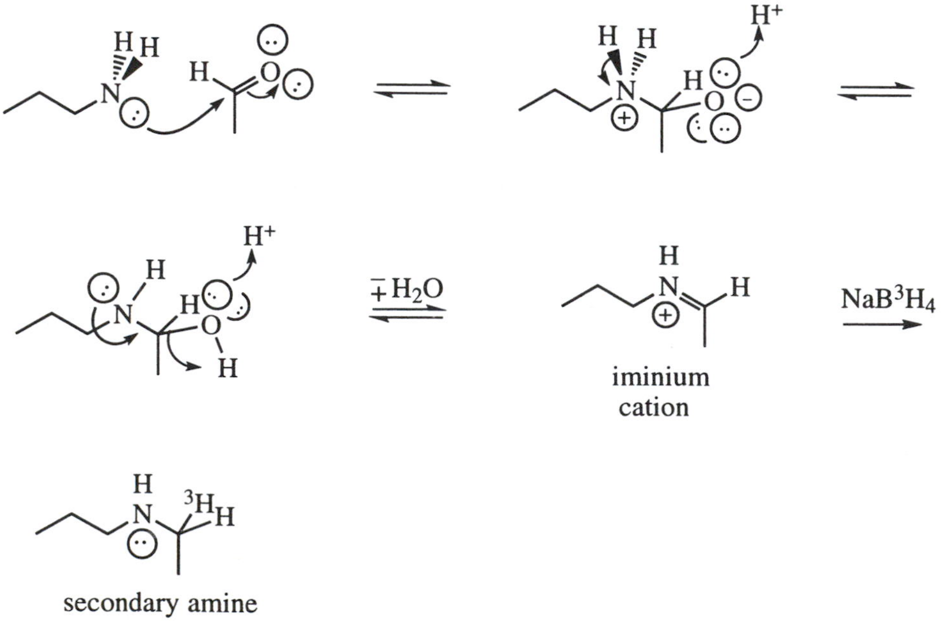

Figure 10–3: Reaction of lysine with an aldehyde to form an iminium cation, which can be reduced to the secondary amine with sodium borohydride ($NaBH_4$). If tritiated sodium borohydride (NaB^3H_4) is used, tritium is incorporated into the secondary amine.

iminium cation

secondary amine

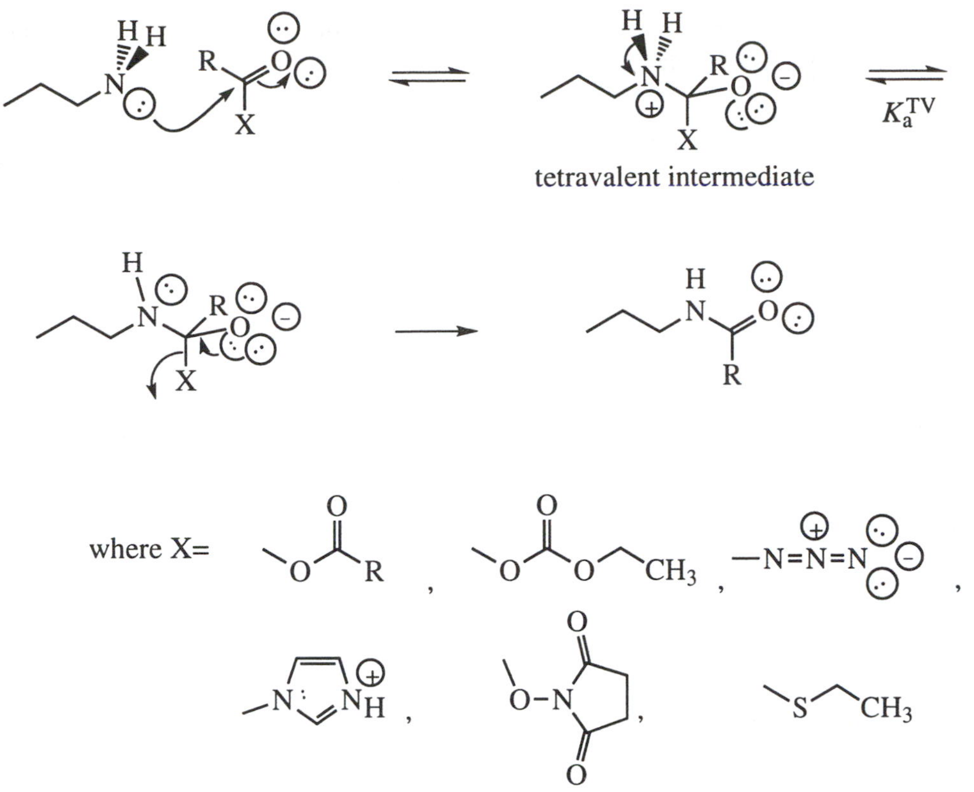

Figure 10–4: Acylation of lysine with any one of a number of acylating agents in which the acyl carbon is activated by attaching an excellent leaving group. In the tetravalent intermediate, the leaving group is expelled, in preference to the nitrogen of the lysine, to produce the lysine amide. The activating groups used to produce the acylating agent from a carboxylic acid are the carboxylic acid itself to form the anhydride, ethyl carbonic acid to form the acyl ethyl carbonate, azide anion to form the acyl azide, imidazole to form the acyl imidazole, N-hydroxysuccinimide to form the N-hydroxysuccinimide ester, or a thiol to form the thioester.

modified form. For example, O-acetyltyrosine can be hydrolyzed back to tyrosine by treatment with hydroxylamine.[31] Often an acylating agent whose structure renders the undesired derivatives particularly unstable can be chosen. For example, the ethoxycarbonyl group is added to tyrosine and histidine as well as lysine when one uses the carbonic acid anhydride diethyl pyrocarbonate

diethylpyrocarbonate

$$\text{H}_3\text{C}-\text{O}-\overset{\text{O}}{\overset{\|}{\text{C}}}-\text{N}-\text{CH}_2 + \text{CO}_2 + \text{HOC}_2\text{H}_5$$

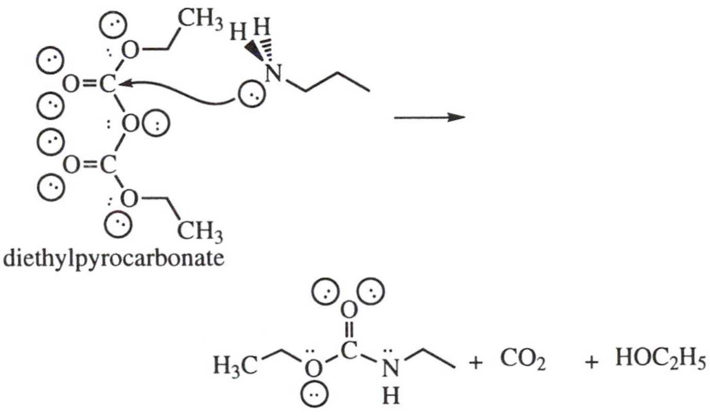

(10–25)

The ethoxycarbonyl group, however, can be removed from the histidine and tyrosine by treatment of the modified protein with hydroxylamine.[29]

Cyclic anhydrides such as succinic anhydride

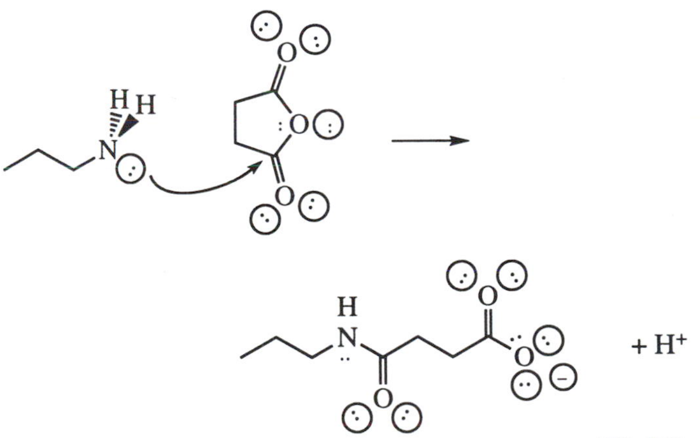

(10–26)

are frequently used to modify the lysines in a protein. They replace the positively charged primary ammonium cation of lysine with a negatively charged carboxylate. The acylation is rendered reversible by the use of maleic anhydride (2,3-dehydrosuccinic anhydride), citraconic anhydride,[32] or 3,4,5,6-tetrahydrophthalic anhydride.[9]

Fluorosulfonic acids are general electrophilic reagents that modify lysine, by N-sulfonation, and tyrosine, by O-sulfonation. The paradigm for these reagents is 5-(dimethylamino)naphthalene-1-sulfonyl fluoride, dansyl fluoride, which is used not only in determinations of the amino terminus of a peptide but also as a fluorescent reagent suitable for the covalent modification of proteins. Presumably, the sulfonyl derivatives of cysteine, histidine, methionine, glutamate, or aspartate, if they form at all, rapidly hydrolyze.

Both 2,4-dinitrofluorobenzene (FDNB) and 2,4,6-trinitrobenzenesulfonate (TNBS)[33] modify lysine by **nucleophilic aromatic substitution**

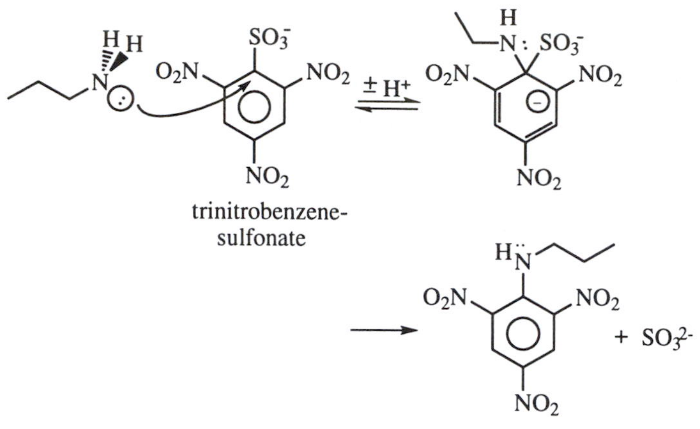

(10-27)

The former compound reacts with every nucleophilic amino acid[34] while the latter can be confined to react with only cysteine[6] and lysine by the proper choice of pH.[33] The derivative formed with cysteine, however, is unstable, so in the end only lysine is modified.

As noted previously, alkylating reagents such as iodoacetamide and other alkyl halides are electrophiles that react with every nucleophilic amino acid to yield stable products. Informed or uninformed manipulation of the pH can affect the distribution of alkylated products. Alkylating reagents can be used to produce stable derivatives of **methionine** specifically. For example, at slightly acidic values of pH, benzyl bromide will alkylate methionine in fumarase quite selectively[35]

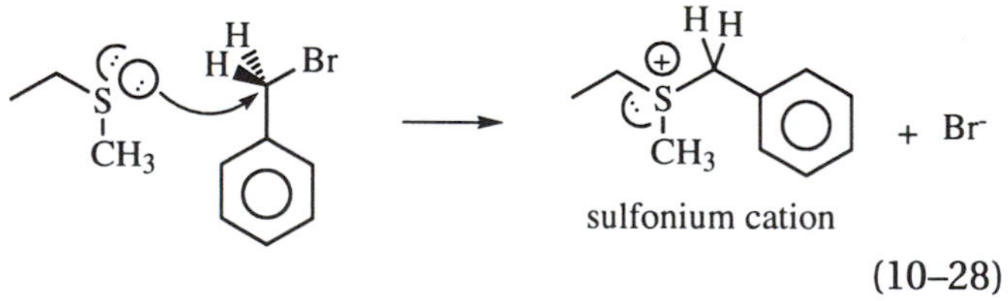

(10-28)

The reagent that has shown the greatest selectivity for **histidine** is **diethyl pyrocarbonate** (Reaction 10-25).[29] Usually this selectivity is obtained by running the reaction slightly below the pK_a for histidine, where the greatest discrimination in favor of histidine, relative to lysine (Reaction 10-25), should be manifested (Figure 10-1). Histidine is also susceptible to **photooxidation** in the presence of dyes such as methylene blue or rose bengal.[36] Under carefully controlled conditions such photooxidation can be confined to histidine,[36,37] but usually many different amino acids are destroyed simultaneously.[36]

One of the most readily modified amino acids in a protein is **cysteine**. At slightly alkaline pH, in the vicinity of its pK_a (Figure 10-1), cysteine is preferentially alkylated by **alkyl halides** such as iodoacetamide and

iodoacetate. **N-Ethylmaleimide** (NEM) is another reagent often used to modify cysteine[38]

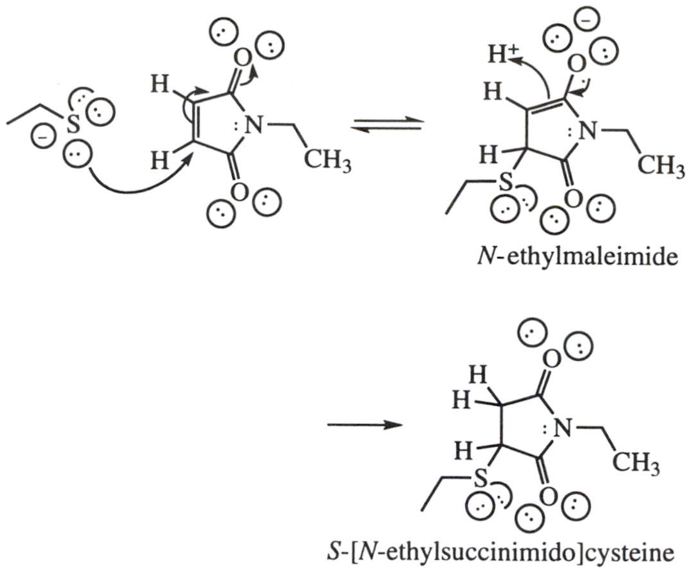

N-ethylmaleimide

S-[N-ethylsuccinimido]cysteine

(10-29)

This reaction is an example of nucleophilic addition to an α,β-unsaturated acyl compound. Lysine also reacts with N-ethylmaleimide to form the analogous N-[N-ethylsuccinimidyl]lysine.[39] 2-Vinylpyridine[40] is highly selective for modification of cysteine in a similar reaction. Organic mercurials, such as p-chloromercuribenzoate (PCMB),[38] are usually specific for cysteine

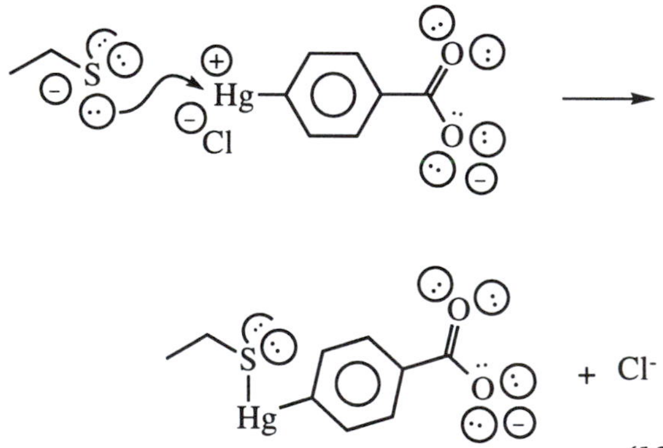

(10-30)

The sulfur–mercury bond, because it is significantly covalent, is particularly stable, but not stable enough to survive upon subsequent digestion of the protein and chromatography of the peptides.

5,5'-Dithiobis(2-nitrobenzoate)

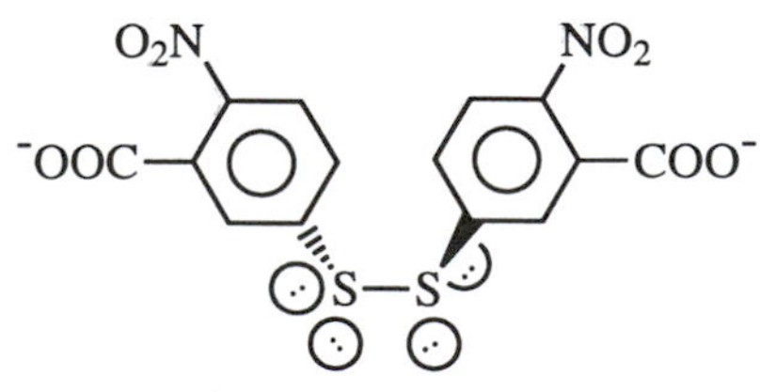

10-1

participates readily in disulfide exchange with a cysteine (Figure 3–23). The reagent contains a disulfide that is reactive[41] because the nitrothiobenzoate dianion is a good leaving group ($pK_a < 5$). This causes the equilibrium to lie in favor of the mixed disulfide but also causes the product, the mixed disulfide between cysteine and 5-thio-2-nitrobenzoate, to be electrophilic. This mixed disulfide in turn can react with a nucleophile to release a second equivalent of the nitrothiobenzoate dianion

nitrothiobenzoate

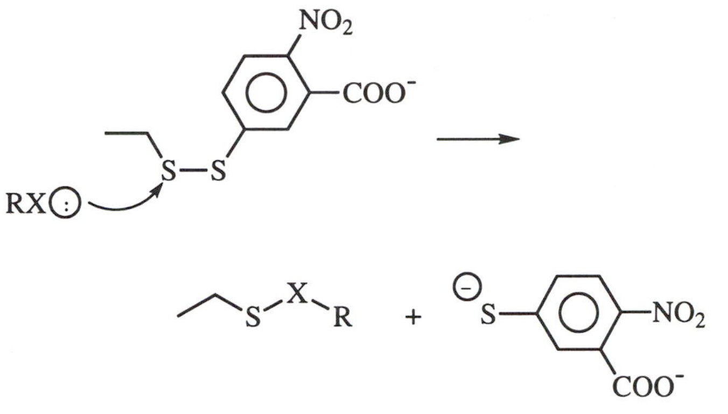

(10–31)

where ⊙X–R is another cysteine or hydroxide anion. The net result of this reaction is to reduce the reagent and oxidize the cysteine, either to cystine or to oxides of cysteine (Figure 2–17). The nitrothiobenzoate dianion released during the initial reaction between cysteine and 5,5′-dithiobis(2-nitrobenzoate) is brightly colored ($\varepsilon_{412} = 13{,}600$ M^{-1} cm^{-1}) and its absorbance can be used to follow the reaction. The situation is complicated, however, by the side reactions releasing the second nitrothiobenzoate (Equation 10–31) and by the reaction of the nitrothiobenzoate itself with oxygen that proceeds with the loss of its absorbance. Nevertheless, the ease with which this reaction can be followed has led to its wide and indiscriminate application to proteins.

Tyrosine is frequently alkylated, acylated, arylated, or sulfonylated inadvertently during modification reactions designed to be restricted to lysines. It can be modified specifically, however, by taking advantage of its elevated susceptibility to **electrophilic aromatic substitution**. As a p-alkylphenol, it is activated toward substitution, which is directed to its ortho positions by the electron-releasing hydroxyl. A simple example of this susceptibility is the facile iodination of tyrosine

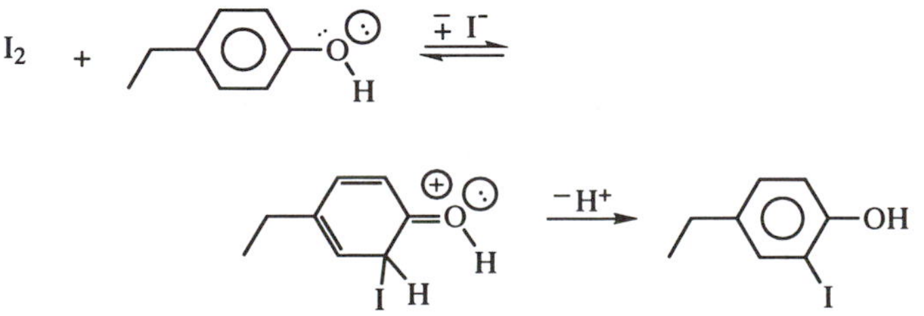

(10–32)

Iodide ion is used as the source of the iodine and it is oxidized to I_2, IOH, or ICl either enzymatically[42] or chemically.[43] Histidine is also iodinated under similar conditions but not so readily as tyrosine.[39] A particularly advantageous method for activating I$^-$ uses N-chlorobenzenesulfonamide attached to polystyrene beads. This reagent produces ICl from I$^-$

(10–33)

The ICl iodinates tyrosines, and the fact that the chlorinating agent is on a solid phase inhibits its direct reaction with the protein.

Diazonium salts also participate in electrophilic aromatic substitution with tyrosine. 5-Diazonium-1-hydrotetrazole[44] is a diazonium salt producing a product with tyrosine that absorbs strongly at 550 nm[39]

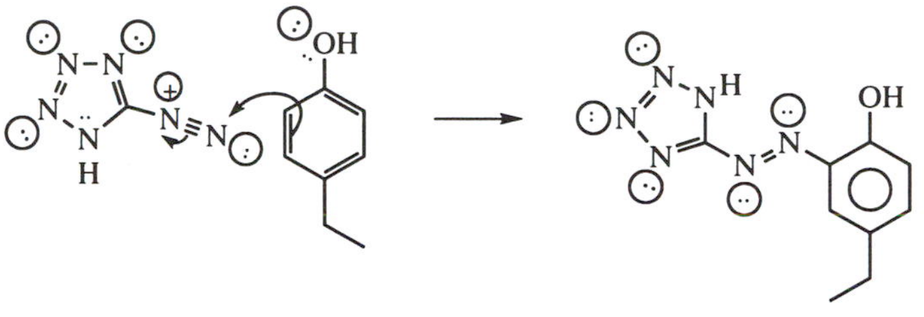

(10–34)

It reacts readily with histidine as well, but at low pH histidine will be mainly protonated, and the imidazolium cation is inert to electrophilic aromatic substitution.

Tetranitromethane is the reagent used most frequently to modify tyrosine

10–2

The nitration that produces the o-nitrotyrosine proceeds by a free radical mechanism.[45] The o-nitrotyrosine produced absorbs strongly at 428 nm as the nitrophenolate anion. It can be reduced to o-aminotyrosine with dithionite.[6] o-Aminotyrosine has a uniquely low pK_a (4.8), and this fact can be exploited to direct further modification to this location in the protein.[46] Unlike O-acylation or O-alkylation, which require the tyrosine to be anionic to react as a nucleophile, the reaction with tetranitromethane proceeds with the un-ionized tyrosine.

Tryptophan is susceptible to electrophilic aromatic substitution because of its similarity to aniline. For example, tritium can be incorporated into tryptophan under strongly acidic conditions[47]

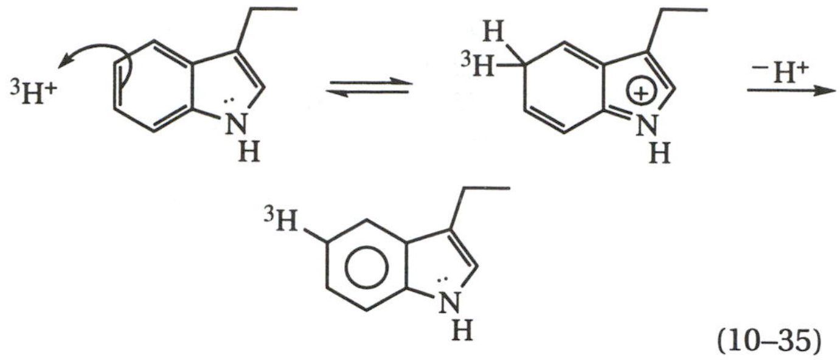

(10–35)

Sulfenyl halides such as 2,4-dinitrobenzenesulfenyl chloride participate in a formal electrophilic aromatic substitution at carbon 2 of tryptophan[48]

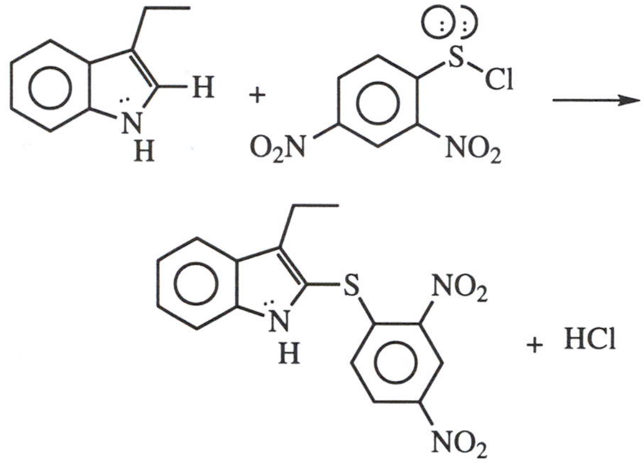

(10–36)

The most peculiar position in tryptophan is the π bond between carbons 2 and 3. This bond displays the properties of an olefin during bromination with mild brominating reagents (Reaction 3–1) by participating in addition rather than substitution. Under mild conditions a relatively inert brominating agent, 2-[(2-nitrophenyl)sulfenyl]-3-methyl-3'-bromoindolenine (BNPS-skatole), oxidized the tryptophan in micrococcal nuclease to the oxindole,[49] presumably through an intermediate halohydrin:

halohydrin   oxindole

(10–37)

Only methionine was oxidized at the same time, and it could be regenerated readily by reduction. The use of addition reactions to this olefin to incorporate nucleophiles other than water might be feasible. This reaction, however, often results in cleavage of the polypeptide at the tryptophan (Reaction 3–1).

Arginine is modified specifically by vicinal diones

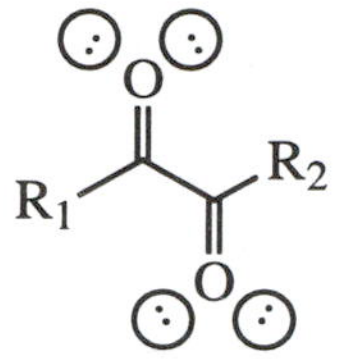

10–3

Although the final products of these reactions are not yet well characterized, it is believed that the initial intermediate is the cyclic adduct 10–4, in which two of the nitrogens of the arginine are covalently attached to the two carbonyl carbons of the dione[50]

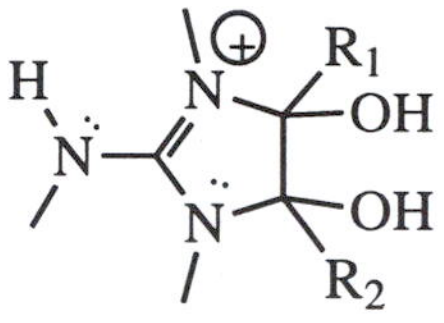

10–4

The initial product of the reaction between 1,2-dioxocyclohexane and arginine is susceptible to periodate cleavage, which produces the product expected from vicinal diol 10–4.[10] Upon treatment with alkali the initial product of the reaction between arginine and 1,2-dioxocyclohexane rearranges to the iminoimidazolidone[51]

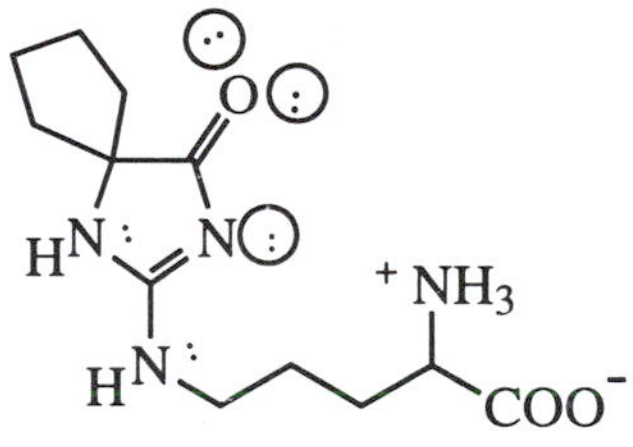

10–5

which can be rationalized as the direct derivative of adduct 10–4.

The modification of arginine by vicinal diones often proceeds beyond adduct 10–4 to products that incorporate additional molecules of the dione. 2,3-Butanedione (10–3, $R_1 = R_2 = CH_3$) under appropriate conditions self-condenses to dimers and trimers that both react with arginine to yield poorly characterized, heterogeneous mixtures of products containing 3 mol of dione for every mole of arginine.[52] Phenyl glyoxal (10–3, R_1 = phenyl, R_2 = H) reacts with arginine to produce a product containing 2 mol of dione for every mole of arginine.[50]

The earliest modifications of arginine, with either benzil (10–3, $R_1 = R_2$ = phenyl)[53] or 1,2-dioxocyclohexane,[51] were performed at alkaline pH (0.2 M NaOH), and the products were quite stable. The conditions, however, were too harsh to avoid destruction of the polypeptide. It was subsequently noted that the addition of borate during the reaction of a protein with 2,3-butanedione accelerated the rate of the reaction at neutral pH and rendered the modification irreversible as long as the borate was present.[54] The product of the reaction of 1,2-dioxocyclohexane and arginine, presumably diol **10–4**, could also be stabilized significantly by the addition of borate.[10] Borate is known to add to vicinal diols, such as sugars, to form cyclic borate diesters

10–6

The addition of borate to render the reaction with diones irreversible under mild conditions has permitted the isolation of modified peptides from proteins modified by 1,2-dioxocyclohexane.[55]

Glutamates and **aspartates** are modified with **carbodiimides** (Figure 10–5).[56] The carbodiimides used can be either very hydrophobic, such as dicyclohexylcarbodiimide (DCCD; $R_1 = R_2$ = cyclohexyl) or hydrophilic and water-soluble, such as N-ethyl-N'-[3-(dimethylamino)propyl] carbodiimide [EDC; $R_1 = C_2H_5$, $R_2 = (CH_3)_2{}^+NHC_3H_6$]. The initial product of the reaction is an O-acylurea[56] in which the acyl carbon of the original glutamate or aspartate has been activated by forming an acyl derivative whose leaving group is the oxygen of an N,N'-dialkylurea ($pK_a = 1$).

Four fates await this central intermediate. If it is buried in a nonnucleophilic environment and sterically constrained, it will remain as the O-acylurea until the protein is unfolded, at which point it will usually hydrolyze back to the unmodified glutamate or aspartate. If it is somewhat buried in a polar environment, but not sterically constrained, the O-acylurea will rearrange to the N-acylurea, which is stable (pathway ① in Figure 10–5). This rearrangement stably incorporates 1 mol of the carbodiimide into a mole of glutamate or aspartate. Dicyclohexylcarbodiimide is often incorporated into a protein in this way. It usually reacts with buried carboxylic acids because it is so hydrophobic, and the buried O-acylurea survives long enough to rearrange; but the reaction rarely proceeds in high yield. If there is a nucleophilic amino acid in the protein (such as a lysine) in the vicinity of the O-acylurea, an intramolecular adduct (such as an amide) between that amino acid and the glutamate or aspartate will form.[57] Usually, however, an external amine, such as the methyl ester of glycine, is added in high concentration to react with the O-acylurea as it is formed and produce the amide between the external amine and the glutamate or aspartate.[56] In this way, a defined covalent modification of the carboxylate can be made. If the external nucleophile is ammonia, glutamates and aspartates are converted to glutamines and asparagines, respectively.[58]

The practical outcome of each of these four fates is

unique. In the first, the native protein is modified by the carbodiimide at glutamate or aspartate but loses the modification upon unfolding. In the second, a stable derivative between the protein and the carbodiimide is formed. In the third, the glutamate or aspartate is stably modified and intramolecularly cross-linked,[57] but neither the carbodiimide nor the external amine is incorporated into the protein. In the fourth, the external amine and not the carbodiimide is incorporated. Often a combination of all of these outcomes occurs, which defies any attempt to quantify the results.

Another reagent used to activate the carboxylates of glutamates and aspartates is N-ethoxycarbonyl-2-ethoxy-1,2-dihydroquinoline (EEDQ). It activates the carboxylate (Figure 10–6) by forming a mixed ethyl carbonic an-hydride ($pK_a^{LG} = 7$), which is somewhat less reactive than an O-acylurea ($pK_a^{LG} = 1$) but capable of either intramolecular or intermolecular reaction with nucleophiles.

Compounds that serve as precursors to nitrenes or carbenes through photolytic reactions are reagents that display even less specificity than alkylating agents in the modification of the amino acids in a protein.

Aryl azides, such as phenyl azides or nitrophenyl azides, are the usual precursors for nitrenes. A nitroaryl azide produces a nitroaryl nitrene upon photolysis

(10–38)

A convenient, widely used reagent for attaching a nitrophenyl azide to other compounds (the R in Reaction 10–38) by nucleophilic aromatic substitution is 4-azido-2-nitrofluorobenzene.

A **nitrene** is a nitrogen whose four valence orbitals are occupied by only six valence electrons and is therefore electron-deficient or electrophilic. In a **singlet** nitrene, three of these orbitals are occupied by pairs of electrons and one orbital is vacant. In theory, a singlet nitrene, because of its vacant orbital, has a higher preference for insertion into nitrogen–hydrogen bonds or oxygen–hydrogen than carbon–hydrogen bonds because atoms of oxygen or nitrogen are electron-rich. A **triplet** nitrene is a diradical in which two of the orbitals on nitrogen are each occupied by only one unpaired electron and the other two are occupied by two pairs of electrons. Theoretically, triplet nitrenes should be able to modify proteins by hydrogen abstraction followed by combination of the two adjacent monoradicals.[59] Because hydrogen is usually more easily abstracted from carbon than from oxygen or nitrogen, triplet nitrenes should abstract hydrogen more readily from carbon–hydrogen bonds than either nitrogen–hydrogen bonds or oxygen–hydrogen bonds.

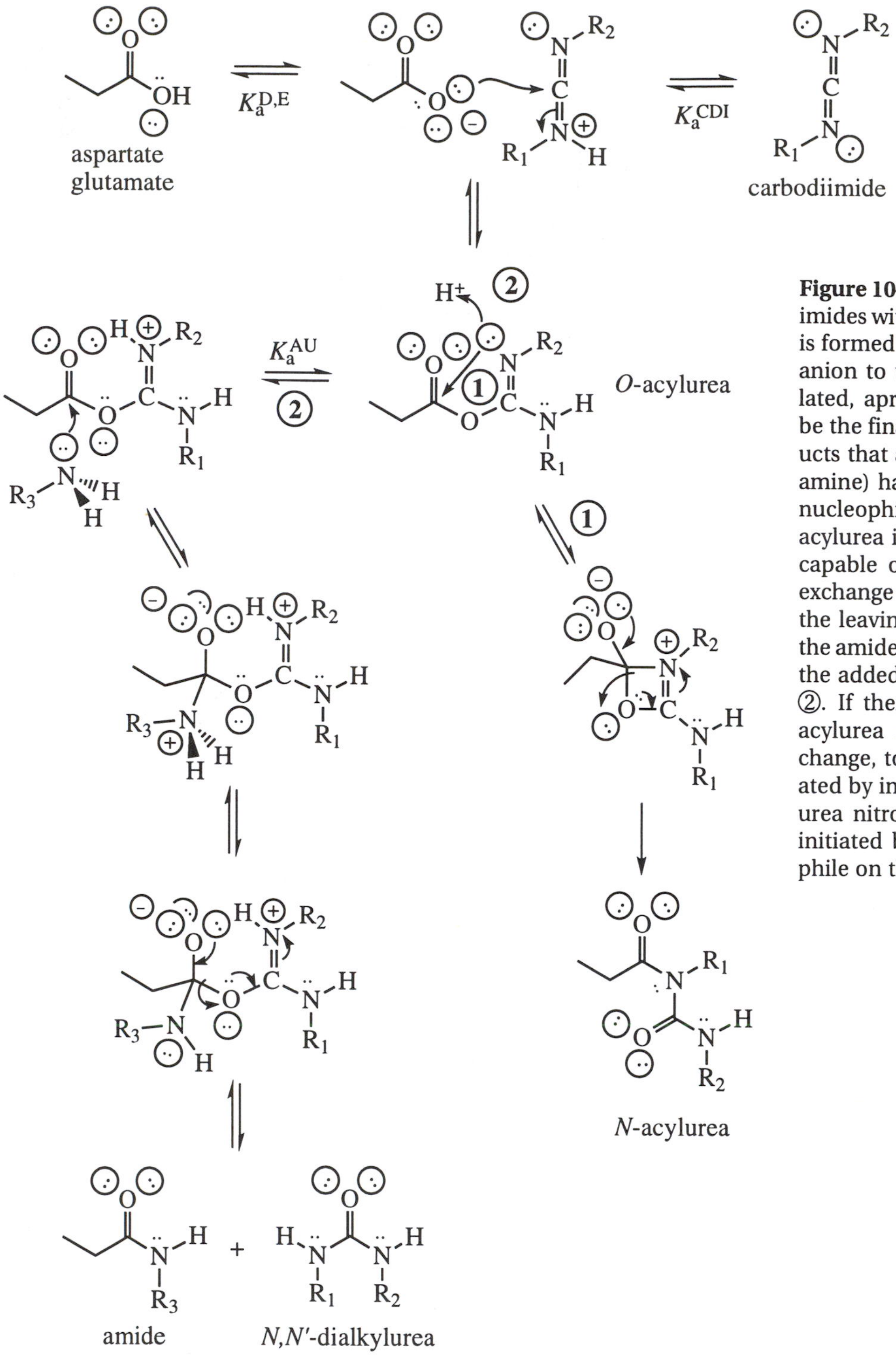

Figure 10–5: Outcomes of the reactions of carbodiimides with aspartate or glutamate. The *O*-acylurea is formed by the direct addition of the carboxylate anion to the protonated carbodiimide. In an isolated, aprotic environment, the *O*-acylurea could be the final product, but there are two other products that are possible. If a nucleophile (usually an amine) has been added, or if there is an adjacent nucleophile in the protein such as a lysine, the *O*-acylurea is an activated carboxylic acid derivative capable of acylating that nucleophile in an acyl exchange reaction to give the *N,N'*-dialkylurea as the leaving group and the acyl derivative (usually the amide) of the glutamate or aspartate with either the added nucleophile or the lysine in the protein ②. If there is no accessible nucleophile, the *O*-acylurea rearranges, by intramolecular acyl exchange, to the *N*-acylurea ①. Pathway ① is initiated by intramolecular attack of the unprotonated urea nitrogen on the acyl carbon; pathway ② is initiated by the attack of the extraneous nucleophile on the activated carboxyl group.

When light is absorbed by an aryl azide, which is a singlet, the excited state is initially a singlet excited state, which must produce a singlet nitrene because N_2 is a singlet molecule. If the excited state lasts long enough, a triplet excited state can be formed by intersystem crossing. The triplet excited state produces a triplet nitrene and singlet N_2. The yield of triplet excited state can also be increased by adding a triplet sensitizer.[60] Singlet nitrene itself can turn into triplet nitrene if it survives long enough. In the absence of a sensitizer, about 10% of the ni-

trene produced by photolysis of phenyl azide is triplet.[61]

Although it is widely believed that aryl nitrenes, such as the phenyl nitrenes or the 3-nitro-4-(alkylamino)phenyl nitrenes usually employed in the modification of proteins, should insert into carbon–hydrogen bonds, a reaction that would require significant yields of the triplet state, the chemistry of such nitrenes belies this belief. In ideal situations, such as the intramolecular insertion in the vapor phase of an aryl nitrene into a tertiary carbon–hydrogen bond four carbons away, a reasonable yield of the *N*-alkylaniline (50%)[62] is obtained. When, however, phenyl nitrene is generated in cyclohexane by photolysis, no insertion (< 30%)[63] into the solvent is observed, and most of the reaction proceeds with either dimerization of the nitrene itself or the production of aniline by two successive hydrogen abstractions by the triplet. Phenyl nitrene generated by photolysis under the same conditions in hydroxylic solvents such as methanol or propanol, however, inserts into those solvents in high yield (80%).[63] The products of the photolytic reactions of 4-substituted phenyl nitrenes with water, methyl alcohol, or diethylamine are the lactams **10–7** (60–90% yield), the 2-methoxy-3-hydroazepines **10–8** (40–80% yield), or the 2-diethylamino-3-hydroazepines **10–9** (90–100% yield), respectively[64]

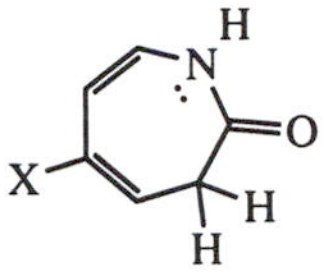

10–7 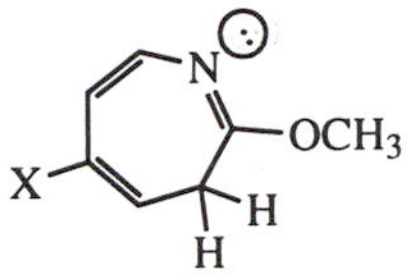   **10–8** 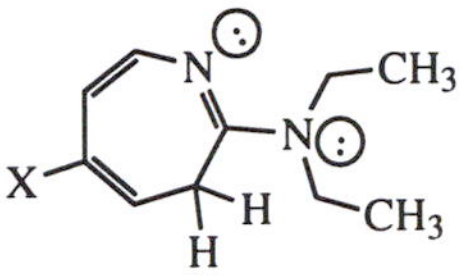   **10–9**

The products of the photolytic reaction of 4-methylamino-3-nitrophenyl nitrene with methanol or 1% diethylamine in methanol are aniline **10–10** (40% yield) and aniline **10–11** (40% yield) or aniline **10–10** (30% yield) and aniline **10–12** (70% yield), respectively[64]

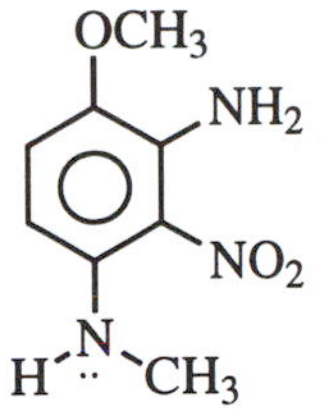

10–10 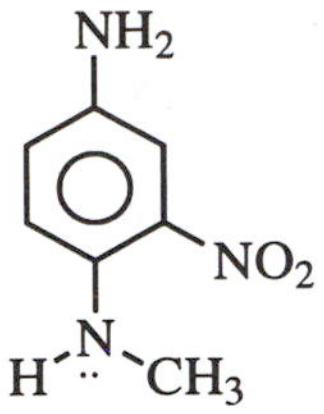   **10–11** 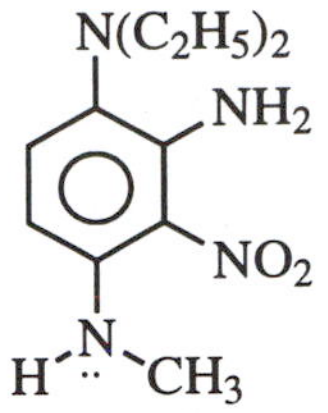   **10–12**

Most of these products can be explained as the results of nucleophilic addition to the two electrophilic species engaged in the following equilibrium

(10–39)

Presumably, the majority of the products from the reaction of an aryl nitrene with a protein result from reaction with nucleophilic functional groups. The identity of the amino acids modified by aryl nitrenes are consistent with these general considerations. Tyrosine, cysteines, and lysines have usually been identified as the reactants, but a leucine, two alanines, and a phenylalanine have also been reported to be modified.[59,61,65]

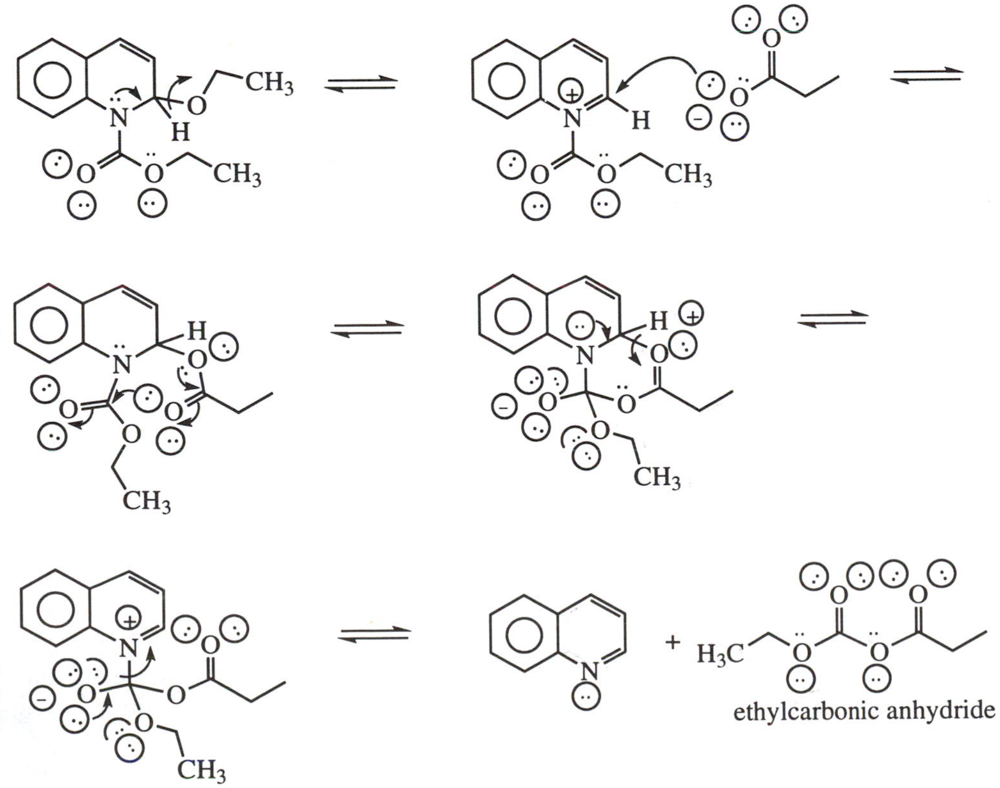

Figure 10–6: Formation of an ethyl carbonic anhydride at a glutamate or aspartate by *N*-ethoxycarbonyl-2-ethoxy-1,2-dihydroquinoline. The 2-ethoxy group is the alcohol of a carbamylether that decomposes in water to the iminium cation. The iminium cation reacts with the carboxylate of glutamate or aspartate. The adduct that is formed undergoes an intramolecular, cyclic rearrangement involving an acyl exchange reaction at one end with quinoline as the leaving group and the expulsion of an ester from its adduct with an iminium cation at the other end.

ethylcarbonic anhydride

Singlet aryl nitrenes can insert intramolecularly into phenyl rings,[60] and this may explain the modification of phenylalanine.

Carbenes, like nitrenes, have only six valence electrons on one atom, but they are distributed around a carbon instead of nitrogen. The carbenes generally used for the modification of proteins are on secondary carbons. They can be singlets or triplets. Again, the singlet is the first product and has a significant preference for insertion into nucleophilic locations such as nitrogen–hydrogen or oxygen–hydrogen bonds[66] and probably reacts before much triplet is formed.

The compounds that have proven to be the most efficient and uncomplicated precursors of carbenes are 1-trifluoromethyl-1-phenyldiazarines[67]

$$(10\text{–}40)$$

The carbene is generated by photolysis. Prior to the advent of these reagents, α-diazoketones, α-diazoacetyl esters, and ethyldiazomalonyl esters were used as precursors of carbenes[59]

$$(10\text{–}41)$$

The first application of a carbene as a reagent for the modification of a protein used a diazoacetyl ester of α-chymotrypsin, and modification of cystine,[68] serine,[69] alanine,[70] and tyrosine[69] was observed. Carbenes have usually been found to display a preference for insertion into oxygen–hydrogen and nitrogen–hydrogen bonds. They have been observed to modify lysines,[59] tyrosines,[66] tryptophans,[71] glutamic acids,[72] and aspartic acids,[66] but incorporation has also been noted into valine[66] and glycine.[59]

Two remarkable illustrations of the preference of carbenes for nitrogen–hydrogen and oxygen–hydrogen bonds can be found in the use of these reagents to modify amino acids found in polypeptides that span membranes of phospholipid. In both instances, several precursors for a carbene were used that should have placed it at different respective locations along the polypeptide crossing the phospholipid bilayer; yet, in each of the two experiments the carbenes reacted with the same amino acid, in one case a tryptophan[71] and in the other case a glutamic acid.[72] Other amino acids with readily abstracted hydrogen at tertiary carbons must have been more accessible to at least some of these carbenes than the tryptophan or the glutamic acid, respectively, into which they eventually inserted. If this was the case, these results demonstrate that carbenes, like nitrenes, are not so promiscuous as they are often thought to be.

The products of the reactions between nitrenes and carbenes and proteins have rarely been characterized. In part, this is due to the low yields encountered in most of these reactions, presumably because of the tendency of the singlet carbene or singlet nitrene and its rearranged products to insert into water.[73]

Site-directed mutation[74,75] produces the covalent modification of a protein by converting one particular amino acid in its sequence into another of the 20 amino acids. It is also possible to delete amino acids from the sequence of a polypeptide or insert extra amino acids at a particular location with this technique. The method requires that the cDNA for the protein of interest has been cloned and that the cloned cDNA can be expressed, or in other words translated into the intact protein. It is also necessary to be able to produce, by this expression, quantities of the protein sufficient for the contemplated experiments. The protein to be studied is produced in the **expression system** by an organism, such as a bacterium, that is usually different from the organism from which the protein was originally isolated, for example, a human; but because the sequence of the polypeptide, as dictated by the sequence of the cDNA, is the same, the protein is the same, unless posttranslational modification is required.

The site-directed mutation is incorporated into the cDNA, and the mutated cDNA is used to direct the production of a covalently modified polypeptide in which one particular amino acid has been deliberately changed. For example, a collection of 13 mutated versions of the lysozyme from T4 bacteriophage, in which Threonine 157 had been changed to 13 of the other 19 amino acids, was produced by site-directed mutation. Each of these 13 different proteins was obtained as a pure crystalline product in quantities sufficient for the production of a map of difference electron density between each of them and the unmutated protein.[76]

A site-directed mutation can be introduced into a particular segment of cDNA by annealing a short piece of synthetic DNA, the mutagenic oligonucleotide, to one of the two strands of the unmutated cDNA to form a short section of double-helical DNA in which one or more of the nucleotide bases are mismatched.[74] The mutagenic oligonucleotide is designed so that the desired mismatches occur in the middle of the short duplex and there are sufficiently long regions of complementary nucleotide sequence on each flank to guarantee a stable and specific duplex. This can be accomplished in the following way.

A restriction fragment of the cDNA encoding the protein of interest and containing the site to be mutated is inserted into the genome of a bacteriophage. The altered bacteriophage produces virus particles containing the enlarged genome on a closed, single-stranded circle of DNA.[77] Closed, single-stranded circles containing the strand of the cDNA complementary to the mutagenic oligonucleotide are selected[75] for hybridization. The mutagenic oligonucleotide is complementary to this single-stranded cDNA except at the central, mismatched positions, chosen to produce the desired change in a particular codon.[76] For example, the deoxyribonucleotide sequence –CTCTACTGCGGGTTTG– occurs in the cDNA encoding the sequence of tyrosyl tRNA synthetase. It encodes the amino acid sequence –LYCGF–, which contains amino acids 33–37 in the sequence of the intact protein. The mutagenic oligonucleotide 5'CAAACCCGCCGTAGAG3' was chemically synthesized.[78] It is complementary to the coding sequence of the unmutated cDNA except at its tenth residue, which is a C instead of the complementary A. When it was annealed to a single-stranded, circular DNA produced from cDNA with the unmutated sequence, it formed a short self-complementary segment of double-stranded DNA in which its C was mismatched with the T of the unmutated sequence. It is this mismatch that eventually produced the mutated cDNA with the sequence –CTCTACGGCGGGTTTG–, encoding the mutated protein sequence, –LYGGF–.

The short mutagenic oligonucleotide sits upon the single-stranded, circular DNA offering a free 3' hydroxyl. This is used to prime the synthesis of DNA by DNA polymerase.[74,75] This enzyme synthesizes a single strand of DNA upon the circular template until it comes around the circle to the 5' end of the mutagenic oligonucleotide. The newly synthesized, single-stranded circle is then closed with DNA ligase to produce a doubly closed, double-stranded circle of DNA, completely complementary except at the designed mismatch. This double-stranded circular DNA is then replicated as a bacteriophage in an appropriate host. Half of the resulting viral DNA should contain the mutated sequence of the segment of cDNA because it is the progeny of the single strand into which the mutagenic oligonucleotide was incorporated originally.

Plaques produced by viruses containing the mutated cDNA are selected,[75] double-stranded DNA is produced from these viruses, and the desired restriction fragment containing the mutation is isolated. The restriction fragment containing the desired mutation can also be produced[79] by the polymerase chain reaction rather than by cloning.

In either case, once the restriction fragment containing the desired mutation has been produced and isolated, it is introduced into the original DNA, and full-length cDNA incorporating the mutation is produced. The mutant protein expressed from this full-length, mutant cDNA should contain the designated substitution. For example, in the case of the mutated tyrosyl-tRNA synthetase, the purified protein had a glycine rather than a cysteine at position 35.[78] That the modification has occurred is usually verified by sequencing the mutated cDNA[74] rather than the protein itself.

In the procedure just described, the initially cloned fragment of the mutated cDNA arises from both the strand of the circular duplex DNA that incorporated the mutagenic oligonucleotide and from the original parental single strand to which the mutagenic oligonucleotide was annealed. This makes the process of identifying the mutated clone more difficult because the original parental cDNA gives rise to unmutated clones. This is particularly troublesome when the mutation causes no change in the behavior of the protein. A way around this difficulty[80] is to use parental cDNA in which thymine has been replaced with uracil, but the new, mutated strand is made with thymine. When the duplex is replicated in normal bacteria, the parental cDNA containing the uracil is destroyed, and most of the clones that survive have the mutated sequence in the cDNA. In this way, the mutants can be selected directly by nucleotide sequencing.

Site-directed mutations can also be produced by insertion of **cassettes** of synthetic double-stranded DNA into a particular cDNA. In this method, preexisting sites for restriction enzymes, or purposely designed sites for restriction enzymes, that flank the region to be mutated are chosen. These sites are designed so that the piece of double-stranded DNA produced by the restriction endonucleases is short and has single-stranded, sticky ends. A double-stranded segment of DNA is synthesized so that it has the appropriate sticky ends and incorporates complementary nucleotide sequences in the middle that perform the desired mutation in the coding sequence. This is the cassette, which is then inserted into the hole in the original cDNA produced by the restriction endonucleases. The advantage of the cassette is that the mutation is produced directly by insertion of synthetic double-stranded DNA. The disadvantage is that two complementary pieces of synthetic single-stranded DNA have to be synthesized. Nevertheless, mutation with cassettes has particular advantages when sets of mutants are prepared in which all of the possible 19 substitutions have been made at a particular location.[81]

A strategic distinction exists between covalent modification with chemical reagents and covalent modification by site-directed mutation. In the former situation, the protein of interest is modified, and the outcome of the reaction is assessed by isolation of modified peptides chromatographically and their identification by sequencing. In the latter case, a particular amino acid in the sequence of the protein is chosen, the modification is performed, and the results are assessed by the effect of the modification on some property of the protein. In the former, the structure of the protein and, in the latter, the intuition of the investigator determines which amino acid or amino acids are modified. Because the intuition of the investigator is usually fallible, the most informative experiments using site-directed mutation have been performed on proteins for which a crystallographic molecular model is already available. Although the reliance on the intuition of the investigator is a disadvantage, a major advantage of site-directed mutation is the absolute specificity of the modification for only one site in the protein.

It is also possible to employ another strategy for studying a protein by performing covalent modification. A particular nucleophilic amino acid in the sequence of the protein is chosen as a target, as is done in site-directed mutation, but its rate of covalent reaction with one or more electrophilic reagents is systematically assessed. If this strategy is employed, the modified amino acid can be rapidly purified, within a particular peptide produced by proteolytic digestion of the modified protein, by affinity adsorption chromatography. Antibodies raised against a synthetic peptide[82] with the same amino- or carboxy-terminal sequence as the peptide containing the modified amino acid can be used to produce an affinity matrix able to capture only the particular peptide containing the modified target.[83,84]

During the covalent modification of a protein, even with as nonspecific a reagent as iodoacetamide,[17] the various types of amino acids do not react as a homogeneous population. This is most readily discerned when the amount of incorporation into a particular type of amino acid is plotted as a function of the duration of the reaction.[17] If the concentration of reagent remains constant, the natural logarithm of the amount of unmodified amino acid should decrease as a linear function of time because the reaction is pseudo-first-order (Equation 10–12). This is usually not observed, and the disappearance of a particular type of amino acid, such as histidine, lysine, or cysteine, as the modification proceeds usually displays nonlinear kinetics (Figure 10–7).[17] This can be ascribed to the fact that each histidine, each lysine, or each cysteine in the sequence of the protein is in a different environment in the folded polypeptide and should react at a unique rate with the reagent. For example, the three histidine residues of α-lactalbumin can be modified by iodoacetamide, but each reacts at a significantly different rate[85] so that the disappearance of total histidine displays heterogeneous kinetics.

It has already been noted that the environment surrounding a particular acidic–basic amino acid shifts its apparent **acid dissociation constant** from the value it would have in an unfolded polypeptide (Table 2–2). Such shifts of the acid dissociation constants from their intrinsic values move each of the inflections of the profiles of log k_i' against pH (Figure 10–1) horizontally to coincide with the altered value of the respective pK_a and simultaneously move the plateau at high pH vertically in response to the Brønsted relationship (Equation 10–14). A unique shift occurs for each amino acid in the protein. An example of such an effect of environment on the nucleophilicity of acid–bases is provided by a pair of cysteine residues, Cysteine 31 and Cysteine 32, in seminal ribonu-

clease.[86] These two adjacent cysteines react more rapidly with 5,5'-dithiobis(2-nitrobenzoate) at pH 7 than do model compounds such as cysteine itself or cysteinyl-cysteine. The synthetic peptide MCCRKM, which incorporates the sequence of seminal ribonuclease around these two cysteines, has the same enhanced reactivity. It was shown that this enhanced reactivity is due to the fact that the cysteines are adjacent to an arginine and a lysine. These cationic amino acids lower the values of the acid dissociation constants for the two cysteines. Although the nucleophilicities of the thiolate anions also decrease accordingly, the Brønsted coefficient β is small (< 0.2). Therefore, the increase in reactivity at pH 7 results from a significant increase in the concentration of the respective thiolate anions, which each have almost the same intrinsic nucleophilicity, k_{Cys}, as a cysteine whose pK_a has not been lowered.

The reaction of acetic anhydride with particular lysines in a protein has been used to monitor their individual acid dissociation constants and nucleophilicities. Because acetic anhydride is rapidly hydrolyzed, the amount of its incorporation at set pH into a particular lysine in the protein when the reaction has reached completion, relative to the amount of its incorporation into an added standard amine, provides a direct measurement of the relative bimolecular rate constant for its reaction with the particular lysine (Equation 10–22) at that pH. If the absolute rate constant for the reaction between acetic anhydride and the standard amine is known, the absolute bimolecular rate constant for the reaction between the lysine of interest and acetic anhydride at that pH can be calculated from the relative rate constant, $k_1 (k_2)^{-1}$.

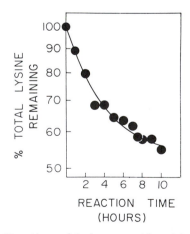

Figure 10–7: Reaction of iodoacetamide with the lysines of glucose-6-phosphate isomerase.[17] Glucose-6-phosphate isomerase (0.12 mM) was mixed with iodoacetamide (49 mM) at pH 8.5 and 40 °C. At the noted times a sample was removed from the solution, the reaction was quenched with 0.3 M 2-mercaptoethanol, and the sample was subjected to total amino acid analysis. The amount of unmodified lysine (percentage of total) is plotted on a logarithmic scale as a function of time. Reprinted with permission from ref 17. Copyright 1970 *Journal of Biological Chemistry*.

The behavior of these absolute rate constants as a function of pH (Figure 10–1) provides the pK_a^{Lys} of the lysine and the bimolecular rate constant for the reaction between its unprotonated conjugate base and acetic anhydride. The Brønsted coefficient β (0.48) and the absolute bimolecular rate constant for an unhindered lysine (1.6×10^5 M^{-1} min^{-1}) of normal pK_a^{Lys} (10.8) with acetic anhydride at 10 °C[15] permit the bimolecular rate constant expected of a lysine with a particular pK_a to be calculated. A comparison of the observed rate constant and the rate constant calculated from the observed pK_a provides an estimate of the **accessibility** of the lysine of interest to the acetic anhydride.

In most instances, the apparent nucleophilicity of a particular amino acid is determined by **steric effects** engendered by neighboring amino acids in the folded polypeptide or a decrease or increase in its nucleophilicity brought about by its participation in intramolecular hydrogen bonds. It has been demonstrated that the reactivity of particular amino acids in the sequence of a protein can provide an indication of their accessibility to the aqueous phase. Tetranitromethane, which is large and quite polar (**10–2**), reacts with the neutral form of tyrosine. At neutral pH, all of the tyrosines in a protein should be un-ionized, and their modification by tetranitromethane should reflect only their accessibility.[14] There are eight tyrosines in human carbonate dehydratase B, and only three of them, Tyrosine 20, Tyrosine 88, and Tyrosine 114, react with tetranitromethane.[14] Subsequent to this assessment, the structure of human carbonate dehydratase B was obtained crystallographically, and in the map of electron density only Tyrosine 20, Tyrosine 88, Tyrosine 114, and Tyrosine 129 were found to be "located on the surface of the molecule."[87] Aspartate 194 in bovine chymotrypsinogen A could not be modified in the native protein with ethyl glycinate and N-ethyl-N′-[3-(dimethylamino)propyl]carbodiimide even under conditions where 13 of its 15 carboxylates did react completely.[88] Aspartate 194 is "buried" in the interior of the crystallographic molecular model of chymotrypsinogen.[89] When fructose-bisphosphate aldolase was modified by methyl acetimidate at high concentrations only 20 of its 30 lysines were modified.[90] The 10 unmodified lysines reacted readily when the protein was unfolded, and it could be shown that these were 10 unique lysines in the sequence of the protein, presumably rendered unreactive by their surroundings in the folded polypeptide.

Site-directed mutation can also be used to monitor the accessibility of particular locations in the amino acid sequence of the protein. As has been pointed out, amino acids in the interior of a molecule of protein are tightly packed and all donors of hydrogen bonds are occupied by conveniently located acceptors. On the surface of a molecule of protein, however, there are few steric restrains and both hydrophilic and hydrophobic amino acids are tolerated. An α helix passes across the surface of the crystallographic molecular model of λ repressor.[91] In this α helix Isoleucine 84 and Methionine 87 are on the face of the α helix directed toward the interior of the protein while Tyrosine 85, Glutamate 86, Tyrosine 88, and Glutamate 89 are on the surface of the α helix that is accessible to the solution. Only isoleucine at position 84 and either methionine or isoleucine at position 87, of all of the 20 amino acids, produces a functional protein, but 10–14 of the 20 amino acids can be substituted at the other four positions and still produce a functional protein. Such an alternating pattern can identify an α helix running along the surface of a protein.

Another covalent modification that has been used to assess the accessibility of particular amino acids in a folded polypeptide is proteolytic cleavage. For a proteolytic enzyme to cleave a peptide bond, the polypeptide at that location must be able to enter the active site of the proteolytic enzyme. This has usually been assumed to require that the susceptible peptide bond be located on a somewhat flexible loop, on the outside surface of the protein, well exposed to the solvent. In the case of chymotrypsinogen A, the proteolytic cleavages of the folded polypeptide (c15 Kin.1) that remove the amino acids between Leucine 13 and Isoleucine 16 and between Tyrosine 146 and Alanine 149 to produce α-chymotrypsin occur within two such loops.[89] In the case of deoxyribonuclease I, however, a less easily explained proteolytic cleavage of the folded polypeptide has been observed. Under the proper set of conditions, chymotrypsin cleaves deoxyribonuclease I completely and exclusively at the peptide bond on the carboxy-terminal side of Tryptophan 178.[92] In the refined crystallographic molecular model of deoxyribonuclease I,[93] Tryptophan 178 is found in the middle of an α helix that is a rigid feature of the structure. This α helix traverses the outer surface of the protein, but Tryptophan 178 is on the side of the α helix pointed toward the interior and itself is inaccessible. There are, however, no more accessible sites in the protein at which chymotrypsin could cleave, and it may be the case that in solution the α helix containing Tryptophan 178 is in equilibrium with a disordered loop.

Covalent cross-linking uses covalent modification to assess the proximity of particular amino acids or polypeptides. A bifunctional reagent[94] used for cross-linking has two electrophilic locations in the same molecule. The electrophiles used are those commonly used in monofunctional reagents for the modification of proteins. They can be identical to each other, or they can be two electrophiles with different specificities. They can be connected by a chain of atoms stably bonded or a chain of atoms containing a bond that can be cleaved as desired to permit later separation and identification of the cross-linked species. An example[95] of the latter class would be the tetraester

10–13

The *N*-hydroxysuccinimide esters at the two ends acylate lysines, and the esters in the middle can be cleaved subsequently with hydroxylamine.

The use of cross-linking to demonstrate nearest-neighbor relationships has been most extensively applied to proteins containing several different polypeptides. The intention is to identify pairs of polypeptides that are readily cross-linked on the assumption that this designates them as neighbors in the structure of the protein. If the cross-linking reactions were carried to completion, all of the polypeptides present would be cross-linked together, the product would be one covalently linked polymer, and this result would state that all of the polypeptides were part of the same protein. If pairs of immediately adjacent neighbors are to be identified, however, such quantitative cross-linking must be avoided, and the reaction must be purposely incomplete so that only intimate contacts are recognized.

The F_1 portion of the mitochondrial H$^+$-transporting ATP synthase is a protein that contains five polypeptides, α, β, γ, δ, and ε. When it was treated with dimethyl 3,3'-dithiobis(propionimidate)[96]

10–14

under conditions producing a low yield of cross-linked products, six new components appeared on polyacrylamide gels run in the presence of sodium dodecyl sulfate. Two-dimensional polyacrylamide gels, on which these six products were separated in the first dimension and reduced and the respective products of the reduction of each component were then separated in the second dimension,[96,97] were used to identify the six products as the covalent dimers α–α, α–γ, β–γ, γ–γ, γ–ε, and δ–ε. It was concluded that these six covalent products reflect six specific noncovalent contacts present in the intact protein.

An even more complex mixture of polypeptides is found in the ribosome, which is a large molecule of protein built from many polypeptides and several ribonucleic acids. A ribosome can be readily dissociated into two components, the 30S subunit and the 50S subunit. Each of these proteins has been examined by cross-linking studies. In this case, the proteins were modified by 2-iminothiolane to convert lysines to thiols[98]

$$(10\text{–}42)$$

These thiols were then oxidized to disulfides to cross-link various lysines on the proteins. The products were identified by two-dimensional gel electrophoresis. With this reagent, and dimethyl 3,3'-dithiobis(propionimidate),[99] covalently cross-linked pairs of polypeptides could be identified among the products from the reactions of the 30S subunit[100] and 15 covalently cross-linked pairs of polypeptides could be identified among the products from the reactions of the 50S subunit.[101]

Intramolecular cross-linking can be used to determine juxtapositions in a single folded polypeptide. Cystines, as a natural cross-link, can provide evidence for the juxtaposition of two segments of polypeptide.[102] The bifunctional reagent 2-(*p*-nitrophenyl)-3-(3-carboxy-4-nitrophenyl)thio-1-propene (**10–15**) can undergo a series of reversible addition–eliminations to form bridges between two nucleophilic amino acids (Figure 10–8), either lysines or cysteines.[103] The reaction is reversible as long as the nitrophenyl group is present to stabilize the carbanion but can be rendered irreversible by reducing the nitro group with dithionite. Therefore, the reagent can be permitted to step around the protein until the most stable cross-link is formed, and this cross-link can then be locked in by reduction. In this way, two pairs of intramolecular cross-links on ribonuclease (c2 Kin.4), one between Lysine 7 and Lysine 37 and the other between Lysine 31 and Lysine 41, could be formed in high yield when only 2 equiv of the reagent was added initially to the protein.

The reagent bromopyruvate is bifunctional by virtue of its carbonyl, which can form an imine with a lysine, and its alkyl bromide, which is an alkylating agent. Bromopyruvate is able to form an imine with Lysine 144 of 2-keto-3-deoxy-6-phosphogluconate aldolase while it simultaneously alkylates Glutamate 56.[104,105] This observation established the proximity of these two amino acids in the folded polypeptide.

Suggested Reading

Buechler, J.A., & Taylor, S.S. (1989) Dicyclohexylcarbodiimide Cross-Links Two Conserved Residues, Asp-184 and Lys-72, at the Active Site of the Catalytic Subunit of c-AMP-Dependent Protein Kinase, *Biochemistry* 28, 2065–2070.

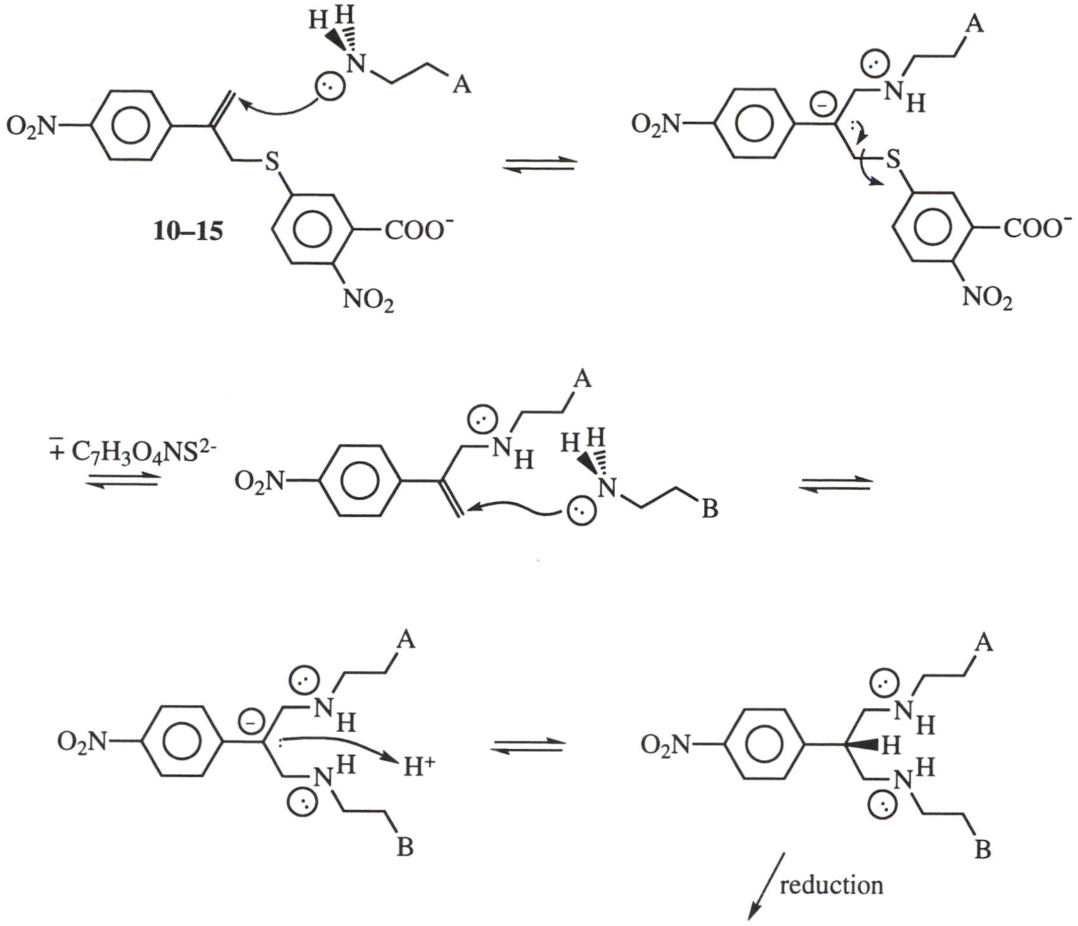

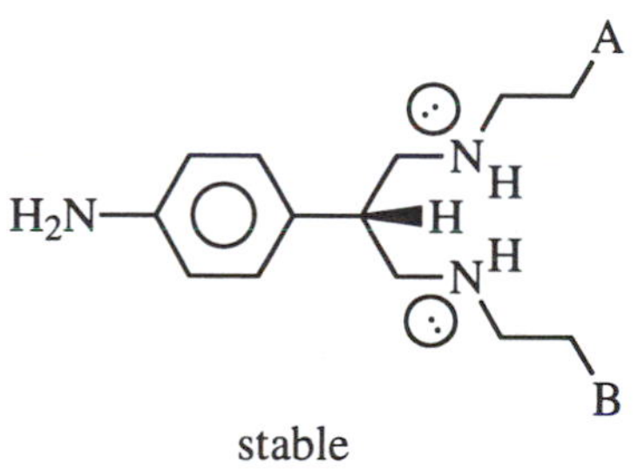

Figure 10–8: Mechanism by which 2-(p-nitrophenyl)-3-(3-carboxy-4-nitrophenyl)thio-1-propene cross-links adjacent lysines. The olefin on the p-nitrostyrene is activated by the electron-withdrawing capacity of the nitro group and participates in reversible, nucleophilic addition–elimination. Because the adduct is symmetric, two nucleophiles are cross-linked reversibly. In the first step of the reaction, the nitrothiobenzoate (Reaction 10–31) is the preferred leaving group from the asymmetric carbanion, but when the carbanion is then formed between two lysines, either can be the leaving group and the reagent can be passed from lysine to lysine over the surface of the protein. The nitrobenzyl carbanion is not very basic, so its protonation is reversible and this allows the cycles of addition and elimination to proceed. When the nitro group is reduced to the amine, the aminobenzyl proton is no longer acidic and the reagent is fixed in place.

Problem 10–1: Mercaptoethanol undergoes the following dissociation

$$HOCH_2CH_2SH \rightleftharpoons HOCH_2CH_2S^- + H^+$$

$$pK_a = 9.50$$

Suppose the total amount of mercaptoethanol in solution is equal to $[SH]_{TOT}$, a quantity you know since you added that much. Suppose, also, that there is a chemical reaction that occurs between only the anion, $HOCH_2CH_2S^-$, and an electrophile, X, and the rate of this reaction is

$$rate = k[HOCH_2CH_2S^-][X]$$

As the pH changes, $[HOCH_2CH_2S^-]$ changes although $[SH]_{TOT}$ is always the same.

(A) Show that

$$rate = k\{f([H^+])\}[SH]_{TOT}[X]$$

(B) What is $f([H^+])$? Give an explicit equation for this function.

(C) Plot log $[k_{observed}\,(k)^{-1}]$ against pH, where $k_{observed} \equiv k\{f([H^+])\}$. Indicate on the plot where pH = pK_a.

(D) At what pH does the rate of the reaction become zero?

(E) By what factor does the rate decrease for each decrease in pH of 1.00 when pH < pK_a?

Problem 10–2: Give mechanisms for the following reactions. In each reaction,

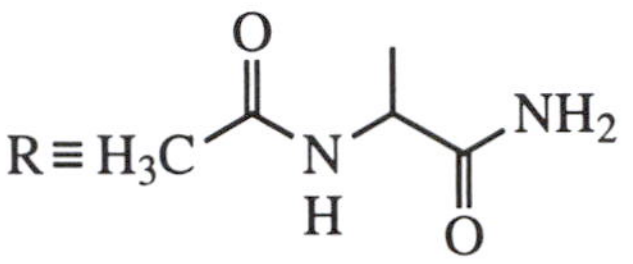

A.

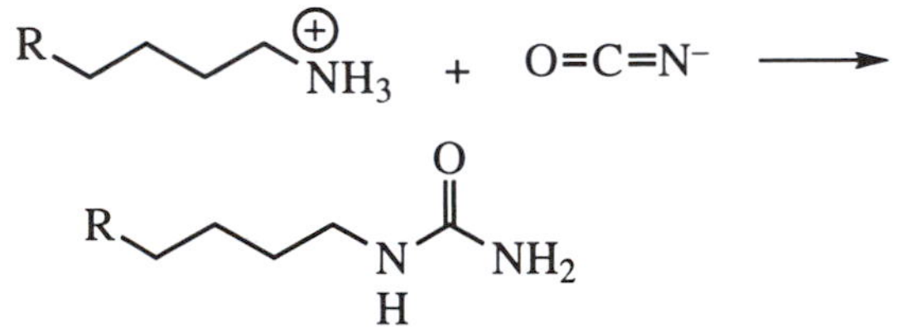

B.

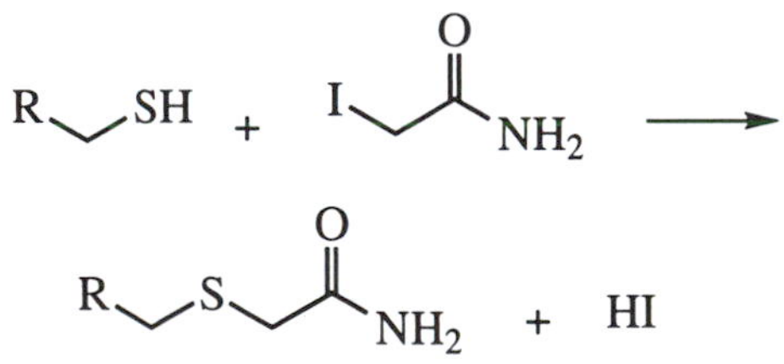

C.

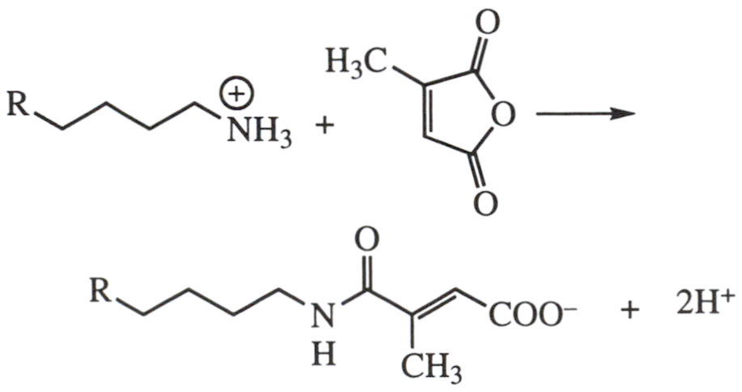

D.

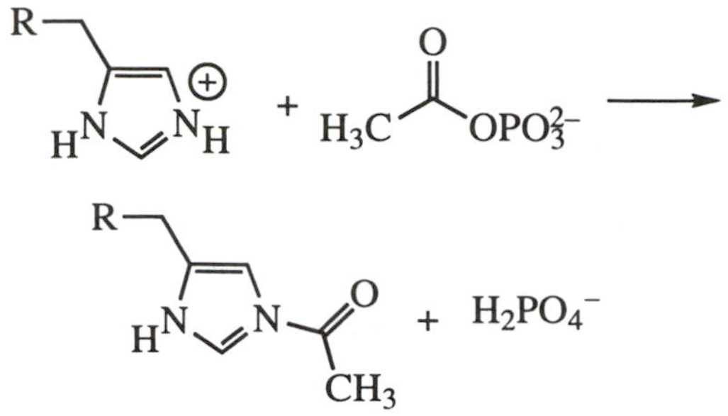

E.

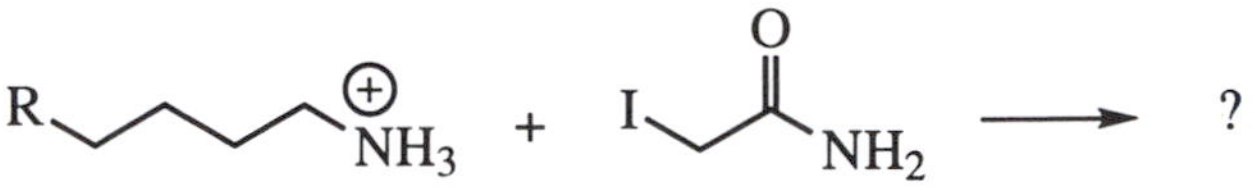

F.

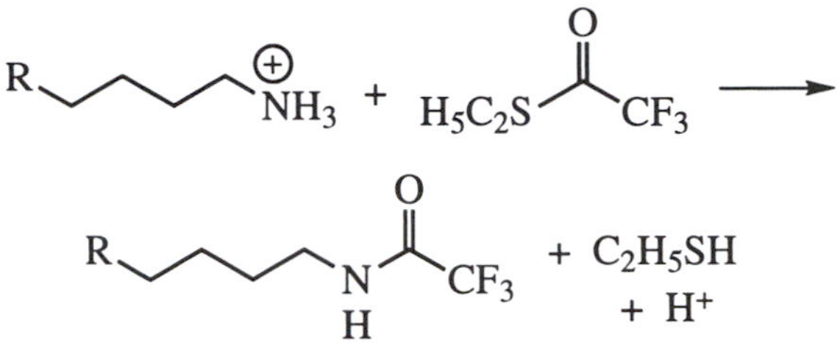

Problem 10–3: Diisopropyl fluorophosphate inhibits serine proteinases by specific phosphorylation of the seryl residue in the active site. Papain is a proteinase that does not have a serine residue in its active site. Nevertheless, it reacts with diisopropyl fluorophosphate with the result that 1 mol of phosphate is bound for every mole of enzyme but without loss of enzymatic activity. The reaction between papain and the reagent was carried out with radioactive diisopropyl [^{32}P]fluorophosphate, the modified protein was digested with chymotrypsin, and the segment of enzyme containing the radioactive label was isolated. It corresponded to amino acids 112–123 in the sequence of papain: ^{112}QVQPYNQGALLY^{123}. What is the most likely site of alkylphosphorylation of papain?

Problem 10–4

(A) Draw the structure of the peptide RDVLMKE in the ionization state in which it would exist at pH 1.4. Indicate all lone pairs.

The peptide was modified with iodoacetamide at pH 1.4, 40 °C, for 20 h. Digestion with carboxypeptidase yielded the full complement of glutamic acid from the resulting peptide. Edman degradation yielded the full complement of arginine. It is possible to estimate the number of charges a peptide bears at a certain pH from its behavior on electrophoresis. This was done for the initial peptide and the product from its reaction with iodoacetamide.

	charge	
pH	original peptide	iodoacetamide product
6.5	0	+1
2.1	+3	+4

(B) Write a stoichiometric mechanism for the reaction that occurred between iodoacetamide and one of the side chains on this peptide.

(C) How would the product of the reaction with iodoacetamide move on a cation-exchange column relative to the unreacted peptide?

Problem 10–5: A peptide has the sequence SVEKCYEKP.

(A) How many charges does the peptide bear at pH 1.9? At pH 5.6?

The peptide was reacted with trimethyloxonium tetrafluoroborate

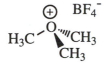

in aqueous solution at pH 6.0. Three methyl groups were covalently attached to the peptide. When the methylated and unmethylated peptides were examined by electrophoresis the following result was observed.

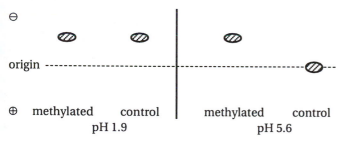

(B) What nucleophiles on the peptide have reacted with the trimethyloxonium cation? Write a mechanism for this reaction. Why is trimethyloxonium so reactive?

References

1. Deisenhofer, J., Epp, O., Miki, K., Huber, R., & Michel, H. (1984) *J. Mol. Biol. 180*, 385–398.
2. Weiss, M.S., Abele, U., Weckesser, J., Welte, W., Schiltz, E., & Schulz, G.E. (1991) *Science 254*, 1627–1630.
3. Xuong, N.H., Freer, S.T., Hamlin, R., Neilsen, C., & Vernon, W. (1978) *Acta Crystallogr. A34*, 289–296.
4. Remington, S., Wiegand, G., & Huber, R. (1982) *J. Mol. Biol. 158*, 111–152.
5. Stammers, D.K., & Muirhead, H. (1975) *J. Mol. Biol. 95*, 213–225.
6. Stark, G.R. (1970) *Adv. Protein Chem. 24*, 261–308.
7. Fuller, G.M., & Doolittle, R.F. (1966) *Biochim. Biophys. Acta 25*, 694–700.
8. Klotz, I.M., & Keresztes-Nagy, S. (1962) *Nature 195*, 900–901.
9. Gibbons, I., & Schachman, H.K. (1976) *Biochemistry 15*, 52–60.
10. Patthy, L., & Smith, E.L. (1975) *J. Biol. Chem. 250*, 557–564.
11. Koshland, D.E., Karkhanis, Y.D., & Latham, H.G. (1964) *J. Am. Chem. Soc. 86*, 1448–1450.
12. Wu, C.W., & Stryer, L. (1972) *Proc. Natl. Acad. Sci. U.S.A. 69*, 1104–1108.
13. Kaplan, H., Stevenson, K.J., & Hartley, B.S. (1971) *Biochem. J. 124*, 289–299.
14. Dorner, F. (1971) *J. Biol. Chem. 246*, 5896–5902.
15. Wofsy, L., & Singer, S.J. (1963) *Biochemistry 2*, 104–116.
16. Gurd, F.R.N. (1967) *Methods Enzymol. 11*, 532–541.
17. Schnakerz, K.D., & Noltmann, E.A. (1970) *J. Biol. Chem. 245*, 6417–6423.
18. Gundlach, H.G., Stein, W.H., & Moore, S. (1959) *J. Biol. Chem. 234*, 1754–1761.
19. Gundlach, H.G., Moore, S., & Stein, W.H. (1959) *J. Biol. Chem. 234*, 1761–1764.
20. Hand, E.S., & Jencks, W.P. (1962) *J. Am. Chem. Soc. 84*, 3505–3514.
21. Makoff, A.J., & Malcolm, A.D.B. (1981) *Biochem. J. 193*, 245–249.
22. Makoff, A.J., & Malcolm, A.D.B. (1980) *Eur. J. Biochem. 106*, 313–320.
23. DiMarchi, R.D., Garner, W.H., Wang, C.C., Hanania, G.I.H., & Gurd, F.R.N. (1978) *Biochemistry 17*, 2822–2828.
24. Means, G.E., & Feeny, R.E. (1968) *Biochemistry 7*, 2192–2201.
25. Rippa, M., Spanio, L., & Pontremoli, S. (1967) *Arch. Biochem. Biophys. 118*, 48–57.
26. Riordan, J.F., & Vallee, B.L. (1967) *Methods Enzymol. 11*, 565–570.
27. Anderson, G.W., Zimmerman, J.E., & Callahan, F.M. (1964) *J. Am. Chem. Soc. 86*, 1839–1842.
28. Serjeant, E.P., & Dempsey, B. (1979) *Ionization Constants of Organic Acids in Aqueous Solution*, Pergamon Press, Oxford, England.
29. Miles, E.W. (1977) *Methods Enzymol. 47*, 431–442.
30. Goldberger, R.F., & Anfinsen, C.B. (1962) *Biochemistry 1*, 401–405.
31. Riordan, J.F., & Vallee, B.L. (1967) *Methods Enzymol. 11*, 570–576.
32. Dixon, H.B.F., & Perham, R.N. (1968) *Biochem. J. 109*, 312–314.
33. Fields, R. (1972) *Methods Enzymol. 25*, 464–468.
34. Henkart, P. (1971) *J. Biol. Chem. 246*, 2711–2713.
35. Rogers, G.A., Shaltiel, N., & Boyer, P.D. (1976) *J. Biol. Chem. 251*, 5711–5717.
36. Westhead, E.W. (1965) *Biochemistry 4*, 2139–2144.
37. Bond, J.S., Francis, S.H., & Park, J.H. (1970) *J. Biol. Chem. 245*, 1041–1053.
38. Riordan, J.F., & Vallee, B.L. (1967) *Methods Enzymol. 11*, 541–548.
39. Cohen, L.A. (1968) *Annu. Rev. Biochem. 37*, 695–726.

40. Banas, T., Gontero, B., Drews, V.L., Johnson, S., Marcus, F., & Kemp, R.G. (1988) *Biochim. Biophys. Acta 957*, 178–184.
41. Ellman, G.L. (1959) *Arch. Biochem. Biophys. 82*, 70–77.
42. Hubbard, A.L., & Cohn, Z.A. (1972) *J. Cell Biol. 55*, 390–405.
43. McConahey, P.J., & Dixon, F.J. (1966) *Int. Arch. Allergy Appl. Immunol. 29*, 185–189.
44. Horinishi, H., Hachimori, Y., Kurihara, K., & Shibata, K. (1964) *Biochim. Biophys. Acta, 86*, 477–489.
45. Bruice, T.C., Gregory, M.J., & Walters, S.L. (1968) *J. Am. Chem. Soc. 90*, 1612–1619.
46. Sokolovsky, M., Riordan, J.F., & Vallee, B.L. (1967) *Biochem. Biophys. Res. Commun. 27*, 20–25.
47. Holt, L.A., Milligan, B., & Rivitt, D.E. (1971) *Biochemistry 10*, 3559–3564.
48. Scoffone, E., Fontana, A., & Rocchi, R. (1968) *Biochemistry 7*, 971–979.
49. Omenn, G.S., Fontana, A., & Anfinsen, C.B. (1970) *J. Biol. Chem. 245*, 1895–1902.
50. Takahashi, K. (1968) *J. Biol. Chem. 243*, 6171–6179.
51. Toi, K., Bynum, E., Norris, E., & Itano, H.A. (1967) *J. Biol. Chem. 242*, 1036–1043.
52. Yankeelov, J.A. (1970) *Biochemistry 9*, 2433–2439.
53. Itano, H.A., & Gottlieb, A.J. (1963) *Biochem. Biophys. Res. Commun. 12*, 405–408.
54. Riordan, J.F. (1973) *Biochemistry 12*, 3915–3923.
55. Patthy, L., Varadi, A., Thész, J., & Kovacs, K. (1979) *Eur. J. Biochem. 99*, 309–313.
56. Hoare, D.G., & Koshland, D.E. (1967) *J. Biol. Chem. 242*, 2447–2453.
57. Buechler, J., & Taylor, S.S. (1989) *Biochemistry 28*, 2065–2070.
58. Lewis, S.D., & Shafer, J.A. (1973) *Biochim. Biophys. Acta, 303*, 284–291.
59. Bayley, H., & Knowles, J.R. (1977) *Methods Enzymol. 46*, 69–114.
60. Iddon, B., Meth-Cohn, O., Scriven, E.F.V., Suschitzky, H.J., & Gallagher, P.T. (1979) *Angew. Chem., Int. Ed. Engl. 18*, 900–917.
61. Garin, J., Boulay, F., Issartel, J.P., Lunardi, J., & Vignais, P.V. (1986) *Biochemistry 25*, 4431–4437.
62. Abramovitch, R.A., & Davis, B.A. (1964) *Chem. Rev. 64*, 149–185.
63. Reiser, A., & Leyshon, L.J. (1971) *J. Am. Chem. Soc. 93*, 4051–4052.
64. Nielsen, P.E., & Buchard, O. (1982) *Photochem. Photobiol. 35*, 317–323.
65. Richards, F.F., Lifter, J., Hew, C., Yoshioka, M., & Konigsberg, W.H. (1974) *Biochemistry 17*, 3572–3575.
66. Westerman, J., Wirtz, K.W.A., Berkhout, T., VanDeenan, L.L.M., Radhakrishnan, R., & Khorana, H.G. (1983) *Eur. J. Biochem. 132*, 441–449.
67. Brunner, J., Senn, H., & Richards, F.M. (1980) *J. Biol. Chem. 255*, 3313–3318.
68. Hexter, C.S., & Westheimer, F.H. (1971) *J. Biol. Chem. 246*, 3934–3938.
69. Hexter, C.S., & Westheimer, F.H. (1971) *J. Biol. Chem. 246*, 3928–3933.
70. Vaughan, R.J., & Westheimer, F.H. (1979) *J. Am. Chem. Soc. 21*, 217–218.
71. Brunner, J., & Richards, F.M. (1980) *J. Biol. Chem. 255*, 3319–3329.
72. Ross, A.H., Radhakrishnan, R., Robson, R.J., & Khorana, H.G. (1982) *J. Biol. Chem. 257*, 4152–4161.
73. Shafer, J., Baronowsky, P., Laursen, R., Finn, F., & Westheimer, F.H. (1966) *J. Biol. Chem. 241*, 421–427.
74. Hutchinson, C.A., Phillips, S., Edgell, M.H., Gillam, S., Jahnke, P., & Smith, M. (1978) *J. Biol. Chem. 253*, 6551–6560.
75. Zoller, M.J., & Smith, M. (1982) *Nucleic Acids Res. 10*, 6487–6500.
76. Alber, T., Dao-pin, S., Wilson, K., Wozniak, J.A., Cook, S.P., & Matthews, B.W. (1987) *Nature 330*, 41–46.
77. Sanger, F., Coulson, A.R., Barrell, B.G., Smith, A.J.H., & Roe, B.A. (1980) *J. Mol. Biol. 143*, 161–178.
78. Wilkinson, A.J., Fersht, A.R., Blow, D.M., & Winter, G. (1983) *Biochemistry 22*, 3581–3586.
79. Landt, O., Grunert, H., & Hahn, U. (1990) *Gene 96*, 125–128.
80. Kunkel, T.A., Roberts, J.D., & Zakour, R.A. (1987) *Methods Enzymol. 154*, 367–382.
81. Reidhaar-Olson, J.F., & Sauer, R.T. (1988) *Science 241*, 53–57.
82. Walter, G., Scheidtmann, K.H., Carbone, A., Laudano, A.P., & Doolittle, R.F. (1980) *Proc. Natl. Acad. Sci. U.S.A. 77*, 5197–5200.
83. Wilchek, M., Bocchini, V., Becker, M., & Givol, D. (1971) *Biochemistry 10*, 2828–2834.
84. Kyte, J., Xu, K.Y., & Bayer, R. (1987) *Biochemistry 26*, 8350–8360.
85. Castellino, F.J., & Hill, R.L. (1970) *J. Biol. Chem. 245*, 417–424.
86. Parente, A., Merrifield, B., Geraci, G., & D'Alessio, G.D. (1985) *Biochemistry 24*, 1098–1104.
87. Kannan, K.K., Notstrand, B., Fridborg, K., Lövgren, S., Ohlsson, A., & Petef, M. (1975) *Proc. Natl. Acad. Sci. U.S.A. 72*, 51–55.
88. Abita, J.P., Maroux, S., Delaage, M., & Lazdunski, M. (1969) *FEBS Lett. 4*, 203–206.
89. Freer, S.T., Kraut, J., Robertus, J.D., Wright, H.T., & Xuong, N.H. (1970) *Biochemistry 9*, 1997–2009.
90. Lambert, J.M., Perham, R.N., & Coggins, J.R. (1977) *Biochem. J. 161*, 63–71.
91. Pabo, C.O., & Lewis, M. (1982) *Nature 298*, 443–447.
92. Hugli, T.E. (1973) *J. Biol. Chem. 248*, 1712–1718.
93. Oefner, C., & Suck, D. (1986) *J. Mol. Biol. 192*, 605–632.
94. Ji, T.H. (1983) *Methods Enzymol. 91*, 580–609.
95. Abdella, P.M., Smith, P.K., & Royer, G.P. (1979) *Biochem. Biophys. Res. Commun. 87*, 734–742.
96. Wang, K., & Richards, F.M. (1974) *J. Biol. Chem. 249*, 8005–8018.
97. Sommer, A., & Traut, R.R. (1974) *Proc. Natl. Acad. Sci. U.S.A. 71*, 3946–3950.
98. Jue, R., Lambert, J.M., Pierce, L.R., & Traut, R.R. (1978) *Biochemistry 17*, 5399–5405.
99. Peretz, H., Towbin, H., & Elson, D. (1976) *Eur. J. Biochem. 63*, 83–92.
100. Sommer, A., & Traut, R.R. (1976) *J. Mol. Biol. 106*, 995–1015.
101. Kenny, J.W., & Traut, R.R. (1979) *J. Mol. Biol. 127*, 243–263.
102. Spackman, D.H., Stein, W.H., & Moore, S. (1960) *J. Biol. Chem. 235*, 648–659.
103. Mitra, S., & Lawton, R.G. (1979) *J. Am. Chem. Soc. 101*, 3097–3110.
104. Meloche, H.P. (1973) *J. Biol. Chem. 248*, 6945–6951.
105. Suzuki, N., & Wood, W.A. (1980) *J. Biol. Chem. 255*, 3427–3435.

CHAPTER 11

Immunochemical Probes of Structure

Immunoglobulins are proteins found, among other locations, in the blood serum of birds and mammals. Immunoglobulins are also called antibodies. In an animal, the function of an immunoglobulin is to recognize a foreign macromolecule by binding to it. The foreign macromolecule is referred to as an antigen. An **antigen** is any foreign macromolecule that elicits, upon its injection into an animal, the production of immunoglobulins capable of binding to it with high specificity. Almost any macromolecule can serve as an antigen. Within the animal, after the antigen has been bound by the immunoglobulin, it is usually destroyed. The important point which should be kept in mind is that the primary biological purpose for a particular immunoglobulin is to distinguish one particular undesirable macromolecule from the myriad of other necessary macromolecules in the animal. Whenever an immunoglobulin recognizes and binds not only to its antigen but also to one or more of the macromolecules normally present in the animal, these indigenous macromolecules are also destroyed in autoimmune processes detrimental to the animal. Therefore, the immune system has evolved to produce immunoglobulins that are highly specific in their recognition of molecular structure. Because no predictions can be made as to what antigens will have to be destroyed during the life of the animal, the immune system must be prepared to make immunoglobulins capable of binding with high specificity to any foreign molecule when it is presented to the animal in an antigenic form.

The serum from any animal contains a wide variety of immunoglobulins, each with its own distinct amino acid sequence and each present in a different concentration from the others. They are the immunoglobulins that have been produced in response to all of the foreign antigens encountered by the particular individual during its peculiar lifetime. These immunoglobulins are present in the serum at a concentration of 10–20 mg mL^{-1}.

The paradigm of the various types of immunoglobulins is **immunoglobulin G** (Figures 7–12 and 11–1). It is composed of two identical heavy polypeptides (n_{aa} = 440–450) and two identical light polypeptides (n_{aa} = 210–220).[1] Each heavy polypeptide is folded into four internally repeating, superposable domains designated V_H, C_H1, C_H2, and C_H3 in the order in which they occur in the sequence of the protein. Each light polypeptide is folded into two internally repeating, superposable domains, V_L and C_L. Each of these six unique domains, each approximately 110 amino acids in length, is superposable in its folded form on each of the other five (Figure 7–12).[2] Two of these $\alpha\beta$ heterodimers are associated across a molecu-

lar 2-fold axis of symmetry to form the entire immunoglobulin.

An immunoglobulin G can be cut into three pieces. When the intact native protein is treated with papain,[3,4] cleavage occurs within the hinged region to the amino-terminal side of the cystines connecting the two heavy polypeptides (Figure 11–1) and two identical Fab fragments (c12 Kin.1) and one Fc fragment are produced. The designation Fab arises from the fact that the fragment contains the antigen-binding site. The designation Fc originally referred to the fact that this fragment could be crystallized. It is now more informative and consistent to consider this the constant fragment. It is because it is constant that it can crystallize. Each of these fragments is a well-behaved, independent, soluble, globular protein. Each contains four of the original 12 internally repeating, superposable domains.

The advantage of using Fab fragments in experiments is that they are **univalent**. Each Fab fragment contains only one binding site for the antigen. An intact molecule of immunoglobulin G is necessarily **bivalent** because, with its molecular rotational axis of symmetry, it must have two binding sites for antigen (Figure 11–1). The fact that two antigens can be bound by intact immunoglobulin G complicates some experiments. A bivalent analog of the Fab fragment can be produced by digesting intact immunoglobulin G with pepsin.[5,6] The pepsin cleaves to the carboxy terminal side of the cystines between the two heavy chains and produces a fragment, $(Fab')_2$, containing two Fab fragments joined by two or more cystines. When reduced, the $(Fab')_2$ fragment dissociates into two Fab' fragments whose C_H1 domains are slightly longer than those of an Fab fragment.

The major immunoglobulins in the serum of a mammal are immunoglobulins G (10–20 mg mL^{-1}), immunoglobulins M (1 mg mL^{-1}), and immunoglobulins A (1 mg mL^{-1}). Each of these immunoglobulins contains light chains that are indistinguishable from one type to the next. It is the heavy chains, present always in equimolar ratio to the light chains, that distinguish one type from the other. **Immunoglobulin M**[7] has a longer heavy polypeptide (n_{aa} = 570–580) than immunoglobulin G by one extra domain, $C_\mu4$, which would be the analog of C_H4, if C_H4 existed. Immunoglobulin M is a pentameric complex of five [(heavy)$_2$(light)$_2$] heterotetramers held together by cystines among themselves. The cystines cross-link pairs of $C_\mu3$ domains to form a pentameric ring of the heterotetramers. **Immunoglobulin A** has a heavy polypeptide only about 30 amino acids (n_{aa} = 470–480)[8] longer than that of immunoglobulin G. Immunoglobulin

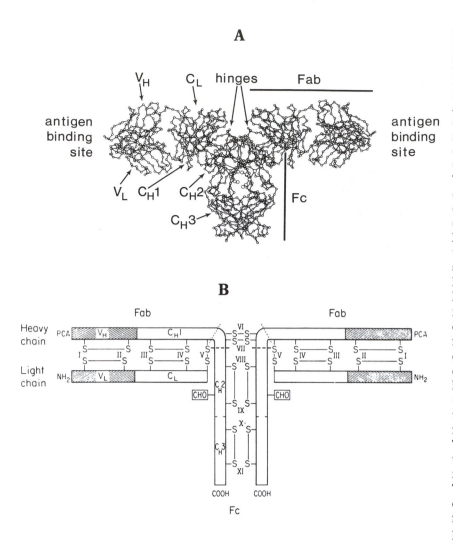

A

B

Figure 11–1: Structure of a molecule of immunoglobulin G. (A) The crystallographic molecular model[2] of immunoglobulin G, which is presented in stereo in Figure 7–12. (Adapted with permission from ref 2. Copyright 1977 National Academy of Sciences.) (B) A diagrammatic representation of the molecule based on the internally repeating domains observed in its amino acid sequences.[1] (Adapted with permission from ref 1. Copyright 1969 National Academy of Sciences.) The molecule is composed of 12 superposable domains all homologous in amino acid sequence, each about 110 amino acids in length. In the center of each domain is a cystine (I, II, III, IV, VIII, IX, X, XI) formed between two structurally adjacent cysteines, 60 amino acids apart in the amino acid sequence. A light chain and a heavy chain are linked by cystine V at the carboxy terminus of the light chain, and heavy chains are linked by two or more cystines (VI and VII) in the hinge region that couples the three arms. The 12 domains are referred to as variable domain, heavy chain (V_H); constant domain 1, heavy chain (C_H1); constant domain 2, heavy chain (C_H2); constant domain 3, heavy chain (C_H3); variable domain, light chain (V_L); and constant domain, light chains (C_L). The three arms are referred to as the antigen-binding fragments (Fab) and the constant fragment (Fc). The binding sites for the antigens are at the tips of the Fab arms and are formed by the variable domains from heavy and light chains, respectively. The light dashed lines indicate where papain cleaves the molecule to produce the Fab and Fc fragments; the heavy dashed lines indicate where pepsin cleaves the molecule to produce the $(Fab')_2$ fragment. The Fab fragments are missing the cystines holding the heavy chains together; the $(Fab')_2$ fragments include the cystines. The hinges at which the cleavages by proteinases occur are indicated in panel A.

A is a mixture of monomeric [(heavy)$_2$(light)$_2$] heterotetramers similar to those of immunoglobulin G and higher oligomers of [(heavy)$_2$(light)$_2$] heterotetramers held together by cystines between their $C_\alpha3$ domains.[9] Both immunoglobulins M and A have a short polypeptide J associated with them that may promote their initial oligomerization even before the intertetrameric cystines are formed.[9] Immunoglobulins M and A have their binding sites for antigens in a similar location to those on immunoglobulins G and distant from the regions ($C_\mu3$, $C_\mu4$, $C_\alpha3$) that account for their distinct oligomeric structures. The only significant differences between these types of immunoglobulins and immunoglobulins G is their size and, hence, their valence. Fab fragments or Fab-like fragments can be produced from each.[7,8,10]

An immunoglobulin of a particular sequence is pro-duced by a colony of lymphocytes, all derived from one single cell that was initially stimulated to divide and manufacture. The colony assumes its identity from its pedigree and not from its situation. All lymphocytes in the colony are descendants of the same cell, but each member of the colony, like all other lymphocytes, is dispersed by the bloodstream and lymphatic system and wanders independently and at random through the animal as it continuously manufactures its particular immunoglobulin. The sole product of the members of a colony is this one immunoglobulin continuously released into the serum and extracellular fluid. Each time a lymphocyte is stimulated to divide and manufacture its particular immunoglobulin against a particular antigen, a new colony is established. Each colony produces molecules of immunoglobulin with the same, unique amino acid sequence, and each particular amino acid sequence endows the ability to bind a particular antigen. Because many (10–100) different lymphocytes are stimulated to divide and manufacture by molecules of the same antigen, even if it is a simple chemical functionality such as a dinitrophenyl substituent, many different colonies continuously produce immunoglobulins after exposure of the animal to the antigen, each of a different amino acid sequence, but all specific for that antigen. Such a set of immunoglobulins, each capable of recognizing a given antigen but each different from the others, is referred to as a **polyclonal** set. The product of the reaction of an intact animal to an antigen is always a polyclonal set of immunoglobulins, which are isolated as a mixture from the serum of the animal.

In a normal animal, the various colonies are stable contributors to the population of immunoglobulins necessary to deal with antigens in the environment. Occasionally, the controls maintaining the stable population of the colony fail, and one lymphocyte begins to multiply malignantly. This uncontrolled cancerous growth causes an enormous increase in the number of lymphocytes producing an immunoglobulin of just one unique sequence and structure. Such a cancer is referred to as a myeloma. The serum of such individuals contains high concentrations of only one type of immunoglobulin. Such **myeloma proteins** are present in sufficient quantities to be purified, crystallized,[2] and sequenced.[1] It is from such proteins that much of our understanding of the structure of immunoglobulins derives.

Myeloma proteins appear by chance as the products of the random malignant transformation of normal lymphocytes. The antigens to which most myelomas are directed are unknown, but a few of the antigens for particular myeloma proteins have been identified. For example, the murine myeloma protein McPC603 can bind pneumococcal polysaccharides.[11] These polysaccharides contain choline, and it was found that the immunoglobulin A designated McPC603 could bind phosphonocholine with high specificity.

The site to which phosphonocholine is bound by immunoglobulin McPC603 has been located by constructing a difference electron density map between the protein in crystals of the Fab fragment of immunoglobulin McPC603 soaked in a solution of phosphonocholine and the protein in unsoaked crystals.[10] The phosphonocholine, bound noncovalently to the protein in the former crystals, appeared as positive electron density in the difference map. A molecular model of phosphonocholine was inserted into this positive electron density to establish its precise location and orientation within the unit cell and hence its location on the surface of the protein. The site to which it was bound, as have been all other binding sites for antigen that have been observed crystallographically, is located in the same place, at the tips of the Fab fragments (c12 Kin.1) most distant from the hinge (Figure 11–1).

The **binding site for antigen** is a region on the surface of the Fab fragment formed from six loops of random meander (c12 Kin.1), three from each polypeptide, heavy and light, and known as the **complementarity-determining regions** of the structure (amino acids designated by open circles in Figure 11–2).[10] Each of these six loops

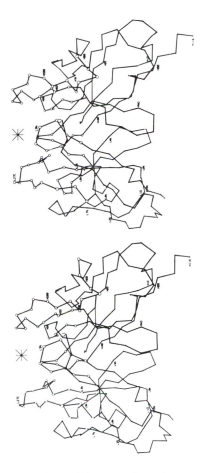

Figure 11–2: Stereo view of the two domains (c12 Kin.1), the heavy-chain variable domain (to the right) and the light-chain variable domain (to the left) forming the site at which phosphonocholine binds in the crystallographic molecular model[10] of the murine myeloma protein McPC603. The immunoglobulin A produced by the murine myeloma cells was digested with pepsin, reduced, and crystallized as the Fab' fragment from ammonium sulfate. A map of electron density was produced by multiple isomorphous replacement. Molecular models were constructed from the known amino acid sequences of the two polypeptides and were inserted into the map of electron density. The figure is an α-carbon trace of the variable domains (the amino-terminal domains) from the two chains. A difference map of electron density was calculated between crystals of the unoccupied Fab' and crystals soaked in phosphonocholine, an antigen bound by this antibody. The phosphonocholine was bound between the two variable domains of the heavy and light chains, respectively (asterisk). The pocket to which it was bound was formed by six loops of random meander, three from each chain (designated by open dots), referred to as the complementarity-determining loops. Reprinted with permission from ref 10. Copyright 1974 National Academy of Sciences.

is one of the connections between the strands of the antiparallel β structure that form the superstructure of the core of the respective domains (Figure 11–2). The amino acid sequences in these six complementarity-determining loops show remarkable variation among different immunoglobulins, and they are referred to as the **hypervariable regions** of the sequences.[12] It is this variety of amino acid sequences that gives the immunoglobulins as a class their ability to provide individual proteins each tailored to bind a particular antigen. The specific sequences in these loops define the specificity of the individual immunoglobulins.

The difficulty with a myeloma protein is that the investigator cannot choose the antigen against which it is directed. Its advantage is that it can be purified to homogeneity and the purified protein is necessarily composed of identical copies of the same molecule, each with an identical ability to recognize the antigen. The advantage of the ability to select the antigen and the advantage of the homogeneity of the product have been combined in the production of **monoclonal immunoglobulins**.[13] In this procedure, lymphocytes from the spleen of a mouse immunized with the antigen of interest are fused with cultured myeloma cells that normally produce a particular myeloma protein. These cultured myeloma cells are immortal cell lines that continuously grow and divide either in flasks in an incubator or as solid tumors in mice. Hybrids, each produced by the fusion of one lymphocyte and one myeloma cell, are selected on the basis of their ability to grow on a particular medium. The hybrids are then reproduced as single colonies of cells. Because each colony arose from one single cell, the cells in a particular colony produce only the immunoglobulin originally secreted by the parental lymphocyte that fused to the myeloma cell and the myeloma protein originally secreted by the myeloma cell. Therefore each colony is the offspring of one lymphocyte in the mouse from which the spleen was taken. If that lymphocyte happened to be producing one of the immunoglobulins directed against the antigen originally injected, its offspring can be cultivated for the production of a homogeneous monoclonal immunoglobulin recognizing that antigen. To identify the colonies secreting monoclonal immunoglobulins against the antigen of interest, each colony is individually screened.

In immunochemistry, there is the immunoglobulin and there is the antigen. An antigen is any macromolecule that elicits the appearance in an animal of immunoglobulins recognizing that antigen. If a macromolecule is injected into an animal and no such immunoglobulin is elicited, the macromolecule cannot be referred to as an antigen. Because the immune system was developed to recognize and destroy foreign organisms such as viruses and bacteria and because other systems are used by animals to eliminate small toxic molecules, antigens are always macromolecules like proteins, never small molecules. This fastidiousness of the immune system can be circumvented by covalently attaching a small molecule to a carrier protein as a **hapten**. This can be done by reactions analogous to those used for covalent modification or cross-linking of proteins. Likewise, many substances capable of modifying proteins cause strong immunological reactions by reacting at random with proteins of the animal and rendering them foreign. Because the immune system has evolved to recognize large oligomeric proteins, such as viral coats, with great efficiency, proteins that have been covalently cross-linked into larger aggregates are more antigenic than they are in their uncross-linked form.[14] A potential antigen is injected into an animal and specific immunoglobulins may or may not appear in the serum. If they do not, there is no way of discovering why they did not.

If a protein is the antigen, each immunoglobulin that does appear in response to the immunization possesses a binding site to which an epitope on the antigen can bind. An **epitope** is the region on the antigenic protein that interacts directly with the binding site on the immunoglobulin. Usually, an epitope is one or more short sequences of amino acids in the antigen that are adjacent to each other in its structure and that associate specifically with the pocket formed by the six complementarity-determining loops. Haptens, although they also associate with this pocket in the respective immunoglobulin, are not referred to as epitopes. If the antigen is a protein, the epitope and the binding site on the immunoglobulin combine noncovalently as if they were two faces forming an interface in an oligomeric protein. Therefore, the epitope can be no larger than the binding site on the immunoglobulin.

The binding site on an immunoglobulin is a flat surface or a depression on the surface of a globular protein formed from the domains V_H and V_L (Figure 11–2). It seems unlikely that a segment of polypeptide on the antigenic protein that resides in a depression on its surface could be recognized as an epitope by the binding site on an immunoglobulin. The surface of picornaviruses such as cold viruses and polioviruses are highly irregular. They are furnished with bosses at the 5-fold rotational axes of symmetry of the icosahedral shell (Figure 9–21). These bosses are separated from each other by deep depressions on the surface of the virus. The epitopes on polioviruses are located mainly on the bosses themselves and a few small protruding segments of polypeptide.[15] It is believed that the crucial regions of the surface of cold viruses that allow them to produce an infection are located in the depressions between the bosses.[16] These locations would be inaccessible to antibodies. Each time the epitopes on the bosses or smaller protrusions of cold viruses sustain a sufficient number of mutations to escape recognition, an antigenically novel but still infectious cold virus arises. This hypothesis explains the large collection of immunologically distinct cold viruses, which all have the same mechanism of infection. The actual machinery of infection lying as it does within the depressions would be protected from being recognized by any immunoglobulin.

When an immunoglobulin is elicited by a hapten attached to a large protein, the immunoglobulins produced against the hapten are believed to bind it well within the depression near the center of the binding site (Figure 11–2). This would explain why the unattached hapten can also be bound efficiently by the immunoglobulin. For example, immunoglobulins raised against a protein whose lysines had been modified by 2,4,6-trinitrobenzenesulfonate (Reaction 10–27) bind N^ε-(2,4,6-trinitrophenyl)lysine with high affinity.[17] Immunoglobulins raised against short peptides covalently attached to other proteins as haptens also can bind the unattached peptides with high affinity. In the crystallographic molecular models of small haptens within the site formed from the hypervariable loops, the hapten is bound in the center of the depression among the loops.[10]

Crystallographic molecular models of complexes between an intact protein antigen and the Fab fragment of the respective monoclonal immunoglobulin are available. One is of a complex between the monomeric protein lysozyme and an Fab fragment from a monoclonal immunoglobulin G specific for this protein (c12 Kin.2);[18] another is of a complex between the tetrameric neuraminidase of influenza virus and four Fab fragments attached at four separate faces that are related by the 4-fold rotational axis of symmetry of the tetramer.[19]

In the crystallographic molecular model of the complex between lysozyme and the Fab fragment (Figure 11–3),[18] the interface between the epitope and the binding site for the antigen on the immunoglobulin, as with any other interface between two folded polypeptides, resembles the interior of a molecule of protein. It is composed of hydrophobic interactions and hydrogen bonds (Figure 11–3B). All six of the complementarity-determining loops of the immunoglobulin are involved, and they associate with two strands of polypeptide from lysozyme, that between Aspartate 18 and Asparagine 27 and that between Lysine 116 and Leucine 129. The two strands forming the epitope are immediately adjacent to each other on the surface of the protein.

The central and most critical amino acid in this epitope is Glutamine 121, the side chain of which occupies a distinct hole on the surface of the Fab fragment, among the six complementarity-determining loops (Figure 11–3). Each of the three amino acids lining the hole for Glutamine 121 is from a different complementarity-determining loop, one from the heavy polypeptide and two from the light polypeptide, and this places the hole in the very center of the binding site on the Fab fragment. If Glutamine 121 is replaced by either a histidine or an asparagine by site-directed mutation, the antigen is no longer bound by the immunoglobulin. In the normal structure of lysozyme, Glutamine 121 is fully exposed to the solvent.

It is often the case that an epitope seems to be focused on a particular amino acid on the surface of a protein. For example, about 30–40% of the polyclonal immunoglobulins raised to human cytochrome c fail to bind to the cytochrome c from *Macaca mulatta*, which differs from the human protein only by the replacement of Isoleucine 58 by a threonine.[20] These same immunoglobulins recognized cytochrome c from the kangaroo that differs from the human at several other locations but does contain Isoleucine 58. No immunoglobulins raised to the cytochrome c from *M. mulatta* failed to recognize the cytochrome c from the human, and this result suggests that when a cytochrome c contains a threonine at position 58, as does the protein from *M. mulatta*, this region on the external surface[21] is not antigenic.[20] The impression left by these observations is that Isoleucine 58 is the key amino acid in this epitope, as is Glutamine 121 in the epitope of lysozyme.

In the crystallographic molecular model of the complex between lysozyme and immunoglobulin, the lysozyme is superposable, within the error of the models, upon its structure in the absence of the immunoglobulin and no obvious differences in the structure between the Fab fragment and other Fab fragments were noted.[18] It was concluded that the fit between antigen and immunoglobulin was entirely complementary and no significant rearrangement of either was required upon the formation of the complex. In the crystallographic molecular model of the complex between the tetrameric neuraminidase and the four Fab fragments, however, the interface between the two domains, V_L and V_H (Figure 11–2), was altered from that seen in crystallographic molecular models of most other free Fab fragments by a rigid-body rotation of about 5–10°, which would rearrange the relative orientation of the heavy polypeptide and the light polypeptide within the binding site by about 0.4 nm.[19] In the center of the epitope on the neuraminidase, the structure in the crystallographic molecular model at one of the more flexible strands of its polypeptide was also distorted during the formation of the complex.

These two crystallographic examples of complexes between antigens and immunoglobulins place in perspective an ambiguity in our understanding of the characteristics of an epitope. At the moment, it is unclear what relationship the amino acids on the surface of the antigen must have to each other to constitute an epitope. In particular, must all of the amino acids in an epitope be directly and covalently attached to each other or can a set of amino acids from distant locations in the primary structure of the antigen but adjacent to each other in the tertiary structure together form an epitope?

The physical state of the antigen within the animal as it is presented to the surface of the lymphocyte that will be stimulated to divide and manufacture the immunoglobulin is not yet precisely understood, but it now seems that only short sequences of amino acids covalently attached to each other can be presented.[22] The immune complex of lysozyme provides an example of

A

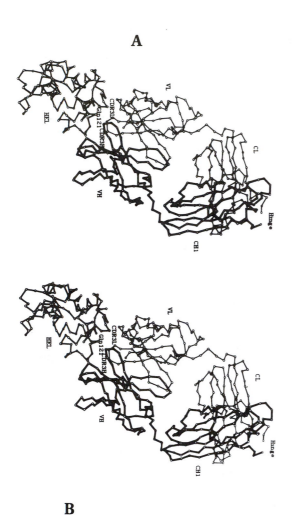

B

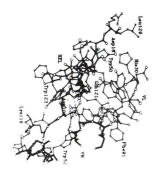

Figure 11–3: Crystallographic molecular model of the binding of hen egg white lysozyme to the Fab fragment of a murine monoclonal immunoglobulin G.[18] A complex was formed between hen egg white lysozyme and the monoclonal Fab fragment, and the complex was crystallized from 20% poly(ethylene glycol). A crystallographic molecular model was prepared from the map of electron density. (A) Drawing of the α-carbon skeleton of the complete molecular model. The Fab fragment (upper right) is drawn with the heavy chain (thick bonds) and light chain (thin bonds) including both domains from each chain (V_H and C_H1 and V_L and C_L, respectively). The hen egg white lysozyme (HEL) is in the lower left. The two epitopic loops of random meander in the lysozyme that are in contact with the complementarity-determining loops of the immunoglobulin are Aspartate 18–Asparagine 27 and Lysine 116–Leucine 129. (B) Detailed view of the interface between these two loops of hen egg white lysozyme (lower left) and the tips of the complementarity-determining loops (upper right) in the same orientation as in panel A. The polypeptide of the lysozyme from Aspartate 116 to Leucine 129 is the most intimately associated and can be traced across the center of the complex. The complementarity-determining loops from the heavy chain are the heaviest lines, those from the light chain are the lightest, and those from the lysozyme are of intermediate thickness. Side chains making contacts are highlighted. Hydrogen bonds are dashed lines. Reprinted with permission from ref 18. Copyright 1986 American Association for the Advancement of Science.

such an epitope because, in the native protein, the sequence between Aspartate 18 and Asparagine 27 is directly and covalently attached to the sequence between Lysine 116 and Leucine 129 by the cystine connecting Cysteine 30 and Cysteine 115. Two sets of observations, one gathered with immunoglobulins raised against myoglobin[23] and one gathered with immunoglobulins raised against lysozyme,[24] have suggested that epitopes are usually composed of amino acids all attached locally to each other by covalent bonds in the native structure.

Several polyclonal mixtures of immunoglobulins were produced, each in a different animal, with sperm whale myoglobin as an antigen. Each of these polyclonal mixtures probably contained more than a hundred different monoclonal immunoglobulins all mixed in varying concentrations. Through an extensive series of experiments,[23] it was determined that the majority of these monoclonal immunoglobulins were directed against only five epitopes on the surface of the molecule of myoglobin. These five epitopes were identified as the segments Lysine 16–Valine 21, Lysine 56–Lysine 62, Alanine 94–Isoleucine 99, Histidine 113–Histidine 119, and Tyrosine 146–Tyrosine 151. Most of these segments form loops of random meander on the surface of the crystallographic molecular model of myoglobin (Figure 11–4).[24] Each was incorporated synthetically into a small peptide and tested for its ability to compete with native myoglobin for binding to the immunoglobulins. When a synthetic peptide inhibits the binding of an antigen by a set of monoclonal antibodies in a polyclonal mixture, it is presumed that the peptide itself is recognized as an antigen and bound at the several binding sites for the antigen present in the polyclonal mixture. In separate competitions,[25-29] the synthetic peptides were able to inhibit the binding to myoglobin of a total of 50–70% of the monoclonal immunoglobulins present in these polyclonal mixtures.

A similar experiment was performed with lysozyme.[24] In this instance, two polyclonal mixtures obtained from two animals appeared to contain immunoglobulins directed against only three epitopes on the protein. Each site included or was adjacent to a cystine in the protein. Only when the protein was digested with its cystines intact could peptides (Figure 11–5) inhibiting the reaction between the polyclonal mixtures and intact lysozyme be isolated.[30] One of these peptides (Figure 11–5A) contained Glycine 22–Asparagine 27 and Lysine 116 of the epitope defined crystallographically at a later date (Figure 11–3). Two of these peptides were either purified or synthesized chemically. In separate experiments,[31,32] they could each compete, respectively, with intact lysozyme for 30% of the monoclonal antibodies in the polyclonal mixtures. It was concluded that the two sets of monoclonal immunoglobulins used in these particular experi-

ments were each directed against a unique epitope, which included the segments of the protein containing these amino acid sequences. When synthetic models of all three of these peptides were attached individually to solid supports for affinity adsorption, the three affinity adsorbents used sequentially could remove greater than 95% of the monoclonal immunoglobulins directed against lysozyme from the mixtures.

All of these experiments, as well as studies of the protein that presents the antigen to the lymphocyte,[22] suggest that direct covalent connections among the amino acids within an epitope are necessary to retain its integrity during the processing of the antigen that occurs within the animal prior to its stimulation of the lymphocyte that will divide and manufacture. If this were the case, an epitope would not include two or more unconnected segments of amino acid sequence that were simply juxtaposed in the native structure of the protein. It is also possible, however, that each of the active peptides in these experiments represented simply the more tightly bound of two or more strands of polypeptide composing the epitope (Figure 11–3), and when that peptide was bound at one side of the binding site for antigen the intact protein could not be bound. In fact, the two synthetic peptides containing Lysine 16–Valine 21 and Histidine 113–Histidine 119 in myoglobin, which are adjacent to each other in native protein (Figure 11–4),

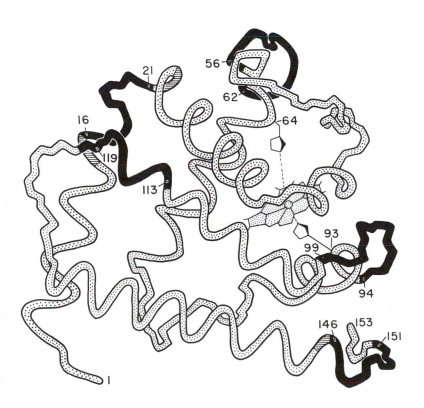

Figure 11–4: Five epitopes on the surface of myoglobin.[24] The five regions were identified by competition[23] between synthetic peptides and myoglobin for polyclonal sets of immunoglobulins raised in goats and rabbits. Synthetic peptides with amino acid sequences identical to these short segments (6 or 7 amino acids) of amino acid sequence from the intact molecule of myoglobin were able to compete for most of the immunoglobulins in various antisera. Each of the five segments identified in this way is highlighted in black on a drawing of the crystallographic molecular model of myoglobin. Reprinted with permission from ref 24. Copyright 1978 Pergamon Press.

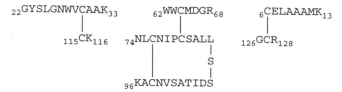

Figure 11–5: Amino acid sequence of the three peptides that are responsible for inhibition (85–89%) of the reaction of native lysozyme with its antisera. Unreduced lysozyme was digested with trypsin to produce a digest that retained antigenic activity. The peptides were mixed with the polyclonal serum and the complexes between immunoglobulins and peptides were isolated by molecular exclusion chromatography. These three peptides were the major components bound by the polyclonal mixture of immunoglobulins raised against the native protein.

each competed for about the same fraction of the immunoglobulins in the polyclonal mixtures.[25,28] Some of these immunoglobulins might recognize both peptides. The strength with which the immunoglobulins bound each of the synthetic peptides from myoglobin was, however, fairly high. Only 10–1000-fold more peptide was required to compete on an equal basis with each intact epitope, and this represents a difference in free energy of association between peptide and intact protein of only 5–17 kJ mol[-1], values which hardly suggest that the peptide occupies only half or less of the binding site. Another possibility is that most of the interactions seen in the crystallographic molecular model (Figure 11–3) provide no favorable contributions to the free energy of association between antigen and immunoglobulins. In this case, the true epitope, that portion providing the strength of the association, would be smaller than it seems. This, however, does not seem to be true of the other paradigm for an immune complex, that between viral neuraminidase and an Fab fragment.

The interface between antigen and immunoglobulin seen in the crystallographic molecular model[19] of the viral neuraminidase and the Fab fragments is formed from at least four strands of polypeptide from distant segments of the amino acid sequence of the neuraminidase. When amino acids in any one of three of these four strands were mutated, the neuraminidase would no longer bind to the immunoglobulin. When amino acids in the fourth strand were mutated, binding affinity was noticeably diminished. All of these results indicate that the epitope on this protein for this immunoglobulin is a large region on the surface comprising all four of these strands. Furthermore, these four strands are all more than seven amino acids away from the nearest cystine and not interconnected even at this distance by cystines.[33] It seems an inescapable conclusion that this epitope would cease to exist if the protein were to unfold in this region. Such an epitope would be referred to as a **conformationally specific epitope**.

Many, if not the majority of, immunoglobulins will recognize their antigens even when the protein is no longer in its native structure. These are **sequence-specific immunoglobulins**. The paradigm of this class would be an immunoglobulin that can be used to isolate by affinity adsorption a peptide comprising its epitope from a digest of its antigen;[30] such an immunoglobulin can recognize its epitope even when it is a formless peptide. Usually such sequence-specific immunoglobulins, which in theory should be the only type that can be produced, are distinguished from the class of conformationally specific immunoglobulins. Although the immunoglobulin specific for neuraminidase that was used in the crystallographic studies seems to be an obvious member of this other, inexplicable class, the evidence for such conformationally specific immunoglobulins is usually anecdotal. A protein is irreversibly unfolded and loses its antigenic properties. To unfold a polypeptide irreversibly, however, it is usually either covalently modified[34] or noncovalently and intermolecularly polymerized by aggregation. Either the epitope could be covalently modified or it could be sterically sequestered within an aggregate of the protein during such uncontrolled reactions. If either of these events occur, it would appear as if the immunoglobulin were conformationally specific and recognized only the native structure of the protein when actually it was sequence specific and recognized a single linear sequence of amino acids that was simply covalently modified or inaccessible. The technical difficulty is to unfold the polypeptide of the antigen to a monodisperse random polymer without doing the same thing to the immunoglobulin, which is also a folded polypeptide. Otherwise, the binding site for the antigen may not be able to find the epitope.

It is probably the case that the distinction between sequence-specific immunoglobulins and conformationally specific immunoglobulins is one of degree. For example, a polyclonal set of immunoglobulins was raised against native staphylococcal nuclease (n_{aa} = 149) and purified on the basis of its ability to recognize a fragment of the intact polypeptide comprising amino acids 99–149. These immunoglobulins could bind the intact, folded polypeptide of staphylococcal nuclease 2×10^4 times more tightly, as judged from the dissociation equilibrium constants, than they could the fragment.[35] The fragment is a monodisperse random polymer. Presumably, the immunoglobulins still recognize the epitope or portions of the epitope after it is unfolded, but with much smaller free energy of dissociation ($\Delta\Delta G°$ = –24 kJ mol[-1]). A difference in dissociation constant of even this magnitude, if the concentrations of immunoglobulin and antigen and the individual dissociation constants are in the appropriate ranges, would be observed as a complete elimination of the ability of antigen to bind to immunoglobulin when it is in fact due only to a finite attenuation of the ability of antigen to bind to immunoglobulin. In any case, the distinction between sequence-specific and conformationally specific immunoglobulins is theoretically and practically ambiguous.

Although immunoglobulins for use in protein chemistry are usually raised by injecting an intact protein into an animal, it is also possible to raise immunoglobulins directed against synthetic peptides with the same amino acid sequence as a segment from a particular protein.[36] The synthetic peptides are attached as haptens to another protein. For example, the amino-terminal amino acid sequence of the large tumor antigen from simian virus is AcMDKVLNR–, where Ac is a posttranslationally added acetyl group. Immunoglobulins raised against a peptide with this sequence were able to recognize and bind exclusively to the intact protein in crude homogenates from animal cells infected with the simian virus.[36] Immunoglobulins directed against a particular peptide can be purified by using an affinity adsorbent to which the peptide is attached.

The difficulty inherent in the use of immunoglobulins raised against a particular amino acid sequence in a protein to study that protein in its native conformation is that the investigator, a fallible judge, rather than the immune system, which is less fallible, has chosen the epitope. A thoughtful consideration of Figures 11–3 and 11–4 leads to the conclusion that a native protein probably does not expose many sequences sufficiently for immunoglobulins to recognize them. If the investigator has chosen the epitope, there is only a small chance that it will be available in the native protein, unless the choice is the safe one of the amino terminus or the carboxy terminus. The segment of sequence against which the immunoglobulin was raised, however, will usually be available in the unfolded polypeptide.

Any solution of a protein, even pristine cytoplasm, freshly drawn serum, or a solution of redissolved crystals, contains some of the irreversibly unfolded polypeptide of any particular protein. Immunoglobulins directed against synthetic peptides necessarily bind preferentially, and perhaps exclusively, to any unfolded protein exposing the sequence of amino acids against which they were raised because the original antigen was a structureless peptide. Yet, most experiments are designed with the requirement that the immunoglobulins recognize and bind to the native protein for the conclusion to be valid. This problem of the antigenicity of the unfolded polypeptide also exists with immunoglobulins raised against a native protein. If partially or fully unfolded forms of the protein are presented to a lymphocyte and if they cause the lymphocyte to divide and manufacture, immunoglobulins recognizing only unfolded forms of the protein could be present in the polyclonal sets resulting from the immunization.

The solution to this problem is analytical. If every molecule of protein in a solution can bind one immunoglobulin at a particular epitope, then the immunoglobulin is recognizing the native protein. If only a small percentage of the molecules of protein in a solution

can bind an immunoglobulin at that epitope, it is only the unfolded polypeptides that are presenting that epitope. When using either monoclonal immunoglobulins or immunoglobulins raised against synthetic peptides, one can assume that each folded polypeptide possesses only one epitope, either exposed or buried. It is also possible to purify by affinity adsorption a subset of polyclonal immunoglobulins recognizing only one epitope from a set of polyclonal immunoglobulins raised against an intact protein.[20] In any experiment relying on the assumption that the immunoglobulins recognize the native protein, it must be demonstrated both that the immunoglobulins bind to only one unique epitope and that every protomer of the antigen is capable of binding one molecule of immunoglobulin. In the absence of such a demonstration, the conclusions can be disregarded.

Immunoglobulins directed against an antigenic protein can effect **immunoprecipitation.** Each immunoglobulin has at least two binding sites for antigen (Figure 11–1). Each antigen usually has more than one epitope (Figure 11–4). If a polyclonal mixture containing monoclonal immunoglobulins, directed against different epitopes on the antigen, is mixed in the proper ratio with that protein, a precipitate forms containing antigen and immunoglobulin cross-linked among themselves. **Complement fixation** is a procedure for detecting the relative concentrations of such immunoprecipitates in a series of samples.[37] It is sensitive to concentrations of immunoprecipitate smaller than can be measured directly. The ratio at which maximum precipitation occurs is known as the **equivalence point**.

If a large excess of antigen is present, each immunoglobulin has its binding sites filled with antigens that are not bound to other immunoglobulins and no precipitate forms. If a large excess of immunoglobulin is present, each antigen is surrounded by immunoglobulins, each with its other binding sites vacant, and no precipitate forms. Such soluble complexes are present under all circumstances, and it is never possible to precipitate directly all of the antigen or all of the immunoglobulins even at equivalence. Finally, the molar ratio between antigenic protein and immunoglobulin in a precipitate gathered at equivalence is a complicated function of the number of epitopes, their relative affinities, and the distribution of monoclonal immunoglobulins in the polyclonal mixture.

The multivalency of an immunoglobulin, essential for immunoprecipitation, complicates most quantitative measurements of the binding of immunoglobulins to antigens, and it is often desirable to use Fab fragments or Fab' fragments, which are simple univalent reagents. For quantitative experiments, the Fab fragments are often rendered radioactive by covalent modification. From their specific radioactivity and the total radioactivity of the immune complex, the moles of Fab fragment bound to the epitopes present in a sample can be determined. If the Fab fragments are from a mixture of polyclonal immunoglobulins all directed against the same epitope or from a monoclonal immunoglobulin, they become reliable and quantitative reagents for protein chemistry.

The reaction between an Fab fragment, Fab, and its unique epitope, Ep, is governed by the reaction

$$Ep + Fab \underset{k_{-1}}{\overset{k_1}{\rightleftharpoons}} Ep \cdot Fab \qquad (11\text{--}1)$$

where $Ep \cdot Fab$ is the **immune complex** (Figure 11–3). The immune complexes between epitopes on proteins and Fab fragments are usually quite strong. This is due to the fact that k_{-1} is often a fairly small rate constant and dissociation of the complex is slow. It is this slow dissociation that allows immunoglobulins to be used as reagents for affinity adsorption. It also permits the complex between Fab fragment and antigen to be separated from uncomplexed Fab fragment or uncomplexed antigen or both. The slow rate constant k_{-1}, however, complicates quantitative analysis of this reaction by decreasing the rate at which it reaches equilibrium.

The amount of immune complex can be determined after the reaction has come to equilibrium. Three or more parameters govern its concentration. These are the total concentration of competent Fab fragments, the total concentration of accessible epitopes, and the dissociation constants between the various Fab fragments and the antigenic site. None of these parameters is known with certainty before measurements are made. The Fab fragments are generally mixtures of competent and incompetent species, and available epitopes may exist on only a small fraction of the molecules of the antigenic protein. It is possible, however, to saturate the epitopes with Fab fragments and determine explicitly the molar concentration of immune complex that forms. This molar concentration will be equal to the total molar concentration of accessible epitopes. Likewise, by saturating the mixture of Fab fragments with antigen, the total molar concentration of competent Fab fragment can be determined. If the immunoglobulin from which the Fab fragments were derived was monoclonal, the dissociation constant for the immune complex can also be determined from the variation of its concentration at equilibrium as the concentration of Fab fragment and antigen are varied.

Unfortunately, most immunochemical assays do not provide the absolute concentration of immune complex but only the relative amount, and absolute concentrations must be estimated indirectly. This is usually done by preadsorption. Either a solution of antigen or a solution of an immunoglobulin with a known molar concentration of epitopes or binding sites is mixed in various ratios with a solution of an immunoglobulin or antigen, respectively, of unknown concentration of binding sites or epitopes. The decrease in the relative concentration of either the unoccupied immunoglobulin or unoccupied antigen brought about by adding the standard solution is then assessed by an immunochemical assay. If the complex between immunoglobulin and antigen is stoichiometric under the chosen conditions, and the complex does not dissociate during the subsequent immunoassay, 1 mol of the antigen or immunoglobulin of unknown concentration should disappear for every mole of immunoglobulin or antigen of known concentration that has been added.

One of the most important uses of immunochemistry is either to identify or to isolate the particular protein to which the antibody is directed. A specific immunoglobulin can be used to stain its antigen among the proteins in a heterogeneous mixture separated by electrophoresis by the procedure of **immunoblotting**.[38-40] The separated proteins are transferred from the polyacrylamide gel onto a membrane of nitrocellulose or poly(vinylidene difluoride),[41] the membrane is soaked in a solution of the specific immunoglobulin, and the immunoglobulins bound to the antigen are tagged with a second immunoglobulin to which either peroxidase[42] or alkaline phosphatase[40] is attached. The peroxidase or alkaline phosphatase produces a colored precipitate at the location of the antigen. In this way, one protein can be picked out of a complex mixture because it is the only protein that is stained.

Antibodies can also be used to isolate a particular protein by **immunoadsorption**. For example, excess specific immunoglobulin G is added to a solution to saturate all of the antigen. Agarose to which protein A from *Staphylococcus aureus*, a protein with a high affinity for immunoglobulin G, has been covalently attached is added, and it adsorbs the immune complexes.[43] The adsorbed antigen is released from the agarose after the other proteins have been rinsed away. Usually, the protein of interest is then identified by sodium dodecyl sulfate–polyacrylamide gel electrophoresis. It is also possible to attach the primary antibody itself covalently to agarose and use this immunoadsorbent to isolate the protein directly by affinity adsorption.[44,45]

Often such procedures are used to identify or isolate a protein for which no assay is available. An example of such a protein is one that is translated from an unidentified reading frame. For such a protein, only the sequence of genomic DNA or complementary DNA is available, but the protein itself has not yet been observed. In such cases, only the amino acid sequence of the protein, read from the nucleic acid sequence, is known. Immunoglobulins can be raised against synthetic peptides with the same amino- or carboxy-terminal sequence as the protein to be identified.[36] Alternatively, the unidentified protein can be tagged by adding a short sequence of amino acids to its carboxy or amino terminus by site-directed mutation.[46] The elongated protein expressed in mutated cells can then be identified[46] or isolated[47] by an immunoglobulin against this extraneous amino acid sequence.

Immunoglobulins have been used in protein chemistry to identify the region on the surface of a protein that contains the antigenic site to which they are directed. The disposition of the various polypeptides within the 30S subunit[48–50] and 50S subunit[51] of ribosomes from *Escherichia coli* has been determined with immunoglobulins directed against the constitutive polypeptides. The 21 unique polypeptides composing the 30S subunit of the ribosome can be catalogued by two-dimensional gel electrophoresis (Figure 11–6).[52,53] They have been separated and individually purified,[54] and their amino acid sequences have been determined.[55] Polyclonal sets of immunoglobulins have been raised against each polypeptide. When immunoglobulins specific for one of these polypeptides are mixed with intact 30S ribosomal subunits and the immune complexes are then prepared for electron microscopy, individual immunoglobulins bound to individual 30S ribosomal subunits or crosslinking two or more 30S ribosomal subunits can be observed (Figure 11–7).[48–50] Because the 30S ribosomal subunit has a characteristic shape (Figure 11–8) the epitopes recognized by these immunoglobulins can be assigned to certain regions of the subunit.

Three classes of polypeptides can be distinguished from these images. Polypeptides in the first class, exemplified by polypeptide S14 (Figure 11–7A), support the binding of only one immunoglobulin at a time to each 30S ribosomal subunit, and each immunoglobulin seems to be bound to about the same location on the subunit. Therefore the epitope or epitopes for polypeptide S14 are all the same or very close to each other. The polypeptides S8, S13, S14, and S20 belong to this class.[49,56] Polypeptides in the second class, exemplified by polypeptides S9 and S10 (Figure 11–7B), can support the binding of two immunoglobulins simultaneously to the same 30S ribosomal subunit, but the two or more epitopes are always close together and in the same region of the subunit. The polypeptides S2, S3, S6, S9, S10, S12, S16, and S17 belong to this class.[49,56] Polypeptides in the third class, exemplified by S4 (Figure 11–7), also can support the binding of two immunoglobulins simultaneously to the same 30S ribosomal subunit, but the epitopes that these polypeptides present to the immunoglobulins are much more widely separated and are even located in different regions of the subunit. The polypeptides S5, S7, S11, S15, S18, and S21 belong to this class.[49,56] It has been concluded that the polypeptides in this last class are severely elongated and meander through the complete native 30S ribosomal subunit, occasionally surfacing to expose an epitope.

Two independent descriptions of the relative locations of the various polypeptides within the 30S ribosomal subunit have been presented, based on the relative distributions of the antigenic sites over its surface.[49,57] In each of these two sets of observations, the location at which each individual immunoglobulin was bound on the 30S ribosomal subunit was ascertained by deciding visually which projection and which orientation of the idealized structure of the 30S ribosomal subunit (Figure 11–8) it represented and deciding visually where the point of contact between the immunoglobulin and the idealized image occurred (drawings in Figure 11–7). Unfortunately, the relative positions for the various polypeptides, assigned independently by the two separate laboratories, do not agree.[56] One of the most peculiar and disturbing findings in both sets of experiments was that the majority of the antigenic sites were located on the smaller lobe of the 30S ribosomal subunit (head in Figure 11–8).[48–50,57] As these immunochemical studies of the 30S ribosomal subunit represent one of the most exhaustive uses of immunoglobulins for topological studies where results were independently checked, the ambiguity of the conclusions is discouraging.

There is, however, uniform agreement on which of the polypeptides within the 30S ribosomal subunit are highly elongated.[56] The conclusion that a subset of the polypeptides is a collection of random meanders is also supported by results from studies of covalent cross-linking, coincident covalent modification, and fluorescence energy transfer between pairs of polypeptides within the 30S ribosomal unit.[58] For example, from results of such experiments, polypeptide S4, thought to be one of the most dispersed, has been found to pass within less than 1.0 nm of polypeptides S3, S5, S7, S9, S12, S13, S15, S16, S17, S19, and S20.

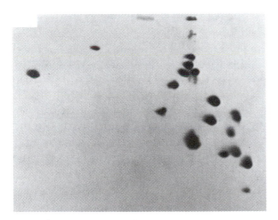

Figure 11–6: Separation of the polypeptides composing the 30S subunit of ribosomes from *E. coli*.[53] Intact ribosomes were isolated from a homogenate of bacteria by centrifugation. They were dissociated into subunits by treatment with $MgCl_2$, and the 30S subunit was isolated by centrifugation through a gradient of sucrose. The protein was extracted from the 30S ribosomal subunit and dissolved in 6 M urea. The individual polypeptides were separated in the first dimension at pH 9.6 and in the second dimension at pH 4.6. Separation in each dimension was performed in 6 M urea. The proteins were stained with amido black. Although only 17 components are observed, three polypeptides coelectrophorese at one spot and two pairs of polypeptides coelectrophorese at two other spots. The total number of polypeptides is 21. Reprinted with permission from ref 53. Copyright 1973 *Journal of Biological Chemistry.*

A

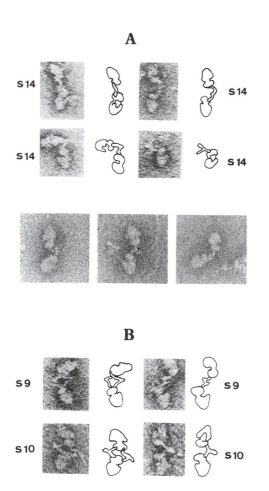

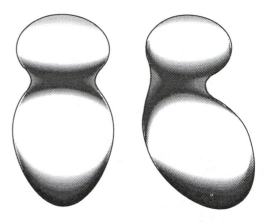

Figure 11–8: Shape of the 30S ribosomal subunit from *E. coli*. The structure is a composite of the shapes defined in two different laboratories[48,49] from projections of the 30S subunit negatively stained with uranyl acetate and observed by electron microscopy. Adapted with permission from refs 48 and 49. Copyright 1974 and 1975 National Academy of Sciences.

Figure 11–7: Electron micrographs[48,49] of immune complexes between polyclonal immunoglobulins G and 30S ribosomal subunits from *E. coli*. Purified 30S subunits were mixed with the polyclonal immunoglobulins raised against the particular purified polypeptide. The complexes that formed were adsorbed to a layer of carbon on a grid for microscopy, negatively stained with uranyl acetate, and observed in the electron microscope. The immunoglobulins G are Y-shaped proteins that connect two globular 30S ribosomal subunits, which can be recognized by their characteristic shapes (Figure 11–8). All images 240000×. (A) Seven images of complexes between 30S ribosomal subunits and immunoglobulins G against polypeptide S14. The upper four images, with accompanying drawings, of complexes formed with different antisera from those used to form the complexes presented in the lower three images were taken in a different laboratory but the results were the same. (B) Two images each of complexes between 30S ribosomal subunits and immunoglobulins G against polypeptides S9 and S10, respectively. Reprinted with permission from refs 48 and 49. Copyright 1974 and 1975 National Academy of Sciences.

Suggested Reading

Amit, A.G., Mariuzza, R.A., Phillips, S.E.V., & Poljak, R.J. (1986) Three-Dimensional Structure of an Antigen–Antibody Complex at 2.8–Å Resolution, *Science 233*, 747–753.

Problem 11–1: A polyclonal set of immunoglobulins was produced against the synthetic peptide –ETYY.

This is the carboxy-terminal sequence of Na⁺/K⁺-transporting adenosinetriphosphatase, a protein embedded in the membranes of animal cells. Immunoglobulins recognizing this peptide were purified by affinity adsorption on a solid phase to which the peptide had been attached. The immunoglobulins were rendered radioactive by reductive methylation (Figure 10–3) with formaldehyde and sodium [³H]borohydride to a final specific radioactivity of $10,760 \text{ cpm nmol}^{-1}$. Equal amounts of this immunoglobulin were mixed with increasing amounts of homogeneous Na⁺/K⁺-transporting adenosinetriphosphatase in its membrane-bound form so that bound and unbound immunoglobulin could be separated by centrifugation after equilibrium had been reached. The amount of bound immunoglobulin increased linearly with the amount of membrane-bound protein added. It was found that each milligram of protein of purified Na⁺/K⁺-transporting adenosinetriphosphatase could bind 670 cpm of radioactive immunoglobulin regardless of the final concentration of immunoglobulin. The asymmetric unit of Na⁺/K⁺-transporting adenosinetriphosphatase is composed of one α polypeptide ($n_{aa} = 1020$) and one β polypeptide ($n_{aa} = 300$). What fraction of the molecules of the adenosinetriphos-phatase displays epitopes?

A monoclonal immunoglobulin was also produced against Na⁺/K⁺-transporting adenosinetriphosphatase. This monoclonal immunoglobulin could bind the synthetic peptide HLLVMKGAPER, which has a sequence identical to a segment of the sequence from the α polypeptide of the enzyme. The relative concentration of binding sites of this immunoglobulin in a solution could be determined by an indirect immunoassay. When samples of a solution of this monoclonal immunoglobulin, at a final concentration of 11 nM in binding sites for antigen, were mixed and brought to equilibrium with increasing concentrations of Na$^+$/K$^+$-transporting adenosinetriphosphatase, the concentration of unoccupied binding sites for antigen decreased. A final concentration of Na$^+$/K$^+$-transporting adenosinetriphosphatase of 300 μg mL^{-1} was required to decrease the concentration of active immunoglobulin by greater than 90% (from 11 nM to less than 1 nM). What fraction of the molecules of Na$^+$/K$^+$-transporting adenosinetriphosphatase displays epitopes?

References

1. Edelman, G.M., Cunningham, B.A., Gall, W.E., Gottlieb, P.D., Rutishauser, V., & Waxdal, M.J. (1969) *Proc. Natl. Acad. Sci. U.S.A. 63*, 78–85.

2. Silverton, E.W., Navia, M.A., & Davies, D.R. (1977) *Proc. Natl. Acad. Sci. U.S.A. 74*, 5140–5144.

3. Mage, M.G. (1980) *Methods Enzymol. 70*, 142–150.

4. Porter, R.R. (1959) *Biochem. J. 73*, 119–126.

5. Nisonoff, A., Wissler, F.C., Lipman, L.N., & Woernley, D.L. (1960) *Arch. Biochem. Biophys. 89*, 230–244.

6. Masson, P.L., Cambiaso, C.L., Collet-Cassart, D.Magnusson, C.M.G., Richards, C.B., and Sindic, C.J.M. (1981) *Methods Enzymol. 74*, 106–139.

7. Putnam, F.W., Florent, G., Paul, C., Shinoda, T., & Shimizu, A. (1973) *Science 182*, 287–291.

8. Torano, A., & Putnam, F.W. (1978) *Proc. Natl. Acad. Sci. U.S.A. 75*, 966–969.

9. Chapuis, R.M., & Koshland, M.E. (1975) *Biochemistry 14*, 1320–1326.

10. Segal, D.M., Padlan, E.A., Cohen, G.H., Rudikoff, S., Potter, M., & Davies, D.R. (1974) *Proc. Natl. Acad. Sci. U.S.A. 71*, 4298–4302.

11. Potter, M. (1972) *Physiol. Rev. 52*, 631–719.

12. Wu, T.T., & Kabat, E.A. (1970) *J. Exp. Med. 132*, 211–250.

13. Köhler, G., & Milstein, C. (1976) *Eur. J. Immunol. 6*, 511–519.

14. Margoliash, E., Nisonoff, A., & Reichlin, M. (1970) *J. Biol. Chem. 245*, 931–939.

15. Hogle, J.M., Chow, M., & Filman, D.J. (1985) *Science 229*, 1358–1365.

16. Rossmann, M.G., Arnold, E., Erickson, J.W., Frankenberger, E.A., Griffith, J.P., Hecht, H.J., Johnson, J.E., Kamer, G., Luo, M., Mosser, A.G., Rueckert, R.R., Sherry, B., & Vriend, G. (1985) *Nature 317*, 145–153.

17. Barisas, B.G., Singer, S.J., & Sturtevant, J.M. (1972) *Biochemistry 11*, 2741–2744.

18. Amit, A.G., Mariuzza, R.A., Phillips, S.E.V., & Poljak, R.J. (1986) *Science 233*, 747–753.

19. Colman, P.M., Laver, W.G., Varghese, J.N., Baker, A.T., Tulloch, P.A., Air, G.M., & Webster, R.G. (1987) *Nature 326*, 358–363.

20. Nisonoff, A., Reichlin, M., & Margoliash, E. (1970) *J. Biol. Chem. 245*, 940–946.

21. Takano, T., Kallai, O.B., Swanson, R., & Dickerson, R.E. (1973) *J. Biol. Chem. 248*, 5234–5255.

22. Silver, M.L., Guo, H., Strominger, J.L., & Wiley, D.C. (1992) *Nature 360*, 367–369.

23. Atassi, M.Z. (1975) *Immunochemistry 12*, 423–438.

24. Atassi, M.Z. (1978) *Immunochemistry 15*, 909–936.

25. Koketsu, J., & Atassi, M.Z. (1974) *Immunochemistry 11*, 1–8.

26. Koketsu, J., & Atassi, M.Z. (1974) *Biochim. Biophys. Acta 342*, 21–29.

27. Pai, R.C., & Atassi, M.Z. (1975) *Immunochemistry 12*, 285–290.

28. Atassi, M.Z., & Pai, R.C. (1975) *Immunochemistry 12*, 735–744.

29. Koketsu, J., & Atassi, M.Z. (1973) *Biochim. Biophys. Acta 328*, 289–302.

30. Atassi, M.Z., Habeeb, A.F.S., & Ando, K. (1973) *Biochim. Biophys. Acta 303*, 203–209.

31. Lee, C.L., & Atassi, M.Z. (1975) *Biochim. Biophys. Acta 405*, 464–474.

32. Atassi, M.Z., Koketsu, J., & Habeeb, A.F.S.A. (1976) *Biochim. Biophys. Acta 420*, 358–375.

33. Varghese, J.N., Laver, W.G., & Colman, P.M. (1983) *Nature 303*, 35–40.

34. Ahern, T.J., & Klibanov, A.M. (1985) *Science 228*, 1280–1284.

35. Sachs, D.H., Schechter, A.N., Eastlake, A., & Anfinsen, C.B. (1972) *Proc. Natl. Acad. Sci. U.S.A. 69*, 3790–3794.

36. Walter, G., Scheidtmann, K.H., Carbone, A., Laudano, A.P., & Doolittle, R.F. (1980) *Proc. Natl. Acad. Sci. U.S.A. 77*, 5197–5200.

37. Levine, L., & VanVunakis, H. (1967) *Methods Enzymol. 11*, 928–936.

38. Renart, J., Reiser, J., & Stark, G. (1979) *Proc. Natl. Acad. Sci. U.S.A. 76*, 3116–3120.

39. Towbin, H., Staehelin, T., & Gordon, J. (1979) *Proc. Natl. Acad. Sci. U.S.A. 76*, 4350–4354.

40. Blake, M.S., Johnston, K.H., Russell-Jones, G.J., & Gotschlich, E.C. (1984) *Anal. Biochem. 136*, 175–179.

41. Pluskal, M.F., Przekop, M.B., Kavonian, M.R., Vecoli, C., & Hicks, D.A. (1986) *BioTechniques 4*, 272–282.

42. Domingo, A., & Marco, R. (1989) *Anal. Biochem. 182*, 176–181.

43. Langone, J.J. (1982) *J. Immunol. Methods 55*, 277–296.

44. Wofsy, L., & Burr, B. (1969) *J. Immunol. 103*, 380–382.

45. Schneider, C., Newman, R.A., Sutherland, D.R., Asser, U., & Greaves, M.F. (1982) *J. Biol. Chem. 257*, 10766–10769.

46. Munro, S., & Pelham, H.R.B. (1984) *EMBO J. 3*, 3087–3093.

47. Munro, S., & Pelham, H.R.B. (1987) *Cell 48*, 899–907.

48. Lake, J.A., Pendergast, M., Kahan, L., & Nomura, M. (1974) *Proc. Natl. Acad. Sci. U.S.A. 71*, 4688–4692.

49. Tischendorf, G.W., Zeichhardt, H., & Stöffler, G. (1975) *Proc. Natl. Acad. Sci. U.S.A. 72*, 4820–4824.

50. Lake, J.A., & Kahan, L. (1975) *J. Mol. Biol. 99*, 631–644.

51. Wittmann, H.G. (1976) *Eur. J. Biochem. 61*, 1–13.

52. Kaltschmidt, E., & Wittmann, H.G. (1970) *Anal. Biochem. 36*, 401–412.

53. Held, W.A., Muzushima, S., & Nomura, M. (1973) *J. Biol. Chem. 248*, 5720–5730.

54. Held, W.A., Mizushima, S., & Nomura, M. (1973) *J. Biol. Chem. 248*, 5720–5730.

55. Wittmann, H.G., Littlechild, J.A., & Wittman-Liebold, B. (1980) in *Ribosomes, Structure, Function, and Genetics* (Chambliss, G., Craven, G.R., Davies, J., Davis, K., Kahan, L., & Nomura, M., Eds.) pp 51–88, University Park Press, Baltimore, MD.

56. Gaffney, P.T., & Craven, G.R. (1980) in *Ribosomes, Structure, Function, and Genetics* (Chambliss, G., Craven, G.R., Davies, J., Davis, K., Kahan, L., & Nomura, M., Eds.) pp 237–253, University Park Press, Baltimore, MD.

57. Winkelman, D.A., Kahan, L., & Lake, J.A. (1982) *Proc. Natl. Acad. Sci. U.S.A. 79*, 5184–5188.

58. Gaffney, P.T., & Craven, G.R. (1978) *Proc. Natl. Acad. Sci. U.S.A. 75*, 3128–3132.

CHAPTER 12

Physical Measurements of Structure

Examples of physical properties used to assess the structure of a protein are its diffusion constant, its sedimentation coefficient, and its intrinsic viscosity, all of which respond to the shape of the macromolecule; its absorption of light, which responds to the environments around particular chromophores, in particular the peptide bonds; its fluorescence, which can be used to measure molecular shape and intramolecular dimensions; and its nuclear magnetic resonance spectrum, which can be used to map spatial relationships among the amino acids in the native structure. These physical properties are derived from measurements made of solutions of the protein. To convert these measurements into the particular physical property associated with the protein, certain assumptions must be made. The measurements themselves, however, usually do not validate these assumptions, and often these assumptions are not independently validated. For example, one assumption usually made is that the solution contains only one protein and that all of the molecules of that protein are identical in their structure both before and during the measurement. If this assumption is not valid, and the solution is heterogeneous, the measured physical parameters cannot be converted into conclusions about the molecular structure of the protein. Therefore, independent monitoring of the state of the protein in the solution by procedures sensitive to heterogeneity is a necessary control whenever physical measurements are made. Other assumptions are made in each of the sets of mathematical equations connecting a particular physical property with a molecular conclusion, and these must be clearly understood to evaluate the conclusions critically.

Shape

A molecule of protein dissolved in an aqueous solution of moderate ionic strength is a compact solid of peculiar shape coated with a layer of water that is more or less fixed upon its surface and that has the effect of smoothing its shape. The available refined molecular models of various proteins are similar enough to each other to provide an accurate mental picture of the boundary between the molecule of protein proper and the liquid solvent of the bulk phase. Crevices on the surface of a molecule of protein are filled with molecules of water. Although they are rapidly exchanging with their neighbors more peripherally located, the presence of these molecules of water can be considered a permanent feature of the molecule of protein. The net effect of these molecules of water is to remove the corrugations on the surface of the compact solid. Between these crevices, over the open surface of a molecule of protein, a large number of molecules of water are situated in locations that are permanently occupied, even though constantly exchanging. The relative locations of these positions becomes less and less fixed the farther from the molecule of protein they are situated until a region is reached where the water is no different from the water in an otherwise identical solution that lacks the protein. This continuous transition between the molecules of water fixed to hydrogen-bond donors and acceptors on the surface of a molecule of protein and the molecules of water in the bulk solvent is characterized by a gradual, rather than an abrupt, decrease of attachment. This is mainly due to the fact that water is a structured solvent that automatically ensnares and is ensnared by a molecule of protein. Therefore, no distinct boundary exists between the macromolecule and the solvent. Nevertheless, the concept of the **hydrodynamic particle**[1] is necessary if specific dimensions are to be extracted from physical measurements of molecules of protein dissolved in free solution.

The hydrodynamic particle is defined as the covalent molecule of protein and any molecules of water or any other small solutes that behave during the measurement as if they were affixed to the molecule of protein. An affixed molecule of water would be identified as a specific location upon the surface of the protein always occupied by one or another molecule of water. If it is assumed that a hydrodynamic particle exists, its mass m_h (in grams) will be

$$m_h = \frac{M_p\left(1+\delta_{H_2O}\right)}{N_{Av}} \qquad (12\text{--}1)$$

where M_p is the molar mass (grams mole^{-1}) of the protein, δ_{H_2O} denotes the grams of water bound for every gram of protein (Table 6–3), and N_{Av} is Avogadro's number (moles^{-1}). It should be recalled that M_p, the molar mass of the covalent structure of the protein, is almost always calculated directly from the amino acid sequences and stoichiometries of its constituent polypeptides and the amount of any attached posttranslational modifications. The volume of the hydrodynamic particle (in centimeters^3) should be

$$V_h = \frac{M_p}{N_{Av}}\left(\bar{v}_p + \delta_{H_2O}\, v^0_{H_2O}\right) \qquad (12\text{--}2)$$

where $\bar{v}_p$ is the partial specific volume of the protein in centimeters^3 gram^{-1} and v_{H2O}^0 is the specific volume of pure water in centimeters^3 gram^{-1}.

If other solutes j are attached to the hydrodynamic particle, Equation 12–1 is expanded by adding a set of terms δ_j, each of which is the grams of solute j for every gram of protein, and Equation 12–2 is expanded by adding a set of terms $\delta_j \bar{v}_j$, where the $\bar{v}_j$ are the partial specific volumes (centimeters^3 gram^{-1}) of the solutes j. An example of bound solutes for which these additional terms are major features of these equations is a case in which the protein has bound detergents or bound lipids.[2]

The concept of the hydrodynamic particle can be appended to the development of the meaning of the **diffusion coefficient** presented in Chapter 1 and the **frictional ratio** presented in Chapter 6. The diffusion coefficient (centimeters^2 second^{-1}) is designated as $D_{20,w}^0$, where the superscript indicates extrapolation to a zero protein concentration and the subscripts indicate a correction to a temperature of 20 °C and to a solvent of pure water. The diffusion coefficient is a measure of f, the **frictional coefficient** (grams second^{-1}) of the hydrodynamic particle in water at 20 °C at infinite dilution

$$f = \frac{k_B T}{D_{20,w}^0} \qquad (12\text{–}3)$$

where k_B is Boltzmann's constant (1.381×10^{-16} erg K^{-1}) and T is the temperature (293.15 K). A diffusion coefficient and a frictional coefficient are particular and intrinsic properties of a given protein in a given solution.

A minimum frictional coefficient for the hydrodynamic particle at infinite dilution can be defined as

$$f_0^{hydro} \equiv 6\pi\eta R_0^{hydro} \qquad (12\text{–}4)$$

where η is the viscosity (grams centimeter^{-1} second^{-1}) of the solvent and the subscript refers to the minimization. The viscosity of pure water at 20 °C is 0.0100 g cm^{-1} s^{-1}. The hydrodynamic radius, R_0^{hydro}, is defined as the radius (centimeters) of a sphere with the same volume as the hydrodynamic particle

$$R_0^{hydro} \equiv \left(\frac{3V_h}{4\pi}\right)^{1/3} \qquad (12\text{–}5)$$

and

$$f_0^{hydro} = 6\pi\eta \left[\frac{3M_p}{4\pi N_{Av}}\left(\bar{v}_p + \delta_{H2O} v_{H2O}^0\right)\right]^{1/3} \qquad (12\text{–}6)$$

The hydrodynamic radius, R_0^{hydro}, the radius of a sphere with the same volume as the hydrodynamic particle, should be distinguished from the apparent radius, or Stokes radius, of the particle, a, the radius of a sphere with the same diffusion coefficient as the particle.

The definition of the minimum frictional coefficient for the molecule of protein, that expected of a hydrated sphere, incorporates the water bound to the protein rather than treating the protein as if it were unhydrated

as was done earlier (Equation 8–33). If δ_{H2O} is 0.3 g (g of protein)$^{-1}$, consistent with the values in Table 6–3, the hydrated effective sphere will have a volume 1.4 times as large as the unhydrated effective sphere if $\bar{v}_p$ is taken as 0.74 cm^3 g^{-1} and $\bar{v}_{H2O}^0$ as 1.00 cm^3 g^{-1}. The frictional coefficient of the hydrated sphere of protein, f_0^{hydro}, should be 1.12 times larger than the frictional coefficient of the unhydrated sphere of protein, f_0^{unh}. This is consistent with the fact that the smallest frictional ratios, f/f_0^{unh}, observed for globular proteins are always greater than or equal to 1.1 when no correction is made for hydration.

The relationship between the frictional ratio (f/f_0) and the shape of a particle has been derived for ellipsoids of revolution, either prolate or oblate.[3] The relationships can be presented graphically (Figure 12–1A). After the frictional ratio f/f_0^{hydro} has been calculated from the observed value of the frictional coefficient f (Equation 12–3) and the calculated value of f_0^{hydro} (Equation 12–6), and a value is assumed for δ_{H2O}, the apparent axial ratio, a/b, of the hydrodynamic particle can be read from the graph.

Molecules of protein are neither prolate nor oblate ellipsoids of revolution, but exact solutions to the hydrodynamic equations are available only for these shapes. Such an approximation, however, may provide some insight into the shape of a particular molecule of protein, especially when the frictional ratio differs greatly from 1. Such a result cannot be explained on the basis of an unexpectedly high degree of hydration and states that the protein of interest is peculiar in its shape. In the particular case where the molecule is thought to resemble a cylindrical rod of length L and diameter d, it has been concluded that the dimensions of the rod can be calculated with use of the frictional ratio to determine the axial ratio of an equivalent prolate ellipsoid, a/b, and then with application of the formula

$$\frac{L}{d} = \left(\frac{3}{2}\right)^{1/2}\frac{a}{b} \qquad (12\text{–}7)$$

The segment of rope formed by triple-helical collagen, usually referred to as a protofilament, is known to be a rod. The molar mass of a triple-helical segment of collagen is 280,000 g mol^{-1}, its partial specific volume[1] is 0.695 cm^3 g^{-1}, and its diffusion coefficient[1] is 0.85×10^{-7} cm^2 s^{-1}. If it is assumed that $\delta_{H2O} = 0.3$ g g^{-1}, then the frictional ratio f/f_0^{hydro} would be 5.3, for which the ratio of L/d would be 212 (Figure 12–1A and Equation 12–7). The volume of the hydrodynamic particle containing a segment of collagen rope, based on the assumption that $\delta_{H2O} = 0.3$, would be 460 nm^3 (Equation 12–2), which would fill a cylinder 285 nm long with a diameter of 1.35 nm. From the dimensions of the triple helix (Figure 9–26) and the length of the polypeptide, a segment of collagen rope is thought to be 300 nm long.

It is also possible to calculate the frictional coefficient for a rigid string of spherical beads.[4] This approximation is useful for molecules of protein such as fibronectin,[5] which is known to be composed of internally repeating domains strung together to make a highly elongated molecule. Each internally repeating domain can be approximated by a sphere of the appropriate radius.

The frictional coefficient of a molecule of protein can also be determined by **sedimentation velocity**. Consider a hydrodynamic particle dissolved in a solution that is submitted to a high centrifugal force in the rotor of an ultracentrifuge. The centrifugal force on the particle is equal to $m_h \omega^2 r$, where ω is the angular velocity (radians second^{-1}) of the rotor, m_h is the mass (grams) of the hydrodynamic particle, and r is the distance (centimeters) the particle is from the axis of the rotor. This force is countered by the buoyant force which is equal to $V_h \rho_{sol} \omega^2 r$, where ρ_{sol} is the density (grams centimeter^{-3}) of the solution displaced by V_h, the volume (centimeters^3) of the hydrodynamic particle. The net force is

$$F = \omega^2 r (m_h - V_h \rho_{sol}) \qquad (12\text{--}8)$$

It is usually assumed, although there is no need to do so, that ρ_{sol} is insignificantly different from ρ_0, the density of water at the appropriate temperature. Because ρ_0 then becomes the reciprocal of $v_{H_2O}^0$ when expressions for m_h (Equation 12–1) and V_h (Equation 12–2) are substituted, the net force is

$$F = \frac{M_p}{N_{Av}} \omega^2 r (1 - \bar{v}_p \rho_0) \qquad (12\text{--}9)$$

The term $(M_p/N_{Av})(1 - \bar{v}_p \rho_0)$ is the **buoyant mass** of a protein.

The net force causes the hydrodynamic particle to accelerate. As it accelerates, the frictional force, which is equal to fu, where f is the frictional coefficient (grams second^{-1}) and u is the velocity (centimeters second^{-1}) of the hydrodynamic particle, increases until it just balances the net centrifugal force. At that point, a steady state is achieved, the forces are equal and opposite, and

$$fu = \frac{M_p}{N_{Av}} \omega^2 r (1 - \bar{v}_p \rho_0) \qquad (12\text{--}10)$$

This equation can be rearranged to give

$$s \equiv \frac{u}{\omega^2 r} = \frac{M_p (1 - \bar{v}_p \rho_0)}{f N_{Av}} \qquad (12\text{--}11)$$

The term on the left, $u\omega^{-2}r^{-1}$, which includes all of the parameters of the instrument, can be directly measured, and it is referred to as the **sedimentation coefficient**, s. The sedimentation coefficient (seconds) is designated as $s_{20,w}^0$, where superscript and subscript have the same meaning as before. Because sedimentation coefficients of proteins are between 10^{-13} and 10^{-11} s, the unit 10^{-13} s is designated as S, the Svedberg. As it is only a function of universal constants and the properties of the molecule of protein itself, the sedimentation coefficient is an intrinsic property. In particular it is, as is the diffusion coefficient,

a direct measurement of the frictional coefficient

$$f = \frac{M_p (1 - \bar{v}_p \rho_0)}{s_{20,w}^0 N_{Av}} \qquad (12\text{--}12)$$

Because the sedimentation coefficient can be measured easily and accurately, it is the most reliable way to measure the frictional coefficient. The molar mass of the protein, however, must be a fixed and known quantity. If the protein is normally engaged in a reaction that changes its molar mass, such as the equilibrium between the dimers and the tetramer of hemoglobin, a molar mass cannot be assigned. If the protein is participating in such a reaction, abnormally large decreases in the sedimentation coefficient will occur as the concentration of the protein is decreased while extrapolating the zero concentration of protein.

Aspartate transcarbamylase from *Escherichia coli* (Figure 9–12) serves as an example of the application of sedimentation analysis to the study of the shape of a hydrodynamic particle. It is a protein of molar mass 310,000 g mol^{-1} (Table 8–1), and its sedimentation coefficient $s_{20,w}^0$ is 11.6×10^{-13} s,[6,7] from which a frictional coefficient f of 11.6×10^{-8} g s^{-1} can be calculated. The minimum frictional coefficient, f_0^{unh}, for the unhydrated protein ($\delta_{H_2O} = 0$), folded as a sphere, would be 8.5×10^{-8} g s^{-1} (Equation 12–6). If hydration is estimated to be 0.3 g g^{-1}, this would give a frictional ratio f/f_0^{hydro} of 1.22. Although the protein does not appear to be so asymmetric (Figure 9–12) as the frictional ratio suggests ($a/b = 5$ for an oblate ellipsoid), the high value might result from the protruding R subunits and an abnormally large amount of bound water within the central cavity between the two C subunits.

It has been demonstrated crystallographically[8] that when aspartate transcarbamylase binds the enzymatic inhibitor N-(phosphonacetyl)-L-aspartate, the disposition of its subunits changes significantly. The net effect of this change in structure is to move the two trimeric C subunits (Figure 9–12) 1.2 nm farther apart. In the process, the water-filled space between the two C subunits widens by the same amount. This change in structure caused by the binding of the inhibitor can be detected as a change in the sedimentation coefficient of aspartate transcarbamylase.[6,7] This change is accurately quantified by difference sedimentation analysis in which the two samples, with and without the inhibitor, are simultaneously monitored in separate cells in the same rotor.[9] The sedimentation coefficient[6,7] decreases by 3.4% upon the change in structure. The crystallographic results demonstrate an expansion of the hydrodynamic particle by about 9% but in only one of its dimensions, coincident with the incorporation of solvent at its center.

The diffusion coefficient and the sedimentation coefficient provide independent determinations of the frictional coefficient of a molecule of protein. The force producing net flux of protein when diffusion is measured

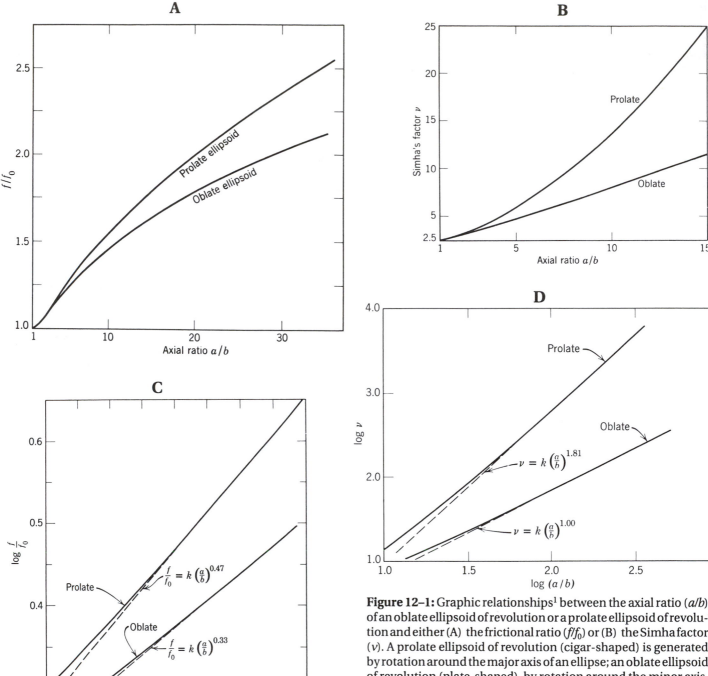

Figure 12–1: Graphic relationships[1] between the axial ratio (a/b) of an oblate ellipsoid of revolution or a prolate ellipsoid of revolution and either (A) the frictional ratio (f/f_0) or (B) the Simha factor (ν). A prolate ellipsoid of revolution (cigar-shaped) is generated by rotation around the major axis of an ellipse; an oblate ellipsoid of revolution (plate-shaped), by rotation around the minor axis. The relationships for smaller values of the axial ratio are given directly (upper panels). For large values of the axial ratio (> 10) the logarithm of the frictional ratio (C) or the logarithm of the Simha factor (D) is given as a function of the logarithm of the axial ratio. The frictional ratio or the Simha factor, determined experimentally, can be converted into an axial ratio with the appropriate graph. When the axial ratios are greater than 100, each of the four curves in the two lower panels (C and D) becomes a straight line to infinity. As a result, values of the frictional ratios or the Simha factors that are greater than those on the graphs can still be converted to values for axial ratios with the use of the slopes of these lines for extrapolation. The slope of the line for frictional ratios of prolate ellipsoids is 0.47; that for oblate ellipsoids, 0.33; the slope of the line for Simha factors for prolate ellipsoids is 1.81; and for oblate ellipsoids, 1.00. Reprinted with permission from ref 1. Copyright 1961 John Wiley.

is the chemical potential, which is unrelated to, as well as being somewhat less concrete of a concept than, centrifugal force. Furthermore, the theoretical derivations of the relationships between the diffusion coefficient and the frictional coefficient and between the sedimentation coefficient and the frictional coefficient are entirely different. Now that the molar masses of proteins can be established independently with precision, it is of interest to compare (Table 12–1) the two frictional coefficients, that calculated from diffusion (f_{diff}) and that calculated from sedimentation (f_{sed}). Depending upon one's prejudice, the agreement between the numbers is either as one expected or quite gratifying. The lack of any systematic deviation verifies the assumption, first made by Einstein, that the same frictional coefficient applies to both diffusion and sedimentation.

The frictional ratios ($f_{\text{av}}/f_0^{\text{hydro}}$), where f_{av} is the average of the two measurements, are close to 1 (1.1–1.2) for most globular proteins (Table 12–1), but even in these instances the frictional ratios of the hydrated particles predict (Figure 12–1A) an axial ratio of greater than 3, which is unrealistic. This may simply reflect the fact that molecules of protein, even when they are almost spherical, are not smooth homogeneous solids but corrugated, heterogeneous macromolecules. Proteins such as fibrinogen and plasminogen, which are known from other observations to be highly asymmetric, have high frictional ratios.

Measurements of **intrinsic viscosity** also provide an evaluation of the shape of the hydrodynamic particle. When a fluid flows through a cylindrical capillary under the appropriate circumstances, laminar flow occurs. The fluid immediately adjacent to the capillary walls is stationary, and the fluid at the center has the highest rate of flow. Each cylindrical lamina between the center and the wall moves with an intermediate velocity that monotonically decreases as the distance from the center increases to a value of zero at the wall. Laminar flow requires that each cylindrical lamina move more slowly than its neighbor toward the center. As such, shear occurs between these lamina throughout the capillary. The surfaces at which shear occurs are all parallel to the axis of the capillary. The more viscous the fluid, the more difficult it will be for these surfaces of shear to move across one another, and the more slowly the fluid will flow through the capillary. The time required for a given volume of fluid to move through a given capillary at a given hydrostatic pressure is directly proportional to η, the **viscosity** (grams centimeter^{-1} second^{-1}) of the fluid.

The addition of macromolecules such as proteins to the fluid in the capillary interrupts the shear that otherwise would occur in the solution in their vicinity and increases the viscosity of the solution. This increase can be expressed in terms of the specific viscosity, η_{sp}, which is defined by

$$\eta_{\text{sp}} \equiv \frac{\eta' - \eta}{\eta} = \frac{\eta'}{\eta} - 1 \tag{12–13}$$

where η' is the viscosity of the solution containing a particular concentration of the protein and η is the viscosity of an otherwise identical solution lacking the protein. The specific viscosity is a positive number because η' is always greater than η.

If the flow through the capillary is driven only by the weight of the fluid, the specific viscosity is readily measured because

$$\frac{\eta'}{\eta} = \frac{t'\rho'}{t\rho} \tag{12–14}$$

where t is the time for a given volume of a solution to flow through the capillary, ρ is the density of the solution, and the primed and unprimed terms refer to the solution of protein and an identical solution minus the protein, respectively.

Table 12–1: Frictional Coefficients from Sedimentation and Diffusion

protein	species	$\bar{v}^a$ (cm^3 g^{-1})	$s_{20,w}^{0\ a}$ (s × 10^{13})	$D_{20,w}^{0\ a}$ (cm^2 s^{-1} × 10^7)	$M_p^{\ b}$ (g mol^{-1})	$f_{\text{sed}}^{\ c}$ (g s^{-1} × 10^8)	$f_{\text{diff}}^{\ d}$ (g s^{-1} × 10^8)	$f_0^{\text{unh}\ e}$ (g s^{-1} × 10^8)	$f_0^{\text{hydro}\ f}$ (g s^{-1} × 10^8)	$f_{\text{av}}/f_0^{\text{hydro}}$
lysozyme	chicken	0.703	1.91	11.20	14,320	3.7	3.6	3.00	3.35	1.09
serum albumin	human	0.735	4.64	6.0	66,500	6.3	6.7	5.07	5.68	1.15
alcohol dehydrogenase	horse	0.750	5.0	6.2	69,700	6.6	6.5	5.42	6.07	1.08
prothrombin	cow	0.700	4.85	6.24	73,000^g	7.5	6.5	5.14	5.77	1.22
aldolase	rabbit	0.742	7.35	4.63	156,900	9.2	8.7	6.77	7.59	1.18
catalase	cow	0.730	11.30	4.10	233,000^h	9.3	9.9	7.69	8.61	1.11
β-galactosidase	E. coli	0.76	15.93	3.12	465,000	11.7	13.0	9.81	10.98	1.12
fibrinogen	human	0.725	7.63	1.98	338,000^i	20.3	20.4	8.68	9.72	2.10
plasminogen	human	0.71	4.30	4.31	103,000^j	11.6	9.4	5.80	6.50	1.61

^{a}From tables in ref 7a. ^{b}From sequence. $^c f_{\text{sed}} = M(1 - \bar{v}\rho)/N_{\text{Av}} s_{20,w}^0$. $^d f_{\text{diff}} = k_B T/D_{20,w}^0$. $^e f_0^{\text{unh}} = 6\pi\eta(3M\bar{v}/4\pi N_{\text{Av}})^{1/3}$. $^f f_0^{\text{hydro}} = 1.12 f_0$. ^{g}10.4 g of CHO (100 g of protein)$^{-1}$. ^{h}One heme subunit^{-1}. ^{i}2 g of CHO(100 g of protein)$^{-1}$. ^{j}17 g of CHO(100 g of protein)$^{-1}$.

The specific viscosity, η_{sp}, is the fractional increase in the viscosity of the solution due to addition of the protein, and it increases monotonically as the concentration of protein is increased. To render this increase an intrinsic property of the protein, regardless of its concentration, the **intrinsic viscosity**, $[\eta]$ (centimeters^3 gram^{-1}), is defined as

$$[\eta] \equiv \lim_{C_p \to 0} \frac{\eta_{sp}}{C_p} \qquad (12\text{--}15)$$

where C_p is the concentration of protein in grams centimeter^{-3}. At low concentrations of protein, η_{sp} should be directly proportional to C_p and $[\eta]$ is simply the slope of the line of η_{sp} plotted against C_p. Neither the specific viscosity nor the intrinsic viscosity is itself a viscosity. The intrinsic viscosity is sometimes called the limiting viscosity number to avoid this confusion.

For macromolecules such as proteins, it can be shown[1] that

$$[\eta] = \nu \frac{V_h N_{Av}}{M_p} \qquad (12\text{--}16)$$

$$[\eta] = \nu \left(\bar{v}_p + \delta_{H_2O} v^0_{H_2O} \right) \qquad (12\text{--}17)$$

where ν is a dimensionless coefficient of proportionality referred to as the Simha factor. As with the frictional ratio, f/f_0^{hydro}, the relationship between the Simha factor ν and shape has been derived for ellipsoids of revolution.[10] The relationships can be presented graphically (Figure 12–1B). If a value for δ_{H_2O} is assumed, ν can be calculated from

$$\nu = \frac{[\eta]}{\bar{v}_p + \delta_{H_2O} v^0_{H_2O}} \qquad (12\text{--}18)$$

and the apparent value of the axial ratio can be read from the graph.

Based on Einstein's calculations, the value of ν for a spherical hydrodynamic particle is 2.5. If $\delta_{H_2O} = 0.3$ g g^{-1}, $\bar{v}_p = 0.7$ cm^3 g^{-1}, and $v^0_{H_2O} = 1$ cm^3 g^{-1}, $[\eta]$ should be 2.5 cm^3 g^{-1} if the hydrodynamic particle is a sphere regardless of its molar mass. What this means is that as long as δ_{H_2O} and $\bar{v}_p$ do not vary, the viscosities observed for a set of solutions of different spherical molecules of protein, each at the same concentration in grams centimeter^{-3}, will be the same regardless of whether the mass is distributed among only a few large spheres because the protein has a large molar mass or is distributed among many small spheres because the protein has a small molar mass. As a result, most globular proteins have intrinsic viscosities between 3.0 and 4.0 cm^3 g^{-1} regardless of their molar masses.

It is with asymmetric molecules of protein that intrinsic viscosity is informative. Collagen has been submitted to viscometry, and its intrinsic viscosity[1] is 1150 cm^3 g^{-1}. If it is assumed that $\delta_{H_2O} = 0.3$ g g^{-1}, the axial ratio (a/b) of

the hydrodynamic particle would be 140 (Figure 12–1B) and the ratio of cylindrical length to diameter (L/d) would be 170 (Equation 12–7). This ratio is that of a cylinder of length 260 nm with a diameter of 1.5 nm and a volume of 460 nm^3.

Another procedure that can provide information about the shape of a molecule of protein is the **scattering of electromagnetic radiation**. In the earlier discussion of light scattering, it was pointed out that the intensity of the scattered light from a solution of protein can depend on the angle at which the measurement is made. This is due to the fact that if the molecule of protein has at least one dimension that is an appreciable fraction of the wavelength of the light, photons scattered from different points in the same molecule of protein will be out of phase, and intramolecular interference will diminish the intensity of the light. This interference increases as the angle at which the scattered light is measured is increased. It can be shown that when the contribution of the virial coefficients to the scattering is eliminated by extrapolating the measurements to zero concentration of protein

$$\lim_{C_p \to 0} \frac{K C_p}{R_\theta} \left(\frac{\partial \tilde{n}}{\partial C_p} \right)_{P,\mu}^2 = \left(\frac{1}{M_p} \right) \left[\frac{1}{P(\theta)} \right] \qquad (12\text{--}19)$$

where K is the optical constant (moles centimeter^{-4}) defined by Equation 8–22, R_θ is the Rayleigh ratio (centimeters^{-1}) calculated from the measurements by Equation 8–24, C_p is the concentration of protein in the units of grams centimeter^{-3}, $(\partial \tilde{n}/\partial C_p)_{P,\mu}$ is the change (centimeters^3 gram^{-1}) in the refractive index of the solution as a function of the concentration of the protein, and M_p is the molar mass (grams mole^{-1}) of the protein. The function $P(\theta)$ is the factor by which the intensity of the scattered light (i_θ) is decreased as a result of the interference[11]

$$P(\theta) = 1 - \frac{16\pi^2 R_G^2}{3\lambda^2} \sin^2 \frac{\theta}{2} + \cdots \qquad (12\text{--}20)$$

where R_G is the radius of gyration (centimeters) and θ is the angle relative to the incident radiation at which the scattered radiation is measured. The value for the wavelength of the light, λ, is its wavelength in the solution

$$\lambda = \frac{\lambda_0}{\tilde{n}} \qquad (12\text{--}21)$$

where λ_0 is its wavelength in a vacuum. Equation 12–20 is an infinite series, but at small values of θ the higher terms become negligible and the approximation

$$\lim_{\theta \to 0} \frac{1}{P(\theta)} = \frac{1}{1 - \frac{16\pi^2 R_G^2}{3\lambda^2} \sin^2 \frac{\theta}{2}} = 1 + \frac{16\pi^2 R_G^2}{3\lambda^2} \sin^2 \frac{\theta}{2} \qquad (12\text{--}22)$$

can be used. In practice

$$\lim_{\substack{C_p \to 0 \\ \theta \to 0}} \frac{C_p}{R_\theta} = \frac{C_p}{R_0}\left(1 + \frac{16\pi^2 R_G^2}{3\lambda^2}\sin^2\frac{\theta}{2}\right) \qquad (12\text{-}23)$$

A plot of the left-hand quotient, extrapolated to zero concentration of protein at each value of θ, against $\sin^2(\theta/2)$, for small values of θ, will be a straight line from the slope and intercept of which a value of R_G, the **radius of gyration**, can be calculated. This equation emphasizes that the interference is independent of the colligative property.

It is the radius of gyration of the protein molecule that is the molecular parameter, obtained from the angular dependence of the intensity of the scattered radiation, that provides information about the shape of the molecule of protein. The radius of gyration of a solid of uniform scattering density, as is usually assumed to be the case for proteins, is defined by the relationship

$$R_G^2 = \frac{\int_{\text{vol}} r^2 dV}{\int_{\text{vol}} dV} \qquad (12\text{-}24)$$

where r is the distance of a volume element dV from the center of mass and the integration is performed over the whole volume of the solid. The advantage of the radius of gyration is that it can be calculated by numerical integration for any hypothetical structure and compared to the value obtained from the measurement. For example, in an elongated protein such as fibronectin, which is known to be constructed from internally repeating domains, the radii of gyration can be calculated for various rigid structures built from a string of spheres representing the individual domains and these calculated values can be compared to the observed values.[12]

The radius of gyration for a single sphere of uniform density is

$$R_G = \left(\frac{3}{5}\right)^{1/2} R_{\text{sph}} \qquad (12\text{-}25)$$

where R_{sph} is the radius of the sphere. The radius of gyration for a cylindrical rod is

$$R_G = \frac{L_r}{12^{1/2}} \qquad (12\text{-}26)$$

where L_r is the length of the rod. The radius of gyration for a prolate ellipsoid[13] is

$$R_G = \left(\frac{2a^2 + b^2}{5}\right)^{1/2} \qquad (12\text{-}27)$$

where a and b are the semi-major and semi-minor axes, respectively.

The effect of the finite size of the molecule of protein on the scattered light is that its intensity, as reflected in R_θ (Equation 8–24), decreases as θ increases, owing to intramolecular interference, but its intensity will decrease significantly only if the term $[16\pi^2 R_G^2 \sin^2(\theta/2)]/3\lambda^2$ in Equation 12–23 is large enough to cause a measurable effect. In practice,[1] this means that at least one dimension of the protein must be greater than $\lambda/20$. The sizes of most molecules of protein are too small for this to be the case when visible light (λ = 300–500 nm in water) is used as the radiation. It is usually necessary to use X-radiation (λ = 0.1–0.2 nm). Unfortunately, this is radiation of such short wavelength that complete intramolecular interference occurs at quite small values of θ (Equation 12–20), and the scattered radiation from the solution of protein becomes equal to that from the solution lacking protein when θ is only 1 or 2 deg. Fortunately, accurate measurements of scattered X-radiation can be made at very low angles.

When the observations of the scattering of X-rays at these low angles are presented, a different convention is used to present $P(\theta)$. Because

$$e^{-x} = 1 - x + \frac{x^2}{2!} - \frac{x^3}{3!} + \cdots \qquad (12\text{-}28)$$

the first two terms in Equation 12–20 are identical to the first two terms in the expansion of $\exp[(4\pi/\lambda)^2 \sin^2(\theta/2)R_G^2/3]$ and at low angles[11]

$$\lim_{\theta \to 0} \ln P(\theta) = -\frac{16\pi^2 R_G^2}{3\lambda^2}\sin^2\frac{\theta}{2} \qquad (12\text{-}29)$$

Because at these low angles none of the terms in Equation 12–19 except R_θ and $P(\theta)$ change as θ is varied, a plot of $\ln i_\theta$ (see Equation 8–24) as a function of $\sin^2(\theta/2)$ will give a straight line with a slope of $-16\pi^2 R_G^2/(3\lambda^2)$. From this slope R_G is readily determined.

The **low-angle scattering of X-rays** has been used in a number of structural studies. For example, from such measurements, colicin E$_3$ was found to have a radius of gyration of 3.4 nm.[13] From its molar mass of 70,000 g mol^{-1} and a partial specific volume of 0.73 cm^3 g^{-1}, it could be calculated that the protein would have a radius of gyration of 2.1 nm if it were a sphere. The measured value, therefore, indicates that the protein is anisotropic. A prolate ellipsoid of the proper volume with an axial ratio of 5 would have the observed radius of gyration. When substrates are added to a solution of yeast phosphoglycerate kinase, there is a decrease of its radius of gyration from 2.33 ± 0.02 to 2.23 ± 0.03 nm as determined by low-angle scattering of X-rays.[14] This protein is built from two structural domains, which can be delineated in its crystallographic molecular model.[15] It had been proposed that these structural domains are actually hinged and close upon the binding of substrates. The change in the radius of gyration expected from such a closing of the domains could be calculated from the crystallographic molecular model of the protein.[14] A reorientation of the two structural domains by about 10–12° around the hinge could produce the change in radius of gyration that is observed. Calmodulin is a small protein that binds calcium and the crystallographic molecular model of which has a peculiar shape, resembling a dumbbell with two globular weights each 2.0 nm across and an α-helical handle 2.4 nm long between them.[16] The radius of gyration for such a structure can be calculated to be 2.1 nm, which agrees well with the value of 2.15 ± 0.02 nm observed in low-angle X-ray scattering.[17] When the calcium is removed from calmodulin, the radius of gyration decreases to 2.06 ± 0.02 nm. This suggests that the molecule changes its shape upon the removal of its calcium.

The difficulty with scattering or hydrodynamic methods for assessing the shape of a molecule of protein is that they provide only one numerical result, either a frictional coefficient, a Simha factor, or a radius of gyration. If the frictional ratio f/f_0^{hydro} is less than 1.15, the Simha factor v is less than 4.0, or the radius of gyration R_G is near a value of $(\frac{3}{5})^{\frac{1}{2}}R_0^{hydro}$, it can be concluded that the protein is globular. If the value of one or more of these parameters for a given protein is significantly greater than the values expected for a sphere, it is usually necessary to conclude that the protein is anisotropic and has one or more extended dimensions. It is clear from the foregoing discussion of some of the results that larger values of these parameters are consistent with an infinite set of particular arrangements of the available mass. The only reason so much is heard about prolate and oblate ellipsoids of revolution is not that molecules of proteins are such geometric solids but that frictional coefficients and radii of gyration can be calculated explicitly for such solids. In using any of the measured parameters in an informative way, other details about the structure of the protein are essential.

One way to observe the shape of a molecule of protein directly is by **electron microscopy**. The three symmetrically protruding regulatory subunits on aspartate transcarbamylase and the hollow, water-filled cavity between its two rotationally symmetric, trimeric C subunits, which together presumably account for its abnormally large frictional coefficient, were first observed in electron micrographs of the protein (Figure 12–2A).[18] Another protein with an abnormally large frictional coefficient is fibrinogen. As no crystallographic molecular model is available, electron micrographs of this elongated molecule of protein are the only views of its shape that are presently available (Figure 12–2B).[19] Whenever results from such electron microscopic studies are presented, a representative field of molecules (Figure 12–2C) should be shown so that the reader can judge what fraction of the molecules of protein on the film of carbon give images that resemble the images chosen for a gallery of "representative" views (Figures 12–2A,B).

Phosphorylase kinase is a protein with an even more

Figure 12–2: Asymmetric molecules of protein viewed by electron microscopy. Solutions of the protein of interest (10 μg mL^{-1}–1.0 mg mL^{-1}) were applied to electron microscopic grids coated with a thin film of carbon[22] supported by a film of either collodion or formvar. The layer of carbon (~5 nm) was ionized so that it was hydrophilic enough to accept the aqueous solution. The molecules of protein were adsorbed to this surface and were then negatively stained with either 2% phosphotungstate or 1–2% uranyl formate. All magnifications are 480000×. (A) Gallery of selected images[18] of aspartate transcarbamylase (Figure 9–12) viewed either along the 3-fold rotational axis of symmetry (left three images) or along one of the 2-fold rotational axes of symmetry (right three images). Reprinted with permission from ref 18. Copyright 1972 American Chemical Society. (B) Gallery of selected images of fibrinogen.[19] The elongated molecule has globular domains at each end. Reprinted with permission from ref 19. Copyright 1981 Academic Press. (C) A field of negatively stained molecules of phosphorylase kinase.[20] This is an accurate representation of the usual situation. Most of the molecules of protein negatively stained on the grid are featureless asymmetric structures. The minority that present a repeating, definable image (indicated by arrowheads) are selected by the microscopist as representative images of the protein and presented in galleries as in panels A and B. In this instance, the shape of the individual images of phosphorylase kinase was so peculiar that digitized optical densities of a large number of the selected images of individual molecules (62) could be sequentially superposed by a computer to obtain an enhanced image (inset). Reprinted with permission from ref 20. Copyright 1985 Academic Press. (D) Gallery of images of Factor V, a protein in the cascade leading to clotting of the blood, that has been activated by treatment with thrombin.[23] The molecule is formed from two almost spherical globular domains, of apparently equal diameters, tethered together. Reprinted with permission from ref 23. Copyright 1986 *Journal of Biological Chemistry.*

peculiar shape than that of either aspartate transcarbamylase or fibrinogen. The shape of the protein was so unusual that a collection of digitized micrographic images of individual molecules (Figure 12–2C) could be easily superposed by a computer and stacked one upon the other.[20] The average of this stack of images could be calculated, and it is presumed that this average represents an enhanced image of the actual molecule (inset, Figure 12–2C). The chalice seen in the enhanced image can be imperfectly discerned in each of the individual selected images (Figure 12–2C), and this fulfills the usual requirement placed upon results of this kind. A similar procedure has been applied to α_2-macroglobulin[21] to obtain enhanced images. This protein has a shape almost as peculiar as that of phosphorylase kinase.

Electron micrographs of activated bovine coagulation Factor Va were instrumental in explaining its unusual behavior upon sedimentation analysis. The protein has a molar mass[24,25] of 170,000 g mol^{-1} and a sedimentation coefficient,[24] $s_{20,w}^0$, of 8.2 × 10^{-13} s, from which a frictional coefficient f = 1.27 × 10^{-7} g s^{-1} can be calculated. The fric-

A

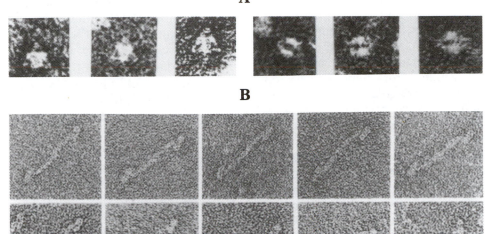

B

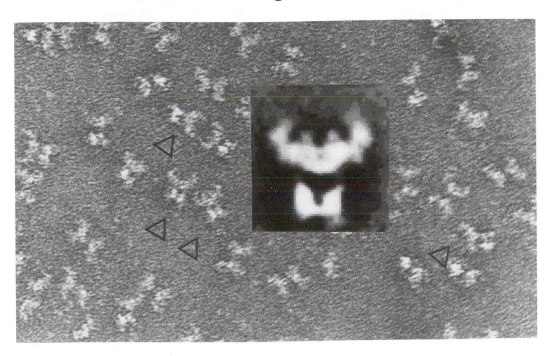

C

D

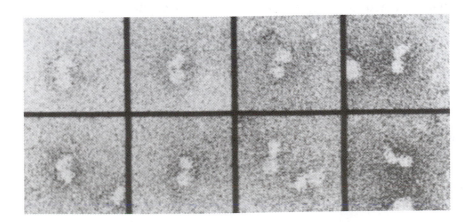

tional ratio f/f_0^{hydro}, based on the assumption that $\delta_{H2O} = 0.3$, would be 1.6. The protein is constructed from two nonidentical polypeptides. These two polypeptides, in turn, are derived by proteolytic posttranslational modification of a much larger precursor, the intact coagulation Factor V, which is constructed from only one very long polypeptide. When activated coagulation Factor Va was observed by electron microscopy, it was found to be two globular domains of protein, very similar in diameter, attached together through a narrow neck (Figure 12–2D).[23] Presumably, the two globular domains represent the two different polypeptides ($n_{aa} = 920$ and 620) produced by the posttranslational modification, separately folded, because in the absence of calcium the two polypeptides separate from each other as independent monomers.[24] The shape seen in the electron micrographs should be responsible for the unusually large frictional ratio of this protein. Although an axial ratio for a prolate ellipsoid was calculated from the earlier results of the sedimentation analysis,[24] it was moot when the electron micrographs became available.[23]

Problem 12–1: Calculate values of f_{sed}, f_{diff}, and f_0^{unh} from $\bar{v}$, $s_{20,w}^0$, $D_{20,w}^0$ and M_p for each protein in Table 12–1. From tabulated values of f_{av} f/f_0^{hydro}, determine a/b for each protein on the assumption that they are prolate ellipsoids of revolution.

Problem 12–2: Human immunoglobulin G is a protein with a molar mass of 167,040 g mol^{-1} and a partial specific volume of 0.739 cm^3 g^{-1}.

(A)   Assume hydration to be 0.3 g of H_2O (g of protein)$^{-1}$ and calculate the minimum frictional coefficient, f_0^{hydro}, for the hydrated hydrodynamic particle at 20 °C in water.

(B)   The standard sedimentation coefficient $s_{20,w}^0$, for immunoglobulin G is 7.0×10^{-13} s, and the standard diffusion coefficient $D_{20,w}^0$ is 4.0×10^{-7} s^{-1}. Calculate the frictional coefficient independently, first from the sedimentation coefficient and then from the diffusion coefficient.

(C)   From the average of these two estimates of the frictional coefficient and from the estimate of the minimum frictional coefficient of the hydrated hydrodynamic particle, estimate the axial ratio a/b upon the assumption that the molecule is a prolate ellipsoid of revolution.

(D)   The shape of human immunoglobulin G is displayed in Figure 7–12. How does this compare with your estimate of its shape?

Problem 12–3: The sedimentation coefficient of human fibrinogen is 7.6×10^{-13} s, its molar mass is 338,000 g mol^{-1}, and its partial specific volume is 0.725 cm^3 g^{-1}.

(A)   Assume $\delta_{H2O} = 0.3$ and determine the volume of the hydrodynamic particle, the frictional co-

efficient of fibrinogen, its frictional ratio, and its axial ratio and dimensions on the basis of the assumption that it is a prolate ellipsoid of revolution or a cylindrical rod.

(B)   The length of the fibrinogen molecule has been determined to be 45 nm by electron microscopy (Figure 12–2B). Calculate the dimensions of a prolate ellipsoid or a cylindrical rod with the same volume as the hydrodynamic particle and a major axis of length 22.5 nm.

(C)   The intrinsic viscosity of fibrinogen is 27 cm^3 g^{-1}. Calculate its Simha factor v and estimate its axial ratio and molecular dimensions on the basis of the assumption that it is a prolate ellipsoid of revolution or a cylindrical rod.

Problem 12–4: Tropomyosin is a protein formed from identical folded polypeptides each 284 amino acids in length ($M = 32,680$ g mol^{-1}). It has a partial specific volume of 0.72 cm^3 g^{-1}. The number average molecular weight in solution was determined by osmotic pressure. Each measurement was extrapolated to $C_p = 0$. The osmotic pressure at 0 °C as a function of ionic strength is presented in the following table.

ionic strength (M)	$\lim_{C_p \to 0} \frac{\Pi}{C_p}$ (mmHg cm^3 g^{-1})	M_n (g mol^{-1})
0.10	126	135,000
0.20	154	
0.27	193	
0.30	236	
0.60	254	
1.10	264	

(A)   Complete the table by filling in the missing values for the number average molar mass. Show in detail the calculation of M_n at ionic strength of 0.10 M.

(B)   How many polypeptides compose the major form of tropomyosin present in solution at high ionic strength? What is its exact molecular weight? Why does the molar mass increase as the ionic strength is lowered?

(C)   The following data were gathered from solutions of tropomyosin at an ionic strength of 1.1 M

η'/η	C_p [g (100 mL)$^{-1}$]
1.210	0.33
1.299	0.44
1.466	0.64
1.588	0.76
1.793	0.96
2.223	1.29
2.972	1.72
4.603	2.14

where η' is the viscosity of the solution of protein, η is the viscosity of the solvent, and C_p is the concentration of protein. Determine the intrinsic viscosity $[\eta]$ at this ionic strength.

(D) Assume that $\delta_{H_2O} = 0.3$ g H_2O (g of protein)$^{-1}$ and calculate v. From v determine the axial ratio for tropomyosin if it were a prolate ellipsoid of revolution by using Figure 12–1B.

(E) From circular dichroism it is known that, at all ionic strengths, tropomyosin is > 90% α-helical. It is a coiled coil in which the two α helices wrap around each other as the strands in a two-stranded rope. The length for each amino acid in an α helix is 0.15 nm. Calculate the length of a molecule of tropomyosin at high ionic strength and, assuming it to be a cylindrical rod, calculate its diameter from its hydrated molecular volume. What is its actual axial ratio? Compare this to the axial ratio obtained from v.

(F) The intrinsic viscosities of solutions of tropomyosin also vary with ionic strength.

$[\eta]_{rel}$	ionic strength (M)
1.00	1.1
1.03	0.6
1.23	0.3
1.75	0.2
2.45	0.1

Plot M_n against $[\eta]_{rel}$ and explain the correlation in terms of structures that could form as the ionic strength is lowered.

Problem 12–5: The definition of Rayleigh's ratio for the scattering of unpolarized light is

$$R_\theta \equiv \frac{r^2 i_\theta}{I_0(1+\cos^2\theta)}$$

where i_θ is the intensity of the scattered light at an angle θ to the incident beam. At very low angles ($\theta < 5 \times 10^{-2}$ rad), $\cos^2\theta \cong 1.00$. In this situation, by Equation 12–19

$$P(\theta) = \lim_{C_p \to 0} \frac{i_\theta}{C_p}\left[\frac{r^2}{2I_0 K M_p}\left(\frac{\partial \tilde{n}}{\partial C_p}\right)\right]^{-2}$$

and

$$\ln P(\theta) = \lim_{C_p \to 0}\left(\ln\frac{i_\theta}{C_p} - A\right)$$

where A is a constant determined by the values of all of the parameters in the brackets. At very low angles, the approximation of Equation 12–29 is also valid, and it follows that

$$\lim_{C_p \to 0} \ln\frac{i_\theta}{C_p} = A - \frac{16\pi^2 R_G^2}{3\lambda^2}\sin^2\frac{\theta}{2}$$

If ω is expressed in radians

$$\sin\omega = \omega - \frac{\omega^3}{3!} + \frac{\omega^5}{5!} + \cdots$$

(A) Show that

$$\lim_{\substack{C_p \to 0 \\ \theta \to 0}} \ln\frac{i_\theta}{C_p} = A - \frac{16\pi^2 R_G^2}{3\lambda^2}\left(\frac{\theta}{2}\right)^2$$

It is more convenient to take a series of measurements at varying scattering angles, θ (in radians), of a single solution of protein than to measure the scattering at a fixed angle for several solutions of protein. Therefore, what is usually done is to determine the slope of $\ln(i_\theta/C_p)$ as a function of θ^2 at various fixed values of C_p and then extrapolate to $C_p = 0$. If, however, C_p is held constant, as is done in such an experiment, then

$$\lim_{\theta \to 0} \ln\frac{i_\theta}{C_p} = \lim_{\theta \to 0}\ln i_\theta + \ln C_p$$

and at constant C_p

$$\lim_{\theta \to 0}\ln i_\theta = A' - \frac{4\pi^2 R_G^2}{3\lambda^2}\theta^2$$

where A' is a constant equal to $A - \ln C_p$.

(B) What should be the slope of the line for a plot of $\ln i_\theta$ against θ^2 at low concentrations of protein and low scattering angle?

Immunity protein is a protein produced by strains of the bacterium *E. coli* producing colicin E$_3$. The intensities of the scattered X-rays ($\lambda_0 = 0.154$ nm; $\tilde{n} = 1.33$; $\lambda = 0.116$ nm) as a function of the square of the scattering angle were measured for a series of solutions of immunity protein[13] (reprinted with permission from ref 13; copyright 1983 *Journal of Biological Chemistry*):

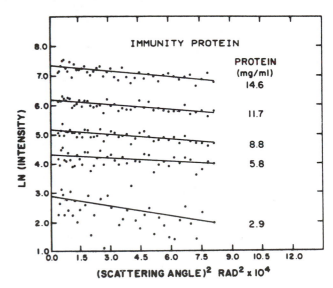

The natural logarithm of the intensity of the scattered X-rays, ln (intensity), is presented as a function of the square of the scattering angle, θ^2, in (radians)$^2 \times 10^4$. A value of 6.0 on the abscissa is equal to 6.0×10^{-4} rad^2. The values for the concentration of protein (C_p in milligrams milliliter^{-1}) are indicated to the right. The slopes of the lines in the plot are

C_p (mg mL^{-1})	slope (rad^{-2})
2.9	1130
5.8	420
8.8	540
11.7	550
14.6	630

(C) Calculate the apparent radius of gyration, R_G^{app}, for each concentration of protein.

(D) Extrapolate to $C_p = 0$ and obtain the actual radius of gyration, R_G.

(E) The molar mass of immunity protein is 9800 g mol^{-1}, and its partial specific volume, $\bar{v}_{imm}$, is 0.73 cm^3 g^{-1}. Calculate its unhydrated volume and the radius a of a sphere with that volume.

(F) What would be the radius of gyration of immunity protein if it were this sphere?

Absorption and Emission of Light

A valence electron in any molecule occupies an atomic orbital or a molecular orbital that has energy levels associated with it (Figure 12–3). These energy levels have discrete magnitudes because of the quantum theory, and the steps between any two energy levels are also of discrete magnitude or quantized. The energy levels that have the smallest steps between them are the **rotational energy levels**. These energy levels correspond to the quantized kinetic and potential energy associated with the rotations around the bond in which a particular electron resides and with the bonds in its vicinity. The steps between successive rotational energy levels are normally 0.5–50 J mol^{-1} in magnitude, corresponding to the energy in a photon of wavelength 2×10^8 to 2×10^6 nm. The energy levels that have the next larger steps between successive stages are the **vibrational energy levels.** These energy levels reflect the quantization of the vibrations of the bond in which a particular electron resides and of neighboring bonds that are coupled to it. The steps between vibrational energy levels are normally 5–50 kJ mol^{-1} in magnitude, corresponding to the energy in a photon of wavelength 2×10^4 to 2×10^3 nm. The energy levels that have the next larger steps among them are the **electronic energy levels** of the molecule. These energy levels are those of the atomic orbital or

molecular orbital in which a particular electron resides and of the vacant atomic orbitals or molecular orbitals that are accessible to it. Steps between two of these electronic energy levels are normally 50–500 kJ mol^{-1} in magnitude corresponding to the energy in a photon of wavelength 2×10^3 to 200 nm.

These three types of energy levels form nested sets (Figure 12–3). Within a given electronic energy level there are a series of vibrational energy levels, and within a given vibrational energy level there are a series of rotational energy levels. In a particular molecule, at a given instant, a discrete set of electronic energy levels will be occupied by the valence electrons to produce the σ bonds, the π molecular orbital systems, and the lone pairs. Within each occupied electronic energy level, a particular vibrational energy level will be occupied. The rotational motions within the molecule will determine which particular rotational energy levels are also occupied. As the differences in energy between electronic energy levels are large, the equilibrium constants governing the occupation of the successive levels at temperatures experienced by living organisms are also large, and only the lowest electronic energy levels are significantly occupied. The electrons in a molecule in the **ground state** are always (> 99.99999%) distributed so as to fill in succession the lowest electronic energy levels. As the differences in energy between vibrational energy levels are significant, the equilibrium constants governing the occupation of the successive levels at temperatures experienced by living organisms are also significant. In the ground state, the vibrational energy levels that are occupied are usually (> 90%) the lowest levels of each particular vibration. Rotational energy levels, however, are widely occupied by the atoms in the molecule.

When a photon encounters an electron in a molecule in such a way that its energy is absorbed by the electron, several consequences can result. If the electron momentarily absorbs the energy of the oscillating electric field and then emits the same photon back again without retaining any of its energy, the direction of the electromagnetic wave is altered so that its new direction of propagation bears no relationship to its incident direction while its wavelength remains the same. This is **elastic scattering,** and it is the phenomenon mainly responsible for X-ray diffraction, low-angle X-ray scattering, and light scattering.

If the excited electron is in a bond that happens to change its vibrational energy level during the brief time the electron is excited, the scattered light will have a wavelength that is shorter or longer than that of the incident light by a difference equivalent to the energy of the transition between the two vibrational energy levels. If the photon is scattered by an electron in the excited state of a vibrational mode, it can carry away the energy of the transition to the ground state and be scattered with higher energy. If it is scattered by an electron in the ground state that enters an excited state during its residence, the photon will provide the energy for this transition and be scattered with lower energy. As a result, the intensity of the scattered light varies symmetrically about the energy of the incident light. This is the **Raman effect**. As with light scattering, the Raman effect is usually measured by sampling the scattered light emitted perpendicular to the incident light. The incident light is from a laser, and it is intense and monochromatic. It is the spectrum of the wavelengths of the scattered light that is determined. Although most of the scattered light is the same wavelength as the incident light, scattered photons of other specific, sharply defined wavelengths are also present and a Raman spectrum of this scattered light provides a catalogue of many of the transitions among the vibrational energy levels present in the molecule. It is customary for investigators using Raman spectroscopy or infrared spectroscopy to present absorption as a function of the inverse of the wavelength, referred to as the

wavenumber (in centimeters^{-1}), which is directly proportional to the energy of the absorption. One advantage of this convention is that the two symmetrical displacements of the Raman effect from the same vibrational mode have the same numerical values when expressed in terms of wavenumber.

If the energy of the light exciting the electron is equal to the difference in energy between the vibrational energy level its bond occupied at the instant it was absorbed and a higher vibrational energy level accessible to it, the light can be absorbed and the vibrational energy of the bond occupied by the electron and those bonds vibrationally coupled to it will increase by that step in energy. Most of the energy absorbed will not be emitted back **radiatively** as light but will be dissipated **nonradiatively** by intermolecular collision or by exciting rotational motions whose differences in energy levels bridge the gap between the vibrational energy level of the

Figure 12–3: Photophysical and photochemical processes of a covalent bond between two atoms.[26] The smooth curve S_0 is the potential energy of the molecular orbital of the covalent bond in which the electron resides as a function of the distance r between the two nuclei. The smooth curve S_1 is the potential energy that would be experienced if an electron were in this unoccupied antibonding molecular orbital between the two atoms as a function of the distance between the two nuclei. When an electron in the occupied molecular orbital, the ground state, absorbs a quantum of electromagnetic energy sufficient to boost its energy high enough to enter the unoccupied molecular orbital, the excited state is created (process **2**). As the excited state relaxes, some of the absorbed energy is lost as heat. When the relaxed excited state emits light as the electron returns to the ground state (process **F**), the quantum of emitted fluorescent light has less energy (longer wavelength) than that of the quantum of light originally absorbed. If the spin of the electron inverts while it is in the excited state (process $S_1 \rightarrow T_1$), the electron enters a triplet excited state. The triplet excited state also relaxes by giving off heat. The phosphorescent light emitted from the relaxed triplet state (process **P**) has even less

energy (even longer wavelength) than the fluorescence from the initial excited state, and the triplet excited state has an even longer lifetime. If enough energy is absorbed by an electron in the ground state to excite it to an energy greater than the energy of the dissociated atoms (process **1**), then the bond can be photolytically cleaved. Each well of potential energy has levels of vibrational energy and levels of rotational energy associated with it (see expanded scale of the potential energy of the ground state to the left). Adapted with permission from ref 26. Copyright 1977 W.A. Benjamin.

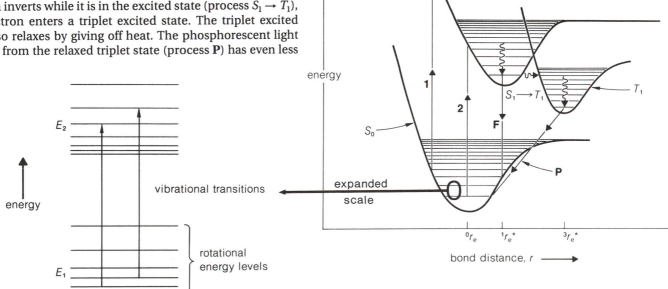

ground state and the vibrational energy level of the excited state (Figure 12–3). Any light emitted radiatively due to a direct transition from the excited state back to the ground state has the same wavelength as the absorbed light but an altered direction and becomes indistinguishable from elastically scattered light. The **absorption of infrared light** (in the range from 2×10^4 to 2×10^3 nm, or 500 to 5000 cm^{-1}) produces transitions among vibrational energy levels, and a spectrum of infrared absorption has discrete maxima whose energies correspond to transitions between pairs of vibrational energy levels. Because the selection rules governing which differences in vibrational energy levels and to what extent these differences absorb infrared light are different from the selection rules governing the Raman effect, infrared absorption spectra provide data that are complementary to Raman scattering spectra.

If the energy of the light exciting the electron is equal to the difference in energy between the molecular orbital or atomic orbital it occupies in the ground state and an unoccupied orbital of higher energy, the light can be absorbed. The electron then enters the unoccupied orbital, and an **electronically excited state** of the molecule is created. Because the excited state differs from the ground state in the distribution of its electrons among molecular and atomic orbitals, it can be thought of as a distinct, albeit similar, molecule. At the very least, the excited state in its most stable structure will have some bond lengths, bond angles, and bond energies that are different from those of the ground state because its bonding differs from that of the ground state. The instant the electron enters the new orbital, however, the molecule has the structure of the ground state. As it relaxes in energy to the most stable structure available to the excited state, the distance in energy between excited state and ground state shortens, and the excited electron loses energy. Because of the overlap required for excitation, the excited electron usually enters the excited state through one of its higher vibrational energy levels, and it simultaneously loses energy by a nonradiative passage to the lowest vibrational energy level. The net result of these relaxations is that the excited electron very rapidly (10^{-13}–10^{-11} s) finds itself at an energy considerably below the energy it had achieved immediately after the light was absorbed.

Because the energy levels of the excited state usually overlap energy levels of the ground state, the electron can reenter the molecular or atomic orbital of the ground state by pursuing a path among the rotational and vibrational energy levels of excited state and ground state. In this case, the energy originally absorbed is dissipated nonradiatively as heat, and only the absorption of the light is detected. The **absorption of ultraviolet and visible light** (in the range between 200 and 2000 nm) produces transitions among electronic energy levels and a spectrum of ultraviolet or visible absorption that has maxima whose energy corresponds to the energies of electronic transitions in the molecule.

If the ground state and the excited state overlap weakly, the excited electron can become trapped in the lowest vibrational energy level of the excited state long enough ($> 10^{-9}$ s) to reenter the ground state with a bang rather than a whimper. The reentry of the excited electron into the ground state with a single step requires that the energy it loses be emitted as a photon and produce either fluorescence or phosphorescence.

If it came from a covalent bond or a lone pair of electrons, at the instant of excitation, the excited electron entering the new orbital has a spin opposite to the spin of the partner it left behind in its previous orbital. As it relaxes into the lowest vibrational energy of the excited state and as the excited state rearranges, the excited electron remains coupled to its old partner, and the excited state remains a **singlet excited state**. From a singlet excited state, the electron can rapidly return to the ground state because the excited electron can readily reenter its old orbital with a spin compatible with the single electron still there. The reentry is **spin-allowed** and rapid ($< 10^{-7}$ s), and the emitted photon is **fluorescence.**

The energy of a quantum of fluorescent light is necessarily less than the energy of the light absorbed by the electron during excitation because of the nonradiative relaxations of the excited state that have occurred. Fluorescent light is light of a longer wavelength (usually visible light) emitted shortly after (within 10^{-7} s) a molecule has absorbed light of a shorter wavelength, usually ultraviolet light. The spectrum of the light absorbed is the **absorption spectrum**; the spectrum of the light emitted is the **emission spectrum**. Fluorescent light, as with scattered light and for the same reasons, is emitted in all directions equally relative to the incident light unless intramolecular interference occurs. It is usually measured perpendicular to the direction of the incident light. It can be measured under continuous excitation, or the excitation can be a flash ($< 10^{-9}$ s in length), and the rate of decay of the fluorescence following the flash can be measured.

If the electronically excited state is structured in such a way that the excited electron can become unpaired with the electron it left behind, it can enter a **triplet excited state** by **intersystem crossing**. The triplet state is usually of lower energy than the singlet state. Once the triplet excited state has been occupied, the electron can return to the ground state only through a **spin-disallowed** process which is very slow (of the order of microseconds). The emitted light, or **phosphorescence**, emerges from the solution over a relatively long period of time and has an even longer wavelength than the rapidly emitted fluorescence. Fluorescence is light emitted from singlet excited states; phosphorescence, from triplet excited states.

Both Raman and infrared spectra of proteins display absorptions of energy resulting from transitions in molecular vibrations. The most obvious and securely assigned absorptions in the infrared or Raman spectrum of a protein arise from excitations of the vibrations of the amides in the polypeptide backbone. A primary amide such as *N*-methylacetamide absorbs infrared energy of wavelength around 6000 nm (1650 cm^{-1}) into its C=O stretching vibration and of wavelength around 8000 nm (1250 cm^{-1}) into a coupled C–N stretching and N–H bending vibration.[27] These two absorbances are referred to as the amide I band and the amide III band, respectively. A solution of protein dissolved in ^2H_2O exhibits a strong, isolated, and characteristic absorbance in the amide I region at 1650 cm^{-1} (Figure 12–4), which is usually the only recognizable absorption in the infrared spectrum.

When a solution of protein is excited with a He–Ne laser, the **Raman spectrum** of the scattered light also displays a maximum with a wavenumber of 1650 cm^{-1} less than the wavenumber of the majority of the scattered light, which has the same wavenumber as the incident light (15,802 cm^{-1}). The amide III maximum at a wavenumber 1250 cm^{-1} less than that of the elastically scattered light and other maxima that can be assigned to vibrational transitions in some of the amino acids, such as phenylalanine, tyrosine, and methionine, are also observed (Figure 12–5).[29]

If the protein contains a functional group that also absorbs the exciting visible light in an electronic transition, as does the heme in hemoglobin,[30] bands in the Raman spectrum resulting from the absorption of energy by vibrations of the atoms within or adjacent to that functional group will be enhanced, and the spectrum is referred to as a **resonance Raman spectrum**.[31] The maxima in a resonance Raman spectrum can often be assigned to vibrations of particular bonds, such as an iron–dioxygen stretching vibration in oxygenated hemoglobin.[30] Otherwise, little information about particular amino acids or particular bonds in a protein is obtained by either infrared or Raman spectroscopy.

When synthetic polyamino acids, such as poly (L-lysine), in aqueous solution are caused to assume α-helical, β, or random structures, the wavelength of maximum absorption, λ_{max}, of the amide I vibrations in their polypeptides assumes characteristic values.[27] For α helix, λ_{max} = 6060 nm (1650 cm^{-1}); for β structure, λ_{max} = 6130 nm (1632 cm^{-1}); and for random coil, λ_{max} = 6040 nm (1656 cm^{-1}). If the contributions of the overlapping absorptions of the individual aromatic amino acids are subtracted from the amide I absorption in the infrared absorption spectrum of a protein to leave only the contribution from the polypeptide,[32] the shape and width of this absorption can be used to obtain an estimate of the relative amounts of these secondary structures in the protein.[33] Circular dichroism, however, is more widely used for this purpose.

Circular dichroism is the consequence of the absorption of visible or ultraviolet light by a chiral solute such as a protein. As such it involves the excitation of

electrons from occupied molecular orbitals into unoccupied molecular orbitals. The most widely used absorptions in spectroscopic studies of proteins by circular dichroism are the electronic absorptions of the amides of the polypeptide backbone between wavelengths of 180 and 240 nm. In this region, two electronic transitions account for the absorption of light.[34] One is a transition ($n \rightarrow \pi^*$) in which an electron leaves one of the lone pairs on the acyl oxygen and enters the vacant antibonding π molecular orbital, π^*, of the amide, which is the orbital of highest energy of the three molecular orbitals of the π molecular orbital system (Figure 2–3). This $n \rightarrow \pi^*$ transition is responsible for the absorption of light at a wavelength of about 220 nm. The other transition ($\pi^\circ \rightarrow \pi^*$) is one in which an electron leaves the highest occupied nonbonding π molecular orbital of the amide (Figure 2–3) and enters the antibonding π molecular orbital. This $\pi^\circ \rightarrow \pi^*$ transition is responsible for the absorption of light at a wavelength of about 200 nm.

Plane-polarized light is light characterized by an electric vector that oscillates within a plane. One way to describe plane-polarized light mathematically is to assume that it is produced by the sum of two electric vectors of equal amplitude emanating from a single point that is traveling at the speed of light in a straight line. While propagating, these two electric vectors spin in opposite directions, clockwise and counterclockwise at the same frequency (in revolutions second^{-1}) as the

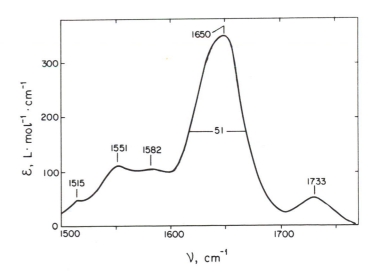

Figure 12–4: Infrared spectrum, in the range between 1500 and 1800 cm^{-1}, of Na$^+$/K$^+$-transporting adenosinetriphosphatase.[28] The protein was purified in its membrane-bound form to obtain a sample of small fragments of membrane in which all of the embedded protein was Na$^+$/K$^+$-transporting ATPase. A suspension of 17 mg mL^{-1} protein in a 3-cm cuvette was used to obtain the spectrum. The extinction coefficient, ε (liters mole^{-1} centimeter^{-1}) is presented as a function of wavenumber, v (centimeters^{-1}). The peak of absorbance at 1650 cm^{-1} is the amide I absorbance. Adapted with permission from ref 28. Copyright 1985 *Journal of Biological Chemistry*.

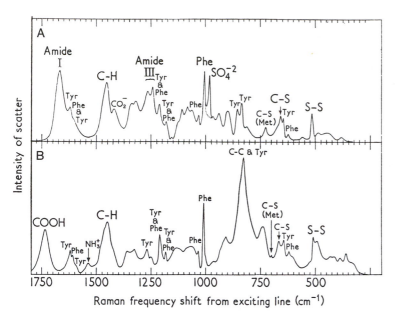

Figure 12–5: Raman spectrum of the intensity of the light scattered from (A) a solution of ribonuclease (c2 Kin.4) (200 mg mL^{-1}) and (B) a solution of amino acids at the same ratio that they are present in ribonuclease.[29] The samples were excited with a He–Ne laser (λ = 632.8 nm), and the intensity of the light scattered was measured as a function of wavenumber (centimeter^{-1}) in the neighborhood of the wavenumber of the elastically scattered light, which had a wavenumber identical to that of the incident light (15,802 cm^{-1}). The intensity of the scattered light is presented as a function of the difference between the wavenumber of the measured light and the wavenumber of the incident and elastically scattered light. As in the infrared spectrum (Figure 12–4), the amide I absorption is the most obvious, but other absorptions that can be assigned to various vibrational modes of the side chains of the amino acids, as well as the partially obscured amide III bond, are clearly seen in the spectrum. The spectrum has high resolution because the exciting light is from a laser. Reprinted with permission from ref 29. Copyright 1970 Academic Press.

frequency of the light (Figure 12–6A). The sum of these two vectors produces an electric vector that oscillates in a plane containing the line of propagation. The two rotating components that consequently spin around the axis defined by the direction the light travels to produce the plane-polarized light are called **circular polarizations**.

When these two circular polarizations encounter a chiral object such as the protein in a solution, they are absorbed and retarded unequally. If the amplitude of one component were decreased more than the amplitude of the other while the two components remained in phase, the plane of polarization of the emerging light would retain the same orientation although its electric vector, which is the sum of the two components, would trace in cross section an ellipse rather than a flat, linear segment (Figure 12–6B). If the phase between the two components were shifted while their amplitudes remained the same, the plane of polarization of the emerging light would rotate by an angle α; however, the electric vector would still trace in cross section a flat, linear segment (Figure 12–6B). The first effect is circular dichroism; the second effect, optical rotation. Both effects are required to occur simultaneously in any circumstance; and, as polarized light is rotated, it necessarily becomes elliptical and vice versa. This obligatory connection permits the spectrum of optical rotation as a function of wavelength to be calculated from the spectrum of circular dichroism as a function of wavelength and vice versa.[35]

The degree to which the emerging light has become elliptical can be measured, and it is expressed as an angle

$$\theta = \tan^{-1}\frac{OB'}{OA'} \qquad (12\text{–}30)$$

where the ratio OB′/OA′ is the ratio of the minor and major axes of the ellipse (Figure 12–6B). The molar ellip-

ticity at a given wavelength λ, $[\theta]_\lambda$, is defined by the relationship

$$[\theta]_\lambda = \frac{\theta}{l[\text{chromophore}]} \qquad (12\text{–}31)$$

where [chromophore] is the molar concentration of the functional group absorbing the light, referred to as the **chromophore**, and l is the path length of the sample chamber. The units of $[\theta]_\lambda$ are usually degrees centimeter^2 decimole^{-1}. An ultraviolet circular dichroic spectrum is a display of the amplitude of $[\theta]_\lambda$ as a function of wavelength.

The optical rotation α (Figure 12–6B) produced by the sample can be registered with a spectropolarimeter. It can also be normalized by the concentration of the chromophore responsible for it to produce the specific rotation $[\alpha]$. A spectrum of the **optical rotatory dispersion** is simply the amplitude of $[\alpha]$ plotted as a function of the wavelength of the polarized light. Because optical rotation arises from a shift in the relative phases of the two circularly polarized components (Figure 12–6B), it is proportional to the derivative with respect to wavelength of the electronic absorption from which it arises. This has the practical disadvantage of both turning one peak into two peaks, a positive and a negative distributed around the wavelength of maximum absorbance, and broadening the signal. In a spectrum of optical rotatory dispersion arising from several maxima of absorbance, the individual components are difficult to resolve.

A circular dichroic spectrum, however, is usually simpler to interpret. Unless excitonic coupling between two chromophores of similar wavelengths of absorption is occurring, the individual bands in a circular dichroic spectrum of a protein are unsplit peaks that coincide with absorption maxima in the absorption spectrum of

A

B

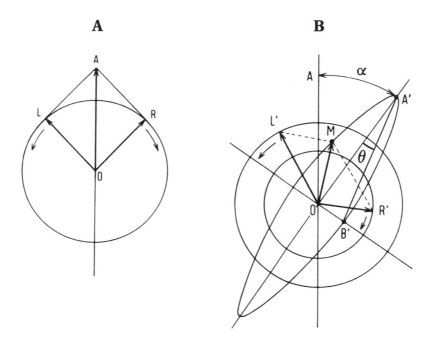

Figure 12–6: Principles of circular dichroism and optical rotation.[35] (A) View down a ray of light plane-polarized in the vertical direction looking from the source. The plane-polarized light can be considered to be the sum of two electric vectors, respectively, of right (R) and left (L) circularly polarized light. The electric vector of the plane-polarized light (A) remains fixed in orientation but oscillates in amplitude. The electric vectors of each of the circular polarizations of the ray remain fixed in amplitude but circle in opposite directions at the same angular velocity. (B) If the index of refraction of a solution for the left circular polarization (L′) of the plane-polarized light is greater than that for the right circular polarization (R′), the left component will have a slower angular velocity than the right component, and the plane of the polarized light (A′) will be rotated to the right by an angle α. If the right circular polarization (R′) of the plane-polarized light is absorbed more than the left circular polarization component (L′), the plane of the polarized light will broaden into an ellipse because when the two electric vectors are at 180° to each other (at B′), they no longer cancel. The ellipse is created by a composite electric vector that rotates to the left if the absorption of the right circular polarization is greater and to the right if the absorption of the left circular polarization is greater. Adapted with permission from ref 35. Copyright 1967 Marcel Dekker.

the same protein. In uncomplicated situations, the circular dichroic spectrum simply registers the optical activity of each chiral contributor to the absorption spectrum. For example, most of the peaks in the absorption spectrum of cytochrome c_1 have only one corresponding negative or positive peak at the same wavelength in its circular dichroic spectrum (Figure 12–7).[36] Because adjacent absorption bands often have different polarities, the circular dichroic spectrum can often reveal details in the absorption spectrum. If excitonic coupling between two or more chromophores is occurring, however, the resulting bands in the circular dichroic spectrum will each be split into two or more components of both positive and

negative amplitude, and this splitting complicates the situation.

A polypeptide folded entirely as an α helix has a circular dichroic spectrum distinct from that of a polypeptide folded entirely in β structure. Both of these spectra are distinct from that of a polypeptide unfolded as a random coil (Figure 12–8).[37] In the spectrum of a polypeptide folded as an α helix, the amide $\pi° \rightarrow \pi^*$ transition at about 200 nm is split into a positive component ($\lambda_{max} = 191$) and a negative component ($\lambda_{max} = 205$). This splitting arises from the fact that each amide is held in the same orientation relative to the axis of the α helix.[38] The $\pi° \rightarrow \pi^*$ transition from a polypeptide in either β structure

A

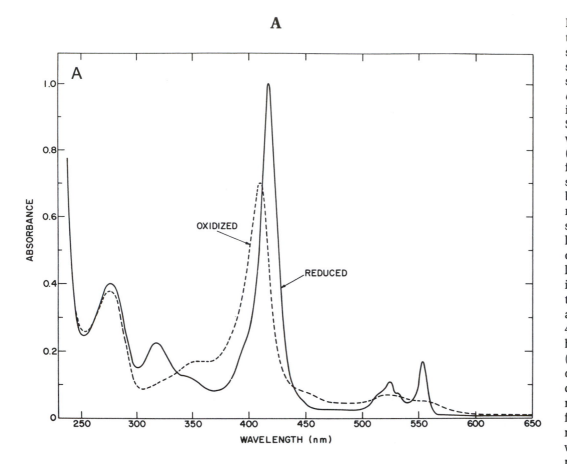

B

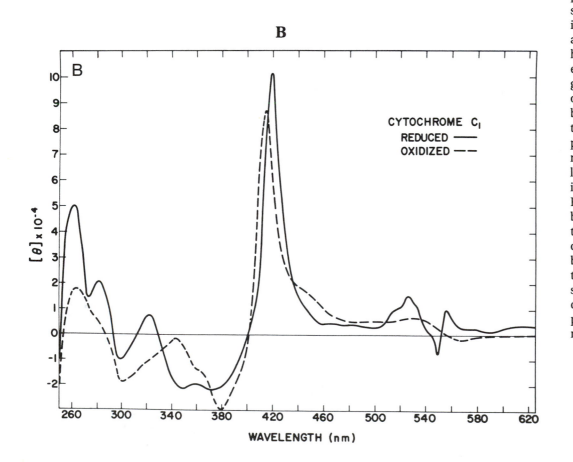

Figure 12–7: Correlation between an optical absorption spectrum (A) and the corresponding circular dichroic spectrum (B).[36] Cytochrome c_1 was purified to homogeneity from bovine heart muscle. Solutions of the hemoprotein were prepared in its oxidized (Fe^{III}) and reduced (Fe^{II}) forms. (A) The optical absorptions of solutions (absorbance) of the oxidized and reduced proteins were measured as a function of wavelength (nanometers). In each case, the absorption at wavelengths greater than 300 nm is due entirely to the heme of the hemoprotein. The intense absorption bands at 400–420 nm are characteristic of hemes. (B) Molar ellipticities ($[\theta] \times 10^{-4}$) of the two solutions of the same two forms of the cytochrome c_1, oxidized and reduced, were measured as a function of wavelength. The molarities of the solutions were expressed as moles of peptide bond in each liter of solution, which is disconcerting since the majority of the absorption arises from the heme. The units for molar ellipticities, therefore, are degrees · centimeter^2 (decimole of peptide bond)$^{-1}$. For each band in the absorption spectrum, each of which has a positive value, there is a corresponding band in the circular dichroic spectrum, which is either positive or negative. For example, the absorption band at 350 nm in the spectrum of the reduced cytochrome c_1 corresponds to a band of negative molar ellipticity in the circular dichroic spectrum, as does the shoulder at 375 nm. Reprinted with permission from ref 36. Copyright 1971 Academic Press.

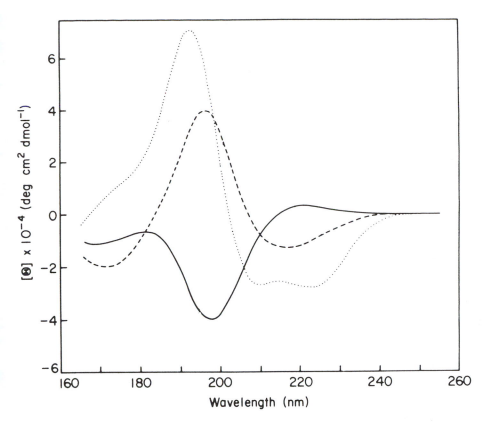

Figure 12–8: Circular dichroic spectra of reference compounds that are used as reference spectra for α helix (dotted line), β structure (dashed line), and random meander (solid line).[37] Molar ellipticity, $[\theta] \times 10^{-4}$ (degree centimeter^2 decimole^{-1}), is presented as a function of wavelength (nanometers). Sperm whale myoglobin dissolved in 0.1 M NaF at pH 7 was used as a polypeptide that is purely α-helical (c3 Kin.3). Poly(Lys-Leu-Lys-Leu) in 0.5 M NaF at pH 7 was used as a polypeptide that is purely β structure. Poly(Pro-Lys-Leu-Lys-Leu) in a salt-free solution is completely structureless because of the prolines and the strong repulsions of the lysines. It is not, however, a typical random coil because of both of these features. Nevertheless, it was used as a model for a polypeptide that is purely random meander. Reprinted with permission from ref 37. Copyright 1980 Academic Press.

or random coil is unsplit. There is also the additional band of negative ellipticity at 225 nm from the $n \to \pi^*$ transition of the peptide bond which, in combination with the band of negative ellipticity at 205 nm, gives the circular dichroic spectrum of the α helix its characteristic double minimum.

The tyrosines, phenylalanines, and tryptophans in a polypeptide also absorb light of wavelength between 180 and 240 nm and have characteristic circular dichroic spectra.[37] The contributions of the tyrosines, phenylalanines, and tryptophans in a protein to its circular dichroic spectrum can be numerically subtracted to reveal the circular dichroic spectrum of the polypeptide alone. Because the polypeptide is the main contributor to the circular dichroic spectrum between wavelengths of 180 and 240 nm, the unit in which the molar ellipticity is usually presented is molarity of peptide bonds.

The experimentally measured circular dichroic spectrum of the folded polypeptide in a protein can always be numerically resolved into three component spectra as similar as possible to those of pure α helix, pure β structure, and pure random coil. If it is assumed that these component spectra, obtained only by numerical analysis, accurately represent the contributions of α helix, β structure, and random meander to the entire spectrum, their relative amplitudes should provide the relative amounts of these three components in the actual molecule of protein.[39]

This simple expectation is diminished by several difficulties. Small peptides that assume particular types of

β turn show different circular dichroic spectra, each unique from that of a random coil, and dissecting out the contribution of each β turn to the spectrum of a particular protein is difficult,[37] if not impossible. A related shortcoming is the confounding of random coils and random meanders. A **random coil** is an unfolded polypeptide continuously changing its structure by rotation around its various covalent bonds. **Random meander** is the path assumed by the backbone of a folded polypeptide that is neither an α helix, a β structure, nor a β turn. Random meander is static and respectively identical in all of the folded polypeptides in a solution of a given protein. The random coil of an unfolded polypeptide used as the standard in circular dichroism, unlike the α helix or β structure used as the standard, bears no relationship to the random meander in a folded polypeptide. The random meander in a particular protein will produce a specific circular dichroic spectrum that is distinct from the common circular dichroic spectrum produced by all random coils and also distinct from the unique spectra produced by random meanders in other proteins.

Nevertheless, a least-squares method has been developed to fit the experimental circular dichroic spectrum of a protein, from which the contributions of tyrosine, tryptophan, and phenylalanine have been subtracted, to a calculated spectrum (Figure 12–9).[37] The parameters of the fitting procedure are the fraction of α helix, the fraction of β structure, the fraction of random meander, and the fraction of each type of β turn. It is assumed that the sum of these fractions is unity, that

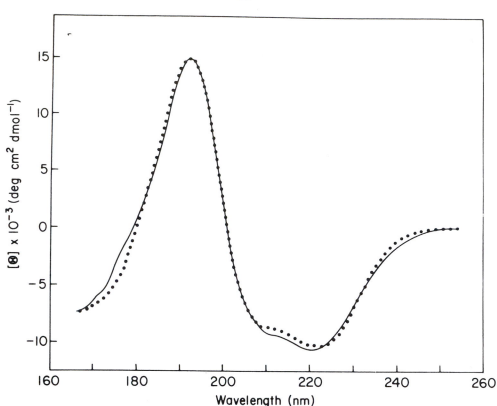

A

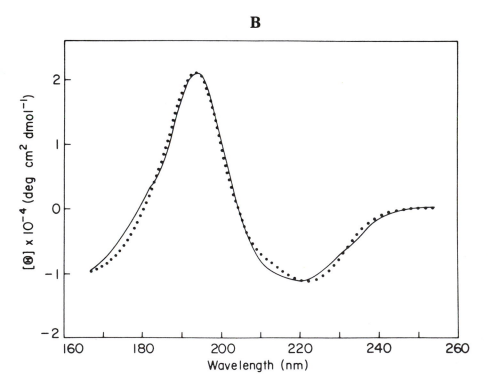

B

Figure 12–9: Circular dichroic spectra[37] of (A) glyceraldehyde-3-phosphate dehydrogenase in 0.1 M NaF, pH 7, and (B) subtilisin (c15 Kin.4) in 0.2 M NaF, pH 7. Molar ellipticities, $[\theta] \times 10^{-4}$ (degree centimeter^2 decimole^{-1}), are presented as a function of wavelength (nm). The spectra were either directly measured (solid lines) or duplicated (dotted lines) by adding together spectra for α helix, β structure, β turn, and random meander (Figure 12–8). In the procedure used to duplicate the experimental spectrum, it was assumed that the proteins only contain α helix, β structure, β turn, and random meander. If f_α, f_β, f_T, and f_{RM} are the fractions of each of these secondary structures, it is assumed that the sum of these four numbers is 1 and that $f_\alpha(\theta_\alpha^\circ) + f\beta(\theta_\beta^\circ) + f_T(\theta_T^\circ) + f_{RM}(\theta_{RM}^\circ)$ is equal to the measured value of θ at every wavelength, where the θ° values are the molar ellipticities of the standard curves (Figure 12–8) at the same wavelength. A least-squares method was used to obtain the best values for f_α, f_β, f_T, and f_{RM}, and these four values were used to construct the calculated curves. Note that f_α, f_β, f_T, and f_{RM} are parameters determined by the structure of the protein and must have the same values for all wavelengths. For the spectrum of glyceraldehyde-3-phosphate dehydrogenase, the best values of f_α, f_β, f_T, and f_{RM} were 0.31, 0.30, 0.22, and 0.17; for the spectrum of subtilisin, 0.30, 0.21, 0.21, and 0.28. Reprinted with permission from ref 37. Copyright 1980 Academic Press.

each point on the experimental spectrum is the sum of the molar ellipticity at that wavelength of the appropriate reference spectrum ($\theta°$) times the fraction for that particular secondary structure, and that the reference spectrum for random meander is that of random coil. The least-squares procedure gives the values for the parameters that produce a calculated curve most closely reproducing the experimental curve. The fractions for each type of secondary structure estimated in this way for a set of proteins agree quite closely with the fractions for each type of secondary structure in the respective crystallographic molecular models of these proteins.

One of the more important and informative uses of circular dichroism is to provide evidence that the structure of the protein has changed under particular circumstances. The change in the structure of aspartate transcarbamylase that occurs upon the binding of its substrates and that is detected both crystallographically and as a change in sedimentation coefficient is also accompanied by significant changes in the circular dichroic spectrum of the protein.[40] Such observations are commonly encountered. This fact increases the concern over the accuracy of secondary structural dissections by numerical analysis of circular dichroic spectra because crystallographic descriptions of structural changes accompanying the binding of substrates rarely involve changes in the content of α helix, β structure, β turns, or random meander or changes in their disposition over the sequence of the folded polypeptide. The changes in the circular dichroic spectrum of Na^+/K^+-transporting adenosinetriphosphatase during a change in its structure caused by the binding of its substrates are consistent with the transformation of 7% of its α helix into β structure.[41] Such a transformation is both unlikely and also inconsistent with the absence of a shift in the amide I absorption in the infrared spectrum (Figure 12–4) under the same circumstances.[28]

Ultraviolet absorption spectra of proteins at wavelengths greater than 240 nm are dominated by the absorption of phenylalanine, tyrosine, and tryptophan. Tryptophan has the largest extinction coefficient ($\varepsilon = 5600$) and longest wavelength of maximum absorbance ($\lambda_{max} = 280$ nm). If one or more of the accessible tyrosines on the protein have been nitrated, their absorption spectrum is shifted into the visible range ($\lambda_{max} = 430$ nm) and their acid dissociation constants are increased ($pK_a = 6.5$), so that at neutral pH they are present mainly as the nitrophenolate, which absorbs strongly.[42]

Either tryptophan or nitrotyrosine can be used as a spectral "reporter" group whose spectrum reflects its environment or a change in its environment.[43] For example, the absorption spectrum of nitrated Tyrosine 115 in micrococcal nuclease indicates that it is in a nonpolar environment in the absence of substrate but a polar environment in the presence of substrates.[42] This change in environment is also reflected in its accessibility to

nitration by tetranitromethane. The change in the structure of aspartate transcarbamylase that occurs on the binding of substrates can be detected by an upfield shift in the wavelength of the absorption of tryptophans in the protein[44] or of nitrated tyrosine residues in the regulatory subunits.[45]

In addition to absorbing ultraviolet light, tryptophan is also both fluorescent and phosphorescent. The wavelength of its maximum fluorescence emission varies from 300 to 350 nm.[46] The fluorescence of indole itself varies between these limits systematically as a function of the polarity of the solvent; the more polar the solvent, the longer the wavelength of the emission. It has been stated[46] that proteins in which tryptophans are more buried display shorter wavelengths of maximum emission. In the case of the absorption spectrum of tryptophan as opposed to the emission spectrum, the situation is, however, reversed. The most buried tryptophans, in the most nonpolar environments, have been found to absorb light of the longest wavelength, on the red edge of the absorption band for tryptophan in the ultraviolet.[47]

If intersystem crossing is not significant, the excited state of a fluorescent functional group such as tryptophan can decay to the ground state by at least four separate pathways[48]

$$\text{excited state} \xrightarrow{k_L} \text{ground state} \qquad (12\text{–}32)$$

$$\text{excited state} \xrightarrow{k_F} \text{ground state} + h\nu \qquad (12\text{–}33)$$

$$\text{excited state} + \text{quencher} \xrightarrow{k_Q}$$
$$\text{ground state} + \text{excited quencher}$$
$$(12\text{–}34)$$

$$\text{excited state} + \text{acceptor} \xrightarrow{k_T}$$
$$\text{ground state} + \text{excited acceptor}$$
$$(12\text{–}35)$$

Equation 12–32 describes a radiationless decay of the energy of excitation through migration among rotational and vibrational energy levels or other piecemeal transfers to its surroundings as heat. Equation 12–33 describes the release of a portion of the energy of excitation as a photon of fluorescent light. Equation 12–34 describes the transfer of the energy of excitation to another molecule, the quencher, which collides with the fluorescent functionality when it is in the excited state. Equation 12–35 describes the radiationless transfer of the energy of excitation through space to a nearby functional group, an acceptor, capable of absorbing the energy.

When neither quencher nor acceptor is present, the **quantum yield**, Q_0, or number of photons appearing as fluorescence for every photon absorbed, is[48]

$$Q_0 = \frac{k_F}{k_L + k_F} = k_F \tau_0 \qquad (12\text{–}36)$$

The time over which 50% of the excited state disappears, or the half-life of the excited state, would be $(\ln 2)\,(k_L + k_F)^{-1}$ but the **fluorescence lifetime**, τ_0, is arbitrarily defined as $(k_L + k_F)^{-1}$.

If a **collisional quencher** is added to the solution, it affects the quantum yield and lifetime of the excited state because every time a molecule of quencher collides with a molecule of excited state, the excitation energy is transferred from the excited state to the quencher. In this circumstance[48]

$$Q_Q = \frac{k_F}{k_L + k_F + k_Q[\text{quencher}]} \qquad (12\text{--}37)$$

and

$$\frac{Q_Q}{Q_0} = \frac{1}{1 + k_Q \tau_0[\text{quencher}]} \qquad (12\text{--}38)$$

The quenched fluorescence lifetime τ_Q is

$$\tau_Q = \frac{\tau_0}{1 + \tau_0 k_Q[\text{quencher}]} \qquad (12\text{--}39)$$

The ratio F_Q/F_0 is the ratio between the fluorescence observed in the presence of the quencher and the fluorescence observed in its absence. This ratio is necessarily equal to Q_Q/Q_0. The ratio F_Q/F_0 can be readily measured with a fluorometer. When F_0/F_Q is plotted as a function of [quencher], a linear relationship is obtained, the slope of which is equal to $k_Q \tau_0$ (Figure 12–10). The bimolecular rate constant k_Q for the collision of the quencher with the fluorescent functional group on a protein can be calculated from the slope of this line and the fluorescent lifetime τ_0 of the unquenched excited state. Phosphorescence, which is simply fluorescence from a triplet state, is subject to the same quenching as fluorescence.

The fluorescence and phosphorescence of tryptophan can be quenched by large inorganic anions, such as I^- and NO_2^-; molecular oxygen; unsaturated amides, such as acrylamide; and ketones, such as 2-oxobutane.[46–49] The bimolecular rate constant k_Q for the quenching reflects the accessibility of the tryptophans to the solvent.[49]

On the basis of the observed rate constants, k_Q, the tryptophans in most proteins can be divided into two classes.[47] The tryptophans in the first class are fully accessible to the aqueous phase, and their fluorescence is readily quenched. The rate constants for the collisions between their singlet excited states and various quenchers are $1–10 \times 10^9$ M^{-1} s^{-1} as expected from a diffusion-controlled process. The fluorescence of the tryptophans in the second class cannot be quenched because no collisions with the quenchers can occur within the lifetime of their excited states. Presumably, this is due to the fact that they are buried. This conclusion follows from the facts that such tryptophans are found in proteins that contain buried tryptophans in their crystallographic molecular models and that such unquenchable tryptophans are preferentially excited by light of a longer wavelength.[47] The phosphorescence of these buried tryp-

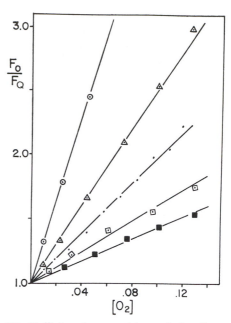

Figure 12–10: Collisional quenching of the fluorescence of tryptophans in several proteins by oxygen.[46] Solutions of the various proteins were placed in cuvettes in a fluorometer and excited with light of wavelength 280 nm. Fluorescence at 90° to the exciting beam was monitored at the wavelength of maximum emission for each protein (325–350 nm). The high concentrations of oxygen (molar) were produced by enclosing the cuvette in a chamber that could be pressurized to 1500 lb in.$^{-2}$ of O_2 gas and allowing the gas to equilibrate with the solution at various pressures. The proteins used were bovine α-chymotrypsin (■), rabbit fructose-bisphosphate aldolase (□), bovine immunoglobulin G (•), and bovine serum albumin (△). A solution of tryptophan (○) was used as an example of a fully exposed side chain. Lines were drawn on the basis of the expectation that F_0/F_Q as a function of [quencher] would be linear (Equation 12–38). Reprinted with permission from ref 46. Copyright 1973 American Chemical Society.

tophans, however, has a sufficiently long lifetime ($\tau_0 \cong 1.4$ s) that it can be quenched. The bimolecular rate constants k_Q for the quenching of the phosphorescence of the tryptophans in the buried class are relatively small ($< 10^6$ M^{-1} s^{-1}), and the quenching registered by these rate constants seems to result from extensive and momentary unfoldings of the folded polypeptide which occasionally provide access to the interior, but only for a short time.[47] There are tryptophans, such as the one tryptophan present in ribonuclease T_1, that seem to have properties intermediate between these two classes.[49]

Oxygen provides an interesting exception to the behavior of quenchers as a group. It is able to quench both accessible and buried tryptophans with almost equal efficiency.[46,50] On the basis of this observation, it has been proposed that oxygen is small enough to insinuate its way through a molecule of protein in liquidlike diffusion among the tightly packed amino acids.

Changes in the accessibility of tryptophans to quenchers dissolved in the aqueous phase have been used to monitor changes in the structure of a protein

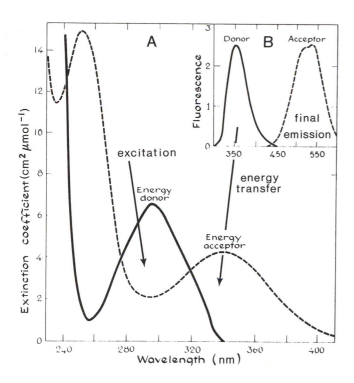

Figure 12–11: Absorption spectra (A) and emission spectra (B) of a typical donor for fluorescent energy transfer, 1-acetyl-4-(1-naphthyl)semicarbazide (solid lines), and a matched acceptor, dansyl-L-prolylhydrazide (dashed lines), both dissolved in ethanol.[52] The measurements of absorbance are made by monitoring the intensity of the light of systematically varied wavelength that passes through each solution. The measurements of emission are made by following the intensity of the light emitted at 90° to the incident beam as a function of wavelength while exciting the chromophore with light of wavelength equal to that of its absorption maximum. The amount of light absorbed is expressed as an extinction coefficient (centimeter^2 micromole^{-1}) as a function of wavelength (nanometers). The amount of light emitted is expressed as fluorescence (in relative units) as a function of wavelength (nanometers). A portion of the excited states would have their energy of emission at 350 nm transferred radiationlessly to the overlapping absorption band of the acceptor, and this transfer would quench the fluorescence of the donor. The transferred energy would be emitted as fluorescence at 540 nm from the excited acceptors. Adapted with permission from ref 52. Copyright 1967 National Academy of Sciences.

caused by the binding of substrates. In the case of succinate–CoA ligase (ADP-forming) from *E. coli*, the binding of ATP to the α subunit of the enzyme causes significant decreases in the accessibility of the tryptophans in the β subunit to acrylamide dissolved in the solution.[51] This suggests that a change in structure propagated throughout the whole protein occurs upon the binding of ATP. The implication that both the α and β subunits change their structure in concert when ATP binds is consistent with the observation that they are intimately associated in the oligomeric structure of the protein (Figure 8–22).

If the absorption spectrum of a nearby chromophore overlaps the emission spectrum of the excited state, it is also possible for the energy of the excited state in excess of the energy of the ground state to be transferred intact through space to this other chromophore. This radiationless **fluorescent energy transfer** discharges the electronically excited state of the functional group that originally absorbed the photon, the **donor**, and produces an electronically excited state in the functional group that receives the energy, the **acceptor**. If this new excited state of the acceptor always returns to its respective ground state by radiationless processes—in other words, if its quantum yield is zero—no fluorescent photon is emitted, and the fluorescence of the donor is quenched. If the acceptor is also a fluorescent functional group, it will release a fluorescent photon, consistent with its quantum yield. The photon released from the acceptor, however, will be of longer wavelength than the one that would have been released from the donor (Figure 12–11) because the initial excited state of the acceptor relaxes to a stable excited state, the energy of which, relative to the ground state of the acceptor, is less than the energy that

was passed from donor to acceptor during the transfer. At the shorter wavelengths in the emission spectrum of the donor, which do not overlap the emission spectrum of the acceptor, the quantum yield Q_A of the fluorescence emitted by the donor in the presence of the acceptor will be less than the quantum yield Q_0 in its absence. The observed fluorescence of the donor will decrease, in the presence of the acceptor, by the ratio Q_A/Q_0.

In general, the acceptor is a functional group that is covalently or noncovalently attached to a specific location on a protein, and the donor is covalently or noncovalently attached to another specific location. The intention of the experiment is to measure the distance between these two locations on the properly folded molecule of protein. For example, if the donor were attached by the specific and stoichiometric covalent modification of a particular amino acid in the sequence of the protein and the acceptor were specifically attached by the unique modification of another amino acid in the sequence, the sole intention of the experiment would be to measure the distance between the donor and acceptor in the folded polypeptide and hence the distance between the two modified amino acids. It turns out that measuring the energy transfer is easy but placing the donors and acceptors at unique and exclusive locations on the protein is difficult. Usually, either an intrinsic donor or an intrinsic acceptor or both, placed by evolution either covalently or noncovalently at a unique location on the protein, is relied upon to circumvent at least half of the difficulty. One of the most widely used class of compounds of this type of evolutionarily positioned donors and acceptors is substrates or analogues of substrates that bind to the active site or ligands that bind to a specific binding site on a protein.

Suppose that a molecule of protein had a fluorescent functional group that could act as a donor covalently attached or noncovalently bound to a particular location in its tertiary structure and a different fluorescent functional group that could act as an acceptor covalently attached or noncovalently bound to a different location. The rate of energy transfer from the donor to the acceptor (Equation 12–35) would be equal to $k_T f_A$[excited donor], where f_A is the fraction of the sites on the protein for acceptor that are occupied and [excited donor] is the molar concentration of the excited donor. As the acceptors are fixed to the protein at a constant fractional occupancy, the pseudo-first-order rate constant for the decay in the concentration of excited donor through energy transfer to acceptor is $k_T f_A$. The efficiency of transfer E_T is defined as

$$E_T \equiv \frac{k_T f_A}{k_L + k_F + k_T f_A} \qquad (12\text{--}40)$$

This is the fraction of the decay of the excited state due to energy transfer, or the ratio between the quanta transferred and the quanta absorbed by the donor.[53] The quantum yield Q_A of the fluorescence in the presence of the acceptor is governed by a relationship analogous to Equation 12–37

$$Q_A = \frac{k_F}{k_L + k_F + k_T f_A} \qquad (12\text{--}41)$$

from which it follows[54] that

$$E_T = 1 - \frac{Q_A}{Q_0} = 1 - \frac{F_A}{F_0} \qquad (12\text{--}42)$$

where F_A/F_0 is the ratio between the fluorescence observed in the presence and that observed in the absence of the acceptor. The lifetime of the excited state τ_A in the presence of acceptor is governed by a relationship analogous to Equation 12–39

$$\tau_A = \frac{\tau_0}{1 + \tau_0 k_T f_A} \qquad (12\text{--}43)$$

from which it follows that

$$E_T = 1 - \frac{\tau_A}{\tau_0} \qquad (12\text{--}44)$$

The efficiency of transfer can be assessed by measuring the decrease in steady-state fluorescence (F_A/F_0) for the decrease in the lifetime of the excited state (τ_A/τ_0) produced by the attachment of the acceptor to the protein in the vicinity of the donor.

The efficiency of energy transfer, E_T, is related to the distance r between the donor and the acceptor by the relationship

$$E_T = \frac{f_A r^{-6}}{f_A r^{-6} + R_0^{-6}} \qquad (12\text{--}45)$$

where R_0 is the distance at which the efficiency of transfer would be 50% when the acceptor site is fully occupied. It has been theoretically calculated[53] that

$$R_0^6 = \frac{9(\ln 10) K^2 Q_0 J}{128 \pi^5 \bar{n}^4 N_{Av}} \qquad (12\text{--}46)$$

where Q_0 is the quantum yield of the donor (dimensionless), $\bar{n}$ is the refractive index of the medium, N_{Av} is Avogadro's number (moles^{-1}), and J is the overlap integral (centimeters^6 mole^{-1})[54] between the fluorescence emission spectrum $I(\lambda)$ of the donor (in relative units) and the extinction coefficient $\varepsilon(\lambda)$ of the acceptor (in liters mole^{-1} centimeters^{-1}) normalized by the total fluorescence of the donor

$$J = \frac{\int I(\lambda)\varepsilon(\lambda)\lambda^4 \, d\lambda}{\int I(\lambda) \, d\lambda} \qquad (12\text{--}47)$$

where λ is the wavelength (in centimeters). The integral J is calculated numerically from the absorption spectrum of the acceptor and the emission spectrum of the donor (Figure 12–11).

The orientation factor K^2 (dimensionless) is the most uncertain parameter in Equation 12–46.[54] It is a function of the orientation of the emission dipole of the donor relative to the absorption dipole of the acceptor. In any given situation, K^2 has a specific numerical value but its value cannot be measured. In the absence of the numerical value for the K^2 that applies to a particular situation, the usual approach is to define upper and lower limits between which it must lie. If both donor and acceptor are free enough to assume all possible relative orientations, K^2 is 2/3.[54] If one of the two is fixed and the other can assume all possible relative orientations, K^2 must have a value between 1/3 and 4/3.[54] If both donor and acceptor are fixed in their relative orientations, for example, both rigidly bound to a molecule of protein, K^2 has a value somewhere between zero and 4.0.

The more freedom the donor or the acceptor has to assume different orientations by rotation around unhindered bonds between it and the protein, the closer the value of K^2 comes to 2/3. An estimate of the orientational freedom of donor or acceptor can be made from its **emission anisotropy**. If either the donor or acceptor is excited with polarized light, the light emitted as fluorescence immediately, that is, before the chromophore has time to reorient, will also be polarized. The polarity of the emitted light will decay over the lifetime of the excited state as it reorients. The rate of this decay and the final residual polarity of its fluorescence,[55] provide an estimate of the orientational freedom of the donor or acceptor. From this emission anisotropy, limits[55,56] can be placed on the possible values for K^2.

Equations 12–45 and 12–46 have been shown to be consistent with the observed transfers of energy between a donor and an acceptor at the two ends of a short

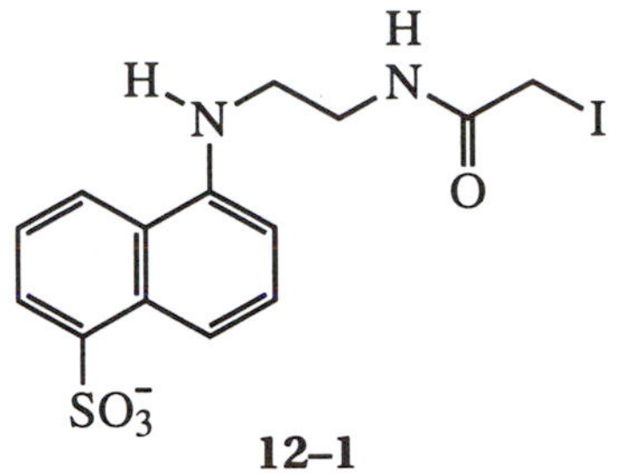

12–1

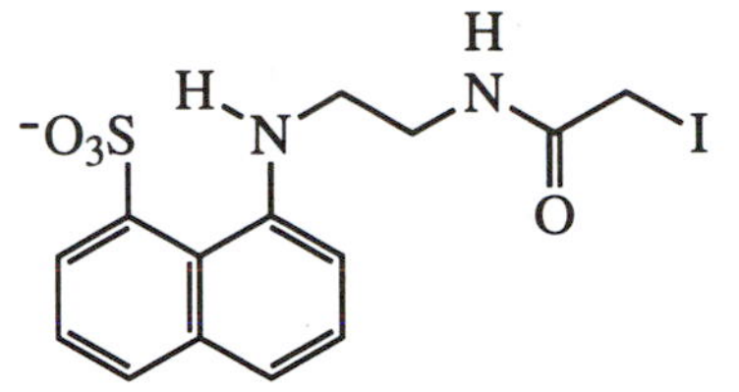

12–2

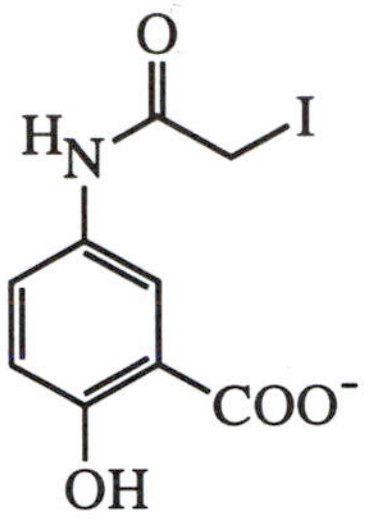

12–3

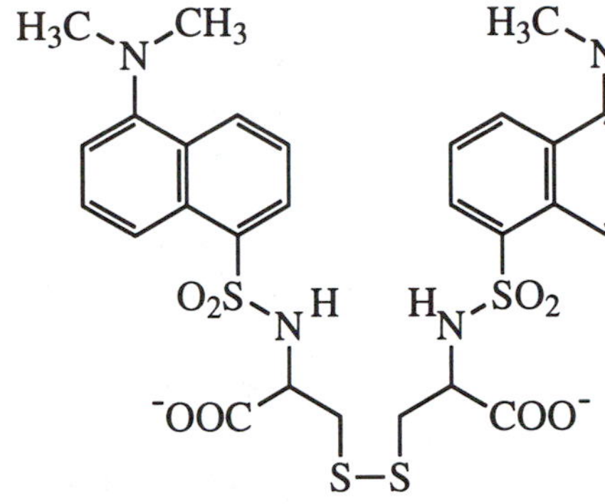

12–4

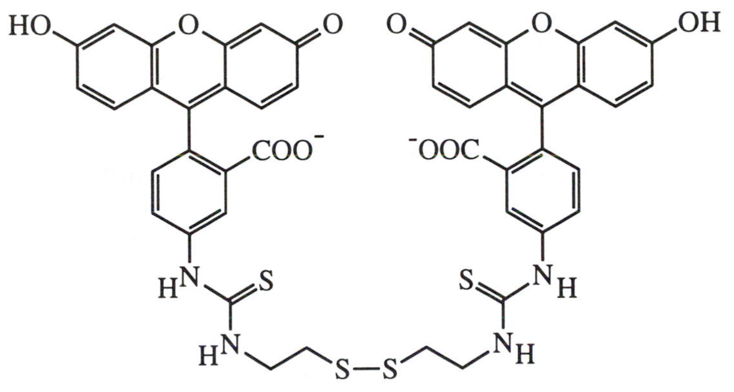

12–5

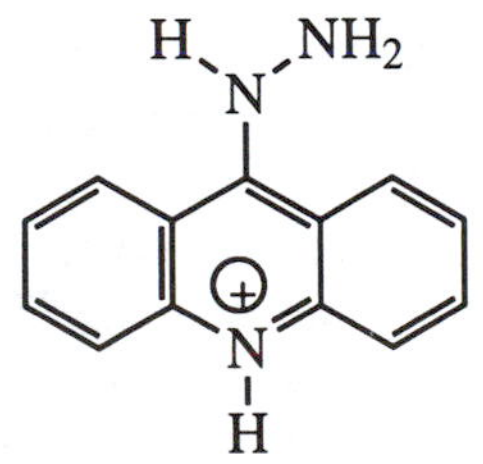

12–6

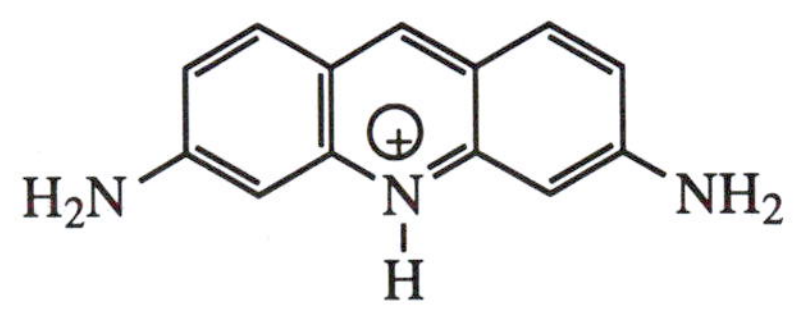

12–7

Figure 12–12: Fluorescent reagents used to modify, covalently or noncovalently, three sites on rhodopsin.[54] *N*-[(Iodoacetamido)ethyl]-1-aminonaphthalene-5-sulfonate anion **12–1** (λ_{abs} = 350 nm; λ_{emit} = 495 nm), *N*-[(iodoacetamido)ethyl]-1-aminonaphthalene-8-sulfonate anion **12–2** (λ_{abs} = 350 nm; λ_{emit} = 495 nm), and 5-(iodoacetamido)salicylate anion **12–3** (λ_{abs} = 323 nm; λ_{emit} = 405 nm) were used to modify a particular cysteine in the protein by alkylation. Bis[1-(dimethylamino)naphthalene-5-sulfonato]-L-cystine **12–4** (λ_{abs} = 350 nm; λ_{emit} = 520 nm) and bis[fluoresceinyl(isothiocarbamido)] cystamine **12–5** (λ_{abs} = 495 nm; λ_{emit} = 518 nm) were used to modify a different cysteine in the protein by disulfide exchange. 9-Hydrazinoacridine **12–6** (λ_{abs} = 440 nm; λ_{emit} = 470 nm) and proflavin **12–7** (λ_{abs} = 470 nm; λ_{emit} = 512 nm) were used as ligands for a particular site on the protein with a high affinity for aromatic cations. All wavelengths (λ) are wavelengths of maximum absorption or maximum emission.

synthetic peptide of polyproline.[52] The distance between donor and acceptor was varied by varying the number of prolines in the peptides to demonstrate that the dependence of efficiency upon distance was as the sixth power. If K^2 was assumed to be $2/3$, the calculated distances between donor and acceptor agreed fairly well (within 25%) with the distances measured from molecular models of these modified peptides.

From the efficiency of energy transfer between donor and acceptor on a molecule of protein, the distance between the emission dipole of the donor and the absorption dipole of the acceptor can be estimated. Rhodopsin contains two different cysteines, one of which, site A, can be modified by derivatives of iodoacetamide and the other of which, site B, reacts readily in disulfide exchange with added dithiols. Rhodopsin also has a site, site C, to which cationic dyes such as acridine or proflavin bind noncovalently. The protein has a covalently attached coenzyme, 11-*cis*-retinal, which can be eliminated at will by bleaching. A series of fluorescent derivatives of iodoacetamide, fluorescent dithiols, or fluorescent cationic dyes (Figure 12–12) were used as donors and acceptors among themselves and as donors to 11-*cis*-retinal to measure the various distances among these locations on the molecule of protein. In each case the efficiency of fluorescence energy transfer was determined from both the ratios of the quantum yields (Equation 12–42) and the ratios of the fluorescent lifetimes (Equation 12–44) of the donor in the presence or absence of the acceptor. A self-consistent set of distances, which were reproducible when different donors or different acceptors were used at the same site, was obtained. The distances were 8 nm from site A to the retinal, 6 nm from site B to the retinal, 5 nm from site C to the retinal, 4 nm from site A to site B, 3 nm from site A to site C, and 3 nm from site B to site C.[54] Several of the fluorescent probes that were used have the aromatic rings responsible for the absorption and fluorescence at a significant distance (~1 nm) from the point at which the reagents were covalently attached to the protein (Figure 12–12), and this, as well as the uncertainty in the actual value of K^2, limited the accuracy of these measurements to at least an uncertainty of ± 1 nm.

Under appropriate circumstances only one lysine of aspartate transcarbamylase, which is in the active site, reacts with pyridoxal phosphate to form an imine that can be reduced to the covalently attached pyridoxamine phosphate with sodium borohydride (Figure 10–3). When such modified catalytic subunits are reconstituted with an excess of both unmodified catalytic subunits and unmodified regulatory subunits, each catalytic subunit containing three pyridoxamine phosphates sits opposite an unmodified catalytic subunit in the hexamer (Figure 9–12). If pyridoxal phosphate is then added, it forms pyridoximine phosphates at the active sites of the unmodified catalytic subunit. As the emission spectrum of pyridoxamine phosphate overlaps the absorption spectrum of pyridoximine phosphate, the distance between the active sites on one catalytic subunit and those on the opposite catalytic subunit could be estimated to be 3 nm from measurements of energy transfer between these two chromophores.[57]

A cysteine on the γ-polypeptide of the chloroplast adenosinetriphosphatase can be modified exclusively with *N*-[4-[7-(diethylamino)-4-methylcoumarin-3-yl]phenyl]maleimide, a fluorescent *N*-arylmaleimide. This covalently attached, fluorescent chromophore could then be used as a donor to 2′(3′)-trinitrophenyladenosine triphosphate bound noncovalently at the active site of the enzyme or to octylrhodamine incorporated into the membrane to which the adenosinetriphosphatase is normally attached. Decreases in both quantum yield and lifetime of the excited state of the donor were used to calculate transfer efficiencies and distances among these locations.[58]

The enzyme transglutaminase was used to exchange the amide nitrogen on one particular glutamine in the sequence of rhodopsin for the primary amine on *N*-dansyl-1,5-diaminopentane. The covalently incorporated dansyl group was shown to have a high rotational mobility by measurement of its emission anisotropy. From these measurements a range was calculated within which the value of K^2 must fall. Using these limits and the measured efficiency of energy transfer between the dansyl group as a donor and the 11-*cis*-retinal of the rhodopsin as an acceptor, the distance between the dansyl group and the 11-*cis*-retinal was calculated to be between 4 and 7 nm.

This last result reemphasizes the problem surrounding the uncertainty in the value for the orientation factor K^2. It is an important example because efforts were made to place limits on the value of K^2 rather than to assume glibly that its value was $2/3$. Yet even in a case where one of the two probes was rapidly reorienting and the other was fixed, which is the situation usually encountered, the uncertainties in the value of K^2 were sufficiently large that the measurements made by fluorescent energy transfer were unfortunately imprecise (±25%). The certainty of the measurement may improve if two very different chromophores attached to the same amino acid, both either as donors or as acceptors, give consistent and independent estimates of the same distance to another specific location.[59]

Suggested Reading

Pober, J.S., Iwanij, V., Reich, E., & Stryer, L. (1978) Transglutaminase-Catalyzed Insertion of a Fluorescent Probe into the Protease-Sensitive Region of Rhodopsin, *Biochemistry* *17*, 2163–2169.

Nuclear Magnetic Resonance and Proton Exchange

Many atomic nuclei display rotational motion known as nuclear spin. Because nuclear spin is quantized, its angular velocities can assume only those magnitudes dictated by spin quantum numbers. Among many other atomic nuclei, those of ^1H, ^{13}C, ^{15}N, ^{19}F, and ^{31}P have only two spin quantum numbers, $+\frac{1}{2}$ and $-\frac{1}{2}$. These dictate two specific angular velocities of the same magnitude but of opposite polarity. As any one of these nuclei is a charged particle by virtue of its protons, either of these angular velocities creates a nuclear magnetic field of the respective polarity aligned with the axis of the nuclear spin. When such a nucleus is placed in an external, homogeneous magnetic field of a given polarity, its axis aligns with the direction of the applied field, and its spin states, because they are of opposite polarity to each other, become different in energy. This difference in energy, ΔE, is directly proportional to the magnitude of the applied magnetic field H

$$\Delta E = \gamma_i h H = h\nu \qquad (12\text{–}48)$$

where h is Planck's constant, ν is the frequency of the electromagnetic energy absorbed, and γ_i is the gyromagnetic ratio, which has a unique value for each atom: ^1H, ^{13}C, ^{15}N, ^{19}F, or ^{31}P. At readily accessible magnetic fields (< 200,000 G), the difference in energy between the two spin states of one of these nuclei is less than 0.5 J mol^{-1}, which is the energy contained in a photon of wavelength greater than 3×10^8 nm and frequency less than 1000 MHz. This is in the radio-frequency range of electromagnetic energy.

Nuclear magnetic resonance spectroscopy measures the same phenomenon as optical spectroscopy. In an applied magnetic field every nucleus of spin ½ has two energy levels. Electromagnetic energy of wavelength somewhere between 2×10^{10} nm (15 MHz) and 3×10^8 nm (1000 MHz) will be absorbed by the sample in the process of exciting nuclei of the spin state with lower energy to the spin state with higher energy. In theory, this absorption of energy could be recorded as a function of wavelength to obtain a spectrum, as is done with an optical absorption spectrum. In practice, it is easier to vary the strength of the magnetic field, H, than the wavelength of the radio-frequency energy. Because the applied magnetic field is directly proportional to the difference in energy levels (Equation 12–48) and the energy of light is directly proportional to its frequency, the spectrum presented is that of absorption as a function of frequency.

There is an additional feature of nuclear magnetic resonance, which is inconsequential for optical spectroscopy. When a population of chemically identical fundamental particles, such as electrons or atomic nuclei, is exposed to electromagnetic radiation of a wavelength equivalent in energy to the difference between two energy levels accessible to the particles, the electromagnetic radiation catalyzes the movement of those particles between these two energy levels. The reason is that the electromagnetic radiation is in **resonance** with transitions between the two energy levels. Energy will be

absorbed during this process only if, at the time of irradiation, the population of particles occupying the lower energy level is greater than the population occupying the higher energy level. The absorption of photons, however, necessarily increases the population in the higher energy level at the expense of the population in the lower energy level. When the populations in the two resonating energy levels become equivalent, absorption can no longer occur and a state of **saturation** is reached.[*]

In the electronic transitions and vibrational transitions of electrons in atomic and molecular orbitals, the energy levels are sufficiently different and relaxation back to the ground state sufficiently rapid that absorption of a particular wavelength of light by a given population of chemically identical electrons rarely displays saturation. In nuclear magnetic resonance, however, the energy difference between the two spin states that can be achieved with the available magnetic fields is so small (< 0.5 J mol^{-1}) that the equilibrium constant K_{sp} between the two spin states of a given nucleus at normal temperatures (300 K) is very close to 1 ($1 < K_{sp} < 1.0002$). This means that the difference in the concentrations of the nuclei in the two spin states at equilibrium in an available magnetic field will be less than 200 parts per million. The difference between the populations in the two energy levels set in resonance is small enough and the relaxation of a nucleus in the level of higher energy back to the level of lower energy is slow enough (≥ seconds^{-1}) that saturation occurs readily. This causes the amplitude of the observed absorption of the electromagnetic energy in nuclear magnetic resonance spectroscopy to be sensitive to the rate of **relaxation** of the populations of individual nuclei from the state of saturation to the state of equilibrium. The faster the population relaxes, the more energy it can absorb. For this relaxation to occur, the excess energy that has been absorbed has to be dissipated. This usually occurs by the transfer of magnetization among the nuclei in the sample.

A solution of molecules contains discrete populations of atomic nuclei in which each and every nucleus is chemically identical. For example, if the naturally present deuterium is ignored, a solution of *p*-xylene

12–8

---

contains one discrete population of hydrogen nuclei, composed from the four protons attached to the phenyl ring in each molecule of *p*-xylene, and another discrete population of hydrogen nuclei, composed from the six protons attached to the methyl groups in each molecule of *p*-xylene. Each of these discrete populations of nuclei in the solution has a corresponding nuclear magnetic resonance absorption associated with it, and this absorption has certain properties.

The **chemical shift** of a nuclear magnetic resonance absorption is the position at which the absorption appears in the spectrum. The chemical shift of the absorption of a particular population of nuclei is determined by the chemical environment of the nuclei that compose the population. The electrons surrounding a given nucleus circulate in response to the applied magnetic field as a current would in a copper coil. This current decreases the local magnetic field experienced by the nucleus and shifts its resonant frequency (Equation 12–48). The chemical shift, δ_i, of a nuclear magnetic resonance absorption i is the normalized difference between the frequency ν_i at which it absorbs and the frequency ν_{std} at which the population of a standard nucleus absorbs radio-frequency energy

$$\delta_i = \frac{\nu_{std} - \nu_i}{\nu_{std}} = \frac{H_{std} - H_i}{H_{std}} \qquad (12\text{–}49)$$

Because a nuclear magnetic resonance spectrum is taken by varying the strength of the magnetic field, the chemical shift is also equivalent to the normalized difference between the magnitude of the magnetic field H_i at which

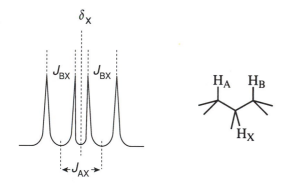

Figure 12–13: Through-bond coupling of proton spins (*J* coupling). The nuclear magnetic resonance of proton H_X would be at the chemical shift δ_X if there were, for example, deuteriums at positions A and B. The proton at position A, because it is spin-up in half the population and spin-down in the other half of the population, increases or decreases the local magnetic field experienced by H_X by a magnitude of J_{AX}. Likewise, the proton at position B increases or decreases the local magnetic field experienced by H_X by a magnitude of J_{BX}. As these shifts are additive and as the distributions of spin-up and spin-down protons at position A and position B are uncorrelated and randomly distributed over the population, four absorptions are observed.

absorption i occurs and the magnitude of the magnetic field H_{std} at which the absorption of the standard occurs. The units used for chemical shift are parts per million (ppm) relative to the absorption of the standard nucleus in a particular reference compound because the local differences in magnetic field are never greater than about 10^{-4} (100 ppm) of the applied field. Chemical shift cannot be expressed in absolute units of energy because the energy difference of a particular absorption varies with the magnitude of the applied field (Equation 12–48). There is, however, a recent, and unfortunate, habit to refer to chemical shift as if it were a frequency, ν, rather than a relative, unitless number. The magnitude of the chemical shift provides chemical information about the disposition of the electrons in the environment of the nucleus, in other words, the molecular structure in its vicinity.

The absorption of a population of identical nuclei is usually split into a series of peaks producing a symmetrical pattern around its mean resonant frequency. This splitting is due to **spin–spin coupling**. It arises from the fact that any adjacent spinning nucleus A, covalently attached to the nucleus X being observed, can act as a small magnet that induces the electrons between it and the nucleus X to circulate. This induced current alters the magnetic field in the vicinity of the nucleus X. Within the whole population of molecules, the various nuclei A assume both of their spin states almost equivalently, but each particular nucleus X in the population can be **spin-coupled** to a particular nucleus A in only one of those spin states. This divides the population of nuclei X into different groups, each group having each of its corresponding nuclei A in only one of its available distributions of spin states.

A simple example of this division of the population of nuclei X into groups, each with a different energy of absorption due to spin–spin coupling, is that of a molecule in which hydrogen X is adjacent to only two other hydrogens A and B to give the pattern A X B. Within the population of protons X, half will be next to protons A with spin +½ and half will be next to protons A with spin −½. This will split the absorption of the population of protons X into two peaks separated in frequency by the coupling constant J_{AX}. The magnitude of J_{AX} is determined by the orientation of nucleus A relative to nucleus X and the number and type of covalent bonds between them. Within each of these two groups of protons X, half of the individuals will be next to protons B with spin +½ and half will be next to protons B with spin −½. Therefore, each of the two peaks will be further split into two peaks separated in frequency by J_{BX}. The final absorption of proton X will have four peaks of equal magnitude centered on δ_X but split by the spin–spin **coupling constants** J_{AX} and J_{BX} (Figure 12–13).

Because spin–spin coupling is a function only of the intrinsic magnetic fields of the neighboring nuclei, coupling constants are not a function of the magnitude of the applied field, and their values (which are invariant differences in energy) are expressed in units of hertz. Because spin–spin coupling is relayed by the electrons in the covalent bonds connecting the nuclei, it decreases in magnitude with the number of bonds separating them. The values of J for the spin–spin coupling of hydrogen nuclei through two bonds can be as large as 15 Hz, and for coupling of hydrogen nuclei through three bonds, as large as 12 Hz, but for coupling of hydrogen nuclei through four bonds, the values are less than 1 Hz.

In addition to being determined by the number of bonds separating the two coupled protons, the value of the spin–spin coupling constant J also depends on the angles at which the two hydrogens are held with respect to each other. With two-bond coupling, the range of values is 10–15 Hz if the bond angle is close to the tetrahedral angle of 109°, but the value of the coupling constant falls to almost zero at 120°. In three-bond coupling, the value of the spin–spin coupling constant depends on the dihedral angle between the two hydrogens along the bond connecting the two neighboring atoms to which they are attached:

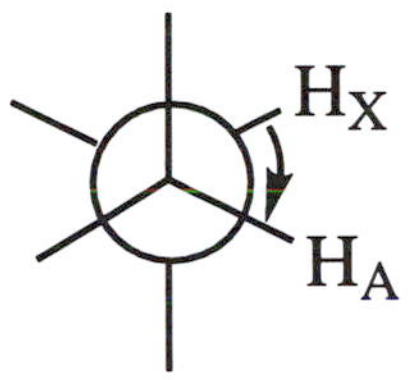

12–9

The coupling constant J_{AX} is at its maximum when the dihedral angle is 0° or 180° and at its minimum when the dihedral angle is 90° or 270°. At these latter angles, the spin–spin coupling constant can be almost zero.

The populations of nuclei, X and A, can also be coupled by a nuclear Overhauser effect. A **nuclear Overhauser effect** of the population of nuclei A on the population of nuclei X is any change in the spin state of the population of nuclei X produced by a change in the spin state of the population of nuclei A. For example, a nuclear Overhauser effect can be either an alteration in the relaxation rate between the two spin states accessible to the population of nuclei X or an alteration in the levels of occupation of the two spin states within the population of nuclei X caused by a change in the spin state of the population of nuclei A. A change in the spin state of the population of one nucleus produced by a change in the spin state of the population of another nucleus results from **dipolar interactions** between the two respective nuclei. Dipolar interactions are functions of, among other things, the distance between the two nuclei, r, and their magnitude is directly proportional to r^{-6}. Fluorescent energy transfer is another example of such a dipolar interaction (Equation 12–45). Because of this inverse dependence on the sixth

power of the distance, nuclear Overhauser effects are significant only if the two nuclei in a particular molecule are close to each other. Because a nuclear Overhauser effect is not transmitted by changes in the static, local magnetic field of nucleus X brought about by nucleus A, there is no requirement that the two nuclei be associated by covalent bonds as there was with spin–spin coupling. For example, nuclear Overhauser effects can indicate that the two nuclei involved are adjacent to each other in the tertiary structure of a protein even though they may be distant from each other in the primary structure.

As nuclear Overhauser effects are manifested only as changes in the spin state of the population of one nucleus under the influence of the spin state of the population of the other, no change occurs in the chemical shift of either nucleus involved in the nuclear Overhauser coupling. Rather, a nuclear Overhauser effect is registered as a change in the amplitude of the absorption for nucleus X. For example, if the rate of relaxation of nucleus X is increased by the change that has occurred in the spin state of the population of nuclei A, then the amplitude of the absorption for nucleus X will increase. If the occupation of the two spin states available to the population of nuclei X is caused to become more equal by the change that has occurred in the spin state of the population of nuclei A, then the amplitude of the absorption for nucleus X will decrease.

In large, relatively rigid molecules such as proteins, nuclear Overhauser effects are usually produced by **saturation transfer** brought about by **spin diffusion**.[60] Spin diffusion can be observed in an experiment analogous to fluorescence energy transfer. The population of nuclei A is irradiated at the radio frequency with which it resonates at the applied magnetic field and with sufficient amplitude to saturate its absorption by equalizing the number of nuclei in its two spin states. The stimulating radiation is then turned off. The population of nuclei A will slowly relax back to its equilibrium distribution by losing the excess energy it has gained. One of the ways the population of nuclei A may relax is by transferring energy to the population of nuclei X, if within the particular molecule nucleus X is close to nucleus A. If the absorption of the population of nuclei X is measured after a time, τ_m, sufficient for some of the excess spins in the population of nuclei A to be transferred by dipolar interaction to some of the nuclei X, the absorption of the population of nuclei X will have decreased relative to its absorption in the absence of saturation transfer because the population of nuclei X will have been moved closer to saturation.

A nuclear Overhauser effect due to saturation transfer can be presented as a difference spectrum. It is the difference between the nuclear magnetic resonance spectra before and after preirradiation of the nuclei A. The absorption of the population of nuclei A is the largest peak in the difference spectrum because it has been saturated by the preirradiation and has disappeared in

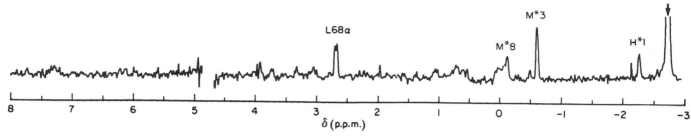

Figure 12–14: Nuclear Overhauser effect arising from saturation transfer from the protons on the δ_2 methyl group of Leucine 68 in tuna cytochrome c (Figure 7–8C) to four other protons in the molecule: H*1, the protons on the methyl group of the adjacent thioether of the heme; M*3, the protons on the δ_1 methyl group of Leucine 68; M*8, the protons on the δ_2 methyl group of Leucine 94; and L68α, the α-proton on Leucine 68.[61] Deionized tuna cytochrome c was dissolved in ^2H_2O, and the pH was adjusted to 7. Two nuclear magnetic resonance spectra were taken, one of the unirradiated solution and the other of the solution that had been irradiated for 0.5 s, immediately prior to acquisition of the spectrum, with energy of a frequency equivalent to that of the absorption ($\delta = -2.73$ ppm) from the protons on the δ_2 methyl group of Leucine 68. The energy used was sufficient to saturate completely the population of protons on the δ_2 methyl group of Leucine 68. Some of this saturation was transferred during the 0.5 s to protons in its vicinity. The fully saturated protons on Leucine 68 will absorb very little energy immediately after the saturating pulse is turned off because there is no bias to the population of spins. Therefore, in the preirradiated spectrum, that absorption is missing. Four other absorptions decrease in intensity following preirradiation but only slightly because only a small fraction of the saturation is transferred to each of them. The spectrum displayed is a difference spectrum (unirradiated–irradiated) of absorption as a function of chemical shift, δ (parts per million). Decreases in absorption following preirradiation are positive peaks in the difference spectrum. The largest peak (the off-scale peak at the arrow) is at the position of the normal absorption from the protons on the δ_2 methyl group of Leucine 68. The four other peaks arise from the small decreases in the absorption of the other four populations of protons. Adapted with permission from ref 61. Copyright 1985 Academic Press.

the latter spectrum. The smaller peaks in the difference spectrum correspond to decreases in the absorptions of populations of other nuclei coupled to nucleus A by a nuclear Overhauser effect.

An example of such a difference spectrum was taken during a spectroscopic study of tuna cytochrome c.[61] The heme in cytochrome c, as it is a large aromatic ring (Structure **2–4**), produces a substantial toroidal magnetic field when its π electrons circulate as a **ring current** in the presence of the applied magnetic field. The δ_2 methyl group of Leucine 68 (Figure 7–8C) resides adjacent to the heme in the region of this local field that is opposed to the applied field, and the chemical shift (–2.7 ppm) of the absorption of its three equivalent protons is even less than that of the reference absorption. This substantial displacement isolates this peak of absorption from the absorptions of the rest of the methyl protons in the protein. When the population of the δ_2 methyl protons on Leucine 68 is saturated by preirradiation at the frequency of its chemical shift, the absorptions of four other populations of protons are found to decrease (Figure 12–14). These were assigned to the protons, in particular hydrogens neighboring the δ_2 methyl group, of Leucine 68 in the crystallographic molecular model of tuna cytochrome c.

Another way that spin diffusion can be registered is to label the nuclear spins of a particular population of protons in a molecule with the frequency ν_2 of their chemical shift at the applied magnetic field and wait a sufficient time, τ_m, for these labeled spins to be transferred by spin diffusion to neighboring protons in a molecule. Those protons that have picked up spins labeled with the frequency ν_2 of the chemical shift can then be identified by the frequency ν_1 of their own chemical shifts. For example, in such an experiment, the proton on the amide nitrogen of Alanine 37 in the proteinase inhibitor IIA was labeled with the frequency ν_2 of its chemical shift. The protons on the amides of Phenylalanine 38 and Cysteine 36, the proton on the α-carbon of Alanine 37, and the protons on the β-carbons of Alanine 37 and Cysteine 36 also became labeled with the frequency ω_2 as time progressed and spin diffusion proceeded outward from the source (Figure 12–15).

The difficulty with a nuclear Overhauser effect mediated by spin diffusion is that it is usually not confined to nuclei immediately adjacent to the source of the diffusing spin but spreads outward from the source in rather complex pathways that cannot be delineated unless the detailed structure of the molecule is known.[60,63] The time between preirradiation and acquisition of the spectrum must be chosen by trial and error to maximize the amount of transfer to immediately adjacent nuclei while minimizing the spread to more distant locations (Figure 12–15). Because the nuclear Overhauser effect results from a dynamic inhomogeneous process, no reliable absolute estimates of particular distances between nuclei can be made. An intuition of relative distances between the protons can be gained, however, by following the changes in the intensity of the nuclear Overhauser effects as a function of the time interval τ_m. If a nuclear Overhauser

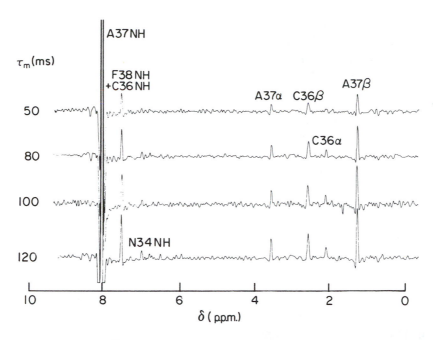

Figure 12–15: Diffusion of saturation beyond the targeted nucleus as a function of the time interval between irradiation and acquisition of a nuclear magnetic resonance spectrum.[62] A 16 mM solution of proteinase inhibitor IIA from bovine seminal plasma (n_{aa} = 57) in ^1H_2O at pH 5.3 and 45 °C was used to obtain the spectra. The sample was irradiated with a short, intense pulse of electromagnetic energy with a frequency equivalent to the resonant frequency of the chemical shift (8.02 ppm) for the proton on the amide nitrogen of Alanine 37 in the amino acid sequence of the protein. At various times, τ_m (in milliseconds), after the saturating pulse, a nuclear magnetic resonance spectrum was acquired. Although not exactly the case, the spectra presented are equivalent to difference spectra between spectra acquired before and after the preirradiation. The intense peak at a chemical shift of 8.02 ppm is from the change in absorption of the proton on the amide nitrogen of Alanine 37. The other peaks are changes in the absorption of protons in the vicinity of the proton on the amide of Alanine 37 to which saturation has diffused during the time τ_m. As the time interval is increased, more and more of the saturation diffuses and the peaks of difference absorption increase in magnitude. The other peaks in the difference spectrum have been identified as arising from the absorptions of the protons on the β-carbon of Alanine 37, the α-carbon of Cysteine 36, the β-carbon of Cysteine 36, the α-carbon of Alanine 37, and the amide protons on Phenylalanine 38 and Cysteine 36. All of these nuclear Overhauser effects, therefore, are localized to only three amino acids in the protein. Adapted with permission from ref 62. Copyright 1985 Academic Press.

effect is observed and it is one that develops early in the progress of spin diffusion, the two nuclei connected by that nuclear Overhauser effect are presumed to be close to each other (< 0.5 nm) in the folded polypeptide.[62] Propinquity, however, is not the only requirement for transfer, and if a nuclear Overhauser effect is not observed, it cannot be concluded that the two nuclei are not in atoms adjacent to each other.

The initial ^1H nuclear magnetic resonance spectra of proteins, dissolved in ^2H_2O, were one-dimensional spectra of absorption as a function of chemical shift. Even a

small protein of 100 amino acids has more than 700 protons in it, most of them unique and most of their absorptions split by spin–spin coupling. It is not surprising, therefore, that such spectra contain, by and large, several broad, unresolved absorptions, each resulting from the overlap of hundreds of individual absorptions.[64] The ranges in which these overlapping absorptions occur can be assigned to particular classes of protons: methyl protons on leucines, isoleucines, valines, alanines, and threonines (δ = 0.9–1.5 ppm); methylene protons (δ = 1.5–3.5 ppm); α-protons on each amino acid (δ = 3.5–

5.5 ppm); the protons on the peripheries of the aromatic rings of tryptophans, phenylalanines, and tyrosines (δ = 6.4–7.4 ppm); and the unexchanged amide protons of the peptide bonds and glutamines and asparagines (δ = 7.0–9.0 ppm).

There are a few distinct individual absorptions that can be identified in the one-dimensional ^{1}H nuclear magnetic resonance spectrum of a protein. The paradigm of this group is the absorption of the proton in the hydrogen on C2 of the imidazole ring of the **histidines**:

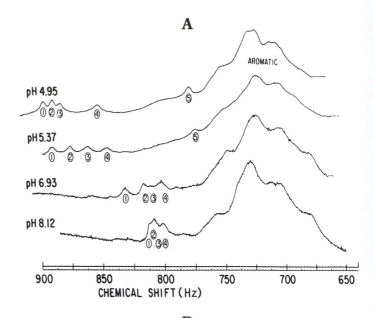

12–10

The π electrons of the aromatic ring are induced to circulate in a ring current by the applied magnetic field. The ring current creates a toroidal magnetic field opposite to the applied field in the center of the ring but reinforcing the applied field at the periphery of the ring. This additional local magnetic field at the periphery causes all protons around aromatic rings to absorb at higher frequencies (Equation 12–48). The nitrogens on either side of carbon 2 of a histidine are electronegative elements that withdraw electrons from carbon 2, decreasing the shielding provided by the σ electrons in the carbon–hydrogen bond and shifting the absorption of the hydrogen on C2 away from the absorptions of the other aromatic hydrogens on phenylalanines, tyrosines, and tryptophans. The absorption of the proton on C2 of the imidazole of a histidine is usually not divided by spin–spin coupling because the adjacent protons on the two nitrogens exchange with deuteriums from the solvent. For all of these reasons, the absorption from this proton on each histidine in a protein appears as a sharp individual peak in the nuclear magnetic resonance spectrum. One of the first nuclear magnetic spectra displaying these absorptions in a native protein was the spectrum for ribonuclease (Figure 12–16A).[65]

As the imidazole of histidine gains a proton during its acid–base reaction, the nitrogens of the conjugate acid become even more electron-withdrawing than those of the conjugate base, and the absorption of the proton on C2 assumes an even larger chemical shift. Because a specific fraction of the imidazoles in the population is the cationic conjugate acid at a particular pH and because the transfer of protons among the individuals in the population is much faster than the time resolution of nuclear magnetic resonance spectroscopy, the absorption of the population of protons on C2 of a given histidine in a protein assumes a chemical shift, δ_H^{obs}, between that of the neutral conjugate base δ_H^B and that of the cationic conjugate acid δ_H^A (Figure 12–16B). The

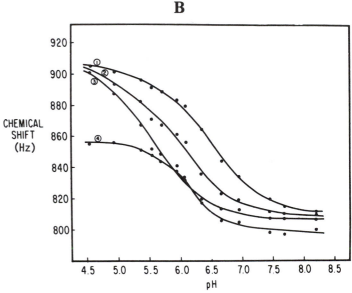

Figure 12–16: (A) Low-resolution (100 MHz) nuclear magnetic resonance spectra covering the region of the unresolved absorptions of the protons of the six tyrosines and three phenylalanines (aromatic) and the four resolved absorptions of the protons on the carbons 2 of the four histidines (numbered 1–4) of ribonuclease A (c2 Kin.4).[65] Absorption is presented as a function of chemical shift (in hertz from the absorption of an internal standard). As the excitation frequency is 100 MHz, 100 Hz of chemical shift is 1 ppm. The sample was a 12 mM solution of ribonuclease A (n_{aa} = 124) in deuterioacetate buffers in ^{2}H$_2$O at various values of p^{2}H (noted on the drawing). Peak 5 is a proton on carbon 4 of one of the histidines. (B) Chemical shift (in hertz from the absorption of the internal standard) of each of the absorptions from protons on the carbons 2 of each of the four histidines in the protein as a function of the pH of the solution. The values were taken directly from a series of spectra including the four displayed in panel A. The same numbering applies. The curves drawn are for simple acid–base titrations with values for pK_a of 6.3 (curve 1), 5.9 (curve 2), 5.6 (curve 3), and 6.1 (curve 4). Reprinted with permission from ref 65. Copyright 1967 National Academy of Sciences.

chemical shift, δ_H^{obs}, assumed is the weighted mean between the two extremes, δ_H^B and δ_H^{HA}

$$\delta_H^{obs} = f_B \delta_H^B + f_{HA} \delta_H^{HA} \qquad (12\text{--}50)$$

where f_B and f_{HA} are the fractions of conjugate base and conjugate acid, respectively, at a particular pH. Therefore, the chemical shift as a function of pH traces the titration curve of a particular histidine in a molecule of protein. The first application of this method was the measurement of the titration curves for the four histidines of ribonuclease (Figure 12–16B).[65]

In small, well-known proteins, such as ribonuclease,[66,67] myoglobin,[68] and carbonic anhydrase,[69] the individual peaks of absorption from individual protons on the carbons 2 can be assigned to specific histidines in the sequence of the protein. Historically, such assignments were performed by determining which absorptions disappeared when a specific histidine was alkylated,[66] by arguing from correlations between the locations of the histidine in the crystallographic molecular model and their values of pK_a, or by comparing the spectra of the proteins from two species in one of which a particular histidine in the sequence has been replaced by another amino acid.[68] Presently, the assignments are usually made by mutating each of the histidines consecutively to another amino acid and observing which of the absorptions disappear in each mutant.[70]

These acid dissociation constants for histidines in native proteins have been used to test computational methods[67,71] for assessing the effect of electric field and dielectric constant on the acid dissociation constant of a particular histidine in a crystallographic molecular model (Table 12–2). The specific assignment of individual absorptions to particular histidines in a protein also permits the behavior of each of those histidines to be monitored by nuclear magnetic resonance spectroscopy as experimental changes are made, such as adding small ligands that are known to bind in the vicinity of particular histidines.[72]

Several other amino acids have protons the absorptions of which can be isolated under certain circumstances. The hydrogens on the indole nitrogens of the tryptophans in a protein can be observed when spectra are gathered in ^1H_2O, to prevent exchange with deuterium, and at low pH, to suppress exchange broadening. Each of the individual absorptions from the six tryptophans in lysozyme has been assigned[73] to the particular position of each tryptophan in the sequence. That from Tryptophan 62 could be assigned because an oxindole (Equation 10–37), which shifts its absorption, can be produced exclusively at this tryptophan. The doublet absorptions from the protons on the methyl groups of threonines appear between 1.0 and 1.4 ppm in a relatively empty area of a nuclear magnetic resonance spectrum of a protein. The eight absorptions from the methyl protons of threonines in human myelin basic protein could be assigned[74] to

Table 12–2: Acid Dissociation Constants Observed for Particular Histidines in Two Proteins by Nuclear Magnetic Resonance and Calculated from Crystallographic Molecular Models

amino acid	observed pK_a	calculated pK_a
ribonuclease[a]		
Histidine 12	5.9	6.3
Histidine 48	6.2	6.5
Histidine 105	6.7	6.6
Histidine 119	6.2	6.7
myoglobin[b,c]		
Histidine 36	7.8	7.7
Histidine 48	6.5	6.5
Histidine 81	6.3	6.2
Histidine 113	5.5	5.3
Histidine 116	6.4	6.4
Histidine 119	5.4	5.6
Histidine 152	6.1	5.9

[a]Reference 67. [b]Reference 68. [c]Reference 71.

their respective locations in the sequence by comparing spectra of the protein from different species that differed in composition of threonine or spectra of large independently folded fragments of the protein.

It is also possible to isolate absorptions from particular amino acids in a protein by enriching them with a magnetic nucleus other than hydrogen. Although this can be done chemically,[75] the method used almost exclusively to enrich a protein in a particular magnetic nucleus is by expressing the protein, after the nucleic acid encoding it has been cloned, in a microorganism.[70] The microorganism used for the expression is one that is auxotrophic for the amino acid chosen for enrichment, and it is grown on an isotopic version of that amino acid. Usually the amino acid added has ^{13}carbon, a nucleus with spin of ½, incorporated at one or more locations in its structure. Because the microorganism is auxotrophic, the [^{13}C]amino acid is incorporated into the expressed protein at each location in the sequence normally occupied by that particular amino acid. In such situations the only prominent features of the ^{13}C nuclear magnetic resonance spectrum are the absorptions of those particular amino acids. For example, when glutathione transferase is expressed in a strain of *E. coli* auxotrophic for histidine and grown on a medium containing L-[2-^{13}C]histidine, the ^{13}C nuclear magnetic resonance spectrum of the purified protein is dominated by the absorptions from its four histidines.[70]

The central difficulty in ^1H nuclear magnetic resonance spectroscopy of even a small protein is that in a one-dimensional spectrum of absorption as a function of chemical shift, the peaks of absorption from the individual protons overlap and cannot be distinguished from each other, let alone assigned. It has been possible, however, to dissect the spectra of small molecules of protein (< 150 amino acids)[76,77] into their individual components by using the techniques of **two-dimensional nuclear magnetic resonance spectroscopy** that have been developed by Ernst and his colleagues.[78,79]

A two-dimensional nuclear magnetic spectrum displays connections between pairs of absorptions in the one-dimensional spectrum of the same sample. These connections are either spin–spin couplings through bonds (*J* couplings) between two populations of protons or nuclear Overhauser couplings through space between two populations of protons. The acronyms coined for these two types of spectra are COSY (correlation spectroscopy) and NOESY (nuclear Overhauser effect spectroscopy), respectively. A two-dimensional nuclear magnetic resonance spectrum (Figure 12–17A) is a presentation of absorption as a function of two values of chemical shift (in parts per million), which can be designated as δ_1 and δ_2. In its three-dimensional representation (Figure 12–17A), it can be thought of as a parallel array of sequential one-dimensional spectra of absorption as a function of chemical shift δ_2. At each fixed value of δ_1, the spectrum traced as a function of δ_2 contains a peak when $\delta_2 = \delta_1$ that is equal in relative amplitude to the relative amplitude of the one-dimensional spectrum at this particular chemical shift. Therefore, the diagonal of the two-dimensional spectrum is the one-dimensional spectrum (Figure 12–17A). At all other values of δ_2 in a particular component spectrum with a fixed value of δ_1, a peak appears when-ever the value of δ_2 becomes equal to the chemical shift of an absorption in the one-dimensional spectrum that is connected by spin–spin coupling to another absorption in the one-dimensional spectrum that has the chemical shift δ_1 of the fixed frequency. Consequently, each off-diagonal peak in a two-dimensional nuclear magnetic resonance spectrum has the same value of the chemical shift δ_2 as a peak buried in the one-dimensional spectrum and the same value of the chemical shift δ_1 as another peak buried elsewhere in the one-dimensional spectrum but connected to the first. The result is that two individual absorptions in the one-dimensional spectrum are simultaneously drawn out of it and placed in isolation of all of the other absorptions otherwise overlapping them. This provides the **resolution**. The **information** provided by an off-diagonal peak is that the two protons responsible for these two now isolated absorptions are connected, either through two or three covalent bonds that mediate spin–spin coupling (Figure 12–17B) or by being close enough in space to participate in a nuclear Overhauser effect (Figure 12–18). Two-dimensional spectra (Figure 12–17A) are most easily read, as maps are most easily read, in topographical projection (Figure 12–17B).

The connections that occur along the polypeptide backbone of a protein alternate between spin–spin couplings and nuclear Overhauser couplings. The proton in the hydrogen on each α-carbon is coupled by spin–spin coupling ($J_{\alpha N}$) to the amide proton on its amino nitrogen and connected by a nuclear Overhauser coupling ($d_{\alpha N}$NOE) to the amide proton of the next amino acid (Figure 12–19). The chemical shift of the proton on the α-carbon of an amino acid is between 3 and 5 ppm, and the chemical shift of the amide proton of the peptide bond is

Figure 12–17: (A) Two-dimensional nuclear magnetic resonance spectrum of a 20 mM solution of bovine basic pancreatic trypsin inhibitor ($n_{aa} = 58$) at pH 4.6 and 80 °C in a 9:1 mixture of ^1H_2O and ^2H_2O.[77] This is a COSY (correlation spectroscopy) spectrum, so the off-diagonal peaks appear at two chemical shifts (δ_1 and δ_2) of two individual absorptions from two particular populations of protons that are spin–spin-coupled through bonds in the molecule (Figure 12–13) and that are necessarily on the same or adjacent atoms. The diagonal running through the two-dimensional spectrum is the one-dimensional spectrum because, where $\delta_1 = \delta_2$, the absorptions of all of the protons are, by the procedure itself, coupled to themselves. Adapted with permission from ref 77. Copyright 1982 Academic Press. (B) A topographic contour map of a two-dimensional COSY spectrum[76] of a 20 mM solution of bovine basic pancreatic trypsin inhibitor in ^2H_2O at p^2H 4.6, and 24 °C. The spectrum is very similar to the spectrum in panel A except that there are fewer absorptions because some of the exchangeable protons have been replaced by deuteriums. The protons on the α-carbons of each amino acid have chemical shifts between 2.5 and 6.0 ppm. The off-diagonal peaks connecting these absorptions with either those of the protons on the respectively adjacent β-carbons ($J_{\alpha\beta}$), which have chemical shifts between 0.5 and 2.5 ppm, or those of the protons on the respectively adjacent amide nitrogens ($J_{\alpha N}$), which have chemical shifts between 6.5 and 10 ppm, are boxed. Two peaks are identified to illustrate such connectivities. The cross-peak Y21 connects the absorptions of the protons on the α-carbon and amide nitrogen of Tyrosine 21. The cross-peak N44 connects absorptions of the protons on the α-carbon and amide nitrogen of Asparagine 44. Note that, in both of these spectra, all off-diagonal peaks are symmetrically displayed. In both spectra, the amplitude of the spin–spin coupling is presented as a function of the two chemical shifts, δ_1 and δ_2 (parts per million). Adapted with permission from ref 76. Copyright 1981 Springer-Verlag.

A

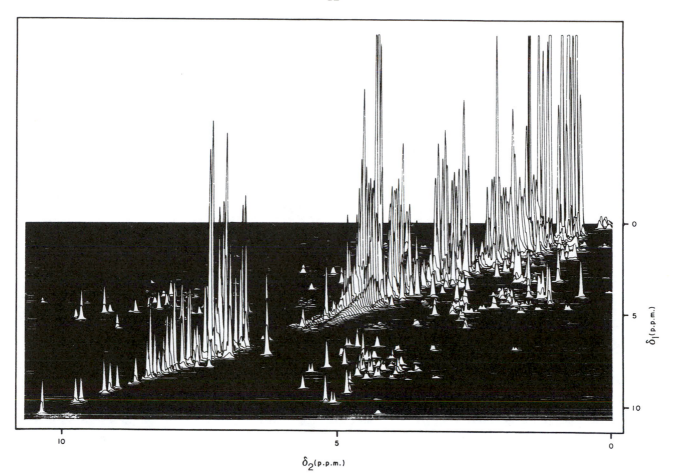

δ_1(p.p.m.)

δ_2(p.p.m.)

B

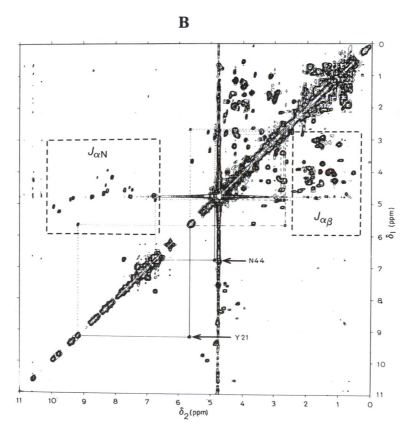

$J_{\alpha N}$

$J_{\alpha\beta}$

N44

Y21

δ_1 (ppm)

δ_2 (ppm)

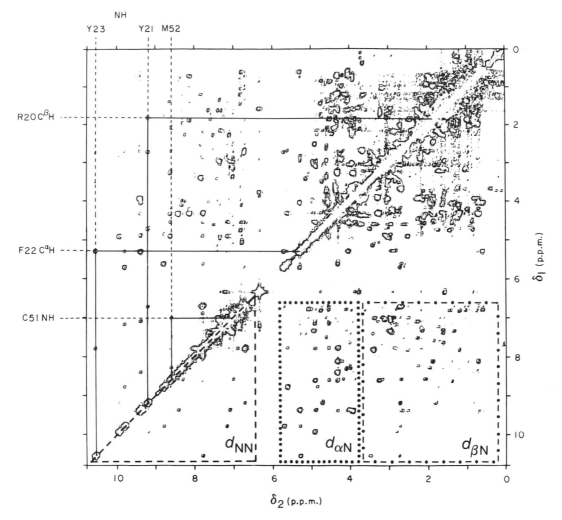

Figure 12–18: Topographic contour map of a two-dimensional NOESY nuclear magnetic resonance spectrum of a 20 mM solution of bovine basic pancreatic trypsin inhibitor in ^2H_2O at p^{2}H 4.6 and 36 °C.[77] The amplitude of the nuclear Overhauser coupling is presented as a function of the two chemical shifts, δ_1 and δ_2 (parts per million). In this spectrum, an off-diagonal peak appears whenever the absorptions of two protons with chemical shifts δ_1 and δ_2, respectively, are connected by a nuclear Overhauser effect. Because most protons on adjacently bonded atoms are close enough to be connected by a nuclear Overhauser effect as well as coupled through the bonds, most if not all of the off-diagonal peaks in the COSY spectrum in Figure 12–17B are also present here. There are, however, far more off-diagonal peaks. The additional ones connect protons on atoms adjacent in space but not connected by covalent bonds. Several regions are highlighted within boxes on the full spectrum: $d_{\alpha N}$, connections between amide protons and protons on α-carbons; d_{NN}, connections between protons on different amide nitrogens; and $d_{\beta N}$, connections between amide protons and protons on β-carbons. Three examples are called out: the nuclear Overhauser connection between the amide proton of Cysteine 51 and the amide proton of Methionine 52; the connection between the proton on the α-carbon of Phenylalanine 22 and the amide proton of Tyrosine 23; and the connection between the proton on the β-carbon of Arginine 20 and the amide proton of Tyrosine 21. Again, the full spectrum is symmetric around the diagonal, so every connection appears twice. Adapted with permission from ref 77. Copyright 1982 Academic Press.

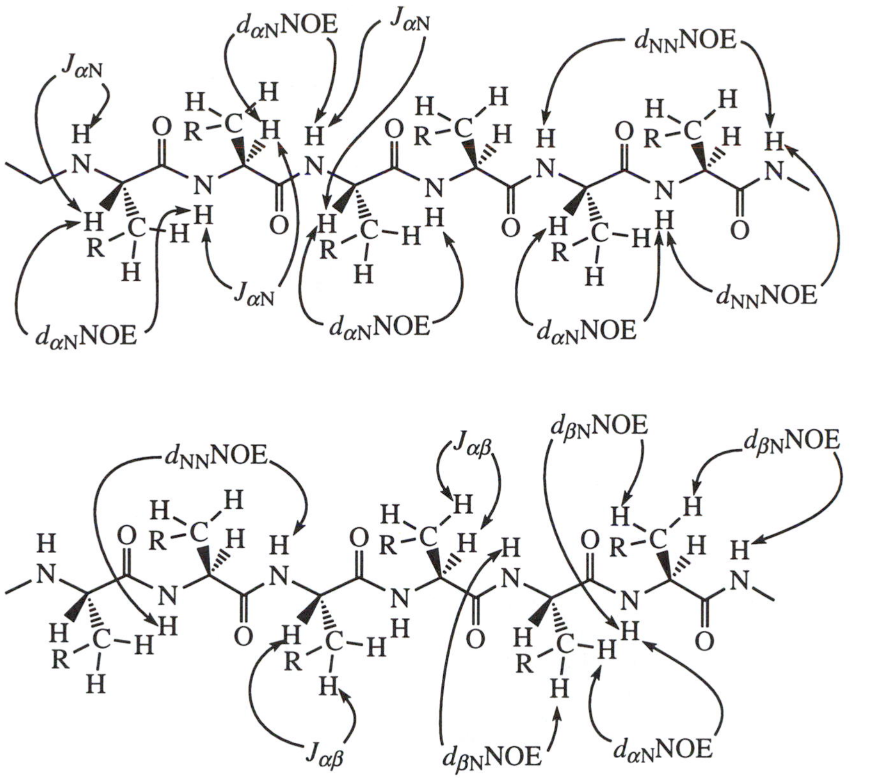

Figure 12–19: Connections highlighted by boxes in Figures 12–17B and 12–18 on two-dimensional COSY or NOESY nuclear magnetic resonance spectra, respectively. Couplings through bonds (J coupling) are identified as $J_{\alpha N}$, coupling between a proton on an α-carbon and a proton on the adjacent amide, or $J_{\alpha\beta}$, coupling between a proton on an α-carbon and a proton on an adjacent β-carbon. Couplings through space by nuclear Overhauser effects are identified as $d_{\alpha N}$NOE, connection between a proton on an α-carbon and the proton on an amide nitrogen of the next amino acid; d_{NN}NOE, connection between two protons on different amide nitrogens; or $d_{\beta N}$NOE, connection between a proton on a β-carbon and a proton on an amide nitrogen. The spectroscopist walks along the polypeptide by recognizing a sequence of alternating $J_{\alpha N}$ and $d_{\alpha N}$NOE connections.

between 7 and 10 ppm. These boundaries define the regions in a COSY spectrum (designated $J_{\alpha N}$ in Figure 12–17B) or in a NOESY spectrum (designated $d_{\alpha N}$ in Figure 12–18) that contain the off-diagonal peaks arising from the spin–spin couplings or nuclear Overhauser effects between protons on the α-carbons and protons on the amides of the peptide bonds.

When a two-dimensional COSY spectrum is gathered in H$_2$O so that the protons remain on the amide nitrogens, almost every pair of protons on the α-carbons and spin–spin-coupled amide protons ($J_{\alpha N}$) produces a peak in the appropriate region (Figure 12–20).[80] Each of these peaks originates from a different amino acid. Each amino acid produces one peak, with the exception of glycine, which usually produces two peaks because it has two diastereotopic protons on its α-carbon, or proline, which produces none because it lacks an amide proton. Each of these peaks connects the absorption of the proton on the α-carbon of an amino acid to the absorption of the proton on the amide nitrogen of the same amino acid and states that the α-carbon and the amide nitrogen producing these two absorptions are immediately adjacent atoms in the polypeptide. Each of these peaks also defines simultaneously the exact value of the chemical shift (δ_1) of the proton on the α-carbon and the exact value of the chemical shift (δ_2) of the proton on the amide nitrogen. Each peak draws two buried absorptions simultaneously out of the one-dimensional spectrum and isolates them from their neighbors. This region of the COSY spectrum catalogues the amino acids present, and one

goal of the analysis is to assign each of these peaks to the amino acid in the sequence of the protein from which it arises. No such assignment can be made in the absence of a complete amino acid sequence of the protein.

The NOESY spectrum from the same region (Figure 12–21) contains the nuclear Overhauser connections ($d_{\alpha N}$NOE) between the protons on the α-carbons of each amino acid and the respective amide protons on the preceding amino acid in the sequence (Figure 12–19). Most of the nuclear Overhauser couplings connecting the proton on each α-carbon with its own amide proton are also in this spectrum, and they can be identified by pairing them with the peaks in the COSY spectrum.

If a COSY spectrum and a NOESY spectrum from the same two regions are mirrored across an arbitrary diagonal, connections along the polypeptide backbone of the protein can be traced systematically (Figure 12–21).[77] From one peak in the COSY spectrum a peak with the same chemical shift (δ_{NH}) as its contributing amide proton is found in the NOESY spectrum by running orthogonal axes through the diagonal. A peak in the COSY spectrum with the same chemical shift ($\delta_{\alpha C}$) as the α-carbon contributing to this NOESY peak is found in the same way. The process is repeated as often as peaks can be correlated. (Follow the spirals in Figure 12–21). This pattern usually ends because a nuclear Overhauser coupling is missing.

If a nuclear Overhauser coupling ($d_{\alpha N}$NOE) between a proton on the α-carbon of one amino acid and the amide proton on the preceding amino acid is missing, there may

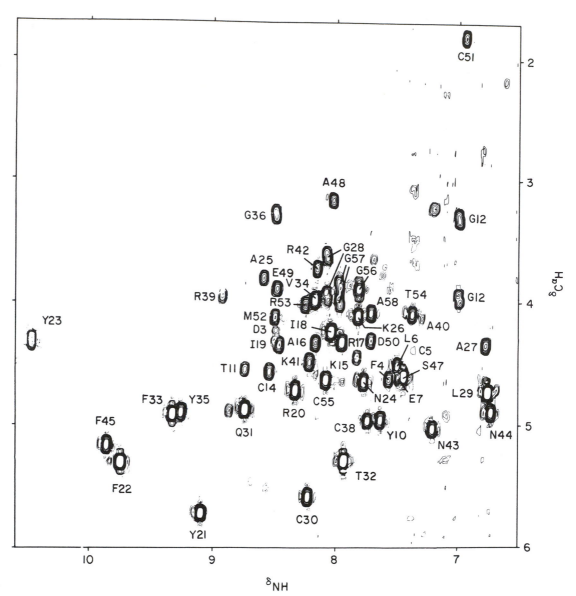

Figure 12–20: Topographic contour map of the two-dimensional COSY nuclear magnetic resonance spectrum[77] of a 20 mM solution of basic pancreatic trypsin inhibitor (n_{aa} = 58) in ^1H_2O at pH 4.6 and 68 °C. Only the off-diagonal region containing 1.7 ppm < δ_1 < 6.0 ppm and 6.6 ppm < δ_2 < 10.6 ppm (see box $J_{\alpha N}$ on full spectrum of Figure 12–17B) is presented. This contains the through-bond spin–spin couplings between the proton and on each α-carbon and the proton on the respective, immediately adjacent amide nitrogen. Every off-diagonal peak in this region of the spectrum has been assigned to one of the amino acids in the sequence of the protein. All of the amino acids are represented, with the exception of the four prolines, Glycine 37, and Arginine 1. Lysine 46, because the chemical shift of its absorption coincides with that of H_2O, is also missing from the spectrum. Each peak represents both the individual absorption from the proton on the α-carbon and the individual absorption from the proton on the amide nitrogen of that amino acid. Both must be present in the one-dimensional spectrum, and each off-diagonal peak gives their respective chemical shifts. Reprinted with permission from ref 77. Copyright 1982 Academic Press.

nevertheless be a coupling (d_{NN}NOE) between the proton on the amide nitrogen of that amino acid and the proton on the amide of the preceding one (found in the region designated d_{NN} in Figure 12–18). The sequential assignments in a short segment of polypeptide in cytochrome c were made entirely from such nuclear Over-hauser couplings (d_{NN}NOE) between consecutive amide hydrogens.[81]

Three-dimensional nuclear magnetic resonance spectra of proteins enriched in ^{13}carbon and ^{15}nitrogen can be used to trace the polypeptide chain in a more systematic fashion.[82] The advantage of this approach is that the connections between atoms can be traced through the entire polypeptide by spin–spin couplings alone. The first step in this approach is to express the cDNA for the

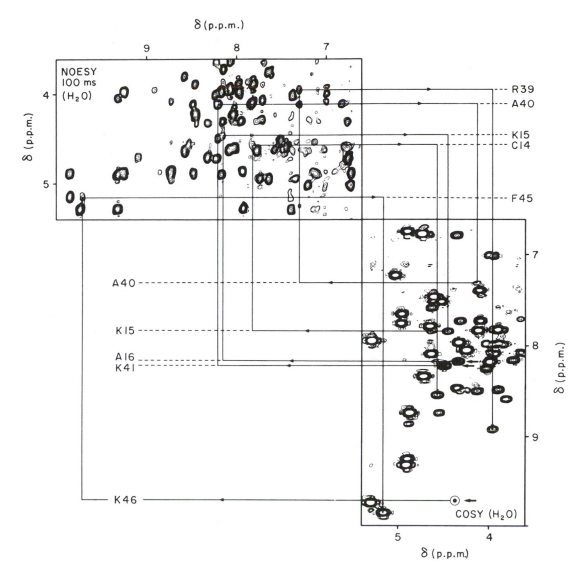

Figure 12–21: Topographic contour map of both a COSY (lower right) and a NOESY (upper left) two-dimensional nuclear magnetic resonance spectrum of the same region of the two-dimensional field from each complete spectrum: 3.5 ppm $< \delta_1 <$ 5.5 ppm and 6.5 ppm $< \delta_2 <$ 10 ppm. (See Figures 12–17B and 12–18 for full spectra and Figure 12–20 for the complete COSY spectrum of the $J_{\alpha N}$ region.)[77] The two spectra are aligned orthogonally so that equivalent peaks of absorption are symmetrically displayed across the diagonal. In this way, any two orthogonal line segments, one horizontal and one vertical, meeting at the diagonal have either identical values of δ_1 or identical values of δ_2 and connect peaks with either the same value of δ_1 or the same value of δ_2, respectively. With these orthogonal line segments, for example, the peak arising from $J_{\alpha N}$ for Arginine 41 is connected with the peak arising from the $d_{\alpha N}$NOE of Arginine 41 and Alanine 40, this latter peak is connected with the peak arising from $J_{\alpha N}$ for Alanine 40, this latter peak is connected with the peak arising from the $d_{\alpha N}$NOE of Alanine 40 and Arginine 39, and so forth. The two spectra presented are both of a 20 mM solution of bovine basic pancreatic trypsin inhibitor at pH 4.6 in ^1H_2O at 68 °C. Adapted with permission from ref 77. Copyright 1982 Academic Press.

protein of interest in *E. coli* grown on $^{15}NH_4Cl$ as a sole nitrogen source and [U-^{13}C]glucose as a sole carbon source. The protein expressed under these circumstances has all (> 95%) of its nitrogen as ^{15}N and all of its carbon as ^{13}C. In this way there are heteronuclear spin–spin couplings among the nuclei of ^1H, ^{13}C, and ^{15}N connecting each pair of atoms along the polypeptide attached to each other by one, two, or three bonds. For example, a given amide nitrogen in a polypeptide is coupled to the adjacent acyl carbon, the amide proton, and the adjacent α-carbon through the respective bonds; to the farther α-carbon, the proton on the closer α-carbon, and the farther acyl carbon through two bonds, respectively; and to the proton on the farther α-carbon, the nitrogen on the farther α-carbon, and the nitrogen beyond the farther acyl carbon through three bonds, respectively (Figure 12–19). In theory each of these spin–spin couplings should produce a peak in a respective COSY spectrum.

The third dimension is added to the spectrum by stepping systematically through the frequencies over which one type of nucleus, the reference nucleus, absorbs and recording two-dimensional spectra of the other nuclei that are coupled to absorptions of the reference nuclei within each step. An example of this strategy is a set of spectra gathered in a study of [U-^{13}C,U-^{15}N] calmodulin.[82] ^{15}Nitrogen at the amide positions was chosen as the reference nucleus, and the range of frequencies (108–132 ppm) over which this nucleus absorbs when it is in an amide was divided into individual steps. In the step containing all of the absorptions between 117.4 and 117.5 ppm, an amide proton spin–spin-coupled to one of the ^{15}nitrogens in this step was also found to be spin–spin-coupled in two-dimensional COSY spectra to a proton on an α-carbon, an α-^{13}carbon, and an acyl ^{13}carbon. These four coupled nuclei were later identified as the amide proton, the proton on the α-carbon, and the α-carbon itself of Lysine 21 and the acyl carbon of Aspartate 20 in calmodulin. In the same step between 117.4 and 117.5 ppm in the ^{15}N range, another amide proton also was coupled to protons on an α-carbon, an α-^{13}carbon, and a acyl ^{13}carbon, which were later identified as the respective nuclei in Leucine 116 and Lysine 115.

^{13}Carbon was then chosen as the reference nucleus. In the step between 58.3 and 58.4 ppm, in the range (45–68 ppm) over which the ^{13}carbons at the α positions would absorb, a proton on an α-carbon spin–spin-coupled to one of the ^{13}carbons in this step was also spin–spin-coupled in two-dimensional spectra to an acyl ^{13}carbon and an amide ^{15}nitrogen. These three nuclei were later identified as the proton on the α-carbon and the acyl carbon of Lysine 21 and the amide nitrogen of Aspartate 22.

These particular three-dimensional spectra connect absorptions of nuclei in Lysine 21 to nuclei in both Aspartate 20 and Aspartate 22 by spin–spin coupling. In this way, spin–spin couplings can be traced down the entire polypeptide. Because so many such connections can be made, each assignment can be double-checked. In addition, because the nuclear magnetic resonance is spread over three dimensions, the resolution is enhanced so that each nucleus in large proteins such as calmodulin ($n_{aa} = 148$) nevertheless can be individually observed and chemical shifts can be assigned to their absorptions. The result of such an assignment is a table of the chemical shifts of each of the hydrogens, carbons, and nitrogens along the polypeptide backbone of the protein. This table, as does the spectrum in Figure 12–20, serves as the reference to which every nucleus in the protein can be connected by spin–spin coupling.

At this point, short segments of the polypeptide or, in the case of three-dimensional heteronuclear spectra, long segments of the polypeptide have been connected but not identified. Either a spin–spin coupling ($J_{\alpha\beta}$ in Figure 12–17B) or a nuclear Overhauser coupling ($d_{\beta N}$NOE in Figure 12–18) or both are used to connect each proton on an α-carbon (Figure 12–20) with the respective proton

or protons on its β-carbon (Figure 12–19). The patterns of the protons connected to the β-carbon identify the amino acid. For example, in a COSY spectrum in which the splitting resulting from spin–spin coupling of each absorption is retained,[83] the region between $\delta_2 = 4$ and 5 ppm and $\delta_1 = 1$ and 2 ppm (Figure 12–22A)[84] contains the spin–spin connections between the protons on the α-carbons of the alanines (δ_2, quartets) and their methyl protons (δ_1, doublets) and the connections between the protons on the β-carbons of the threonines (δ_2, quartets) and their methyl protons (δ_1, doublets). Also, in a COSY spectrum the connections between the proton on the α-carbon of each valine (δ_2, 4–5.5 ppm, doublets) and the connections between each of these protons on the β-carbons and its six γ-protons (δ_2, 0–1 ppm, pair of doublets) can also be made (Figure 12–22B).[85] Patterns like these establish the identity of many of the amino acids attached to each of the protons on the α-carbons responsible for the peaks in the $J_{\alpha N}$ region of the COSY spectrum containing connections between proton α-carbons and amide protons (Figure 12–20). These assignments establish the amino acid sequence of each segment of polypeptide that has been identified in the nuclear magnetic spectrum; and, because the amino acid sequence of the protein is known, these partial sequences position each segment in the overall sequence. Eventually, a significant fraction of the peaks in the COSY and NOESY spectra have been assigned to particular protons, and to ^{13}carbons and ^{15}nitrogens in the case of heteronuclear spectra, in particular amino acids in the sequence of the protein (Figure 12–20).

To apply these **assignments** in the extraction of **information** about the structure of the folded polypeptide in solution, nuclear Overhauser couplings between pairs of protons distant from each other in the primary structure of the polypeptide are assembled. Each of these connections is established by a peak that has two chemical shifts, each identical, respectively, to the chemical shift of two other peaks known to originate from two protons whose position in the covalent structure of the protein has been assigned. For example, in the region of a NOESY spectrum (Figure 12–23) containing connections between pairs of protons on α-carbons (4–5.5 ppm), a peak that has both the chemical shift (δ_1) known to be that of a proton on a particular α-carbon in the sequence and the chemical shift (δ_2) known to be that of a proton on another α-carbon distant from the first in the sequence establishes a nuclear Overhauser connection between these two protons.[86] It can be assumed that the two protons responsible for this peak are close to each other (< 0.5 nm) in the folded polypeptide.

As with the process of tracing connections along the polypeptide chain, three-dimensional heteronuclear nuclear magnetic resonance spectra can also be used to increase the resolution of a proton NOESY spectrum.[87] As with a three-dimensional COSY spectrum, a reference nucleus is chosen, for example, ^{15}nitrogen within the

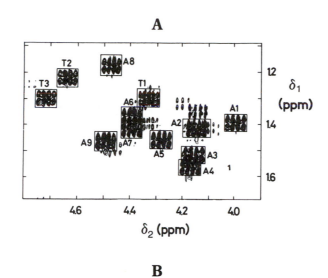

A

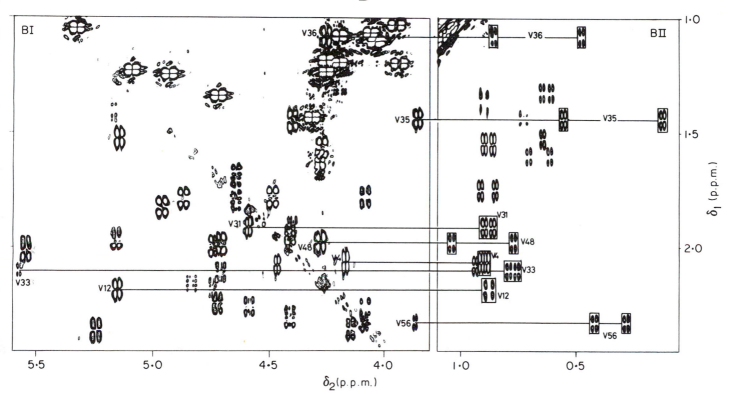

B

Figure 12–22: Patterns of spin–spin couplings and chemical shifts characteristic of the absorptions from protons on the side chains of alanine and threonine (A)[84] or valine (B).[85] (A) The absorption of the proton on the α-carbon of alanine (δ_2) is a quartet, and the absorption of its β-methyl protons (1 ppm $< \delta_1 <$ 2 ppm) is a doublet (δ_1). Any α-proton in the $J_{\alpha N}$ box (Figure 12–20) that has the same chemical shift (δ_2) as a 4 × 2 pattern (with 1 ppm $< \delta_1 <$ 2 ppm) must be the α-proton on an alanine. The absorption of the proton on the β-carbon of a threonine (δ = 4–5 ppm because of the OH) is a quartet of doublets (δ_2), and the resonance of the three protons on its methyl group (1 ppm $< \delta_1 <$ 2 ppm) is a doublet (δ_1). Any proton on an α-carbon in the $J_{\alpha N}$ box (Figure 12–20) that has the same chemical shift ($J_{\alpha\beta}$) as a proton on a β-carbon (with 4 ppm $< \delta_2 <$ 5 ppm) that in turn has the same chemical shift (δ_2) as a 4 × 2 pattern (with 1 ppm $< \delta_1 <$ 2 ppm) must be the

proton on the α-carbon of a threonine. Adapted with permission from ref 84. Copyright 1985 Springer-Verlag. (B) The resonance of the proton on the α-carbon (δ_2, left) of a valine is a doublet, the absorption of the proton on its β-carbon (δ_1) is a doublet of unresolved quartets, and the absorptions of the protons on each of its methyl groups (0 ppm $< \delta_2 <$ 1 ppm), each of which has a different chemical shift, are doublets. Any proton on an α-carbon in the $J_{\alpha N}$ box (Figure 12–20) that has the same chemical shift (δ_2, left) as a 2 × 2 pattern (with 1.5 ppm $< \delta_1 <$ 2.5 ppm) that in turn has the same chemical shift (δ_1) as a pair of 2 × 2 patterns (0 ppm $< \delta_2 <$ 1 ppm) must be the proton on the α-carbon of a valine. The spectra are of a 5 mM solution of metallothionein-2 in ^2H_2O at pH 7.0 and 24 °C (A) and a 7 mM solution of Tendamistat in ^2H_2O at pH 3.2 and 50 °C (B). Adapted with permission from ref 85. Copyright 1986 Academic Press.

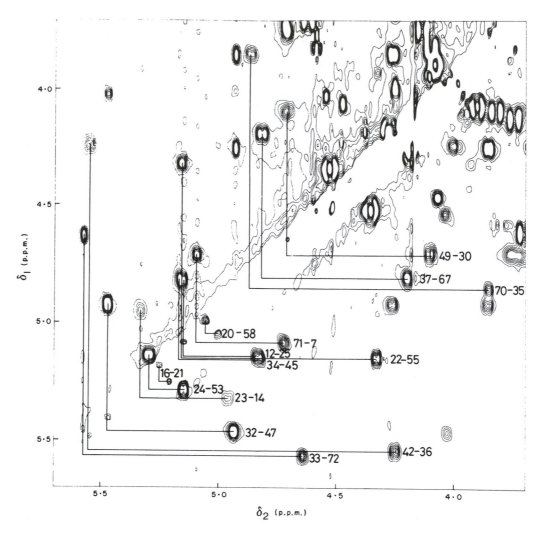

Figure 12–23: Topographic contour map[86] of a two-dimensional NOESY nuclear magnetic resonance spectrum of a 7 mM solution of Tendamistat in ^2H_2O at pH 3.2 and 70 °C. The region of the complete spectrum shown (3.7 ppm $< \delta_1 <$ 5.7 ppm; 3.7 ppm $< \delta_2 <$ 5.7 ppm) is symmetrical across the diagonal, so that each peak is mirrored. (The lines do not indicate connectivities as they do in Figure 12–21.) This is the region for the chemical shifts of the protons on the α-carbons of the individual amino acids in the protein. The precise chemical shift of the resonance of each α-carbon in the protein has been determined by the assignment of all of the peaks in the $J_{\alpha N}$ box of the COSY spectrum (Figure 12–20). The peaks displayed in the region represented here (see Figures 12–17B and 12–18 to be-come oriented) arise from nuclear Overhauser effects between two protons on two different α-carbons adjacent to each other in space. Each peak has two chemical shifts (δ_1 and δ_2) that are respectively identical to the chemical shifts of two particular protons already assigned to the α-carbons of particular amino acids in the sequence of the protein. Consequently, these peaks identify protons on α-carbons close together in the native structure of the protein. For example, the proton on the α-carbon of Threonine 32 must be less than 0.5 nm from the proton on the α-carbon of Alanine 47 in the native structure, that of Glutamine 22 must be less than 0.5 nm from that of Threonine 55, and so forth. Adapted with permission from ref 86. Copyright 1985 Academic Press.

amides of the polypeptide or ^{13}carbon at the α positions. At this stage in the process, however, the individual chemical shift of each of the reference nuclei in the polypeptide is known, so that the reference nucleus at a particular location in the amino acid sequence, for example, the amide ^{15}nitrogen of Glycine 132 (δ = 108.3 ppm)[82] in [U-^{13}C, U-^{15}N]calmodulin, can be chosen by its chemical shift. In this way only nuclear Overhauser effects involving protons coupled to the reference nucleus contribute to the particular two-dimensional NOESY spectrum. For example, if an amide ^{15}nitrogen in a particular amino acid is chosen as the reference nucleus of the third dimension, the corresponding two-dimensional NOESY spectrum will display nuclear Overhauser couplings involving only the amide proton, the proton on the α-carbon, and the proton on the β-carbon of that amino acid and the proton on the α-carbon of the preceding amino acid. In this way the various nuclear Overhauser effects are distributed over many NOESY spectra, each one corresponding to one or two particular amide ^{15}nitrogens or α-^{13}carbons.

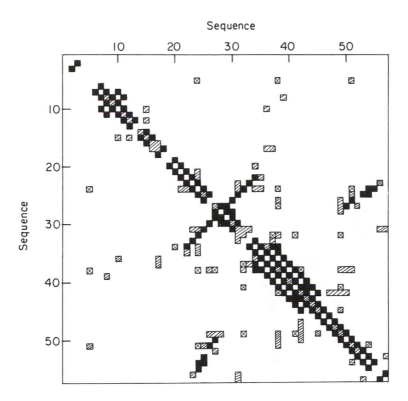

Sequence

Sequence

Sequence

Figure 12–24: Diagonal plot of the nuclear Overhauser effects observed between different amino acids in the amino acid sequence of proteinase inhibitor IIA from bovine seminal plasma.[62] Each nuclear Overhauser effect was established by the existence of a peak in a two-dimensional NOESY nuclear magnetic spectrum of the protein that had two chemical shifts (δ_1 and δ_2) identical to the respective chemical shifts of particular protons on the two different amino acids. The two axes are the numbering of the amino acids in the sequence of the protein, and a square represents a connection between the two positions in the sequence. Filled squares represent nuclear Overhauser effects between protons on α-carbons or amide protons of both amino acids; hatched squares, between a proton on an α-carbon or amide proton on one amino acid and a proton on the side chain of the other; squares with a cross, between protons on the side chains of both amino acids. Reprinted with permission from ref 62. Copyright 1985 Academic Press.

The nuclear Overhauser connections established by such assignments can be presented as a diagonal plot (Figure 12–24).[62] In such plots, patterns of connections can be distinguished. An α helix (Asparagine 34–Lysine 44 in proteinase inhibitor IIA; Figure 12–24) shows nuclear Overhauser connections[88,89] between the amide proton of one of its amino acids and the proton on the α-carbon three residues ahead of it in the sequence (Figure 4–17). Antiparallel β sheets produce diagonal features perpendicular to the main diagonal in the plot (Figures 12–24 and 12–25),[86,90] and parallel β sheets produce diagonal features parallel to the main diagonal but displaced from it.

Once the patterns of the secondary structures have been recognized, connections can be made between these individual structural elements, usually by the nuclear Overhauser couplings between pairs of protons on side chains distant from each other in the amino acid sequence (the widely scattered connections in Figures 12–24 and 12–25A). In theory, it is possible to extract from the rate of spin diffusion between two hydrogen nuclei an estimate of the distance between them. In practice, the rate of spin diffusion is too dependent on the character of the unique surroundings around each nucleus to obtain reliable estimates of distances. Consequently, the observation of a nuclear Overhauser effect between two hydrogen nuclei can be taken to mean only that they are less than about 0.5 nm apart in the native structure of the protein. In effect, the existence of a nuclear Overhauser effect allows the investigator to connect the two hydrogens in a molecular model of the covalent structure of the polypeptide with a rubber band that is elastic enough to stretch to a distance equivalent to about 0.5 nm. If enough of these rubber bands can be inserted into a molecular model of the polypeptide and the regions of the amino acid sequence identified as β structure and α helix (Figures 12–24 and 12–25) are locked into these secondary structures, the model will snap into a conformation that should resemble the native folded conformation of the polypeptide. In this way, a low-resolution molecular model of the protein can be built. For example, a molecular model of the protein Tendamistat ($n_{aa} = 74$) could be built that incorporated two antiparallel β sheets (Figure 12–25B) and in which all pairs of hydrogens displaying

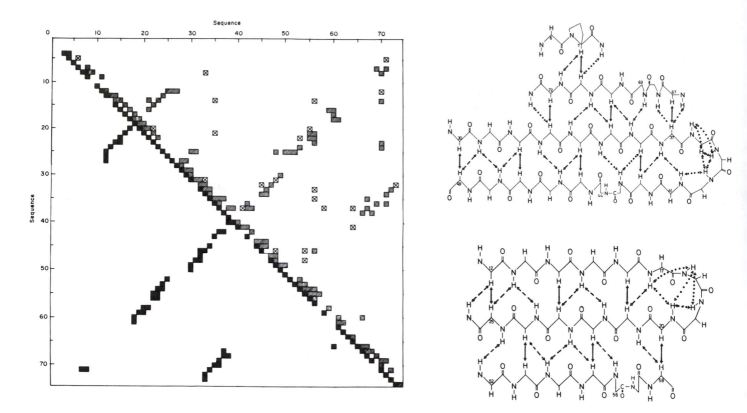

Figure 12–25: (A) Diagonal plot of the nuclear Overhauser effects observed between protons on different amino acids in the amino acid sequence of Tendamistat.[90] The symbols for the different squares are as in Figure 12–24. The three strong diagonal features perpendicular to the main diagonal, seen in the nuclear Overhauser effects between protons on the backbone of the polypeptide (lower left) and between protons on the backbone and the side chains (upper right), are indicative of β structure. Reprinted with permission from ref 90. Copyright 1986 Academic Press. (B) These orthogonal, diagonal patterns are consistent with the existence of two β sheets in the native structure of the protein.[86] Those two β sheets are drawn out schematically, and the nuclear Overhauser effects observed in the various two-dimensional NOESY nuclear magnetic resonance spectra are designated in the two structures by double arrows. Filled arrows indicate nuclear Overhauser effects between protons on carbon; dotted arrows are nuclear Overhauser effects observed only in ^1H_2O, not in ^2H_2O; and dashed arrows are nuclear Overhauser effects that are observed in ^2H_2O even though they involve one or two exchangeable protons. Protons that exchange in ^2H_2O must be more exposed in the native structure of the protein than those that do not. Reprinted with permission from ref 86. Copyright 1985 Academic Press.

nuclear Overhauser coupling were close to each other.[90] The nuclear magnetic molecular model resembled the crystallographic molecular model,[91] which was published after the nuclear magnetic molecular model had been built.

Because the nuclear magnetic molecular model is based on a relatively small number of connections of imprecise distance other than those within the characteristic elements of the secondary structure (Figures 12–24 and 12–25A), the atomic details of the structure of the protein cannot be established. It is usually possible, however, to obtain a poorly resolved description of the folding of the polypeptide of a small protein. In regions that can be assigned as either α-helical or β structure, the molecular model is most precise, but only because these hydrogen-bonded features can be constructed with a high degree of certainty. It is often the case that in regions where the polypeptide assumes random meander, it is

impossible to assign a unique configuration from the available connections provided by the nuclear magnetic resonance spectra of the protein. It is customary to display many different configurations for such undefined regions. This conveys the erroneous impression that in the actual molecule these regions are flexible rather than the fact that no reliable information exists about these regions and therefore their structure is unknown. In a small protein with a high fraction of regular secondary structure, an accurate representation of the actual molecule can often be derived. Rather than supplanting crystallography, however, the real contribution of two-dimensional nuclear magnetic resonance is to permit the observation of the behavior of individual hydrogen atoms in the structure of a protein while it is in solution. An example of such an application is the observation of **amide proton exchange**.

The amide protons in the peptide bonds of a molecule

of protein are subject to exchange with protons or deuterons of the water. Because the concentrations of water, protons, and hydroxide ions are constant at a particular pH, the exchange is a first-order process with a first-order rate constant. The exchange of an amide proton for a deuteron when a peptide is dissolved in ^2H_2O can be followed by observing the decrease in the nuclear magnetic absorptions of the protons on the amides as a function of time.[92] The first-order rate constants for this process, k_{obs}, display both **specific acid** and **specific base catalysis** (Figure 12–26)

$$k_{obs} = k_{D^+}[D^+] + k_{OD^-}[OD^-] \qquad (12\text{–}51)$$

where D stands for deuterium. At the minimum of the relationship (Figure 12–26), acid catalysis and base catalysis are of equal magnitude and the domination of one over the other inverts. The mechanisms of these catalyses are known to be

$$(^2H \text{ for } {}^1H)$$

$$(12\text{–}52)$$

and

$$(^2H \text{ for } {}^1H)$$

$$(12\text{–}53)$$

respectively.[93,92] The rate constants, k_{D^+} and k_{OD^-}, have been tabulated for the amide nitrogens on either side of specific amino acids. The rate constants for acid catalysis, k_{D^+}, vary between 1×10^2 and 5×10^3 M^{-1} min^{-1} and those for base catalysis, k_{OD^-}, vary between 1×10^9 and 1×10^{11} M^{-1} min^{-1}. Therefore, at pH 7, hydrogen exchange is dominated by the base-catalyzed reaction. From the tabulated rates of exchange in derivatives of the amino acids (Figure 12–26), the unhindered rate of exchange for any amide in a protein can be calculated.[92]

When a protein such as myoglobin is incubated in tritiated water for an extended period of time at moderate temperature (37 °C), most if not all of its amide protons reach equilibrium with the protons and tritons in the water. If the tritiated water is then replaced with untritiated water by molecular exclusion chromatography and the tritium remaining on the protein is assessed as a function of time at low temperature (0 °C), a population of tritiated amides that lose their tritiums very

slowly can be distinguished (Figure 12–27).[94] These amides exchange far more slowly than would be expected from the rates observed with small peptides in solution.

It has usually been assumed that the protons on peptide bonds that exchange very slowly with hydrogen isotopes in the solvent are protons participating in stable hydrogen bonds in the folded polypeptide.[95] That the number of slowly exchanging protons in myoglobin (Figure 12–27) is about equal to the number of peptide amide protons participating in buried hydrogen bonds in the crystallographic molecular model (approximately 120) is consistent with this assumption.[94] In the case of lysozyme, the exchange of protons for deuterons at the amides of the polypeptide could be followed directly by observing the decrease in the absorbance of the amide II vibration in the infrared spectrum relative to the absorbance of the amide I vibration. The agreement between the number of slowly exchanging amide protons (44 moles for every mole of lysozyme) and the number of buried hydrogen bonds involving the amide hydrogens of the polypeptide in the crystallographic molecular model is also quite close.[96] Therefore, proton exchange can be used to estimate the fraction of the amide protons in the polypeptide participating in buried internal hydrogen bonds in the native structure.

The difficulty with such global measurements of amide proton exchange is that the identity of the amides in the polypeptide which display slow exchange is not established. Proton exchange can be followed, however, by two-dimensional nuclear magnetic resonance spectroscopy, and in this way the rate of exchange of protons on individual amides that have been assigned to specific amino acids in the sequence of the protein can be followed. The region of the COSY spectrum that registers the spin–spin couplings ($J_{\alpha N}$) between the protons on α-carbons and the adjacent amide protons (Figure 12–20) is used. In this region of the spectrum of cytochrome c, the peaks arising from Lysine 5–Alanine 15 have been assigned. These amino acids form an external α helix in the crystallographic molecular model (Figure 7–8C). When cytochrome c is transferred to ^2H_2O while this region of the spectrum is monitored, peaks disappear as the amide protons are replaced by deuterons (Figure 12–28).

The individual rate constants of these first-order proton exchanges are usually presented relative to the rate constants that each of these amide hydrogens would have had in a small unstructured peptide of the same sequence (Figure 12–26). These slowing factors are the number of times slower the rate of exchange is in the folded protein relative to the unhindered peptide. They can be quite small. For example, in lysozyme there are amide hydrogens in the core of the protein that exchange their protons 10^6 times more slowly than the same amides would in a small peptide.[96] There is a group of protons that all have this same small slowing factor. This group of amides in the center of the protein may be exposed to the solvent only during large cooperative unfoldings of a

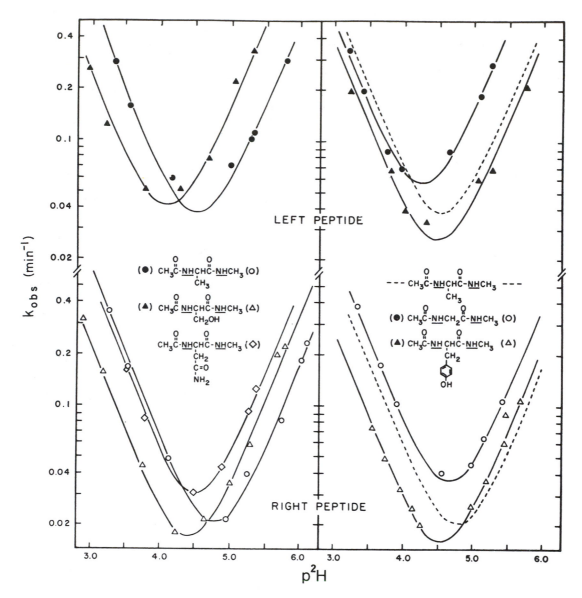

Figure 12–26: Magnitude of the observed first-order rate constants for the exchange of the amide proton on the amino-terminal side (filled symbols) or carboxy-terminal side (open symbols) in the N^{α}-acetyl-N-methyl amides of alanine ($\bullet$, $\circ$, left panel), serine ($\blacktriangle$, $\triangle$, left panel), asparagine ($\diamondsuit$, left panel), glycine ($\bullet$, $\circ$, right panel), and tyrosine ($\blacktriangle$, $\triangle$, right panel).[92] Solutions of each model compound were prepared in $^{2}H_{2}O$ at the noted $p^{2}H$ and immediately introduced into a nuclear magnetic resonance spectrometer. The two amide protons in each compound produced by spin–spin coupling the usual splitting into a doublet of the resonance of the proton or protons on the adjacent carbons. As the proton on each nitrogen exchanged for a deuteron, the respective doublet was converted into a singlet. The areas of doublet and singlet were measured as a function of time, and it was observed that the doublet was converted to the singlet in a first-order process. The rate of this process was converted to an observed first-order rate constant, k_{0bs} (minute^{-1}), and its logarithm is plotted as a function of $p^{2}H$. Adapted with permission from ref 92. Copyright 1972 American Chemical Society.

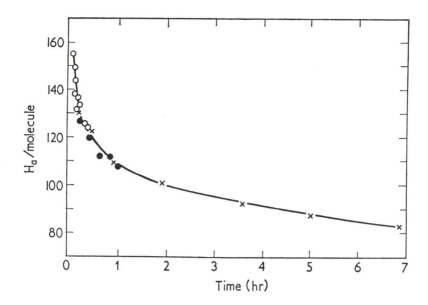

Figure 12–27: Exchange of tritium[94] from myoglobin equilibrated with ^3H_2O and then transferred to ^1H_2O at pH 5, 0 °C. Sperm whale myoglobin ($n_{aa} = 153$) was incubated at 37 °C and pH 9 with ^3H_2O until equilibrium was reached (20 h) at all of its amides. The solution was cooled to 0 °C, and the protein was rapidly transferred by molecular exclusion chromatography to ^1H_2O. The loss of ^3H from the protein (H_a/molecule) was followed as a function of time (hours). Over these time intervals of preequilibrium and separation, only protons on amides and more acidic acid–bases on the protein should acquire tritons, and only the amides should retain any tritium at the first time point. Myoglobin contains 162 amide protons, and this number agrees favorably with the 155 tritons found on the protein at the earliest time after equilibration with ^3H_2O. Reprinted with permission from ref 94. Copyright 1969 Academic Press.

considerable fraction of the protein that expose many amide hydrogens simultaneously but only for a short time before the polypeptide snaps shut again. If this is the case, regions of lysozyme may spend 10^{-6} of their life in an unfolded state.

The individual amide protons of basic pancreatic trypsin inhibitor, however, have rather heterogeneous rate constants of exchange.[98] This suggests that only local vibrational modes producing local unfoldings are responsible, at least in this protein, for performing the disconnection of a particular hydrogen bond required to expose the amide to the solvent so that its proton can exchange. Whether the motions responsible for proton exchange are local, regional, or global, the recurring observation that all or almost all of the amide protons, at least of small proteins, eventually exchange means that a molecule of protein is continuously breathing or unfolding throughout its lifetime.

It has been observed that the binding of substrate to micrococcal nuclease or the binding of heme to apomyoglobin increases the number of slowly exchanging amide hydrogens in these proteins. It may be that this is due to a damping of local or regional fluctuations caused by a tightening of the structure of the protein when the ligand binds. Such a tightening of the structure of a protein can also be reflected in a decrease in susceptibility to proteolytic digestion.

It is also possible to observe proton exchange by neutron diffraction. Crystals of protein are routinely prepared for neutron diffraction by soaking them in ^2H_2O to replace as many protons as possible with deuterons, which scatter neutrons more strongly. Molecules of either trypsin[99] or ribonuclease[100] that had been soaking within crystals in ^2H_2O for periods of 1 year retained hydrogen on 54 or 28 of their amides, respectively. All of the sites that remained unexchanged were in the interior of the folded polypeptide. These were mainly on the central strands of β sheets and at the centers of α helices. Most of the sites retaining hydrogen after 1 year were not even partially exchanged with deuterium, while those sites that had deuteriums were almost fully exchanged. The location of these sites and their occurrence in regions with very little thermal motion suggest that in a crystal, only local motions are responsible for what ex-

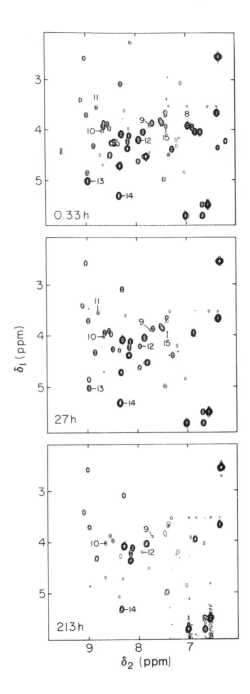

Figure 12–28: Topographic contour maps of two-dimensional COSY nuclear magnetic resonance spectra of horse heart cytochrome c in the region (2 ppm < δ_1 < 6 ppm; 6 ppm < δ_2 < 10 ppm) in which the peaks connecting the protons on the α-carbons and the adjacent amide protons ($J_{\alpha N}$, Figure 12–19) are located as a function of the time (0.33, 27, or 213 h) that the protein was in ^2H_2O at 20 °C.[97] An 8 mM solution of cytochrome c in ^1H_2O was passed rapidly over a molecular exclusion chromatographic medium equilibrated with ^2H_2O at p^{2}H 7.0 to exchange all of the ^1H_2O in the solvent for ^2H_2O. At the noted times, the p^{2}H of a sample from the solution was dropped to 5.0 to slow the exchange (Figure 12–26) and a two-dimensional COSY nuclear magnetic resonance spectrum was obtained over a period of 6 h at 20 °C. In the panels, the amplitude of the spin–spin coupling is presented as a function of the two chemical shifts, δ_1 and δ_2 (parts per million), at the three different times of equilibration. While the cytochrome c was in the ^2H_2O at p^{2}H 7.0, each peak associated with the spin–spin coupling between the α-proton and the amide proton of each amino acid decreased in magnitude in a first-order process with time as its amide proton was replaced with a deuteron. A first-order rate constant could be assigned to each peak. In earlier studies, the peaks arising from the amide protons on the amino acids from Lysine 5 to Alanine 15 in the amino acid sequence of cytochrome c, which form three turns of α helix across the surface of the native structure of the protein (Figure 7–8C), had been identified in the two-dimensional spectrum (numbers in the three spectra). The peaks corresponding to the amide protons of Lysine 5, Glycine 6, and Lysine 7 had already disappeared at the earliest time (0.33 h), but first-order rate constants for the exchange of each of the other amide protons could be measured. Adapted with permission from ref 97. Copyright 1986 American Chemical Society.

change takes place. Large regional unfolding or entire unfolding of the polypeptide cannot occur in a crystal, and these observations also suggest that the proton exchange at deep locations in a protein in free solution result from such global configurational fluctuations.

Suggested Reading

Wand, A.J., Roder, H., & Englander, S.W. (1986) Two-dimensional ^{1}H NMR Studies of Cytochrome c: Hydrogen Exchange in the N-Terminal Helix, *Biochemistry 25*, 1107–1114.

Problem 12–6: This figure is a two-dimensional nuclear magnetic resonance spectrum of the cytochrome c-551 from *Pseudomonas aeruginosa* (adapted with permission from ref 101; copyright 1990 American Chemical Society). In the spectrum, the off-diagonal peaks arise entirely from nuclear Overhauser enhancements ($\tau_m = 150$ ms).

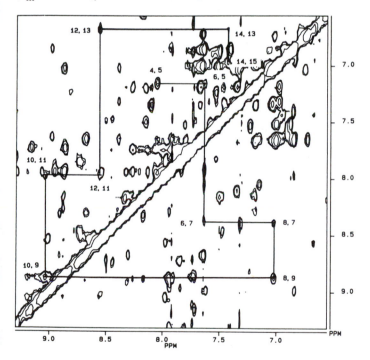

The corresponding peaks on the two sides across the diagonal result from the same nuclear Overhauser effects. You should convince yourself that the patterns are symmetric across the diagonal.

(A) What hydrogens in a protein produce off-diagonal peaks in this region of the spectrum?

(B) The peaks connected by the horizontal and vertical lines have been identified with a particular subset of these hydrogens. Each of the peaks connected by these lines is produced by a nuclear Overhauser enhancement between two hydrogens. Draw a polypeptide in the extended conformation as in Figure 12–19. On your drawing show with double arrows only the connec-

tions that give rise to those peaks that are highlighted by the horizontal and vertical lines.

(C) What do the horizontal and vertical lines indicate about these peaks? Why are the peaks labeled with pairs of numbers that increase consecutively?

(D) What other information was used to assign the numbers to the particular peaks?

References

1. Tanford, C. (1961) *Physical Chemistry of Macromolecules*, pp 275–456, John Wiley, New York .
2. Reynolds, J.A., & Tanford, C. (1976) *Proc. Natl. Acad. Sci. U.S.A. 73*, 4467–4470.
3. Perrin, F. (1936) *J. Phys. Radium 7*, 1–11.
4. de la Torre, J.G., & Bloomfield, V. (1977) *Biopolymers 16*, 1747–1763.
5. Rocco, M., Carson, M., Hantgan, R., McDonagh, J., & Hermans, J. (1983) *J. Biol. Chem. 258*, 14545–14549.
6. Gerhart, J.C., & Schachman, H.K. (1968) *Biochemistry 7*, 538–552.
7. Howlett, G.J., & Schachman, H.K. (1977) *Biochemistry 16*, 5077–5083.
7a. Smith, M.H. (1970) in *Handbook of Biochemistry and Selected Data for Molecular Biology* (Sober, H., Ed.) pp C–3 to C–12, CRC Press, Cleveland.
8. Krause, K.L., Volz, K.W., & Lipscomb, W.N. (1985) *Proc. Natl. Acad. Sci. U.S.A. 82*, 1643–1647.
9. Kirschner, M.W., & Schachman, H.K. (1971) *Biochemistry 10*, 1919–1926.
10. Simha, R. (1940) *J. Phys. Chem. 44*, 25–34.
11. vanHolde, K.E. (1985) *Physical Biochemistry*, 2nd Ed., Prentice-Hall Inc., Englewood Cliffs.
12. Rocco, M., Infusini, E., Daga, M.G., Gogioso, L., & Cuniberti, C. (1987) *EMBO J. 6*, 2343–2349.
13. Levinson, B.L., Pickover, C.A., & Richards, F.M. (1983) *J. Biol. Chem 258*, 10967–10972.
14. Pickover, C.A., McKay, D.B., Engelman, D.M., & Steitz, T.A. (1979) *J. Biol. Chem. 254*, 11323–11329.
15. Bryant, T.N., Watson, H.C., & Wendell, P.L. (1974) *Nature 247*, 14–17.
16. Babu, Y.S., Sack, J.S., Greenough, T.J., Bugg, C.E., Means, A.R., & Cook, W.J. (1985) *Nature 315*, 37–40.
17. Seaton, B.A., Head, J.F., Engelman, D.M., & Richards, F.M. (1985) *Biochemistry 24*, 6740–6743.
18. Richards, K.E., & Williams, R.C. (1972) *Biochemistry 11*, 3393–3395.
19. Williams, R.C. (1981) *J. Mol. Biol. 150*, 399–408.
20. Schramm, H.J., & Jennissen, H.P. (1985) *J. Mol. Biol. 181*, 503–516.
21. Boisset, N., Taveau, J., Pochon, F., Barray, M., Delain, E., & Lamy, J.N. (1991) *J. Struct. Biol. 106*, 31–41.
22. Valentine, R.C., Shapiro, B.M., & Stadtman, E.R. (1968) *Biochemistry 7*, 2143–2152.
23. Dahlbäck, B. (1986) *J. Biol. Chem. 261*, 9495–9501.
24. Lane, T.M., Johnson, A.E., Esmon, C.T., & Yphantis, D.A. (1984) *Biochemistry 23*, 1339–1348.
25. Suzuki, K., Dahlbäck, B., & Stenflo, J. (1982) *J. Biol. Chem. 257*, 6556–6564.

26. Roberts, J.D., & Caserio, M.C. (1977) *Basic Principles of Organic Chemistry*, 2nd Ed., W.A. Benjamin, Menlo Park, CA.

27. Susi, H. (1972) *Methods Enzymol. 26*, 455–472.

28. Chetverin, A.B., & Brazhnikov, E.V. (1985) *J. Biol. Chem. 260*, 7817–7819.

29. Lord, R.C., & Yu, N.T. (1970) *J. Mol. Biol. 51*, 203–213.

30. Duff, L.L., Appelman, E.H., Shriver, D.F., & Klotz, I.M. (1979) *Biochem. Biophys. Res. Commun. 90*, 1098–1103.

31. Carey, P.R., Schneider, H., & Bernstein, H.J. (1972) *Biochem. Biophys. Res. Commun. 47*, 588–595.

32. Chirgadze, Y.N., Brazhnikov, E.V., & Nevskaya, N.A. (1976) *J. Mol. Biol. 102*, 781–792.

33. Chirgadze, Y.N., & Brazhnikov, E.V. (1974) *Biopolymers 13*, 1701–1712.

34. Holzwarth, G., & Doty, P. (1965) *J. Am. Chem. Soc. 87*, 218–228.

35. Beychok, S. (1967) *Poly-α-Amino Acids: Protein Models for Conformational Studies* (Fasman, G.D., Ed.) pp 293–337, Marcel Dekker, New York.

36. Yu, C.A., Yong, F.C., Yu, L., & King, T.E. (1971) *Biochem. Biophys. Res. Commun. 45*, 508–513.

37. Brahms, S., & Brahms, J. (1980) *J. Mol. Biol. 138*, 149–178.

38. Moffitt, W. (1956) *Proc. Natl. Acad. Sci. U.S.A. 42*, 736–746.

39. Greenfield, N., & Fasman, G.D. (1969) *Biochemistry 8*, 4108–4116.

40. Griffin, J.H., Rosenbusch, J.P., Weber, K., & Blout, E.R. (1972) *J. Biol. Chem. 247*, 6482–6490.

41. Gresalfi, T.J., & Wallace, B.A. (1984) *J. Biol. Chem. 259*, 2622–2628.

42. Cuatracasas, P., Fuchs, S., & Anfinsen, C.B. (1968) *J. Biol. Chem. 243*, 4787–4798.

43. Kirtley, M.E., & Koshland, D.E. (1972) *Methods Enzymol. 26*, 578–601.

44. Jacobson, G.R., & Stark, G.R. (1973) *J. Biol. Chem. 248*, 8003–8014.

45. Wang, C.M., Yang, Y., Hu, C., & Schachman, H.K. (1981) *J. Biol. Chem. 256*, 7028–7034.

46. Lakowicz, J.R., & Weber, G. (1973) *Biochemistry 12*, 4171–4179.

47. Calhoun, D.B., Vanderkooi, J.M., & Englander, S.W. (1983) *Biochemistry 22*, 1533–1539.

48. Lehrer, S.S. (1971) *Biochemistry 10*, 3254–3263.

49. Eftink, M.R., & Ghiron, C.A. (1975) *Proc. Natl. Acad. Sci. U.S.A. 72*, 3290–3294.

50. Calhoun, D.B., Vanderkooi, J.M., Woodrow, G.V., & Englander, S.W. (1983) *Biochemistry 22*, 1526–1532.

51. Prasad, A.R.S., Nishimura, J.S., & Horwitz, P.M. (1983) *Biochemistry 22*, 4272–4275.

52. Stryer, L., & Haugland, R.P. (1967) *Proc. Natl. Acad. Sci. U.S.A. 58*, 719–726.

53. Latt, S.A., Cheung, H.T., & Blout, E.R. (1965) *J. Am. Chem. Soc. 87*, 995–1003.

54. Wu, C.W., & Stryer, L. (1972) *Proc. Natl. Acad. Sci. U.S.A. 69*, 1104–1108.

55. Pober, J.S., Iwanij, V., Reich, E., & Stryer, L. (1978) *Biochemistry 17*, 2163–2169.

56. Berman, H.A., Yguerabide, J., & Taylor, P. (1980) *Biochemistry 19*, 2226–2235.

57. Hahn, L.H.E., & Hammes, G.G. (1978) *Biochemistry 17*, 2423–2429.

58. Snyder, B., & Hammes, G. (1984) *Biochemistry 23*, 5787–5795.

59. Craig, D.W., & Hammes, G.G. (1980) *Biochemistry 19*, 330–334.

60. Gordon, S.L. & Wüthrich, K. (1978) *J. Am. Chem. Soc. 100*, 7095–7096.

61. Williams, G., Moore, G.R., Porteus, R., Robinson, M.N., Soffe, N., & Williams, R.J.P. (1985) *J. Mol. Biol. 183*, 409–428.

62. Williamson, M.P., Havel, T., & Wüthrich, K. (1985) *J. Mol. Biol. 182*, 295–315.

63. Dubs, A., Wagner, G., & Wüthrich, K. (1979) *Biochim. Biophys. Acta 577*, 177–194.

64. Mandel, M. (1965) *J. Biol. Chem. 240*, 1586–1592.

65. Meadows, D.H., Markley, J.L., Cohen, J.S., & Jardetsky, O. (1967) *Proc. Natl. Acad. Sci. U.S.A. 58*, 1307–1313.

66. Meadows, D.H., Jardetsky, O., Epand, R.M., Ruterjans, H.H., & Scheraga, H.A. (1968) *Proc. Natl. Acad. Sci. U.S.A. 60*, 766–772.

67. Matthews, J.B., & Richards, F.M. (1982) *Biochemistry 21*, 4989–4999.

68. Botelho, L.H., & Gurd, F.R.N. (1978) *Biochemistry 17*, 5188–5196.

69. Pesando, J.M. (1975) *Biochemistry 14*, 675–688.

70. Zhang, P., Graminski, G.F., & Armstrong, R.N. (1991) *J. Biol. Chem. 266*, 19475–19479.

71. Botelho, L.H., Friend, S.H., Matthew, J.B., Lehman, L.D., Hanania, G.I.H., & Gurd, F.R.N. (1978) *Biochemistry 17*, 5197–5205.

72. Meadows, D.H., & Jardetsky, O. (1968) *Proc. Natl. Acad. Sci. U.S.A. 61*, 406–413.

73. Glickson, J.D., Phillips, W.D., & Rupley, J.A. (1971) *J. Am. Chem. Soc. 93*, 4031–4038.

74. Mendz, G.L., Moore, W.J., & Martenson, R.G. (1983) *Biochim. Biophys. Acta 742*, 215–223.

75. Jones, W.C., Rothgeb, T.M., & Gurd, F.R.N. (1976) *J. Biol. Chem. 251*, 7452–7460.

76. Wagner, G., Kumar, A., & Wüthrich, K. (1981) *Eur. J. Biochem. 114*, 375–384.

77. Wagner, G., & Wüthrich, K. (1982) *J. Mol. Biol. 155*, 347–366.

78. Aue, W.P., Bartholdi, E., & Ernst, R.R. (1976) *J. Chem. Phys. 64*, 2229–2246.

79. Jeener, J., Meier, B.H., Bachmann, P., & Ernst, R.R. (1979) *J. Chem. Phys. 71*, 4546–4553.

80. Strop, P., Wider, G., & Wüthrich, K. (1983) *J. Mol. Biol. 166*, 641–667.

81. Wand, A.J., & Englander, S.W. (1986) *Biochemistry 25*, 1100–1106.

82. Ikura, M., Kay, L.E., & Bax, A. (1990) *Biochemistry 29*, 4659–4667.

83. Piantini, O.W., Sorensen, O.W., & Ernst, R.R. (1982) *J. Am. Chem. Soc. 104*, 6800–6801.

84. Neuhaus, D., Wagner, G., Vasak, M., Kagi, J.H.R., & Wüthrich, K. (1985) *Eur. J. Biochem. 151*, 257–273.

85. Kline, A.D., & Wüthrich, K. (1986) *J. Mol. Biol. 192*, 869–890.

86. Kline, A.D., & Wüthrich, K. (1985) *J. Mol. Biol. 183*, 503–507.

87. Ikura, M., Spera, S., Barbato, G., Kay, L.E., Krinks, M., & Bax, A. (1991) *Biochemistry 30*, 9216–9228.

88. Wagner, G., Neuhaus, D., Worgotter, E., Vasak, M., Kagi, J.H.R., & Wüthrich, K. (1986) *J. Mol. Biol. 187*, 125–129.

89. Williamson, M.P., Havel, T.F., & Wüthrich, K. (1985) *J. Mol. Biol. 182*, 295–315.

90. Kline, A.D., Braun, W., & Wüthrich, K. (1986) *J. Mol. Biol. 189*, 377–382.

91. Pflugrath, J.W., Wiegand, G., & Huber, R. (1986) *J. Mol. Biol. 189*, 383–386.

92. Molday, R.S., Englander, S.W., & Kallen, R.G. (1972) *Biochemistry 11*, 150–158.

93. Perrin, C.L., Lollo, C.P., & Johnston, E.R. (1984) *J. Am. Chem. Soc. 106*, 2749–2753.

94. Englander, S.W., & Staley, R. (1969) *J. Mol. Biol. 45*, 277–295.

95. Hvidt, A., & Linderstrom-Lang, K. (1954) *Biochim. Biophys. Acta 14*, 574–575.

96. Nakanishi, M., Tsuboi, M., & Ikegami, A. (1972) *J. Mol. Biol. 70*, 351–361.

97. Wand, A.J., Roder, H., & Englander, S.W. (1986) *Biochemistry 25*, 1107–1114.

98. Wagner, G., Stassinopoulou, C.I., & Wüthrich, K. (1984) *Eur. J. Biochem. 145*, 431–436.

99. Kossiakoff, A.A. (1982) *Nature 296*, 713–721.

100. Wlodawer, A., & Sjoln, L. (1982) *Proc. Natl. Acad. Sci. U.S.A. 79*, 1418–1422.

101. Chau, M., Cai, M., & Timkovich, R. (1990) *Biochemistry 29*, 5076–5087.

CHAPTER 13

Folding and Assembly

Each polypeptide begins its existence by emerging, amino terminus foremost, from a ribosome. Its initial amino acid sequence is the complete translation of the sequence in which the codons are arranged between the start codon and the stop codon on the messenger RNA. At some point in its early history, the polypeptide folds to assume its native structure. The **native structure** of a polypeptide is the one conformation or the small set of conformations in which it will spend the remainder of its lifetime and in which it is capable of performing its role within or on behalf of the living organism. The native structure is the conformation or conformations of the polypeptide represented by the crystallographic molecular model. It is also referred to as the **folded state**. On the basis of the easily verified existence and identity of the native structure, a denatured state of a polypeptide can be defined as its antonym. A **denatured state** of a polypeptide is any conformation or set of conformations of the polypeptide that is not or does not contain the conformation or conformations of the native structure. As it emerges from the ribosome, the nascent polypeptide is in a denatured state.

The native structure of the polypeptide can undergo posttranslational modification, it can combine with several other identically folded polypeptides of the same sequence or several other folded polypeptides of a different sequence and structure, or it can enter a helical polymeric protein as one of the protomers. The product of these steps is the **mature native structure** of the protein encountered in the living tissue. The order in which these later processes occurs cannot be predicted, but all of them usually follow the folding of the unadorned polypeptide because it is usually the folded polypeptide that contains the information controlling them.

Accordingly, the steps in the maturation of a protein can be divided into folding, posttranslational modification, and assembly. **Folding** is any process by which the polypeptide initially in a denatured state, for example, its conformation as it emerges from the ribosome, assumes the folded native structure. **Assembly** is the process by which individual folded polypeptides associate to form oligomeric or polymeric proteins.

Thermodynamics of Folding

A polypeptide is a polymer of amino acids (Structure 2–8). It is known from studies of polymers in general that their conformational behavior depends critically on the solvent in which they are dissolved.[1] If the monomers are miscible with the solvent, the polymer is free to expand and expose all of its monomers to the solvent without penalty. Such a solvent is referred to as a good solvent. When a polypeptide is dissolved in a good solvent, rotation about each bond between amide nitrogen and α-carbon and between α-carbon and acyl carbon is permitted, within the confines of the clashes represented in the Ramachandran plot (Figure 6–3) and within the requirement that no two atoms anywhere in the polypeptide can occupy the same space at the same time. As with any other unconfined organic molecule in solution, the conformation of a polypeptide in a good solvent is continuously changing as these rotations occur at random. Such a protean polymer is referred to as a **random coil**. This term incorporates unavoidably the uncontrolled and continuous motion of this process. A random coil is a special type of denatured state. The **unfolded state** of a polypeptide is a state in which the polypeptide is either a random coil or so slightly constrained in its motion as to be indistinguishable in its behavior from a random coil.

Unfortunately, there are very few good solvents, if there are any, for naturally occurring polypeptides. This is due to the fact that naturally occurring polypeptides are created to fold. To fold, they must be composed of hydrophobic and hydrophilic amino acids, placed in a particular sequence, but in a sequence almost impossible to distinguish from a random array. There is almost no solvent in which this mixture of amino acids is miscible. In particular, water, although it is a good solvent for the hydrophilic amino acids, is a bad solvent for the hydrophobic amino acids.

A **bad solvent** is a solvent in which the monomers of a polymer are only sparingly soluble. In a bad solvent, a polymer contracts to decrease the exposure of its monomers. The hydrophobic effect is a force that seeks to minimize the exposure of a hydrophobic solute to water. Consequently, at neutral pH and an ionic strength of 0.2 M, which are the conditions under which most proteins are found, water is a bad solvent for a polypeptide. If it were not, proteins would not fold. This fundamental contradiction means that, under most conditions, if a polypeptide is not in its native conformation, it will also not be a random coil.

This presents a subtle experimental problem. If one is studying the folding of a polypeptide, both the native structure and the denatured state must be well-defined. The only denatured state of a polypeptide that can be defined with sufficient accuracy is a random coil. Consequently, the folding of polypeptides is most informatively studied if the process that is monitored is the isomerization between the random coil and the native structure, even though this may not be what occurs in a cell.

A polypeptide will fold only if the free energy of the native structure is less than the free energy of all accessible denatured states. Because of this requirement, for example, a nascent polypeptide cannot fold until it is long enough for the native structure to contain a large enough collection of noncovalent interactions to overcome the unfavorable loss of entropy that must always accompany folding. It is also the case that a polypeptide which has undergone extensive covalent posttranslational modification after it originally folded may not be able to fold again after it has been returned to a denatured state. The usually cited example of such a situation is the protein proinsulin. Proinsulin can be unfolded and its cystines reduced to cysteine. The protein will then refold spontaneously to its native structure, and the proper cystines will re-form under oxidizing conditions.[2] Insulin, however, which is a posttranslationally modified fragment of proinsulin, missing 25 amino acids from the middle of the polypeptide, does not refold spontaneously after it has been unfolded and reduced and can be refolded only with subterfuge and in low yield. The only fact that seems to be inescapable is that, at some point in its lifetime, a polypeptide has a covalent structure capable of folding to produce either the mature native structure directly or an initial native structure, which is modified subsequently but retains its basic folded conformation.

A polypeptide that has not been modified so extensively as to render the mature native structure higher in free energy than the random coil or higher in free energy than any denatured state will, under the the proper circumstances, spontaneously refold to its mature native structure after it has been purposely turned into a random coil. Most of our understanding of the folding of polypeptides has been derived from the study of such reactions. Their existence states that all of the information necessary to achieve the proper native structure resides in the amino acid sequence of the polypeptide.

The chemical reaction that encompasses the process of protein folding can be presented as the equilibrium

$$U \underset{k_U}{\overset{k_F}{\rightleftharpoons}} F \qquad (13\text{--}1)$$

where F is the polypeptide folded in its native structure and U is the unfolded state. The rate constants k_F and k_U are composite rate constants including any kinetic steps between the two extremes. This reaction can be referred to as folding–unfolding to avoid any bias, but the thermodynamic properties such as the equilibrium constant and the changes in free energy, enthalpy, entropy, and heat capacity should be referred to as those for folding to indicate the direction to which they refer. This is the chemical reaction usually referred to when the terms *protein folding* or *denaturation* are used.

Because polypeptides folded in their native structure are by design reasonably stable at physiological tem-

peratures and ranges of pH, the concentration of the unfolded state under normal circumstances is immeasurably low, and neither the equilibrium constant nor the thermodynamic changes associated with folding can be measured in such a situation. The only solution to this problem is to shift the equilibrium by introducing unnatural perturbations. These perturbations are lowering the pH, raising the temperature, adding high concentrations of urea or guanidinium chloride, or some combination of these treatments. Under each of these circumstances, the intention is to decrease the magnitude of the equilibrium constant for folding sufficiently so that it can be measured and the reaction can be studied.

In any meaningful measurement of the properties of folding–unfolding, the conditions must be such that the reaction remains reversible. This requirement cannot be avoided. When a concentrated solution of ovalbumin and lysozyme, otherwise known as the white of an egg, is heated, the polypeptides unfold but then rapidly coagulate among themselves to form a white, intractable, gelatinous precipitate. In the unfolded state, otherwise buried hydrophobic amino acids on these polypeptides all become simultaneously exposed to the solution and intermolecular polymerization, rather than individual intramolecular isomerizations, takes place. There is little doubt that a significant portion if not the majority of the changes in enthalpy and entropy proceeding during such a process are those of the coagulation, which is of only marginal interest.

In all studies of protein folding, the first result presented should demonstrate the complete reversibility of the reaction. Few folding–unfoldings perturbed solely by increases in temperature are reversible at neutral pH.[3] At low pH (pH 2–3), however, most thermally perturbed folding–unfoldings, even though they would proceed with coagulation at higher pH, often become reversible.[4,5] Presumably, this is due to the fact that coagulation is prevented by charge repulsion among the denatured polypeptides. It is usually observed that a polypeptide denatured thermally and reversibly at low pH will coagulate visibly and irreversibly as the pH is increased, and often the onset of this coagulation is found to occur abruptly within a very narrow range of pH.[6] Folding–unfoldings perturbed by the addition of urea or guanidinium chloride usually remain reversible.

When proteins whose cystines have been reduced to cysteines are dissolved in solutions of **guanidinium chloride (13–1)** or **urea (13–2)**

13–1

13–2

at concentrations of 6 or 8 M, respectively, they become completely unfolded, and their constituent polypeptides become random coils. There are several ways to demonstrate this fact.[7] The molecular weights of the proteins determined from the colligative properties of these solutions are those of the constitutive polypeptides rather than the oligomers. The intrinsic viscosities of proteins dissolved in these solutions range from 15 to 100 $cm^3 g^{-1}$ even though the intrinsic viscosities of the native proteins are between 3 and 5 $cm^3 g^{-1}$. Furthermore, within a set of proteins the intrinsic viscosities are correlated to the length of the constituent polypeptides by a relationship that agrees with theoretical expectation for the behavior of random coils. The optical rotatory dispersion spectra and circular dichroic spectra of proteins in such solutions are those theoretically expected from a polypeptide lacking any regular secondary structure, even if the spectra of the active proteins indicate that they are predominantly α helix and β structure. The acid–base titration curves of proteins dissolved in these solutions lose the normally observed shifts in intrinsic pK_a brought about by the electrostatic features of the native structure and become simple sums of the constituent intrinsic acid–base titrations of the constituent amino acids. All of the tyrosines in the protein display ultraviolet, spectrophotometric acid–base titrations with expected intrinsic values of pK_a. The rates of amide hydrogen exchange become very rapid when proteins are dissolved in these solutions, and no evidence for a class of slowly exchanging amide hydrogens is usually found. The ultraviolet spectra between 270 and 300 nm of proteins dissolved in these solutions are simple summations of the spectra of phenylalanine, tyrosine, and tryptophan and display none of the spectral shifts of the native structures.[8]

A solution of guanidinium chloride at 6 M seems to produce the most complete unfolding to the random coil.[7] The values of the various physical parameters for the same protein dissolved in 8 M urea rather than 6 M guanidinium chloride are slightly but significantly displaced in the direction of a folded structure. Proteins dissolved in solutions of low pH that have been heated until no further change in optical rotation occurs will still display further changes when guanidinium chloride is added, and this suggests that reversible thermal denaturation does not produce an unfolded state.

Solutions of either guanidinium chloride or urea promote the unfolding of a polypeptide by increasing the stability of the random coil.[9-12] This is due to favorable changes in the solvation both of the side chains of the amino acids and of the polypeptide backbone brought about by these solutes. From measurements of the solubility of various amino acids, as well as diglycine and triglycine, in solutions of either urea[9] or guanidinium chloride,[10] the standard free energies of transfer of the side chains of the amino acids, as well as the peptide bond, between water and solutions of urea or guanidinium chloride have been estimated (Table 13–1). These standard free energies of transfer were derived from the differences between the solubilities of each of the amino acids and the solubilities of glycine in water, 7 M urea, or 5 M guanidinium chloride. To arrive at these estimates, it was assumed that the differences between the standard

Table 13–1: Estimates of the Standard Free Energy of Transfer of Various Side Chains of the Amino Acids between Water and Solutions of Urea or Guanidinium Chloride

amino acid side chain	$\Delta G^\circ_{H_2O \to 7\,M\,urea}$ (kJ mol^{-1})			$\Delta G^\circ_{H_2O \to 5\,M\,guanCl}$ (kJ mol^{-1})		
	amino acid[a]	alkane model[b]	N-acetyl ethyl ester[c]	amino acid[a]	alkane model[b]	N-acetyl ethyl ester[c]
leucine	−1.1	−1.0	+0.1	−1.8	−1.3	−0.1
phenylalanine	−2.2	−1.9	−0.7	−2.8	−2.5	−1.3
tryptophan	−3.2	−3.2		−4.6	−4.0	
methionine	−1.5			−2.0		
threonine	−0.4			−0.5		
tyrosine	−2.8			−2.9		
histidine	−1.0			−1.7		
asparagine	−1.6			−2.4		
glutamine	−0.8			−1.4		
peptide bond	−0.8		−0.5	−1.3		−0.8

[a]Difference between solubility of glycine and the appropriate amino acid.[9,10] [b]Free energy of transfer of isobutane, toluene, or skatole.[12] [c]Difference in free energy of transfer of ethyl N-acetylglycinate and N-acetyl ethyl ester of the amino acid.[11]

free energies of solution of glycine and each of the other amino acids in the various solutions of denaturants would give the standard free energy of transfer of each of the side chains.

The values obtained for leucine, phenylalanine, and tryptophan agreed quite closely with direct measurements of the standard free energies of transfer for isobutane, toluene, and skatole as models of the respective side chains (Table 13–1).[12] The N-acetyl ethyl esters of leucine and phenylalanine, however, were found to have standard free energies of transfer, relative to ethyl N-acetylglycinate, that were much less negative (Table 13–1).[11] It may be premature to attach any significance to the absolute numerical values of these various estimates.

It has been uniformly observed, however, that the free energies of transfer of both hydrophobic solutes and neutral hydrophilic solutes such as peptides between water and either 7 M urea or 5 M guanidinium chloride have negative values. As a result, the functional groups, both hydrophobic and hydrophilic, that are exposed to the solution only in the random coil are more favorably solvated in solutions of urea or guanidinium chloride than they would be in water. This stabilization of the random coil increases monotonically with the concentration of denaturant.[9–11] At some concentration of the denaturant, which differs for each protein, the unfolded polypeptide becomes more stable than the folded polypeptide because, while the unfolded polypeptide exposes far more of its side chains and peptide bonds to the solvent than does the folded polypeptide, it is also more disordered.

The effects of guanidinium chloride on an unfolded polypeptide can also be described in terms of **preferential solvation**.[13] The parameter that assesses preferential solvation is $(\partial m_s / \partial m_p)_{T, \mu_{H_2O}, \mu_s}$ where m_s is the moles of a particular solute, in this case guanidinium chloride, m_p is the moles of protein added to the solution, and μ_{H_2O} and μ_s indicate that the chemical potentials of both the water and the solute remain constant as the moles of protein are varied. A positive value of $(\partial m_s / \partial m_p)_{T, \mu_{H_2O}, \mu_s}$ states that the moles of solute in the solution must be increased whenever the moles of protein are increased to maintain constant chemical potential. Therefore, the solute prefers to interact with the protein rather than with the solvent, and this interaction renders the protein more soluble. This situation is referred to as **salting-in** of the protein by the solute. When the concentration of guanidinium chloride is 6 M, $(\partial m_s / \partial m_p)_{T, \mu_{H_2O}, \mu_s}$ is a positive number whose magnitude is directly proportional to the length of the random coil with a slope of 1 mol of guanidinium chloride for every 2 mol of amino acid.[14] This preferential solvation is generally explained as preferential binding of the denaturant to the random coil,[7] but no such specific molecular explanation is required.

Negative preferential solvations are also displayed by some solutes.[15,16] A negative value of $(\partial m_s / \partial m_p)_{T, \mu_{H_2O}, \mu_s}$

states that molecules of solute must be removed from the solution as protein is added to maintain a constant chemical potential. Such solutes are the solutes excluded from the layers of hydration around a protein. Examples are sulfates, polyhydroxylic alcohols, and sugars. These are said to exhibit a **salting-out** property, and they render a protein less soluble. Because the random coil displays a higher surface area than a folded polypeptide, solutes that salt out proteins or cause them to precipitate from the solution simultaneously destabilize the random coil relative to the native structure and consequently stabilize the native structure. From these considerations it can be seen that the effects of urea and guanidinium chloride are only one extreme of the spectrum of preferential solvation.

The preferential solvation of hydrophobic side chains by urea and guanidinium chloride makes a major contribution to the stabilization of the random coil (Table 13–1). The estimated free energies of transfer for hydrophobic side chains between water and solutions of urea and guanidinium chloride are correlated with their surface areas.[17] This suggests that urea and guanidinium chloride cause the solution to become more like a usual organic solvent in its properties (Figure 5–22). Perhaps this effect results from the stable introduction of the nonpolar π clouds of the denaturants (Figure 2–16) into the solution. N-Alkyl-, N,N'-dialkyl-, and N,N,N' N' -tetraalkylureas are even more effective at increasing the solubility of naphthalene, indole, and ethyl N-acetyltryptophanate in water than is urea itself.[18,19] This also suggests that it is nonpolar noncovalent interactions between urea and the hydrophobic amino acids that explain its ability to solvate them preferentially.[18] The fact that methylurea, dimethylurea, and tetramethylurea are each in turn increasingly better denaturants of proteins than urea itself[19] and the fact that some alkylureas are better denaturants than even guanidinium chloride[20] suggest that the preferential solvation by urea of the hydrophobic functionalities revealed during the formation of the random coil, rather than its ability as a donor or acceptor of hydrogen bonds, is the major feature of its ability to promote unfolding.

The decrease in the magnitude of the equilibrium constant for Equation 13–1 brought about by heat is presumably due to increases in thermal motion that always shift reactions in favor of more disordered states. The decrease in the magnitude of the equilibrium constant brought about by lowering the pH is due to protonations of amino acids or acyl oxygens of the peptide bonds that cannot be protonated in the native structure owing to their burial but can be protonated in the denatured state.[21] When either the temperature is increased or the pH is lowered, hydrophobic clusters (Figure 6–21) in otherwise unfolded polypeptides should remain associated. This would account for the incomplete unfolding observed in these situations.

When a physical property of a protein such as its

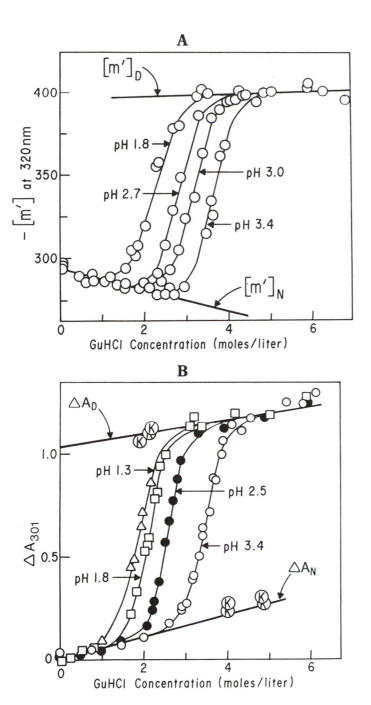

Figure 13–1: Effect of the molar concentration of guanidinium chloride, GuHCl (moles/liter), on the optical rotation at 320 nm (A) or the absorbance at 301 nm (B) of a solution of hen egg lysozyme[23] at various values of pH at 25 °C. The optical rotation is presented as the mean residue rotation, $[m']_{320}$, which normalizes the rotation (in degrees) to the molar concentration of total amino acids in the solution [(degrees centimeter^2 mole^{-1}) × 10^{-3}]. The effect on the absorbance at 301 nm is expressed as the difference in absorbance (ΔA_{301}) between the protein in the absence of guanidinium chloride and in its presence. The values of pH noted for each curve are the values measured at the midpoint of the transition. Values of pH increased by about 0.1 unit as the concentration of guanidinium chloride was increased within the zone of the transition. At 6 M guanidinium chloride at all values of pH, the protein is a random coil cross-linked by its cystines. The upper lines marked $[m']_D$ and ΔA_D are thought to be the mean residue rotation or absorbance, respectively, of the random coil at all concentrations of guanidinium chloride, estimated by extrapolation of the measured values at high concentrations. The points marked K on the upper line in panel B are the immediate absorbances of solutions rapidly diluted from 6 M guanidinium chloride to the noted concentrations, and they fall on the line as expected. The lower lines marked $[m']_N$ and ΔA_N are extrapolations of the points at low concentrations of guanidinium chloride and give the values of the mean residue rotation and absorbance, respectively, of the native structure as a function of concentration. The points marked K on the lower line in panel B are the immediate absorbances of solutions of the native protein rapidly mixed with a concentrated solution of guanidinium chloride to produce the noted concentrations, and they fall on the line as expected. Adapted with permission from ref 23. Copyright 1969 American Chemical Society.

intrinsic viscosity, optical rotation, molar ellipticity, absorption of ultraviolet light, electrophoretic mobility,[17] nuclear magnetic resonance absorptions,[22] or sedimentation velocity is measured as a function of temperature, pH, or the concentration of urea or guanidinium chloride, changes indicative of a shift in the value of the equilibrium constant K_{FU} for folding (Equation 13–1) are observed (Figure 13–1).[23] Each experimental point in Figure 13–1 represents a solution of lysozyme at a particular temperature, pH, and concentration of guanidinium chloride. For each point the solution has been allowed to reach equilibrium, which is assumed to be the state after all changes in optical rotation or ultra-

violet absorption have ceased. The same equilibrium value is reached whether the folded protein or the unfolded protein is the initial state, and this demonstrates that a reversible process is being monitored. That the changes in the measured parameter cease at intermediate values between the extremes is also consistent with the conclusion that equilibrium has been reached.

Below certain concentrations of guanidinium chloride, there is a monotonic smooth increase in ultraviolet absorption or decrease in optical rotation that is established immediately upon mixing of the solution. Even in samples of native protein that eventually will unfold, the immediate change in the ultraviolet absorption ΔA_N upon addition of guanidinium chloride falls upon this baseline. This baseline traces the perturbation in the optical rotation, $[m']_N$, or ultraviolet absorption, ΔA_N, of the fully folded structure due only to addition of the denaturant in the absence of any unfolding.

Above certain concentrations of guanidinium chloride there is a monotonic, smooth increase both in optical rotation, $[m']_D$, and in absorption, ΔA_D, presumed to reflect the effect of increasing the concentration of denaturant on these properties of the unfolded polypeptide. If fully unfolded lysozyme is diluted into the range of concentrations of guanidinium chloride where it will fold, the immediate absorbance ΔA_D of the solution before folding occurs falls on the line of the upper limit.

At intermediate concentrations of guanidinium chloride, the observed optical rotations and ultraviolet absorbances fall between the extremes of the fully native and fully unfolded protein. It is in this region that the equilibrium constant between the native structure and the unfolded state can be measured. Assume for the moment that at all concentrations of guanidinium chloride the solution contains an equilibrium mixture of only fully native protein and random coil. This is referred to as the **two-state assumption**.[24] At high enough concentrations of guanidinium chloride, the concentration of random coil becomes sufficiently large that it contributes significantly to the optical rotation or ultraviolet absorption. At this point, the isomerization of the folding–unfolding continuously interconverts measurable quantities of native structure and unfolded polypeptide in equilibrium with each other. As the concentration of guanidinium chloride is increased further, a greater fraction of the

protein is in the unfolded state until, finally, insignificant amounts of native protein are present at equilibrium. As the pH is lowered and its particular perturbation of the equilibrium in favor of the random coil is increased, the perturbation required from the guanidinium chloride decreases. As the pH is lowered, the perturbation in favor of the random coil required from an elevation of the temperature also decreases (Figure 13–2).[25]

If the two-state assumption is made, the observed optical rotation at each concentration of guanidinium chloride in Figure 13–1A

$$[m']_{obs} = f_F[m']_F + f_U[m']_U \qquad (13\text{–}2)$$

where $[m']_F$ is the optical rotation of the fully folded native structure, $[m']_U$ is the optical rotation of the unfolded state, f_F is the fraction of the protein in the folded state, and f_U is the fraction of the protein in the unfolded state, all at that concentration of guanidinium chloride. Because $[m']_D$, and $[m']_N$ at each concentration of guanidinium chloride are known (Figure 13–1) and because $f_F + f_U = 1.0$, f_F and f_U at each concentration can be calculated. From f_F and f_U, the equilibrium constant for folding ($K_{FU} = f_F/f_U$) can be determined for that concentration of guanidinium chloride, temperature, and pH. The same analysis can also be applied to the behavior of the ultraviolet absorption (Figures 13–1B and 13–2).

It is only if the two-state assumption is made that values for the equilibrium constants and standard changes in free energy, enthalpy, entropy, and heat capacity can be calculated. If there were one or more significant, stable intermediates, then Equation 13–2 would have to be expanded

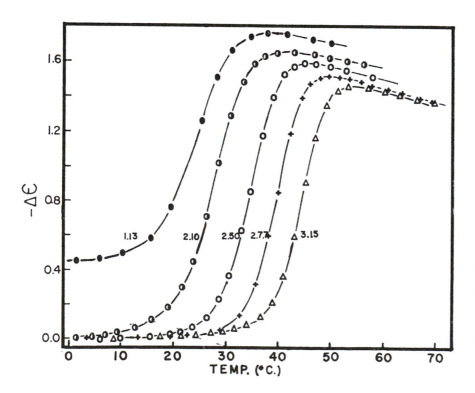

Figure 13–2: Change in the extinction coefficient ($\Delta \varepsilon$) as a function of temperature for five separate solutions of ribonuclease A at different values of pH: 1.13 (●), 2.10 (◐), 2.50 (○), 2.77 (+), and 3.15 (△).[25] The difference in extinction coefficient (deciliters gram^{-1}) is the difference between the extinction coefficient at 287 nm for ribonuclease at pH 7, 25 °C, and the samples at the noted pH and temperature. The changes in extinction coefficient over the ranges monitored were fully reversible. The samples were buffered with 40 mM glycine (pH 2.77 and 3.15) or by the protein itself (pH < 2.7). For each point on the curves, the solution was brought to the noted temperature and the absorbance was noted only after it no longer changed. As the pH of the solution decreased, a lower temperature was required to produce the transition between native structure and denatured state. Reprinted with permission from ref 25. Copyright 1967 American Chemical Society.

$$[m']_{obs} = f_F[m']_F + f_U[m']_U + \sum_i f_i[m']_i \qquad (13\text{-}3)$$

and the situation would become one of too many unknowns for the number of independent simultaneous equations. As a result, one is left with the choice between drawing a conclusion if the assumption is made and saying nothing if the assumption is not made. There are several observations which suggest that, in many cases, the two-state assumption is valid.

A point on the curve between the two extremes of fully native and fully unfolded protein should be, if the two-state assumption is valid, an equilibrium mixture of only fully native protein and the random coil. Suppose that either the purely native structure in the absence of guanidinium chloride or the purely random coil in a concentrated solution of guanidinium chloride is transferred into a solution at a pH and concentration of guanidinium chloride that, by the two-state assumption, supports an equilibrium mixture containing only the native structure and the random coil (intermediate ranges in Figure 13–1).

This reaction is a special case of a general kinetic category referred to as an **approach to equilibrium**. The approach to equilibrium of either the unfolding or the folding polypeptide, respectively, should be governed only by the two rate constants k_F and k_U (Equation 13–1) if the two-state assumption is valid. Either the random coil or the native structure, respectively, should be the exclusive product produced as the equilibrium is established. The rate at which the concentration of either species, F or U, in Equation 13–1 changes is

$$\frac{d[F]}{dt} = -\frac{d[U]}{dt} = k_F[U] - k_U[F] \qquad (13\text{-}4)$$

Because the concentration of total protein, $[protein]_{TOT}$, remains constant, it follows that

$$[U] + [F] = [protein]_{TOT} = [U]_{eq} + [F]_{eq} \qquad (13\text{-}5)$$

where $[F]_{eq}$ and $[U]_{eq}$ are the concentrations of native structure and random coil at equilbrium. Combining Equations 13–4 and 13–5 and focusing on the concentration that is decreasing, arbitrarily chosen to be the denatured form for the following derivation

$$-\frac{d[U]}{dt} = (k_F + k_U)\{[U] - [U]_{eq}\} - k_U[F]_{eq} + k_F[U]_{eq}$$
$$(13\text{-}6)$$

At equilibrium no further changes occur in the concentrations of either the native structure or the random coil and

$$k_U[F]_{eq} = k_F[U]_{eq} \qquad (13\text{-}7)$$

Parenthetically, this means that

$$K_{FU} = \frac{[F]_{eq}}{[U]_{eq}} = \frac{k_F}{k_U} \qquad (13\text{-}8)$$

To return to the derivation

$$-\frac{d[U]}{dt} = -\frac{d([U] - [U]_{eq})}{dt} = (k_F + k_U)\{[U] - [U]_{eq}\} \qquad (13\text{-}9)$$

because $[U]_{eq}$ is a constant. Equation 13–9 is a simple first-order differential equation in the variable $([U] - [U]_{eq})$ and describes a first-order process in this variable. If $\ln([U] - [U]_{eq})$ is plotted as a function of t, linear behavior should be observed and the slope of the line should be $(k_F + k_U)$. If the derivation just presented had been based on [F] instead of [U], the variable would have been $([F] - [F]_{eq})$ rather than $([U] - [U]_{eq})$ but the rate constant of the approach to equilibrium would still be $(k_F + k_U)$.

The absorbance of the solution, A, is equal to

$$A = \varepsilon_F[F] + \varepsilon_U[U] \qquad (13\text{-}10)$$

where ε_F and ε_U are the extinction coefficients of the native structure and the random coil. When Equation 13–10 is combined with Equation 13–5

$$A = (\varepsilon_U - \varepsilon_F)[U] + \varepsilon_F[protein]_{TOT} \qquad (13\text{-}11)$$

The difference between the absorbance at any time, A, and the absorbance at infinite time A_∞, when $[U] = [U]_{eq}$ is

$$A - A_\infty = (\varepsilon_U - \varepsilon_F)\{[U] - [U]_{eq}\} \qquad (13\text{-}12)$$

and, because $[U] - [U]_{eq} > 0$

$$\ln\{[U] - [U]_{eq}\} = \ln|A - A_\infty| - \ln|\varepsilon_U - \varepsilon_F| \qquad (13\text{-}13)$$

Because $\varepsilon_U - \varepsilon_F$ is a constant, a plot of $\ln|A - A_\infty|$ against t will be linear with a slope of $(k_F + k_U)$.

This derivation for the approach to equilibrium as monitored by a physical property is completely general and can be applied to any situation where the equilibrium can be described by two first-order rate constants, forward and reverse. In the particular case of the approach to equilibrium when either the native structure or the random coil is shifted to various conditions where it must partially unfold or partially fold, respectively, every plot of $\ln|A - A_\infty|$ should be a linear relationship. That they are in the case of the folding–unfolding of lysozyme in guanidinium chloride at low pH (Figure 13–3)[23] is consistent with the two-state assumption that was applied at several steps in the preceding derivation.

Usually conditions are chosen so that the approaches to equilibrium of the folding–unfolding reaction in the region of the transition between fully native and fully denatured protein are simple first-order processes. Under other conditions of temperature, pH, or concentration of denaturant, however, either folding or unfolding or both may not be first-order processes. For example, the equilbrium constant for folding of myoglobin at 25 °C is shifted into the measurable range at pH 4.2. At this pH both the folding and unfolding of the polypeptide are first-order processes.[26] They both proceed with clean isosbestic points in the Soret region of the visible spectrum, and this also indicates that both folding and unfolding at this pH are two-state processes. At higher or lower values of pH, however, the kinetics of both the folding and unfolding reactions become complex.

If a point on the curve between the two extremes in Figure 13–1 represents a mixture containing only the

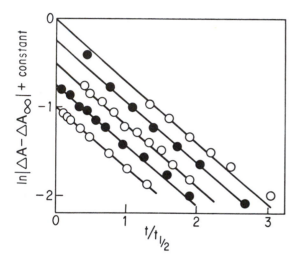

Figure 13–3: Kinetic behavior[23] of the approach to equilibrium when either fully native (○) or fully unfolded (●) lysozyme was shifted into the transition region in which an equilibrium between the native and unfolded states exists. In each experiment, a solution of lysozyme in either buffered 6 M guanidinium chloride (fully unfolded) or in an aqueous buffer in the absence of guanidinium chloride was mixed rapidly with the appropriate volume of either buffered water or buffered 6 M guanidinium chloride, respectively, to achieve the desired pH and concentration of guanidinium chloride at 25 °C. The final values of pH and guanidinium chloride concentration, chosen to be in the transition zone (see Figure 13–1), were, from top to bottom, pH 1.3 and 3.20 M guanidinium chloride ($k_{obs} = 4 \times 10^{-2}\,s^{-1}$); pH 2.65 and 1.92 M guanidinium chloride ($k_{obs} = 2.2 \times 10^{-2}\,s^{-1}$); pH 2.64 and 2.50 M guanidinium chloride ($k_{obs} = 6 \times 10^{-3}\,s^{-1}$); pH 2.44 and 2.58 M guanidinium chloride ($k_{obs} = 4 \times 10^{-3}\,s^{-1}$); and pH 4.66 and 4.01 M guanidinium chloride ($k_{obs} = 5 \times 10^{-4}\,s^{-1}$). The difference in absorbance of the solutions at 301 nm, relative to the absorbance of the fully native protein (Figure 13–1B), was followed as a function of time. The initial points of the lines were arbitrarily offset by a constant to space the lines on the field. The time was normalized for each plot by dividing each time point (t) by the half-time ($t_{1/2}$) of the first-order reaction. The linearity of the courses is the point that is emphasized in this way. Adapted with permission from ref 23. Copyright 1969 American Chemical Society.

native structure and the random coil, then Equation 13–2 with appropriate substitutions should describe the behavior of every physical parameter measured. In this equation, the values for the fraction of native structure, f_F, and the fraction of random coil, f_U, must be the same at a particular pH and a given concentration of guanidinium chloride (Figure 13–1) regardless of the physical property used to follow folding–unfolding. It was shown that both the optical rotation at 320 nm and the absorbance at 301 nm of the solutions of lysozyme satisfied this criterion.[23] An even more convincing demonstration of the coincidence of physical measurements was presented for the thermal denaturation of ribonuclease A (Figure 13–2). When the fraction of denatured ribonuclease, f_U, as a function of temperature at pH 2.10 was determined by intrinsic viscosity, optical rotation, and ultraviolet absorption, all of these properties gave the same result (Figure 13–4).[27] If even one intermediate were present, the values of all three of these physical properties for this intermediate would have to assume the same fractional deviation relative to the two extremes of each respective property. It seems unlikely that an intermediate could exist, for example, that had an intrinsic viscosity a third of the way between the intrinsic viscosities of the native protein and the random coil and an optical rotation also a third of the way between those of the native protein and the random coil and an ultraviolet absorption also a third of the way between that of the native protein and the random coil. In effect, this test is formally equivalent to the observation of an isosbestic point in a series of spectra, and this has always been accepted as evidence for a transition between only two states.

The standard enthalpy change of folding can be calculated from the behavior of K_{FU} as a function of temperature.[28] In the case of the folding–unfolding of chymotrypsinogen as a function of temperature, this enthalpy change agreed with that measured directly in a calorimeter,[28] and this agreement validates the two-state assumption used to calculate K_{FU}, at least for this transition.

A series of replacements of Threonine 157 in the lysozyme of bacteriophage T4 was made by site-directed mutation. The reversible folding–unfolding of each of these mutant proteins at pH 2.0 was followed by varying the temperature.[4] Although the changes in the standard free energy of folding produced by these mutations were not estimated directly[29] from the available values of K_{FU}, they were estimated by an indirect method based on the two-state assumption.[30] When the hydroxyl of threonine was replaced by a methyl group (valine), the standard free energy change of folding at pH 2.0 and 45 °C was 7 kJ mol^{-1}. This agrees well with the difference in standard free energy between a hydroxyl and a methyl group placed in an axial position on cyclohexane (5 kJ mol^{-1}).[31] Therefore, the standard free energy change for the steric effect observed in the protein, the estimation of which

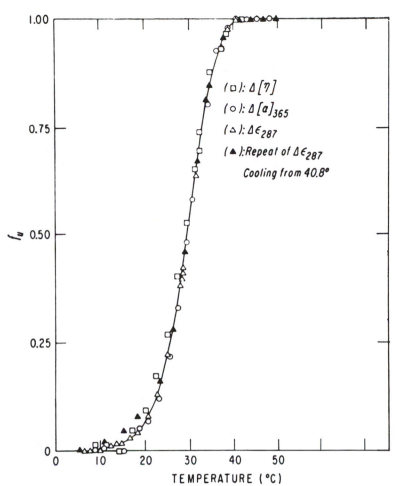

Figure 13–4: Change in f_U, calculated by Equation 13–2, for ribonuclease A at pH 2.10 (Figure 13–2) as a function of temperature.[27] The variation in three properties of a solution of ribonuclease A at pH 2.10 and ionic strength of 20 mM were measured as a function of temperature. These three properties were the absorbance at 287 nm ($\triangle$), the intrinsic viscosity ($\square$), and the optical rotation ($[\alpha]_{365}$) at 365 nm ($\bigcirc$). In each case, the direct observations were first plotted as a function of temperature as in Figure 13–2. The behavior of the physical property for the native structure and the fully denatured state as a function of temperature were estimated by linear extrapolation as in Figure 13–1. The fraction of the denatured state f_U was then calculated from each separate curve by Equation 13–2. The values of f_U determined by each physical property were then presented together as a function of temperature. To demonstrate that the process being followed was reversible, the absorbance at 287 nm was followed a second time ($\blacktriangle$) after a sample was heated to 40.8 °C for 16 h and then cooled. Adapted with permission from ref 27. Copyright 1965 American Chemical Society.

In the graph legend:

($\square$): $\Delta [\eta]$

($\bigcirc$): $\Delta [a]_{365}$

($\triangle$): $\Delta \epsilon_{287}$

($\blacktriangle$): Repeat of $\Delta \epsilon_{287}$

Cooling from 40.8°

was based on the two-state assumption, agrees with an established measurement of a similar steric effect. The steric effect observed when an ethyl group (isoleucine) replaced the hydroxyl in the protein (12 kJ mol⁻¹) was much larger than that for the same change at the axial position of cyclohexane (6 kJ mol⁻¹), but a molecule of protein, because it surrounds the amino acid rather than simply providing 1,3-diaxial hydrogens, should be more sensitive to this latter change. A similar mutation in myoglobin through which Valine 28 (a methyl) was replaced with an isoleucine (an ethyl) showed a much smaller difference in the standard free energy change of folding (0.8 kJ mol⁻¹),[32] presumably because the environment is not so sterically constrained.

In folding–unfolding reactions that are perturbed by lowering the pH at a constant temperature, the perturbation of the equilibrium constant is caused by the fact that the denatured state can bind more protons than the native structure because buried acid–bases become accessible. The difference in protons bound before and after unfolding can be measured directly as an uptake of protons. In the case of the folding–unfolding of myoglobin, the number of protons observed to be taken up during the transition (6.0 mol mol⁻¹) agrees closely with

the numerical value (5.6 ± 0.2 mol mol⁻¹) that can reproduce analytically the behavior of the folding–unfolding as a function of pH at fixed temperatures between 0 and 60 °C. As the analytical function used to reproduce the behavior was based on the two-state assumption, the agreement between observed and calculated values validates this assumption.[21]

All of the foregoing observations suggest that the two-state assumption is valid for at least some of the folding–unfolding isomerizations observed experimentally. In those instances, thermodynamic parameters for the equilibrium between folded and unfolded states can be measured.

The **standard enthalpy change** of folding can be calculated from the dependence of the equilibrium constant, K_{FU}, on temperature. From the van't Hoff relationship

$$\left[\frac{\partial \ln K_{FU}}{\partial \left(\frac{1}{T} \right)} \right] = - \frac{\Delta H^{\circ}_{FU}}{R} \qquad (13\text{–}14)$$

if a folding–unfolding is followed as the temperature is varied and the value of the natural logarithm of the

A

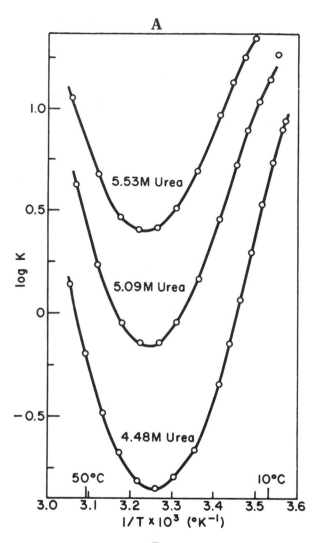

B

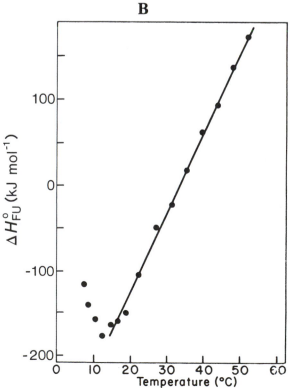

equilibrium constant is plotted as a function of T^{-1}, the slope of the relationship will be directly proportional to the standard enthalpy change. When the folding–unfolding of β-lactoglobulin was rendered reversible and kinetically first-order in both directions by adding appropriate concentrations of urea and adjusting the pH to 3, the equilibrium constant for folding could be measured for temperatures between 10 and 50 °C. The behavior of ln K_{FU} as a function of T^{-1} (Figure 13–5A)[33] demonstrates that the standard enthalpy change for the reaction, ΔH°_{FU}, is not constant but varies with temperature.

When the standard enthalpy change, ΔH°_{FU}, is plotted against the temperature (Figure 13–5B),[33] the slope of the relationship observed is the **standard heat capacity change**, ΔC°_P, of folding

$$\left(\frac{\partial \Delta H^{\circ}_{FU}}{\partial T} \right)_P = \Delta C^{\circ}_P \qquad (13\text{–}15)$$

The standard heat capacity change[33] for the folding of the polypeptide of β-lactoglobulin (n_{aa} = 162) between 5.5 and 4.4 M urea, pH 2.5 and 3.2, and 15 and 50 °C is -8700 ± 700 J K^{-1} mol^{-1}, or -54 J K^{-1} (mol of amino acid)$^{-1}$. The measured values for the standard heat capacity change for the folding of simple proteins with a single polypeptide are -55 ± 10 J K^{-1} (mol of amino acid)$^{-1}$, regardless of the perturbation used to shift the equilibrium.[33-35]

Unlike the standard entropy and enthalpy changes that vary considerably from situation to situation, this

Figure 13–5: Determination of ΔH°_{FU} and ΔC°_P for the folding of β-lactoglobulin in solutions of urea between pH 2.5 and 3.5. A series of measurements of the optical rotation at 365 nm of solutions of β-lactoglobulin as a function of the concentration of urea (as in Figure 13–1) were made at various temperatures. From the lines extrapolated over the pre- and posttransition regions of these smooth curves, the optical rotation of the native structure and the unfolded state, respectively, of β-lactoglobulin at the particular temperature and any concentration of urea could be estimated. From these estimates, it was possible to plot the optical rotations of the native structure and the unfolded state at a particular concentration of urea as a function of temperature. Solutions were then prepared at 4.48, 5.09, and 5.53 M urea, concentrations at which equilibrium constants could be measured over the temperature range 10–50 °C. For each of these solutions, from the optical rotation at a given temperature and the estimated optical rotations of the native structure and the unfolded state at that temperature and concentration of urea, the equilibrium constant K_{FU} could be calculated (Equation 13–2). (A) The logarithm of K_{FU} (log K) when plotted as a function of the inverse of the temperature [$1/T \times 10^3$ (kelvin^{-1})] gives a curve, from the slope of which ΔH°_{FU} can be calculated (Equation 13–15) at any particular temperature. (B) When ΔH°_{FU} (kilojoules mole^{-1}) is plotted as a function of the temperature T (°C), the slope of the line is the standard heat capacity change, ΔC°_P, for folding. Reprinted with permission from ref 33. Copyright 1968 American Chemical Society.

uniformly smaller heat capacity of the native structure relative to the random coil seems to be a fundamental property of folding–unfolding. It should arise both from the decrease in heat capacity that occurs when hydrophobic amino acids are transferred from the aqueous phase into the interior of the molecule of protein and from the decrease in heat capacity that occurs when vibrations and rotations along the polypeptide become more hindered after it is folded.[34] Its magnitude is consistent with the decrease in standard heat capacity (–200 to –400 J K^{-1} mol^{-1}) observed for the transfer of alkanes and arenes from water into an organic phase (Table 5–6) if it is recalled that hydrophobic amino acids make up only a fraction (30%) of the amino acids in a polypeptide and that many of them remain accessible to water in the native structure after the protein has folded. These considerations suggest that the characteristic decrease in standard heat capacity [–55 J K^{-1} (mol of amino acid)$^{-1}$] associated with the folding of a polypeptide is one of the few signatures of the hydrophobic effect resulting from the segregation of hydrophobic amino acids from the solvent in the interior of the folded native structure. It is usually assumed that the hydrophobic effect is the only noncovalent force that can contribute favorably to the standard free energy of folding.

It has been pointed out that there is a tendency for this characteristic change in standard heat capacity of folding to decrease in magnitude as the frequency of disulfides in the polypeptide increases.[35] This probably does not indicate that disulfides provide positive changes in standard heat capacity of folding but simply that the folded state, gaining stability by the introduction of the disulfides, requires less of a contribution from the hydrophobic effect to have the proper standard free energy of folding.

The fact that the changes in standard heat capacity of folding for all proteins are significant negative numbers dictates that the change in standard enthalpy of folding, ΔH°_{FU}, must vary significantly with the temperature (Figure 13–5B) and must pass through a value of zero at some temperature. This causes the standard free energy change of the reaction to pass through a minimum at that same temperature (Figure 13–5A). From these considerations it follows that if the negative change in heat capacity is an intrinsic property of folding–unfolding, each protein must have a characteristic temperature of maximum stability.[33] These temperatures of maximum stability vary from less than 0 °C to more than 35 °C.

It is also possible to shift the equilibrium constant for folding in the direction of denaturation by applying pressure to a solution of protein at low pH (pH 2–4).[36,37] This observation requires that the solvated denatured state have a smaller volume than the solvated native structure because

$$\left(\frac{\partial \Delta G^{\circ}_{FU}}{\partial P} \right)_{T} = \Delta V^{\circ}_{FU} \qquad (13-16)$$

where ΔV°_{FU} is the **standard volume change** during the reaction. The volume changes observed for folding are quite small even at low pH. At pH 2.0 and 0 °C, the volume change for folding of ribonuclease[36] is 47 cm^3 mol^{-1}; and for chymotrypsinogen,[37] 14 cm^3 mol^{-1}. The volume changes decrease as the pH is raised and may have a negative sign at pH 7. As with the changes in enthalpy and heat capacity, it is not the change in volume that is the characteristic property but the change in compressibility.[37] The **change in compressibility** of folding, $\Delta \beta$, where

$$\Delta \beta_{FU} = \left(\frac{\partial \Delta V^{\circ}_{FU}}{\partial P} \right)_{T} \qquad (13-17)$$

is invariant with pH,[36] and it has a value[37] of 1–2 × 10^{-10} cm^6 J^{-1} (mol of amino acids)$^{-1}$. Its positive value indicates that the solvated denatured state is more compressible than the native structure. This is not a surprising result because the adiabatic compressibilities of native proteins are very small, 10-fold smaller than those of organic liquids and 2-fold smaller than those of amorphous organic solids.[38] The greater compressibility of the denatured state is probably due to its more fluid structure, but it is also possible that the hydrophobic functional groups revealed in the denatured state increase the structure of the water surrounding them and thereby increase its compressibility. This increase in the structure of water, if it is significant, would resemble the increase caused by decreasing the temperature, and decreasing the temperature of liquid water increases its compressibility (Figure 5–5).

High pressures also are able to dissociate multimeric proteins into monomers, reversibly, without causing denaturation, even at neutral pH. The volume change is small; in the case of enolase at 10 °C and pH 7.4, $\Delta V^{\circ} = 0.025$ cm^3 (mol of amino acid)$^{-1}$. Presumably the individual volume changes occur only at the faces of the subunits that are exposed during the dissociation.[39]

To be able to measure the equilibrium constant K_{FU} for folding, it must be decreased significantly by a combination of rather unphysiological perturbations. It would be of interest to be able to estimate its value at pH 7 and 25 °C in the absence of urea or guanidinium chloride. This is generally accomplished by extrapolation[29] from realms of pH, temperature, and concentrations of urea and guanidinium chloride where measurements can be made. This extrapolation is accomplished by deriving theoretical equations to describe the behavior of K_{FU} as a function of the perturbation, showing that the theoretical equations reproduce the behavior of K_{FU} in the range in which measurements are possible, and then extrapolating the value of K_{FU} beyond the available data to the conditions of physiological values of pH and temperature. Various equations have been derived for extrapolating to small or zero concentrations of urea and guanidinium chloride[40,41] and from acidic to neutral

Table 13–2: Standard Free Energies of Folding in the Absence of Guanidinium Chloride or Urea, ΔG_{FU}° (H_2O), at 25 °C

protein	pH	perturbation extrapolated	$\Delta G_{FU}^{\circ}(H_2O)$ (kJ mol^{-1})	ref
ribonuclease	6	guanCl	−50	44
	6	guanCl	−40	44
	6.6	urea	−30	40,45
	6.6	guanCl	−40	40,45
lysozyme	6	guanCl	−60	44
	6	guanCl	−50	44
	7	guanCl, pH	−50	46
	7	guanCl, pH	−80	46
α-chymotrypsin	4.3	urea	−30	40,45
	4.3	guanCl	−30	40,45
	4.3	guanCl	−50	40
	4.3	guanCl	−40	45
[(phenylmethyl)sulfonyl]- α-chymotrypsin	4.0	guanCl	−40	42
	4.0	urea	−40	42
	4.0	1,3-dimethylurea	−40	42
	6.0	urea, guanCl	−50	47
myoglobin	7	guanCl, pH	−50	43
	6.0	guanCl	−40	48
	6.0	guanCl	−50	48
cytochrome c	6.5	guanCl	−50	49

pH.[41] Most of these equations plot the observed standard free energies of folding as linear functions of some other variable to perform the extrapolations. It is also possible to obtain nonlinear least-squares fits of theoretical equations to plots of the directly observed changes of a physical property as a function of denaturant (as it is presented in Figure 13–1).[42] Extrapolation both from high concentrations of guanidinium chloride and from low pH can also be performed simultaneously.[43] Unfortunately, each theoretical curve, although it is successful at reproducing the behavior in the measurable regions, deviates from the other theoretical curves beyond the measurable regions. Nevertheless, some representative values for extrapolated **standard free energy changes of folding** ($\Delta G_{FU}^{\circ} = -RT \ln K_{FU}$) under physiologically relevant conditions in the absence of denaturants have been assembled in Table 13–2. Each value for a particular protein is the result of a different extrapolation, often from the same experimental data.

The remarkable feature of this tabulation is that the standard free energies of folding for at least these five proteins are very similar and fall between −30 and −60 kJ mol^{-1}. These are not large standard free energy changes when the magnitude and the number of the individual noncovalent interactions involved in the process are considered. They are clearly the sums of a large number of positive and negative terms which cancel each other to produce small negative numbers. That they are all negative is merely the result of evolution by natural selection. Whether or not the small range that the values assume is also a consequence of natural selection is not known.

With these values for the standard free energy changes (Table 13–2), the values for the equilibrium constants for the folding of these native proteins should be between 10^5 and 10^{10}. If this is the case, 10^{-5}–10^{-10} of the lifetime of each of these proteins is spent in the unfolded state under normal circumstances. It has already been noted that slowing factors for the hydrogen exchange of deeply buried amide hydrogens are 10^6 or greater. This is consistent with the possibility that the most deeply buried amide hydrogens are exchanged only during the brief periods when the native structure has become reversibly unfolded. The requirement that a protein must unfold and refold during its lifetime may be viewed as a con-sequence of the need to fold in the first place (Equation 13–1) and the inescapable dictates of microscopic reversibility. If the rate constant k_F for the spontaneous refolding of a recently unfolded polypeptide[50] is of the order of 10 s^{-1} at 25 °C and the equilibrium constant K_{FU} for folding is of the order of 10^8, then the rate constant for the unfolding of a native protein to the random coil ($k_U = k_F/K_{FU}$) is of the order of 10^{-7} s^{-1}. This would state that a protein has a 50% chance of unfolding to the random coil every 100 days at 25 °C. This is not a major problem in the life of a protein.

The constant fragment of the light chain of immunoglobulin G (Figure 11–1) is a small protein, the folding–unfolding of which as a function of the concentration of guanidinium chloride has been documented.[51] The protein contains a deeply buried cystine which is readily reduced by dithiothreitol when it is unfolded in solutions of guanidinium chloride. The rate of its reduction in the random coil in the absence of guanidinium chloride can be estimated by extrapolation of its rate of reduction at various concentrations of guanidinium chloride. The actual rate of its reduction in the absence of guanidinium chloride when the protein is folded is much slower than the extrapolated value. If it is assumed that the reduction of this cystine in the absence of guanidinium chloride occurs only when the native protein is briefly and reversibly a random coil, the difference in rates of reduction between native structure and random coil is consistent with a value of standard free energy of folding of -26 kJ mol^{-1}. This is close to the value (-30 kJ mol^{-1}) obtained by extrapolation from ranges of guanidinium chloride concentrations in which the equilibrium constant can be measured.

If the reaction determining the unique native structure of a folded polypeptide is the equilibrium that exists between that native structure and the denatured state under physiological conditions, then the individual contributions to the overall free energy change for this isomerization would be the most critical factors in determining its outcome rather than any kinetic features. These thermodynamic forces, by the very nature of the process, must be noncovalent interactions. Neither hydrogen bonds nor ionic interactions can provide any net favorable free energy for the formation of the native structure of a protein in aqueous solution. Therefore, by exclusion and perhaps for the lack of a better candidate, the hydrophobic effect has attracted the most attention in discussions of the folding of a polypeptide.[52]

Felicitously, this has drawn attention to the significant role of the aqueous phase in this process.[53] The hydrophobic effect is simply that free energy, arising from the strong cohesion of the solvent, that drives molecules lacking any favorable interactions with the water molecules themselves from the aqueous phase. In the case of the formation of the native structure from the random coil, the hydrophobic effect participates in the reaction because hydrophobic side chains, which are exposed to water in the random coil, are removed to the interior of the protein during the folding of the native structure.[54] This transfer appears to provide the only favorable free energy available to drive the reaction to completion. The more aversion water has for a given amino acid side chain, the more free energy is gained when that residue or a portion of it ends up inside the native structure.

Conversely, and of equal importance, it is also the case that the more attraction water has for a functional group on an amino acid side chain, the more free energy is lost when that functional group is removed from water during the folding process. This point becomes clear upon examination of the data in Table 5–8 when it is realized that most of the free energies of transfer for the side chains of the amino acids from water to the condensed vapor are actually unfavorable, many by a considerable amount. This is due to the fact that water participates in strong interactions with hydrogen-bond donors and acceptors and to the fact that when charged side chains are withdrawn from water they must be neutralized first. One of the major free energy deficits in the folding of a protein results from the requirement to unsolvate those hydrophilic functional groups destined for the interior. The removal of even neutral hydrogen-bond donors from water, even though they may always find an acceptor in the interior of the protein, is a significantly unfavorable transfer.[55] For all of these reasons, there is a significant probability that free energy will increase whenever a hydrophilic amino acid is removed from water during the folding process.

Both the hydrophobicity and the hydrophilicity of a given sequence of amino acids affect the outcome of the equilibrium between a denatured state and the native structure. Although one or the other of these two properties is often emphasized to make a particular point, neither is more important than the other. For example, it is often stated that the interior of a folded polypeptide is formed from its hydrophobic sequences, but it is seldom pointed out that the particular interior of the protein is also selected because hydrophilic sequences are difficult to bury. To a first approximation, the native structure of a molecule of protein will be the structure that, within the topological constraints of the polypeptide, permits the removal of the greatest amount of hydrophobic surface area and the smallest number of hydrophilic positions from exposure to water.

This does not mean that the interior of a protein will be filled with only hydrophobic amino acids and its surface paved with only hydrophilic amino acids.[54] Because proteins are the product of evolution by natural selection, and because it appears that structures of only marginal stability rather than maximum stability have been selected, the interior is only enriched in hydrophobic amino acids rather than being formed exclusively from hydrophobic amino acids. Because the replacement of a hydrophilic amino acid on the surface by a hydrophobic amino acid on the surface has no effect on the equilibrium between the denatured state and the native structure, the only selective advantage to having hydrophilic amino acids outside of the active sites on the surface of a protein is to keep it in solution.

Although the standard free energies of transfer (Table 5–8) provide a scale with which to judge the hydropathy of particular amino acids,[56] the actual values of the standard free energies of transfer cannot be used to provide estimates of the free energy of transfer of an amino acid in a random coil to the interior of a protein.[57] The fact that the amino acids are covalently incorporated into a polymer, rather than being free in solution, has

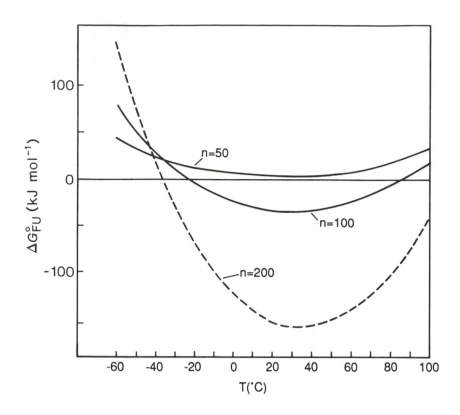

Figure 13–6: Estimated standard free energy of folding, ΔG°_{FU} (kilojoules mole^{-1}), for polypeptides of various fixed lengths ($n = 50$, 100, and 200).[60] The polypeptides were given configurational entropies in the random coil that were calculated from considerations of excluded volume. The poly-peptides were composed in such a way that half (0.5) of their amino acids were hydrophobic, each with a hydrophobic effect of -8 kJ mol^{-1} at 25 °C. The values of ΔC°_P for the individual hydrophobic effects were set at 250 J K^{-1} mol^{-1}. The standard change in free energy for the collapse of the polypeptide to a condensed state was first cal-culated and then the free energy change for the rearrangement of the condensed state to maximize the segregation of the hydrophobic amino acids from the solvent. Calculated values for the overall standard free energy of folding, ΔG°_{FU}(kilojoules mole^{-1}), as a function of temperature, $T(°C)$, are presented. Adapted with permission from ref 60. Copyright 1989 American Chemical Society.

significant effects on their configurational entropies. It has already been noted that neither a liquid solvent nor the vapor phase can adequately mimic the solid interior of a protein. The fact that the surface of a protein is irregular means that most of the amino acids in the folded polypeptide are neither fully accessible nor fully buried (Figure 6–18), and transfer between bulk solvents is irrelevant to this situation.

Aside from the unfavorable standard free energy of transfer associated with the removal of hydrophilic functional groups into the interior of the folded polypeptide, the other major deficit that must be overcome during the folding process is the **configurational entropy** of the random coil. This is the positive, intrinsic entropy that arises from the fact that the random coil can assume a large number of different configurations. It represents a deficit during the folding of the polypeptide because the native structure, to a first approximation, assumes only one or a few configurations (Figure 6–4). Therefore, when the random coil becomes the native structure its configurational entropy almost disappears.

At first glance, it seems that the configurational entropy of the random coil, dictated by the sum over all of its states, should be very large because each amino acid has at least the two dihedral angles, ψ and ϕ (Figure 6–1), each of which can assume a number of values as dictated by the Ramachandran plot (Figure 6–4). This initial intuition, however, neglects a rather substantial consideration referred to as excluded volume.[58] The term **excluded volume** designates the qualification that every configuration of the random coil in which two or more atoms would occupy the same space at the same time is

impossible and thus cannot contribute to the configurational entropy. This is the consequence of the steric effects that produce the Ramachandran plot itself operating over the whole polymer rather than just between neighboring amino acids. Excluded volume makes a large contribution to diminishing the configurational entropy of the random coil. For a polypeptide 100 amino acids in length, a set of configurations could be generated by randomly assigning values to the dihedral angles ψ and ϕ within their allowed ranges. The number of these configurations that would not superpose two or more atoms in the polypeptide has been estimated to be only 10^{-44} of the total number of randomly generated configurations.[57]

Even though a consideration of excluded volume decreases the number of configurations available to the random coil remarkably, there are still a large number of configurations that are accessible. Only one or a small number of these configurations is the compact native structure of the folded polypeptide. For the native structure to be stable relative to the random coil, the configurational entropy resulting from the sum over all of the allowed denatured configurations must be overcome.

It has been shown that if the attractive noncovalent free energies between a subpopulation of the monomers along a polymer are significantly more negative than their individual free energies of solvation, the polymer should spontaneously collapse to a globular form.[59] Because the constraints of excluded volume are even more extreme in this compact, globular form, the number of accessible configurations should be much smaller. Because the hydrophobic effect is the only interaction capable of producing net favorable free energies of association between amino acids, it is generally assumed that the attractive noncovalent force that would perform the condensation leading to the globular state of a polypeptide is the hydrophobic effect exerted by the subpopulation of the amino acids in a polypeptide that are hydrophobic. This view of the folding of a polypeptide could be called the **condensation model**. Its central feature is that the collapse of the random coil to the condensed state, caused by the fact that water is a bad solvent for a polypeptide, decreases the configurational entropy of the polypeptide dramatically and narrows the search for the native structure to a much smaller number of accessible conformations.

On the basis of this model, folding–unfolding can be treated theoretically as a process in which the unfavorable loss of the configurational entropy of the random coil is balanced only by the favorable removal of hydrophobic amino acids from contact with the aqueous phase.[57,60] The statistical treatment of the random coil developed by Flory,[61–63] which takes account of excluded volume and the solvation of the monomeric units, can be expanded[57] to include hydrophobic forces both among the monomers in the condensed state[63] and among monomers fortuitously juxtaposed in the random coil,[57] as well as the much smaller, but still significant, configurational entropy of the condensed polypeptide before it assumes the unique native structure. The process of folding is divided into two imaginary steps,[57] not necessarily related to the actual steps. These imaginary steps are the condensation of the random coil to a globular structure excluding water and the reconfiguration of the polymer in this condensed state to maximize the exposure of hydrophilic groups and minimize the exposure of the hydrophobic groups to the water. With reasonable values both for the hydrophobic force of the average hydrophobic amino acid (-8 kJ mol^{-1}) and for the fraction of the amino acids in the polypeptide that are hydrophobic (0.50), the formation of a unique globular state from a random coil should proceed with net negative standard free energy change for polypeptides greater than about 70 amino acids in length (Figure 13–6).[60] Polypeptides less than about 70 amino acids in length would not fold because they could not bury a large enough number of hydrophobic amino acids to overcome the configurational entropy of their random coils.

One of the more remarkable features of the process of folding a polypeptide to produce a monomeric protein is that the amount of accessible surface area buried during the transformation is only a function of the length of the polypeptide.[64,65] This results from the obvious fact that the accessible surface area of the random coil must be directly proportional to its length and the unexpected fact that the amount of accessible surface area in the native structure of a monomeric protein happens also to be a function of only the length of its polypeptide. The fraction of the buried surface area that is contributed by hydrophobic functional groups is quite constant (0.61 ±

0.04),[64] and the change in configurational entropy is a function of only the length of a polypeptide. Taken together, all of these observations suggest that it is the total favorable free energy change contributed by the hydrophobic effect that is the property that has been constrained by natural selection to become more negative as the length of the polypeptide increases. In this way, there is an enforced balance between the favorable change in free energy of the hydrophobic effect and the unfavorable change in configurational free energy to maintain the total standard free energy change for the folding of the polypeptide forming a monomeric protein within an acceptable range.

It has also been shown, however, that the same relationship between final accessible surface area in the native structure and the total length of the constitutive polypeptides applies to oligomeric proteins[66] as well as monomeric proteins. Because there are significant differences in the fraction of the surface area buried between dimers and tetramers when they assemble from their isolated subunits and because monomeric proteins have no interfaces, the invariance of this relationship suggests that the correlation for all proteins is actually between the amount of buried hydrophobic surface area and total number of amino acids regardless of the number of polypeptides in a particular protein. If this is the case, the reason for this correlation becomes unclear because the change in configurational entropy should not be directly related to the total number of amino acids in an oligomeric protein but should also be a function of the number of subunits.

The presence of **cystines** in folded polypeptides (c2 Kin.5) makes a contribution to the standard free energy change of the isomerization between random coil and native protein. It takes little imagination at the present level of our understanding of the molecular structures of folded polypeptides to realize that the polypeptide must first fold before the thereby juxtaposed cysteines can be oxidized to cystines. It was, however, an informative insight into the sequence of events in protein folding when this was proven by White[67] and Anfinsen et al.[68] who showed that the fully reduced random coil of ribonuclease could be refolded under oxidizing conditions to produce fully active enzyme that contained each of the four properly paired cystines. It was also demonstrated that ribonuclease which had been oxidized while it was denatured so that a random array of intramolecular and intermolecular cystines had formed would remain inactive under physiological conditions until 2-mercaptoethanol was added to catalyze disulfide interchange (Figure 3–23) and permit the spontaneous regeneration of the native protein.[69]

Because the folded native structure is a prerequisite for the formation of a proper cystine, it necessarily follows that the cystine itself cannot change the intrinsic configurational entropy of the folded protein and can change its intrinsic enthalpy only by the standard enthalpy of formation of the cystine. It has been demonstrated, however, that the standard enthalpy of formation for cystine in the random coil is about the same as that estimated for the standard enthalpy of formation of cystine in the native structure.[70] Consequently, the incorporation of a cystine cannot affect the standard enthalpy change of folding either.

A cystine between two cysteines that are adjacent in the native structure increases the value of the equilibrium constant K_{FU} for the folding and decreases the standard free energy change ΔG°_{FU} for the reaction because it decreases the configurational entropy of the random coil. This can be demonstrated experimentally by introducing a specific cross-link between two adjacent amino acids into the native structure and determining its effect on the folding–unfolding of the protein. Glutamate 35 and the enol tautomer of the oxindole (Reaction 10–37) of Tryptophan 108 in lysozyme form an ester

13–3

This ester introduces an intramolecular cross-link between these two amino acids[71] that are adjacent to each other in the crystallographic molecular model of the protein. The cross-linked lysozyme has a standard free energy of folding, ΔG°_{FU}, at pH 2 in 2 M guanidinium chloride at 62 °C that is 22 kJ mol^{-1} less than that of un-cross-linked lysozyme.[70] The decrease in the configurational entropy of the random coil that should occur upon the introduction of such a cross-link can be calculated[72,73] to be −50 J K^{-1} mol^{-1}, which should give a decrease in the standard free energy change of 17 kJ mol^{-1}. This agrees satisfactorily with the decrease in free energy observed.[70] Lysine 7 and Lysine 41 in ribonuclease (c2 Kin.4) can be cross-linked specifically by 2-(p-nitrophenyl)-3-(3-carboxy-4-nitrophenyl)thio-1-propene (Figure 10–8).[74] The difference in the standard free energy change of folding between cross-linked and un-cross-linked ribonuclease at pH 2.0 and 40 °C[75] is −21 kJ mol^{-1}. This agrees closely with the calculated difference in configurational entropy (−69 J K^{-1} mol^{-1}) between a polymer of the length of ribonuclease with four cross-links, situated where the cystines are situated in ribonuclease, and the same polymer with an additional cross-link between amino acids 7 and 41.[75]

When the results of a series of such experiments were tabulated,[76] it was found that the changes in standard free energy of folding produced by adding a covalent cross-link to the random coil varied between −10 and −30 kJ mol^{-1}. The greater the distance in the amino acid

sequence of the polypeptide between the two amino acids cross-linked, the greater the magnitude of the effect. This correlation cannot be explained unless it is assumed that the major effect of the cross-link is on the free energy of the unfolded state. The tabulated values for the changes in standard free energy of folding agree approximately with theoretical predictions based on polymer theory.[72,73] It has recently been pointed out, however, that the number of amino acids in the polypeptide between the two that are cross-linked has a greater effect on the magnitude of the change in standard free energy of folding than these theories would predict.[77]

The equilibrium constant for the folding of the constant fragment of the light chain of immunoglobulin (c 12 Kin.1) at pH 7.5 and 25 °C in solutions of guanidinium chloride is decreased when its single cystine is reduced. All of the change could be accounted for by the fact that the rate constant for folding (k_F in Equation 13–1) of the random coil with the cystine was 100-fold greater than the rate constant for the random coil without the cystine.[78] This is consistent with the conclusion that the intact, correct cystine decreases the configurational entropy of only the random coil, while retaining access to the properly folded structure, and permits the random coil to fold more rapidly. Whether or not the proper cystine was present had no effect on the rate constant of unfolding (k_U in Equation 13–1).

If the standard free energy change, ΔG°_{FU}, for the folding of a protein under physiological conditions is rather small (−30 to −60 kJ mol^{-1}), perhaps for cause, and if the favorable contribution to the standard free energy change arises from the sum of the individual hydrophobic effects of the constituent amino acids, the balance should be easily upset. An example of the effect of such an alteration on the change in standard free energy for the folding of a protein is a series of site-directed mutations performed on the ribonuclease (n_{aa} = 110) from *Bacillus amyloliquefaciens*.[29] When either Isoleucine 96 or Isoleucine 88, two buried hydrophobic amino acids in the crystallographic molecular model, was replaced with valine, which is one methylene group shorter, ΔG°_{FU} increased from −40 to −35 kJ mol^{-1}. When Isoleucine 96, Isoleucine 88, or Leucine 14, another buried amino acid, was replaced with alanine, which is two methylenes and a methyl group shorter, ΔG°_{FU} increased from −40 to −23 kJ mol^{-1}. These are fairly small changes in the covalent structure of the polypeptide, but they significantly destabilized the folded structure relative to the unfolded random coil.

Such observations should also provide an estimate of the magnitude of the hydrophobic effect on the folding of a protein. From a number of such studies,[78a] it has been estimated that replacement of isoleucine by valine results in a change in standard free energy of folding of +6 ± 3 kJ mol^{-1}, replacement of isoleucine or leucine by alanine results in a change of +17 ± 6 kJ mol^{-1}, and replacement of valine by alanine results in a change of

$+12 \pm 4$ kJ mol^{-1}. These results suggest that the hydrophobic effect contributes favorably to protein folding. In these three types of mutation, the amount of hydrophobic surface area exposed to water in the unfolded state is less than in the wild-type protein, and the standard free energy of folding, ΔG_{FU}°, is missing -6, -17, or -12 kJ mol^{-1}, respectively.

Unfortunately, the magnitudes of these effects are too large. The difference in standard free energy between isoleucine and valine for transfer from water to 1-octanol[79] is only $+3$ kJ mol^{-1}, that between leucine or isoleucine and alanine is only $+8$ kJ mol^{-1}, and that between valine and alanine is only $+5$ kJ mol^{-1}. Most likely, the reason that these replacements destabilize the protein by free energies larger than the differences in their hydrophobic effects is that when isoleucine is converted to valine or when isoleucine, valine, or leucine is converted to alanine, empty space would be created in the protein unless the native structure rearranged to fill it. Such a rearrangement must itself be destabilizing and contribute the observed unfavorable increases in standard free energy of folding in excess of that predicted from the hydrophobic effect alone. The large standard deviations of the observed values are also consistent with this explanation. A similar but less ambiguous conclusion can be drawn from two mutations in the lysozyme from bacteriophage T4. When Threonine 157 was replaced[4] by serine, ΔG_{FU}° increased by $+3$ kJ mol^{-1}; and, when Threonine 157 was replaced with alanine, ΔG_{FU}° increased by $+6$ kJ mol^{-1}. The equal increases in standard free energy of folding observed when each functional group was deleted in turn suggests that the effect is almost entirely steric and results in large part from the creation of vacant space in the interior of the protein.

It follows from a number of similar observations that even if the sequence of a polypeptide lacks only a small amount of the necessary **information** or contains only a small amount of misinformation, it may be incapable of folding. Whenever the sequence of a protein is changed, for example, by site-directed mutation, the possibility exists that the mutant will not fold, for reasons that will never be learned. Likewise, incomplete polypeptides often lack sufficient information to fold properly. A form of the polypeptide of bovine ribonuclease ($n_{aa} = 124$) that is missing the last six amino acids is unable to produce a folded protein with enzymatic activity, and what structure it does have at 20 °C is eliminated by heating to only 40 °C at pH 7.5 in the absence of denaturants.[80] This truncated polypeptide is also susceptible to proteolytic degradation, unlike the intact native protein. When the last 23 amino acid residues, which form only a small number of contacts with the bulk of the folded polypeptide, are removed from the polypeptide of micrococcal nuclease ($n_{aa} = 149$), the polypeptide produced is a random coil by the criteria of circular dichroism, optical rotation, and ultraviolet absorption.[81] It is also readily digested by trypsin, unlike the native enzyme. Its residual enzymatic

activity of 0.1%, which is an intrinsic property of the shortened polypeptide,[82] suggests that it can still fold properly to form an active enzyme but that the equilibrium constant for the folding–unfolding is displaced heavily ($K_{FU} = 10^{-3}$) in the direction of the random coil.

Another set of examples of the fact that a polypeptide can fold only when all the necessary information is present are proteins that are posttranslationally modified during their natural maturation. In many instances, the polypeptide that folds to produce the native structure is longer than the final product because the initial folded form is clipped, and the smaller piece resulting from the cleavage of the polypeptide dissociates. For example, subtilisin from *Bacillus subtilis* folds naturally when it is a polypeptide 352 amino acids in length. After it folds, it is posttranslationally modified. During this process, the peptide bond after Tyrosine 77 is cleaved, and the first 77 amino acids of the polypeptide are lost. If the mature enzymatically active form of the protein ($n_{aa} = 275$) is unfolded, it will not refold; but, if the full-length polypeptide ($n_{aa} = 352$) is unfolded in 6 M guanidinium chloride, it readily refolds to produce the native structure.[83] Only when the complete amino acid sequence of the longer polypeptide is intact is there sufficient information to produce the mature form. Once folded and posttranslationally modified, the mature protein is stable and biologically competent, as long as it is not unfolded.

There are many other examples of proteins that lose portions of their polypeptide, usually from the amino terminus, after they have folded. This is so common that the term **proprotein** is used to designate the longer polypeptide that folds, with the implication that the cleaved, mature native structure is designated as the protein. Examples of this designation are proinsulin, proalbumin, and prothrombin.

Fragments of a polypeptide, each lacking sufficient information to fold separately, can sometimes cooperate to produce the proper native structure. The first example of this was the ability of the amino-terminal fragment of ribonuclease (Lysine 1–Alanine 20), which is almost structureless in isolation,[85] to reassume its native structure as an α helix when combined with the remainder of the polypeptide (Serine 21–Valine 124).[86] Both the fragment Alanine 1–Arginine 126 and the fragment Glycine 49–Glutamine 149 of micrococcal nuclease ($n_{aa} = 149$) are structureless in isolation.[81,84] When they are mixed together, however, they combine with each other to form two different native structures that both appear to be properly folded but together have only 10% of the nuclease activity of the native enzyme. Each product contains a globular region that includes the complete amino acid sequence of one molecule of micrococcal nuclease with redundant tails protruding from it (Figure 13–7). That the tails were structureless and protruding at the proposed locations was demonstrated by shaving them away with trypsin.[84] These experiments and those with the fragment Alanine 1–Arginine 126 in isolation suggest

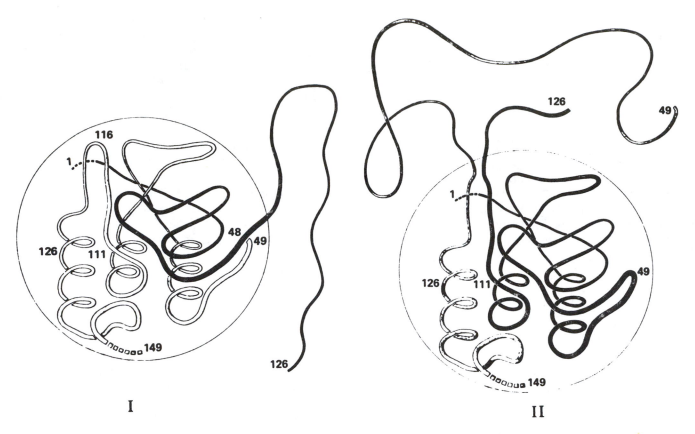

I   II

Figure 13–7: Schematic representation of two forms of enzymatically active micrococcal nuclease formed alternatively from the same two fragments of the complete polypeptide.[84] Micrococcal nuclease (n_{aa} = 149) was trifluoroacetylated and digested with trypsin and a long fragment of polypeptide containing Alanine 1–Arginine 126 was purified from the digest and the trifluoroacetyl groups were removed. This gave fragment (1–126). Micrococcal nuclease was cleaved by cyanogen bromide at, among others, Methionine 99. From this digest, the fragment containing Valine 99–Glutamine 149 was purified. This was fragment (99–149). Fragment (1–126) was mixed with fragment (99–149) in 50 mM NH_4HCO_3, pH 8 at 25 °C. A complex between these two fragments could be purified from this mixture by cation-exchange chromatography and molecular exclusion chromatography. The complex had 10% of the enzymatic activity of the native enzyme. The presence of the two complexes (I and II) presented diagrammatically in the figure in this purified material could be demonstrated by digesting the mixture with trypsin, which was able to shave away the tail beyond Lysine 48 in complex I and the tails between Lysine 49 and Lysine 116 and beyond Lysine 116 in complex II. The schematic drawing of the structure within the globular region was based on the crystallographic molecular model of this protein in its complete state. Reprinted with permission from ref 84. Copyright 1971 *Journal of Biological Chemistry.*

that almost the complete sequence of micrococcal nuclease is required to provide the information necessary to form the properly folded polypeptide. As long as one copy of this information, either in pieces or with redundancy, is provided, the polypeptide or polypeptides can fold spontaneously to produce one copy of the native structure.

It has been suggested that if a protein has domains, each domain might be required to fold as it emerged from the ribosome during biosynthesis before the next emerged. The logic behind this speculation is that a latter domain, in its unfolded state, might interfere with the folding of a former domain and vice versa. This would imply that the complete sequence of the polypeptide is missing some of the information necessary to refold, namely, which region of its sequence emerged first from

the ribosome. There is no evidence in favor of this conjecture. In fact, proteins containing two or more domains undergo reversible folding–unfolding as readily as proteins with only one domain. For example, an Fab fragment of an immunoglobulin G (Figure 11–1), which contains four internally superposable domains, can be unfolded in 6 M guanidinium chloride and its cystines cleaved by disulfide interchange. The reduced, unfolded polypeptides will refold, recombine, and form Fab fragments with the proper physical properties, such as sedimentation coefficient and the capacity to recognize antigen.[87,88] The yield of binding sites for antigen is low (10–15%), but it has been shown that this is due to the fact that the Fab fragments were from polyclonal mixtures and incorrect matching of $V_H C_{H1}$ polypeptides with $V_L C_L$ polypeptides occurs during renaturation.[89] Intact plasminogen, a protein containing reversibly folding domains, can be reversibly unfolded and refolded thermally at low pH, and each domain regains its structure independently during the cycle even though some of the others are unfolded at the same time.[5]

To this point, all of the thermodynamic analyses of the results of measurements of folding–unfolding have been based on the two-state assumption. Because this is the least ambiguous way to extract thermodynamic information from the experiments, the assumption has its appeal. The temptation even arises to apply it to irreversible denaturations or situations where it is known that more than two states exist in equilibrium with each other. There are numerous instances in which the criteria that must be satisfied for the reaction at least to be consistent with the two-state assumption fail. In the folding–unfolding of yeast 3-phosphoglycerate kinase, the shift in the equilibrium constant for folding upon addition of guanidinium chloride is fully reversible, but the measurements of the fraction of native structure, f_F, and the fraction of unfolded state, f_U, made by following changes in the fluorescence emission of tryptophan at 345 nm upon excitation at 300 nm do not agree with those made by following circular dichroic ellipticity at 222 nm.[90] Phosphoglycerate kinase contains two structural domains, thought to be hinged, and one explanation for the discrepancy would be that the two domains are independently folding domains, able to refold independently and properly even in each other's presence. Both tryptophans are in the carboxy-terminal structural domain of phosphoglycerate kinase. If the folding–unfolding of this domain required a higher concentration of guanidinium chloride to shift its equilibrium constant to a value close to 1.0 than the concentration required by the amino-terminal domain, the differences between fluorescence and circular dichroism could be quantitatively explained.[90] Disagreement among measurements of elastic light scattering, sedimentation velocity, circular dichroism, fluorescence, and enzymatic activity, as well as the heterogeneity of the changes observed with several of these physical properties, when the folding–unfolding of pyruvate kinase was followed in solutions of urea and guanidinium chloride suggests that this protein also contains independently folding domains.[91] In this case, dissociation of the subunits of the tetramer had to be distinguished from the folding–unfolding of the individual polypeptides.

The postulation of independently folding domains may be reasonable in circumstances where there are crystallographic observations consistent with their existence, as with phosphoglycerate kinase or pyruvate kinase, but behavior suggesting the existence of stable intermediates other than the fully native protein and the random coil is often observed when there is no reason to suspect that the protein has domains. For example, the shift in the equilibrium constant of the folding of carbonate dehydratase, the crystallographic molecular model of which contains no domains, has been followed by changes

in circular dichroic ellipticity at 269 nm, ultraviolet absorption at 290 nm, and optical rotation at 400 nm as a function of the concentration of guanidinium chloride. All three physical parameters detected a different transition (Figure 13–8A).[92] Furthermore, the kinetics of the refolding of the random coil was not a homogeneous, first-order process. It was concluded that one or more stable conformers of the polypeptide of carbonate dehydratase, other than the native structure and the random coil, are present in solutions of guanidinium chloride between 2 and 3 M in concentration. The properties of these other conformers are distinct from those of either the native structure or the random coil.

Similar results have been observed for α-lactalbumin (Figure 13–8B),[93] cytochrome c,[94] and myoglobin[95] under various conditions and for carbonate dehydratase[96] under circumstances other than in solutions of guanidinium chloride. These several observations have been used as evidence that a particular thermodynamically stable intermediate in the folding–unfolding of a protein can be defined. This intermediate is referred to as the molten globular state.[96,97]

The **molten globular state** is a state of a polypeptide in which the polymer has collapsed to a globular particle from the expanded random coil but remains fluid with a constantly changing conformation rather than achieving the single conformation that is the native structure. It has already been noted that water is a bad solvent for a polypeptide, and observations on synthetic polymers in bad solvents suggest that such a collapse is to be expected for a polypeptide in water. In such a condensed state, the configurational entropy of the random coil would be significantly reduced, and only a relatively small number of conformations that avoid the problems of excluded volume should be accessible.[59] Many of these conformations should display α helices and β structure that form spontaneously.[98] Under conditions that differ significantly from those in the living system in which the particular polypeptide has evolved to fold, the native structure may no longer be the most stable of these condensed conformations of the polypeptide, and a number of these structured conformations may be equally stable. Peculiar conditions such as low pH or the presence of denaturants, however, would be necessary to prevent the polypeptide from assuming the one native structure, as it would do normally, and they would trap it in the molten globular state. If this is actually what is occurring, a molten globular state could be an intermediate in the normal pathway of folding.

It is argued that the stable intermediates detected in the folding–unfolding of several proteins under several circumstances are all examples of this molten globular state. It should be kept in mind, however, that measurements of the same set of physical properties have not yet been made on all of the proteins under all of the various sets of conditions, and the assignment of the same state to all of them may be mistaken.

A

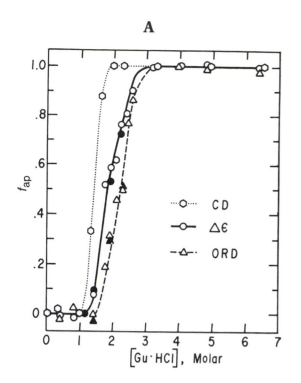

Figure 13–8: Changes in the apparent fraction of denatured protein (f_{ap}) as a function of the molar concentration of guanidinium chloride for (A) bovine carbonate dehydratase B at pH 7.0, 25 °C[92] and (B) bovine α-lactalbumin at pH 6.7, 25 °C.[93] The physical properties monitored were, for carbonate dehydratase (A), the molar ellipticity (CD) at 269 nm (O), the absorbance ($\Delta\varepsilon$) at 292 nm (O), and optical rotation (ORD) at 400 nm ($\triangle$). Filled symbols are measurements of solutions that were first exposed to a high concentration of guanidinium chloride and then diluted to the noted concentration. Panel A adapted with permission from ref 92. Copyright 1973 *Journal of Biological Chemistry*. For α-lactalbumin (B), the properties measured were the molar ellipticities at 270 nm (O), 296 nm ($\triangle$), and 222 nm (●). Panel B reprinted with permission from ref 93. Copyright 1976 Academic Press. In all cases, measurements of these physical properties were made on a series of solutions of varying guanidinium chloride, GuHCl, concentration. The direct results were plotted as in Figure 13–1 for each set of observations. Lines were drawn for the behavior of fully native and fully unfolded protein as a function of the concentration of guanidinium chloride, and the apparent fraction of denatured protein was estimated from the position of each data point relative to these lines. These estimated values of the apparent fraction of denatured protein are plotted in each panel as a function of the concentration of guanidinium chloride.

B

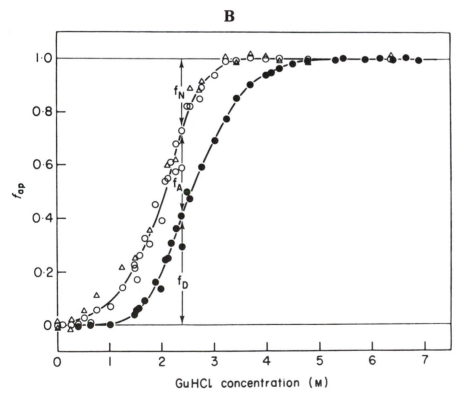

Intermediates believed to be molten globular states have been detected under several circumstances. They have been observed for α-lactalbumin[99] below pH 4.5 at concentrations of guanidinium chloride below 2.5 M; for α-lactalbumin,[100a] stripped of bound Ca^{2+}, at pH 8 and guanidinium chloride concentrations between 0.5 and 2.0 M; for cytochrome c[94] below pH 3 at chloride concentrations greater than 0.1 M; and for carbonate dehydratase[96] at temperatures below 60 °C and values of pH less than 3.5. These are all unphysiological conditions, but proteins have evolved to be entirely in their native states under physiological conditions. Consequently, it would not be surprising to find stable intermediates in the normal process of folding only under such peculiar conditions.

Various physical measurements have been made of these stable intermediates categorized as molten globular states. The majority of the circular dichroic absorption between 200 and 240 nm seen in the native conformation is retained in the intermediate state, and this suggests that the state retains α helices and β structure.[93,100] The circular dichroic absorptions between 260 and 290 nm, however, are largely lost, and this must result from the loss of unique asymmetric environments around tryptophans, tyrosines, and phenylalanines in the intermediate state. The complex nuclear magnetic resonance spectrum of the native state becomes much simpler and more like that of the random coil upon formation of the intermediate state,[101] as would be expected if the unique environments around each amino acid had been lost and each side chain now sampled continuously a broad range of changing environments. The fluorescence intensities of the tryptophans in cytochrome c, which are quenched by the nearby heme in the native state but are fully fluorescent in the random coil, remain quenched in the intermediate state.[101] The slow hydrogen exchange observed for deeply buried peptide bonds in the native state increases by a factor of about 1000 upon the transition to the intermediate state, even though the same amino acids remain less accessible.[95] This suggests that the same elements of secondary structure remain at the same locations in the amino acid sequence of the polypeptide but open up 1000-fold more often. The intrinsic viscosity, rotational relaxation times, and diffusion coefficients of these intermediate states are indistinguishable from those of the corresponding native states but are very different from those of the random coil.[100,101] These observations demonstrate that these intermediates are condensed, globular structures like that of the native state.

It is thought that the molten globular state represents the random coil that has collapsed to a globular state because of the hydrophobic effect, even though it cannot exist as the unique conformation that is the native state. In this regard, it is interesting that the majority (85%) of the change in heat capacity between the random coil and the native structure of α-lactalbumin is experienced in the transition between the random coil and the intermediate that has been characterized as the molten globular state.[100a] Large changes in heat capacity are a hallmark of the hydrophobic effect.

If the molten globular state does define all of these intermediates and if such states resemble intermediates on the normal kinetic pathway between the random coil and the native state, then the condensation model for the folding of a polypeptide may be an accurate rendition of the process. In this description of folding, the random coil spontaneously collapses under the influence of the hydrophobic force to form a condensed structure that would be a molten globule. This molten globule would fluidly sample the limited number of conformations available to the condensed polymer until the native conformation, the conformation of lowest free energy, was encountered.

The alternative to the condensation model for the folding of a polypeptide could be referred to as the **nucleation model**. In this view of the process, a short segment of the polypeptide would spontaneously assume a metastable configuration similar to the configuration of that short segment in the complete native structure. This nucleus for folding would resemble the configuration of the native structure in both its secondary and tertiary interactions, and it would represent the most independently stable region of the native structure. From this nucleus, folding would rapidly spread to produce the entire native structure. Evidence for this proposal comes from the study of short segments of polypeptide that can assume structured states other than the random coil.

Although almost all short segments of polypeptide have proven to be structureless, recently a few examples have been found that assume a structured state. Two peptides from bovine pancreatic trypsin inhibitor (n_{aa} = 58), Asparagine 43–Alanine 58 and Arginine 20–Phenylalanine 33, were chemically synthesized and joined by forming the cystine between Cysteine 51 and Cysteine 30 that occurs naturally in the native protein. This covalent complex, containing only half of the covalent structure of the full-length protein, nevertheless formed a structure[102] that seemed to resemble the structure of this region in the crystallographic molecular model of the complete protein. This inference was reached by examining circular dichroic spectra and two-dimensional nuclear magnetic resonance spectra. There were, however, some drawbacks to the interpretation. The effects of temperature on the nuclear magnetic resonance spectra and the circular dichroic spectra did not seem to be consistent with a two-state equilibrium between a unique folded conformation and a random denatured state. Furthermore, although the nuclear Overhauser effects between the protons on the α-carbons and the amide protons expected of the antiparallel sheet of β structure (Figure 12–25) within this region of the crystallographic molecular model of the complete protein were clearly observed, the nuclear Overhauser effects between the α-protons and the amide protons expected from the α helix (Figure 12–24) that is also in this region were not. Nevertheless, it was proposed that this segment of the protein could represent the nucleus that initiates the folding of the full-length polypeptide. The short, stable α-helical segments of polypeptide discussed earlier (Chapter 5) have also been proposed as models of nucleation points in protein folding.[103]

Suggested Reading

Salahuddin, A., & Tanford, C. (1970) Thermodynamics of the Denaturation of Ribonuclease by Guanidine Hydrochloride, *Biochemistry 9*, 1342–1347.

Taniuchi, H., & Anfinsen, C.B. (1971) Simultaneous Formation of Two Alternative Enzymatically Active Structures by Complementation of Two Overlapping Fragments of Staphylococcal Nuclease, *J. Biol. Chem. 246*, 2291–2301.

Problem 13–1: As urea is added to a solution containing a protein in its native structure, the protein usually begins to unfold when the concentration of urea rises above 4–5 M. We shall assume for a moment that this is due to the ability of urea to stabilize the unfolded state; prove this assumption, and then decide how urea is able to do this.

Consider an amino acid residue that is located in the interior of a protein and cannot see the solvent when the protein is folded. In considering this interior residue, the following series of equilibria govern the unfolding process.

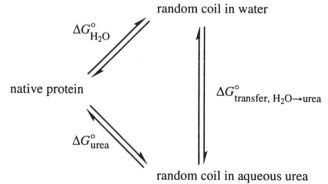

The standard free energy changes $\Delta G^\circ_{H_2O}$ and ΔG°_{urea} are standard free energy changes that occur as the amino acid is transferred from the interior of the protein either into pure water or a solution of urea, respectively, as the protein unfolds to a random coil, and $\Delta G^\circ_{transfer\ H_2O \to urea}$ is the standard free energy change involved in transferring an amino acid, which is exposed during unfolding, from water into a urea solution.

(A) How are these three ΔG°s related? What sign must each carry to explain the unfolding caused by urea?

The following is a table of the solubilities of a series of amino acids in solutions of several concentrations of urea at 25.1 °C.

solute	\multicolumn{5}{c}{solubilities [g(100 g of solvent)$^{-1}$] at urea concentration of}				
	0	2 M	4 M	6 M	8M
glycine	25.1	22.7	20.4	17.5	15.00
alanine	16.7	15.3	13.7	12.1	10.60
leucine	2.16	2.37	2.34	2.29	2.25
phenylalanine	2.80	3.42	3.94	4.33	4.67
tryptophan	1.38	1.98	2.65	3.31	3.95
methionine	5.59	6.19	6.74	7.00	6.99
threonine	9.80	9.56	9.07	8.31	7.41
tyrosine	0.0451	0.0600	0.0732	0.0870	0.0986
histidine	4.33	4.66	4.70	4.46	4.23
glutamine	4.30	4.49	4.49	4.30	4.02
asparagine	2.51	2.89	⌣.08	3.22	3.32

By calculating $\Delta G^\circ_{transfer\ H_2O \to urea}$ for each model compound and subtracting $\Delta G^\circ_{transfer\ H_2O \to urea}$ for glycine (or ethyl acetate in the case of *N*-acetyl-tetraglycine ethyl ester), the $\Delta G^\circ_{transfer\ H_2O \to urea}$ for only the side chains and the peptide group can be calculated. These are tabulated below. Note that the negative of the standard free energy of transfer is tabulated.

side chain	\multicolumn{8}{c}{$-\Delta G^\circ_{transfer\ H_2O \to urea}$ (cal mol^{-1})}							
	\multicolumn{4}{c}{urea}				\multicolumn{4}{c}{guanCl}			
	2 M	4 M	6 M	8 M	1 M	2 M	4 M	6 M
peptide[a]	49	86	118	130	83	134	207	245
Ala	0	−15	−10	−10	10	20	30	45
Val[b]	60	85	125	160	85	115	195	265
Leu	110	155	225	295	150	210	355	480
Ile[b]	100	140	205	265	135	190	320	430
Met	115	225	325	415	150	245	400	535
Phe	180	330	470	600	215	355	580	775
Tyr	225	395	580	735	235	385	605	770
Trp	270	505	730	920	400	630	980	1,235
Pro[b]	75	105	155	200	100	140	240	320
Thr	40	60	90	115	65	90	120	125
His	100	160	205	255	180	285	385	420
Asn	135	225	330	430	200	320	490	645
Gln	80	130	190	230	135	215	315	360

[a]Based on solubility studies of *N*-acetyltetraglycine ethyl ester and ethyl acetate.
[b]The values for these side chains are estimates based on results for the other side chains and on results at a single denaturant concentration.

(B) Plot $\Delta G^\circ_{transfer\ H_2O \to urea}$ against [urea] for each of these side chains. Determine the slopes of these lines to yield values for $(\partial \Delta G^\circ_{transfer} / \partial [urea])$ in joules (mole of side chain)$^{-1}$ [liter (mole of urea)$^{-1}$].

(C) How do these numbers correlate with your expectations in part A? Explain why the protein unfolds when [urea] rises above a certain critical level.

It is now necessary to determine what types of interactions are affected by urea. The accessible surface area of each of these side chains has been calculated from molecular models by a computer.

model	surface area of side chain (nm^2)
Ala	0.21
Val	0.48
Thr	0.51
Leu	0.67
Met	0.90
Phe	0.93
Tyr	1.10
Trp	1.34
Asn	0.60
Gln	0.89
His	0.83
peptide	0.72

(D) Plot $(\partial \Delta G^\circ_{transfer} / \partial [urea])$ against accessible surface area, labeling each point on your curve to keep track of the side chain it represents.

(E) From this display of the data alone, which noncovalent forces would you decide were affected by urea? Discuss both hydrophobic forces and hydrogen bonds.

(F) From the correlations you have made, determine a quantity joules (mole of side chain)$^{-1}$ [liter (mole of urea)$^{-1}$] nanometer^{-2} and calculate a quantity joules (mole of side chain)$^{-1}$ nanometer^{-2} at 5 M urea to compare with −9 kJ mol^{-1} nm^{-2}, the approximate relationship between the hydrophobic effect and accessible surface area. At 5 M urea, the value of joules (mole of side chain)$^{-1}$ nanometer^{-2} is much greater than −9 kJ mol^{-1} nm^{-2}. How is urea able to unfold a protein?

A denaturation curve can be employed to obtain a value of the equilibrium constant for the transformation between the native structure and the random coil at different concentrations of a denaturant. From each of these equilibrium constants, ΔG°_{FU} for the reaction can be calculated. The figure at the top of the next column is an example of the relationship between ΔG°_{FU} and denaturant concentration for the denaturation of lysozyme by urea and guanidinium chloride.

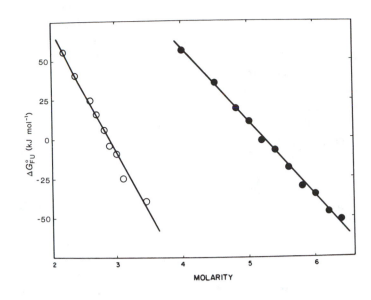

Apparent standard free energy of folding, ΔG°_{FU}, of lysozyme as a function of the molar concentrations of urea (●) or guanidinium chloride (○) at pH 2.9.[104] The apparent free energy of folding is zero at the concentration of denaturant at which [U] = [F]. Adapted with permission from ref 104. Copyright 1974 *Journal of Biological Chemistry*.

The slopes of these two lines ($\partial \Delta G^\circ / \partial [denaturant]$) are relative measures of the effectiveness of the two denaturants. By fitting a straight line to these data it is possible to obtain, by extrapolation, the value for ΔG° when [denaturant] = 0, which is referred to as ΔG°_{FU} (H$_2$O). The following table gathers the results from four separate proteins, where [guanCl]$_{1/2}$ or [urea]$_{1/2}$ is the concentration of denaturant when [U] = [F] and $\Delta G^\circ_{UF} = 0$.

(G) From an examination of the figure and an understanding of where the two points tabulated fall upon each curve, calculate $m = \partial \Delta G^\circ / \partial [denaturant]$ for each combination of protein and denaturant and complete columns 4 and 7 of the table below.

(H) Calculate the quantity

$$R_{prot} = \frac{(\partial \Delta G^\circ / \partial [guanidinium])}{(\partial \Delta G^\circ / \partial [urea])}$$

for each protein. This number will serve as a quantitative estimate of the relative effectiveness of the two denaturants.

protein	guanidinium chloride [guanCl]$_{1/2}$ (M)	ΔG°_{FU} (H$_2$O) (kJ mol^{-1})	m	urea [urea]$_{1/2}$ (M)	ΔG°_{FU} (H$_2$O) (kJ mol^{-1})	m
ribonuclease	3.01	39		6.96	32	
lysozyme	3.07	24		5.21	24	
chymotrypsin	1.90	32		4.04	35	
goat β-lactoglobulin	3.23	52		5.01	44	

(I) In part D you calculated a quantity $\partial(\partial\Delta G_{transfer}^{\circ}/\partial[urea])/\partial(surface\ area)$. Using the same methods, calculate a value for $\partial(\partial\Delta G_{transfer}^{\circ}/\partial[guanidinium])/\partial(surface\ area)$ from the data in the table at the bottom of page 467.

(J) Calculate the quantity $\partial(\partial\Delta G_{transfer}^{\circ}/\partial[guanidinium])/\partial(\partial\Delta G_{transfer}^{\circ}/\partial[urea]) = R_{transfer}$.

It can be seen by comparing $R_{transfer}$ and R_{prot} that although the relative abilities of guanidinium and urea to stabilize nonpolar surface area are in the right direction, they cannot explain the entire difference between these two denaturants which is oberved when a protein is unfolding.

(K) By a critical examination of the structures of the amino acid side chains and the peptide backbone, estimate the ratio between the number of hydrogen-bond donors and acceptors which become exposed when a protein unfolds. What is the ratio of donors to acceptors in H_2O, urea, and guanidinium?

(L) What solution will more readily satisfy all of the donors and acceptors that are exposed on unfolding?

(M) With these considerations in mind, use selected $\partial\Delta G_{transfer}^{\circ}/\partial[denaturant])$ values to explain why guanidinium is a much better denaturant than one would predict from $R_{transfer}$ alone.

(N) How do these pK_a values strengthen your case? $pK_a(guanidinium) = 14$, $pK_a(urea) = 18$.

Problem 13–2: Ribonuclease is a protein containing four cystines and 124 amino acids. Ribonuclease was added to a series of solutions containing different concentrations of guanidinium chloride and the reduced specific rotation $[\alpha']$ was measured at 365 nm when each of the solutions had come to equilibiribum.

[guanCl]	$-[\alpha']_{365}$	[guanCl]	$-[\alpha']_{365}$
0.00	239.0	3.19	289.9
1.23	238.5	3.35	300.8
1.98	239.3	3.52	302.9
2.43	247.3	3.89	306.4
2.55	247.6	4.42	306.5
2.79	257.8	5.03	306.3
3.03	276.6	5.40	306.8

The reduced specific rotation is a spectral indicator of the conformation of the peptide backbone of the protein. Make a plot of these data.

(A) What is $[\alpha']$ of native ribonuclease at 4 M guanidinium chloride?

(B) What is $[\alpha']$ of unfolded ribonuclease at 1 M guanidinium chloride?

The total reduced specific rotation is the sum of the reduced specific rotations due to each species present in the solution. In the case of ribonuclease it has been proven that only native and fully random coil structures are present at any [guanCl].

(C) Calculate K_{FU} for the equilibrium of Equation 13–1 for each [guanCl] and enter your values into a table.

(D) Plot $\ln K_{FU}$ against [guanCl] and determine, by extrapolation, the standard free energy of folding for ribonuclease in pure water, $\Delta G_{FU}^{\circ}(H_2O)$.

Problem 13–3: In the region of the nuclear magnetic resonance spectrum of ribonuclease between 8.0 and 9.0 ppm, the only absorptions present are those from the carbons 2 of the imidazole rings of the histidines. The figure presents this region of the nuclear magnetic resonance spectrum of ribonuclease and the changes that occur when guanidinium chloride is added to the sample at the noted concentration.

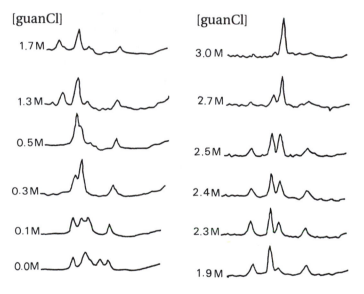

In the native protein (0.0 M guanidinium chloride), four absorptions from the protons on carbons 2 are observed. They have been assigned to Histidines 48, 119, 12, and 105, the four histidines of ribonuclease.

(A) Why does each absorption have a unique position in the spectrum of native ribonuclease?

(B) Why is there only one absorption from the protons on the carbons 2, which integrates as four protons from the protein, when it is dissolved in 3.0 M guanidinium chloride?

(C) Between 0.0 and 1.7 M guanidinium chloride the resonances shift around, but above 1.7 M the four absorptions simply disappear as the one absorption appears. What process is the spectrometer monitoring between 1.7 and 3.0 M guanidinium chloride?

(D) What would be the position of the absorption from the protons on the carbons 2 of N^α-acetyl-histidine ethyl ester in 3.0 M guanidinium chloride?

Problem 13–4:

(A) Below are listed several thermodynamic parameters that are involved in the process of protein folding. Which one is most affected by the steric constraints described in the Ramachandran plot?

(a) Hydrophobic standard free energy change

(b) Donor–acceptor standard free energy change

(c) Standard enthalpy hydrogen bond formation

(d) Electrostatic standard free energy change

(e) Configurational entropy of random coil

(f) Configurational entropy of native structure

Suppose proteins were held together by imine linkages rather than peptide bonds

(B) What effect would this have on the parameter you have chosen above?

(C) How would the value of ΔG°_{FU} be affected by this change?

Problem 13–5:
Micrococcal nuclease can be isolated from the extracellular medium of a culture of *Staphylococcus aureus*. It cleaves the phosphodiester bonds of nucleic acids. The protein is a single polypeptide of 149 amino acids containing no disulfides. The sequence of the protein is

ATSTKKLHKEPATLIKAIDGDTVKLMYKGQPMT
FRLLLVDTPETKHPKKGVEKYGPEASAFTKKMV
ENAKKIEVEFNKGQRTDKYGRGLAYIYADGKM
VNEALVRQGLAKVAYVYKPNNTHEQLLRKSEAQA
KKEKLNIWSENDADSGQ

The protein was modified with ethyl thiotrifluoroacetate, which reacts specifically with lysines.

The trifluoroacetylated nuclease was digested with trypsin and aliquots were removed at various times. The amino-terminal amino acids present in each of these samples were determined by labeling the free amino termini with fluorodinitrobenzene and identifying the labeled amino acids by thin layer chromatography of an acid hydrolysate. The results were as follows:

digestion time (min)	labeled amino termini (mole/mole of nuclease)				
	Ala	Lys	Leu	Glu	Thr
0	0.9	< 0.01	< 0.01	< 0.01	< 0.01
15	0.9	0.6	0.2	0.1	0.03
30	0.8	0.7	0.3	0.1	0.06
45	0.9	0.8	0.5	0.2	0.1

(A) What bonds are being split by the trypsin? Why were the lysines trifluoroacetylated?

(B) The trifluoroacetylated micrococcal nuclease was digested with trypsin for 8 min and the trypsin was rapidly inactivated. All the trifluoroacetyl groups were then removed. What are the major products of this tryptic digestion?

The products were first separated by gel filtration, which gave a high molecular weight fraction which contained nuclease activity and a peptide fraction from which a peptide, referred to as T peptide, was purified. It had the following amino acid composition:

amino acid	nmol (g of peptide)$^{-1}$
Lys	1.68
Asx	1.54
Ser	1.22
Glx	2.10
Gly	0.40
Ala	1.18
Ileu	0.44
Leu	0.39

What is T peptide?

The high molecular weight fraction still contained intact active nuclease, as well as a large fragment. This material was passed over a column containing a solid phase to which was covalently attached deoxythymidine, a molecule that binds to active micrococcal nuclease. None of the activity emerged from the column. The material that did emerge was further purified to yield a protein referred to as T protein.

(C) When T protein was digested with carboxypeptidase, arginine and leucine were released in a 1:2 molar ratio. What is T protein?

When micrococcal nuclease was added to 6 M guanidinium chloride, it became a structureless random coil. When the guanidinium chloride was removed, the native enzyme refolded completely, as judged by optical properties and enzymatic activity. Therefore, nuclease will reversibly fold and unfold; the native structure is the most stable in the absence of guanidinium chloride.

T Protein is completely structureless under all conditions. When T protein and T peptide were mixed together, no enzymatic activity appeared and the mixture remained structureless as judged by optical properties.

(D)   Discuss this observation in terms of the information that is required for the proper folding of a polypeptide.

Kinetics of Folding and Unfolding

When a protein is either dissolved in a solution of urea or guanidinium chloride or heated at low pH, the polypeptide unfolds and forms either a random coil or a denatured state reminiscent of a random coil, respectively. When the urea or guanidinium chloride concentration is rapidly lowered by dilution into a buffered aqueous solution or the pH is raised by mixing with a strongly buffered aqueous solution, the protein usually refolds in high yield to its native structure with full biological or enzymatic activity, if proper conditions are chosen.[103] Just as irreversible coagulations of the fully denatured protein must be avoided, coagulation of the unfolded protein after the denaturant is removed must be avoided. Before appreciable folding has occurred, the hydrophobic amino acids on all of the unfolded molecules are exposed simultaneously to the solution and no longer solvated by the denaturant. Intermolecular aggregation can compete with intramolecular folding. This is prevented by diluting the protein sufficiently to favor the intramolecular over the intermolecular pathway.

Many of the kinetic observations of the folding of polypeptides during the transition from a denatured state to the native structure have been performed with ribonuclease. In one sense, it is unfortunate that it is unfolded ribonuclease with its cystines intact that has been used in these studies. In the unfolded state, the presence of the correctly paired cystines brings the cross-linked random coil closer to the native structure both thermodynamically and kinetically. Nevertheless, several important insights into the kinetics of folding, in particular, the role of proline isomerization, have been gained in studies of the folding–unfolding of ribonuclease.

When ribonuclease is added to 5 M guanidinium chloride at pH 2.3 the unfolding of the polypeptide is very rapid (< 10 s), and it produces a cross-linked random coil.[50] If the solution containing this unfolded polypeptide is diluted immediately (within 15 s) to 1.3 M guanidinium chloride and pH 6.4, at 25 °C, all of the polypeptide (> 95%)[50] refolds to the native structure, capable of full enzymatic activity,[105] in an uncomplicated first-order relaxation[50] with a rate constant of 10 s^{-1}. This result demonstrates that the cross-linked random coil that has just been native ribonuclease can refold extremely rapidly, and with no obvious complications, back into native ribonuclease.

If unfolded ribonuclease is allowed to sit as a cross-linked random coil in a solution of guanidinium chloride, slow changes occur over 10 min. These changes are independent of pH (for pH ≥ 1.9) or the concentration of guanidinium chloride, and they produce an equilibrium mixture of at least two types of cross-linked random coil. This is ascertained by following the kinetics of refolding. During this equilibration the kinetics of refolding become dissociated into two phases, one with the same rate constant as that of the initially produced random coil and one that is much slower. These two phases in the refolding reaction are explained by assuming that a portion of the initially produced cross-linked random coil that refolds rapidly, U_F, has isomerized to a random coil that refolds slowly, U_S. If this is the case, the ratio between the concentration of isomer U_F and the concentration of isomer U_S at equilibrium is 1:4,[105] and the rate constant for the approach to the equilibrium between U_F and U_S is 5×10^{-3} s^{-1} at 10 °C.[50]

Very similar behavior has been observed in studies of the kinetics of folding–unfolding of cytochrome c (Figure 13–9).[106,107] If native cytochrome c is mixed with guanidinium chloride at a final concentration of 2.5 M, the unfolding (forward) reaction displays two phases: the rapid production of one isomer of the random coil, analogous to isomer U_F of ribonuclease, and the slow isomerization of this initial isomer into another, analogous to isomer U_S of ribonuclease. When the equilibrium mixture of these two isomers of the random coil in 4.0 M guanidinium chloride is diluted to 2.5 M guanidinium chloride, the folding (reverse) reaction displays the two phases corresponding to the rapid folding of isomer U_F and the slow folding of isomer U_S. This behavior and that of ribonuclease is consistent with the kinetic mechanism

$$F \underset{k_{-1}}{\overset{k_1}{\rightleftharpoons}} U_F \underset{k_{-2}}{\overset{k_2}{\rightleftharpoons}} U_S \qquad (13\text{--}18)$$
$$\underset{\text{fast}}{} \qquad \underset{\text{slow}}{}$$

In the case of cytochrome c, the rate constants k_1 and k_{-1} are both dependent on the concentration of guanidinium chloride,[106] but the apparent first-order rate constant ($k_2 + k_{-2}$, see Equation 13–9) for the approach to the equilibrium between U_F and U_S is independent of the concentration of guanidinium chloride[106] and has a value of 0.06 ± 0.01 s^{-1} at 25 °C.

It has been pointed out[108] that chymotrypsin and chymotrypsinogen, and trypsin and trypsinogen, in the form of random coils, are also equilibrium mixtures of rapidly folding and slowly folding isomers and that the rate constants for the production of the native structure from the slowly folding isomers at 25 °C are all very similar, between 0.1 and 0.02 s^{-1}. This has led to the proposal that the common mechanism explaining all of the slowly folding isomers of these random coils is that they contain peptide bonds on the amino-terminal side of particular proline residues which are in the wrong

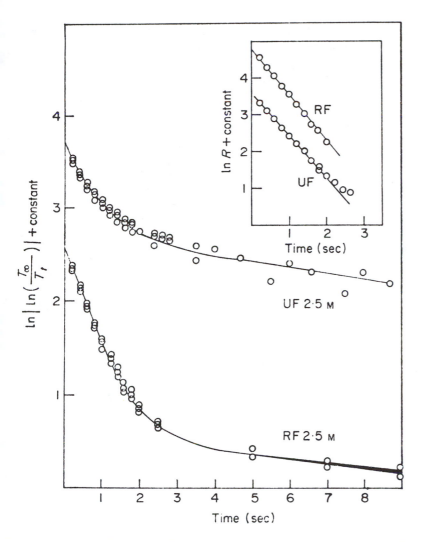

Figure 13–9: Kinetics of folding (RF, bottom trace) and unfolding (UF, top trace) of cytochrome *c* in 2.5 M guanidinium chloride at pH 6.5 and 25 °C.[106] Cytochrome *c*, dissolved in either a buffered aqueous solution (for unfolding) or a buffered solution at 4 M guanidinium chloride (for folding), was mixed rapidly with concentrated guanidinium chloride or water, respectively, to achieve the same final concentration of guanidinium chloride (2.5 M) and protein (70 μg mL^{-1}). The transmittance, *T*, of each solution was followed at 400 nm, the region in which the covalently attached heme of the cytochrome *c* absorbs. The natural logarithm of the transmittance, ln *T*, is directly proportional to the molar concentration of an absorbing species, and the natural logarithm of the difference between ln *T* at a given time, ln T_t, and ln *T* at equilibrium, ln T_∞ (Equation 13–22), is plotted as a function of time (seconds). On such a plot, a homogeneous first-order reaction would display linear behavior (Figure 13–3). Both folding (RF) and unfolding (UF), however, displayed two distinct first-order relaxations, a fast relaxation ($k_{F,obs}^{fast}$ = 1.2 s^{-1} and $k_{U,obs}^{fast}$ = 1.2 s^{-1}; see inset to figure) and a slow relaxation ($k_{F,obs}^{slow}$ = 0.06 s^{-1} and $k_{F,obs}^{slow}$ = 0.06 s^{-1}). All four of these rate constants should describe approaches to equilibrium (Equation 13–22). If only three species (N, U_F, and U_S) are in equilibrium with each other, the observed forward and reverse rate constants for the same process should be identical, as they are. Reprinted with permission from ref 106. Copyright 1973 Academic Press.

geometric isomer for the native structure and that it is the rate of **proline isomerization** that governs the folding of the slowly folding isomers.

In the crystallographic molecular models of proteins, about 25% of the peptide bonds on the amino-terminal sides of proline are *cis* peptide bonds and the rest are *trans* peptide bonds (Figure 2–9). These are geometric isomers. The peptide bond on the amino-terminal side of a proline at a particular position in the amino acid sequence of a particular protein either will be *cis* in every molecule of native structure or it will be *trans* in every molecule of the native structure. In the random coil, however, every proline is free to adopt either geometric isomer, and the *cis* and *trans* isomers slowly come to equilibrium.

The rate constant of the approach to this equilibrium between *cis* and *trans* isomers in dipeptides containing proline is between 2×10^{-3} and 5×10^{-3} s^{-1} at 25 °C above pH 3.[108] This is in the range of the rate constants for the slow isomerizations required before the fully native structure can form from slowly folding random coils. In dipeptides, the ratio between *cis* and *trans* isomers of proline varies with pH but is between 1:9 and 4:6. The more

prolines that have to be one or the other isomer before the random coil can fold to the native structure, the more random coils with at least one incorrect isomer of proline will be present in the solution, and the greater will be the proportion of slowly folding random coils at equilbrium.

In the crystallographic molecular model of ribonuclease, the peptide bond on the amino-terminal side of Proline 93 is a *cis* peptide bond. The amount of this peptide bond in the *cis* form in the random coil can be monitored by its insensitivity to proteolytic cleavage by prolidase.[109] In 8.5 M urea at 10 °C, 70% of this peptide bond is *cis*. When the urea is diluted to 0.3 M, the 30% of this peptide bond that is *trans* slowly and completely reverts to the *cis* isomer with a rate constant of 0.01 s^{-1}, as the polypeptide folds. Under these conditions, 30% of the random coil refolds slowly and 30% of the activity of the enzyme is regained slowly, both with a rate constant of 0.01 s^{-1}, as the polypeptide folds. Tyrosine 92 resides at the other end of this peptide bond, and it is the fluorescence from this tyrosine that monitors the slow isomerization that produces fully native enzyme during the refolding of the cross-linked random coil after a decrease in the concentration of guanidinium chloride.[110] It is presumed that the slow process monitored by the fluorescence of Tyrosine 92 is the state of isomerization of the peptide bond between Tyrosine 92 and Proline 93.

There are several properties of the transition between rapidly folding and slowly folding forms of the random coil of ribonuclease consistent with the isomerization being due entirely to isomerization of peptide bonds on the amino-terminal sides of prolines in the sequence. The rate constant for the approach to the equilibrium between rapidly folding and slowly folding forms of the random coil[111] and the formation of the native structure from the slowly folding random coil[110] are both increased by strong acid, as are the rate constants for the cis–trans isomerization in dipeptides of proline.[108] The enthalpy of activation for the equilibrium between rapidly folding and slowly folding forms of the random coil is between 75 and 90 kJ mol^{-1}, either at low pH or in 5 M guanidinium chloride,[111,112] which compares favorably to the values for the enthalpy of activation (80–90 kJ mol^{-1}) for the cis–trans isomerization of dipeptides of proline.[108] The slow phases of folding of other proteins also show similar enthalpies of activation (70–90 kJ mol^{-1}).[108] The rates of the approach to cis–trans equilibrium of neither dipeptides of proline nor the rapidly folding and slowly folding forms of the random coil are affected by the concentration of guanidinium chloride.[113]

The amide bonds on the amino-terminal sides of only certain prolines are critical for proper folding. Ribonucleases from some species have prolines at position 15 or 17 in their amino acid sequence, yet the folding of these polypeptides is indistinguishable from those that have no proline at either of these positions.[114] The fact that some of the prolines in the amino acid sequence of a polypeptide can be irrelevant to the kinetics of refolding would explain why the number of kinetically detectable, slowly folding intermediates is usually less than might be expected from the proline content of a polypeptide.

When the equilibrium mixture of rapidly folding and slowly folding random coils of ribonuclease is diluted into conditions favorable to folding, some of the slowly folding random coils assume nativelike conformations in which the critical peptide bonds on the amino-terminal sides of prolines are nevertheless the incorrect isomer. One intermediate, I_1, is sufficiently folded to trap almost 20 amide hydrogens in stable hydrogen bonds,[115] and another, I_N, is compactly folded by several criteria,[116] including its insensitivity to digestion by pepsin.[117] These intermediates, however, differ from the native structure and can be distinguished from it by having the incorrect isomers at particular prolines.[110]

Similar results were obtained in studies of the refolding of ribonuclease T_1 from Aspergillus oryzae.[118,119] In this case, only 3.5% of the unfolded polypeptides at equilibrium were able to fold rapidly (< 1 s). The remaining 96.5% of the unfolded polypeptides required minutes to hours to fold. They could be divided into three populations: two with one incorrect proline isomer each and one with two incorrect proline isomers. The complex kinetics of folding of the slow forms could be explained as the result of rate-limiting isomerizations of the prolines to the proper isomers followed by rapid folding.

All of these observations leave the impression that protein folding is a rather clumsy affair that becomes slower and more prone to inefficiency as more and more critical prolines accumulate in longer and longer polypeptides. This, however, may be an inaccurate picture of the situation within the cell at the locations where the folding of polypeptides to form their native structures actually occur. A number of enzymes, known as **peptidyl prolyl isomerases**, have been described that are able to catalyze the cis–trans isomerization of proline when it is either in short peptides or in folding polypeptides. The first enzyme with this ability that was purified to homogeneity[120,121] was assayed by its ability to catalyze the cis–trans isomerization in N-glutaryl-Ala-Ala-Pro-Phe 4-nitrophenylanilide.[122] This particular enzyme is able to increase the rates of the slowest phases in the refolding of ribonuclease A,[123] the S protein of ribonuclease A,[124] the light chain of immunoglobulin,[125] and type III collagen.[126]

In all of these instances the increase in the rate of folding observed was relatively unremarkable (less than a factor of 10) even at high concentrations of the peptidyl prolyl isomerase. It has been found, however, that there are a number of distinct enzymes responsible for this enzymatic activity, even within the same cell,[127,128] and it has been proposed that the slow isomerization of proline during the folding of polypeptides has been exploited as a regulatory mechanism. It is possible that polypeptides are prevented from folding by the problem of the isomerization of several critical prolines until they come in contact with the proper peptidyl prolyl isomerase, which catalyzes the isomerization of those prolines only at the proper time or proper place. If this is the case, the acceleration of the rate of folding by the peptidyl prolyl isomerase to which the particular polypeptide was matched might be much more remarkable than simply an order of magnitude.

A proper appreciation of the effects of proline isomerization on the rate at which a polypeptide folds returns the argument to the **rapidly folding form** of ribonuclease A. If this represents a form of the cross-linked random coil in which all slow and essential geometric isomerizations have reached the particular isomer required in the native structure and if the rapidly folding random coil is truly a random coil with no secondary structure, then the folding of a random coil to the native structure is a rapid process. It is through a kinetic examination of these rapid processes, uncomplicated by proline isomerization, that intermediates in the folding process will be identified and defined.

As a result of this realization, techniques that can follow the folding process over very short time periods are being applied to an evaluation of protein folding with the hope of detecting **kinetic intermediates in folding**.

For example, it is possible to follow circular dichroic ellipticity as a function of time in a stopped-flow, rapid mixing device.[129] By following the ellipticity at 222.5 and 420 nm, it was possible to detect an intermediate in the folding of cytochrome c. This intermediate displayed the ellipticity of the native structure at 222.5, which indicated that it had a content of α helix equivalent to that of the native structure, but displayed an ellipticity almost equivalent to that of the unfolded state at 420 nm, which suggested that the asymmetric environment of the native structure surrounding the heme had not yet formed. This intermediate formed in a time interval too short (< 20 ms) for the apparatus to resolve and then isomerized to the native structure in a biphasic process (k_1 = 8 s^{-1}; k_2 = 0.8 s^{-1} at 25 °C). Both of these rate constants seem too fast for proline isomerizations.

It is also possible to monitor the folding of polypeptides by following, as a function of time, the removal of particular amide protons on the backbone from access to the solvent. In such an experiment, fully deuterated protein as a random coil in a solution of urea or guanidinium chloride is rapidly diluted from the high concentration of the denaturant into buffered ^2H_2O and folding commences. At various times after folding begins the solution is diluted a second time into ^1H_2O at high pH, and all accessible amide deuterons exchange over a short time period (Figure 12–26). The protein is then rapidly and completely folded at a low pH to prevent further exchange. The final folded product is then examined by two dimensional nuclear magnetic resonance spectroscopy to determine which amide nitrogens have gained protons. Those that have were accessible when the pulse of ^1H_2O was applied. In this way, the sequestration of particular amide nitrogens from the solvent can be followed as the protein folds.

Rapid exchange has been used to study the folding of ribonuclease A with its cystines intact,[130] cytochrome c,[131] and the ribonuclease from B. amyloliquefaciens.[132] The amide protons that could be monitored in the rapidly folding form of ribonuclease A all were sequestered from solvent in concert, in agreement with the first-order behavior of the folding of this form of the unfolded polypeptide. With cytochrome c, however, the amide protons in the amino-terminal and carboxy-terminal α helices, which are adjacent to each other in the crystallographic molecular model, became sequestered at early times (10 ms), but the amide protons in the rest of the molecule became sequestered only at later times (1 s). In the ribonuclease from B. amyloliquefaciens, the amide protons in α helices and sheets of β structure in the crystallographic molecular model of the native protein became sequestered from the solution very rapidly (< 2 ms) in the rapidly folding form (25%) of the protein and at the same rate (20 s^{-1} at 25 °C) in a slowly folding form (75%) of the protein. As a group, these amide protons seemed to be sequestered in concert during the folding process. Amide protons, however, on the few

amino acids that could be followed and that were not involved in α helices or sheets of β structure in the crystallographic molecular model were sequestered more slowly still (5 s^{-1} at 25 °C).

The results of observations with the rapidly folding form of ribonuclease A suggest that its folding is a rapid cooperative process without intermediates. The results with cytochrome c suggest that the two ends of the polypeptide form α helices that nucleate the folding of the remainder. Unfortunately, in this instance, the effects of proline isomerization have not been defined.[131] The results obtained with ribonuclease from B. amyloliquefaciens suggest that in this protein all of the proper α helices and sheets of β structure form initially and the final tertiary structure snaps into place somewhat later. At the moment, these results suggest that the folding of each protein may be unique and that no one mechanism may describe the folding of polypeptides to form the native structure.

All of these experimental observations are relevant to the problem of predicting the native structure of a protein from its amino acid sequence. If the native structure is actually the conformation of the polypeptide, among all possible conformations, that has the lowest free energy, then the problem of predicting structure is a thermodynamic problem. If the native structure, however, is only a local minimum of free energy in which the polypeptide is trapped as the polypeptide condenses to a globular state, then the problem of predicting structure is a kinetic one, and the pathway through which folding proceeds must be predicted. The approach to the problem would be very different if it were thermodynamic or kinetic.

Now that the thermodynamics and kinetics of folding–unfolding have been discussed, a proper appreciation of the **thermal inactivation** of a protein is possible. The thermal inactivation of a protein is studied in the following manner. A solution of protein, usually at neutral pH, is divided into equal portions which are then each heated at a different temperature for the same length of time or each heated at the same temperature for a different length of time. The solution is then cooled and the biological or enzymatic activity of the protein is measured. The experiment is easy to perform, and results are always obtained.

As the temperature is raised, the equilibrium constant between native structure and random coil shifts in the direction of the random coil. Both the rate at which the protein unfolds, governed by k_U in Equation 13–1, and the time it spends in the unfolded state, governed by K_{FU}, increase. At some elevated temperature either the concentration of unfolded polypeptides is high enough that intermolecular coagulation occurs among them or covalent modifications of the protein[133] become rapid enough that inactivation proceeds through reactions on individual polypeptides. Both of these processes lead to an irreversible loss of the competent random coil as either a precipitate or an accumulation of covalently modified polypeptides. The random coils that survive these irreversible processes may or may not refold upon cooling and regain enzymatic activity. It is wonderful that so many thermal inactivations are presented as if they were informative.

Suggested Reading

Schmid, F.X., Grafl, R., Wrba, A., & Beintema, J.J. (1986) Role of Proline Peptide Bond Isomerization in Unfolding and Refolding of Ribonuclease, *Proc. Natl. Acad. Sci. U.S.A. 83*, 872–876.

Assembly of Oligomeric Proteins

When oligomeric proteins are dissolved in solutions of guanidinium chloride, they dissociate into their constitutive polypeptides and unfold to random coils. When the guanidinium chloride is removed from the solution, the random coils refold and the refolded monomers reassociate to form the native structure. For example, the pyrophosphatase from *Escherichia coli* is an α_6 hexamer in its native structure. When dissolved in 5 M guanidinium chloride at pH 7, it dissociates into single α polypeptides ($n_{aa} = 175$), as judged by sedimentation equilibrium, that are random coils, as judged by their sedimentation coefficient ($s^0_{20} = 0.59$ S) and intrinsic viscosity ($[\eta] = 22$ cm^3 g^{-1}). When the guanidinium chloride is removed by dialysis, 80–90% of the enzymatic activity slowly returns, and the protein that results is an α_6 hexamer indistinguishable in sedimentation coefficient, optical rotatory dispersion, or ultraviolet absorption spectrum from the original native enzyme.[134] The native protein contains no cystine, but it does contain cysteine. Renaturation is successful only if an external thiol is added to prevent adventitious intermolecular and intramolecular formation of cystine, a reaction that interferes with proper refolding.

The steps in the assembly of an oligomeric protein can be followed by **quantitative cross-linking**. Phosphoglycerate mutase from yeast is an α_4 tetramer (Figure 9–9). When it is dissolved in 4 M guanidinium chloride, it dissociates into random coils of the α polypeptide as judged by circular dichroism (Figure 13–10A). When the solution is diluted 40-fold to 0.1 M guanidinium chloride, greater than 80% of the ellipticity of the native structure is regained in less than 30 s (Figure 13–10B). At this point greater than 80% of the protein is still monomeric. The appearance of dimers and tetramers as a function of time could be followed by quantitative cross-linking to catalogue the species present at each point (Figure 13–10C–E).[135] No trimers were observed, as would be expected. From an examination of the progress of the reaction, it could be concluded that the dimer was the initial oligomer, which, as it built up in concentration, dimerized to produce tetramer. Although the circular dichroism of the sample changed insignificantly as the reaction progressed

(Figure 13–10B), the intrinsic fluorescence of the protein increased in concert with the oligomerization. Both the rate of the oligomerization (Figure 13–10C–E) and the rate of the increase in fluorescence (except for a small immediate increase of 20% that was invariant) were dependent on the absolute concentration of the protein. The kinetic behavior of both the oligomerization and the increase in fluorescence could be accounted for quantitatively[135] by the mechanism

$$4\alpha_U \xrightarrow{k_F} 4\alpha_F \underset{k_{-1}}{\overset{k_1}{\rightleftharpoons}} 2(\alpha_F)_2 \xrightarrow{k_2} [(\alpha_F)_2]_2 \quad (13\text{–}19)$$

wherein all unfolded polypeptides, α_U, have folded after the first 30 s; the folded monomer, α_F, regains 20% of the fluorescence of the native structure; and both the dimer and tetramer have the full fluorescence of the native structure. That the folded monomer and reassembled dimer can be digested with trypsin while the reassembled tetramer cannot[136] suggests that the monomer and the dimer are loosely folded. That both the monomer and the dimer possess some enzymatic activity[136] suggests that they are properly folded. The tetramer is produced in quantitative yield with full enzymatic activity at these concentrations of protein (< 50 µg mL^{-1}).[136]

The **assembly of a dimer** from its dissociated random coils is an even simpler reaction. Porcine mitochondrial malate dehydrogenase is an α_2 dimer that can be reversibly unfolded in several different ways. After random coils, α_U, of the α polypeptide are transferred to a solution at neutral pH, coincident with the dilution of the denaturant, the reappearance of enzymatic activity shows the same time course regardless of the mode of the original denaturation (Figure 13–11).[137] The time course displays two phases, a lag followed by an increase. The increase in activity has a second-order dependence on the concentration of protein. The lag is unaffected by the concentration of protein and is a first-order process. The results can be explained quantitatively with the following mechanism

$$2\alpha_U \xrightarrow{k_1} 2\alpha_F \xrightarrow{k_2} (\alpha_F)_2 \quad (13\text{–}20)$$

if only the dimer, $(\alpha_F)_2$, and not the folded monomer, α_F, is enzymatically active. At pH 7.6 and 20 °C, $k_1 = 6 \times 10^{-4}$ s^{-1} and $k_2 = 3 \times 10^4$ M^{-1} s^{-1}. The value of k_1 is too small to be the rate constant for the rapid refolding of a polypeptide with the correct proline isomers, and almost all of the random coil, α_U, must be slowly refolding. If the folding of the random coil of malate dehydrogenase must pass through a rapidly folding form (Equation 13–18), the isomerizations producing it must be first-order processes and therefore independent of the concentration of protein.

The **assembly of a trimer** (Figure 9–11) is somewhat more complicated than that of either a tetramer or a dimer because the addition of the third subunit to the dimer is quite different from the initial combination of two monomers to form the dimer.

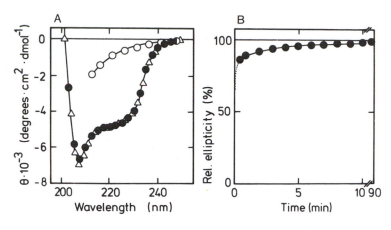

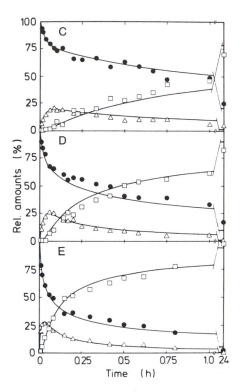

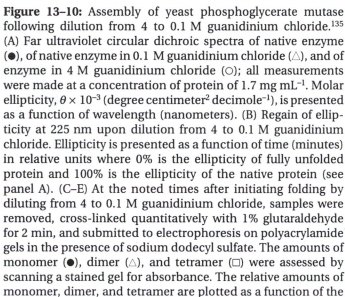

Figure 13–10: Assembly of yeast phosphoglycerate mutase following dilution from 4 to 0.1 M guanidinium chloride.[135] (A) Far ultraviolet circular dichroic spectra of native enzyme (●), of native enzyme in 0.1 M guanidinium chloride (△), and of enzyme in 4 M guanidinium chloride (○); all measurements were made at a concentration of protein of 1.7 mg mL^{-1}. Molar ellipticity, $\theta \times 10^{-3}$ (degree centimeter^2 decimole^{-1}), is presented as a function of wavelength (nanometers). (B) Regain of ellipticity at 225 nm upon dilution from 4 to 0.1 M guanidinium chloride. Ellipticity is presented as a function of time (minutes) in relative units where 0% is the ellipticity of fully unfolded protein and 100% is the ellipticity of the native protein (see panel A). (C–E) At the noted times after initiating folding by diluting from 4 to 0.1 M guanidinium chloride, samples were removed, cross-linked quantitatively with 1% glutaraldehyde for 2 min, and submitted to electrophoresis on polyacrylamide gels in the presence of sodium dodecyl sulfate. The amounts of monomer (●), dimer (△), and tetramer (□) were assessed by scanning a stained gel for absorbance. The relative amounts of monomer, dimer, and tetramer are plotted as a function of the

time (hours) between dilution of the guanidinium chloride and the addition of the glutaraldehyde. The final concentrations of protein were (C) 10.6 μg mL^{-1}, (D) 20.6 μg mL^{-1}, and (E) 36.9 μg mL^{-1}. The curves were drawn in all three panels on the basis of Equation 13–9 with $k_1 = 6.25 \times 10^3$ M^{-1} s^{-1}; $k_{-1} = 6.0 \times 10^{-3}$ s^{-1}; and $k_2 = 2.75 \times 10^4$ M^{-1} s^{-1}. The temperature for all of these experiments was 20 °C and they were run at pH 7.5. Reprinted with permission from ref 135. Copyright 1983 *Journal of Biological Chemistry*.

The catalytic subunit of aspartate transcarbamylase (Figure 9–12) is an α_3 trimer. The assembly of trimers of the catalytic subunit from random coils of the α polypeptide is a first-order process with no evident intermediates and a rate constant of 2×10^{-4} s^{-1} at 0 °C.[138] It seems that a slow isomerization of the partially folded α polypeptide is the rate-limiting step in the assembly from random coils. To circumvent the barrier presented by this isomerization to the observation of intermediates in the process, native α_3 trimer was dissociated into globular rather than unfolded α monomers with thiocyanate (S=C=N$^-$),[139] which is a milder denaturant than either urea or guanidinium chloride. It is an anion that salts in protein as does urea or guanidinium but not so vigorously. The enzymatically inactive α monomers retain most of the circular dichroic ellipticity and ultraviolet absorption of the native α_3 trimers and have an f/f_0 of 1.27.[139] These globular α monomers reassemble readily after the dilution of the thiocynate.

When the assembly was followed by quantitative cross-linking, α monomers turned directly into α_3 trimers with

no evidence for the formation of any α_2 dimers (< 3%). The appearance of α_3 trimers was coincident with the return of enzymatic activity. Both of these processes, however, were strictly second-order in the concentration of α monomer;[139] in other words

$$\frac{d[\alpha_3 \text{ trimer}]}{dt} = k[\alpha \text{ monomer}]^2 \qquad (13\text{–}21)$$

This result means that the transition state in the rate-limiting step of the assembly must contain only two folded α subunits. A mechanism consistent with both of these results is

$$2\alpha \xrightarrow{k_1} \alpha_2 \qquad (13\text{–}22)$$

$$\alpha_2 + \alpha \xrightarrow{k_2} \alpha_3 \qquad (13\text{–}23)$$

where $k_2 \gg k_1$. The explanation for this behavior is reasonable. When the third monomer adds to the dimer, two interfaces form simultaneously, and this reaction could have a much lower free energy of activation than the formation of the dimer itself. Because Reaction 13–23

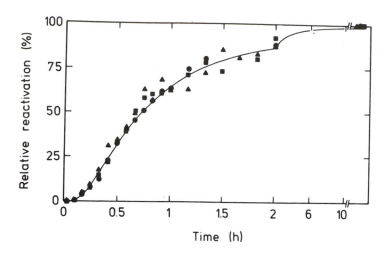

Figure 13–11: Reassembly and reactivation of porcine mitochondrial malate dehydrogenase at a concentration of 60 nM and at pH 7.6 and 20 °C after dilution from various denaturants.[137] The protein was unfolded at pH 2.3 (●), in 6 M guanidinium chloride (■), or in 6 M urea (▲). After it was fully unfolded and dissociated into its separate polypeptides in each instance, it was diluted to initiate refolding and reassembly. Samples were removed at the noted times and assayed for enzymatic activity; the enzymatic activity is presented in relative units with 0% being the immediately observed enzymatic activity (< 4% of the final) and 100% being the enzymatic activity after full reactivation (> 24 h). The curve was calculated for the mechanism of Equation 13–20 with $k_1 = 6.5 \times 10^{-4}$ s^{-1} and $k_2 = 3 \times 10^4$ M^{-1} s^{-1}. Reprinted with permission from ref 137. Copyright 1979 American Chemical Society.

is so much faster than Reaction 13–22, no α_2 dimer accumulates. Reaction 13–22, however, a bimolecular reaction, is the rate-limiting step.

When homooligomeric proteins are assembled from random coils, the observations are always consistent with the first step in the process being the folding of the random coil to a globular structure. In many instances this structure is loosely folded. For example, it may be sensitive to proteolytic cleavage. This globular monomer either combines directly with other globular monomers to form the oligomer, or it undergoes an isomerization before it is competent to assemble. These isomerizations may be *cis–trans* isomerizations of critical prolines, but this possibility has not been demonstrated. The competent monomers then assemble in simple, reasonable **bimolecular steps** to form the enzymatically active oligomer. The rate constants for these bimolecular steps are between 10^4 and 10^6 M^{-1} s^{-1} at 25 °C,[135,137,139] several orders of magnitude below diffusion-controlled rates for the collision of molecules of this size. Therefore, they proceed with significant free energies of activation.

Whether or not enzymatic activity is displayed by the folded intermediates in this process seems to be a property of the individual protein. Both α monomer and α_2 dimer of phosphoglycerate mutase have enzymatic activity.[136] Fumarase, an α_4 tetramer, can be denatured to random coils and reassembled to an α_2 dimer. The α_2 dimer is enzymatically inactive until it assembles to the α_4 tetramer.[140] When fructose-bisphosphate aldolase, an α_4 tetramer, is denatured to random coils which are then transferred to a solution at pH 5.5, the coils fold to form α monomers that have the sedimentation coefficient of a globular protein of their length and the circular dichroic spectra and ultraviolet spectra of the native protein. Their enzymatic activity cannot be determined because these α monomers oligomerize too rapidly to α_4 tetramers when mixed with substrates.[141]

The assembly of **heterooligomers** constructed from several copies of each of two or more different polypeptides is somewhat more complex than that of

homooligomers. When assembly reactions of this type have been studied, the reactants employed are the globular, homooligomeric subunits, such as catalytic subunits and regulatory subunits of aspartate transcarbamylase (Figure 9–12). It is generally assumed, in the absence of any evidence, that under physiological circumstances the homooligomeric subunits assemble first and then combine to form the heterooligomer. Only those proteins formed from two or more polypeptides translated from different messenger RNAs are of interest. Proteins containing different polypeptides arising from the posttranslational cleavage of identical larger polypeptides fold and assemble as simple homooligomers before the posttranslational modification occurs.

Tryptophan synthase is one of the simplest examples of the assembly of a heterooligomer.[142] This protein is an $\alpha_2\beta_2$ heterotetramer. When it is dissociated into its components, the products are α monomers and β_2 dimers, and both can be obtained in globular, folded states. When α monomer is mixed with excess β_2 dimer, the major product is the complex $\alpha\beta_2^*$. It forms in a reaction whose kinetics are consistent with the mechanism

$$\alpha + \beta_2 \underset{k_D}{\overset{k_R}{\rightleftharpoons}} \alpha\beta_2 \underset{k_b}{\overset{k_f}{\rightleftharpoons}} \alpha\beta_2^* \qquad (13\text{–}24)$$

where the complex $\alpha\beta_2^*$ is an isomerized form of the initial intermediate $\alpha\beta_2$. The rate constants k_R, k_D, k_f, and k_b for this process at 25 °C are 1×10^6 M^{-1} s^{-1}, 3 s^{-1}, 6 s^{-1}, and 1×10^{-3} s^{-1}, respectively. When excess α monomer is then added to the $\alpha\beta_2^*$ complex, the next step in the assembly has kinetic behavior consistent with the mechanism

$$\alpha + \alpha\beta_2^* \underset{k_D'}{\overset{k_R'}{\rightleftharpoons}} \alpha_2\beta_2^* \underset{k_b'}{\overset{k_f'}{\rightleftharpoons}} \alpha_2\beta_2^{**} \qquad (13\text{–}25)$$

the rate constants k_R', k_D', k_f', and k_b' for this process at 25 °C are 1.6×10^6 M^{-1} s^{-1}, 26 s^{-1}, 16 s^{-1}, and 2×10^{-3} s^{-1}, respectively.

Each time an interface forms between an α monomer

and one of the two β monomers in the β_2 dimer of tryptophan synthase, an isomerization of the structure of either the participating β monomer or the conjoined α monomer, or both, occurs, producing the asterisked conformer. The isomerizations producing the conformers $\alpha\beta_2^*$ or $\alpha\beta_2^{**}$ are too rapid to be isomerizations of prolines and presumably represent rearrangements of the structures similar to the changes that permit the tetramer of phosphoglycerate mutase to resist proteolytic degradation or that permit enzymatically inactive subunits to regain activity after oligomers such as malate dehydrogenase and fumarase reach the native stoichiometry. The equilibrium constant ($K_{isom} = k_f/k_b$; Equation 13–8) for the isomerization following the addition of the first α monomer ($K_{isom} = 6000$) is the same as that for the addition of the second α monomer ($K_{isom}' = 8000$), indicating that the same local adjustments are occurring after each α monomer adds in turn.

The overall dissociation constant ^1K_d for the first step in the assembly of tryptophan synthase (Equation 13–24) is defined by the relationship

$$^1K_d = \frac{2[\alpha]_{eq}[\beta_2]_{eq}}{[\alpha\beta_2^*]_{eq}} \qquad (13\text{–}26)$$

where the factor of 2 is a **statistical correction** that

accounts for the fact that each β_2 dimer has two faces for an α monomer. Because, at equilibrium (Equation 13–8),

$$k_R[\alpha]_{eq}[\beta_2]_{eq} = k_D[\alpha\beta_2]_{eq} \qquad (13\text{–}27)$$

and

$$k_f[\alpha\beta_2]_{eq} = k_b[\alpha\beta_2^*]_{eq} \qquad (13\text{–}28)$$

it follows that

$$^1K_d = \frac{2k_bk_D}{k_fk_R} = 1\times10^{-9}\ \text{M} \qquad (13\text{–}29)$$

The overall dissociation constant ^2K_d for the second reaction (Equation 13–25) is defined by the relationship

$$^2K_d = \frac{[\alpha]_{eq}[\alpha\beta_2^*]_{eq}}{2[\alpha_2\beta_2^{**}]_{eq}} = \frac{k_b'k_D'}{2k_f'k_R'} = 1\times10^{-9}\ \text{M} \qquad (13\text{–}30)$$

where the factor 2 in the denominator is a statistical correction which states that each $\alpha_2\beta_2^{**}$ tetramer has two α monomers that can dissociate. In this reaction, therefore, $^1K_d = {}^2K_d$ and each α monomer seems to add independently to a face on the β_2 dimer. This may not reflect exactly the true state of affairs, however, because the rate constant k_D' for the dissociation of an α monomer from the $\alpha_2\beta_2^*$ complex is almost 10-fold the rate constant k_D

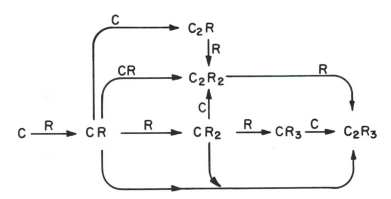

Figure 13–12: Intermediates in the assembly of aspartate transcarbamylase.[143] Catalytic (C) or regulatory (R) subunits that had been made radioactive by iodinating their tyrosines with ^{125}I were mixed with excess unradioactive regulatory or C subunits, respectively, to initiate assembly. At various times, the assembly reaction was quenched with either excess C subunit or excess succinylated C subunit to cap off partially formed complexes and scavenge all unreacted R subunit. From examining the specific radioactivity of complexes separated by electrophoresis, an estimate of the relative concentrations of all of the intermediates in the process of assembly at each time point could be made. The changes in these relative concentrations with time was used to formulate the assembly diagram displayed in the figure. Four of the steps in this process are rapidly reversible: $C + R \rightleftharpoons CR$, $CR + R \rightleftharpoons CR_2$, $CR_2 + R \rightleftharpoons CR_3$, and $CR + C \rightleftharpoons C_2R$. In contrast, processes forming the complexes C_2R_2 and C_2R_3 are essentially irreversible because these complexes are so stable. Reprinted with permission from ref 143. Copyright 1980 *Journal of Biological Chemistry.*

for the dissociation of an α monomer from the $\alpha\beta_2$ complex. This suggests that there are steric effects expelling a second monomer from the crowded complex $\alpha_2\beta_2^*$. Presumably these effects are relaxed after the rearrangement to the complex $\alpha_2\beta_2^{**}$ has occurred.

Aspartate transcarbamylase is constructed from two catalytic C subunits that are α_3 trimers and three regulatory R subunits that are β_2 dimers. From its crystallographic molecular model (Figure 9–12), it is clear that only certain steps are possible in its assembly from separated C subunits and R subunits (Figure 13–12). The intermediates that appear during the assembly of the intact $(\alpha_3)_2(\beta_2)_3$ heterododecamer, or $C_2 R_3$ heteropentamer, have been followed by quenching the assembly of radioactive catalytic or regulatory subunits and their unradioactive complements with large excesses of unradioactive catalytic or succinylated catalytic subunits, respectively.[143] The specific radioactivity of the various mosaic oligomers, which can be separated from each other by electrophoresis, permits the concentrations of the various intermediates at the time the reaction was quenched to be calculated.

When a limiting concentration of C subunit is mixed with various excesses of R subunit, equilibrium mixtures of the intermediates CR, CR_2, and CR_3 are formed. Subsequent addition of excess C subunit causes CR_3 to be trapped as CR_3C, the intact native protein, and CR_2 to be trapped as CR_2C.[143] When excess C subunit is mixed with a limiting concentration of R subunit, the only two products other than unreacted C subunit are CR_3C and CR_2C with the former in the majority.[144] If any CRC is formed it must disproportionate rapidly to 2C and CR_2C. The complex CR_2C can be isolated as a stable protein.[145] When it is combined with R subunit, it produces CR_3C in a clean bimolecular reaction.[144] In these experiments, most of the intermediates in the general scheme (Figure 13–12) have been directly observed, and rate constants and equilibrium constants for their interconversion have been established.[146] Most of the steps in the scheme seem to occur simultaneously, and different pathways become more or less important as concentrations of the subunits are changed.

The pyruvate dehydrogenase complex of E. coli is composed of three different polypeptide chains, α, β, and γ. The protein can be resolved into three independent components. These are the dihydrolipoamide transacetylase core, the pyruvate decarboxylase subunits, and the dihydrolipoamide dehydrogenase subunits (Reactions 9–4, 9–5, and 9–6). The dihydrolipoamide transacetylase core is an octahedral α_{24} oligomer (Figure 9–23), pyruvate decarboxylase is a β_2 dimer, and lipoamide dehydrogenase is a γ_2 dimer. No association can be detected between the β_2 dimers of pyruvate decarboxylase and the γ_2 dimers of lipoamide dehydrogenase.[147] Therefore, the α_{24} oligomer of the dihydrolipoamide transacetylase serves as the point of attachment of the other components.

Unlike the closely related dihydrolipoamide trans-succinylase from α-ketoglutarate dehydrogenase, which can associate with only six β_2 dimers of α-ketoglutarate decarboxylase at saturation because of a steric effect,[148] the empty α_{24} oligomer of the dihydrolipoamide transacetylase from E. coli can associate with up to 24 β_2 dimers of pyruvate decarboxylase.[147,149] Presumably, in the saturated complex, one of the two faces on each of the 24 β_2 dimers occupies one of the 24 equivalent faces of the octahedral α_{24} oligomer with no steric hindrance. The empty α_{24} oligomer of the dihydrolipoamide transacetylase can also associate with as many as 20 γ_2 dimers of dihydrolipoamide dehydrogenase in the absence of pyruvate decarboxylase.[149]

When both β_2 dimers of pyruvate decarboxylase and γ_2 dimers of dihydrolipoamide dehydrogenase are added together to the dihydrolipoamide transacetylase core, substoichiometric amounts of each are bound,[149] presumably because of steric crowding. Certainly the native protein, which has an average of about 12 γ_2 dimers of dihydrolipoamide dehydrogenase and an average of somewhat less than 24 β_2 dimers of pyruvate decarboxylase,[147] appears to be a crowded structure (Figure 13–13). When a preformed complex containing an average of 12 γ_2 dimers of dihydrolipoamide dehydrogenase for each α_{24} oligomer of dihydrolipoamide transacetylase is mixed with increasing amounts of pyruvate decarboxylase, about 22 β_2 dimers of pyruvate decarboxylase bind to the α_{24} oligomers at saturation, and the overall enzymatic activity increases in direct proportion to the number bound.[147] All of these results suggest that β_2 dimers of pyruvate decarboxylase and γ_2 dimers of dihydrolipoamide dehydrogenase add at random to the respective faces on the α_{24} oligomer of dihydrolipoamide transacetylase, at least under the circumstances of these experiments, until there is no more room left around the core. What is not clear is whether the dimers of dihydrolipoamide dehydrogenase and pyruvate decarboxylase add at random to the core during normal assembly in the cell until no more can fit or there is some ordered sequence that determines the final stoichiometry.

A ribosome from E. coli is assembled from two ribonucleoproteins, the 30S subunit and the 50S subunit. These subunits associate with and dissociate from each other during the normal cycle of protein biosynthesis, so they can be considered as separate oligomeric proteins. The 30S subunit, being the smaller, is the more extensively understood. It is composed of a single strand of 16S ribosomal RNA (n_{nuc} = 1541 nucleotides)[151] and 21 polypeptides. When the ribosomal RNA and the polypeptides are mixed together, they spontaneously reassemble in high yield to form 30S subunits that are fully competent to participate in protein biosynthesis.[152] The assembly of the intact 30S subunit from its components (Figure 13–14) proceeds through an explicit sequence of steps beginning with the binding of a few of the polypeptides to the 16S ribosomal RNA itself.[153]

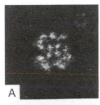

Figure 13–13: Electron micrographs of (A) the pyruvate dehydrogenase complex from *E. coli* and (B) the dihydrolipoamide transacetylase core from the same protein.[150] Both specimens were adsorbed onto a thin, supported layer of amorphous carbon on an electron microscopic grid and negatively stained with sodium methylphosphotungstate. Magnification is 300000×. The complete complex was purified directly from a homogenate of the bacteria; the transacetylase core was prepared from the complete complex by stripping away dihydrolipo-amide dehydrogenase and pyruvate carboxylase. Reprinted with permission from ref 150. Copyright 1971 Cold Spring Harbor Laboratory.

Figure 13–14: Assembly diagram for the 30S subunit of the ribosome from *E. coli*.[153] The sequence of events was determined by mixing, in various combinations, the 21 purified polypeptides with the purified 16S RNA and assaying for formation of a complex or complexes among the components. For example, only polypeptides S15, S17, S4, and S7 would bind alone to 16S ribosomal RNA. Polypeptide S20 will form a complex with 16S RNA only when polypeptides S4 and S8 are already bound. Polypeptide S13 binds to 16S RNA only when polypeptides S4, S8, and S20 are bound, and so forth. Reprinted with permission from ref 153. Copyright 1974 *Journal of Biological Chemistry.*

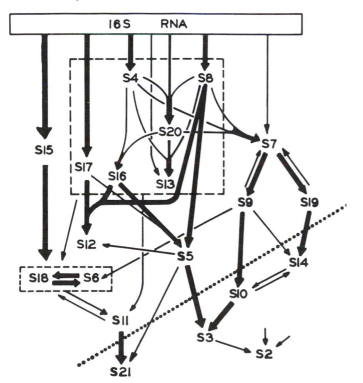

The diagram suggests that the RNA is the core of the 30S subunit upon which is assembled, in layers, a shell of protein. Under the conditions of assembly, the 16S ribosomal RNA assumes a compact globular structure which may result in part from the presence of intramolecular double-helical segments. These are believed to form throughout the molecule from about 60 pairs of complementary sequences 4–10 base pairs in length that are found spread uniformly throughout the overall sequence of the nucleic acid.[154] These segments of double-helical secondary structure in the form of both double-helical hairpins and double helices formed from two distant segments of the sequence would associate among themselves to form a unique globular structure as do the segments of secondary structure in a molecule of protein or the four double-helical hairpins in the folded polynucleotide of a transfer RNA. In animal ribosomes, and presumably also in the 30S subunit of a ribosome from *E. coli*, the RNA forms the compact center of the final ribonucleoprotein,[155] a location consistent with its role in the initiation of assembly. The folded 16S ribosomal RNA has specific binding sites for each of the polypeptides that can initially associate with it or for consecutive sets of polypeptides that can associate with it. The regions of the RNA forming these binding sites have been identified by submitting complexes between the 16S ribosomal RNA and various combinations of the ribosomal polypeptides to digestion with T_1 ribonuclease or pancreatic ribonuclease and identifying the segments of the RNA protected from digestion by the bound polypeptides.[154,156]

As the assembly progresses, the binding to the 16S ribosomal RNA of the polypeptides earlier in the sequence of events or the binding of polypeptides to complexes between the 16S ribosomal RNA and other polypeptides creates sites to which polypeptides later in the sequence of events can attach (Figure 13–14). If a polypeptide is added to the mixture before all of the polypeptides that must precede it have been added, it will not bind to the partially assembled 30S subunit. An assembly map, necessarily of greater complexity but describing a similar hierarchically ordered process, has been drawn for the assembly of the 50S ribosomal subunit of *E. coli* from 23S ribosomal RNA, 5S ribosomal RNA, and 31 polypeptides.[157]

It has already been noted that some of the polypeptides in the assembled 30S subunit seem to be highly elongated because they present antigenic sites over broad ranges upon the surface of the ribonucleoprotein. One explanation for this observation is that each of these polypeptides enters the 30S subunit during its assembly as a random coil which takes up a widely distributed but precisely defined track passing through the 30S subunit. Along this track the polypeptide would assume α helices and β structure, but there would be no requirement that these segments of secondary structures associate with other segments from the same polypeptide. The atomic details of the structure of the protein in a 30S subunit can be no different from that of any other protein; but, as with any other protein, segments of polypeptide associate with each other regardless of their location in the amino acid sequence or regardless of whether they are on the same polypeptide or on different polypeptides.

One polypeptide that seems to be almost structureless before it associates with the 16S ribosomal RNA is polypeptide S4 (n_{aa} = 203). The nuclear magnetic resonance spectrum of polypeptide S4 under the conditions in which assembly takes place is almost indistinguishable from its spectrum in 8 M urea, which is the spectrum of the sum of the amino acids from which it is composed.[158] When a polypeptide folds, the chemical shifts of the nuclear magnetic absorptions of each amino acid move away from their standard values as each amino acid enters its unique environment in the folded structure, but such shifts are not observed in polypeptide S4. This result indicates that polypeptide S4 cannot assume a unique native structure. The circular dichroic spectrum[158] and frictional ratio (f/f_0 = 1.7), however, are not those of a fully random coil (f/f_0 = 2.4 for n_{aa} = 245),[159] and an explanation of these results and those from nuclear magnetic resonance spectroscopy would be that the polypeptide in solution is rapidly passing through an array of loosely folded conformations, none of which is unique. As it binds to the 16S ribosomal RNA, its final structure could easily be dictated solely by the noncovalent interactions in which it participates as it spreads over the surface of the folded polynucleotide. This would explain how polypeptide S4 alone is able to protect a larger portion of the RNA from digestion than can any other.[154]

The most peculiar of the ribosomal polypeptides in this regard is polypeptide L7 (n_{aa} = 120) from the 50S ribosomal subunit of *E. coli*. By sedimentation equilibrium the polypeptide, under the conditions suitable for assembly, is an α_2 dimer[160] whose total length would be 240 amino acids, yet it has the intrinsic viscosity ($[\eta]$ = 28 cm^3 g^{-1})[161] of a random coil 240 amino acids long.[7] It is difficult to reconcile a random coil with a noncovalent dimer. To confuse the situation further, the circular dichroic spectrum of polypeptide L7 suggests that the majority of it is α-helical.[161]

Some ribosomal polypeptides seem to enter the assembling 30S subunit as folded globular proteins. For example, polypeptide S17 (n_{aa} = 83) under the conditions of assembly has both the frictional ratio (f/f_0 = 1.24; calculated from its sedimentation coefficient) and the intrinsic viscosity ($[\eta]$ = 4.2 cm^3 g^{-1}) of a globular protein and the molar mass, as determined by sedimentation equilibrium, of a monomer.[162] Other polypeptides, however, seem to be elongated but compact proteins under the conditions of assembly. For example, polypeptides S3, S5, S6, and S7 have frictional ratios f/f_0 = 1.4–1.6 that are too large to be those of globular proteins.[159,162] It

seems clear that the original image of folded globular polypeptides adding sequentially to the enlarging ribosomal subunit is unrealistic. Polypeptides meandering through the structure are probably interspersed with tightly folded neighbors in the usual mess that results from evolution by natural selection.

Suggested Reading

Hermann, R., Rudolf, R., Jaenicke, R., Price, N.C., & Scobbie, A. (1983) The Reconstitution of Denatured Phosphoglycerate Mutase, *J. Biol. Chem.* **258**, 11014–11019.

Problem 13–6: Phosphofructokinase is a protein found in all organisms. It has been purified from rabbit liver and rabbit muscle. In both cases, it is a tetramer composed of four identical subunits, each formed from folded polypeptides 780 amino acids long. The proteins from these two organs, however, differ from each other in electrophoretic mobility. This results from a small number of amino acid substitutions between them. When a mixture of rabbit muscle and rabbit liver phosphofructokinase was incubated at 8 °C in 6 mM sodium citrate, pH 6.9, for 3 days, a new protein of intermediate electrophoretic mobility was formed from the other two. This new protein is also a tetramer. Below is a diagram of the electrophoresis patterns of three samples, each of which has been incubated for 3 days at 8 °C in 6 mM sodium citrate, pH 6.9.

	direction of electrophoresis →		
	⊖		⊕
muscle	□		
liver			□
muscle + liver	□	□	□

(A) What types of amino acid substitutions in the sequence of phosphofructokinase could explain the observed difference in electrophoretic mobility between the muscle protein and the liver protein? Give several possibilities.

(B) From how many of what kind of subunits is the new intermediate protein formed?

(C) Explain how the new protein was formed during the incubation. Why is it the only new protein which appears when there are three possibilities? What property of these tetramers does this experiment illustrate?

Problem 13–7: A protein is reversibly denatured by guanidinium chloride. The extent of denaturation is measured by the loss of the biological activity, and it is complete in 5 M guanidinium chloride. The rate of unfolding is measured by making a 10^{-4} M solution of enzyme 5 M in guanidinium chloride. The rate of renaturation is measured by diluting a concentrated solution of denatured enzyme in 5 M guanidinium chloride with a large volume of buffer so that the concentration of guanidinium chloride is negligible and the enzyme concentration is 10^{-4} M. The following data were obtained at 27 °C.

time (h)	denaturation (fractional activity)	time (h)	renaturation (fractional activity)
0.1	0.82	0.5	0.20
0.2	0.67	1.0	0.37
0.3	0.54	1.5	0.50
0.4	0.45	2.0	0.60
0.5	0.36	2.5	0.68
0.7	0.24	3.0	0.75
1.0	0.13	48.0	1.00

Determine the order of these reactions and the value of the rate constants. What mechanisms for these processes are compatible with the kinetics, in view of the fact that the protein is composed of four subunits?

Assembly of Helical Polymeric Proteins

There are two classes into which helical polymeric proteins can be divided when the question of assembly is considered, those that assemble irreversibly and those that assemble reversibly. Those that assemble irreversibly polymerize initially by the noncovalent assembly of monomeric subunits, but the polymers are often strengthened secondarily by the formation of naturally occurring covalent cross-links between adjacent subunits. Those that assemble reversibly require that reversibility for their biological role, and their assembly is not an approach to equilibrium but the result of a steady state.

Examples of helical polymeric proteins that assemble irreversibly are collagen (Figure 9–26), intermediate filaments (Figure 9–29), thick filaments, and fibrin.

The thick fibers of **fibrin** that form clots of blood are readily observed in scanning electron micrographs. They are lateral aggregates of thinner cables of fibrin; these cables of fibrin are assembled irreversibly from a protomeric unit known as **fibrinogen** (Figures 13–15A and 12–2), which is a freely soluble $(\alpha\beta\gamma)_2$ heterohexamer.[163] Each molecule of fibrinogen is constructed from two $\alpha\beta\gamma$ heterotrimers arrayed about a 2-fold rotational axis of symmetry. Each of the two $\alpha\beta\gamma$ heterotrimers contains a long segment of rope (112 amino acids) in which the three strands are wound around each other in a triple coiled coil (Figure 9–28). At the two ends of each rope are globular domains known as the terminal domain and the central domain, respectively. A terminal domain is composed of the folded carboxy-terminal regions of the β polypeptide and the γ polypeptide; the central domain is composed of the folded amino-terminal regions of all three polypeptides. Two $\alpha\beta\gamma$ heterotrimers are associated at their central domains around

the 2-fold rotational axis of symmetry to produce the molecule of fibrinogen.

Fibrinogen is a knotted segment of rope with a 2-fold rotational axis of symmetry centered on the knot, and it is ready to assemble into a helical cable. The infinite helix defined by a smooth curve is a geometric structure with 2-fold rotational axes of symmetry intersecting every one of its points, so the creation of an infinite helical polymer from a monomer with a molecular 2-fold axis of symmetry produces a structure with a 2-fold rotational axis of symmetry at each molecular 2-fold rotational axis of symmetry. When this polymer is finite, the two ends are identical.

Fibrinogen does not assemble into a helical cable until four short peptides, two fibrinopeptides A and two fibrinopeptides B, are removed from the amino-terminal ends of the two α polypeptides and the two β polypeptides to produce **fibrin monomers**. The fibrinopeptides have sequences that vary extensively among species and seem to satisfy only the requirement that they be polar and structureless. Immediately beyond the points of the proteolytic cleavages that remove the four fibrinopeptides, the amino acid sequences of the α and β polypeptides become highly conserved among species.[165] The α polypeptide of a fibrin monomer has the amino-terminal sequence GPRAlk–, where Alk is the alkyl group of valine, leucine, or isoleucine; and the β polypeptide of a fibrin monomer has the amino-terminal sequence GHRP–. A synthetic peptide of the sequence GPR can inhibit completely the polymerization of fibrin monomers to fibrin polymer.[165] It does so by competing with the amino termini of the α polypeptides, which are in the central domain, for a binding site on the terminal domains. These results suggest that the helical cable is assembled by the noncovalent binding of the central domain and terminal domain of one molecule of the fibrin monomer to the terminal domain and central domain, respectively, of another fibrin monomer to form a rotationally symmetric, doubly bonded dimer. The dimer would be elongated to the polymer by adding other monomers, dimers, or oligomers through steps each creating identical interfaces (Figure 13–15B). Each of the consecutive, individual noncovalent interactions holding the helical polymer together is between the proteolytically exposed amino terminus of an α polypeptide on a central domain and the binding site for its amino-terminal sequence, GPR–, on one of the terminal domains of another fibrin monomer. Electron micrographs of intermediates in the polymerization of fibrin monomers are consistent with this structural proposal (Figure 13–15C).[164]

The helical cable that forms as the assembly proceeds has a thickness, as determined by light scattering,[166] of about two fibrin monomers (Figure 13–15B,C) and is referred to as a **protofibril**. When fibrin monomers are created instantly by adding a large excess of the necessary proteolytic enzyme to a solution of fibrinogen, the initial rate of formation of protofibrils is bimolecular in

A

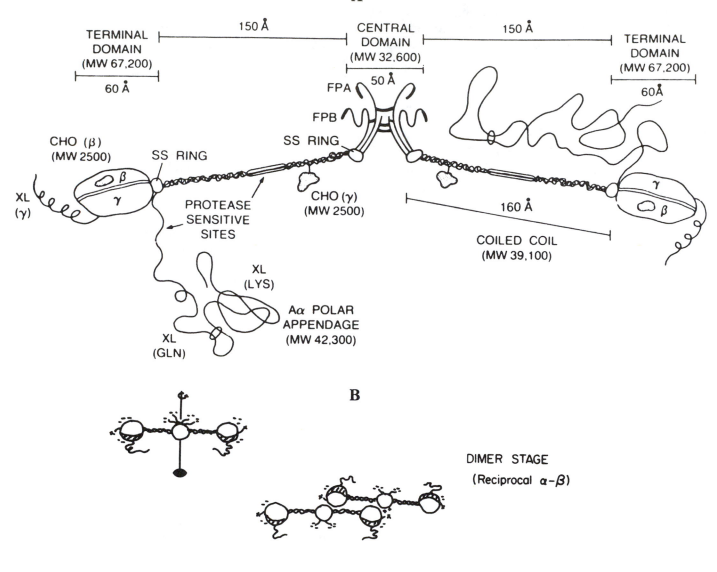

TERMINAL DOMAIN (MW 67,200)

60 Å

150 Å

CENTRAL DOMAIN (MW 32,600)

50 Å

150 Å

TERMINAL DOMAIN (MW 67,200)

60 Å

FPA

FPB

SS RING

CHO (β) (MW 2500)

SS RING

XL (γ)

β

γ

PROTEASE SENSITIVE SITES

CHO (γ) (MW 2500)

160 Å

COILED COIL (MW 39,100)

γ

β

XL (LYS)

XL (GLN)

Aα POLAR APPENDAGE (MW 42,300)

B

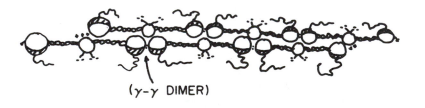

DIMER STAGE (Reciprocal α–β)

POLYMER STAGE

(γ–γ DIMER)

C

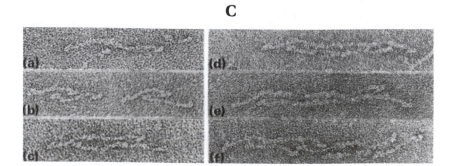

(a) (d)

(b) (e)

(c) (f)

Figure 13–15: Assembly of fibrin from fibrinogen. (A) Schematic drawing of fibrinogen showing the central domain with amino termini of all six chains, the connecting coiled coils, the two terminal domains, and the α chain protuberances.[163] The linear arrangement of the three major domains and their noted dimensions reflect results obtained from electron microscopy (Figure 12–2). The lengths of the coiled coils (16 nm) are based on calculations for a triple-helical coiled coil (Figure 9–28) 112 residues in length and agree closely with results from electron microscopy. Fibrinopeptides A and B are noted at the amino-terminal ends of the α and β chains. The two sets of polypeptide chains are held together by three intersubunit cystines, two between the γ chains and one between the α chains. There are four carbohydrate clusters (CHO), each about 14 monosaccharides in length, located on the γ chains near the central domain and on the β chains on each terminal domain. Primary cross-linking sites (XL) are situated very near the carboxy termini of the γ chains. On the α chain, the two cross-linking glutamine acceptor sites (Gln) are about 200 residues from the five potential donor lysine sites (Lys). The length of the human α polypeptide is 610 amino acids; the human β polypeptide, 461 amino acids; and the human γ polypeptide, 411 amino acids. (B) Schematic drawing of the initial polymerization events in fibrin formation. Removal of fibrinopeptides exposes sites on the central domain that can then interact with complementary sites on terminal domains of other molecules. An additional set of contacts (end-to-end) comes into play upon addition of the third molecule. Panels A and B reprinted with permission from ref 163. Copyright 1984 Annual Reviews Inc. (C) Thrombin was added to a solution of bovine fibrinogen (0.3 mg mL^{-1}) to initiate polymerization. At short times after initiation of polymerization (about 10 min), the macroscopic clot was removed and samples of the clear solution were placed on hydrophilic films of carbon supported by a network of formvar. The adsorbed complexes of fibrin monomers were negatively stained with 1.0% uranyl acetate.[164] Magnification 290000×. The complexes of fibrin monomers presented in the gallery are (a) a dimer, (b) two dimers, (c) a tetramer, (d) a pentamer, (e) a hexamer, and (f) a heptamer. Panel C reprinted with permission from ref 164. Copyright 1981 Academic Press.

the concentration of fibrin monomers (Figure 13–16A) and shows no evidence of a lag. The time required for half of the final light scattering to be established is inversely proportional to the initial concentration of fibrin monomers (Figure 13–16B).[164] Because light scattering is not proportional to bulk concentration of polymer, this is simply the time required for a particular fraction α of the polymer to form.

It can be shown[61] that such behavior is that expected from a polymerization in which end-to-end connections among monomers, dimers, and oligomers form at random with no initiation required and in which each connection has the same rate constant of formation regardless of the lengths of the two participants, including unconnected monomers. If the assumption is made that the degree of polymerization is equal to the total molar concentration of connected interfaces between monomers, [interfaces], and it is realized that the molar con-

centration of central faces is always equal to the molar concentration of terminal faces, then

$$\frac{d[\text{interfaces}]}{dt} = k[\text{faces}]^2 \tag{13–31}$$

where [faces] is the total molar concentration at any time, t, of open faces, either terminal or central, at the ends of monomers, dimers, and oligomers (Figure 13–15B). The initial rate of polymer formation will be second-order in the concentration of fibrin monomer because $[\text{faces}]_0 = 2[\text{monomer}]_0$, where $[\text{faces}]_0$ and $[\text{monomer}]_0$ are the initial concentrations of faces and monomers, respectively. Because one interface is formed from two of each kind of face, central and terminal (Figure 13–15),

$$2\frac{d[\text{interfaces}]}{dt} = -\frac{d[\text{faces}]}{dt} \tag{13–32}$$

Upon combination of Equations 13–31 and 13–32 and rearrangement

$$-\frac{d[\text{faces}]}{[\text{faces}]^2} = 2k\,dt \tag{13–33}$$

Upon integration between $t = 0$ and t

$$\frac{1}{[\text{faces}]} - \frac{1}{[\text{faces}]_0} = 2kt \tag{13–34}$$

Choose any time, t_α, at which a particular fraction α of the polymer has formed. Because that fraction of the faces has disappeared, $[\text{faces}] = (1-\alpha)[\text{faces}]_0$. When this equality is inserted into Equation 13–34

$$\frac{1}{[\text{faces}]_0} = 2\left(\frac{1-\alpha}{\alpha}\right)kt_\alpha \tag{13–35}$$

It follows that the time at which any particular fraction, α, of the polymer has formed will be inversely proportional to the initial concentration of monomer, as was observed. The kinetic mechanism just described, based on the assumption that the combination of two faces is independent of whether they are faces on monomers, dimers, or oligomers, is completely consistent with the proposed molecular mechanism (Figure 13–15B).

As protofibrils form, they begin to associate laterally to form fibers of fibrin. This lateral association can be prevented by raising the ionic strength. This is the way in which lateral association is inhibited so that the kinetics of only the assembly of protofibrils can be followed (Figure 13–16). The rate of lateral association is also decreased remarkably when fibrin monomers are formed with a proteolytic enzyme that can remove only the fibrinopeptides A and not the fibrinopeptides B, even though fibrin monomers formed in this way assemble normally into protofibrils at undiminished rates.[166] This suggests that the separate binding site on the terminal domain for the synthetic peptide containing the amino

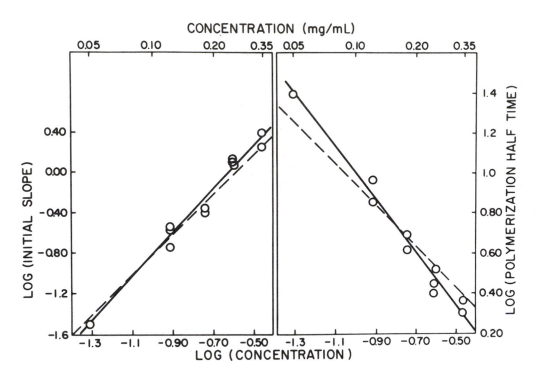

Figure 13–16: Dependence of the initial rate of fibrin polymerization (A) and the time required to reach half of the total increase in light scattering (B) on the initial concentration of fibrin monomer.[166] Solutions of fibrinogen (0.05–0.35 mg mL⁻¹ final concentration) were mixed rapidly (< 0.5 s) with solutions of thrombin at concentrations sufficient to remove the fibrinopeptides from the molecules of fibrinogen immediately (< 0.5 s). The polymerization of the fibrin was then followed by monitoring the increase in light scattering of the solution at 633 nm as a function of time (seconds). The initial rate of increase in light scattering (arbitrary units) was determined for each trace as well as the time (2–25 s) required for the light scattering to reach half its final value. The logarithm of the initial rate, log (initial slope), and the logarithm of the time required to reach half of the final value, log (polymerization half time), are presented, respectively, as a function of the logarithm of the concentration (milligram milliliter⁻¹) of fibrinogen, log (concentration). The polymerization was performed at 23 °C in 0.5 M NaCl. The high ionic strength prevented side-to-side aggregation of the polymers so that only elongation of protofibrils was occurring during the measurements. Adapted with permission from ref 166. Copyright 1979 *Journal of Biological Chemistry.*

terminus of the β polypeptide of the fibrin monomer, GHRP,[165] is involved in the lateral association of the protofibrils and that the amino terminus of a β polypeptide in one protofibril would associate with the binding site on a terminal domain of a fibrin monomer in another protofibril to mediate side-to-side associations of the cables.

The final step in the irreversible polymerization of fibrin is the covalent cross-linking of the fibrin monomers among themselves by the enzyme transglutaminase. This enzyme catalyzes replacement of the ammonia in a glutamine on one monomer with the ε-amine of a lysine on another monomer

$$(13\text{–}36)$$

Two symmetrically disposed pairs of lysines and glutamines near the carboxy termini of two γ subunits are cross-linked in this way to produce a covalent dimer of γ polypeptides with the following structure[167]

– G Q Q H Q L G G A K Q A G D VCOOH

HOOCV D G A Q K A G G L Q H Q Q G –

13–4

The rotationally symmetric juxtaposition of these two sequences from the γ polypeptides of different fibrin monomers in the protofibril is consistent with the structural model in which there is a 2-fold rotational axis of symmetry at this location (Figure 13–15C). Amide cross-links are also formed among α polypeptides juxtaposed during the lateral association of the protofibrils.[168] The pairing in these cross-links reflects the side-by-side orientation of the α polypeptides in different protofibrils in the final fiber of fibrin.

Examples of helical polymeric proteins that assemble reversibly are actin (Figure 9–1) and microtubules.

Microtubules are hollow threaded cylinders of indefinite length whose overall radius is 12 nm and whose hollow center has a radius of 6 nm. These helical polymeric proteins are often assembled and disassembled during the life of a cell, unlike fibrin that is irreversibly assembled and only eliminated by proteolytic disintegration. The protomer from which microtubules are assembled is **tubulin**. It is an $\alpha\beta$ heterodimer.[169,170] The α polypeptide ($n_{aa} = 450$)[171] and the β polypeptide ($n_{aa} = 445$)[172] are homologous in sequence [41% identity with 1.1 gaps (100 amino acids)$^{-1}$].[172] It necessarily follows that the two polypeptides share a common ancestor and that their native structures are superposable and indistinguishable from each other at low resolution.

When viewed in the electron microscope, by negative staining, microtubules appear to be constructed from a tubular bundle of 13 indistinguishable rows of protein (Figure 13–17A). Each row is parallel to all of the others and parallel to the axis of the microtubule. At the end of a microtubule these rows can be frayed into 13 threads.[173] From this observation it can be concluded that the interfaces forming the rows are stronger than the interfaces between them.[174]

Upon the microtubule there is a helical surface lattice.[174] This lattice has been identified by image reconstruction of electron micrographic images of intact microtubules. The reciprocal lattice of the Fourier transform of digitized images from electron micrographs has been assigned to that of the Fourier transform of a triply threaded, left-handed ($n = -3$) screw whose three helices are spaced at 4.0-nm intervals along the axis (Figure 13–17B). These three helices are helices of individual, single folded polypeptides, where folded α polypeptides are indistinguishable from folded β polypeptides. A set of $(8.0 \text{ nm})^{-1}$ reflections also appear in the Fourier transform of the images, and these reflections results from an octuply threaded, left-handed ($n = -8$) screw. The eight helices in this pattern represent helices of $\alpha\beta$ heterodimers. Because the strongest interfaces of the structure are contained within the rows parallel to the axis of the microtubule, it is assumed that the $\alpha\beta$ heterodimers are aligned with their internal interfaces within the rows rather than between the rows. An array of $\alpha\beta$ dimers consistent with these assignments is shown in Figure 13–17B. Because the basic lattice is built on 13 parallel rows, the three left-handed helices ($n = -3$) of monomers with spacing of 4.0 nm create 10 right-handed helices ($n = 10$) with spacing of 4.0 nm, and the eight left-handed helices ($n = -8$) of $\alpha\beta$ heterodimers with spacing of 8.0 nm create five right-handed helices ($n = 5$) of $\alpha\beta$ dimers with spacing of 8.0 nm.

The two ends of a microtubule are different, and a microtubule is **polar**. This has been verified by the observation that growth at one end of a microtubule during assembly is slower than growth at the other end.[175] In the arrangement depicted in Figure 13–17B, one end would display only folded α polypeptides; and the other end, only folded β polypeptides.

When purified tubulin is brought to the proper conditions of temperature and pH and is mixed with the proper substrates, it spontaneously polymerizes to form microtubules. The process can be divided into two phases, initiation and elongation.[176] **Initiation** is the sequence of events that leads to the formation of a large enough oligomer of tubulin protomers to act as an origin from which a microtubule can then grow by the repetitive addition of $\alpha\beta$ dimers of tubulin. **Elongation** is the addition of $\alpha\beta$ dimers of tubulin to one end of the growing microtubule in such a way that each successive addition is formally equivalent regardless of the length of the microtubule. Therefore, elongation begins only when the end of a cylinder in every way identical to the end of a long microtubule has been created in the solution. Until such a stage is reached, the steps in the assembly of a microtubule are steps in the process of initiation.

Under all experimental conditions, initiation of microtubules is a complicated process involving a number of intermediates of peculiar structure.[177,178] This is not surprising because the steps between a set of unassociated $\alpha\beta$ dimers of tubulin in free solution and an oligomer of those dimers large enough to offer the end of a cylinder equivalent to the end of the cylinder in an elongated microtubule are not easy to imagine. In living cells, however, all of the microtubules in each cell originate at only one precise location.[179] In cells containing centrioles, it is the centriole that serves as the origin. In cells lacking centrioles, the point of origin is associated with structures resembling centrioles. Isolated centrioles are able to initiate the formation of microtubules readily[180] and do so at lower concentrations of free tubulin than are necessary for spontaneous initiation.[179] From these observations, it becomes clear that initiation in simple experimental situations in the absence of centrioles is an adventitious process for which the protein was never designed because all initiation in a cell occurs at the centriole.

Elongation of microtubules in the absence of the complications of spontaneous initiation can be instituted immediately by adding short segments of preformed microtubules, referred to as seeds, which have already passed through the steps of spontaneous initiation, to solutions containing high enough concentrations of unpolymerized tubulin to support the elongation of the seeds.[176] Preparations of tubulin of high purity are reasonably unsusceptible to spontaneous intitiation[179] and readily support the elongation of seeds in a reaction that assumes its maximum rate immediately after the seeds are added to the solution.

Consider a polymerization in which monomers, such as $\alpha\beta$ heterodimers of tubulin, are successively adding to

A

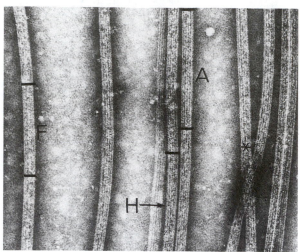

B

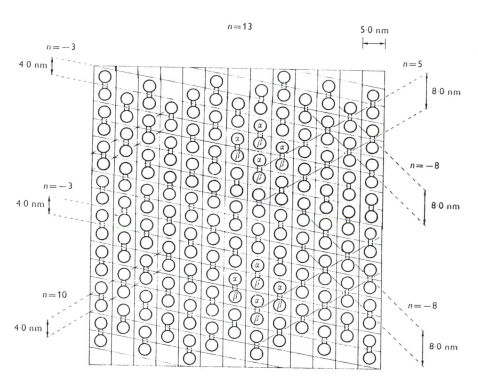

one end of a polymer, such as a microtubule, by the reaction

$$\text{polymer}_n + \text{monomer} \underset{k_{-1}}{\overset{k_1}{\rightleftharpoons}} \text{polymer}_{n+1} \qquad (13\text{--}37)$$

where polymer_n is a polymer composed of n protomeric units and polymer_{n+1} is a polymer composed of $(n + 1)$ protomeric units. Dissociated from the polymer in free solution, the individual units are referred to as **monomers**. The $\alpha\beta$ heterodimer of tubulin in free solution is

the monomer of the microtubule. It is referred to as monomeric tubulin. Within the polymer, the individual units are referred to as **protomers**. $\alpha\beta$ Heterodimers of tubulin within a microtubule are referred to as protomeric tubulin. When the reaction has come to equilibrium, the dissociation constant, $^{n}K_{d}^{\text{poly}}$, of a monomer from the end of a polymer $n+1$ units in length is

$$^{n}K_{d}^{\text{poly}} = \frac{[\text{polymer}_n][\text{monomer}]}{[\text{polymer}_{n+1}]} = \frac{k_{-1}}{k_1} \qquad (13\text{--}38)$$

Figure 13–17: Helical lattice of $\alpha\beta$ heterodimers of tubulin producing a microtubule.[174] *Trichonympha*, flagellated protists from the gut of the termite *Zootermopsis angusticollis*, were mixed with 1% phosphotungtate, pH 7, and applied to a film of amorphous carbon supported by a network of collodion. The excess negative stain was drained, the preparation was dried, and the negatively stained specimens were examined in the electron microscope (A). Although flagella, over most of their length, are composed of pairs of microtubules in which two microtubules are fused to each other, at the distal end of a flagellum these pairs dwindle to single microtubules. The electron micrograph is of a group of these individual, unpaired microtubules. Regions marked F, H, and A were chosen for image enhancement. The optical density of each of these images was digitized with an optical densitometer and the Fourier transform of the digitized image was calculated. From examination of this calculated Fourier transform and from optical diffraction patterns of the images themselves, the reflections arising from the helical array of protomers were identified and indexed. These reflections defined the helical lattice in which the $\alpha\beta$ heterodimers of tubulin are arrayed in a microtuble. (B) A schematic diagram of that lattice is presented. The cylinder on which the lattice is arrayed was cut along one of the lattice lines parallel to its axis and the cylindrical surface was flattened onto the page. The $\alpha\beta$ dimers of tubulin are aligned in 13 rows parallel to the axis of the cylinder (see panel A). Individual folded polypeptides, if no distinction is made between α and β, lie on a three-stranded, left-handed helix ($n = -3$) and a 10-stranded, right-handed helix ($n = 10$). $\alpha\beta$ Heterodimers lie on a five-stranded right-handed helix ($n = 5$) and an eight-stranded left-handed helix ($n = -8$). Reprinted with permission from ref 174. Copyright 1974 Biochemical Society.

(see Equation 13–8). When all values of n are large enough so that only elongation is being considered, $^nK_d^{poly}$ has the same value, $^nK_d^{poly}$, for all values of n because the ends of all of the polymers, for example, the ends of all of the microtubules (Figure 13–17) regardless of length, are indistinguishable.

Two features of the solution of a polymer can be separately defined. The **bulk concentration** of polymer, [polymer]$_b$, or bulk concentration of microtubules, [microtubule]$_b$, is equal to the molar concentration of protomers, or of protomeric tubulin, that are incorporated into polymers. The bulk concentration is directly proportional to the total length of polymer in the solution. In the case of microtubules the bulk concentration is usually determined by light scattering, which is proportional to the total length of microtubule present. Normally, a significant fraction of the light is scattered by these solutions, and the absorbance is measured rather than the amplitude of the scattered light. The **number concentration** of polymer, [polymer]$_n$, or number concentration of microtubules, [microtubule]$_n$, is equal to the molar concentration of individual intact molecules of polymer or microtubules. The number concentration of microtubules is measured by quantitative electron mi-

croscopy.[181] Individual microtubules in a field are counted, and their density is related to their density in the original solution. The number concentration of polymer is equal to the concentration of either end of the polymer, either [+ ends] or [– ends], when the two ends are separately distinguished.

Assume for the moment that elongation is occurring at only one end of the polymer, or at one end of the microtubule, and that

$$\frac{d[\text{polymer}]_b}{dt} = k_1[\text{ends}][\text{monomer}] - k_{-1}[\text{ends}]$$

$$(13\text{–}39)$$

If the reaction has been initiated by adding seeds to a solution of protomer, and if no spontaneous initiation occurs, the molar concentration of ends, [ends], at which elongation is proceeding must remain constant, and the initial rate of bulk polymer formation is defined by the relationship

$$\left(\frac{d[\text{polymer}]_b}{dt}\right)_0 = [\text{ends}]_0(k_1[\text{monomer}]_0 - k_{-1})$$

$$(13\text{–}40)$$

where the subscripts indicate initial quantities. It has been shown, in agreement with this equation, that the initial rate at which the bulk concentration of microtubles increases is directly proportional to the volume of a particular solution of seeds added to a series of reactions (Figure 13–18).[176]

The initial rate of formation of bulk polymer should also be directly proportional to the term $(k_1[\text{monomer}]_0 - k_{-1})$, and a plot of initial rate against initial concentration of monomer should be linear and pass through zero when $k_1[\text{monomer}]_0 = k_{-1}$. Immediately after addition of seeds, the bulk concentration of polymer [polymer]$_b$ should increase when $k_1[\text{monomer}]_0 > k_{-1}$ and decrease when $k_1[\text{monomer}]_0 < k_{-1}$. When the bulk concentration of polymer is decreasing rather than increasing, individual molecules of polymer are **depolymerizing** by shedding monomer from their ends and decreasing in length. The initial concentration of monomer at which neither elongation nor depolymerization should occur is referred to as the **critical concentration**. If only one molecule of polymer, or one microtubule, were being observed, its initial   rate of elongation (in monomers second^{-1}) should be equal to $(k_1[\text{monomer}]_0 - k_{-1})$. When $k_1[\text{monomer}]_0 > k_{-1}$, that molecule of polymer should elongate; and when $k_1[\text{monomer}]_0 < k_{-1}$, it should depolymerize.

Seeds were added to a series of solutions containing increasing concentrations of pure monomeric tubulin. At various times after the addition, samples were taken from each mixture. They were quenched by quantitative cross-linking with glutaraldehyde, sedimented onto a specimen grid, and examined in the electron microscope. The seeds that were used were fragments of

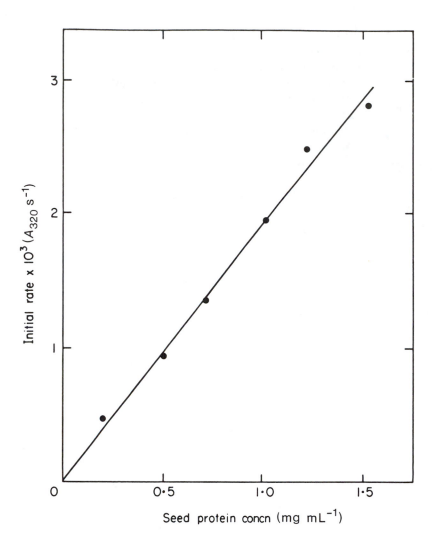

Figure 13–18: Initial rate of tubulin polymerization as a function of the initial concentration of fragmented whole microtubules.[176] A preparation of microtubules (7.1 mg mL^{-1}) that had been polymerized separately by spontaneous initiation were sheared by passing them through a 22-gauge needle to produce fragments about 1 μm in length. The fragments were immediately added to a solution of unpolymerized monomeric tubulin at a concentration high enough (1.8 mg mL^{-1}) to elongate the fragments. The solution was 0.1 mM MgCl$_2$, 0.5 mM GTP, and pH 6.9. The elongation of the fragments was followed by the increase in absorbance at 320 nm due to light scattering. The initial rate of increase in absorbance (A_{320} second^{-1}) is plotted against the initial concentration (milligrams milliliter^{-1}) of sheared microtubules added to initiate elongation. Adapted with permission from ref 176. Copyright 1977 Academic Press.

axonemes, which are naturally occurring bundles of microtubules whose polarity can be determined visually. The two ends are arbitrarily termed the *plus end* and the *minus end*. The initial rate of increase in the length of the microtubules projecting from each end of the axoneme could be measured in this way and plotted against the initial concentration of tubulin (Figure 13–19).[181] The initial rate of elongation is linearly related to the initial concentration of monomeric tubulin, [monomer]$_0$, as predicted by Equation 13–40, but k_{-1} is too small to be estimated accurately.

When elongation occurs because seeds are added to a solution of monomer above its critical concentration, the free concentration of monomer should decrease as polymer is formed until equilibrium is reached and further elongation ceases. Assume for the moment that only one end of each polymer can elongate; then, from Equation 13–39

$$\frac{d[polymer]_b}{dt} = -\frac{d[monomer]}{dt} = [ends](k_1[monomer] - k_{-1})$$

$$(13\text{–}41)$$

When this relationship is rearranged and integrated between $t = 0$ and t

$$\ln(k_1[monomer] - k_{-1}) = \ln(k_1[monomer]_0 - k_{-1}) - k_1[ends]t$$

$$(13\text{–}42)$$

when t = ∞ and [monomer] = [monomer]$_{eq}$, k_1[monomer]$_{eq}$ = k_{-1}. Therefore, the concentration of monomer at equilibrium should be equal to the critical concentration at which no elongation occurs when seeds are added to a solution of monomer.

A microtubule, however, has two ends. Perform the following imaginary reaction. At equilibrium, remove an $\alpha\beta$ heterodimer from the plus end and add an $\alpha\beta$ heterodimer to the minus end. Because of its helical symmetry (Figure 13–17B), the microtubule after this exchange is the same as the microtubule before the exchange. Because free energy is a state function, the free energy change for this reaction must be zero, and the standard free energy of dissociation of a molecule of tubulin from the plus end must be equal to the standard free energy of dissociation of a molecule of tubulin from

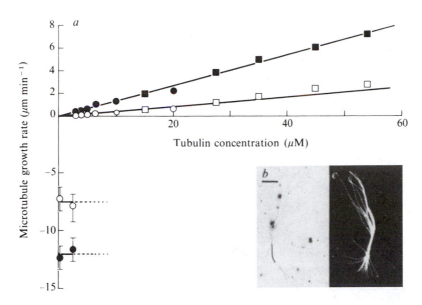

Figure 13–19: Rate of elongation (micrometers minute^{-1}) and rate of depolymerization (–micrometers minute^{-1}) of microtubules from the plus end (●,■) and the minus end (○,□) of fragments of axonemes (A).[181] Fragments of axonemes from *Tetrahymena* were added to final concentrations of 10^7 mL^{-1} to solutions of monomeric tubulin at the noted concentrations (in micromolar). Each of these solutions was 1 mM MgCl$_2$, 1 mM EDTA, and 1 mM GTP, pH 6.8. At a series of time points, aliquots were withdrawn from these solutions and rapidly fixed with glutaraldehyde, and the products were sedimented onto electron microscopic grids or glass cover slips. The elongated axonemes were examined, following negative staining, in the electron microscope (B) or, following staining with fluorescent, anti-tubulin immunoglobulin G, in a fluorescence microscope (C). Rates of growth were calculated from measurements of the length of the microtubules grown from the ends of the axoneme as a function of time. At concentrations of monomeric tubulin greater than 3 μM, microtubules would grow from axonemes and the rate of growth was linearly related to the concentration of unpolymerized tubulin. At concentrations of monomeric tubulin below 3 μM, no growth would occur. If microtubules, however, were grown on axonemes to 20 μm on the plus end and 7.5 μm on the minus end at a high monomeric tubulin concentration and the concentration of monomeric tubulin was then dropped to one of the noted concentrations less than 3 μM, the microtubules would begin to depolymerize at the noted rates (points with minus values of rates), which were zero-order functions of the concentration of unpolymerized tubulin. The scale bar in panel B is 4.5 μm, and the scale of panel C is the same as that of panel B. The circles in panel A are measurements made by electron microscopy and the squares are measurements made by immunofluorescence. Reprinted with permission from *Nature*, ref 181. Copyright 1984 Macmillan Magazines Limited.

the minus end. Therefore, the dissociation constants, K_d^{tub}, for tubulin at either end must be the same. Because $K_d^{tub} = k_{-1}/k_1 = $ [tubulin]$_{eq}$ (Equation 13–38), the concentration of monomeric tubulin in equilibrium with either end, the critical concentration, is the same. Therefore, a microtubule should not elongate at one end while it is depolymerizing at the other.

If monomeric tubulin adds to a population of elongating microtubules until the concentrations of monomeric tubulin and microtubules reach equilibrium with each other, the total bulk concentration of microtubule is linearly related to the initial concentration of monomeric tubulin, and the line relating these two variables intersects the abscissa at the value of the critical concentra-

tion (Figure 13–20).[182] This necessarily follows from the facts that the amount of tubulin incorporated into microtubules when equilibrium is reached is equal to [monomer]$_0$ – [monomer]$_{eq}$; that [monomer]$_{eq}$, the critical concentration, is the same for each point; and that when [monomer]$_0$ equals the critical concentration, no increase in [polymer]$_b$ can occur.

One of the ingredients that is essential for the elongation of microtubules is GTP[183] or GDP.[182] Guanosine triphosphate is bound by monomeric tubulin to one site on each folded polypeptide in the $\alpha\beta$ heterodimer.[184] The two molecules of GTP that are bound are slowly hydrolyzed to GDP in a reaction initiated by the addition of the monomeric (GTP)$_2$·tubulin to an elongating microtu-

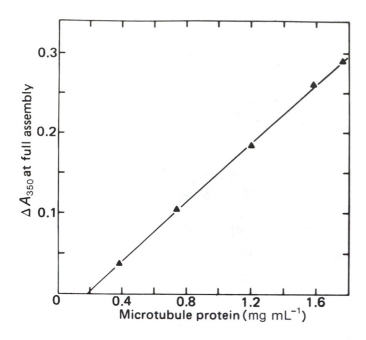

Figure 13–20: Demonstration of the critical concentration of unpolymerized $\alpha\beta$ heterodimers of tubulin in the formation of microtubules in a solution of GDP.[182] $\alpha\beta$ Heterodimers of tubulin at a concentration of 1.75 mg mL^{-1} were assembled into microtubules at 30 °C in the presence of 0.1 mM GTP. After 60 min, when steady state had been reached, samples were removed from the solution and diluted to the noted final concentration of microtubule protein (milligrams milliliter^{-1}) into solutions prepared so that the final concentration of GDP would be 2.0 mM. When equilibrium had been achieved (25 min), the final concentration of polymer was determined by measuring the absorbance at 350 nm (A_{320}) due to light scattering. Under these circumstances, the critical concentration of microtubule protein, below which no polymer remains, is 0.19 mg mL^{-1}. Above 0.19 mg mL^{-1} the amount of polymer at equilibrium increases linearly with the total concentration of tubulin. Adapted with permission from ref 182. Copyright 1979 National Academy of Sciences.

bule, and the two molecules of GDP that are formed remain trapped at the two binding sites while the protomeric $(GDP)_2$·tubulin is within the microtubule.[185,186] This peculiar feature of elongation causes an elongating microtubule to have a short region near its end in which all the tubulin has bound GTP because it has recently been added. Proximal to this cap of protomeric $(GTP)_2$·tubulin, however, the frequency of protomeric $(GDP)_2$·tubulin increases until the main body of the microtubule is reached, which is entirely formed from protomeric $(GDP)_2$·tubulin. When a protomer of $(GDP)_2$·tubulin leaves the microtubule, the GDP rapidly dissociates from it and is then replaced by GTP from the solution. It is the reinstatement of GTP on the $\alpha\beta$ heterodimer of tubulin that renders it again capable of adding to the end of a microtubule during elongation.

The critical concentration for monomeric $(GTP)_2$·tubulin (Figure 13–19) is at least 30 times lower[179,182] than the critical concentration for $(GDP)_2$·tubulin (Figure 13–20). As microtubules do not elongate in the absence of a nucleotide, the critical concentration for monomeric tubulin whose nucleotide site is unoccupied must be much higher than that for monomeric $(GDP)_2$·tubulin.

The rate constants for the four elongation reactions can be distinguished separately by the following convention. Those for elongation at the positive end can be identified with a plus sign (+); those for elongation at the negative end, with a minus sign (–); those for $(GTP)_2$·tubulin, with the letter T; and those for $(GDP)_2$·tubulin, with the letter D. From the results in Figure 13–19, $k_1^{T+} = 3.8 \times 10^6$ M^{-1} s^{-1} and $k_1^{T-} = 1.2 \times 10^6$ M^{-1} s^{-1} at 37 °C. Both k_{-1}^{T+} and k_{-1}^{T-} are too small (≤ 1 s^{-1}) to be measured with any accuracy.[181] Under most circumstances in which elongation occurs in the presence of GTP, it is governed entirely by k_1^{T+} and k_1^{T-}.

Even though the dissociation constants for $(GTP)_2$·tubulin to both ends of the microtubule must be the same, the two rate constants are not the same because the microtubule is a polar structure. It happens that a microtubule elongates about 3-fold more rapidly from its plus end than from its minus end.

When microtubules depolymerize, for example, upon dilution, the short cap of protomeric $(GTP)_2$·tubulin near the end that is formed from recently added monomeric $(GTP)_2$·tubulin is rapidly lost, and depolymerization proceeds by the dissociation of monomers of $(GDP)_2$·tubulin. If the dilution has been great enough so that the concentration of monomeric $(GTP)_2$·tubulin is below its apparent critical concentration, ends cannot be recapped and only k_{-1}^{D+} and k_{-1}^{D-} govern the rates of depolymerization. The values for these rate constants for depolymerization upon dilution,[181] under the same conditions (Figure 13–19) in which k_1^{T+} and k_1^{T-} were measured, are $k_{-1}^{D+} = 340$ s^{-1} and $k_{-1}^{D-} = 210$ s^{-1} at 37 °C. At concentrations just above the apparent critical concentration of tubulin,[179] these rates of depolymerization from uncapped ends containing protomeric $(GDP)_2$·tubulin are about 10-fold greater than the rates of elongation from ends capped with protomeric $(GTP)_2$·tubulin. As a result, a microtubule that is depolymerizing from an uncapped end will rapidly disappear even if it is elongating at its other end.

Depolymerization of microtubules under conditions where k_{-1} is much greater than k_1[monomer] proceeds initially as a zero-order reaction (Figure 13–21).[176,181] This is because the rate of depolymerization is defined by the equation

$$\frac{d[\text{polymer}]_b}{dt} = k_{-1}^{D+}[\text{+ends}] + k_{-1}^{D-}[\text{–ends}] \qquad (13\text{–}43)$$

As long as [ends] remains constant, the reaction remains

zero-order. As [ends] begins to decrease, when more and more microtubules cease to exist, the rate of depolymerization decreases.[176] From electron micrographs of samples removed at various times from a population of depolymerizing microtubules, the decrease in the rate constant of depolymerization at the longer times could be quantitatively correlated to the decrease in the number concentration of microtubules and hence the molar concentration of ends.[187]

The purpose for the peculiar role of GTP and GDP in tubulin polymerization seems to be the maintenance of the origin of the network of microtubules in the cell.[181] When a microtubule breaks into two pieces for any reason, the break will almost always occur in the region where protomeric $(GDP)_2$·tubulin is located. The broken tubule will rapidly depolymerize from its broken ends, and the broken end attached to the centriole and the fragment unattached to the centriole will disappear. This prevents broken pieces from initiating microtubules unattached to centrioles. As the centriole can initiate microtubules at concentrations lower than those at which they initiate spontaneously, all microtubules end up originating at the centriole.

The system, however, creates a peculiar steady state. At all times and at random, regions of protomeric $(GDP)_2$·tubulin are overtaking elongating ends of individual tubules and switching those tubules from ones that are elongating to ones that are depolymerizing (Figure 13–22). The monomeric tubulin released picks up new molecules of GTP and reenters a tubule that happens by chance to be outdistancing its protomeric $(GDP)_2$·tubulin. The centriole has a finite capacity to initiate microtubules, but sites become empty as the result of depolymerization; new microtubules, the elongating ends of which are again outracing their destruction, are initiated at those empty sites.

When seeds are introduced into a solution of tubulin, elongation will commence on most of the ends, but some will be uncapped and begin to depolymerize immediately. Others will be uncapped as the reaction proceeds; some of these will depolymerize awhile and then be recapped, if the concentration of monomeric $(GTP)_2$·tubulin is high enough, and some of them will disappear. If the concentration of monomeric $(GTP)_2$·tubulin is less than a certain **apparent critical concentration**, the boundary of protomeric $(GDP)_2$·tubulin, whose rate of propagation is independent of the concentration of monomeric $(GTP)_2$·tubulin, will move along the microtubule faster than the rate of its elongation, which is dependent on the concentration of monomeric $(GTP)_2$·tubulin, and the microtubules will be switched (Figure 13–22) and begin to depolymerize with the rate constants k_{-1}^{D+} and k_{-1}^{D-}. At concentrations of tubulin in excess of this apparent critical concentration, elongation will be faster

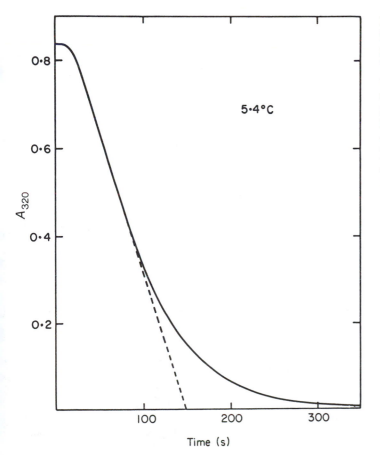

Figure 13–21: Kinetics of the depolymerization of microtubules.[176] Tubulin (3.5 mg mL^{-1}) was polymerized at 30 °C in 0.1 mM MgCl$_2$ and 0.5 mM GTP at pH 6.9 to steady state (30 min). When the temperature of a solution of microtubules is dropped, the microtubules depolymerize. When the temperature of this solution was brought to 5 °C, the depolymerization could be followed by the decrease in the absorbance at 320 nm. After a lag coinciding with the time necessary to lower the temperature, the depolymerization followed zero-order kinetics (dotted line) until the number of microtubules in the solution began to decrease. Adapted with permission from ref 176. Copyright 1977 Academic Press.

Phase transition

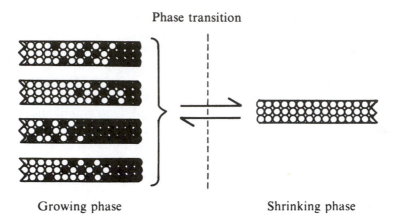

Growing phase   Shrinking phase

Figure 13–22: Schematic model describing the role of GTP in the elongation and polymerization of microtubules. The dark circles represent protomers of tubulin to which GTP is bound, and the open circles represent protomers of tubulin on which the GTP has hydrolyzed to GDP. As long as the tubule is elongating at a significant rate, the end of the microtubule is occupied by protomers of $(GTP)_2 \cdot$ tubulin and the end has a low critical concentration (Figure 13–19). In this state, it will not depolymerize, and this is the growing phase. If elongation slows down because the concentration of monomeric $(GTP)_2 \cdot$ tubulin decreases, at a certain point the spreading wave of GTP hydrolysis will reach the end of the microtubule. The end of the microtubule will then be occupied mostly by protomeric $(GDP)_2 \cdot$ tubulin and the end will then have a much higher critical concentration (Figure 13–20). It has passed into the shrinking phase and will rapidly depolymerize. Reprinted with permission from *Nature*, ref 181. Copyright 1984 Macmillan Magazines Limited.

than the rate at which the boundary of protomeric $(GDP)_2 \cdot$ tubulin can move, and the microtubules will elongate with rate constants k_1^{T+} and k_1^{T-} as if they contained only $(GTP)_2 \cdot$ tubulin. At the apparent critical concentration, the rate of depolymerization from uncapped ends will be equal in magnitude to the rate of elongation from capped ends. This apparent critical concentration is not the reflection of an equilibrium process as is the actual critical concentration (Figure 13–20) but the result of a complex combination of rate constants producing a steady state, and it should not be confused with a real critical concentration. The steady state is maintained by the continuous hydrolysis of GTP, and as long as GTP is available, equilibrium cannot be reached.

The features of this steady state can explain a number of peculiar observations. When a uniform population of microtubules elongating rapidly at high concentrations of monomeric $(GTP)_2 \cdot$ tubulin is diluted to a concentration slightly below the apparent critical concentration, the bulk concentration of microtubules [microtubule]$_b$ immediately begins to decrease. This decrease, however, is entirely due to a decrease in the number concentration of microtubules rather than the mean length of the remaining microtubules. The microtubules that remain are still slowly elongating because they are still in the

race.[181] Those that have disappeared have lost the race with the boundary of protomeric $(GDP)_2 \cdot$ tubulin at one of their ends. When microtubules grown from seeds reach a steady-state bulk concentration, this concentration is maintained by a decrease in the number of microtubules and an elongation of those that remain (Figure 13–23).[181] Those that are still in the race are keeping ahead at the expense of the losers.

This last observation explains the phenomenon of treadmilling.[188,189] A population of microtubules is elongated from seeds until the steady state is reached and [microtubule]$_b$ has become constant. Either [³H]tubulin[190] or [³H]GTP[186] is then added to the solution, and [³H]tubulin or ([³H]GTP)$_2 \cdot$ tubulin, respectively, is incorporated into the microtubules. The amount of [³H]tubulin or [³H]GTP trapped within the population of microtubules increases with time. This is due to incorporation of the mono-meric [³H]tubulin or monomeric ([³H]GTP)$_2 \cdot$ tubulin into those microtubules that are still in the race and are still elongating. It had been mistakenly thought that the incorporation of [³H]GDP was due to ([³H]GTP)$\cdot$ tubulin entering a microtubule at an elongating end while the other end of the same tubule was depolymerizing at the same rate that elongation was occurring.[188] If both ends of the same tubule, however, are capped with protomeric $(GTP)_2 \cdot$tubulin, both must be elongating because both must have the same critical concentration. If one end is uncapped, it will depolymerize about 10 times faster than the other end can elongate and that microtubule will quickly disappear.

As time progresses, the rate of incorporation of [³H]tubulin into microtubules at steady state decreases.[190] This is not due to radioactive [³H]tubulin passing through the microtubules and beginning to exit the other end. It is due to the fact that as time progresses, the microtubules that have lost the race at one of their ends and that are forced to depolymerize completely contain more and more [³H]tubulin within themselves, and more and more [³H]tubulin is released each time a microtubule depolymerizes. This release of [³H]tubulin competes more and more successfully with the incorporation of [³H]tubulin into the fewer and fewer elongating ends.

If microtubules are elongated to steady state, exposed to [³H]GTP for 1 h, and then transferred to a solution containing unradioactive GTP, the incorporated [³H]GDP is lost very slowly.[188] This is not due to the fact that it must traverse the length of its particular microtubule before it can be released at the depolymerizing end but is due to the fact that it can be released only when its microtubule depolymerizes completely. In such an experiment, the trapped [³H]GDP measures the rate at which entire microtubules are lost, or, in other words, the decrease in the number concentration of microtubules $[\text{microtubule}]_n$, because, at the end of the one-hour pulse, each existing microtubule contains the same amount of [³H]GDP.

Colchicine and podophyllotoxin are inhibitors of tubulin elongation. They bind to free $\alpha\beta$ heterodimers of tubulin which then have a higher affinity for an elongating end of a microtubule. After the toxin·tubulin complex enters an elongating end, however, it is no longer able to elongate further[191] because it is capped by colchicine or podophyllotoxin. When podophyllotoxin is added to microtubules grown to steady state from seeds, the bulk concentration of microtubule $[\text{microtubule}]_b$ decreases in two phases. One phase has the rate constant $(k_{-1}^D{}^+ + k_{-1}^D{}^-)$ of normal depolymerization, and the other phase is much slower. The rapid phase is the depolymerization from the ends that lose the race with the boundary of $(GDP)_2 \cdot$tubulin and begin to depolymerize before they can be capped by podophyllotoxin. The slow phase is the slow depolymerization of microtubules that have become capped at both ends by podophyllotoxin before either end can begin to depolymerize. If fresh tubulin is added with the podophyllotoxin, the magnitude of the rapid phase is decreased as expected.

The assembly of **actin** into **thin filaments** (Figure 9–1) is similar to that of tubulin into microtubules. Thin filaments of actin are necessarily polar. As with a microtubule, a thin filament elongates at both ends, but it elongates at one end more rapidly (about 7 times more) than at the other.[192,193] The end that elongates more rapidly is the "barbed" end of a thin filament when it is decorated with fragments of myosin. This end is the unattached end of a thin filament within the cell, away from which a thick filament of myosin always slides, and should be the only end at which a thin filament would normally elongate during its assembly in the cell.

The assembly of monomeric actin to form polymeric thin filaments requires that ATP be present.[194] The ATP binds to the actin, the ATP·actin complex is incorporated into the thin filament, and the ATP is then hydrolyzed slowly to ADP. The ADP·actin complex is unable to polymerize at concentrations that readily promote the polymerization of ATP·actin,[194] and it follows that the critical concentration of ADP·actin is much higher than that of ATP·actin. Actin also displays behavior that has been described incorrectly as treadmilling[189] but presumably is due to disproportionation at the steady state.

Thin filaments, like microtubules, must always be attached at one of their ends to some structure in the cell because it is their function to pull upon these structures as does a chain upon a stump. If there were an unattached thin filament in the cell, thick filaments of myosin, pulling upon that thin filament, would not be performing any noticeable work, and the thick filaments would be hydrolyzing ATP uselessly as they drew a loose thin filament past themselves. Presumably, the same mechanism as that for eliminating unattached microtubules applies to unattached thin filaments. The point of attachment of thin filaments of actin to other structures in the cell is associated with the protein vinculin,[195,196] which is engaged in the structures that presumably initiate the assembly of thin filaments as centrioles initiate the assembly of microtubules.

Unlike microtubules, however, thin filaments of actin are often sculpted to precisely regulated lengths and shapes for certain purposes. There are a number of proteins responsible for this function. Gelsolin is a protein found in both cytoplasm[197] and serum.[198] It binds to monomeric actin, and the gelsolin·actin complex binds to the barbed end of an elongating thin filament of actin and blocks further polymerization.[199] As the barbed end is the elongating end that would be unattached in the cell under normal circumstances, complexes of gelsolin and actin are thought to terminate elongation and create stable actin filaments of defined lengths attached at their end of initiation to vinculin and capped at their other end with gelsolin. This would turn a protean steady state into a collection of defined filaments. It is possible that a protein such as gelsolin also caps microtubules.

Fimbrin[200] and villin[201] are two proteins that have been purified from intestinal brush border. Fimbrin binds to thin filaments of actin and gathers them into thick bundles[200] of tens to hundreds of parallel strands. Villin combines the roles of both gelsolin and fimbrin. It can cap elongating thin filaments at their barbed ends but can also cause them to form bundles.[202] A detachable domain constituting most of the molecule of villin contains amino acid sequences that are homologous to

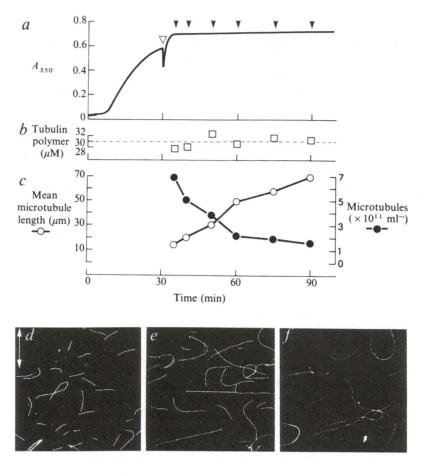

Figure 13–23: Changes in the distribution of microtubule lengths as a function of time at steady state. (*a*) Absorbance at 350 nm of a 59 μM solution of tubulin at 1 mM $MgCl_2$, 1 mM EDTA, and 1 mM GTP, pH 6.8 and 37 °C, was followed as a function of time. At 8 min, fragments of preformed microtubules were added to initiate elongation. At the point marked by the open triangle, the sample was sheared by drawing it through a 22-gauge needle. Following shearing, the concentration of microtubules rapidly (< 5 min) reached steady state. At the times noted by the arrows, samples were withdrawn, rapidly fixed with glutaraldehyde, and sedimented onto cover slips. The microtubules were stained with fluorescent anti-tubulin immunoglobulin and examined by fluorescence microscopy (*d–f*). The lengths of the microtubules were measured on photomicrographs, and the mean length was calculated. Because all of the microtubules in the solution end up on the cover slip, the number of microtubules milliliter^{-1} in the original samples could be calculated from the number of microtubules centimeter^{-2} on the cover slips. (*b*) The product of mean length and number of microtubules milliliter^{-1} could be converted to the bulk molar concentration of polymerized tubulin, which, in agreement with the constant absorbance of the solution, remained fixed. (*c*) This fixed molar concentration of polymerized tubulin, however, was maintained while the number of microtubules decreased (●) and the mean length of the remaining microtubules increased (○). The photomicrographs of microtubules are for the time points of (*d*) 35 min, (*e*) 50 min, and (*f*) 90 min. The scale in (*d*) is 30 μm. Adapted with permission from ref 181. Copyright 1984 *Nature*, Macmillan Magazines Limited.

regions of the amino acid sequence of gelsolin.[203] This observation would explain its ability to cap thin filaments of actin. The other detachable domain of villin, comprising the last 80 amino acids in its amino acid sequence, can be removed by digestion with the proteolytic enzyme from *S. aureus* strain V8.[204] This domain by itself is able to straighten sinuous thin filaments by binding to them at a stoichiometry of 1 mol (mol of actin)$^{-1}$, and this binding inhibits the bundling ability of intact villin.[202] Presumably this domain contains only one of the two or more binding sites for actin in each molecule of villin that permit it to bundle thin filaments. Gelsolin, fimbrin, and villin are representative members of a large set of proteins that control the assembly of thin filaments and determine the ultimate pattern they display within a cell.

Thick filaments of myosin are the oligomeric proteins that pull upon the thin filaments of actin. **Thick filaments** are helical polymeric proteins irreversibly but noncovalently assembled from a monomer known as **myosin**. Myosin is composed from two identical α polypeptides ($n_{aa} = 1940$) and several shorter polypeptides ($n_{aa} = 150–200$). The α polypeptides are entwined around each other to form a two-stranded, α-helical coiled coil (Figure 9–27) 150 nm in length[205] that has two globular, detachable domains known as heads, one formed from each α polypeptide at one of its ends (Figure 13–24A).[205,206] The shorter polypeptides are incorporated into these globular heads.

The individual coiled coils of the myosin molecules are segments of rope that are assembled into a helical cable from which the myosin heads protrude (Figure 13–24B).[207] The segments of rope add to the elongating cable at each of its two ends with opposite orientation. In each direction along the cable, the molecules of myosin add so that the empty ends of their segments of rope point toward the middle of the cable and the ends of the segments of rope to which the myosin heads are attached point away from the middle (Figure 13–24C).[209] The absence of myosin heads where the segments of rope pointing in opposite directions overlap in the middle of the cable creates a bare zone 150 nm in length.[207] Because of this pattern of assembly, thick filaments (Figure 13–24B),[207] unlike thin filaments, which have two distinct ends, have two ends that are identical but of opposite orientation. This necessarily produces a 2-fold rotational axis of pseudosymmetry normal to the axis of the thick filament in the center of the bare zone (Figure 13–24C).

Upon the surface of the thick filament distal to the bare zone, the myosin heads are arranged in a helical surface lattice, reflecting the underlying helical symmetry of the cable (Figure 13–24D). This helical surface lattice is septuply threaded, and myosin heads protrude from each of the seven constituent helices at intervals that are vertically in register[207] to create horizontal rings, or crowns, each spaced at 14.4-nm intervals.[207,210] The seven globular protrusions around each crown are each single myosin heads because each crown accounts for

the total molecular mass of about 3.5 molecules of myosin.[210,211] Because seven is an odd number, no two adjacent globular heads in a crown can be from the same molecule of myosin. The pair of heads from the same molecule of myosin must be adjacent to each other within the same helix. Therefore, along each of the seven helices the heads must alternate in the pattern lower head of myosin$_i$, upper head of myosin$_i$, lower head of myosin$_{i+1}$, upper head of myosin$_{i+1}$, and so forth. If upper heads are in register across the seven helices and lower heads are in register, crowns of lower heads and crowns of upper heads would alternate along the thick filament. Such an alternating pattern has been observed.[207]

Thick filaments of myosin assemble irreversibly and spontaneously from monomers of myosin[209] but do not become helical polymeric proteins of indefinite length such as fibrin, tubulin, or actin. Their final length is between 1000 and 3000 nm when they are polymerized under experimental situations[212] or between 1600 and 2000 nm when they are polymerized within a contractile tissue such as skeletal muscle (Figure 13–24B).[207,213]

When monomeric actin is induced to polymerize in the presence of thick filaments of myosin, thin filaments of actin form around each thick filament of myosin.[212] The thin filaments are positioned around the thick filaments at seven evenly spaced intervals. Presumably, the assembly of this 7-fold array is dictated by the underlying seven helices of the myosin heads. The pitch of the seven thin filaments of this actin, however, is much steeper than that of the seven primary helices on the surface of the thick filament, and the assembled thin filaments of actin are almost parallel to the axis of the thick filament. This alignment could be explained if the thick filaments of actin were in contact only with every other crown or every fourth crown along the thick filament and stepped up one helix for each contact (Figure 13–24D).

The measured rise for each subunit in a thin filament of actin is 2.8 nm and the measured rotation for each subunit is 166°. For a thin filament to span two crowns and step up one helix (Figure 13–24D) would require 29.7 nm, the distance covered by 11 subunits of actin if the rise for each subunit in a thin filament were actually 2.7 nm rather than 2.8 nm. The eleventh protomer further along a thin filament would be pointed in exactly the same direction toward the thick filament as a subunit of actin already attached to a myosin head in that thick filament if the rotation for each subunit of actin in a thin filament were 164° instead of 166°. A similarly successful but not so remarkable fit of the dimensions can be made if the thin filament of actin makes contact only with every fourth crown and steps up one helix. The coincidences between the dimensions of the thin filament and the thick filament are reminders that the more primordial of the two served as the template for the evolution of the other.

With the assembly of helical polymeric proteins such as microtubules, thick filaments of myosin, and sculpted

A   **B**   **C**   **D**

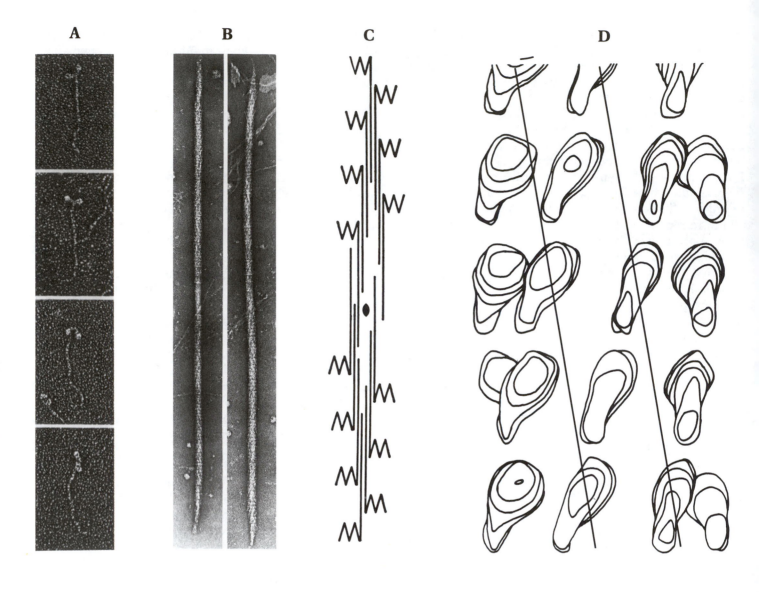

thin filaments of actin and the assembly of oligomeric proteins such as ribosomes, protein chemistry enters the microscopic realm and becomes cell biology. Another set of striking microscopic cellular features that are of importance in cell biology are membranes.

Suggested Reading

Mitchison, T., & Kirschner, M. (1984) Microtubule Assembly Nucleated by Isolated Centrosomes, and Dynamic Instability of Microtubule Growth, *Nature 312*, 232–242.

Problem 13–8: The rate constants for microtubule assembly listed in this table were obtained from the data in Figure 13–19.

(A) Describe which rate constant was derived from which aspect of this figure. Which rate constants are in error and why?

(B) If the concentration of monomeric tubulin were 15 μM, the steady-state critical concentration, at what rate would a tubule capped with GTP·tubulin be elongating at its plus end?

(C) If the minus end of this tubule were capped with GDP·tubulin, at what rate would it be depolymerizing?

Problem 13–9: Make a xerographic copy of Figure 13–17B. Cut out the surface lattice and roll the paper into a cylinder. Follow the various helices over the cylinder and identify how many individual strands each of them has.

rate constant	value	rate constant	value
k_1^{T+}	3.8×10^6 M^{-1} s^{-1}	k_{-1}^{T-}	1.1 s^{-1}
k_1^{T-}	1.2×10^6 M^{-1} s^{-1}	k_{-1}^{D+}	340 s^{-1}
$^-k_1^{T+}$	0.4 s^{-1}	k_{-1}^{D-}	210 s^{-1}

Figure 13–24: Structures of myosin and thick filaments. (A) Gallery of electron micrographs of myosin molecules.[205] Myosin from thick filaments in rabbit skeletal muscle, which had been disassembled at high ionic strength (0.5 M KCl), was purified by precipitation at low ionic strength, ammonium sulfate precipitation, and anion-exchange chromatography. A solution of purified myosin at 50 μg mL^{-1} in 0.6 M ammonium formate was sprayed onto the surface of freshly cleaved mica and the liquid and salt were evaporated from the surface. The adsorbed molecules of myosin were coated with platinum as the mica was rotated in a beam of platinum vapor. The specimen was then viewed in an electron microscope. Magnification 120000×. Panel A reprinted with permission from ref 205. Copyright 1978 Academic Press. (B) Thick filaments from scallop (*Placopectin magellanicus*) muscle.[207] Strips of scallop muscle were chopped finely and homogenized in a solution containing MgATP to dissociate thick and thin filaments. A drop of this homogenate was placed on a carbon film and negatively stained with 3% uranyl acetate. Thick filaments were located in the specimen in the electron microscope and photographs were taken. Magnification 60000×. Panel B adapted with permission from ref 207. Copyright 1983 Academic Press. (C) Diagrammatic representation of the way in which molecules of myosin are assembled to form a thick filament.[208] Each line segment is an individual molecule of myosin; the globular head (panel A) is represented by a W and the tail by a line. All of the tails point toward the center of the filament, and at the center the orientation reverses. Because the heads are directed distally, the bare zone (the smooth central portion of each of the thick filaments in panel B) has no heads protruding from it. The structure assembled in this way has a 2-fold rotational axis of pseudosymmetry at its center. Panel C reprinted with permission from ref 208. Copyright 1969 American Association for the Advancement of Science. (D) Helical surface lattice of globular myosin heads upon a thick filament.[207] The optical density of electron micrographs, such as those in panel B, was digitized and the Fourier transform of the digitized optical density was calculated. Discrete reflections arising from longitudinal spacings of 14.5 and 29.0 nm along the thick filament and helical spacings of 48.0 nm (panel B) were observed in the Fourier transform. Reflections on the reciprocal helical lattice in the Fourier transform were selected and used to calculate a three-dimensional distribution of electron scattering density. The image presented is of the front four helical strands of the seven-start right-handed helical lattice on one thick filament. This helical pattern produces the strong reflections of a 48.0-nm helical repeat. The globular heads are arranged in circular disks 14.5 nm wide and seven heads around that are stacked helically to produce the lattice. The lines drawn in the figure indicate the probable orientation of two thin filaments relative to the surface lattice of the thick filament. Panel D reprinted with permission from ref 207. Copyright 1983 Academic Press.

References

1. Dill, K.A., & Shortle, D. (1991) *Annu. Rev. Biochem. 60*, 795–826.
2. Steiner, D.F., & Clark, J.L. (1968) *Proc. Natl. Acad. Sci. U.S.A. 60*, 622–629.
3. Edge, V., Allewell, N.M., & Sturtevant, J.M. (1985) *Biochemistry 24*, 5899–5906.
4. Alber, T., Dao-pin, S., Wilson, K., Wozniak, J.A., Cook, S.P., & Matthews, B.W. (1987) *Nature 330*, 41–48.
5. Novokhatany, V.V., Kudinov, S.A., & Privalov, P.L. (1984) *J. Mol. Biol. 179*, 215–232.
6. Flanagan, M.T., & Hesketh, T.R. (1974) *Eur. J. Biochem. 44*, 251–259.
7. Tanford, C. (1968) *Adv. Protein Chem. 23*, 122–283.
8. Edelhoch, H. (1967) *Biochemistry 6*, 1948–1954.
9. Nozaki, Y., & Tanford, C. (1963) *J. Biol. Chem. 238*, 4074–4081.
10. Nozaki, Y., & Tanford, C. (1970) *J. Biol. Chem. 245*, 1648–1652.
11. Nandi, P.K., & Robinson, D.R. (1984) *Biochemistry 23*, 6661–6668.
12. Wetlaufer, D.B., Malik, S., Stoller, L., & Coffin, R.L. (1964) *J. Am. Chem. Soc. 86*, 508–514.
13. Arakawa, T., & Timasheff, S.N. (1984) *Biochemistry 23*, 5924–5929.
14. Lee, J.C., & Timasheff, S.N. (1974) *Biochemistry 13*, 257–265.
15. Arakawa, T., & Timasheff, S.N. (1982) *Biochemistry 21*, 6545–6552.
16. Arakawa, T., & Timasheff, S.N. (1982) *Biochemistry 21*, 6536–6544.
17. Creighton, T.E. (1979) *J. Mol. Biol. 129*, 235–264.
18. Roseman, M., & Jencks, W.P. (1975) *J. Am. Chem. Soc. 97*, 631–639.
19. Herskovits, T.T., Jaillet, H., & Gadegbeku, B. (1970) *J. Biol. Chem. 245*, 4544–4550.
20. Pace, C.N., & Marshall, H.F. (1980) *Arch. Biochem. Biophys. 199*, 270–276.
21. Hermans, J., & Acampora, G. (1967) *J. Am. Chem. Soc. 89*, 1547–1552.
22. Benz, F.W., & Roberts, G.C.K. (1975) *J. Mol. Biol. 91*, 367–387.
23. Aune, K.C., & Tanford, C. (1969) *Biochemistry 8*, 4579–4585.
24. Lumry, R., Biltonen, R., & Brandts, J.F. (1966) *Biopolymers 4*, 917.
25. Brandts, J.F., & Hunt, L. (1967) *J. Am. Chem. Soc. 89*, 4826–4838.
26. Shen, L.L., & Hermans, J. (1972) *Biochemistry 11*, 1836–1841.
27. Ginsburg, A., & Carroll, W.R. (1965) *Biochemistry 4*, 2159–2174.
28. Jackson, W.M., & Brandts, J.F. (1970) *Biochemistry 9*, 2294–2301.
29. Kellis, J.T., Nyberg, K., & Fersht, A.R. (1989) *Biochemistry 28*, 4914–4922.
30. Becktel, W.J., & Schellman, J.A. (1987) *Biopolymers 26*, 1859–1877.
31. Roberts, J.D., & Caserio, M.C. (1977) *Basic Principles of Organic Chemistry*, 2nd ed., p 457, W.A. Benjamin, Menlo Park, CA.

32. Flanagan, M.A., Garcia-Moreno, B., Friend, S.H., Feldmann, R.J., Scouloudi, H., & Gurd, F.R.N. (1983) *Biochemistry 22*, 6027–6037.
33. Pace, N.C., & Tanford, C. (1968) *Biochemistry 7*, 198–208.
34. Sturtevant, J.M. (1977) *Proc. Natl. Acad. Sci. U.S.A. 74*, 2236–2240.
35. Doig, A.J., & Williams, D.H. (1991) *J. Mol. Biol. 217*, 389–398.
36. Brandts, J.F., Oliveira, R.J., & Westort, C. (1970) *Biochemistry 9*, 1038–1047.
37. Hawley, S.A. (1971) *Biochemistry 10*, 2436–2442.
38. Gavish, B., Gratton, E., & Hardy, C.J. (1983) *Proc. Natl. Acad. Sci. U.S.A. 80*, 750–754.
39. Paladini, A.A., & Weber, G. (1981) *Biochemistry 20*, 2587–2593.
40. Pace, C.N. (1975) *CRC Crit. Rev. Biochem. 3*, 1–43.
41. Tanford, C. (1970) *Adv. Protein Chem. 24*, 1–95.
42. Santoro, M.M., & Bolen, D.W. (1988) *Biochemistry 27*, 8063–8068.
43. Puett, D. (1973) *J. Biol. Chem. 248*, 4623–4634.
44. Salahuddin, A., & Tanford, C. (1970) *Biochemistry 9*, 1342–1347.
45. Greene, R.F., & Pace, C.N. (1974) *J. Biol. Chem. 249*, 5388–5393.
46. Aune, K.C., & Tanford, C. (1969) *Biochemistry 8*, 4586–4590.
47. Bolen, D.W., & Santoro, M.M. (1988) *Biochemistry 27*, 8069–8074.
48. Pace, C.N., & Vanderburg, K.E. (1979) *Biochemistry 18*, 288–292.
49. Knapp, J.A., & Pace, C.N. (1974) *Biochemistry 13*, 1289–1294.
50. Rehage, A., & Schmid, F.X. (1982) *Biochemistry 21*, 1499–1505.
51. Goto, Y., & Hamaguchi, K. (1982) *J. Mol. Biol. 156*, 891–910.
52. Kauzman, W. (1959) *Adv. Protein Chem. 14*, 1–63.
53. Kyte, J., & Doolittle, R.F. (1982) *J. Mol. Biol. 157*, 105–132.
54. Chothia, C. (1976) *J. Mol. Biol. 105*, 1–14.
55. Dill, K.A. (1990) *Biochemistry 29*, 7133–7155.
56. Nozaki, Y., & Tanford, C. (1971) *J. Biol. Chem. 246*, 2211–2217.
57. Dill, K.A. (1985) *Biochemistry 24*, 1501–1509.
58. Flory, P.J. (1949) *J. Chem. Phys. 17*, 303–310.
59. Lau, K.F., & Dill, K.A. (1989) *Macromolecules 22*, 3896–3997.
60. Dill, K.A., Alonso, D.O.V., & Hutchinson, K. (1989) *Biochemistry 28*, 5439–5449.
61. Flory, P.J. (1953) *Principles of Polymer Chemistry*, Cornell Univesity Press, Ithaca, NY.
62. Flory, P.J., & Fisk, S. (1966) *J. Chem. Phys. 44*, 2243–2248.
63. Sanchez, I.C. (1979) *Macromolecules 12*, 980–988.
64. Miller, S., Janin, J., Lesk, A.M., & Chothia, C. (1987) *J. Mol. Biol. 196*, 641–656.
65. Chothia, C. (1975) *Nature 254*, 304–308.
66. Miller, S., Lesk, A.M., Janin, J., & Chothia, C. (1987) *Nature 328*, 834–836.
67. White, F.H. (1961) *J. Biol. Chem. 236*, 1353–1360.
68. Anfinsen, C.B., Haber, E., Sela, M., & White, F.H. (1961) *Proc. Natl. Acad. Sci. U.S.A. 47*, 1309–1314.
69. Haber, E., & Anfinsen, C.B. (1962) *J. Biol. Chem. 237*, 1839–1844.
70. Johnson, R.E., Adams, P., & Rupley, J.A. (1978) *Biochemistry 17*, 1479–1483.
71. Imoto, T., & Rupley, J.A. (1973) *J. Mol. Biol. 80*, 657–667.
72. Flory, P.J. (1956) *J. Am. Chem. Soc. 78*, 5222–5235.
73. Schellman, J.A. (1955) *C.R. Trav. Lab. Carlsberg, Ser. Chim. 29*, 230–259.
74. Mitra, S., & Lawton, R.G. (1979) *J. Am. Chem. Soc. 101*, 3097–3110.
75. Lin, S.H., Konishi, Y., Denton, M.E., & Scheraga, H.A. (1984) *Biochemistry 23*, 5504–5512.
76. Pace, C.N., Grimsley, G.R., Thomson, J.A., & Barnett, B.J. (1988) *J. Biol. Chem. 263*, 11820–11825.
77. Chan, H.S., & Dill, K.A. (1989) *J. Chem. Phys. 90*, 492–509.
78. Goto, Y., & Hamaguchi, K. (1982) *J. Mol. Biol. 156*, 911–926.
78a. Jackson, S.E., Moracci, M., el Masry, N., Johnson, C.M., & Fersht, A.R. (1993) *Biochemistry 32*, 11259–11269.
79. Fauchere, J., & Pliska, V. (1983) *Eur. J. Med. Chem. Chim. Ther. 18*, 369–375.
80. Lin, M.C. (1970) *J. Biol. Chem. 245*, 6726–6731.
81. Taniuchi, H., & Anfinsen, C.B. (1969) *J. Biol. Chem. 244*, 3864–3875.
82. Sachs, D.H., Schechter, A.N., Eastlake, A., & Anfinsen, C.B. (1974) *Nature 251*, 242–244.
83. Ikemura, H., & Inouye, M. (1988) *J. Biol. Chem. 263*, 12959–12963.
84. Taniuchi, H., & Anfinsen, C.B. (1971) *J. Biol. Chem. 246*, 2291–2301.
85. Klee, W.A. (1968) *Biochemistry 7*, 2731–2736.
86. Wyckoff, H.W., Hardman, K.D., Allewell, N.M., Inagami, T., Johnson, L.N., & Richards, F.M. (1967) *J. Biol. Chem. 242*, 3984–3988.
87. Haer, E. (1964) *Proc. Natl. Acad. Sci. U.S.A. 52*, 1099–1106.
88. Whitney, P.L., & Tanford, C. (1965) *Proc. Natl. Acad. Sci. U.S.A. 53*, 524–532.
89. Painter, R.G., Sage, H.J., & Tanford, C. (1972) *Biochemistry 11*, 1338–1345.
90. Adams, B., Burgess, R.J., & Pain, R.H. (1985) *Eur. J. Biochem. 152*, 715–720.
91. Doster, W., & Hess, B. (1981) *Biochemistry 20*, 772–780.
92. Wong, K., & Tanford, C. (1973) *J. Biol. Chem. 248*, 8518–8523.
93. Kuwajima, K., Nitta, K., Yoneyama, M., & Sugai, S. (1976) *J. Mol. Biol. 106*, 359–373.
94. Stellwagen, E., & Babul, J. (1975) *Biochemistry 14*, 5135–5140.
95. Hughson, F.M., Wright, P.E., & Baldwin, R.L. (1990) *Science 249*, 1544–1548.
96. Brazhnikov, E.V., Chirgadze, Y.N., Dolgikh, D.A., & Ptitsyn, O.B. (1985) *Biopolymers 24*, 1899–1907.
97. Ptitsyn, O.B. (1987) *J. Protein Chem. 6*, 273–293.
98. Chan, H.S., & Dill, K.A. (1990) *Proc. Natl. Acad. Sci. U.S.A. 87*, 6388–6392.
99. Kuwajima, K. (1977) *J. Mol. Biol. 114*, 241–258.
100. Dolgikh, D.A., Gilmanshin, R.I., Brazhnikov, E.V., Bychkova, V.E., Semisotnov, G.V., Venyaminov, S.Y., & Ptitsyn, O.B. (1981) *FEBS Lett. 136*, 311–315.
100a. Xie, D., Bhakuni, V., & Friere, E. (1991) *Biochemistry 30*, 10673–10678.
101. Ohgushi, M., & Wada, A. (1983) *FEBS Lett. 164*, 21–24.
102. Oas, T.G., & Kim, P.S. (1988) *Nature 336*, 42–48.
103. Baldwin, R.L. (1989) *Trends Biochem. Sci. 14*, 291–294.

104. Greene, R.F., & Pace, C.N. (1974) *J. Biol. Chem. 249*, 5388–5393.

105. Garel, J.R., & Baldwin, R.L. (1973) *Proc. Natl. Acad. Sci. U.S.A. 70*, 3347–3351.

106. Ikai, A., Fish, W.W., & Tanford, C. (1973) *J. Mol. Biol. 73*, 165–184.

107. Ikai, A., & Tanford, C. (1971) *Nature 230*, 100–102.

108. Brandts, J.F., Halvorson, H.R., & Brennan, M. (1975) *Biochemistry 14*, 4953–4963.

109. Li, L.N., & Brandts, J.F. (1983) *Biochemistry 22*, 559–563.

110. Schmid, F.X., Grafl, R., Wrba, A., & Beintema, J.J. (1986) *Proc. Natl. Acad. Sci. U.S.A. 83*, 872–876.

111. Schmid, F.X., & Baldwin, R.L. (1978) *Proc. Natl. Acad. Sci. U.S.A. 75*, 4764–4768.

112. Henkens, R.W., Gerber, A., Cooper, M.R., & Herzog, W.R. (1980) *J. Biol. Chem. 255*, 7075–7078.

113. Schmid, F.X., & Baldwin, R.L. (1979) *J. Mol. Biol. 133*, 285–287.

114. Krebs, H., Schmid, F.X., & Jaenicke, R. (1983) *J. Mol. Biol. 169*, 619–635.

115. Kim, P.S., & Baldwin, R.L. (1980) *Biochemistry 19*, 6124–6129.

116. Schmid, F.X., & Blaschek, H. (1981) *Eur. J. Biochem. 114*, 111–117.

117. Schmid, F.X., & Blaschek, H. (1984) *Biochemistry 23*, 2128–2133.

118. Kiefhaber, T., Quaas, R., Hahn, U., & Schmid, F.X. (1990) *Biochemistry 29*, 3053–3061.

119. Kiefhaber, T., Quass, R., Hahn, U., & Schmid, F.X. (1990) *Biochemistry 29*, 3061–3070.

120. Fischer, G., Wittmann-Liebold, B., Lang, K., Kiefhaber, T., & Schmid, F.X. (1989) *Nature 337*, 476–478.

121. Takahashi, N., Hayano, T., & Suzuki, M. (1989) *Nature 337*, 473–475.

122. Fischer, G., Bang, H., & Mech, C. (1984) *Biomed. Biochim. Acta 43*, 1101–1111.

123. Fischer, G., & Bang, H. (1985) *Biochim. Biophys. Acta 828*, 39–42.

124. Lang, K., Schmid, F.X., & Fischer, G. (1987) *Nature 329*, 268–270.

125. Lang, K., & Schmid, F.X. (1988) *Nature 331*, 453–455.

126. Bachinger, H.P. (1987) *J. Biol. Chem. 262*, 17144–17148.

127. Siekierka, J.J., Hung, S.H.Y., Poe, M., Lin, C.S., & Sigal, N.H. (1989) *Nature 341*, 755–757.

128. Harding, M.W., Galat, A., Uehling, D.E., & Schreiber, S.L. (1989) *Nature 341*, 758–760.

129. Kuwajima, K., Yamaya, H., Miwa, S., Sugai, S., & Nagamura, T. (1987) *FEBS Lett. 221*, 115–118.

130. Udgaonkar, J.B., & Baldwin, R.L. (1988) *Nature 335*, 694–699.

131. Roder, H., Elöve, G.A., & Englander, S.W. (198) *Nature 335*, 700–704.

132. Bycroft, M., Matouschek, A., Kellis, J.T., Serrano, L., & Fersht, A.R. (1990) *Nature 346*, 488–490.

133. Ahern, T.J., & Klibanov, A.M. (1985) *Science 228*, 1280–1284.

134. Wong, S.C.K., Burton, P.M., & Josse, J. (1970) *J. Biol. Chem. 245*, 4353–4357.

135. Hermann, R., Rudolf, R., Jaenicke, R., Price, N.C., & Scobbie, A. (1983) *J. Biol. Chem. 258*, 11014–11019.

136. Hermann, R., Jaenicke, R., & Price, N.C. (1985) *Biochemistry 24*, 1817–1821.

137. Jaenicke, R., Rudolph, R., & Heider, I. (1979) *Biochemistry 18*, 1217–1223.

138. Burns, D.L., & Schachman, H.K. (1982) *J. Biol. Chem. 257*, 8648–8654.

139. Burns, D.L., & Schachman, H.K. (1982) *J. Biol. Chem. 257*, 8638–8647.

140. Yamamoto, S., & Murachi, T. (1979) *Eur. J. Biochem. 93*, 189–195.

141. Vimard, C., Orsini, G., & Goldberg, M.E. (1975) *Eur. J. Biochem. 51*, 521–527.

142. Lane, A.N., Paul, C.H., & Kirschner, K. (1984) *EMBO J. 3*, 279–287.

143. Bothwell, M.A., & Schachman, H.K. (1980) *J. Biol. Chem. 255*, 1962–1970.

144. Yang, Y.R., Syvanen, J.M., Nagel, G.M., & Schachman, H.K. (1974) *Proc. Natl. Acad. Sci. U.S.A. 71*, 918–922.

145. Jacobsen, G.R., & Stark, G.R. (1973) *J. Biol. Chem. 248*, 8003–8014.

146. Bothwell, M.A., & Schachman, H.K. (1980) *J. Biol. Chem. 255*, 1971–1977.

147. Bates, D.L., Danson, M.J., Hale, G., Hooper, E.A., & Perham, R.N. (1977) *Nature 268*, 313–316.

148. Wagenknecht, T., Francis, N., & DeRosier, D.J. (1983) *J. Mol. Biol. 165*, 523–541.

149. Reed, L.J., Pettit, F.H., Eley, M.H., Hamilton, L., Collins, J.H., & Oliver, R.M. (1975) *Proc. Natl. Acad. Sci. U.S.A. 72*, 3068–3072.

150. DeRosier, D.J., & Oliver, R.M. (1971) *Cold Spring Harbor Symp. Quant. Biol. 36*, 199–203.

151. Brosius, J., Palmer, M.L., Kennedy, P.J., & Noller, H.F. (1978) *Proc. Natl. Acad. Sci. U.S.A. 75*, 4801–4805.

152. Held, W.A., Mizushima, S., & Nomura, M. (1973) *J. Biol. Chem. 248*, 5720–5730.

153. Held, W.A., Ballou, B., Mizushima, S., & Nomura, M. (1974) *J. Biol. Chem. 249*, 3103–3111.

154. Woese, C.R., Gutell, R., Gupta, R., & Noller, H.F. (1983) *Microbiol. Rev. 47*, 621–669.

155. Kühlbrandt, W., & Unwin, P.N.T. (1982) *J. Mol. Biol. 156*, 431–448.

156. Ungewickell, E., Garrett, R., Ehresmann, C., Stiegler, P., & Fellner, P. (1975) *Eur. J. Biochem. 51*, 165–180.

157. Herold, M., & Nierhaus, K.H. (1987) *J. Biol. Chem. 262*, 8826–8833.

158. Morrison, C.A., Garret, R.A., & Bradbury, E.M. (1977) *Eur. J. Biochem. 78*, 153–159.

159. Rohde, M.E., O'Brien, S., Cooper, S., & Aune, K.C. (1975) *Biochemistry 14*, 1079–1087.

160. Wong, K.P., & Paradies, H.H. (1974) *Biochem. Biophys. Res. Commun. 61*, 178–184.

161. Luer, C.A., & Wong, K.P. (1979) *Biochemistry 18*, 2019–2027.

162. Franz, A., Georgalis, Y., & Giri, L. (1979) *Biochim. Biophys. Acta, 578*, 365–371.

163. Doolittle, R.F. (1984) *Annu. Rev. Biochem. 53*, 195–229.

164. Williams, R. (1981) *J. Mol. Biol. 150*, 399–408.

165. Laudano, A.P., & Doolittle, R.F. (1980) *Biochemistry 19*, 1013–1019.

166. Hantgan, R.R., & Hermans, J. (1979) *J. Biol. Chem. 254*, 11272–11281.

167. Chen, R., & Doolittle, R.F. (1971) *Biochemistry 10*, 4486–4491.

168. Doolittle, R.F., Cassman, K.G., Cottrell, B.A., & Friezner, S.J. (1977) *Biochemistry 16*, 1715–1719.

169. Ludueña, R.F., Shooter, E.M., & Wilson, L. (1977) *J. Biol. Chem. 252*, 7006–7014.

170. Feit, H., Slusarek, L., & Shelanski, M.L. (1971) *Proc. Natl. Acad. Sci. U.S.A. 68*, 2028–2031.

171. Ponstingl, H., Krauhs, E., Little, M., & Kempf, T. (1981) *Proc. Natl. Acad. Sci. U.S.A. 78*, 2757–2761.

172. Krauhs, E., Little, M., Kempf, T., Hofer-Warbinek, R., Ade, W., & Ponstingl, H. (1981) *Proc. Natl. Acad. Sci. U.S.A. 78*, 4156–4160.

173. Kirschner, M.W., Williams, R.C., Weingarten, M., & Gerhart, J.C. (1974) *Proc. Natl. Acad. Sci. U.S.A. 71*, 1159–1163.

174. Amos, L.A., & Klug, A. (1974) *J. Cell Sci. 14*, 523–549.

175. Bergen, L.G., & Borisy, G.G. (1980) *J. Cell Biol. 84*, 141–149.

176. Johnson, K.A., & Borisy, G.G. (1977) *J. Mol. Biol. 117*, 1–31.

177. Kirschner, M.W., Honig, L.S., & Williams, R.C. (1975) *J. Mol. Biol. 99*, 263–276.

178. Scheele, R.B., & Borisy, G. (1978) *J. Biol. Chem. 253*, 2846–2851.

179. Mitchison, T., & Kirschner, M. (1984) *Nature 312*, 232–237.

180. Bergen, L.G., Kuriyama, R., & Borisy, G.G. (1980) *J. Cell Biol. 84*, 151–159.

181. Mitchison, T., & Kirschner, M. (1984) *Nature 312*, 237–242.

182. Karr, T.L., Podrasky, A.E., & Purich, D.L. (1979) *Proc. Natl. Acad. Sci. U.S.A. 76*, 5475–5479.

183. Weisenberg, R.C. (1972) *Science 177*, 1104–1105.

184. Weisenberg, R.C., Borisy, G.G., & Taylor, E.W. (1968) *Biochemistry 7*, 4466–4479.

185. Weisenberg, R.C, Deery, W.J., & Dickinson, P.J. (1976) *Biochemistry 15*, 4248–4254.

186. Margolis, R.L. (1981) *Proc. Natl. Acad. Sci. U.S.A. 78*, 1586–1590.

187. Karr, T.L., Kristofferson, D., & Purich, D.L. (1980) *J. Biol. Chem. 255*, 8560–8566.

188. Margolis, R.L., & Wilson, L. (1978) *Cell 13*, 1–8.

189. Wegner, A. (1976) *J. Mol. Biol. 108*, 139–150.

190. Cote, R.H., & Borisy, G.G. (1981) *J. Mol. Biol. 150*, 577–602.

191. Margolis, R.L., & Wilson, L. (1977) *Proc. Natl. Acad. Sci. U.S.A. 74*, 3466–3470.

192. Woodrum, D.T., Rich, S.A., & Pollard, T.D. (1975) *J. Cell Biol. 67*, 231–237.

193. Pollard, T.D., & Mooseker, M.S. (1981) *J. Cell Biol. 99*, 654–659.

194. Straub, F.B., & Feuer, G. (1950) *Biochim. Biophys. Acta 4*, 455–470.

195. Geiger, B., Tokuyasu, K.T., Dutton, A., & Singer, S.J. (1980) *Proc. Natl. Acad. Sci. U.S.A. 77*, 4127–4131.

196. Small, J.V. (1985) *EMBO J. 4*, 45–49.

197. Yin, H.L., & Stossel, T.P. (1980) *J. Biol. Chem. 255*, 9490–9493.

198. Harris, D.A., & Schwartz, J.H. (1981) *Proc. Natl. Acad. Sci. U.S.A. 78*, 6798–6802.

199. Yin, H.L., Hartwig, J.H., Maruyama, K., & Stossel, T.P. (1981) *J. Biol. Chem. 256*, 9693–9697.

200. Glenney, J.R., Kaulfus, P., Matsudaira, P., & Weber, K. (1981) *J. Biol. Chem. 256*, 9283–9288.

201. Bretscher, A., & Weber, K. (1979) *Proc. Natl. Acad. Sci. U.S.A. 76*, 2321–2325.

202. Glenney, J.R., Geisler, N., Kaulfus, P., & Weber, K. (1981) *J. Biol. Chem. 256*, 8156–8161.

203. Kwiatkowski, D.J., Stossel, T.P., Orkin, S.H., Mole, J.E., Colten, H.R., & Yin, H.L. (1986) *Nature 323*, 455–458.

204. Hesterberg, L.K., & Weber, K. (1983) *J. Biol. Chem. 258*, 365–369.

205. Elliot, A., & Offer, G. (1978) *J. Mol. Biol. 123*, 505–519.

206. Slayter, H.S., & Lowey, S. (1967) *Proc. Natl. Acad. Sci. U.S.A. 58*, 1611–1618.

207. Vibert, P., & Craig, R. (1983) *J. Mol. Biol. 165*, 303–320.

208. Huxley, H.E. (1969) *Science 164*, 1356–1366.

209. Huxley, H.E. (1963) *J. Mol. Biol. 7*, 281–308.

210. Knight, P.J., Erickson, M.A., Rodgers, M.E., Beer, M., & Wiggins, J.W. (1986) *J. Mol. Biol. 189*, 167–177.

211. Morimoto, K., & Harrington, W.F. (1974) *J. Mol. Biol. 83*, 83–97.

212. Hayashi, T., Silver, R.B., Ip, W., Cayer, M.L., & Smith, D.S. (1977) *J. Mol. Biol. 111*, 159–171.

213. Morimoto, K., & Harrington, W.F. (1973) *J. Mol. Biol. 77*, 165–175.

CHAPTER 14

Membranes

Implicit in the cellular theory is the existence of a boundary between the cytoplasm of a cell and its surroundings, be they seawater or the extracellular fluid in a highly organized tissue. The boundary is a defined physical structure known as the **plasma membrane**, and it is a thin, continuous, closed bag defining the boundary of the cell. In electron micrographs of thin sections through a cell, the plasma membrane appears as a continuous closed curve that marks the perimeter of the cytoplasm.

In many microorganisms, such as algae, fungi, and bacteria, the plasma membrane is surrounded on its outer surface by an outer membrane or cell wall. The cells of higher plants are also surrounded by a thick cell wall. Usually, the cells of animals, when they are located in organized tissues, are surrounded by networks of collagen and mucopolysaccharide. All of these integuments encasing these various cells are tough polymeric materials that provide support and security for the plasma membrane, which is the formal boundary between the cytoplasm and the environment, between the living and the inert.

When a thin section of a eukaryotic cell is examined in the electron microscope, the most striking feature of the image is the collection of curves that represent systems of intracellular membranes cut in cross section. These intracellular membranes are the **endoplasmic reticulum**, the **Golgi apparatus**, and the membranes of the **mitochondria**, the **nucleus**, the **lysosomes**, the **peroxisomes**, the **chloroplasts**, the **endosomes**, and the **vacuoles** of the cell. Each of these structures at any instant is a closed, continuous, often highly irregular bag enclosing its respective volume of fluid, which is isolated by its membrane from the cytoplasm of the cell. The aqueous solution of proteins found within any one of these bags is unique from that in the cytoplasm surrounding it and is characteristic of the particular organelle. The membranes creating each of these organelles have the same structure as the plasma membrane, although each is distinct in its chemical composition. Because almost every membrane in the cell separates the cytoplasm from another space, the two sides of a membrane are defined relative to the cytoplasm, and they are referred to as **cytoplasmic** and **extracytoplasmic**, respectively. Among the exceptions to this situation are the inner membranes of the mitochondria and the thylakoids of chloroplasts.

Each of the membranes composing a cell can be purified from the homogenate of a eukaryotic tissue by **cell fractionation**.[1] Originally these purifications were performed in water,[1] but it was subsequently noted that the organelles retained their appearance more successfully in concentrated solutions of sucrose.[2] Tissues are usually homogenized in a solution of 0.25 M sucrose, and the membranes are isolated by centrifugation on gradients composed of solutions of increasing concentrations of sucrose, which is a solute that stabilizes proteins by salting in, or of other solutes that change the density or osmolarity of the solution. During homogenization, the mitochondria, lysosomes, peroxisomes, chloroplasts, nuclei, and Golgi apparatus remain intact and can be identified by their characteristic morphology.[3,4] Plasma membranes and endoplasmic reticulum are disintegrated by the homogenization and become rounded fragments, often goblet-shaped or vesicular in morphology, and these fragments are known as **microsomes**.[5] Microsomes of rough endoplasmic reticulum are readily identified by their adherent ribosomes.[3]

The various membranes and intact organelles suspended in the homogenate of a eukaryotic cell differ in their size and shape, their ratio of protein, lipid, and carbohydrate, and their composition of fixed acid–bases. Therefore, they can be separated from each other by differences in their sedimentation coefficients and their buoyant densities (Figure 14–1)[6] or their net charges at a particular pH. Homogenates are often submitted to sequential centrifugations at different centrifugal forces for different durations. Such **differential centrifugations**[1] separate only crudely on the basis of sedimentation coefficient because large differences in sedimentation coefficient between two organelles are necessary if one of the organelles is to form a pellet exclusively at one centrifugal force and duration while the other forms a pellet exclusively at a higher force or longer duration. **Rate sedimentation**[7] is a technique in which a narrow band of sample is layered onto a gradient that changes only gradually in sucrose concentration, and the components are separated owing to their differences in sedimentation coefficient as they move through the gradient under the influence of a centrifugal field. Rate sedimentation provides much higher resolution than differential centrifugation. Unfortunately, the population of a given organelle in a tissue usually has a significant variation (Figure 14–1)[6] in its sedimentation coefficient, and additional steps that separate membranes by other independent properties are often required for complete purification. During **isopycnic centrifugation**[4] the sample is layered onto a much steeper gradient of density formed usually either by varying sucrose concentration or by varying Ficoll concentration.[8] Ficoll is a polysaccharide that does not have the high osmolarity of sucrose. The gradients are submitted to centrifugation until all of the components have traveled to their respective buoyant densities, at which point they cease to move. The various membranes suspended in a homogenate can also be separated on the basis of their charge by **free-flow electrophoresis**.[9] In certain instances, where the fixed anionic functionalities on a particular type of membrane are properly oriented, these membranes can be precipitated exclusively with divalent cations such as magnesium or calcium.[10]

During purification, the various classes of membranes can be followed by assaying for particular **marker enzymes**. Each type of organelle has at least one enzymatic activity that is almost exclusively confined to it and can be used as a measure of its concentration.[3,4,11,12] The ability of certain membranes to bind very specific ligands has also been used to follow their purification. For example, the plasma membranes of animal cells bind the protein wheat germ agglutinin, the peptide hormone insulin, and the toxin from *Vibrio cholerae* with high specificity; the binding of any one of these three ligands can be used to identify plasma membranes.[13] The final identification of the purified suspension of membranes as the organelle of interest, however, must always be made by examining thin sections of pellets of the purified material by electron microscopy.

All of these procedures have been used to develop methods for the purification of mitochondria,[4] peroxisomes,[4] lysosomes,[4] Golgi apparatuses,[12] rough endoplasmic reticulum,[14] and chloroplasts.[15] Ironically, it is the plasma membrane of most cells that is the most

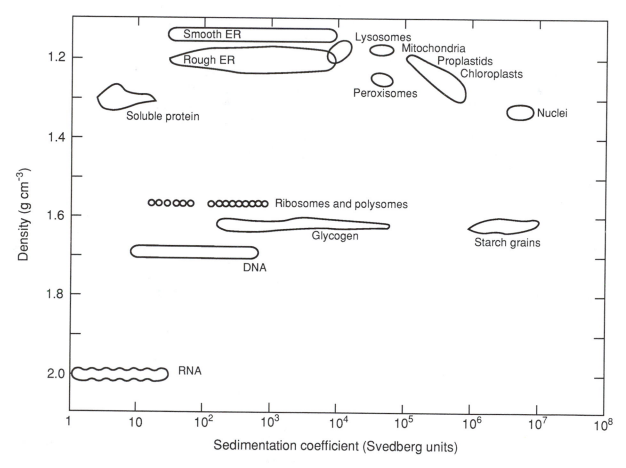

Figure 14–1: Buoyant density and sedimentation coefficient of the organelles and fragments of membrane found in a homogenate of a eukaryotic cell. Each of the boundaries surrounds the distribution of buoyant densities (grams centimeter^{-3}) and sedimentation coefficients of the particular organelle or fragment of membrane. Uniform organelles such as mitochondria, lysosomes, nuclei, and peroxisomes have relatively tight distributions of density and sedimentation coefficient. Smooth endoplasmic reticulum (ER), plasma membrane, and rough endoplasmic reticulum (ER), because they are heterogeneous fragments of membrane broken from much larger continuous structures, have fairly uniform buoyant densities but a wide range of sedimentation coefficients. The diagram illustrates that each of these classes of particles occupies a unique region in the two-dimensional space, and this allows each class of membranes to be isolated by a method exploiting differences in sedimentation coefficient in combination with a method exploiting differences in buoyant density. Usually, the boundaries for plasma membrane and smooth endoplasmic reticulum coincide, so these two types of membrane fragments are difficult to separate. Adapted with permission from ref 6. Copyright 1974 Academic Press.

difficult membrane to purify because upon homogenization it fragments and becomes very similar to the much more abundant fragments of smooth endoplasmic reticulum. For this reason, the plasma membrane of the erythrocyte, a cell lacking any other membranes, has often been used as a model for an animal plasma membrane. Plasma membranes, however, have been purified from a number of tissues, including liver,[16] kidney,[9] adipose tissue,[8] and brain,[17] and from cells grown in tissue culture, such as mouse fibroblasts (L-cells)[18] and HeLa cells.[19] Plasma membranes from fungi[20] or bacteria, such as *Escherichia coli*,[21] are prepared from spheroplasts, which are individual cells that have been enzymatically stripped of their outer membranes or cell walls. The spheroplasts are ruptured and the smooth plasma membranes are isolated from the homogenate.

Any of the membranes comprising an organelle or derived from a larger cellular structure can be freed from the soluble proteins it encloses by lysis and sedimentation. Such a membrane is constituted from lipids, carbohydrate, and proteins. All of the carbohydrate is covalently attached to protein in the form of glycoprotein (Chapter 3) or to lipid in the form of glycolipid. The component lipids and proteins are both heterogeneous mixtures, and the fraction of the mass that is protein can vary up to 75%. The basic structure upon which biological membranes are based is a **bilayer** of amphipathic lipids in which some neutral lipid is dissolved.

Suggested Reading

Price, C.A. (1974) Plant Cell Fractionation, *Methods Enzymol.* 31, 501–519.

The Bilayer

The basic structural element of a biological membrane is a bilayer of **phospholipids** (Figure 14–2). The most prevalent phospholipids are phosphatidylcholine (Figure 14–2A) and phosphatidylethanolamine (Figure 14–2B). Phosphatidylcholine and phosphatidylethanolamine are constructed from a molecule of *sn*-glycerol 3-phosphate

14–1

A molecule of either ethanolamine (Figure 14–2B) or trimethylethanolammonium (choline; Figure 14–2A) is esterified to the phosphate, forming a phosphate diester. The two remaining hydroxyls of the glycerol 3-phosphate are acylated with a pair of fatty acids. Phosphatidylcholine and phosphatidylethanolamine are **amphipathic**

lipids. One end of each of the molecules, the end containing the phosphate diester and the choline or ethanolamine, is hydrophilic. The other end, the end containing the linear hydrocarbon of the fatty acids, is hydrophobic. Such **linear hydrocarbons** are the most hydrophobic functional groups among biological molecules (Table 5–6).[22]

All phospholipids have two fatty acids attached to the glycerol in ester linkages. The fatty acids found in naturally occurring phospholipids seem to be chosen almost at random from the mixtures of fatty *S*-acylcoenzymes A produced by the particular organism in which the membrane is located. Saturated fatty acids are esterified mainly to carbon 1 of the *sn*-glycerol 3-phosphate, and the unsaturated fatty acids are esterified mainly to carbon 2 (Figure 14–2), so that the amphipathic lipids end up with a roughly equal ratio of saturated and unsaturated linear hydrocarbons, inescapably intermixed.

The **saturated fatty acids** are mainly linear carboxylic acids that vary in length from 12 to 24 carbons. The most frequently encountered saturated fatty acids in biological membranes are palmitic acid (*n*-hexadecanoic acid; Figure 14–2C,F) and stearic acid (*n*-octadecanoic acid; Figure 14–2B,D). The most stable conformation of a linear hydrocarbon is all-*trans* (**14–2**), but the introduction of a *gauche* (**14–3**) conformation at one of the carbon–carbon bonds requires only about 4 kJ mol^{-1} if the hydrocarbon is unhindered. Therefore, at 25 °C about 15% of the unhindered carbon–carbon single bonds in a fatty acid should be *gauche*. The *gauche* conformation, however, transiently introduces an elbow with an angle of about 109° into the chain

14–2

14–3

A second *gauche* conformation can reorient the chain to its original direction. **Unsaturated fatty acids** contain one or more carbon–carbon double bonds. A *trans* double bond would not put an elbow in the hydrocarbon, but almost every carbon–carbon double bond in naturally occurring fatty acids is *cis*. One *cis* double bond introduces a permanent elbow with an angle of about 120–130° into an otherwise unsaturated linear hydrocarbon

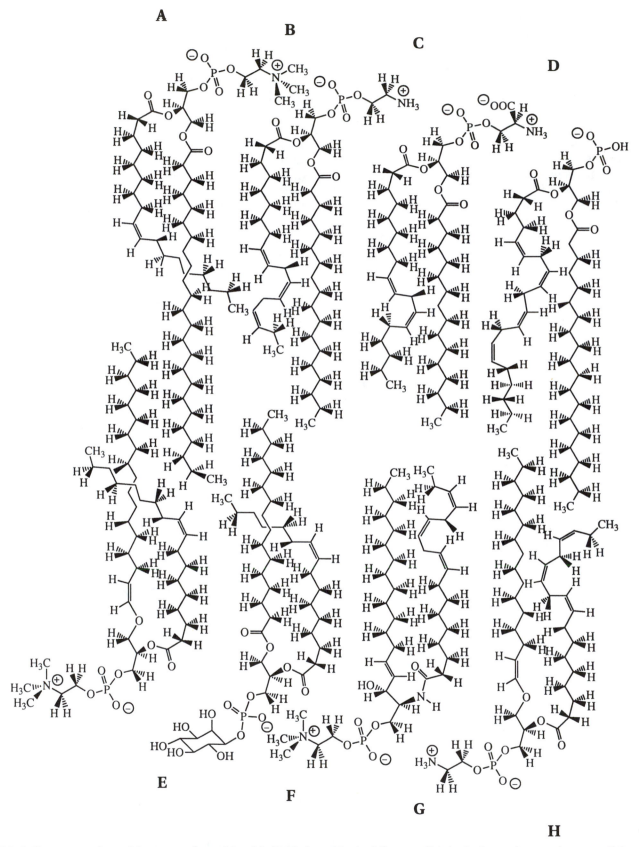

Figure 14–2: Representatives of the types of amphipathic lipids found in the bilayers of biological membranes drawn as if they were in a bilayer. (A) 1-Lignoceryl-2-oleylphosphatidylcholine. (B) 1-Stearyl-2-linolenylphosphatidylethanolamine. (C) 1-Palmityl-2-linoleylphosphatidylserine. (D) 1-Arachidyl-2-arachidonylphosphatidic acid. (E) 1-Octadec-1′-enyl-2-oleylglycero-3-phosphocholine (a plasmalogen). (F) 1-Palmityl-2-palmitoleylphosphatedylinositol. (G) N-Linolenylsphingosine-1-phosphocholine (a sphingomyelin). (H) 1-Hexadec-1′-enyl-2-linolenyl-3-phosphoethanolamine (a plasmalogen).

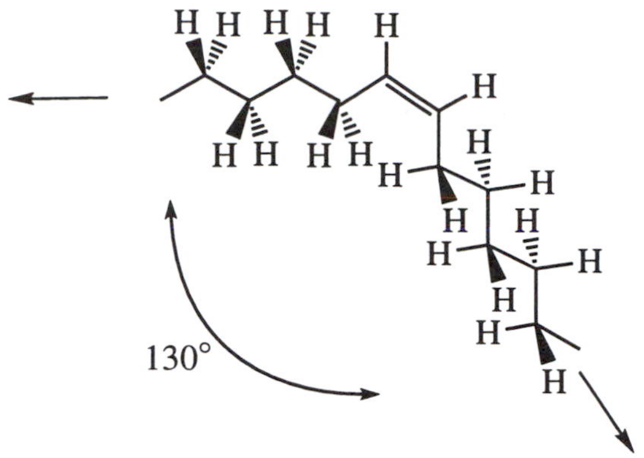

14-4

In the most stable conformation, one of the hydrogens on one of the two methylenes adjacent to the double bond fits between the two hydrogens of the other. The most common unsaturated fatty acids with only one carbon–carbon double bond are palmitoleic acid (*cis*-hexadec-9-enoic acid) and oleic acid (*cis*-octadec-9-enoic acid; Figure 14–2A,E), and the latter predominates.

When an unsaturated fatty acid contains two or three double bonds, they are spaced three carbons apart with an unsaturated carbon between them. This prevents the double bonds from conjugating with each other and becoming a rigid planar structure. Two *cis* double bonds spaced in this way produce a sinuous curve in the alkyl chain but do not change its ultimate direction

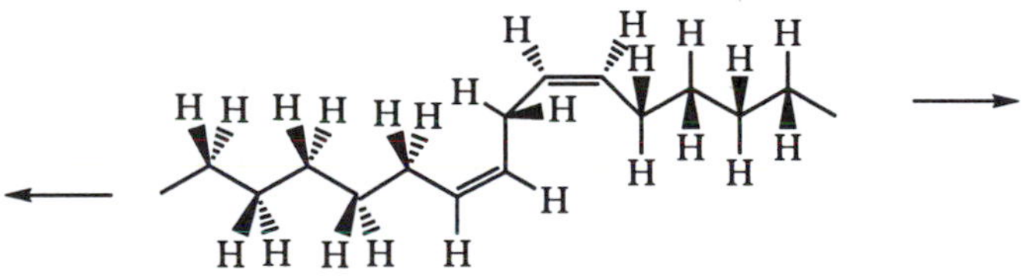

14-5

They do, however, shorten its ultimate length by the equivalent of two carbon atoms, and the volume lost at the end is expressed as a bulge at the location of the unsaturation. The most common unsaturated fatty acid with two double bonds is linoleic acid (*cis,cis*-octadeca-9,12-dienoic acid; Figure 14–2C). Three *cis* double bonds produce an even longer sinuous curve that does place a permanent elbow in the alkyl chain

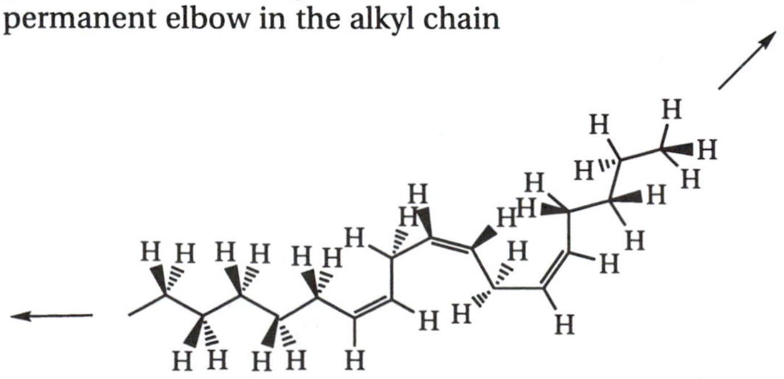

14-6

The most common fatty acid with three carbon–carbon double bonds is linolenic acid in its two geometric isomers, α-linolenic acid (*cis,cis,cis*-octadeca-9,12,15-trienoic acid; Figure 14–2B,H) and γ-linolenic acid (*cis,cis,cis*-octadeca-6,12,15-trienoic acid; Figure 14–2F) Arachidonic acid (*cis,cis,cis,cis*-eicosa-5,8,11,14-tetraenoic acid; Figure 14–2D) is a less common unsaturated fatty acid that serves as the precursor to prostaglandins, and its frequency is regulated for that purpose. Almost all of the unsaturation commences at carbon 9 in the usual mixture of the various fatty acids, so the portion of the hydrocarbon closest to the glycerol is fully saturated and that farthest away is unsaturated and geometrically disordered (Figure 14–2).

The majority of the polar functional groups esterified to the phosphate of *sn*-glycerol 3-phosphate in naturally occurring phospholipids are based on ethanolamine

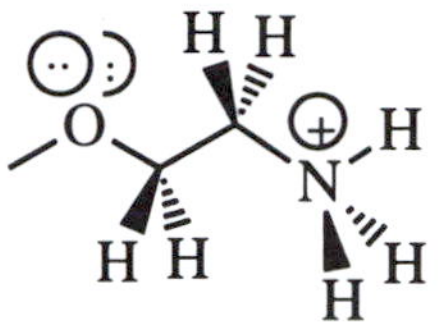

14-7

In **phosphatidylethanolamine** (Figure 14–2B), the ethanolamine is unaltered. In **phosphatidylcholine** (Figure 14–2A), the ethanolamine is triply methylated on nitrogen. In **phosphatidylserine** (Figure 14–2C) the ethanolamine is carboxylated. Two phospholipids not based on ethanolamine are **phosphatidic acid** (Figure 14–2D), which lacks an ester on phosphate, and **phosphatidylinositol** (Figure 14–2F). Phosphatidylinositol is a minor phospholipid present at less than 5% in membranes. It provides the anchor for phosphatidylinositol-linked proteins (Figure 3–19). Because it is an intermediate in the biosynthesis of the second messenger inositol triphosphate, its levels are independently regulated.

There are several other phospholipids that are variations on the theme. The **plasmalogens** (Figure 14–2E,H) have an enol ether at carbon 1 of the *sn*-glycerol 3-phosphate rather than an acylated oxygen. Upon treatment with BF$_3$ in methanol, the enol ether is released as the dimethyl acetal of a fatty aldehyde.[23] The enol ethers are usually derived from *n*-hexadecanal (Figure 14–2H) or *n*-octadecanal (Figure 14–2E), but minor amounts of the derivatives of unsaturated fatty aldehydes are also present.[18] Either ethanolamine (Figure 14–2E) or choline (Figure 14–2G) is esterified, respectively, to the *sn*-glycerol 3-phosphate of plasmalogens. **Sphingomyelins** have a primary alkene of 15 carbons replacing one of the hydrogens on carbon 1 of the *sn*-glycerol 3-phosphate, the oxygen on carbon 1 is not acylated, and a nitrogen replaces the oxygen on carbon 2 and forms an amide with a fatty acid rather than an ester (Figure 14–2G). Upon saponification, sphingosine (2-amino-3-hydroxyoctadec-

Table 14–1: Composition of Amphipathic Lipids in Plasma Membranes from Human Erythrocytes and Murine Fibroblasts (L-Cells)

	% total lipid	
lipid	erythrocytes[a]	L-cells[b]
amphipathic lipid		
phosphatidylcholine	16	23
sphingomyelin	16	14
phosphatidylethanolamine	16	9
phosphatidic acid	2	9
phosphatidylserine	8	3
phosphatidylinositol	2	3
choline plasmalogen	2	3
ethanolamine plasmalogen	NR[c]	2
ganglioside	4	NR
neutral lipid		
cholesterol	27	20
triglyceride	0	13

[a]Ref 24. [b]Ref 18. [c]Not reported.

4-en-1-ol) is released. Sphingosine is the fundamental skeleton on which sphingomyelins are constructed. A **cerebroside** is a sphingomyelin in which the phosphocholine is replaced with a **monosaccharide** in glycosidic linkage to the oxygen on carbon 1 of the sphingosine. When an **oligosaccharide** is so situated the sphingolipid is referred to as a **ceramide oligosaccharide**, and when that oligosaccharide contains sialic acid, the sphingolipid is referred to as a **ganglioside**. There are also other glycolipids in which the phosphoethanolamine of phosphatidylethanolamine is replaced with a monosaccharide or an oligosaccharide in glycosidic linkage with carbon 3 of the glycerol. In all of these variations, the physical properties of the amphipathic lipids, which are the reflection of its long chain, its alkyl and alkenyl substituents, and its hydrophilic terminus, are insignificantly altered.

The composition (Table 14–1) of the lipids in plasma membranes from human erythrocytes[24] or from murine fibroblasts (L-cells) grown in tissue culture[18] are typical of plasma membranes from animal cells. The distribution of fatty acids among the various phospholipids from the plasma membranes of murine fibroblasts (Table 14–2) illustrates the heterogeneity of the collection. Membranes from fungi, such as the yeast *Saccharomyces cerevisiae*, have a similar ratio of phosphatidylcholine to

Table 14–2: Fatty Acid Composition of Major Phospholipids and Neutral Lipid of Surface Membranes of L-Cells[a]

major fatty acids	total neutral lipid	phosphatidyl-serine	sphingo-myelin	phosphatidyl-choline	phosphatidylethanol-amine	phosphatidyl-inositol	phospha-tidic acid
14:0	4	0.5	trace	1	1	0.6	
16:0	13	8	3	31	29	24	41
16:1	1	1		2	8	9	5
18:0	46	8	3	20	10	13	15
18:1	23	1	2	5	14	16	2
18:2	1	2	0.3		2	3	1
18:3	4	49	60	27	25	20	27
22:0		2	1		0.3	1	1
20:4	4		0.4	0.2		trace	trace
24:0	3	28	30	14	11	13	8
unsaturated fatty acids	33	53	63	34	48	48	35
long-chain fatty acids[b]	7	30	32	14	11	15	9
polyunsaturated fatty acids	9	51	61	27	27	22	28

[a]Data represent the average composition of fatty acids present in the lipids of two membrane preparations. Plasma membrane neutral lipid and phospholipid were separated by silicic acid chromatography. Phospholipids were further separated by two-dimensional thin-layer chromatography, and the lipids were visualized by spraying with bromthymol blue. Iodine was not used in order to avoid possible losses of unsaturated fatty acids. Fatty acid methyl esters were prepared and analyzed. Data are from ref 18. [b]Long-chain fatty acids are defined as fatty acids containing 20 or more carbon atoms.

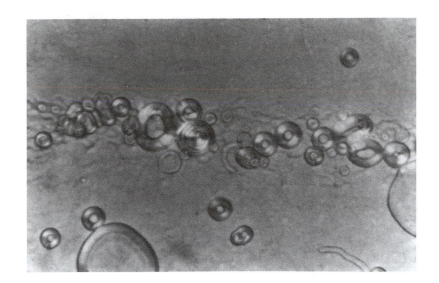

Figure 14–3: Spherical multibilayers of phosphatidylcholine.[30] Phosphatidylcholine was prepared from egg yolks by purification by chromatography on alumina and silicic acid with chloroform in methanol. The pure solid phosphatidylcholine was suspended in 0.15 M sodium chloride and the resulting suspensions were examined in a polarizing microscope. The structures observed are small spheres of phospholipid. The pattern of alternating dark and light sectors around the wall of the sphere within the plane of the page results from the fact that each sphere is formed from many concentric spherical shells, nested each within the other. Each of these spherical shell is a single bilayer of phosphatidylcholine. Reprinted with permission from ref 30. Copyright 1965 Academic Press.

phosphatidylethanolamine but greater amounts of both phosphatidylinositol and phosphatidylserine. Fungi lack plasmalogens and sphingolipids.[25]

The plasma membranes of prokaryotes, such as the bacterium *E. coli*,[26,27] are composed mainly of phosphatidylethanolamine (70%) but also contain small amounts of phosphatidic acid and phosphatidylserine as well as two unusual phospholipids, phosphatidylglycerol, where a glycerol is esterified to the phosphate of phosphatidic acid, and diphosphatidylglycerol, where a single molecule of glycerol is esterified at its two ends with two respective phosphatidic acids. These latter two phospholipids are found exclusively in prokaryotes and mitochondria. The lipids of the plasma membranes from the bacterium *Mycoplasma laidlawii*, however, have high percentages (45%) of the glycolipids 3-[*O*-α-D-glucopyranosyl]-1,2-diacyl-*sn*-glycerol and 3-[*O*-α-D-glucopyranosyl-(1,2)-*O*-α-D-glucopyranosyl]-1,2-diacyl-*sn*-glycerol.[28] This bacterium also contains phosphatidylglucose, a homologue of phosphatidylinositol. The composition of the lipids varies among the various types of membranes in an animal cell: plasma membrane, endoplasmic reticulum, and mitochondria.[29] Mitochondria have less neutral lipid, sphingomyelin, and phosphatidylserine and more phosphatidylethanolamine than plasma membranes but about the same amount of phosphatidylcholine and plasmalogen.

Phosphatidylcholine, purified from a natural animal source such as eggs, spontaneously forms bilayers when it is suspended in water.[30] Initially, these bilayers are gathered in sets of nested spherical shells known as multibilayer vesicles (Figure 14–3).[30] Each of the shells in a multibilayer vesicle is a thin, closed, continuous bag. If such a suspension is submitted to sonication, the multibilayers eventually fragment and become small, unilamellar spherical vesicles.[31] A flat planar membrane,

which is a single bilayer of phospholipids, can be formed across a small circular hole separating two aqueous compartments.[32,33] Oriented bilayers of phosphatidylcholine can also be produced by evaporating a solution of the phospholipid in chloroform–methanol onto a flat surface such as mica and hydrating it with moist helium.[34] In each of these forms it is believed that the basic structural element, the bilayer, is the same and that the bilayer is a thin (4.0–5.0 nm) plastic film of phospholipid that can assume all of these different forms.

When bilayers of egg phosphatidylcholine that are stacked upon mica as flat, planar, parallel sheets are placed in a beam of X-radiation, they produce a diffraction pattern (Figure 14–4A) that is characterized by a set of sharp meridional arcs and two broad symmetrical equatorial reflections.[34] The meridional arcs arise from diffraction by the planes stacked parallel to the orienting surface. The diffraction pattern of the meridional arcs can be transformed into a distribution of electron density normal to the orienting surface (Figure 14–4B). Because the stack of flat bilayers is a regularly repeating structure, the recurring variation of electron density in this dimension produces the diffraction pattern.

The repeating pattern in the properly phased Fourier transform of the meridional diffraction pattern consists of two regions of high electron density symmetrically sandwiching a region of low electron density (Figure 14–4B). This sandwich is the bilayer of phosphatidylcholine. The two regions with the highest electron density on either surface of the bilayer have been assigned to the glycerol, the phosphate, and the choline of the phosphatidylcholine (Figure 14–2). These functional groups, because they contain oxygen, nitrogen, and phosphorus, must have high electron density. The central region of the bilayer has been assigned to the hydrocarbon of the fatty acids. The width of this bilayer of phosphatidylcholine

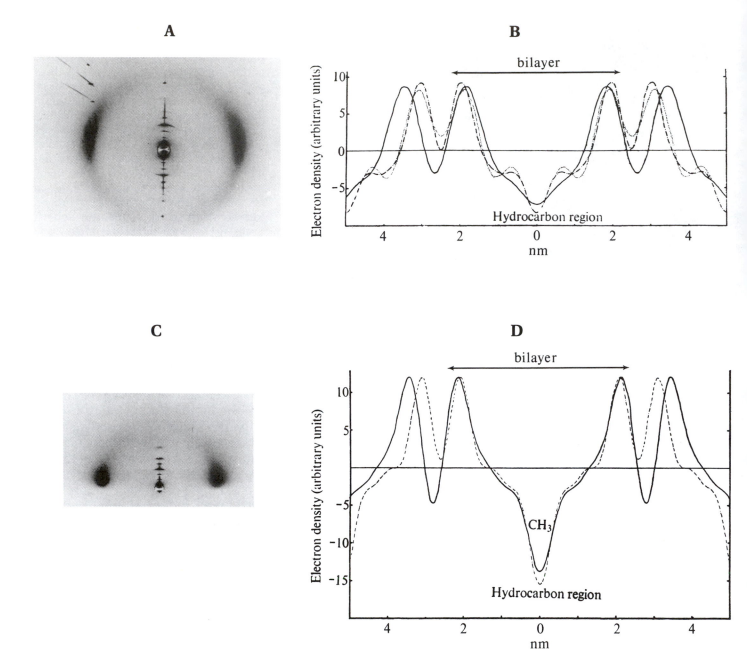

A

B

C

D

Figure 14–4: Diffraction patterns (A and C) and computed Fourier transform (B and D) of multilayers of pure phosphatidylcholine (A and B) from egg yolk or an equimolar mixture of phosphatidylcholine and cholesterol (C and D). Lipids dissolved in chloroform in methanol were smeared on mica sheets, and the solvent was evaporated under a stream of moist helium. The multilayers that resulted were submitted to diffraction in a beam of X-rays. The two symmetrically displayed, diffuse but intense diffractions on the equator (0.46 nm) are from the spacings of the linear hydrocarbons of the phospholipids oriented normal to the plan of the specimen. The sharp reflections on the meridian (the vertical axis of the pattern) are the reflections arising from the set of planes produced by the stacking of the bilayers. With the appropriate choice of phase, the amplitudes of the meridional reflections can be submitted to Fourier transform to obtain a profile of electron density along an axis normal to the plane of the specimen (B and D). Presumably this axis is normal to the flat sheets producing the multilayer. The electron density in arbitrary units is presented as a function of the distance (nanometers) from the center of the bilayer. In the figure the profile of electron density is given for lipids equilibrated with moist helium of 57% relative humidity (dashed lines, 13% water content) or 100% humidity (solid line, 22% water content). It was the expectation of the investigators that the width and structure of the bilayer (within the double arrow) would remain constant while the distance between bilayers would increase as the water content increased. This expectation was used to assign the phases so its fulfillment is inconsequential. Adapted with permission from *Nature,* ref 34. Copyright 1971 Macmillan Magazines Limited.

from eggs at 23 °C is 3.7 nm from maximum to maximum and the width of the hydrocarbon is about 2.8 nm.

When the amount of water in a sample of hydrated natural phosphatidylcholine from eggs is systematically increased from 10% to 45%, changes in the structure of the bilayers occur.[35,36] The width of the bilayers decrease from 4.5 to 3.8 nm,[35] and the cross-sectional area for each phospholipid in one of the two monolayers increases[35,36] from 0.55 to 0.68 nm². It is thought that these changes are responses to the steric effects coincident to the hydration of the hydrophilic functional groups on the external surfaces. As hydration increases it pushes apart the adjacent molecules of phosphatidylcholine in each monolayer of the bilayer and produces the observed changes. Above a certain level of hydration (> 40% water), when all hydrophilic groups are fully hydrated, the thickness of the bilayer no longer decreases.

Because it is heterogenous, naturally occurring phosphatidylcholine will not crystallize. Synthetically prepared di-*n*-tetradecanoylphosphatidylcholine, however, has been crystallized, and a crystallographic molecular model has been prepared (Figure 14–5).[37] The crystalline material consists of stacks of bilayers whose distribution of electron density along an axis normal to the planes of the bilayers (vertical axis in Figure 14–5) would resemble the distribution seen in Figure 14–4B. Because the fatty acids are homogenous they have solidified into a crystalline array that is hexagonally packed. Di-*n*-dodecanoylphosphatidylethanolamine crystallizes from acetic acid in bilayers, and the vertical hexagonal packing of the fatty acids in its crystallographic molecular model is readily discerned from a view normal to the plane of one of the bilayers (Figure 14–6).[38] Such crystallographic models provide a starting point for a discussion of the structure of a bilayer of lipids. It must be remembered, however, that they represent homogeneous lipids in which the alkane is fully saturated and solid.

Four types of conformation at the glycerol backbone of phospholipids have been observed in crystallographic molecular models (Figure 14–7),[39] and there is evidence from nuclear magnetic resonance spectra that in the liquid state, the phospholipids fluctuate among these conformations. One interesting aspect of these structures is that in the conformations represented by ditetradecanoylphosphatidylcholine (DTPC) and didodecanoyl-*N*,*N*-dimethylphosphatidylethanolamine (DDPEM₂) the acyl carbon of the fatty acid on carbon 3 of the glycerol is buried more deeply in the bilayer than the acyl carbon of the fatty acid on carbon 2, while in the two conformations represented by didodecanoylphosphatidic acid (DDPA) and ditetradecanoylphosphatidylglycerol (DTPG) it is the acyl carbon of the fatty acid on carbon 2 that is buried more deeply. This means that, in the rapid transitions among these conformations that occur in bilayers of liquid phospholipid, the linear hydrocarbons slide back and forth past each other in a direction normal to the plane of the bilayer. As these sliding movements

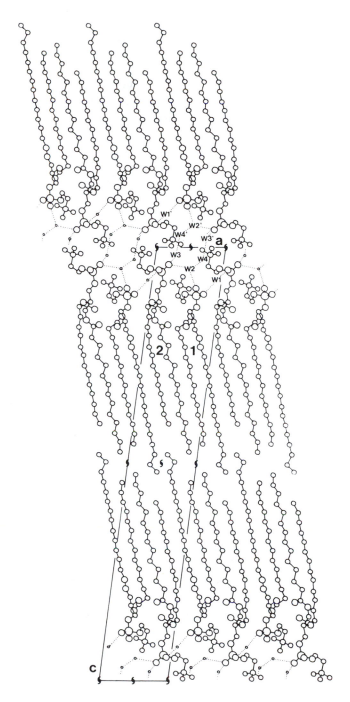

Figure 14–5: Arrangement of the molecules of phosphatidylcholine in the crystallographic molecular model of di-*n*-tetradecanoylphosphatidylcholine dihydrate.[37] The asymmetric unit in the unit cell of the *P*2₁ space group is formed from two molecules of di-*n*-tetradecanoylphosphatidylcholine (1 and 2). The unit cell is outlined, the axes **a** and **c** are designated, and the 2-fold screw axes of symmetry normal to the plane of the page are designated. Because the crystal is the dihydrate, grown from a mixture of ether, ethanol, and water, each asymmetric unit has four water molecules, labeled W1, W2, W3, and W4, associated with it in the hydrophilic region between the bilayers. This corresponds to a water content of 5%. Reprinted with permission from *Nature*, ref 37. Copyright 1979 Macmillan Magazines Limited.

occur, each of the acyl oxygens on the two fatty acids comes in turn to the surface of the bilayer (Figure 14–7).

In these sliding fluctuations, the positions of the phosphate and the charged nitrogen on phosphatidylcholine must average to about the same mean location relative to the surface of the bilayer. This follows from the fact that vesicles of phosphatidylcholine have zero electrophoretic mobility, which demonstrates that on the average the negative charges on their phosphates must reside in the same plane parallel to the surface of the bilayer as the positive charges on their cholines.[40] The dielectric properties of bilayers of phosphatidylcholine are also consistent with this disposition.

The bilayers observed in crystallographic molecular models are solids in which the hydrocarbon is frozen; the bilayers of the mixture of phospholipids purified from a natural source or the bilayer present in a biological membrane is liquid. The transition between solid and liquid resembles the melting of an organic solid, and it can be observed in any bilayer of phospholipids that can be solidified by solidifying the bilayer and then raising the temperature gradually to melt it. As phospholipids from most natural sources remain fluid even at low temperatures, the transition is usually followed either in homogeneous, synthetic phospholipids or in biological membranes whose composition is highly enriched in one particular fatty acid. For example, when the bacterium *M. laidlawii* is grown on medium supplemented with a chosen fatty acid, up to 70% of the fatty acids in its membranes are the supplemented fatty acid.[41]

The transition between solid and liquid can be observed by X-ray diffraction. In a solid bilayer the alkanes of the amphipathic lipids are in a hexagonal array (Figure 14–6), and the spacing between the linear, all-*trans* alkanes produces a strong sharp equatorial reflection at $(0.415 \text{ nm})^{-1}$ characteristic of crystalline paraffins.[42] The cross-sectional area of such an array of paraffin hydrocarbons is 0.40 nm^2 for every two alkyl chains,[42] and this agrees closely with the cross-sectional areas of 0.39–0.41 nm^2 for one complete molecule of phospholipid in each monolayer of the crystallographic molecular models.[37,38,43] The width of a solid bilayer of phospholipids with only saturated fatty acids is consistent[42] with the width of two layers of slightly tilted ($\leq 20°$) alkanes of the appropriate length in the all-*trans* configuration (Figure 14–7).

When a solid bilayer is melted to a liquid bilayer by raising the temperature, its width decreases by 0.5–1.0 nm.[42,44,45] If at the same time the hydrocarbon is expanding to the extent that paraffins expand as they become liquid, the cross-sectional area for each phospholipid in one of the monolayers of the bilayer must increase[42,44] to 0.55–0.70 nm^2. This expansion of the cross-sectional area of the bilayer, among other factors, reflects the establishment of the normal disorder of the liquid state of a paraffin. In this state, *gauche* conformations, which necessarily shorten the distance that can be cov-

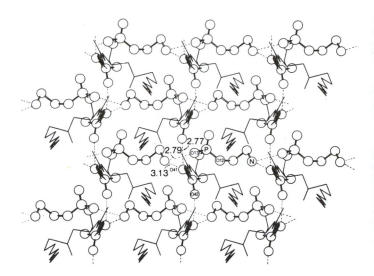

Figure 14–6: View of the crystallographic molecular model of didecanoylphosphatidylethanolamine looking down on the hydrophilic surface of the bilayer. Only one monolayer of the bilayer is shown in the drawing. The phosphorus and nitrogen of one of the molecules of phospholipid are labeled. Oxygen 13 of each phosphate forms a hydrogen bond with the ammonium group of an adjacent ethanolamine. The didecanoylphosphatidylethanolamine was crystallized from glacial acetic acid and in the crystals there was one molecule of acetic acid for each molecule of phospholipid. Oxygen 41 of each acetic acid forms a hydrogen bond with the ammonium group of an adjacent ethanolamine and a hydrogen bond with Oxygen 42 of an adjacent molecule of acetic acid. Because this is a projection onto the plane of the membrane, atoms are allowed to overlap in the drawing, making it somewhat complicated. Nevertheless, the fact that the linear hydrocarbons are all normal to the surfaces of the monolayer is apparent, and the drawing gives a representation of the view from above the surface of a bilayer. The numbers are the bond lengths for the various hydrogen bonds in angstroms. Reprinted with permission from ref 38. Copyright 1974 National Academy of Sciences.

ered by a hydrocarbon, become common features that lead to the narrowing of the bilayer.

In such molten bilayers of fully saturated phospholipid, the hydrocarbons remain oriented preferentially with their long axis aligned with an axis normal to the plane of the bilayer. This conformation has been demonstrated by neutron diffraction of oriented bilayers of dipalmitylphosphatidylcholine in which different carbons along the palmitates have been labeled with deuterium, an atom that scatters neutrons strongly. In the solid phase at low hydration, where all of the hydrocarbons are in hexagonal array normal to the plane of the membrane, the location of the deuteriums in the distribution of scattering density is easily distinguished (Figure 14–8).[45] In such solid bilayers, the deuteriums appear at the expected distances from the center (Table 14–3). When

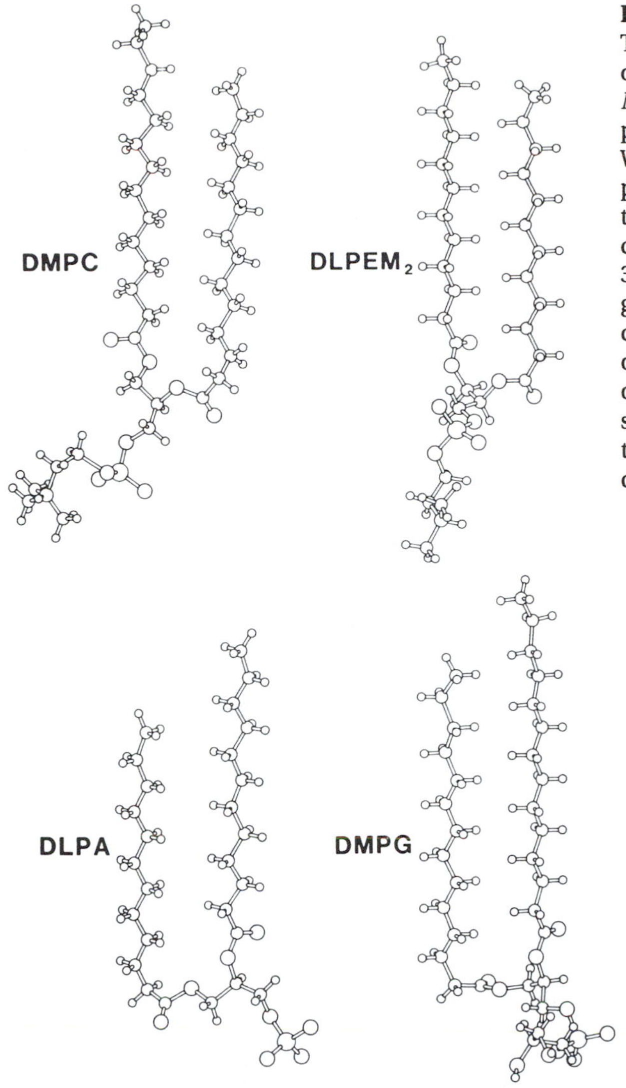

Figure 14–7: Four conformations available to a phospholipid in a bilayer.[39] These four drawings are taken from the crystallographic molecular models of crystalline ditetradecanoylphosphatidylcholine (DTPC), didodecanoyl-*N,N*-dimethylphosphatidylethanolamine (DDPEM$_2$), didodecanoylphosphatidic acid (DDPA), and ditetradecanoylphosphatidylglycerol (DTPG). Within each structure the fatty acids are the same length, so the relative positions of the two fatty acids are most readily ascertained by looking at their ends. Within each structure the fatty acid to the right is the one on carbon 2 of the glycerol, and the fatty acid to the left is the one on carbon 3. In the upper two conformations, the fatty acid on carbon 3 of the glycerol is deeper in the bilayer than that on carbon 2, and in the lower two conformations the fatty acid on carbon 2 is deeper in the bilayer than that on carbon 3. In the upper two structures the acyl oxygen of the fatty acid on carbon 2 of the glycerol is at the surface of the bilayer; in the lower two structures the acyl oxygen of the fatty acid on carbon 3 is at the surface of the bilayer. Reprinted with permission from ref 39. Copyright 1988 American Chemical Society.

bilayers of the various dipalmitylphosphatidylcholines are melted, the deuteriums in the fluid hydrocarbon remain similarly distributed, but each moves closer to the center. This rearrangement is consistent with the narrowing of the bilayer that occurs upon melting and would reflect the appearance of *gauche* conformations (**14–3**) that necessarily decrease mean perpendicular displacements all along the hydrocarbon.

The transition between solid and liquid in a bilayer composed of mixtures of various homogeneous, synthetic phospholipids has also been studied. When a suspension of bilayers composed of only one phospholipid such as dipalmitylphosphatidylcholine is melted, a single sharp transition that occurs completely over 2–3 °C is observed. It can be monitored in a calorimeter as the absorption of heat resulting from the heat of fusion.[46] When two phospholipids are mixed, the transition occurs over a broader range of temperatures somewhere between the temperatures of the transitions of the sepa-

rated components.[47–49] At temperatures in the range over which the transition is occurring, regions of fluid phase are in equilibrium with regions of solid phase laterally separated from each other in the plane of the bilayer.[47] In many instances, the two component phospholipids are not miscible with each other as solids and a separate solid phase of one or the other remains laterally isolated in the bilayer.[47] For example, mixtures of up to 40% dipalmitylphosphatidylethanolamine in dimyristylphosphatidylcholine seem to contain significant regions of unmixed dimyristylphosphatidylcholine in the solid phase.[47] These results suggest that regions of immiscible, unmelted phospholipid might form in natural bilayers under certain circumstances. The heterogeneous mixtures of phospholipids normally found in natural circumstances, however, appear to form bilayers that are fully liquid and fully miscible at all physiological temperatures.

The fluid bilayers in normal biological membranes and in bilayers formed spontaneously from the amphipathic lipids extracted from normal biological membranes have been studied by following the **electron spin resonance** of probes incorporated into them. Fatty acyl nitroxide **14–8**[50] is an example of such a probe

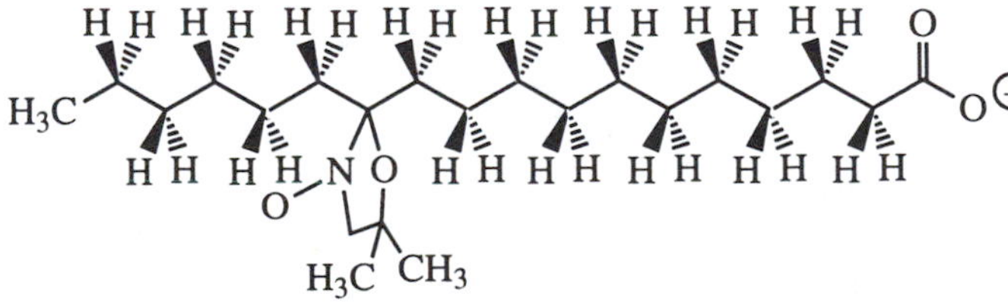

14–8

The nitroxide is in a five-membered ring and has an unpaired electron which remains a stable free radical because it is sterically hindered from reacting by the four vicinal alkyl substituents

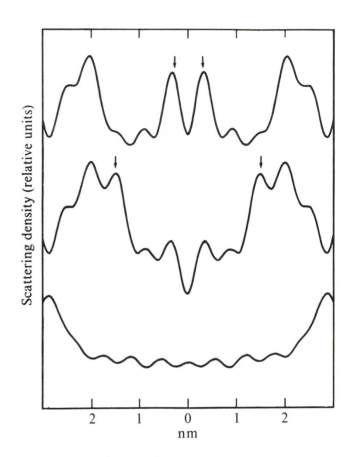

14–9

The unpaired electron is located mainly over the nitrogen[51] in a *p* orbital whose principal axis is aligned parallel to the axis of the all-*trans* linear hydrocarbon.[52]

Electron spin resonance is analogous in theory to nuclear magnetic resonance. An unpaired electron has a spin. Upon application of an external magnetic field, the two degenerate energy levels of the spinning electron are split into two distinct energy levels, one for the spin aligned with the field and one for the spin opposed to the magnetic field. At normally employed magnetic fields the resonant frequency between these two energy levels is in the microwave region of the electromagnetic spectrum (10^{10} Hz). The magnetic field, H_{TOT}, sensed by a given electron is the sum of the applied magnetic field, H_{app}, and any local magnetic fields, H_{loc}, created by neighboring nuclei.

In the dimethyl cyclic nitroxide **14–9**, the nucleus that dominates the local magnetic field is that of the nitrogen (^{14}N), which has a spin quantum number of 1 and is therefore quadripolar. The nitrogen nucleus can assume spins of +1, 0, and −1 with equal probability, because the

Figure 14–8: Distribution of neutron scattering density across bilayers in which hydrogen has been replaced by deuterium at specific locations in the di-(15,15-dideuteriopalmityl)phosphatidylcholine (A), di-(5,5-dideuteriopalmityl)phosphatidylcholine (B), or dipalmitylphosphatidylcholine hydrated with ^2H_2O were prepared on quartz slides. The multilayers were brought to 20 °C, which is below the melting point of dipalmitylphosphatidylcholine under these circumstances, and allowed to diffract neutrons. From the meridional reflections and appropriate phases, a distribution of neutron scattering density (relative units) normal to the plane of the membranes as a function of distance from the center (nanometers) could be calculated by Fourier transformation. The positions of the deuterated carbons in the first two samples are clearly observed (arrows). The position of the ^2H_2O in the third sample was defined by a difference map of neutron scattering density between a specimen hydrated with H_2O and one hydrated with ^2H_2O (C). Reprinted with permission from *Nature*, ref 45. Copyright 1978 Macmillan Magazines Limited.

Table 14–3: Distance of Various Carbons from the Center of a Bilayer[a]

carbon deuterated	distance from center (nm)	
	solid bilayer	fluid bilayer
C15	0.20 ± 0.1	0.19 ± 0.1
C14	0.36 ± 0.1	0.36 ± 0.1
C9	0.94 ± 0.1	0.81 ± 0.1
C5	1.21 ± 0.2	1.05 ± 0.1
C4	1.53 ± 0.2	1.22 ± 0.1

[a]Determined by neutron diffraction of bilayers of dipalmitylphosphatidylcholine selectively deuteriated at those positions. From ref 45.

distribution among its energy levels is insignificantly affected by the applied magnetic field. As a result, the local magnetic field created by the nitrogen nucleus assumes three values, one of which is zero. The spectrum consists of a central absorption, arising from unpaired electrons coupled to nitrogen nuclei of spin quantum number 0 and from unpaired electrons not coupled to any nucleus, and two peaks of **hyperfine absorption** on either side of the central peak, arising from unpaired electrons coupled to nitrogen nuclei of spin quantum number +1 and −1. The hyperfine absorptions are of variable magnitude and split different distances from the central absorption depending on the quality of the coupling between the nitrogen and the electron, but the central absorption will always be fixed because it is at the position where H_{loc} is zero. Information about environment, rotational diffusion, and isotropy is contained in the hyperfine absorptions.

Electron spin resonance spectra are presented as derivatives of the absorption spectra, so peaks of absorbance appear as pairs of positive and negative deflections representing the positive slope of the rising phase and the negative slope of the declining phase of the absorption itself. The spectrum of the tetramethyl nitroxide **14–10** freely tumbling in aqueous solution displays three sharp peaks of equal intensity (Figure 14–9A)[51] reflecting the full coupling of the electron and the nitrogen nucleus and the rapid isotropic motion of the molecule

14–10

When fatty acyl nitroxide **14–8** is placed in an oriented multilayer of phosphatidylcholine from eggs, several differences from the isotropic spectrum can be noted (Figures 14–9B,C).[53] The absorbances are much broader because the motion of the nitroxide is not so rapid. The

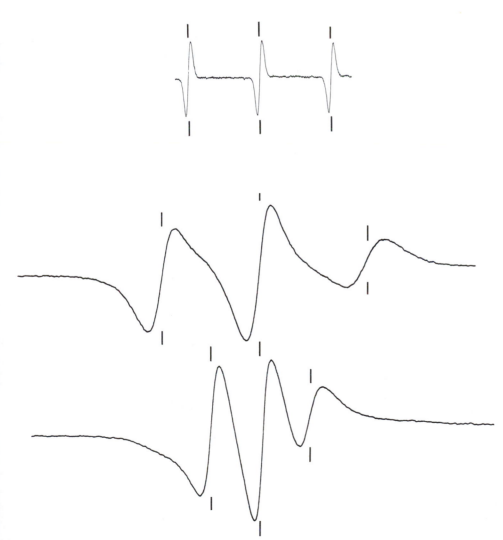

Figure 14–9: Electron spin resonance spectra of 2,2,5,5-tetraalkylazacyclopentyl oxides. (A) 2,2,5,5-Tetramethyl-3-(aminocarbonyl)azacyclopentane oxide (**14–10**) dissolved in water.[51] (B,C) 2-(10-Carboxydecyl)-2-hexyl-4,4-dimethyl-3-azatetrahydrofuran *N*-oxide (**14–8**) incorporated into multilayers of egg phosphatidylcholine, formed on a glass cover slip, with the magnetic field oriented perpendicular (B) or parallel (C) to the plane of the cover slip.[53] Nitroxide **14–8** and phosphatidylcholine were dissolved in chloroform and the mixture was evaporated to dryness. Water was added and a portion of the opalescent suspension that resulted was spread on a cover slip. The water was evaporated at 39 °C to produce the multilayers oriented by the plane of the cover slip. In all of the spectra, the derivative of the absorbance is presented on the vertical axis. The microwave frequency was held at a constant value while the strength of the magnetic field (in gauss) was changed. The magnetic field strength is the variable on the horizontal axis. The distance between the two end resonances in panel A is 28 gauss. The vertical lines mark the positions of maximum absorption of microwave energy. Reprinted with permission from refs 51 and 53. Copyright 1965 and 1969 National Academy of Sciences.

spectra also have become anisotropic because the motion of the nitroxide has become anisotropic. The easiest way to demonstrate this is to take two spectra from the specimen oriented so that the magnetic field is either normal to the planes of the multilayers (Figure 14–9B) or parallel to the planes of the multilayers (Figure 14–9C). In the one case the splitting of the peaks is larger than that of the isotropic spectrum, and in the other it is smaller. From theoretical simulations of these spectra, it could be concluded that in the bilayers of phosphatidylcholine, fatty acyl nitroxide **14–8** is oriented with the axis of its hydrocarbon perpendicular to the planes of the multilayers and rotates exclusively or almost exclusively about this axis. These conclusions are consistent with expectations based on the structure of a phospholipid bilayer (Figure 14–2) and the amphipathic structure of fatty acyl nitroxide **14–8**. When such nitroxides are incorporated into multilamellar vesicles of phosphatidylcholine suspended in water (Figure 14–8), the spectrum that results is a composite of the perpendicular and parallel spectra seen in Figures 14–9, panels B and C, respectively.[54]

The full coupling of the electron to the nitrogen is expressed in aqueous solution (Figure 14–9A) because the electronic structure in which the radical occupies a *p* orbital over nitrogen, requiring a separation of charge, can be readily solvated by the water. In nonpolar environments such as the center of the bilayer, the nitroxide can shift its hybridization to form an ethenyl molecular orbital system

14–11

— bonding orbital
--- antibonding orbital

composed from one of the lone pairs on oxygen and the radical. The unpaired electron occupies the antibonding molecular orbital and spends some of its time over oxygen, which is diamagnetic. This decreases the coupling to the paramagnetic nitrogen, and the hyperfine components decrease accordingly in intensity.

The molecular motion of the linear hydrocarbons in a fluid bilayer of phospholipid increases in proceeding from the acyl carbon to the center.[54] A series of phosphatidylcholines were synthesized in which the acyl groups on carbon 1 of the *sn*-glycerol 3-phosphate were derived from either palmitic acid or stearic acid, and the acyl groups on carbon 2 were derivatives of palmitic acid or stearic acid, respectively, on which the cyclic dimethyl nitroxide **14–9** was positioned at the fifth, eighth, twelfth, and sixteenth carbon.[54] An order parameter, *S*, can be

defined, which is a number that quantifies the confinement of the rotational motion of these cyclic nitroxides to one particular axis. When $S = 1$, the ring has a fixed orientation in space; and when $S = 0$, its rotational motion is isotropic (Figure 14–9A). When the various labeled phospholipids were incorporated into multilayers of egg phosphatidylcholine, the order parameter *S* was observed to decrease as the cyclic nitroxide was situated farther from the acyl carbon. For cyclic nitroxides at the fifth, eighth, twelfth, and sixteenth carbon of the labeled fatty acid, the order parameters *S* were 0.68, 0.50, 0.33, and 0.16, respectively.[55]

A more extensive determination of the variation in the order parameter with position along the hydrocarbon has been made by deuterium nuclear magnetic resonance.[56] A series of synthetic phospholipids were prepared which contained either palmitic acid at both carbon 1 and carbon 2 or palmitic acid at carbon 1 and oleic acid at carbon 2 of the *sn*-glycerol 3-phosphate. In each member of the series a deuterium atom was placed synthetically at a specific carbon in the palmitic acids. Because a deuterium nucleus, like a nitrogen nucleus, is quadripolar, the nuclear magnetic resonance spectrum of deuterium can also be used to estimate an order parameter *S* for the degree of anisotropy experienced by the carbon to which it is attached.[43,57] Order parameters *S* for dipalmitylphosphatidylcholine, 1-palmityl-2-oleyl-phosphatidylcholine, and dipalmitylphosphatidylserine, gathered from bilayers of these phospholipids held at temperatures an equivalent distance above each of their melting points, have been presented as a function of the carbon on which the deuterium was located (Figure 14–10).[58] The confinement experienced by a carbon in the liquid hydrocarbon of the bilayer decreases as the distance from the acyl carbon increases. Carbons at the very core of the bilayer are able to assume almost every orientation, while carbons near the periphery are confined in their orientations. These more complete data are in substantial agreement with more limited results obtained by using electron spin resonance.[54]

These observations relate to a stereochemical paradox in the structure of a bilayer of amphipathic lipids from natural sources. Beyond the eighth carbons of the fatty acids attached to carbon 2 of either the *sn*-glycerol 3-phosphates or the sphingosines, within the **core of the hydrocarbon**, the carbon–carbon double bonds begin (Figure 14–2). This has two consequences. First, permanent elbows inconsistent with a straight alkane necessarily disrupt the alignments of the hydrocarbons of the fatty acids. Second, the average length for each carbon is decreased by the multiple double bonds. Both the disorder introduced by single and triple *cis* double bonds and the shortening of the chains caused by all of the double bonds necessarily increase the mean cross-sectional area parallel to the plane of the bilayer for each hydrocarbon in this region. Added to this effect is the disorder that naturally occurs farther away from the surfaces of the

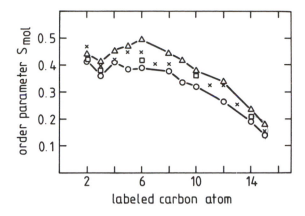

Figure 14–10: Variation of order parameter, S_{mol}, from deuterium nuclear magnetic resonance as a function of the position of the deuterium along the hydrocarbon of 1,2-di(deuteriopalmityl) phosphatidylcholine (O) or 1-(deuteriopalmityl)-2-oleylphosphatidylcholine (△).[58] A series of selectively deuterated palmitic acids were synthesized, each with deuterium at a different carbon along the chain. From these deuteriopalmitic acids a series of dipalmitylphosphatidylcholines was synthesized, each with two palmitic acids in which deuterium occupied the same position. These lipids were separately suspended in water at a temperature 19 °C above their melting points (41 °C) and deuterium nuclear magnetic resonance spectra were recorded, from which order parameters, S_{mol}, were calculated. Samples of each of these dipalmitylphosphatidylcholines were digested with phospholipase A, and the resulting 2-lysophosphatidylcholines were esterified with oleic acid to produce a series of 1-(deuteriopalmityl)-2-oleylphosphatidylcholines in each of which deuterium occupied a different position in the palmitic acid. Deuterium nuclear magnetic resonance spectra were taken of suspensions of these phospholipids at a temperature 16 °C above their melting points (–5 °C). A series of palmitic acids selectively deuterated at specific positions were separately fed to *M. laidlawii* bacteria to enrich (70%) the membranes of these cells in the added fatty acid, and the order parameter for samples of each of these selectively deuterated membranes were also determined (×). Reprinted with permission from ref 58. Copyright 1978 Elsevier Science Publishers.

bilayer even in saturated phospholipids (Figure 14–10). Before the eighth carbons of the acyl groups on carbon 2 of the *sn*-glycerol 3-phosphates and sphingosines, however, essentially all of the hydrocarbons are linear, saturated, and more ordered (Figure 14–10). In a bilayer composed only of phospholipids and sphingomyelins, these **peripheral regions of the hydrocarbon** are nevertheless required to have the same mean cross-sectional area for each hydrocarbon as exists in the core. One way the mean cross-sectional area could be increased would be the incorporation of a high frequency of *gauche* conformations in the hydrocarbon in these peripheral regions, but this would be inconsistent with the universal

observation that these regions are the more oriented than the regions nearer the core (Figure 14–10). It is also inconsistent with the distribution of electron density in bilayers of naturally occurring phosphatidylcholine (Figure 14–4B) because the two peripheral regions of the hydrocarbon have the highest electron density, and the core has the lowest electron density. The two symmetrical shoulders of intermediate electron density within the peripheral regions of the hydrocarbons are thought to be real features of the structure of the bilayer rather than artifacts of the sinusoidal transform.[34] All of these considerations require that the solution to this paradox incorporate a large cross-sectional area, low density, and high disorder at the core of the hydrocarbon and a large cross-sectional area, high density, and low disorder at the two symmetrically peripheral regions of the hydrocarbon in a bilayer.

One solution to this paradox would be that the linear, saturated hydrocarbons in the two peripheral regions are tilted.[55] Tilting the hydrocarbon increases its cross-sectional area parallel to the plane of the bilayer while retaining the density of the condensed, hexagonal array. This possibility has been examined with a series of synthetic phosphatidylcholines that each had dimethyl cyclic nitroxide **14–9** attached at a different carbon in the saturated fatty acyl group on carbon 2 of their *sn*-glycerol 3-phosphates. These labeled phosphatidylcholines were incorporated into bilayers of egg phosphatidylcholine. When the dimethyl cyclic nitroxide was on the fifth or the eighth carbon, its principal axis was tilted 30° relative to the plane of the bilayer, but when it was on the twelfth or the sixteenth carbon, its principal axis was oriented normal to the plane of the bilayer.[55] Tilted hydrocarbon chains have been directly observed in crystallographic molecular models of bilayers of amphipathic lipids.[43] A tilting of the alkyl chains in the two peripheral regions would also explain the increase in cross-sectional area and decrease in width that occurs upon the melting of bilayers of homogeneous phospholipids.[42] The only implausible feature of this explanation is that all of the alkyl chains in a fluid bilayer would have to tilt in the same direction for it to be successful.

Such a coordinated tilting of the hydrocarbon over large areas of a solid bilayer has been observed by X-ray diffraction. In bilayers of dipalmitylphosphatidylcholine and distearylphosphatidylcholine below the temperature at which they melt, the equatorial reflection in the X-ray diffraction pattern that arises from the aligned hydrocarbon chains displays the fine structure of a sharp reflection superimposed upon a broader reflection.[59] This distribution of reflected intensity has been shown to arise from hydrocarbons tilted relative to the axis normal to the bilayer, and the degree of tilt can be calculated from the diffraction pattern. As the degree of hydration in these solids was increased from 6% to 30%, the tilt of the hydrocarbons increased from 17° to 40°. Presumably the steric effects of the hydration force the hydrophilic functional groups of the phospholipids to take up a greater surface area for each molecule of phospholipid, and the linear hydrocarbons adjust to the required increase in their cross-sectional area by tilting.

It has also been observed that dimethyl cyclic nitroxide 14–9, attached at the fifth carbon of a fatty acid on carbon 2 of the sn-glycerol 3-phosphate of dipalmitylphosphatidylcholine, appeared to be in a more polar environment in bilayers of natural phospholipids than nitroxides attached farther down the fatty acid.[60] If the hydrocarbons were tilted in this region, their surfaces should be more exposed to the aqueous phase. These observations, however, illustrate a difficulty in interpreting many of the physical studies on bilayers. The bilayers used in these experiments were vesicles prepared by sonication. It has been shown by nuclear magnetic resonance spectroscopy that vesicles of small diameter (30–90 nm) prepared by sonication display anomalous physical properties because of their high curvature.[61] It has been pointed out that these anomalies arise from the fact that the high curvature unavoidably forces a portion of the hydrocarbon adjacent to the hydrophilic functional groups of the phospholipids to occupy locations on the outer surface of the vesicle in contact with the water,[62] and this would explain why the environment of the nitroxide in this situation appears to be so polar. Naturally occurring bilayers, however, rarely have such high curvature. The tension within such small vesicles produced by sonication seems to be significant. When the kinetic barrier is overcome by adding appropriate catalysts, small vesicles of phospholipid spontaneously fuse among themselves to produce much larger single-walled structures.[63]

The paradox of the cross-sectional areas has been phrased in terms of the structure of naturally occurring bilayers because this is the most critical situation. Natural phospholipids have unsaturated fatty acids that necessarily disrupt the core of their bilayers. Consistent with this stereochemical consequence of the cis double bonds, the most obvious discontinuity in the plot of the order parameter S against position (Figure 14–10) occurs after the sixth carbon of the palmitate on 1-palmityl-2-oleylphosphatidylcholine (see Figure 14–2A). A similar, although not so abrupt, discontinuity, however, also seems to be present in the plot for bilayers formed from fully saturated phospholipids such as dipalmitylphosphatidylcholine. Even in this case, the disorder increases most precipitously beyond the eighth carbon.

Cholesterol plays the major role in overcoming the stereochemical paradox posed by the high disorder and low density in the core and the low disorder and high density in the peripheral regions of the hydrocarbon in a bilayer of amphipathic lipids. Eukaryotic membranes contain significant quantities of steroids. In animal membranes, cholesterol is the major steroid

14–12

It accounts for about 20–30% of the mass of the lipids in a membrane, and the molar ratios of cholesterol to phospholipid[18] vary between 0.5 and 1.3. Each molecule of cholesterol is confined to one or the other surface of the bilayer,[64] presumably because its hydroxyl is hydrogen bonded to water. Its long axis is aligned normal to the bilayer.[53] The nuclear magnetic resonance spectrum of [^{1}H]cholesterol incorporated into bilayers of [^{2}H]dipalmitylphosphatidylcholine is consistent with the confinement of its fused rings to the more ordered peripheral regions of the hydrocarbon and the incorporation of its isoprenoid tail into the more fluid core.[65] Along its long axis, a molecule of cholesterol has a van der Waals cross-sectional area in a Corey–Pauling–Koltun model of 0.25 nm^2 for the first 1.0 nm and then abruptly, at its isoprenoid tail, the van der Waals cross-sectional area decreases to 0.12 nm^2 for its last 0.8 nm.[66] It has been proposed that the widest portion of the cholesterol occupies the space between the saturated hydrocarbons in the peripheral regions of the hydrocarbon in a natural bilayer and permits them to straighten their posture and assume a fully extended all-trans configuration normal to the plane of the bilayer. This stereochemical function for cholesterol would explain why its addition to bilayers of natural phosphatidylcholine from eggs decreases their fluidity but its addition to bilayers of dipalmitylphosphatidylcholine increases their fluidity.[67]

Measurements by X-ray diffraction from bilayers formed from mixtures of natural phosphatidylcholine from eggs and cholesterol also support this structural proposal. When the distribution of electron density in oriented bilayers of cholesterol and phosphatidylcholine is compared to that of bilayers of phosphatidylcholine alone, an increase in electron density occurs in the peripheral regions rather than in the core of the hydrocarbon (compare panel B with panel D in Figure 14–4).[34] Unlike bilayers of pure phosphatidylcholine from eggs, in which the alignment of the linear hydrocarbon with an axis normal to the plane of the bilayer is poor and decreases as hydration is increased, bilayers of an equimolar mixture of phosphatidylcholine and cholesterol have their linear hydrocarbons closely aligned with the normal axis, and this alignment does not change as hydration is changed (compare the equatorial reflections in panels A and C of Figure 14–4).[34]

As cholesterol is added to a bilayer of natural phosphatidylcholine from eggs at a constant concentration of water, the width of the bilayer increases linearly[68] with the concentration of cholesterol until it reaches a maximum width at a molar ratio of 0.50 mol of cholesterol (mol of phospholipid)$^{-1}$. At the maximum, the width of the bilayer has increased by 19%.[68] At the same time, however, the cross-sectional area for each molecule of phosphatidylcholine in a monolayer of the bilayer decreases from 0.62 to 0.48 nm^2, if it is assumed that each molecule of cholesterol contributes 0.37 nm^2 to the surface area.[68] The minimum value of 0.48 nm^2 is not far from the value of 0.40 nm^2 for the perpendicular cross-sectional area of a pair of hexagonally arrayed linear hydrocarbons. Because the molecules of cholesterol are spacing the molecules of phosphatidylcholine, the phosphocholines are more widely separated, and the steric effects of their hydration are no longer significant.

The addition of cholesterol to bilayers of homogeneous, saturated synthetic phospholipids broadens their melting point and at high enough molar ratios [> 0.5 mol of cholesterol (mol of phospholipid)$^{-1}$] eliminates it completely, causing the lipid to remain fluid.[46,69,70] There is evidence from calorimetric studies,[46] X-ray diffraction,[71] and pressure–area functions of monolayers[70] that complementary interactions occur between cholesterol and phospholipids causing them to segregate into phases of specific molar compositions. These phases have volumes that are smaller than the sum of the volumes of the separate components[70] and have much broader phase transitions.[46] It has been proposed that the molar ratios in such phases are two phospholipids for each cholesterol in the case of dipalmitylphosphatidylcholine,[71] but that the ratio would vary depending on the adjustment of volumes within the bilayer that is required of the cholesterol.[70]

The structure of the bilayer of amphipathic lipids in a membrane from a eukaryotic cell is faithfully reproduced by bilayers formed from an equimolar mixture of cholesterol and natural phosphatidylcholine. The distribution of electron density for such a bilayer (Figure 14–4D) displays three regions. First, the two symmetrical boundaries of high electron density, formed by the hydrophilic functional groups, sandwich the hydrocarbon and provide interfaces compatible with the water on either side. The distance between the two maxima of electron density that designate these interfaces is 4.2 nm. In a natural membrane, these surfaces are irregular (Figure 14–6) and are formed by the interdigitation of the phosphocholines phosphoethanolamines, phosphoserines, and phosphoinositols. These surfaces are constantly changing in appearance owing to the fluid state of the bilayer. Second, the two peripheral regions of the hydrocarbon formed by the first eight saturated carbons of the fatty acyl chains (Figure 14–2) and the fused rings of the cholesterol (**14–12**) have a lower electron density (Figure 14–4D) only

because they are hydrocarbon. They are densely packed, with the hydrocarbon predominantly in its fully extended configuration aligned normal to the plane of the bilayer and appropriately spaced and supported by the fused rings of the cholesterol. In one of the monolayers of a bilayer formed from an equimolar mixture of cholesterol and phosphatidylcholine from eggs, the cross-sectional area for each phospholipid is 0.48 nm^2, if the cross-sectional area for each cholesterol is 0.37 nm^2. The width of each of these two symmetrically displayed peripheral regions of the hydrocarbon is about 0.9 nm. They commence at the level of the two acyl carbons attached to the glycerol and extend into the bilayer to the level at which the unsaturation of the fatty acids and the isoprenoid tail of the cholesterol commence. Third, within these boundaries, the central core of the bilayer contains the unsaturated hydrocarbon of the fatty acids and the disordered hydrocarbons. It has the lowest electron density (Figure 14–4D) because the disorder of the hydrocarbon in this region increases the frequency at which vacant space is encountered. It is believed that the hydrocarbon in the core of the bilayer, if not the hydrocarbon in the peripheral regions, has most of the properties of liquid paraffin. The width of the central core, about 1.2 nm, brings the width of the entire sheet of hydrocarbon to 3.0 nm.

When an amphipathic lipid, such as egg phosphatidylcholine, is spread at an interface between air and water, it forms a monolayer with its hydrophilic functional groups directed toward the water and its hydrophobic hydrocarbon directed toward the air. The area for each phospholipid in this monolayer is a function of the surface pressure. This pressure is exerted mechanically by decreasing systematically the area of the surface. The surface pressure is measured with a torsion balance as the surface area is decreased. Above a certain pressure, or in other words below a certain area, the monolayer becomes so compressed that phospholipid molecules leave it and form small patches of bilayer adhering to the monolayer. Before this breakdown, however, the area for each molecule of phospholipid at the interface is a smooth inverse function of the pressure (Figure 14–11).[72] The explanation for this behavior is that at low pressure, the tendency of the hydrocarbon to be maximally disordered causes the monolayer to have a large surface area, which is about 3-fold greater than that of hexagonally packed hydrocarbon oriented normal to the interface. The molecules of phospholipid, at zero pressure, do not lie flat upon the surface, presumably because this would bring all of their hydrophobic hydrocarbon into contact with water. The observed surface area at zero pressure is a balance between the entropy that would spread the lipids and the hydrophobic effect that would contract them. As the surface is compressed and its free energy is thereby increased, the hydrocarbons become more and more aligned, and of lower and lower entropy. The surface area of a molecule of phosphatidylcholine in a normal fluid bilayer of pure egg phosphatidylcholine is about 0.7 nm², which corresponds to a surface pressure of about 37 dyn cm⁻¹ in a monolayer at an air–water interface.

The same measurements can be made on a monolayer of egg phosphatidylcholine at an interface between an alkane, such as *n*-hexadecane, and water (Figure 14–11).[73] At each surface area the monolayer in this situation has a greater surface pressure than when it is backed by air. The reason for this is that, because of van der Waals forces, liquid alkane is more compatible with the hydrocarbon side of the monolayer than is air; and, when the monolayer is backed by alkane, there is not an interfacial free energy causing it spontaneously to contract and minimize its surface area at the side of the hydrocarbon–air interface as well as at the aqueous interface. The pressure needed to compress this monolayer to 0.7 nm² (molecule of phosphatidylcholine)⁻¹ is 44 dyn cm⁻¹.

A droplet of alkane in water has a surface tension of about 50 dyn cm⁻¹, which reflects the free energy of the hydrophobic effect. As amphipathic lipid is added to a droplet of alkane, its surface tension rapidly decreases and reaches zero before the mole fraction of the amphipathic lipid reaches 1.[74] When its surface tension reaches zero, the surface area of the droplet will begin to expand indefinitely, and at a mole fraction of amphipathic lipid equal to 1, it will be a bilayer.[75] That the surface tension of a bilayer of amphipathic lipid is zero has been verified experimentally.[73] The initial surface tension of the droplet of alkane is a direct measurement of the cohesive force of the hydrophobic effect. In the bilayer, this cohesive force is still in operation, trying to minimize its surface area.

From these various considerations, it follows that a bilayer of amphipathic lipids immersed in an aqueous solution represents a compromise among a number of forces. The hydrocarbon of the fatty acyl groups is hydrophobic and is withdrawn as successfully as possible from contact with the water by the cohesive force of the hydrophobic effect. The most successful stereochemical solution to this problem would be all-*trans*-alkane oriented normal to the surface in hexagonal array. The presence of *cis* double bonds, the steric effects of hydration, and the entropy of the liquid state defeat this solution (Figures 14–10 and 14–11) and cause the cross-sectional area for each amphipathic lipid to be greater than the minimum value of 0.40 nm² for hexagonal packing. This stereochemically enforced spreading of the bilayer must expose some of the hydrocarbon in the peripheral regions to water.[62] The hydrophilic functional groups of the amphipathic lipids are facing the aqueous phase. If they were buried or sterically excluded from contact with water, their free energies of hydration would be lost, which would be unfavorable (Figure 5–9). In the compromise among the various forces, a bilayer of phosphatidylcholine ends up with about half the surface area (0.7 nm²) for each molecule in one of its two monolayers as that for a molecule in a monolayer of phosphatidylcholine at an air–water interface (Figure 14–11) at zero surface pressure. Presumably, this difference between a bilayer and a monolayer arises from the greater cohesion, due to van der Waals forces, that can be established within an interior of liquid hydrocarbon as opposed to a thin layer of hydrocarbon at an interface with air.

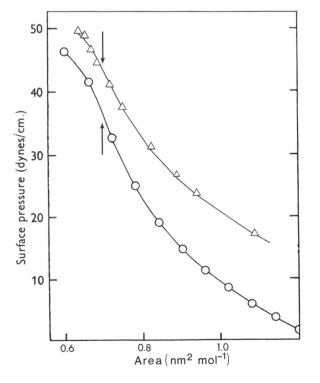

Figure 14–11: Relationship of surface pressure (dynes centimeter⁻¹) and area [nanometers (mole of phosphatidylcholine)⁻¹] for monolayers of egg phosphatidylcholine at an interface between air and water[72] (○) or between *n*-hexadecane and water (△).[73] Phospholipid was spread at the interfaces from a solution in *n*-hexane; and, in the case of the interface with air, the hexane evaporated immediately leaving behind the monolayer of lipid. In each case, a monolayer of egg phosphatidylcholine was produced. The area of the interface could be varied by movable boundaries, and the interfacial pressure was measured directly by a torsion balance. Areas for a molecule of phospholipid were calculated on the assumption that all of the phospholipid added to the system was incorporated into the monolayer. Arrows mark areas of 0.7 nm² mol⁻¹. Adapted with permission from refs 72 and 73. Copyright 1960 Biochemical Society and 1971 Springer-Verlag.

A bilayer represents one example from a spectrum of different structures that can be formed by various **amphipathic compounds** such as amphipathic lipids, soaps, and detergents. An amphipathic compound usually contains a hydrocarbon that is covalently attached at one of its ends to one or more hydrophilic functional groups. When an amphipathic compound is added to an aqueous solution, it forms noncovalent, multimolecular complexes referred to as micelles or bilayers. In these complexes, all of the hydrophilic functional groups of the constituent molecules reside on the surface at the interfaces with the aqueous phase so that they can be hydrated by the water. The hydrocarbon occupies the interior of the complex sequestered from the aqueous phase by the hydrophobic effect. The molar volume of the hydrocarbon in the complex is determined simply by the partial molar volume of the hydrocarbon from which it is composed.

The final molar surface area at the interface of the complex with the aqueous phase is determined by the balance between two opposing forces.[22,76] The hydrophilic functional groups have an inescapable atomic cross-sectional area for their covalent structure that forces them to be spaced at least a minimum distance apart on the surface of the complex. This spacing is increased by the layers of hydration that are noncovalently associated with each hydrophilic functional group and any mutual electrostatic repulsion driving them apart. The farther apart the hydrophilic functional groups are spaced to relieve these repulsive forces, the more of the hydrocarbon to which they are covalently attached is drawn out to the surface to come in contact with water. This exposure of the hydrocarbon to water is resisted by the hydrophobic effect. The balance between the intermolecular repulsion among the hydrophilic functional groups and the hydrophobic effect determines the ultimate molar surface area of the complex.

There is a third geometric constraint on the complex. Because every hydrocarbon has a hydrophilic functional group covalently attached to it and every hydrophilic functional group must remain in contact with the aqueous phase, no carbon in the interior of the complex can be located more than a maximum distance from the aqueous phase. If the hydrocarbon were fully extended in the all-*trans* conformation, that maximum distance would be the maximum length of the amphipathic molecule. A certain amount of the hydrocarbon, however, having been dragged out of the interior by the repulsive forces among the hydrophilic functional groups, is required to occupy the surface of the complex, and the hydrocarbon in the interior is rarely fully extended because it is fluid and because it must mix to fill the interior. Therefore, the maximum distance any carbon can be from the aqueous phase is significantly less than the maximum length of the fully extended hydrocarbon found in the amphipathic molecule. One dimension of the complex formed from molecules of the amphipathic compound must always be less than or equal to twice this maximum distance. If it were not, the complex would contain a region farther from the aqueous phase than any matter can be located.

The shape and size of the complex that an amphipathic compound can form is dictated by this **maximum dimension**. The shapes available are a sphere whose radius is less than or equal to the maximum dimension; an ellipsoid of revolution, prolate or oblate, whose minor axis is less than or equal to the maximum dimension; a cylindrical rod of indefinite length whose diameter is less than or equal to the maximum dimension; a bilayer of indefinite area whose width is less than or equal to the maximum dimension; or rods, ellipsoids of revolution, or spheres of water embedded uniformly in a volume otherwise filled with the amphipathic compound and spaced such that no distance between two adjacent surfaces of these aqueous inclusions is greater than the maximum dimension.

The choice among these different alternatives in a given situation is determined by the ratio between the molar surface area, which is determined independently by the repulsion among the hydrophilic functional groups and the hydrophobic effect, and the molar volume, which is determined by the molecular structure of the particular amphipathic compound in the particular aqueous solution. The hydrocarbon of the compound must also be able sterically to fill the volume allotted to it by a particular shape;[62] certain volumes are too anisotropic to be filled by real hydrocarbon, which is made from atoms of carbon and hydrogen joined by covalent bonds of precise bond angle and bond length and around which only particular rotations are permitted, even though these same volumes can be filled with imaginary hydrocarbon, which is a uniform plastic fluid that can fill any shape drawn on a sheet of paper. For example, when one of the fatty acyl chains is removed from phosphatidylcholine to form lysophosphatidylcholine, the product forms ellipsoidal micelles[22] rather than bilayers because the internal volume of the narrow bilayer that would be dictated by the appropriate molar surface area and molar volume cannot be filled uniformly by the hydrocarbon available, but an elliptical micelle, with its smaller and more isotropic volume for each unit of surface area, can be filled readily. Dodecyl sulfate in 0.3 M lithium chloride forms spherical micelles because the electrostatic repulsion among the sulfates is sufficient to produce the largest possible ratio of molar surface area to molar volume and the linear, fully saturated hydrocarbon is flexible enough to fill the appropriate spherical volume uniformly.[77] A heterogeneous mixture of phospholipids extracted from mammalian brain, when hydrated at 37 °C to a low content of water ($\leq 20\%$), forms a complex in which parallel cylinders of water spaced 4.5 nm apart are embedded in a volume otherwise filled with phospholipid.[78,79] In this case, the poorly hydrated hydrophilic functional groups of the phospholipids produce a molar surface area that when combined with the unavoidable molar volume of the phospholipids, would produce a bilayer wider than the maximum dimension permitted these phospholipids. The fact that the particular phospholipids synthesized by living organisms form bilayers spontaneously rather than one of these other structures is as much a result of evolution by natural selection as the fact that the polypeptides synthesized by living organisms fold.

Suggested Reading

Levine, Y.K., & Wilkins, M.H.F. (1971) Structure of Oriented Lipid Bilayers, *Nature New Biol. 230*, 69–72.

Hubbell, W.L., & McConnell, H.M. (1971) Molecular Motion in Spin-Labeled Phospholipids and Membranes, *J. Am. Chem. Soc. 93*, 314–326.

The Proteins

Even a homogeneous suspension of biological membranes, highly purified by a series of centrifugations, contains a diverse collection of proteins. These proteins fall into several categories. All membranes when they are present in the cell are closed continuous sacs, usually containing solutions of soluble proteins. Even if the final suspension of purified membranes has been submitted to lysis and centrifugation, some of these **entrapped soluble proteins** may still be enclosed in small vesicles of membrane and contaminate the preparation. For example, it is difficult to obtain from erythrocytes a suspension of plasma membranes completely devoid of hemoglobin. Other proteins are bound to membranes only through interfaces with more firmly attached proteins. These **peripheral membrane-bound proteins**[80] make no contact with the bilayer. They can be removed from the membrane by mild treatments that are normally used to dissociate the subunits of multimeric proteins but that are not able to dissolve the bilayer itself and release any of the proteins physically embedded within it. Examples of such treatments are increasing or decreasing either the pH or the ionic strength, removing divalent cations, using mild denaturants, or some combination of these treatments.[81-83]

Anchored membrane-bound proteins are proteins that make firm contact with the bilayer itself, but only through a short segment of their polypeptide at one of their termini, the sole purpose of which is to attach them to the bilayer by being immersed in the hydrocarbon. This segment of polypeptide embedded in the bilayer is not engaged in the native structure of the globular domain to which it is attached and can often be removed proteolytically to render the globular, detachable domain a perfectly soluble protein that displays all of the functions of the membrane-bound form. There is a family of proteins, exemplified by the receptor for epidermal growth factor,[84,85] that also contain only one short hydrophobic

segment of their polypeptide embedded within the bilayer, but this segment is in the middle of the amino acid sequence.[86-88] The role of these proteins is to transmit across the membrane to their cytoplasmic domains the information that a circulating hormone is bound at their extracytoplasmic domains. Upon receipt of the information, a protein tyrosine kinase activity, catalyzed by the cytoplasmic domain, is activated. Neither domain by itself is capable of displaying hormone-dependent protein tyrosine kinase activity,[87] so in this case, the single membrane-spanning segment performs a greater role than merely anchoring the protein in the membrane.

Phosphatidylinositol-linked (PI-linked) membrane-bound proteins are a class of proteins that are located on the extracellular surfaces of protozoa or animal cells and that are connected firmly to the plasma membrane by covalent attachment to phosphatidylinositol in the bilayer (Figure 3–19).[89] This is a heterogeneous collection of proteins, and their respective functions are unrelated to each other, so their common mode of attachment is probably fortuitous. Immediately after they have been synthesized, these proteins are anchored temporarily in the membrane by a carboxy-terminal segment of their polypeptide that is rich in hydrophobic amino acids. The ultimate carboxy terminus of the posttranslationally modified protein is a glycine, alanine, cysteine, serine, or asparagine 15–30 amino acids in from the initial carboxy terminus. Through an enzymatically catalyzed transamidation, the carboxy terminus of this amino acid is transferred from the carboxy-terminal segment of 15–30 amino acids to which it was originally attached to the amine of the ethanolamine phosphate connected through the oligosaccharide to the phosphatidylinositol.[90] A protein belonging to this class can be identified by its ability to be released from the surface of the cell by a phospholipase C specific for phosphatidylinositol.[91]

Finally, there is the class of proteins the polypeptides of which span the bilayer several times and present structured surfaces on both sides of it. A significant portion of the native structure of the polypeptide or polypeptides of such **integral membrane-bound proteins**[80] is within the hydrocarbon. Integral membrane-bound proteins can never be detached in a functional form from the bilayer by proteolytic cleavage and usually lose their native structure or precipitate from solution or both when the bilayer is dissolved by detergents. As with any protein, the native structure of an integral membrane-bound protein is determined by the solvent in which it is dissolved. This solvent is a sheet of liquid paraffin 3.0 nm wide possessing covalently attached anions and zwitterions at both of its surfaces and in contact on each of its surfaces with an aqueous solution. Most integral membrane-bound proteins are responsible for the transport of polar, water-soluble metabolites across the bilayer, a function that requires the protein to have a significant portion of its mass within the hydrocarbon.

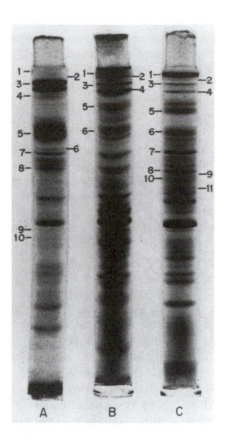

Figure 14–12: Polyacrylamide gels displaying the collection of polypeptides found in the plasma membranes of rat erythrocytes (A), the plasma membranes of cells from rat liver (B), and the portion of the plasma membrane of kidney cells referred to as the brush border (C). Purified membranes from these various tissues were dissolved in a solution of sodium dodecyl sulfate, which denatured and coated each polypeptide with a layer of dodecyl sulfate. The polypeptides were then separated by electrophoresis on polyacrylamide gels cast in a solution of sodium dodecyl sulfate. Each band represents a different polypeptide. The numbers indicate polypeptides that are glycosylated. Reprinted with permission from ref 92. Copyright 1971 *Journal of Biological Chemistry.*

When a suspension of purified membranes is examined by electrophoresis in solutions of dodecyl sulfate, a large collection of different polypeptides, each present at its own characteristic concentration, is observed (Figure 14–12). Each of these polypeptides is a component of one of the many native proteins bound to the membranes, and the protein of which it is a part is responsible for a particular biochemical function. Therefore, a biological membrane, although often much less complex, resembles cytoplasm in being a heterogeneous mixture of a large number of different proteins each present at a different concentration and each with a specific function. Many of the functions performed by membrane-bound proteins have been identified, and biochemical assays have been developed for determining their presence and their concentration. In many cases, the protein responsible for one of these functions has been identified, purified, cloned, and sequenced.

The **purification** of the particular membrane-bound protein identified by a biochemical assay proceeds in two stages. In the first stage, a biological source is chosen that contains the highest possible concentration of the protein. This involves assaying the biochemical activity in different tissues from different species or trying to increase the concentration of the protein in its active form by genetic manipulation of a microorganism or cultured eukaryotic cells. Membranes that contain the protein in high concentration are then separated by the procedures of cell fractionation from other membranes in the biological source that contain little or none of the protein. These **purified membranes** are lysed to release the entrapped soluble proteins and submitted to treatments that release peripheral membrane-bound proteins without inactivating or releasing the protein of interest. The product of these manipulations is a suspension of membranes to which are bound the protein of interest in the highest possible concentration. At the completion of this first stage, the membrane-bound protein being purified may be essentially homogeneous. For example, fragments of membrane whose only protein (90%)[93] is sodium and potassium ion-activated adenosinetriphosphatase can be purified[94] from a region of the mammalian kidney, the only function of which is to transport sodium and potassium. In the kidney these membranes are paved with this protein. Many of the membrane-bound proteins that have been purified to homogeneity are those that are already present at high density in such suspensions of appropriately purified and extracted membranes.

The second state in the purification of a membrane-bound protein is to **dissolve the membranes** without unfolding the protein and then purify the dissolved protein as if it were a soluble protein. A nonionic or zwitterionic detergent is used to dissolve the membranes, because nonionic or zwitterionic detergents do not bind tightly to the polypeptide and thus unfold the native structure of a protein.[95,96]

Nonionic or zwitterionic detergents are amphipathic compounds that have neutral or weakly zwitterionic functionalities attached to one end of their hydrocarbon. A common class of nonionic detergents (Brij series) is synthesized from linear, saturated alcohols produced commercially by the reduction of linear fatty acids 12–20 carbons in length. Ethylene oxide is polymerized at random to the hydroxyl of one of these alcohols to form the amphipathic compound

14–13

The length of the alcohol ($n + 2$) is defined, but the number of ethylene oxide units ($m + 2$) is a random set within a certain range. Another widely used class of nonionic detergents (Triton series) has a p-alkylphenol attached to the poly(ethylene oxide) rather than an alcohol

14–14

Mildly zwitterionic detergents such as the N-oxides of linear dimethylalkyl amines have also been used

14–15

All of these detergents, as they are amphipathic compounds, form elliptical micelles in aqueous solution in the absence of added membranes.

A nonionic detergent is able to form mixed micelles with phospholipids and cholesterol and thereby dissolve membranes.[97–99] If the removal of the bilayer by this process does not unfold the protein being purified, the nonionic detergent forms either a torus or a micelle surrounding the segments of the protein formerly embedded in the membrane and replacing the hydrocarbon of the bilayer with the hydrocarbon of the detergent. Ideally, this arrangement presents the hydrophilic functional groups of the detergent to the aqueous phase while its hydrocarbon supports the portions of the protein formerly embedded in the membrane (Figure 14–13). Such a fantasy is realized only fortuitously and infrequently with integral membrane-bound proteins (Figure 14–13B).

Many anchored membrane-bound proteins have been dissolved with nonionic detergents as biologically active proteins and purified to homogeneity by the normal methods of chromatography or affinity adsorption. A few examples of such anchored membrane-bound proteins are cytochrome b_5,[100,101] HLA histocompatibility antigens,[102] HLA-linked B-cell antigen,[103] dipeptidylpeptidase IV,[104,105] brush border aminopeptidase,[106] sucrase–isomaltase,[107] and the hemagglutinin of influenza virus.[108] In each of these cases, the protein can be purified by dissolving the membranes with a nonionic detergent and performing the steps of the chromatography or affinity adsorption in solutions of nonionic detergent. The main functional portion of an anchored membrane-bound protein can sometimes be released from the membrane as a **water-soluble, detachable** domain by a proteolytic cleavage that cuts away the **embedded anchor** of the intact polypeptide. For this approach to be feasible, the embedded anchor must be located either at the amino terminus[104,106,107] or at the carboxy terminus of the uncleaved protein,[101,108] so that its removal alters none of the biological activity of the protein, only its attachment to the bilayer. When anchored membrane-bound proteins are purified intact in the presence of nonionic detergent, they will recombine with bilayers of phospholipid when the detergent is removed[101,107,109] and become anchored again. When they are removed from the membrane by proteolytic cleavage, the detached domains have no affinity for bilayers.

Many of these proteolytically released detachable domains have been crystallized, and crystallographic molecular models have been constructed from the maps of electron density.[110,111] These crystallographic molecular models are indistinguishable from those of normal water-soluble proteins. The terminal region of the polypeptide at which the cleavage releasing the detachable domain from the membrane occurred is usually disordered and featureless in the map of electron density. From all of these observations it can be concluded that an anchored membrane-bound protein is simply a water-soluble protein that is leashed to the bilayer by a flexible segment of its polypeptide attached in turn to the embedded anchor.

A membrane-bound protein known as glycophorin is found in the plasma membrane of erythrocytes. The protein is 131 amino acids long, and its embedded anchor[112] is located to the carboxy-terminal side of Glutamate 72. This embedded anchor spans the plasma membrane.[113] The 35 carboxy-terminal amino acids on the cytoplasmic side of the membrane are rich in proline and seem to be structureless and functionless. The function of glycophorin is to serve as the source of most of the polysaccharide on the extracytoplasmic surface of the erythrocyte, and the protein is about 60% carbohydrate by weight.[112] This carbohydrate is entirely linked as oligosaccharides[114] to the extracytoplasmic amino-terminal portion of the protein through 15 O-glycosidic linkages to threonines and serines and one *N*-glycosidic linkage.[112] If it is assumed that the last 59 amino acids in the sequence of glycophorin serve only as the embedded anchor, then this protein also is an anchored membrane-bound protein, even though the embedded anchor is almost as long as the functional region and cannot be separated from it by proteolytic cleavage.

The embedded anchor left behind in the bilayer when an anchored membrane-bound protein is released by proteolytic digestion always has at least one segment of sequence composed of very hydrophobic amino acids. The entire stretch of polypeptide left behind in the membrane may be quite long, but the length of each **hydrophobic segment** is usually only about 20 amino acids. 3-Hydroxy-3-methylglutaryl coenzyme A reductase, which is an anchored membrane-bound protein by virtue of the fact that a catalytically active detachable domain can be removed from the membrane by proteolytic cleavage, has an embedded anchor almost 400 amino acids in length containing seven hydrophobic segments of 20

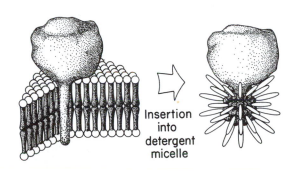

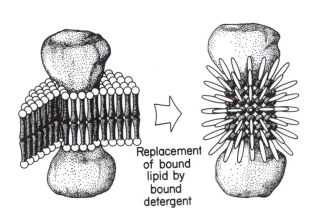

Figure 14–13: Diagrammatic representations of the mechanism by which a nonionic detergent dissolves an anchored membrane-bound protein (A) or an integral membrane-bound protein (B) by the insertion of the anchor into a micelle of the detergent or by replacement of the hydrocarbon of the bilayer with the hydrocarbon of the detergent, respectively. Figure courtesy of Steven Clarke, Department of Chemistry and Biochemistry, University of California at Los Angeles.

amino acids each,[115] but most anchored membrane-bound proteins contain only one hydrophobic segment. Anchored membrane-bound proteins such as glycophorin that do not happen to have a loop of their polypeptide capable of being cleaved by a proteolytic enzyme to release a detachable domain also possess such hydrophobic segments. The amino acid sequence of the hydrophobic segment from cytochrome b_5 is –WWTN-WVIPAISAVVVALMY–,[116] that from one of the HLA histocompatibility antigens is –VPIVGIVAGLVLLVAV-VTGAVVAAVMW–,[117] that from sucrase–isomaltase is –LIVLFVIVFIIAIALIAVLA–,[118] that from the hemagglutinin of influenza virus is –WILWISFAISCFLLCVVLGFIM-WAS–,[119] and that from glycophorin is –ITLIIFGVMAG-VIGTILLISYGI–.[120] Such hydrophobic segments from anchored membrane-bound proteins are the most hydrophobic sequences of their length found in any protein.[121] These sequences are flanked at both ends by regions containing normal or above normal frequencies of neutral and charged hydrophilic amino acids.

Presumably, it is these hydrophobic segments of anchored membrane-bound proteins that are embedded in the hydrocarbon of the bilayer. As most of the anchors span the bilayer only once, this places the two termini of the intact polypeptide composing the protein on opposite sides of the membrane.[122] That each of these membrane-spanning segments of amino acid sequence is completely surrounded by liquid hydrocarbon explains their extreme hydrophobicity. The short hydrophilic segments that end up in most of these proteins on the opposite side of the membrane from the globular, detachable domains could act simply as barbs that cannot be pulled through the bilayer, but other roles, such as the relay of information across the membrane, have been proposed for some of them, in analogy with the membrane-spanning segment of epidermal growth factor receptor.

The hydrophobic segments of the polypeptide that serve to attach anchored membrane-bound proteins to the bilayer spontaneously assume an **α helix** over most or all of their length, as judged by circular dichroic spectra, when they are incorporated into micelles of detergent[123,124] or bilayers of phospholipid.[125] It is believed that they also adopt α helices within the hydrocarbon of the normal biological membranes in which they are found.

An α helix is the logical structure for a segment of polypeptide to assume when it is immersed in liquid hydrocarbon in the total absence of water or any other donor or acceptor of hydrogen bonds. Within itself, an α helix satisfies all of the hydrogen-bond donors on the polypeptide (Figure 4–17). It has already been noted that it is the donors which are the substoichiometric participants in the overall balance of hydrogen bonds in a solution of protein and that every donor must be paired with an acceptor. Serines and threonines occur fairly often in these fully engulfed membrane-spanning segments, but the donors on their hydroxyls can be automatically satisfied by the intramolecular hydrogen bonds, in which these two amino acids usually participate, with the empty lone pairs on the acyl oxygens of the amino acids three or four residues ahead of them in an α helix (Figure 6–6B). Within the hydrocarbon of the bilayer these lone pairs, partners to the lone pairs on the same oxygens that are already participating in hydrogen bonds within the α helix, will always be empty and available to

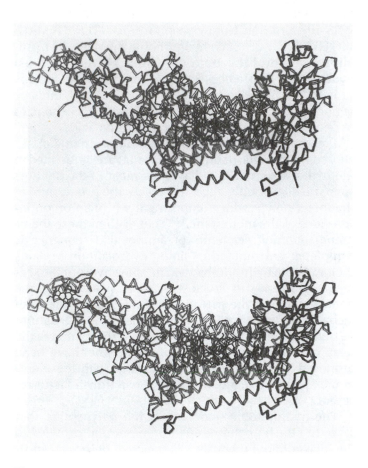

Figure 14–14: Drawing of the α-carbon skeleton of the crystallographic molecular model (c13 Kin.2) of the photosynthetic reaction center from the photosynthetic bacterium *Rhodopseudomonas viridis*. A purified preparation of photosynthetic membranes was isolated from broken cells by differential centrifugation followed by isopycnic centrifugation. The membranes were dissolved in *N*-dodecyl-*N,N*-dimethylamine *N*-oxide (**14–15**) and submitted to molecular exclusion chromatography in a solution of the same detergent. Fractions containing reaction center were identified by its characteristic absorbance at 830 nm. These fractions were pooled and crystals of reaction center were produced in concentrated solutions of ammonium sulfate. These crystals were used for X-ray diffraction. To produce the drawing, the positions of the α-carbons of the polypeptide in the crystallographic molecular model were connected with rectangular links. Reprinted with permission from *Nature*, ref 127. Copyright 1985 Macmillan Magazines Limited.

act as acceptors to unoccupied donors. The presence of such intrahelical hydrogen bonds may introduce bends in the α helix appropriate to its function. Therefore, all of the hydrogen-bond donors in a hydrophobic segment can be satisfied intramolecularly with the notable exception of those on tyrosines and tryptophans. These two amino acids, however, usually occur at the ends of a hydrophobic segment in an anchored membrane-bound protein where their lone hydrogen-bond donors can remain in contact with water as they do almost always in a crystallographic molecular model of a soluble protein.

The width of the hydrocarbon in a bilayer of naturally occurring amphipathic lipids and cholesterol is 3.0 nm (Figure 14–4D). As the rise for each amino acid in an α helix is 0.15 nm, it should require 20 amino acids to span the hydrocarbon. The fact that the lengths of the hydrophobic segments of anchored membrane-bound proteins are usually about 20 amino acids is further support for the proposal that these hydrophobic segments are α-helical in their normal situation.

The belief that an α helix is the structure most compatible with the environment in which the membrane-spanning segments of a membrane-bound protein are located is also confirmed by what little is known about the detailed structure of integral membrane-bound proteins. The **photosynthetic reaction center** is an integral membrane-bound protein located in membranes of photosynthetic bacteria. Membranes containing high concentrations of the protein were purified by differential centrifugation and isopycnic centrifugation and dissolved in a solution of zwitterionic detergent **14–15**. The protein was purified by molecular exclusion chromatography and could then be crystallized.[126] A crystallographic molecular model (c13 Kin.2) of the photosynthetic reaction center has been constructed from the map of electron density (Figure 14–14).[127] The protein has 11 membrane-spanning segments. Each traverses the hydrocarbon of the bilayer in one unbroken α helix. The amino acid sequences of these 11 α helices[128] are –SLGVLSLFSGLMWFFTIGIWFWYNA–, –LKEGGLWLIASFFMFVAVWSWWGRTYLRAQA–, –AWAFLSAIWLWMVLGFIRPILM–, –PFHGLSIAFLYGSALLFAMHGATILAV–, –MEGIHRWAIWMAVLVTLTGGIGILL–, –GFFGVATFFAALGIILIAWSAVL–, –GGLWQIITICATGAFVSWALREVEICRKL–, –HIPFAFAILAYLTLVLFRPVM–, –PAHMIAISFFFTNALALALHGALVLSAA–, –GTLGIHRLGLLLSLSAVFFSALCMII–, and –IAQLVWYAQWLVIWTVVLLYLRREDR–.[129–131] In each of these amino acid sequences there is a region 20 amino acids in length that contains no amino acids that are charged at neutral pH with the exception of the arginines to the carboxy-terminal ends of the third and the eighth α helices. These two α helices, however, have their amino-terminal ends a short way into the bilayer and emerge at the other side of the bilayer just before these arginines. The reason that these 11 α helices are embedded in the bilayer is that they are composed of hydrophobic amino acids and the bilayer is an organic solvent into which the side chains of these amino acids have dissolved.

These 11 hydrophobic segments of amino acid sequence are similar to the five listed from the anchored membrane-bound proteins but differ in flavor. The hydrogen-bond donors, tryptophan and tyrosine, are more uniformly distributed over the length of these segments; the neutral but hydrophilic hydrogen-bond donors and acceptors glutamine, asparagine, and histidine now occasionally appear; and the frequency with which glycine is encountered is greater. Each of these subtle changes indicates that these amino acid sequences are from a bundle of α helices gathered together as a protein rather than from individual α helices spanning the harsh environment of the membrane as isolated entities. The α helices still present branched alkane to the hydrocarbon of the bilayer, but among themselves the interactions are more complex.

The increased frequency of donors and acceptors of hydrogen bonds in these amino acid sequences results in part from the formation of hydrogen bonds among adjacent α helices. When a hydrogen-bond donor and acceptor enter a hydrogen bond during the folding of a polypeptide in aqueous solution, the standard enthalpy change for the reaction is zero because the reaction proceeds with no net change in the number of hydrogen bonds (Equation 5–19). A hydrogen-bond donor or acceptor in the middle of an otherwise hydrophobic segment spanning a membrane, however, is held within the hydrocarbon by the α helix. The price of withdrawing the hydrogen-bond donors and acceptors from the water and stripping them of their hydration has already been paid by the hydrophobic effect that immersed the membrane-spanning segment in the first place. When a hydrogen-bond donor and acceptor form a hydrogen bond while aligning these α helices within the hydrocarbon during the formation of the native structure of the protein, the standard enthalpy change for the reaction is –12 to –20 kJ mol^{-1} (Table 5–2). Once the membrane-spanning α helix has become inserted in the bilayer, however, the hydrophobic effect is no longer in operation. The importance of hydrogen bonding and the importance of the hydrophobic effect are reversed within the hydrocarbon of bilayer. A few hydrogen-bond donors and acceptors can be responsible for significant, favorable standard enthalpy of formation, yet no favorable change in standard free energy occurs when two hydrophobic surfaces are juxtaposed.

Several of the buried hydrogen-bond donors and acceptors in the membrane-spanning segments from reaction center are engaged in hydrogen bonds with the several coenzymes in the middle of the protein. For example, the histidines in the third positions of both the fourth- and the ninth-listed amino acid sequences are ligands to the magnesiums of the bacteriochlorophylls, and the tryptophan in the eighteenth position of the

eighth-listed sequence forms a hydrogen bond with a bacteriopheophytin.[132]

A map of electron scattering density is available for **bacteriorhodopsin**, which is another integral membrane-bound protein. Bacteriorhodopsin, when present in its native purple membrane in the bacterium *Halobacterium halobium*, is already in a **two-dimensional crystalline array** within the bilayer.[133] The space group of this two-dimensional lattice is $P3$, and the asymmetric units related by the 3-fold rotational axes of symmetry within the unit cell are individual molecules of bacteriorhodopsin. Although bacteriorhodopsin is one of the few membrane-bound proteins that is naturally crystalline, most integral membrane-bound proteins can be induced to crystallize in two dimensions.[134] Such crystals can be embedded in a glass of negative stain (Figure 14–15),[135] a glass of glucose,[136] or a glass of amorphous ice.[137]

A two-dimensional crystalline array of an integral membrane-bound protein is a three-dimensional distribution of electron scattering density, $\theta(x,y,z)$, that is periodic in the two dimensions of the bilayer. The electrons that it scatters in an electron microscope will form an electron diffraction pattern (Figure 14–16). This diffraction pattern does not arise from reflections generated by sets of parallel planes running through a three-dimensional lattice (Figure 4–8) but from reflections generated by sets of parallel lines running through the two-dimensional lattice (Figure 4–4) of the projection of the three-dimensional array on a plane normal to the beam of electrons. Each reflection has an amplitude, an index, and a phase, but the phases cannot be measured.

The indexed set of amplitudes and phases of the electron diffraction pattern from such an array are the amplitudes and phases of the Fourier transform of the projection of the three-dimensional distribution of electron scattering density of the array upon the plane normal to the axis of the beam (Equation 9–10). Therefore, they are the amplitudes and phases of a central section through the three-dimensional Fourier transform of the three-dimensional distribution of electron scattering density (Equation 9–9).

An electron micrograph of a two-dimensional crystalline array of a membrane-bound protein (Figure 14–15) is a projection of the natural logarithm of the three-dimensional electron scattering density $\theta(x,y,z)$ of the array upon a plane normal to the axis of the beam of electrons (Equation 9–10). From the digitized distribution of contrast on the electron micrograph, the amplitudes and phases of the central section of the Fourier transform of the three-dimensional electron scattering density can be calculated by a computer (Equation 9–9). Because the array is a two-dimensional crystal, the central section through its Fourier transform is a lattice of spots. Each spot in the transform calculated by the computer from the digitized micrograph corresponds to one of the reflections in the electron diffraction pattern (Figure 14–16) and has the same phase and the same relative amplitude

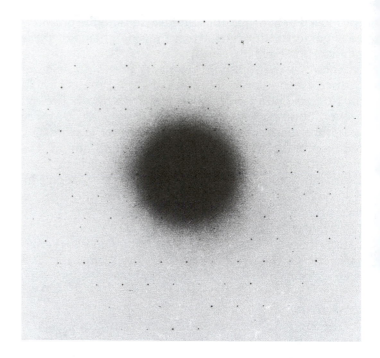

Figure 14–15: Electron micrograph of a two-dimensional crystalline array of cytochrome oxidase.[138] Beef heart mitochondria were sonicated, and the resulting fragments of membrane were separated by differential centrifugation. Fragments rich in cytochrome oxidase were extracted with Triton X-114 and then Triton X-100 to dissolve away other proteins. The purified particulate material was composed of fragments of membrane in which the only protein was cytochrome oxidase. These fragments of membrane were attached to carbon films and embedded in a glass of the negative stain uranyl acetate. Upon examination in the electron microscope, it was found that in many of the fragments the cytochrome oxidase had crystallized into two-dimensional arrays. The upper view is one of these arrays viewed normal to the electron beam. The lower view is the same array tilted on a horizontal axis so that its plane was at an angle of 36° to the electron beam. Reprinted with permission from ref 138. Copyright 1977 Academic Press.

Figure 14–16: Electron diffraction pattern produced by a two-dimensional crystalline array of bacteriorhodopsin. Fragments of membrane containing crystalline arrays of the protein were attached to a film of carbon and embedded in a glass of glucose. The specimen was centered in the beam of electrons of an electron microscope and a photographic plate was used to record the reflections of the diffraction pattern. The reflections emerge from the specimen at characteristic angles determined by the lattice, and they are recorded on a piece of film at an exact distance from the specimen. The dark central spot is the majority of the electrons that passed through the specimen undeflected. The sharp dots of varying intensity are the reflections themselves. Reprinted with permission from ref 136. Copyright 1975 Academic Press.

as it does. This permits the phases to be estimated from the micrograph and the amplitudes to be measured from the electron diffraction pattern.

The Fourier transform of the periodic three-dimensional distribution of electron density that is a three-dimensional crystal of protein is a three-dimensional lattice of spots in reciprocal space. Each spot has an amplitude and a phase. Each reflection in the diffraction of X-radiation from the crystal represents one of these spots. The Fourier transform of the three-dimensional distribution of scattering density in a crystalline array of a membrane-bound protein, which is periodic in only two dimensions, is a lattice of parallel lines in reciprocal space. Each of these parallel lines has an amplitude and a phase that vary periodically along its length. If the variations of the amplitudes and phases along each of these lattice lines could be measured, the three-di-

mensional distribution of electron scattering density in the unit cell of the crystalline array could be calculated by Fourier transformation of this set of functions.

If a crystalline array is tilted in the electron beam (Figure 14–15, lower panel), a projection along an axis tilted relative to the axis normal to the plane of the array is recorded on the micrograph, and the amplitudes and phases of the electron diffraction pattern, which are now the amplitudes and phases of the Fourier transform of this new projection, have changed. Each of the micrographs and electron diffraction patterns in a series in which the specimen is systematically tilted represents a different central section through the lattice of lines in the Fourier transform of the three-dimensional array of scattering density.[139] If enough of these central sections are gathered, the amplitudes and phases of the Fourier transform within certain ranges along the lattice lines can be gathered (Figure 14–17).[140] The amplitudes can be gathered from either electron diffraction patterns or Fourier transforms of the digitized contrast on electron micrographs, but the phases can be obtained only from Fourier transforms of the digitized distributions of contrast on the electron micrographs. Because the amplitudes and phases are inaccurate (Figure 14–17), reconstructions from such Fourier transforms have usually been performed only when the space group of the crystal has rotational symmetry, such as the P3 space group of bacteriorhodopsin. The Fourier transform of a lattice with rotational symmetry contains sets of identical amplitudes and phases along sets of symmetrically related lattice lines, and these identities serve as an internal check upon the reality of the measurements.

When the amplitudes of the electron diffraction patterns of crystalline arrays of bacteriorhodopsin embedded in a glass of glucose and tilted at various angles were combined with the phases from the digitized distributions of contrast in electron micrographs taken at the same angles of tilt (Figure 14–17), a map of the three-dimensional electron scattering density within the asymmetric unit in the unit cell of the two-dimensional crystal of the protein could be calculated.[140,141] At low resolution, when only reflections from lattice lines with spacing out to 0.7 nm were included in the calculation,[140] seven rods of scattering density aligned roughly perpendicular to the plane of the array within the membrane were observed (Figure 14–18A). The details of the scattering density fade away at the two ends of the molecule above and below the bilayer because the specimen could be tilted only between –57° and +57° and the lattice lines arising from the unseen portions of the protein were outside the regions that could be sampled with these tilts. Nevertheless, the conclusion drawn from these observations was that the portion of bacteriorhodopsin passing through the bilayer, as with the portion of photosynthetic reaction center that passes through the bilayer, is a bundle of α helices.

Within the amino acid sequence of bacteriorhodopsin[142,143] there are seven hydrophobic segments, each about 20 amino acids in length.[121] These seven hydrophobic segments are similar in their composition of amino acids to the 11 hydrophobic segments of amino acid sequence in photosynthetic reaction center, with the exception that aspartic acids and glutamic acids are scattered among the α helices of bacteriorhodopsin but are mostly absent from those of the reaction center. These acidic amino acids may simply be neutral hydrogen-bond donors and acceptors, but there is evidence from site-directed mutation that some of them are involved in the ability of bacteriorhodopsin to transport protons.[144]

When reflections from lattice lines with spacing out to 0.35 nm were included in the calculation of the map of electron scattering density,[141] the same seven rods were observed, but the rods became more detailed at their ends so that some of the connections between the rods could be observed and, more importantly, protrusions of electron scattering density appeared along the rods (Figure 14–18B). These protrusions represent the side chains of the amino acids in the sequence of the protein. Their appearance at the expected regular intervals along each of the seven rods verified that each rod represented an α helix. From the few connections observed and the pattern of the largest protrusions along each rod (representing the aromatic amino acids), a molecular model of the polypeptide, built with the known amino acid sequence, could be unambiguously placed into the map of electron scattering density.

The seven hydrophobic segments in the amino acid sequence of the protein occupied the rods of scattering density representing the seven α helices that span the bilayer in the native protein. In the crystallographic molecular model, the first hydrophobic segment in the amino acid sequence occupies the rod of electron scattering density at 9 o'clock in the cross section presented in Figure 14–18B, and each successive hydrophobic segment in the amino acid sequence occupies a rod of scattering density one position in a clockwise direction from the preceding one. The seven membrane-spanning segments have the sequences: –WIWLALGTALMGLGTLYFLV–, –FYAITTLVPAIFTMYL-SMLL–, –ARYADWLFTTPLLLLDLALLV–, –ILAIVGA-DGIMIGTGLVGAL–, –VWWAVSTAAMLYILYVLFFGF–, –KVLRNVTVVLWSAYPVVWLI–, and –IETLLFMVLDVSA-KVGFGLI–. The lysine in the middle of the last segment is covalently linked to the coenzyme, retinal.

In the structures of both photosynthetic reaction center (Figure 14–14) and bacteriorhodopsin (Figure 14–18), the α helices are not perfectly normal to the plane of the bilayer but are tilted 10–35° from the normal.[128,140] Presumably, this is the result of the fact that each is forced to be tilted relative to its neighbors. The dihedral angle Ω (Figure 6–22) between two almost parallel, adjacent α helices should be about +27° if the maximum interdigitation is to occur between the amino acids within the interface between them (Figure 9–27).[145] Presumably the tilting of the α helices within each bundle reflects a compromise between this stereochemical imperative and the requirement to produce a tight bundle of almost perpendicular rods in the bilayer.

It is generally believed that, although the structure of an integral membrane-bound protein on either side of the bilayer is that of a typical globular protein with a mixture of α helices, β structure, and random meander, all packed as unpredictably as is usually observed, the portions within the bilayer are such bundles of several α helices, each oriented roughly normal to the plane of the bilayer. This belief is based on the structures of photosynthetic reaction center and bacteriorhodopsin. It is based on the fact that the hydrophobic segments of anchored membrane-bound proteins spontaneously form α helices. It is based on the high frequency with which

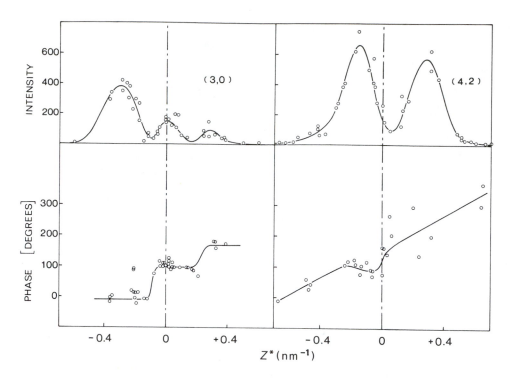

Figure 14–17: Variation of intensity and phase along two of the lattice lines [(3,0) and (4,2)] in the Fourier transform of the three-dimensional distribution of electron scattering density within the two-dimensional crystal of bacteriorhodopsin.[140] The intensities are the intensities of the spots on electron diffraction patterns of the array (Figure 14–16) tilted at various angles. The phases were determined from Fourier transforms of the distribution of contrast on electron micrographs of the specimens tilted at various angles. The lattice lines that occur in the Fourier transform of the electron micrograph are the same as the lattice lines upon which the reflections of the electron diffraction lie because the electron diffraction pattern is the same Fourier transform of the same array. Each data point on each graph represents a measurement from a different electron diffraction pattern or a different electron micrograph, respectively, from a specimen at a different angle of tilt. As the angle of tilt is varied, different positions along the lattice lines are sampled. The phase or intensity is presented as a function of the distance along the lattice line, Z^* (nanometer^{-1}). Reprinted with permission from *Nature*, ref 140. Copyright 1975 Macmillan Magazines Limited.

hydrophobic segments about 20 amino acids in length appear in the amino acid sequences of integral membrane-bound proteins. It is based on the fact that an α helix automatically occupies all of the donors of hydrogen bonds on the polypeptide backbone before the segment is incorporated into the bilayer. And it is based on the continuing emotional appeal of the α helix.

There is, however, an alternative way to occupy all of the donors of hydrogen bonds on the polypeptide backbone, and this is illustrated by the structure of porin. Porin is an integral membrane-bound protein found in high abundance in the outer membranes of Gram-negative bacteria.[146] It is a trimer of three folded polypeptides 300–340 amino acids in length[147,148] depending on the species. The protein is known[149] to form water-filled channels through the outer membrane large enough to pass molecules of molar mass 600 g mol^{-1}. The protein

has been purified after dissolving it in solutions of anionic[146] or neutral detergents,[148] and has been crystallized in three dimensions from these solutions. A crystallographic molecular model (c13 Kin.3) of the protein has been built (Figure 14–19) from crystallographic studies with X-ray diffraction.[148]

Each of the three subunits of porin (Figure 14–19A) is a 16-stranded barrel of antiparallel β structure.[150] In the middle of each subunit is a large, hydrophilically lined, water-filled channel (Figure 14–19B), presumably the pore through which solutes pass across the bacterial outer membrane. About 12 strands of β structure from each barrel are in contact with the hydrocarbon of the bilayer, and this outer surface is lined with hydrophobic amino acids (Figure 14–19B). The remaining four strands are responsible for the interfaces among the three subunits (Figure 14–19A). The peptide backbone of the 12

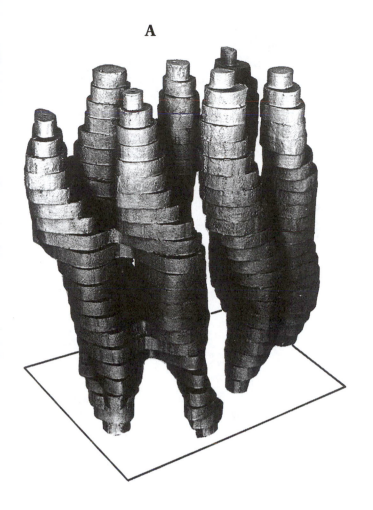

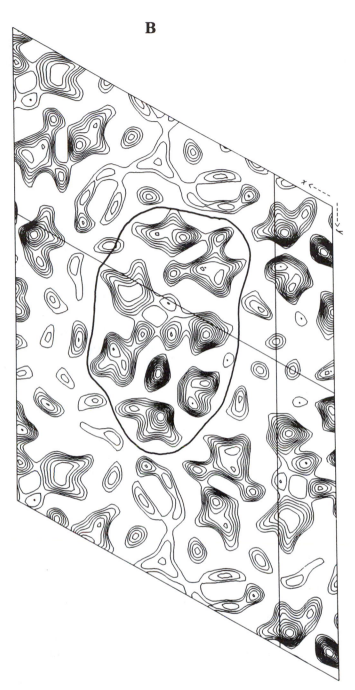

Figure 14–18: Maps of electron scattering density for bacteriorhodopsin (c13 Kin.1). Large sheets of purple membrane were applied to a thin film of amorphous carbon. The sheets of membrane were embedded in a glass of hydrated glucose. Electron diffraction patterns and electron micrographs were gathered at various angles of tilt between –60° and +60°. The variations of phase along the lattice lines of the Fourier transform of the two-dimensional array were estimated from the Fourier transforms of the digitized electron micrographs and the variations of amplitude were estimated from the same Fourier transforms and the electron diffraction patterns. (A) Map of electron scattering density at low resolution (lattice spacings to 0.7 nm). Sections through the three-dimensional map of electron scattering density were made at 0.2-nm intervals and traced onto balsa wood of the proper thickness, and the cutouts were stacked one on top of the other to create a three-dimen-

sional reconstruction. Panel A reprinted with permission from *Nature*, ref 140. Copyright 1975 Macmillan Magazines Limited. (B) Slice through the map of electron scattering density at much higher resolution (lattice spacings to 0.35 nm). The cross sections of the seven rods are seen within the circumscribed boundaries of one molecule. The intense feature at the lower center of the molecule of protein is the β-ionone ring of the retinal. Protrusions can be seen emerging from the seven α helices cut in cross section. These are side chains of amino acids. For example, the protrusion at 9 o'clock on the upper right α helix has been identified as Tyrosine 57; the protrusion at 9 o'clock on the right middle α helix, Tryptophan 86; the protrusion at 2 o'clock on the lower left α helix, Tyrosine 185; and the protrusion at 10 o'clock on the bottom α helix, Tryptophan 138. Panel B reprinted with permission from ref 141. Copyright 1990 Academic Press.

strands in contact with the hydrocarbon is folded in such a way that all of the donors of hydrogen bonds are occupied in the typical arrangement of an antiparallel β sheet (Figure 4–17).

This last feature allows this structure to fold stably. It is an arrangement available to a large β barrel such as the one found in porin because the curvature of this sheet (roughly 20° strand^{-1}) is gradual enough that properly aligned hydrogen bonds can be formed. The other β barrel that has been observed frequently in crystallographic molecular models is the one seen in triosephosphate isomerase (Figure 9–1), and other members of this evolutionarily related family.[151] It has already been noted, however, that this eight-stranded barrel is flattened (c2 Kin.8 and c5 Kin.1), causing the two ends of high curvature to have fewer intact hydrogen bonds than the two flattened sheets of low curvature.[152] This suggests that to be capable of immersion in a bilayer, a β barrel must have enough strands to form a structure with a complete set of unstrained hydrogen bonds. This suggestion is reinforced by the crystallographic molecular model of the B subunit of cholera toxin,[153] a protein that can insert into animal cell membranes. This rotationally symmetric pentamer includes a β barrel composed from 35 strands of β structure. Such large β barrels are probably not very common. The crystallographic molecular model of porin has renewed interest in the possibility that sheets of β structure may be found spanning the bilayer in integral membrane-bound proteins. The problem with this proposal, however, is the need to occupy all of the donors of hydrogen bonds within the two strands of polypeptide backbone at the two ends of the sheet before it is plunged into the hydrocarbon.

A membrane-spanning bundle of α helices in an integral membrane-bound protein is immersed in the bilayer and surrounded on all sides by the hydrocarbon of the amphipathic and neutral lipids. The molecules of lipid whose behavior is affected at a given instant by the presence of the protein can be formally distinguished from those molecules of lipid that behave as if they were in an unadulterated bilayer of amphipathic lipids. This distinction resembles in its ambiguity the distinction between water of hydration associated with a protein (Table 6–3) and water in the bulk solution. In the case of water of hydration, there is a gradual diminution of the influence of the protein the farther a particular molecule of water is from its surface, but a water molecule several shells from the protein may still be influenced by it because of the nets of hydrogen bonds that ensnare both water and protein. Likewise, a molecule of amphipathic lipid somewhat distant from the protein may be marginally influenced by it when one or two of its methylenes strike against the surface of the protein as the linear hydrocarbon writhes within the liquid paraffin, but molecules of amphipathic lipid embracing the protein should be more severely affected. Networks of hydrogen bonds among the hydrophilic functionalities of the phospholipids and sphingomyelins (Figure 14–6) may also spread the influence of the protein beyond its immediate vicinity. In this context, a class of **lipids in the boundary layer**,[154] surrounding the protein and under its influence, has been defined operationally just as water of hydration has been defined operationally. Just as in the case of waters of hydration, a single numerical value for moles of lipid in the boundary layer (mole of protein)$^{-1}$ is measured.

When 1-palmityl-2-stearylphosphatidylcholine, to which dimethyl cyclic nitroxide **14–9** is attached at the fourteenth carbon of the stearyl group,

14–16

is incorporated into bilayers of natural phosphatidylcholine containing an integral membrane-bound protein such as Ca^{2+}-transporting adenosinetriphosphatase (Ca^{2+}-ATPase), the electron spin resonance spectrum that is observed (Figure 14–20A)[155] can be decomposed into two spectra (Figures 14–20B,C) of which it is the sum.[154,156] One of the component spectra is the same as that of nitroxyl phosphatidylcholine **14–16** in pure vesicles of liquid phosphatidylcholine (Figure 14–20C), and one is that of nitroxyl phosphatidylcholine **14–16** when its motion is restricted (Figure 14–20B). As nitroxyl phosphatidylcholine **14–16** is fully saturated and has a fairly prominent protrusion on the fourteenth carbon of its stearyl substituent, it would not be surprising if its motion were restricted when it was immediately adjacent to a molecule of protein. This expectation and other observations[157] led to the conclusion that at any instant there are two sets of nitroxyl phosphatidylcholines **14–16** present in these bilayers, one set constituted by molecules of restricted mobility located in the boundary layer immediately adjacent to the protein and the other constituted by molecules of unrestricted mobility in the bulk bilayer. The class adjacent to the protein is used to monitor that location.

As the ratio between egg phosphatidylcholine and Ca^{2+}-ATPase in these membranes is increased, the fraction of restricted nitroxyl phosphatidylcholine **14–16** decreases. This decrease results from a competition between unlabeled molecules of egg phosphatidylcholine and molecules of nitroxyl phosphatidylcholine **14–16** for positions in the boundary layer adjacent to the protein and the increase in the concentration of the former. From the numerical values of the fraction of restricted nitroxyl phosphatidylcholine **14–16** as a function of the molar

Figure 14–19: Crystallographic molecular model (c13 Kin.3) of the porin from *Rhodobacter capsulatus*.[148] Fragments of the outer membrane of *R. capsulatus* were dissolved in a solution of the detergent *N*-dodecyl-*N,N*-dimethylamine *N*-oxide, and the porin was purified by salt extraction and molecular exclusion chromatography in solutions of the same detergent. From the final solutions of the purified protein, three-dimensional crystals suitable for X-ray crystallography could be grown. A refined crystallographic molecular model ($R = 0.22$ at 0.18 nm) was built. (A) Tracing of the α-carbon positions in the molecular model for the entire trimer (3×301 amino acids) viewed at an angle of 40° to the 3-fold rotational axis of symmetry normal to the bilayer. The wall of each of the three channels can be divided into a segment exposed to the hydrocarbon of the bilayer with a height of 4.0 nm (the width of the bilayer) and a segment participating in the interfaces among the three subunits, where the height of the wall drops to only 2.0 nm. The shorter wall is possible in the center of the structure because no contact is made with the bilayer at this location. (B) Complete molecular model (all non-hydrogen atoms) of a single subunit, viewed along an axis parallel to the 3-fold rotational axis of symmetry (indicated by the triangle in the figure). The taller outer wall of the subunit, coated on its outside surface with hydrophobic amino acids (above), which is responsible for holding back the bilayer, can be distinguished from the shorter inner wall (below), which forms the interfaces. The pore is readily distinguished in this view. Reprinted with permission from ref 148. Copyright 1991 Elsevier Science Publishers.

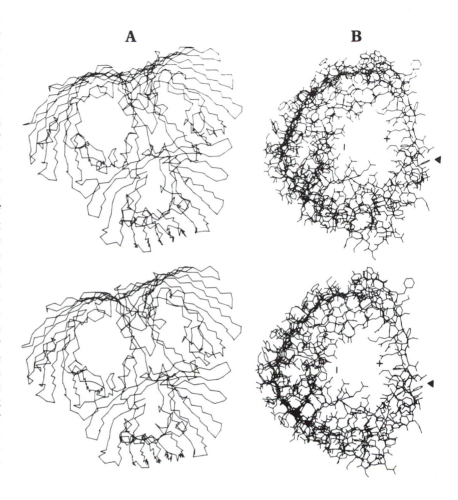

A   B

ratio between egg phosphatidylcholine and protein, the ratio between the affinities of nitroxyl phosphatidylcholine **14–16** and egg phosphatidylcholine for positions adjacent to protein can be estimated, and the number of molecules of egg phosphatidylcholine occupying positions adjacent to the protein can be calculated.[158] In the case of Ca^{2+}-ATPase at 25 °C,[155,156] the egg phosphatidylcholine and nitroxyl phosphatidylcholine **14–16** have equal affinity for positions around the protein and the number of positions is 22 mol (mol of protein)$^{-1}$. In other words, there are, on the average, 22 molecules of phospholipid in the boundary layer around a molecule of Ca^{2+}-ATPase, each of which can exchange with no bias for a molecule of nitroxyl phosphatidylcholine **14–16**, and these 22 molecules of natural, unlabeled phosphatidylcholine occupy positions at which a molecule of nitroxyl phosphatidylcholine **14–16** would be restricted in its motion by the protein. Therefore, the definition of lipid in the boundary layer, like the definition of water of hydration, is directly related to a particular experimental situation.

The ability of cholesterol to occupy positions in the boundary layer adjacent to an integral membrane-bound protein has also been examined.[155] 17-Spiroxylcholesterol nitroxide

14–17

was used as the probe. Its electron spin resonance spectrum could also be decomposed into mobile and restricted components. Spectral analysis of its behavior showed that this probe could displace any one of the 22 mol of egg phosphatidylcholine (mol of Ca^{2+}-ATPase)$^{-1}$ acting as boundary lipid but that its affinity for these 22 positions was about two-thirds the affinity of natural phosphatidylcholine. In agreement with this observation, unlabeled cholesterol was found to be able to com-

Figure 14–20: Decomposition of the electron spin resonance spectrum of nitroxyl phosphatidylcholine **14–16** in vesicles of phospholipid in which Ca²⁺-ATPase has been incorporated.[155] Ca²⁺-Transporting ATPase, from which all indigenous phospholipid had been removed, was incorporated into vesicles of egg phosphatidylcholine into which nitroxyl phosphatidylcholine **14–16** was also incorporated. In the final suspension of vesicles containing the protein, the molar ratios of Ca²⁺-ATPase (molar mass = 115,000 g mol⁻¹) to egg phosphatidylcholine to phosphatidylcholine nitroxide were 1:55:0.4. The nitroxide was acting as a probe present in dilute concentration within a solvent of egg phosphatidylcholine. The electron spin resonance spectrum of this probe (A) could be decomposed into two component spectra. One component (C), which accounted for 54% of the spins producing spectrum A, had the same spectrum as the probe dissolved in egg phosphatidylcholine in the absence of protein (F). The other component (B), which accounted for 46% of the spins producing spectrum A, had the same spectrum as the probe in a viscous bilayer composed of dipalmitylphosphatidylcholine and palmityloleylphosphatidylcholine at a ratio of 4:1 (E). The effect of the severe immobilization of the probe by the viscous bilayer resembles the effect of the protein on the probe (compare spectra B and E). A summation of reference spectra E and F, at a molar ratio of 0.46 to 0.54, produced theoretical spectrum D, which is identical to experimental spectrum A. All spectra are the amplitude of the first derivative of the adsorption of the microwave energy as a function of the strength of the magnetic field (gauss). Reprinted with permission from ref 155. Copyright 1984 American Chemical Society.

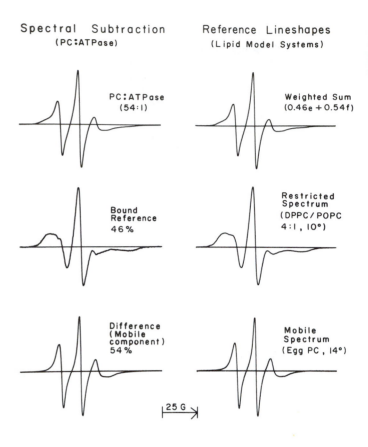

Spectral Subtraction (PC:ATPase)

Reference Lineshapes (Lipid Model Systems)

PC:ATPase (54:1)

Weighted Sum (0.46e + 0.54f)

Bound Reference 46%

Restricted Spectrum (DPPC/POPC 4:1, 10°)

Difference (Mobile component) 54%

Mobile Spectrum (Egg PC, 14°)

25 G

pete for restricted positions occupied by nitroxyl phosphatidylcholine **14–16** but not so effectively as natural phosphatidylcholine. It was proposed that the difference in affinity between cholesterol and phosphatidylcholine reflects the ability of the flexible hydrocarbon of egg phosphatidylcholine to adapt more readily than can the fused rings of cholesterol to the irregular surface of Ca²⁺-ATPase.

It is not necessarily the case, when a position in the boundary layer is occupied by a natural phospholipid rather than nitroxyl phosphatidylcholine **14–16**, that the motion of that natural phospholipid is noticeably affected. The amino acids presented by the membrane-spanning bundle of α helices to the linear hydrocarbon of the phospholipid are themselves branched hydrocarbons that are free to rotate fluidly, and the hydrocarbon of phosphatidylcholine may experience little change as it enters or leaves these positions. When Ca²⁺-ATPase incorporated into vesicles of dioleylphosphatidylcholine in which either the second carbons or the ninth and tenth carbons on the two fatty acids were labeled with deuterium rather than a nitroxide, the effect of the protein on the motion of these lipids could be followed by nuclear magnetic resonance spectroscopy.[159] The protein was present at a ratio of about 1 mol (100 mol of phospholipid)⁻¹. Significant increases in the anisotropy at the

ninth and tenth carbons were detected when the protein was added to the bilayer but not at the second carbon. The increase in anisotropy observed, even with the assumption that only 20–30% of the lipid was in the boundary layer, was only about 10–20% for the lipids in these positions. This small increase may indicate that the protein does not force the hydrocarbon in this region to assume conformations much more irregular than those it would normally assume. The deuterium spin–lattice relaxation times, which are measures of the fluidity of the hydrocarbon, were barely altered by the presence of the protein, and no evidence for two separate sets of phospholipids remaining distinct over time intervals greater than 5 msec was observed. All of these results demonstrate that when the lipids in the boundary layer are dioleylphosphatidylcholines rather than nitroxyl phosphatidylcholines **14–16**, negligible restriction on their molecular motion is imposed by membrane-spanning segments of the protein relative to their molecular motion in an unadulterated bilayer. Neither of these probes, however, resembles the heterogeneous mixture of phospholipids in a natural membrane any more closely than the other.

The general problem of identifying within the amino acid sequence of an integral membrane-bound protein, for which a crystallographic molecular model is unavailable, those segments of 20 amino acids in length that span the bilayer as α helices, rather than simply spanning the interior of the globular protein on either side of the bilayer, is supposed to have both a computational solution and an experimental solution. The computational solution[121,160] is based on one or the other of the scales of numerical values for the hydropathies of the amino acids. These values are averaged over the amino acid sequence of a given segment. If the mean numerical value of the hydropathy of the 20 amino acids in a given segment is greater than a certain magnitude, where hydrophobic amino acids have positive values and hydrophilic amino acids have negative values on the scale chosen, then there is a high probability that that segment spans the membrane within the native structure of the folded polypeptide. This approach is based on the assumptions that the change in free energy for the insertion of an α helix into a 3.0-nm layer of hydrocarbon from an aqueous phase is directly related to the partition of model solutes for its amino acid side chains between water and hydrocarbon,[160] that scales of hydropathy regardless of their origin ultimately reflect the free energy of this partition, that the hydrocarbon of the bilayer is more nonpolar than the interior of any protein, and that a longer stretch of polypeptide, 20 amino acids, is required to span a bilayer than is required to span the interior of a molecule of protein.[121] The 11 amino acid sequences known to span the membrane in photosynthetic reaction center (Figure 14–14) were designated as the only membrane-spanning α helices in the native structure by one of these computational algorithms[129–131] before the crystallographic molecular model became available, and every hydrophobic segment of the amino acid sequence of this protein ultimately observed to span the membrane had been so designated. This result might be taken as an indication of the reliability of these predictions.

In another instance, however, an entire set of assignments based on mean hydropathy seems to have failed. Cytochrome P-450 is a membrane-bound protein from the endoplasmic reticulum of liver cells. A cytochrome P-450 is also found in the cytoplasm of a pseudomonad able to metabolize camphor.[161] This bacterial protein is water-soluble and has been crystallized. It is believed that the membrane-bound protein from the liver and the soluble protein from the bacterium share homologies in their amino acid sequences,[162] and hence must have superposable structures. This result and the ease with which the membrane-bound protein can be dissolved in solutions of nonionic detergents and purified as if it were a soluble protein[163] suggest that the protein from the liver is an anchored membrane-bound protein. In fact, the membrane-bound protein from the liver has an extremely hydrophobic segment at its amino terminus, MEFSLLLLLAFLAGLLLLLF–,[164,165] the sequence of which is that of a typical hydrophobic segment in an embedded anchor. It seems that cytochrome P-450 from the liver is a normal soluble protein leashed to the membrane through this hydrophobic anchor. Yet at least one of the computational algorithms designates five other segments of the sequence of this protein as membrane-spanning.[166] If cytochrome P-450 from the liver is an anchored membrane-bound protein rather than an integral membrane-bound protein, these must be simply hydrophobic amino acid sequences that span the soluble, globular domain of the protein and not the bilayer.

The experimental solution to the problem of identifying a membrane-spanning segment relies upon covalent modification of the protein from within the liquid hydrocarbon of the bilayer. Because only poorly nucleophilic amino acids are found within membrane-spanning segments, nitrenes or carbenes have been universally used as reagents for their selective modification. The precursor of a nitrene or carbene is incorporated into a hydrophobic molecule that partitions preferentially into the hydrocarbon of the bilayer of amphipathic lipids surrounding the membrane-spanning segments of the protein. The nitrene or carbene is generated from the precursor by photolysis, and it inserts into the membrane-spanning segments of the polypeptide, albeit in very low yield. The intact polypeptides that are susceptible to the modification can be identified by electrophoresis in solutions of dodecyl sulfate,[167–169] the regions of the polypeptides that have been modified can be identified by isolating and identifying peptides containing them,[170,171] and the particular amino acids modified can be identified by submitting the peptides to sequencing.[172–174a]

The reagents that have been used are precursors of carbenes or nitrenes attached to two different types of carriers. 1-Tritiospiro[adamantane-4,3'-diazirine] (14–18),[167] 5-[125I]iodonaphthyl azide (14–19),[169] and 3-(trifluoromethyl)-3-(m-[125I]iodophenyl)diazirine (14–20)[175] are examples of hydrophobic solutes that can diffuse freely through the liquid hydrocarbon of the bilayer.

14–18   **14–19**   **14–10**

Precursors of carbenes[172,173,176,177] and nitrenes[172,174b] have also been incorporated covalently into phospholipids that can then be incorporated into bilayers of amphipathic lipids surrounding membrane-bound proteins. An example would be diazirinyl phospholipid **14–21**.

14–21

In these derivatives of phospholipids, the precursor of the carbene or the nitrene can be incorporated into the fatty acyl chains, as in diazirinyl phospholipid **14–21**, or it can be incorporated into the hydrophilic functionality esterified to the phosphate of the phospholipid.[177] In the former case, amino acids within the hydrocarbon are modified; and, in the latter case, amino acids at the two ends of the membrane-spanning segments are modified.[173]

Originally it was thought that by varying the position along the hydrocarbon of the fatty acid at which the carbene was located within a phospholipid, amino acids within the membrane-spanning segment that were located at different depths within the bilayer could be identified. Unfortunately, carbenes show a significant preference for insertion into nitrogen–hydrogen and oxygen–hydrogen bonds over carbon–hydrogen bonds, and this preference usually directs all of the carbenes in the liquid hydrocarbon, regardless of their mean depth in the bilayer, to the same one or two most susceptible amino acids in each membrane-spanning segment.[172,173] Therefore, there is no obvious advantage to the derivatives of the phospholipids over the simpler hydrophobic precursors other than their elegance.

Often the incorporation observed with such hydrophobic reagents is consistent with the identification of a membrane-spanning segment based on its mean hydropathy. Both glycophorin,[173] which spans the membrane once, presumably with its only hydrophobic segment, and subunit IV of cytochrome oxidase,[170] which also contains only one hydrophobic segment 20 amino acids in length, have been modified by either a carbene or a nitrene, respectively, incorporated into a phospholipid. The large majority of the incorporation on each case was located in a peptide 68 or 49 amino acids in length, respectively, that contained the hydrophobic segment of 20 amino acids picked out by the computational algorithms. When (iodophenyl)diazirine **14–20** was used to modify bacteriorhodopsin, incorporation also was found to occur in a region of the amino acid sequence containing a segment that had been identified computationally as spanning the membrane.[174b]

When the adamantyldiazirine **14–18** was used to modify Na⁺/K⁺-transporting ATPase, however, and the long tryptic peptides of the intact polypeptide that were modified by the reagent were isolated and identified,[171] it was found that substantial amounts of adamantylidene had been incorporated into three tryptic peptides that did not contain hydrophobic segments designated computationally as membrane-spanning. These three peptides contained segments 20 amino acids in length that were hydrophobic but not sufficiently hydrophobic to be picked out by their mean hydropathy. Their sequences[178] are –VNFPVDNLCFVGLISMIDPP–, –QIGMIQALGGFFTYF-VIMAE–, and –PTWWFCAFPSLLIFVYDEV–.[171] Such segments, if they do span the bilayer, could be centrally located in a large bundle of α helices, and they could be responsible for the ability of this protein to transport sodium and potassium cations across the membrane. Three of the segments, however, of the amino acid sequence of this protein designated as membrane-spanning by their high mean hydropathy were also within tryptic peptides modified by adamantyldiazirine **14–18**.

Both the experience with cytochrome P-450, where hydrophobic segments that do not seem to span the bilayers were designated as hydrophobic enough to do so, and the experience with Na⁺/K⁺-transporting ATPase, where hydrophobic segments that may span the bilayer were not hydrophobic enough to be designated as doing so, suggest that there is no reliable method for making this designation by inspection. Integral membrane-bound proteins may contain a set of membrane-spanning segments whose hydrophobicity is so remarkable that they can be designated as traversing the bilayer without too much doubt. Many integral membrane-bound proteins, however, may also also contain another set of membrane-spanning segments that are hydrophobic but not so hydrophobic as to be distinguishable from those segments elsewhere within the amino sequence that merely span the globular portions of the protein on either side of the membrane. In many instances, it may be these less hydrophobic membrane-spanning segments, impossible to identify by inspection, that are most intimately involved with the function of the integral membrane-bound protein, particularly if it catalyzes the transport of a hydrophilic solute across the bilayer. This makes their identification even more desirable. If one end of such a doubtful segment can be shown to be located on the cytoplasmic surface of the protein and the other end on the extracytoplasmic surface of the protein, then it can be concluded that it spans the bilayer.

The identification of the side of a membrane, cytoplasmic or extracytoplasmic, upon which a particular amino acid or peptide within the sequence of an integral membrane-bound protein is located can be made with oriented, sealed structures and an impermeant reagent. The most reliable oriented, sealed structures are intact cells or intact organelles such as undamaged mitochondria or lysosomes.[179] The intact cells that have been used most frequently for such studies are erythrocytes,[180] which present only the extracytoplasmic surfaces of their membrane-bound proteins to the solution surrounding them. These cells are even more ideal for this purpose because sealed inside-out vesicles that present only the cytoplasmic surfaces of their membrane-bound proteins to the solution can be prepared from intact erythrocytes.[82] Intact animal cells grown in tissue culture[181] or spheroplasts of microorganisms have also been used in such experiments. Mitochondria as they are usually prepared contain both an outer membrane, which is the porous cellular membrane isolating these organelles from direct contact with the cytoplasm, and an inner membrane, which is the boundary of the functional mitochondrion. The outer membrane can be removed[182] to produce sealed, unwrapped mitochondria that present the cytoplasmic surfaces of their membrane-bound proteins to the external solution.[183] Sealed, inside-out vesicles, which present the extracytoplasmic surfaces of their membrane-bound proteins to the solution, can be prepared from unwrapped mitochondria.[184]

Sealed vesicles often form spontaneously from fragments of the constituent membranes during homogenization of a tissue. As these structures are adventitious, they are not necessarily sealed to all hydrophilic solutes. For example, vesicles of plasma membrane can be isolated from the electric organ of *Torpedo californica* that are sealed to large solutes such as proteins,[185] but only a minority of them are sealed to small solutes such as the cations of alkali metals.[186] Occasionally, however, a suspension of homogeneously and tightly sealed vesicles, in which all of the proteins are oriented as they were when the membrane containing them was in the cell, can be prepared from a homogenate.[187]

It is also possible to incorporate purified membrane-bound proteins into vesicles of purified phospholipid.[188] Many different procedures have been developed for such a **reconstitution**, but the goal of each of them, which is realized only imperfectly, is to create a suspension of tightly sealed, unilamellar vesicles of phospholipid in which are embedded the membrane-bound protein as it had been in the original cellular membrane from which it was purified. It is usually the case that during a reconstitution the membrane-bound protein inserts at random in either of the two possible orientations, cytoplasmic surface directed outward or extracytoplasmic surface directed outward. If only one of these two surfaces is susceptible to proteolytic digestion when the protein is in its native structure, as is often the case, digestion of the reconstituted vesicles will nick only those molecules of protein exposing that surface, and intact polypeptides, derived exclusively from molecules of protein inserted in the opposite orientation, can be purified by electrophoresis or molecular exclusion chromatography performed in solutions of dodecyl sulfate.[189,190]

Many impermeant reagents have been used to modify such sealed structures. Because a bilayer contains a continuous sheet of hydrocarbon 3.0 nm wide, charged solutes or solutes with large numbers of donors and acceptors for hydrogen bonds cannot pass through it. An **impermeant reagent** is such a hydrophilic solute that also contains an electrophilic functionality appropriate for the modification of proteins. Diazotized *p*-[35S]sulfanilic acid (**14–22**),[191] *N*-formyl-[35S]sulfinylmethionylmethylphosphate (**14–23**),[192] isethionyl[14C]acetimidate (**14–24**)[193]

14–22   **14–23**   **14–24**

and pyridoxal phosphate[194] and sodium borohydride (Figure 10–3) are impermeant reagents that have been used to modify only the surface of a protein presented to the external solution in a suspension of sealed membranes. For example, both intact mitochondria and sealed inside-out vesicles of mitochondria were separately modified with pyridoxal phosphate and sodium borohydride, and labeled ATP–ADP carrier, an integral membrane-bound protein, was purified from each sample. Thermolytic peptides containing the labeled lysines were isolated from the protein and identified by their amino-terminal sequences. It was found that Lysine 146 was modified by pyridoxal phosphate and sodium borohydride in the protein from the sealed inside-out vesicles, while Lysines 95, 198, 205, 259, and 267 were modified by pyridoxal phosphate and sodium borohydride in the protein from intact mitochondria.[195] Lysine 146 was assigned to the extracytoplasmic surface of the protein; and the other lysines, to the cytoplasmic surface.

Cytochrome oxidase, another integral membrane-bound protein, was isolated by immunoprecipitation from intact mitochondria or sealed inside-out vesicles of mitochondria, both of which had been modified separately by diazotized *p*-[35S]sulfanilic acid. Only subunit II of cytochrome oxidase incorporated radioactivity when intact mitocondria were labeled, but subunits II, III, IV, V, and VII all incorporated radioactivity when sealed inside-out vesicles were labeled.[196] These results demonstrate that subunit II spans the membrane while subunits III, IV, V, and VII present surfaces to the extracytoplasmic space within the mitochondrion.

The tryptic peptide HLLVMKGAPER, the amino acid sequence containing Lysine 501 from Na+/K+-transporting adenosinetriphosphatase, could be isolated from digests of the intact protein by immuno-adsorption with immunoglobulins G directed against its carboxy terminus. Lysine 501 would not incorporate pyridoxal phosphate when the protein was present in sealed vesicles that presented only the extracytoplasmic surface of the protein to the solution but readily incorporated pyridoxal phosphate when the vesicles were opened by adding the surfactant saponin.[197] Saponin, by combining with the cholesterol in the bilayer, is able to form large (8.0-nm) holes in natural membranes[198] without significantly altering the membrane-bound proteins. The results of these experiments demonstrated that Lysine 501 is located on the cytoplasmic surface of sodium and potassium ion-activated adenosinetriphosphatase.

Molecules of soluble proteins such as proteinases or immunoglobulins G are also impermeant reagents. Each of the proteinases pronase,[199] chymotrypsin,[200] and papain[201] is able to cleave native membrane-bound anion carrier (n_{aa} = 911),[202] an integral membrane-spanning protein of human erythrocytes, within the short segment of the polypeptide, –QDHPLQKTYNYNVLMVPKPW-QGPLP–, between Glutamine 545 and Proline 568. Papain cleaves after Glutamine 550, and chymotrypsin cleaves after Tyrosine 553.[202] These cleavages occur quantitatively when any one of the proteinases is added to the extracytoplasmic solution in which intact erythrocytes are suspended. These results demonstrate that this short segment of amino acids is fully exposed on the extracytoplasmic surface of the intact protein.

Two different monoclonal immunoglobulins G raised against purified acetylcholine receptor, an integral membrane-bound protein from the electric organ of *T. californica*, recognize as an antigen the synthetic peptide KAEEYILKKPRSELMFEEC, which has an amino acid sequence from the interior of one of the polypeptides composing the protein. Presumably, the natural antigenic sites on the intact protein are composed of sequences from this region. It could be shown that these monoclonal immunoglobulins G were bound only at the cytoplasmic surfaces of membranes containing this protein.[203] From this result, it was concluded that this sequence in the native structure of acetylcholine receptor is exposed on the cytoplasmic surface of the protein.

The enzyme transglutaminase (Reaction 13–36) has also been used to catalyze the modification of exposed glutamines on the surfaces of membrane-bound proteins with fluorescent primary amines.[204]

The assumption behind all of these experiments is that membrane-bound proteins are inserted into natural membranes such that every copy of the same protein is oriented in the same direction relative to the cytoplasm. The earliest observations addressing this point explicitly confirmed this assumption of **vectorial insertion**.[80,180,205] For example, nucleophilic amino acids in 10 of the thermolytic peptides on the peptide map of a digest of anion carrier from erythrocytes could not be modified with methionine sulfone **14–23** when the native protein was in intact sealed erythrocytes, even though they could be readily modified when the erythrocytes were broken open.[180] The explanation of this observation is that every

copy of anion carrier is oriented the same way in the membrane, each presenting the same unique surface to the cytoplasmic space of the cell, and that surface is inaccessible to the reagent in an intact cell. Since these early studies, many examples of vectorial insertion have been verified, and no example of a membrane-bound protein whose copies are oriented at random in a natural membrane has been described.

A property related to the vectorial insertion of every protein in a biological membrane is the asymmetric distribution of the polysaccharides on membrane-bound **glycoproteins** and **glycolipids**. Almost all[206] of the saccharide bound to the plasma membrane of an animal cell is located upon its extracytoplasmic surface.[207,208] This feature is a corollary of the fact that almost no[209] glycoproteins are found in cytoplasm, only in extracytoplasmic spaces.

Many examples of **membrane-bound oligomeric proteins** have been observed. Cytochrome oxidase is the paradigm of an oligomeric membrane-bound protein constructed from several unrelated polypeptides. It is composed of at least seven[196] polypeptides varying in length from 510[210] to less than 100 amino acids.[211] Each is separately translated, and the entire complex assembles from one copy of each of the separate subunits. It is, however, when the subunits of an oligomeric membrane-bound protein are identical in sequence, and hence in structure, or homologous in sequence, and hence folded into superposable structures, that **rotational symmetry** comes into play.

Oligomeric proteins constructed from identical subunits or superposable subunits, with the educational exception of hexokinase (Figure 9–2), always incorporate rotational axes of symmetry or pseudosymmetry, respectively, into their structures. In a bilayer of amphipathic lipids, all of the folded polypeptides either of an oligomeric, anchored membrane-bound protein or of an oligomeric, integral membrane-bound protein with the same or homologous amino acid sequences are inserted so that they point in the same direction. Because the same or the homologous hydrophobic segments of their common or homologous sequence span the bilayer, all folded polypeptides of the same sequence or of homologous sequences float at the same depth and have the same orientation. These inescapable requirements placed upon the common structure of the subunits of oligomeric membrane-bound proteins force any rotational axis of symmetry or pseudosymmetry relating the individual subunits in the protein to be normal to the plane of the bilayer. Therefore, a membrane-bound oligomeric protein can have only one rotational axis of symmetry or pseudosymmetry, and that axis will be normal to the plane of the membrane.

Closed structures with rotational axes of symmetry are even more exclusive necessities for oligomeric membrane-bound proteins than they are for soluble proteins. A screw axis of symmetry is incompatible with the vectorial two-dimensional distribution of identical or homologous subunits enforced by the bilayer, and helical polymeric fibers are not available structures. The only interfaces that could propagate a polymer of indefinite length in a membrane would have to form from complementary faces arrayed at precisely 180° across from each other on each subunit, or the row of subunits would eventually come around upon itself to form a broken or unbroken ring. During evolution by natural selection, therefore, every time two complementary faces appear at random on the surface of one subunit of a membrane-bound protein such that they can produce a series of interfaces joining several of the subunits in an oligomer, either an incomplete ring or a complete ring of an integral number of subunits will always form. A complete ring, because no pair of complementary faces remains unassociated, is a more stable structure than an incomplete ring. If the angle between the two complementary faces on a single subunit is an integral quotient of 360°, where the integer is greater than or equal to 3, complete rings containing that number of subunits will form. Furthermore, as the bilayer orients the subunits in the normal direction and provides structural support, single rings with a larger number of subunits than those seen with soluble proteins are possible. Because only one rotational axis of symmetry normal to the bilayer is available, magic numbers such as four or six, applicable to soluble oligomeric proteins having sets of perpendicular rotational axes of symmetry, are irrelevant to membrane-bound oligomeric proteins. For example, unlike a soluble protein such as hemoglobin, a membrane-bound protein cannot be tetrameric under one set of conditions and dimeric under another,[212] nor can it even be a tetramer like that of hemoglobin.

It is not surprising that the oligomeric membrane-bound proteins whose structures have been directly observed are all assembled around single rotational axes of symmetry normal to the plane of the bilayer. The distinction between anchored and integral membrane-bound proteins is irrelevant because both are constrained by the same requirements. Thus both bacteriorhodopsin (Figure 14–21),[136] an integral membrane-bound protein, and the hemagglutinin of influenza virus,[108] an anchored membrane-bound protein, have three identical subunits arrayed around a 3-fold rotational axis of symmetry normal to the plane of the membrane. Photosynthetic reaction center, an integral membrane-bound protein containing two different polypeptides with homologous amino acid sequences,[129,130] has those two folded polypeptides arrayed around a 2-fold rotational axis of pseudosymmetry.[127] Neuraminidase of influenza virus, an anchored membrane-bound protein, has four identical subunits arrayed around a 4-fold rotational axis of symmetry.[213] Acetylcholine receptor, an integral membrane-bound protein, has five subunits of homologous amino acid sequence arrayed around a 5-fold rotational axis of pseudosymmetry.[214] Gap junction connexon, an integral membrane-bound protein, has six identical subunits arrayed around a 6-fold rotational axis of symmetry.[215]

It is of interest that each gap junction connexon, which is a ring of six subunits in the plasma membrane of an animal cell, is associated by a 2-fold rotational axis of symmetry to another gap junction connexon in the membrane of an adjacent cell to produce a dodecamer that is a dimer of hexamers (Figure 9–15), one hexamer in each plasma membrane; and, within the plasma membrane of a given cell, the gap junction connexons are in crystalline arrays of indefinite surface area.[216] Another hexamer of identical subunits arrayed around a 6-fold rotational axis of symmetry normal to the plane of the bilayer[217] is the major constituent of the luminal plasma membrane of urinary bladder, and this protein is also present in the cell in a crystalline array.[135] These observations suggest that an unsupported hexamer may be too flexible to be stable on its own. Seven-fold or higher rotational axes of symmetry should be as probable in membrane-bound proteins as lower multiplicities, but they have not yet been observed. They may be too flexible in isolation to be stable, while they are not permitted by their symmetry to crystallize and form stable arrays.

Now that the properties of membrane-spanning α helices, lipid in the boundary layer, cytoplasmic and extracytoplasmic surfaces, and rotational symmetry have been discussed, the structures of particular integral membrane-bound proteins can be appreciated. Anion carrier, cytochrome oxidase, acetylcholine receptor, and proteins catalyzing active transport can serve as typical examples.

Anion carrier is an integral membrane-bound protein in the plasma membrane of the erythrocyte that is responsible for the transport of anions such as chloride, bicarbonate, or phosphate,[218] and it is the integral membrane-bound protein present in the highest concentration in this plasma membrane. It is composed of a single polypeptide 840–930 amino acids long[202,219] that bears covalently attached carbohydrate[218] and spans the bilayer in its native structure.[180] Human anion carrier ($n_{aa} = 911$) has a detachable domain on the cytoplasmic side of the protein that can be released by trypsin from fragments of membrane[220] and that constitutes the first 360 amino acids from the amino terminus of the folded polypeptide.[202] After cleavage from the membrane, the detached domain is freely water-soluble. It is an α_2 dimer,[221] which may explain why the entire protein is a dimer when dissolved from the membrane with nonionic detergent.[222] In the membrane, the 2-fold rotational axis relating the two identical subunits would be normal to the plane of the bilayer.

The domain 550 amino acids long, left in the membrane after the amino-terminal domain has been detached by proteolytic digestion and washed away, is still able to transport anions as rapidly as does the intact protein.[223] Its amino acid sequence contains at least 13 individual hydrophobic segments, 20 amino acids in length, that are

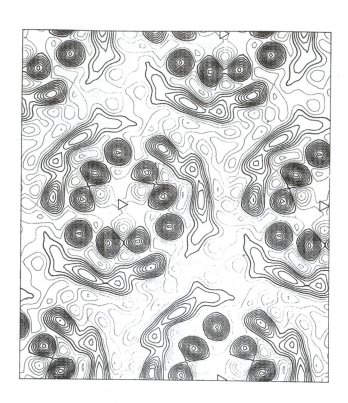

Figure 14–21: Projection of the map of electron scattering density of bacteriorhodopsin onto the plane of the membrane.[136] Amplitudes of the reflections were obtained from an electron diffraction pattern from purple membrane embedded in glucose (Figure 14–16), and phases were obtained from Fourier transforms of a digitized electron micrograph of an identical specimen. In each case, the specimens were oriented so that the plane of the membrane was normal to the beam of electrons. As a result, the Fourier transform of the amplitudes and phases of the indexed reflections is the projection of the electron scattering density on the plane of the membrane. Amplitudes and phases of the reflections arising from sets of parallel lines with spacings as small as 0.7 nm were included in the Fourier transform. The projection is that of the structure with the same resolution as the one in Figure 14–18A. The individual molecules of bacteriorhodopsin are arranged about 3-fold rotational axes of symmetry normal to the plane of the membrane (triangles) to form trimers in the membrane. The view is in the same orientation as that in Figure 14–18B, but Figure 14–18B is at higher resolution and is a slice through a three-dimensional map rather than a projection. Reprinted with permission from ref 136. Copyright 1975 Academic Press.

candidates for spanning the membrane (Figure 14–22).[219] The formal amino terminus of this embedded domain in the human protein,[224] Glycine 361, must be on the cytoplasmic surface of the protein because cleavage by trypsin at the α-amide of Glycine 361 to release the detachable domain occurs at that surface. Lysine 430 can be modified from the extracytoplasmic space by formylation followed by reduction with sodium borohydride, which, under the appropriate conditions, is impermeant to intact erythrocytes.[225] Therefore, the hydrophobic segment between Glutamine 404 and Glycine 428 (segment *a* in Figure 14–22) must span the membrane.

In the region between Lysine 542 and Proline 568 (between segments *d* and *e* in Figure 14–22) there is a loop of polypeptide in the native structure of the protein that is susceptible to cleavage by pronase,[199] chymotrypsin,[226] pepsin,[227] and papain[201] from the extracellular surface. Therefore, the polypeptide must span the membrane, if it does at all, an even number of times between Lysine 430 and Lysine 542. There are four hydrophobic segments 20 amino acids in length in this region (segments *b–d* in Figure 14–22).

Tyrosine 596 is not iodinated by iodide and lactoperoxidase when the lactoperoxidase is present only at the extracytoplasmic side of the membrane, and Asparagine 593 does not bear an *N*-linked polysaccharide, even though it is in the proper sequence to do so. These negative results suggest that this part of the amino acid sequence is on the cytoplasmic surface of the membrane and that the hydrophobic segment between Asparagine 569 and Arginine 589 (segment *e* in Figure 14–22) spans the membrane.[227]

Tyrosine 682 is iodinated in the presence of iodide and extracytoplasmic lactoperoxidase,[227] and the peptide bond between Threonine 629 and Glutamine 630 is cleaved by extracytoplasmic papain[201] in intact erythrocytes. Therefore, if segment *e* spans the membrane, the hydrophobic segment between Lysine 600 and Aspartate 621 (segment *f* in Figure 14–22) should also span the membrane.

Trypsin can cleave the polypeptide in the native protein at Lysine 743, but only from the cytoplasmic surface.[228] Therefore, only one of the two hydrophobic segments (segment *g* or segment *h* in Figure 14–22) between Glutamine 630 and Lysine 743 can span the membrane. The carboxy terminus of anion carrier when it is in inside-out vesicles but not when it is in right-side-out vesicles can bind an immunoglobulin raised against its amino acid sequence.[229] Therefore, the polypeptide must span the membrane an even number of times or not at all between Lysine 743 and the carboxy terminus. There are four potential membrane-spanning segments (in segments *i* and *j* in Figure 14–22) between these two points.

There are several candidates for the role of spanning the membrane still to be examined. Nevertheless, these experiments provide an example of the systematic use of impermeant reagents for assessing the topology of the polypeptide in an integral membrane-bound protein.

There are seven integral membrane-bound proteins that catalyze the active transport of inorganic cations across cellular membranes at the expense of the hydrolysis of MgATP. These are Na⁺/K⁺-transporting adenosine-triphosphatase (Na⁺/K⁺-ATPase) from animal plasma membranes, Ca²⁺-transporting ATPase (ER Ca²⁺-ATPase)

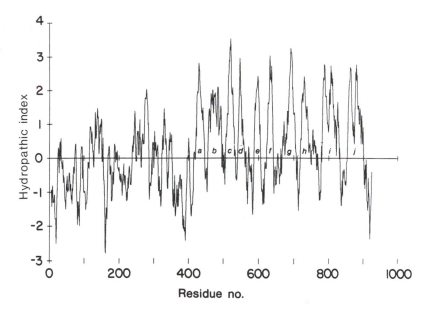

Figure 14–22: Plot of the distribution of hydropathy over the amino acid sequence of murine anion carrier (n_{aa} = 929).[219] Each amino acid in the sequence is assigned its numerical value of hydropathy in the scale of Kyte and Doolittle.[121] A moving average with a span of 11 amino acids is calculated from this sequence of numbers. The numerical value (hydropathic index) assigned to each residue in the amino acid sequence is the average of the segment of 11 amino acids of which it is the central amino acid. Positive values indicate hydrophobic locations; negative values, hydrophilic locations. Ten long hydrophobic segments are found within the last 520 amino acids of the sequence (*a–j*). Segments *b*, *i*, and *j* are long enough (> 40 amino acids) to contain two membrane-spanning segments. The amino-terminal cytoplasmic domain in the murine protein is 15 amino acids longer than the one in the human protein, so each potential membrane-spanning segment is 15–20 amino acids farther along in the sequence of the murine protein. Adapted with permission from *Nature*, ref 219. Copyright 1985 Macmillan Magazines Limited.

from animal endoplasmic reticulum, calmodulin-regulated Ca^{2+}-ATPase from animal plasma membranes, H^+/K^+-transporting ATPase from the luminal plasma membranes of gastric mucosa, K^+-transporting ATPase from bacterial plasma membranes, H^+-transporting ATPase from fungal plasma membranes, and H^+-transporting ATPase from plant plasma membranes. Each of these seven proteins has a long polypeptide, designated the α polypeptide, which is responsible for the catalysis of active transport. All of the seven α polypeptides are homologous in sequence,[230–234] and therefore each must fold to assume a native structure that is superposable upon the native structure of all of the others. Renal Na^+/K^+-ATPase and gastric H^+/K^+-transporting ATPase are constructed from two polypeptides present in equimolar ratio,[235] but the smaller of the two, β (n_{aa} = 300),[236,237] does not seem to participate in the catalysis of the active transport of Na^+ and K^+ or of H^+ and K^+ because several of the other members of the superfamily lack a β subunit and yet still transport cations effectively. The larger polypeptide, α, from each of these proteins is the functional unit homologous to the single polypeptides present in the other proteins.

Both Na^+/K^+-ATPase[238,239] and ER Ca^{2+}-ATPase[240] have been induced to crystallize in two dimensions. Both proteins crystallize in lattices of each the *P*1 or the *P*2 space group. In the *P*1 space group,[239] each unit cell contains an $\alpha\beta$ protomer or an α monomer, respectively, each representing a single folded α polypeptide. Because both proteins can crystallize within the membrane as monomers, it necessarily follows that these proteins are monomers in the normal uncrystalline state and that the $(\alpha\beta)_2$ heterotetramers or α_2 dimers present in the P2 crystals form during crystallization. This follows from the fact that oligomeric proteins do not dissociate during

crystallization but often associate to produce unit cells of higher symmetry than the proteins possessed in solution. The reason for this is that crystallization is necessarily performed under conditions promoting association. Both the $\alpha\beta$ protomer of Na^+/K^+-ATPase[241] and the α monomer of Ca^{2+}-ATPase[242] have full enzymatic activity, observations also suggesting that the proteins are an $\alpha\beta$ protomer within the plasma membrane and an α monomer within the endoplasmic reticulum, respectively.

Low-resolution three-dimensional maps of electron scattering density have been produced by analyzing two-dimensional crystals of ER Ca^{2+}-ATPase[243] and Na^+/K^+-ATPase.[244] In both cases the majority of the mass of the protein is located on the cytoplasmic surface of the membrane. Because the polypeptide composing these proteins is so long, the globular portion on the cytoplasmic side of the membrane outside of the bilayer proper is quite large, about 6 nm in diameter.[239] The amount of the protein that is within the bilayer, however, cannot be determined from these images of very low resolution.

Some of the hydrophobic segments in the amino acid sequence of the α polypeptides of Na^+/K^+-ATPase and Ca^{2+}-ATPase that span the bilayer have also been identified. There are 11 hydrophobic segments in the amino acid sequence of the α polypeptide of Na^+/K^+-ATPase that are candidates for spanning the membrane, either because they have high mean hydropathies or because they have been labeled with hydrophobic reagents that modify proteins from within the bilayer.[245] These 11 segments all have their homologues in the amino acid sequences of the other proteins catalyzing active transport.[231,246] Because all of these proteins share a common ancestor, decisions about the disposition of one of these hydrophobic segments in one of the seven proteins will be applicable to the homologous hydrophobic segments in all of the proteins.

The first two hydrophobic segments in the α polypeptide of Na$^+$/K$^+$-ATPase, between Glutamine 88 and Glutamine 111 and between Asparagine 122 and Serine 140, together form a hairpin of α helices that spans the membrane twice. This follows from the facts that [^{125}I]iodonaphthyl azide labels heavily[247] a portion of the α polypeptide lying on the amino-terminal side of a tryptic cleavage occurring at Arginine 262 and that both this tryptic cleavage and another tryptic cleavage at Lysine 30 can take place only when trypsin has access to the cytoplasmic surface of the protein.[248] Trypsin is also able to cleave Ca^{2+}-ATPase from the cytoplasmic surface of intact endoplasmic reticulum,[249] at the arginine of its polypeptide[246] in a position in its amino acid sequence homologous to Asparagine 228 in the amino acid sequence of the α polypeptide of Na$^+$/K$^+$-ATPase.[250] The amino-terminal fragment generated by this cleavage is modified by a hydrophobic nitrene precursor when Ca^{2+}-ATPase is in the membrane in its native structure.[251]

The third and fourth hydrophobic segments, between Histidine 283 and Glutamate 307 and between Glutamate 312 and Lysine 342, in the amino acid sequence of the α polypeptide of Na$^+$/K$^+$-ATPase together seem to form another hairpin that spans the membrane twice. This follows from the facts that these two segments are present in a large tryptic peptide labeled with adamantanyl-diazirine 14–18,[171] that they both have a very high mean hydropathy, and that Aspartate 369 is located in the active site of the protein[252] on its cytoplasmic surface. Therefore, at least four of the 11 candidates for spanning the membrane do so. The carboxy termini of H$^+$-transporting ATPase from fungi[253] and calmodulin-regulated Ca^{2+}-ATPase from animal plasma membranes[254] are on the cytoplasmic surfaces of these proteins, so the carboxy termini of all of the members of the superfamily must be cytoplasmic. As the amino terminus is on the cytoplasmic surface, these proteins can span the membrane only an even number of times, and the membrane-spanning portion of the transport ATPases should be a bundle of 4–10 α helices.

If there were seven membrane-spanning α helices bundled as are the α helices in bacteriorhodopsin (Figure 14–18), the cross section of the membrane-spanning portion of the transport ATPases would resemble the projection of the bundle of α helices in bacteriorhodopsin (Figure 14–21). Each phospholipid in one monolayer of a bilayer of natural phosphatidylcholine contributes 0.7 nm^2 to its surface area, which is 0.35 nm^2 for each of its two linear hydrocarbons. This cross section for each hydrocarbon can be represented by a circle 0.64 nm in diameter, which would form a hexagonal array of circles that has an area of 0.35 nm^2 for each circle. Twenty circles of this diameter are able to surround the perimeter of the projection of the bundle of seven α helices of bacteriorhodopsin (Figure 14–23B). If there were only six α helices spanning the bilayer in Ca^{2+}-ATPase, about 19

such circles could fit around the perimeter of its projection (Figure 14–23A); and if there were 10 α helices spanning the bilayer, about 23 such circles could fit around the perimeter of its projection (Figure 14–23C). These geometric estimates agree satisfactorily with the observed values of 22 mol of lipid in the boundary layer (mol of Ca^{2+}-ATPase)$^{-1}$ in bilayers of natural phosphatidylcholine.[155,156] Each phosphatidylcholine contributes two hydrocarbons, but there are two monolayers for each bilayer. Therefore, the number of boundary lipids measured agrees with a structure passing through the bilayer composed of 6–10 α helices.

It has been estimated from the behavior of cyclic nitroxides incorporated into fatty acids that 36 of the phospholipids around an $\alpha\beta$ protomer of Na$^+$/K$^+$-ATPase serve as lipid in the boundary layer.[255] The somewhat larger value for this protein is thought to reflect the presence of the β polypeptide, which has one hydrophobic segment in its amino acid sequence[236] that probably spans the membrane in addition to the 6–10 segments contributed by the α polypeptide.

Cytochrome oxidase is an integral membrane-bound protein located in the mitochondrial membrane. It transfers electrons from cytochrome c to molecular oxygen and at the same time catalyzes the transport of protons. The protein is composed of at least seven different polypeptides,[256,257] referred to as subunits. If they are present in equimolar ratio, the complete protein should contain a total of about 1500 amino acids.

The structure of the protein has been probed both with phospholipids containing precursors of nitrenes and with impermeant reagents. In its purified form, cytochrome oxidase is present in open fragments of membrane, but it can be reconstituted into sealed vesicles of phospholipid[258] in which almost all (> 80%) of the molecules of protein are oriented with their cytoplasmic faces to the external solution.[259] Unwrapped mitochondria, in which the cytoplasmic face of the protein is presented to the external solution, or inside-out vesicles of mitochondria, in which the extracytoplasmic face of the protein is presented to the external solution, can also be used in these experiments. The labeled protein can be purified by immunoprecipitation,[256] or the individual subunits can be identified immunochemically within the electrophoretic catalogue on a polyacrylamide gel of mitochondria dissolved in dodecyl sulfate.[257]

Subunit I of cytochrome oxidase is the longest ($n_{aa} = 510$)[260] and most hydrophobic of the polypeptides composing the protein. Its amino acid sequence contains at least nine hydrophobic segments,[121] and the polypeptide in the native protein is heavily labeled by phospholipids containing nitrene precursors.[261] It is poorly labeled by charged impermeant reagents but can be labeled from its cytoplasmic surface by diazotized p-[^{35}S]sulfanilic acid in sealed reconstituted vesicles.[259]

Subunit II of cytochrome oxidase ($n_{aa} = 230$)[260] has two hydrophobic segments in the 57 amino acids be-

B   A   C

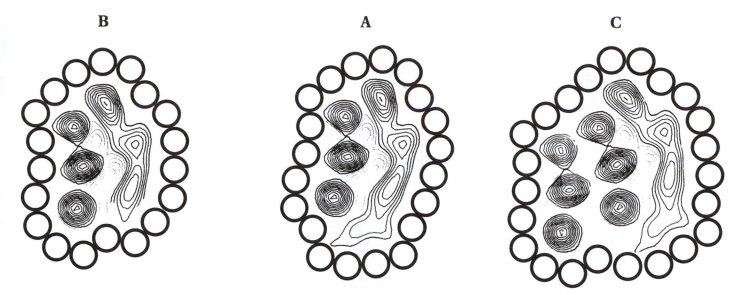

Figure 14–23: Spacing of lipid in the boundary layer around bundles of α helices. (A) The projection of a single molecule of bacteriorhodopsin on the plane of the membrane (Figure 14–21) can be surrounded by 20 circles 0.64 nm in diameter, each repre- senting the cross section of a linear alkane of a phospholipid. (B) If one α helix is removed from the projection, the number of circles accommodated is only 19. (C) If three α helices are added to the projection, 23 circles are required to surround it.

tween Histidine 24 and Arginine 82. This region, includ- ing Histidine 24, could be modified by a phospholipid that contains a nitrene precursor directly attached to carbon 2 of its *sn*-glycerol 3-phosphate.[262] Therefore, at least the first hydrophobic segment spans the mem- brane, and it has been decided that the other does also.[262] Almost all of the mass of subunit II, however, would be on only one surface of cytochrome oxidase.[259] Nevertheless, subunit II has amino acids available for modification by diazotized p-[35S]sulfanilic acid both from the cytoplasmic surface of unwrapped mitochondria[256] and reconstituted vesicles[259] and from the extracytoplasmic surface of in- side-out vesicles of mitochondria.[256] It is also accessible to digestion by both pronase and subtilisin in both un- wrapped mitochondria and inside-out vesicles of mito- chondria.[257] If both hydrophobic segments do span the bilayer as a hairpin, and they are the only membrane- spanning segments, the short segment between Lysine 47 and Glutamate 58 must account for all of the availabil- ity of this subunit on one of the two surfaces of the membrane.

Subunit III of cytochrome oxidase ($n_{aa} = 260$)[260] has at least three hydrophobic segments.[121] Because chymo- trypsin could cleave the polypeptide in a region between Tryptophan 34 and Phenylalanine 37 as well as in a region between Tryptophan 99 and Tryptophan 116 when it had access only to the cytoplasmic surface of cyto- chrome oxidase in reconstituted vesicles,[259] it has been concluded that the protein must span the membrane not at all or twice between these two points. There is, how- ever, only one obvious hydrophobic segment in this interval, that between Arginine 79 and Histidine 103. Subunit III is heavily labeled when cytochrome oxidase

in fragments of membrane is modified with a phospho- lipid bearing a nitrene precursor.[261] Glutamate 90, within the hydrophobic segment between Arginine 79 and His- tidine 103, is modified by dicyclohexylcarbodiimide,[263] which is a hydrophobic carbodiimide, and this result suggests that this hydrophobic segment is in the bilayer. The polypeptide of subunit III is accessible at the extracytoplasmic surface of inside-out vesicles of mito- chondria to diazotized p-[35S]sulfanilic acid[256] and at both the extracytoplasmic surface of inside-out vesicles and the cytoplasmic surface of unwrapped mitochondria to cleavage by pronase and subtilisin.[257] From all of these observations it has been concluded that subunit III spans the membrane several times.

Subunit IV ($n_{aa} = 150$),[264] however, seems to span the bilayer only once.[170] A hydrophobic segment is located between Threonine 80 and Tryptophan 98. This segment is within the amino acid sequence of a large tryptic fragment, 49 amino acids in length. This peptide contains all of the labeling when cytochrome oxidase in fragments of membrane is modified with a phospholipid containing a nitrene precursor.[170] No other prominent hydrophobic segment is present in the amino acid sequence. The peptide bond of Lysine 7 of subunit IV is susceptible to cleavage when intact cytochrome oxidase is digested with trypsin from its extracytoplasmic surface in inside- out vesicles of mitochondria[265] and only when sealed vesicles, in which the protein is oriented with its cyto- plasmic surface outward, are opened with nonionic detergent.[259] The polypeptide of subunit IV is also sus- ceptible to digestion by pronase in unwrapped mito- chondria,[257] which only expose the cytoplasmic surface of cytochrome oxidase. Both of these results together

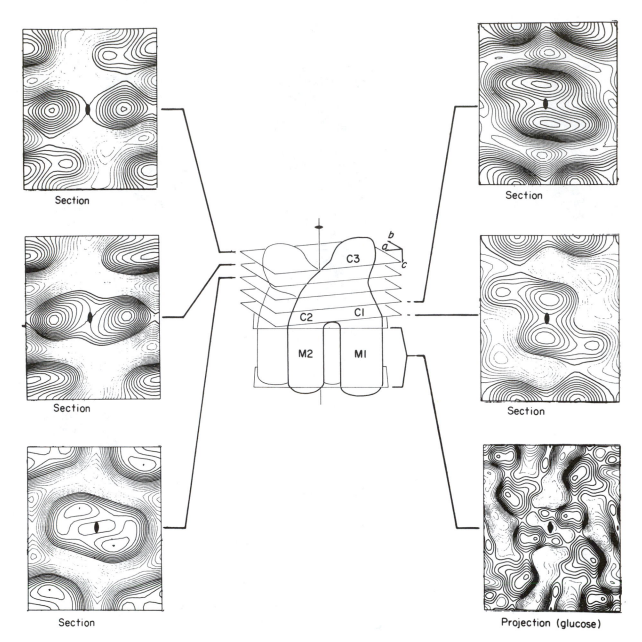

Figure 14–24: Low-resolution map of electron scattering density of cytochrome oxidase.[268] Fragments of natural mitochondrial membranes, in which the only protein was cytochrome oxidase, were embedded in a glass of uranyl acetate (Figure 14–15) or a glass of glucose on grids for electron microscopy. Fragments containing crystalline arrays were chosen for observation, and a series of micrographs were taken at various angles of tilt. The crystalline areas on the electron micrographs were divided into 20-μm squares and the mean absorbance in each square was determined. The digitized images that resulted were submitted to Fourier transformation, and the amplitudes and phases of the indexed maxima in these transforms were tabulated. The variations in amplitude and phase along the lattice lines of the complete Fourier transform were estimated from the collection of Fourier transforms of the images at various angles of tilt (Figure 14–17). From these variations, the distribution of electron scattering density through a unit cell could be calculated by Fourier transformation. Five sections through that map of electron scattering density on the outside of the bilayer are presented, each parallel to the plane of the bilayer. The distances separating the successive planes are each 1 nm, and the position of the sections in the overall molecule is indicated in the central drawing. The projection of the electron scattering density within the bilayer is also presented. In this crystal form, the molecule of cytochrome oxidase is a dimer, through the center of which passes a crystallographic 2-fold rotational axis of symmetry relating the two protomers. Each protomer is constructed from about seven different polypeptides. Reprinted with permission from ref 268. Copyright 1982 Academic Press.

demonstrate that subunit IV does span the membrane in cytochrome oxidase, with its amino terminus on the extracytoplasmic side.

Subunit V (n_{aa} = 110)[266] is a polypeptide containing no hydrophobic segments. It is labeled by neither 5-[^{125}I]iodonaphthyl azide[267] nor phospholipids containing precursors of nitrenes,[261] and it does not seem to span the bilayer. The various additional subunits of cytochrome oxidase are all equal to or smaller than subunit V in length. Each contains one or two hydrophobic segments, but it is unclear which of these short polypeptides are subunits of the protein and which are merely carried along with it as contaminants during its purification.

Two-dimensional crystals of cytochrome oxidase have been prepared, and from micrographs of tilted specimens a three-dimensional map of scattering density for the protein has been computed (Figure 14–24).[268] The map for the portions of the protein above the bilayer in the figure was computed from images of the protein embedded in uranyl acetate, and the projection of the portions of cytochrome oxidase within the bilayer was computed from images of the protein embedded in a glass of glucose. The space group was $P22_12_1$ in these crystals, and the protein was present as a dimer arranged around a crystallographic 2-fold rotational axis of symmetry. In another crystalline form, the protein crystallizes in the $P12_1$ space group[269] as single monomers of the same structure as the monomers in Figure 14–24.

Cytochrome oxidase is peculiar in that it appears to have two independent membrane-spanning bundles (M1 and M2 in Figure 14–24), one about half the volume of the other. There are about 16 unique hydrophobic segments that are candidates for spanning the membrane within the amino acid sequences of the subunits.[121] If bundles M1 and M2 had 10 and 6 membrane-spanning α helices, respectively, packed as are the helices in bacteriorhodopsin (Figure 14–23) but into shapes resembling those in the projection for cytochrome oxidase within the membrane (Figure 14–24), their perimeters would support about 19 and 23 linear hydrocarbons, respectively, as lipids in the boundary layer in each monolayer (Figure 14–23A,C). Accordingly, the two monolayers of the bilayer should contain about 43 mol of phospholipid in the boundary layer (mol of cytochrome oxidase).$^{-1}$ This estimate compares favorably with the 47 ± 5 mol lipid in the boundary layer (mol of cytochrome oxidase)$^{-1}$ observed with nitroxyl phosphatidylcholine **14–16** as the probe in reconstituted vesicles containing increasing ratios of dioleoylphosphatidylcholine competing for boundary positions.[157]

Acetylcholine receptor is a membrane-spanning protein found at the neuromuscular junction. It is responsible for the postsynaptic currents of sodium and potassium that arise upon the release of acetylcholine into the synapse between nerve and muscle. Although it was first isolated from the electric organ of the electric eel,[270] another rich source of acetylcholine receptor is the electric organ of the torpedo ray.[271,272] The protein purified from this source is homologous in its sequence to the protein present at the neuromuscular junctions in mammals,[273] so there is no doubt that the elasmobranch and mammalian proteins are structurally indistinguishable.

Acetylcholine receptor is constructed from only[274,275] four unique polypeptides,[271,272] designated α, β, γ, and δ on the basis of their electrophoretic mobilities. All four polypeptides are glycoproteins,[276,277] and it is assumed that the carbohydrate is attached to their extracytoplasmic surfaces as is normally the case. The α polypeptide from *T. californica* is 437 residues long; the β polypeptide, 469 residues long; the γ polypeptide, 489 residues long; and the δ polypeptide, 501 residues long.[278–280] The molar ratio of the polypeptides in acetylcholine receptor is $\alpha_2\beta\gamma\delta$.[281]

Probably the most important observation pertinent to the molecular structure of this protein is that all four polypeptides are homologous in sequence.[282] Although this was originally established from the amino-terminal sequences of the separated polypeptides, it has been confirmed and amplified by the elucidation of the complete sequence of each.[278–280] All four can be readily aligned, and in the six pairwise comparisons the percent identity averages around 40%.[279] There is no question that these four proteins were derived from a common ancestor and that all four of them assume the same unique superposable tertiary structure upon folding. Any structural feature that is established for one of them must be true for all of them.

This consideration and an understanding of the constraints placed on assembling a homopolymer within a membrane require that acetylcholine receptor be a pentamer whose five superposable subunits would be disposed around a 5-fold rotational axis of pseudosymmetry normal to the plane of the bilayer. This expectation has been confirmed by a determination of the three-dimensional structure of acetylcholine receptor by Fourier analysis of electron micrographs of the protein embedded in amorphous ice (Figure 14–25).[214,283] The pentameric toroid is shaped like a pentagonally faceted shot glass, with its mouth directed extracytoplasmically. The cylindrical shot glass has an outer diameter of 8 nm and an inner diameter of 3 nm and is 8 nm deep. The bottom of the glass is about 5 nm thick, and there is a depression on the underside of the bottom 3 nm wide and about 2 nm deep. The 3 nm between the bottom of the inside of the glass and the top of the lower depression is a featureless solid that must contain the channel that opens to permit the ions to pass across the membrane.

The first 211–225 amino acids from the amino terminus of each polypeptide are located on the extracytoplasmic surface of the protein and form most if not all of the wall of the shot glass. These regions are homologous in length as well as in sequence, and this explains the uniformity of the mouth. Beyond 211–225 amino acids in the respective sequences, the first potentially membrane-

A

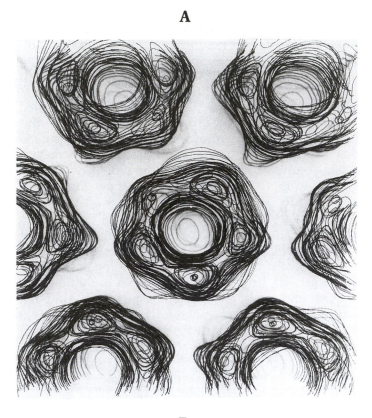

B

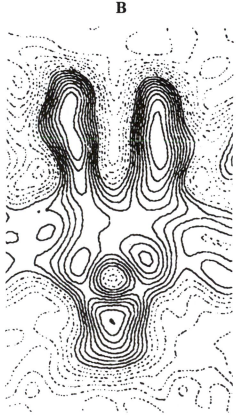

Figure 14–25: Map of electron scattering density of acetylcholine receptor embedded in a glass of amorphous ice.[214,283] Membranes enriched in acetylcholine receptor were prepared from electric organs of *Torpedo marmorata* by differential centrifugation. The membranes were resuspended in 0.1 M tris(hydroxymethyl)aminomethane hydrochloride, pH 6.8, and allowed to stand at 10 °C for 1 month, at which time long (< 1 μm) cylindrical tubes 70 nm in diameter had formed. These tubes were cylindrical, helical, crystalline arrays of molecules of acetylcholine receptor. The asymmetric unit in the helical array was a dimer of identical acetylcholine receptors. The asymmetric units formed the rows in a five-stranded right-handed helix in one dimension and the rows of a 15-stranded left-handed helix in the other dimension of the surface lattice. These tubes were embedded in a thin layer of amorphous ice on a film of carbon on an electron microscopic grid. Digitized electron micrographs were submitted to Fourier transformation. The layer lines of the resulting diffraction pattern were indexed, and variations in phase and amplitude along the layer lines were measured from these diffraction patterns. These functions were then submitted to Fourier–Bessel inversion to obtain a three-dimensional map of electron scattering density for the tube. (A) A view perpendicular to the surface of the tube of this map of scattering density. The image was made by stacking about 20 successive sheets of clear plastic of the appropriate thickness, each with a cross section of the map traced upon it. The successive sections chosen were 0.5 nm apart. Reprinted with permission from *Nature*, ref 214. Copyright 1985 Macmillan Magazines Limited. (B) A cross section through the center of a molecule of acetylcholine receptor in a plane normal to the axis of the tube. The blocklike structure at the bottom of the image is thought to be a protein other than acetylcholine receptor. The bilayer is to the right and left. There is a deep cylindrical depression on the upper, extracytoplasmic surface of the protein and a small shallow depression on the lower, cytoplasmic surface of the protein. The five subunits arrayed about a 5-fold rotational axis of pseudosymmetry produce a thick cylindrical pipe about 8 nm in diameter and 8 nm in height with a wall 3 nm thick extending out from the cytoplasmic surface of the membrane. Reprinted with permission from ref 283. Copyright 1990 Rockefeller University Press.

spanning, hydrophobic segments in each polypeptide commence. The first three hydrophobic segments in each polypeptide occur within lengths of only 90 amino acids in the respective amino acid sequences, and each of the three is thought to span the membrane,[280] which would bring each of the polypeptides to the cytoplasmic surface of the protein at the completion of its respective third segment. This membrane-spanning region containing the three hydrophobic segments from each polypeptide has the most highly conserved amino acid sequences among the four polypeptides of any region in the protein.[279] Results obtained with site-directed mutation[284,285] and direct chemical modification[286–288] implicate the second of these three hydrophobic segments from each subunit as a participant in forming the gated channel for cations. It is thought that this channel is formed from a pseudo-symmetric ring of five membrane-spanning α helices, each one the second membrane-spanning segment in the respective subunit.[289]

Up to the point 299–313 amino acids from the amino termini, the four aligned polypeptides have remained almost in register throughout their sequence. The next region contains hydrophilic segments 109, 132, 141, and 142 amino acids in length, respectively. These regions, each from a different subunit, have been shown to be located on the cytoplasmic surface of native acetylcholine receptor.[203,290] It is in these cytoplasmic regions that the differences among the polypeptides in both length and amino acid sequence are most apparent. Following this hypervariable cytoplasmic region in each of the four polypeptides is a fourth hydrophobic segment that spans the bilayer to bring the carboxy termini to the extracytoplasmic surface.[290]

The membrane-spanning portion of the three-dimensional map of electron scattering density (Figure 14–25) has a cross-sectional area[283] of 18 nm^2. This is consistent with the existence of four membrane-spanning α helices in each subunit. The number of boundary lipids around a molecule of acetylcholine receptor has been estimated to be 40 ± 7 using nitroxyl derivatives of stearic acid, 17-hydroxyandrostan-3-one, phosphatidylcholine, phosphatidylethanolamine, phosphatidylserine, and phosphatidic acid.[291] Each of the six probes gave the same number of boundary positions but differed in their affinity for the protein. The nitroxyl derivative of 17-hydroxyandrostan-3-one, a homologue of cholesterol, had 4 times the affinity for acetylcholine receptor as the nitroxyl derivative of phosphatidylcholine. This preference for steroid displayed by acetylcholine receptor is opposite to the preference for phosphatidylcholine displayed by Ca^{2+}-ATPase.[155] From a low-resolution image of acetylcholine receptor, it was calculated that 43 boundary lipids should occupy its perimeter.[291]

Suggested Reading

Henderson, R., Baldwin, J.M., Ceska, T.A., Zemlin, F., Beckman, E., & Downing, K.H. (1990) Model for the Structure of Bacteriorhodopsin Based on High-Resolution Cryomicroscopy, *J. Mol. Biol. 213*, 899–929.

Problem 14–1: Pick out potential candidates for membrane-spanning α helices from the following amino acid sequence of an integral membrane-bound protein.

MNWTGLYTLLSGVNRHSTAIGRVWLSVIFIFRIMVLVVA
AESVWGDEKSSFICNTLQPGSNSVCYDQFFPISHVRLWS
KQLILVSTPALLVAMHVAHQQHIEKKMLRLEGHGDP
LHLEEVKRHKVHISGTLWWTYVISVVFRLLFEAVFM
YVFYLLYPGYAMVRLVKCDVYPCPNTVDCFVSRPTEKTV
FTVFMLAASGICIILNVAEVVYLIIRACARRAQRRSNPPSR
KGSGFGHRLSPEYKQNEINKLLSEQDGSLKDILRRSP
GTGAGLAEKSDRCSAC

Problem 14–2: Membranes containing only cytochrome oxidase can be isolated from mitochondria. By successive

acetone extractions it is possible to deplete these membranes of their phospholipids and obtain preparations with different ratios of protein to phospholipid. The protein in these preparations retains its native conformation at all times. The following spin label was incorporated into the membranes containing different amounts of phospholipid.

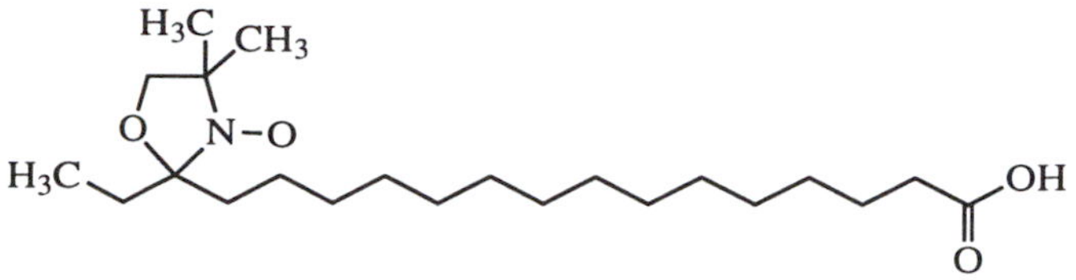

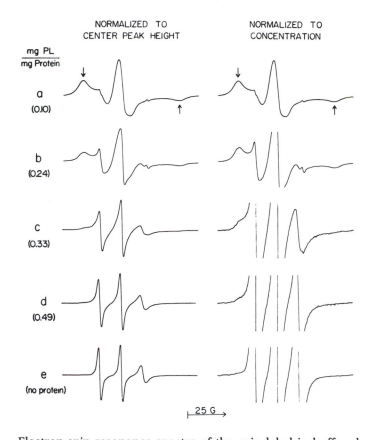

2-ethyl-2-(14-carboxytetradecyl)-4,4-dimethyloxazolidine *N*-oxide

The ratio of protein to spin label was held constant. The ESR spectra of each preparation was taken with the following result:

Electron spin resonance spectra of the spin label in buffered aqueous dispersions of membranous cytochrome oxidase with various lipid contents. Ratio of spin label to protein remained constant. The lipid to protein ratio expressed as milligrams of lipid (milligram of protein)$^{-1}$ is indicated at the far left. Left, spectra normalized to the center-line height; right, the same spectra normalized to give equivalent values after two integrations. Therefore, the right column represents constant concentration of spin label. Reprinted with permission from ref 292. Copyright 1973 National Academy of Sciences.

The left-hand column gives an idea of spectral shape; the right-hand column, amplitude of the signal. Spectrum *a* is that of a highly immobilized probe; spectrum *e*, of a highly mobile one. Explain these observations.

Problem 14–3: Label each of the following reagents as a hydrophobic reagent for modifying membrane-spanning segments of a protein or as an impermeant reagent. Indicate the reactive position in each reagent with an arrow, and circle the portion of the molecule that renders

it hydrophobic or impermeant, respectively.

Problem 14–4: For what purpose were the following reagents synthesized and used to study membrane-bound proteins?

Phenyl [^{35}S]isothiocyanate modified the lysine in the segment –PNTALLSLVLMAGTFFFAMMLRKF–. Where is this lysine probably located relative to the bilayer?

Problem 14–5:

(A) The density of protein is 1.35 g cm^{-3}. If one polypeptide of anion carrier were coiled so as to form a hard sphere, what would be its diameter? Compare this diameter to the width of a phospholipid bilayer.

(B) *N*-Formyl-[^{35}S]sulfinylmethionylmethylphosphate reacts indiscriminately with lysine residues on the external surfaces of protein molecules to form a derivative of the ε-amino group that is radioactive. This reagent cannot pass through the erythrocyte membrane because of its polar character. Write the mechanism of this modification of lysine.

Intact erythrocytes were mixed with this reagent, and the reaction was allowed to proceed for 10 min. The cells were then washed three times with buffer. The anion carrier protein was purified from these cells and was found to be radioactive. This radioactive protein was cleaved with the proteolytic enzyme thermolysin, and the digest was spread on a two-dimensional chromatogram. The chromatogram was placed over a sheet of photographic film and set aside for several days. The film was developed and radioactive peptides were located visually. This figure is a diagrammatic representation of the spots observed on this film.

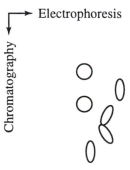

(C) Why was thermolysin used rather than trypsin?

This experiment was repeated with erythrocytes that had been broken open instead of intact erythrocytes. The distribution of radioactive peptides is shown in the tracing below.

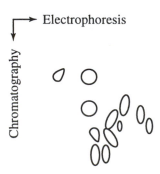

You should convince yourself that each spot on the peptide maps corresponds to a unique lysine residue on a surface of the anion carrier. You should also understand that each erythrocyte contains 3×10^5 copies of the anion carrier in its membrane.

(D) What two fundamental chemical properties of integral membrane-bound proteins are demonstrated by this experiment? How?

(E) How does the surface area of the anion carrier exposed to the exterior of the cell compare to the surface area exposed to the cytoplasm?

Problem 14–6: Rhodopsin is a protein that is firmly embedded in the membranes of a vertebrate rod. If the sacs of membrane known as disks are purified from these rods, the only protein they contain in significant quantity is rhodopsin, an integral membrane-bound protein. Purified disks were dissolved in a solution of nonionic detergent and mixed with excess phospholipid in the same detergent. When the detergent was slowly removed, small unilamellar vesicles, 40–70 nm in diameter, form spontaneously. The rhodopsin molecules ended up embedded in the membranes of the vesicle. Spectral measurements demonstrated that the tertiary structure of the rhodopsin in the vesicles is the same as it was in the disk.

(A) Papain, a proteolytic enzyme, cleaves native rhodopsin at only one position in its entire sequence to yield two fragments from the original molecule of protein. In disk membranes, papain cleaves every rhodopsin molecule. In the reconstituted vesicles, it can cleave only 65% of the rhodopsin molecules. Draw a diagram of a reconstituted vesicle with bilayers and rhodopsin molecules and explain why 35% of the rhodopsin is resistant to cleavage.

(B) Reconstituted vesicles were labeled with lactoperoxidase, an enzyme that cannot pass through a bilayer. Lactoperoxidase produces a very reactive iodinating agent that inserts ^{125}iodine into any tyrosine residue on the outside surface of the vesicle. Those rhodopsin molecules that cannot be cleaved by papain are nevertheless labeled by lactoperoxidase. What does this experiment demonstrate about the rhodopsin molecule? Why?

The Fluid Mosaic [80]

A membrane within a living cell often differs in shape and extension from the same membrane purified from a homogenate of that cell. Small organelles such as mitochondria, chloroplasts, and lysosomes remain intact and are not visibly or functionally altered during gentle disruption of the cell and purification by centrifugation. The endoplasmic reticulum is rather plastic and fluid even within the cytoplasm, a property reflected in the fact that upon homogenization, it readily disintegrates into small microsomes. The plasma membrane, however, in its natural state in an intact cell, is required to remain at all times a continuous enclosure surrounding the cytoplasm even while it must maintain a total surface area much greater than the membranes of any of the stable organelles it contains. When the cell is homogenized, the plasma membrane, like the endoplasmic reticulum, also disintegrates into small microsomes, but much more reluctantly.

This reluctance seems to result from the fact that plasma membranes are skins stretched and pinned upon a frame. In bacteria and fungi the frames are the outer membranes and cell walls on the extracytoplasmic surface for, when these integuments are digested away, a fragile, naked spheroplast remains that is easily disintegrated. In animal cells, however, there is often a frame or **cytoskeleton** on the cytoplasmic side of the plasma membrane upon which the membrane is stretched and to which it is pinned. A limited number of integral membrane-bound proteins function as the pins. These pins connect the continuous bilayer of the plasma membrane to the frame at random points scattered over its surface but do not noticeably affect the physical properties of the bilayer. The support provided by the cytoskeleton, the cell wall, or the outer membrane allows a plasma membrane to remain unbroken over its entire surface area even though it is a thin, fragile fluid film that disintegrates when it is removed from the frame.

Every membrane in a living cell, regardless of its total surface area and shape, is an individual, **isolated solution**. In each case, the solvent is a fluid bilayer of the appropriate surface area and shape, and the solutes are anchored membrane-bound proteins and integral membrane-bound proteins. The bilayer is a film of liquid paraffin 3.0 nm in width sandwiched between thin hydrophilic lamellae 0.5–1.0 nm wide. The bilayer is isotropic in the two dimensions of the surface defined by its center. Because every membrane in a cell is a closed sac, this surface is finite, continuous, and unbounded. Each protein floats upon the sheet of the bilayer at an unvarying draught; its membrane-spanning α helices cannot move up or down in the bilayer owing to the hydrophobic effect. These proteins, however, with the interesting exception of the pins upon the frame, are free to diffuse in the two unbounded dimensions of the bilayer.

That the solvent from which biological membranes are composed is a **bilayer of amphipathic lipids** has been demonstrated in many ways. The diffraction of X-radiation by biological membranes is mainly from the bilayer they contain. A myelinated nerve is a bundle of parallel axons each coated with a tubular spiral of myelin, which is the plasma membrane of a Schwann cell wrapped around and around the axon. Therefore, all of the plasma membranes of all of the Schwann cells are cylindrically oriented. A myelinated nerve diffracts X-radiation. From the diffraction pattern, a radial distribution of electron density normal to the axis of an axon can be computed,[293] and it is indistinguishable from that of multibilayers formed from an equimolar mixture of phosphatidylcholine and cholesterol (Figure 14–4D). Vesicles of biological membranes such as plasma membranes from erythrocytes, plasma membranes from the bacterium *M. laidlawii*, or endoplasmic reticulum from skeletal muscle diffract X-radiation in a circular pattern that can be thought of as the diffraction pattern of oriented bilayers (Figure 14–4A,C) spun around its center so that its equatorial and meridional reflections form circles. These circular diffraction patterns from biological membranes have the same periodicity as the diffraction patterns from vesicles containing only the lipids from the respective membranes,[294] and the amplitudes and periodicities of these diffraction patterns can be explained if they are assumed to arise from shells with distributions of electron density identical to those of bilayers of cholesterol and phosphatidylcholine (Figure 14–4D).[295]

When spin-labeled probes are incorporated into biological membranes, they behave almost as if they were incorporated into bilayers of pure phospholipid. In oriented biological membranes such as flattened endoplasmic reticulum,[296] nerves,[297] or oriented erythrocytes,[297] fatty acids containing dimethyl nitroxide **14–9** at various locations along the hydrocarbon incorporate with their long axes perpendicular to the surface of the membrane as they do in bilayers of phospholipid, and they display anisotropic motion resembling that displayed in oriented bilayers of pure amphipathic lipids. When 2,2,6,6-tetramethylpiperidine *N*-oxide

14–25

is incorporated into vesicles of endoplasmic reticulum from skeletal muscle, its spectrum is the same as its spectrum in vesicles of pure amphipathic lipid, and the absolute amplitude of its absorbance can be used to show that at least 85% of the amphipathic lipid in the endoplasmic reticulum is present as an unperturbed bilayer.[298]

Various microorganisms, such as fatty acid auxotrophs of *E. coli*, can be forced to incorporate high percentages of specific fatty acids into their plasma membranes. In preparations of these native membranes with a more homogeneous lipid composition, transitions between solid bilayers, with their hydrocarbons packed in hexagonal array, and liquid bilayers, with their hydrocarbons in the disordered fluid state, can be detected by X-ray

diffraction.[299] The transitions observed with such biological membranes are very similar to those observed when pure bilayers of the lipids extracted from these membranes undergo the same solid to liquid transformation.

These transitions can also be monitored by fluorescent probes. Fluorescent molecules such as *N*-phenyl-1-naphthylamine[300] are hydrophobic enough to partition preferentially into the hydrocarbon of the bilayer and register the transition between solid and liquid by changes in the intensity of their emission of fluorescence. In vesicles of the amphipathic lipids purified from bacteria of *E. coli* whose membranes were enriched in various fatty acids, in the intact native membranes purified from these cells, and in the whole cells themselves, the fluorescent probes detected the same phase transitions.[301] Quantitative analysis of these results showed that at least 80% of the amphipathic lipid in the native membranes was in the form of a bilayer indistinguishable in its phase transitions from a bilayer of the purified lipids.[300]

The molar surface area of the phospholipids in the bilayer of an intact biological membrane can be estimated by using phospholipases as probes. Phospholipases are enzymes that catalyze the hydrolysis of either the carboxylic esters or phosphate esters in a phospholipid. Their ability to catalyze these hydrolyses in monolayers of phospholipid depends on the molar surface area of the monolayer (Figure 14–11) and is abruptly inhibited below a characteristic molar surface area.[72] From the ability of several phospholipases that differed in these characteristic molar surface areas to catalyze the hydrolysis of phospholipids in the plasma membrane of the erythrocyte, it was shown that the molar surface area in this natural bilayer is equivalent[302] to the molar surface area of a monolayer of phospholipid at a surface pressure between 30 and 35 dyn cm^{-1}. This is in a region of the pressure–area curve for a monolayer of phosphatidylcholine (Figure 14–11) that would correspond to a surface area of 0.72–0.75 nm^2 (molecule of phosphatidylcholine)$^{-1}$, which is close to the usual surface area of hydrated egg phosphatidylcholine in a homogeneous bilayer.

It has already been noted that in a biological membrane each membrane-bound protein has its particular **vectorial orientation**. This orientation is maintained through the lifetime of a molecule of that protein by its inability to rotate even once 180° around any axis parallel to the surface of the membrane. For this to occur, the hydrophilic surfaces of the protein on the two sides of the membrane would have to pass through the 3.0 nm of hydrocarbon within the bilayer. This would require that the hydrogen bonds ensnaring these surfaces in the lattices of the liquid water (Figure 6–32) would all have to break simultaneously to permit the protein to capsize. Apparently, this cannot be accomplished.

It is also the case that the phospholipids in a cellular membrane maintain a vectorial orientation.[303] These vectorial orientations of the phospholipids have been demonstrated by submitting sealed, oriented biological membranes to digestion by phospholipases under nonlytic conditions[304] or to modification by impermeant reagents,[303,305] or by exchanging the phospholipids accessible on their outer surface with radioactive phospholipids in other vesicles in a reaction catalyzed by phospholipid exchange proteins.[306] A phospholipid exchange protein carries a specific phospholipid tightly bound to itself,[307] which it is able to exchange for another phospholipid of the same type at the external surface of a sealed membrane. If small vesicles bearing one type of [^{32}P]phospholipid, for example, [^{32}P]phosphatidylcholine, are mixed with a suspension of sealed native membranes in the presence of the phospholipid exchange protein specific for that phospholipid, the phospholipids of that type, for example, the phosphatidylcholines, in the outer monolayers of the membranes will rapidly reach the same specific radioactivity as that of the outer monolayers of the added radioactive vesicles.[308] From this specific radioactivity, the fraction of that type of phospholipid present in the outer monolayers of the membranes can be calculated.

With use of one or the other of these procedures, the distributions of the various phospholipids across the bilayers of various biological membranes have been determined. In each case, the total moles of phospholipid in one monolayer always equals, within experimental error, the total moles in the other monolayer of the membrane, but the distribution of each type between the monolayers is biased (Table 14–4). Phosphatidylethanolamine and phosphatidylserine tend to be concentrated on the cytoplasmic surfaces of plasma membranes. There is an interesting inversion in the case of unwrapped mitochondria in which phosphatidylethanolamine is concentrated on the extracytoplasmic surface, perhaps as a vestige of the ancestry of the mitochondrion, which is thought to have arisen from a prokaryotic symbiote. Sphingomyelin is enriched in the extracytoplasmic surface of plasma membranes. Phosphatidylcholine, in animals, or phosphatidylglycerol, in bacteria, seems simply to make up the differences between the two monolayers.

It is unclear whether or not cholesterol is asymmetrically distributed across biological membranes. Two independent measurements of cholesterol distribution in human erythroctyes found an equal ratio[311b] or a ratio of about 2-fold in favor of the extracytoplasmic surface.[312] The membrane of influenza virus, derived directly from the plasma membrane of the host, has cholesterol evenly distributed between its two faces.[313]

The question that these vectorial distributions raise is whether they are maintained kinetically or thermodynamically. In other words, is there a barrier to equilibration of these lipids across the bilayer after the asymmetry is established biosynthetically, or are there gradients of free energy unique to each that produce these distributions at equilibrium? These gradients of free energy would be created by other, more rigidly established vectorial asymmetries such as those of the proteins and polysaccharides across the bilayer or solutes on the two sides of the membranes. Because the largest ratio of a given phospholipid between the two faces of a membrane may be only about a factor of 10, the free energy required

Table 14–4: Asymmetric Distribution of Phospholipid and Sphingomyelin between the Two Sides of a Biological Membrane[a]

| | phospholipid (% of total) | | | | | | | | | | | |
| | extracytoplasmic monolayer | | | | | | cytoplasmic monolayer | | | | | |
membrane	PC[b]	PE	PS	SM	PG	DPG	PC	PE	PS	SM	PG	DPG
plasma membranes												
human erythrocyte[304]	20	5	< 1	20			10	25	10	5		
rat erythrocyte[309]	30	5	1[c]	10			15	20	15[c]	< 1		
human erythrocyte[305]	ND[d]	5	< 1	ND			ND	25	15	ND		
Bacillus megaterium[305]		25			25[e]			50			< 5[e]	
disks from rod outer												
segment[310]	ND	10	5	ND			ND	30	5	ND		
unwrapped bovine												
mitochondria[311a]	10	20				15	30	10				5

[a]Numbers are mole percentages based on the amount of total phospholipid in the membrane. [b]PC, phosphatidylcholine; PE, phosphatidylethanolamine; SM, sphingomyelin; PG, phosphatidylglycerol; DPG, diphosphatidylglycerol. [c]PS + PI. [d]ND, not determined. [e]By difference.

to maintain the bias would be only about 6 kJ mol⁻¹.

In sonicated vesicles of purified phospholipids, the rate at which a phospholipid, labeled in its hydrophilic functionality with a tetramethyl cyclic nitroxide, can pass from the external monolayer to the internal monolayer is very slow. The time required for the distribution to come halfway to equilibrium at 30 °C was measured to be 6 h, but there was evidence that oxidation of the phospholipids was adventitiously accelerating the rate.[314] When an exchange protein specific for phosphatidylcholine was used to exchange the phosphatidylcholine in unilamellar vesicles of pure phosphatidylcholine, a very slowly exchanging component ($t_{1/2} \geq 10$ days at 37 °C) was observed that accounted for about 40% of the total phosphatidylcholine. This slow component was assigned to phosphatidylcholine on the inner monolayer that had to transfer to the outer monolayer before it could exchange.[308] Both of these results suggest that there is a kinetic barrier to this transfer in bilayers of pure phospholipid.

In biological membranes, however, the transfer of phospholipids between the two monolayers of the bilayer seems to be much more rapid. [³²P]Phosphatidylcholine was observed to transfer from the extracytoplasmic monolayer to the cytoplasmic monolayer in an erythrocyte with a half-time for equilibration of 1–2 h at 37 °C.[309] The same phospholipid labeled in its hydrophilic functional group with the tetramethyl cyclic nitroxide, which equilibrated so slowly in vesicles of pure phospholipid, could equilibrate in vesicles of membrane from the electric organ of *Electrophorus electricus* with a half-time of less than 10 min at 15 °C.[315] In *Bacillus megaterium*, newly synthesized phosphatidylethanolamine in the cytoplasmic monolayer reaches equilibrium between the two monolayers within 30 min at 24 °C.[316] All of these results suggest that in living cells there is not a kinetic barrier to the distribution of phospholipids between the two monolayers as there is to the rotation of a molecule of protein and that the asymmetries persisting over the life of the cell are distributions at equilibrium.

Although rotational diffusion of a molecule of an integral membrane-bound protein about any axis parallel to the surface of the membrane does not occur and rotational diffusion of a molecule of phospholipid about any axis parallel to the surface of the membrane is slow, integral membrane-bound proteins, anchored membrane-bound proteins, and phospholipids all display one degree of **rotational diffusion** about axes normal to the surface of the membrane and two degrees of **translational diffusion** along axes parallel to the surface of the membrane. These diffusional degrees of freedom, prescribed by the two-dimensional nature of the solvent, are the only ones available to these molecules.

The rapid translational diffusion of proteins over the two dimensions of a plasma membrane was strikingly demonstrated by an experiment involving the fusion of two cells.[317] Immunoglobulins G specific for antigens on the surface of mouse (c11D) and human (VA-2) cells, respectively, were produced. These immunoglobulins were covalently modified with different fluorescent reagents, one that fluoresced green and one that fluoresced orange, respectively. The addition of the former immunoglobulins to mouse cells rendered proteins in their plasma membranes green, and the addition of the latter immunoglobulins to human cells rendered proteins in their plasma membranes orange. When a mouse cell was fused with a human cell, the hybrid was initially stained green at one side and orange at the other, but the two sets of antigenic proteins rapidly diffused over the plasma membrane of the hybrid until, within 40 min, they were each uniformly distributed. This intermixing resulted from two-dimensional translational diffusion.

In a three-dimensional, isotropic solvent, such as an aqueous solution, the **translational diffusion coefficient**, D_T, is the proportionality constant (Equation 1–53) between the flux, J, of a certain solute across a unit area, in a plane normal to its direction of net flux, and the gradient of its concentration, c, along the direction of net flux, x:

$$J = -D_T \left(\frac{\partial C}{\partial x} \right)_t \tag{14–1}$$

The units on J are moles centimeter⁻² second⁻¹, and on the gradient of concentration, they are (moles centimeter⁻³) centimeter⁻¹. Therefore, the units of D_T are centimeters² second⁻¹. The mean square displacement, $\overline{r^2}$, that the molecules of the solute will experience while in Brownian motion over a given time interval, t, is related to the diffusion coefficient by

$$\overline{r^2} = 4 D_T t \tag{14–2}$$

Molecules of the solute will also experience Brownian rotational motion as well as Brownian translational motion, and a **rotational diffusion coefficient**, D_R, for this rotational motion can be defined, in analogy with Equation 14–2, as

$$D_R = \frac{\overline{\theta^2}}{2t} \tag{14–3}$$

where $\overline{\theta^2}$ is the mean square angular rotation experienced in time t. If θ is expressed in radians, the units on D_R are radians second⁻¹. The relationship between the translational diffusion coefficient and the frictional coefficient, ^3f_T, for translational motion in three dimensions is

$$^3f_T = \frac{k_B T}{D_T} \tag{14–4}$$

where k_B is Boltzmann's constant and T is the temperature. An analogous relationship can be written for rotational diffusion

$$^3f_R = \frac{k_B T}{D_R} \qquad (14\text{--}5)$$

where ^3f_R is the frictional coefficient for rotational motion in three dimensions. If the molecules of the solute were hard spheres of radius a in a solvent of viscosity η (Equation 1–57), then

$$^3f_T = 6\pi\eta a \qquad (14\text{--}6)$$

and

$$^3f_R = 8\pi\eta a^3 \qquad (14\text{--}7)$$

Diffusion in a two-dimensional solvent, isotropic in those two dimensions, can be treated in parallel. The two-dimensional translational diffusion coefficient, ^2D_T, is the proportionality constant (Equation 14–1) between the flux of a substance across a unit width on a line normal to its direction of net flux, in moles centimeter^{-1} second^{-1}, and the gradient of its concentration along the direction of net flux, in (moles centimeter^{-2}) centimeter^{-1}. The units of ^2D_T are centimeter^2 second^{-1}. The mean square displacement is still governed by Equation 14–2. Each molecule of the solute will experience Brownian rotational motion about one axis, that normal to the surface of the two-dimensional solvent, and the one-dimensional rotational diffusion coefficient ^1D_R is defined by Equation 14–3. The frictional coefficient, ^2f_T, for translation in a two-dimensional solvent is defined by an equation analogous to Equation 14–4, and the frictional coefficient for one-dimensional rotational diffusion, ^2f_R, is defined by an equation analogous to Equation 14–5, with the substitution of ^2f_T and ^2f_R for ^3f_T and ^3f_R, respectively.

For right cylinders of radius a that are either the size of membrane-bound proteins or the size of phospholipids, undergoing two-dimensional diffusion in a sheet of liquid paraffin with a width, h, of about 3 nm and a viscosity, η_H, of approximately 1 P, which is sandwiched on both sides by water with a viscosity, η_W, of 0.01 P, at 298 K

$$^2f_T \cong 4\pi\eta_H h \left[\ln\frac{\eta_H h}{\eta_W a} - \gamma_E \right]^{-1} \qquad (14\text{--}8)$$

where γ_E is Euler's constant (0.5772).[318] This relationship is not precise because the equations for slow viscous flow used to derive Equation 14–6 in three dimensions have no exact solution in two dimensions. The equations for deriving the frictional coefficient, ^1f_R, for rotational motion about an axis normal to the surface of the solvent, of a right cylinder of any radius a in a two-dimensional solvent do have an exact solution, which is

$$^1f_R = 4\pi\eta_H a^2 h \qquad (14\text{--}9)$$

When translational or rotational diffusion of a membrane-bound protein in two dimensions is evaluated, it is usually assumed that the protein can be treated as an equivalent right circular cylinder of **Stokes radius** a.

Unlike Equations 14–6 and 14–7, Equations 14–8 and 14–9 cannot be used quantitatively to determine the Stokes radius a or the shape of a molecule of protein or lipid in a bilayer of phospholipids. It is not possible to determine independently the value of η_H experienced by the diffusing solute because a bilayer is not an isotropic sheet of liquid hydrocarbon (Figure 14–2). It is also not known what value should be used for h, the width of the bilayer. These equations, however, can be used qualitatively to show that the diffusion coefficients measured are in reasonable agreement with the sizes of the diffusing solutes, the width of a bilayer, and the expected viscosity of the liquid hydrocarbon.

Large multilamellar vesicles of dimyristoylphosphatidylcholine, 25–50 μm across, can be prepared with various concentrations of bacteriorhodopsin.[319] Bacteriorhodopsin contains a chromophore, retinal, that is fluorescent, and the distribution of the protein over the membranes of a vesicle can be monitored in a microscope by the distribution of fluorescence. When a circular area 5 μm in diameter in the middle of a vesicle is submitted to intense irradiation by a laser, the retinal within the circle is photolytically bleached. After bleaching, the molecules of bacteriorhodopsin in the circular area are no longer fluorescent, but those surrounding the circle still are. As translational diffusion takes place, the circle slowly fills with fluorescent molecules entering from the perimeter, and the bleached circle gradually disappears. From such **fluorescence photobleaching recovery**, the translational diffusion coefficient D_T of the unbleached bacteriorhodopsin moving into the circle can be calculated.[320]

Measurements were made of the two-dimensional translational diffusion coefficient, ^2D_T, for bacteriorhodopsin at several different surface concentrations and temperatures.[319] The values varied between 4×10^{-8} and 0.1×10^{-8} cm^2 s^{-1}. As the surface concentration of bacteriorhodopsin was decreased from 1 mol (30 mol of phospholipid)$^{-1}$ to 1 mol (240 mol of phospholipid)$^{-1}$, the translational diffusion coefficient increased from 0.15×10^{-8} to 3.4×10^{-8} cm^2 s^{-1}, presumably because of a decrease in the viscosity of the bilayer due to the decrease in the concentration of protein. Because the diffusion coefficient was still increasing significantly at the lowest surface concentration at which measurements could be made, the translational diffusion coefficient at zero density, $^2D_T^0$, could not be determined accurately by extrapolation. At the lowest densities of protein examined at 30 °C, the translational diffusion coefficient of bacteriorhodopsin, 3.4×10^{-8} cm^2 s^{-1}, is that of a cylinder of protein with a Stokes radius, a, equal to 2.0 nm, in a bilayer with a width, h, of 4.5 nm, the viscosity of whose hydrocarbon, η_H, is 1.1 P. The broadest dimension of the bundle of α helices in a single molecule of bacteriorhodopsin is 4.0 nm (Figure 14–23B), the width of a bilayer of natural phospholipid and cholesterol, including the hydrophilic functional groups, is 5 nm

(Figure 14–4D), the viscosity of motor oil at 30 °C is between 1 and 2 P, and the viscosity of vegetable oil at 30 °C is about 0.5 P. Because realistic numerical values for these three parameters can be used to calculate, by Equations 14–8 and 14–5, a diffusion coefficient equal to the one that is measured, it appears that bacteriorhodopsin, at least when it is in bilayers of dimyristoylphosphatidylcholine, is diffusing predictably within the two-dimensional solvent formed by those bilayers. Its diffusion is a random walk driven by thermal energy through an isotropic, viscous medium just as is the diffusion of a soluble protein through an isotropic aqueous solution.

The rotational diffusion coefficient of bacteriorhodopsin in dimyristoylphosphatidylcholine has been measured from the rate at which the anisotropy of its fluorescence decays over the lifetime of the excited state.[321] The rotational diffusion coefficient, ^1D_R, as opposed to the translational diffusion coefficient, was affected little by the density of the protein, and at 30 °C it was equal to $7 \times 10^4 \ s^{-1}$. If the Stokes radius of a cylindrical bacteriorhodopsin, a, is taken as 2.0 nm and the width of the bilayer, h, as 4.5 nm, then the viscosity of the bilayer, η_H, sensed by the rotating protein (Equation 14–9), is 4 P.

Translational diffusion coefficients have been measured for other integral membrane-bound proteins. For rhodopsin at 20 °C, densely packed in the disks in photoreceptors, the translational diffusion coefficient[322] is $0.3 \times 10^{-8} \ cm^2 \ s^{-1}$ and the rotational diffusion coefficient[323] is $5 \times 10^4 \ s^{-1}$. For anion carrier at a surface concentration of 1 mol (200 mol of phospholipid)$^{-1}$, the translational diffusion coefficient is $1.6 \times 10^{-8} \ cm^2 \ s^{-1}$ at 30 °C.[324] The translational diffusion coefficients of bovine rhodopsin, Ca^{2+}-ATPase from endoplasmic reticulum of skeletal muscle, and acetylcholine receptor have all been determined by fluorescence photobleaching recovery at several temperatures in reconstituted membranes at high dilution [< 1 mole of protein (3000 mol of phospholipid)$^{-1}$].[322] The values for the translational diffusion coefficients at 25 °C are between 1.4×10^{-8} and $2 \times 10^{-8} \ cm^2 \ s^{-1}$ for all three proteins, and they are indistinguishable within the ranges of their standard deviations.

The translational diffusion coefficients of integral membrane-bound proteins vary by more than a factor of 10 as the concentration of the protein in the membrane is varied, yet they vary insignificantly with variations in the Stokes radius, a, of the protein because of the logarithmic dependence of the frictional coefficient on this parameter (Equation 14–8). This is one reason why translational diffusion coefficients cannot be used to provide any insight into the cross-sectional areas of these proteins in the bilayer.

At high dilution at 25 °C, the translational diffusion coefficients of all integral membrane-bound proteins are about $(2–3) \times 10^{-8} \ cm^2 \ s^{-1}$. This means that after 1 s the average value of the square of the distance that a protein will be situated from the position it occupied initially will be 10 μm^2. In a densely packed plasma membrane, however, the translational diffusion coefficients are significantly less. For example, a value of $0.02 \times 10^{-8} \ cm^2 \ s^{-1}$ has been determined for randomly labeled proteins in the plasma membranes of L-6 cells.[325] In this situation, the mean square displacement for one of these proteins after 1 s will be only 0.1 μm^2. The diameter of a normal eukaryotic cell is about 20 μm. Therefore, it should take about 100 min for a protein with this low value for its translational diffusion coefficient to spread over the plasma membrane of a cell of this size if the protein were added at only one point on its surface.

The phospholipids in a membrane also display translational diffusion. This can be followed by using phospholipids, such as phosphatidylethanolamine, that have been modified by fluorescent reagents, in the case of phosphatidylethanolamine at its primary amine. The translational diffusion coefficients for such a fluorescent lipid in bilayers of various phospholipids are between 6×10^{-8} and $9 \times 10^{-8} \ cm^2 \ s^{-1}$ at 25 °C.[326] These values compare favorably with those calculated from spin-exchange among molecules of a nitroxylphosphatidylcholine in vesicles of various natural phospholipids, which are about $10 \times 10^{-8} \ cm^2 \ s^{-1}$ at 45 °C.[327] The translational diffusion coefficients for lipids are not much greater than those for proteins even though their Stokes radii are much smaller than those of proteins. This again is due to the logarithmic dependence of translational diffusion coefficient on the Stokes radius of the equivalent cylinder (Equation 14–8). In natural membranes of endoplasmic reticulum from skeletal muscle, the translational diffusion coefficients for a nitroxylphosphatidylcholine is approximately the same as it is in vesicles of the pure lipid extracted from these membranes.[327]

Associated with the bilayer of a membrane is a **microviscosity** experienced by small hydrophobic solutes dissolved in the liquid hydrocarbon. This microviscosity is generally estimated from the polarization retained in the fluorescence of a hydrophobic, fluorescent solute such as 2-methylanthracene[328] after its excitation with polarized light. A hydrophobic solute is used so that most of the molecules of the solute in the sample have been incorporated into the bilayer. The more rapidly the solute is reorienting within the bilayer during the lifetime of the excited state, the greater will be the loss of polarization. This loss of polarization can be calibrated by the behavior of the fluorescent solute in hydrocarbon solvents of known macroscopic viscosity.[328] In bilayers of pure natural phospholipid, microviscosities of 1–2 P have been observed at 25 °C.[329,330] The addition of cholesterol, however, increases the microviscosity by a factor of 5–10.[329] The microviscosity determined in vesicles formed from only the lipids in a biological membrane is almost the same as that determined for the complete biological membranes from which the lipids were extracted,[330] and addition of an integral membrane protein to vesicles of pure phospholipid affects the microviscosity only slightly.[331] These results suggest that the regions within the bilayer occupied by the fluorescent solutes used to monitor this property are mainly the bulk lipid between the molecules of protein. Because it is the rotational diffusion of the probe that senses the microviscosity rather than its translational diffusion, the presence of protein, which provides obstacles mainly to translation, has only a small effect. Nevertheless, the microviscosities determined are in the same range as the viscosities that seem to be controlling the translational and rotational diffusion of molecules of protein at low surface concentrations in bilayers of phospholipids.

Epidermal growth factor receptor is an integral membrane-bound protein that depends on its ability to diffuse over a bilayer of phospholipids to accomplish its function. Epidermal growth factor is a circulating polypeptide hormone that stimulates the growth of cells from a variety of tissues. Human epidermal growth factor receptor is a transmembrane glycoprotein composed from a folded polypeptide 1186 amino acids in length.[86] It is a member of a group of structurally and functionally related receptors for growth hormones on the cell surface characterized by an intrinsic activity for protein tyrosine kinase.[332] The initial response in the cascade leading to the mitosis caused by the binding of epidermal growth factor to epidermal growth factor receptor is the activation of this tyrosine kinase.

The amino acid sequence of human epidermal growth factor receptor[86] can be divided into two domains of about equal size. The extracytoplasmic half of the protein (620 amino acids) contains the binding site for epidermal growth factor.[333] The cytoplasmic half of the protein (540 amino acids) contains the active site for protein tyrosine kinase.[334] Both of these domains have been produced independently, in their entirety, and they are well-behaved soluble proteins with the appropriate functions.[333,334] In the intact native polypeptide the short segment between these two domains is composed of 23 hydrophobic amino acids, 15 of which are leucine, isoleucine, valine, methionine, or phenylalanine. There is no doubt that this segment spans the membrane connecting the extracytoplasmic domain to the cytoplasmic domain in the complete native protein. It has been shown that mutations, insertions, or deletions in this membrane-spanning segment have little effect on activation of the protein tyrosine kinase;[335,336] hence, its role in the activation of the protein tyrosine kinase activity must not involve any severe structural requirements, short of spanning the bilayer.

There have been a number of reports, however, implicating the dimerization of epidermal growth factor receptor in the activation of protein tyrosine kinase. Chemical cross-linking between monomers is observed but only after epidermal growth factor has been bound;[337,338]

bivalent immunoglobulins against epidermal growth factor receptor can activate its tyrosine kinase;[339,340] and mutant forms of epidermal growth factor receptor can suppress the activation of wild-type epidermal growth factor receptor.

It has been possible to follow the dimerization of monomeric epidermal growth factor receptor as a function of time by quantitative cross-linking,[341] exactly as the dimerization of malate dehydrogenase was followed by quantitative cross-linking (Figure 13–11). After epidermal growth factor was added to epidermal growth factor receptor dissolved in a solution of Triton X-100, the initially monomeric protein dimerized in a reaction that could be shown to be second-order in the concentration of protein. When the activation of protein tyrosine kinase activity was followed in the same preparations, it was found that it was also a process second-order in the concentration of epidermal growth factor receptor and that the second-order rate constants for dimerization and the activation of protein tyrosine kinase were identical. It follows that the rate-limiting step in the activation of the enzyme is dimerization of the protein. In the plasma membrane, this dimerization must result from collisions between monomers of epidermal growth factor receptor as they diffuse in two dimensions across the surface of the bilayer.

There is another set of membrane-bound proteins that also relies on translational diffusion over the surface of the plasma membrane to fulfill its biological function. This is the **adenylate cyclase system**. The role of this set of proteins is also to respond to the presence of a hormone in the medium surrounding the cell. Binding of the hormone to the extracytoplasmic surface of a particular protein in the plasma membrane either activates or inhibits adenylate cyclase, which is the enzymatic activity of an active site at the cytoplasmic surface of a different protein in the same plasma membrane. This active site is responsible for the reaction

$$MgATP \rightleftharpoons MgP_2O_7^{2-} + cyclic\ AMP \qquad (14\text{–}10)$$

If the hormone increases the rate of production of cyclic AMP, it is stimulatory; if it decreases the production, it is inhibitory.

The hormone initiates the process by binding to one of a set of **receptors**, each of which is a membrane-bound protein. A typical example of one of these receptors is β-adrenergic receptor. This protein has the binding site to which β-adrenergic agonists bind as the first step in the stimulation of adenylate cyclase. β-Adrenergic receptor has been purified to homogeneity from microsomes of plasma membrane from hamster lung that have been dissolved in the nonionic detergent digitonin.[342] The ability of the receptor to bind hormone was used as an assay, and the purification involved affinity adsorption to a solid phase to which a β-adrenergic antagonist had been attached (Table 1–2). The purified receptor is a membrane-spanning glycoprotein. The nucleotide sequence of cDNA encoding β-adrenergic receptor has provided its amino acid sequence.[343,344] The polypeptide composing the protein from the hamster is 418 amino acids in length and contains at least six hydrophobic segments, each of 20 amino acids, that are candidates to be α helices spanning the membrane. On the basis of such observations, it is usually assumed that all of the members of this class of receptors are integral membrane-bound proteins.

The **adenylate cyclase** itself is a much larger protein. When the mammalian adenylate cyclase is dissolved from membranes with the detergent Lubrol PX, it has an apparent molecular mass indicating that it contains between 1500 and 2000 amino acids.[345,346] The amino acid sequence of the adenylate cyclase present in the fungus S. cerevisiae has been obtained from cDNA encoding it, by taking advantage of a mutation in its gene to identify the cDNA.[347] The polypeptide encoded by this cDNA is responsible for the adenylate cyclase activity in these eukaryotic microorganisms and has a length of 2026 amino acids, which would account for most or all of the molecular mass of the intact protein in mammalian plasma membranes. Within the sequence of 2026 amino acids, however, there is no obvious hydrophobic segment 20 amino acids in length that might serve the role of an α helix that spans the membrane.[347] There is, however, within one region of the sequence a remarkable set of at least 14 consecutive, internally repeating sequences. They can be aligned by their repeating patterns of leucines, prolines, asparagines, isoleucines, and valines. The sequence that repeats consecutively is -PXXαXXLXX-LXXLXLXXNXαXXα-, where α can be leucine, isoleucine, or valine and X is usually a hydrophilic or charged amino acid. This pattern is neither the heptad repeat of an amphipathic helix nor the alternating pattern of a β sheet hydrophobic on one face and hydrophilic on the other. Furthermore, an identical pattern occurs in eight consecutive, internally repeating sequences in the α_2 glycoprotein from human serum, which is a perfectly water-soluble protein.[348] Nevertheless, the catalytic subunit of the adenylate system in mammalian cells seems to have hydrophobic surfaces suitable for spanning the membrane. This decision is based on its ability to bind between 50 and 90 mol of nonionic detergent (mol of polypeptide)$^{-1}$ when the protein is dissolved from the membrane.[345,346]

The final component in the overall adenylate cyclase system is a **guanine nucleotide-binding protein** or G-protein. There are several types of G-proteins present in the same membrane, one type mediating the stimulation of adenylate cyclase, another type mediating its inhibition, and other types with other roles in other systems.

The stimulation of adenylate cyclase by hormones requires the constant presence of GTP under physiological conditions[349] owing to the requirement that the associated stimulatory G-protein have GTP bound to it before the active site on adenylate cyclase can produce cyclic AMP. The stimulatory G-protein involved in this process binds GTP tightly[350] and under the appropriate circumstances catalyzes a slow hydrolysis of the GTP to GDP and inorganic phosphate.[351] Both of these properties are reminiscent of the binding and hydrolysis of GTP performed by tubulin. The rate of hydrolysis of GTP at the active site of a stimulatory G-protein is enhanced by the binding of hormone to the receptor,[352,353] and this is expressed as a hormone-activated guanosine triphosphatase.

The slow hydrolysis of GTP bound to the stimulatory G-protein is a timing device to terminate the activity of adenylate cyclase when hormone is no longer present at the extracytoplasmic surface of the cell.[351,354] When a hormone is abruptly removed from the binding site of a receptor by adding an antagonist to the solution in which the membranes are suspended, the adenylate cyclase activity decays slowly over a minute.[355] The rate at which the adenylate cyclase activity decays is equal to the rate at which GTP is hydrolyzed within the active site of a G-protein when it is coupled to adenylate cyclase.[354] If this hydrolysis of GTP is blocked by cholera toxin, a very specific inhibitor of this process, the adenylate cyclase no longer decays with time. The decrease in the rate of the GTPase activity as a function of the concentration of cholera toxin parallels the decrease in the fraction of the adenylate cyclase that is turned off.[351] The hydrolysis of the guanine nucleotide can also be prevented by the use of a chemical analogue of GTP that cannot be hydrolyzed. These analogues, like cholera toxin, produce adenylate cyclase activity that does not turn off when hormone is driven from the receptor.[351]

After the GTP is hydrolyzed, GDP remains tightly bound to the G-protein, preventing the binding of GTP if the hormone receptors are unoccupied. The binding of hormone to the receptor accelerates the dissociation of this bound GDP.[356]

Therefore, the overall sequence of steps is the following. Guanosine triphosphate binds to an empty G-protein, and this permits cyclic AMP to be produced at the active site on the catalytic subunit as long as the GTP remains unhydrolyzed. Following hydrolysis, the tightly bound GDP prevents the G-protein from activating the catalytic subunit. When that GDP is released in response to hormone bound on the receptor, the empty site can again bind GTP.

Guanine nucleotide-binding proteins are judged to be membrane-bound on the basis of their strong attachment to plasma membranes before those membranes have been dissolved by a neutral or weakly ionic detergent such as cholate.[357] They are usually purified in the continuous presence of such detergents, and this is assumed to mean that they were once integral or anchored membrane-bound proteins, although it need not have been so.

The purified G-proteins are each constructed from three different polypeptides, α, β, and γ. The α polypeptides from all of the various G-proteins are highly homologous in sequence (averaging about 60% identity in pairwise comparisons),[358] but they differ significantly in length (n_{aa} = 310–400)[358] because of three regions in which long gaps occur. The α polypeptide of the stimulatory G-protein from bovine adrenal glands is 394 amino acids long.[359] Each of the α polypeptides from other G-proteins contains one hydrophobic segment homologous to that between Arginine 232 and Arginine 258 in the bovine adrenal α polypeptide. This hydrophobic segment is a candidate for spanning the membrane.[358] Its sequence, –WIQCFNDVTAIIFVVASSSYNMVI, however, is not reminiscent of the hydrophobic segments in anchored membrane-bound proteins because it contains too many donors and acceptors of hydrogen bonds. The β polypeptide of the G-protein from bovine adrenal glands is 340 amino acids long,[360] and the γ polypeptide from bovine retinal G-protein is 69 amino acids long.[361,362] Neither the β nor γ polypeptides have any obvious hydrophobic segments. Therefore, it is unclear how G-proteins are bound to the membrane.

One resolution of this dilemma would be that each G-protein is tightly bound as a peripheral membrane-bound protein to a molecule of adenylate cyclase and remains associated with it as long as the membranes are intact and dissociates from it only after the membranes have been dissolved with detergents. Indeed, a tightly associated complex of G-protein and adenylate cyclase has been identified when plasma membranes are dissolved in solutions of one particular nonionic detergent,[363] but this does not mean the complex cannot dissociate under other normally occurring conditions. Whether two proteins are associated or dissociated when they are dissolved in solutions of nonionic detergent says nothing about whether they are associated or dissociated in the intact plasma membrane because the magnitudes of two-dimensional surface concentrations are completely different from those of three-dimensional concentrations. This point is especially relevant to the adenylate cyclase system, where associations and dissociations are known to occur on the surface of the membrane during its normal function.

The fact that the individual components of the adenylate cyclase system can be transferred among themselves by diffusion in the two dimensions of the plasma membrane was first demonstrated in a fusion experiment. Turkey erythrocytes, the adenylate cyclase of which responds to β-adrenergic agonists, were treated with N-ethylmaleimide to inactivate the active sites of adenylate cyclase completely. When these cells were fused with Friend erythroleukemia cells, which have the adenylate cyclase of the adenylate cyclase system but no β-adrenergic receptor, the adenylate cyclase activity in membrane fragments prepared from these hybrids had become responsive to β-adrenergic agonists.[364] Therefore, the β-adrenergic receptor of the turkey erythrocytes had become coupled to the adenylate cyclase of the Friend cells following fusion. Because only fragments of membrane were present in the final enzymatic assay, it was concluded that the proteins involved had diffused over the intact plasma membranes of the cells and associated among themselves to couple the binding of hormone to the catalytic activity.

In a cell responsive to a particular hormone, receptors for hormones are not permanently associated with molecules of adenylate cyclase. When increasing fractions of the hormone receptors in a membrane are destroyed by covalent modification, the final activities of adenylate cyclase achieved following addition of the hormone remain the same (Figure 14–26A).[365] This observation is inconsistent with a permanent association between receptors and catalytic subunits because a permanently inactivated receptor does not produce a permanently unresponsive catalytic subunit.

There is a step between the binding of hormone to the receptor and the increase in the enzymatic activity at the active site of adenylate cyclase that requires the protein to which the hormone binds to collide with the protein responsible for the catalytic activity as they diffuse independently over the plasma membrane. At 25 °C, the adenylate cyclase system in turkey erythrocytes takes several minutes to reach full enzymatic activity after hormone is added to a suspension of plasma membranes (Figure 14–26B).[365] As the fraction of the receptors occupied by hormone is increased, the lag preceding the expression of full activity decreases (Figure 14–26B). The relationship between the rate at which full activity is established, the final activity of the enzyme, and the fraction of the receptors occupied by hormone, which was determined by a direct measurement of bound hormone in separate experiments, is consistent[365] with the kinetic mechanism

$$H + X \overset{K_d^H}{\rightleftharpoons} H \cdot X + Y \xrightarrow{k_D} Z \qquad (14\text{–}11)$$

where K_D^H is the dissociation constant for hormone, $H \cdot X$ is a protein containing the hormone receptor, Y is a protein containing the active site for adenylate cyclase in its inactive form, and Z is a protein containing the active site in its active form. When a kinetic mechanism such as Reaction 14–11 is written, it can be said that it is consistent only with the observations, not that it is the kinetic mechanism. The actual kinetic mechanism can be much more complicated than the provisional kinetic mechanism and involve many more steps. It is just that the experiments as designed detect only the steps indicated in the provisional mechanism, and the other steps do not affect the reaction under these conditions. The experiments do, however, isolate and identify the step dependent on the collision of $H \cdot X$ and Y and governing the rate constant k_D.

A

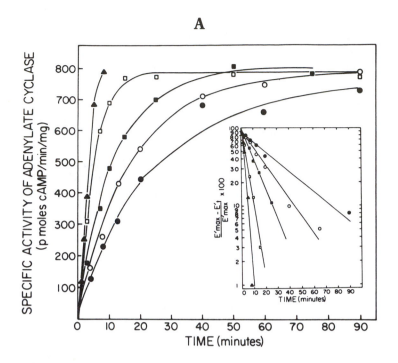

B

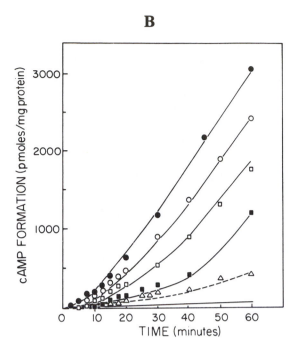

Figure 14–26: Rate of activation of adenylate cyclase as a function of the relative concentration of functional, occupied β-adrenergic receptor.[365] (A) Samples of plasma membranes from turkey erythrocytes were exposed to various concentrations ($\blacktriangle$, none; $\square$, 17 μM; $\blacksquare$, 43 μM; $\bigcirc$, 100 μM; and $\bullet$, 230 μM) of N-[2-hydroxy-3-(1-naphthyloxy)propyl]-N'-(bromoacetyl)ethylenediamine, a specific, covalent, irreversible inhibitor of β-adrenergic receptor, to destroy functionally various fractions of the receptor. Equivalent amounts of each sample (1.4 mg mL^{-1}) were mixed at 25 °C with saturating concentrations of the specific hormone, epinephrine, and guanosine β,γ-imidotriphosphate, an analogue of GTP that cannot be hydrolyzed by G-proteins. At various times, samples were removed, and the specific activity (picomoles of cAMP minute^{-1} milligram^{-1}) of the adenylate cyclase was determined. As time progressed, the specific activity of the membranes increased monotonically. Inset: Data in panel A are replotted on a semilogarithmic field to show that the approach to maximum velocity is a first-order process. (B) Equivalent samples of plasma membranes from turkey erythrocytes (1.4 mg mL^{-1}) were mixed in several tubes with MgATP at 25 °C and hormone-activated adenylate cyclase was initiated by adding guanosine β,γ-imidotriphosphate and various concentrations of the hormone epinephrine ($\triangle$, 0.5 μM; $\blacksquare$, 1.0 μM; $\square$, 3.0 μM; $\bigcirc$, 6 μM; and $\bullet$, 15 μM). At the indicated times, samples were removed from each tube and the total accumulations of cyclic-3′,5′-adenosine monophosphate were assessed. Panel B represents the integrated accumulation of cyclic-3′,5′-adenosine monophosphate; and panel A, the rate at which it is accumulating at any time. Reprinted with permission from ref 365. Copyright 1978 American Chemical Society.

The rate constant k_D for the activation of adenylate cyclase represents that of a bimolecular reaction between two independent proteins H·X and Y diffusing over the membrane. When receptor is saturated with hormone, all of the protein X is in the liganded form, H·X, and the reaction becomes dependent only on the rate constant k_D. Under these circumstances, the rate of activation of adenylate cyclase activity, E, is a first-order reaction in the concentration of the unactivated protein Y containing the inactive active site (inset to Figure 14–26A) where

$$\frac{[Y]}{[Y]_{TOT}} = \frac{[Z]_{TOT} - [Z]}{[Z]_{TOT}} = \frac{E_{max} - E}{E_{max}} \qquad (14\text{–}12)$$

The subscripts indicate total concentration of unactivated catalytic protein Y at the beginning of the reaction, total concentration of activated catalytic protein Z at the end of the reaction, and maximum adenylate cyclase activity

reached after full activation, and the assumption made is that $[Y]_{TOT} = [Z]_{TOT}$. When the concentration of the complex between hormone and receptor at saturation [H·X] is decreased systematically by decreasing the concentration of the competent hormone receptor by a specific covalent modification, the rate of production of the protein Z, containing the activated active site, decreases in direct proportion to the decrease in the concentration of occupied receptors at saturation (Figure 14–26A).[365] These two observations demonstrate that the reaction governed by k_D is a bimolecular reaction which is unimolecular in Y and unimolecular in H·X.

When the microviscosity of the membrane, as judged by the residual polarization of a hydrophobic fluorescent molecule, is decreased by adding *cis*-vaccenic acid, the rate of production of Z at saturating concentrations of hormone increases monotonically.[365] When the viscosity of a plasma membrane is increased by removing some of its bulk phospholipid, the ability of the binding of hormone to the receptor to activate adenylate cyclase is significantly inhibited.[366]

If it is assumed that every collision between H·X and Y produces an active Z, in other words, that the reaction is diffusion-controlled, and that the concentration of H·X is equal to the concentration of occupied receptors, a lower limit to the two-dimensional translational diffusion coefficient of the complex H·X necessary for collision between H·X and Y to explain the rate of production of Z can be calculated. The lower limit on the translational diffusion coefficient of H·X is only 0.005×10^{-8} cm^2 s^{-1}, which is less than the expected translational diffusion coefficient for an integral membrane-bound protein even in a crowded plasma membrane. Therefore the production of Z by collision of two proteins, H·X and Y, is reasonable.

At the moment, the compositions of protein X and protein Y in Reaction 14–11 are unresolved. Together they must contain one molecule of hormone receptor, one molecule of G-protein, and one molecule of adenylate cyclase, because it is the G-protein that connects the binding of hormone to the activation of adenylate cyclase activity. When the adenylate cyclase system is depleted of G-protein by affinity adsorption, no coupling between hormone binding and adenylate cyclase activity can occur until it is added back.[367] The membranes of Cyc$^-$549 lymphoma cells contain both a receptor for β-adrenergic agonists and a catalytic subunit but lack a stimulatory G-protein and cannot display hormone-activated adenylate cyclase activity. When homogeneous stimulatory G-protein is added to these membranes, hormone-activated adenylate cyclase activity appears.[357] Therefore, the G-protein must be involved in the collision producing the activation of adenylate cyclase as well as the hormone receptor and catalytic subunit.

The G-protein can bind to either the hormone receptor or adenylate cyclase under appropriate circumstances. It has already been noted that a tight complex between one adenylate cyclase and one G-protein has been identified in solutions derived from plasma membranes dissolved with nonionic detergents.[363] A complex between one β-adrenergic receptor and one G-protein has also been identified in solutions derived from plasma membranes dissolved in nonionic detergent, but only when they are pretreated with β-adrenergic agonists.[368] In addition, the incorporation of purified homogeneous G-protein and purified homogeneous β-adrenergic receptor together into phospholipid vesicles causes the receptor to have a higher affinity for agonist, by 12 kJ mol^{-1}, than it does in the absence of G-protein.[369] This effect must result from the formation of a complex between hormone receptor and G-protein in these membranes, but it is eliminated by the addition of GTP, which binds to the G-protein. Therefore, the question that has yet to be answered is whether a complex between an occupied hormone receptor and a G-protein collides with a catalytic subunit or an occupied hormone receptor collides with a complex between a G-protein and a catalytic subunit during the rate-limiting step in the activation of adenylate cyclase. There seems little doubt, however, that diffusion of the various components over the surface of the bilayer is essential to the process. Presumably, this permits the same catalytic protein to service a number of different hormone receptors.

The plasma membrane of a cell is a fluid mosaic of proteins diffusing in two dimensions through a solvent of liquid phospholipid and cholesterol that is stretched and pinned upon a cytoskeleton. Although most of the proteins are free to diffuse, the proteins pinning the plasma membrane to the cytoskeleton do not. In the short run, the membrane is attached at these points but fluid everywhere else. When the frame collapses and, as a result, the pins cluster rather than remaining spread out, the unsupported plasma membrane in the abandoned regions slowly fragments into microsomes.[370]

The cytoskeleton in an erythrocyte is composed of the proteins actin[371] and spectrin. Spectrin[372] was originally identified by Rosenthal, Kregenow, and Moses[373] as a protein composing a fuzzy network on the cytoplasmic side of the plasma membrane of an erythrocyte. It was isolated by extraction of plasma membranes from erythrocytes at low ionic strength in the presence of a chelating agent for multivalent cations.[373] The protein is composed of polypeptides about 2000 amino acids in length[374] and is an $\alpha\beta$ heterodimer or $(\alpha\beta)_2$ heterotetramer depending upon the conditions.[375] It has been highly purified from the initial low ionic strength extract,[376] and the purified dimer has a high intrinsic viscosity ($[\eta] = 140$ cm^3 g^{-1}),[376] indicating that it is elongated. The two polypeptides of the $\alpha\beta$ heterodimer are homologous in sequence, and each contains internally repeating domains (Figure 7–16).[377] In electron micrographs, the $\alpha\beta$ heterodimer appears as a flexible two-stranded segment of rope 97 nm long (Figure 14–27B). The two strands are not held together along their entire length and tend to splay. An $(\alpha\beta)_2$ heterotetramer can also form from two dimers associating end to end (Figure 14–27A).

When erythrocytes are added to a solution of nonionic detergent, the plasma membrane dissolves and leaves behind a cytoskeleton that has the shape of an erythrocyte but is a basket rather than a bag.[378] The cytoskeleton exposed by this treatment contains mainly spectrin, actin, and a protein referred to as band 4.1. Band 4.1 is a globular protein that can bind to an $\alpha\beta$ heterodimer of spectrin near the other end of the rope from that which combines with another $\alpha\beta$ heterodimer to form the $(\alpha\beta)_2$ heterotetramer.[379,380] When purified actin, band 4.1, and spectrin are mixed together with ATP, they spontaneously form a macroscopic gel made up of filaments of actin cross-linked by spectrin tetramers.[381] This presumably is analogous to the meshwork seen in the cytoskeleton.

The pins attaching the plasma membrane of an erythrocyte to the cytoskeleton are formed from two proteins, ankyrin and anion carrier. Ankyrin is a monomer con-

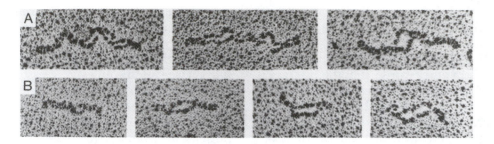

Figure 14–27: Electron micrographic images of individual molecules of spectrin as the $(\alpha\beta)_2$ heterotetramer (A) or $\alpha\beta$ heterodimer (B). Erythrocytes were washed and lysed, and the resulting plasma membranes were washed to remove the hemoglobin. The purified plasma membranes were extracted with 0.1 mM EDTA, pH 8, at 0 °C for 40 h, and the membranes were then removed by centrifugation. The released spectrin was purified from the extract by molecular exclusion chromatography. Solutions containing the spectrin were brought to 70% glycerol and sprayed onto freshly cleaved mica. The decorated surface of the mica was then sprayed at an angle of 9° to the surface with a mixture of platinum and carbon vaporized by electric discharge. The spray was applied while the sample was rapidly rotating about an axis normal to the surface. This causes the molecules to be surrounded by drifts of platinum, which is electron-dense. These drifts produce an outline of the molecules of protein. The film was then transferred from the mica to a grid for electron microscopy. The molecules are represented by the tortuous, elongated outlines. Magnification 170000×. Reprinted with permission from ref 382. Copyright 1979 Academic Press.

structed from one single polypeptide about 1800 amino acids in length.[379,383] It has a fairly high frictional ratio ($f/f_0 = 1.46$),[383] and in electron micrographs it appears as a cluster of three to five globular domains.[379] In keeping with this structure, it has a detachable domain ($n_{aa} = 650$) that contains the binding site specific for spectrin.[384] Ankyrin binds tightly to spectrin near the end of the rope that associates to form the $(\alpha\beta)_2$ heterotetramer, the opposite end from that to which band 4.1 binds.[379]

Intact ankyrin can also bind to the amino-terminal, detachable domain of anion carrier in a simple bimolecular reaction ($K_d = 10^{-8}$ M) as well as to intact anion carrier in solutions of nonionic detergents.[385] A freely soluble heterodimer containing one polypeptide of ankyrin and one polypeptide from anion carrier can be purified in solutions of nonionic detergent.[386] Because intact ankyrin binds tightly to both anion carrier, which is an integral membrane-bound protein, and spectrin, which is incorporated into the cytoskeleton, it can link the cytoskeleton to the plasma membrane. As there are far fewer molecules of ankyrin in an erythrocyte than molecules of anion carrier, only a minority of the molecules of anion carrier, presumably chosen at random, are linked to the cytoskeleton. In the short run, these fixed molecules of anion carrier are the stationary points around which flow the proteins and lipids of the plasma membrane. In the long run, as the cell changes shape and size, these points of attachment may also rearrange fluidly to accommodate the alterations.

Suggested Reading

Peters, R., & Cherry, R.J. (1982) Lateral and Rotational Diffusion of Bacteriorhodopsin in Lipid Bilayers: Experimental Test of the Saffman-Delbrück Equations, *Proc. Natl. Acad. Sci. U.S.A.* 79, 4317–4321.

Tolkovsky, A.M., & Levitski, A. (1978) Mode of Coupling between the β-Adrenergic Receptor and Adenylate Cyclase in Turkey Erythrocytes, *Biochemistry 17*, 3795–3810.

Problem 14–7: Cytochrome b_5 is a protein that is firmly attached to the membranes of the endoplasmic reticulum. The amino acid sequence of the protein from the rat is

```
  1 A E Q S D K D V K Y Y T L E E I Q K H K D S K S T W V I L H 30
 31 H K V Y D L T K F L E E H P G G E E V L R E Q A G G D A T E 60
 61 N F E D V G H S T D A R E L S K T Y I I G E L H P D D R S K 90
 91 I A K P S E T L I T T V E S N S S W W T N W V I P A I S A L 120
121 V V A L M Y R L Y M A E D *
```

When endoplasmic reticulum is treated with trypsin or pancreatic lipase contaminated with trypsin, only the peptide bond following Lysine 90 is cleaved, and a protein containing the first 90 amino acids of cytochrome b_5 falls off the membrane. This soluble protein can be purified and crystallized and its structure has been determined by X-ray crystallography. This protein will be referred to as cytochrome b_5(trypsin) or cytochrome b_5(lipase), respectively.

(A) Explain these observations in terms of the distribution of specific amino acids in the sequence of the protein.

(B)   Draw a representation of the complete cytochrome b_5 molecule attached to the membrane.

It is possible to release, by the use of a detergent, intact (133 amino acids) cytochrome b_5 from the endoplasmic reticulum membrane and to purify this protein. It will not crystallize. This protein will be referred to as cytochrome b_5(detergent).

(C)   When various amounts of cytochrome b_5(detergent) or cytochrome b_5(lipase) are mixed with endoplasmic reticulum membranes and incubated for 18 h at 2 °C, and the membranes are then washed extensively, the detergent form attaches to the membranes while the lipase form does not (see figure below). Explain this observation.

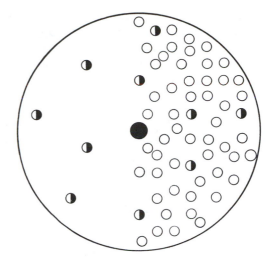

Schematic representation of the spatial relationships between phospholipid and cytochrome b_5 reductase on the surface of a microsomal vesicle. The surface areas used for these components were phospholipid, 0.63 nm²; cytochrome b_5, 4 nm²; and cytochrome b_5 reductase, 10 nm². The entire area (diameter = 56 nm) would include 4000 molecules of phospholipid. The endogenous concentrations of the two proteins would place approximately 10 molecules of cytochrome b_5 (◑) and approximately 1 molecule of reductase (●) in this area, assuming a random distribution on the outer surface of the membrane. The left sector indicates the molecular density with only endogenous cytochrome b_5, and the right sector, the density with a 10-fold molar excess of cytochrome b_5 (detergent) (○). Reprinted with permission from ref 387. Copyright 1972 *Journal of Biological Chemistry.*

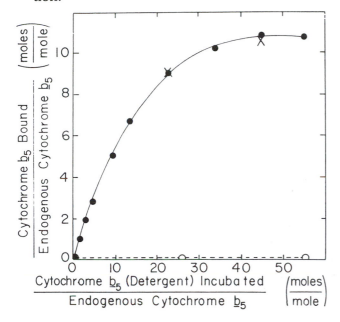

Binding of cytochrome b_5(detergent) (●) or cytochrome b_5(lipase) (○) to microsomes from rabbit liver. The results are expressed as the moles of exogenous cytochrome b_5 bound to the microsomes for every mole of endogenous cytochrome b_5 in the liver microsomes as a function of the moles of exogenous cytochrome b_5 added for every mole of endogenous cytochrome b_5 present. The points marked with × were samples that were washed further in 0.5 M NaCl, pH 8.0, to remove any loosely bound cytochrome b_5. Reprinted with permission from ref 387. Copyright 1972 *Journal of Biological Chemistry.*

There is an enzyme, also attached to the endoplasmic reticulum, known as cytochrome b_5 reductase, that catalyzes the following reaction:

$$\text{cytochrome } b_{5ox} \rightleftharpoons \text{cytochrome } b_{5red}$$

The native endoplasmic reticulum membrane contains one molecule of cytochrome b_5 reductase and 10 molecules of cytochrome b_5 for every 2.5×10^5 Å² (see figure top of next column).

When all of these cytochrome b_5 molecules are in the oxidized form and the reaction is initiated by adding reducing equivalents, the time required for the reduction of half of all the cytochrome b_5 molecules by the reductase is 0.47 s for the unenriched membranes and 0.13 s for the enriched membranes.

(D)   What ability must cytochrome b_5 possess in order to participate as a substrate in this reaction?

(E)   Why is the $t_{1/2}$ shorter in the case of the enriched membranes?

(F)   The concentration of cytochrome b_5 in the solution during the reduction experiments with the enriched membranes just described was about 2×10^{-6} M and the rate of the reaction catalyzed by the reductase was independent of the concentration of membranes suspended in the solution. In order to observe the same turnover rate, however, when cytochrome b_5(trypsin) was the substrate, its concentration had to be 5×10^{-5} M and the rate of the reaction catalyzed by the enzyme depended on the concentration of cytochrome b_5(trypsin) in the solution. Explain these observations.

Problem 14–8: Calculate the translational diffusion coefficients at 20 °C for integral membrane-bound proteins, the Stokes radii, a, of whose bundles of α helices are 1.0, 2.0, 4.0, and 5.0 nm. Assume that the viscosity of the bilayer is 1 P and the width of the bilayer is 5 nm.

Problem 14–9: The vertebrate rod is a cell in the retina responsible for registering light rays in the visual process. The one end of the cell that performs this task is called the outer segment. It is a cylinder filled with disks, which are flattened circular, closed sacs pinched off from the plasma membrane.

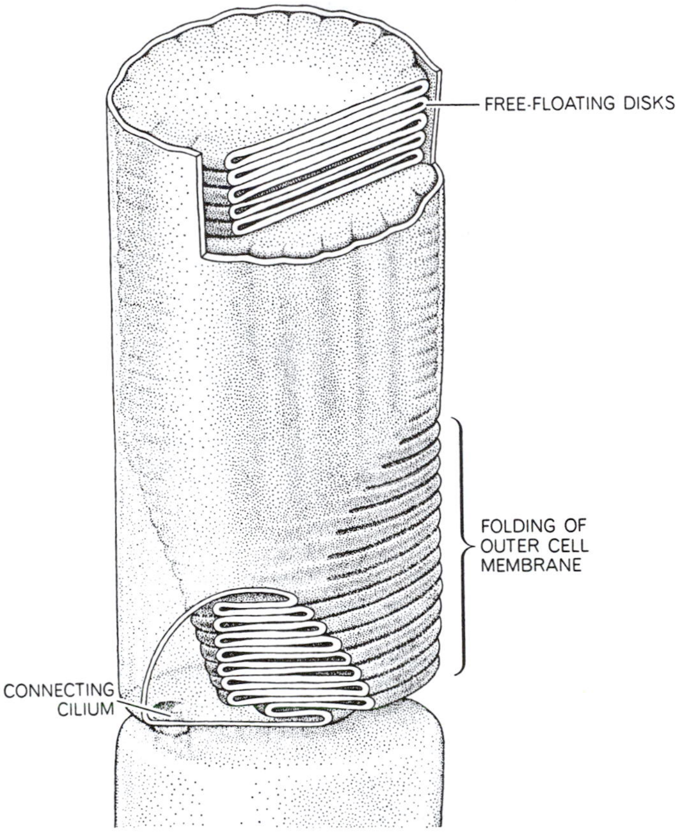
FREE-FLOATING DISKS
FOLDING OF OUTER CELL MEMBRANE
CONNECTING CILIUM

Diagram showing frog rod outer segment. The lamellar membranous structure of the rod outer segments consists of a stack of sacs, except near the base, where it consists of infoldings of the cell plasma membrane. Reprinted with permission from ref 388. Copyright 1970 *Scientific American.*

The disks are stacked in the rod as poker chips are stacked in a rack. In this way, greater than 90% of the membrane from which the disk is made lies normal to the cylindrical axis of the rod outer segment.

Rhodopsin is the only protein dissolved in the disk membrane. It has been shown that rhodopsin spans the membrane.

(A) In an outline of the cross section of a disk draw a schematic diagram of the molecular structure of the disk membrane. Include rhodopsin molecules, labeled with an arrow indicating direction of insertion, and phospholipids.

Rhodopsin is a protein to which is attached, by an imine linkage, one molecule of 11-*cis*-retinal. When 11-*cis*-retinal is exposed to intense light it isomerizes to all-*trans*-retinal. This process is known as bleaching and results in a color change from orange to white.

(B) What does the following experiment demonstrate about the properties of a protein such as rhodopsin in a biological membrane? Reprinted with permission from *Nature*, ref 322. Copyright 1974 Macmillan Magazines Limited.

Rod outer segments were obtained by gently shaking retinas dissected under dim red light from the eyes of frog (*Rana catesbeiana*) and mudpuppy (*Necturus maculosus*) which had been dark-adapted for more than 10 h. The rods were shaken into a microchamber containing a standard Ringer solution and examined in a Shimadzu 50L microspectrophotometer (MSP) fitted with a high quanted efficiency photomultiploer (Hamamatsu type R375). Single rods which appeared intact and which lay flat on the bottom of the chamber were selected for observation, and all observations were completed within 30 min after the rods were isolated. The rhodopsin in isolated rods, once bleached, does not regenerate, hence dim red light was used for selection, focusing, and alignment.

The measuring beam of the MSP was limited by an aperature and a condensing lens to form a rectangle about 2 × 20 μm in cross section. The long axis of the rectangle was aligned with the long axis of the rod and a simple motor-driven "alternator" optically shifted the rectangular measuring beam back and forth between the two sides of the rod. Thus the absorbance of the rhodopsin on each side of the rod could be compared directly. The wavelength of the measuring beam was set at the obsorption peak of the visual pigment: 500 nm for frog, 530 nm for mudpuppy.

With suitable alignment and focusing the absorbance was essentially equal on both sides of the unbleached rods, as shown by the first pair of measurements on an unbleached rod at the beginning of each of the two recordings in the figure below. The alternator was then stopped momentarily and the intensity of the measuring beam was increased about 1000-fold to bleach some pigment on one side of the rod. The exponential decrease iin absorbance during the bleach was recorded, and then the intensity was dropped to the original level and the alternator turned on again. The figure below shows that immediately after the bleach the absorbance on the unbleached side was little changed, but there was a marked drop in absorbance on the bleached side. Within the next few seconds, however, the absorbance of the unbleached side decreased while that on the bleached side increase, and within less than 1 min the absorbance of the two sides became equal, reaching a final level midway between that of the each side immediately after the bleach.

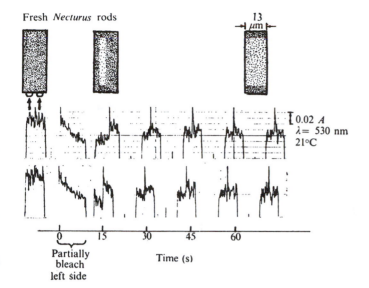

The diagrams of a rod depict the pigment distribution corresponding in time with the absorbance measurements shown below. The arrows indicate the location on the rod at which each absorbance measurement was made. Recordings made from two different rods are shown to give an indication of the repeatability of the measurements. In each experiment the chart recorder was run continuously, as shown by the time base. The alternator also ran continuously except during the bleach. The records thus consist of a repeated pattern in which absorbance measurements were made first on the left side, then the right side of the rod. Between each pair of measurements baseline measurements were also made to ensure that no drifts occurred (for clarity, these were omitted from the figure). The spikes on the traces were caused by switching transients in the alternator. The diameter of a disk is essentially equal to the width of the rod, and the width is measured in the MSP after completing each experiment.

References

1. Claude, A. (1941) *Cold Spring Harbor Symp. Quant. Biol.* 9, 263–271.
2. Hogeboom, G.H., Schneider, W.C., & Pallade, G.E. (1948) *J. Biol. Chem. 172*, 619–636.
3. Fleischer, S., & Kervina, M. (1974) *Methods Enzymol. 31*, 6–41.
4. Leighton, F., Poole, B., Beaufay, H., Baudhuin, P., Coffey, J.W., Fowler, S., & DeDuve, C. (1968) *J. Cell Biol. 37*, 482–513.
5. Palade, G.E., & Siekevitz, P. (1956) *J. Cell Biol. 2*, 171–200.
6. Price, C.A. (1974) *Methods Enzymol. 31*, 501–519.
7. Schneider, W.C., Hogeboom, G.H., & Striebich, M.J. (1953) *Cancer Res. 13*, 617–632.
8. McKeel, D.W., & Jarett, L. (1970) *J. Cell Biol. 44*, 417–432.
9. Heidrich, H.G., Kinne, R., Kinne-Saffran, E., & Hannig, K. (1972) *J. Cell Biol. 54*, 232–245.
10. Booth, A.G., & Kenny, A.J. (1974) *Biochem. J. 142*, 575–581.
11. DePierre, J.W., & Karnovsky, M.L. (1973) *J. Cell Biol. 56*, 275–303.
12. Fleischer, B., & Zambrano, F. (1974) *J. Biol. Chem. 249*, 5995–6003.
13. Chang, K.J., Bennett, V., & Cuatracasas, P. (1975) *J. Biol. Chem. 250*, 488–500.
14. Dallner, G. (1963) *Acta Pathol. Microbiol. Scand. Suppl. 166*, 1–94.
15. Nobel, P.S. (1974) *Methods Enzymol. 31*, 600–606.
16. Neville, D.M. (1960) *J. Biophys. Biochem. Cytol. 8*, 413–422.
17. Morgan, I.G., Wolfe, L.S., Mandel, P., & Gombos, G. (1971) *Biochim. Biophys. Acta 241*, 737–751.
18. Weinstein, D.B., Marsh, J.B., Glick, M.C., & Warren, L. (1969) *J. Biol. Chem. 244*, 4103–4111.
19. Atkinson, P.H., & Summers, P.F. (1971) *J. Biol. Chem. 246*, 5162–5175.
20. Wiley, W.R. (1974) *Methods Enzymol. 31*, 609–626.
21. Kaback, H.R. (1968) *J. Biol. Chem. 243*, 3711–3724.
22. Tanford, C. (1973) *The Hydrophobic Effect: Formation of Micelles and Biological Membranes*, John Wiley & Sons, New York.
23. Farquhar, J.W. (1962) *J. Lipid Res. 3*, 21–30.
24. Rouser, G., Nelson, G.J., Fleischer, S., & Simon, G. (1968) in *Biological Membranes, Physical Fact and Function* (Chapman, D., Ed.) pp 5–69, Academic Press, London.
25. Erwin, J.A. (1973) *Lipids and Biomembranes of Eukaryotic Microorganisms*, Academic Press, New York.
26. Kaback, H.R. (1971) *Methods Enzymol. 22*, 99–120.
27. Cronan, J.E., & Vagelos, P.R. (1972) *Biochim. Biophys. Acta 265*, 25–60.
28. Shaw, N., Smith, P.F., & Koostra, W.L. (1968) *Biochem. J. 107*, 329–333.
29. Fiehn, W., Peter, J.B., Mead, J.F., & Gan-Elepano, M. (1971) *J. Biol. Chem. 246*, 5617–5620.
30. Bangham, A.D., Standish, M.M., & Watkins, J.C. (1965) *J. Mol. Biol. 13*, 238–252.
31. Huang, C. (1969) *Biochemistry 8*, 344–352.
32. Mueller, P., Rudin, D.O., Tien, H.T., & Wescott, W.C. (1962) *Nature 194*, 979–980.
33. Montal, M., & Mueller, P. (1972) *Proc. Natl. Acad. Sci. U.S.A. 69*, 3561–3566.
34. Levine, Y.K., & Wilkins, M.H.F. (1971) *Nature New Biol. 230*, 69–72.
35. Reiss-Husson, F. (1967) *J. Mol. Biol. 25*, 363–382.
36. Small, D.M. (1967) *J. Lipid Res. 8*, 551–557.
37. Pearson, R.H., & Pascher, I. (1979) *Nature 281*, 499–501.
38. Hitchcock, P.B., Mason, R., Thomas, K.M., & Shipley, G.G. (1974) *Proc. Natl. Acad. Sci. U.S.A. 71*, 3036–3040.
39. Hauser, H., Pascher, I., & Sundell, S. (1988) *Biochemistry 27*, 9166–9174.
40. Hanai, T., Haydon, D.A., & Taylor, J. (1965) *J. Theor. Biol. 9*, 278–296.
41. McElhaney, R.N., & Tourtellotte, M.E. (1969) *Science 164*, 433–434.
42. Engelman, D.M. (1971) *J. Mol. Biol. 58*, 153–165.
43. Pascher, I., & Sundell, S. (1981) *J. Mol. Biol. 153*, 791–806.
44. Chapman, D., Williams, R.M., & Ladbrooke, D.B. (1967) *Chem. Phys. Lipids 1*, 445–475.
45. Buldt, G., Gally, H.U., Seelig, A., Seelig, J., & Zaccai, G. (1978) *Nature 271*, 182–184.

46. Mabrey, S., Mateo, P.L., & Sturtevant, J.M. (1978) *Biochemistry 17*, 2464–2468.

47. Shimshick, E.J., & McConnell, H.M. (1973) *Biochemistry 12*, 2351–2360.

48. Oldfield, E., & Chapman, D. (1972) *FEBS Lett. 23*, 285–296.

49. Chapman, D., Urbina, J., & Keough, K.M. (1974) *J. Biol. Chem. 249*, 2512–2521.

50. Keana, J.W., Keana, S.B., & Beetham, D. (1967) *J. Am. Chem. Soc. 89*, 3055–3056.

51. Stone, T.J., Buckman, T., Nordio, P.L., & McConnell, H.M. (1965) *Proc. Natl. Acad. Sci. U.S.A. 54*, 1010–1017.

52. McConnell, H.M., & McFarland, B.G. (1970) *Q. Rev. Biophys. 3*, 91–136.

53. Libertini, L.J., Waggoner, A.S., Jost, P.C., & Griffith, O.H. (1969) *Proc. Natl. Acad. Sci. U.S.A. 64*, 13–19.

54. Hubbell, W.L., & McConnell, H.M. (1971) *J. Am. Chem. Soc. 93*, 314–326.

55. McFarland, B.G., & McConnell, H.M. (1971) *Proc. Natl. Acad. Sci. U.S.A. 68*, 1274–1278.

56. Seelig, A., & Seelig, J. (1974) *Biochemistry 13*, 4839–4845.

57. Seelig, J., & Niedererger, W. (1974) *J. Am. Chem. Soc. 96*, 2069–2072.

58. Seelig, J., & Browning, J.L. (1978) *FEBS Lett. 92*, 41–44.

59. Tardieu, A., Luzzati, V., & Reman, F.C. (1973) *J. Mol. Biol. 75*, 711–733.

60. Griffith, O.H., Dehlinger, P.J., & Van, S.P. (1974) *J. Membr. Biol. 15*, 159–192.

61. Sheetz, M.P., & Chan, S.I. (1972) *Biochemistry 11*, 4573–4581.

62. Dill, K.A., & Flory, P.J. (1981) *Proc. Natl. Acad. Sci. U.S.A. 78*, 676–680.

63. Papahadjopoulos, D., Vail, W.J., Jacobson, K., & Poste, G. (1975) *Biochim. Biophys. Acta 394*, 483–491.

64. Lenard, J., & Rothman, J.E. (1976) *Proc. Natl. Acad. Sci. U.S.A. 73*, 391–395.

65. Kroon, P.A., Kainosho, M., & Chan, S.I. (1975) *Nature 256*, 582–584.

66. Rothman, J.E., & Engelman, D.M. (1972) *Nature New Biol. 237*, 42–44.

67. Oldfield, E., & Chapman, D. (1971) *Biochem. Biophys. Res. Commun. 43*, 610–616.

68. Lecuyer, H., & Dervichian, D.G. (1969) *J. Mol. Biol. 45*, 39–57.

69. Shimshick, E.J., & McConnell, H.M. (1973) *Biochem. Biophys. Res. Commun. 53*, 446–451.

70. DeKriyff, B., Demel, R.A., Slotboom, A.J., VanDeenan, L.L.M., & Rosenthal, A.F. (1973) *Biochim. Biophys. Acta 307*, 1–19.

71. Engelman, D.M., & Rothman, J.E. (1972) *J. Biol. Chem. 247*, 3694–3697.

72. Bangham, A.D., & Dawson, R.M.C. (1960) *Biochem. J. 75*, 133–138.

73. Fettiplace, R., Andrews, D.M., & Haydon, D.A. (1971) *J. Membr. Biol. 5*, 277–296.

74. Haydon, D.A., & Taylor, F.H. (1960) *Philos. Trans. R. Soc. London, A 252*, 225–248.

75. Haydon, D.A., & Taylor, J. (1963) *J. Theor. Biol. 4*, 281–296.

76. Tanford, C. (1974) *Proc. Natl. Acad. Sci. U.S.A. 71*, 1811–1815.

77. Dill, K.A., Koppel, D.E., Cantor, R.S., Dill, J.D., Bendedouch, D., & Chen, S.H. (1984) *Nature 309*, 42–45.

78. Luzzati, V., & Husson, F. (1962) *J. Cell Biol. 12*, 207–219.

79. Stoeckenius, W. (1962) *J. Cell Biol. 12*, 221–229.

80. Singer, S.J., & Nicholson, G.L. (1972) *Science 175*, 720–731.

81. Rosenberg, S.A., & Guidotti, G. (1969) *J. Biol. Chem. 244*, 5118–5124.

82. Steck, T.L., Weinstein, R.S., Straus, J.H., & Wallach, D.F.H. (1970) *Science 168*, 255–257.

83. Steck, T.L., & Yu, J. (1973) *J. Supramol. Struct. 1*, 220–232.

84. Carpenter, G. (1987) *Annu. Rev. Biochem. 56*, 881–914.

85. Yarden, Y., & Ullrich, A., (1988) *Annu. Rev. Biochem. 57*, 443–478.

86. Ullrich, A., Coussens, L., Hayflick, J.S., Dull, T.J., Gray, A., Tam, A.W., Lee, J., Yarden, Y., Libermann, T.A., Schlessinger, J., Downward, J., Mayes, E.L.V., Whittle, N., Waterfield, M.D., & Seeburg, P.H. (1984) *Nature 309*, 418–425.

87. Wedegaertner, P.B., & Gill, G.N. (1989) *J. Biol. Chem. 264*, 11346–11353.

88. Lax, I., Mitra, A.K., Ravera, C., Hurwitz, D.R., Rubinstein, M., Ullrich, A., Stroud, R.M., & Schlessinger, J. (1991) *J. Biol. Chem. 266*, 13828–13833.

89. Ferguson, M.A.J., & Williams, A.F. (1988) *Annu. Rev. Biochem. 57*, 285–320.

90. Micanovic, R., Kodulka, K., Gerber, L.D., & Udenfriend, S. (1990) *Proc. Natl. Acad. Sci. U.S.A. 87*, 7939–7943.

91. Ikezawa, H., Yamanegi, M., Taguchi, R., Miyashita, T., & Ohyabu, T. (1976) *Biochim. Biophys. Acta 450*, 154–164.

92. Glossmann, H., & Neville, D.M. (1971) *J. Biol. Chem. 246*, 6339–6346.

93. Zampighi, G., Kyte, J., & Freytag, W. (1984) *J. Cell Biol. 98*, 1851–1864.

94. Jørgensen, P.L. (1974) *Biochim. Biophys. Acta 356*, 36–52.

95. Makino, S., Reynolds, J.A., & Tanford, C. (1973) *J. Biol. Chem. 248*, 4926–4932.

96. Clarke, S. (1975) *J. Biol. Chem. 250*, 5459–5469.

97. Ribiero, A.A., & Dennis, E.A. (1973) *Biochim. Biophys. Acta 332*, 26–35.

98. Simons, K., Helenius, A., & Garoff, H. (1973) *J. Mol. Biol. 80*, 119–133.

99. Helenius, A., & Söderlund, H. (1973) *Biochim. Biophys. Acta 307*, 287–300.

100. Spatz, L., & Strittmatter, P. (1971) *Proc. Natl. Acad. Sci. U.S.A. 68*, 1042–1046.

101. Strittmatter, P., Rogers, M.J., & Spatz, L. (1972) *J. Biol. Chem. 247*, 7188–7194.

102. Springer, T.A., Strominger, J.L., & Mann, D. (1974) *Proc. Natl. Acad. Sci. U.S.A. 71*, 1539–1543.

103. Springer, T.A., Kaufman, J.F., Terhorst, C., & Strominger, J.L. (1977) *Nature 268*, 213–217.

104. MacNair, R.D., & Kenny, A.J. (1979) *Biochem. J. 179*, 379–395.

105. Kenney, J., Booth, A.G., George, S.G., Ingram, J., Kershaw, D., Wood, E.J., & Young, A.R. (1976) *Biochem. J. 155*, 169–182.

106. Maroux, S., & Louvard, D. (1976) *Biochim. Biophys. Acta 419*, 189–195.

107. Brunner, J., Hauser, H., Braun, H., Wilson, K.J., Wacker, H., O'Neill, B., & Semenza, G. (1979) *J. Biol. Chem. 254*, 1821–1828.

108. Skehel, J.J., & Waterfield, M.D. (1975) *Proc. Natl. Acad. Sci. U.S.A. 72*, 93–97.

109. Pattus, F., Verger, R., & Desnuelle, P. (1976) *Biochem. Biophys. Res. Commun. 69*, 718–723.

110. Wilson, I.A., Skehel, J.J., & Wiley, D.C. (1981) *Nature 289*, 366–373.

111. Mathews, F.S., Argos, P., & Levine, M. (1971) *Cold Spring Harbor Symp. Quant. Biol. 36*, 387–395.

112. Tomita, M., & Marchesi, V.T. (1975) *Proc. Natl. Acad. Sci. U.S.A. 72*, 2964–2968.

113. Bretscher, M.S. (1971) *Nature New Biol. 231*, 229–232.

114. Thomas, D.B., & Winzler, R.J. (1969) *J. Biol. Chem. 244*, 5943–5946.

115. Chin, D.J., Gil, G., Russell, D.W., Liscum, L., Luskey, K.L., Basu, S.K., Okayama, H., Berg, P., Goldstein, J.L., & Brown, M.S. (1984) *Nature 308*, 613–617.

116. Ozols, J., & Gerard, C. (1977) *J. Biol. Chem. 252*, 8549–8553.

117. Malissen, M., Malissen, B., & Jordan, B.R. (1982) *Proc. Natl. Acad. Sci. U.S.A. 79*, 893–897.

118. Frank, G., Brunner, J., Hauser, H., Wacker, H., Semenza, G., & Zuber, H. (1978) *FEBS Lett. 96*, 183–188.

119. Verhoeyen, M., Fang, R., Jou, W.M., Devos, R., Huylebroeck, D., Saman, E., & Fiers, W. (1980) *Nature 286*, 771–776.

120. Tomita, M., Furthmayr, H., & Marchesi, V.T. (1978) *Biochemistry 17*, 4756–4770.

121. Kyte, J., & Doolittle, R.F. (1982) *J. Mol. Biol. 157*, 105–132.

122. Louvard, D., Semeriva, M., & Maroux, S. (1976) *J. Mol. Biol. 106*, 1023–1035.

123. Schulte, T.H., & Marchesi, V.T. (1979) *Biochemistry 18*, 275–279.

124. Visser, L., Robinson, N.C., & Tanford, C. (1975) *Biochemistry 14*, 1194–1199.

125. Spiess, M., Brunner, J., & Semenza, G. (1982) *J. Biol. Chem. 257*, 2370–2377.

126. Michel, H. (1982) *J. Mol. Biol. 158*, 567–572.

127. Deisenhofer, J., Epp, O., Miki, K., Huber, R., & Michel, H. (1985) *Nature 318*, 618–624.

128. Allen, J.P., Feher, G., Yeates, T.U., Komiya, H., & Rees, D.C. (1987) *Proc. Natl. Acad. Sci. U.S.A. 84*, 6162–6166.

129. Williams, J.C., Steiner, L.A., Ogden, R.C., Simon, M.I., & Feher, G. (1983) *Proc. Natl. Acad. Sci. U.S.A. 80*, 6505–6509.

130. Williams, J.C., Steiner, L.A., Feher, G., & Simon, M.I. (1984) *Proc. Natl. Acad. Sci. U.S.A. 81*, 7303–7307.

131. Michel, H., Weyer, K.A., Gruenberg, H., & Lottspeich, F. (1985) *EMBO J. 4*, 1667–1672.

132. Michel, H., Epp, O., & Deisenhofer, J. (1986) *EMBO J. 5*, 2445–2451.

133. Blaurock, A.E., & Stoeckenius, W. (1971) *Nature New Biol. 233*, 152–155.

134. Henderson, R., Capaldi, R.A., & Leigh, J.S. (1977) *J. Mol. Biol. 112*, 631–648.

135. Vergara, J., Longley, W., & Robertson, J.D. (1969) *J. Mol. Biol. 46*, 593–596.

136. Unwin, P.N.T., & Henderson, R. (1975) *J. Mol. Biol. 94*, 425–440.

137. Unwin, P.N.T., & Ennis, P.D. (1984) *Nature 307*, 609–613.

138. Henderson, R., Capaldi, R.A., & Leigh, J.S. (1977) *J. Mol. Biol. 112*, 631–648.

139. Crowther, R.A., DeRosier, D.J., & Klug, A. (1970) *Proc. R. Soc. London A 317*, 319–340.

140. Henderson, R., & Unwin, P.N.T. (1975) *Nature 257*, 28–32.

141. Henderson, R., Baldwin, J.M., Ceska, T.A., Zemlin, F., Beckmann, E., & Downing, K.H. (1990) *J. Mol. Biol. 213*, 899–929.

142. Khorana, H.G., Gerber, G.E., Herlihy, W.C., Gray, C.P., Anderegg, R.J., Nihei, K., & Biemann, K. (1979) *Proc. Natl. Acad. Sci. U.S.A. 76*, 5046–5050.

143. Ovchinnikov, Y., Abdulaev, N.F., Feigina, M.Y., Kiselev, A.V., & Lobanov, N.A. (1979) *FEBS Lett. 100*, 219–224.

144. Mogi, T., Stern, L.J., Marti, J., Chao, B.H., & Khorana, H.G. (1988) *Proc. Natl. Acad. Sci. U.S.A. 85*, 4148–4152.

145. Crick, F.H.C. (1953) *Acta Crystallogr. 6*, 689–697.

146. Rosenbusch, J.P. (1974) *J. Biol. Chem. 24*, 8019–8029.

147. Chen, R., Kramer, C., Schmidmayr, W., Chen-Schmeisser, U., & Henning, U. (1982) *Biochem. J. 203*, 33–43.

148. Weiss, M.S., Kreusch, A., Schiltz, E., Nestel, U., Welte, W., Weckesser, J., & Schulz, G.E. (1991) *FEBS Lett. 280*, 379–382.

149. Benz, R., & Bauer, K. (1988) *Eur. J. Biochem. 176*, 1–19.

150. Kleffel, B., Garavito, R.M., Baumeister, W., & Rosenbusch, J.P. (1985) *EMBO J. 4*, 1589–1592.

151. Farber, G.K., & Petsko, G.A. (1990) *Trends Biochem. Sci. 15*, 228–234.

152. Banner, D.W., Bloomer, A.C., Petsko, G.A., Phillips, D.C., Pogson, C.I., Wilson, I.A., Corran, P.H., Furth, A.J., Milman, J.D., Offord, R.E., Priddle, J.D., & Wale, S.G. (1975) *Nature 255*, 609–614.

153. Sixman, T.K., Pronk, S.E., Kalk, K.H., Wartna, E.S., vanZanten, B.A.M., Witholt, B., & Hol, W.G.J. (1991) *Nature 351*, 371–377.

154. Jost, P.C., Griffith, O.H., Capaldi, R.A., & Vanderkooi, G. (1973) *Proc. Natl. Acad. Sci. U.S.A. 70*, 480–484.

155. Silvius, J.R., McMillan, D.A., Saley, N.D., Jost, P.C., & Griffith, O.H. (1984) *Biochemistry 23*, 538–547.

156. East, J.M., Melville, D., & Lee, A.G. (1985) *Biochemistry 24*, 2615–2623.

157. Griffith, O.H., McMillen, D.A., Keana, J.F.W., & Jost, P.C. (1986) *Biochemistry 25*, 574–584.

158. Brotherus, J., Griffith, O.H., Brotherus, M.O., Jost, P.C., Silvius, J.R., & Hokin, L.E. (1981) *Biochemistry 20*, 5261–5267.

159. Seelig, J., Tamm, L., Hymel, L., & Fleischer, S. (1981) *Biochemistry 20*, 3922–3932.

160. Engelman, D.M., Steitz, T.A., & Goldman, A. (1986) *Annu. Rev. Biophys. Biophys. Chem. 15*, 321–353.

161. Haniu, M., Armes, L.G., Tanaka, M., Yasunobu, K.T., Shastry, B.S., Wagner, G.C., & Gunsalus, I.C. (1982) *Biochem. Biophys. Res. Commun. 105*, 889–894.

162. Gonzalez, F.J., Nerbert, D.W., Hardwick, J.P., & Kasper, C.B. (1985) *J. Biol. Chem. 260*, 7435–7441.

163. Haugen, D.A., & Coon, M.J. (1976) *J. Biol. Chem. 251*, 7929–7939.

164. Heineman, F.S., & Ozols, J. (1982) *J. Biol. Chem. 257*, 14988–14999.

165. Haugen, D.A., Armes, L.G., Yasunobu, K.T., & Coon, M.J. (1977) *Biochem. Biophys. Res. Commun. 77*, 967–973.

166. Eisenberg, D., Schwartz, E., Komaromy, M., & Wall, R. (1984) *J. Mol. Biol. 179*, 125–142.

167. Bayley, H., & Knowles, J.R. (1980) *Biochemistry 19*, 3883–3892.

168. Prochaska, L., Bisson, R., & Capaldi, R. (1980) *Biochemistry 19*, 3174–3179.

169. Bercovici, T., & Gitler, C. (1978) *Biochemistry 17*, 1484–1489.

170. Malatesta, F., Darley-Usmar, V., deJong, C., Prochaska, L., Bisson, R., Capaldi, R.A., Steffans, G.C.M., & Buse, G. (1983) *Biochemistry 22*, 4405–4411.

171. Nicholas, R.A. (1984) *Biochemistry 23*, 888–898.

172. Brunner, J., & Richards, F.M. (1980) *J. Biol. Chem. 255*, 3319–3329.

173. Ross, A.H., Radhakrishnan, R., Robson, R.J., & Khorana, H.G. (1982) *J. Biol. Chem. 257*, 4152–4161.

174a. Bisson, R., & Montecucco, C. (1981) *Biochem. J. 193*, 757–763.

174b. Brunner, J., Franzusoff, A.J., Lüscher, B., Zugliani, C., & Semenza, G. (1985) *Biochemistry 24*, 5422–5430.

175. Brunner, J., & Semenza, G. (1981) *Biochemistry 20*, 7174–7182.

176. Brunner, J., Spiess, M. Aggeler, R., Huber, P., & Semenza, G. (1983) *Biochemistry 22*, 3812–3820.

177. Burnett, B.K., Robson, R.J., Takagaki, Y., Radhakrishnan, R., & Khorana, H.G. (1985) *Biochim. Biophys. Acta 815*, 57–67.

178. Shull, G.E., Schwartz, A., & Lingrel, J.B. (1985) *Nature 316*, 691–695.

179. Schneider, D.L., Burnside, J., Gorga, F.R., & Nettleton, C.J. (1978) *Biochem. J. 176*, 75–82.

180. Bretscher, M.S. (1971) *J. Mol. Biol. 59*, 351–357.

181. Evans, R.M., & Fink, L.M. (1977) *Proc. Natl. Acad. Sci. U.S.A. 74*, 5341–5344.

182. Schnaitman, C., & Greenawalt, J.W. (1968) *J. Cell Biol. 38*, 158–175.

183. Clarke, S. (1976) *J. Biol. Chem. 251*, 1354–1363.

184. Hackenbrock, C.R., & Hammon, K.M. (1975) *J. Biol. Chem. 250*, 9185–9197.

185. St. John, P.A., Froehner, S.C., Goodenough, D.A., & Cohen, J.B. (1982) *J. Cell Biol. 92*, 333–342.

186. Karpen, J.W., & Hess, G.P. (1986) *Biochemistry 25*, 1777–1785.

187. Forbush, B. (1982) *J. Biol. Chem. 257*, 12678–12684.

188. Kagawa, Y., & Racker, E. (1971) *J. Biol. Chem. 246*, 5477–5487.

189. Fung, B.K.K., & Hubbell, W.L. (1978) *Biochemistry 17*, 4403–4410.

190. O'Connell, M.A. (1982) *Biochemistry 21*, 5984–5991.

191. Berg, H.C. (1969) *Biochim. Biophys. Acta 183*, 65–78.

192. Bretscher, M.S. (1971) *J. Mol. Biol. 58*, 775–781.

193. Whiteley, N.M., & Berg, H.C. (1974) *J. Mol. Biol. 87*, 541–561.

194. Cabantchik, I.Z., Balshin, M., Breur, W., & Rothstein, A. (1975) *J. Biol. Chem. 250*, 5130–5136.

195. Bogner, W., Aquila, H., & Klingenberg, M. (1982) *FEBS Lett. 146*, 259–261.

196. Ludwig, B., Downer, N.W., & Capaldi, R.A. (1979) *Biochemistry 18*, 1401–1407.

197. Kyte, J., Xu, K., & Bayer, R. (1987) *Biochemistry 26*, 8350–8360.

198. Bangham, A.D., & Horne, R.W. (1962) *Nature 196*, 952–953.

199. Bender, W.W., Garan, H., & Berg, H.C. (1971) *J. Mol. Biol. 58*, 783–797.

200. Drickamer, L.K. (1976) *J. Biol. Chem. 251*, 5115–5123.

201. Jennings, M.L., Adams-Lackey, M., & Denney, G.H. (1984) *J. Biol. Chem. 259*, 4652–4660.

202. Tanner, M.J.A., Martin, P.G., & High, S. (1988) *Biochem. J. 256*, 703–712.

203. LaRochelle, W.J., Wray, B.E., Sealock, R., & Froehner, S.C. (1985) *J. Cell Biol. 100*, 684–691.

204. Dutton, A., & Singer, S.J. (1975) *Proc. Natl. Acad. Sci. U.S.A. 72*, 2568–2571.

205. Cone, R.A. (1972) *Nature New Biol. 236*, 39–43.

206. Torres, C., & Hart, G.W. (1984) *J. Biol. Chem. 259*, 3308–3317.

207. Nicolson, G.L., & Singer, S.J. (1971) *Proc. Natl. Acad. Sci. U.S.A. 68*, 942–945.

208. Nicolson, G.L., & Singer, S.J. (1974) *J. Cell Biol. 60*, 236–248.

209. Schindler, M., Hogan, M., Miller, R., & DeGaetano, D. (1987) *J. Biol. Chem. 262*, 1254–1260.

210. Bonitz, S.G., Coruzzi, G., Thalenfeld, B.E., & Tzagoloff, A. (1980) *J. Biol. Chem. 255*, 11927–11941.

211. Buse, G., & Steffens, G.J. (1978) *Hoppe-Seyler's Z. Physiol. Chem. 359*, 1005–1009.

212. Kaul, R.K., Murthy, S.N.P., Reddy, A.G., Steck, T.L., & Kohler, H. (1983) *J. Biol. Chem. 258*, 7981–7990.

213. Varghese, J.N., Laver, W.G., & Colman, P.M. (1983) *Nature 303*, 35–40.

214. Brisson, A., & Unwin, P.N.T. (1985) *Nature 315*, 474–477.

215. Unwin, P.N.T., & Zampighi, G. (1980) *Nature 283*, 545–549.

216. Robertson, J.D. (1963) *J. Cell Biol. 19*, 201–221.

217. Brisson, A., & Wade, R.H. (1983) *J. Mol. Biol. 166*, 21–36.

218. Ho, M.K., & Guidotti, G. (1975) *J. Biol. Chem. 250*, 675–683.

219. Kopito, R.R., & Lodish, H.F. (1985) *Nature 316*, 234–238.

220. Steck, T.L., Ramos, B., & Strapazon, E. (1976) *Biochemistry 15*, 1154–1161.

221. Appell, K.C., & Low, P.S. (1981) *J. Biol. Chem. 250*, 11104–11111.

222. Yu, J., & Steck, T.L. (1975) *J. Biol. Chem. 250*, 9176–9184.

223. Grinstein, S., Ship, S., & Rothstein, A. (1978) *Biochim. Biophys. Acta. 507*, 294–304.

224. Mawby, W.J., & Findlay, J.B.C. (1982) *Biochem. J. 205*, 465–475.

225. Jennings, M.L., & Nicknish, J.S. (1984) *Biochemistry 23*, 6432–6436.

226. Drickamer, L.K. (1976) *J. Biol. Chem. 251*, 5115–5123.

227. Brock, C.J., Tanner, M.J.A., & Kempf, C. (1983) *Biochem. J. 213*, 577–586.

228. Jennings, M.L., Anderson, M.D., & Monaghan, R. (1986) *J. Biol. Chem. 261*, 9002–9010.

229. Lieberman, D.M., & Reithmeier, R.A.F. (1988) *J. Biol. Chem. 263*, 10022–10028.

230. Kyte, J. (1981) *Nature 292*, 201–204.

231. Serrano, R.S., Kielland-Brandt, M.C., & Fink, G.R. (1986) *Nature 319*, 689–693.

232. Walderhaug, M.O., Post, R.L., Saccomani, G., Leonard, R.T., & Briskin, D.P. (1985) *J. Biol. Chem. 260*, 3852–3859.

233. Niggli, V., Penniston, J.T., & Carafoli, E. (1979) *J. Biol. Chem. 254*, 9955–9958.

234. Filoteo, A.G., Gorski, J.P., & Penniston, J.T. (1987) *J. Biol. Chem. 262*, 6526–6530.

235. Craig, W.S., & Kyte, J. (1980) *J. Biol. Chem. 255*, 6262–6269.

236. Shull, G.E., Lane, L., & Lingrel, J.B. (1986) *Nature 321*, 429–431.

237. Reuben, M.A., Lasater, L.S., & Sachs, G. (1990) *Proc. Natl. Acad. Sci. U.S.A. 87*, 6767–6771.

238. Herbert, H., Jorgensen, P.L., Skriver, E., & Maunsbach, A.B. (1982) *Biochim. Biophys. Acta 689*, 571–574.

239. Zampighi, G., Kyte, J., & Freytag, W. (1984) *J. Cell Biol. 98*, 1851–1864.

240. Dux, L., Taylor, K.A., Ting-Beall, H.P., & Martonosi, A. (1985) *J. Biol. Chem. 260*, 11730–11743.

241. Craig, W.S. (1982) *Biochemistry 21*, 5707–5717.

242. Møller, J.V., Lind, K.E., & Andersen, J.P. (1980) *J. Biol. Chem. 255*, 1912–1920.

243. Taylor, K.A., Dux, L., & Martonosi, A. (1986) *J. Mol. Biol. 187*, 417–427.

244. Hebert, H., Skriver, E., Söderholm, M., & Maunsbach, A. (1988) *J. Ultrastruct. Mol. Struct. Res. 100*, 86–93.

245. Kyte, J. (1987) in *Perspectives of Biological Energy Transduction* (Mukahota, Y., Morales, M.F., & Fleischer, S., Eds.) pp 231–240, Academic Press, Tokyo.

246. MacLennan, D.H., Brandl, C.J., Korczak, B., & Green, N.M. (1985) *Nature 316*, 696–700.

247. Jorgensen, P.L., Karlish, S.J.D., & Gitler, C. (1982) *J. Biol. Chem. 257*, 7435–7442.

248. Castro, J., & Farley, R.A. (1979) *J. Biol. Chem. 254*, 2221–2228.

249. Stewart, P.S., MacLennan, D.H., & Shamoo, A.E. (1976) *J. Biol. Chem. 251*, 712–719.

250. Shull, G.E., Schwartz, A., & Lingrel, J.B. (1985) *Nature 316*, 691–695.

251. Allen, G., Trinnaman, B.J., & Green, N.M. (1980) *Biochem. J. 187*, 591–596.

252. Bastide, F., Meissner, G., Fleischer, S., & Post, R.L. (1973) *J. Biol. Chem. 248*, 8385–8391.

253. Mandala, S.M., & Slayman, C.W. (1989) *J. Biol. Chem. 264*, 16276–16281.

254. James, P., Maeda, M., Fischer, R., Verma, A.K., Krebs, J., Penniston, J.T., & Carafoli, E. (1988) *J. Biol. Chem. 263*, 2905–2910.

255. Esmann, M., Hideg, K., & Marsh, D. (1988) *Prog. Clin. Biol. Res. 268A*, 191–196.

256. Ludwig, B., Downer, N.W., & Capaldi, R.A. (1979) *Biochemistry 18*, 1401–1407.

257. Jarausch, J., & Kadenbach, B. (1985) *Eur. J. Biochem. 146*, 219–225.

258. Carroll, R.C., & Racker, E. (1977) *J. Biol. Chem. 252*, 6981–6990.

259. Zhang, Y.Z., Georgevich, G., & Capaldi, R. (1984) *Biochemistry 23*, 5616–5621.

260. Anderson, S., deBruijn, M.H.L., Coulson, A.R., Eperon, I.C., Sanger, F., & Young, I.G. (1982) *J. Mol. Biol. 156*, 683–717.

261. Prochaska, L., Bisson, R., & Capaldi, R.A. (1980) *Biochemistry 19*, 3174–3179.

262. Bisson, R., Steffens, G.C.M., & Buse, G. (1982) *J. Biol. Chem. 257*, 6716–6720.

263. Prochaska, L., Bisson, R., Capaldi, R.A., Steffens, G.C.M., & Buse, G.(1980) *Biochim. Biophys. Acta 637*, 360–373.

264. Sacher, R., Steffins, G.J., & Buse, G. (1979) *Hoppe-Zeyler's Z. Physiol. Chem. 360*, 1385–1392.

265. Malatesta, F., & Capaldi, R. (1982) *Biochem. Biophys. Res. Commun. 109*, 1180–1185.

266. Tanaka, M., Haniu, M., Yasunobu, K.T., Yu, C.A., Yu, L., Wei, Y.H., & King, T.E. (1979) *J. Biol. Chem. 254*, 3879–3885.

267. Cerletti, N., & Schatz, G. (1979) *J. Biol. Chem. 254*, 7746–7751.

268. Deatherage, J.F., Henderson, R., & Capaldi, R.A. (1982) *J. Mol. Biol. 158*, 501–514.

269. Fuller, S.D., Capaldi, R.A., & Henderson, R. (1979) *J. Mol. Biol. 134*, 305–327.

270. Lindstrom, J., & Patrick, J. (1974) in *Synaptic Transmission and Neuronal Interaction* (Bennett, M.V.L., Ed.) pp 191–216, Raven Press, New York.

271. Weill, C.L., McNamee, M.G., & Karlin, A. (1974) *Biochem. Biophys. Res. Commun. 61*, 997–1003.

272. Reed, K., Vandlen, R., Bode, J., Duguid, J., & Raftery, M.A. (1975) *Arch. Biochem. Biophys. 167*, 138–144.

273. Boulter, J., Luyten, W., Evans, K., Mason, P., Ballivet, M., Goldman, D., Stengelin, S., Martin, G., Heinemann, S., & Patrick, J. (1985) *J. Neurosci. 5*, 2545–2552.

274. Neubig, R.R., Krodel, E.K., Boyd, N.D., & Cohen, J.B. (1979) *Proc. Natl. Acad. Sci. U.S.A. 76*, 690–694.

275. Moore, H.P.H., Hartig, P.R., & Raftery, M.A. (1979) *Proc. Natl. Acad. Sci. U.S.A. 76*, 6265–6269.

276. Vandlen, R.L., Wu, W.C.S., Eisenach, J.C., & Raftery, M.A. (1979) *Biochemistry 18*, 1845–1854.

277. Kellaris, K.V., & Ware, D.K. (1989) *Biochemistry 28*, 3469–3482.

278. Noda, M., Takahashi, H., Tanabe, T., Toyosato, M., Furutani, Y., Hirose, T., Asai, M., Inayama, S., Miyata, T., & Numa, S. (1982) *Nature 299*, 793–797.

279. Noda, M., Takahashi, H., Tanabe, T., Toyosato, M., Kikyotani, S., Furatani, Y., Hirose, T., Takashima, H., Inayama, S., Miyata, T., & Numa, S. (1983) *Nature 302*, 528–531.

280. Claudio, T., Ballivet, M., Patrick, J., & Heinemann, S. (1983) *Proc. Natl. Acad. Sci. U.S.A. 80*, 1111–1115.

281. Reynolds, J.A., & Karlin, A. (1978) *Biochemistry 17*, 2035–2038.

282. Raftery, M.A., Hunkapiller, M.W., Strader, C.D., & Hood, L.E. (1980) *Science 208*, 1454–1457.

283. Toyoshima, C., & Unwin, P.N.T. (1990) *J. Cell Biol. 111*, 2623–2635.

284. Imoto, K., Busch, C., Sakmann, B., Mishina, M., Konno, T., Nakai, J., Bujo, H., Mori, Y., Fukuda, K., & Numa, S. (1988) *Nature 335*, 645–648.

285. Leonard, R.J., Labarca, C.G., Charnet, P., Davidson, N., & Lester, H.A. (1988) *Science 242*, 1578–1581.

286. Hucho, F., Oberthur, W., & Lottspeich, F. (1986) *FEBS Lett. 205*, 137–142.

287. Oberthur, W., Muhn, P., Baumann, H., Lottspeich, F., Wittman-Leibold, B., & Hucho, F. (1986) *EMBO J. 5*, 1815–1819.

288. Revah, F., Galzi, J., Giraudat, J., Haumont, P., Lederer, F., & Changeux, J. (1990) *Proc. Natl. Acad. Sci. U.S.A. 87*, 4675–4679.

289. Kersh, G.J., Tomich, J.M., & Montal, M. (1989) *Biochem. Biophys. Res. Commun. 162*, 352–356.

290. Dwyer, B.P. (1991) *Biochemistry 30*, 4105–4112.

291. Ellena, J.F., Blazing, M.A., & McNamee, M.G. (1983) *Biochemistry 22*, 5523–5535.

292. Jost, P., Griffith, O.H., Capaldi, R.A., & Vanderkooi, G. (1973) *Proc. Natl. Acad. Sci. U.S.A. 70*, 480–484.

293. Caspar, D.L.D., & Kirschner, D.A. (1971) *Nature New Biol. 231*, 46–52.

294. Wilkins, M.H.F., Blaurock, A.E., & Engelman, D.M. (1971) *Nature New Biol. 230*, 72–76.

295. Engelman, D.M. (1971) *J. Mol. Biol. 58*, 153–165.

296. Eletr, S., & Inesi, G. (1972) *Biochim. Biophys. Acta 282*, 174–179.

297. Hubbell, W.L., & McConnell, H.M. (1969) *Proc. Natl. Acad. Sci. U.S.A. 64*, 20–27.

298. McConnell, H.M., Wright, K.L., & McFarland, B.G. (1972) *Biochem. Biophys. Res. Commun. 47*, 273–281.

299. Esfahani, M., Limbrick, A.R., Knutton, S., Oka, T., & Wakil, S.J. (1971) *Proc. Natl. Acad. Sci. U.S.A. 68*, 3180–3184.

300. Träuble, H., & Overath, P. (1973) *Biochim. Biophys. Acta 307*, 491–512.

301. Overath, P., & Träuble, H. (1973) *Biochemistry 12*, 2625–2634.

302. Demel, R.A., Geurts Van Kessel, W.S.M., Zwall, R.F.A., Roelofson, B., & VanDeenan, L.L.M. (1975) *Biochim. Biophys. Acta 406*, 97–107.

303. Bretscher, M.S. (1972) *Nature New Biol. 236*, 11–12.

304. Verkleij, A.J., Zwaal, R.F.A., Roelofsen, B., Comfurius, P., Kastelijn, D., & VanDeenan, L.L.M. (1973) *Biochim. Biophys. Acta 323*, 178–193.

305. Gordesky, S.E., Marinetti, G.V., & Love, R. (1975) *J. Membrane Biol. 20*, 111–132.

306. Bloj, B., & Zilversmit, D.B. (1976) *Biochemistry 15*, 1277–1283.

307. Kamp, H.H., Wirtz, K.W.A., & VanDeenan, L.L.M. (1973) *Biochim. Biophys. Acta 318*, 313–325.

308. Johnson, L.W., Hughes, M.E., & Zilversmit, D.B. (1975) *Biochim. Biophys. Acta 375*, 176–185.

309. Renooij, W., VanGolde, L.M.G., Zwaal, R.F.A., & VanDeenan, L.L.M. (1976) *Eur. J. Biochem. 61*, 53–58.

310. Rothman, J.E., & Kennedy, E.P. (1977) *J. Mol. Biol. 110*, 603–618.

311a. Krebs, J.J.R., Hauser, H., & Carafoli, E. (1979) *J. Biol. Chem. 254*, 5308–5316.

311b. Blau, L., & Bittman, R. (1978) *J. Biol. Chem. 253*, 8366–8368.

312. Fisher, K.A. (1976) *Proc. Natl. Acad. Sci. U.S.A. 73*, 173–177.

313. Lenard, J., & Rothman, J.E. (1976) *Proc. Natl. Acad. Sci. U.S.A. 73*, 391–395.

314. Kornberg, R.D., & McConnell, H.M. (1971) *Biochemistry 10*, 1111–1120.

315. McNamee, M.G., & McConnell, H.M. (1973) *Biochemistry 12*, 2951–2957.

316. Rothman, J.E., & Kennedy, E.P. (1977) *Proc. Natl. Acad. Sci. U.S.A. 74*, 1821–1825.

317. Frye, C.D., & Ediden, M. (1970) *J. Cell. Sci. 7*, 313–336.

318. Saffman, P.G., & Delbrück, M. (1975) *Proc. Natl. Acad. Sci. U.S.A. 72*, 3111–3113.

319. Peters, R., & Cherry, R.J. (1982) *Proc. Natl. Acad. Sci. U.S.A. 79*, 4317–4321.

320. Axelrod, D., Koppel, D.E., Schlessinger, J., Elson, E., & Webb, W.W. (1976) *Biophys. J. 16*, 1055–1069.

321. Cherry, R.J., & Godfrey, R.E. (1981) *Biophys. J. 36*, 257–276.

322. Poo, M., & Cone, R.A. (1974) *Nature 247*, 438–441.

323. Cone, R.A. (1972) *Nature New Biol. 236*, 39–43.

324. Chang, C.H., Takeuchi, H., Ito, T., Machida, K., & Ohnishi, S. (1981) *J. Biochem. (Tokyo) 90*, 997–1004.

325. Schlessinger, J., Axelrod, D., Koppel, D.E., Webb, W.W., & Elson, E.L. (1977) *Science 195*, 307–309.

326. Vaz, W.L.C., Criado, M., Madeira, V.M.C., Schoellmann, G., & Jovin, T.M. (1982) *Biochemistry 21*, 5608–5612.

327. Scandella, C.J., Devaux, P., & McConnell, H.M. (1972) *Proc. Natl. Acad. Sci. U.S.A. 69*, 2056–2060.

328. Shinitsky, M., Dianoux, A.C., Gitler, C., & Weber, G. (1971) *Biochemistry 10*, 2106–2113.

329. Cogan, U., Shinitsky, M., Weber, G., & Nishida, T. (1973) *Biochemistry 12*, 521–527.

330. Shinitzky, M., & Inbar, M. (1974) *J. Mol. Biol. 85*, 603–615.

331. Kinoshita, K., Kawato, S., Ikegami, A., Yoshida, S., & Orii, Y. (1981) *Biochim. Biophys. Acta 647*, 7–17.

332. Yarden, Y., & Ullrich, A. (1988) *Annu. Rev. Biochem. 57*, 443–478.

333. Lax, I., Mitra, A.K., Ravera, C., Hurwitz, D.R., Rubinstein, M., Ullrich, A., Stroud, R.M., & Schlessinger, J. (1991) *J. Biol. Chem. 266*, 13828–13833.

334. Wedegaertner, P.B., & Gill, G.N. (1989) *J. Biol. Chem. 264*, 11346–11353.

335. Kashles, O., Szapary, D., Bellot, F., Ullrich, A., Schlessinger, J., & Schmidt, A. (1988) *Proc. Natl. Acad. Sci. U.S.A. 85*, 9567–9571.

336. Carpenter, C.D., Ingraham, H.A., Cochet, C., Walton, G.M., Lazar, C.S., Sowadski, J.M., Rosenfeld, M.G., & Gill, G.N. (1991) *J. Biol. Chem. 266*, 5750–5755.

337. Cochet, C., Kashles, O., Chambaz, E.M., Borrello, I., King, C.R., & Schlessinger, J. (1988) *J. Biol. Chem. 263*, 3290–3295.

338. Fanger, B.O., Stephens, J.E., & Staros, J.V. (1989) *FASEB J. 3*, 71–75.

339. Yarden, Y., & Schlessinger, J. (1987) *Biochemistry 26*, 1434–1442.

340. Spaargaren, M., Defize, L.H.K., Boonstra, J., & deLaat, S.W. (1991) *J. Biol. Chem. 266*, 1733–1739.

341. Canals, F. (1992) *Biochemistry 31*, 4493–4501.

342. Benovic, J.L., Schorr, R.G.L., Caron, M.G., & Lefkowitz, R.J. (1984) *Biochemistry 23*, 4510–4518.

343. Dixon, R.A.F., Kobilka, B.K., Strader, D.J., Benovic, J.L., Dohlman, H.G., Frielle, T., Bolanowski, M.A., Bennett, C.D., Rands, E., Diehl, R.E., Mumford, R.A., Slater, E.E., Sigal, I.S., Caron, M.G., Lefkowitz, R.J., & Strader, C.D. (1986) *Nature 321*, 75–79.

344. Yarden, Y., Rodriguez, H., Wong, S.K.F., Brandt, D.R., May, D.C., Burnier, J., Harkins, R.N., Chen, E.Y., Ramachandran, J., Ullrich, A., & Roos, E.M. (1986) *Proc. Natl. Acad. Sci. U.S.A. 83*, 6795–6799.

345. Neer, E.J. (1974) *J. Biol. Chem. 249*, 6527–6531.

346. Haga, T., Haga, K., & Gilman, A.G. (1977) *J. Biol. Chem. 252*, 5776–5782.

347. Kataoka, T., Broek, D., & Wigler, M. (1985) *Cell 43*, 493–505.

348. Takahashi, N., Takahashi, Y., & Putnam, F.W. (1985) *Proc. Natl. Acad. Sci. U.S.A. 82*, 1906–1910.

349. Rodbell, M., Birnbaumer, L., Pohl, S.L., & Krans, H.M.J. (1971) *J. Biol. Chem. 246*, 1877–1882.

350. Pfeuffer, T. (1979) *FEBS Lett. 101*, 85–89.

351. Cassel, D., & Selinger, Z. (1977) *Proc. Natl. Acad. Sci. U.S.A. 74*, 3307–3311.

352. Cassel, D., & Selinger, Z. (1976) *Biochim. Biophys. Acta 452*, 538–551.

353. Pike, L.J., & Lefkowitz, R.J. (1980) *J. Biol. Chem. 255*, 6860–6867.

354. Cassel, D., Levkovitz, H., & Selinger, Z. (1977) *J. Cyclic Nucleotide Res. 3*, 393–406.

355. Cassel, D., Eckstein, F., Lowe, M., & Selinger, Z. (1979) *J. Biol. Chem. 254*, 9835–9838.

356. Cassel, D., & Selinger, Z. (1978) *Proc. Natl. Acad. Sci. U.S.A. 75*, 4155–4159.

357. Sternweis, P.C., Northrup, J.K., Smigel, M.D., & Gilman, A.G. (1981) *J. Biol. Chem. 256*, 11517–11526.

358. Itoh, H. Kozasa, T., Nagata, S., Nakamura, S., Katada, T., Ui, M., Iwai, S., Ohtsuka, E., Kawasaki, H., Suzuki, K., & Kaziro, Y. (1986) *Proc. Natl. Acad. Sci. U.S.A. 83*, 3776–3780.

359. Robishaw, J.D., Russell, D.W., Harris, B.A., Smigel, M.D., & Gilman, A.G. (1986) *Proc. Natl. Acad. Sci. U.S.A. 83*, 1251–1255.

360. Fong, H.K.W., Amatruda, T.T., Birren, B.W., & Simon, M.I. (1987) *Proc. Natl. Acad. Sci. U.S.A. 84*, 3792–3796.

361. McConnell, D.G., Kohnken, R.E., & Smith, A.J. (1984) *Fed. Proc. 43*, 1585.

362. Yatsunami, K., Pandya, B.V., Oprian, D.D., & Khorana, H.G. (1985) *Proc. Natl. Acad. Sci. U.S.A. 82*, 1936–1940.

363. Arad, H., Rosenbusch, J.P., & Levitski, A. (1984) *Proc. Natl. Acad. Sci. U.S.A. 81*, 6579–6583.

364. Orly, J., & Schramm, M. (1976) *Proc. Natl. Acad. Sci. U.S.A. 73*, 4410–4414.

365. Tolkovsky, A.M., & Levitski, A. (1978) *Biochemistry 17*, 3795–3810.

366. Kazarov, A.R., Rosenkranz, A.A., & Sobolev, A.S. (1986) *Biokhimiia (Moscow) 51*, 355–363.

367. Pfeuffer, T. (1977) *J. Biol. Chem. 252*, 7224–7234.

368. Limbird, L.E., Gill, D.M., & Lefkowitz, R.J. (1980) *Proc. Natl. Acad. Sci. U.S.A. 77*, 775–779.

369. Cerione, R.A., Codina, J., Benovic, J.L., Lefkowitz, R.J., Birnbaumer, L., & Caron, M.G. (1984) *Biochemistry 23*, 4519–4525.

370. Elgsaeter, A., Schotton, D.M., & Branton, D. (1976) *Biochim. Biophys. Acta 426*, 101–122.

371. Sheetz, M.P., Painter, R.G., & Singer, S.J. (1976) *Biochemistry 15*, 4486–4492.

372. Marchesi, V.T., & Steers, E. (1968) *Science 159*, 203–204.

373. Rosenthal, A.S., Kregenow, F.M., & Moses, H.L. (1971) *Biochim. Biophys. Acta 196*, 254–262.

374. Gwynne, J.T., & Tanford, C. (1970) *J. Biol. Chem. 245*, 3269–3273.

375. Kam, Z., Josephs, R., Eisenberg, H., & Gratzer, W.B. (1977) *Biochemistry 16*, 5568–5572.

376. Clarke, M. (1971) *Biochem. Biophys. Res. Commun. 45*, 1063–1070.

377. Speicher, D.W., & Marchesi, V.T. (1984) *Nature 311*, 177–180.

378. Steck, T.L. (1974) *J. Cell Biol. 62*, 1–19.

379. Tyler, J.M., Hargreaves, W.R., & Branton, D. (1979) *Proc. Natl. Acad. Sci. U.S.A. 76*, 5192–5196.

380. Tyler, J.M., Reinhardt, B.N., & Branton, D. (1980) *J. Biol. Chem. 255*, 7034–7039.

381. Ungewickell, E., Bennett, P.M., Calvert, R., Ohanian, V., & Gratzer, W.B. (1979) *Nature 280*, 811–814.

382. Shotton, D.M., Burke, B., & Branton, D. (1979) *J. Mol. Biol. 131*, 303–329.

383. Bennett, V., & Stenbuck, P.J. (1980) *J. Biol. Chem. 255*, 2540–2548.

384. Bennett, V. (1978) *J. Biol. Chem. 253*, 2292–2299.

385. Bennett, V., & Stenbuck, P.J. (1980) *J. Biol. Chem. 255*, 6424–6432.

386. Bennett, V. (1982) *Biochim. Biophys. Acta 689*, 475–484.

387. Strittmatter, P., Rogers, M.G., & Spatz, L. (1972) *J. Biol. Chem. 247*, 7188–7194.

388. Young, R. (1970) *Sci. Am. 223* (4), 81–91.

Index

Page numbers with an F refer to a figure; page numbers with a T refer to a table.